Beijing 2002
August 20–28

Proceedings of the
International
Congress of
Mathematicians

Vol.III : Invited Lectures

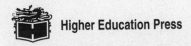

Higher Education Press

International Congress of Mathematicians (2002, Beijing)
Proceedings of the International Congress of Mathematicians
August 20–28, 2002, Beijing

Editor: LI Tatsien (LI Daqian)
Department of Mathematics, Fudan University
Shanghai 200433, China   Email: `dqli@fudan.edu.cn`
Editorial Assistants: Cai Zhijie, Lu Fang, Xue Mi, Zhou Chunlian

This volume is the second part of the collection of manuscripts of the lectures given by the invited speakers of the ICM2002. The first part of this collection is published in Volume II.

The manuscripts of the invited lectures are ordered by sections and, in each section, alphabetically by author's names, except in Section 18, where a special arrangement is made. In case of several authors for one manuscript, the name of invited speaker is written in boldface type. The electronic version of this work will be published on the international Math ArXiv with the address

`http://front.math.ucdavis.edu/`

©2002 Higher Education Press
55 Shatan Houjie, Beijing 100009, China
`http://www.hep.com.cn`   `http://www.hep.edu.cn`

Copy Editors: Li Rui, Li Yanfu, Wang Yu

ISBN 7-04-008690-5   Set of 3 Volumes

Contents

Section 12. Ordinary Differential Equations and Dynamical Systems

Section 13. Mathematical Physics

Section 14. Combinatorics

Section 15. Mathematical Aspects of Computer Science

Section 16. Numerical Analysis and Scientific Computing

Section 17. Application of Mathematics in the Sciences

Section 18. Mathematics Education and Popularization of Mathematics

Section 19. History of Mathematics

Section 10. Probability and Statistics

ICM 2002 · Vol. III · 3–14

Aging and Spin-glass Dynamics

Gérard Ben Arous*

Abstract

We survey the recent mathematical results about aging in certain simple disordered models. We start by the Bouchaud trap model. We then survey the results obtained for simple models of spin-glass dynamics, like the REM (the Random Energy Model, which is well approximated by the Bouchaud model on the complete graph), then the spherical Sherrington-Kirkpatrick model. We will insist on the differences in phenomenology for different types of aging in different time scales and different models. This talk is based on joint works with A.Bovier, J.Cerny, A.Dembo, V.Gayrard, A.Guionnet, as well as works by C.Newman, R.Fontes, M.Isopi, D.Stein.

2000 Mathematics Subject Classification: 60H10, 60K35, 82C44, 82C31, 20J05.
Keywords and Phrases: Aging, Spin Glass, Random Media, Trapping models, Statistical mechanics.

1. Introduction

Aging is an interesting long-time property of dynamics in complex disordered media, and in particular in certain random media. A system ages when its decorrelation properties are age-dependent: the older the system the longer it takes to forget its past. Aging has been heavily studied both experimentally, numerically and theoretically by physicists, in particular in the context of spin-glass dynamics, but the mathematical litterature is still rather sparse. Our interest in aging stemmed from the study of dynamics of mean field spin glasses models, and more precisely of Langevin dynamics for the Sherrington Kirkpatrick model. This remains the ultimate goal, far from being achievable. But we will survey some of the partial progress which has been made both in short time scales (for the spherical SK model, see section 4) or much longer time scales (for the REM dynamics, see section 3). We begin with aging the Bouchaud trap model on various graphs (see

---

*Department of Mathematics, Ecole Polytechnique Federale de Lausanne, 1005 Lausanne, Switzerland. E-mail: gerard.benarous@epfl.ch

section 2)even though it is not directly related to Spin Glasses, because the mechanism of aging there is close to the one for the REM. This talk is based on joint works with A.Bovier, J.Cerny, A.Dembo, V.Gayrard, A.Guionnet, as well as works by C.Newman, R.Fontes, M.Isopi.

2. Bouchaud's random trap model

This model is simple model of a random walk trapped (or rather slowed down) by random wells. It is nevertheless quite rich. It has been introduced by Bouchaud and coworkers (see [9] for a beautiful general survey) as an ansatz to understand "activated" dynamics of spin glasses.

2.1. The random energy landscape

Consider $G = (V, B)$ a graph where V is the set of vertices and B the set of bonds. We introduce a "random energy landscape" on V, a collection $(E_x)_{x\in V}$ of i.i.d non-negative random variables, indexed by the vertices of the graph and exponentially distributed (with mean one) , i.e

$$P(E_x > a) = e^{-a}.$$

From these random variables, we define a random measure τ on V by:

$$\tau(x) = e^{\beta E_x}.$$

Here $\beta > 0$ is an inverse temperature parameter. We will denote by $\alpha = \frac{1}{\beta}$ and concentrate on the low temperature phase, i.e $\alpha < 1$.

The physicists would see the set of vertices V as a set of " favourable valleys" for the configurations of a much more complex system (for instance a spin-glass), $-E_x$ as the energy of the bottom of the valley x. Since these energies are usually the extreme values of some other random landscape (see below the discussion about the REM), the hypothesis that they are exponential random variables is reasonable. The sites x where E_x is large are seen as " very favourable valleys", or equivalently as very deep traps, where the system should stay for a long time, in any sensible definition of the dynamics.

2.2. Bouchaud's random walk

We consider the continuous-time Markov Chain $X(t)$ on the set V of vertices, whose jump rates are given, when x and y are neighbours on the graph, by:

$$w_{x,y} = \nu e^{-\beta((1-a)E_x - aE_y)}$$

and $w_{x,y} = 0$ if x and y are not neighbours.

Here $\nu > 0$ is a time-scale parameter, often set to 1, and $a \in [0, 1]$ is a symmetry index.

One can also write

$$w_{x,y} = \nu\tau(x)^{-(1-a)}\tau(y)^a.$$

These jump rates satisfy the detailed balance equation:

$$\sum_{y \in V} w_{x,y} \tau(x) = \sum_{y \in V} w_{y,x} \tau(y).$$

So that the measure τ is reversible for the Markov Chain X, whatever the value of the parameter a.

Notice nevertheless that the case where $a = 0$ is simpler. Then the jump rate $w_{x,y} = \frac{\nu}{\tau(x)}$ depends only on the random landscape at site x. The Markov Chain X is simply obtained by a random time-change from the simple random walk on the graph. This case is always easier to handle, and will be called the Random Hopping Times dynamics (RHT dynamics). In this case, $\tau(x)$ is simply the mean time spent at x.

Remark The important feature of the random variables $\tau(x)$ is the fact that they are heavy-tailed:

$$P(\tau(x) > a) = \frac{1}{a^{\alpha}}$$

in particular their expectation is infinite, when $\alpha < 1$. The whole model could in fact be directly defined in terms of the $\tau(x)$ rather than from the energies $(-E_x)$, and then one could assume that the $\tau(x)$ are i.i.d and in the domain of attraction of an α-stable law.

2.3. Aging and two-point functions

The natural question about Bouchaud's Markov Chain is to study its long-time behaviour, either in the "quenched" regime, i.e almost surely in the randomness of the energy landscape, or in the "annealed" regime, i.e after averaging in this randomness. One would expect a dynamical phase transition, between the high temperature phase where $\alpha > 1$ and the low temperature phase where $\alpha < 1$. A general idea about the low temperature phase is that the system spends most of its time in very deep traps: more precisely, that by time t, the chain has explored a large part of the space and has found traps of depth depending on t. So that , at age t, with high probability the system sits waiting in a deep trap, whose depth is t-dependent, for a time thus depending on t, before being able to get out and find another deep trap. The beautiful idea put forward by the physics litterature is "to think in the two-times plane", i.e to consider the evolution of the system between two large times, generally denoted t_w (like waiting time, t_w is the age of the system) and $t_w + t$ (t is then the duration of the observation of the system), and to let both t_w and t tend to infinity. The next step proposed by physicists is to choose appropriate two-point functions, i.e functions of the evolution of the system in the time interval $(t_w, t_w + t)$, in order to measure how much the system forgets its past in this time interval. The simplest such two-point function is the quenched probability that the system is in the same state at times t_w and $t_w + t$:

$$R^{\omega}(t_w, t_w + t) = P^{\omega}(X(t_w) = X(t_w + t)).$$

Here the probability is quenched, i.e conditioned on the random medium, and the superscript ω denotes this randomness of medium, i.e the i.i.d collection of energies. One will also consider the annealed version of this two-point function:

$$R(t_w, t_w + t) = \langle R^\omega(t_w, t_w + t) \rangle$$

where $\langle \ , \ \rangle$ denotes the expectation w.r.t the medium. One also considers often another two-point function, i.e the quenched probability that the system has not jumped at all in the time interval $(t_w, t_w + t)$:

$$\Pi^\omega(t_w, t_w + t) = P^\omega(X(t_w) = X(t_w + s), \forall s < t)$$

or its annealed conterpart

$$\Pi(t_w, t_w + t) = \langle \Pi^\omega(t_w, t_w + t) \rangle.$$

Finding an "agime regime" is then proving that when t is too small as a function of t_w (typically $t = o(t_w^\gamma)$ for some "aging exponent" γ), then such a two-point function is close to one, and when it is large enough (typically $t >> t_w^\gamma$) it is close to zero. Naturally it is even more desirable to find the limit of these two-point functions for the critical regime (typically $t = Ct_w^\gamma$).

This program is now understood rigorously for a few important graphs, which we will now review. We will review the case of Z^d, first for $d = 1$ after the work of Fontes-Isopi-Newman ([16],[17])and more recently Cerny (see [4],[14]), then the case of $d = 2$ (see [5]) (for $d > 2$ see [14]). Then we treat the case of the complete graph on M points when M tends to infinity. This is in fact the original Bouchaud model, which was introduced as an ansatz for the dynamics of the Random Energy Model. We will then survey the recent results on these dynamics of the REM ([1],[2],[3]) which will be the first results really pertaining to the topic of spin glass dynamics.

2.4.   Two aging regimes for Bouchaud's model on Z

Bouchaud's model on Z has been first studied by Fontes-Isopi-Newman, when $a = 0$, i.e for the Random Hopping Times dynamics. To understand aging in dimension $d = 1$, it is important to introduce a limiting object which will play the role of the random medium:

Definition 1 *The random speed measure ρ.*

Let (x_i, v_i) be a Poisson Point Process on $R \times (0, \infty)$, with intensity measure $\alpha v^{-(1+\alpha)} dx dv$. We define a random measure ρ on R by

$$\rho = \sum_i v_i \delta_{x_i}.$$

Definition 2 *The FIN (Fontes-Isopi-Newman) singular diffusion $Z(s)$.*

Let $W(t)$ be a standard one-dimensional Brownian Motion, and $l(t,y)$ its local time at y. Define the random time-change:

$$\phi^\rho(t) = \int l(t, y) \rho(dy)$$

and its inverse
$$\psi^\rho(t) = inf(s, \phi^\rho(s) = t).$$
Then the FIN singular diffusion is $Z(s) = W(\psi^\rho(t))$.

Notice that the Random speed measure and the FIN singular diffusion are entirely independent of the symmetry parameter a, but depend only on the temperature parameter α.

Then the following (annealed) aging result has been proved in [16] for $a = 0$, and in [4] for general a's.

Theorem 1 *For any $\alpha < 1$ and any $a \in [0,1]$, the following limit exists*
$$\lim_{t_w \to \infty} R(t_w, (1+\theta)t_w) = f(\theta).$$

Moreover the function f can be computed using the singular diffusion Z:
$$f(\theta) = \langle P(Z(1+\theta) = Z(1)) \rangle.$$

This result shows that the two-point function R exhibits an aging regime, $t = \theta t_w$, independently of a. To be able to feel the influence of a, one should use the other (annealed) two-point function Π, which exhibits another aging regime (see [4]). Let us introduce some notation: Denote by F the annealed distribution function of the important r.v $\rho(Z(1))$
$$F(u) = \langle P(\rho(Z(1)) \leq u) \rangle.$$

Here the brackets denote the average w.r.t the environment i.e the randomness of the measure ρ. Let g_a be the Laplace transform of the r.v $\tau(0)^a$
$$g_a(\lambda) = E[e^{-\lambda \tau(0)^a}]$$
and C the constant given by $C = 2^{a-1}[E(\tau(0)^{-2a})]^{1-a}$.

Theorem 2 *For any $\alpha < 1$ and any $a \in [0,1]$, the following limit exists*
$$\lim_{t_w \to \infty} \Pi(t_w, t_w + \theta t_w^\gamma) = q_a(\theta)$$

where $\gamma = \frac{1-a}{1+a}$ and this limit can be computed explicitly
$$q_a(\theta) = \int_0^\infty g_a^2(C\theta u^{a-1}) \, dF(u).$$

In particular, when $a = 0$,
$$q_0(\theta) = \int_0^\infty e^{-\frac{\theta}{u}} \, dF'(u).$$

These two results show that the Markov Chain essentially succeeds in leaving the site it has reached at age t_w only after a time t of the order of t_w^γ, but that for time scales between t_w^γ and t_w it will jump out of the site reached at age t_w but will not find an other trap deep enough and so will be attracted back to the trap it reached at age t_w.

2.5. Aging for Bouchaud's model on Z^2, the RHT case

In dimension 2, i.e on Z^2, the only case studied is the case where a=0, the RHT case. An aging regime is exibited (in [5])for both quenched two-point functions

Theorem 3 *The following limits exists almost surely in the environment ω,*

$$\lim_{t_w \to \infty} R^\omega(t_w, t_w + \theta t_w) = h(\theta),$$

$$\lim_{t_w \to \infty} \Pi^\omega(t_w, t_w + \theta \frac{t_w}{\ln t_w}) = k(\theta).$$

The functions h and k satisfy

$$\lim_{\theta \to 0} h(\theta) = \lim_{\theta \to 0} k(\theta) = 1$$

and

$$\lim_{\theta \to \infty} h(\theta) = \lim_{\theta \to \infty} k(\theta) = 0.$$

In fact the functions h and k can be computed explicitly easily using arcsine laws for stable processes. For instance

$$h(\theta) = \frac{\sin \alpha \pi}{\pi} \int_0^{1+\theta} u^{\alpha-1}(1-u)^{-\alpha} \, du.$$

Here again there is a difference between the subaging regime (i.e $t = \theta \frac{t_w}{\ln(t_w)}$ for Π))and aging regime(i.e $t = \theta t_w$ for R), but much slighter than in dimension 1. Indeed it is naturally much more difficult to visit a trap again after leaving it.

There is another important difference between $d=1$ and $d=2$, noticed by [16], namely there is localisation in dimension 1 and not in dimension 2. More precisely, in $d=1$

$$\limsup_{t \to \infty} \sup_{x \in Z} \langle P^\omega(X(t) = x) \rangle > 0.$$

And this property is wrong in dimension 2.

2.6. Aging for Bouchaud's model on a large complete graph

Consider now the case where G is the complete graph on M points. We will also study here only the case the RHT case, where a=0. We consider Bouchaud's Markov Chain X on G, started from the uniform measure. Then it is easy to see that the times of jump form a renewal process and that the two-point function Π_M^ω is the solution of a renewal equation. This renewal process converges, when M tends to ∞, to a heavy-tailed renewal process we now introduce.

Let

$$F_\infty(t) = 1 - \alpha \int_1^\infty e^{-\frac{t}{x}} x^{-(1+\alpha)} \, dy.$$

Consider $\Pi_\infty(t_w, t_w + t)$ the unique solution of the renewal equation

$$\Pi_\infty(t_w, t_w + t) = 1 - F_\infty(t_w + t) + \int_0^t \Pi_\infty(t_w - u, t_w - u + t) \, dF_\infty(u).$$

Theorem 4 (see [10], [3]) *Almost surely in the environment, for all t_w and t, $\Pi_M^\omega(t_w, t_w + t)$ converges to $\Pi_\infty(t_w, t_w + t)$.*

It is easy to see that the limiting two-point function $\Pi_\infty(t_w, t_w + t)$ shows aging. Let

$$H(\theta) = \frac{1}{\pi cosec(\frac{\pi}{\alpha})} \int_\theta^\infty \frac{1}{(1 + x)x^\alpha \, dx}.$$

Obviously $H(\theta) \sim 1 - C_\alpha \theta^{1-\alpha}$ when θ tends to 0, and $H(\theta) \sim C'_\alpha \theta^{-\alpha}$ when θ tends to ∞ .

Theorem 5

$$\lim_{t_w \to \infty} \Pi_\infty(t_w, t_w + \theta t_w) = H(\theta).$$

So that

$$\lim_{t_w \to \infty} \lim_{M \to \infty} \Pi_M^\omega(t_w, t_w + \theta t_w) = H(\theta).$$

We will see in the next section how this simple result gives an approximation (in a weak sense) for the much more complex problem of aging for the RHT dynamics for the REM.

3.   Aging for the Random Energy Model

We report here on the joint work with Anton Bovier and Veronique Gayrard, see [1],[2], [3]. This work uses heavily the general analysis of metastability for disordered mean-field models in [11], [12]. Let us first, following Derrida, define the REM, often called the simplest model of a spin glass. A spin configuration σ is a vertex of the hypercube $S_N = \{-1, 1\}^N$. As in Bouchaud model on the graph S_N we consider a collection of i.i.d random variables $(E_\sigma)_{\sigma \in S_N}$ indexed by the vertices of S_N. But we will here assume that the distribution of the E_σ 's is standard Gaussian. We then define the Gibbs measure $\mu_{\beta,N}$ on S_N by setting:

$$\mu_{\beta,N}(\sigma) = \frac{e^{\beta\sqrt{N}E_\sigma}}{Z_{\beta,N}}$$

where $Z_{\beta,N}$ is the normalizing partition function. The statics of this model are well understood (see[15], and [13]). It is well known that the REM exhibits a static phase transition at $\beta_c = \sqrt{2 \ln 2}$. For $\beta > \beta_c$ the Gibbs measure gives ,asymptotically when N tends to ∞, positive mass to the configurations σ where the extreme values of the order statistics of the i.i.d $N(0, 1)$ sample $(E_\sigma)_{\sigma \in S_N}$ are reached, i.e if we order the spin configurations according to the magnitude of their (-) energies:

$$E_{\sigma(1)} \geq E_{\sigma(2)} \geq ... \geq E_{\sigma(2^N)}.$$

Then for any fixed k, the mass $\mu_{\beta,N}(\sigma^{(k)}$ will converge to some positive random variable. In fact the whole collection of masses $\mu_{\beta,N}(\sigma^{(k)})$ will converge to a point process, called Ruelle's point process. Consider for any E, the set (which we have called the top in [2] and [3], and should really be called the bottom) of configurations with energies below a certain threshold $u_N(E)$.

$$T_N(E) = \{\sigma \in S_N, E_\sigma \geq u_N(E)\}.$$

We will choose here the natural threshold for extreme of standard Gaussian i.i.d rv's, i.e

$$u_N(E) = \sqrt{2N \ln 2} + \frac{E}{\sqrt{2N \ln 2}} - \frac{\ln(N \ln 2) + \ln(4\pi)}{2\sqrt{2N \ln 2}}.$$

Now we define discrete-time dynamics on S_N by the transition probabilities

$$p_N(\sigma, \eta) = \begin{cases} \frac{1}{N} e^{-\beta \sqrt{N} E_\sigma^+}, & \text{if } \sigma \text{ and } \eta \text{ differ by a spin-flip}; \\ 1 - e^{-\beta \sqrt{N} E_\sigma^+}, & \text{if } \sigma = \eta; \\ 0, & \text{otherwise.} \end{cases} \tag{3.1}$$

Notice we have here truncated the negative values of the E_σ's , this truncation is technical and irrelevant. We could truncate much less drastically. Anyway the states with very negative E's wont be seen on the time scales we are interested in.

Then the idea defended by Bouchaud is that the motion of these REM dynamics when seen only on the deepest traps $T_N(E)$ should be close to the dynamics of the Bouchaud model on the complete graph for large M. This is true only to some extent. It is true that, if one conditions on the size of the top $T_N(E)$ to be M, then the sequence of visited points in $T_N(E)$ has asymptotically, when N tends to ∞, to the standard random walk on the complete graph with M points. Nevertheless Bouchaud's picture would be completely correct if the process observed on the top would really be Markovian, which is not the case, due to a lack of time scales separation between the top and its complement. Nevertheless it is remarkable that in a weak asymptotic form Bouchaud's prediction about aging is correct. Let us consider the following natural two-point function:

$$\Pi_N(n, m) = \frac{1}{|T_N(E)|} \sum_{\sigma \in T_N(E)} \Pi_\sigma(n, m)$$

where $\Pi_\sigma(n, m)$ is the quenched probability that the process starting at time 0 in state σ does not jump during the time interval $(n, n+m)$ from one state in the top $T_N(E)$ to another such state.

Theorem 6 *Let $\beta > \beta_c$ Then there is a sequence $c_{N,E} \sim e^{\beta \sqrt{N} u_N(E)}$, such that for any $\epsilon > 0$*

$$\lim_{t_w \to \infty} \lim_{E \to -\infty} \lim_{N \to \infty} P(|\frac{\Pi_N(c_{N,E} t_w, c_{N,E}(t_w + t))}{H(\frac{t}{t_w})}) - 1| \geq \epsilon) = 0.$$

Remark 1 The rescaling of the time by the factor $c_{N,E}$ shows that Bouchaud's trap model is a good approximation of the REM dynamics for the very large time asymptotics, on the last time scale before equilibrium is reached. This is to be contrasted with the other model of glassy dynamics for spin glasses, as advocated by Parisi, Mezard or Cugliandolo-Kurchan where the infinite volume limit is taken before the large time limit. We will see in the next section an example of such a very different aging phenomenon in a much shorter time-scale.

Remark 2 The REM is the first model on which this study of aging on very long times scales (activated dynamics)has been rigorously achieved. This phenomenon should be present in many more models, like the Generalized Random Energy Model, which is the next achievable goal, or even in much harder problems like the p-spin models for large enough p's. The tools developped in [11][12] should be of prime relevance.

4.   Aging for the spherical Sherrington-Kirkpatrick model

Studying spin glass dynamics for the Sherrington Kirkpatrick model might seem premature, since statics are notoriously far from fully understood, as opposed to the REM. Nevertheless, following Sompolinski and Zippelius, a mathematical study of the Langevin dynamics has been undertaken in the recent years jointly with A.Guionnet (see [7],[8],[19]). The output of this line of research has been to prove convergence and large deviation results for the empirical measure on path space as well as averaged and quenched propagation of chaos. The same problem has been solved by M. Grunwald for discrete spins and Glauber dynamics, see [18]. The law of the limiting dynamics (the self consistent single spin dynamics) is characterized in various equivalent ways, from a variational problem to a non-Markovian implicit stochastic differential equation, none of which being yet amenable, for the moment, to a serious understanding.

The Sherrington-Kirkpatrick Hamiltonian is given by

$$H_J^N(x) = \frac{1}{\sqrt{N}} \sum_{i,j=1}^{N} J_{ij} x_i x_j$$

with a $N \times N$ random matrix $\mathbf{J} = (J_{ij})_{1 \leq i,j \leq N}$ of centered i.i.d standard Gaussian random variables. The Langevin dynamics for this model are described by the stochastic differential system :

$$dx_t^j = dB_t^j - U'(x_t^j)dt - \frac{\beta}{\sqrt{N}} \sum_{1 \leq i \leq N} J_{ji} x_t^i dt, \qquad (4.2)$$

where B is a N-dimensional Brownian motion, and U a smooth potential growing fast enough to infinity. It was proved in [8],[19] that, for any time $T > 0$, the empirical measure on path space

$$\mu_N := \frac{1}{N} \sum_{i=1}^{N} \delta_{x_{[0,T]}^i}$$

converges almost surely towards a non Markovian limit law $Q_{\mu_0}^T$, when the initial condition is a "deep quench", i.e when $x_0 = \{x_0^j, 1 \leq j \leq N\}$ are i.i.d. $Q_{\mu_0}^T$ is called the self consistent single spin dynamics in the physics litterature. It is the law of a self-consistent non Markovian process, which is very hard to study. One expects that the long time behaviour of this process shows an interesting dynamical phase transition, and in particular exhibits aging. A consequence of the convergence stated above is that the limit of the empirical covariance exists, and is simply the autocovariance of the law $Q_{\mu_0}^T$:

$$C(t_w, t_w + t) := \lim_{N \to \infty} \frac{1}{N} \sum_{i=1}^{N} x_{t_w}^i x_{t_w+t}^i = \int x_{t_w} x_{t_w+t} \, dQ_{\mu_0}^T(x).$$

Unfortunately it is not possible to find a simple, autonomous equation satisfied by C, or even by C and the so-called response function R. This is a very hard open problem. But in the physics litterature (mainly in the work of Cugliandolo and Kurchan) one find that the same program is tractable and gives a very rich picture of aging for a large class of models, i.e the spherical p-spin models.

We report here on the joint work with A.Dembo and A.Guionnet [6], on the simplest of such models, i.e the Spherical SK, or spherical p-spin model with p=2. The general case of $p > 2$ is harder and will be our next step in the near future. In this work, we study the Langevin dynamics for a spherical version of the Sherrington Kirkpatrick (SSK) spin glass model.

More precisely, we shall consider the following stochastic differential system

$$du_t^i = \beta \sum_{j=1}^{N} J_{ij} u_t^j dt - f'(\frac{1}{N} \sum_{j=1}^{N} (u_t^j)^2) u_t^i dt + dW_t^i \qquad (4.3)$$

where f' is a uniformly Lipschitz, bounded below function on R^+ such that $f(x)/x \to \infty$ as $x \to \infty$ and $(W^i)_{1 \leq i \leq N}$ is an N-dimensional Brownian motion, independent of $\{J_{i,j}\}$ and of the initial data $\{u_0^i\}$.

The term containing f is a Lagrange multiplier used to implement a "soft" spherical constraint.

Here again the empirical covariance admits a limit

$$C(t_w, t_w + t) := \lim_{N \to \infty} \frac{1}{N} \sum_{i=1}^{N} x_{t_w}^i x_{t_w+t}^i.$$

But now, as opposed to the true SK model, this limiting two point function is easily computable from an autonomous renewal equation. Indeed the induced rotational symmetry of the spherical model reduces the dynamics in question to an N-dimensional coupled system of Ornstein-Uhlenbeck processes whose random drift parameters are the eigenvalues of a GOE random matrices.

Theorem 7 *There exists a critical β_c such that: When starting from i.i.d initial conditions: If $\beta < \beta_c$, then*

$$C(t_w, t_w + t) \leq C_\beta \exp(-\delta_\beta |t - s|)$$

for some $\delta_\beta > 0$, $C_\beta < \infty$ and all (t_w, t).

If $\beta = \beta_c$, then $C(t_w, t_w + t) \to 0$ as $t \to \infty$. If t_w/t is bounded, then the decay is polynomial $t^{-1/2}$, and otherwise it behaves like $\frac{t_w^{1/2}}{t_w + t}$.

If $\beta > \beta_c$ then the following limit exists

$$\lim_{t_w \to \infty} C(t_w, t_w + \theta t_w) = f(\theta).$$

Moreover if $t \gg t_w \gg 1$, then $C(t_w, t_w + t) \frac{t}{t_w}^{3/4}$ is bounded away from zero and infinity. In particular, the convergence of $C(t_w, t_w + t)$ to zero occurs if and only if $\frac{t}{t_w} \to \infty$.

In these much shorter time scales than for the former sections, the aging phenomenon we exhibit here is quite different than the one shown in the REM. Here the system has no time to cross any barrier, or to explore and find deep wells, it simply goes down one well, in very high dimension. In some of these very many directions (corresponding to the top eigenvectors of the random matrix $\mathbf{J}$), the curvature of the well is so weak that the corresponding coordinates of the system are not tightly bound and are very slow to equilibrate. These very slow components are responsible for the aging phenomenon here.

References

[1] G.Ben Arous, A.Bovier, V.Gayrard: Aging in the random energy model, volume 88, issue 8 of Physical Review Letters, 2002.

[2] G.Ben Arous, A.Bovier, V.Gayrard: Glauber Dynamics of the Random Energy Model, Part I, Metastable Motion on the extreme states, to appear in Communications in Mathematical Physics.

[3] G.Ben Arous, A.Bovier, V.Gayrard: Glauber Dynamics of the Random Energy Model, Part II, Aging below the critical temperature, to appear in Communications in Mathematical Physics.

[4] G.Ben Arous, J.Cerny: Bouchaud's model exhibits two aging regimes in dimension one. Preprint.

[5] G.Ben Arous, J.Cerny: T.Mountford, Aging for Bouchaud's model in dimension 2. Preprint.

[6] G.Ben Arous, A.Dembo, A.Guionnet : Aging of spherical spin glasses, Probability Theory and Related Fields, 120, (2001), 1–67.

[7] G.Ben Arous, A.Guionnet : Large deviations for Langevin Spin Glass Dynamics, Probability Theory and Related Fields, 102, (1995), 455–509.

[8] G.Ben Arous, A.Guionnet :Symmetric Langevin Spin Glass Dynamics, Annals of Probability, 25, (1997), 1367–1422.

[9] J.P Bouchaud,L. Cugliandolo, J.Kurchan, M.Mezard: Out of equilibrium dynamics in spin glasses and other glassy systems, Spin Glass dynamics and Random Fields, A.P Young editor,World Scientific, Singapore(1998).

[10] J.P Bouchaud, D.Dean: Aging on Parisi tree, Journal of Physics I (france), 5, 265 (1995).

[11] A.Bovier, M.Eckhoff, V.Gayrard, M.Klein: Metastability in stochastic dynamics of disordered mean field models,Probability Theory and Related Fields 120 (2001), 1–67.

[12] A.Bovier, M.Eckhoff, V.Gayrard, M.Klein:Metastability and low lying spectra in reversible Markov Chains, to appear in Comunications of Mathematical Physics.

[13] A.Bovier, I.Kurkova, M.Lowe: Fluctuations of the free energy in the REM and the p-spin SK models, Annals of Probability, 2001.

[14] J.Cerny: PhD thesis, EPFL, October 2002.

[15] B.Derrida, Random Energy Model: Limit of a family of disordered models, Physics review Letters,45,(1980),79–82.

[16] L.R.G Fontes, M.Isopi, C.Newman: Random walks with strongly inhomogeneous rates and singular diffusions: convergence, localization, and aging in one dimension, preprint.

[17] L.R.G Fontes, M.Isopi, C.Newman: Chaotic time dependance in a disordered spin system, Probability Theory and Related Fields 115 (1999), vol3, 417–443.

[18] M.Grunwald: Sanov Results for Glauber spin Glass dynamics, Probability Theory and Related Fields, 106, (1996), 187–232.

[19] A.Guionnet: Annealed and Quenched Propagation of Chaos for langevin Spin glass dynamics, Probability Theory and Related Fields, 109, (1997), 183–215.

Some Aspects of Additive Coalescents

Jean Bertoin*

Abstract

We present some aspects of the so-called additive coalescence, with a focus on its connections with random trees, Brownian excursion, certain bridges with exchangeable increments, Lévy processes, and sticky particle systems.

2000 Mathematics Subject Classification: 60K35, 60G51, 60G09.
Keywords and Phrases: Additive coalescence, Lévy processes, Ballistic aggregation.

1.   Additive coalescence in the finite setting

The additive coalescence is a simple Markovian model for random aggregation that arises for instance in the study of droplet formation in clouds [13, 11], gravitational clustering in the universe [19], phase transition for parking [10], ... As long as only finitely many clusters are involved, it can be described as follows. A typical configuration is a finite sequence $x_1 \geq \ldots \geq x_n > 0$ with $\sum_1^n x_i = 1$, which may be thought of as the ranked sequence of masses of clusters in a universe with unit total mass. Each pair of clusters, say with masses x and y, merges as a single cluster with mass $x + y$ at rate $K(x,y) = x + y$, independently of the other pairs in the system. This means that to each pair (i,j) of indices with $1 \leq i < j \leq n$, we associate an exponential variable $e(i,j)$ with parameter $x_i + x_j$, such that to different pairs correspond independent variables. If the minimum $\gamma_1 := \min_{1 \leq i < j \leq n} e(i,j)$ of these variables is reached, say for the pair (i_0, j_0), i.e. $\gamma_1 = e(i_0, j_0)$, then at time γ_1, we replace the clusters with labels i_0 and j_0 by a single cluster with mass $x_{i_0} + x_{j_0}$. Then the system keeps evolving with the same dynamics until it is reduced to a single cluster.

An additive coalescent $(X(t), t \geq 0)$ started from a finite number n of masses is a Markov chain in continuous times for which the sequence of jump times $\gamma_1 < \ldots < \gamma_{n-1}$ has a simple structure. Specifically, the increments between consecutive coalescence times, $\gamma_1, \gamma_2 - \gamma_1, \ldots, \gamma_{n-1} - \gamma_{n-2}$, are independent exponential variables with parameters $n - 1, n - 2, \ldots, 1$, and are independent of the state chain

*Laboratoire de Probabilités et Modèles Aléatoires and Institut universitaire de France, Université Paris 6, 175, rue du Chevaleret, F-75013 Paris, France. E-mail: jbe@ccr.jussieu.fr

$(X(\gamma_k), k = 0, \ldots, n-1)$. This elementary observation enables one to focus on the chain of states, and to derive simple relations with other random processes, as we shall now see.

Pitman [17] pointed at the following connection with random trees. Pick a tree $\tau^{(n)}$ at random, uniformly amongst the n^{n-2} trees on n labelled vertices. Enumerate its $n-1$ edges at random and decide that they are all closed at the initial time. At time $k = 1, \ldots, n-1$, open the k-th edge, and say that two vertices belong to the same sub-tree if all the edges of the path connecting those two vertices are open at time k. If we denote by $nY^{(n)}(k)$ the ranked sizes of sub-trees at time k, then the chain $Y^{(n)} = (Y^{(n)}(k), k = 0, \ldots n-1)$ has the same law as the chain of states $(X^{(n)}(\gamma_k), k = 0, \ldots n-1)$ of the additive coalescent started from the initial configuration $(1/n, \ldots, 1/n)$.

Forest derived at time 10 from a tree with 14 vertices

Second, we lift from [8] a different construction which is closely related to hashing with linear probing, cf. [10]. We view the initial configuration $x_1 \geq \ldots \geq x_n > 0$ as the ranked jumps of some bridge $b = (b(u), 0 \leq u \leq 1)$ with exchangeable increments. That is we introduce $U_1, \ldots, U_n$, n independent and uniformly distributed variables and define

$$b(u) = \sum_{i=1}^{n} x_i \left(\mathbf{1}_{\{u \geq U_i\}} - u \right), \qquad 0 \leq u \leq 1. \tag{1.1}$$

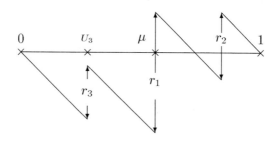

bridge b

Next, we consider a path transformation (see the picture below) that has been introduced by Takács [21] and used by Vervaat [23] to change a Brownian bridge into a normalized Brownian excursion. Specifically, we set

$$\epsilon(u) = b(u + \mu \, [\mathrm{mod} \, 1]) - b(\mu-), \qquad 0 \leq u \leq 1, \tag{1.2}$$

where μ stands for the location of the infimum of the bridge b.

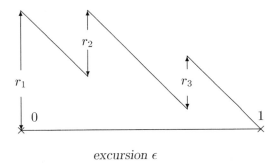

excursion ϵ

Finally, for every $t \geq 0$, call t-interval any maximal interval $[a, b[\subseteq [0, 1]$ on which

$$tu - \epsilon(u) < \max\left\{(tv - \epsilon(v))^+, 0 \leq v \leq u\right\}, \qquad \text{for all } u \in [a, b].$$

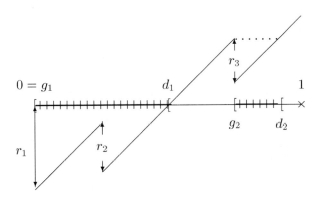

graph of $v \to tv - \epsilon(v)$ and t-intervals (hatched)

It is easy to see that the t-intervals get finer as t increases and tend to reduce to the jump times of ϵ when $t \to \infty$. Denote by $F(t)$ the ranked sequence of the sums of the jumps made by ϵ on each t-interval, and by $0 < \delta_1 < \ldots < \delta_{n-1}$ the jump times of $F(\cdot)$. Then the chain $(F(\delta_{n-k-1}), k = 0, \ldots n - 1)$ has the same law as the chain of states $(X(\gamma_k), k = 0, \ldots n - 1)$ of the additive coalescent started from the initial configuration $(x_1, \ldots, x_n)$.

2.   Standard and other eternal coalescents

Dealing with a finite number of clusters may be useful to give a simple description of the dynamics, however it is a rather inconvenient restriction in practice. In fact, it is much more natural to work with the infinite simplex

$$\mathcal{S}^{\downarrow} = \left\{ x = (x_1, x_2, \ldots) : x_i \geq 0 \text{ and } \sum_{i=1}^{\infty} x_i = 1 \right\}$$

endowed with the uniform distance. In this direction, Evans and Pitman [12] have shown that the semigroup of the additive coalescence enjoys the Feller property on $\mathcal{S}^{\downarrow}$. Approximating a general configuration $x \in \mathcal{S}^{\downarrow}$ by configurations with a finite number of clusters then enables us to view the additive coalescence as a Markovian evolution on $\mathcal{S}^{\downarrow}$. It is interesting in this setting to consider asymptotics when the coalescent starts with a large number of small clusters, which we shall now discuss.

Evans and Pitman [12] have first observed that the so-called *standard additive coalescent* $(X^{(\infty)}(t), -\infty < t < \infty)$ arises at the limit as $n \to \infty$ of the additive coalescent process $(X^{(n)}(t), -\frac{1}{2}\log n \leq t < \infty)$ started at time $-\frac{1}{2}\log n$ with n clusters, each with mass $1/n$. This limit theorem is perhaps better understood if we recall the connection with the uniform random tree $\tau^{(n)}$ on n vertices that was presented in the previous section. Indeed, if one puts a mass $1/n$ at each vertex and let each edge have length $n^{-1/2}$, then $\tau^{(n)}$ converges weakly as $n \to \infty$ to the so-called *continuum random tree* $\tau^{(\infty)}$; see Aldous [1]. More precisely, $\tau^{(\infty)}$ is a compact metric space endowed with a probability measure (arising as the limit of the masses on vertices) which is concentrated on the leaves of the tree, and a skeleton equipped with a length measure which is used to define the distance between leaves. This suggests that the standard additive coalescent might be constructed as follows: as time passes, one creates a continuum random forest by logging the continuum random tree along its skeleton and consider the ranked sequence of masses of the subtrees. This yields a fragmentation process, and the standard additive coalescent is finally obtained by time-reversing this fragmentation process. Aldous and Pitman [3] have made this construction rigorous; more precisely they showed that the tree $\tau^{(\infty)}$ has to be cut at points that appear according to a Poisson point process on the skeleton with intensity given by the length measure. This representation yields a number of explicit statistics for the standard additive coalescent. For instance, for every $t \in \mathbf{R}$, the distribution of $X_t^{(\infty)}$ is given by that of the ranked sequence $\xi_1 \geq \xi_2 \geq \ldots$ of the atoms of a Poisson measure on $]0, \infty[$ with intensity $e^{-t}(2\pi x^3)^{-1/2}dx$ and conditioned by $\xi_1 + \cdots = 1$.

The continuum random tree bears remarkable connections with the *Brownian excursion* (cf. for instance Le Gall [14]), and one naturally expects that the standard additive coalescent could also be constructed from the latter. This is indeed feasible (see [7] and also [10]) although its does not seem obvious to relate the following construction with that based on the continuum random tree. Specifically, let $(\epsilon(s), 0 \leq s \leq 1)$ be a Brownian excursion with unit duration, and for every $t \geq 0$, consider the random open set

$$G(t) = \left\{ s \in [0,1] : ts - \epsilon(s) < \max_{0 \leq u \leq s} (tu - \epsilon(u)) \right\}. \qquad (2.1)$$

Then $G(t)$ decreases as t increases, and if we denote by $F(t)$ the ranked sequence of the lengths of its intervals components (which of course are related to the so-called

t-intervals of the preceding section), then $(F(\mathrm{e}^{-t}), -\infty < t < \infty)$ is a standard additive coalescent.

More generally, Aldous and Pitman [4] have characterized all the processes that may arise as the limit of additive coalescents started with a large number of small clusters. They are referred to as *eternal* additive coalescents as these processes are indexed by times in $]-\infty, \infty[$. They can be constructed by the same procedure as in [3] after replacing the continuum random tree $\tau^{(\infty)}$ by a so-called inhomogeneous continuum random tree.

An alternative construction was proposed in [8] and [15]. Specifically, one may replace the standard Brownian excursion ϵ by that obtained by the Takács-Vervaat transformation (1.2) where $(b(s), 0 \leq s \leq 1)$ is now a bridge with exchangeable increments, no positive jumps and infinite variation (which arises as the limit of elementary bridges of the type (1.1), see Kallenberg [16]). The ranked sequence $F(t)$ of the lengths of the interval components of $G(t)$ defined by (2.1) then yields a fragmentation process, and by time-reversal, $(F(\mathrm{e}^{-t}), -\infty < t < \infty)$ is an eternal additive coalescent.

Roughly, this construction can be viewed as the limit of that presented in Section 1 when the additive coalescent starts from a finite number of clusters.

3. Eternal coagulation and certain Lévy processes

A long time before the notion of stochastic coalescence was introduced, Smoluchowski [20] considered a family of differential equations to model the evolution in the hydrodynamic limit of a particle system in which particles coagulate pairwise as time passes. It bears natural connections with the stochastic coalescence; we refer to the survey by Aldous [2] for detailed explanations, physical motivations, references ... Typically, we are given a symmetric kernel $K :]0, \infty[\times]0, \infty[\to [0, \infty[$ that specifies the rate at which two particles coagulate as a function of their masses. Here, we take of course $K(x, y) = x + y$. If we represent the density of particles with mass $\mathrm{d}x$ at time t by a measure $\mu_t(\mathrm{d}x)$ on $]0, \infty[$, then

$$\frac{\mathrm{d}}{\mathrm{d}t}\langle \mu_t, f \rangle = \frac{1}{2} \int_{]0,\infty[\times]0,\infty[} (f(x+y) - f(x) - f(y))\,(x+y)\mu_t(\mathrm{d}x)\mu_t(\mathrm{d}y)\,, \quad (3.1)$$

where f a test function and $\langle \mu_t, f \rangle = \int f(x)\mu_t(\mathrm{d}x)$. Motivated by the preceding section, we are interested in the eternal solutions of (3.1), in the sense that the time parameter t is real (possibly negative). It is proven in [9] that every eternal solution $(\mu_t)_{t \in \mathbf{R}}$ subject to the normalizing condition $\int x\mu_t(\mathrm{d}x) = 1$ (i.e. the total mass of the system is 1), can be constructed as follows.

First, define the function

$$\Psi_{\sigma^2,\Lambda}(q) = \frac{1}{2}\sigma^2 q^2 + \int_{]0,\infty[} \left(\mathrm{e}^{-qx} - 1 + qx\right)\Lambda(\mathrm{d}x)\,, \qquad q \geq 0, \quad (3.2)$$

where $\sigma^2 > 0$ and Λ is a measure on $]0, \infty[$ with $\int (x \wedge x^2)\Lambda(\mathrm{d}x) < \infty$. We further impose that either $\sigma^2 > 0$ or $\int x\Lambda(\mathrm{d}x) = \infty$. Next, let $\Phi(\cdot, s)$ be the inverse of the bijection $q \to \Psi_{\sigma^2,\Lambda}(sq) + q$. One can check that $\Phi(q, e^t)$ can be expressed in the form

$$\Phi(q, e^t) = \int_{]0,\infty[} (1 - e^{-qx}) \mu_t(dx), \qquad q \geq 0, \tag{3.3}$$

where $(\mu_t)_{t \in \mathbf{R}}$ is then an eternal solution to Smoluchowski's coagulation equation. For instance, when $\sigma^2 = 1$ and $\Lambda = 0$, ξ is a standard Brownian motion and we recover the well-known solution

$$\mu_t(dx) = \frac{e^{-t}}{\sqrt{2\pi x^3}} \exp\left(-\frac{xe^{-2t}}{2}\right) dx, \qquad t \in \mathbf{R}, \, x > 0.$$

This invites a probabilistic interpretation. Indeed, (3.2) is a special kind of *Lévy-Khintchine formula*; see section VII.1 in [5]. More precisely, there exists a *Lévy process with no positive jumps*, $\xi = (\xi_r, r \geq 0)$, such that

$$\mathbf{E}\left(\exp\left(q\xi_r\right)\right) = \exp\left(r\Psi_{\sigma^2,\Lambda}(q)\right), \qquad q \geq 0.$$

It is then well-known (e.g. Theorem VII.1 in [5]) that the first passage process

$$T_x^{(s)} := \inf\left\{r \geq 0 : s\xi_r + r > x\right\}, \qquad x \geq 0$$

is a subordinator with

$$\mathbf{E}\left(\exp\left(-qT_x^{(s)}\right)\right) = \exp\left(-x\Phi(q, s)\right), \qquad q, x \geq 0,$$

where the Laplace exponent $\Phi(\cdot, s)$ is the inverse bijection of $q \to \Psi_{\sigma^2,\Lambda}(sq) + q$. Thus (3.3) can be interpreted as the Lévy-Khintchine formula for $\Phi(\cdot, s)$, and we conclude that the eternal solution μ_t can be identified as the Lévy measure of the subordinator $T^{(s)}$ for $s = e^t$.

This probabilistic interpretation also points at a simple random model for aggregation of intervals. Indeed, the closed range $\mathcal{T}^{(s)} = \left\{T_x^{(s)}, x \geq 0\right\}^{cl}$ of $T^{(s)}$ induces a partition of $[0, \infty[$ into a family of random disjoint open intervals, namely the interval components of $G(s) = [0, \infty[\backslash \mathcal{T}^{(s)}$. We now make the key observation that

$$\mathcal{T}^{(s)} \subseteq \mathcal{T}^{(s')} \qquad \text{for } 0 < s' < s, \tag{3.4}$$

because an instant at which $r \to s\xi_r + r$ reaches a new maximum is always also an instant at which $r \to s'\xi_r + r$ reaches a new maximum. Roughly, (3.4) means that the random partitions get coarser as the parameter s increases; and therefore they induce a process in which intervals aggregate. The latter is closely related to a special class of eternal additive coalescents, and has been studied in [7, 18] in the Brownian case, and in [15] in the general case.

4.   Sticky particle systems

Sticky particle systems evolve according to the dynamics of completely inelastic collisions with conservation of mass and momentum, which are also known as the dynamics of ballistic aggregation. This means that the velocity of particles only changes in case of collision, and in that case, a heavier cluster merges at the location of the shock with mass and momentum given by the sum of the masses and momenta of the clusters involved. This has been proposed as a model for the formation of large scale structures in the universe; see the survey article [22]. We now have two quite different dynamics for clustering: on the one hand the ballistic aggregation which is deterministic, and on the other hand the additive coalescence which is random and may appear much more elementary as it does not take into account significant physical parameters such as distances between clusters and the relative velocities. Nonetheless, there is a striking connection between the two when randomness is introduced in the deterministic model, as we shall now see.

We henceforth focus on dimension one and assume that at the initial time, particles are infinitesimal (i.e. fluid) and uniformly distributed on the line. The evolution of the sticky particle system can then be completely analyzed in terms of the entropy solution to a single PDE, the transport equation

$$\partial_t u + u \partial_x u = 0 \,. \tag{4.1}$$

Here $u(x,t)$ represents the velocity of the particle located at x at time t, and the entropy condition imposes that for every fixed $t > 0$, the function $u(\cdot, t)$ has only discontinuities of the first kind and no positive jumps (the latter restriction accounts for the total inelasticity of collisions). Provided that the initial velocity $u(\cdot, 0)$ satisfies some very mild hypothesis on its rate of growth, there is a unique weak solution to the equation (4.1) which fulfills the entropy condition, and which can be given explicitly in terms of $u(\cdot, 0)$.

We assume that the initial velocities in the particle system are random, and more precisely

$$u(r,0) = 0 \text{ for } r < 0 \quad \text{and} \quad (u(r,0), r \geq 0) \overset{\mathcal{L}}{=} (\xi_r, r \geq 0) \,,$$

where ξ denotes the Lévy process with no positive jumps which was used in the preceding section.

Roughly, the dynamics of sticky particles are not only deterministic, but also induce a loss of information as time goes by, in the sense that the initial state of the system entirely determines the state at time $t > 0$, but cannot be completely recovered from the latter. In this direction, let us observe the system at some fixed time $t > 0$, i.e. we know the locations, masses and velocities of the clusters at this time. Let us pick a cluster located in $[0, \infty[$, using for this only the information available at time t (for instance, we may choose the heaviest cluster located at time t in $[0, 1]$). We shall work conditionally on the mass of this cluster, and for simplicity, let us assume it has unit mass. For every $r \in]0, t[$, denote by $M(r) = (m_1(r), m_2(r), \ldots)$ the ranked sequence of masses of clusters at time r which, by time t have aggregated to form the cluster we picked, so $M(r)$ can be viewed as a random variable with values in $\mathcal{S}^\downarrow$. Then the time-changed processes

$$M\left(t\left(1 - \frac{t}{t + e^s}\right)\right), \qquad -\infty < s < \infty$$

is an eternal additive coalescent. This was established in [6] in the case of Brownian initial velocity; and the recent developments on eternal additive coalescents made in [4, 8, 15] show that the arguments also applies for Lévy type initial velocities.

References

[1] D. J. Aldous. The continuum random tree III. *Ann. Probab.* **21** (1993), 248–289.

[2] D. J. Aldous. Deterministic and stochastic models for coalescence (aggregation, coagulation): a review of the mean-field theory for probabilists. *Bernoulli* **5** (1999), 3–48.

[3] D. J. Aldous and J. Pitman. The standard additive coalescent. *Ann. Probab.* **26** (1998), 1703–1726.

[4] D. J. Aldous and J. Pitman. Inhomogeneous continuum random trees and the entrance boundary of the additive coalescent. *Probab. Theory Related Fields* **118** (2000), 455–482.

[5] J. Bertoin. *Lévy processes.* Cambridge University Press, Cambridge, 1996.

[6] J. Bertoin. Clustering statistics for sticky particles with Brownian initial velocity. *J. Math. Pures Appl.* **79** (2000), 173–194.

[7] J. Bertoin. A fragmentation process connected to Brownian motion. *Probab. Theory Relat. Fields* **117** (2000), 289–301.

[8] J. Bertoin. Eternal additive coalescents and certain bridges with exchangeable increments. *Ann. Probab.* **29** (2001), 344–360.

[9] J. Bertoin. Eternal solutions to Smoluchowski's coagulation equation with additive kernel and their probabilistic interpretations. *Ann. Appl. Probab.* (2002) (To appear).

[10] P. Chassaing and G. Louchard. Phase transition for parking blocks, Brownian excursions and coalescence. *Random Structures Algorithms.* (2002) (To appear).

[11] R. L. Drake. A general mathematical survey of the coagulation equation. In G. M. Hidy and J. R. Brock (eds): *Topics in current aerosol research, part 2.* International Reviews in Aerosol Physics and Chemistry, 201–376, Pergammon, 1972.

[12] S. N. Evans and J. Pitman (1998). Construction of Markovian coalescents. *Ann. Inst. H. Poincaré, Probabilités Statistiques* **34**, 339–383.

[13] A. M. Golovin (1963). The solution of the coagulation equation for cloud droplets in a rising air current. *Izv. Geophys. Ser.* **5**, 482–487.

[14] J.-F. Le Gall (1993). The uniform random tree in a Brownian excursion. *Probab. Theory Relat. Fields* **96**, 369–384.

[15] G. Miermont (2001). Ordered additive coalescent and fragmentations associated to Lévy processes with no positive jumps. *Electr. J. Prob.* **6** paper 14, 1–33. http://math.washington.edu/ ejpecp/ejp6contents.html

[16] O. Kallenberg (1973). Canonical representations and convergence criteria for processes with interchangeable increments. *Z. Wahrscheinlichkeitstheorie verw. Gebiete* **27**, 23–36.

[17] J. Pitman (1999). Coalescent random forests. *J. Comb. Theory A.* **85**, 165–193.

[18] J. Schweinsberg (2001). Applications of the continuous-time ballot theorem to Brownian motion and related processes. *Stochastic Process. Appl.* **95**, 151–176.

[19] R. K. Sheth and J. Pitman (1997). Coagulation and branching processes models of gravitational clustering. *Mon. Not. R. Astron. Soc.*

[20] M. von Smoluchowski (1916). Drei Vorträge über Diffusion, Brownsche Bewegung und Koagulation von Kolloidteilchen. *Physik. Z.* **17**, 557–585.

[21] L. Takács (1966). *Combinatorial methods in the theory of stochastic processes.* Wiley, New York.

[22] M. Vergassola, B. Dubrulle, U. Frisch and A. Noullez (1994). Burgers' equation, devil's staircases and the mass distribution function for large-scale structures. *Astron. Astrophys.* **289**, 325–356.

[23] W. Vervaat (1979). A relation between Brownian bridge and Brownian excursion. *Ann. Probab.* **7**, 141–149.

Localization-Delocalization Phenomena
for Random Interfaces

Erwin Bolthausen[*]

Abstract

We consider d-dimensional random surface models which for $d = 1$ are
the standard (tied-down) random walks (considered as a random "string").
In higher dimensions, the one-dimensional (discrete) time parameter of the
random walk is replaced by the d-dimensional lattice $\mathbb{Z}^d$, or a finite subset
of it. The random surface is represented by real-valued random variables ϕ_i,
where $i \in \mathbb{Z}^d$. A class of natural generalizations of the standard random walk
are gradient models whose laws are (formally) expressed as

$$P\left(d\phi\right) = \frac{1}{Z}\exp\left[-\sum_{|i-j|=1} V\left(\phi_i - \phi_j\right)\right]\prod_i d\phi_i,$$

$V : \mathbb{R} \to \mathbb{R}^+$, convex, and with some growth conditions.

Such surfaces have been introduced in theoretical physics as (simplified)
models for random interfaces separating different phases. Of particular interest are localization-delocalization phenomena, for instance for a surface interacting with a wall by attracting or repulsive interactions, or both together.
Another example are so-called heteropolymers which have a noise-induced
interaction.

Recently, there had been developments of new probabilistic tools for such
problems. Among them are:

- Random walk representations of Helffer-Sjöstrand type,
- Multiscale analysis,
- Connections with random trapping problems and large deviations.

We give a survey of some of these developments.

2000 Mathematics Subject Classification: 60.

1. Introduction

[*]Institut für Mathematik, University of Zürich, Winterthurerstrasse 190, 8057 ZÜRICH,
Switzerland. E-mail: eb@amath.unizh.ch

Gradient models are an important class of random interfaces and random surfaces. In the mathematical physics literature they are often called "effective interface models". The (discrete) random surface is described by random variables $(\phi_x)_{x \in V}$, where V is $\mathbb{Z}^d$ or a subset of it. The ϕ_x itself are either $\mathbb{Z}$-valued or $\mathbb{R}$-valued. We will mainly concentrate on the latter situation which is easier in some respects. If V is a finite subset of $\mathbb{Z}^d$, the law P_V of $\phi = (\phi_x)_{x \in V}$ is described via a Hamiltonian

$$H_V(\phi) \stackrel{\text{def}}{=} \frac{1}{2} \sum_{x,y \in V} p(y-x) U(\phi_x - \phi_y) + \sum_{x \in V, y \notin V} p(y-x) U(\phi_x), \qquad (1)$$

where $U : \mathbb{R} \to \mathbb{R}^+$ is symmetric and convex, and $p : \mathbb{Z}^d \to [0,1]$ is a symmetric probability distribution on $\mathbb{Z}^d$. The above choice of the Hamiltonian corresponds to 0 boundary conditions. Of course, one can consider more general ones, where the second summand is replaced by $\sum_{x \in V, y \notin V} p(y-x) U(\phi_x - \psi_y)$, ψ being a configuration outside V. We will be mainly interested in the nearest neighbor case $p(x) = 1/2d$, for $|x| = 1$, and $p(x) = 0$ otherwise, but more general conditions can also be considered. We always assume that the matrix $\mathcal{Q} = (q_{ij})$ given by

$$q_{ij} \stackrel{\text{def}}{=} \sum_x x_i x_j p(x) \qquad (2)$$

is positive definite, and that p has exponentially decaying tails. Furthermore, the random walk $(\eta_t)_{t \in \mathbb{N}}$ with transition probabilities p is assumed to be irreducible. The Hamiltonian defines a probability distribution on $\mathbb{R}^V$ by

$$P_V(d\phi) \stackrel{\text{def}}{=} \frac{1}{Z_V} \exp\left[-H_V(\phi)\right] \prod_{x \in V} d\phi_x, \qquad (3)$$

where $d\phi_x$ denotes the Lebesgue measure. Z_V is the norming constant

$$Z_V \stackrel{\text{def}}{=} \int_{\mathbb{R}^V} \exp\left[-H_V(\phi)\right] \prod_{x \in V} d\phi_x. \qquad (4)$$

In the one-dimensional case $d = 1$, P_V is the law of a tied down random walk: Let ξ_i, $i \geq 1$, be i.i.d. random variables with the density const $\times e^{-U(x)}$. If $V = \{1, \ldots, n\}$, then P_V is the law of the sequence $\left(\sum_{j=1}^i \xi_j\right)_{1 \leq i \leq n}$, conditioned on $\sum_{j=1}^{n+1} \xi_j = 0$.

A special case is the *harmonic* one with $U(x) = x^2/2$. Then P_V is a Gaussian measure on $\mathbb{R}^V$ which is centered for 0-boundary conditions. We usually write P_V^{harm} in this case. The law is therefore given by its covariances

$$\gamma_V(x,y) \stackrel{\text{def}}{=} \int \phi_x \phi_y \, dP_V^{\text{harm}}.$$

These covariances have a random walk representation: If V is a finite set then

$$\gamma_V(x,y) = \mathbb{E}_x\left(\sum_{s=0}^{\tau_V - 1} 1_y(\eta_s)\right), \qquad (5)$$

where $(\eta_s)_{s \geq 0}$ under $\mathbb{P}_x$ is a discrete time random walk on $\mathbb{Z}^d$ starting at x and with transition probabilities $\mathbb{P}_x(\eta_1 = y) = p(y - x)$. τ_V is the first exit time from V. As a consequence of this representation one sees that the thermodynamic limit

$$P_\infty^{\mathrm{harm}} \overset{\text{def}}{=} \lim_{n \to \infty} P_{V_n}^{\mathrm{harm}}, \ V_n \overset{\text{def}}{=} \{-n, -n+1, \ldots, n\}^d \tag{6}$$

exists for $d \geq 3$. P_∞^{harm} is the centered Gaussian measure on $\mathbb{R}^{\mathbb{Z}^d}$ whose covariances are given by the Green's function of the random walk. It is important to notice that this random field has slowly decaying correlations:

$$\gamma_\infty(x, y) \approx \frac{\mathrm{const}}{|x - y|^{d-2}}, \ |x - y| \to \infty.$$

For $d = 2$, the thermodynamic limit does not exist, and in fact

$$E_{V_n}^{\mathrm{harm}}(\phi_0^2) \approx \mathrm{const} \times \log n, \ n \to \infty.$$

For $d = 1$, the variance grows of course like n in the bulk. The harmonic surface is therefore localized for $d \geq 3$, but not for $d = 1, 2$.

Many of these properties carry over to non-harmonic cases with a convex and symmetric interaction function U in (1). Of particular importance is that there is a generalization of the representation (5), the Helffer-Sjöstrand representation, see [26]. The random walk (η_s) has to be replaced by a random walk in a dynamically changing random environment. Using this representation, many of the results for the harmonic case can be generalized to the case of a convex U, although often not in a quantitatively as precise form as in the harmonic case. For a probabilistic description of the Helffer-Sjöstrand representation, see [20].

The main topic of this paper are effects arising from interactions of the random surface (ϕ_x) with a "wall". The simplest case of such a wall is the configuration $\phi \equiv 0$. There are many type of interactions which had been considered in the literature, both in physics and in mathematics. The simplest one is a local attraction of the surface to this wall. It turns out that an arbitrary weak attraction localizes the random field in a strong sense, and in all dimensions. This will be discussed in a precise way in Section 2.. Interesting localization-delocalization phenomena may occur when mixed attractive and repulsive interactions are present, with phase transitions depending on the parameters regulating the strength of the interactions. Naturally, these phenomena are best understood for the one-dimensional case. A simple example is the following one, which is discussed in details in [25]: Let $\phi_0 = 0, \phi_1, \ldots, \phi_{2n-1}, \phi_{2n} = 0$ be a discrete time $\mathbb{Z}$-valued, and tied-down, simple random walk, i.e. P_n is simply the uniform distribution on all such paths which satisfy $|\phi_x - \phi_{x-1}| = 1$. Introducing an arbitrary pinning to the wall in the form

$$\hat{P}_{n,\beta}(\phi) = \frac{1}{\hat{Z}_{n,\beta}} \exp\left[\beta \sum_{x=1}^{2n-1} 1_0(\phi_x)\right] P_n(\phi), \ \beta > 0$$

strongly localizes the "random string", i.e. $\sup_{n,x} \hat{E}_{n,\beta}(\phi_x^2) < \infty$ holds for all $\beta > 0$. Furthermore, the correlations $\hat{E}_{n,\beta}(\phi_x \phi_y)$ are exponentially decaying in $|x - y|$, uniformly in n. These facts are easily checked.

On the other hand, if the string is confined to be on one side of the wall, the situation is completely different. Let $\Omega_{2n}^+ \overset{\text{def}}{=} \{\phi : \phi_x \geq 0,\ 1 \leq x \leq 2n - 1\}$, and $\hat{P}_{n,\beta}^+ (\cdot) \overset{\text{def}}{=} \hat{P}_{n,\beta} (\cdot \mid \Omega_{2n}^+)$. Then there is a critical $\beta_c > 0$ such that the above localization property holds for $\beta > \beta_c$, but not for $\beta < \beta_c$, where the path measure converges, after Brownian rescaling, to the Brownian excursion. For a proof of this so called "wetting transition", see [25]. More precise information has been obtained recently in this one-dimensional situation in [28].

There are similar phase transitions for more complicated models. Some of them will be discussed in Section 3. and Section 4.. We begin in the next section by discussing the pinning effect alone mainly in the difficult two-dimensional case.

2.   Pinning of two-dimensional gradient fields

We consider now a gradient field (3), but we modify it by introducing an attractive local pinning to the wall $\{\phi \equiv 0\}$. This is often done by modifying the Hamiltonian in the following way: Let $\psi : \mathbb{R} \to \mathbb{R}^-$ be symmetric and with compact support. Then we put

$$H_{V,\psi} (\phi) = H_V (\phi) + \sum_{x \in V} \psi (\phi_x).$$ (7)

Evidently, the corresponding finite volume Gibbs measure favours surfaces which have the tendency to stick close to the wall. It should be emphasized that this is a much weaker attraction than in a so-called massive field, where one takes ψ to be convex, for instance $\psi (x) = x^2$. A formally slightly easier model can be obtained by not changing the Hamiltonian, but replacing the Lebesgue measure as the reference measure by a mixture of the Lebesgue measure and a Dirac measure at 0. This corresponds to the following probability measure on $\mathbb{R}^V$:

$$\hat{P}_{V,\varepsilon} (d\phi) \overset{\text{def}}{=} \frac{1}{\hat{Z}_{V,\varepsilon}} \exp\left[-H_V (\phi)\right] \prod_{x \in V} (d\phi_x + \varepsilon \delta_0 (d\phi_x)),\ \varepsilon > 0.$$ (8)

This measure can be obtained from measures defined by the Hamiltonian (7) via an appropriate limiting procedure. The nice feature of (8) is that $\hat{P}_{V,\varepsilon}$ can trivially be expanded into a mixture of "free" measures: We just have to expand out the product:

$$\hat{P}_{V,\varepsilon} (d\phi) = \sum_{A \subset V} \varepsilon^{|V \setminus A|} \frac{Z_A}{\hat{Z}_{V,\varepsilon}} \frac{1}{Z_A} \exp\left[-H_V (\phi)\right] \prod_{x \in A} d\phi_x \prod_{x \in V \setminus A} \delta_0 (d\phi_x)$$ (9)

$$= \sum_{A \subset V} \varepsilon^{|V \setminus A|} \frac{Z_A}{\hat{Z}_{V,\varepsilon}} P_A (d\phi),$$

where P_A is the measure defined by (3), extended by 0 outside A. Remark that

$$\nu_{V,\varepsilon} (A) \overset{\text{def}}{=} \varepsilon^{|V \setminus A|} \frac{Z_A}{\hat{Z}_{V,\varepsilon}}$$

defines a probability distribution on the set of subsets of V. Therefore, we have represented $\hat{P}_{V,\varepsilon}$ as a mixture of free measures P_A. It should be remarked that similar but technically more involved expansions are possible also in the case of the Hamiltonian (7). The case of $\psi(x) = -a1_{[-b,b]}(x)$ is discussed in [10]. Probably, more general cases could be handled with the help of the Brydges-Fröhlich-Spencer random walk representation (see [15]), but the results presented here have not been derived in this more general case. For the sake of simplicity, we stick here to the δ-pinning case (8).

The above representation easily leads to a representation of the covariances of the pinned field. This is particularly simple in the harmonic case $U(x) = x^2/2$, where one gets

$$\int \phi_x \phi_y \hat{P}_{V,\varepsilon}^{\mathrm{harm}}(d\phi) = \sum_{A \subset V} \nu_{V,\varepsilon}(A) \, \mathbb{E}_x \left(\sum_{s=0}^{\tau_A-1} 1_y(\eta_s) \right).$$

The problem is therefore reduced to a problem of a random walk among random traps: The distribution $\nu_{V,\varepsilon}$ defines a random trapping configuration, let's denote it by $\mathcal{A}$, i.e. $P_{\mathrm{trap}}\left(\mathcal{A} = V^c \cup (V \backslash A)\right) \overset{\mathrm{def}}{=} \nu_{V,\varepsilon}(A)$, and the covariances of our pinned measure are given in terms of the discrete Green's function among these random traps which are killing the random walk when it enters one of these traps. A difficult point is a precise analysis of the distribution of $\mathcal{A}$, and a crucial step is a comparison with Bernoulli measures. The two-dimensional case is the most difficult one. In three and more dimensions, a comparison of the distribution of $\mathcal{A}$ with a Bernoulli measure is quite easy.

It turns out that the pinning localizes the field in a strong sense. First of all, the variance of the variables stay bounded as $V \uparrow \mathbb{Z}^d$. Secondly, there is exponential decay of the covariances, uniformly in V. Results of this type have a long history. For $d \geq 3$, and for the harmonic case with pinning of the type (7), the localization has been obtained in [15]. In [24], boundedness of the absolute first moment has been proved for $d = 2$, but no exponential decay of the correlations. The first proof of exponential decay of correlations in the two-dimensional case has been obtained in [7] for the harmonic case. One drawback of the method used there was that it uses reflection positivity, which holds only under restrictive assumptions on p. Also, periodic boundary conditions are required, and so the results are not directly valid for the 0-boundary case. A satisfactory approach had then been obtained in [21] and [27]. The quantitatively precise results presented here are from [10], where the critical exponents for the depinning transition $\varepsilon \to 0$ have been derived, including the correct log-corrections for $d = 2$.

We define the mass $m_\varepsilon(x)$, $x \in S^{d-1}$, by

$$m_\varepsilon(x) \overset{\mathrm{def}}{=} - \lim_{k \to \infty} \frac{1}{k} \log \lim_{V \uparrow \mathbb{Z}^d} \hat{E}_{V,\varepsilon}\left(\phi_0 \phi_{[kx]}\right).$$

The most precise results we have are for the harmonic case:

Theorem 1   *a) If $d = 2$, then for small enough ε :*

$$\left| \lim_{V \uparrow \mathbb{Z}^d} \hat{E}_{V,\varepsilon}^{\mathrm{harm}}\left(\phi_0^2\right) - \frac{|\log \varepsilon|}{2\pi\sqrt{\det Q}} \right| \leq \mathrm{const} \times \log|\log \varepsilon|$$

b) *If $d = 2$, then for all $x \in S^{d-1}$ and small enough ε :*

$$\text{const} \times \frac{\sqrt{\varepsilon}}{|\log \varepsilon|^{3/4}} \leq m_\varepsilon^{\text{harm}}(x) \leq \text{const} \times \frac{\sqrt{\varepsilon}}{|\log \varepsilon|^{3/4}}.$$

c) *If $d \geq 3$, then for all $x \in S^{d-1}$ and small enough ε :*

$$\text{const} \times \sqrt{\varepsilon} \leq m_\varepsilon^{\text{harm}}(x) \leq \text{const} \times \sqrt{\varepsilon}.$$

The constants depend on the dimension d and p only.

The proof of the results depends on a comparison of the laws of the trapping configurations with Bernoulli measures. This is particularly delicate in $d = 2$. The following result is the key comparison of the distribution of traps with Bernoulli measures. We formulate it only in the harmonic case. Somewhat weaker results are proved in [10] also for the anharmonic situation.

Theorem 2 *Let $\mathcal{A}_{\varepsilon,V}$ be a random subset of V with $P(\mathcal{A}_{\varepsilon,V} = V \backslash A) = \nu_{V,\varepsilon}(A)$. Assume $d = 2$, and $U(x) = x^2/2$.*

a) *Let $\alpha > 0$. There exists $\varepsilon_0 > 0$ and $C(\alpha) > 0$ such that for $\varepsilon \leq \varepsilon_0$, any finite set $V \subset \mathbb{Z}^d$, and any $B \subset V$ with $\text{dist}(B, V^c) > \varepsilon^{-\alpha}$, one has the estimate*

$$P(\mathcal{A}_{\varepsilon,V} \cap B = \emptyset) \geq \left(1 - C(\alpha) \frac{\varepsilon}{\sqrt{|\log \varepsilon|}}\right)^{|B|}.$$

b) *There exist $C > 0$ and $\varepsilon_0 > 0$ such that for $\varepsilon \leq \varepsilon_0$, any finite set $V \subset \mathbb{Z}^d$, and all $B \subset V$, one has*

$$P(\mathcal{A}_{\varepsilon,V} \cap B = \emptyset) \leq \left(1 - C \frac{\varepsilon}{\sqrt{|\log \varepsilon|}}\right)^{|B|}.$$

The case of dimension $d \geq 3$ is simpler and somewhat better estimates can be obtained. With the help of the above theorem and the random walk representation (5), a comparison can be made, relating the quantities in Theorem 1 to random trapping problems for Bernoulli traps. For instance, when investigating the variance, we get

$$\hat{E}_{V,\varepsilon}^{\text{harm}}(\phi_0^2) = E_{\text{traps}} \mathbb{E}_0 \left(\sum_{t=0}^{\tau-1} 1_0(\eta_t)\right) = E_{\text{traps}} \sum_{t=0}^{\infty} p_t(0) \mathbb{P}_{0,0}^{(t)}(A \cap \eta_{[0,t]} = \emptyset),$$

where $\mathcal{A}$ is the random set of points with traps, as introduced above, τ is the first entrance time into this trapping set and $\eta_{[0,t]}$ is the set of points visited by the random walk between time 0 and t. $\mathbb{P}_{0,0}^{(t)}$ refers to a random walk bridge from 0 to 0 in time t, and $p_t(x)$, $x \in \mathbb{Z}^d$ are the transition probabilities of the random walk. With the help of Theorem 2, the right hand side can be estimated in terms of a

Bernoulli trapping problem. If the traps are Bernoulli on $\mathbb{Z}^d$, with probability ρ that a trap is present at a given site, then (in the $V \uparrow \mathbb{Z}^d$ limit)

$$E_{\text{traps}} \sum_{t=0}^{\infty} p_t (0) \, \mathbb{P}_{0,0}^{(t)} \left(A \cap \eta_{[0,t]} = \emptyset \right) = \sum_{t=0}^{\infty} p_t (0) \, \mathbb{E}_{0,0}^{(t)} \exp \left[\left| \eta_{[0,t]} \right| \log (1 - \rho) \right].$$

There are classical results about the right hand side, due to Donsker and Varadhan [23], Sznitman [30], and most recently in [1] investigating such questions. The classical Donsker-Varadhan result is not sharp enough to prove the results of Theorem 1, but a modification of the arguments in [1] is exactly what is needed. The following result is a discrete but somewhat weaker version of one of the main results in [1].

Proposition 3 *Assume $d = 2$. There exists a function $\mathbb{R}^+ \ni a \to r(a) \in \mathbb{R}^+$, satisfying $\lim_{a \to 0} r(a) = \infty$, such that*

$$\mathbb{P}_{0,0}^{(t)} \left(\left| \eta_{[0,t]} \right| \leq a \frac{t}{\log t} \right) \leq t^{-r(a)}$$

for large enough t.

(In [1], a variational formula for $r(a)$ is given, in the continuous Wiener sausage case.) This proposition and Theorem 2 lead to the appropriate variance estimates in Theorem 1 a).

For the anharmonic case, the results are less precise, but we still get the correct leading order dependence of the variance on the pinning parameter ε. Assume that there is a $C > 0$ such that

$$1/C \leq U''(x) \leq C, \ \forall x.$$

Under this condition we have the following result:

Theorem 4 *Assume $d = 2$. There exists a constant D, depending on p, such that*

$$\frac{1}{D} \left| \log \varepsilon \right| \leq \sup_V \hat{E}_{V,\varepsilon} \left(\phi_0^2 \right) \leq D \left| \log \varepsilon \right|.$$

The upper bound is in [21] and [27], and the lower bound is in [10].

3.   Entropic repulsion and the wetting transition

In view of the example of Fisher discussed shortly in Section 1. it is natural to ask similar question for higher-dimensional interfaces. The first task is to investigate the effect of a wall on the random surface without the presence of a pinning effect. There are different ways to take the presence of a wall into account. We have mainly worked with a "hard wall", i.e. where the measure is simply conditioned on the event $\Omega_V^+ \stackrel{\text{def}}{=} \{ \phi : \phi_x \geq 0 \ \forall x \in V \}$. There are other possibilities, for instance by introducing a "soft wall". This means that the Hamiltonian (1) is changed by adding $\sum_{x \in V} f(\phi_x)$, where $f : \mathbb{R} \to \mathbb{R}$ satisfies $\lim_{x \to \infty} f(x) = 0$, $\lim_{x \to -\infty} f(x) = \infty$. We will only work with a hard wall here, and consider the conditional law for the random field $P\left(\cdot | \Omega_V^+\right)$.

What is the effect of the presence of the wall on the surface? The crucial point is that the surface has local fluctuations, which push the interface away from the wall. On the other hand, the long-range correlations give the surface a certain global stiffness. In order to understand what is going on, consider first the case where there are no such long-range correlations, in the extreme case, where the ϕ_x are just i.i.d. random variables. In that case, evidently nothing interesting happens: The variables are individually conditioned to stay positive. In particular, $E\left(\phi_x | \Omega_V^+\right)$ stays bounded for $V \uparrow \mathbb{Z}^d$. This picture remains the same for fields with rapidly decaying correlations. However, gradient fields behave differently, and so do interfaces in more realistic statistical physics models. As the surface has some global stiffness, the energetically best way for the surface to leave some room for the local fluctuations is to move away from the wall in some global sense. This effect is called "entropic repulsion" and is well known in the physics literature.

The first mathematically rigorous treatment of entropic repulsion appeared in the paper by Bricmont, Fröhlich and El Mellouki [14]. In a series of papers [4], [17], [18], and [6], sharp quantitative results have been derived, the most accurate ones for the harmonic case.

In most of these and related questions, the two-dimensional case is the most difficult but also the most interesting one. In fact, interfaces in the "real world" are mostly two-dimensional.

We first present the results for $d \geq 3$. For gradient non-Gaussian models, some results in the same spirit have been obtained in [18], but they are not as precise as the ones obtained in the Gaussian model. The case where one starts with the field P_∞ (which exists for $d \geq 3$) is somewhat easier than the field on the finite box V_n with zero boundary condition. In the latter case, there are some boundary effects complicating the situation without changing it substantially. This is investigated in [17]. Despite the fact that we consider P_∞, we consider the wall only on a finite box, i.e., we consider $P_\infty\left(\cdot | \Omega_{V_n}^+\right)$, and we are interested in what happens as $n \to \infty$. We usually write Ω_n^+ for $\Omega_{V_n}^+$. Our first task is to get information about $P_\infty\left(\Omega_n^+\right)$. The following results are proved only for the case of nearest neighbor interactions, i.e. when $p(x) = 1/2d$ for $|x| = 1$.

Theorem 5 *Let* $d \geq 3$*. Then*

a)

$$P_\infty^{\mathrm{harm}}\left(\Omega_n^+\right) = \exp\left[-2\Gamma(0)\mathrm{cap}\,(V)\,n^{d-2}\log n\,(1 + o(1))\right],$$

where $V = [-1, 1]^d$*,* $\mathrm{cap}(A)$ *denotes the Newtonian capacity of* A

$$\mathrm{cap}(A) \overset{\mathrm{def}}{=} \inf\left\{\|\nabla f\|^2 : f \geq 1_A\right\},$$

and $\Gamma(0) = \gamma_\infty(0, 0)$ *is the variance of* ϕ_0 *under* P_∞^{harm}*.*

b)

$$E_\infty^{\text{harm}}\left(\phi_0|\Omega_n^+\right) = 2\sqrt{\Gamma(0)\log n}(1 + o(1)).$$

c)

$$\mathcal{L}_{P_\infty^{\text{harm}}(\cdot|\Omega_n^+)}\left(\left(\phi_x - E_\infty(\phi_x|\Omega_n^+)\right)_{x\in\mathbb{Z}^d}\right) \to P_\infty^{\text{harm}} \text{ weakly,}$$

as $n \to \infty$, where $\mathcal{L}_{P_\infty^{\text{harm}}(\cdot|\Omega_n^+)}$ denotes the law of the field under the conditioned measure.

Part b) gives the exact rate at which the random surface escapes to infinity, while part c) states that the effect of the entropic repulsion essentially is only this shifting: after subtraction of the shift by the expectation, the surface looks as it does without the wall. However, there is some subtlety in this picture. From the theorem in particular part c), one might conclude that $\lim_{n\to\infty} P_\infty^{\text{harm}} \theta^{-1}_{2\sqrt{\Gamma(0)\log n}}(\Omega_n^+) = 1$, where $\theta_a : \mathbb{R}^{\mathbb{Z}^d} \to \mathbb{R}^{\mathbb{Z}^d}$ is the shift mapping $\theta_a\left((\phi_x)_{x\in\mathbb{Z}^d}\right) = (\phi_x + a)_{x\in\mathbb{Z}^d}$. But this is not the case. In fact $P_\infty^{\text{harm}} \theta^{-1}_{2\sqrt{\Gamma(0)\log n}}(\Omega_n^+)$ converges rapidly to 0. As part c) states only the weak convergence, this is no contradiction. Parts a) and b) of Theorem 5 had been proved in [4], part c) in [18].

We come now to the two-dimensional case which is considerably more delicate than the higher dimensional one. We again consider only the harmonic case. We write P_n for P_{V_n}. If the lattice is two-dimensional, a thermodynamic limit of the measures P_n does not exist as the variance blows up. $P_n^{\text{harm}}(\Omega_n^+)$ is of order $\exp[-cn]$, as has been shown in [17]. As remarked above, this is mainly a boundary effect and is not really relevant for the phenomenon of the entropic repulsion. To copy somehow the procedure of the case $d \geq 3$, we consider a subset $D \subset V = [-1,1]^2$ which has a nice boundary and a positive distance from the boundary of V. To be specific, just think of taking $D \stackrel{\text{def}}{=} \lambda V$ for some $\lambda < 1$. Then let $D_n \stackrel{\text{def}}{=} nD \cap \mathbb{Z}^2$ and $\Omega_{D_n}^+ \stackrel{\text{def}}{=} \{\phi_x \geq 0,\ x \in D_n\}$. In contrast to $P_n(\Omega_n^+)$, $P_n\left(\Omega_{D_n}^+\right)$ decays much slower, but still faster than any polynomial rate. In [6] we proved the following result:

Theorem 6 *Assume $d = 2$ and let $g \stackrel{\text{def}}{=} 1/2\pi$.*

a)

$$\lim_{n\to\infty} \frac{1}{(\log n)^2} \log P_n^{\text{harm}}(\Omega_{D_n}^+) = -2g\text{cap}_V(D),$$

where $\text{cap}_V(D)$ is the relative capacity of D with respect to V:

$$\text{cap}_V(D) \stackrel{\text{def}}{=} \inf\left\{\|\nabla f\|_2^2 : f \in H_0^1(V),\ f \geq 1 \text{ on } D\right\}.$$

Here, $H_0^1(V)$ is the Sobolev space of (weakly) differentiable functions f with square integrable gradient and $f|_{\partial V} = 0$.

b) For any $\varepsilon > 0$

$$\lim_{n\to\infty}\sup_{x\in D_n} P_n^{\text{harm}}\left(|\phi_x - 2\sqrt{g}\log n| \geq \varepsilon\log n|\Omega_{D_n}^+\right) = 0.$$

This corresponds to parts a) and b) of Theorem 5. Part c) does not make sense here as P_∞^{harm} does not exist. Remark that under the unconditional law P_n^{harm}, $|\phi_x|$ is typically of order $\sqrt{\log n}$ in the bulk.

Roughly speaking, the delicacy in the two-dimensional case is coming from the fact that the relevant "spikes" responsible for the repulsion are thicker than in the higher dimensional case, where essentially just very local spikes are responsible for the effect. This makes necessary to apply a multiscale analysis separating the scales of the spikes.

It is well-known that the two-dimensional harmonic field has much similarity with a hierarchical field defined in the following way: We call a sequence $\alpha = \alpha_1 \alpha_2 \ldots \alpha_m$, $\alpha_i \in \{0, 1\}$ a *binary string*. $\ell(\alpha) = m$ is the length. $\emptyset$ is the empty string of length 0. We write T for the set of all such strings of finite length, and $T_m \subset T$ for the set of strings of length m. If $\alpha \in T_m$, $0 \le k \le m$, we write $[\alpha]_k$ for the truncation at level k :

$$[\alpha_1 \alpha_2 \ldots \alpha_m]_k \stackrel{\text{def}}{=} \alpha_1 \alpha_2 \ldots \alpha_k.$$

If $\alpha, \beta \in T_m$ we define the hierarchical distance

$$d_H(\alpha, \beta) \stackrel{\text{def}}{=} m - \max \{k \le m : [\alpha]_k = [\beta]_k\}.$$

We consider the following family $(X_\alpha)_{\alpha \in T_m}$ of centered Gaussian random variables by

$$\text{cov}(X_\alpha, X_\beta) = \gamma(m - d_H(\alpha, \beta)), \tag{10}$$

with a parameter $\gamma > 0$. We argue now that there is much similarity between the two dimension harmonic field $(\phi_x)_{x \in D_n}$ and the field $(X_\alpha)_{\alpha \in T_m}$. To see this, we first match the number of variables, i.e. put $2^m = |D_n|$. As $|D_n|$ is of order n^2, this just means that $m \sim 2 \log n / \log 2$. Then we should also match the variances, i.e. take $\gamma = g/2 \log 2$. For the free field (ϕ_x), it is known that $\text{cov}(\phi_x, \phi_y)$ behaves like $g(\log n) / \log |x - y|$, if x, y are not too close to the boundary. This follows from the random walk representation. Comparing this with (10), we see that for any number $s \in (0, g)$

$$\# \{y \in D_n : \text{cov}(\phi_x, \phi_y) \le s \log n\} \sim \# \{\beta \in T_m : \text{cov}(X_\alpha, X_\beta) \le s \log n\} \tag{11}$$

in first order, for any $x \in D_n$, $\alpha \in T_m$. Therefore, the two fields have roughly the same covariance structure. The hierarchical field is much simpler and is very well investigated (see e.g. [2], [12], [22]), and the entropic repulsion is much easier to discuss than for the harmonic field. The approach to prove Theorem 6 consist in introducing a hierarchical structure in the (ϕ_x)-field with the help of successive conditionings on a hierarchy of scales, and then adapt the methods from the purely hierarchical case.

We come now back to the question of a wetting transition, as discussed in the one-dimensional case by Michel Fisher [25]. One is interested in the behavior of $\hat{P}_{V,\varepsilon}(\cdot \mid \Omega_V^+)$ for large V, where $\hat{P}_{V,\varepsilon}$ is the pinned measure introduced in (8). Unfortunately, we are not able to describe this path measure. The simplest way to discuss the wetting transition is in terms of free energy considerations. For this we expand $\hat{P}_{V,\varepsilon}\left(\Omega_V^+\right)$ (see (9)):

$$\hat{P}_{V,\varepsilon}\left(\Omega_V^+\right) = \sum_{A\subset V} \varepsilon^{|V\setminus A|} \frac{Z_A}{\hat{Z}_{V,\varepsilon}} P_A\left(\Omega_V^+\right).$$

It is plausible, that pinning "wins" over entropic repulsion, if this sum is much larger than the contribution to the sum coming from subsets A having essentially no pinning sites, i.e. $A \approx V$. It is therefore natural to consider the quantity

$$p_+\left(\varepsilon\right) \overset{\text{def}}{=} \lim_{V\uparrow\mathbb{Z}^d} \frac{1}{|V|} \log \frac{\hat{Z}_{V,\varepsilon}\hat{P}_{V,\varepsilon}\left(\Omega_V^+\right)}{Z_V P_V\left(\Omega_V^+\right)} = \lim_{V\uparrow\mathbb{Z}^d} \frac{1}{|V|} \log \frac{\hat{Z}_{V,\varepsilon}\hat{P}_{V,\varepsilon}\left(\Omega_V^+\right)}{Z_V}.$$

The limit is easily seen to exist. It is also not difficult to see that $p_+\left(\varepsilon\right) > 0$ for large enough $\varepsilon > 0$, and in any dimension (see [9]). Similar to the discrete random walk case in [25], the Gaussian model has a wetting transition, too, for $d = 1$: There exists an $\varepsilon_{\text{crit}} > 0$, such that $p_+\left(\varepsilon\right) = 0$ for $\varepsilon < \varepsilon_{\text{crit}}$. This is easy to see for $d = 1$. For the harmonic model, there is remarkably no such transition for $d \geq 3$, but for $d = 2$ there is a wetting transition.

Theorem 7 [5] *For $d \geq 3$, $p_+^{\text{harm}}\left(\varepsilon\right) > 0$ for all $\varepsilon > 0$.*

Theorem 8 [16] *For $d = 2$, there exists $\varepsilon_{\text{crit}}^{\text{harm}} > 0$, such that $p_+^{\text{harm}}\left(\varepsilon\right) = 0$ for $\varepsilon < \varepsilon_{\text{crit}}^{\text{harm}}$.*

Remarkably, too, Caputo and Velenik have proved that such a wetting transition exists for $d \geq 3$ for some non-harmonic models, e.g. for $U\left(x\right) = |x|$.

There are many open questions concerning this wetting transition, which is very poorly understood (mathematically). For instance, the methods discussed in Section 2. do not apply, and we are not able to prove that in the pinning dominated region $p_+\left(\varepsilon\right) > 0$, the measure is pathwise localized, i.e. that

$$\sup_V \sup_{x\in V} \hat{P}_{V,\varepsilon}\left(\phi_x^2 \mid \Omega_V^+\right) < \infty,$$

which certainly should be expected. To discuss the nature of the transition (first order or second order?) is probably even much more delicate.

4. Localization-delocalization transitions for one-dimensional copolymers

We stick here to the standard simple random walk case where P_n simply is the uniform distribution on the set of paths $\phi_0 = 0, \phi_1, \ldots, \phi_n \in \mathbb{Z}$, satisfying $|\phi_i - \phi_{i-1}| = 1$, $1 \leq i \leq n$. There is not much difference when considering more general random walks, or the tied-down situation, but most of the published results are for the simple random walk. An interesting case of a mixed attractive-repulsive interaction is given in the following way. Regard the above random walk as a (very simplified) model of a polymer chain imbedded in two liquids, say water and oil. The water is at the bottom, say at points $(i, j) \in \mathbb{N} \times \mathbb{Z}$, $j \leq 0$, and the oil above at $j > 0$. The polymer chain is attached with one end at the interface between the two liquids, and interacts with them in the following way: To each "node" (i, ϕ_i) of the polymer chain, we attach a value $\sigma_i \in \mathbb{R}$ which is < 0 if the node is water-repellent, and > 0 if it is oil-repellent. The overall effect is described by the Hamiltonian

$$H_{n,\sigma}(\phi) \stackrel{\text{def}}{=} \sum_{i=1}^{n} \sigma_i \operatorname{sign}(\phi_i),$$

where we put $\operatorname{sign}(0) \stackrel{\text{def}}{=} 0$. With this Hamiltonian, we define the σ-dependant path measure

$$P_{n,\beta,\sigma}(\phi) \stackrel{\text{def}}{=} \frac{1}{Z_{n,\beta,\sigma}} \exp\left[-\beta H_{n,\sigma}(\phi)\right],$$

where $\beta > 0$ is a parameter governing the strength of the interaction. We assume that the σ_i change sign either in a periodic way or randomly. There may be two competing effects. The polymer chain may try to follow the preferences described by the σ's as closely as possible in which case the path evidently would have to stay close to the oil-water-interface and gets localized. On the other hand, this strategy may be entropically too costly, in particular if there is no balance between oil-repellence and water-repellence. We will always assume that

$$h \stackrel{\text{def}}{=} \lim_{n \to \infty} \frac{1}{n} \sum_{j=1}^{n} \sigma_j,$$

exists, and we assume it to be ≥ 0. (The case $h \leq 0$ can be treated symmetrically). It turns out that typically, there is a non-trivial curve in the (β, h)-plane which separates the localized from the delocalized region. This phase separation line is quite model dependent, but the behavior near $(0, 0)$ appears to be much more universal but it is completely different depending whether the σ_i are random or periodic.

The first rigorous results in this direction had been obtained by Sinai [29] who proved the following result in the balanced case (i.e. $h = 0$). Let $\mathbb{P}$ be the symmetric Bernoulli-measure on $\{-1, 1\}^{\mathbb{N}}$.

Theorem 9 *Let $\beta > 0$. There exist constants C and $\rho(\beta) > 0$, and for $\mathbb{P}$-almost all $\sigma = (\sigma_i)_{i \geq 1}$, there exists a sequence $(R_n(\sigma))_{n \in \mathbb{N}}$ of natural numbers such that*

$$P_{n,\beta,\sigma}(|\phi_n| \geq r) \leq C \exp\left[-\rho(\beta) r\right]$$

for $r \geq R_n(\sigma)$. The sequence (R_n) is stochastically bounded, i.e.

$$\lim_{m \to \infty} \sup_{n} \mathbb{P}(R_n \geq m) = 0.$$

In a paper with Frank den Hollander [8] we proved that there is a localization-delocalization transition in the random non-balanced case. This transition is discussed in this paper in terms of the free energy. To describe the results, let

$\sigma_i = \pm 1 + h$ with probabilities $1/2$, and independently, $h \geq 0$. One strategy of the path could be just to stay on the negative side all the time, i.e. $\phi_i < 0$ for all $i \leq n$. This leads to a trivial lower bound of the free energy

$$f(\beta, h) \overset{\text{def}}{=} \lim_{n \to \infty} \frac{1}{n} \log Z_{n,\beta,\sigma}$$

which is easily seen to exist, and is non-random:

$$Z_{n,\beta,\sigma} \geq E_n \left(\exp\left[-\beta H_{n,\sigma}(\phi) \right] 1_{\{\phi_i < 0, \, \forall i \leq n\}} \right)$$
$$= \exp\left[\beta \sum_{i=1}^{n} \sigma_i \right] P(\phi_i < 0, \, \forall i \leq n).$$

¿From this we get

$$f(\beta, h) \geq \beta h.$$

It is quite plausible that localization dominates in the case where there is a strict inequality, and that delocalization holds if $f(\beta, h) = \beta h$.

Theorem 10 *There exists a positive, continuous, and increasing function $\beta \to h^*(\beta)$ such that*

$$f(\beta, h) > \beta h \text{ for } 0 \leq h < h^*(\beta), \tag{12}$$
$$f(\beta, h) = \beta h \text{ for } h > h^*(\beta). \tag{13}$$

The function $\beta \to h^*(\beta)$ has a positive tangent at $\beta = 0$.

The phase separating function h^* is certainly very much model dependent, but we expect that the tangent at 0 is model independent, and would be the same for any random law of the σ-sequence which has variance 1 and a expectation h, and has exponentially decaying tails, but this is not proved in [8]. In physics literature, there are non-rigorous arguments claiming that the tangent is 1, but we neither have been able to prove or disprove it, yet. We prove that the tangent at 0 can be described in terms of a phase separation line for a continuous model, where the random walk is replaced by a Brownian motion, and the random environment σ is replaced by (biased) white noise. In this case, the phase separation line is a straight line, and we prove that this line is the tangent at 0 of our model. It should be remarked that the $(\beta, h) \approx (0, 0)$ situation, cannot be handled by simple perturbation techniques.

A natural question is if in the localized region $f(\beta, h) > \beta h$ the path measure is really localized in the sense described in the paper of Sinai. This is indeed the case and has been proved by Biskup and den Hollander [3]. One might also wonder if in the localized region $f(\beta, h) = \beta h$ or at least in the interior of it, the path measure is really delocalized, which should mean, that it converges, after Brownian rescaling, to the limit of a random walk conditioned to stay negative, which is the negative of the Brownian meander. This seems to be a rather difficult question and has not been answered, yet.

The positive tangent is essentially tied to the randomness of the sequence. For the periodic case, the situation is different, as has recently been proved in [11]:

Theorem 11 *Let $\sigma_i = \omega_i + h$, where $\omega_i \in \{-1, 1\}$ is periodic, i.e. such that there exists T with $\omega_{i+2T} = \omega_i$ for all i, and $\sum_{i=1}^{2T} \omega_i = 0$. Then there is a function h^* such that (12) and (13) hold. In this case*

$$C = \lim_{\beta \to 0} \frac{h^*(\beta)}{\beta^3}$$

exists and is positive.

In this paper an expression for C in terms of a variational problem is derived, where the exact nature of the periodic sequence enters.

References

[1] van den Berg, M., Bolthausen, E. and den Hollander, F.: *Moderate deviations for the Wiener sausage.* Annals of Mathematics **153**(2001), 355–406.

[2] Biggins, J.D.: *Chernoff's theorem in the branching random walk.* J. Appl. Prob. **14**(1977), 630–636.

[3] Biskup, M. and den Hollander, F.: *A heteropolymer near a linear interface.* Ann. Appl. Probab. **9**(1999), 668–687.

[4] Bolthausen, E., Deuschel, J. D., and Zeitouni, O.: *Entropic repulsion for the lattice free field,* Comm. Math. Phys. **170**(1995), 417–443.

[5] Bolthausen, E., Deuschel, J.D., and Zeitouni, O. *Absence of a wetting transition for lattice free fields in dimensions three and larger.* J. Math. Phys. **41**(2000), 1211–1223.

[6] Bolthausen, E., Deuschel, J.D., and Giacomin, G.: *Entropic repulsion for the two-dimensional lattice free field.* Annals of Probability **29**(2001), 1670–1692.

[7] Bolthausen, E., and Brydges, D.: *Gaussian surface pinned by a weak potential.* IMS Lecture Notes Vol. **36**(2001), 134–149.

[8] Bolthausen, E. and den Hollander, F.: *Localization transition for a polymer near an interface.* Ann. Probability **25**(1997), 1334–1365.

[9] Bolthausen, E. and Ioffe, D.: *Harmonic crystal on the wall: a microscopic approach,* Comm. Math. Phys. **187**(1997), 523–566.

[10] Bolthausen, E. and Velenik, Y.: *Critical behavior of the massless free field at the depinning transition.* Comm. Math. Phys. **233**(2001), 161–203.

[11] Bolthausen, E. and Giacomin, G.: *On the critical delocalization-localization line for periodic copolymers at interfaces.* Preprint.

[12] Bramson, M.: *Maximal displacement of branching Brownian motion.* Comm. Pure Appl. Math. **31**(1978), 531–581.

[13] Brezin, E., Halperin, and Leibler, S.: *Critical wetting in three dimensions.* Phys. Rev. Lett. **50**(1983), 1387.

[14] Bricmont, J., Fröhlich, J., and El Mellouki, A.: *Random surfaces in statistical mechanics: Roughening, rounding, wetting.* J. Stat. Phys. **42**(1986), 743.

[15] Brydges, D.C., Fröhlich, J., and Spencer, T.: *The random walk representation of classical spin systems and correlation inequalities.* Commun. Math. Phys., **83**(1982), 123–150.

[16] Caputo, P., and Velenik, I.: *A note on wetting transition for gradient fields.* Stoch. Proc. Appl. **87**(2000), 107–113.

[17] Deuschel, J.D.: *Entropic repulsion of the lattice free field.* II. The 0-boundary case. Comm. Math. Phys. **181**(1996), 647–665.

[18] Deuschel, J. D., and Giacomin G.: *Entropic repulsion for the free field: path-wise characterization in* $d \geq 3$, Comm. Math. Phys. **206**(1999), 447–462.

[19] Deuschel, J.D., and Giacomin, G.: *Entropic repulsion for massless fields*, Stoch. Process. Appl. **89**(2000), 333–354.

[20] Deuschel, J.D., Giacomin, G., and Ioffe, D.: *Large deviations and concentration properties for* $\nabla \phi$ *interface models.* Prob. Theory Rel. Fields **117**(2000), 49–111.

[21] Deuschel, J.D., and Velenik, Y.: *Non-Gaussian surface pinned by a weak potential.* Prob. Theory Rel. Fields **116**(2000), 359–377.

[22] Derrida, B. and Spohn, H.: *Polymers on disordered trees, spin glasses, and travelling waves*, J. Stat. Phys. **51**(1988), 817–840.

[23] Donsker, M. and Varadhan, S.R.S.: *On the number of distinct sites visited by a random walk.* Comm. Pure Appl. Math. **32**(1979), 721–747.

[24] Dunlop, F., Magnen, J., Rivasseau, V., and Roche, Ph.: *Pinning of an interface by a weak potential.* J. Statist. Phys. **66**(1992), 71–98.

[25] Fisher, M.: *Walks, walls, wetting and melting.* J. Stat. Phys. **34**(1984), 667–729.

[26] Helffer, B., and Sjöstrand, J.: *On the correlation for Kac-like models in the convex case.* J. Statist. Phys. **74**(1994), 349–409.

[27] Ioffe, D., and Velenik, I.: *A note on the decay of correlations under* δ-*pinning.* Prob. Theory Rel. Fields **116**(2000), 379–389.

[28] Isozaki, Y. and Yoshida, N.: *Weakly pinned random walk on the wall: pathwise descriptions of the phase transition.* Stoch. Proc. Appl. **96**(2001), 261–284.

[29] Sinai, Ya. G.: *A random walk with a random potential.* Theory Probab. Appl. **38**(1993), 382–385.

[30] Sznitman, A.-S.: *Brownian Motion, Obstacles, and Random Media.* Springer, Heidelberg 1998.

ICM 2002 · Vol. III · 41–52

Ergodic Convergence Rates of Markov Processes—Eigenvalues, Inequalities and Ergodic Theory

Abstract

This paper consists of four parts. In the first part, we explain what eigenvalues we are interested in and show the difficulties of the study on the first (non-trivial) eigenvalue through examples. In the second part, we present some (dual) variational formulas and explicit bounds for the first eigenvalue of Laplacian on Riemannian manifolds or Jacobi matrices (Markov chains). Here, a probabilistic approach—the coupling methods is adopted. In the third part, we introduce recent lower bounds of several basic inequalities; these are based on a generalization of Cheeger's approach which comes from Riemannian geometry. In the last part, a diagram of nine different types of ergodicity and a table of explicit criteria for them are presented. These criteria are motivated by the weighted Hardy inequality which comes from Harmonic analysis.

2000 Mathematics Subject Classification: 35P15, 47A75, 49R50, 60J99.
Keywords and Phrases: Eigenvalue, Variational formula, Inequality, Convergence rate, Ergodic theory, Markov process.

I. Introduction

We will start by explaining what eigenvalues we are interested in.

1.1 Definition. *Consider a birth-death process with a state space* $E = \{0, 1, 2, \cdots, n\}$ $(n \leqslant \infty)$ *and an intensity matrix* $Q = (q_{ij})$: $q_{k,k-1} = a_k > 0 \, (1 \leqslant k \leqslant n)$, $q_{k,k+1} = b_k > 0 \, (0 \leqslant k \leqslant n - 1)$, $q_{k,k} = -(a_k + b_k)$, *and* $q_{ij} = 0$ *for other* $i \neq j$.

Since the sum of each row equals 0, we have $Q\mathbb{1} = 0 = 0 \cdot \mathbb{1}$. This means that the Q-matrix has an eigenvalue 0 with an eigenvector $\mathbb{1}$. Next, consider the finite case of $n < \infty$. Then, the eigenvalues of $-Q$ are discrete: $0 = \lambda_0 < \lambda_1 \leqslant \cdots \leqslant \lambda_n$. We are interested in the first (non-trivial) eigenvalue $\lambda_1 = \lambda_1 - \lambda_0$ (also called spectral gap of Q). In the infinite case ($n = \infty$), λ_1 can be 0. Certainly, one can consider a self-adjoint elliptic operator in $\mathbb{R}^d$, the Laplacian Δ on manifolds, or an infinite-dimensional operator as in the study of interacting particle systems.

---

*Department of Mathematics, Beijing Normal University, Beijing 100875, China. E-mail: mfchen@bnu.edu.cn, Home page: http://www.bnu.edu.cn/~chenmf/main_eng.htm

1.2 Difficulties. To get a concrete feeling about the difficulties of this topic, let us first look at the following examples with a finite state space. When $E = \{0,1\}$, it is trivial that $\lambda_1 = a_1 + b_0$. The result is nice because when either a_1 or b_0 increases, so does λ_1. When $E = \{0,1,2\}$, we have four parameters b_0, b_1, a_1, a_2 and $\lambda_1 = 2^{-1}\left[a_1 + a_2 + b_0 + b_1 - \sqrt{(a_1 - a_2 + b_0 - b_1)^2 + 4a_1 b_1}\right]$. When $E = \{0,1,2,3\}$, we have six parameters: $b_0, b_1, b_2, a_1, a_2, a_3$. In this case, the expression for λ_1 is too lengthy to write. The roles of the parameters are inter-related in a complicated manner. Clearly, it is impossible to compute λ_1 explicitly when the size of the matrix is greater than five.

Next, consider the infinite state space $E = \{0,1,2,\cdots\}$. Denote the eigenfunction of λ_1 by g and the degree of g by $D(g)$ when g is polynomial. Three examples of the perturbation of λ_1 and $D(g)$ are listed in Table 1.1.

$b_i (i \geqslant 0)$	$a_i (i \geqslant 1)$	λ_1	$D(g)$
$i + c (c > 0)$	$2i$	1	1
$i + 1$	$2i + 3$	2	2
$i + 1$	$2i + (4 + \sqrt{2})$	3	3

Table 1.1 Three examples of the perturbation of λ_1 and $D(g)$

The first line is the well known linear model for which $\lambda_1 = 1$, independent of the constant $c > 0$, and g is linear. Keeping the same birth rate, $b_i = i + 1$, changes the death rate a_i from $2i$ to $2i + 3$ (resp. $2i + 4 + \sqrt{2}$), which leads to the change of λ_1 from one to two (resp. three). More surprisingly, the eigenfunction g is changed from linear to quadratic (resp. triple). For the other values of a_i between $2i$, $2i + 3$ and $2i + 4 + \sqrt{2}$, λ_1 is unknown since g is non-polynomial. As seen from these examples, the first eigenvalue is very sensitive. Hence, in general, it is very hard to estimate λ_1.

In the next section, we find that this topic is studied extensively in Riemannian geometry.

II. New variational formula for the first eigenvalue

2.1 Story of estimating λ_1 in geometry. At first, we recall the study of λ_1 in geometry.

Consider Laplacian Δ on a compact Riemannian manifold (M, g), where g is the Riemannian metric. The spectrum of Δ is discrete: $\cdots \leqslant -\lambda_2 \leqslant -\lambda_1 < -\lambda_0 = 0$ (may be repeated). Estimating these eigenvalues λ_k (especially λ_1) is very important in modern geometry. As far as we know, five books, excluding those books on general spectral theory, have been devoted to this topic: Chavel (1984), Bérard (1986), Schoen and Yau (1988), Li (1993) and Ma (1993). For a manifold M, denote its dimension, diameter and the lower bound of Ricci curvature by d, D, and K ($\text{Ricci}_M \geqslant Kg$), respectively. We are interested in estimating λ_1 in terms of these three geometric quantities. It is relatively easy to obtain an upper bound by applying a test function $f \in C^1(M)$ to the classical variational formula:

$$\lambda_1 = \inf \left\{ \int_M \|\nabla f\|^2 \mathrm{d}x : f \in C^1(M), \int f \mathrm{d}x = 0, \int f^2 \mathrm{d}x = 1 \right\}, \qquad (2.0)$$

where "$\mathrm{d}x$" is the Riemannian volume element. To obtain the lower bound, however, is much harder. In Table 2.1, we list eight of the strongest lower bounds that have been derived in the past, using various sophisticated methods.

Author(s)	Lower bound	
A. Lichnerowicz (1958)	$\dfrac{d}{d-1} K, \quad K \geqslant 0.$	(2.1)
P. H. Bérard, G. Besson & S. Gallot (1985)	$d \left\{ \dfrac{\int_0^{\pi/2} \cos^{d-1} t \mathrm{d}t}{\int_0^{D/2} \cos^{d-1} t \mathrm{d}t} \right\}^{2/d}, \quad K = d-1 > 0.$	(2.2)
P. Li & S. T. Yau (1980)	$\dfrac{\pi^2}{2D^2}, \quad K \geqslant 0.$	(2.3)
J. Q. Zhong & H. C. Yang (1984)	$\dfrac{\pi^2}{D^2}, \quad K \geqslant 0.$	(2.4)
P. Li & S. T. Yau (1980)	$\dfrac{1}{D^2(d-1)\exp\left[1 + \sqrt{1 + 16\alpha^2}\right]}, \quad K \leqslant 0.$	(2.5)
K. R. Cai (1991)	$\dfrac{\pi^2}{D^2} + K, \quad K \leqslant 0.$	(2.6)
H. C. Yang (1989) & F. Jia (1991)	$\dfrac{\pi^2}{D^2} e^{-\alpha}, \quad \text{if } d \geqslant 5, \quad K \leqslant 0.$	(2.7)
H. C. Yang (1989) & F. Jia (1991)	$\dfrac{\pi^2}{2D^2} e^{-\alpha'}, \quad \text{if } 2 \leqslant d \leqslant 4, \quad K \leqslant 0,$	(2.8)

Table 2.1 Eight lower bounds of λ_1

In Table 2.1, the two parameters α and α' are defined as $\alpha = D\sqrt{|K|(d-1)}/2$ and $\alpha' = D\sqrt{|K|((d-1) \vee 2)}/2$. Among these estimates, five ((2.1), (2.2), (2.4), (2.6) and (2.7)) are sharp. The first two are sharp for the unit sphere in two or higher dimensions but fail for the unit circle; the fourth, the sixth, and the seventh are all sharp for the unit circle. As seen from this table, the picture is now very complete, due to the efforts of many geometers in the past 40 years. Our original starting point is to learn from the geometers and to study their methods, especially the recent new developments. In the next section, we will show that one can go in the opposite direction, i.e., studying the first eigenvalue by using probabilistic methods. Exceeding our expectations, we find a general formula for the lower bound.

2.2 New variational formula. Before stating our new variational formula, we introduce two notations:

$$C(r) = \cosh^{d-1}\left[\frac{r}{2}\sqrt{\frac{-K}{d-1}} \right], \ r \in (0, D). \qquad \mathscr{F} = \{ f \in C[0, D] : f > 0 \text{ on } (0, D) \}.$$

Here, we have used all the three quantities: the dimension d, the diameter D, and the lower bound K of Ricci curvature.

Theorem 2.1[General formula] (Chen & Wang (1997a)).

$$\lambda_1 \geqslant \sup_{f \in \mathcal{F}} \inf_{r \in (0,D)} \frac{4f(r)}{\int_0^r C(s)^{-1}\mathrm{d}s \int_s^D C(u)f(u)\mathrm{d}u} =: \xi_1. \tag{2.9}$$

The new variational formula has its essential value in estimating the lower bound. It is a dual of the classical variational formula in the sense that "inf" in (2.0) is replaced by "sup" in (2.9). The classical formula can be traced to Lord S. J. W. Rayleigh (1877) and E. Fischer (1905). Noticing that these two formulas (2.0) and (2.9) look very different, which explains that why such a formula (2.9) has never appeared before. This formula can produce many new lower bounds. For instance, the one corresponding to the trivial function $f \equiv 1$ is non-trivial in geometry. Applying the general formula to the test functions $\sin(\alpha r)$ and $\cosh^{d-1}(\alpha r)\sin(\beta r)$ with $\alpha = D\sqrt{|K|(d-1)}/2$ and $\beta = \pi/(2D)$, we obtain the following:

Corollary 2.2 (Chen & Wang (1997a)).

$$\lambda_1 \geqslant \frac{dK}{d-1}\left\{1 - \cos^d\left[\frac{D}{2}\sqrt{\frac{K}{d-1}}\right]\right\}^{-1}, \quad d > 1, \quad K \geqslant 0, \tag{2.10}$$

$$\lambda_1 \geqslant \frac{\pi^2}{D^2}\sqrt{1 - \frac{2D^2 K}{\pi^4}}\cosh^{1-d}\left[\frac{D}{2}\sqrt{\frac{-K}{d-1}}\right], \quad d > 1, K \leqslant 0. \tag{2.11}$$

Applying this formula to some very complicated test functions, we can prove the following result:

Corollary 2.3 (Chen, Scacciatelli and Yao (2002)).

$$\lambda_1 \geqslant \pi^2/D^2 + K/2, \qquad K \in \mathbb{R}. \tag{2.12}$$

The corollaries improve all the estimates (2.1)—(2.8). Especially, (2.10) improves (2.1) and (2.2), (2.11) improves (2.7) and (2.8), and (2.12) improves (2.3) and (2.6). Moreover, the linear approximation in (2.12) is optimal in the sense that the coefficient $1/2$ of K is exact.

A test function is indeed a mimic of the eigenfunction, so it should be chosen appropriately in order to obtain good estimates. A question arises naturally: does there exist a single representative test function such that we can avoid the task of choosing a different test function each time? The answer is seemingly negative since we have already seen that the eigenvalue and the eigenfunction are both very sensitive. Surprisingly, the answer is affirmative. The representative test function, though very tricky to find, has a rather simple form: $f(r) = \sqrt{\int_0^r C(s)^{-1}\mathrm{d}s}$. This is motivated from the study of the weighted Hardy inequality, a powerful tool in harmonic analysis (cf. Muckenhoupt (1972), Opic and Kufner (1990)).

Corollary 2.4 (Chen (2000)). *For the lower bound ξ_1 of λ_1 given in Theorem 2.1, we have*

$$4\delta^{-1} \geqslant \xi_1 \geqslant \delta^{-1}, \qquad \text{where} \tag{2.13}$$

$$\delta = \sup_{r \in (0,D)} \left(\int_0^r C(s)^{-1}\mathrm{d}s\right)\left(\int_r^D C(s)\mathrm{d}s\right), \qquad C(s) = \cosh^{d-1}\left[\frac{s}{2}\sqrt{\frac{-K}{d-1}}\right].$$

Theorem 2.1 and its corollaries are also valid for manifolds with a convex boundary endowed with the Neumann boundary condition. In this case, the estimates (2.1)—(2.8) are conjectured by the geometers to be correct. However, only the Lichnerowicz's estimate (2.1) was proven by J. F. Escobar in 1990. The others in (2.2)—(2.8) and furthermore in (2.10)—(2.13) are all new in geometry.

On the one hand, the proof of this theorem is quite straightforward, based on the coupling introduced by Kendall (1986) and Cranston (1991). On the other hand, the derivation of this general formula requires much effort. The key point is to find a way to mimic the eigenfunctions. For more details, refer to Chen (1997).

Applying similar proof techniques to general Markov processes, we also obtain variational formulas for non-compact manifolds, elliptic operators in $\mathbb{R}^d$ (Chen and Wang (1997b)), and Markov chains (Chen (1996)). It is more difficult to derive the variational formulas for the elliptic operators and Markov chains due to the presence of infinite parameters in these cases. In contrast, there are only three parameters (d, D, and K) in the geometric case. In fact, formula (2.9) is a particular example of our general formula (which is complete in dimensional one) for elliptic operators.

To conclude this part, we return to the matrix case introduced at the beginning of the paper.

2.3 Birth-death processes. Let $b_i > 0 (i \geqslant 0)$ and $a_i > 0 (i \geqslant 1)$ be the birth and death rates, respectively. Define $\mu_0 = 1$, $\mu_i = b_0 \cdots b_{i-1}/a_1 \cdots a_i$ ($i \geqslant 1$). Assume that the process is non-explosive:

$$\sum_{k=0}^{\infty} (b_k \mu_k)^{-1} \sum_{i=0}^{k} \mu_i = \infty \qquad \text{and moreover} \qquad \mu = \sum_i \mu_i < \infty. \qquad (2.14)$$

The corresponding Dirichlet form is $D(f) = \sum_i \pi_i b_i (f_{i+1} - f_i)^2$, $\mathcal{D}(D) = \{f \in L^2(\pi) : D(f) < \infty\}$. Here and in what follows, only the diagonal elements $D(f)$ are written, but the non-diagonal elements can be computed from the diagonal ones by using the quadrilateral role. We then have the classical formula $\lambda_1 = \{D(f) : \pi(f) = 0, \pi(f^2) - 1\}$. Define $\mathcal{F}' = \{f : f_0 = 0, \text{there exists } k : 1 \leqslant k \leqslant \infty \text{ so that } f_i = f_{i \wedge k}$ and f is strictly increasing in $[0, k]\}$, $\mathcal{F}'' = \{f : f_0 = 0, f \text{ is strictly increasing}\}$, and $I_i(f) = [\mu_i b_i (f_{i+1} - f_i)]^{-1} \sum_{j \geqslant i+1} \mu_j f_j$. Let $\bar{f} = f - \pi(f)$. Then we have the following results:

Theorem 2.5 (Chen (1996, 2000, 2001))[1]. *Under* (2.14), *we have*

(1) *Dual variational formula.* $\inf\limits_{f \in \mathcal{F}'} \sup\limits_{i \geqslant 1} I_i(\bar{f})^{-1} = \lambda_1 = \sup\limits_{f \in \mathcal{F}''} \inf\limits_{i \geqslant 0} I_i(\bar{f})^{-1}.$

(2) *Explicit estimate.* $\mu \delta^{-1} \geqslant \lambda_1 \geqslant (4\delta)^{-1}$, *where* $\delta = \sup\limits_{i \geqslant 1} \sum\limits_{j \leqslant i-1} (\mu_j b_j)^{-1} \sum\limits_{j \geqslant i} \mu_j.$

(3) *Approximation procedure. There exist explicit sequences* η_n' *and* η_n'' *such that* $\eta_n'^{-1} \geqslant \lambda_1 \geqslant \eta_n''^{-1} \geqslant (4\delta)^{-1}.$

Here the word "dual" means that the upper and lower bounds are interchangeable if one exchanges "sup" and "inf". With slight modifications, this result is also valid for finite matrices, refer to Chen (1999).

---

[1] Due to the limitation of the space, the most of the author's papers during 1993–2001 are not listed in References, the readers are urged to refer to [11].

III. Basic inequalities and new forms of Cheeger's constants

3.1 Basic inequalities. We now go to a more general setup. Let $(E, \mathcal{E}, \pi)$ be a probability space satisfying $\{(x, x) : x \in E\} \in \mathcal{E} \times \mathcal{E}$. Denote by $L^p(\pi)$ the usual real L^p-space with norm $\|\cdot\|_p$. Write $\|\cdot\| = \|\cdot\|_2$.

For a given Dirichlet form $(D, \mathcal{D}(D))$, the classical variational formula for the first eigenvalue λ_1 can be rewritten in the form of (3.1) below with an optimal constant $C = \lambda_1^{-1}$. From this point of view, it is natural to study other inequalities. Two additional basic inequalities appear in (3.2) and (3.3) below.

$$\textit{Poincaré inequality}: \qquad \mathrm{Var}(f) \leqslant CD(f), \qquad f \in L^2(\pi), \qquad (3.1)$$

$$\textit{Logarithmic Sobolev inequality}: \int f^2 \log \frac{f^2}{\|f\|^2} \mathrm{d}\pi \leqslant CD(f), \ f \in L^2(\pi), \ (3.2)$$

$$\textit{Nash inequality}: \qquad \mathrm{Var}(f) \leqslant CD(f)^{1/p} \|f\|_1^{2/q}, \ f \in L^2(\pi), \qquad (3.3)$$

where $\mathrm{Var}(f) = \pi(f^2) - \pi(f)^2$, $\pi(f) = \int f \mathrm{d}\pi$, $p \in (1, \infty)$ and $1/p + 1/q = 1$. The last two inequalities are due to Gross (1976) and Nath (1958), respectively.

Our main object is a symmetric (not necessarily Dirichlet) form $(D, \mathcal{D}(D))$ on $L^2(\pi)$, corresponding to an integral operator (or symmetric kernel) on $(E, \mathcal{E})$:

$$D(f) = \frac{1}{2} \int_{E \times E} J(\mathrm{d}x, \mathrm{d}y)[f(y) - f(x)]^2, \quad \mathcal{D}(D) = \{f \in L^2(\pi) : D(f) < \infty\}, \ (3.4)$$

where J is a non-negative, symmetric measure having no charge on the diagonal set $\{(x, x) : x \in E\}$. A typical example is the reversible jump process with a q-pair $(q(x), q(x, \mathrm{d}y))$ and a reversible measure π. Then $J(\mathrm{d}x, \mathrm{d}y) = \pi(\mathrm{d}x)q(x, \mathrm{d}y)$.

For the remainder of this part, we restrict our discussions to the symmetric form of (3.4).

3.2 Status of the research. An important topic in this research area is to study under what conditions on the symmetric measure J do the above inequalities hold. In contrast with the probabilistic method used in Part (II), here we adopt a generalization of Cheeger's method (1970), which comes from Riemannian geometry. Naturally, we define $\lambda_1 := \inf\{D(f) : \pi(f) = 0, \|f\| = 1\}$. For bounded jump processes, the fundamental known result is the following:

Theorem 3.1 (Lawler & Sokal (1988)). $\lambda_1 \geqslant \dfrac{k^2}{2M}$, *where*

$$k = \inf_{\pi(A) \in (0,1)} \frac{\int_A \pi(\mathrm{d}x)q(x, A^c)}{\pi(A) \wedge \pi(A^c)} \qquad and \qquad M = \sup_{x \in E} q(x).$$

In the past years, the theorem has been collected into six books: Chen (1992), Sinclair (1993), Chung (1997), Saloff-Coste (1997), Colin de Verdière (1998), Aldous, D. G. & Fill, J. A. (1994–). From the titles of the books, one can see a wide range of the applications. However, this result fails for the unbounded operator.

Thus, it has been a challenging open problem in the past ten years to handle the unbounded case.

As for the logarithmic Sobolev inequality, there have been a large number of publications in the past twenty years for differential operators. (For a survey, see Bakry (1992) or Gross (1993)). Still, there are very limited results for integral operators.

3.3 New results. Since the symmetric measure can be unbounded, we choose a symmetric, non-negative function $r(x, y)$ such that

$$J^{(\alpha)}(\mathrm{d}x, \mathrm{d}y) := I_{\{r(x,y)^\alpha > 0\}} \frac{J(\mathrm{d}x, \mathrm{d}y)}{r(x, y)^\alpha} \quad (\alpha > 0) \quad \text{satisfies} \quad \frac{J^{(1)}(\mathrm{d}x, E)}{\pi(\mathrm{d}x)} \leqslant 1, \, \pi\text{-a.s.}$$

For convenience, we use the convention $J^{(0)} = J$. Corresponding to the three inequalities above, we introduce the following new forms of Cheeger's constants.

Inequality	Constant $k^{(\alpha)}$	
Poincaré	$\displaystyle\inf_{\pi(A)\in(0,1)} \frac{J^{(\alpha)}(A \times A^c)}{\pi(A) \wedge \pi(A^c)}$	(Chen & Wang(1998))
Log. Sobolev	$\displaystyle\lim_{r\to 0}\inf_{\pi(A)\in(0,r]} \frac{J^{(\alpha)}(A \times A^c)}{\pi(A)\sqrt{\log[e + \pi(A)^{-1}]}}$	(Wang (2001a))
Log. Sobolev	$\displaystyle\lim_{\delta\to\infty}\inf_{\pi(A)>0} \frac{J^{(\alpha)}(A \times A^c) + \delta\pi(A)}{\pi(A)\sqrt{1 - \log\pi(A)}}$	(Chen (2000))
Nash	$\displaystyle\inf_{\pi(A)\in(0,1)} \frac{J^{(\alpha)}(A \times A^c)}{[\pi(A) \wedge \pi(A^c)]^{(2q-3)/(2q-2)}}$	(Chen (1999))

Table 3.1   New forms of Cheeger's constants

Our main result can be easily stated as follows.

Theorem 3.2. $k^{(1/2)} > 0 \Longrightarrow$ *the corresponding inequality holds.*

In other words, we use $J^{(1/2)}$ and $J^{(1)}$ to handle the unbounded J. The first two kernels come from the use of Schwarz inequality. This result is proven in four papers quoted in Table (3.1). In these papers, some estimates which are sharp or qualitatively sharp for the upper or lower bounds are also presented.

IV. New picture of ergodic theory and explicit criteria

4.1 Importance of the inequalities. Let $(P_t)_{t\geqslant 0}$ be the semigroup determined by a Dirichlet form $(D, \mathcal{D}(D))$. Then, various applications of the inequalities are based on the following results:

Theorem 4.1 (Liggett (1989), Gross (1976) and Chen (1999)).

(1) Poincaré *inequality* $\Longleftrightarrow \|P_t f - \pi(f)\|^2 = \mathrm{Var}(P_t f) \leqslant \mathrm{Var}(f)\exp[-2\lambda_1 t]$.
(2) *Logarithmic Sobolev inequality* $\Longrightarrow$ *exponential convergence in entropy:* $\mathrm{Ent}(P_t f) \leqslant \mathrm{Ent}(f)\exp[-2\sigma t]$, *where* $\mathrm{Ent}(f) = \pi(f \log f) - \pi(f)\log\|f\|_1$.
(3) *Nash inequality* $\Longleftrightarrow \mathrm{Var}(P_t f) \leqslant C\|f\|_1/t^{1-q}$.

In the context of diffusions, one can replace "$\Longrightarrow$" by "$\Longleftrightarrow$" in part (2). There-fore, the above inequalities describe some type of L^2-ergodicity for the semigroup $(P_t)_{t\geqslant 0}$. These inequalities have become powerful tools in the study on infinite-dimensional mathematics (phase transitions, for instance) and the effectiveness of random algorithms.

4.2 Three traditional types of ergodicity. The following three types of ergo-dicity are well known for Markov processes.

Ordinary ergodicity : $\qquad \lim\limits_{t\to\infty} \|p_t(x,\cdot) - \pi\|_{\mathrm{Var}} = 0$

Exponential ergodicity : $\qquad \|p_t(x,\cdot) - \pi\|_{\mathrm{Var}} \leqslant C(x)e^{-\alpha t}$ for some $\alpha > 0$

Strong ergodicity : $\qquad \lim\limits_{t\to\infty} \sup\limits_{x} \|p_t(x,\cdot) - \pi\|_{\mathrm{Var}} = 0$

$\qquad\qquad \Longleftrightarrow \lim\limits_{t\to\infty} e^{\beta t} \sup\limits_{x} \|p_t(x,\cdot) - \pi\|_{\mathrm{Var}} = 0$ for some $\beta > 0$

where $p_t(x, \mathrm{d}y)$ is the transition function of the Markov process and $\|\cdot\|_{\mathrm{Var}}$ is the total variation norm. They obey the following implications:

Strong ergodicity $\Longrightarrow$ Exponential ergodicity $\Longrightarrow$ Ordinary ergodicity.

It is natural to ask the following question. does there exist any relation between the above inequalities and the three traditional types of ergodicity?

4.3 New picture of ergodic theory.

Theorem 4.2 (Chen (1999), ...). *For reversible Markov processes with densities, we have the diagram shown in Figure 4.1.*

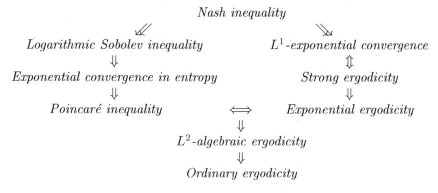

Figure 4.1   Diagram of nine types of ergodicity

In Figure 4.1, L^2-algebraic ergodicity means that $\mathrm{Var}(P_t f) \leqslant CV(f)t^{1-q}$ $(t > 0)$ holds for some V having the properties (cf. Liggett (1991)): V is homogeneous of degree two (in the sense that $V(cf + d) = c^2 V(f)$ for any constants c and d) and $V(f) < \infty$ for all functions f with finite support.

The diagram is complete in the following sense: each single-side implication can not be replaced by double-sides one. Moreover, strong ergodicity and logarithmic Sobolev inequality (resp. exponential convergence in entropy) are not comparable. With exception of the equivalences, all the implications in the diagram are suitable for more general Markov processes. Clearly, the diagram extends the ergodic theory of Markov processes.

The diagram was presented in Chen (1999), originally for Markov chains only. Recently, the equivalence of L^1-exponential convergence and strong ergodicity was mainly proven by Y. H. Mao. A counter-example of diffusion was constructed by Wang (2001b) to show that strong ergodicity does not imply exponential convergence in entropy. For other references and a detailed proof of the diagram, refer to Chen (1999).

4.4 Explicit criteria for several types of ergodicity. As an application of the diagram in Figure 4.1, we obtain a criterion for the exponential ergodicity of birth-death processes, as listed in Table 4.2. To achieve this, we use the equivalence of exponential ergodicity and Poincaré inequality, as well as the explicit criterion for Poincaré inequality given in part (3) of Theorem 2.5. This solves a long standing open problem in the study of Markov chains (cf. Anderson (1991), §6.6 and Chen (1992), §4.4).

Next, it is natural to look for some criteria for other types of ergodicity. To do so, we consider only the one-dimensional case. Here we focus on the birth-death processes since the one-dimensional diffusion processes are in parallel. The criterion for strong ergodicity was obtained recently by Zhang, Lin and Hou (2000), and extended by Zhang (2001), using a different approach, to a larger class of Markov chains. The criteria for logarithmic Sobolev, Nash inequalities, and the discrete spectrum (no continuous spectrum and all eigenvalues have finite multiplicity) were obtained by Bobkov and Götze (1999) and Mao (2000, 2002a,b), respectively, based on the weighted Hardy inequality (see also Miclo (1999), Wang (2000), Gong and Wang (2002)). It is understood now the results can also be deduced from generalizations of the variational formulas discussed in this paper (cf. Chen (2001b)). Finally, we summarize these results in Theorem 4.3 and Table 4.2. The table is arranged in such an order that the property in the latter line is stronger than the property in the former line. The only exception is that even though the strong ergodicity is often stronger than the logarithmic Sobolev inequality, they are not comparable in general, as mentioned in Part III.

Theorem 4.3 (Chen (2001a)). *For birth-death processes with birth rates $b_i (i \geqslant 0)$ and death rates $a_i (i \geqslant 1)$, ten criteria are listed in Table 4.2. Recall the sequence (μ_i) defined in Part II and set $\mu[i,k] = \sum_{i \leqslant j \leqslant k} \mu_j$. The notion "$(*)$ & $\cdots$" appeared in Table 4.2 means that one requires the uniqueness condition in the first line plus the condition "$\cdots$". The notion "(ε)" in the last line means that there is still a small room $(1 < q \leqslant 2)$ left from completeness.*

Property	Criterion
Uniqueness	$\displaystyle\sum_{n\geqslant 0}\frac{1}{\mu_n b_n}\mu[0,n]=\infty \quad (*)$
Recurrence	$\displaystyle\sum_{n\geqslant 0}\frac{1}{\mu_n b_n}=\infty$
Ergodicity	$(*)\ \&\ \mu[0,\infty)<\infty$
Exponential ergodicity L^2-exp. convergence	$\displaystyle(*)\ \&\ \sup_{n\geqslant 1}\mu[n,\infty)\sum_{j\leqslant n-1}\frac{1}{\mu_j b_j}<\infty$
Discrete spectrum	$\displaystyle(*)\ \&\ \lim_{n\to\infty}\sup_{k\geqslant n+1}\mu[k,\infty)\sum_{n\leqslant j\leqslant k-1}\frac{1}{\mu_j b_j}=0$
Log. Sobolev inequality	$\displaystyle(*)\ \&\ \sup_{n\geqslant 1}\mu[n,\infty)\log[\mu[n,\infty)^{-1}]\sum_{j\leqslant n-1}\frac{1}{\mu_j b_j}<\infty$
Strong ergodicity L^1-exp. convergence	$\displaystyle(*)\ \&\ \sum_{n\geqslant 0}\frac{1}{\mu_n b_n}\mu[n+1,\infty)=\sum_{n\geqslant 1}\mu_n\sum_{j\leqslant n-1}\frac{1}{\mu_j b_j}<\infty$
Nash inequality	$\displaystyle(*)\ \&\ \sup_{n\geqslant 1}\mu[n,\infty)^{(q-2)/(q-1)}\sum_{j\leqslant n-1}\frac{1}{\mu_j b_j}<\infty\ (\varepsilon)$

Table 4.2   Ten criteria for birth-death processes

References

[1] Anderson, W. J. (1991), *Continuous-Time Markov Chains*, Springer Series in Statistics.

[2] Aldous, D. G. & Fill, J. A. (1994–), *Reversible Markov Chains and Random Walks on Graphs,* www. stat.Berkeley.edu/users/aldous/book.html.

[3] Bakry, D. (1992), *L'hypercontractivité et son utilisation en théorie des semi-groupes*, LNM, Springer-Verlag, **1581**.

[4] Bérard, P. H. (1986), *Spectral Geometry: Direct and Inverse Problem*, LNM. vol. 1207, Springer-Verlag.

[5] Bérard, P. H., Besson, G., Gallot, S. (1985), *Sur une inégalité isopérimétrique qui généralise celle de Paul Lévy-Gromov*, Invent. Math. 80: 295–308.

[6] Bobkov, S. G. and Götze, F. (1999), *Exponential integrability and transportation cost related to logarithmic Sobolev inequalities*, J. Funct. Anal. 163, 1–28.

[7] Cai, K. R. (1991), *Estimate on lower bound of the first eigenvalue of a compact Riemannian manifold*, Chin. Ann. of Math. 12(B):3, 267–271.

[8] Chavel, I. (1984), *Eigenvalues in Riemannian Geometry*, Academic Press.

[9] Cheeger, J. (1970), *A lower bound for the smallest eigenvalue of the Laplacian, Problems in analysis, a symposium in honor of S. Bochner*, 195–199, Princeton U. Press, Princeton.

[10] Chen, M. F. (1992), *From Markov Chains to Non-Equilibrium Particle Systems*, World Scientific.

[11] Chen, M. F. (1993–2001), *Ergodic Convergence Rates of Markov Processes* (Collection of papers): www.bnu.edu.cn/~chenmf/main_eng.htm.

[12] Chen, M. F. (2001a), *Explicit criteria for several types of ergodicity*, Chin. J. Appl. Prob. Stat. 17:2, 1–8.

[13] Chen, M. F. (2001b), *Variational formulas of Poincaré-type inequalities in Banach spaces of functions on the line*, to appear in Acta Math. Sin. Eng. Ver.

[14] Chen, M. F., Scacciatelli, E. and Yao, Liang (2002), *Linear approximation of the first eigenvalue on compact manifolds*, Sci. Sin. (A) 45:4, 450–461.

[15] Chen, M. F. and Wang, F. Y. (1997a), *General formula for lower bound of the first eigenvalue on Riemannian manifolds*, Sci. Sin. 40:4, 384–394.

[16] Chen, M. F. and Wang, F. Y. (1997b), *Estimation of spectral gap for elliptic operators*, Trans. Amer. Math. Soc. 349, 1239-1267.

[17] Chen, M. F. and Wang, F. Y. (1998), *Cheeger's inequalities for general symmetric forms and existence criteria for spectral gap*, Abstract. Chin. Sci. Bulletin 43:18, 1516–1519. Ann. Prob. 2000, 28:1, 235–257.

[18] Chung, F. R. K. (1997), *Spectral Graph Theory*, CBMS, **92**, AMS, Providence, Rhode Island.

[19] Colin de Verdière, Y. (1998), *Spectres de Graphes*, Publ. Soc. Math. France.

[20] Cranston, M. (1991), *Gradient estimates on manifolds using coupling*, J. Funct. Anal. 99, 110–124.

[21] Escobar, J. F. (1990), *Uniqueness theorems on conformal deformation of metrics, Sobolev inequalities, and an eigenvalue estimate*, Comm. Pure and Appl. Math. XLIII: 857–883.

[22] Gong, F. Z. Wang, F. Y.(2002), *Functional inequalities for uniformly integrable semigroups and application to essential spectrums*, Forum Math. 14, 293 313.

[23] Gross, L. (1976), *Logarithmic Sobolev inequalities*, Amer. J. Math. 97, 1061–1083.

[24] Gross, L. (1993), *Logarithmic Sobolev inequalities and contractivity properties of semigroups*, LNM **1563**, Springer-Verlag.

[25] Jerrum, M. R. and Sinclair, A. J. (1989), *Approximating the permanent*, SIAM J. Comput.18, 1149–1178.

[26] Jia, F. (1991), *Estimate of the first eigenvalue of a compact Riemannian manifold with Ricci curvature bounded below by a negative constant* (In Chinese), Chin. Ann. Math. 12(A):4, 496–502.

[27] Kendall, W. (1986), *Nonnegative Ricci curvature and the Brownian coupling property*, Stochastics 19, 111–129.

[28] Lawler, G. F. and Sokal, A. D.(1988), *Bounds on the L^2 spectrum for Markov chain and Markov processes: a generalization of Cheeger's inequality*, Trans. Amer. Math. Soc.309, 557–580.

[29] Li, P. (1993), *Lecture Notes on Geometric Analysis*, Seoul National U., Korea.

[30] Li, P. and Yau, S. T. (1980), *Estimates of eigenvalue of a compact Riemannian manifold*, Ann. Math. Soc. Proc. Symp. Pure Math. 36, 205–240.

[31] Lichnerowicz, A. (1958), *Geometrie des Groupes des Transformationes*, Dunod.

[32] Liggett, T. M. (1989), *Exponential L_2 convergence of attractive reversible nearest particle systems*, Ann. Prob. 17, 403–432.

[33] Liggett, T. M. (1991), *L_2 rates of convergence for attractive reversible nearest particle systems: the critical case*, Ann. Prob. 19:3, 935–959.

[34] Ma, C. Y. (1993), *The Spectrum of Riemannian Manifolds* (In Chinese), Press of Nanjing U., Nanjing.

[35] Mao, Y. H. (2000), *On empty essential spectrum for Markov processes in dimension one*, preprint.

[36] Mao, Y. H. (2002a), *The logarithmic Sobolev inequalities for birth-death process and diffusion process on the line*, Chin. J. Appl. Prob. Statis., 18(1), 94–100.

[37] Mao, Y. H. (2002b), *Nash inequalities for Markov processes in dimension one*, Acta. Math. Sin. Eng. Ser., 18(1), 147–156.

[38] Miclo, L. (1999), *An example of application of discrete Hardy's inequalities*, Markov Processes Relat. Fields 5, 319–330.

[39] Muckenhoupt, B. (1972), *Hardy's inequality with weights*, Studia Math. XLIV, 31–38.

[40] Nash, J. (1958), *Continuity of solutions of parabolic and elliptic equations*, Amer. J. Math., 80, 931–954.

[41] Opic, B. and Kufner, A. (1990), *Hardy-type Inequalities*, Longman, New York.

[42] Saloff-Coste, L. (1997), *Lectures on finite Markov chains*, LNM **1665**, 301–413, Springer-Verlag.

[43] Schoen, R. and Yau, S. T. (1988), *Differential Geometry* (In Chinese), Science Press, Beijing, China.

[44] Sinclair, A. (1993), *Algorithms for Random Generation and Counting: A Markov Chain Approach*, Birkhäuser.

[45] Wang, F. Y. (2000), *Functional inequalities for empty essential spectrum*, J. Funct. Anal. 170, 219–245.

[46] Wang, F. Y. (2001a), *Sobolev type inequalities for general symmetric forms*, Proc. Amer. Math. Soc. 128:12, 3675–3682.

[47] Wang, F. Y. (2001b), *Convergence rates of Markov semigroups in probability distances*, preprint.

[48] Yang, H. C. (1989), *Estimate of the first eigenvalue of a compact Riemannian manifold with Ricci curvature bounded below by a negative constant* (In Chinese), Sci. Sin.(A) 32:7, 698–700.

[49] Zhang, H. J., Lin, X. and Hou, Z. T. (2000), *Uniformly polynomial convergence for standard transition functions*, In "*Birth-death Processes*" by Hou, Z. T. et al (2000), Hunan Sci. Press, Hunan.

[50] Zhang, Y. H. (2001), *Strong ergodicity for continuous-time Markov chains*, J. Appl. Prob. 38, 270–277.

[51] Zhong, J. Q. and Yang, H. C. (1984), *Estimates of the first eigenvalue of a compact Riemannian manifolds*, Sci. Sin. 27:12, 1251–1265.

ICM 2002 · Vol. III · 53–62

Toeplitz Determinants, Random Growth and Determinantal Processes

K. Johansson*

Abstract

We summarize some of the recent developments which link certain problems in combinatorial theory related to random growth to random matrix theory.

2000 Mathematics Subject Classification: 60C05.
Keywords and Phrases: Random matrices, Toeplitz determinants, Determinantal processes, Random Growth, Random permutations.

1.   Introduction

Let σ be a permutation from S_N. We say that $\sigma(i_1), \ldots, \sigma(i_m)$, $i_1 < \cdots < i_m$, is an *increasing subsequence* of σ if $\sigma(i_1) < \cdots < \sigma(i_m)$. The number m is the length of the subsequence. The length of the longest increasing subsequence in σ is denoted by $\ell_N(\sigma)$. If we pick σ from S_N uniformly at random $\ell_N(\sigma)$ becomes a random variable. *Ulam's problem*, [29], is the study of the asymptotic properties as $N \to \infty$ of this random variable in particular its mean. It turns out that there is a surprisingly rich mathematical structure around this problem as we hope will be clear from the presentation below. It has been known for some time that $\mathbb{E}[\ell_N] \sim 2\sqrt{N}$ as $N \to \infty$, [30], [16]. We refer to [2] for some background to the problem. A Poissonized version of the problem can be obtained by letting N be an independent Poisson random variable with mean α. This gives a random variable $L(\alpha)$ with distribution

$$\mathbb{P}[L(\alpha) \leq n] = \sum_{N=0}^{\infty} \frac{e^{-\alpha} \alpha^N}{N!} \mathbb{P}[\ell_N \leq n]. \tag{1.1}$$

Since $\mathbb{P}[\ell_N \leq n]$ is a decreasing function of N, [9], asymptotics of the left hand side of (1.1) can be used to obtain asymptotics of $\mathbb{P}[\ell_N \leq n]$ (de-Poissonization).

---

*Department of Mathematics, Royal Institute of Technology, SE-100 44 Stockholm, Sweden. E-mail: kurtj@math.kth.se

The random variable $L(\alpha)$ can be realized geometrically using Hammersley's picture, [8]. Consider a Poisson process with intensity 1 in the square $[0, \gamma]^2$, $\gamma = \sqrt{\alpha}$. An *up/right path* is a sequence of Poisson points $(x_1, y_1), \ldots, (x_m, y_m)$ in the square such that $x_i < x_{i+1}$ and $y_i < y_{i+1}$, $i = 1, \ldots, m - 1$. The maximal number of points in an up/right path has the same distribution as $L(\alpha)$. A sequence of points realizing this maximum is called a *maximal path*. It is expected from heuristic arguments, see below, that the standard deviation of $L(\alpha)$ should be of order $\gamma^{1/3} = \alpha^{1/6}$. The proof that this is true, [3], and that we can also understand the law of the fluctuations is the main recent result that will be discussed below. Also, the deviations of a maximal path from the diagonal $x = y$ should be of order $\gamma^{2/3}$. This last statement is proved in [11].

A generalization of the random variable $L(\alpha)$ can be defined in the following way. Let $w(i, j)$, $(i, j) \in \mathbb{Z}_+^2$, be independent geometric random variables with parameter q. An *up/right path* π from $(1, 1)$ to (M, N) is a sequence $(1, 1) = (i_1, j_1), (i_2, j_2), \ldots, (i_m, j_m) = (M, N)$, $m = M + N - 1$, such that either $i_{r+1} - i_r = 1$ and $j_{r+1} = j_r$, or $i_{r+1} = i_r$ and $j_{r+1} - j_r = 1$. Set

$$G(M, N) = \max_\pi \sum_{(i,j) \in \pi} w(i, j), \qquad (1.2)$$

where the maximum is taken over all up/right paths π from $(1, 1)$ to (M, N). Alternatively, we can define $G(M, N)$ recursively by

$$G(M, N) = \max(G(M - 1, N), G(M, N - 1)) + w(M, N). \qquad (1.3)$$

Some thought shows that if we let $q = \alpha/N^2$ then $G(N, N)$ converges in distribution to $L(\alpha)$ as $N \to \infty$, [12], so we can view $G(N, N)$ as a generalization of $L(\alpha)$. We can think of (1.2) as a directed last-passage site percolation problem. Since all paths π have the same length, if $w(i, j)$ were a bounded random variable we could relate (1.2) to the corresponding first-passage site percolation problem, with a min instead of a max. in (1.2). The random variable $G(M, N)$ connects with many different problems, a corner growth model, zero-temperature directed polymers, totally asymmetric simple exclusion processes and domino tilings of the Aztec diamond, see [10], [12] and references therein. It is also related to another growth model, the *(discrete) polynuclear growth (PNG) model*, [15], [21] defined as follows. Let $h(x, t) \in \mathbb{N}$ denote the height above $x \in \mathbb{Z}$ at time $t \in \mathbb{N}$. The growth model is defined by the recursion

$$h(x, t + 1) = \max(h(x - 1, t), h(x, t), h(x + 1, t)) + a(x, t), \qquad (1.4)$$

where $a(x, t)$, $(x, t) \in \mathbb{Z} \times \mathbb{N}$, are independent random variables. If we assume that $a(x, t) = 0$ whenever $x - t$ is even, and that the distribution of $a(x, t)$ is geometric with parameter q, then setting $w(i, j) = a(i - j, i + j - 1)$, we obtain $G(i, j) = h(i - j, i + j - 1)$. The growth model (1.4) has some relation to the so called Kardar-Parisi-Zhang (KPZ) equation, [15], and is expected to fall within the so called KPZ-universality class. The exponents $1/3$ and $2/3$ discussed above are the conjectured exponents for $1 + 1$-dimensional growth models in this class.

2. Orthogonal polynomial ensembles

Consider a probability (density) on Ω^N, $\Omega = \mathbb{R}$, $\mathbb{Z}$, $\mathbb{N}$ or $\{0, 1, \ldots, M\}$ of the form

$$u_N(x) = \frac{1}{Z_N} \Delta_N(x)^2 \prod_{j=1}^{M} w(x_j), \qquad (2.5)$$

where $\Delta_N(x) = \prod_{1 \le i < j \le N} (x_i - x_j)$ is the Vandermonde determinant, $w(x)$ is some non-negative weight function on Ω and Z_N is a normalization constant. We call such a probability an *orthogonal polynomial ensemble*. We can think of this as a finite point process on Ω. Let $d\mu$ be Lebesgue or counting measure on Ω and let $p_n(x) = \kappa_n x^n + \ldots$ be the normalized orthogonal polynomials with respect to the measure $w(x)d\mu(x)$ on Ω. The correlation functions $\rho_{m,N}(x_1, \ldots, x_m)$ of the point process are given by determinants, we have a so called *determinantal point process*, [24]. In fact,

$$\rho_{m,N}(x_1, \ldots, x_m) = \frac{N!}{(N-m)!} \int_{\Omega^{N-m}} u_N(x) d\mu(x_{m+1}) \ldots d\mu(x_N) \qquad (2.6)$$

$$= \det(K_N(x_i, x_j))_{1 \le i,j \le m},$$

where the kernel K_N is given by

$$K_N(x,y) = \frac{\kappa_{N-1}}{\kappa_N} \frac{p_N(x)p_{N-1}(y) - p_{N-1}(x)p_N(y)}{x - y} (w(x)w(y))^{1/2}. \qquad (2.7)$$

A computation shows that for bounded $f : \Omega \to \mathbb{C}$,

$$\mathbb{E}[\prod_{j=1}^{N} (1 + f(x_j))] = \sum_{k=0}^{N} \frac{1}{k!} \int_{\Omega^k} \prod_{j=1}^{k} f(x_j) \det(K_N(x_i, x_j))_{1 \le i,j \le k} d^k \mu(x) \qquad (2.8)$$

$$= \det(I + f K_N)_{L^2(\Omega, d\mu)},$$

where the last determinant is the Fredholm determinant of the integral operator on $L^2(\Omega, d\mu)$ with kernel $f(x)K_N(x, y)$. In particular, we can compute hole or gap probabilities, e.g. the probability of having no particle in an interval $I \subseteq \Omega$ by taking $f = -\chi_I$, minus the characteristic function of the interval I. If $x_{\max} = \max x_j$ denotes the position of the rightmost particle it follows that

$$\mathbb{P}[x_{\max} \le a] = \det(I - K_N)_{L^2((a,\infty), d\mu)}. \qquad (2.9)$$

As $N \to \infty$ we can obtain limiting determinantal processes on $\mathbb{R}$ or $\mathbb{Z}$ with kernel K, i.e. the probability (density) of finding particles at $x_1, \ldots, x_m$ is given by $\det(K(x_i, x_m))_{1 \le i,j \le m}$. We will be interested in the limit process around the rightmost particle. This is typically given by the *Airy kernel*, $x, y \in \mathbb{R}$,

$$A(x,y) = \frac{\mathrm{Ai}\,(x)\mathrm{Ai}\,'(y) - \mathrm{Ai}\,'(x)\mathrm{Ai}\,(y)}{x - y}. \qquad (2.10)$$

This limit, the *Airy point process* has a rightmost particle almost surely and its position has the distribution function

$$F_2(\xi) = \det(I - A)_{L^2(\xi,\infty)}, \qquad (2.11)$$

known as the *Tracy-Widom distribution*, [26]. Hole probabilities as functions of the endpoints of the intervals satisfy systems of differential equations, [27], [1]. For example, we have

$$F_2(\xi) = \exp(-\int_\xi^\infty (x - \xi)u(x)^2 dx), \qquad (2.12)$$

where u solves the Painlevé-II equation $u'' = xu + 2u^3$ with boundary condition $u(x) \sim Ai(x)$ as $x \to \infty$.

An example of a measure of the form (2.5) comes from the *Gaussian Unitary Ensemble (GUE)* of random matrices. The GUE is a Gaussian measure on the space $\mathcal{H}_N \cong \mathbb{R}^{N^2}$ of all $N \times N$ Hermitian matrices. It is defined by $d\mu_{\text{GUE},N}(M) = Z_N^{-1} \exp(-\text{tr } M^2)dM$, where dM is the Lebesgue measure on $\mathcal{H}_N$ and Z_N is a normalization constant. The corresponding eigenvalue measure has the form (2.5) with $w(x) = \exp(-x^2)$ and $\Omega = \mathbb{R}$, [17]. Hence the p_n:s are multiples of the ordinary Hermite polynomials. The largest eigenvalue $x_{\max}$ will lie around $\sqrt{2N}$. This is related to the fact that the largest zero of p_N lies around $\sqrt{2N}$. The local asymptotics of $p_N(x) \exp(-x^2/2)$ around this point, $x = \sqrt{2N} + \xi/N^{1/6}\sqrt{2}$, is given by the Airy function, $Ai(\xi)$. This asymptotics, some estimates, (2.9) and (2.11) give the following result,

$$\mathbb{P}_{\text{GUE},N}[\frac{\sqrt{2N}x_{\max} - 2N}{N^{1/3}} \leq \xi] \to F_2(\xi) \qquad (2.13)$$

as $N \to \infty$.

3.   Some theorems

The previous section may seem unrelated to the first but as the next theorems will show the problem of understanding the distribution of $L(\alpha)$ and $G(M,N)$ fits nicely into the machinery of sect. 2.

Theorem 3.1. [10]. *Take* $\Omega = \mathbb{N}$, $M \geq N$ *and* $w(x) = \binom{M-N+x}{x}$ *in (2.5). Then* $G(M,N)$ *is distributed exactly as* $x_{\max}$.

The corresponding orthogonal polynomials are the Meixner polynomials, a classical family of discrete orthogonal polynomials, and we refer to the measure obtained as the *Meixner ensemble*. It is an example of a *discrete orthogonal polynomial ensemble*, [12]. By computing the appropriate Airy asymptotics of the Meixner polynomials we can use (2.9) to prove the next theorem.

Theorem 3.2. [10]. *Let* $\gamma \geq 1$ *be fixed and set* $\omega(\gamma,q) = (1-q)^{-1}(1+\sqrt{q\gamma})^2 - 1$ *and* $\sigma(\gamma,q) = (1-q)^{-1}(q/\gamma)^{1/6}(\sqrt{\gamma} + \sqrt{q})^{2/3}(1 + \sqrt{q\gamma})^{2/3}$. *Then,*

$$\lim_{N \to \infty} \mathbb{P}[\frac{G([\gamma N], N) - \omega(\gamma,q)N}{\sigma(\gamma,q)N^{1/3}} \leq \xi] = F_2(\xi). \qquad (3.14)$$

Thus $G([\gamma N], N)$ fluctuates like the largest eigenvalue of a GUE matrix.

As discussed above by setting $q = \alpha/N^2$ we can obtain $L(\alpha)$ as a limit of $G(N, N)$ as $N \to \infty$. By taking this limit in theorem 3.1, and using the fact that the measure has determinantal correlation functions, we see that $L(\alpha)$ behaves like the rightmost particle in a determinantal point process on $\mathbb{Z}$ given by the *discrete Bessel kernel*, [12], [4],

$$B^\alpha(x, y) = \sqrt{\alpha} \frac{J_x(2\sqrt{\alpha})J_{y+1}(2\sqrt{\alpha}) - J_{x+1}(2\sqrt{\alpha})J_y(2\sqrt{\alpha})}{x - y}, \qquad (3.15)$$

$x, y \in \mathbb{Z}$. This gives

$$\mathbb{P}[L(\alpha) \le n] = \det(I - B^\alpha)_{\ell^2(\{n, n+1, \dots\})}. \qquad (3.16)$$

Once we have this formula we see that all we need is the classical asymptotic formula $\alpha^{1/6} J_{2\sqrt{\alpha} + \xi \alpha^{1/6}}(2\sqrt{\alpha} \to \mathrm{Ai}(\xi)$ as $\alpha \to \infty$ uniformly in compact intervals, and some estimates of the Bessel functions in order to get a limit theorem for $L(\alpha)$:

Theorem 3.3. [3]. *As $\alpha \to \infty$,*

$$\mathbb{P}\left[\frac{L(\alpha) - 2\sqrt{\alpha}}{\alpha^{1/6}} \le \xi\right] \to F_2(\xi). \qquad (3.17)$$

Note the similarity with (2.13), just replace N by $\sqrt{\alpha}$. This result was first proved in [3] by another method, see below. De-poissonizing we get a limit theorem for $\ell_N(\pi)$, see [3].

4. Rewriting Toeplitz determinants

The *Toeplitz determinant* of order n with generating function $f \in L^1(\mathbb{T})$ is defined by

$$D_n(f) = \det(\hat{f}_{i-j})_{1 \le i, j \le n}, \qquad (4.18)$$

where $\hat{f}_k = (2\pi)^{-1} \int_{-\pi}^{\pi} f(e^{i\theta}) e^{-ik\theta} \, d\theta$ are the complex Fourier coefficients of f. Consider the generating function

$$f(z) = \prod_{\ell=1}^{M} (1 + \frac{a_\ell}{z})(1 + b_\ell z), \qquad (4.19)$$

where a_ℓ, b_ℓ are complex numbers. The elementary symmetric polynomial $e_m(a)$, $a = (a_1, \dots, a_M)$ is defined by $\prod_{j=1}^{M}(1 + a_j z) = \sum_{|m| < \infty} e_m(a) z^m$. A straightforward computation shows that when f is given by (4.19) then

$$\hat{f}_{i-j} = \sum_{m=0}^{\infty} e_{m-j}(a) e_{m-k}(b).$$

Insert this into the definition (4.18) and use the Heine identity,

$$\frac{1}{n!} \int_{\Omega^n} \det(\phi_i(x_j))_{1\leq i,j\leq n} \det(\psi_i(x_j))_{1\leq i,j\leq n} d^n\mu(x) \qquad (4.21)$$

$$= \det(\int_\Omega \phi_i(x)\psi_j(x)d\mu(x))_{1\leq i,j\leq n},$$

to see that

$$D_n(f) = \sum_{m_1>m_2>\cdots>m_n\geq 0} \det(e_{m_i-j}(a))_{1\leq i,j\leq n} \det(e_{m_i-j}(b))_{1\leq i,j\leq n}. \qquad (4.22)$$

Here we have removed the $n!$ by ordering the variables. These determinants are again symmetric polynomials, the so called Schur polynomials. Let $\lambda = (\lambda_1, \lambda_2, \dots)$ be a partition and let $\lambda' = (\lambda'_1, \lambda'_2, \dots)$ be the conjugate partition, [23]. Set $m_i = \lambda'_i + n - i$, $i = 1, \dots, n$ and $\lambda'_i = 0$ if $i > n$, so that λ' has at most n parts, $\ell(\lambda') \leq n$, which means that $\lambda_1 \leq n$. Then the Schur polynomial $s_\lambda(a)$ is given by

$$s_\lambda(a) = \det(e_{\lambda'_i-i+j}(a))_{1\leq i,j\leq n} = \det(e_{m_i-j}(a))_{1\leq i,j\leq n}, \qquad (4.23)$$

the Jacobi-Trudi identity. Hence,

$$D_n(f) = \sum_{\lambda\,;\,\lambda_1\leq n} s_\lambda(a)s_\lambda(b), \qquad (4.24)$$

and we have derived Gessel's formula, [7]. If we let $n \to \infty$ in the right hand side we obtain $\prod_{i,j=1}^M (1 - a_ib_j)^{-1}$ by the Cauchy identity, [23]. In the case when all $a_i, b_j \in [0,1]$, $s_\lambda(a)s_\lambda(b) \geq 0$, and we can think of

$$\prod_{i,j=1}^M (1 - a_ib_j)s_\lambda(a)s_\lambda(b) \qquad (4.25)$$

as a probability measure on all partitions λ with at most n parts, the *Schur measure*, [19]. In this formula we can insert the combinatorial definition of the Schur polynomial, [23],

$$s_\lambda(a) = \sum_{T\,:\,\text{sh}\,(T)=\lambda} a_1^{m_1(T)} \dots a_M^{m_M(T)}, \qquad (4.26)$$

where the sum is over all semi-standard Young tableaux T, [23], with shape λ, and $m_i(T)$ is the number of i:s in T.

A connection with the random variables in section 1 is now provided by the Robinson-Schensted-Knuth (RSK) correspondence, [23]. This correspondence maps an $M \times M$ integer matrix to a pair of semi-standard Young tableaux (T, S) with entries from $\{1, 2, \dots, M\}$. If we let the random variables $w(i,j)$ be independent geometric with parameter a_ib_j then the RSK-correspondence maps the measure we get on the integer matrix $(w(i,j))_{1\leq i,j\leq M}$ to the Schur measure (4.25). Also, the RSK-correspondence is such that $G(M, M) = \lambda_1$, the length of the first row. If we put $a_j = 0$ for $N < j \leq M$ and $a_i = b_i = \sqrt{q}$ for $1 \leq i \leq N$ in the Schur measure and set $x_j = \lambda_j + N - j$, $1 \leq j \leq N$, we obtain the result in Theorem 3.1.

In the limit $M = N \to \infty$, $q = \alpha/N^2$, in which case $G(N, N)$ converges to $L(\alpha)$, the Schur measure converges to the so called *Plancherel measure* on partitions, [30], [12]. In the variables $\lambda_i - i$ this measure is a determinantal point process on $\mathbb{Z}$ given by the kernel B^α, (3.15). This result was obtained independently in [4], which also gives a descrption in terms of different coordinates. See also [18] for a more direct geometric relation between GUE and the Plancherel measure. In this limit the Toeplitz determinant formula (4.24) gives

$$\mathbb{P}[L(\alpha) \leq n] = e^{-\alpha} D_n(e^{2\sqrt{\alpha}\cos\theta}). \tag{4.27}$$

This variant of Gessel's formula was the starting point for the original proof of Theorem 3.3 in [3]. The right hand side of (4.27) can be expressed in terms of the leading coefficients of the orthogonal polynomials on $\mathbb{T}$ with respect to the weight $\exp(2\sqrt{\alpha}\cos\theta)$. These orthogonal polynomials in turn can be obtained as a solution to a matrix-valued Riemann-Hilbert problem (RHP), and the asymptotics of this RHP as $\alpha \to \infty$ can be analyzes using the powerful asymptotic techniques developed by Deift and Zhou, [6]. This approach leads to the formula (2.12) for the limiting distribution.

Write $f = \exp(g)$ and insert the definition of the Fourier coefficients into the definition (4.18). By the Heine identity we obtain an integral formula for the Toeplitz determinant,

$$D_n(f) = \frac{1}{(2\pi)^n n!} \int_{[-\pi,\pi]^n} \prod_{1 \leq \mu < \nu \leq n} |e^{i\theta_\mu} - e^{i\theta_\nu}|^2 \prod_{\mu=1}^n e^{g(e^{i\theta_\mu})} d^n\theta \tag{4.28}$$

$$= \int_{U(n)} e^{\operatorname{tr} g(U)} dU.$$

In the last integral dU denotes normalized Haar measure on the unitary group $U(n)$ and the identity is the Weyl integration formula. The limit of (4.27) as $\alpha \to \infty$ is then a so called *double scaling limit* in a unitary matrix model, [20]. The formula (4.27) can also be obtained by considering the integral over the unitary group, see [22].

Another way to obtain the Schur measure is via families of non-intersecting paths which result from a multi-layer PNG model, [13]. The determinants in the measure then come from the Karlin-McGregor theorem or the Lindström-Gessel-Viennot method.

5. A curiosity

Non-intersecting paths can also be used to describe certain tilings, e.g. domino tilings and tilings of a hexagon by rhombi. By looking at intersections with appropriate lines one can obtain discrete orthogonal polynomial ensembles. In the case of tilings of a hexagon by rhombi, which correspond to boxed planar partitions, [5], the Hahn ensemble, i.e. (2.5) with $\Omega = \{0, \ldots, M\}$ and a weight giving the Hahn polynomials, is obtained, [13]. The computation leading to this result also gives a proof of the classical MacMahon formula, [25], for the number of boxed planar partitions in an abc cube, i.e. the number of rhombus tilings of an abc-hexagon. In terms of Schur polynomials the result is

$$\sum_{\mu \,;\, \ell(\mu) \le c} s_{\mu'}(1^a) s_{\mu'}(1^b) = \prod_{i=1}^{a} \prod_{j=1}^{b} \prod_{k=1}^{c} \frac{i+j+k-1}{i+j+k-2}, \tag{5.29}$$

where the right hand side is MacMahon's formula. (Here 1^a means $(1, \ldots, 1)$ with a components.) Comparing this formula with the formula (4.24) we find

$$(-1)^{a+b} D_n(\prod_{\ell=1}^{a}(1 - e^{-i\theta}) \prod_{\ell=1}^{b}(1 - e^{i\theta})) = \prod_{i=1}^{a} \prod_{j=1}^{b} \prod_{k=1}^{c} \frac{i+j+k-1}{i+j+k-2}. \tag{5.30}$$

It has been conjectured by Keating and Snaith, [14], that the following result should hold for the moments of Riemann's ζ-function on the critical line,

$$\lim_{T \to \infty} \frac{1}{(\log T)^{k^2}} \frac{1}{T} \int_0^T |\zeta(1/2 + it)|^{2k} dt = f_{\mathrm{CUE}}(k) a(k), \tag{5.31}$$

where $a(k)$ is a constant depending on the primes,

$$f_{\mathrm{CUE}}(k) = \lim_{n \to \infty} \frac{1}{n^{k^2}} \int_{U(n)} |Z(U, \theta)|^{2k} dU = \prod_{j=0}^{k-1} \frac{j!}{(j+k)!} \tag{5.32}$$

and $Z(U, \theta) = \det(I - U e^{-i\theta})$ is the characteristic polynomial of the unitary matrix U. If we take $a = b = k$ in (5.30) and use (4.28) we find

$$\int_{U(n)} |Z(U, \theta)|^{2k} dU = \prod_{i=1}^{k} \prod_{j=1}^{k} \prod_{\ell=1}^{n} \frac{i+j+\ell-1}{i+j+\ell-2} = \prod_{j=0}^{n-1} \frac{j!(j+2k)!}{(j+k)!^2} \tag{5.33}$$

as computed in [14] by different methods. Letting $n \to \infty$ we obtain the last expression in (5.32). Hence, we see that the formula (5.33) has a curious combinatorial interpretation via MacMahon's formula.

References

[1] M. Adler, T. Shiota & P. van Moerbeke, *Random matrices, vertex operators and the Virasoro algebra*, Phys. Lett., **A 208** (1995), 67–78.

[2] D. Aldous & P. Diaconis, *Longest increasing subsequences: From patience sorting to the Baik-Deift-Johansson theorem*, Bull. AMS, **36** (1999), 199–213.

[3] J. Baik, P. A. Deift & K. Johansson, *On the distribution of the length of the longest increasing subsequence in a random permutation*, J. Amer. Math. Soc., **12** (1999), 1119–1178.

[4] A. Borodin, A. Okounkov & G. Olshanski, *Asymptotics of Plancherel measures for symmetric groups,* J. Amer. Math. Soc. **13** (2000), 481–515.

[5] H. Cohn, M. Larsen & J. Propp, *The shape of a typical boxed plane partition,* New York J. of Math., **4** (1998), 137–165.

[6] P. Deift, T. Kriecherbauer, K. T.-R. McLaughlin, S. Venakides & X. Zhou, *Uniform asymptotics for orthogonal polynomials,* Proceedings of the International Congress of Mathematicians, Vol. III (Berlin, 1998), Doc. Math. Extra Vol. III (1998), 491–501.

[7] I. M. Gessel, *Symmetric functions and P-recursiveness,* J. Combin. Theory Ser. A, **53** (1990), 257–285.

[8] J. M. Hammersley, *A few seedlings of research,* in *Proc. Sixth Berkeley Symp. Math. Statist. and Probability,* Volume 1, University of California Press (1972), 345–394.

[9] K. Johansson, *The longest increasing subsequence in a random permutation and a unitary random matrix model,* Math. Res. Lett., **5** (1998), 63–82.

[10] K. Johansson, *Shape fluctuations and random matrices,* Commun. Math. Phys., **209** (2000), 437–476.

[11] K. Johansson, *Transversal fluctuations for increasing subsequences on the plane,* Probab. Theory Related Fields, **116** (2000), 445–456.

[12] K. Johansson, *Discrete orthogonal polynomial ensembles and the Plancherel measure,* Annals of Math., **153** (2001), 259–296.

[13] K. Johansson, *Non-intersecting paths, random tilings and random matrices,* Probab. Th. Rel. Fields (to appear).

[14] J. P. Keating & N. C. Snaith, *Random Matrix Theory and $\zeta(1/2+it)$,* Commun. Math. Phys., **214** (2000), 57–89.

[15] J. Krug & H. Spohn, *Kinetic Roughening of Growing Interfaces,* in *Solids far from Equilibrium: Growth, Morphology and Defects,* C Godrèche ed., Cambridge University Press (1992),479–582.

[16] B. F. Logan & L. A. Shepp, *A variational problem for random Young tableaux,* Advances in Math., **26** (1977), 206–222.

[17] M. L. Mehta, *Random Matrices,* 2nd ed., Academic Press, San Diego 1991.

[18] A. Okounkov, *Random matrices and random permutations,* Internat. Math. Res. Notices, no.20 (2000), 1043–1095.

[19] A. Okounkov, *Infinite wedge and random partitions,* Selecta Math. (N.S.), **7** (2001), 57–81.

[20] V. Periwal & D. Shevitz, *Unitary-Matrix Models as Exactly Solvable String Theories,* Phys. Rev. Lett., **64** (1990), 1326–1329.

[21] M. Prähofer & H. Spohn, *Statistical Self-Similarity of One-Dimensional Growth Processes,* Physica A, **279** (2000), 342.

[22] E. Rains, *Increasing subsequences and the classical groups,* Elcctr. J. of Combinatorics, **5(1)** (1998), R12.

[23] B. Sagan, *The Symmetric Group,* Brooks/Cole Publ. Comp., 1991.

[24] A. Soshnikov, *Determinantal random point fields,* Russian Math. Surveys, **55** (2000), 923–975.

[25] R. P. Stanley, *Enumerative Combinatorics,* Vol. 2, Cambridge University Press

(1999).

[26] C. A. Tracy & H. Widom, *Level Spacing Distributions and the Airy Kernel,* Commun. Math. Phys., **159** (1994), 151–174.

[27] C. A. Tracy & H. Widom, *Fredholm determinants, differential equations and matrix models,* Commun. Math. Phys., **163** (1995), 33–72.

[28] C. A. Tracy & H. Widom, *Correlation Functions, Cluster Functions, and Spacing Distributions for Random Matrices,* J. Statist. Phys., **92** (1998), 809–835.

[29] S. M. Ulam, *Monte Carlo calculations in problems of mathematical physics,* in *Modern Mathematics for the Engineers,* E. F. Beckenbach ed., McGraw-Hill (1961), 261–281.

[30] A. Vershik & S. Kerov, *Asymptotics of the Plancherel measure of the symmetric group and the limiting form of Young tables,* Soviet Math. Dokl., **18** (1977), 527–531.

Conformal Invariance, Universality, and the Dimension of the Brownian Frontier

G. Lawler*

Abstract

This paper describes joint work with Oded Schramm and Wendelin Werner establishing the values of the planar Brownian intersection exponents from which one derives the Hausdorff dimension of certain exceptional sets of planar Brownian motion. In particular, we proof a conjecture of Mandelbrot that the dimension of the frontier is 4/3. The proof uses a universality principle for conformally invariant measures and a new process, the stochastic Loewner evolution (SLE), introduced by Schramm. These ideas can be used to study other planar lattice models from statistical physics at criticality. I discuss applications to critical percolation on the triangular lattice, loop-erased random walk, and self-avoiding walk.

2000 Mathematics Subject Classification: 60J65, 60K35.
Keywords and Phrases: Brownian motion, Critical exponents, Conformal invariance, Stochastic Loewner evolution.

1. Exceptional sets for planar Brownian motion

Let B_t be a standard Brownian motion taking values in $\mathbb{R}^2 = \mathbb{C}$ and let $B[s,t]$ denote the random set $B[s,t] = \{B_r : s \leq r \leq t\}$. For $0 \leq t \leq 1$, we say that B_t is a

- *cut point* for $B[0,1]$ if $B[0,t) \cap B(t,1] = \emptyset$;
- *frontier point* for $B[0,1]$ if B_t is on the boundary of the unbounded component of $\mathbb{C} \setminus B[0,1]$;
- *pioneer point* for $B[0,1]$ if B_t is on the boundary of the unbounded component of $\mathbb{C} \setminus B[0,t]$, i.e., if B_t is a frontier point for $B[0,t]$.

I will discuss the following result proved by Oded Schramm, Wendelin Werner, and myself.

---

*Department of Mathematics, Cornell University, Malott Hall, Ithaca, NY 14853-4201 and Department of Mathematics, Duke University, Durham, NC 27708-0320, USA. E-mail: lawler@math.cornell.edu

Theorem 1. [17, 18, 20] *If B_t is a standard Brownian motion in $\mathbb{R}^2 = \mathbb{C}$, then with probability one,*

$$\dim_h [\text{ cut points for } B[0,1]] = 3/4,$$

$$\dim_h [\text{ frontier points for } B[0,1]] = 4/3,$$

$$\dim_h [\text{ pioneer points for } B[0,1]] = 7/4,$$

where $\dim_h$ denotes Hausdorff dimension.

Mandelbrot [27] first gave the conjecture for the Brownian frontier, basing his conjecture on numerical simulation and then noting that simulations of the frontier resembled simulations of self-avoiding walks. It is conjectured that the scaling limit of planar self-avoiding walks has paths of dimension 4/3. Duplantier and Kwon [5] used nonrigorous conformal field theory techniques to make the above conjectures for the cut points and pioneer points. More precisely, they made conjectures about certain exponents called the Brownian or simple random walk *intersection exponents*. More recently, Duplaniter [6] has given other nonrigorous arguments for the conjectures using quantum gravity.

To prove Theorem 1, it suffices to find the values of the Brownian intersection exponents. In fact, before Theorem 1 had been proved, it had been established [11, 12, 13] that the Hausdorff dimensions of the set of cut points, frontier points, and pioneer points were $2 - \eta_1, 2 - \eta_2$, and $2 - \eta_3$, respectively, where η_1, η_2, η_3 are defined by saying that as $\epsilon \to 0+$,

$$\mathbf{P}\{B[0, \tfrac{1}{2} - \epsilon^2] \cap B[\tfrac{1}{2} + \epsilon^2, 1] = \emptyset\} \approx \epsilon^{\eta_1},$$

$$\mathbf{P}\{B[0, \tfrac{1}{2} - \epsilon^2] \cup B[\tfrac{1}{2} + \epsilon^2, 1] \text{ does not disconnect } B_{1/2} \text{ from infinity}\} \approx \epsilon^{\eta_2},$$

$$\mathbf{P}\{B[\epsilon^2, 1] \text{ does not disconnect } 0 \text{ from infinity}\} \approx \epsilon^{\eta_3}.$$

It had also been established [3, 16] that the analogous exponents for simple random walk are the same as for Brownian motion.

There are two main ideas in the proof. The first is a one parameter family of conformally invariant processes developed by Oded Schramm [30] which he named the Stochastic Loewner evolution (*SLE*). The second is the idea of "universality" which states roughly that all conformally invariant measures that satisfy a certain "locality" or "restriction" property must have the same exponents as Brownian motion (see [26]). In this paper, I will define *SLE* and give some of its properties; describe how analysis of *SLE* leads to finding the Brownian intersection exponents; and finally describe some other planar lattice models in statistical physics at criticality that can be understood using *SLE*.

2. Stochastic Loewner evolution

I will give a brief introduction to the stochastic Loewner evolution (*SLE*); for more details, see [29, 17, 18, 15, 28]. Let W_t denote a standard one dimensional

Brownian motion. If $\kappa \geq 0$ and z is in the upper half plane $\mathbb{H} = \{w \in \mathbb{C} : \Im(w) > 0\}$, let $g_t(z)$ be the solution to the Loewner differential equation

$$\partial_t g_t(z) = \frac{2}{g_t(z) - \sqrt{\kappa}\, W_t}, \quad g_0(z) = z. \tag{2.1}$$

For each $z \in \mathbb{H}$, the solution $g_t(z)$ is defined up to a time $T_z \in (0, \infty]$. Let $H_t = \{z : T_z > t\}$. Then g_t is the unique conformal transformation of H_t onto $\mathbb{H}$ with $g_t(z) - z = o(1)$ as $z \to \infty$. In fact,

$$g_t(z) = z + \frac{2t}{z} + O(\frac{1}{|z|^2}), \quad z \to \infty.$$

It is easy to show that the maps g_t are well defined. It has been shown [28, 22] that there is a (random) continuous path $\gamma : [0, \infty) \to \overline{\mathbb{H}}$ such that H_t is the unbounded component of $\mathbb{H} \setminus \gamma[0, t]$ and $g_t(\gamma(t)) = \sqrt{\kappa}\, W_t$. The conformal maps g_t or the corresponding paths $\gamma(t)$ are called the *chordal stochastic Loewner evolution with parameter κ (chordal SLE_κ)*. It is easy to check that the distribution of SLE_κ is invariant (modulo time change) under dilations $z \mapsto rz$. Using this, we can use conformal transformations to define chordal SLE_κ connecting two distinct boundary points of any simply connected domain. This gives a family of probability measures on curves (modulo reparametrization) on such domains that is invariant under conformal transformation.

Chordal SLE_κ can also be considered as the only probability distributions on continuous curves (modulo reparametrization) $\gamma : [0, \infty) \to \overline{\mathbb{H}}$ with the following properties.

- $\gamma(0) = 0, \gamma(t) \to \infty$ as $t \to \infty$, and $\gamma(t) \in \partial H_t$ for all $t \in [0, \infty)$, where H_t is the unbounded component of $\mathbb{H} \setminus \gamma[0, t]$.
- Let $h_t : H_t \to \mathbb{H}$ be the unique conformal transformation with $h_t(\gamma(t)) = 0, h_t(\infty) = \infty, h'_t(\infty) = 1$. Then the conditional distribution of $\hat{\gamma}(s) := h_t \circ \gamma(s + t), 0 \leq s < \infty$, given $\gamma[0, t]$ is the same as the original distribution.
- The measure is invariant under $x + iy \mapsto -x + iy$.

There is a similar process called *radial SLE_κ* on the unit disk. Let W_t be as above, and for z in the unit disk $\mathbb{D}$, consider the equation

$$\partial_t g_t(z) = g_t(z) \frac{e^{i\sqrt{\kappa} W_t} + g_t(z)}{e^{i\sqrt{\kappa} W_t} - g_t(z)}, \quad g_0(z) = z.$$

Let U_t be the set of $z \in \mathbb{D}$ for which $g_t(z)$ is defined. It can be shown that there is a random path $\gamma : [0, \infty) \to \overline{\mathbb{D}}$, such that U_t is the component of $\mathbb{D} \setminus \gamma[0, t]$ containing the origin; $g_t(\gamma(t)) = e^{i\sqrt{\kappa} W_t}$; and g_t is a conformal transformation of U_t onto $\mathbb{D}$ with $g_t(0) = 0, g'_t(0) = e^t$. We can define radial SLE_κ connecting any boundary point to any interior point of a simply connected domain by conformal transformation.

The qualitative behavior of the paths γ varies considerably as κ varies, although chordal and radial SLE_κ for the same κ are qualitatively similar. The

Hausdorff dimension of $\gamma[0,t]$ for chordal or radial SLE_κ is conjectured to be $\min\{1+(\kappa/8),2\}$. This has been proved for $\kappa = 8/3, 6$, see [2], and for other κ it is a rigorous upper bound [28]. For $0 \le \kappa \le 4$, the paths γ are simple (no self-intersections) and $\gamma(0,\infty)$ is a subset of $\mathbb{H}$ or $\mathbb{D}$. For $\kappa > 4$, the paths have double points and hit $\partial\mathbb{H}$ or $\partial\mathbb{D}$ infinitely often. If $\kappa \ge 8$, the paths are space filling.

Investigation of SLE_κ requires studying the behavior of SLE_κ under conformal maps. Suppose A is a compact subset of $\overline{\mathbb{H}}$ not containing the origin such that $\overline{A \cap \mathbb{H}} = A$ and $\mathbb{H} \setminus A$ is simply connected. Let Φ denote the conformal transformation of $\mathbb{H} \setminus A$ onto $\mathbb{H}$ with $\Phi(0) = 0, \Phi(\infty) = \infty, \Phi'(\infty) = 1$. Let γ denote a chordal SLE_κ starting at the origin, and let T be the first time t that $A \cap \mathbb{H} \not\subset H_t$. For $t < T$, let $\tilde\gamma(t) = \Phi \circ \gamma(t)$. Let $\tilde g_t$ be the conformal transformation of the unbounded component of $\mathbb{H} \setminus \tilde\gamma[0,t]$ onto $\mathbb{H}$ with $\tilde g_t(z) - z = o(1)$ as $z \to \infty$; define $a(t)$ by $\tilde g(z) - z \sim a(t) z^{-1}$. Then $\tilde g(t)$ satisfies the modified Loewner equation

$$\partial_t \tilde g(t) = \frac{\partial_t a}{\tilde g_t(z) - \tilde W_t}, \quad \tilde g_0(z) = z,$$

for some $\tilde W_t$. In fact $\tilde W_t = \tilde g_t \circ \Phi \circ g_t^{-1}(\sqrt\kappa\, W_t)$. Using the Loewner differential equation and Itô's formula, we can write $\tilde W_t$ as a local semimartingale, $d\tilde W_t = b(t)\,dt + \sqrt{\kappa\,\partial_t a/2}\,dW_t$; here $b(t)$ and $a(t)$ are random depending on $W_s, 0 \le s \le t$. For $\kappa = 6$, and only $\kappa = 6$, the drift term $b(t)$ disappears and hence $\tilde W_t$ is a time change of Brownian motion.

Locality property for SLE_6. [17] *If $\kappa = 6$, $\tilde\gamma(t), 0 \le t < T$, has the same distribution as a time change of SLE_6.*

For other values of κ, the image $\tilde\gamma(t), t < T$, has a distribution that is absolutely continuous with respect to that of (a time change of) SLE_κ. This follows from Girsanov's theorem (see, e..g, [1, Theorem I.6.4]) that states roughly that Brownian motions with the same variance but different drifts give rise to absolutely continuous measures on paths. Similarly, radial SLE_κ can be obtained from chordal SLE_κ by considering its image under a map taking $\mathbb{H}$ to $\mathbb{D}$. For all values of κ we get absolutely continuous measures (which is why radial SLE_κ is qualitatively the same as chordal SLE_κ), but for $\kappa = 6$ we get a special relationship [18, Theorem 4.1].

One of the reasons that SLE_κ is useful is that "crossing probabilities" and "critical exponents" for the process can be calculated. The basic idea is to relate an event about the planar path γ to an event about the driving process $\sqrt\kappa W_t$ and then to use standard methods of stochastic calculus to relate this to solutions of partial differential equations. As an example, consider chordal SLE_6 in the upper half plane $\mathbb{H}$ going from $x \in (0,1)$ to infinity. Let T be the first time t that $\gamma(t) \in (-\infty, 0] \cap [1, \infty)$; since $\kappa > 4$, $T < \infty$ with probability one. Let $\mathcal{E}$ be the event that $\gamma(T) \in (-\infty, 0]$, and let H_T be the unbounded component of $\mathbb{H} \setminus \gamma[0,T]$. Let y_1, y_2 be the minimum and maximum of $\gamma[0,T] \cap \mathbb{R}$; on the event $\mathcal{E}$, $y_1 \le 0$ and $x \le y_2 < 1$. Let $\mathcal{L}$ denote the π-*extremal distance* between $(-\infty, y_1]$ and $[y_2, 1]$ in H_T, i.e., the number $\mathcal{L}$ such that H_T can be mapped conformally onto $[0, \mathcal{L}] \times [0, \pi]$ in a way that $(-\infty, y_1]$ and $[y_2, 1]$ are mapped onto the vertical boundaries. In order to relate SLE_6 to intersection exponents for Brownian motion one needs to understand the behavior of $\mathbf{E}^x[1_{\mathcal{E}} \exp\{-\lambda\mathcal{L}\}]$ as $x \to 1-$ for $\lambda \geq 0$. It is not hard to show that this quantity is closely related to $\mathbf{E}^x[1_{\mathcal{E}} \; g_T'(1)^\lambda]$. If we differentiate (2.1) with respect to z we get an equation for $\partial_t g_t'(1)$, and standard techniques of stochastic calculus can be applied to give a differential equation for the function $r(x, \lambda) = \mathbf{E}^x[1_{\mathcal{E}} \; g_T'(1)^\lambda]$. We get an exact solution in terms of hypergeometric functions [17, Theorem 3.2]. If $\lambda = 0$, so that $r(x) = \mathbf{P}^x[\mathcal{E}]$, we get the formula given by Cardy [4] for crossing probabilities of percolation clusters (see §3.2.).

3. Applications

3.1. Brownian motion

As already mentioned, computation of dimensions for many exceptional sets for Brownian motion reduces to finding the Brownian intersection exponents. These exponents, which can be defined in terms of crossing probabilities for non-intersecting paths, were studied in [25, 26]. In these papers, relations were given between different exponents and a "universality" principle was shown for conformally invariant processes satisfying an additional hypothesis (the term completely conformally invariant was used there). Heuristic arguments indicated that self-avoiding walks and percolation should also satisfy this hypothesis. Unfortunately, from a rigorous standpoint, we had only reduced a hard problem, computing the Brownian intersection exponents, to the even harder problem of showing conformal invariance and computing the exponents for self-avoiding walks or critical percolation.

At the same time Schramm [29] was completing his beautiful construction of SLE_κ and conjecturing that SLE_6 gave the boundaries of critical percolation clusters. While he was unable to prove that critical percolation has a conformally invariant limit, he was able to conclude that if the limit was conformally invariant then it must be SLE_6. The identification $\kappa = 6$ was determined from rigorous "crossing probabilities" for SLE_κ; only $\kappa = 6$ was consistent with Cardy's formula (see §3.2.) or even the simple fact that a square should have crossing probability $1/2$.

Since both Brownian motion and SLE_6 were conjectured to be related to the scaling limit of critical percolation, it was natural to try to use SLE_6 to prove results about Brownian motion (and, as mentioned before, the Hausdorff dimension of exceptional sets on the path); see [17, 18, 20, 19]. There were two major parts of the proof. First, the locality property for SLE_6 was formulated and proved; this allowed ideas as in [26] to show that the exponents of SLE_6 can be used to find the exponents for Brownian motion. Second, the exponents for SLE_6 had to be computed. The basic idea is discussed at the end of the last section. What makes SLE so powerful is that it reduces problems about a two-dimensional process to analysis of a one-dimensional stochastic differential equation (and hence a partial differential equation in one space variable).

The universality in these papers was in terms of exponents. We now know that the paths of planar Brownian motion and SLE_6 are even more closely related.

The "hull" generated by an SLE_6 is the same as the hull generated by a Brownian motion with oblique reflection (see [33]). In particular, the frontiers (outer boundaries) of the two processes have the same dimension. There are now direct proofs that the Hausdorff dimension of the frontier of SLE_6 is 4/3 ([2]) and this stronger universality principle implies the same holds for Brownian paths.

3.2. Critical percolation

Suppose each vertex of the planar triangular lattice is colored independently white or black, with the probability of a white being 1/2. This is called *critical percolation* (on the triangular lattice). Let D be a simply connected domain in $\mathbb{C} = \mathbb{R}^2$ and let A_1, A_2 be disjoint nontrivial connected arcs on ∂D. Consider the limit as $\delta \to 0$ of the probability that in critical percolation on a lattice with mesh size δ that there is a connected set of white vertices in D connecting A_1, A_2. It has long been believed that this limit, $p(A_1, A_2; D)$, exists and is strictly between 0 and 1. (Note: if the probability of a white vertex is p, then $p(A_1, A_2; D)$ is 0 for $p < 1/2$ and 1 for $p > 1/2$. One of the features of *critical* percolation is the fact that this quantity is strictly between 0 and 1.) Moreover, it has been conjectured that $p(A_1, A_2; D)$ is a conformal invariant [4, 10] . It is also believed that this limit does not depend on the nature of the lattice; for example, critical bond percolation in $\mathbb{Z}^2$ (each bond is colored white or black independently with probability 1/2) should give the same limit.

Cardy [4] used nonrigorous methods from conformal field theory to find an exact formula for $p(A_1, A_2; D)$; his calculations were done for $D = \mathbb{H}$ and the formula involves hypergeometric functions. Carleson noted that the formula was much nicer if one chooses D to be an equilateral triangle of side length 1; A_1, one of the sides; and A_2, a line segment of length x with one endpoint on the vertex opposite A_1. In this case, Cardy's formula is $p(A_1, A_2; D) = x$. Schramm [29] went further and, *assuming existence and conformal invariance of the limit*, showed that the limiting boundary between black and white clusters can be given in terms of SLE_6. If A_3 denotes the third side of the triangle (so that $A_3 \cap A_2$ is a single point), we can consider the limiting cluster formed by taking all the white vertices that are connected by a path of white vertices to A_3. In the limit, the outer boundary of this "hull" has the same distribution as the outer boundary of the hull of chordal SLE_6 going from the vertex $A_3 \cap A_1$ to the vertex $A_3 \cap A_2$. The identification with SLE comes from the conformal invariance assumption; Schramm determined the value $\kappa = 6$ from a particular crossing probability, but we now understand this in terms of the locality property which scaling limits of these boundary curves can be seen to satisfy. Cardy's formula (and generalizations) were computed for SLE_κ in [17].

Recently Smirnov [31] made a major breakthrough by proving conformal invariance and Cardy's formula for the limit of critical percolation in the triangular lattice. As a corollary, the identification of the limit with SLE_6 has become a theorem. This has also led to rigorous proofs of a number of critical exponents for the lattice model [21, 30, 32]. The basic strategy is to compute the exponent for SLE_6 and to then to use Smirnov's result to relate this exponent to lattice percolation.

It is an open problem to show that critical percolation on other planar lattices, e.g., bond percolation on the square lattice, has the same limiting behavior.

3.3. Loop-erased random walk

Loop-erased random walk (LERW) in a finite set $A \subset \mathbb{Z}^2$ starting at $0 \in A$ is the measure on self-avoiding paths obtained from starting a simple random walk at the origin, stopping at the first time that it leaves A, and erasing loops chronologically from the path. It can also be defined as a nonMarkov chain which at each time n chooses a new step using probabilities weighted by the probability that simple random walk starting at the new point avoids the path up to that point (see, e.g., [14]). It is also related to uniform spanning trees; if one choose a spanning tree uniformly among all spanning trees of A, considered as a graph with appropriate boundary conditions, then the distribution of the unique self-avoiding path from the origin to the boundary is the same as LERW. Wilson gave a beautiful algorithm to generate uniform spanning trees using LERW [34].

One can hope to define a scaling limit of planar LERW on a domain connecting an interior point to a boundary point by taking LERW on finer and finer grids and taking the limit. There are a number of reasons to believe that this limit is conformally invariant. For example. the limit of simple random walk (Brownian motion) is conformally invariant and the ordering of points used in the loop-erasing procedure is not changed under conformal maps. Also, certain crossing probabilities for LERW can be given by determinants of probabilities for simple random walk (see [7]), and hence these quantities are conformally invariant. Kenyon [9] used a conformal invariance argument (using a determinant relation from a related domino tiling model) to prove that the growth exponent for LERW is 5/4; roughly, this says it takes about $r^{5/4}$ steps for a LERW to travel distance r.

Schramm [29] showed that under the assumption of conformal invariance, the scaling limit of LERW must be radial SLE_2. He used conformal invariance and a natural Markovian-type property of LERW to conclude that it must be an SLE_κ, and then he used Kenyon's result to determine κ. Recently, Schramm, Werner, and I [22] proved that the scaling limit of loop-erased random walk is SLE_2.

There is another path obtained from the uniform spanning tree that has been called the uniform spanning tree Peano curve. This path, which lies on the dual lattice, encodes the entire tree (not just the path from the origin to the boundary). A similar, although somewhat more involved, argument can be used to show that this process converges to the space-filling curve SLE_8 [22].

3.4. Self-avoiding walk

A *self-avoiding walk (SAW)* in the lattice $\mathbb{Z}^2$ is a nearest neighbor walk with no self-intersections. The problem of the SAW is to understand the uniform measure on all such walks of a given length (or sometimes the measure that assigns weight a^n to all walks of length n). It is still an open problem to prove there is a limiting distribution; it is believed that such a limit in conformally invariant (see [23] for precise statements). However, if the conjectures hold there is only one possible limit, $SLE_{8/3}$.

The conformal invariance property leads one to conclude that the limit must be an SLE_κ and $\kappa \leq 4$ is needed in order to have a measure on simple paths. The property that $SLE_{8/3}$ has that is not held by SLE_κ for other $\kappa \leq 4$ is the *restriction property*. The restriction property is similar to, but not the same, as the locality property. Let A be a compact subset and Φ the transformation as in §2. Then [24] if $\gamma[0, \infty)$ is an $SLE_{8/3}$ path from 0 to ∞, the distribution of $\Phi \circ \gamma$ given the event $\{\gamma[0, \infty) \cap A = \emptyset\}$ is the same as (a time change of) $SLE_{8/3}$. In fact, the probability that $\{\gamma[0, \infty) \cap A = \emptyset\}$ is $\Phi'(0)^{5/8}$.

If the scaling limit of SAW has a conformally invariant limit then one can show easily that the limit satisfies the restriction property. Hence, the only candidate for the limit (assuming a conformally invariant scaling limit) is $SLE_{8/3}$. The conjectures for critical exponents for SAW can be interpreted in terms of rigorous properties of $SLE_{8/3}$ (see [23]). For example, the Hausdorff dimension of $SLE_{8/3}$ paths is 4/3 [2, 24]; this gives strong evidence that the limit of SAWs should give paths of dimension 4/3. Monte Carlo simulations [8] support the conjecture that the limit of SAW is $SLE_{8/3}$.

Acknowledgment. Oded Schramm and Wendelin Werner should be considered co-authors of this paper since this describes joint work. I thank both of them for an exciting collaboration.

References

[1] R. Bass, *Probabilistic Techniques in Analysis*, Springer-Verlag, 1995.

[2] V. Beffara, Hausdorff dimensions for SLE_6, preprint.

[3] K. Burdzy & G. Lawler, Non-intersection exponents for random walk and Brownian motion. Part II: Estimates and applications to a random fractal, *Ann. Probab*, 18 (1990), 981–1009.

[4] J. Cardy, Critical percolation in finite geometries, *J. Phys. A*, 25 (1992), L201–L206.

[5] B. Duplantier & K.-H. Kwon, Conformal invariance and intersections of random walks, *Phys. Rev. Lett.*, 61, 2514–2517.

[6] B. Duplantier, Random walks and quantum gravity in two dimensions, *Phys. Rev. Let.* , 81 (1998), 5489–5492.

[7] S. Fomin, Loop-erased walks and total positivity, *Trans. Amer. Math. Soc.*, 353 (2001), 3563–3583.

[8] T. Kennedy, Monte Carlo tests of SLE predictions for the 2D self-avoiding walk, *Phys. Rev. Lett.* 88 (2002), 130601

[9] R. Kenyon, The asymptotic determinant of the discrete Laplacian, *Acta Math.*, 185 (2000), 239–286.

[10] R. Langlands, P. Pouliot, & Y. Saint-Aubin, Conformal invariance in two-dimensional percolation, *Bull. Amer. Math. Soc. (N.S.)* 90 (1994), 1–61.

[11] G. Lawler, Hausdorff dimension of cut points for Brownian motion, *Electronic J. Probab.*, 1 (1996), paper no. 2.

[12] G. Lawler, The dimension of the frontier of planar Brownian motion, *Electronic Comm. Probab.*, 1 (1996), paper no. 5.

[13] G. Lawler, Geometric and fractal properties of Brownian motion and random walk paths in two and three dimensions, in *Random Walks, Budapest 1998*, Bolyai Mathematical Studies, 9 (1999), 210–258.

[14] G. Lawler, Loop-erased random walk, *Perplexing Problems in Probability*, Birkhäuser (1999), 197–217.

[15] G. Lawler, An introduction to the stochastic Loewner evolution, preprint.

[16] G. Lawler & E. Puckette, The intersection exponent for simple random walk, *Combinatorics, Probab., and Computing*, 9 (2000), 441–464.

[17] G. Lawler, O. Schramm, & W. Werner, Values of Brownian intersection exponents I: Half-plane exponents, *Acta. Math.*, 187 (2001), 237–273.

[18] G. Lawler, O. Schramm, & W. Werner, Values of Brownian intersection exponents II: Plane exponents, *Acta. Math.*, 187 (2001), 275–308.

[19] G. Lawler, O. Schramm, & W. Werner, Values of Brownian intersection exponents III: Two-sided exponents, *Ann. Inst. Henri Poincaré*, 38 (2002), 109–123.

[20] G. Lawler, O. Schramm, & W. Werner, Analyticity of intersection exponents for planar Brownian motion, *Acta. Math.*, to appear.

[21] G. Lawler, O. Schramm, & W. Werner, One arm exponent for critical 2D percolation, *Electronic J. Probab.*, 7 (2002), paper no. 2.

[22] G. Lawler, O. Schramm, & W. Werner, Conformal invariance of planar loop-erased random walk and uniform spanning trees, preprint.

[23] G. Lawler, O. Schramm, & W. Werner, On the scaling limit of planar self-avoiding walk, preprint.

[24] G. Lawler, O. Schramm, & W. Werner, Conformal restriction properties: the chordal case, in preparation.

[25] G. Lawler & W. Werner, Intersection exponents for planar Brownian motion, *Annals of Probab.*, 27 (1999), 1601–1642.

[26] G. Lawler & W. Werner, Universality for conformally invariant intersection exponents, J. European Math. Soc. 2 (2000), 291–328.

[27] B. Mandelbrot, *The Fractal Geometry of Nature*, Freeman, 1982.

[28] S. Rohde & O. Schramm, Basic properties of SLE, preprint.

[29] O. Schramm, Scaling limits of loop-erased random walks and uniform spanning trees, *Israel J. Math*, 118 (2001), 221–288.

[30] O. Schramm, A percolation formula, *Electronic Comm. Probab.*, 8 (2001), paper no. 12.

[31] S. Smirnov, Critical percolation in the plane: Conformal invariance, Cardy's formula, scaling limits, *C. R. Acad. Sci. Paris. Sr. I Math.*, 333 (2001), 239–244.

[32] S. Smirnov & W. Werner, Critical exponents for two-dimensional percolation, *Math. Res. Lett.*, to appear.

[33] W. Werner, Critical exponents, conformal invariance and Brownian motion, *Proceedings of the 3rd Europ. Congress Math.*, Prog. Math 202 (2001), 87–103.

[34] D. Wilson, Generating random spanning trees more quickly than the cover time, *Proceedings of the Twenty-eighth Annual ACM Symposium on the Theory of Computing*, ACM (1996), 296–303.

Brownian Intersections, Cover Times and Thick Points via Trees

Yuval Peres*

Abstract

There is a close connection between intersections of Brownian motion paths and percolation on trees. Recently, ideas from probability on trees were an important component of the multifractal analysis of Brownian occupation measure, in joint work with A. Dembo, J. Rosen and O. Zeitouni. As a consequence, we proved two conjectures about simple random walk in two dimensions: The first, due to Erdős and Taylor (1960), involves the number of visits to the most visited lattice site in the first n steps of the walk. The second, due to Aldous (1989), concerns the number of steps it takes a simple random walk to cover all points of the n by n lattice torus. The goal of the lecture is to relate how methods from probability on trees can be applied to random walks and Brownian motion in Euclidean space.

2000 Mathematics Subject Classification: 60J15.
Keywords and Phrases: Random walk, Cover time, Thick point, Lattice, Brownian motion, Percolation, Tree.

1. Introduction

In [18], the author showed that long-range intersection probabilities for random walks, Brownian motion paths and Wiener sausages in Euclidean space, can be estimated up to constant factors by survival probabilities of percolation processes on trees.

More recently, several long-standing problems involving cover times and "thick points" for random walks in two dimensions were solved in joint works [9, 10] of A. Dembo, J. Rosen, O. Zeitouni and the author. These solutions were motivated by powerful analogies with corresponding problems on trees, but these analogies were not discussed explicitly in the research papers cited. The goal of the present note is to describe the tree problems and solutions, that correspond to the problems studied in [9, 10].

*Departments of Statistics & Mathematics, University of California, Berkeley CA, USA. E-mail: peres@stat.berkeley.edu

The *cover time* for a random walk on a finite graph is the number of steps it takes the random walk to visit all vertices. The cover time has been studied intensively by probabilists, combinatorialists, statistical physicists and computer scientists, with a variety of motivations; see, e.g., [7, 16, 2, 8, 17]. The problem of determining the expected cover time $\mathcal{T}_n$ for the n by n lattice torus $\mathbf{Z}_n^2$, was posed by Wilf [22] and Aldous [1]. In [9] we proved the following conjecture of Aldous [1].

Theorem 1 *If $\mathcal{T}_n$ denotes the time it takes for the simple random walk in $\mathbf{Z}_n^2$ to completely cover $\mathbf{Z}_n^2$, then*

$$\lim_{n\to\infty} \frac{\mathcal{T}_n}{(n\log n)^2} = \frac{4}{\pi} \quad \textit{in probability.} \tag{1.1}$$

The first step toward proving Theorem 1, was to find a sufficiently robust proof for the asymptotics of the cover time of finite b-ary trees. These asymptotics were originally determined by Aldous in [4], but his elegant recursive method was quite sensitive and did not adapt to the approximate tree structure that can be found in Euclidean space. Cover times on trees are discussed in the next section.

Turning to a different but related topic, Erdős and Taylor (1960) posed a problem about simple random walks in $\mathbf{Z}^2$: *How many times does the walk revisit the most frequently visited site in the first n steps?*

Theorem 2 ([10]) *Denote by $T_n(x)$ the number of visits of planar simple random walk to $x \in \mathbf{Z}^2$ by time n, and let $T_n^* := \max_{x\in\mathbf{Z}^2} T_n(x)$. Then*

$$\lim_{n\to\infty} \frac{T_n^*}{(\log n)^2} = \frac{1}{\pi} \quad a.s.. \tag{1.2}$$

This was conjectured by Erdős and Taylor [11, (3.11)]. After D. Aldous heard one of us describing this result, he pointed us to his cover time conjecture, and this eventually led to Theorem 1. Although the proofs of that theorem and of Theorem 2 differ in important technical points, they follow the same basic pattern:

(i) Formulate a suitable tree-analog and find a "robust" proof.
(ii) Establish a Brownian version using excursion counts.
(iii) Deduce the lattice result via strong approximation a-la [12].

2.   Cover times for trees

Let Γ_k denote the balanced b-ary tree of height k, which has $n_k = (b^{k+1} - 1)/(b - 1)$ vertices, and $n_k - 1$ edges.

Theorem 3 (Aldous [4]) *Denote by $\mathcal{C}_k$ the time it takes for simple random walk in Γ_k, started at the root, to cover Γ_k. Then*

$$\lim_{k\to\infty} \frac{\mathbf{E}\mathcal{C}_k}{n_k k^2} = 2\log(b). \tag{2.1}$$

Remark The expected hitting time from one vertex to another is bounded by the commute time, which equals the effective resistance times twice the number of edges (see, e.g., [5]). Therefore the expected hitting time between two vertices in Γ_k is at most $4kn_k$. From a general result in [3], it follows that also

$$\lim_{k\to\infty} \frac{C_k}{n_k k^2} = 2\log(b) \quad \text{in probability.} \tag{2.2}$$

Proof of theorem 3 Denote by $\mathcal{C}_k^+$ the time it takes the walk to cover and return to the root, and by R_k the number of returns to the root until time $\mathcal{C}_k^+$. By the remark preceding the proof, $\mathbf{E}\mathcal{C}_k^+ - \mathbf{E}\mathcal{C}_k \le 4kn_k$, so to prove the theorem it suffices to establish that

$$\lim_{k\to\infty} \frac{\mathbf{E}\mathcal{C}_k^+}{n_k k^2} = 2\log(b) \,. \tag{2.3}$$

The expected time to return to the root is the reciprocal of the root's stationary probability $b/(2n_k - 2)$, so by Wald's lemma

$$\mathbf{E}(\mathcal{C}_k^+) = \frac{2n_k - 2}{b}\mathbf{E}(R_k) \,. \tag{2.4}$$

Thus the theorem reduces to showing

$$\lim_{k\to\infty} \frac{\mathbf{E}R_k}{k^2} = b\log(b) \,. \tag{2.5}$$

We start by reproducing the straightforward proof of the upper bound. Denote by R_v the number of returns to the root of Γ_k until the first visit to v, and observe that R_k is the maximum of R_v over all leaves v at level k. At each visit to the root, the chance to hit a specific leaf v before returning to the root is $1/bk$, whence

$$\mathbf{P}[R_v > rbk^2] \le (1 - \frac{1}{bk})^{rbk^2} \le e^{-rk} \,.$$

Summing over all leaves, we infer that

$$\mathbf{P}[R_k > rbk^2] < \min\{1, b^k e^{-rk}\} \,. \tag{2.6}$$

Integrating over $r > 0$,
$$\mathbf{E}[R_k] \le bk^2(\log b + 1/k) \,. \tag{2.7}$$

This yields the upper bound in (2.5). To prove a lower bound, Aldous [4] uses a delicate recursion, and an embedded branching process argument. Here we will give the shortest argument we know, which only involves an embedded branching process. Given $\lambda < \log b$, our next goal is to show that

$$\mathbf{P}[R_k > \lambda bk^2] \to 1 \text{ as } k \to \infty \,. \tag{2.8}$$

Let T_λ be the number of steps until the root is visited λbk^2 times.

Fix $r \in (\lambda, \log b)$, and choose ℓ large, depending on r. Let v be a vertex at level $k - (j+1)\ell$ of Γ_k, and suppose that w is a descendant of v at level $k - j\ell$.

Observe that the expected number of visits to v by time T_λ is $\lambda(b+1)k^2$, and the expected number of excursions between v and w by time T_λ is $\lambda k^2/\ell$.

Say that w is "special" if the number of excursions from v to w by time T_λ is at most $r\ell j^2$. Note that vertices close to the root (i.e., at level $k - j\ell$ where $r\ell^2 j^2 > \lambda k^2$) are special with high probability, because $r > \lambda$. If $k > (j+2)\ell$, then every visit to v is equally likely to start an excursion to w as to the ancestor of v at distance ℓ from v. Thus, if v is special then w is special with probability at least $\mathbf{P}[X < r\ell j^2]$, where X has binomial law with parameters $r\ell(j^2 + (j+1)^2)$ and $1/2$. By the central limit theorem, as j grows, $\mathbf{P}[X < r\ell j^2] \to \mathbf{P}(Z > (2r\ell)^{1/2})$, where Z is standard normal. Since $r < \log b$, we find that $\mathbf{P}(Z > (2r\ell)^{1/2}] > b^{-\ell}$, if ℓ is large enough. Therefore, special vertices considered at jumps of 2ℓ levels (to ensure the required independence) dominate a supercritical branching process; the survival probability tends to 1 as $k \to \infty$, because vertices near the root are almost guaranteed to be special. This establishes (2.8). It follows that $\mathbf{E}(R_k) > \lambda b k^2$ for large k, and since $\lambda < \log b$ is arbitrary, this completes the proof of (2.5) and the theorem.

Remark The argument above is quite robust: it readily extends to family trees of Galton watson trees with mean offspring $b > 1$. With a little more work, using the notion of quasi-Bernoulli percolation (see [13] or [19]), it can be extended to the first k levels of any tree Γ that has growth and branching number both equal to $b > 1$. The most robust argument, the truncated second moment method used in [9], is too technical to include here.

3.   From trees to Euclidean space

The following "dictionary" was offered in [18] to illustrate the reduction of certain intersection problems from Euclidean space to trees:

Problem in Euclidean space	**Corresponding problem on trees**
• How many (independent) Brownian paths in $\mathbf{R}^d$ can intersect?	• Which branching processes can have an infinite line of descent?
• What is the probability that several random walk paths, started at random in a cube of side-length 2^k, will intersect?	• What is the probability that a branching process survives for at least k generations?
• Which sets in $\mathbf{R}^3$ contain double points of Brownian motion?	• Which trees percolate at a fixed threshold p?
• What is the Hausdorff dimension of the intersection of a fixed set in $\mathbf{R}^d$ with one or two Brownian paths?	• What is the dimension of a percolation cluster on a general tree?

The Brownian analogs of Theorems 1 and 2, respectively, are given below. Throughout, denote by $D(x, \epsilon)$ the disk of radius ϵ centered at x.

Theorem 4 ([9]) *For Brownian motion $w_{\mathbf{T}}(\cdot)$ in the two-dimensional torus $\mathbf{T}^2$,* *consider the hitting time of a disk,*

$$\mathcal{T}(x, \epsilon) = \inf\{t > 0 \mid X_t \in D(x, \epsilon)\},$$

and the ϵ-covering time,

$$\mathcal{C}_\epsilon = \sup_{x \in \mathbf{T}^2} \mathcal{T}(x, \epsilon)$$

which is the amount of time needed for the Wiener sausage of radius ϵ to completely cover $\mathbf{T}^2$. Then

$$\lim_{\epsilon \to 0} \frac{\mathcal{C}_\epsilon}{(\log \epsilon)^2} = \frac{2}{\pi} \qquad a.s. \qquad (3.1)$$

Theorem 5 ([10]) *Denote by μ_w the occupation measure for a planar Brownian motion $w(\cdot)$ run for unit time. Then*

$$\limsup_{\epsilon \to 0} \sup_{x \in \mathbf{R}^2} \frac{\mu_w(D(x, \epsilon))}{\epsilon^2 \left(\log \frac{1}{\epsilon}\right)^2} = 2, \qquad a.s. \qquad (3.2)$$

(This was conjectured by Perkins and Taylor [20].)

The basic approach used to prove these results, which goes back to Ray, [21], is to control occupation times using excursions between concentric discs. The approximate tree structure that is (implicitly) used arises by considering discs of the same radius r around different centers and varying r; for fixed centers x, y, and "most" radii r (on a logarithmic scale) the discs $D(x, r)$ and $D(y, r)$ are either well-separated (if $r << |x - y|$) or almost coincide (if $r >> |x - y|$).

References

[1] D. Aldous, *Probability approximations via the Poisson clumping heuristic*, Applied Mathematical Sciences **77**, Springer-Verlag, New York, 1989.

[2] D. Aldous, *An introduction to covering problems for random walks on graphs*, J. Theoret. Probab. 2 (1989), 87–89.

[3] D. Aldous, Threshold limits for cover times, *J. Theoret. Probab.* 4 (1991), 197–211.

[4] D. Aldous, Random walk covering of some special trees, *J. Math. Anal. Appl.—* 157 (1991), 271–283.

[5] D. Aldous and J. Fill, *Reversible Markov Chains and Random Walks on Graphs*, monograph in preparation, draft available at
http://oz.Berkeley.EDU/users/aldous/book.html

[6] N. Alon and J. Spencer, *The Probabilistic Method*, Second Edition, Wiley, 2000.

[7] A. Broder, *Universal sequences and graph cover times. A short survey.* Sequences (Naples/Positano, 1988), 109–122, Springer, New York, 1990.

[8] M. J. A. M. Brummelhuis and H. J. Hilhorst, *Covering of a finite lattice by a random walk* Phys. A 176 (1991), no. 3, 387–408.

[9] A. Dembo, Y. Peres, J. Rosen and O. Zeitouni, *Cover Times for Brownian Motion and Random Walks in two dimensions*, submitted.

[10] A. Dembo, Y. Peres, J. Rosen and O. Zeitouni, *Thick points for planar Brownian motion and the Erdős-Taylor conjecture on random walk*, Acta Math. 186 (2001), 239–270.

[11] P. Erdős and S. J. Taylor, Some problems concerning the structure of random walk paths, *Acta Sci. Hung.* 11 (1960), 137–162.

[12] J. Komlós, P. Major and G. Tusnády, An approximation of partial sums of independent RV's, and the sample DF. I, *Zeits. Wahr. verw. Gebiete*, 32 (1975), 111–131.

[13] R. Lyons, Random walks and percolation on trees, *Annals Probab.* 18 (1990), 931–958.

[14] R. Lyons and R. Pemantle, Random walks in a random environment and first-passage percolation on trees, *Ann. Probab.* 20 (1992), 125–136.

[15] P. Matthews, Covering problems for Brownian motion on spheres, *Ann. Probab.* 16 (1988), 189–199.

[16] M. Mihail and C. H. Papadimitriou, On the random walk method for protocol testing, *Computer aided verification* (Stanford, CA), 132–141, Lecture Notes in Comput. Sci. 818, Springer, Berlin, 1994.

[17] A. M. Nemirovsky, M. D. Coutinho-Filho, Lattice covering time in D dimensions: theory and mean field approximation, *Current problems in statistical mechanics* (Washington, DC, 1991). *Phys. A* 177 (1991), 233–240.

[18] Y. Peres, Intersection-equivalence of Brownian paths and certain branching processes, *Commun. Math. Phys.* 177 (1996), 417–434.

[19] Y. Peres, Probability on trees: an introductory climb, Lectures on probability theory and statistics (Saint-Flour, 1997), 193–280, *Lecture Notes in Math.* 1717, Springer, Berlin, (1999).

[20] E. A. Perkins and S. J. Taylor, Uniform measure results for the image of subsets under Brownian motion, *Probab. Theory Related Fields* 76 (1987), 257–289.

[21] D. Ray, Sojourn times and the exact Hausdorff measure of the sample path for planar Brownian motion, *Trans. Amer. Math. Soc.* 106 (1963), 436–444.

[22] H. S. Wilf, The editor's corner: the white screen problem, *Amer. Math. Monthly* 96 (1989), 704–707.

Renormalization, Large Deviations and Phase Separation in Ising and Percolation Models*

Ágoston Pisztora†

Abstract

Phase separation is a fairly common physical phenomenon with examples including the formation of water droplets from humid air (fog, rain), the separation of a crystalline structure from an isotropic material such as a liquid or even the formation of the sizzling gas bubbles when a soda can is opened.

It was recognized long ago (at least on a phenomenological level) that systems exhibiting several phases in equilibrium can be described with an appropriate variational principle: the phases arrange themselves in such a way that the energy associated with the phase boundaries is minimal. Typically this leads to an almost deterministic behavior and the phase boundaries are fairly regular. However, when looked at from a microscopic point of view, the system consists of a bunch of erratically moving molecules with relatively strong short-range interaction and the simplicity of the above macroscopic description looks more than miraculous. Indeed, when starting from the molecular level, there are many more questions to be asked and understood: which are the phases which we will see? why do only those occur? why are the phase boundaries sharp? how should we find (define) the energy associated with the interfaces? Only then can we ask the question: why does the system minimize this energy?

It is only in the last decade that a mathematically satisfactory understanding of this phenomenon has been achieved. The main goal of the talk is to present the current state of affairs focusing thereby on results obtained in joint works with Raphael Cerf. The connection to fields of mathematics other than probability theory or statistical mechanics will be highlighted; namely, to geometric measure theory and to the calculus of variations.

2000 Mathematics Subject Classification: 60K35, 82B, 60F10.
Keywords and Phrases: Ising model, Potts model, Random cluster model, FK percolation, Phase separation, Phase coexistence, Large deviations, Double bubble, Minimal surfaces.

---

*The author thanks the following institutions for their hospitality and/or financial support: Center for Nonlinear Analysis at CMU, Forschungsinstitut at ETHZ, Université Paris Sud and IMPA, Brazil. The author has been supported by NSF Grant DMS-0072217.

†Department of Mathematical Sciences, Carnegie Mellon University, Pittsburgh, PA 15213, USA. E-mail: pisztora@andrew.cmu.edu

1. Introduction

Although phase separation is a fairly common physical phenomenon (examples will be given further below) its mathematically satisfactory understanding, even in the simplest models, has not been achieved until the last decade. In order to uncover the mechanism leading to the separation of various phases in an initially homogeneous material, in particular to explain why and what kind of phases will occur and which shapes they take, one has to work with a microscopic description of the system, often at a molecular scale. Materials on such scales however tend to behave 'chaotic' and this strongly motivates (if not compels) the use of a probabilistic approach. In this approach, the system is modeled by randomly moving (in equilibrium theory randomly located) particles which interact with each other according to some simple, typically short range, mechanism. The goal is then to derive the large scale behavior of the system based only on the specification of the local interaction.

The difficulty of the analysis stems from the fact that the interaction, although local, might be strong enough (depending on some parameter such as the temperature) to cause a subtle spatial propagation of stochastic dependence across the entire system. As a consequence, one has to leave the familiar realm of classical probability theory whose focus has been laid on the large scale effects of randomness arising from independent (or weakly dependent) sources. Instead, we have to deal with a strongly dependent system and it is exactly this strong dependence which causes a highly interesting cooperative behavior of the particles on the macroscopic level which can, in certain cases, be observed as phase separation.

The problem of phase separation and related issues have been a driving force behind developing, and a benchmark for testing various new techniques. Postponing historical remarks until section 2.1, let me highlight here only those ones which play an essential role in our approach [10, 11] achieved in collaboration with Raphael Cerf. The basic framework is *(abstract) large deviation theory*, see e.g. [37], whose power contributed substantially (admittedly rather to my own surprize) to the success of this approach. To have sufficient control of the underlying model, in our case the Ising-Potts model, we employ *spatial renormalization techniques*, as developed in [34], in conjunction with the Fortuin-Kasteleyn percolation representation of the Potts model, [21]. Finally, tools from geometric measure theory a la Caccioppoli and De Giorgi will be employed to handle the geometric difficulties associated with three or higher dimensions, in a similar fashion as was done in [9].

The goal of this article is to present some recent developments in the equilibrium theory of coexisting phases in the framework of the Ising-Potts model. The presentation will be based mainly on the results contained in [34, 10, 11]. It will include a description of the underlying physical phenomenon, of the corresponding mathematical model and its motivation and, following the statement of the main results, comments on the proofs will be included. In order to address an audience broader than usual, I will try to use as little formalism as possible.

1.1. Phase separation: examples and phenomenology

Perhaps the most common and well known example of phase separation is the development of fog and later rain from humid air. When warm humid air is cooled down so much that its *relative* humidity at the new cooler temperature would exceed 100% (i.e. it would become over-saturated) the excess amount of water precipitates first in the form of very small droplets which we might observe as fog. The system at this time is not in equilibrium, rather in a so called metastable state. After waiting very long time or simply dropping the temperature further down, the droplets grow bigger and ultimately fall to the ground due to gravitation in the form of common rain droplets. In this example phase separation occurred since from a single homogeneous phase (warm humid air) two new phases have been formed: a cooler mixture of water and air (note: with 100% relative humidity (saturated) at the new lower temperature) plus a certain amount of water, more precisely a saturated solution of air in water, in the form of macroscopic droplets. In fact, in the absence of gravitation and after very long time, only a huge droplet of fluid would levitate in a gas (both saturated solutions of air in water and water in air, respectively). The opposite situation (water majority, air minority) may also occur. Consider the following familiar example; think of a bottle of champagne when opened. Here the change of temperature is replaced by a change of pressure but the phenomenon is similar with the roles exchanged; at the new lower pressure the liquid is not able to dissolve the same amonunt of carbon dioxide, hence this latter precipitates in the form of small bubbles (droplets), etc.

The phenomenological theory explains this type of phenomenon as follows. At any temperature there are saturation densities of air/water and water/air mixtures and only saturated solutions will coexist in equilibrium. They also determine the volumes of the two coexisting phases. Moreover the phases arrange themselves in a way so as to minimize the so-called *surface energy*, associated with the interface between the phases. In fact, it is supposed to exist a (in general direction-dependent) scalar quantity τ, called the *surface tension*, whose surface integral along the interface gives the surface energy. The surface tension, as well as the saturation densities, have to be measured experimentally. It is implicitly assumed that the interface is 'surfacelike' and regular enough so that the integral along the surface makes sense. The prediction which can be made is that the shape of the phases in equilibrium is just a solution of the variational principle. By the classical isoperimetric theorem, in the isotropic case the solution is just a sphere, hence the occurrence of bubbles and droplets. In the non isotropic case the corresponding variational problem is called the *Wulff problem*. The solution is known to be explicitly given [38, 13, 19, 20, 36] by rescaling appropriately the so called *Wulff crystal:*

$$\mathcal{W}_\tau = \left\{ x \in \mathbb{R}^d \; ; \; x \cdot \nu \leq \tau(\nu) \text{ for all unit vectors } \nu \right\}.$$

It is worth noting that the same arguments are used to describe macroscopic crystal shapes as well.

1.2.   The mathematical model and the goals of the analysis

The next step is to find a model which is simple enough to be analyzed by rigorous methods yet rich enough to exhibit the phenomenon we want to study.

In order to accommodate a multitude of phases we consider a finite number (q) possible types of particles (called colors or spins). Physics suggests to choose a short range interaction of "ferromagnetic" character which means that particles of identical type prefer to stay together and/or they repel particles of different types. For simplicity we assume that the interaction distinguishes only between identical and different types, otherwise it is invariant under permutation of colors. There is a standard model of statistical mechanics, called the (ferromagnetic) *q-states Potts model*, which corresponds exactly to these specifications. We consider the closed unit cube $\Omega \in \mathbb{R}^d$, $d \geq 3$ (modeling the container of the mixture of particles) overlapped by the rescaled integer lattice $\mathbb{Z}_n^d = \mathbb{Z}^d/n$. We define $\Omega_n = \Omega \cap \mathbb{Z}_n^d$, and denote by $\partial^{in}\Omega_n$ the internal vertex boundary of Ω_n. At each lattice point x there is a unique particle σ_x of one of the types $1, 2, ..., q$. The energy $H(\sigma)$ of a configuration $\sigma = (\sigma_x)_{x \in \Omega_n}$ can be chosen to be the number of nearest neighbor pairs of different types of particles corresponding to nearest neighbor repulsion. According to the Gibbs formula, the probability of observing a configuration σ is proportional to $e^{-\beta H(\sigma)}$, where $\beta = 1/T$ is the 'inverse temperature' which adjusts the interaction strength. (High T = large disorder = *relatively* small interaction, etc.) Note that we use a static description of the equilibrium system. It corresponds to a snap-shot of the system at a given time and the task is to understand the 'typical' picture we will see.

A *restricted ensemble* is a collection of certain feasible configurations. For instance, in the situation of the water/air mixture every configuration with a fixed number of water and air particles is possible, and this collection forms our restricted ensemble. It turns out that the direct study of this particular ensemble is extremely difficult and it is a crucial idea (discovered long ago) to go over to a larger, more natural ensemble, namely to that without any restrictions on the particle numbers. Then, the restricted system can be regarded as a very rare event $=: G_n$ in the large ensemble and conditional probabilities can be used to describe the restricted system. The events G_n are often in the large deviations regime, and it is from here that large deviations theory enters the analysis in an essential way.

The unrestricted system is usually referred to as the Potts model with free boundary conditions. In the case $q = 2$, it is equivalent to the classical Ising model. The Gibbs formula and the energy uniquely determines the probability measure in this (and in every) ensemble. We can introduce mixed boundary conditions as follows. Divide the boundary $\Gamma = \partial\Omega$ of the 'container' into $q + 1$ parts indexed by $\Gamma^0, \Gamma^1, ...\Gamma^q$. The parts can be fairly general but the $(d-1)$-dimensional Hausdorff measure of their relative boundaries has to be zero. We set for $n \in \mathbb{N}$ and $i = 0, \ldots, q$,

$$\Gamma_n^i = \left\{ x \in \Gamma_n ; \ d_\infty(x, \Gamma^i) < 1/n \text{ and } \forall j < i, \ d_\infty(x, \Gamma^j) \geq 1/n \right\} \quad i = 0, \ldots, q$$

where d_∞ denotes the distance corresponding to the max-norm. We use the sequence of $q + 1$-tuples of sets $\gamma(n) = (\Gamma_n^0, \ldots, \Gamma_n^q)$ to specify boundary conditions by imagining that all particles in Γ_n^j are of type j for $j = 1, ..., q$ and none occupies Γ_n^0. This defines a restricted ensemble and the corresponding probability measure is denoted by $\mu_n = \mu_n^{\gamma(n), \beta, q}$. The choice of b.c.s $\gamma(n)$ is understood to be fixed.

Let's consider the Ising-Potts model in an $n \times n$ lattice box $B(n)$ with boundary conditions $j \in \{1, 2, ..., q\}$ at a fixed inverse temperature β with the corresponding probability measure $\mu_{B(n)}^{(j),\beta}$. It is well known that, as $n \to \infty$, a unique, translation invariant infinite volume measure $\mu_{\infty}^{(j),\beta}$ emerges as the weak limit of the sequence $(\mu_{B(n)}^{(j),\beta})_{n \geq 1}$. We can define the *order parameter* $\theta = \theta(\beta, q, d)$ as the *excess density* of the dominant color, the excess is measured from the symmetric value $1/q$. The model exhibits phase transition in the sense that in dimensions $d \geq 2$ there exists a critical value $0 < \beta_c(d) < \infty$ such that for $\beta < \beta_c$, $\theta(\beta) := \mu_{\infty}^{(j),\beta}[\sigma_0 = j] - 1/q = 0$ but for $\beta > \beta_c$, $\theta(\beta) > 0$, i.e., when the interaction becomes strong enough, the influence of the (arbitrarily) far away boundary still propagates all the way through the inside of the volume and creates a majority of j-type particles. (Note that in the Ising model ($q = 2$), the spontaneous magnetization m^* is equal to θ.) The probability measures $\mu_{\infty}^{(j),\beta}$, $j = 1, 2, ..., q$, describe in mathematical terms what we call 'pure' phases and which correspond to the saturated solutions in the initial example.

Having chosen our model, let us formulate the goals of the analysis. Clearly, the main goal is to verify the predictions of the phenomenological theory, namely, that on the macroscopic scale the phases will be arranged according to some solution of the variational principle corresponding to minimal surface energy. First, however, the participating phases have to be found and identified. Moreover, as pointed out earlier, the previous statement contains a couple of implicit assumptions, such as the existence of the surface tension, the absence of transitional states (where one phase would smoothly go over into another one) the regularity of the interface boundaries, etc., all of which have to be justified from a microscopic point of view.

1.3. Connection to minimal surfaces

In this section a partially informal discussion of some examples will be presented with the aim to make the close relation to minimal surfaces transparent.

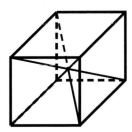 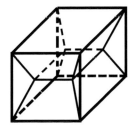

figure 1

Consider the Potts model with $q = 6$ colors (states) in a three dimensional box with boundary condition i on the i-th face of the box. Naively, one might expect that all phases will try to occupy the region closest to the corresponding piece of the boundary, which would lead to a phase partition consisting of symmetric and pyramid-like regions, as can be seen in figure 1, left. However, at least in the case when the surface tension is isotropic (which is presumably the case in the limit $T \uparrow T_c$), there exists a better configuration with lower total surface free energy. Recall that in this case our desired interface is simply a minimal surface spanned by the edges of the box. A picture of the well known solution to this problem can be seen in figure 1, right. In order to be able to discuss this example at temperatures $0 < T < T_c$, we have to make certain assumptions about the surface tension τ. We assume that the sharp simplex inequality holds, that the value of τ is minimal in axis directions and that τ increases as the normal vector moves from say $(0, 0, 1)$ to $(1, 1, 1)$. (Although these assumptions are very plausible, none of them has been proved in dimensions $d \geq 3$). Under these hypotheses, we conjecture that the phase partition at moderate subcritical temperatures looks like in figure 2, left. In the limit $T \downarrow 0$, only two phases survive, as shown in figure 2, right. At $T = 0$, there is no reason for the middle plane to stay centered, in fact, any horizontal plane is equally likely.

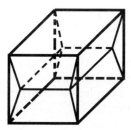 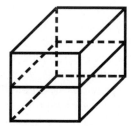

figure 2

In the next example we consider the three dimensional Ising model with free boundary conditions below T_c, conditioned on the event that the average magnetization is positive and does not exceed $m^* - \varepsilon$, where ε is a sufficiently small positive number and m^* denotes the spontaneous magnetization. It is natural to conjecture that a minimizer of the corresponding variational problem is a droplet attached symmetrically to one of the corners of the box.

The single bubble sitting in one of the corners is filled with the minus phase and in the rest of the box we see the plus phase. The size of the bubble is determined by ε and its internal boundary coincides with the corresponding piece of the surface of the Wulff crystal.

Another Wulff-type problem arises by conditioning the q-states Potts model (with say $q \geq 4$) to have a moderate excess of colors 2 and 3 while imposing 1-boundary conditions on the entire box. In this case it is conceivable that a so-called "double bubble" is created, consisting of two adjacent macroscopic droplets filled with the (pure) phases 2 and 3, respectively. The double bubble is swimming in the phase 1 which fills the rest of the box. Of course, we might have an excess of color 4 as well; in this case a further bubble will presumably appear which will be attached to the previous two bubbles.

For related variational questions concerning soap films and immiscible fluids, see [30].

In fact, by studying questions concerning phase boundaries we are very quickly confronted with the theory of minimal surfaces, such as the Plateau problem, corresponding to anisotropic surface measures. Let Ω be a bounded open set in $\mathbb{R}^3$ with smooth boundary and let γ be a Jordan curve drawn on $\partial\Omega$ which separates $\partial\Omega$ into two disjoint relatively open sets Γ^+ and Γ^-. Typical configurations in the Ising model on a fine grid in Ω with plus b.c.s on Γ^+ and minus b.c.s on Γ^- will exhibit two phases separated with an interface close to a minimal surface which is a global solution of the following Plateau type problem:

$$\text{minimize } \int_S \tau(\nu_S(x)) \, d\mathcal{H}^{d-1}(x) \ : \ S \text{ is a surface in } \Omega \text{ spanned by } \gamma$$

where $\nu_S(x)$ is the normal vector to S at x. We remark that it is conjectured that, as the temperature approaches T_c from below, the surface tension τ becomes more and more isotropic and it is conceivable that the solution of the above minimization problem approaches the solution of the classical (isotropic) Plateau problem.

1.4. Further background

There is a beautiful and extremely useful way to decompose the Ising-Potts model into a certain bond percolation model, called FK-percolation and some simple 'coloring' procedure discovered by Fortuin and Kasteleyn [21]. Consider the q-state Potts model with mixed boundary conditions γ in a finite lattice box B at inverse temperature β and set $p = 1 - e^{-\beta}$. Consider a bond percolation model (called FK percolation) specified by the following formula:

$$\Phi_B^{\gamma,\beta,q}[\eta] = P_p[\eta] \, q^{\#(\eta)} / Z_B^{\gamma,\beta,q}$$

where η is a bond configuration with the property that there is no open connection between differently colored boundary parts, P_p is the usual Bernoulli measure with parameter p, $\#(\eta)$ denotes the number of clusters (conn. components) in the configuration η with the rule that identically colored boundary parts (and their connected components) count as one single cluster. Finally $Z_B^{\gamma,\beta,q}$ is the appropriate normalizing constant.

In the second step we assign colors to every cluster (and their sites) as follows: the boundary pieces inherit the color of the boundary condition, the remaining clusters will get one of the colors $1, 2, ..., q$ with probability $1/q$ each independently from each other. The distribution of the coloring of the sites corresponds exactly to the Potts model. Note that in the case of free or constant b.c.s. there is no constraint prohibiting open connections, and indeed these measures, denoted by $\Phi_B^{f,\beta,q}$ (free b.c.s.) and $\Phi_B^{w,\beta,q}$ ("wired" b.c.s.) behave very similar to regular Bernoulli percolation. Their thermodynamic limits $\Phi_\infty^{f,\beta,q}$ and $\Phi_\infty^{w,\beta,q}$ exist as the box size tends to infinity and we can define the percolation probability as usual $\theta^*(p) = \Phi_\infty^{*,p}[0 \leftrightarrow \infty]$, for $*$ free or wired. It is easy to check that the order parameter $\theta(\beta)$ of the Ising-Potts model agrees with $\theta^w(\beta)$, and correspondingly the FK model exhibits a percolation phase transition. Further it is known [23], that $\theta^w(\beta) = \theta^f(\beta)$ for all but at most countably many values of β and it is conjectured that the equality is valid for all values except possibly the critical point $\beta_c = \beta_c(q)$. Moreover this condition is equivalent to the equality of the thermodynamical limits: $\Phi_\infty^{w,\beta,q} = \Phi_\infty^{f,\beta,q}$. We define the set of "regular" inverse temperatures by

$$\mathcal{U}(q) = \left\{ \beta > 0 \,;\, \theta^w(\beta) = \theta^f(\beta) \right\}.$$

Although the status of the bonds are dependent, their correlation tends fast to zero as the distance between them becomes large and this holds for both the sub and supercritical phase of FK percolation. This property is of crucial importance in our large deviation analysis since in the original Potts model there exists no corresponding asymptotic independence when $\beta > \beta_c$.

Our results are valid above the so called *slab-threshold* $\widehat{\beta}_c = \widehat{\beta}_c(q,d)$, introduced in [34]. This threshold is conjectured to agree with the critical point and at least in the case of percolation ($q = 1$) this have been proved by Grimmett and Marstrand [24]. It is possible to characterize this threshold as the smallest value such that when β exceeds it, it is possible to find $\alpha > 0$ and $L \geq 1$ such that at least in the center of the of slabs $S(L,n) = [-L,L] \times [-n,n]^{d-1} \cap \mathbb{Z}^d$ there is "uniform long range order", i.e.,

$$\inf_{n\geq 1} \inf_{x,y\in S(L,\alpha n)} \Phi_{S(L,n)}^{f,\beta}\left[x \leftrightarrow y \right] > 0.$$

It has been proved in [34] that above $\widehat{\beta}_c$, α always can be chosen to be one, guaranteeing a strictly positive probability (uniformly in n) for connections within a sufficiently (depending on β) thick slab. This property is crucial for establishing the basic properties of supercritical FK percolation; the existence of a unique crossing cluster in a box, its omnipresence, the concentration of its density around the percolation probability, the exponential tail decay of the diameter of other clusters in the box, etc. These properties are then used to establish a renormalization scheme which is essential for the large deviation analysis of this and the Ising-Potts model.

2.   The results

2.1.   Historical remarks

Before we start with the presentation of our results we give a brief summary of the previous work on this subject. As we have already mentioned, large deviations theory plays an important role in this context and not surprisingly the first efforts were devoted to the study of large deviations of the empirical magnetization in the Ising model, i.e., the average value of the spins in a large box. A *volume order* large deviation principle (LDP) has been established for the Ising model by various authors: Comets, Ellis, Föllmer, Orey, Olla [12, 15, 17, 32]. The corresponding rate function has been found to vanish in $[-m^*, m^*]$ where m^* denotes the spontaneous magnetization. In fact, it was suspected that the correct order of decay is exponential to *surface order*. Indeed, Schonmann [35] found a proof of this conjecture, valid for any dimensions and low enough temperatures and Chayes, Chayes and

Schonmann extended the result for the supercritical $\beta > \beta_c$ regime in the two dimensional case. Föllmer and Ort [18] investigated this phenomenon on the level of empirical measures. Finally, inspired by the work of Kesten and Zhang [29] on related questions in percolation, Pisztora [34] established surface order upper bounds for the remaining dimensions $d \geq 3$ above the slab-threshold $\widehat{\beta}_c$, introduced in the same work, which is conjectured to agree with the critical point β_c. In that work a renormalization scheme has been developed for supercritical Fortuin-Kasteleyn percolation (or random cluster model) in conjunction with a stochastic domination argument (generalized and improved in [31]) which allows to control the renormalized process, and so, the original one.

The monograph of Dobrushin, Kotecký and Shlosman [14] opened the way to the rigorous study of the phase separation phenomenon creating thereby an immense interest and activity which lasts up to the present time. Their analysis, which provided the first mathematical proof of phase separation, had been performed in the context of the Ising model. The main tool of their work is the cluster expansion, which, on the one hand allowed the derivation of results much finer than necessary to verify the Wulff construction, on the other hand it restricted the validity of the results to two dimensions and low temperatures. Significant improvements of these results in two dimensions have been derived by Pfister [33], Alexander, Chayes and Chayes [4] (treating percolation), Alexander [3], Ioffe [26, 27]. Finally Ioffe and Schonmann [28] extended the results of [14] up to T_c.

The next challenge was to analyze phase separation for short range models in higher dimensions. The additional difficulties came mainly from two sources. First, new techniques have to be developed to avoid the use of perturbative methods (such as the cluster expansion) which severely limit the applicability of the arguments and methods which are specific to two dimensions only (duality). Second, the emerging geometry is far more complex than in two dimensions and this requires the use of new tools and ideas. The complexity of the geometry causes problems also within the probabilistic analysis (for instance the lack of the skeleton technique for surfaces) and even the correct formulation of the results is far from obvious ("hairs").

The first issue has been resolved by the application of the aforementioned renormalization technology from [34]. Renormalization arguments lie at the heart of the proof of much of the intermediate steps (for instance exponential tightness, decoupling) and even in the remaining parts they play an important role usually in combination with geometric arguments (interface lemma, etc.).

To handle the geometric difficulties, the use of appropriate tools from geometric measure theory has been introduced in the works of Alberti, Bellettini, Bodineau, Buttà, Cassandro, Presutti [2, 5, 6] and by Cerf [9]. In these works also a novel and very general large deviation framework have been proposed to tackle the problem. In fact, this framework turned out to be crucial for the success of the entire approach. It is the work of Cerf [9] in which the first complete analysis of phase separation in a three dimensional model have been achieved, namely the asymptotic analysis of the shape of a large finite cluster (Wulff problem) in percolation.

The results presented in this article have been derived in the works [34, 9, 10, 11]. It should be mentioned that in an independent work [7] Bodineau carried out an analysis of the Wulff problem in the Ising model with conclusions slightly weaker than the results appearing in [10].

Finally, for current developments in the field we refer the reader to the preprint [8] and the references therein.

2.2. Statement of the results

Range of validity of the results. Our results for the Ising-Potts models hold in the region: $d \geq 3$, $q \in \mathbb{N} \setminus \{0,1\}$, $\beta > \widehat{\beta}_c(q,d)$, $\beta \in \mathcal{U}(q,d)$.

At this point it is natural to comment on the case of two dimensions. Although most of our results should hold for $d = 2$, there are several points in the proofs which would require a significant change, making the proofs even longer. The main reason, however, for not to treat the two dimensional case is that the natural topology for the LDP-s in $d = 2$ is not the one we use (which is based on the distance dist_{L^1}) but a topology based on the Hausdorff distance.

Surface tension. From FK percolation we can extract a direction dependent surface tension $\tau(\nu) = \tau(p,q,d,\nu)$, cf [10]. For a unit vector ν, let A be a unit hypersquare orthogonal to ν, let $\mathrm{cyl}\,A$ be the cylinder $A + \mathbb{R}\nu$, then $\tau(\nu)$ is equal to the limit

$$\lim_{n \to \infty} -\frac{1}{n^{d-1}} \log \Phi_\infty^{p,q} \left(\begin{array}{l} \text{inside } n\,\mathrm{cyl}\,A \text{ there exists a finite set of closed edges } E \\ \text{cutting } n\,\mathrm{cyl}\,A \text{ in at least 2 unbounded components and} \\ \text{the edges of } E \text{ at distance less than } 2d \text{ from the boundary} \\ \text{of } n\,\mathrm{cyl}\,A \text{ are at distance less than } 2d \text{ from } nA \end{array} \right)$$

The function τ satisfies the weak simplex inequality, is continuous, uniformly bounded away from zero and infinity and invariant under the isometries which leave $\mathbb{Z}^d$ invariant (see section 4 in [10] for details).

Identification of the phases. The typical picture which emerges from the Potts model with mixed b.c.s. at the macroscopic level is a partition of Ω in maximal q phases corresponding to the dominant color in that phase. The individual phases need not be connected. In order to identify the phases we choose first a sequence of *test events* which we regard as characteristic for that phase. More specifically, for $j = 1, 2, ..., q$ we select events $E_n^{(j)}$ defined on a $n \times n$ lattice box such that $\mu_\infty^{(j),\beta,q}[E_n^{(j)}] \to 1$ as $n \to \infty$. We may also assume that $E_n^{(j)} \cap E_n^{(j)} = \emptyset$ for $j \neq i$. For instance, one natural choice is to require that the densities of the different colors in the box do not deviate more than some small fraction from their expected value. This will guarantee that the right mixture of colors occurs which is typical for that particular pure phase. Alternately, we may request that the empirical measure defined by the given configuration is close to the restriction of $\mu_\infty^{(j),\beta,q}$ to $\Lambda(n)$ with respect some appropriate distance between probability measures, etc.

For $x \in \mathbb{R}^d$ and $r > 0$ we define the box $\Lambda(x,r)$ by

$$\Lambda(x,r) = \left\{ y \in \mathbb{R}^d \; ; \; -r/2 < y_i - x_i \leq r/2, \; i = 1, \ldots, d \right\}$$

and we introduce an *intermediate length scale* represented by a fixed function f :

$\mathbb{N} \to \mathbb{N}$ satisfying

$$\lim_{n\to\infty} n/f(n)^{d-1} = \lim_{n\to\infty} f(n)/\log n = \infty. \qquad (1)$$

Given a configuration in Ω_n, we say that the point $x \in \Omega$ belongs to the phase j, if the event $E_n^{(j)}$ occurs in the box $\Lambda(x, f(n)/n)$. For $j = 1, 2, .., q$, we denote by $A_n^{(j)}$ the set of points in Ω belonging to the phase j and set $A_n^{(0)} = \Omega \setminus \cup_j A_n^{(j)}$ (indefinite phase).

The random partition of Ω, $\vec{A}_n = (A_n^0, A_n^1, \ldots, A_n^q)$, is called the *empirical phase partition*. Our first result shows that up to super-surface order large deviations, the region of indefinite phase A_n^0 has negligible density, i.e., the pure phases fill out the entire volume.

Theorem 2.1 *Let $d \geq 3$, $q \in \mathbb{N} \setminus \{0, 1\}$, $\beta > \widehat{\beta}_c$, $\beta \in \mathcal{U}(q, d)$. For $\delta > 0$,*

$$\limsup_{n\to\infty} \frac{1}{n^{d-1}} \log \mu_n \big[\mathrm{vol}\,(A_n^0) > \delta \big] = -\infty.$$

Although Theorem 2.1 guarantees that the pure phases fill out the entire volume (up to negligible density) but it does not exclude the possibility that the connected components of the pure phases are very small. For instance, they could have a diameter not much larger than our fixed intermediate scale (in which case they would be invisible on a macroscopic scale.) If this happened, the total area of the phase boundaries would be exceedingly high. Before stating our next result, which will exclude this possibility, we introduce some geometric tools.

We define a (pseudo) metric, denoted by dist_{L^1}, on the set $\mathcal{B}(\Omega)$ of the Borel subsets of Ω by setting

$$\forall A_1, A_2 \in \mathcal{B}(\Omega) \qquad \mathrm{dist}_{L^1}(A_1, A_2) = \mathrm{vol}\,(A_1 \Delta A_2). \qquad (2)$$

We consider then the *space of phase partitions* $P(\Omega, q)$ consisting of $q + 1$-tuples $(A^0, A^1, \ldots, A^q)$ of Borel subsets of Ω forming a partition of Ω. We endow $P(\Omega, q)$ with the following metric:

$$\mathrm{dist}_P\Big((A^0, \ldots, A^q), (B^0, \ldots, B^q)\Big) = \sum_{i=0,\ldots,q} \mathrm{dist}_{L^1}(A^i, B^i).$$

In order to define the surface energy $\mathcal{I}$ of a phase partition $\vec{A}_n$, we recall some notions and facts from the theory of sets of finite perimeter, introduced initially by Caccioppoli and subsequently developed by De Giorgi, see for instance [22, 16]. The perimeter of a Borel set E of $\mathbb{R}^d$ is defined as

$$\mathcal{P}(E) = \sup \Big\{ \int_E \mathrm{div}\, f(x)\, dx : f \in C_0^\infty(\mathbb{R}^d, B(1)) \Big\}$$

where $C_0^\infty(\mathbb{R}^d, B(1))$ is the set of the compactly supported C^∞ vector functions from $\mathbb{R}^d$ to the unit ball $B(1)$ and div is the usual divergence operator. The set E is of finite perimeter if $\mathcal{P}(E)$ is finite. A unit vector ν is called the measure theoretic exterior normal to E at x if

$$\lim_{r \to 0} r^{-d} \operatorname{vol}(B_-(x,r,\nu) \setminus E) = 0, \quad \lim_{r \to 0} r^{-d} \operatorname{vol}(B_+(x,r,\nu) \cap E) = 0.$$

Let E be a set of finite perimeter. Then there exists a certain subset of the topological boundary of E, called the *reduced boundary*, denoted by $\partial^* E$, with the same $d-1$ dimensional Hausdorff measure as ∂E, such that at each $x \in \partial^* E$ there is a measure theoretic exterior normal to E at x. For practical (measure theoretic) purposes, the reduced boundary represents the boundary of any set of finite perimeter, for instance, the following generalization of Gauss Theorem holds: For any vector function f in $C_0^1(\mathbb{R}^d, \mathbb{R}^d)$,

$$\int_E \operatorname{div} f(x)\, dx = \int_{\partial^* E} f(x) \cdot \nu_E(x)\, \mathcal{H}^{d-1}(dx).$$

(For more on this see e.g. the appendix in [10] and the references there.)

The *surface energy* $\mathcal{I}$ of a phase partition $(A_n^0 A_n^1, \ldots, A_n^q) \in P(\Omega, q)$ is defined as follows:

- for any $(A^0, A^1, \ldots, A^q)$ such that either $A^0 \neq \emptyset$ or one set among $A^1, \ldots, A^q$ has not finite perimeter, we set $\mathcal{I}(A^0, \ldots, A^q) = \infty$,
- for any $(A^0, A^1, \ldots, A^q)$ with $A^0 = \emptyset$ and $A^1, \ldots, A^q$ having finite perimeter, we set

$$\mathcal{I}(A^0, \ldots, A^q) = \sum_{i=1,\ldots,q} \frac{1}{2} \int_{\partial^* A^i \cap \Omega} \tau(\nu_{A_i}(x))\, d\mathcal{H}^{d-1}(x)$$

$$+ \sum_{\substack{i,j=1,\ldots,q \\ i \neq j}} \int_{\partial^* A^i \cap \Gamma^j} \tau(\nu_{A_i}(x))\, d\mathcal{H}^{d-1}(x).$$

Note that $\mathcal{I}$ depends on τ and the boundary conditions $\gamma = (\Gamma^1, \ldots, \Gamma^q)$. The first term in the above formula corresponds to the interfaces present in Ω, while the second term corresponds to the interfaces between the elements of the phase partition and the boundary Γ. It is natural to define the *perimeter of the phase partition* by using the same formula with τ replaced by constant one. Since τ is uniformly bounded away from zero and infinity at any temperature, the surface energy can be bounded by a multiple of the perimeter and vice versa. It is known that the surface energy $\mathcal{I}$ and the perimeter $\mathcal{P}$ are lower semi continuous and their level sets of the form $\mathcal{I}^{-1}[0, K]$ are compact on the space $(\mathcal{B}(\mathbb{R}^d), \operatorname{dist}_{L^1})$.

The next result states that up to surface order large deviations (and the constant can be made arbitrarily large by adjusting the bound K below) the empirical phase partition will be (arbitrarily) close to the set of phase partitions with perimeter not exceeding K.

Theorem 2.2 Let $d \geq 3$, $q \in \mathbb{N} \setminus \{0,1\}$, $\beta > \widehat{\beta}_c$, $\beta \in \mathcal{U}(q,d)$. For $\delta > 0$,

$$\limsup_{n \to \infty} \frac{1}{n^{d-1}} \log \mu_n\left[\operatorname{dist}_P(\vec{A}_n, \mathcal{I}^{-1}[0, K]) > \delta\right] \leq -c\,K.$$

Our fundamental result is a large deviation principle (LDP) for the empirical phase partition $(A_n^0, A_n^1, \ldots, A_n^q)$.

Theorem 2.3 *The sequence* $(\vec{A}_n)_{n \in \mathbb{N}} = ((A_n^0, A_n^1, \ldots, A_n^q))_{n \in \mathbb{N}}$ *of the empirical phase partitions of* Ω *satisfies a LDP in* $(P(\Omega, q), \mathrm{dist}_P)$ *with respect to* μ_n *with speed* n^{d-1} *and rate function* $\mathcal{I} - \min_{P(\Omega, q)} \mathcal{I}$, *i.e., for any Borel subset* $\mathbb{E}$ *of* $P(\Omega, q)$,

$$- \inf_{\overset{o}{\mathbb{E}}} \mathcal{I} + \min_{P(\Omega, q)} \mathcal{I} \leq \liminf_{n \to \infty} \frac{1}{n^{d-1}} \log \mu_n \left[\vec{A}_n \in \mathbb{E} \right]$$

$$\leq \limsup_{n \to \infty} \frac{1}{n^{d-1}} \log \mu_n \left[\vec{A}_n \in \mathbb{E} \right]$$

$$\leq - \inf_{\overline{\mathbb{E}}} \mathcal{I} + \min_{P(\Omega, q)} \mathcal{I}.$$

Note that the constant $\min_{P(\Omega, q)} \mathcal{I}$ is always finite. Every large deviation result includes a (weak) law of large numbers; here the corresponding statement is as follows: Asymptotically, the empirical phase partition will be concentrated in an (arbitrarily) small neighborhood of the set of partitions minimizing the surface energy. In other words, on the macroscopic level, the typical phase partition will coincide with some of the minimizers of the variational problem, in agreement with the phenomenological prediction. Of course, the LDP states much more than this, in particular we will be able to extract similar statements for restricted ensembles. Recall that imposing mixed boundary conditions is not the only way to force the system to exhibit coexisting phases. In the Wulff problem in the Ising model context, for instance, a restricted ensemble is studied which is characterized by an artificial excess of say minus spins in the plus phase. Technically this can be achieved by conditioning the system to have a magnetization larger than the spontaneous magnetization while imposing plus b.c.s.

The next result describes the large deviation behavior of the phase partition in a large class of restricted ensembles. Although it is a rather straightforward generalization of Theorem 2.3, we state it separately because of its physical relevance.

Let $(G_n)_{n \geq 1}$ be a sequence of events, i.e., sets of spin configurations, satisfying the following two conditions: first there exists a Borel subset $\mathbb{G}$ of $P(\Omega, q)$ such that the sequence of events $(G_n)_{n \in \mathbb{N}}$ and $(\{\vec{A}_n \in \mathbb{G}\})_{n \in \mathbb{N}}$ are exponentially equivalent, i.e.,

$$\limsup_{n \to \infty} \frac{1}{n^{d-1}} \log \mu_n \left[G_n \bigtriangleup \{ \vec{A}_n \in \mathbb{G} \} \right] = -\infty, \tag{3}$$

where $\bigtriangleup$ denotes the symmetric difference. Second, the following limit exists and is finite:

$$\mathcal{I}_G = \lim_{n \to \infty} \frac{1}{n^{d-1}} \log \mu_n [G_n] > -\infty. \tag{4}$$

The sequence of events $(G_n)_{n \geq 1}$ determines a restricted (conditional) ensemble. Note that if

$$\inf_{\overset{o}{\mathbb{G}}} \mathcal{I} = \inf_{\overline{\mathbb{G}}} \mathcal{I} > -\infty, \tag{5}$$

then Theorem 2.3 implies that (4) is satisfied, with $\mathcal{I}_G = \inf_G \mathcal{I}$.

Theorem 2.4 *Assume that the sequence* $(G_n)_{n\geq 1}$ *satisfies (3) and (4) and define for each* $n \geq 1$ *the conditional measures*

$$\mu_n^G = \mu_n(\,\cdot\,|G_n).$$

Then the sequence $\left(\vec{A}_n\right)_{n\geq 1}$ *of the empirical phase partitions of* Ω *satisfies a LDP in* $(P(\Omega, q), \mathrm{dist}_P)$ *with respect to* μ_n^G *with speed* n^{d-1} *and rate function* $\mathcal{I} - \mathcal{I}_G$, *i.e., for any Borel subset* $\mathbb{E}$ *of* $P(\Omega, q)$,

$$
\begin{aligned}
-\inf_{\overset{\circ}{\mathbb{E}} \cap G} \mathcal{I} + \mathcal{I}_G \;\; &\leq \;\; \liminf_{n\to\infty} \frac{1}{n^{d-1}} \log \mu_n^G\left[\vec{A}_n \in \mathbb{E}\right] \\[4pt]
&\leq \;\; \limsup_{n\to\infty} \frac{1}{n^{d-1}} \log \mu_n^G\left[\vec{A}_n \in \mathbb{E}\right] \\[4pt]
&\leq \;\; -\inf_{\overline{\mathbb{E}} \cap G} \mathcal{I} + \mathcal{I}_G.
\end{aligned}
$$

Theorem 2.4 gives a rigorous verification of the basic assumption underlying the phenomenological theory, namely, that in a given ensemble, the typical configurations are those minimizing the surface free energy. A general compactness argument implies the existence of at least one such minimizer. However, in most examples one cannot say much about the minimizers themselves. (One notable exception is the Wulff problem.) The difficulty stems from the fact that the surface tension τ is anisotropic and almost no quantitative information about its magnitude is available. Moreover, the corresponding variational problems are extremely hard even in the isotropic case and the (few) resolved questions represent the state of the art in the calculus of variations. For instance, a famous conjecture related to the symmetric double-bubble in the three dimensional case with isotropic surface energy (perimeter) has only been resolved recently [25] and the asymmetric case remains unresolved (even in the isotropic case).

We show next how Theorem 2.4 can be applied to the Wulff and multiple bubble problem. We take pure boundary conditions with color 1, that is, $\Gamma^1 = \Gamma$, $\Gamma^2 = \cdots = \Gamma^q = \emptyset$. Let $s_2, \cdots, s_q$ be $q - 1$ real numbers larger than or equal to $(1 - \theta)/q$. We set

$$\forall i \in \{2, \ldots, q\} \qquad v_i = \mathrm{vol}\,(\Omega)\theta^{-1}(s_i - (1 - \theta)/q).$$

We define next the events

$$\forall n \in \mathbb{N} \qquad G_n = \{\forall i \in \{2, \ldots, q\} \quad S_n(i) \geq s_i\}$$

and the collection of phase partitions

$$\mathbb{G}(v_2, \ldots, v_q) = \{\,\vec{A} = (A_0, A_1, \ldots, A_q) \in P(\Omega, q) : \mathrm{vol}\,(A_2) \geq v_2, \ldots, \mathrm{vol}\,(A_q) \geq v_q\,\}.$$

It can be shown that the sequences of events

$$(G_n)_{n\in\mathbb{N}} \qquad \text{and} \qquad (\vec{A}_n \in \mathbb{G}(v_2, \ldots, v_q))_{n\in\mathbb{N}}$$

are exponentially equivalent, i.e., they satisfy the condition (3). In order to ensure condition (5), we suppose that the minimum of the surface energy $\mathcal{I}$ over $\mathbb{G}(v_2, \ldots, v_q)$ is reached with a phase partition having no interfaces on the boundary Γ. More precisely, we suppose that the following assumption is fulfilled.

Assumption. The region Ω and the real numbers $v_2, \ldots, v_q$ are such that there exists $\vec{A}^* = (A_0^*, A_1^*, \ldots, A_q^*)$ in $\mathbb{G}(v_2, \ldots, v_q)$ such that

$$\mathcal{I}(\vec{A}^*) = \min \{ \mathcal{I}(\vec{A}); \vec{A} \in \mathbb{G}(v_2, \ldots, v_q) \}$$
$$\forall i \in \{ 2, \ldots, q \} \qquad d_2(A_i^*, \Gamma) > 0.$$

We expect that this assumption is fulfilled provided the real numbers $v_2, \ldots, v_q$ are sufficiently small (or equivalently, $s_2, \ldots, s_q$ are sufficiently close to $(1 - \theta)/q$), depending on the region Ω. This is for instance the case when $q = 2$. Indeed, let $\mathcal{W}_\tau$ be the Wulff crystal associated to τ. We know that $\mathcal{W}_\tau$ is, up to dilatations and translations, the unique solution to the anisotropic isoperimetric problem associated to τ. For v_2 sufficiently small, a dilated Wulff crystal $x_0 + \alpha_0 \mathcal{W}_\tau$ of volume v_2 fits into Ω without touching Γ, and the phase partition $\vec{A}^* = (\emptyset, \Omega \setminus (x_0 + \alpha_0 \mathcal{W}_\tau), x_0 + \alpha_0 \mathcal{W}_\tau)$ answers the problem. In the case $q > 2$, we expect that a minimizing phase partition corresponds to a multiple bubble having $q - 1$ components.

Under the above assumption, we claim that the collection of phase partitions $\mathbb{G}(v_2, \ldots, v_q)$ satisfies (5). For $\lambda > 1$, we define

$$\vec{A}^*(\lambda) = \left(\emptyset, \Omega \setminus \bigcup_{2 \leq i \leq q} \lambda A_i^*, \lambda A_2^*, \ldots, \lambda A_q^* \right).$$

Since by hypothesis the sets $A_2^*, \ldots, A_q^*$ are at positive distance from Γ, for λ larger than 1 and sufficiently close to 1, the phase partition $\vec{A}^*(\lambda)$ satisfies

$$\vec{A}^*(\lambda) \in \mathbb{G}(\lambda^d v_2, \ldots, \lambda^d v_q) \subset \overset{\circ}{\mathbb{G}}(v_2, \ldots, v_q)$$

and moreover $\mathcal{I}(\vec{A}^*(\lambda)) = \lambda^{d-1} \mathcal{I}(\vec{A}^*)$. Sending λ to 1, and remarking that $\mathbb{G}(v_2, \ldots, v_q)$ is closed, we see that $\mathbb{G}(v_2, \ldots, v_q)$ satisfies (5). Thus we can apply Theorem 2.4 with the sequence of events $(G_n)_{n \in \mathbb{N}}$, thereby obtaining a LDP and a weak law of large numbers for the conditional measures $\mu_n^G = \mu_n(\cdot | G_n)$. In the particular case $q = 2$, we obtain again the main result (Wulff problem) of our previous paper [10]. In the more challenging situations $q > 2$, the unresolved questions concerning the macroscopic behavior of such systems belong to the realm of the calculus of variations.

References

[1] Almgren, F.J., Jr.: Existence and regularity almost everywhere of solutions to elliptic variational problems with constraints. Mem. Amer. Math. Soc. 4 no. 165 (1976).

[2] Alberti, G., Bellettini, G., Cassandro, M., Presutti, E.: Surface tension in Ising systems with Kac potentials. J. Stat. Phys. 82, 743–796 (1996).

[3] Alexander, K.S.: Stability of the Wulff minimum and fluctuations in shape for large finite clusters in two–dimensional percolation. Probab. Theory Relat. Fields 91, 507–532 (1992).

[4] Alexander, K.S., Chayes, J.T., Chayes, L.: The Wulff construction and asymptotics of the finite cluster distribution for two-dimensional Bernoulli percolation. Comm. Math. Phys. 131, 1–50 (1990).

[5] Benois, O., Bodineau, T., Buttà, P., Presutti, E.: On the validity of van der Waals theory of surface tension. Markov Process. Rel. Fields 3, 175–198 (1997).

[6] Benois, O., Bodineau, T., Presutti, E.: Large deviations in the van der Waals limit. Stochastic Process. Appl. 75, 89–104 (1998).

[7] Bodineau, T.: The Wulff construction in three and more dimensions. Commun. Math. Phys. 207 no.1, 197–229 (1999).

[8] Bodineau, T., Presutti, E.: Surface tension and Wulff theory for a lattice model without spin flip symmetry. Preprint (2002).

[9] Cerf, R.: Large deviations for three dimensional supercritical percolation. Astérisque 267 (2000).

[10] Cerf, R., Pisztora, A.: On the Wulff crystal in the Ising model. Ann. Probab. 28 no. 3, 945–1015 (2000).

[11] Cerf, R., Pisztora, A.: Phase coexistence in Ising, Potts and percolation models. Ann. I. H. Poincare, PR 37, 6 (2001) 643–724.

[12] Comets, F.: Grandes deviations pour des champs de Gibbs sur $\mathbb{Z}^d$. C.R.Acad.Sci. Ser. I, 303, 511–513 (1986).

[13] Dinghas, A.: Uber einen geometrischen Satz von Wulff für die Gleichgewichtsform von Kristallen. Z. Kristallogr. 105, 304–314 (1944).

[14] Dobrushin, R.L., Kotecký, R., Shlosman, S.B.: Wulff construction: a global shape from local interaction. AMS translations series, Providence (Rhode Island) (1992).

[15] Ellis, R.S.: Entropy, Large Deviations and Statistical Mechanics. Springer (1986).

[16] Evans, L.C., Gariepy, R.F.: Measure theory and fine properties of functions. Studies in Advanced Mathematics, Boca Raton: CRC Press (1992).

[17] Föllmer, H., Orey, S.: Large deviations for the empirical field of a Gibbs measure. Ann. Probab. Vol. 16, 961–977 (1988).

[18] Föllmer, H., Ort, M.: Large deviations and surface entropy for Markov fields. Asterisque 157–158, 173–190 (1988).

[19] Fonseca, I.: The Wulff theorem revisited. Proc. R. Soc. Lond. Ser. A 432 No.1884, 125–145 (1991).

[20] Fonseca, I., Müller, S.: A uniqueness proof for the Wulff theorem. Proc. R. Soc. Edinb. Sect. A 119 No.1/2, 125–136 (1991).

[21] Fortuin C.M., and Kasteleyn, P.W.: On the random cluster model. I. Physica 57, 536–564 (1972).

[22] Giusti, E.: Minimal surfaces and functions of bounded variation. Birkhäuser (1984).

[23] Grimmett, G.R.: The stochastic random-cluster process and the uniqueness of random-cluster measures. Ann. Probab. 23, 1461–1510 (1995).

[24] Grimmett, G.R., Marstrand J.M.: The supercritical phase of percolation is well behaved. Proc. R. Soc. Lond. Ser. A 430, 439–457 (1990).

[25] Hass, J.; Hutchings, M.; Schlafly, R.: The double bubble conjecture. Electron. Res. Announc. Amer. Math. Soc. 1 (1995), no. 3, 98–102 (electronic).

[26] Ioffe, D.: Large deviations for the 2D Ising model: a lower bound without cluster expansions. J. Stat. Phys. 74, 411–432 (1993).

[27] Ioffe, D.: Exact large deviation bounds up to T_c for the Ising model in two dimensions. Probab. Theory Relat. Fields 102, 313–330 (1995).

[28] Ioffe, D., Schonmann, R.: Dobrushin-Kotecký-Shlosman Theorem up to the critical temperature. Comm. Math. Phys. 199, 117–167 (1998).

[29] Kesten, H., Zhang, Y.: The probability of a large finite cluster in supercritical Bernoulli percolation. Ann. Probab. Vol. 18, 537–555 (1990).

[30] Lawlor, G., Morgan, F.: Paired calibrations applied to soap films, immiscible fluids, and surfaces or networks minimizing other norms. Pacific Journal of Mathematics Vol. 166 No. 1, 55–82 (1994).

[31] Liggett, T.M., Schonmann, R.H., Stacey, A.M.: Domination by product measures. Ann. Probab. Vol.25, 71–95 (1997).

di $\mathbb{R}^n$ con varietà differenziabili Bollettino U.M.I. (4) 10, 532–544 (1974).

[32] Olla, S.: Large deviations for Gibbs random fields. Probab. Theory Relat. Fields, Vol. 77, 395–409 (1988).

[33] Pfister, C.E.: Large deviations and phase separation in the two-dimensional Ising model. Helv. Phys. Acta 64, 953–1054 (1991).

[34] Pisztora, A.: Surface order large deviations for Ising, Potts and percolation models. Prob. Theor. Rel. Fields 104, 427–466 (1996).

[35] Schonmann, R.H.: Second order large deviation estimates for ferromagnetic systems in the phase coexistence region. Comm. Math. Phys. 112, 409–422 (1987).

[36] Taylor, J.E.: Crystalline variational problems. Bull. Am. Math. Soc. 84 no.4, 568–588 (1978).

[37] Varadhan, S.R.S.: Large deviations and applications. CBMS-NSF Regional Conference Series in Applied Mathematics, 46. SIAM, Philadelphia, PA. (1984).

[38] Wulff, G.: Zur Frage der Geschwindigkeit des Wachstums und der Auflösung der Kristallflächen. Z. Kristallogr. 34, 449–530 (1901).

ICM 2002 · Vol. III · 97–106

Biological Sequence Analysis

T. P. Speed*

Abstract

This talk will review a little over a decade's research on applying certain stochastic models to biological sequence analysis. The models themselves have a longer history, going back over 30 years, although many novel variants have arisen since that time. The function of the models in biological sequence analysis is to summarize the information concerning what is known as a motif or a domain in bioinformatics, and to provide a tool for discovering instances of that motif or domain in a separate sequence segment. We will introduce the motif models in stages, beginning from very simple, non-stochastic versions, progressively becoming more complex, until we reach modern profile HMMs for motifs. A second example will come from gene finding using sequence data from one or two species, where generalized HMMs or generalized pair HMMs have proved to be very effective.

2000 Mathematics Subject Classification: 60J20, 92C40.
Keywords and Phrases: Motif, Regular expression, Profile, Hidden Markov model.

1.   Introduction

DNA (deoxyribonucleic acid), RNA (ribonucleic acid), and proteins are macro-molecules which are unbranched polymers built up from smaller units. In the case of DNA these units are the 4 nucleotide residues A (adenine), C (cytosine), G (guanine) and T (thymine) while for RNA the units are the 4 nucleotide residues A, C, G and U (uracil). For proteins the units are the 20 amino acid residues A (alanine), C (cysteine) D (aspartic acid), E (glutamic acid), F (phenylalanine), G (glycine), H (histidine), I (isoleucine), K (lysine), L (leucine), M (methionine), N (asparagine), P (proline), Q (glutamine), R (arginine), S (serine), T (threonine), V (valine), W (tryptophan) and Y (tyrosine). To a considerable extent, the chemical properties of DNA, RNA and protein molecules are encoded in the linear sequence of these basic units: their primary structure.

---

*Department of Statistics, University of California, Berkeley, CA 94720, USA; Division of Genetics and Bioinformatics, Walter and Eliza Hall Institute of Medical Research, VIC 3050, Australia. E-mail: terry@stat.berkeley.edu

The use of statistics to study linear sequences of biomolecular units can be descriptive or it can be predictive. A very wide range of statistical techniques has been used in this context, and while statistical models can be extremely useful, the underlying stochastic mechanisms should never be taken literally. A model or method can break down at any time without notice. Further, biological confirmation of predictions is almost always necessary.

The statistics of biological sequences can be global or it can be local. For example, we might consider the global base composition of genomes: *E. coli* has 25% A, 25% C, 25% G, 25% T, while *P. falciparum* has 82%A+T. At the very local, the triple ATG is the near universal motif indicating the start of translation in DNA coding sequence. A major role of statistics in this context is to characterize individual sequences or classes of biological sequences using probability models, and to make use of these models to identify them against a background of other sequences. Needless to say, the models and the tools vary greatly in complexity.

Extensive use is made in biological sequence analysis of the notions of motif or domain in proteins, and site in DNA. We shall use these terms interchangeably to describe the recurring elements of interest to us. It is important to note that while we focus on the sequence characteristics of motifs, domains or sites, in practice they also embody (biochemical) structural significance.

2.   Deterministic models

The C2H2 (cysteine-cysteine histidine-histidine) zinc-finger DNA binding domain is composed of 25-30 amino acid residues including two conserved cysteines and two conserved histidines spaced in a particular way, with some restrictions on the residues in between and nearby. Of course the arrangement reflects the three-dimensional molecular structure into which the amino-acid sequence folds, for it is the structure which has the real biochemical significance, see Figure 1, which was obtained from `http://www.rcsb.org/pdb/`. An example of this motif is the 27-

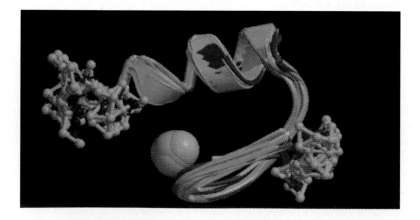

Figure 1: A C2H2 zinc finger DNA binding domain

letter sequence known as 1ZNF, this being a Protein Data Bank identifier for the structure XFIN-31 of *X. laevis*. Its amino acid sequence is

```
1ZNF:   XYKCGLCERSFVEKSALSRHQRVHKNX
```

Note the presence of the two *C*s separated by 2 other residues, and the two *H*s separated by 3 other residues. Here and elsewhere, X denotes an arbitrary amino acid residue. A popular and useful summary description of C2H2 zinc fingers which clearly includes our example, is the regular expression

$$C - X(2,4) - C - X(3) - [LIVMFYWC] - X(8) - H - X(3,5) - H$$

where $X(m)$ denotes a sequence of n unspecified amino acids, while $X(m,n)$ denotes from m to n such, and the brackets enclose mutually exclusive alternatives. There is a richer set of notation for *regular expressions* of this kind, but for our purposes it is enough to note that this representation is essentially deterministic, with uncertainty included only through mutually exclusive possibilities (e.g. length or residue) which are not otherwise distinguished.

Simple and efficient algorithms exist for searching query sequences of residues to find every instance of the regular expression above. In so doing with sequence in which all instances of the motif are known, we may identify some sub-sequences of the query sequence which are not C2H2 zinc finger DNA binding domains, i.e. which are false positives, and we may miss some sub-sequences which are C2H2 zinc fingers, i.e. which are false negatives. Thus we have essentially deterministic descriptions and search algorithms for the C2H2 motifs using regular expressions. Their performance can be described by the frequency of false positives and false negatives, equivalently, their complements, the specificity and sensitivity of the regular expression. We do not have space for an extensive bibliography, so for more on regular expressions and on most of the other concepts we introduce below, see [2].

3. Regular expressions can be limiting

Most protein binding sites are characterized by some degree of sequence specificity, but seeking a consensus DNA sequence is often an inadequate way to recognize their motifs. Simply listing the alternatives seen at a position may not be very informative, but keeping track of the frequencies with which the different alternatives appear can be very valuable. Thus position-specific nucleotide or amino acid distributions came to represent the variability in DNA or protein motif composition. This is just the set of marginal distribution of letters at each position. Rather than present an extensive tabulation of frequencies for our C2H2 zinc finger example, we present a pictorial representation: a sequence logo coming from http://blocks.fhcrc.org.

Sequence logos are scaled representation of position-specific nucleotide or amino acid distributions. The overall height at a given position is proportional to information content, which is a constant minus the entropy of the distribution at that

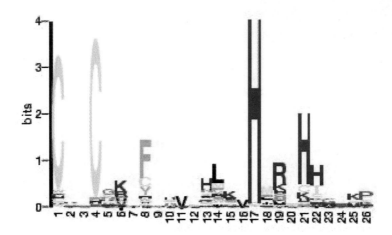

Figure 2: Sequence logo for C2H2 zinc finger

position. The proportions of each nucleotide or amino acid at a position are in relation to their observed frequency at that position, with the most frequent on top, the next most frequent below, etc.

4. Profiles

It is convenient for our present purposes to define a profile as a set of position-specific distributions describing a motif. (Traditionally the term has been used for the derived scores.) How would we use a set of such distributions to search a query sequence for instances of the motif? The answer from bioinformatics is that we *score* the query sequence, and for suitably large scores, declare that a candidate subsequence is an instance of our motif.

There are a number of approaches for deriving profile scores, but the easiest to explain here is this: scores are *log-likelihood ratio test statistics*, for discriminating between a probability model M for the motif and a model B for the background. The model M will be the direct product of the position-specific distributions, (i.e. the independent but not identical distribution model), while the background model B will be the direct product of a set of relevant background frequencies (i.e. the independent and identical distribution model). Thus, if f_{al} is the frequency of residue a at position l of the motif, and f_a background frequency of the same residue, then the profile score assigned to residue a at position l in a possible instance of the motif will be $s_{al} = \log f_{al}/f_a$. These scores are then summed across the positions in the motif, and compared to a suitably defined threshold. Note that proper setting of the threshold requires a set of data in which all instances of the motif are known. The false positive and false negative rate could then be determined for various thresholds, and a suitable choice made.

We briefly discuss variants of the log-likelihood ratio scores. In many contexts, it will matter little whether a position is occupied by a leucine (L) rather than an isoleucine (I), as each can evolve in time to or from the other rather more readily than from other residues. Thus it might make sense to modify the scores to take this and similar evolutionary patterns into account. Indeed the first use of profiles involved scores of this kind, using the position specific amino acid distribution of an alignment of instances of the motif and entries from what are known as PAM matrices, which embody patterns of molecular evolution. In addition, the background distribution of residues may be modelled more detailed manner, e.g. using the so-called Dirichlet mixture models.

It is also possible to include position-specific scores for insertion and deletion of residues, relative to a consensus pattern. When these are used, the scoring becomes a little more subtle, as the problem is then quite analogous to pairwise sequence alignment, but with position dependent scoring parameters for matches, mismatches, insertions and deletions.

We summarise this section by noting that probability has entered into our description through the use of frequencies, and scores based on them, but so far we do not have global statistical models, at least not ones embodying insertions and deletions, on which we base our estimation and testing. These are all part of the use of profile HMMs, but first we introduce HMMs.

5.   Hidden Markov models

Hidden Markov models (HMMs) are processes $(S_t, O_t), t = 1, \ldots, T$, where S_t is the hidden state and O_t the observation at time t. Their probabilistic evolution is constrained by the equations

$$
\begin{aligned}
pr(S_t | S_{t-1}, O_{t-1}, S_{t-2}, O_{t-2}, \ldots) &= pr(S_t | S_{t-1}), \\
pr(O_t | S_{t-1}, O_{t-1}, S_{t-2}, O_{t-2}, \ldots) &= pr(O_t | S_t, S_{t-1}).
\end{aligned}
$$

The definitions and basic facts concerning HMMs were laid out in a series of beautiful papers by L. E. Baum and colleagues around 1970, see [2] for references. Much of their formulation has been used almost unchanged to this day. Many variants are now used. For example, the distribution of O may not depend on previous S, or it may also depend on previous O values,

$$
\begin{aligned}
pr(O_t | S_t, S_{t-1}, O_{t-1}, \ldots) &= pr(O_t | S_t), \quad \text{or} \\
pr(O_t | S_t, S_{t-1}, O_{t-1}, \ldots) &= pr(O_t | S_t, S_{t-1}, O_{t-1}).
\end{aligned}
$$

Most importantly for us below, the times of S and O may be decoupled, permitting the observation corresponding to state time t to be a string whose length and composition depends on S_t (and possibly S_{t-1} and part or all of the previous observations). This is called a hidden semi-Markov or generalized hidden Markov model.

Early applications of HMMs were to finance, but these were never published, to speech recognition, and to modelling ion channels. In the mid-late 1980s HMMs entered genetics and molecular biology, where they are now firmly entrenched. One of the major reasons for the success of HMMs as stochastic models is the fact that although they are substantial generalizations of Markov chains, there are elegant dynamic programming algorithms which permit full likelihood calculations in many cases of interest. Specifically, there are algorithms which permit the efficient calculation of a) $pr(sequence|M)$, where $sequence$ is a sequence of observations and M is an HMM; b) the maximum over $states$ of $pr(states|sequence, M)$, where $states$ is the unobserved state sequence underlying the observation $sequence$; and c) the maximum likelihood estimates of parameters in M based on the observation $sequence$. Step c) is carried out by an iterative procedure which in the case of independent states was later termed the EM algorithm.

6.   Profile HMMs

In a landmark paper A. Krogh, D. Haussler and co-workers introduced profile HMMs into bioinformatics. An illustrative form of their profile HMM architecture is given in Figure 3. There we depict the underlying state space of the hidden

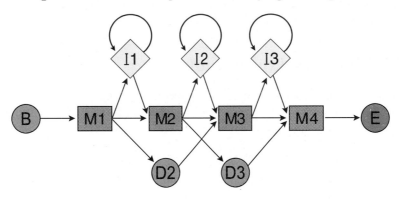

Figure 3: State space of a simple profile HMM

Markov chain of a profile HMM of length 4, with M denoting $match$ states, I $insert$ states and D $delete$ states, while B and E are $begin$ and end states, respectively. Encircled states (D, B and E) do not emit observations, while each of the match and insert states will have position-specific observation or emission distributions. Finally, each arrow will have associated transition probabilities, with the expectation being that the horizontal transition probabilities are typically near unity. This the chain proceeds from left to right, and if it remains within match states, its output will be an amino acid sequence of length 4. Deviation to the insert or delete states will modify the output accordingly. The similarity with a direct product of a sequence of position-specific distributions should be unmistakeable. The profile HMMs in use now have considerably more features, while sharing the basic M, I and D architecture.

Why was the introduction of the HMM formalism such an advance? The answer is simple: it permitted the construction and application of profiles to be conducted entirely within a formal statistical framework, and that really helped. Instances of the motif embodied in an HMM could be identified by calculating $pr(sequence|M)/pr(sequence|B)$ as was done with profiles, using the algorithm for problem a) in X above. Instances of the motif could be aligned to the HMM by calculating the most probable state sequence giving rise to the motif sequence, in essence finding the most probable sequence of matches, insertions, deletions which align the given sequence to the others which gave rise to the HMM, cf. problem b) above. And finally, the parameters in the HMMs could be estimated from data comprising known instances of the motif by using maximum likelihood, an important step for many reasons, one being that it put insert and delete scores on precisely the same footing as match and mismatch scores. Although the estimation of HMM parameters is easiest if the example sequences are properly aligned, the EM algorithm (problem c) above) does not require aligned sequences.

In the years since the introduction of profile HMMs, they have been become the standard approach to representing motifs and protein domains. The database Pfam (`http://pfam.wustl.edu`) now has 3,849 hidden Markov models (May 2002) representing recognized protein or DNA domains or motifs. Profile HMMs have essentially replaced the use of regular expressions and the original profiles for searching other databases to find novel instances of a motif, for finding a motif or domain match to an input sequence, and for aligning a motif or domain to a an existing family. There is considerable evidence that the HMM-based searches are more powerful than the older profile based ones, though they are slower computationally, and at times that is an important consideration.

7. Finding genes in DNA sequence

Identifying genes in DNA sequence is one of the most challenging, interesting and important problems in bioinformatics today. With so many genomes being sequenced so rapidly, and the experimental verification of genes lagging far behind, it is necessary to rely on computationally derived genes in order to make immediate use of the sequence.

What is a gene? Most readers will have heard of the famous *central dogma* of molecular biology, in which the hereditary material of an organism resides in its genome, usually DNA, and where genes are expressed in a two-stage process: first DNA is *transcribed* into a messenger RNA (mRNA) sequence, and later a processed form of this sequence is *translated* into an amino acid sequence, i.e. a protein. In general the transcribed sequence is longer than the translated portion: parts called introns (intervening sequence) are removed, leaving exons (expressed sequence), of which only some are expressed, while the rest remain untranslated. The translated sequence comes in triples called codons, beginning and ending with a unique start (ATG) and one of three stop (TAA, TAG, TGA) codons. There are also characteristic intron-exon boundaries called splice donor and acceptor sites, and a variety of other motifs: promoters, transcription start sites, polyA sites, branching sites, and so on.

All of the foregoing have statistical characterizations, and in principle they can all help identify genes in long otherwise unannotated DNA sequence segments. To get an idea of the magnitude of the task with the human genome, consider the following facts about human gene sequences [5]: the coding regions comprise about 1.5% of the entire genome; the average gene length is about 27,000 bp (base pair); the average total coding region is 1,340 bp; and the average intron length is about 3,300 bp. Further, only about 8% of genes have a single exon. We see that the information in human genes is very dispersed along the genome, and that in general the parts of primary interest, the coding exons, are a relatively small fraction of the gene, on average about $\frac{1}{20}$.

8. Generalized HMMs for finding genes

The HMMs which are effective in finding genes are the generalized HMMs (GHMMs) described in section 5. above. Space does not permit our giving an adequate description here, so we simply outline the architecture of Genscan [1] one of the most widely used human genefinders. States represent the gene features we mentioned above: exon, intron, and of course intergenic region, and a variety of other features (promotor, untranslated region, polyA site, and so on. Output observations embody state-dependent nucleotide composition, dependence, and specific signal features (such as stop codons). In a GHMM the state *duration* needs to be modelled, as well as two other important features of genes in DNA: the *reading frame*, which corresponds to the triples along the mRNA sequence which are sequentially translated, and the *strand*, as DNA is double stranded, and genes can be on either strand, i.e. they can point in either direction. These features can be seen in Figure 4, which was kindly supplied by Lior Pachter.

The output of a GHMM genefinder after processing a genomic segment is broadly similar to that from a profile HMM after processing an amino acid sequence: the most probable state sequence given an observation sequence is a best gene annotation of that sequence, and a variety of probabilities can be calculated to indicate the support in the observation sequence for various specific gene features.

9. Comparative sequence analysis using HMMs

The large number of sequenced genomes now available, and the observation that functionally important regions are evolutionarily conserved, has led to efforts to incorporate conservation into the models and methods of biological sequence analysis. Pair HMMs were introduced in [2] as a way of including alignment problems under the HMM framework, and recently [4] they were combined with GHMMs (forming GPHMMs) to carry out alignment and genefinding with homologous segments of the mouse and human genomes. Use of the program SLAM on the whole

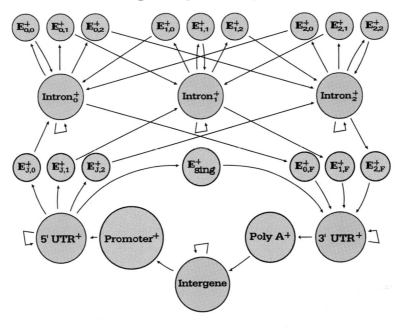

Figure 4: Forward half of the Genscan GHMM state space

mouse genome (http://bio.math.berkeley.edu/slam/mouse/) demonstrated the value of GPHMMs in this context.

10. Challenges in biological sequence analysis

The first challenge is to understand the biology well enough to begin biological sequence analysis. This part will frequently involve collaborations with biologists. With HMMs, GHMMs and GPHMMs, designing the underlying architecture, and carrying out the modelling for the components parts, e.g. for splice sites in genefinding GHMM is perhaps the next major challenge. Undoubtedly the hardest and most important task of all is the implementation: coding up the algorithms and making it all work with error-prone and incomplete sequence data. Finally, it is usually a real challenge to find good data sets for calibrating and evaluating the algorithms, and for carrying out studies of competing algorithms.

For a recent example of this process, which is a model of its kind, see [3]. There an HMM is presented for the so-called σ^A recognition sites, which involve two DNA motifs separated by a variable number of base pairs. In addition to the examples mentioned so far, there are many more HMMs in the bioinformatics literature, see p. 79 of [2] for ones published before 1998.

11. Closing remarks

In this short survey of biological sequence analysis, I have simply touched on some of the major ideas. A much more comprehensive treatment of material covered here can be found in the book [2], whose title not coincidentally is the same as that of this paper. Many important ideas from biological sequence analysis have not been mentioned here, including molecular evolution and phylogenetic inference, and the use of stochastic context-free grammars, a form of generalization of HMMs suited to the analysis of RNA sequence data.

At this Congress I have talked (and am now writing) on the research of others, in an area in which my own contributions have been negligible. I chose to do so upon being honoured by the invitation to speak at this Congress because I believe this topic – HMMs – to be one of the great success stories of applying mathematics to bioinformatics. In my view it is the one most worthy of a wider mathematical audience. I hope that the fact that there are many others better suited than me to speak on this topic will not prevent readers from appreciating it and following it up through the bibliography.

I owe what understanding I have of this field to collaborations and discussions with a number of people, and I would like to acknowledge them here. Firstly, Tony Wirth, Simon Cawley and Mauro Delorenzi, with whom I have worked on HHMMs. Next, it has been an honour and pleasure to observe from close by the development of SLAM, by Simon Cawley, Lior Pachter and Marina Alexandersson. Finally I'd like to thank Xiaoyue Zhou and Ken Simpson for their kind help to me when I was preparing my talk and this paper.

References

[1] C. Burge & S. Karlin, Prediction of complete gene structures in human genomic DNA, *J. Mol. Biol.* 268 (1997) 78–94.

[2] R. Durbin, S. Eddy, A. Krogh & G. Mitchison, *Biological Sequence Analysis. Probabilistic models of proteins and nucleic acids*, Cambridge University Press, 1998.

[3] H. Jarmer, T. S. Larsen, A. Krogh, H. H. Saxild, S. Brunak & S. Knudsen, Sigma A recognition sites in the *Bacillus subtilis* genome, *Microbiology* 147 (2001), 2417–2424.

[4] L. Pachter, M. Alexandersson & S. Cawley, Applications of generalized pair hidden Markov models to alignment and gene finding problems, *J. Comp. Biol.* 9 (2002), 389–399.

[5] The Genome Sequencing Consortium, Initial sequencing and analysis of the human genome, *Nature* 409 (2001), 860–921.

Estimates for the Strong Approximation in Multidimensional Central Limit Theorem[*]

A. Yu. Zaitsev[†]

Abstract

In a recent paper the author obtained optimal bounds for the strong Gaussian approximation of sums of independent $\mathbf{R}^d$-valued random vectors with finite exponential moments. The results may be considered as generalizations of well-known results of Komlós–Major–Tusnády and Sakhanenko. The dependence of constants on the dimension d and on distributions of summands is given explicitly. Some related problems are discussed.

2000 Mathematics Subject Classification: 60F05, 60F15, 60F17.
Keywords and Phrases: Strong approximation, Prokhorov distance, Central limit theorem, Sums of independent random vectors.

1. Introduction

Let $X_1, \ldots, X_n, \ldots$ be mean zero independent $\mathbf{R}^d$-valued random vectors and $\mathbb{D}_n = \operatorname{cov} S_n$ the covariance operator of the sum $S_n = \sum_{i=1}^n X_i$. By the Central Limit Theorem, under some simple moment conditions the distribution of normalized sums $\mathbb{D}_n^{-1/2} S_n$ is close to the standard Gaussian distribution. The invariance principle states that, in a sense, the distribution of the *whole* sequence $\mathbb{D}_n^{-1/2} S_1, \ldots, \mathbb{D}_n^{-1/2} S_n, \ldots$ is close to the distribution of the sequence $\mathbb{D}_n^{-1/2} T_1, \ldots,$ $\mathbb{D}_n^{-1/2} T_n, \ldots$, where $T_n = \sum_{i=1}^n Y_i$ and $Y_1, \ldots, Y_n, \ldots$ is a corresponding sequence of independent Gaussian random vectors (this means that Y_i has the same mean and the same covariance operator as X_i, $i = 1, \ldots, n, \ldots$).

We consider here the problem of strong approximation which is more delicate than that of estimating the closeness of distributions. It is required to construct on a probability space a sequence of independent random vectors $X_1, \ldots, X_n$ (with

[*]Research partially supported by Russian Foundation of Basic Research (RFBR) Grant 02-01–00265, and by RFBR-DFG Grant 99-01–04027.

[†]St. Petersburg Branch of the Steklov Mathematical Institute, Fontanka 27, St. Petersburg 191011, Russia. E-mail: zaitsev@pdmi.ras.ru

given distributions) and a corresponding sequence of independent Gaussian random vectors $Y_1, \ldots, Y_n$ so that the quantity

$$\Delta(X, Y) = \max_{1 \leq k \leq n} \left\| \sum_{i=1}^{k} X_i - \sum_{i=1}^{k} Y_i \right\|$$

would be so small as possible with large probability. Here $\| \cdot \|$ is the Euclidean norm. It is clear that the vectors even with the same distributions can be very far one from another.

In some sense this problem is one of the most important in probability approximations because many well-known probability theorems can be considered as consequences of results about strong approximation of sequences of sums by corresponding Gaussian sequences. This is related to the law of iterated logarithm, to several theorems about large deviations, to the estimates for the rate of convergence of the Prokhorov distance in the invariance principles (Prokhorov [19], Skorokhod [26], Borovkov [4]), as well as to the Strassen-type approximations (Strassen [28], see, for example, Csörgő and Hall [8]).

The rate for strong approximation in the one-dimensional invariance principle was studied by many authors (see, e.g., Prokhorov [19], Skorokhod [26], Borovkov [4], Csörgő and Révész [6] and the bibliography in Csörgő and Révész [7], Csörgő and Hall [8], Shao [20]). Skorokhod [26] developed a method of construction of close sequences of sequential sums of independent random variables on the same probability space. For a long time the best rates of approximation were obtained by this method, known now as the Skorokhod embedding. However, Komlós, Major and Tusnády (KMT) [17] elaborated a new, more powerful method of dyadic approximation. With the help of this method they obtained optimal rates of Gaussian approximation for sequences of independent identically distributed random variables.

We restrict ourselves on the most important case, where the summands have finite exponential moments. Sakhanenko [24] generalized and essentially sharpened KMT results in the case of non-identically distributed random variables. He considered the following class of one-dimensional distributions:

$$\mathcal{S}_1(\tau) = \left\{ \mathcal{L}(\xi) : \; \mathbf{E}\, \xi = 0, \; \mathbf{E}\, |\xi|^3 \exp\left(\tau^{-1} |\xi|\right) \leq \tau \, \mathbf{E}\, |\xi|^2 \right\}$$

(the distribution of a random vector ξ will be denoted by $\mathcal{L}(\xi)$). His main result is formulated as follows.

Theorem 1 (Sakhanenko [24]). *Suppose that $\tau > 0$, and $\xi_1, \ldots, \xi_n$ are independent random variables with $\mathcal{L}(\xi_j) \in \mathcal{S}_1(\tau)$, $j = 1, \ldots, n$. Then one can construct on a probability space a sequence of independent random variables $X_1, \ldots, X_n$ and a sequence of independent Gaussian random variables $Y_1, \ldots, Y_n$ so that $\mathcal{L}(X_j) = \mathcal{L}(\xi_j)$, $\mathbf{E}\, Y_j = 0$, $\mathbf{E}\, Y_j^2 = \mathbf{E}\, X_j^2$, $j = 1, \ldots, n$, and*

$$\mathbf{E} \exp\left(c\, \Delta(X, Y)/\tau\right) \leq 1 + B/\tau, \tag{1.1}$$

where c is an absolute constant and $B^2 = \mathbf{E}\, \xi_1^2 + \cdots + \mathbf{E}\, \xi_n^2$.

KMT [17] supposed that $\xi, \xi_1, \ldots, \xi_n$ are identically distributed and $\mathbf{E}\, e^{\langle h,\xi \rangle} < \infty$, for $h \in V$, where $V \subset \mathbf{R}^d$ is some neighborhood of zero. The KMT (1975–76) result follows from Theorem 1. It is easy to see that there exists $\tau(F)$ such that $F = \mathcal{L}(\xi_j) \in \mathcal{S}_1(\tau(F))$. Applying the Chebyshev inequality, we observe that (1.1) imply that

$$\mathbf{P}\,(c_1\,\Delta(X,Y)/\tau(F) \geq x) \leq \exp\left(\log\left(1 + \sqrt{n\,\mathbf{E}\,\xi^2}/\tau(F) \right) - x \right), \quad x > 0. \quad (1.2)$$

Inequality (1.2) provides more information than the original KMT formulation which contains unspecified constants depending on F. In (1.2) the dependence of constants on the distribution F is written out in an explicit form. The quantity $\tau(F)$ can be easily calculated or estimated for any concrete distribution F.

The first attempts to extend the KMT and Sakhanenko approximations to the multidimensional case (see Berkes and Philipp [3], Philipp [18], Berger [2], Einmahl [10, 11]) had a partial success only. Comparatively recently U. Einmahl [12] obtained multidimensional analogs of KMT results which are close to optimal. Zaitsev [33, 34] removed an unnecessary logarithmic factor from the result of Einmahl [12] and obtained multidimensional analogs of KMT results (see Theorem 2 below). In Theorem 2 the random vectors are, generally speaking, non-identically distributed. However, they have the same identity covariance operator $\mathbb{I}$. Therefore, the problem of obtaining an adequate multidimensional generalization of the main result of Sakhanenko [24] remained open. This generalization is given in Theorem 3 below.

2. Main results

For formulations of results we need some notations. Let $\mathcal{A}_d(\tau)$, $\tau \geq 0$, $d \in \mathbf{N}$, denote classes of d-dimensional distributions, introduced in Zaitsev [29], see as well Zaitsev [33–35]. The class $\mathcal{A}_d(\tau)$ (with a fixed $\tau \geq 0$) consists of d-dimensional distributions F for which the function $\varphi(z) = \varphi(F, z) = \log \int_{\mathbf{R}^d} e^{\langle z, x \rangle} F\{dx\}$ ($\varphi(0) = 0$) is defined and analytic for $\|z\|\, \tau < 1$, $z \in \mathbf{C}^d$, and $\left| d_u d_v^2\, \varphi(z) \right| \leq \|u\|\tau \, \langle \mathbb{D}v, v \rangle$ for all $u, v \in \mathbf{R}^d$ and $\|z\|\,\tau < 1$, where $\mathbb{D} = \operatorname{cov} F$, the covariance operator corresponding to F, and $d_u\varphi$ is the derivative of the function φ in direction u.

Theorem 2 (Zaitsev, [33, 34]). *Suppose that $\tau \geq 1$, $\alpha > 0$ and $\xi_1, \ldots, \xi_n$ are random vectors with distributions $\mathcal{L}(\xi_k) \in \mathcal{A}_d(\tau)$, $\mathbf{E}\,\xi_k = 0$, $\operatorname{cov}\xi_k = \mathbb{I}$, $k = 1, \ldots, n$. Then one can construct on a probability space a sequence of independent random vectors $X_1, \ldots, X_n$ and a sequence of independent Gaussian random vectors $Y_1, \ldots, Y_n$ so that*

$$\mathcal{L}(X_k) = \mathcal{L}(\xi_k), \quad \mathbf{E}\,Y_k = 0, \quad \operatorname{cov}\mathcal{L}\,(Y_k) = \mathbb{I}, \qquad k = 1, \ldots, n,$$

and

$$\mathbf{E} \exp\left(\frac{c_1(\alpha)\,\Delta(X,Y)}{\tau d^{7/2} \log^* d} \right) \leq \exp\left(c_2(\alpha)\, d^{9/4+\alpha} \log^*\left(n/\tau^2 \right) \right),$$

where $c_1(\alpha)$, $c_2(\alpha)$ are positive quantities depending on α only and $\log^* b = \max\{1, \log b\}$, for $b > 0$.

Corollary 1. *In the conditions of Theorem 2 for all $x \geq 0$ the following inequality is valid*

$$\mathbf{P}\left\{\Delta(X,Y) \geq \frac{c_2(\alpha)\,\tau d^{23/4+\alpha} \log^* d \, \log^*\left(n/\tau^2\right)}{c_1(\alpha)} + x\right\} \leq \exp\left(-\frac{c_1(\alpha)\,x}{\tau d^{7/2}\log^* d}\right).$$

It is easy to see that if $V \subset \mathbf{R}^d$ is some neighborhood of zero and $\mathbf{E}\,e^{\langle h, \xi \rangle} < \infty$, for $h \in V$, then $F = \mathcal{L}(\xi) \in \mathcal{A}_d(c(F))$. Below we list some simple and useful properties of classes $\mathcal{A}_d(\tau)$ which are essential in the proof of Theorem 2. Theorem 2 implies in one-dimensional case Sakhanenko's Theorem 1 for identically distributed random variables with finite exponential moments as well as the result of KMT [17].

Corollary 2. *Suppose that a random vector ξ has finite exponential moments $\mathbf{E}\,e^{\langle h, \xi \rangle}$, for $h \in V$, where $V \subset \mathbf{R}^d$ is some neighborhood of zero. Then one can construct on a probability space a sequence of independent random vectors $X_1, X_2, \ldots$ and a sequence of independent Gaussian random vectors $Y_1, Y_2, \ldots$ so that*

$$\mathcal{L}(X_k) = \mathcal{L}(\xi), \quad \mathbf{E}\,Y_k = 0, \quad \operatorname{cov} Y_k = \operatorname{cov} \xi, \qquad k = 1, 2, \ldots,$$

and

$$\sum_{k=1}^{n} X_k - \sum_{k=1}^{n} Y_k = O(\log n) \qquad \text{a.s..}$$

As it is noted in KMT [17], from the results of Bártfai [1] that the rate of approximation in Corollary 2 is the best possible for non-Gaussian vectors ξ. An analog of Corollary 2 was obtained by Einmahl [12] under additional smoothness-type restrictions on the distribution $\mathcal{L}(\xi)$. The following statement is a sharpening of Corollary 2.

Corollary 3 (Zaitsev [36]). *Suppose that a random vector ξ has the distribution such that $\mathcal{L}(\mathbb{D}^{-1/2}\xi) \in \mathcal{A}_d(\tau)$, where $\mathbb{D} = \operatorname{cov}\mathcal{L}(\xi)$ is a reversible operator. Let σ^2, $\sigma > 0$, be the maximal eigenvalue of $\mathbb{D}$. Then for any $\alpha > 0$ there exists a construction from Corollary 2 such that*

$$\mathbf{P}\left\{\limsup_{n \to \infty} \frac{1}{\log n}\left\|\sum_{k=1}^{n} X_k - \sum_{k=1}^{n} Y_k\right\| \leq c_3(\alpha)\,\sigma\,\tau\,d^{23/4+\alpha}\log^* d\right\} = 1 \qquad (2.1)$$

with $c_3(\alpha)$ depending on α only.

In Theorems 2 and Corollary 3 we consider the case $\tau \geq 1$. The case of small τ was investigated by Götze and Zaitsev [16]. It is shown that under additional smoothness-type restrictions on the distribution $\mathcal{L}(\xi)$ the expression in the right-hand side of the inequality in (2.1) can be arbitrarily small if the parameter τ is small enough. It is clear that the statements of Theorem 2 and Corollary 3 becomes stronger for small τ. In Götze and Zaitsev [16] one can find simple examples in which the sufficiently complicated smoothness condition is satisfied. The approximation is better in the case when summands have smooth distributions which are close to Gaussian ones (see inequalities (3.1) and (3.2) below).

The following Theorem 3 is a generalization of Theorem 2 to the case of multivariate random variables. In one-dimensional situation, Theorem 3 implies Theorem 1.

Theorem 3 (Zaitsev [35]). *Suppose that* $\alpha > 0$, $\tau \geq 1$, *and* $\xi_1, \ldots, \xi_n$ *are independent random vectors with* $\mathbf{E}\,\xi_j = 0$, $j = 1, \ldots, n$. *Assume that there exists a strictly increasing sequence of non-negative integers* $m_0 = 0$, $m_1, \ldots, m_s = n$ *satisfying the following conditions. Write*

$$\zeta_k = \xi_{m_{k-1}+1} + \cdots + \xi_{m_k}, \qquad k = 1, \ldots, s,$$

and suppose that (for all $k = 1, \ldots, s$*)* $\mathcal{L}(\zeta_k) \in \mathcal{A}_d(\tau)$, $\operatorname{cov} \zeta_k = \mathbb{B}_k$ *and, for all* $u \in \mathbf{R}^d$,

$$c_4 \left\| u \right\|^2 \leq \langle \mathbb{B}_k u, u \rangle \leq c_5 \left\| u \right\|^2 \tag{2.2}$$

with some constants c_4 *and* c_5. *Then one can construct on a probability space a sequence of independent random vectors* $X_1, \ldots, X_n$ *and a corresponding sequence of independent Gaussian random vectors* $Y_1, \ldots, Y_n$ *so that* $\mathcal{L}(X_j) = \mathcal{L}(\xi_j)$, $\mathbf{E}\,Y_j = 0$, $\operatorname{cov} \mathcal{L}(Y_j) = \operatorname{cov} \mathcal{L}(X_j)$, $j = 1, \ldots, n$, *and*

$$\mathbf{E} \exp\left(\frac{a_1 \, \Delta(X, Y)}{\tau d^{9/2} \log^* d} \right) \leq \exp\left(a_2 \, d^{3+\alpha} \log^*(s/\tau^2) \right),$$

where a_1, a_2 *are positive quantities depending only on* α, c_4, c_5.

3. Properties of classes $\mathcal{A}_d(\tau)$

Let us consider elementary properties of classes $\mathcal{A}_d(\tau)$ which are essentially used in the proof of Theorems 2 and 3, see Zaitsev [29, 31, 33–35]. It is easy to see that $\tau_1 < \tau_2$ implies $\mathcal{A}_d(\tau_1) \subset \mathcal{A}_d(\tau_2)$. Moreover, the class $\mathcal{A}_d(\tau)$ is closed with respect to convolution: if $F_1, F_2 \in \mathcal{A}_d(\tau)$, then $F_1 F_2 = F_1 * F_2 \in \mathcal{A}_d(\tau)$. Products of measures are understood in the convolution sense. Note that the condition $\mathcal{L}(\zeta_k) \in \mathcal{A}_d(\tau)$ in Theorem 3 is satisfied if $\mathcal{L}(\xi_j) \in \mathcal{A}_d(\tau)$, for $j = 1, \ldots, n$.

Let $\tau \geq 0$, $F = \mathcal{L}(\xi) \in \mathcal{A}_d(\tau)$, $y \in \mathbf{R}^m$, and $\mathbb{A} : \mathbf{R}^d \to \mathbf{R}^m$ is a linear operator. Then

$$\mathcal{L}(\mathbb{A}\xi + y) \in \mathcal{A}_m\left(\|\mathbb{A}\| \, \tau \right), \qquad \text{where} \quad \|\mathbb{A}\| = \sup_{x \in \mathbf{R}^d, \, \|x\| \leq 1} \|\mathbb{A}x\|.$$

Suppose that $\tau \geq 0$, $F_k = \mathcal{L}\left(\xi^{(k)} \right) \in \mathcal{A}_{d_k}(\tau)$, and the vectors $\xi^{(k)}$, $k = 1, 2$, are independent. Let $\xi \in \mathbf{R}^{d_1 + d_2}$ be the vector with the first d_1 coordinates coinciding with those of $\xi^{(1)}$ and with the last d_2 coordinates coinciding with those of $\xi^{(2)}$. Then $F = \mathcal{L}(\xi) \in \mathcal{A}_{d_1 + d_2}(\tau)$.

The classes $\mathcal{A}_d(\tau)$ are closely connected with other naturally defined classes of multidimensional distributions. From the definition of $\mathcal{A}_d(\tau)$ it follows that if

$\mathcal{L}(\xi) \in \mathcal{A}_d(\tau)$ then the vector ξ has finite exponential moments $\mathbf{E}\, e^{\langle h, \xi \rangle} < \infty$, for $h \in \mathbf{R}^d$, $\|h\|\,\tau < 1$. This leads to exponential estimates for the tails of distributions.

The condition $\mathcal{L}(\xi) \in \mathcal{A}_1(\tau)$ is equivalent to Statulevičius' [27] conditions on the rate of increasing of cumulants γ_m of the random variable ξ:

$$|\gamma_m| \le \frac{1}{2} m!\, \tau^{m-2} \gamma_2, \qquad m = 3, 4, \ldots.$$

This equivalence means that if one of these conditions is satisfied with parameter τ, then the second is valid with parameter $c\tau$, where c denotes an absolute constant. However, the condition $\mathcal{L}(\xi) \in \mathcal{A}_d(\tau)$ differs essentially from other multidimensional analogs of Statulevičius' conditions, considered by Rudzkis [23] and Saulis [25].

Zaitsev [30] considered classes of distributions

$$\begin{aligned}
\mathcal{B}_d(\tau) &= \left\{ F = \mathcal{L}(\xi) : \mathbf{E}\,\xi = 0, \ \left| \mathbf{E}\, \langle \xi, v \rangle^2 \langle \xi, u \rangle^{m-2} \right| \right. \\
&\le \left. \frac{1}{2} m!\, \tau^{m-2} \|u\|^{m-2}\, \mathbf{E}\, \langle \xi, v \rangle^2 \quad \text{for all } u, v \in \mathbf{R}^d, \ m = 3, 4, \ldots \right\}
\end{aligned}$$

satisfying multidimensional analogs of the Bernstein inequality condition. Sakhanenko's condition $\mathcal{L}(\xi) \in \mathcal{S}_1(\tau)$ is equivalent to the condition $\mathcal{L}(\xi) \in \mathcal{B}_1(\tau)$. Note that if $F\left\{ \{x \in \mathbf{R}^d : \|x\| \le \tau\} \right\} = 1$ then $F \in \mathcal{B}_d(\tau)$.

Let us formulate a relation between classes $\mathcal{A}_d(\tau)$ and $\mathcal{B}_d(\tau)$. Denote by $\sigma^2(F)$ the maximal eigenvalue of the covariance operator of a distribution F. Then

a) If $F = \mathcal{L}(\xi) \in \mathcal{B}_d(\tau)$, then $\sigma^2(F) \le 12\,\tau^2$, $\mathbf{E}\,\xi = 0$ and $F \in \mathcal{A}_d(c\tau)$.
b) If $F = \mathcal{L}(\xi) \in \mathcal{A}_d(\tau)$, $\sigma^2(F) \le \tau^2$ and $\mathbf{E}\,\xi = 0$, then $F \in \mathcal{B}_d(c\tau)$.

If F is an infinitely divisible distributions with spectral measure concentrated on the ball $\{x \in \mathbf{R}^d : \|x\| \le \tau\}$ then $F \in \mathcal{A}_d(c\tau)$, where c is an absolute constant. It is obvious that the class $\mathcal{A}_d(0)$ coincides with the class of all d-dimensional Gaussian distributions. The following inequality was proved in Zaitsev [29] and can be considered as an estimate of stability of this characterization:

$$\text{if } F \in \mathcal{A}_d(\tau), \text{ then } \pi\left(F, \Phi(F)\right) \le c\, d^2 \tau \log^*(\tau^{-1}); \tag{3.1}$$

where $\pi(\cdot, \cdot)$ is the Prokhorov distance and $\Phi(F)$ denotes the Gaussian distribution whose mean and covariance operator coincide with those of F. The Prokhorov distance between distributions F, G may be defined by means of the formula

$$\pi(F, G) = \inf\left\{ \lambda : \pi(F, G, \lambda) \le \lambda \right\},$$

where

$$\pi(F, G, \lambda) = \sup_X \max\left\{ F\{X\} - G\{X^\lambda\}, G\{X\} - F\{X^\lambda\} \right\}, \quad \lambda > 0,$$

and $X^\lambda = \{y \in \mathbf{R}^d : \inf_{x \in X} \|x - y\| < \lambda\}$ is the λ-neighborhood of the Borel set X. Moreover, in Zaitsev [29] it was established that

$$\pi(F, \Phi(F), \lambda) \le c\, d^2 \exp\left(-\frac{\lambda}{c\, d^2 \tau} \right). \tag{3.2}$$

It is very essential (and important) that the inequality (3.2) is proved for all $\tau > 0$ and for arbitrary $\mathrm{cov}F$, in contrast to Theorems 2 and 3, where $\tau \geq 1$ and covariance operators satisfy condition (2.2). The question about the necessity of condition (2.2) in Theorems 2 and 3 remains open. In Zaitsev [30] inequalities (3.1) and (3.2) were proved for convolutions of distributions from $\mathcal{B}_d(\tau)$

By the Strassen–Dudley theorem (see Dudley [9]) coupled with inequality (3.2), one can construct on a probability space the random vectors ξ and η with $\mathcal{L}(\xi) = F$ and $\mathcal{L}(\eta) = \Phi(F)$ so that

$$\mathbf{P}\left\{\|\xi - \eta\| > \lambda\right\} \leq c\,d^2 \exp\left(-\frac{\lambda}{c\,d^2\tau}\right). \tag{3.3}$$

For convolutions of bounded measures, this fact was used by Rio [21], Einmahl and Mason [13], Bovier and Mason [5], Gentz and Löwe [15], Einmahl and Kuelbs [14].

The scheme of the proof of Theorems 2 and 3 is very close to that of the main results of Sakhanenko [24] and Einmahl [12]. We suppose that the Gaussian vectors $Y_1, \ldots, Y_n$, $n = 2^N$, are already constructed and construct the independent vectors which are bounded with probability one, have sufficiently smooth distributions and the same moments of the first, second and third orders as the needed independent random vectors $X_1, \ldots, X_n$. For the construction we use the dyadic scheme proposed by KMT [17]. Firstly we construct the sum of 2^N summands using the Rosenblatt [22] quantile transform for conditional distributions (see Einmahl [12]). Then we construct blocks of $2^{N-1}, 2^{N-2}, \ldots, 1$ summands. The rate of approximation is estimated using the fact that, for smooth summands distributions, the corresponding conditional distribution are smooth and close to Gaussian ones. Then we construct the vectors $X_1, \ldots, X_n$ in several steps. After each step the number of X_k which are not constructed becomes smaller in 2^p times, where p is a suitably chosen positive integer. In each step we begin with already constructed vectors which are bounded with probability one and have sufficiently smooth distributions and the needed moments up to the third order. Then we construct the vectors such that, in each block of 2^p summands, only the first vector has the initial bounded smooth distribution. The rest $2^p - 1$ vectors have the needed distributions $\mathcal{L}(\xi_k)$. These $2^p - 1$ vectors from each block will be chosen as X_k and will be not involved in the next steps of the procedure. The coincidence of third moments will allow us to use more precise estimates of the closeness of quantiles of conditional distributions contained in Zaitsev [32]. In the estimation of closeness of random vectors in the steps of the procedure described above, we use essentially properties of classes $\mathcal{A}_d(\tau)$.

4. Infinitely divisible approximation

Let us finally mention a result about strong approximation of sums of independent random vectors by infinitely divisible distributions. Theorem 4 below follows from the main result of Zaitsev [32] coupled with the Strassen–Dudley theorem. Inequality (4.1) can be considered as a generalization of inequality (3.3) to convolution of distribution with unbounded supports.

Theorem 4. *Let d-dimensional probability distributions F_i, $i = 1, \ldots, n$, be represented as mixtures of d-dimensional probability distributions U_i and V_i:*

$$F_i = (1 - p_i)U_i + p_i V_i,$$

where

$$0 \leq p_i \leq 1, \qquad \int x\, U_i\{dx\} = 0, \qquad U_i\left\{\{x \in \mathbf{R}^d : \|x\| \leq \tau\}\right\} = 1,$$

and V_i are arbitrary distributions. Then for any fixed $\lambda > 0$ one can construct on the same probability space the random vectors ξ and η so that

$$\mathbf{P}\left\{\|\xi - \eta\| > \lambda\right\} \leq c(d)\left(\max_{1 \leq i \leq n} p_i + \exp\left(-\frac{\lambda}{c(d)\tau}\right)\right) + \sum_{i=1}^{n} p_i^2 \qquad (4.1)$$

and

$$\mathcal{L}(\xi) = \prod_{i=1}^{n} F_i, \qquad \mathcal{L}(\xi) = \prod_{i=1}^{n} e(F_i),$$

where $c(d)$ depends on only and $e(F_i)$ denotes the compound Poisson infinitely divisible distribution with characteristic function $\exp(\widehat{F}_i(t) - 1)$, where $\widehat{F}_i(t) = \int e^{itx} F_i\{dx\}$. If the distributions V_i are identical, the term $\sum_{i=1}^{n} p_i^2$ in (4.1) can be omitted.

References

[1] P. Bártfai, Die Bestimmung der zu einem wiederkehrenden Prozess gehörenden Verteilungfunktion aus den mit Fehlern behafteten Daten einer einzigen Realisation, *Studia Sci. Math. Hungar.*, 1 (1966), 161–168.

[2] E. Berger, *Fast sichere Approximation von Partialsummen unabhängiger und stationärer ergodischer Folgen von Zufallsvectoren* Dissertation, Universität Göttingen, 1982.

[3] I. Berkes, & W. Philipp, Approximation theorems for independent and weakly dependent random vectors. *Ann. Probab.*, 7 (1979), 29–54.

[4] A. A. Borovkov, On the rate of convergence in the invariance principle, *Theor. Probab. Appl.*, 18 (1973), 207–225.

[5] A. Bovier & D. Mason, *Extreme value behaviour in the Hopfield model,* Preprint 1998.

[6] M. Csörgő & P. Révész, A new method to prove Strassen type laws of invariance principle. I; II, *Z. Wahrscheinlichkeitstheor. verw. Geb.* 31 (1975), 255–259; 261–269.

[7] M. Csörgő & P. Révész, *Strong approximations in probability and statistics,* New York, Academic Press, 1981.

[8] S. Csörgő & P. Hall, The Komlós–Major–Tusnády approximations and their applications, *Austral. J. Statist.* 26 (1984), 189–218.

[9] R. M. Dudley, *Real analysis and probability,* Pacific Grove, California: Wadsworth & Brooks/Cole, 1989.

[10] U. Einmahl, A useful estimate in the multidimensional invariance principle, *Probab. Theor. Rel. Fields,* 76 (1987), 81–101.

[11] U. Einmahl, Strong invariance principles for partial sums of independent random vectors, *Ann. Probab.* 15 (1987), 1419–1440.

[12] U. Einmahl, Extensions of results of Komlós, Major and Tusnády to the multivariate case, *J. Multivar. Anal.,* 28 (1989), 20–68.

[13] U. Einmahl & D. M. Mason, Gaussian approximation of local empirical processes indexed by functions, *Probab. Theor. Rel. Fields,* 107 (1997), 283–311.

[14] U. Einmahl & J. Kuelbs, *Cluster sets for a generalized law of iterated logarthm in Banach spaces,* Preprint, 1989, 1–25.

[15] B. Gentz & M. Löwe, *Fluctuations in the Hopfield model at the critical temperature,* Preprint 98-003 EURANDOM, Eindhoven Institute of Technology 1998, 1–20.

[16] F. Götze & A. Yu. Zaitsev, Multidimensional Hungarian construction for vectors with almost Gaussian smooth distributions, *In: Asymptotic Methods in Probability and Statsistics (N. Balakrishnan, I. Ibragimov, V. Nevzorov eds.),* Birkhäuser, Boston, 2001, 101-132.

[17] J. Komlós, P. Major & G. Tusnády, An approximation of partial sums of independent RV'-s and the sample DF. I; II, *Z. Wahrscheinlichkeitstheor. verw. Geb.* 32 (1975) 111–131; 34 (1976), 34–58.

[18] W. Philipp, Almost sure invariance principles for sums of B-valued random variables, *Lect. Notes in Math.* 709 (1979), 171–193.

[19] Yu. V. Prokhorov, Convergence of random processes and limit theorem of probability theory, *Theor. Probab. Appl.,* 1 (1956), 157–214.

[20] Qi-Man Shao, Strong approximation theorems for independent random variables and their applications, *J. Multivar. Anal.,* 52 (1995), 107–130.

[21] E. Rio, Vitesses de convergence dans le principe d'invariance faible pour la fonction de répartition empirique multivariée. *C. R. Acad. Sci. Paris Sér. I Math.,* 322 (1996), 2, 169–172.

[22] M. Rosenblatt, Remarks on a multivariate transformation, *Ann. Math. Statist.,* 23 (1952), 470–472.

[23] R. Rudzkis, Probabilities of large deviations of random vectors, *Lithuanian Math. J.,* 23 (1983), 113–120.

[24] A. I. Sakhanenko, Rate of convergence in the invariance principles for variables with exponential moments that are not identically distributed, *In: Trudy Inst. Mat. SO AN SSSR 3,* Nauka, Novosibirsk, 1984, 4–49 (in Russian).

[25] L. Saulis, Large deviations for random vectors for certain classes of sets. I, *Lithuanian Math. J.,* 23 (1983), 308–317.

[26] A. V. Skorokhod, *Studies in the theory of random processes,* Univ. Kiev, Kiev, 1961 (in Russian); Engl. transl.: Addison–Wesley Reading, Mass., 1965.

[27] V. A. Statulevičius, On large deviations *Z. Wahrscheinlichkeitstheor. verw. Geb.,* 62 (1966), 133–144.

[28] V. Strassen, An invariance principle for the law of iterated logarithm *Z. Wahrscheinlichkeitstheor. verw. Geb.,* 3 (1964), 211–226.

[29] A. Yu. Zaitsev, Estimates of the Lévy–Prokhorov distance in the multivariate

central limit theorem for random variables with finite exponential moments, *Theor. Probab. Appl.*, 31 (1986), 203–220.

[30] A. Yu. Zaitsev, On the Gaussian approximation of convolutions under multidimensional analogues of S. N. Bernstein inequality conditions, *Probab. Theor. Rel. Fields*, 74 (1987), 535–566.

[31] A. Yu. Zaitsev, On the connection between two classes of probability distributions, *In: Rings and modulus. Limit theorems of probability theory. Vol. 2*, Leningrad University Press, Leningrad, 1988, 153–158.

[32] A. Yu. Zaitsev, Multidimensional version of the second uniform limit theorem of Kolmogorov, *Theor. Probab. Appl.*, 34 (1989), 108–128.

[33] A. Yu. Zaitsev, Estimates for quantiles of smooth conditional distributions and multidimensional invariance principle, *Siberian Math. J.*, 37 (1996), 807–831 (in Russian).

[34] A. Yu. Zaitsev, Multidimensional version of the results of Komlós, Major and Tusnády for vectors with finite exponential moments, *ESAIM : Probability and Statistics*, 2 (1998), 41–108.

[35] A. Yu. Zaitsev, Multidimensional version of the results of Sakhanenko in the invariance principle for vectors with finite exponential moments. I; II; III, *Theor. Probab. Appl.*, 45 (2000), 718–738; 46 (2001), 535-561; 744-769.

[36] A. Yu. Zaitsev, On the strong Gaussian approximation in multidimensional case, *Ann. de l'I.S.U.P. Publications de l'Institut de Statistique de l'Université de Paris*, 45 (2001), 2–3, 3–7.

ICM 2002 · Vol. III · 117–127

Random Walks in Random Environments

Ofer Zeitouni*

Abstract

Random walks in random environments (RWRE's) have been a source of surprising phenomena and challenging problems since they began to be studied in the 70's. Hitting times and, more recently, certain regeneration structures, have played a major role in our understanding of RWRE's. We review these and provide some hints on current research directions and challenges.

2000 Mathematics Subject Classification: 60K37, 82C44.
Keywords and Phrases: Random walks, Random environment, Regeneration.

1. Introduction

Let S denote the 2d-dimensional simplex, set $\Omega = S^{Z^d}$, and let $\omega(z, \cdot) = \{\omega(z, z + e)\}_{e \in Z^d, |e|=1}$ denote the coordinate of $\omega \in \Omega$ corresponding to $z \in Z^d$. Ω is an "environment" for an inhomogeneous nearest neighbor random walk (RWRE) started at x with *quenched* transition probabilities $P_\omega(X_{n+1} = z + e | X_n = z) = \omega(x, x + e)$ $(e \in \mathbb{Z}^d, |e| = 1)$, whose law is denoted P_ω^x. In the RWRE model, the environment is random, of law P, which is always assumed stationary and ergodic. We also assume here that the environment is *elliptic*, that is there exists an $\epsilon > 0$ such that P-a.s., $\omega(x, x + e) \geq \epsilon$ for all $x, e \in \mathbb{Z}^d, |e| = 1$. Finally, we denote by $\mathbb{P}$ the *annealed* law of the RWRE started at 0, that is the law of $\{X_n\}$ under the measure $P \times P_\omega^0$.

The RWRE model has a natural physical motivation and interpretation in terms of transport in random media. Mathematically, and especially for $d > 1$, it leads to the analysis of irreverservible, inhomogeneous Markov chains, to which standard tools of homogenization theory do not apply well. Further, unusual phenomena, such as sub-diffusive behavior, polynomial decay of probabilities of large deviations, and trapping effects, arise, already in the one dimensional model.

When $d = 1$, we write $\omega_x = \omega(x, x + 1)$, $\rho_x = \omega_x/(1 - \omega_x)$, and $u = E_P \log \rho_0$. The following reveals some of the surprising phenomena associated with the RWRE:

---

*Departments of Electrical Engineering and of Mathematics, Technion, Haifa 32000, Israel. E-mail: zeitouni@ee.technion.ac.il

Theorem 1.1 (Transience, recurrence, limit speed, $d = 1$) *(a) With* $\text{sign}(0)$ *$= 1$, it holds that* $\mathbb{P}$*-a.s.,*

$$\limsup_{n \to \infty} \kappa X_n = \text{sign}(\kappa u)\infty\,, \quad \kappa = \pm 1\,.$$

Further, there is a v *such that*

$$\lim_{n \to \infty} \frac{X_n}{n} = v\,, \quad \mathbb{P} - \text{a.s.}\,, \tag{1.2}$$

$v > 0$ *if* $\sum_{i=1}^{\infty} E_P(\prod_{j=0}^{i} \rho_{-j}) < \infty$, $v < 0$ *if* $\sum_{i=1}^{\infty} E_P(\prod_{j=0}^{i} \rho_{-j}^{-1}) < \infty$, *and* $v = 0$
if both these conditions do not hold.
(b) If P *is a product measure then*

$$v = \begin{cases} \frac{1 - E_P(\rho_0)}{1 + E_P(\rho_0)}\,, & E_P(\rho_0) < 1, \\ -\frac{1 - E_P(\rho_0^{-1})}{1 + E_P(\rho_0^{-1})}\,, & E_P(\rho_0^{-1}) < 1, \\ 0\,, & \text{else.} \end{cases} \tag{1.3}$$

Theorem 1.1 is essentially due to [25], see [29] for a proof in the general ergodic setup. The surprising features of the RWRE model alluded to above can be appreciated if one notes, already for a product measure P, that the RWRE can be transient with zero speed v. Further, if P is a product measure and $v_0(\omega)$ denotes the speed of a (biased) simple random walk with probability of jump to the right equal, at any site, to ω_0, then Jensen's inequality reveals that $|v| \leq |E_P(v_0(\omega))|$, with examples of strict inequality readily available.

The reason for this behavior is that the RWRE spends a large time in small traps. This is very well understood in the case $d = 1$, to which the next section is devoted. We introduce there certain hitting times, show how they yield precise information on the RWRE, and describe the analysis of these hitting times. Understanding the behavior of the RWRE when $d > 1$ is a major challenging problem, on which much progress has been done in recent years, but for which many embarrassing open questions remain. We give a glimpse of what is involved in Section 3., where we introduce certain *regeneration* times, and show their usefulness in a variety of situations. Here is a particularly simple setup where law of large numbers (and CLT's, although we do not emphasize that here) are available:

Theorem 1.4 *Assume* P *is a product measure,* $d \geq 6$, *and* $\omega(x, x + e) = \eta > 0$
for $e = \pm e_i, i = 1, \dots, 5$. *Then there exists a deterministic constant* v *such that*
$X_n/n \to v$, $\mathbb{P}$*-a.s..*

2.   The one-dimensional case

Recursions

Let us begin with a sketch of the proof of Theorem 1.1. The transience and recurrence criterion is proved by noting that conditioned on the environment ω, the

Markov chain X_n is reversible. More explicitly, fix an interval $[-m_-, m_+]$ encircling the origin and for z in that interval, define

$$\mathcal{V}_{m_-, m_+, \omega}(z) := P_\omega^z(\{X_n\} \text{ hits } -m_- \text{ before hitting } m_+).$$

Then,

$$\mathcal{V}_{m_-, m_+, \omega}(z) = \frac{\sum\limits_{i=z+1}^{m_+} \prod\limits_{j=z+1}^{i-1} \rho_j}{\sum\limits_{i=z+1}^{m_+} \prod\limits_{j=z+1}^{i-1} \rho_j + \sum\limits_{i=-m_-+1}^{z} \left(\prod\limits_{j=i}^{z} \rho_j^{-1} \right)}, \tag{2.1}$$

from which the conclusion follows. The proof of the LLN is more instructive: define the hitting times $T_n = \min\{t > 0 : X_t = T_n\}$, and set $\tau_i = T_{i+1} - T_i$. Suppose that $\limsup_{n \to \infty} X_n/n = \infty$. One checks that τ_i is an ergodic sequence, hence $T_n/n \to \mathbb{E}(\tau_0)$ $\mathbb{P}$-a.s., which in turns implies that $X_n/n \to 1/\mathbb{E}(\tau_0)$, $\mathbb{P}$-a.s.. But,

$$\tau_0 = \mathbf{1}_{\{X_1=1\}} + \mathbf{1}_{\{X_1=-1\}}(1 + \tau_{-1}' + \tau_0'),$$

where τ_{-1}' (τ_0') denote the first hitting time of 0 (1) for the random walk X_n after it hits -1. Hence, taking P_ω^0 expectations, and noting that $\{E_{P_\omega^i}(\tau_i)\}_i$ are, P-a.s., either all finite or all infinite,

$$E_{P_\omega^0}(\tau_0) = \frac{1}{\omega_0} + \rho_0 E_{P_\omega^{-1}}(\tau_{-1}). \tag{2.2}$$

When P is a product measure, ρ_0 and $E_{P_\omega^{-1}}(\tau_{-1})$ are P-independent, and taking expectations results with $\mathbb{E}(\tau_0) = (1 + E_P(\rho_0))/(1 - E_P(\rho_0))$ if the right hand side is positive and ∞ otherwise, from which (1.3) follows. The ergodic case is obtained by iterating the relation (2.2).

The hitting times T_n are also the beginning of the study of limit laws for X_n. To appreciate this in the case of product measures P with $E_P(\log \rho_0) < 0$ (i.e., when the RWRE is transient to $+\infty$), one first observes that from the above recursions,

$$\mathbb{E}(\tau_0^r) < \infty \iff E_P(\rho_0^r) < 1.$$

Defining $s = \max\{r : E_P(\rho_0^r) < 1\}$, one then expects that $(X_n - vn)$, suitably rescaled, possesses a limit law, with s-dependent scaling. This is indeed the case: for $s > 2$, it is not hard to check that one obtains a central limit theorem with scaling $\sqrt{n}$ (this holds true in fact for ergodic environments under appropriate mixing assumptions and with a suitable definition of the parameter s, see [29]). For $s \in (0,1) \cup (1,2)$, one obtains in the i.i.d. environment case a Stable(s) limit law with scaling $n^{1/s}$ (the cases $s = 1$ or $s = 2$ can also be handled but involve logarithmic factors in the scaling and the deterministic shift). In particular, for $s < 2$ the walk is *sub-diffusive*. We omit the details, referring to [16] for the proof, except to say that the extension to ergodic environments of many of these results has recently been carried out, see [23].

Traps

The unusual behavior of one dimensional RWRE is due to the existence of traps in the environment. This is exhibited most dramatically when one tries to evaluate the probability of slowdown of the RWRE. Assume that P is a product measure, X_n is transient to $+\infty$ with positive speed v (this means that $s > 1$ by Theorem 1.1), and that $s < \infty$ (which means that $P(\omega_0 < 1/2) > 0$). One then has:

Theorem 2.3 ([8, 11]) *For any* $w \in [0, v)$, $\eta > 0$, *and* $\delta > 0$ *small enough,*

$$\lim_{n \to \infty} \frac{\log \mathbb{P}\left(\frac{X_n}{n} \in (w - \delta, w + \delta)\right)}{\log n} = 1 - s, \tag{2.4}$$

$$\liminf_{n \to \infty} \frac{1}{n^{1-1/s+\eta}} \log P^0\left(\frac{X_n}{n} \in (w - \delta, w + \delta)\right) = 0, \quad P - a.s., \tag{2.5}$$

and

$$\limsup_{n \to \infty} \frac{1}{n^{1-1/s-\eta}} \log P^0\left(\frac{X_n}{n} \in (w - \delta, w + \delta)\right) = -\infty, \quad P - a.s.. \tag{2.6}$$

(Extensions of Theorem 2.3 to the mixing environment setup are presented in [29]. There are also precise asymptotics available in the case $s = \infty$ and $P(\omega_0 = 1/2) > 0$, see [20, 21]).

One immediately notes the difference in scaling between the annealed and quenched slowdown estimates in Theorem 2.3. These are due to the fact that, under the quenched measure, traps are almost surely of a maximal given size, determined by P, whereas under the annealed measure $\mathbb{P}$ one can create, at some cost in probability, larger traps.

To demonstrate the role of traps in the RWRE model, let us exhibit, for $w = 0$, a lower bound that captures the correct behavior in the annealed setup, and that forms the basis for the proof of the more general statement. Indeed, $\{X_n \leq \delta\} \subset \{T_{n\delta} \geq n\}$. Fixing $R_k = R_k(\omega) := k^{-1} \sum_{i=1}^{k} \log \rho_i$, it holds that R_k satisfies a large deviation principle with rate function $J(y) = \sup_\lambda(\lambda y - \log E_P(\rho_0^\lambda))$, and it is not hard to check that $s = \min_{y \geq 0} y^{-1} J(y)$. Fixing a y such that $J(y)/y \leq s + \eta$, and $k = \log n/y$, one checks that the probability that there exists in $[0, \delta n]$ a point z with $R_k \circ \theta^z \omega \geq y$ is at least $n^{1-s-\eta}$. But, the probability that the RWRE does not cross such a segment by time n is, due to (2.1), bounded away from 0 uniformly in n. This yields the claimed lower bound in the annealed case. In the quenched case, one has to work with traps of size almost $k = \log n/sy$ for which $kR_k \geq y$, which occur with probability 1 eventually, and use (2.1) to compute the probability of an atypical slowdown inside such a trap. The fluctuations in the length of these typical traps is the reason why the slowdown probability is believed, for P-a.e. ω, to fluctuate with n, in the sense that

$$\liminf_{n \to \infty} \frac{1}{n^{1-1/s}} \log P_\omega^0\left(\frac{X_n}{n} \in (-\delta, \delta)\right) = -\infty, \quad P - a.s.,$$

while it is known that

$$\limsup_{n \to \infty} \frac{1}{n^{1-1/s}} \log P_\omega^0 \left(\frac{X_n}{n} \in (-\delta, \delta) \right) = 0 , \quad P - a.s..$$

This has been demonstrated rigorously in some particular cases, see [10].

The role of traps, and the difference they produce between the quenched and annealed regimes, is dramatic also in the scale of large deviations. Roughly, the exponential (in n) rate of decay of the probability of atypical events differ between the quenched and annealed regime:

Theorem 2.7 *The random variables X_n/n satisfy, for P-a.e. realization of the environment ω, a large deviations principle (LDP) under P_ω^0 with a deterministic rate function $I_P(\cdot)$. Under the annealed measure $\mathbb{P}$, they satisfy a LDP with rate function*

$$I(w) = \inf_{Q \in \mathcal{M}_1^e} \left(h(Q|P) + I_Q(w) \right), \tag{2.8}$$

where $h(Q|P)$ is the specific entropy of Q with respect to P and $\mathcal{M}_1^e$ denotes the space of stationary ergodic measures on Ω.

Theorem 2.7 means that to create an annealed large deviation, one may first "modify" the environment (at a certain exponential cost) and then apply the quenched LDP in the new environment. We refer to [13] (quenched) and [3, 7] for proofs and generalizations to non i.i.d. environments. We also note that Theorem 2.7 stands in sharp contrast to what happens for random walks on Galton-Watson trees, where the growth of the tree creates enough variability in the (quenched) environment to make the annealed and quenched LDP's identical, see [6].

Sinai's recurrent walk and aging

When $E_P(\log \rho_0) = 0$, traps stop being local, and the whole environment becomes a diffused trap. The walk spends most of its time "at the bottom of the trap", and as time evolves it is harder and harder for the RWRE to move. This is the phenomenum of *aging*, captured in the following theorem:

Theorem 2.9 *There exists a random variable B^n, depending on the environment only, such that*

$$\mathbb{P} \left(\left| \frac{X_n}{(\log n)^2} - B^n \right| > \eta \right) \xrightarrow[n \to \infty]{} 0 .$$

Further, for $h > 1$,

$$\lim_{\eta \to 0} \lim_{n \to \infty} \mathbb{P} \left(\frac{|X_{n^h} - X_n|}{(\log n)^2} < \eta \right) = \frac{1}{h^2} \left[\frac{5}{3} - \frac{2}{3} e^{-(h-1)} \right] . \tag{2.10}$$

The first part of Theorem 2.9 is due to Sinai [24], with Kesten [15] providing the evaluation of the limiting law of B^n. The second part is implicit in [12], we refer to [5] and [29] for the proof and references.

3.   Multi-dimensional RWRE

Homogenization

Two special features simplify the analysis of the RWRE in the one-dimensional case: first, for every realization of the environment, the RWRE is a reversible Markov chain. This gave transience and recurrence criteria. Then, the location of the walk at the hitting times T_n is deterministic, leading to stationarity and mixing properties of the sequence $\{\tau_i\}$ and to a relatively simple analysis of their tail properties. Both these features are lost for $d > 1$.

A (by now standard) approach to homogenization problems is to consider the *environment viewed from the particle*. More precisely, with θ^x denoting the $\mathbb{Z}^d$ shift by x, the process $\omega_n = \theta^{X_n}\omega$ is a Markov chain with state-space Ω. Whenever the invariant measure of this chain is absolutely continuous with respect to P, law of large numbers and CLT's can be deduced, see [17]. For reversible situations, e.g. in the "random conductance model" [19], the invariant measure of the chain $\{\omega_n\}$ is known explicitly. In the non-reversible RWRE model, this approach has had limited consequences: one needs to establish absolute continuity of the invariant measure without knowing it explicitly. This was done in [18] for balanced environments, i.e. whenever $\omega(x, x+e) = \omega(x, x-e)$ P-a.s. for all $e \in \mathbb{Z}^d, |e| = 1$, by developing a-priori estimates on the invariant measure., valid for *every* realization of the environment. Apart from that (and the very recent [22]), this approach has not been very useful in the study of RWRE's.

Regeneration

We focus here on another approach based on analogs of hitting times. Throughout, fix a direction $\ell \in \mathbb{Z}^d$, and consider the process $Z_n = X_n \cdot \ell$. Define the events $A_{\pm\ell} = \{Z_n \to_{n\to\infty} \pm\infty\}$. Then, with P a product measure, one shows that $\mathbb{P}(A_\ell \cup A_{-\ell}) \in \{0, 1\}$, [14]. We sketch a proof: Call a time t *fresh* if $Z_t > Z_n, \forall n < t$, and for any fresh time t, define the return time $D_t = \min\{n > t : Z_n < Z_t\}$, calling t a regeneration time if $D_t = \infty$. Then, $\mathbb{P}(A_\ell) > 0$ implies by the Markov property that $\mathbb{P}(A_\ell \cap \{D_0 = \infty\}) > 0$. Similarly, on A_ℓ, each fresh time has a bounded away from zero probability to be a regeneration time. One deduces that $\mathbb{P}(\exists \text{ a regeneration time}|A_\ell) = 1$. In particular, on $A_{\pm\ell}$, Z_n changes signs only finitely many times. If $\mathbb{P}(A_\ell \cup A_{-\ell}) < 1$ then with positive probability, Z_n visits a finite centered interval infinitely often, and hence it must change signs infinitely many times. But this implies that $\mathbb{P}(A_\ell \cup A_{-\ell}) = 0$.

The proof above can be extended to non-product P-s having good mixing properties using, due to the uniform ellipticity, a coupling with simple nearest neighbor random walk. This is done as follows: Set $W = \{0\} \cup \{\pm e_i\}_{i=1}^d$. Define the measure

$$\overline{\mathbb{P}} = P \otimes Q_\epsilon \otimes \overline{P}^0_{\omega,\varepsilon} \quad \text{on} \quad \left(\Omega \times W^{\mathbb{N}} \times (\mathbb{Z}^d)^{\mathbb{N}}\right)$$

in the following way: Q_ϵ is a product measure, such that with $\varepsilon = (\epsilon_1, \epsilon_2, \ldots)$ denoting an element of $W^{\mathbb{N}}$, $Q_\epsilon(\epsilon_1 = \pm e_i) = \epsilon/2, i = 1, \cdots, d$, $Q_\epsilon(\epsilon_1 = 0) = 1 - \epsilon d$. For each fixed ω, ε, $\overline{P}^0_{\omega,\varepsilon}$ is the law of the Markov chain $\{X_n\}$ with state space $\mathbb{Z}^d$, such that $X_0 = 0$ and, for each $e \in W$, $e \neq 0$,

$$\overline{P}^0_{\omega,\varepsilon}(X_{n+1} = z + e | X_n = z) = \mathbf{1}_{\{\epsilon_{n+1}=e\}} + \frac{\mathbf{1}_{\{\epsilon_{n+1}=0\}}}{1 - d\epsilon}[\omega(z, z + e) - \epsilon/2].$$

It is not hard to check that the law of $\{X_n\}$ under $\overline{\mathbb{P}}$ coincides with its law under $\mathbb{P}$, while its law under $Q_\epsilon \otimes \overline{P}^0_{\omega,\varepsilon}$ coincides with its law under P^0_ω. Now, one introduces modified regeneration times $D_t^{(L)}$ by requiring that after the fresh time t, the "ε" coin was used for L steps in the direction ℓ: more precisely, requiring that $\epsilon_{t+i} = u_i, i = 1, \ldots, L$ for some fixed sequence $u_i \in \mathbb{Z}^d, |u_i| = 1, u_i \cdot \ell > 0$ such that $\sum_{i=1}^L u_i \cdot \ell \geq L/2$. This, for large L, introduces enough decoupling to carry through the proof, see [29, Section 3.1]. We can now state the:

Embarrassing Problem 1 Prove that $\mathbb{P}(A_\ell) \in \{0, 1\}$.

For $d = 2$, and P i.i.d., this was shown in [31], where counter examples using non uniformly elliptic, ergodic P's are also provided. The case $d > 2$, even for P i.i.d., remains open.

Embarrassing Problem 2 Find transience and recurrence criteria for the RWRE under $\mathbb{P}$.

The most promising approach so far toward Problem 2 uses regeneration times. Write $0 \leq d_1 < d_2 < \ldots$ for the ordered sequence of regeneration times, assuming that $\mathbb{P}(A_\ell) = 1$. The name regeneration time is justified by the following property, which for simplicity we state in the case $\ell = e_1$:

Theorem 3.1 ([28]) *For P a product measure, the sequence*

$$\{\{\omega_z\}_{z \cdot \ell \in [Z_{d_i}, Z_{(d_{i+1}-1)})}, \{X_t\}_{t \in [d_i, d_{i+1})}\}_{i=2,3,\ldots}$$

is i.i.d..

From this statement, it is then not hard to deduce that once $\mathbb{E}(d_2 - d_1) < \infty$, a law of large numbers results, with a non-zero limiting velocity. Sufficient conditions for transience put forward in [14] turn out to fall in this class, see [28]. More recently, Sznitman has introduced a condition that ensures both a LLN and a CLT: *Sznitman's T' condition:* $\mathbb{P}(A_\ell) = 1$ *and , for some $c > 0$ and all $\gamma < 1$,*

$$\mathbb{E}(\exp(c \sup_{0 \leq n < d_1} |X_n|^\gamma)) < \infty.$$

A remarkable fact about Sznitman's T' condition is that he was able to derive, using renormalization techniques, a (rather complicated) criterion, depending on the restriction of P to finite boxes, to check it. Further, Sznitman's T' condition implies a good control on d_1, and in particular that d_1 possesses all moments, which is the key to the LLN and CLT statements:

$$\mathbb{E}\left(\exp(\log d_1)^\delta\right) < \infty, \forall \delta < 2d/(d+1).$$

For these, and related, facts see [27]. This leads one to the

Challenging Problem 3 Do there exist non-ballistic RWRE's for $d > 1$ satisfying that $\mathbb{P}(A_\ell) = 1$ for some ℓ?

For $d = 1$, the answer is affirmative, as we saw, as soon as $E_P \log \rho_0 < 0$ but $s < 1$. For $d > 1$, one suspects that the answer is negative, and in fact one may suspect that $\mathbb{P}(A_\ell) = 1$ implies Sznitman's condition T'. The reason for the striking difference is that for $d > 1$, it is much harder to force the walk to visit large traps.

It is worthwhile to note that the modified regeneration times $\{D_t^{(L)}\}$ can be used to deduce the LLN for a class of mixing environments. We refer to [4] for details. At present, the question of CLT's in such a general set up remains open.

Cut points

Regeneration times are less useful if the walk is not ballistic. Special cases of non-ballistic models have been analyzed in the above mentioned [18], and using a heavy renormalization analysis, in [2] for the case of symmetric, low disorder, i.i.d. P. In both cases, LLN's with zero speed and CLT's are provided. We now introduce, for another special class of models, a different class of times that are not regeneration times but provide enough decoupling to lead to useful consequences.

The setup is similar to that in Theorem 1.4, that is we assume that $d \geq 6$ and that the RWRE, in its first 5 coordinate, performs a deterministic random walk: For $i = 1, \ldots, 5$, $\omega(x, x \pm e_i) = q_{\pm i}$, for some deterministic $q_{\pm i}$, $P - a.s..$ Set $S = \sum_{i=1}^5 (q_i + q_{-i})$, let $\{R_n\}_{n \in \mathbb{Z}}$ denote a (biased) simple random walk in $\mathbb{Z}^5$ with transition probabilities $q_{\pm i}/S$, and fix a sequence of independent Bernoulli random variable with $P(I_0 = 1) = S$, letting $U_n = \sum_{i=0}^{n-1} I_i$. Denote by X_n^1 the first 5 components of X_n and by X_n^2 the remaining components. Then, for every realization ω, the RWRE X_n can be constructed as the Markov chain with $X_n^1 = R_{U_n}$ and transition probabilities

$$\overline{P}_\omega^0(X_{n+1}^2 = z | X_n) = \begin{cases} 1, & X_n^2 = z, I_n = 1, \\ \omega(X_n, (X_n^1, z))/(1 - S), & I_n = 0. \end{cases}$$

Introduce now, for the walk R_n, cut times c_i as those times where the past and future of the path R_n do not intersect. More precisely, with $\mathcal{P}_I = \{X_n\}_{n \in I}$,

$$c_1 = \min\{t \geq 0 : \mathcal{P}_{(-\infty,t)} \cap \mathcal{P}_{[t,\infty)} = \emptyset\}, c_{i+1} = \min\{t > c_i : \mathcal{P}_{(-\infty,t)} \cap \mathcal{P}_{[t,\infty)} = \emptyset\}.$$

The cut-points sequence depends on the ordinary random walk R_n only. In particular, because that walk evolves in $\mathbb{Z}^5$, it follows, as in [9], that there are infinitely many cut points, and moreover that they have a positive density. Further, the increments $X_{c_{i+1}}^2 - X_{c_i}^2$ depend on disjoint parts of the environment. Therefore, conditioned on $\{R_n, I_n\}$, they are independent if P is a product measure, and they possess good mixing properties if P has good mixing properties. From here, the statement of Theorem 1.4 is not too far. We refer the reader to [1], where this and CLT statements (with 5 replaced by a larger integer) are proved. An amusing consequence of [1] is that for $d > 5$, one may construct ballistic RWRE's with, in the notations of Section 2., $E_P(v_0(\omega)) = 0$!

Challenging Problem 4 Construct cut points for "true" non-ballistic RWRE's.

The challenge here is to construct cut points and prove that their density is positive, without imposing a-priori that certain components of the walk evolve independently of the environment.

Large deviations

We conclude the discussion of multi-dimensional RWRE's by mentioning large deviations for this model. Call a RWRE *nestling* if $\text{co} \operatorname{supp} Q$, where Q denotes the law of $\sum_{e \in \mathbb{Z}^d : |e| = 1} e \omega(0, e)$. In words, an RWRE is nestling if by combining local drifts one can arrange for zero drift. One has then:

Theorem 3.2 ([30]) *Assume P is a product nestling measure. Then, for P-almost every ω, X_n/n satisfies a LDP under P_ω^0 with deterministic rate function.*

The proof of Theorem 3.2 involves hitting times: let T_y denote the first hitting time of $y \in \mathbb{Z}^d$. One then checks, using the subaddititve ergodic theorem, that

$$\Lambda(y, \lambda) := \lim_{n \to \infty} n^{-1} \log E_\omega^0 (\exp(-\lambda T_{ny}) \mathbf{1}_{\{T_{ny} < \infty\}})$$

exists and is deterministic, for $\lambda \geq 0$. In the nestling regime, where slowdown has sub-exponential decay rate due to the existence of traps much as for $d = 1$, this and concentration of measure estimates are enough to yield the LDP. But:

Embarrassing Problem 5 Prove the quenched LDP for non-nestling RWRE's.

A priori, non nestling walks should have been easier to handle than nestling walks due to good control on the tail of regeneration times!

Challenging Problem 6 Derive an annealed LDP for the RWRE, and relate the rate function to the quenched one.

One does not expect a relation as simple as in Theorem 2.7, because the RWRE can avoid traps by contouring them, and to change the environment in a way that surely modifies the behavior of the walk by time n has probability which seems to decay at an exponential rate faster than n. This puts the muti-dimensional RWRE in an intermediate position between the one-dimensional RWRE and walks on Galton-Watson trees [6]. We also note that certain estimates on large deviations for RWRE's, without matching constants, appear in [26].

References

[1] E. Bolthausen, A. S. Sznitman and O. Zeitouni, Cut points and diffusive random walks in random environments, *preprint* (2002). http://www-ee.technion.ac.il/~zeitouni/ps/BSZ4.ps

[2] J. Bricmont and A. Kupiainen, Random walks in asymmetric random environments, *Comm. Math. Phys* 142 (1991), 345–420.

[3] F. Comets, N. Gantert and O. Zeitouni, Quenched, annealed and functional large deviations for one dimensional random walk in random environment, *Prob. Th. Rel. Fields* 118 (2000), 65–114.

[4] F. Comets and O. Zeitouni, A law of large numbers for random walks in random mixing environments, math.PR/0205296 (2002).

[5] A. Dembo, A. Guionnet and O. Zeitouni, Aging properties of Sinai's random walk in random environment, math.PR/0105215 (2001).

[6] A. Dembo, N. Gantert, Y. Peres and O. Zeitouni, Large deviations for random walks on Galton-Watson trees: averaging and uncertainty, *Prob. Th. Rel. Fields* 122 (2002), 241–288.

[7] A. Dembo, N. Gantert and O. Zeitouni, Large deviations for random walk in random environment with holding times, *preprint* (2002).

[8] A. Dembo, Y. Peres and O. Zeitouni, Tail estimates for one-dimensional random walk in random environment, *Comm. Math. Physics* 181 (1996), 667–684.

[9] P. Erdös and S. J. Taylor, Some intersection properties of random walks paths, *Acta Math. Acad. Sci. Hungar.* 11 (1960), 231–248.

[10] N. Gantert, Subexponential tail asymptotics for a random walk with randomly placed one-way nodes, to appear, *Ann. Inst. Henri Poincaré* (2002).

[11] N. Gantert and O. Zeitouni, Quenched sub-exponential tail estimates for one-dimensional random walk in random environment, *Comm. Math. Physics* 194 (1998), 177–190.

[12] A. O. Golosov, On limiting distributions for a random walk in a critical one dimensional random environment, *Comm. Moscow Math. Soc.* 199 (1985), 199–200.

[13] A. Greven and F. den Hollander, Large deviations for a random walk in random environment, *Annals Probab.* 22 (1994), 1381–1428.

[14] S. A. Kalikow, Generalized random walks in random environment, *Annals Probab.* 9 (1981), 753–768.

[15] H. Kesten, The limit distribution of Sinai's random walk in random environment, *Physica* 138A (1986), 299–309.

[16] H. Kesten, M. V. Kozlov and F. Spitzer, A limit law for random walk in a random environment, *Comp. Math.* 30 (1975), 145–168.

[17] S. M. Kozlov, The method of averaging and walks in inhomogeneous environments, *Russian Math. Surveys* **40** (1985), 73–145.

[18] G.F. Lawler, Weak convergence of a random walk in a random environment, *Comm. Math. Phys.* **87** (1982), 81–87.

[19] A. De Masi, P. A. Ferrari, S. Goldstein and W. D. Wick, An invariance principle for reversible Markov processes. Applications to random motions in random environments, *J. Stat. Phys.* 55 (1989), 787–855.

[20] A. Pisztora and T. Povel, Large deviation principle for random walk in a quenched random environment in the low speed regime, *Annals Probab.* 27 (1999), 1389–1413.

[21] A. Pisztora, T. Povel and O. Zeitouni, Precise large deviation estimates for a one-dimensional random walk in a random environment, *Prob. Th. Rel. Fields* 113 (1999), 191–219.

[22] F. Rassoul-Agha, A law of large numbers for random walks in mixing random environment, *Preprint* (2002).

[23] A. Roitershtein, Ph.D. thesis, *Dept. of Mathematics, Technion* (Forthcoming).

[24] Ya. G. Sinai, The limiting behavior of a one-dimensional random walk in random environment, *Theor. Prob. and Appl.* 27 (1982), 256–268.

[25] F. Solomon, Random walks in random environments, *Annals Probab.* 3 (1975), 1–31.

[26] A. S. Sznitman, Slowdown estimates and central limit theorem for random walks in random environment, *JEMS* **2** (2000), 93–143.

[27] A. S. Sznitman, An effective criterion for ballistic behavior of random walks in

random environment, *Prob. Th. Rel. Fields* 122 (2002), 509–544.

[28] A. S. Sznitman and M. Zerner, A law of large numbers for random walks in random environment, *Annals Probab.* 27 (1999), 1851–1869.

[29] O. Zeitouni, Lecture notes on RWRE, notes from the St.-Flour summer school in probability, 2001. Available at http://wwww-ee.technion.ac.il/~zeitouni/ps/notes1.ps

[30] M. P. W. Zerner, Lyapounov exponents and quenched large deviations for multidimensional random walk in random environment, *Annals Probab.* 26 (1998), 1446–1476.

[31] M. P. W. Zerner and F. Merkl, A zero-one law for planar random walks in random environment, *Annals Probab.* **29** (2001), 1716–1732.

Section 11. Partial Differential Equations

ICM 2002 · Vol. III · 131–140

Optimal Transport Maps in Monge-Kantorovich Problem

L. Ambrosio*

Abstract

In the first part of the paper we briefly decribe the classical problem, raised by Monge in 1781, of optimal transportation of mass. We discuss also Kantorovich's weak solution of the problem, which leads to general existence results, to a dual formulation, and to necessary and sufficient optimality conditions.

In the second part we describe some recent progress on the problem of the existence of optimal transport maps. We show that in several cases optimal transport maps can be obtained by a singular perturbation technique based on the theory of Γ-convergence, which yields as a byproduct existence and stability results for classical Monge solutions.

2000 Mathematics Subject Classification: 49K, 49J, 49Q20.
Keywords and Phrases: Optimal transport maps, Optimal plans, Wasserstein distance, c-monotonicity, Γ-convergence, Transport density.

1. The optimal transport problem and its weak formulation

In 1781, G.Monge raised in [26] the problem of transporting a given distribution of matter (a pile of sand for instance) into another (an excavation for instance) in such a way that the work done is minimal. Denoting by h_0, $h_1 : \mathbf{R}^2 \to [0, +\infty)$ the Borel functions describing the initial and final distribution of matter, there is obviously a compatibility condition, that the total mass is the same:

$$\int_{\mathbf{R}^2} h_0(x)\, dx = \int_{\mathbf{R}^2} h_1(y)\, dy. \tag{1.1}$$

Assuming with no loss of generality that the total mass is 1, we say that a Borel map $\psi : \mathbf{R}^2 \to \mathbf{R}^2$ is a *transport* if a local version of the balance of mass condition

---

*Scuola Normale Superiore, Piazza Cavalieri 7, 56126 Pisa, Italy. E-mail: luigi@ambrosio.sns.it

holds, namely

$$\int_{\psi^{-1}(E)} h_0(x)\, dx = \int_E h_1(y)\, dy \qquad \text{for any } E \subset \mathbf{R}^2 \text{ Borel.} \qquad (1.2)$$

Then, the Monge problem consists in minimizing the work of transportation in the class of transports, i.e.

$$\inf \left\{ \int_{\mathbf{R}^2} |\psi(x) - x| h_0(x)\, dx \; : \; \psi \text{ transport} \right\}. \qquad (1.3)$$

The Monge transport problem can be easily generalized in many directions, and all these generalizations have proved to be quite useful:

• General measurable spaces X, Y, with measurable maps $\psi : X \to Y$;

• General probability measures μ in X and ν in Y. In this case the local balance of mass condition (1.2) reads as follows:

$$\nu(E) = \mu(\psi^{-1}(E)) \qquad \text{for any } E \subset Y \text{ measurable.} \qquad (1.4)$$

This means that the push-forward operator $\psi_\#$ induced by ψ, mapping probability measures in X into probability measures in Y, maps μ into ν.

• General cost functions: a measurable map $c : X \times Y \to [0, +\infty]$. In this case the cost to be minimized is

$$W(\psi) := \int_X c\,(x, \psi(x))\; d\mu(x).$$

Even in Euclidean spaces, the problem of existence of optimal transport maps is far from being trivial, mainly due to the non-linearity with respect to ψ of the condition $\psi_\#\mu = \nu$. In particular the class of transports is not closed with respect to any reasonable weak topology. Furthermore, it is easy to build examples where the Monge problem is ill-posed simply because there is no transport map: this happens for instance when μ is a Dirac mass and ν is not a Dirac mass.

In order to overcome these difficulties, in 1942 L.V.Kantorovich proposed in [21] a notion of weak solution of the transport problem. He suggested to look for *plans* instead of transports, i.e. probability measures γ in $X \times Y$ whose marginals are μ and ν. Formally this means that $\pi_{X\#}\gamma = \mu$ and $\pi_{Y\#}\gamma = \nu$, where $\pi_X : X \times Y \to X$ and $\pi_Y : X \times Y \to Y$ are the canonical projections. Denoting by $\Pi(\mu, \nu)$ the class of plans, he wrote the following minimization problem

$$\min \left\{ \int_{X \times Y} c(x, y)\, d\gamma \; : \; \gamma \in \Pi(\mu, \nu) \right\}. \qquad (1.5)$$

Notice that $\Pi(\mu, \nu)$ is not empty, as the product $\mu \otimes \nu$ has μ and ν as marginals. Due to the convexity of the new constraint $\gamma \in \Pi(\mu, \nu)$ it turns out that weak topologies can be effectively used to provide existence of solutions to (1.5): this happens for instance whenever X and Y are Polish spaces and c is lower semicontinuous (see for instance [28]). Notice also that, by convexity of the energy, the infimum is attained on a *extremal* element of $\Pi(\mu, \nu)$.

The connection between the Kantorovich formulation of the transport problem and Monge's original one can be seen noticing that any transport map ψ induces a planning γ, defined by $(Id \times \psi)_{\#}\mu$. This planning is concentrated on the graph of ψ in $X \times Y$ and it is easy to show that the converse holds, i.e. whenever γ is concentrated on a graph, then γ is induced by a transport map. Since any transport induces a planning with the same cost, it turns out that

$$\inf (1.3) \geq \min (1.5).$$

Moreover, by approximating any plan by plans induced by transports, it can be shown that equality holds under fairly general assumptions (see for instance [3]). Therefore we can really consider the Kantorovich formulation of the transport problem as a weak formulation of the original problem.

If all extremal points of $\Pi(\mu,\nu)$ were induced by transports one would get existence of transport maps directly from the Kantorovich formulation. It is not difficult to show that plannings γ induced by transports are extremal in $\Pi(\mu,\nu)$. The converse holds in some very particular cases, but unfortunately it is not true in general. It turns out that the existence of optimal transport maps depends not only on the geometry of $\Pi(\mu,\nu)$, but also (in a quite sensible way) on the choice of the cost function c.

2. Existence of optimal transport maps

In this section we focus on the problem of the existence of optimal transport maps in the sense of Monge. Before discussing in detail in the next sections the two model cases in which the cost function is the square of a distance or a distance (we refer to [19] for the case of *concave* functions of the distance, not discussed here), it is better to give an informal description of the tools by now available for proving the existence of optimal transport maps.

Strategy A (Dual formulation). This strategy is based on the duality formula

$$\min (MK) = \sup \left\{ \int_X h \, d\mu + \int_Y k \, d\nu \right\}, \tag{2.6}$$

where the supremum runs among all pairs $(h, k) \in L^1(\mu) \times L^1(\nu)$ such that $h(x) + k(y) \leq c(x, y)$. The duality approach to the (MK) problem was developed by Kantorovich, and then extended to more general cost functions (see [22]). The transport map is obtained from an optimal pair (h, k) in the dual formulation by making a first variation. This strategy for proving the existence of an optimal transport map goes back to the papers [18] and [11].

Strategy B (Cyclical monotonicity). In some situations the necessary (and sufficient) minimality conditions for the primal problem, based upon the so-called c-cyclical monotonicity ([32], [28], [29]) yield that *any* optimal Kantorovich solution γ is concentrated on a graph Γ (i.e. for μ-a.e. x there exists a *unique* y such that $(x, y) \in \Gamma$) and therefore is induced by a transport ψ.

This happens for instance when $c(x, y) = H(x - y)$, with H *strictly* convex in $\mathbf{R}^n$. This approach is pursued in the papers [19], [30].

Strategy C (Singular perturbation with strictly convex costs). One can try to get an optimal transport map by making the cost strictly convex through a perturbation and then passing to the limit (see [12] and Theorem 4.1, Theorem 4.2 below). The main difficulty is to show (strong) convergence at the level of the transport maps and not only at the level of transport plans.

Strategy D (Reduction to a lower dimensional problem). This strategy has been initiated by V.N.Sudakov in [33]. It consists in writing (typically through a disintegration) μ and ν as the superposition of measures concentrated on lower dimensional sets and in solving the lower dimensional transport problems, trying in the end to "glue" all the partial transport maps into a single transport map. This strategy is discussed in detail in [3] and used, together with a "variational" decomposition, in [5]. The simplest case is when the lower dimensional problems are 1-dimensional, since the solution of the 1-dimensional transport problem is simply given by an increasing rearrangement, at least for convex functions of the distance (see for instance [2], [28], [35]).

Strategies A and B are basically equivalent and yield existence and uniqueness at the same time: the first one could be preferable for someone, as a very small measure-theoretic apparatus is involved. On the other hand, it strongly depends on the existence of maximizing pairs in the dual formulation, and this existence issue can be more subtle than the existence issue for the primal problem (see [28] and the discussion in [3]). For this reason it seems that the second strategy can work for more general classes of cost functions.

Strategies C and D have been devised to deal with situations where the cost function is convex but not strictly convex. Also these two strategies are closely related, as the strictly convex perturbation often leads to an effective dimension reduction of the problem (see for instance [5]).

3.   cost=distance^2

In this section we consider the case when $X = Y$ and the cost function c is proportional to the square of a distance d. For convenience we normalize c so that $c = d^2/2$. The first result in the Euclidean space $\mathbf{R}^n$ has been discovered independently by many authors Y.Brenier [8], [9], S.T.Rachev and L.R.üschendorf [27], [29], and C.Smith and M.Knott [31].

Theorem 3.1 *Assume that μ is absolutely continuous with respect to $\mathcal{L}^n$ and that μ and ν have finite second order moments. Then there exists a unique optimal transport map ψ. Moreover ψ is the gradient of a convex function.*

In this case the proof comes from the fact that both strategies A and B yield that the displacement $x - \psi(x)$ is the gradient of a c-concave function, i.e. a function representable as

$$h(x) = \inf_{(y,t) \in I} c(x, y) + t \qquad \forall x \in \mathbf{R}^n$$

for a suitable non-empty set $I \subset Y \times \mathbf{R}$. The concept of c-concavity [29] has been extensively used to develop a very general duality theory for the (MK) problem, based on (2.6). In this special Euclidean situation it is immediate to realize that c-concavity of h is equivalent to concavity (in the classical sense) of $h - \frac{1}{2}|x|^2$, hence

$$\psi(x) = x - \nabla h(x) = \nabla \left[\frac{1}{2}|x|^2 - h(x) \right]$$

is the gradient of a convex function. Finally, notice that the assumption on μ can be sharpened (see [19]), assuming for instance that $\mu(B) = 0$ whenever B has finite $\mathcal{H}^{n-1}$-measure. This is due to the fact that the non-differentiability set of a concave function is σ-finite with respect to $\mathcal{H}^{n-1}$ (see for instance [1]). Also the assumption about second order moments can be relaxed, assuming only that the infimum of the (MK) problem with data μ, ν is finite.

The following result, due to R.Mc Cann [25], is much more recent.

Theorem 3.2 *Assume that M is a C^3, complete Riemannian manifold with no boundary and d is the Riemannian distance. If μ, ν have finite second order moments and μ is absolutely continuous with respect to vol_M there exists a unique optimal transport map ψ.*

Moreover there exists a c-concave potential $h : M \to \mathbf{R}$ such that

$$\psi(x) = \exp_x \left(-\nabla h(x) \right) \quad vol_M\text{-a.e.}.$$

This Riemannian extension of Theorem 3.1 is non trivial, due to the fact that d^2 is not smooth in the large. The proof uses some semiconcavity estimates for d^2 and the fact that d^2 is C^2 for x close to y (this is where the C^3 assumption on M is needed). It is interesting to notice that the results of [24] (where the eikonal equation is read in local coordinates), based on the theory of viscosity solutions — see in particular Theorem 5.3 of [23] — allow to push Mc Cann's technique up to C^2 manifolds.

Can we go beyond Riemannian manifolds in the existence theory? A model case is given by stratified Carnot groups endowed with the Carnot-Carathéodory metric d_{CC}, as these spaces arise in a very natural way as limits of Riemannian manifolds with respect to the Gromov-Hausdorff convergence (see [20]). At this moment a general strategy is still missing, but some preliminary investigations in the Heisenberg group H_n show that positive results analogous to the Riemannian ones can be expected. The following result is proved in [6]:

Theorem 3.3 *If $n = 1, 2$ and μ is a probability measure in H_n absolutely continuous with respect to $\mathcal{L}^{2n+1}$, then:*
(a) there exists a unique optimal transport map ψ, deriving from a c-concave potential h;
(b) If $d_p \uparrow d_{CC}$ are Riemannian left invariant metrics then Mc Cann's optimal transport maps ψ_p relative to $c_p = d_p^2/2$ converge in measure to ψ as $p \to \infty$.

The restriction to H_n, $n \leq 2$, arises from the fact that so far we have been able to carry on some explicit computations only for $n \leq 2$. We expect that this restriction could be removed. The proof of (b) is not direct, as Mc Cann's exponential representation $\psi_p = \exp_x^p(-\nabla^p h_p)$ "degenerates" as $p \to \infty$, because the injectivity radius of the approximating manifolds tends to 0. This is due to the fact that in CC metric spaces geodesics exist but are not unique, not even in the small.

Finally, if we replace c by the square of the Korányi norm (related to the fundamental solution of the Kohn sub-Laplacian), namely

$$\tilde{c}(x,y) := \frac{1}{2}\|y^{-1}x\|^2 \quad \text{with} \quad \|(z,t)\| := \sqrt[4]{|z|^4 + t^2}$$

(here we identify H_n with $\mathbf{C}^n \times \mathbf{R}$) then we are still able to prove existence in any Heisenberg group H_n. The proof uses some fine properties of BV functions on sub-Riemannian groups [4]. However, we can't hope for a Riemannian approximation result, as the Korányi norm induces a metric d_K which is not geodesic. It turns out that the geodesic metric associated to d_K is a constant multiple of d_{CC}.

4. cost=distance

In this section we consider the case when $X = Y$ and the cost function c is a distance. In this case both strategies A and B give only a partial information about the location of y, for given x. In particular it is not true that any optimal Kantorovich plan γ is induced by a transport map. Indeed, if the first order moments of μ and ν are finite, the dual formulation provides us with a maximizing pair $(h,k) = (u,-u)$, with $u : X \to \mathbf{R}$ 1-Lipschitz. If $X = \mathbf{R}^n$ and the distance is induced by a norm $\|\cdot\|$, this provides the implication

$$(x,y) \in \operatorname{spt}\gamma \quad \implies \quad y \in \{x - s\xi : \xi \in (du(x))^*, \ s \geq 0\} \qquad (4.7)$$

at any differentiability point of u. Here we consider the natural duality map between covectors and vectors given by

$$L^* := \{\xi \in \mathbf{R}^n : L(\xi) = \|L\|_* \text{ and } \|\xi\| = 1\}.$$

The most favourable case is when the norm is strictly convex (e.g. the Euclidean norm): in this situation the $*$ operator is single-valued and we recover from (4.7) an information on the direction of transportation, i.e. $(du(x))^*$, but not on the length of transportation. If the norm is not strictly convex (e.g. the l_1 or l_∞ norm) then even the information on the direction of transportation, encoded in $(du(x))^*$, is partial.

The first attempt to bypass these difficulties came with the work of V.N.Sudakov [33], who claimed to have a solution for any distance cost function induced by a norm. Sudakov's approach is based on a clever decomposition of the space $\mathbf{R}^n$ in affine regions with variable dimension where the Kantorovich dual potential u associated to the transport problem is an affine function. His strategy is to solve the transport problem in any of these regions, eventually getting an optimal transport map just by gluing all these transport maps. An essential ingredient in his proof is Proposition 78, where he states that, if $\mu << \mathcal{L}^n$, then the conditional measures induced by the decomposition are absolutely continuous with respect to the Lebesgue measure (of the correct dimension). However, it turns out that this property is not true in general even for the simplest decomposition, i.e. the decomposition in segments: G.Alberti, B.Kirchheim and D.Preiss found an example of a compact faily of pairwise disjoint open segments in $\mathbf{R}^3$ such that the family M of their midpoints has strictly positive Lebesgue measure (the construction is a variant of previous examples due to A.S.Besicovitch and D.G.Larman, see also [2] and [5]). In this case, choosing $\mu = \mathcal{L}^3 \llcorner M$, the conditional measures induced by the decomposition are Dirac masses. Therefore it is clear that this kind of counterexamples should be ruled out by some kind of additional "regularity" property of the decomposition. In this way the Sudakov strategy would be fully rigorous. As noticed in [5], this regularity comes for free only in the case $n = 2$, using the fact that transport rays do not cross in their interior.

Several years later, L.C.Evans and W.Gangbo made a remarkable progress in [15], showing by differential methods the existence of a transport map, under the assumption that $\operatorname{spt}\mu \cap \operatorname{spt}\nu = \emptyset$, that the two measures are absolutely continuous with respect to $\mathcal{L}^n$ and that their densities are Lipschitz functions with compact support. The missing piece of information about the length of transportation is recovered by a p-laplacian approximation

$$-\operatorname{div}\left(|\nabla u|^{p-2}\nabla u\right) = \mu - \nu, \qquad u \in H_0^1(B_R), \qquad R \gg 1$$

obtaining in the limit as $p \to +\infty$ a nonnegative function $a \in L^\infty(\mathbf{R}^n)$ and a 1-Lipschitz function u solving

$$-\operatorname{div}(a\nabla u) = \mu - \nu, \qquad |\nabla u| = 1 \ \mathcal{L}^n\text{-a.e. on } \{a > 0\}.$$

The diffusion coefficient a in the PDE above plays a special role in the theory. Indeed, one can show (see [2]) that the measure $\sigma := a\mathcal{L}^n$, the so-called *transport density*, can be represented in several different way, and in particular as

$$\sigma(B) = \int \mathcal{H}^1\left(B \cap [x, y]\right) d\gamma(x, y) \qquad \forall B \subset \mathbf{R}^n \text{ Borel} \tag{4.8}$$

for some optimal planning γ. Notice that the total mass of σ is $\int |x - y|\, d\gamma$, the total work done and the meaning of $\sigma(B)$ is the work done *within* B during the transport process. This representation of the transport density has been introduced by G.Bouchitté and G.Buttazzo in [7], who showed that the a constant multiple of the transport density is a solution of their so-called mass optimization problem. Later, in [2], it was shown that there is actually a 1-1 correspondence between solutions of the mass optimization problem and transport densities, defined as in (4.8).

One can also show ([2], [13], [16], [14]) that σ is unique (unlike γ) if either μ or ν are absolutely continuous. Moreover, the nonlinear operator mapping $(\mu, \nu) \in L^1 \times L^1$ into $a \in L^1$ maps $L^p \times L^p$ into L^p for $1 \le p \le \infty$.

Coming back to the problem of the existence of optimal transport maps with Euclidean distance $|x - y|$ (or, more generally, with a distance induced by a C^2 and uniformly convex norm), the first existence results for general absolutely continuous measures μ, ν with compact support have been independently obtained by

L.Caffarelli, M.Feldman and R.Mc Cann in [12] and by N.Trudinger and L.Wang in [34]. Afterwards, the author estabilished in [2] the existence of an optimal transport map assuming only that the initial measure μ is absolutely continuous, and the results of [12] and [34] have been extended to a Riemannian setting in [17]. All these proofs involve basically a Sudakov decomposition in transport rays, but the technical implementation of the idea is different from paper to paper: for instance in [12] a local change of variable is made, so that transport rays become parallel and Fubini theorem, in place of abstract disintegration theorems for measures, can be used. The proof in [3], instead, uses the co-area formula to show that absolute continuity with respect to Lebesgue measure is stable under disintegration.

The following result [3] is a slight improvement of [12], where existence of an optimal transport map was estabilished but not the stability property. The result holds under regularity and uniform convexity assumptions for the norm $\|\cdot\|$.

Theorem 4.1 *Let μ,ν be with compact support, with $\mu << \mathcal{L}^n$, and let ψ_ϵ be the unique optimal transport maps relative to the costs $c_\epsilon(x,y) := \|x - y\|^{1+\epsilon}$. Then ψ_ϵ converge as $\epsilon \downarrow 0$ to an optimal transport map ψ for $c(x,y) = \|x - y\|$.*

The proof is based only the fact that any plan γ_0, limit of some sequence of plans $(Id \times \psi_{\epsilon_i})$, is not only optimal for the (MK) problem, but also for the *secondary* one

$$\min_{\gamma \in \Pi_1(\mu,\nu)} \int_{\mathbf{R}^n \times \mathbf{R}^n} \|x - y\| \ln(\|x - y\|) \, d\gamma, \qquad (4.9)$$

where $\Pi_1(\mu,\nu)$ denotes the class of all optimal plannings for the Kantorovich problem (the entropy function in (4.9) comes from the Taylor expansion of c_ϵ around $\epsilon = 0$). It turns out that this additional minimality property selects a unique plan induced by a transport ψ and, a posteriori, ψ is the same map built in [12]. A class of counterexamples built in [3] shows that the absolute continuity assumption on μ cannot be weakened, unlike the strictly convex case.

This "variational" procedure seems to select extremal elements of $\Pi(\mu,\nu)$ in a very effective way. This phenomenon is apparent in view of the following result [5], which holds for all "crystalline" norms $\|\cdot\|$ (i.e. norms whose unit sphere is contained in finitely many hyperplanes).

Theorem 4.2 *Let μ,ν be as in Theorem 4.1 and let ψ_ϵ be the unique optimal transport maps relative to the costs*

$$c_\epsilon(x,y) := \|x - y\| + \epsilon|x - y| + \epsilon^2|x - y| \ln |x - y|.$$

Then ψ_ϵ converge as $\epsilon \downarrow 0$ to an optimal transport map ψ for $c(x,y) = \|x - y\|$.

In this case a secondary and a ternary variational problem are involved, and we show that the latter has a unique solution which is also induced by a transport.

Some borderline cases between "crystalline" norms and "Euclidean" norms apparently can't be attacked by any of the existing techniques. In particular the existence of optimal transport maps for the cost induced by a general norm in $\mathbf{R}^n$, $n \geq 3$, is still open.

References

[1] G.ALBERTI & L.AMBROSIO: *A geometric approach to monotone functions in* $\mathbf{R}^n$. Math. Z., **230** (1999), 259–316.

[2] L.AMBROSIO: *Lecture Notes on the Optimal Transport Problems.* Notes of a CIME Course given in Madeira (2000), to be published in the CIME Springer Lecture Notes (see also http://cvgmt.sns.it).

[3] L.AMBROSIO & A.PRATELLI: *Existence and stability results in the* L^1 *theory of optimal transportation.* Notes of a CIME Course given in Martina Franca (2002), to be published in the CIME Springer Lecture Notes (see also http://cvgmt.sns.it).

[4] L.AMBROSIO & V.MAGNANI: *Weak differentiability of BV functions on sub-Riemannian groups.* Submitted to Math. Z.

[5] L.AMBROSIO, B.KIRCHHEIM & A.PRATELLI: *Existence of optimal transports with crystalline norms.* In preparation.

[6] L.AMBROSIO & S.RIGOT: *Optimal mass transportation in the Heisenberg group.* In preparation.

[7] G.BOUCHITTÉ & G.BUTTAZZO: *Characterization of optimal shapes and masses through Monge-Kantorovich equation.* J. Eur. Math. Soc., **3** (2001), 139–168.

[8] Y.BRENIER: *Décomposition polaire et réarrangement monotone des champs de vecteurs.* C.R. Acad. Sci. Paris, Sér I Math., **305** (1987), 805–808.

[9] Y.BRENIER: *Polar factorization and monotone rearrangement of vector-valued functions.* Comm. Pure Appl. Math., **44** (1991), 375–417.

[10] L.CAFFARELLI: *Allocation maps with general cost functions.* Lecture Notes in Pure and Appl. Math., **177** (1996), 29–35.

[11] L.CAFFARELLI: *Boundary regularity of maps with a convex potential.* Commun. Pure Appl. Math., **45** (1992), 1141–1151.

[12] L.CAFFARELLI, M.FELDMAN & R.J.McCANN: *Constructing optimal maps for Monge's transport problem as a limit of strictly convex costs.* J. Amer. Math. Soc., **15** (2002), 1–26.

[13] L.DE PASCALE & A.PRATELLI: *Regularity properties for Monge transport density and for solutions of some shape optimization problem.* Calc. Var., **14** (2002), 249–274.

[14] L.DE PASCALE, L.C. EVANS & A.PRATELLI: In preparation.

[15] L.C.EVANS & W.GANGBO: *Differential Equation Methods for the Monge-Kantorovich Mass Transfer Problem.* Memoirs AMS, **653**, 1999.

[16] M.FELDMAN & R.McCANN: *Uniqueness and transport density in Monge's mass transportation problem.* 2000, to appear on Calc. Var.

[17] M.FELDMAN & R.McCANN: *Monge's transport problem on a Riemannian manifold.* Trans. Amer. Mat. Soc., **354** (2002), 1667–1697.

[18] W.GANGBO: *An elementary proof of the polar factorization theorem for functions.* Arch. Rat. Mech. Anal., **128** (1994), 381–399.

[19] W.GANGBO & R.J.McCANN: *The geometry of optimal transportation.* Acta Math., **177** (1996), 113–161.

[20] M.GROMOV: *Carnot-Carathéodory spaces seen from within.* In *Subriemannian*

Geometry, Progress in Mathematics, **144**, ed. by A.Bellaiche and J.Risler, Birkhäuser, Basel, 1996.

[21] L.V.KANTOROVICH: *On the transfer of masses.* Dokl. Akad. Nauk. SSSR, **37** (1942), 227–229.

[22] H.G.KELLERER: *Duality theorems for marginal problems.* Z. Wahrsch. Verv. Gebiete, **67** (1984) 399–432.

[23] P.L.LIONS: *Generalized solutions of Hamilton-Jacobi equations*, Research Notes in Math., **69**, Pitman (1982).

[24] C.MANTEGAZZA & A.MENNUCCI: *Hamilton-Jacobi equations and distance functions on Riemannian manifolds.* (1999) App. Math. Optimization, to appear (see also http://cvgmt.sns.it).

[25] R.MCCANN: *Polar factorization of maps on Riemannian manifolds.* Geom. Funct. Anal., **11** (2001), 589–608.

[26] G.MONGE: *Memoire sur la Theorie des Déblais et des Remblais.* Histoire de l'Acad. des Sciences de Paris, 1781.

[27] S.T.RACHEV & L.RÜSCHENDORF: *A characterization of random variables with minimum L^2 distance.* J. Multivariate Anal., **32** (1990), 48–54.

[28] S.T.RACHEV & L.RÜSCHENDORF: *Mass transportation problems.* Vol I: Theory, Vol. II: Applications. Probability and its applications, Springer, 1998.

[29] L.RÜSCHENDORF: *Fréchet bounds and their applications.* In: G. Dall'Aglio et al. Editors, Advances in Probability distributions with given marginals, **67** (1991) Math. Appl., 151–187.

[30] L.RÜSCHENDORF: *Optimal solutions of multivariate coupling problems.* Appl. Math. (Warsaw), **23** (1995), 325–338.

[31] C.SMITH & M.KNOTT: *On the optimal transportation of distributions.* J. Optim. Theory Appl., **52** (1987), 323–329.

[32] C.SMITH & M.KNOTT: *On Hoeffding-Fréchet bounds and cyclic monotone relations.* J. Multivariate Anal., **40** (1992), 328–334.

[33] V.N.SUDAKOV: *Geometric problems in the theory of infinite dimensional distributions.* Proc. Steklov Inst. Math., **141** (1979), 1–178.

[34] N.S.TRUDINGER & X.J.WANG: *On the Monge mass transfer problem.* Calc. Var. PDE, **13** (2001), 19–31.

[35] C.VILLANI: *Topics in mass transportation.* Forthcoming book by AMS.

ICM 2002 · Vol. III · 141–153

Quasilinear Wave Equations and Microlocal Analysis

Hajer Bahouri[*]   Jean-Yves Chemin[†]

Abstract

In this text, we shall give an outline of some recent results (see [3] [4] and [5]) of local wellposedness for two types of quasilinear wave equations for initial data less regular than what is required by the energy method. To go below the regularity prescribed by the classical theory of strictly hyperbolic equations, we have to use the particular properties of the wave equation. The result concerning the first kind of equations must be understood as a Strichartz estimate for wave operators whose coefficients are only Lipschitz while the result concerning the second type of equations is reduced to the proof of a bilinear estimate for the product of two solutions for wave operators whose coefficients are not very regular. The purpose of this talk is to emphasise the importance of ideas coming from microlocal analysis to prove such results.

The method known to prove Strichartz estimates uses a representation eventually approximate of the solution. In the case of the wave equation, the approximation used is the one coming from the Lax method, namely the one connected to the geometrical optics. But it seems impossible, in the framework of the quasilinear wave equations, to construct a suitable approximation of the solution on some interval $[0, T]$, since the associate Hamilton-Jacobi equation develop singularities (it is the caustic phenomenon) at a time connected with the frequency size. We have then to microlocalize, which means to localize in frequencies, and then to work on time interval whose size depend on the frequency considered. It is the alliance of geometric optics and harmonic analysis which allow to establish a quasilinear Strichartz estimate and to go below this minimal regularity in the case of the first kind of equations.

To study the second kind of equations, we are confronted to an additional problem: Contrary to the constant case, the support of the Fourier transform is not preserved by the flow of the variable coefficient wave equation. To overcome this difficulty, we show that the relevant information in the variable case is the concept of microlocalized function due to J.M.Bony [11]. The proof that for solutions of variable coefficient operators, microlocalization properties propagate nicely along the Hamiltonian flows related to the wave operator is the key point in the proof of the result in the second case.

[*]Université de Tunis, Département de Mathématiques, 1060 Tunis, Tunisia. E-mail: Hajer.Bahouri@fst.rnu.tn

[†]Centre de Mathématiques École polytechnique, 91 128 Palaiseau Cedex, France. E-mail: chemin@math.polytechnique.fr

Hajer Bahouri   Jean-Yves Chemin

2000 Mathematics Subject Classification: 35J10.
Keywords and Phrases: Quasilinear wave equations, Paradifferential calculus, Microlocalazed functions.

1. Introduction

In this paper, our interest is to prove local solvability for quasilinear wave equations of the type

$$(E) \begin{cases} \partial_t^2 u - \Delta u - g(u) \cdot \nabla^2 u &= Q(\partial u, \partial u) \\ (u, \partial_t u)_{|t=0} &= (u_0, u_1) \end{cases}$$

where g is a smooth function vanishing at 0 with value in K such that $\mathrm{Id} + K$ is a convex compact subset of the set of positive symmetric matrices and Q is a quadratic form on $\mathbb{R}^{d+1}$. Our interest proceeds also for cubic quasilinear wave equations of the type

$$(EC) \begin{cases} \partial_t^2 u - \Delta u - \displaystyle\sum_{1 \le j,k \le d} g^{j,k} \partial_j \partial_k u &= \displaystyle\sum_{1 \le j,k \le d} \widetilde{Q}_{j,k}(\partial g^{j,k}, \partial u) \\ \Delta g^{j,k} &= Q_{j,k}(\partial u, \partial u) \\ (u, \partial_t u)_{|t=0} &= (u_0, u_1) \end{cases}$$

where $Q_{j,k}$ and $\widetilde{Q}_{j,k}$ are quadratic forms on $\mathbb{R}^{d+1}$ and where all the quadratic forms are supposed to be smooth functions of u.

The basic tool to prove local solvability for such equations is the following energy estimate, also valid for the symmetric systems

$$\|\partial u(t, .)\|_{H^{s-1}} \le \|\partial u(0, .)\|_{H^{s-1}} e^{C \int_0^T \|\partial g(\tau, .)\|_{L^\infty} d\tau}. \tag{1}$$

So thanks to classical arguments, local solvability derives easily from the control of the quantity

$$\int_0^T \|\partial g(\tau, .)\|_{L^\infty} d\tau.$$

In the framework of the equation (E), the control of this key quantity requires initial data (u_0, u_1) in $H^s \times H^{s-1}$ for $s > \frac{d}{2} + 1$ while in the framework of (EC) (with small data, which makes sense in this case) it only requires initial data (u_0, u_1) in $H^{\frac{d}{2}+\frac{1}{2}} \times H^{\frac{d}{2}-\frac{1}{2}}$.

The goal of this paper is to go below this regularity for the initial data. Let us first have a look at the scaling properties of equations (E) and (EC). If u is a solution of (E) or (EC), then $u_\lambda(t, x) \stackrel{\text{def}}{=} u(\lambda t, \lambda x)$ is also a solution of the same equation. The space which is invariant under this scaling for the couple (u_0, u_1) is $\dot{H}^{\frac{d}{2}} \times \dot{H}^{\frac{d}{2}-1}$. So the results given by the classical energy estimate appear to require more regularity than the scaling in the two cases.

In fact, the energy methods despise the particular properties of the wave equation. It is on the impulse of the pioneer work of S. Klainerman (see [19]) that a vast series of works have been attached to improve the span life time of regular solutions of quasilinear wave equations using the Lorentz invariance. Let us notice the results of S. Alinhac (see [1] and [2]), of L. Hörmander (see [14]), of F. John (see [15]), of F. John and S. Klainerman (see [16]), of S. Klainerman (see [20]) and of J-M.Delort ([12]) concerning the Klein-Gordon equation.

In this talk, we shall limit our self to the question of minimal regularity. Concerning this subject, the only case studied is the semilinear case, which means the case of the equation (E) with $g \equiv 0$. As it has been shown by S. Klainerman and M.Machedon (see [21] and [22]) we can, when the quadratic form Q verifies a structure condition known by "null condition", nearly reach the space invariant by scaling. For any quadratic form Q, we have the following theorem, proved by G. Ponce and T. Sideris in [27]

Theorem 1.1 *Let us define $\underline{s}_d$ by*

$$\underline{s}_d \stackrel{def}{=} \frac{d}{2} + \frac{1}{2} \quad if \quad d \geq 3 \quad and \quad \underline{s}_2 = \frac{7}{4}.$$

Let (u_0, u_1) be a Cauchy data in $H^s \times H^{s-1}$ with $s > \underline{s}_d$ then there exists a time T such that there exists a unique solution u of the equation (E) such that

$$u \in L^\infty([0,T]; H^s) \cap Lip([0,T]; H^{s-1}) \quad and \quad \partial u \in L^2([0,T]; L^\infty).$$

The proof of this result lies on specific properties of the wave equation, namely the following Strichartz estimate

$$\|\partial u\|_{L^2_T(L^\infty)} \leq C \left(\|\partial u(0,.)\|_{H^{s-1}} + \|\Box u\|_{L^1_T(H^{s-1})} \right),$$
$$\text{for } d \geq 3 \text{ and } s > \frac{d}{2} + \frac{1}{2}. \tag{2}$$

Indeed, if we couple it with the standard energy estimate

$$\|\partial u\|_{L^\infty_T(H^{s-1})} \leq C \left(\|\partial u(0,.)\|_{H^{s-1}} + \|\Box u\|_{L^1_T(H^{s-1})} \right)$$

we obtain, owing to the tame estimates and the Cauchy-Schwarz inequality

$$\|\partial u\|_{L^2_T(L^\infty)} + \|\partial u\|_{L^\infty_T(H^{s-1})}$$
$$\leq C \left(\|\partial u(0,.)\|_{H^{s-1}} + T^{\frac{1}{2}} \|\partial u(t,.)\|_{L^2_T(L^\infty)} \|\partial u\|_{L^\infty_T H^{s-1}} \right),$$

which ensures by the theory of evolution equations the local solvability for $T \leq \frac{C}{\|\partial u(0,.)\|^2_{H^{s-1}}}$.

In other respects, in [26], H. Linblad shows that for $d = 3$ the above result is optimum, which means that the problem (E) with $g = 0$ is not wellposed in H^2. Let us also notice that the same kind of result is also true on the Heisenberg group (see [9]).

The authors (see [3] and [4]) adjust a method followed by D. Tataru (see [32]) based on microlocal analysis to improve the minimal regularity for the equation (E) in the quasilinear case. Let us recall this result

Theorem 1.2 *If $d \geq 3$, let (u_0, u_1) be in $H^s \times H^{s-1}$ for $s > s_d$ with $s_d = \dfrac{d}{2} + \dfrac{1}{2} + \dfrac{1}{6}$.
Then, a positive time T exists such that a unique solution u of the equation (E)
exists such that*
$$\partial u \in C([0,T]; H^{s-1}) \cap L^2([0,T]; L^\infty).$$

Remarks

- This theorem has been proved with $1/4$ instead than $1/6$ in [3] and then improved a little bit in [4] and proved with $1/6$ by D. Tataru in [32]. Strichartz estimates for quasilinear equations are the key point of the proofs.
- Let us notice that the improvement with $1/6$ of D. Tataru in [32] is due to a different manner of counting the intervals where microlocal estimates are true.
- Recently, S. Klainerman and I. Rodnianski in [24] have obtained a better index in dimension 3. Their proof is based on very different methods.
- Let us notice that we have also improved the minimal regularity in dimension 2, but the gain is only of $\frac{1}{8}$ derivative, this is explained by the mean dispersif effect in this dimension already known for the constant case.

The analogous theorem in the case of equation (EC) is the following

Theorem 1.3 *If $d \geq 4$, let (u_0, u_1) be in $H^s \times H^{s-1}$ with $s > \frac{d}{2} + \frac{1}{6}$ such
that $\|\gamma\|_{\dot{H}^{\frac{d}{2}-1}}$ is small enough. Then, a positive times T exists such that a unique
solution u of (EC) exists such that*
$$\partial u \in C([0,T]; H^{s-1}) \cap L_T^2(\dot{B}_{4,2}^{\frac{d}{4}-\frac{1}{2}}), \quad for \quad d \geq 5,$$

and
$$\partial u \in C([0,T]; H^{s-1}) \cap L_T^2(\dot{B}_{6,2}^{\frac{1}{6}}) \quad and \quad \partial g \in L_T^1(L^\infty) \quad for \quad d = 4.$$

Remarks

- The case when $d \geq 5$ can be treated only with Strichartz estimates simply because if ∂u belongs to $L_T^2(\dot{B}_{4,2}^{\frac{d}{4}-\frac{1}{2}})$ then ∂g is in $L_T^1(L^\infty)$.
- The case when $d = 4$ requires bilinear estimates. This fact appears in the statement of Theorem 1.3 through the following phenomenon: The fact that ∂u is in $L_T^2(\dot{B}_{6,2}^{\frac{1}{6}})$ does not imply that the time derivative of g belongs to $L_T^1(L^\infty)$. Of course this condition is crucial in particular to get the basic energy estimate. But we have been unable to exhibit a Banach space $\mathcal{B}$ which contains the solution u and such that if a function a is in $\mathcal{B}$, then $\partial \Delta^{-1}(a^2)$ belongs to $L_T^1(L^\infty)$.
- For technical obstructions, this theorem is limited to the dimensions $d \geq 4$.

2. Quasilinear Strichartz estimates

Following the process of G. Ponce and T. Sideris in [27], we reduce the proof of the theorem 1.2 to the following a priori estimate

Theorem 2.1 *If $d \geq 3$, a constant C exists such that, for any regular solution u of the equation (E), if*

$$T^{\frac{1}{2}+(s-s_d)}\left(\|\gamma\|_{H^{\frac{d}{2}-\frac{1}{2}+(s-s_d)}} + T^{\frac{1}{6}}\|\gamma\|_{H^{s-1}}\right) \leq C, \quad with \quad s > s_d$$

then we have

$$\|\partial u\|_{L^2_T(L^\infty)} \leq C\left(\|\gamma\|_{H^{s-1}} + \|Q(\partial u, \partial u)\|_{L^1_T(H^{s-1})}\right).$$

The estimate in hand must be understood as a Strichartz estimate for wave equations with variable coefficients and not very regular. The Strichartz estimates have a long history begining with Segal's work [28] for the wave equation with constant coefficients. After the fundamental work of Strichartz [30], it was developed by diverse authors, we refer to the synthesis article of Ginibre and Velo [13] to which it is advisable to add the recents works of Keel-Tao [18] consecrated to some limited cases and of Bahouri, Gérard and Xu [8] for the wave equation on the Heisenberg group. For Strichartz estimates with C^∞ coefficients, we refer to the result of L. Kapitanski (see [17]). The article of H. Smith (see [29]) constitutes an important step in the study of Strichartz estimates for operators with coefficients not very regular since it proves Strichartz estimates with coefficients only $C^{1,1}$.

We shall now explain how to establish this quasilinear Strichartz estimate, showing where are the difficulties and what are the essential ideas which allow to overcome them. The method known to prove these estimates uses a representation, eventually approximate, but always explicit of the solution. In the case of the wave equation, the approximation used is the one coming from the Lax method, namely the one connected to the geometrical optics. To make such a method work in the framework of quasilinear wave equations requires a "regularization" of the coefficients also in time. This leads to the following iterative scheme introduced in [4]. Let us define the sequence $(u^{(n)})_{n\in\mathbb{N}}$ by the first term $u^{(0)}$ satisfying

$$\left\{\begin{array}{rcl} \partial_t^2 u^{(0)} - \Delta u^{(0)} & = & 0 \\ (u^{(0)}, \partial_t u^{(0)})_{|t=0} & = & (S_0 u_0, S_0 u_1), \end{array}\right.$$

and by the following induction

$$(E_n)\left\{\begin{array}{rcl} \partial_t^2 u^{(n+1)} - \Delta u^{(n+1)} - g_{n,T} \cdot \nabla^2 u^{(n+1)} & = & 0 \\ (u^{(n+1)}, \partial_t u^{(n+1)})_{|t=0} & = & (S_{n+1} u_0, S_{n+1} u_1) \end{array}\right.$$

where $g_{n,T} \overset{\text{def}}{=} \theta(T^{-1})g_n$ with $g_n \overset{\text{def}}{=} g(u^n)$ and θ a function of $\mathcal{D}(]-1,1[)$ whose value is 1 near 0 and where S_n is a frequencies truncated operator which only conserves the frequencies lower than $C2^{n-1}$. Let us introduce some notations which will be used all along this section. If $s = s_d + \alpha$ where α is a small positive number, let us define

$$N_T^\alpha(\gamma) \overset{\text{def}}{=} \|\gamma\|_{H^{\frac{d}{2}-\frac{1}{2}+\alpha}} + T^{\frac{1}{6}}\|\gamma\|_{H^{s-1}}.$$

The assertion we have to prove by induction, for $T^{\frac{1}{2}+\alpha}N_T^\alpha(\gamma)$ small enough, is the following: If $d \geq 3$,

$$(\mathcal{P}_n)\left\{\begin{array}{rcl} \|\partial u^{(n)}\|_{L^2_T(L^\infty)} & \leq & C_\alpha T^\alpha N_T^\alpha(\gamma) \\ \|\partial u^{(n)}\|_{L^\infty_T(H^{s-1})} & \leq & C\|\gamma\|_{H^{s-1}}. \end{array}\right.$$

For this, we shall transform the equation (E_n) into a paradifferential equation, more precisely an equation of the type

$$(E_q) \qquad \partial_t^2 u_q^{(n+1)} - \Delta u_q^{(n+1)} - (S_q g_{n,T}) \nabla^2 u_q^{(n+1)} = \widetilde{R}_q(n)$$

where the term $\widetilde{R}_q(n)$ is a remainder term estimated as agreed and where u_q denotes the part of u which is relative to the frequencies of size 2^q.

This transformation of the equation, which is the classical paralinearization defined by J.-M. Bony in [10] is here not sufficient, since it is well known that the paradifferential operators defined in [10] belong to a bad class of pseudodifferential operators (class $S_{1,1}$ of Hörmander), class in particular devoid of any asymptotic calculus, which forbids of course to envisage any approximate method of type "Lax method".

The idea is as in [25] to truncate more in the frequencies of the metric g and to transform the equation (E_q) on the following equation (EPM_q)

$$(EPM_q) \qquad \partial_t^2 u_q^{(n+1)} - \Delta u_q^{(n+1)} - (S_{\delta q} g_{n,T}) \nabla^2 u_q^{(n+1)} = R_q(n);$$

where $0 < \delta < 1$ and $S_{\delta q}$ is a frequencies truncated operator which conserves only the frequencies smaller than $CT^{-(1-\delta)} 2^{\delta q - 1}$. We can interpret it as a localization in the pseudodifferential calculus sense $(1, \delta)$ of Hörmander.

This localization allows us to construct an approximation of the solution but engenders a loss in the remainder $R_q(n)$. This approximation is on the form

$$\int e^{i\Phi_q(t,x,\xi)} \sigma_q(t,x,\xi) \widehat{\gamma}(\xi) d\xi$$

where Φ_q is a solution of the Hamilton-Jacobi equation and σ_q is a symbol calculated by resolving a sequence of transport equations; it is about a classical method. But, on account of the the caustic phenomenon, this approximation is microlocal, which means valid only a time interval whose length depends on the size of the frequencies we work with.

Nevertheless, following the classical method, we prove microlocal Strichartz estimates

$$\|\partial u_q^{(n+1)}\|_{L_{I_q}^2(L^\infty)} \le C_\beta (2^q T)^\beta 2^{q(\frac{d-1}{2})} \left(\|\gamma_q\|_{L^2} + \|R_q(n)\|_{L_{I_q}^1(L^2)} \right) \qquad (3)$$

for any positive β, where $\gamma_q \overset{\text{def}}{=} (\nabla(u_0)_q, (u_1)_q)$ and I_q satisfies

$$\|\nabla^2 G_\delta^{(n)}\|_{L_{I_q}^1(L^\infty)} \le \epsilon \qquad (4)$$

where $G_\delta^{(n)} \overset{\text{def}}{=} S_{\delta q} g(u^{(n)})$ and

$$|I_q| \le T(2^q T)^{1-2\delta - \epsilon}. \qquad (5)$$

The condition (4) is imposed by the Hamilton-Jacobi equation while the condition (5) is required by the asymptotic calculus to turn out the "Lax method".

Finally to prove the complete estimate, the method we used consists in a decomposition of the interval $[0, T]$ on subintervals I_q on which the above microlocalized estimates are true. The key point is a careful counting of the number of such intervals, for this we shall use here D. Tataru's version of the method we introduced in [3].

The idea consists to seize at the opportunity of this decomposition to compensate the loss on the remainder. To do so, we impose on the interval I_q the supplementary condition

$$\|R_q(n)\|_{L^1_{I_q}(L^2)} \leq \lambda \|R_q(n)\|_{L^1_T(L^2)} \tag{6}$$

where the parameter λ is to be determined in the interval $[0, 1]$. This constraint joint to the conditions (4) and (5) leads by optimization to the best choice

$$\lambda = (2^q T)^{-\frac{\delta}{2}}, \qquad \delta = \frac{2}{3},$$

and allows to conclude that the number N of such intervals is less than $C(2^q T)^{\frac{1}{3}+\epsilon}$. If we denote by $(I_{q,\ell})_{1\leq\ell\leq N}$ the partition of the interval $[0, T]$ on such intervals, we can write thanks to (3),

$$\|\partial u_q^{(n+1)}\|^2_{L^2_T(L^\infty)} \leq C_\beta \sum_{\ell=1}^{N} (2^q T)^{2\beta} 2^{2q(\frac{d-1}{2})} \left(\|\gamma_q\|_{L^2} + \|R_q(n)\|_{L^1_{I_{q,\ell}}(L^2)} \right)^2$$

$$\leq C_\beta \sum_{\ell=1}^{N} (2^q T)^{2\beta} 2^{2q(\frac{d-1}{2})} \left(\|\gamma_q\|_{L^2} + (2^q T)^{-\frac{1}{3}} \|R_q(n)\|_{L^1_T(L^2)} \right)^2.$$

As N is less than $C(2^q T)^{\frac{1}{3}+\epsilon}$, we obtain

$$\|\partial u_q^{(n+1)}\|_{L^2_T(L^\infty)} \leq C_\beta (2^q T)^\beta 2^{q(\frac{d-1}{2})} (2^q T)^{\frac{1}{6}+\frac{\epsilon}{2}}$$
$$\cdot \left(\|\gamma_q\|_{L^2} + (2^q T)^{-\frac{1}{3}} \|R_q(n)\|_{L^1_T(L^2)} \right). \tag{7}$$

Now as the loss in the remainder $R_q(n)$ is of order $2^{q(1-\delta)}$ and more precisely, for $N^\alpha_T(\gamma)$ small enough, we have

$$\|R_q(n)\|_{L^1_T(L^2)} \leq c_q C 2^{-q(\frac{d-1}{2})} (2^q T)^{-(s-\frac{d}{2}-\frac{3}{2}+\delta)} T^{s-\frac{d}{2}-\frac{1}{2}} \|\gamma\|_{H^{s-1}}$$
$$\cdot \left(1 + T^{\frac{1}{2}} \|\partial u^{n+1}\|_{L^2_T(L^\infty)} \right) \tag{8}$$

where $(c_q) \in \ell^2$. We deduce, owed to the choice of δ that

$$\|\partial u_q^{(n+1)}\|_{L^2_T(L^\infty)} \leq c_q C (2^q T)^{-(\alpha-\frac{\epsilon}{2}-\beta)} T^{s-\frac{d}{2}-\frac{1}{2}} \|\gamma\|_{H^{s-1}} \left(1 + T^{\frac{1}{2}} \|\partial u^{n+1}\|_{L^2_T(L^\infty)} \right)$$

which implies the result by summation.

3. Quasilinear bilinear estimates

The method used here is not without any interaction with the one used to prove the theorem 1.2. As in the case of equation (E), the basic fact is the control of

$$\int_0^T \|\partial g(\tau,\cdot)\|_{L^\infty} d\tau,$$

and the proof of the theorem 1.3 follows from the following a priori estimate

Theorem 3.1 *If $d \geq 4$, a constant C exists such that, for any regular solution u of the equation (EC), if $\|\gamma\|_{\dot{H}^{\frac{d}{2}-1}}$ is small enough and*

$$T^{\frac{1}{6}+(s-\frac{d}{2}-\frac{1}{6})}\|\gamma\|_{H^{s-1}} \leq C, \quad with \quad s > \frac{d}{2} + \frac{1}{6}$$

then we have

$$\|\partial \Delta^{-1} Q(\partial u, \partial u)\|_{L^1_T(L^\infty)} \leq C\|\gamma\|^2_{H^{s-1}}.$$

This is the quasilinear version of the following bilinear estimate owed to D. Tataru and S. Klainerman (see [23])

Proposition 3.1 *Let u be a solution of $\partial_t^2 u - \Delta u = 0$ and $(\partial u)_{|t=0} = \gamma$. Then, if $d \geq 4$,*

$$\|\partial \Delta^{-1} Q(\partial u, \partial u)\|_{L^1_T(L^\infty)} \leq C_{\epsilon,T}\|\gamma\|^2_{\frac{d}{2}-1+\epsilon}.$$

Remark We find a gain of one derivative from the regularity of the initial data compared with the product laws and a gain of half a derivative about the regularity of the initial data compared with purely Strichartz methods.

To explain the basic ideas of bilinear estimates, let us first consider the case of constant coefficients. As $\partial_t \Delta^{-1}\big(\partial_j u(t,\cdot)\partial_k u(t,\cdot)\big) = \Delta^{-1}\big(\partial_t \partial_j u \partial_k u(t,\cdot)\big) + \Delta^{-1}\big(\partial_j u \partial_t \partial_k u(t,\cdot)\big)$, we have to control expression of the type

$$\int_0^T \|\Delta^{-1}\big(\partial_t \partial_j u \partial_k u(t,\cdot)\big)\|_{L^\infty} dt.$$

For this we introduce Bony's decomposition which consists in writing

$$ab = \sum_q S_{q-1} a \Delta_q b + \sum_q S_{q-1} b \Delta_q a + \sum_{\substack{-1 \leq j \leq 1 \\ q}} \Delta_q a \Delta_{q-j} b.$$

When $d \geq 4$, we have $\|\partial^k u_q\|_{L^2_T(L^\infty)} \leq C 2^{q(\frac{d}{2}-\frac{1}{2}+k-1)}\|\gamma_q\|_{L^2}$, then it is easy to prove that

$$\left\|\Delta^{-1}\Big(\sum_q S_{q-1}\partial^2 u \partial u_q\Big)\right\|_{L^1_T(L^\infty)} \leq C\|\gamma\|^2_{\frac{d}{2}-1}.$$

The symmetric term can be treated exactly along the same lines. The remainder term

$$\Delta^{-1}\Big(\sum_{\substack{-1 \leq j \leq 1 \\ q}} \partial^2 u_q \partial u_{q-j}\Big) \tag{9}$$

is much more difficult to treat in particular in dimension 4. The idea introduced by D. Tataru and S. Klainerman (see [23]) consists to treat this term using precised Strichartz estimates and interaction lemma.

The precised Strichartz estimates are described by the following proposition.

Proposition 3.2 *Let $\mathcal{C}$ be a ring of $\mathbb{R}^d$. If $d \geq 3$, a constant C exists such that for any T and any $h \leq 1$, if $\operatorname{Supp} \widehat{u}_j$ are included in a ball of radius h and in the ring $\mathcal{C}$, we have*

$$\|u\|_{L^2_T(L^\infty)} \leq C\big(h^{d-2}\log(e+T)\big)^{\frac{1}{2}}\big(\|u_0\|_{L^2} + \|u_1\|_{L^2}\big),$$

where u denotes the solution of $\partial_t^2 u - \Delta u = 0$ and $\partial_t^j u_{|t=0} = u_j$.

As usual it is deduced with the $TT^\star$ argument from the following dispersive inequality.

Lemma 3.1 *A constant C exists such that if u_0 and u_1 are functions in $L^1(\mathbb{R}^d)$ such that*

$$\operatorname{Supp}(\widehat{u}_j) \subset \mathcal{C} \quad and \quad \max\{\delta(\operatorname{Supp}(\widehat{u}_0)), \delta(\operatorname{Supp}(\widehat{u}_1))\} \leq h,$$

then, for any $\widetilde{d}$ between 0 and $d-1$, we have

$$\|u(t,\cdot)\|_{L^\infty} \leq \frac{Ch^{d-\widetilde{d}}}{t^{\frac{\widetilde{d}}{2}}}\big(\|u_0\|_{L^1} + \|u_1\|_{L^1}\big),$$

where u denotes the solution of $\partial_t^2 u - \Delta u = 0$ and $\partial_t^j u_{|t=0} = u_j$.

This inequality is proved in [23] in the case $\widetilde{d} = d-1$. The general case is obtained by interpolation with the classical Sobolev embedding.

Let us now show how to take account the interactions of the solutions to control the accumulation of frequencies at the origin in the study of the remainder term.

Lemma 3.2 *"Interaction Lemma" There exists a constant C such that if v_1 and v_2 are two solutions of $\partial_t^2 v_j - \Delta v_j = 0$ satisfying $(\partial v_j)_{|t=0} = \gamma_j$ with $\operatorname{Supp}(\widehat{\gamma}_j) \subset \mathcal{C}$ we have for, $0 < h < 1$*

$$\|\chi(h^{-1}D)(\partial^2 v_1 \partial v_2)\|_{L^1_T(L^\infty)} \leq Ch^{d-2}\log(e+T)\|\gamma_1\|_{L^2}\|\gamma_2\|_{L^2}, \tag{10}$$

where χ is a radial function in $\mathcal{D}$ which is equal to 1 near the origin.

Let us define $(\phi_\nu)_{1\leq\nu\leq N_h}$ a partition of unity of the ring $\mathcal{C}$ such that $\operatorname{Supp} \phi_\nu \subset B(\xi_\nu, h)$. Then, using the fact that the support of the Fourier transform of the product of two functions is included in the sum of the support of their Fourier transform, a family of functions $(\widetilde{\phi}_\nu)_{1\leq\nu\leq N_h}$ exists such that $\operatorname{Supp} \widetilde{\phi}_\nu \subset B(-\xi_\nu, 2h)$ and

$$\chi(h^{-1}D)(\partial^2 v_1 \partial v_2) = \sum_{\nu=1}^{N_h} \chi(h^{-1}D)\big(\partial^2 \widetilde{\phi}_\nu(D)v_1 \partial \phi_\nu(D)v_2\big). \tag{11}$$

Applying Proposition 3.2 gives

$$\|\chi(h^{-1}D)(\partial^2 v_1 \partial v_2)\|_{L^1_T(L^\infty)} \leq Ch^{d-2}\log(e+T)\sum_{\nu=1}^{N_h}\|\widetilde{\phi}_\nu(D)\gamma_1\|_{L^2}\|\phi_\nu(D)\gamma_2\|_{L^2}.$$

The Cauchy Schwarz inequality implies that

$$\|\chi(h^{-1}D)(\partial^2 v_1 \partial v_2)\|_{L^1_T(L^\infty)}$$

$$\leq \quad Ch^{d-2}\log(e+T)\left(\sum_{\nu=1}^{N_h}\|\widetilde{\phi}_\nu(D)\gamma_1\|_{L^2}^2\right)^{\frac{1}{2}}\left(\sum_{\nu=1}^{N_h}\|\phi_\nu(D)\gamma_2\|_{L^2}^2\right)^{\frac{1}{2}}.$$

The almost orthogonality of $(\widetilde{\phi}_\nu(D)\gamma_1)_{1\leq\nu\leq N_h}$ and $(\phi_\nu(D)\gamma_2)_{1\leq\nu\leq N_h}$ implies (11) and leads then to the estimate of the remainder term (10) by rescaling.

To establish the theorem 3.1, we shall follow the steps of the proof of the theorem 2.1 which consists owed to the gluing method to reduce the problem to the proof of "microlocal" bilinear estimates. The generalization of the precised Strichartz estimates to the framework of the equation (EC) doesn't cost more than the generalization of the Strichartz estimates to the framework of the equation (E), the supplementary difficulty to study the equation (EC) lies in the generalization of the interaction lemma. The preservation of the support of the Fourier transform by the flow of the wave equation is the crucial point in the proof of this lemma. The defect of this property in the case of the variable coefficients constitutes the additional major problem in the proof of the theorem 3.1.

To palliate this difficulty, we have used a finer localization in phase space. This localization is given by the concept of microlocalized function near a point $X = (x,\xi)$ of the cotangent space $T^\star\mathbb{R}^d$ (the cotangent space of $\mathbb{R}^d$). More precisely, if we consider the positive quadratic form g on $T^\star\mathbb{R}^d$ defined by

$$g(dy^2, d\eta^2) \stackrel{\text{def}}{=} \frac{dy^2}{K^2} + \frac{d\eta^2}{h^2} \quad \text{with} \quad \lambda \stackrel{\text{def}}{=} Kh \geq 1$$

a function u in $L^2(\mathbb{R}^d)$ is said to be microlocalized in $X_0 = (x_0, \xi_0)$ a point of $T^\star\mathbb{R}^d$ if

$$\mathcal{M}_{X_0,N}^{C_0,r}(u) \stackrel{\text{def}}{=} \sup_{g(X-X_0)^{\frac{1}{2}}\geq C_0 r} \lambda^{2N} g(X-X_0)^N \sup_{\substack{\varphi\in\mathcal{D}(B_g(X,r))\\ \|\varphi\|_{k_N,g}\leq 1}} \|\varphi^D u\|_{L^2}$$

are finite, where $B_g(X,r)$ denotes the g-ball of center X and radius r, the operator φ^D is defined by

$$(\varphi^D u)(x) = (2\pi)^{-d}\int_{T^\star\mathbb{R}^d} e^{i(x-y|\xi)}\varphi(y,\xi)u(y)dyd\xi,$$

and

$$\|\varphi\|_{j,g} \stackrel{\text{def}}{=} \sup_{\substack{k\leq j\\ X\in T^\star\mathbb{R}^d}} \sup_{\substack{(T_\ell)_{1\leq\ell\leq k}\\ g(T_\ell)\leq 1}} |D^k\varphi(X)(T_1,\cdot,T_k)|.$$

This notion due to J.-M.Bony ([11]) means that the function u is concentrated in space near the point x_0 and in frequency near the point ξ_0 and behaves well against the product, namely, we show that if

$$g(\check{Y}_1 - Y_2)^{\frac{1}{2}} \geq C_0 r,$$

where $\check{Y} \stackrel{\text{def}}{=} (y, -\eta)$ if $Y = (y, \eta)$ then for any N, we have

$$\left\| \chi(h^{-1}D)(\varphi_1^D u_1 \varphi_2^D u_2) \right\|_{L^1}$$
$$\leq \quad C_N \|\varphi_1\|_{k_N,g} \|\varphi_2\|_{k_N,g} \left(1 + \lambda^2 g(\check{Y}_1 - Y_2)\right)^{-N} \|u_1\|_{L^2} \|u_2\|_{L^2}$$

where $\varphi_i \in \mathcal{D}(B_g(Y_i, r))$.

This study of the interaction between two typical examples of microlocalized functions allows as in (11) to concentrate the bilinear estimate on real interaction.

Anyway, the choice of the localization metric g is essential and it is crucial to impose that the size of the g-balls is preserved by Hamiltonian flow which leads to the only choice $K = C|2^q I_q|h$ thanks to the properties of the solution of the associate Hamilton Jacobi equation.

The key point in the generalization of the bilinear estimate is the proof that for solutions of a variable coefficients wave equation, microlocalization properties propagate nicely along the Hamiltonian flows related to the wave operator; this point follows from the choice of the metric used to localize in the cotangent space of $\mathbb{R}^d$.

Finally to end the proof of the microlocal bilinear estimate, the strategy consists to decompose the Cauchy data using unity partition whose elements are supported in g-balls and then to apply the product and the propagation theorems to concentrate on real interaction (see the proof in the constant coefficient case). Because of the fact that interaction in the product and propagation of microlocalization are badly related, we need at this step to recourse to a second microlocalization, which means that we have to decompose again the interval on which we work on sub intervals where the Hamiltonian flow is nearly constant .

References

[1] S. Alinhac, Blow up of small data solutions for a class of quasilinear wave equations in two space dimensions I, *Annals of Mathematics*, **149**, 1999, 97–127.

[2] S. Alinhac, Blow up of small data solutions for a class of quasilinear wave equations in two space dimensions II, *Acta Mathematica*, **182**, 1999, 1–23.

[3] H. Bahouri and J.-Y. Chemin, Équations d'ondes quasilinéaires et inégalités de Strichartz, *American Journal of Mathematics*, **121**, 1999, 1337–1377.

[4] H. Bahouri and J.-Y. Chemin, Équations d'ondes quasilinéaires et effet dispersif, *International Mathematical Research News*, **21**, 1999, 1141–1178.

[5] H. Bahouri and J.-Y. Chemin, Microlocal analysis, bilinear estimates and cubic quasilinear wave equation, *Prépublication Mathématiques de l École Polytechnique, 2002* .

[6] H. Bahouri and P. Gérard, High Frequency Approximation of Solutions to Critical Nonlinear Wave Equations, American Journal of Mathematics 121(1999), 131–175.

[7] H. Bahouri and J. Shatah, Global estimate for the critical semilinear wave equation, Annales de l'Institut Henri Poincaré, Analyse non linéaire, **15**, N 6 (1998) 783–789.

[8] H. Bahouri, P. Gérard and C.-J. Xu, Espaces de Besov et estimations de Strichartz géneralisées sur le groupe de Heisenberg, *Journal d'Analyse Mathématique*, **82**, 2000, 93–118.

[9] H. Bahouri and I. Gallagher, Paraproduit sur le groupe de Heisenberg et applications, *Revista Matematica Iberoamericana* **17**, 2001, 69–105.

[10] J.-M. Bony, Calcul symbolique et propagation des singularités pour les équations aux dérivées partielles non linéaires, *Annales de l'École Normale Supérieure*, **14**, 1981, 209–246.

[11] J.-M. Bony, personnal communication.

[12] J-M. Delort, Temps d' éxistence pour l'équation de Klein-Gordon semi-linéaire á données petites périodiques, *American Journal of Mathematics*, **120**, 1998, 663–689.

[13] J. Ginibre and G. Velo, Generalized Strichartz inequalities for the wave equation, *Journal of Functional Analysis*, **133**, 1995, 50–68.

[14] L. Hörmander, *Lectures on Nonlinear Hyperbolic Differential Equations*, Mathématiques et Applications, **26**, Springer, 1996.

[15] F. John, Existence for large times of strict solutions of non linear wave equations in three space dimensions, *Communications in Pure and Applied Mathematics*, **40**, 1987, 79–109.

[16] F. John and S. Klainerman, Almost global existence existence to non linear wave equations in three space dimensions, *Communications in Pure and Applied Mathematics*, **37**, 1984, 443–455.

[17] L. Kapitanski, Some generalization of the Strichartz-Brenner inequality, *Leningrad Mathematical Journal*, **1**, 1990, 693–721.

[18] M.Keel and T.Tao, Endpoint Strichartz estimates,American Journal of Mathematics, **120** (1988), 955–980.

[19] S. Klainerman, Global existence for nonlinear wave equations, *Communications in Pure and Applied Mathematics*, **33**, 1980, 43–101.

[20] S. Klainerman, Uniform decay estimates and the Lorentz invariance of the classical wave equation *Communications in Pure and Applied Mathematics*, **38**, 1985, 321–332.

[21] S. Klainerman and M. Machedon, On the regularity properties of a model problem relates to wave maps, *Duke Mathematical Journal*, **87**, 1997, 553–589.

[22] S. Klainerman and M. Machedon, Estimates for null forms and the spaces $H_{s,\delta}$, *International Mathematical Research News*, **15**, 1998, 756–774.

[23] S. Klainerman and D. Tataru, On the optimal local regularity for the Yang-Mills equations in $\mathbf{R}^{4+1}$, Journal of the American Mathematical Society, **12**, 1999, 93–116.

[24] S. Klainerman and I.Rodnianski, Improved local well posedness for quasilinear

wave equations in dimension three, to appear in *Duke Mathematical Journal*

[25] G. Lebeau, Singularités de solutions d'équations d'ondes semi-linéaires, *Annales Scientifiques de l'École Normale Supérieure*, **25**, 1992, 201–231.

[26] H. Lindblad, A sharp counterexample to local existence of low regularity solutions to non linear wave equations, *Duke Mathematical Journal*, **72**, 1993, 503–539.

[27] G. Ponce and T. Sideris, Local regularity of non linear wave equations in three space dimensions, *Communications in Partial Differential Equations*, **18**, 1993, 169–177.

[28] I. E. Segal, Space-time decay for solutions of wave equations, *Adv. Math.*, **22** (1976), 304–311.

[29] H. Smith, A parametrix construction for wave equation with $C^{1,1}$ coefficients, *Annales de l'Institut Fourier*, **48**, 1998, 797–835.

[30] R. S. Strichartz, Restriction of Fourier transform to quadratic surfaces and decay of solutions of wave equations, *Duke Math.*, 44 (1977), 705–774.

[31] D. Tataru, Local and global results for wave maps I, *Communications in Partial Differential Equations*, **23**, 1998, 1781–1793.

[32] D. Tataru, Strichartz estimates for second order hyperbolic operators with nonsmooth coefficients III, *preprint*.

Some New Developments of Realization of Surfaces into R^3

Jiaxing Hong*

Abstract

This paper intends to give a brief survey of the developments on realization of surfaces into R^3 in the last decade. As far as the local isometric embedding is concerned, some results related to the Schlaffli-Yau conjecture are reviewed. As for the realization of surfaces in the large, some developments on Weyl problem for positive curvature and an existence result for realization of complete negatively curved surfaces into R^3, closely related to Hilbert-Efimov theorem, are mentioned. Besides, a few results for two kind of boundary value problems for realization of positive disks into R^3 are introduced.

2000 Mathematics Subject Classification: 53A05, 53C21, 35J60.
Keywords and Phrases: Isometric embedding, Hilbert-Efimov theorem, Boundary value problem.

Given a smooth n-dimensional Riemannian manifold (M^n, g), can we find a map

$$\phi : M^n \longrightarrow R^p \text{ such that } \phi^* h = g$$

where h is the standard metric in R^p? This is a long standing problem in Differential Geometry. The map ϕ is called isometric embedding or isometric immersion if ϕ is embedding or immersion. There are several very nice surveys. For example, for known results before 1970, particularly obtained by Russian mathematicians, see [GR], [G] and for results of $n = 2$, see [Y3] [Y4]. Whereas, what development has been made during the passed decades? In higher dimensional cases, the most important one is to improve the Nash's theorem so that (M^n, g) has isometric embedding in a Euclidean space of much lower dimension than that given by Nash and meanwhile, to use the contraction mapping principle in place of the theorem of the complicated hard implicit functions, see [H] or [GUN]. In contrast to higher dimensional cases it seems that problems of two dimensional Riemannian manifolds embedded into R^3 has attracted much more attention of mathematicians, particularly in the last decade. The present paper is devoted to a survey on the

*Institute of Mathematics, Fudan University, Shanghai 200433, China. E-mail: jxhong@fudan.ac.cn

developments of isometric embedding (or immersion) of surfaces into R^3 in the last decade. Of course the results mentioned in this survey are by no means exhaustive and depend a lot on the author's taste.

Smoothly and isometrically immersing a surface (M, g) into R^3 is equivalent to finding three smooth $(C^s, s \geq 1)$ functions $x_\alpha : M \mapsto R^1$, $\alpha = 1, 2, 3$ such that

$$g = dx_1^2 + dx_2^2 + dx_3^2. \tag{1}$$

Although as far as the formulation of (1) be concerned, C^1 regularity is enough and [K] gives a very nice result in this category. In order to see the role of the curvature of surfaces in the problem considered here we prefer to assume $s \geq 2$ throughout the present paper. In the sequel, sometimes we use x, y, z to denote x_1, x_2, x_3. In a local coordinates near a point $p \in M$, the metric g is of the form $g = g_{ij} du^i du^j$. Then (1) can be written as follows

$$\frac{\partial x_\alpha}{\partial u^i} \frac{\partial x_\alpha}{\partial u^j} = g_{ij} \ i, j = 1, 2, \tag{2}$$

(2) is a system composed of three differential equations of first order and hence, this is a determine system. We say that $\vec{r} = (x_1, x_2, x_3)$ is a local smooth (C^s) isometric embedding in R^3 of the given surface if $\vec{r} = (x_1, x_2, x_3)$ is a smooth (C^s) solution to (2) in a neighbourhood of the point p and that $\vec{r}$ is a global smooth (C^s) isometric embedding (immersion) in R^3 if $\vec{r} = (x_1, x_2, x_3)$ is a smooth (C^s) solution to (1) on M and, meanwhile, is an embedding (immersion) into R^3. This survey consists of three parts. The first and the second part include some recent developments in local isometric embedding and global isometric embedding respectively. The third part contains some developments on boundary value problems for realization of positive disks into R^3.

Local isometric embedding. With the aid of the Cauchy-Kowalevsky theorem Cartan and Janet proved that any n-dimensional analytic metric always admits a local analytic isometric embedding in R^{s_n} with $s_n = n(n+1)/2$. In the smooth category Gromov in [GR] proved that any n-dimensional C^∞ metric always admits a local smooth isometric embedding in R^{s_n+n}. As $n = 2$, $s_n = 5$ and from (2) the present result looks far away from the optimal in the smooth category. On the other hand, [P] proves that any smooth surface always has a local smooth isometric embedding in R^4. In [Y2, No.22] and also in [Y1, No. 54] Yau posed to prove that any smooth surface always has a local smooth isometric embedding in R^3. In this direction, it is Lin who first made important breakthrough and his results in [LC1] and [LC2] state

Theorem 1. (C. S. Lin) *(1) Any C^s, $s > 10$ nonnegatively curved metric always admits a local C^{s-6} isometric embedding in R^3.*

(2) If g is a C^s, $s > 6$ metric and if its curvature K satisfies

$$K(p) = 0 \ and \ dK(p) \neq 0$$

then it admits a C^{s-3} isometric embedding in R^3 near p.

By means of Lin's technique the problem for local isometric embedding related to nonpositive curvature metric are also solvable in the following cases.

(3) In [IW], K nonpositive in a neighbourhood of a point p and $d^2K(p) \neq 0$,

(4) In [HO1], $K = h^{2q}K_1$ where K_1, h are smooth functions, $K_1(p) < 0$, $dh(p) \neq 0$ and q is an integer.

In what follows let us simply explain what the technique ones use while attacking the problem of local isometric embedding. Suppose that $\vec{r} = (x, y, z)$ is a smooth solution to (2) in a neighbourhood of the point in question previously. By the Gauss equations we have, in a local coordinate system,

$$\vec{r}_{ij} = \Gamma_{ij}^k \vec{r}_k + \Omega_{ij}\vec{n} \text{ or } \nabla_{ij}\vec{r} = \Omega_{ij}\vec{n}, i, j = 1, 2 \tag{3}$$

where subscripts i,j and ∇_{ij} denote Euclidean and covariant derivatives respectively and Ω_{ij} the coefficients of the second fundamental form, Γ_{ij}^k the Christoffel symbols with respect to the metric and $\vec{n}$ the unit normal to $\vec{r}$. For each unit constant vector, for instance, the unit vector $\vec{k}$ of the z axis, taking the scale product of $\vec{k}$ and (3) and using the Gauss equations one can get

$$\det(\nabla_{ij}z) = K \det(g_{ij})(\vec{n}, \vec{k})^2. \tag{4}$$

Notice that

$$(\vec{n}, \vec{k})^2 = 1 - \left(\frac{(\vec{r}_1 \times \vec{r}_2) \times \vec{k}}{|\vec{r}_1 \times \vec{r}_2|} \right)^2$$

$$= 1 - g^{ij}z_i z_j = 1 - |\nabla z|^2.$$

Inserting the last expression into (4) we deduce the Darboux equation

$$F(z) = \det(\nabla_{ij}z) - K \det(g_{ij})(1 - |\nabla z|^2) = 0. \tag{5}$$

Obviously each component of $\vec{r}$ does satisfy the Darboux equation. Conversely, for each smooth solution z to (5) satisfying $|\nabla z| < 1$, $\hat{g} = g - dz^2$ is a smooth flat metric. Therefore in simply connected domain Ω we can always find a smooth mapping (x, y): $\Omega \mapsto R^2$ such that $dx^2 + dy^2 = g - dz^2$. So the realization of a given metric into R^3 is equivalent to finding a smooth (or C^s) solution to the Darboux equation with a subsidiary condition $|\nabla z| < 1$. If K is positive or negative at the point considered, then (5) is elliptic or hyperbolic Monge-Ampere equation and the local solvability is well known for both of them. But if K vanishes at this point, the situation is very complicated and so far there has been no standard way to deal with such kind of Monge-Ampere equation. Indeed, its linearized operator

$$L_z\xi = \lim_{t \longrightarrow 0} \frac{F(z + t\xi) - F(z)}{t} = F^{ij}\nabla_{ij}\xi + 2K(\nabla z, \nabla \xi) \tag{6}$$

where $F^{ij} = \partial \det(\nabla_{lk} z)/\partial \nabla_{ij} z$. It is easy to see that the type of this linear differential operator completely depends on K. When K vanishes at the point considered, (6) may be degenerate elliptic, hyperbolic or mixed type and its local solvability is not clear. Using a regularized operator instead of (6) and the Nash-Moser procedure Lin succeeded in proving Theorem 1. But it is still not clear whether there is obstruction for the local isometric embedding in smooth category even for the nonpositive curvature metric. Some results [EG] on linear degenerate hyperbolic operators of second order which have no local solvability should be noticed. Anyway, to author's knowledge the problem for local isometric embedding of surfaces into R^3 is still open !!

Global isometric embedding. The first result on global isometric embedding of complete surfaces in R^3 is due to Weyl and Lewy for analytic metric and to Nirenberg and Pogorelov for smooth metric.

Theorem. (Weyl-Lewy Nirenberg-Pogorelov) *Any analytic (smooth) positive curvature metric defined on S^2 always admits an analytic (a smooth) isometric embedding in R^3.*

For noncompact case, for example, a complete smooth positive curvature metric defined on R^2, this problem was solved by two Russian mathematicians. Olovjanisnikov first found the weak isometric embedding based on the Aleksandrov's theory on convex surfaces and Pogorelov proved the weak solution smooth if the metric smooth.

Theorem. (Olovjanisnikov-Pogorelov) *Any smooth complete positive curvature metric defined on R^2 admits a smooth isometric embedding in R^3.*

The next natural development is to consider the realization in R^3 of nonnegatively curved surfaces. Recently [GL] and [HZ] independently obtained the following result

Theorem 2. (Guan-Li, Hong-Zuily) *Any C^4 nonnegative curvature metric defined on S^2 always admits a $C^{1.1}$ isometric embedding in R^3.*

Olovjanisnikov-Pogorelov's result on complete positively curved plane is also extended to the nonnegatively curved case, see [HO3].

Theorem 3. *Any complete C^4 nonnegative curvature metric defined on R^2 always admits a $C^{1.1}$ isometric embedding in R^3. Moreover it is smooth where the metric is smooth and the curvature positive.*

In this direction a special case is also obtained in [AM].

As far as the regularity of isometric embedding be concerned, ones are interested in the following question. Can we improve the regularity of the isometric embedding obtained in Theorem 2 and Theorem 3 if the metric is smooth ? It is very interesting that [IA] gives a $C^{2.1}$ convex surface which is not C^3 continuous but realizes analytic metric on S^2 with positive curvature except one point. On the other hand, Pogorelov gave a $C^{2,1}$ geodesic disk with nonnegative curvature not even admitting a C^2 local isometric embedding in R^3 at the center of this disk. Therefore a natural open question is : Does there exist a $C^{2 \cdot \alpha}$ $(0 < \alpha < 1)$ isometric embedding in R^3 for any sufficiently smooth (even analytic) nonnegatively curved sphere or plane?

The Hilbert theorem is one of the most important theorems in 3- Euclidean space. This theorem as well as Efimov's generalization in [EF1] provide a negative answer for the problem of realization of complete negatively curved surfaces into R^3.

Theorem. (Hilbert-Efimov) *Any complete surface with negative constant curvature (with curvature bounded above by a negative constant) has no C^2 isometric immersion in R^3.*

Another result [EF2] also due to Efimov should be mentioned.

Theorem. (Efimov) *Let M be a smooth complete negatively curved surface with curvature K subject to*

$$\sup_M |K|, \ \sup_M grad(\frac{1}{\sqrt{|K|}}) \leq C \tag{7}$$

for some constant C. Then M has no C^2 isometric immersion in R^3

Evidently, Efimov's second result yields a necessary condition for a complete negatively curved surface to embed isometrically in R^3,

$$\sup_M |\nabla \frac{1}{\sqrt{|K|}}| = \infty \text{ if } \sup_M |K| < \infty. \tag{8}$$

Yau posed the following question [Y1, No.57].
Find a nontrivial sufficient condition for a complete negatively
curved surface to embed isometrically in R^3.
He also pointed out that such a nontrivial condition might be the rate of decay of the curvature at infinity. Recently some development in this direction has been made in [HO2]. Let M be a simply connected noncompact complete surface of negative curvature K. By the Hadamard theorem $exp.T_p(M) \longrightarrow M$ is a global diffeomorphism for each point $p \in M$ which induces a global geodesic polar coordinates (ρ, θ) on M centered at p.

Theorem 4. *Suppose that*
(a) for some $\delta > 0$, $\rho^{2+\delta}|K|$ is decreasing in ρ outside a compact set and that
(b) $\partial_\theta^i \ln |K|$, $i = 1, 2$ and $\rho \partial_\rho \partial_\theta \ln |K|$ bounded on M.
Then M admits a smooth isometric immersion in R^3.

Remark 1. If $M \in C^{s.1}(s \geq 4)$ and other assumptions in Theorem 4 are fulfilled, then it admits a $C^{s-1.1}$ isometric immersion in R^3.

Remark 2. The assumption (a) implies the rate of decay of the curvature at the infinity

$$|K| \le \frac{A}{\rho^{2+\delta}}, \text{ for a positive constant } A. \tag{9}$$

Such a condition on the decay of the curvature at the infinity is nearly sharp for the existence since if $\delta = 0$, there might be no existence. Consider a radius symmetric surface (R^2, g) with the Gaussian curvature

$$K = -\frac{A}{1 + \rho^2} \text{ for some positive constant } A \tag{10}$$

where ρ is the distance function from some point. Evidently

$$\sup_{R^2} |\nabla \frac{1}{\sqrt{|K|}}| = \frac{1}{\sqrt{A}} < \infty.$$

Therefore Efimov's second theorem tells us that such a complete negatively curved surface (R^2, g) has no any C^2 isometric immersion in R^3 for arbitrary positive constant A. The arguments of [EF1] and [EF2] are very genuine but ones expect a more analysis proof for Efimov's results in [EF1]. Anyway, a result in [EF3] also by Efimov which is much easier understood, shows that the negatively curved surface mentioned above has no C^2 isometric immersion if $A > 3$ in (10).

Boundary value problems for isometric embedding in R^3. Recently, the study on boundary value problems for isometric embedding of surface in R^3 has attracted much attention of mathematicians. Various formulations of such questions can be found in [Y4]. To author's knowledge this field has not been extensively studied. Let g be a smooth positive curvature metric defined on the closed unit disk $\bar{D}$. Throughout the present paper we always call such surfaces $(\bar{D}, g)$ positive disk. According to the classification in [Y4] there are two kinds of boundary value problems: one is Dirichlet problem and another is Neumann problem. Let us first consider the Dirichlet problem.

Given a smooth positive disk $(\bar{D}, g)$ and a complete smooth surface $\Sigma \subset R^3$, can we find an isometric embedding $\vec{r}$ of $(\bar{D}, g)$ in R^3 such that $\vec{r}(\partial D) \subset \Sigma$?

This problem is also raised in [PO1]. As a first step, one can consider a simple case.

Assume that Σ is a plane. Give a complete description of all isometric embedding of $(\bar{D}, g)$ satisfying $\vec{r}(\partial D) \subset \Sigma$.

Assume that $\Sigma : \{z = 0\}$. Then we are faced with the following boundary value problem.

D: To find an isometric embedding $\vec{r} = (x, y, z)$ of the given positive disk $(\bar{D}, g)$ such that $z(\partial D) = 0$.

It is easy to see that there is some obstruction for the existence of solutions to the above boundary value problem. Suppose that $\vec{r} = (x, y, z) \in C^2(\bar{D})$ is a solution to the above boundary value problem. Obviously the intersection $\vec{r}(\partial D)$ of $\vec{r}$ and the plane $\{z = 0\}$ is a C^2 planar convex curve. Denoting its curvature by $\tilde{k}$ we have

$$\tilde{k}^2 = k_g^2 + k_n^2 \tag{11}$$

where k_g and k_n are respectively the geodesic curvature and normal curvature of $\vec{r}(\partial D)$. k_n is positive everywhere since $(\bar{D}, g)$ has positive curvature. Notice that the total curvature of $\vec{r}(\partial D)$, as a planar curve, equals 2π. Hence

$$\int_{\vec{r}(\partial D)} |k_g| ds < 2\pi. \tag{12}$$

This is a necessary condition in order that the above boundary value problem D can be solvable. Indeed, this necessary condition is not sufficient for the solvability. In [HO4] there is a smooth positive disk satisfying (12) but not admitting any C^2 solution to the problem D. Furthermore this counter example also shows that too many changes of the sign of the geodesic curvature of the boundary will make the problem D unsolvable. So we distinguish two cases

$$\text{Case a: } k_g > 0 \text{ on } \partial D \text{ and Case b: } k_g < 0 \text{ on } \partial D. \tag{13}$$

It should be pointed out that Pogorelov gave the first solution to the problem D in [PO1] which states that

Theorem. (Pogorelov) *Let $(\bar{D}, g)$ be a smooth positive disk. Then the boundary value problem admits a solution $\vec{r} \in C^\infty(D) \cap C^{0.1}(\bar{D})$ provided that the geodesic curvature of ∂D with respect to the metric g is nonnegative.*

Pogorelov only obtained a local smooth solution, namely, a solution smooth inside. One wonder that under what conditions the problem D always admits a global smooth solution, namely, a solution smooth up to the boundary. Such global smooth solutions are obtained by [DE] for Case a if there exists a global C^2 subsolution ψ for the Darboux equation (5) in the unit disk D, vanishing on ∂D and with $|\nabla \psi|$ strictly less than 1 on $\bar{D}$. Recently, [HO5] removes this technique requirement.

Theorem 5. *For Case a the problem D always admits a unique solution in $C^\infty(\bar{D})$ if any one of the following assumptions is satisfied (1) $K > 0$ on $\bar{D}$, (2) $K > 0$ in D and $K = 0 \neq |dK|$ on ∂D.*

Obviously, the necessary condition (12) is always satisfied for Case a. As for Case b it looks rather complicated since there are some smooth convex surfaces in R^3 which are of no infinitesimal rigidity. The presence of such convex surfaces makes us fail to prove the existence of the problem D of Case b by means of the standard method of continuity. Thus the solvability of the problem D for Case b is still open !

Let us consider a spherical crown, in the spherical coordinates,

$$\Sigma_{\theta_*} = \{(\sin\theta\cos\phi, \sin\theta\sin\phi, -\cos\theta)|0 \le \phi \le 2\pi, 0 \le \theta \le \theta_*\}$$

and $\theta = 0$ stands for the South pole. Σ_{θ_*} is the isometric embedding of the metric $g = d\theta^2 + \sin^2\theta d\phi^2, 0 \le \theta \le \theta_*$. If $\theta_* > \frac{\pi}{2}$, then Σ_{θ_*} contains the below hemisphere and the geodesic curvature of its boundary is negative. We have in [HO4]

Theorem 6. *There is a countable set* $\Lambda = \{\theta_1, \theta_2, ...\theta_n, ...\} \subset (\pi/2, \pi)$ *with a limit point* $\pi/2$ *such that* Σ_{θ_*} *is not infinitesimally rigid if* $\theta_* \in \Lambda$.

In what follows we proceed to discuss the Neumann problem for realization of surfaces into R^3. The formulation is as follows. Give a smooth positive disk $(\bar{D}, g)$ and a positive function $h \in C^\infty(\partial D)$, The Neumann problem (later, called the problem N) is as follows.

$$N : \text{ Find a surface } \vec{r} : \bar{D} \mapsto R^3 \text{ such that } d\vec{r}^2 = g$$

with the prescribed mean curvature h on $\vec{r}(\partial D)$.   (14)

Let us first introduce an invariant related to umbilical points of surfaces in R^3. Suppose that the given metric is of the form

$$g = Edx^2 + 2Fdxdy + Gdy^2, (x, y) \in \bar{D}.$$ (15)

Let $\vec{r}$ be a smooth isometric immersion of $(\bar{D}, g)$ with the second fundamental form

$$II = Ldx^2 + 2Mdxdy + Ndy^2 \text{ for } (x, y) \in \bar{D}.$$ (16)

Definition. *If* $\vec{r}$ *is of no umbilical points on* ∂D, *with*

$$\sigma = (EM - FL) + \sqrt{-1}(GL - EN)$$

the winding number of σ *on* ∂D *is called the index of the umbilical points of the surface* $\vec{r}$ *and denoted by* $Index(\vec{r})$.

Obviously, this definition makes sense since p is an umbilical point if and only if $\sigma(p) = 0$. Moreover, the definition of the index of the umbilical points is coordinate-free and hence, an invariant of describing umbilical points of surfaces. Such invariance comes from that of a differential form. Indeed, assume that in some orthonomal frame, the induced metric and the second fundamental form of a given surface $\vec{r}$ in R^3 are of the form

$$g = \omega_1^2 + \omega_2^2 \text{ and } II = h_{ij}\omega_i\omega_j$$

respectively. As is well known,

$$[(h_{11} - h_{22}) + 2ih_{12}](\omega_1^2 + \omega_2^2)$$

is an invariant differential form. So the index of umbilical points is nothing else but

$$Index(\vec{r}) = Index\,\{2h_{12} + i(h_{11} - h_{22})\}$$

if no umbilical point on ∂D occurs.

The boundary value problem for realization of positive disks into R^3 seems to have some obstruction. Indeed, even if no imposing any restriction on the boundary the problem of realization of positive disks into R^3 is not always solvable. For details, refer to Gromov's counter example [GR] which contains an analytic positive disk not admitting any C^2 isometric immersion in R^3. Therefore for the Neumann boundary value problem the following hypothesis is natural. Assume that

$$(\bar{D}, g) \text{ admits a } C^2 \text{ isometric immersion } \vec{r}_0 \text{ in } R^3. \tag{17}$$

We have in [HO6]

Theorem 7. *If $(\bar{D}, g)$ is a smooth positive disk satisfying (17) then for any non-negative integer n and arbitrary $(n+1)$ distinct points $p_0 \in \partial D$, $p_1, ..., p_n \in D$, the problem N admits two and only two solutions $\vec{r}$ in $C^\infty(\bar{D}, R^3)$ with prescribed mean curvature h on ∂D and moreover,*

one principal direction at p_0 is tangent to ∂D,

$$Index(\vec{r}) = n \text{ and } H(p_k) = H_0(p_k), k = 1, ..., n$$

where H and H_0 are respectively the mean curvature of $\vec{r}$ and $\vec{r}_0$ provided that

$$\frac{h}{\sqrt{K}} - 1 > 4 \max_{\partial D}\left[\frac{H_0}{\sqrt{K}} - 1\right] \quad on \ \partial D. \tag{18}$$

It is worth pointing out two extreme cases. The first one involves the existence. Suppose that the given positive disk $(\bar{D}, g)$ is of positive constant curvature. Then it is easy to see that this positive disk admits a priori smooth isometric embedding $\vec{r}_0$ in R^3 which is a simply connected region of the sphere. Under the present circumstance $\vec{r}_0$ is totally umbilical and hence, the right hand side of (18) vanishes. Therefore if $(\bar{D}, g)$ is of constant curvature and $\sqrt{K} < h \in C^\infty(\partial D)$, then the problem N is always solvable for each nonnegative integer n and arbitrary $(n+1)$ distinct points $p_0 \in \partial D, p_1, , ..., p_n \in D$.

The second extreme case involves the nonexistence. If the given positive disk is radius symmetric, i.e., $g = dr^2 + G^2(r)d\theta^2$ $0 \leq r \leq 1$ where $G \in C^\infty([0,1])$ and $G(0) = 0$, $G'(0) = 1$, $G > 0$ as $r > 0$. Then if $G_r > -1$, $(\bar{D}, g)$ has such a priori smooth isometric embedding in R^3,

$$\vec{r}_0 : x = G(r)\cos\theta, y = G(r)\sin\theta, z = -\int_r^1 \sqrt{1 - G_r^2}dr. \tag{19}$$

With its mean curvature $H_0 = H_0(r)$, if $H_0(1) > \sqrt{K(1)}$, then [HO6] proves that for arbitrary $h \in C^\infty(\partial D)$ satisfying $\sqrt{K(1)} \leq h < H_0(1)$ the problem N has no any C^2 solution. Of course, if $h > 4H_0(1) - 3\sqrt{K(1)}$, by Theorem 7 the problem N always admits two and only two smooth solutions for any nonnegative integer n and arbitrary $n+1$ points $p_0 \in \partial D$, $p_1, ..., p_j \in D$.

References

[AL] Aleksandrov, A.D., Intrinsic geometry of convex surfaces, Moscow, 1948, German transl., Academie Verlag Berlin, 1955.

[AM] Amano, K., Global isometric embedding of a Riemannian 2-manifold with nonnegative curvature into a Euclidean 3-space, J.Diff.Geometry, **35**(1991), 49–83.

[DE] Delanoe, Ph., Relations globalement regulieres de disques strictement convexes dans les espaces d'Euclide et de Minkowski par la methode de Weingarten, Ann.Sci.Ec Norm.Super.,IV. **21**(1988), 637–652.

[EF1] Efimov, N.V., Generalization of singularities on surfaces of negative curvature, Mat. Sb., 64(1964), 286–320.

[EF2] Efimov, N.V., A Criterion of homemorphism for some mappings and its applications to the theory of surfaces, Mat. Sb., 76(1968), 499–512.

[EF3] Efimov, N.V., Surfaces with slowly varying negative curvature, Russ. Math. Survey, 21(1966), 1–55.

[EG] Egorov, Y., Sur un exemple d'equation lineaire hyperbolique du second ordre n'ayant pas de solution, Journees Equations aux Derivees Partielles, Saint-Jean-de-Monts, 1992.

[G] Gromov. M., Partial differential relations, Springer-Verlage, Berlin Heidelberg, 1986.

[GR] Gromov, M. and Rohklin, V.A., Embeddings and Immersions in Riemannian Geometry, Russian Mathematical Surveys, 25(1970), 1–57.

[GUN] Gunther, M., Isometric embeddings of Riemannian manifolds, Proceeding of the International Congress of Mathematicians, Kyoto, Japan, 1990.

[GL] Guan, P. and Li, Y., The weyl problem with nonnegative Gauss curvature, J. Diff. Geometry, 39(1994), 331–342.

[H] Hormander, L., On the Nash-Moser implicit function theorem, Ann. Acad.Sci of Fenn., 10, 255–259.

[HI] Hilbert, D., Uber Plachen von konstanter Gausscher krummung, Trans. Amer. Math. Soc., 2(1901),87–99.

[HO1] Hong, J.X., Cauchy problem for degenerate hyperbolic Monge-Ampere equations, J.Partial Diff. Equations, 4(1991), 1–18.

[HO2] Hong, J.X., Realization in R^3 of complete Riemannian manifolds with negative curvature, Communications Anal. Geom., 1(1993), 487–514.

[HO3] Hong, J.X., Isometric embedding in R^3 of complete noncompact nonnegatively curved surfaces, Manuscripta Math., 94(1997), 271–286.

[HO4] Hong, J.X., Recent developments of realization of surfaces in R^3,AMS/IP, Studies in Advanced Mathematics, Vol.20, (2001), 47–62.

[HO5] Hong, J.X., Darboux equations and isometric embedding of Riemannian manifolds with nonnegative curvature in R^3, Chin. Ann. of math., 2(1999), 123–136.

[HO6] Hong, J.X., Positive disks with prescribed mean curvature on the boundary,

Asian J. Math., Vol.5 (2001), 473-492.

[HZ] Hong, J.X. and Zuily, C., Isometric embedding of the 2-sphere with non negative curvature in R^3, Math.Z., 219(1995), 323–334.

[IA] Iaia, J.A., Isometric embedding of surfaces with nonnegative curvature in R^3, Duke Math. J., 67(1992) 423–459.

[IW] Iwasaki, N., The stronger hyperbolic equation and its applications, Proc. Sym. Pure Math., 45(1986), 525–528.

[K] Kuiper, N.H., On C^1-isometric embeddings, I, II, Indag. Math., (1955), 545–556, 683–689.

[LC1] Lin, C.S., The local isometric embedding in R^3 of 2-dimensional Riemannian manifolds with nonnegative curvature, J. Diff. Geometry, 21(1985), 213–230.

[LC2] Lin, C.S., The local isometric embedding in R^3 of two dimensional Riemannian manifolds with Gaussian curvature changing sign clearly, Comm. Pure Appl.Math., 39(1986), 307–326.

[NI1] Nirenberg, L., The Weyl and Minkowski problems in Differential Geometry in the large, Comm. Pure Appl.Math., 6(1953), 337–394.

[NI2] Nirenberg, L., In Nonlinear problems, editor by R.E.Langer, University of Wisconsin Press, Madison, 1963, 177–193.

[P] Poznjak, E., Isometric immersion of 2-dimensional metrics into Euclidean spaces, Uspechy 28(1970), 47–76.

[PO1] Pogorelov, A.V., Extrinsic geometry of convex surfaces (transl. Math. Monogr. Vol 35), Providence, RI, Am.Math.Soc., 1973.

[PO2] Pogorelov, A.V., An example of a two dimensional Riemannian metric not admitting a local realization in R^3, Dokl. Akad. Nauk. USSR., 198(1971), 42–43.

[Y1] Yau, S.T., Problem Section, Seminar on Differential Geometry, Princeton University Press, 1982.

[Y2] Yau, S.T., Open problems in Geometry, Chern-A great Geometer on the Twentieth Century, International Press, 1992.

[Y3] Yau, S.T., Lecture on Differential Geometry, in Berkeley, 1977.

[Y4] Yau, S.T., Review on Geometry and Analysis, Asian J. of Math., 4(2000), 235–278.

ICM 2002 · Vol. III · 167–176

p-Laplacian Type Equations Involving Measures

T. Kilpeläinen*

Abstract

This is a survey on problems involving equations $-\operatorname{div}\mathcal{A}(x, \nabla u) = \mu$, where μ is a Radon measure and $\mathcal{A}: \mathbf{R}^n \times \mathbf{R}^n \to \mathbf{R}^n$ verifies Leray-Lions type conditions. We shall discuss a potential theoretic approach when the measure is nonnegative. Existence and uniqueness, and different concepts of solutions are discussed for general signed measures.

2000 Mathematics Subject Classification: 35J60, 31C45.
Keywords and Phrases: p-Laplacian, Quasilinear equations with measures, Nonlinear potential theory, Uniqueness.

1. Introduction

Throughout this paper we let Ω be an open set in $\mathbf{R}^n$ and $1 < p < \infty$ a fixed number. We shall consider equations

$$-\operatorname{div}\mathcal{A}(x, \nabla u) = \mu, \tag{1.1}$$

where μ is a Radon measure. We suppose that the mapping $\mathcal{A}: \mathbf{R}^n \times \mathbf{R}^n \to \mathbf{R}^n$, $(x, \xi) \mapsto \mathcal{A}(x, \xi)$, is measurable in x and continuous in ξ and that it verifies the structural conditions:

$$\mathcal{A}(x, \xi) \cdot \xi \geq \lambda |\xi|^p, \quad |\mathcal{A}(x, \xi)| \leq \Lambda |\xi|^{p-1}, \quad \text{and}$$
$$(\mathcal{A}(x, \xi) - \mathcal{A}(x, \zeta)) \cdot (\xi - \zeta) > 0, \tag{1.2}$$

for a.e. $x \in \mathbf{R}^n$ and all $\xi \neq \zeta \in \mathbf{R}^n$. A prime example of the operators is the p-Laplacian

$$-\Delta_p u = -\operatorname{div}(|\nabla u|^{p-2}\nabla u).$$

In Section 2 we discuss how nonlinear potential theory is related to equations like (1.1); it corresponds to nonnegative measures. Then in Section 3 we discuss the existence and uniqueness for (1.1) with general measures.

---

*Department of Mathematics & Statistics, University of Jyväskylä, PO Box 35, FIN-40014 Jyväskylä, Finland. E-mail: terok@math.jyu.fi

2. Potential theoretic approach

A continuous solution $u \in W^{1,p}_{\text{loc}}(\Omega)$ of $-\operatorname{div} \mathcal{A}(x, \nabla u) = 0$ is called $\mathcal{A}$-*harmonic* in Ω. An $\mathcal{A}$-*superharmonic* function in Ω is a lower semicontinuous function $u \colon \Omega \to \mathbf{R} \cup \{\infty\}$ that is not identically ∞ in any component of Ω and that obeys the following comparison property: *for each open $D \subset\subset \Omega$ and each $h \in C(\bar{D})$, $\mathcal{A}$-harmonic in D, the inequality $u \geq h$ on ∂D implies $u \geq h$ in D.*

For $k > 0$ and $s \in \mathbf{R}$, let

$$T_k(s) = \max\left(-k, \min(s, k)\right)$$

be the truncation operator. Then we have

2.1. Theorem. [12, 26, 15] *If u is $\mathcal{A}$-superharmonic in Ω, then*

$$T_k(u) \in W^{1,p}_{\text{loc}}(\Omega) \quad \text{for all } k > 0\,.$$

This enables us to show that $\mathcal{A}$-superharmonic functions have a "gradient": Suppose that a function u that is finite a.e. has the property that $T_k(u) \in W^{1,p}_{\text{loc}}(\Omega)$ for all $k > 0$. Then we define the (weak) *gradient* of u as

$$\nabla u(x) = DT_k(u) \quad \text{if } |u(x)| < k\,.$$

Here $DT_k(u)$ is the distributional gradient of the Sobolev function $T_k(u) \in W^{1,p}_{\text{loc}}(\Omega)$. Then ∇u is well defined. Observe that if ∇u is locally integrable, then it is the distributional derivative of u. However, it may happen that u or ∇u fails to be locally integrable and so ∇u is not always distributional derivative, see [15], [8]; this is a real issue only for $p < 2 - 1/n$.

2.2. Theorem. [26, 12, 15, 10, 1] *Suppose that u is $\mathcal{A}$-superharmonic in Ω.*

i) *If $1 < p < n$, then*

$$u \in \text{weak} - L^{n(p-1)/(n-p)}_{\text{loc}}(\Omega) \quad and \quad \nabla u \in \text{weak} - L^{n(p-1)/(n-1)}_{\text{loc}}(\Omega).$$

ii) *If $p = n$, then u is locally in BMO and hence $u \in L^q_{\text{loc}}(\Omega)$ for all $q > 0$; moreover*

$$\nabla u \in \text{weak} - L^n_{\text{loc}}(\Omega).$$

iii) *If $p > n$, then*

$$u \in W^{1,p}_{\text{loc}}(\Omega)$$

and hence u is (Hölder) continuous.

If $p > n$ $\mathcal{A}$-superharmonic functions are continuous and locally in $W^{1,p}$, moreover Radon measures then are in the dual of Sobolev space $W^{1,p}$. This makes the cases $p > n$ very special and quite simple by the classical results of Leray and Lions. Henceforth we shall be concerned mainly with cases $1 < p \le n$.

For $\mathcal{A}$-superharmonic u we have by Theorem 2.2 that $|\nabla u|^{p-1}$ is locally integrable. So the distribution

$$- \operatorname{div} \mathcal{A}(x, \nabla u)(\varphi) := \int_\Omega \mathcal{A}(x, \nabla u) \cdot \nabla \varphi \, dx \,, \quad \varphi \in C_0^\infty(\Omega) \,,$$

is well defined. $\mathcal{A}$-superharmonic functions give rise to equation (1.1).

2.3. Theorem. [18] *If u is $\mathcal{A}$-superharmonic in Ω, then $- \operatorname{div} \mathcal{A}(x, \nabla u)$ is represented by a nonnegative Radon measure μ.*

As to the existence we have:

2.4. Theorem. [18, 3] *Given a finite nonnegative Radon measure μ on bounded Ω, there is an $\mathcal{A}$-superharmonic function u in Ω such that*

$$\begin{cases} - \operatorname{div} \mathcal{A}(x, \nabla u) = \mu \\ T_k(u) \in W_0^{1,p}(\Omega) \text{ for all } k > 0 \,. \end{cases} \tag{2.5}$$

If Ω is undounded, then there also is an $\mathcal{A}$-superharmonic solution to (1.1). This is rather easily seen if $1 < p < n$. The case $p \ge n$ requires a more careful analysis, see [9, 17].

In light of Theorem 2.2 we have that a solution u to (2.5) satisfies

$$u \in W_0^{1,q}(\Omega) \quad \text{for all } q < \frac{n(p-1)}{n-1}$$

(for $p > 2 - 1/n$). One naturally asks if such a solution is unique. Unfortunately this is not the case in general (except for $p > n$). To see this consider the function

$$u(x) = \begin{cases} |x|^{\frac{p-n}{p-1}} - 1 & \text{if } 1 < p < n, \\ -\log |x| & \text{if } p = n \,. \end{cases}$$

In the p-Laplacian cases, u is then a p-superharmonic solution to (2.5) with $\mu=0$ in the punctured ball $\Omega = B(0,1) \setminus \{0\}$, but $v \equiv 0$ is another solution. This is rather artificial example but there are more severe ones. The question of uniqueness is a real issue to which we shall return in Section 3 below.

Regularity and estimates

In the classical potential theory the uniqueness can be solved by the aid of the Riesz decomposition theorem which states that superharmonic functions are sums of a potential and a harmonic function. No such decomposition is available in the nonlinear world. However, this lack can be compensated for to an extent by estimating in terms of the *Wolff potential* of the measure μ,

$$\mathbf{W}_\mu^{1,p}(x_0, r) = \int_0^r \left(\frac{\mu(B(x_0, t))}{t^{n-p}} \right)^{\frac{1}{p-1}} \frac{dt}{t} \,.$$

2.6. Theorem. [19] *Let u be a nonnegative $\mathcal{A}$-superharmonic function in $B(x_0, 3r)$ with $\mu = -\operatorname{div}\mathcal{A}(x, \nabla u)$. Then*

$$c_1 \mathbf{W}_\mu^{1,p}(x_0, r) \leq u(x_0) \leq c_2 \mathbf{W}_\mu^{1,p}(x_0, 2r) + c_3 \inf_{B(x_0,r)} u\,,$$

where $c_j = c_j(n, p, \lambda, \Lambda) > 0$.

Theorem 2.6 was discovered by the author with Malý [18, 19] and later generalized for equations depending also on u by Malý and Ziemer [30]. Mikkonen [32] worked out the argument for weighted operators, this was later written up in a metric space setup in [2]. Recently a totally different proof that works for quasilinear subelliptic operators was found by Trudinger and Wang [39]. Labutin [22] gave a generalization for k-Hessian operators.

As the first major application of the potential estimate in 2.6 the author and Malý established the necessity of the Wiener test for the regularity for the Dirichlet problem: We say that $x_0 \in \partial\Omega$ is an $\mathcal{A}$-*regular* boundary point of bounded Ω if

$$\lim_{x \to x_0} u(x) = \varphi(x_0)$$

whenever $\varphi \in C^\infty(\mathbf{R}^n)$ and u is $\mathcal{A}$-harmonic in Ω with $u - \varphi \in W_0^{1,p}(\Omega)$. Then it turns out that regularity is independent of the particular operator and it depends only on its type p. More precisely, define the p-*capacity* of the set E as

$$\operatorname{cap}_p(E) := \inf \int_{\mathbf{R}^n} |v|^p + |\nabla v|^p \, dx\,,$$

where the infimum is taken over all $v \in W^{1,p}(\mathbf{R}^n)$ such that $v \geq 1$ on an open neighborhood of E. Then

2.7. Theorem. [31, 19] *A boundary point $x_0 \in \partial\Omega$ is regular if and only if*

$$\int_0^1 \left(\frac{\operatorname{cap}_p(\complement\Omega \cap B(x_0, t))}{t^{n-p}} \right)^{\frac{1}{p-1}} \frac{dt}{t} = \infty\,.$$

Maz'ya [31] introduced the Wiener type test in 2.7 and proved its sufficiency. Gariepy and Ziemer [11] generalized the result by a different argument. Lindqvist and Martio [27] proved the necessity for $p > n - 1$; the general case was treated in [19]. For generalizations see [30, 32, 39, 22].

Theorem 2.6 can also be used to characterize singular solutions and (Hölder) continuity of $\mathcal{A}$-superharmonic functions (see [17, 19, 21]). Recall that $\mathcal{A}$-harmonic functions are locally Hölder continuous: there is a constant $\varkappa \in (0, 1]$ depending only on the structure such that

$$\operatorname{osc}(u, B(x, r)) \leq c(\frac{r}{R})^\varkappa \operatorname{osc}(u, B(x, R)) \tag{2.8}$$

whenever u is $\mathcal{A}$-harmonic in $B(x,R)$, $r < R$ [37, 15]. For the p-Laplacian $\varkappa = 1$. We have

2.9. Theorem. [21] *Suppose that $u \in W^{1,p}_{\text{loc}}(\Omega)$ is $\mathcal{A}$-superharmonic with $\mu = -\operatorname{div}\mathcal{A}(x,\nabla u)$. If u is α-Hölder continuous, then*

$$\mu\big(B(x,r)\big) \leq c\,r^{n-p+\alpha(p-1)} \qquad whenever \qquad B(x,2r) \subset \Omega.$$

The converse holds if $\alpha < \varkappa$.

The case $\alpha = \varkappa$ is quite different and it is not yet well understood.

Rakotoson and Ziemer [36] proved that the condition of Theorem 2.9 for the measure gives the Hölder continuity of the solution with some exponent. Lieberman [25] showed that for smooth operators like the p-Laplacian the condition $\mu\big(B(x,r)\big) \leq c\,r^{n-1+\varepsilon}$ for some $\varepsilon > 0$ implies that the solution is in $C^{1,\beta}$.

Theorem 2.9 can be employed to establish the following removability result which is due to Carleson [5] in the Laplacian case.

2.10. Theorem. [21] *Suppose that $u \in C^{0,\alpha}(\Omega)$ is $\mathcal{A}$-harmonic in $\Omega \setminus E$. If E is of $n-p+\alpha(p-1)$ Hausdorff measure zero, then u is $\mathcal{A}$-harmonic in Ω.*

If E is of positive $n-p+\alpha(p-1)$ Hausdorff measure and $\alpha < \varkappa$, then there is $u \in C^{0,\alpha}(\Omega)$ that is $\mathcal{A}$-harmonic in $\Omega \setminus E$ but not in the whole Ω.

3. General Radon measures

In this section we let μ be any signed Radon measure and consider equation (1.1). More specifically, we shall discuss the problem

$$\begin{cases} -\operatorname{div}\mathcal{A}(x,\nabla u) = \mu \\ u = 0 \text{ on } \partial\Omega \end{cases} \tag{3.1}$$

where Ω is a bounded domain in $\mathbf{R}^n$. Here the equation is understood in the distributional sense, i.e.

$$\int_\Omega \mathcal{A}(x,\nabla u) \cdot \nabla\varphi\,dx = \int_\Omega \varphi\,d\mu\,, \qquad \varphi \in C^\infty_0(\Omega),$$

where we, of course, assume that $x \mapsto \mathcal{A}(x,\nabla u)$ is locally integrable. The boundary values $u = 0$ are assumed in a weak Sobolev space sense. The existence of the solution to this problem is known:

3.2. Theorem. [3, 8, 9, 10] *For each Radon measure μ of finite total variation, there is a solution u to (3.1) such that*

i) *the truncations*

$$T_k(u) \in W^{1,p}_0(\Omega) \text{ for all } k > 0\,,$$

ii)

$$u \in \text{weak} - L^{n(p-1)/(n-p)}(\Omega) \text{ if } 1 < p < n \qquad and \qquad u \in \text{BMO} \text{ if } p = n\,,$$

iii)

$$\nabla u \in \text{weak} - L^{n(p-1)/(n-1)}(\Omega).$$

We next discuss the uniqueness of such a solution. There are examples of mappings $\mathcal{A}$ for which there is a solution u of equation $\mathcal{A}(x, \nabla u) = 0$ such that u satisfies ii) and iii) of Theorem 3.2, but fails to be $\mathcal{A}$-harmonic, see [33, 38, 16, 29]. See also the nonuniqueness example in an irregular domain after Theorem 2.4 above.

There are various approaches trying to treat the uniqueness problem by attaching additional attributes to the solution. To formulate these we need to recall a decomposition of measures. The p-capacity, defined in Section 2, is an outer measure. Hence the usual proof of the Lebesgue decomposition theorem gives us that any Radon measure μ can be decomposed as

$$\mu = \mu_0 + \mu_s,$$

where μ_0 and μ_s are Radon measures such that μ_0 is *absolutely continuous* with respect to the p-capacity (i.e., $\mu_0(E) = 0$ whenever $\operatorname{cap}_p(E) = 0$) and μ_s is *singular* with respect to the p-capacity (i.e., there is a Borel set B such that $\operatorname{cap}_p(B) = 0$ and $\mu_s(E \setminus B) = 0$ for all E).

Let u be a solution to (3.1) described in Theorem 3.2. We say that

- u is an *entropy solution* of (3.1) if u is Borel measurable and

$$\int_\Omega \mathcal{A}(x, \nabla u) \cdot \nabla T_k(u - \varphi)\, dx \leq \int_\Omega T_k(u - \varphi)\, d\mu$$

for all $\varphi \in C_0^\infty(\Omega)$ and $k > 0$.

- u is a *renormalized solution* of (3.1) if for all $h \in W^{1,\infty}(\mathbf{R})$ such that h' has compact support we have

$$\int_\Omega \mathcal{A}(x, \nabla u) \cdot \nabla u h'(u)\varphi\, dx + \int_\Omega \mathcal{A}(x, \nabla u) \cdot h(u)\nabla\varphi\, dx$$

$$= \int_\Omega h(u)\varphi\, d\mu_0 + h(\infty) \int_\Omega \varphi\, d\mu_s^+ - h(-\infty) \int_\Omega \varphi\, d\mu_s^-$$

whenever $\varphi \in W^{1,r}(\Omega) \cap L^\infty(\Omega)$ with $r > n$ is such that $h(u)\varphi \in W_0^{1,p}(\Omega)$; here

$$h(\infty) = \lim_{t\to\infty} h(t), \quad h(-\infty) = \lim_{t\to-\infty} h(t),$$

and μ_s^+ and μ_s^- are the positive and negative parts of the singular measure μ_s.

Observe that we assume here that u satisfies equation (3.1) in the distributional sense. The entropy condition in this context was first used in [1]; the renormalized solution was introduced by Lions and Murat [28] and in the refined form in [7, 8]. The artificial function we had as a counterexample for the uniqueness (after Theorem 2.4) is not entropy nor renormalized solution. Existence is known; most existence proofs follow a similar idea to that in [3]. The uniqueness can be established for certain measures since one can use truncated solutions as test functions.

3.3. Theorem. [4, 20, 8, 40] *Suppose that μ is a finite Radon measure. Then there is a renormalized and an entropy solution of (3.1). Moreover such a solution is unique if μ is absolutely continuous with respect to the p-capacity.*

The uniqueness with $\mu \in L^1$ was proved in [1] and [28] see also [33], [34]. For measures absolutely continuous wrt p-capacity, the uniqueness is established e.g. in [4, 20, 8, 40].

A renormalized solution is always an entropy solution, whence the concepts coincide at least if μ is absolutely continuous wrt p-capacity; see [8, 6].

In case of a general measure the uniqueness of renormalized solution appears to be an open problem. There are some partial results: Assume that the following strong monotoneity assumption holds. For all $0 \neq \xi, \eta \in \mathbf{R}^n$

$$\left(\mathcal{A}(x, \xi) - \mathcal{A}(x, \eta)\right) \cdot (\xi - \eta) \geq \begin{cases} \beta|\xi - \eta|^p & \text{if } p \geq 2, \\ \beta\dfrac{|\xi - \eta|^2}{(|\xi| + |\eta|)^{2-p}} & \text{if } p < 2. \end{cases} \quad (3.4)$$

Assume also the Hölder continuity:

$$|\mathcal{A}(x, \xi) - \mathcal{A}(x, \eta)| \leq \begin{cases} \gamma\big(b(x) + |\xi| + |\eta|\big)^{p-2}|\xi - \eta|^2 & \text{if } p \geq 2 \\ \gamma|\xi - \eta|^{p-1} & \text{if } p < 2, \end{cases} \quad (3.5)$$

where $b \in L^p$ is nonnegative. For instance, the p-Laplacian satisfies these assumptions. Then we have:

3.6. Theorem. [7, 8, 14] *Suppose the additional assumptions (3.4) and (3.5) hold. If u and v are two renormalized solutions of (3.1) with measure μ such that either $\nabla u - \nabla v \in L^p(\Omega)$ or $u - v$ is bounded from one side, then $u = v$.*

Rakotoson [35] proved that a continuous renormalized solution is unique in smooth domains; continuity requires the measure be rather special.

Borderline case $p = n$

We close this paper by considering the special case when $p = n$. Then the uniqueness can be reached:

3.7. Theorem. [41, 13, 10] *Suppose that $\mathcal{A}$ verifies additional assumption (3.4) with $p = n$ and that Ω is bounded and regular. For each Radon measure μ of finite total variation, there is a unique solution u to (3.1) such that*

i) *the truncations*

$$T_k(u) \in W_0^{1,n}(\Omega) \text{ for all } k > 0,$$

ii) *u is in BMO, and*

$$\nabla u \in \text{weak} - L^n(\Omega).$$

The regularity of Ω refers to the fact that the complement of Ω needs to be thick enough to exclude counterexamples we had in Section 2. Zhong [41] formulated a weak condition for this by requiring that the complement of Ω is *uniformly p-thick*, i.e., $\mathrm{cap}_p(\complement\Omega \cap B(x,r)) \approx r^{n-p}$ for all small $r > 0$; see [23, 15, 32] for more information about uniform thickness.

In fact stronger uniqueness properties than in Theorem 3.7 hold: by using a Hodge decomposition argument Greco, Iwaniec, and Sbordone [13] proved that for p-Laplacian the solution is unique in the *grand Sobolev space* $W^{1,n)}(\Omega)$, i.e.

$$u \in \bigcap_{q<n} W_0^{1,q}(\Omega) \quad \text{and} \quad \sup_{\varepsilon>0} \varepsilon \int_\Omega |\nabla u|^{n-\varepsilon}\, dx < \infty \,.$$

The regularity $\nabla u \in \text{weak} - L^n(\Omega)$ in Theorem 3.7 (proved in [10]) is better than $u \in W^{1,n)}(\Omega)$.

By using a maximal function argument similar to that introduced by Lewis [24], Zhong [41] proved a stronger uniqueness result: the solution is unique in

$$u \in \bigcap_{q<n} W_0^{1,q}(\Omega) \,.$$

Even stronger result appeared in [10]: there is a $\varepsilon > 0$ depending on the structural assumptions and on Ω such that any solution in $W_0^{1,n-\varepsilon}(\Omega)$ is actually the unique solution declared in Theorem 3.7. A similar result follows from the estimates in [13].

References

[1] P. Bénilan, L. Boccardo, T. Gallouët, R. Gariepy, M. Pierre, & J. L. Vázquez, *An L^1-theory of existence and uniqueness of solutions of nonlinear elliptic equations*, Ann. Scuola Norm. Sup. Pisa Cl. Sci. (4) **22** (1995), 241–273.

[2] J. Björn, P. MacManus & N. Shanmugalingam, *Fat sets and pointwise boundary estimates for p-harmonic functions in metric spaces*, J. Anal. Math. **85** (2001), 339–369.

[3] L. Boccardo & T. Gallouët, *Nonlinear elliptic equations with right-hand side measures*, Comm. Partial Diff. Eq. **17** (1992), 641–655.

[4] L. Boccardo, T. Gallouët, & L. Orsina, *Existence and uniqueness of entropy solutions for nonlinear elliptic equations with measure data*, Ann. Inst. H. Poincaré Anal. Non Linéaire **13** (1996), 539–551.

[5] L. Carleson, *Selected Problems on Exceptional Sets*, Van Nostrand, 1967.

[6] G. Dal Maso & A. Malusa, *Some properties of reachable solutions of nonlinear elliptic equations with measure data*, Ann. Scuola Norm. Sup. Pisa Cl. Sci. (4) **25** (1997), 375–396.

[7] G. Dal Maso, F. Murat, L. Orsina, & A. Prignet, *Definition and existence of renormalized solutions of elliptic equations with general measure data*, C. R. Acad. Sci. Paris Sér. I Math. **325** (1997), 481–486.

[8] G. Dal Maso, F. Murat, L. Orsina, & A. Prignet, *Renormalized solutions of elliptic equations with general measure data*, Ann. Scuola Norm. Sup. Pisa Cl. Sci. (4) **28** (1999), 741–808.

[9] G. Dolzmann, N. Hungerbühler, & S. Müller, *Non-linear elliptic systems with measure-valued right hand side*, Math. Z. **226** (1997), 545–574.

[10] G. Dolzmann, N. Hungerbühler, & S. Müller, *Uniqueness and maximal regularity for nonlinear elliptic systems of n-Laplace type with measure valued right hand side*, J. Reine Angew. Math. **520** (2000), 1–35.

[11] R. Gariepy & W. P. Ziemer, *A regularity condition at the boundary for solutions of quasilinear elliptic equations*, Arch. Rat. Mech. Anal. **67** (1977), 25–39.

[12] S. Granlund, P. Lindqvist, & O. Martio, *Conformally invariant variational integrals*, Trans. Amer. Math. Soc. **277** (1983), 43–73.

[13] L. Greco, T. Iwaniec, & C. Sbordone, *Inverting the p-harmonic operator*, Manuscripta Math. **92** (1997), 249–258.

[14] O. Guibé, *Remarks on the uniqueness of comparable renormalized solutions of elliptic equations with measure data*, Ann. Mat. Pura Appl. (4) **180** (2002), 441–449.

[15] J. Heinonen, T. Kilpeläinen, & O. Martio, *Nonlinear Potential Theory of Degenerate Elliptic Equations*, Oxford University Press, Oxford, 1993.

[16] T. Kilpeläinen, *Nonlinear potential theory and PDEs*, Potential Analysis **3** (1994), 107–118.

[17] T. Kilpeläinen, *Singular solutions to p-Laplacian type equations*, Ark. Mat. **37** (1999), 275–289.

[18] T. Kilpeläinen & J. Malý, *Degenerate elliptic equations with measure data and nonlinear potentials*, Ann. Scuola Norm. Sup. Pisa Cl. Science, Ser. IV **19** (1992), 591–613.

[19] T. Kilpeläinen & J. Malý, *The Wiener test and potential estimates for quasilinear elliptic equations*, Acta Math. **172** (1994), 137–161.

[20] T. Kilpeläinen & X. Xu, *On the uniqueness problem for quasilinear elliptic equations involving measures*, Rev. Mat. Iberoamer. **12** (1996), 461–475.

[21] T. Kilpeläinen & X. Zhong, *Removable sets for continuous solutions of quasilinear elliptic equations*, Proc. Amer. Math. Soc. **130** (2002), 1681–1688.

[22] D. Labutin, *Potential estimates for a class of fully nonlinear elliptic equations*, Duke Math. J. **111** (2002), 1–49.

[23] J. L. Lewis, *Uniformly fat sets*, Trans. Amer. Math. Soc. **308** (1988), 177–196.

[24] J. L. Lewis, *On very weak solutions of certain elliptic systems*, Comm. Partial Diff. Eq. **18** (1993), 1515–1537.

[25] G. M. Lieberman, *Sharp forms of estimates for subsolutions and supersolutions of quasilinear elliptic equations involving measures*, Comm. Partial Diff. Eq. **18** (1993), 1191–1212.

[26] P. Lindqvist, *On the definition and properties of p-superharmonic functions*, J. Reine Angew. Math. **365** (1986), 67–79.

[27] P. Lindqvist & O. Martio, *Two theorems of N. Wiener for solutions of quasi-linear elliptic equations*, Acta Math. **155** (1985), 153–171.

[28] P.L. Lions & F. Murat, *Solutions renormalisées d'équations elliptiques non linéaires*, (to appear).

[29] J. Malý, *Examples of weak minimizers with continuous singularities*, Exposition. Math. **13** (1995), 446–454.

[30] J. Malý & W. P. Ziemer, *Fine Regularity of Solutions of Elliptic Partial Differential Equations*, Mathematical Surveys and Monographs, 51, Amer. Math. Soc., Providence, RI, 1997.

[31] V. G. Maz'ya, *On the continuity at a boundary point of solutions of quasi-linear elliptic equations (English translation)*, Vestnik Leningrad Univ. Math. **3** (1976), 225–242, Original in *Vestnik Leningrad. Univ.* **25** (1970), 42–55 (in Russian).

[32] P. Mikkonen, *On the Wolff potential and quasilinear elliptic equations involving measures*, Ann. Acad. Sci. Fenn. Ser. A I. Math. Dissertationes **104** (1996), 1–71.

[33] A. Prignet, *Remarks on existence and uniqueness of solutions of elliptic problems with right-hand side measures*, Rend. Mat. Appl. **15** (1995), 321–337.

[34] J.-M. Rakotoson, *Uniqueness of renormalized solutions in a T-set for the L^1-data problem and the link between various formulations*, Indiana Univ. Math. J. **43** (1994), 685–702.

[35] J.-M. Rakotoson, *Properties of solutions of quasilinear equations in a T-set when the datum is a Radon measure*, Indiana Univ. Math. J. **46** (1997), 247–297.

[36] J.-M. Rakotoson & W. P. Ziemer, *Local behavior of solutions of quasilinear elliptic equations with general structure*, Trans. Amer. Math. Soc. **319** (1990), 747–764.

[37] J. Serrin, *Local behavior of solutions to quasi-linear equations*, Acta Math. **111** (1964), 247–302.

[38] J. Serrin, *Pathological solutions of elliptic differential equations*, Ann. Scuola Norm. Sup. Pisa (3) **18** (1964), 385–387.

[39] N. S. Trudinger & X.J. Wang, *On the Weak continuity of elliptic operators and applications to potential theory*, Amer. J. Math. **124** (2002), 369–410.

[40] N. S. Trudinger & X.J. Wang, *Dirichlet problems for quasilinear elliptic equations with measure data*, preprint.

[41] X. Zhong, *On nonhomogeneous quasilinear elliptic equations*, Ann. Acad. Sci. Fenn. Math. Diss. **117** (1998).

ICM 2002 · Vol. III · 177–184

On Some Conformally Invariant Fully Nonlinear Equations

YanYan Li*

Abstract

We will report some results concerning the Yamabe problem and the Niren-berg problem. Related topics will also be discussed. Such studies have led to new results on some conformally invariant fully nonlinear equations arising from geometry. We will also present these results which include some Liou-ville type theorems, Harnack type inequalities, existence and compactness of solutions to some nonlinear version of the Yamabe problem.

2000 Mathematics Subject Classification: 35, 58.
Keywords and Phrases: Conformally invariant, Fully nonlinear, Yamabe problem, Liouville type theorem.

In this talk, we present some recent joint work with Aobing Li [15] on some conformally invariant fully nonlinear equations.

For $n \geq 3$, consider

$$-\Delta u = \frac{n-2}{2} u^{\frac{n+2}{n-2}}, \qquad \text{on} \quad R^n. \tag{1}$$

The celebrated Liouville type theorem of Caffarelli, Gidas and Spruck ([3]) asserts that positive C^2 solutions of (1) are of the form

$$u(x) = (2n)^{\frac{n-2}{4}} \left(\frac{a}{1 + a^2 |x - \bar{x}|^2} \right)^{\frac{n-2}{2}},$$

where $a > 0$ and $\bar{x} \in R^n$. Under an additional decay hypothesis $u(x) = O(|x|^{2-n})$, the result was proved by Obata ([20]) and Gidas, Ni and Nirenberg ([8]).

Let ψ be a Möbius transformation.

$$\left(u_\psi^{-\frac{n+2}{n-2}} \Delta u_\psi \right) = \left(u^{-\frac{n+2}{n-2}} \Delta u \right) \circ \psi, \qquad \text{on } R^n,$$

---

*Department of Mathematics, Rutgers University, 110 Frelinghuysen Rd., Piscataway, NJ 08854, USA. E-mail: yyli@math.rutgers.edu

where $u_\psi := |J_\psi|^{\frac{n-2}{2n}} (u \circ \psi)$ and J_ψ denotes the Jacobian of ψ. In particular, if u is a positive solution of (1), so is u_ψ.

We call a fully nonlinear operator $H(x, u, \nabla u, \nabla^2 u)$ conformally invariant on R^n if for any Möbius transformation ψ and any positive function $u \in C^2(R^n)$

$$H(\cdot, u_\psi, \nabla u_\psi, \nabla^2 u_\psi) \equiv H(\cdot, u, \nabla u, \nabla^2 u) \circ \psi. \qquad (2)$$

We showed in [15] that $H(x, u, \nabla u, \nabla^2 u)$ is conformally invariant if and only if

$$H(x, u, \nabla u, \nabla^2 u) \equiv F(A^u),$$

where

$$A^u := -\frac{2}{n-2} u^{-\frac{n+2}{n-2}} \nabla^2 u + \frac{2n}{(n-2)^2} u^{-\frac{2n}{n-2}} \nabla u \otimes \nabla u - \frac{2}{(n-2)^2} u^{-\frac{2n}{n-2}} |\nabla u|^2 I, \quad (3)$$

and F is invariant under orthogonal conjugations.

Let U be an open subset of $n \times n$ symmetric matrices which is invariant under orthogonal conjugations (i.e. $O^{-1}UO = U$ for all orthogonal matrices O) and has the property that $U \cap \{M + tN \mid 0 < t < \infty\}$ is convex for any $n \times n$ symmetric matrix M and any $n \times n$ positive definite symmetric matrix N.

Let $F \in C^1(U)$ be invariant under orthogonal conjugation and be elliptic, i.e.

$$(F_{ij}(M)) > 0, \qquad \forall\, M \in U,$$

where $F_{ij}(M) := \frac{\partial F}{\partial M_{ij}}(M)$.

The following theorem extends the result of Obata and Gidas, Ni and Nirenberg to all conformally invariant operators of elliptic type.

Theorem 1 ([15]) *For $n \geq 3$, let U and F be as above, and let $u \in C^2(R^n)$ be a positive solution of*

$$F(A^u) = 1, \qquad on\ R^n.$$

Assume that u is regular at infinity, i.e., $|x|^{2-n} u(x/|x|^2)$ can be extended to a positive C^2 function near the origin. Then for some $\bar{x} \in R^n$ and for some positive constants a and b,

$$u(x) \equiv \left(\frac{a}{1 + b^2 |x - \bar{x}|^2} \right)^{\frac{n-2}{2}}, \qquad \forall\, x \in R^n.$$

Remark 1 In fact, as established in [15], the conclusion of the above theorem still holds when replacing the assumption $u \in C^2(R^n)$ by a weaker assumption that $u \in C^2(R^n \setminus \{0\})$, u can be extended to a positive continuous function near the origin, and $\lim_{x \to 0}(|x||\nabla u(x)|) = 0$.

Theorem 1 indicates that behavior of solutions to conformally invariant equations is very rigid. Thus we expect some good theories for conformally invariant uniformly elliptic fully nonlinear equations. Let F be C^∞ functions defined on $n \times n$

real symmetric matrices, and let F be invariant under orthogonal conjugations. We assume that for some constants $0 < \lambda \le \Lambda < \infty$,

$$\lambda I \le (F_{ij}(M)) \le \Lambda I, \qquad \text{for all } n \times n \text{ real symmetric matrices.}$$

We raise the following

Question 1 Let F be as above, and let B_1 be a unit ball in R^n and $a > 0$ be some constant. Are there some positive constants α and C, depending only on F, a and n such that for any positive C^∞ solution u of

$$F\left(A^u\right) = 0, \qquad \text{in } B_1$$

satisfying

$$\min_{\overline{B_1}} u \ge a, \qquad \|u\|_{C^2(B_1)} \le \frac{1}{a},$$

we have

$$\|u\|_{C^{2,\alpha}(B_{\frac{1}{2}})} \le C?$$

Other interesting questions include to understand behavior near an isolated singularities of a solution in a punctured disc of this subclass of uniformly elliptic equations and to establish some removable singularity results.

Let (M, g) be an n−dimensional smooth Riemannian manifold without boundary, consider the Schouten tensor

$$A_g = \frac{1}{n-2}\left(Ric_g - \frac{R_g}{2(n-1)}g\right),$$

where Ric_g and R_g denote respectively the Ricci tensor and the scalar curvature associated with g.

For $1 \le k \le n$, let

$$\sigma_k(\lambda) = \sum_{1 \le i_1 < \cdots < i_k \le n} \lambda_{i_1} \cdots \lambda_{i_k}, \qquad \lambda = (\lambda_1, \cdots, \lambda_n) \in R^n,$$

denote the k−th symmetric function, and let Γ_k denote the connected component of $\{\lambda \in R^n \mid \sigma_k(\lambda) > 0\}$ containing the positive cone $\{\lambda \in R^n \mid \lambda_1, \cdots, \lambda_n > 0\}$.

It is known (see, e.g., [2]) that Γ_k is a convex cone with its vertex at the origin,

$$\Gamma_n \subset \cdots \subset \Gamma_2 \subset \Gamma_1,$$

$$\frac{\partial \sigma_k}{\partial \lambda_i} > 0 \quad \text{in } \Gamma_k, \ 1 \le i \le n,$$

and

$$\sigma_k^{\frac{1}{k}} \text{ is concave in } \Gamma_k.$$

Fully nonlinear elliptic equations involving $\sigma_k(D^2 u)$ have been investigated in the classical and pioneering paper of Caffarelli, Gidas and Nirenberg [2]. For extensive studies and outstanding results on such equations, see, e.g., Guan and Spruck [10], Trudinger [25], Trudinger and Wang [26], and the references therein. On Riemannian manifolds of nonnegative curvature, Li studied in [17] equations

$$\sigma_k^{\frac{1}{k}}\left(\lambda(\nabla_g^2 u + g)\right) = \psi(x, u), \tag{4}$$

where $\lambda(\nabla_g^2 u + g)$ denotes eigenvalues of $\nabla_g^2 u + g$ with respect to g. On general Riemannian manifolds, Viaclovsky introduced and systematically studied in [28] and [27] equations

$$\sigma_k^{\frac{1}{k}}\left(\lambda(A_g)\right) = \psi(x, u), \tag{5}$$

where $\lambda(A_g)$ denotes the eigenvalues of A_g with respect to g. On 4−dimensional general Riemannian manifolds, remarkable results on (5) for $k = 2$ were obtained by Chang, Gursky and Yang in [4] and [5], which include Liouville type theorems, existence and compactness of solutions, as well as applications to topology. On the other hand, works on the Yamabe equation by Caffarelli, Gidas and Spruck ([3]), Schoen ([22] and [23]), Li and Zhu ([19]), and Li and Zhang ([18]), have played an important role in our approach to the study of (5) as developed in [15].

Consider

$$\sigma_k\left(\lambda(A_g)\right) = 1, \tag{6}$$

together with

$$\lambda(A_g) \in \Gamma_k. \tag{7}$$

Let $g_1 = u^{\frac{4}{n-2}} g_0$ be a conformal change of metrics, then (see, e.g., [28]),

$$A_{g_1} = -\frac{2}{n-2} u^{-1} \nabla_{g_0}^2 u + \frac{2n}{(n-2)^2} u^{-2} \nabla_{g_0} u \otimes \nabla_{g_0} u - \frac{2}{(n-2)^2} u^{-2} |\nabla_{g_0} u|_{g_0}^2 g_0 + A_{g_0}.$$

Let $g = u^{\frac{4}{n-2}} g_{flat}$, where g_{flat} denotes the Euclidean metric on R^n. Then by the above transformation formula,

$$A_g = u^{\frac{4}{n-2}} A_{ij}^u dx^i dx^j,$$

where A^u is given by (3).

Equations (6) and (7) take the form

$$\sigma_k(\lambda(A^u)) = 1, \qquad \text{on} \quad R^n, \tag{8}$$

and

$$\lambda(A^u) \in \Gamma_k, \qquad \text{on} \quad R^n. \tag{9}$$

Our next result extends the Liouville type theorem of Caffarelli, Gidas and Spruck to all σ_k, $1 \le k \le n$. For $k = 1$, equation (8) is (1).

Theorem 2 ([15]) *For $n \geq 3$ and $1 \leq k \leq n$, let $u \in C^2(R^n)$ be a positive solution of (8) satisfying (9). Then for some $a > 0$ and $\bar{x} \in R^n$,*

$$u(x) = c(n,k) \left(\frac{a}{1 + a^2 |x - \bar{x}|^2} \right)^{\frac{n-2}{2}}, \qquad \forall \, x \in R^n, \tag{10}$$

where $c(n,k) = 2^{(n-2)/4} \binom{n}{k}^{(n-2)/4k}$.

The case $k = 2$ and $n = 4$ was obtained by Chang, Gursky and Yang ([5]). More recently, they ([6]) have independently established the result for $k = 2$ and $n = 5$, and they also established the result for $k = 2$ and $n \geq 6$ under an additional hypothesis $\int_{R^n} u^{\frac{2n}{n-2}} < \infty$. Under an additional hypothesis that $\frac{1}{|x|^{n-2}} u(\frac{x}{|x|^2})$ can be extended to a C^2 positive function near $x = 0$, the case $2 \leq k \leq n$ was obtained by Viaclovsky ([28], [29]). As mentioned above, the case $k = 1$ was obtained by Caffarelli, Gidas and Spruck, while under an additional hypothesis that $\frac{1}{|x|^{n-2}} u(\frac{x}{|x|^2})$ is bounded near $x = 0$, the case $k = 1$ was obtained by Obata, and by Gidas, Ni and Nirenberg,

The methods of Chang, Gursky and Yang in [5] and [6] include an ingenious way of using the Obata technique which, as they pointed out, allows the possibility to be generalized to establish the uniqueness of solutions on general Einstein manifolds. Our proof of Theorem 2 is very different from that of [5] and [6]. A crucial ingredient in our proof is the following Harnack type inequality.

Theorem 3 ([15]) *For $n \geq 3$, $1 \leq k \leq n$, and $R > 0$, let $B_{3R} \subset R^n$ be a ball of radius $3R$ and let $u \in C^2(B_{3R})$ be a positive solution of*

$$\sigma_k(A^u) = 1, \qquad in \; B_{3R}, \tag{11}$$

satisfying

$$\lambda(A^u) \in \Gamma_k, \qquad in \; B_{3R}. \tag{12}$$

Then

$$(\max_{\overline{B}_R} u)(\min_{\overline{B}_{2R}} u) \leq C(n) R^{2-n}. \tag{13}$$

The above Harnack type inequality for $k = 1$ was obtained by Schoen ([23]) based on the Liouville type theorem of Caffarelli, Gidas and Spruck. An important step toward our proof of Theorem 3 was taken in an earlier work of Li and Zhang ([18]), where they gave a different proof of Schoen's Harnack type inequality without using the Liouville type theorem.

Our next result concerns existence and compactness of solutions.

Theorem 4 ([15]) *For $n \geq 3$ and $1 \leq k \leq n$, let (M, g) be an n-dimensional smooth compact locally conformally flat Riemannian manifold without boundary satisfying*

$$\lambda(A_g) \in \Gamma_k, \qquad on \; M.$$

*Then there exists some smooth positive function u on M such that $\hat{g} = u^{\frac{4}{n-2}}g$
satisfies*

$$\lambda(A_{\hat{g}}) \in \Gamma_k, \quad \sigma_k(\lambda(A_{\hat{g}})) = 1, \qquad on \ M. \qquad (14)$$

*Moreover, if (M, g) is not conformally diffeomorphic to the standard $n-$sphere, all
solutions of the above satisfy, for all $m \geq 0$, that*

$$\|u\|_{C^m(M,g)} + \|u^{-1}\|_{C^m(M,g)} \leq C,$$

where C depends only on (M, g) and m.

For $k = 1$, it is the Yamabe problem for locally conformally flat manifolds with
positive Yamabe invariants, and the result is due to Schoen ([21]-[22]). The Yamabe
problem was solved through the work of Yamabe [30], Trudinger [24], Aubin [1],
and Schoen [21]. For $k = 2$ and $n = 4$, the result was proved without the locally
conformally flatness hypothesis by Chang, Gursky and Yang [5]. For $k = n$, the
existence result was established by Viaclovsky [27] for a class of manifolds which
are not necessarily locally conformally flat. For $k \neq \frac{n}{2}$, the result is independently
obtained by Guan and Wang in [12] using a heat flow method. More recently, Guan,
Viaclovsky and Wang [9] have proved that $\lambda(A_g) \in \Gamma_k$ for $k \geq \frac{n}{2}$ implies the positivity of the Ricci tensor, and therefore, by classical results, (M, g) is conformally
covered by S^n and the existence and compactness results in this case follow easily.

Our proof of Theorem 1, different from the ones in [20], [8], [3], [28] and [29],
is in the spirit of the new proof of the Liouville type theorem of Caffarelli, Gidas
and Spruck given by Li and Zhu in [19]. We also make use of the substantial
simplifications of Li and Zhang in [18] to the proof in [19]. The proof is along the
line of the pioneering work of Gidas, Ni and Nirenberg [8], which in particular does
not need the kind of divergence structure needed for the method of Obata [20] and
therefore can be applied in much more generality.

In our proofs blow up arguments are used, which require local derivative estimates of solutions. For σ_1 (the Yamabe equation), such estimates follow from
standard elliptic theories. Guan and Wang [11] established local gradient and second derivative estimates for σ_k, $k \geq 2$. Global gradient and second derivative
estimates for σ_k were obtained by Viaclovsky [27]. For the related equation (4)
on manifolds of nonnegative curvature, global gradient and second derivative esti-
mates were obtained by Li in [17]. By the concavity of $\sigma_k^{\frac{1}{k}}$, $C^{2,\alpha}$ estimates hold
due to the classical work of Evans [7] and Krylov [14]. For the proof of the exis-
tence part of Theorem 4, we introduce a homotopy $\sigma_k(t\lambda + (1-t)\sigma_1(\lambda))$, defined
on $(\Gamma_k)_t = \{\lambda \in R^n \mid t\lambda + (1-t)\sigma_1(\lambda) \in \Gamma_k\}$, which establishes a natural link
between (14) and the Yamabe problem. We extend the local estimates in [11] for
σ_k to $\sigma_k(t\lambda + (1-t)\sigma_1(\lambda))$, with estimates uniform in $0 \leq t \leq 1$. The compactness
results as stated in Theorem 4 were established in [15] along the homotopy. The
compactness results for the Yamabe problem was established by Schoen [22]. We
gave a different proof which does not rely on the Liouville type theorem, which allows us to establish existence results for more general f than σ_k for which Liouville
type theorems are not available. The existence results follow from the compactness
results with the help of the degree theory for second order fully nonlinear elliptic

operators ([16]) as well as the degree counting formula for the Yamabe problem ([22]).

The first step in our proof of the Liouville type Theorem 2 is to establish the Harnack type inequality (Theorem 3), from which we obtain sharp asymptotic behavior at infinity of an entire solution. Then we establish Theorem 2 by distinguishing into two cases. In the case $k > \frac{n}{2}$, Theorem 2 is proved by using the sharp asymptotic behavior of an entire solution and Theorem 1-Remark 1, together with a result of Trudinger and Wang ([26]). In the case $1 \le k \le \frac{n}{2}$, Theorem 2 is proved by the sharp asymptotic behavior of an entire solution together with the Obata type integral formula of Viaclovsky ([28]). For the second case, divergence structure of the equation is used.

Theorem 2, Theorem 3 and Theorem 4 are established for more general nonlinear f than σ_k in [15], including those for which no divergence structure is available.

References

[1] T. Aubin, Équations différentielles non linéaires et probléme de Yamabe concernant la courbure scalaire, J. Math. Pures Appl., 55 (1976), 269–296.

[2] L. Caffarelli, L. Nirenberg and J. Spruck, The Dirichlet problem for nonlinear second-order elliptic equations, III: Functions of the eigenvalues of the Hessian. Acta Math., 155 (1985), 261–301.

[3] L. Caffarelli, B. Gidas and J. Spruck, Asymptotic symmetry and local behavior of semilinear elliptic equations with critical Sobolev growth, Comm. Pure Appl. Math., 42 (1989), 271–297.

[4] S.Y. A. Chang, M. Gursky and P. Yang, An equation of Monge-Ampere type in conformal geometry, and four-manifolds of positive Ricci curvature, Ann. of Math., to appear.

[5] S.Y. A. Chang, M. Gursky and P. Yang, An a priori estimate for a fully nonlinear equation on four-manifolds, preprint.

[6] S.Y. A. Chang, M. Gursky and P. Yang, Entire solutions of a fully nonlinear equation, preprint.

[7] L.C. Evans, Classical solutions of fully nonlinear, convex, second-order elliptic equations, Comm. Pure Appl. Math., 35 (1982), 333–363.

[8] B. Gidas, W.M. Ni and L. Nirenberg, Symmetry and related properties via the maximum principle, Comm. Math. Phys., 68 (1979), 209–243.

[9] P. Guan, J. Viaclovsky and G. Wang, Some properties of the Schouten tensor and applications to conformal geometry, preprint.

[10] B. Guan and J. Spruck, Boundary-value problems on S^n for surfaces of constant Gauss curvature, Ann. of Math., 138 (1993), 601–624.

[11] P. Guan and G. Wang, Local estimates for a class of fully nonlinear equations arising from conformal geometry, preprint.

[12] P. Guan and G. Wang, A fully nonlinear conformal flow on locally conformally flat manifolds, preprint.

[13] M. J. Gursky and J. Viaclovsky, Fully nonlinear equations on Riemannian manifolds with negative curvature, preprint.

[14] N.V. Krylov, Boundedly inhomogeneous elliptic and parabolic equation in a domain, Izv. Akad. Nauk SSSR, 47 (1983), 75–108.

[15] A. Li and Y.Y. Li, On some conformally invariant fully nonlinear equations, preprint.

[16] Y.Y. Li, Degree theory for second order nonlinear elliptic operators and its applications, Comm. in Partial Differential Equations, 14 (1989), 1541–1578.

[17] Y.Y. Li, Some existence results of fully nonlinear elliptic equations of Monge-Ampere type, Comm. Pure Appl. Math., 43 (1990), 233–271.

[18] Y.Y. Li and L. Zhang, Liouville type theorems and Harnack type inequalities for semilinear elliptic equations, Journal d'Analyse Mathematique, to appear.

[19] Y.Y. Li and M. Zhu, Uniqueness theorems through the method of moving spheres, Duke Math. J., 80 (1995), 383–417.

[20] M. Obata, The conjecture on conformal transformations of Riemannian manifolds, J. Diff. Geom., 6 (1971), 247–258.

[21] R. Schoen, Conformal deformation of a Riemannian metric to constant scalar curvature, J. Diff. Geom., 20 (1984), 479–495.

[22] R. Schoen, On the number of constant scalar curvature metrics in a conformal class, Differential Geometry: A symposium in honor of Manfredo Do Carmo (H.B. Lawson and K. Tenenblat, eds), Wiley, 1991, 311–320.

[23] R. Schoen, Courses at Stanford University, 1988, and New York University, 1989.

[24] N. Trudinger, Remarks concerning the conformal deformation of Riemannian structures on compact manifolds, Ann. Scuola Norm. Sup. Cl. Sci., (3) 22 (1968), 265–274.

[25] N.S. Trudinger, On the Dirichlet problem for Hessian equations, Acta Math., 175 (1995), 151–164.

[26] N.S. Trudinger and X. Wang, Hessian measures II, Ann. of Math., 150 (1999), 579–604.

[27] J. Viaclovsky, Estimates and existence results for some fully nonlinear elliptic equations on Riemannian manifolds, Comm. Anal. Geom., to appear.

[28] J. Viaclovsky, Conformal geometry, contact geometry, and the calculus of variations, Duke Math. J., 101 (2000), 283–316.

[29] J. Viaclovsky, Conformally invariant Monge-Ampere equations: global solutions, Trans. Amer. Math. Soc., 352 (2000), 4371–4379.

[30] H. Yamabe, On a deformation of Riemannian structures on compact manifolds, Osaka Math. J., 12 (1960), 21–37.

ICM 2002 · Vol. III · 185–188

Shock Waves

Tai-Ping Liu*

Abstract

Shock wave theory was first studied for gas dynamics, for which shocks appear as compression waves. A shock wave is characterized as a sharp transition, even discontinuity in the flow. In fact, shocks appear in many different physical situation and represent strong nonlinearity of the physical processes. Important progresses have been made on shock wave theory in recent years. We will survey the topics for which much more remain to be made. These include the effects of reactions, dissipations and relaxation, shock waves for interacting particles and Boltzmann equation, and multi-dimensional gas flows.

2000 Mathematics Subject Classification: 35.

1. Introduction

The most basic equations for shock wave theory are the systems of hyperbolic conservation laws

$$u_t + \nabla_x \cdot f(u) = 0,$$

where $x \in R^m$ is the space variables and $u \in R^n$ is the basic dependence variables. Such a system represents basic physical model for which $u = u(x, t)$ is the density of conserved physical quantities and the flux $f(u)$ is assumed to be a function of u. More complete system of partial differential equations takes the form

$$u_t + \nabla_x \cdot f(u) = \nabla_x \cdot (B(u, \varepsilon)\nabla_x u) + g(u, x, t),$$

with $B(u, \varepsilon)$ the viscosity matrix and ε the viscosity parameters, and $g(u, x, t)$ the sources. Other evolutionary equations which carry shock waves include the interacting particles system, Boltzmann equation, and discrete systems. Discrete systems can appear as difference approximations to hyperbolic conservation laws. In all these systems, shock waves yield rich phenomena and also present serious mathematical difficulties due to their strong nonlinear character.

---

*Institute of Mathematics, Academia Sinica, Taipei, China. Department of Mathematics, Stanford University, Stanford, CA 94305, USA. E-mail: tpliu@math.sinica.edu.tw

2.   Hyperbolic conservation laws

Much has been done for hyperbolic conservation laws in one space dimension

$$u_t + f(u)_x = 0, \ x \in R^1,$$

see [16], [4], [11], and the article by Bressan in this volume. Because the solutions in general contain discontinuous shock waves, the system provides impetus for the introduction of new ideas, such as the Glimm functional, and is a good testing ground for new techniques, such as the theory of compensated compactness, in nonlinear analysis.

3.   Viscous conservation laws

Physical models of the form of viscous conservation laws are not uniformly parabolic, but hyperbolic-parabolic. Basic study of the dissipation of solutions for such a system has been done using the energy method, see [6]. Study of nonlinear waves for these systems has been initiated, [9], [14], however, much more remains to be done. The difficulty lies in the nonlinear couplings due to both the nonlinearity of the flux $f(u)$, which is the topic of consideration for hyperbolic conservation laws, as well as that of the viscosity matrix $B(u,\varepsilon)$. For instance, the study of zero dissipation limit $\varepsilon \to 0$, see Bressan's article, has been done only for the artificial viscosity matrix $B(u,\varepsilon) = \varepsilon I$.

4.   Conservation laws with sources

Sources added to conservation laws may represents geometric effects, chemical reactions, or relaxation effects. Thus there should be no unified theory for it. When the source represents the geometric effects, such the multi-dimensional spherical waves , hyperbolic conservation laws takes the form

$$u_t + f(u)_r = \frac{m-1}{r}h(u), \ r^2 = \sum_{i=1}^{m}(x_i)^2.$$

There has stablizing and destablizing effects, such as in the nozzle flows, [8]. The chemical effects occur in the combustions. There is complicated, still mostly not understood, phenomena on the rich behaviour of combustions. One interesting problem is the transition from the detonations to deflagrations, where the combined effects of dissipation, compression and chemical energy gives rise to new wave behaviour. Viscous effects are important on the qualitative behaviour of nonlinear waves when the hyperbolic system is not strictly hyperbolic, see [10]. Relaxation, such as for the kinetic models and thermal non-equilibrium in general, is interesting because of the rich coupling of dissipation, dispersion and hyperbolicity, [2].

5.   Discrete conservation laws

Conservative finite differences to the hyperbolic conservation laws, in one space dimension, take the form:

$$u^{n+1}(x) - u^n(x) = \frac{\Delta t}{\Delta x}(F[u^n](x + \frac{\Delta x}{2}) - F[u^n](x - \frac{\Delta x}{2})).$$

It has been shown for a class of two conservation laws and dissipative schemes, such as Lax-Friedrichs and Godunov scheme that the numerical solutions converge to the exact solutions of the conservation laws, [5]. On the other hand, qualitative studies on the nonlinear waves for difference schemes indicate rich behaviour. In particular, the shock waves depend sensitively on the its C-F-L speeds, [15], [12].

6. Multi-dimensional gas flows

Shock wave theory originated from the study of the Euler equations in gas dynamics. The classical book [3] is still important and mostly updated on multi-dimensional gas flows. Because of its great difficulty, the study of multi-dimensional shocks concentrate on flows with certain self-similarity property. One such problem is the Riemann problem, with initial value consisting of finite many constant states. In that case, the solutions are function of x/t, not general function of (x, t), see [7]. See also [18] for other self-similar solutions. However, unlike single space case, multi-dimensional Riemann solutions do not represent general scattering data, and are quite difficult to study. It is more feasible to consider flows with shocks and solid boundary, e.g. [1] [8].

7. Boltzmann equation

The Boltzmann equation

$$f_t + \xi \nabla_x f = Q(f, f)$$

contains much more information than the gas dynamics equations. Nevertheless, the shock waves for all these equations have the same Rankine-Hugoniot relation at the far states. The difference is on the transition layer. There is beginning an effort to make use of the techniques for the conservation laws to study the Boltzmann shocks, [13]. This is a line of research quite different from the intensive current efforts on the incompressible limits of the Boltzmann equation.

8. Interacting particle systems

Interacting particle systems is even closer to the first physical principles than the Boltzmann equation. There is the long-standing problem, the Zermelo paradox, in passing from the reversible particle systems to the irreversible systems such as the Boltzmann equation, and the Euler equations with shocks. This is fine for particle system with random noises. However, except for scalar models, so far the derivation of Euler equations from the particle systems has been done only for solutions with no shocks, [17].

References

[1] Chen, S., Existence of stationary supersonic flows past a pointed body. Arch. Ration.Mech.Anal., 156(2001), No.2, 141–181.

[2] Chen, G.-Q., Levermore, D., and Liu, T.-P., Hyperbolic conservation laws with stiff relaxation terms and entropy. Comm. Pure Appl. Math., 47(1994), no. 6, 787–830.

[3] Courant, R. and Friedrichs, K. O., Supersonic Flows and hock Waves. Springer, 1948.

[4] Dafermos, C.M., Hyperbolic Conservation Laws in Continuum Physics. Springer-Verlag, 2000.

[5] Ding, X., Chen, G., and Luo, P., Convergence of the fractional step Lax-Friedrichs scheme and Godunov scheme for the isentropic system of gas dynamics. Comm. math. Phys., 121(1989), no. 1, 63–84.

[6] Kawashima, S., Large-time behaviour of solutions to hyperbolic-parabolic systems of conservation laws and applications. Proc. Roy. Soc. Edinburgh Sect. A 106(1987), no.1-2, 169–194.

[7] Li, J., Zhang, T. and Yang, S., The two-dimensional Riemann problems in gas dynamics. Pitman Mornographs and surveys in Pure and Applied Mathematics, 98. Longman, Harlow, 1998.

[8] Lien, W.-C., and Liu, T.-P., Nonlinear stability of a self-similar 3-dimensional gas flow. Comm. Math. Phys., 204(1999), no. 3, 525–549.

[9] Liu, T.-P., Hyperbolic and Viscous Conservation Laws. CBMS-NSF Regional Conference Series in Applied Mathematics, 72. SIAM, Philadelphia, PA, 2000.

[10] Liu, T.-P., Zero dissipation and stability of shocks. Methods Appl. Anal., 5(1998), no.1, 81–94.

[11] Liu, T.-P., and Yang, T., Weak solutions of general systems of hyperbolic conservation laws. Comm. Math. Phys., (to appear.)

[12] Liu, T.-P., and Yu, S.-H., Continuum shock profiles for discrete conservation laws, II. Stability. Comm. Pure Appl. Math., 52(1999), no. 9, 1047–1073.

[13] Liu, T.-P., and Yu, S.-H., Boltzmann equation: Micro-macro decompositions and positivity of shock profiles. (Preprint.)

[14] liu, T.-P., and Zeng, Y., Large time behaviour of solutions for general quasilinear hyerbolic-parabolic systems of conservation laws. Mem. Amer. Math. Soc., 125(1997), no. 599.

[15] Serre, D., Discrete shock profiles and their stability. Hyperbolic porblems: theory, numerics, applications, Vol II (1998) (Zurich) 853, Internat. Ser. Numer. Math., 130, Birkhauser, Basel, 1999.

[16] Smoller, J., Shock Waves and Reaction-Diffusion Equations, Springer-Verlag, 1994.

[17] Varadhan, S.R.S., Lectures on hydrodynamics scaling. Hydrodynamic limits and related topics (Toronto, ON, 1998), 3–40, Fields Inst. Comm. 27, Amer. Math. Soc., Providence, RI, 2000.

[18] Zheng, Yuxi, Systems of Conservation Laws. Two-Dimensional Riemann Problems. Progress in Nonlinear Differential Equations and their Applications, 38. Birkhauser Boston, 2001.

The Wiener Test for Higher Order Elliptic Equations

Vladimir Maz'ya*

1. Introduction. Wiener's criterion for the regularity of a boundary point with respect to the Dirichlet problem for the Laplace equation [W] has been extended to various classes of elliptic and parabolic partial differential equations. They include linear divergence and nondivergence equations with discontinuous coefficients, equations with degenerate quadratic form, quasilinear and fully nonlinear equations, as well as equations on Riemannian manifolds, graphs, groups, and metric spaces (see [LSW], [FJK], [DMM], [LM], [KM], [MZ], [AH], [Lab], [TW] to mention only a few). A common feature of these equations is that all of them are of second order, and Wiener type characterizations for higher order equations have been unknown so far. Indeed, the increase of the order results in the loss of the maximum principle, Harnack's inequality, barrier techniques, and level truncation arguments, which are ingredients in different proofs related to the Wiener test for the second order equations.

In the present work we extend Wiener's result to elliptic differential operators $L(\partial)$ of order $2m$ in the Euclidean space $\mathbf{R}^n$ with constant real coefficients

$$L(\partial) = (-1)^m \sum_{|\alpha|=|\beta|=m} a_{\alpha\beta} \partial^{\alpha+\beta}.$$

We assume without loss of generality that $a_{\alpha\beta} = a_{\beta\alpha}$ and $(-1)^m L(\xi) > 0$ for all nonzero $\xi \in \mathbf{R}^n$. In fact, the results can be extended to equations with variable (for example, Hölder continuous) coefficients in divergence form but we leave aside this generalization to make exposition more lucid.

We use the notation ∂ for the gradient $(\partial_{x_1}, \ldots, \partial_{x_n})$, where ∂_{x_k} is the partial derivative with respect to x_k. By Ω we denote an open set in $\mathbf{R}^n$ and by $B_\rho(y)$ the ball $\{x \in \mathbf{R}^n : |x - y| < \rho\}$, where $y \in \mathbf{R}^n$. We write B_ρ instead of $B_\rho(O)$.

Consider the Dirichlet problem

$$L(\partial)u = f, \quad f \in C_0^\infty(\Omega), \quad u \in \mathring{H}^m(\Omega), \tag{1}$$

where we use the standard notation $C_0^\infty(\Omega)$ for the space of infinitely differentiable functions in $\mathbf{R}^n$ with compact support in Ω as well as $\mathring{H}^m(\Omega)$ for the completion of $C_0^\infty(\Omega)$ in the energy norm.

*Linköpings Universitet, Mathematiska Institutionen, 58183 Linköping, Sweden. E-mail: vlmaz@mail.liu.se

We call the point $O \in \partial\Omega$ regular with respect to $L(\partial)$ if for any $f \in C_0^\infty(\Omega)$ the solution of (1) satisfies

$$\lim_{\Omega \ni x \to O} u(x) = 0. \tag{2}$$

For $n = 2, 3, \ldots, 2m - 1$ the regularity is a consequence of the Sobolev imbedding theorem. Therefore, we suppose that $n \geq 2m$. In the case $m = 1$ the above definition of regularity is equivalent to that given by Wiener.

The following result coincides with Wiener's criterion in the case $n = 2$ and $m = 1$.

Theorem 1 *Let $2m = n$. Then O is regular with respect to $L(\partial)$ if and only if*

$$\int_0^1 C_{2m}(B_\rho \backslash \Omega)\rho^{-1}d\rho = \infty. \tag{3}$$

Here and elsewhere C_{2m} is the potential-theoretic Bessel capacity of order $2m$ (see [AHed]). If $n = 2m$ and O belongs to a continuum contained in the complement of Ω, condition (3) holds.

The case $n > 2m$ is more delicate because no result of Wiener's type is valid for all operators $L(\partial)$ (see [MN]). To be more precise, even the vertex of a cone can be irregular with respect to $L(\partial)$ if the fundamental solution of $L(\partial)$:

$$F(x) = F(x/|x|)|x|^{2m-n}, \quad x \in \mathbf{R}^n \backslash O, \tag{4}$$

changes sign. Examples of operators $L(\partial)$ with this property were given in [MN] and [D]. For instance, according to [MN] the vertex of a sufficiently thin 8-dimensional cone K is irregular with respect to the operator

$$L(\partial)u := 10\partial_{x_8}^4 u + \Delta^2 u, \quad u \in \overset{\circ}{H}{}^2(\mathbf{R}^8 \backslash K).$$

In the sequel, Wiener's type characterization of regularity for $n > 2m$ is given for a subclass of the operators $L(\partial)$ called *positive with the weight F*. This means that for all real-valued $u \in C_0^\infty(\mathbf{R}^n \backslash O)$,

$$\int_{\mathbf{R}^n} L(\partial)u(x) \cdot u(x)F(x)\, dx \geq c \sum_{k=1}^m \int_{\mathbf{R}^n} |\nabla_k u(x)|^2 |x|^{2k-n} dx, \tag{5}$$

where ∇_k is the gradient of order k, i.e. $\nabla_k = \{\partial^\alpha\}$ with $|\alpha| = k$.

The positivity of the left-hand side in (5) is equivalent to the inequality

$$\int_{\mathbf{R}^n} \int_{\mathbf{R}^n} \frac{L(\xi) + L(\eta)}{L(\xi - \eta)} f(\xi) f(\eta) d\xi d\eta > 0$$

for all non-zero $f \in C_0^\infty(\mathbf{R}^n)$.

Theorem 2 *Let $n > 2m$ and let $L(\partial)$ be positive with weight F. Then O is regular with respect to $L(\partial)$ if and only if*

$$\int_0^1 C_{2m}(B_\rho \backslash \Omega)\rho^{2m-n-1}d\rho = \infty. \tag{6}$$

Note that in direct analogy with the case of the Laplacian we could say, in Theorems 1 and 2, that O is irregular with respect to $L(\partial)$ if and only if the set $\mathbf{R}^n \setminus \Omega$ is $2m$-thin in the sense of linear potential theory [L], [AHed].

Let, for example, the exterior of Ω contain the region

$$\{ x : 0 < x_n < 1, \quad (x_1^2 + \ldots + x_{n-1}^2)^{1/2} < f(x_n) \},$$

where f is an increasing function such that $f(0) = f'(0) = 0$. Then the point O satisfies (6) if and only if

$$\int_0^1 |\log f(\tau)|^{-1} \tau^{-1} d\tau = \infty \quad \text{for} \ \ n = 2m + 1$$

and

$$\int_0^1 f(\tau) \tau^{2m-n} d\tau = \infty \qquad \text{for} \ \ n \geq 2m + 2.$$

Since, obviously, the operator $L(\partial)$ of the second order is positive with the weight F, Wiener's result for $n > 2$ is contained in Theorem 2.

We note that the pointwise positivity of F follows from (5), but the converse is not true. In particular, the m-harmonic operator with $2m < n$ satisfies (5) if and only if $n = 5, 6, 7$ for $m = 2$ and $n = 2m + 1, 2m + 2$ for $m > 2$ (see [M2], where the proof of sufficiency of (6) is given for $(-\Delta)^m$ with m and n as above, and also [E] dealing with the sufficiency for noninteger powers of the Laplacian in the intervals $(0, 1)$ and $[n/2 - 1, n/2)$).

We state some auxiliary assertions of independent interest which concern the so called L-capacitary potential U_K of the compact set $K \subset \mathbf{R}^n$, $n > 2m$, i.e. the solution of the variational problem

$$\inf \{ \int_{\mathbf{R}^n} L(\partial) u \cdot u \, dx : \ u \in C_0^\infty(\mathbf{R}^n), u = 1 \text{ in vicinity of } K \}.$$

These assertions are used in the proof of necessity in Theorem 2.

By the m-harmonic capacity $\text{cap}_m(K)$ of a compact set K we mean

$$\inf \left\{ \sum_{|\alpha|=m} \frac{m!}{\alpha!} \|\partial^\alpha u\|_{L_2(\mathbf{R}^n)}^2 : \ u \in C_0^\infty(\Omega), \ u = 1 \text{ in vicinity of } K \right\}. \tag{7}$$

Lemma 1 *Let $\Omega = \mathbf{R}^n$, $2m < n$. For all $y \in \mathbf{R}^n \setminus K$*

$$U_K(y) = 2^{-1} U_K(y)^2$$

$$+ \int_{\mathbf{R}^n} \sum_{m \geq j \geq 1} \sum_{|\mu|=|\nu|=j} \partial^\mu U_K(x) \cdot \partial^\nu U_K(x) \cdot \mathcal{P}_{\mu\nu}(\partial) F(x - y) \, dx, \tag{8}$$

where $\mathcal{P}_{\mu\nu}(\zeta)$ are homogeneous polynomials of degree $2(m - j)$, $\mathcal{P}_{\mu\nu} = \mathcal{P}_{\nu\mu}$ and $\mathcal{P}_{\alpha\beta}(\zeta) = a_{\alpha\beta}$ for $|\alpha| = |\beta| = m$.

Corollary 1 *Let $\Omega = \mathbf{R}^n$ and $2m < n$. For all $y \in \mathbf{R}^n \backslash K$ there holds the estimate*

$$|\nabla_j U_K(y)| \leq c_j \operatorname{dist}(y, K)^{2m-n-j} \operatorname{cap}_m K, \tag{9}$$

where $j = 0, 1, 2, \ldots$ and c_j does not depend on K and y.

By $\mathcal{M}$ we denote the Hardy-Littlewood maximal operator.

Corollary 2 *Let $2m < n$ and let $0 < \theta < 1$. Also let K be a compact subset of $\overline{B}_\rho \backslash B_{\theta\rho}$. Then the L-capacitary potential U_K satisfies*

$$\mathcal{M}\nabla_l U_K(0) \leq c_\theta\, \rho^{2m-l-n} \operatorname{cap}_m K, \tag{10}$$

where $l = 0, 1, \ldots, m$ and c_θ does not depend on K and ρ.

Let $L(\partial)$ be positive with the weight F. Then identity (8) implies that the L-capacitary potential of a compact set K with positive m-harmonic capacity satisfies

$$0 < U_K(x) < 2 \quad \text{on } \mathbf{R}^n \backslash K. \tag{11}$$

In general, the bound 2 in (11) cannot be replaced by 1.

Proposition 1 *If $L = \Delta^{2m}$, then there exists a compact set K such that*

$$(U_K - 1)\big|_{\mathbf{R}^n \backslash K}$$

changes sign in any neighbourhood of a point of K.

We give a lower pointwise estimate for U_K stated in terms of capacity (compare with the upper estimate (9)).

Proposition 2 *Let $n > 2m$ and let $L(\partial)$ be positive with the weight F. If K is a compact subset of B_d and $y \in \mathbf{R}^n \backslash K$, then*

$$U_K(y) \geq c\, (|y| + d)^{2m-n} \operatorname{cap}_m K.$$

Sufficiency in Theorem 2 follows from the next assertion which is of interest in itself.

Lemma 2 *Let $2m < n$ and let $L(\partial)$ be positive with the weight F. Also let $u \in \mathring{H}^m(\Omega)$ satisfy $L(\partial)u = 0$ on $\Omega \cap B_{2R}$. Then, for all $\rho \in (0, R)$,*

$$\sup\{|u(p)|^2 : p \in \Omega \cap B_\rho\} + \int_{\Omega \cap B_\rho} \sum_{k=1}^m \frac{|\nabla_k u(x)|^2}{|x|^{n-2k}}\, dx$$

$$\leq c_1 M_R(u) \exp\left(-c_2 \int_\rho^R \operatorname{cap}_m(\bar{B}_\tau \backslash \Omega) \frac{d\tau}{\tau^{n-2m+1}}\right), \tag{12}$$

where c_1 and c_2 are positive constants, and

$$M_R(u) = R^{-n} \int_{\Omega \cap (B_{2R} \backslash B_R)} |u(x)|^2 dx.$$

The present work gives answers to some questions posed in [M2]. I present several simply formulated unsolved problems.

1. Is it possible to replace the positivity of $L(\partial)$ with the weight $F(x)$ by the positivity of $F(x)$ in Theorem 2?

A particular case of this problem is the following one.

2. Does Theorem 2 hold for the operator $(-\Delta)^m$, where

$$n \geq 8, \quad m = 2 \quad \text{and} \quad n \geq 2m + 3, \; m > 2 \; ? \tag{13}$$

The next problem concerns Green's function G_m of the Dirichlet problem for $(-\Delta)^m$ in an arbitrary domain Ω.

3. Prove or disprove the estimate

$$|G_m(x,y)| \leq \frac{c(m,n)}{|x-y|^{n-2m}}, \tag{14}$$

where $c(m,n)$ is independent of Ω and m and n are the same as in (13).

For $n = 5, 6, 7$, $m = 2$ and $n = 2m + 1$, $2m + 2$, $m > 2$ estimate (13) was proved in [M3]. In the sequel, by u we denote a solution in $\overset{\circ}{H}{}^m(\Omega)$ of the equation

$$(-\Delta)^m u = f \quad \text{in} \quad \Omega. \tag{15}$$

Clearly, (14) leads to the following estimate of the maximum modulus of u

$$\|u\|_{L_\infty} \leq c(m,n,\text{mes}_n\Omega)\|f\|_{L_p(\Omega)},$$

where $p > n/2m$. However, the validity of this estimate for the same n and m as in (13) is an open problem. Moreover, the following questions arise.

4. Let $m = 2$, $n \geq 8$, and let Ω be an arbitrary bounded domain. Is u uniformly bounded in Ω for any $f \in C_0^\infty(\Omega)$?

5. Let $m > 2$ and $n \geq 2m + 3$. Also, let $\partial\Omega$ have a conic singularity. Is u uniformly bounded in Ω for any $f \in C_0^\infty(\Omega)$?

For $m = 2$, the affirmative answer to the last question is given in [MP].

I formulate two related open problems.

6. Let $m = 2$ and $n = 2$. Is u Lipschitz up to the boundary of an arbitrary bounded domain, for any $f \in C_0^\infty(\Omega)$?

7. Let $m = 2$ and $n \geq 3$. Does u belong to the class $C^{1,1}(\Omega)$ for any $f \in C_0^\infty(\Omega)$ if Ω is convex?

According to [KoM], the last is true in the two-dimensional case.

I conclude with the following variant of the Phragmén Lindelöf principle (see [M3]).

Proposition 3 *Let either $n = 5, 6, 7$, $m = 2$ or $n = 2m + 1$, $2m + 2$, $m > 2$. Further, let $\eta u \in \overset{\circ}{H}{}^m(\Omega)$ for all $\eta \in C^\infty(\mathbf{R}^n)$, $\eta = 0$ near O. If*

$$\Delta^m u = 0 \quad \text{on} \quad \Omega \cap B_1,$$

then either $u \in \mathring{H}^m(\Omega)$ *and*

$$\limsup_{\rho \to 0} \sup_{B_\rho \cap \Omega} |u(x)| \exp\Big(c \int_\rho^1 \mathrm{cap}_m(\bar{B}_\rho \setminus \Omega) \frac{d\rho}{\rho} \Big) < \infty$$

or

$$\liminf_{\rho \to 0} \rho^{n-2m} M_\rho(u) \exp\Big(-c \int_\rho^1 \mathrm{cap}_m(\bar{B}_\rho \setminus \Omega) \frac{d\rho}{\rho} \Big) > 0.$$

It would be interesting to extend this assertion to other values of n and m.

References

[AH]   Adams, D.R. & Herd, A., The necessity of the Wiener test for some semi-linear elliptic PDE. Ind. Univ. Math. J., **41** (1992), 109–124.

[AHed]   Adams, D.R. & Hedberg, L.I., Function Spaces and Potential Theory. Springer-Verlag, Berlin, 1996.

[DMM]   Dal Maso, G. & Mosco, U., Wiener criteria and energy decay for relaxed Dirichlet problems. Arch. Rational Mech. Anal., **95** (1986), 345–387.

[D]   Davies, E.B., Limits in L^p regularity of self-adjoint elliptic operators. J. Diff. Equations, **135**:1 (1997), 83–102.

[E]   Eilertsen, S., On weighted positivity and the Wiener regularity of a boundary point for the fractional Laplacian. Ark. för Mat., **38**:1 (2000), 53–75.

[FJK]   Fabes, E.G., Jerison, D. & Kenig, C., The Wiener test for degenerate elliptic equations. Ann. Inst. Fourier (Grenoble), **32** (1982), 151–182.

[KM]   Kilpeläinen, T. & Malý, J., The Wiener test and potential estimates for quasi-linear elliptic equations. Acta Math., **172** (1994), 137–161.

[KoM]   Kozlov, V.A. & Maz'ya, V., Boundary behavior of solutions to linear and nonlinear elliptic equations in plane convex domains. Math. Research Letters, **8** (2001), 1–5.

[L]   Landkof, N.S., Foundations of Modern Potential Theory. Nauka, Moscow, 1966 (Russian). English translation: Springer-Verlag, Berlin, 1972.

[Lab]   Labutin, D.A. Potential estimates for a class of fully nonlinear elliptic equations. Duke Math. J., **111** (2002), 1–49.

[LM]   Lindqvist, P. & Martio, O., Two theorems of N. Wiener for solutions of quasilinear elliptic equations, Acta Math., **155** (1985), 153–171.

[LSW]   Littman, W., Stampacchia, G. & Weinberger, H.F., Regular points for elliptic equations with discontinuous coefficients. Ann. Scuola Norm. Sup. Pisa, Ser. III, **17** (1963), 43–77.

[MZ]   Malý, J. & Ziemer, W.P. Regularity of Solutions of Elliptic Partial Differential Equations. Mathematical Surveys and Monographs **51**, AMS, Providence, RI, 1997.

[M1]   Maz'ya, V., The Dirichlet problem for elliptic equations of arbitrary order in unbounded regions. Dokl. Akad. Nauk SSSR, **150** (1963), 1221–1224 (Russian). English translation in Soviet Math., **4** (1963), 860–863.

[M2]   Maz'ya, V., Unsolved problems connected with the Wiener criterion. In The Legacy of Norbert Wiener: A Centennial Symposium, Proc., Cambridge, Massachusetts, Amer. Math. Soc., 199–208,1994.

[M3]   Maz'ya, V., On Wiener's type regularity of a boundary point for higher order elliptic equations. Nonlinear Analysis, Function Spaces and Applications VI, 119–155. Proceedings of the Spring School held in Prague, May 31–June 6, 1998. Prague, 1999.

[MN]   Maz'ya, V. & Nazarov, S., The apex of a cone can be irregular in Wiener's sense for a fourth-order elliptic equation. Mat. Zametki, **39**:1 (1986), 24–28 (Russian). English translation in Math. Notes, **39** (1986), 14–16.

[MP]   Maz'ya, V. & Plamenevskii, B. A., On the maximum principle for the biharmonic equation in a domain with conical points. Izv. Vyssh. Uchebn. Zaved. Mat. (1981) no. 2, 52–59 (Russian). English translation in Soviet Math. (Iz. VUZ), **25** (1981), 61–70.

[TW]   Trudinger, N.S. & Wang, X.-J., On the weak continuity of elliptic operators and applications to potential theory. Amer. J. Math., **124** (2002), 369–410.

[W]   Wiener, N., The Dirichlet problem. J. Math. and Phys., **3** (1924), 127–146. Reprinted in Norbert Wiener: Collected Works with Commentaries, vol. 1, 394–413, MIT Press, Cambridge, Massachusetts, 1976.

ICM 2002 · Vol. III · 197–208

Bubbling and Regularity Issues in Geometric Non-linear Analysis

T. Rivière[*]

Abstract

Numerous elliptic and parabolic variational problems arising in physics and geometry (Ginzburg-Landau equations, harmonic maps, Yang-Mills fields, Omega-instantons, Yamabe equations, geometric flows in general...) possess a critical dimension in which an invariance group (similitudes, conformal groups) acts. This common feature generates, in all these different situations, the same non-linear effect. One observes a strict splitting in space between an almost linear regime and a dominantly non-linear regime which has two major characteristics : it requires a quantized amount of energy and arises along rectifiable objects of special geometric interest (geodesics, minimal surfaces, J-holomorphic curves, special lagrangian manifolds, mean-curvature flows...).

2000 Mathematics Subject Classification: 35D10, 35J20, 35J60, 49Q20, 58E15, 58E20.
Keywords and Phrases: Harmonic maps, Yang-Mills fields, Ginzburg-Landau vortices.

1. Energy quantization phenomena for harmonic maps and Yang-Mills Fields

1.1. The archetype of energy quantization: the ε-regularity

Let B^m be the flat m-dimensional ball and N^n be a compact, without boundary Riemannian manifold. By the Nash embedding Theorem, we may assume $N \subset \mathbb{R}^k$. Let $W^{1,2}(B^m, N^n)$ be the maps in $W^{1,2}(B^m, \mathbb{R}^k)$ that take values almost everywhere in N^n. A map $u \in W^{1,2}(B^m, N^n)$ is called stationary harmonic if it is a critical point for the Dirichlet energy

$$E(u) = \int_{B^m} |\nabla u|^2 \, dx$$

---

[*]D-Math., ETH-Zentrum, CH-8092 Zürich, Switzerland. E-mail: riviere@math.ethz.ch

for both perturbations in the target (of the form $\pi_N(u+t\phi)$ for any ϕ in $C_0^\infty(B^m, \mathbb{R}^k)$ where $\pi_N(y)$ is the nearest neighbor of y in N) and in the domain (of the form $u \circ (id + tX)$ for any vector field X in $C_0^\infty(B^m, \mathbb{R}^m)$). There are relations between these two conditions, in particular a smooth weakly harmonic map u (i.e. critical for perturbations in the target) is automatically stationary (i.e. critical for perturbations in the domain). This is not true in the general case : there exists weakly harmonic maps which are not stationary (see [HLP]) and that can even be nowhere continuous (see [Ri1]). On the one hand, as a consequence of being weakly harmonic, one has the Euler Lagrange equation (*harmonic map equation*)

$$\Delta u + A(u)(\nabla u, \nabla u) = 0 \quad , \tag{1.1}$$

where $A(u)$ is the second fundamental form of N embedded in $\mathbb{R}^k$. Stationarity, on the other hand, implies the *monotonicity formula* saying that for any point x_0 in B^m the *density of energy* $r^{2-m} \int_{B_r(x_0)} |\nabla u|^2$ is an increasing function of r. Among stationary harmonic maps are the *minimizing harmonic maps* (minimizing E for their boundary datas). In [ScU] R. Schoen and K. Uhlenbeck proved that there exists $\varepsilon(m, N) > 0$ such that, for any minimizing harmonic map, u in $W^{1,2}(B^m, N^n)$ and any ball $B_r(x_0) \subset B^m$ the following holds

$$\frac{1}{r^{n-2}} \int_{B_r(x_0)} |\nabla u|^2 \, dx \leq \varepsilon_0 \quad \Longrightarrow \quad \|\nabla u\|_{L^\infty(B_{r/2}(x_0))} \leq \frac{C}{r}$$

where C only depends on m and N^n. In other words, there exists a number $\varepsilon > 0$, depending only m and N, making a strict splitting between the *almost linear regime* for minimizing harmonic maps in which derivatives of the maps are under control and the *totally non-linear regime* where the map may be singular. Since this theorem in the early eighties, $\varepsilon-$regularity results has been found in various problems of geometric analysis (minimal surfaces, geometric flows, Yang-Mills Fields...etc). In particular it was a natural question whether the above result was true for arbitrary stationary harmonic maps. This was done 10 years later by L.C. Evans ([Ev]) in the case where N^n is a standard sphere. Evans' result was extended to general target by F. Bethuel in [Be]. Evans benefitted from a deeper understanding of the non-linearity in (1.1) developed in Hélein's proof that any weakly harmonic map from a two dimensional domain is C^∞ (see [He] and [CLMS]). This fact will be discussed in the next subsection.

Combining the $\varepsilon-$regularity result with a classical Federer-Ziemer covering argument yields the following upper-bound on the size of the singular set of stationary harmonic maps. For a stationary harmonic map u let $\text{Sing}\, u$ be the complement of the largest open subset where u is C^∞. Then

$$\mathcal{H}^{m-2}(\text{Sing}\, u) = 0 \quad , \tag{1.2}$$

where $\mathcal{H}^{m-2}$ denotes the $m - 2-$dimensional Hausdorff measure. For stationary maps and general targets, this is the best estimate available. For minimizing harmonic maps Schoen and Uhlenbeck proved the optimal result : $\dim(\text{Sing}\, u) \leq m-3$. The reason for this improvement is the striking fact that weakly converging *minimizing harmonic maps* in $W^{1,2}(B^m, N)$ are in fact strongly converging in this space.

This feature is very specific to the minimizing map case. When we blow up the map at a singular point strong convergence then implies roughly that the "singular set of the limit is not smaller than the limit of the singular set". *The Federer dimension reduction argument* from the theory of minimal surfaces then gives by induction the result. Here one sees the strong connection between understanding the singular set and compactness properties of sequences of solutions and bubblings as presented in the next subsection.

1.2.   Bubblings of harmonic maps

The following result of J.Sacks and K.Uhlenbeck [SaU] is perhaps the earliest example of energy quantization in non-linear analysis. Their aim was to extend the work of Eells and Sampson [ES] to targets of not necessarily negative sectional curvatures, that is : find in a given 2-homotopy class of an arbitrary riemannian manifold N^n a more "natural" representant that would minimize the area and the Dirichlet energy E in the given homotopy class . This raises the question of compactness and the possible reasons for the lack of compactness for harmonic maps from a 2-sphere (or even more generally a Riemann surface) into N^n. We should mention not only the original contribution of J.Sacks and K.Uhlenbeck that focused on minimizing sequences but also the later works by J.Jost [Jo] for critical points in general, by M. Struwe [St] for it's heat flow version and the more recent contributions by T.Parker [Pa], W. Ding and G. Tian [DiT] and F.H.Lin and C. Wang [LW]. The following result has influenced deeply the non-linear analysis of the eighties, from the concentration-compactness of P.L.Lions to the compactness of J-holomorphic curves by M.Gromov and the analysis of self-dual instantons on 4-manifolds by S.K. Donaldson and K.Uhlenbeck.

Theorem 1. [SaU], [Jo] *Let u_n be a sequence of weakly harmonic maps from a surface Σ into a closed manifold N^n having a uniformly bounded energy. Then a subsequence $u_{n'}$ weakly converges in $W^{1,2}(\Sigma, N^n)$ to a harmonic map u into N^n. Moreover, there exist finitely many points $\{a_1 \cdots a_k\}$ in Σ such that the convergence is strong in $\Sigma \setminus \{a_1 \cdots a_k\}$ and the following holds*

$$|\nabla u_{n'}|^2 \; dvol_\Sigma \rightharpoonup |\nabla u|^2 \; dvol_\Sigma + \sum_{j=1}^{k} m_j \delta_{a_j} \qquad in \ Radon \ measure$$

where $m_j = \sum_{i=1}^{P_j} E(\phi_i^j)$ and ϕ_i^j are nonconstant harmonic 2-spheres of N^n (harmonic maps from S^2 into N^n).

The loss of energy during the weak convergence is not only concentrated at points but is also quantized : the amount is given by a sum of energies of harmonic 2-spheres of N^n, the *bubbles*, that might sometimes be even explicitly known (for instance if $N = S^2$, $E(\phi_i^j) \in 8\pi \mathbb{Z}$). The striking fact in this result is that it excludes the possibility of losing energy in the neck between u and the bubbles or between the bubbles themselves. This *no-neck property* is quite surprising. It is a-priori conceivable, for instance, that, in a tiny annulus surrounding a blow-up point, an axially symmetric harmonic map into N, that is a portion of geodesic of arbitrarily small length in N, breaking the quantization of the energy, would appear. The *no-neck property* disappears if instead of exact solutions we consider in general Palais-Smale sequences for E in general (see [Pa]).

Only relatively recently a first breakthrough was made by F.H. Lin in the attempt of extending Sacks Uhlenbeck result beyond the conformal dimension.

Theorem 2. [Li] *Let u_n be a sequence of stationary harmonic maps from B^m into a closed manifold N^n having a uniformly bounded energy, then there exists a subsequence $u_{n'}$ weakly converging in $W^{1,2}(B^m, N^n)$ to a map u and there exists a $m-2$ rectifiable subset K of B^m such that the convergence is strong in $B^m \setminus K$ and moreover*

$$|\nabla u_{n'}|^2 \, dvol_\Sigma \rightharpoonup |\nabla u|^2 \, dvol_\Sigma + f(x) \, \mathcal{H}^{m-2} \lfloor K \qquad \text{in Radon measure}$$

where f is a measurable positive function of K.

This result, establishing the regularity of the blow-up set of weakly converging stationary harmonic map, is related to the resolution by D.Preiss [Pr] of the Besicovitch conjecture on measures admitting densities. Given a positive Borel regular measure μ such that there exists an integer k for which μ-a.e. the density $\lim_{r \to 0} \mu(B_r)/r^k$ exists and is positive, it is proved in [Pr] that there is a rectifiable k-dimensional rectifiable set K such that $\mu(B^m \setminus K) = 0$. In the present situation calling $\mu_n = |\nabla u_n|^2 \, dvol_\Sigma$ converging to the Radon measure $\mu = |\nabla u|^2 \, dvol_\Sigma + \nu$, from the monotonicity formula one deduces easily that the defect measure ν fulfills the assumptions of Preiss theorem for $k = m - 2$ and the rectifiability follows at once. It has to be noted that the original proof of Theorem 2 in [Li] is self-contained and avoids D. Preiss above result.

Beyond the regularity of the defect measure ν, the question remained whether the whole picture established in the conformal dimension could be extended to higher dimensions and if the *no neck property* still holds. F.H. Lin and the author brought the following answer to that question.

Theorem 3. [LR2] *Let f be the function in the previous theorem, if $N^n = (S^n, g_{stand})$ or if there is a uniform bound on $\|u_{n'}\|_{W^{2,1}(B^m)}$, then, for $\mathcal{H}^{m-2}$ a.e. x of K, there exists a finite family of harmonic 2-spheres in N^n $(\phi_x^j)_{j=1\cdots P_x}$ such that*

$$f(x) = \sum_{j=1}^{P_x} E(\phi_x^j).$$

The proof of this *no neck property* in higher dimension is of different nature from the one provided previously in dimension 2 where the use of objects relevant to the conformal dimension only, such as the Hopf differential, was essential. The idea in higher dimension was to develop a technique of slicing, averaging method combined with estimates in Lorentz spaces $L^{2,\infty} - L^{2,1}$. This technique seems to be quite general as it have some genericity and permitted to solve problems of apparently different nature as we shall expose in the next section. The requirement of the $W^{2,1}$ bound for the case of a general target seems to be technical and should be removed. In the particular case when the target is the round sphere it was proved in [He] and [CLMS] that the non-linearity $A(u)(\nabla u, \nabla u)$ in equation (1.1) is in the Hardy space $\mathcal{H}^1_{loc}$ which immediately implies the desired $W^{2,1}$ bound for the maps

u. Whether this fact and this $W^{2,1}$ bound can be extended to general targets is still unknown.

The 2 previous results suggest to view the loss of compactness in higher dimension as being exactly the one happening in the conformal 2-dimensional case in the plane normal to K, locally invariant in the remaining $m-2$ dimensions tangent to K. This understanding of the loss of compactness through creation of 2-bubbles for stationary harmonic maps has consequences in regularity theory. Indeed, in the case where N^n admits no harmonic 2-spheres (take for instance N^n a surface of positive genus), no blow-up can arise ; weakly converging stationary harmonic maps are strongly converging and the *Federer dimension reduction argument* discussed at the end of the previous subsection combined with the analysis in [Si] may be applied to improve the bound of the size of the singular set to $\dim \text{Sing} \, u \leq m-4$ (see [Li]).

1.3. High dimensional gauge theory

The work of S.K. Donaldson and R. Thomas [DoT] has given a new boost in the motivation for developing the non-linear analysis of high dimensional gauge theory.

Bubbling of Yang-Mills Fields

Theorems 1,2,3 and their proofs are transposable to many others geometric variational problems (such has Yang-Mills fields, Yamabe metrics...) having a given conformal invariant dimension p (p=2 for harmonic maps, 4 for YM...etc). For instance in [Ti], G.Tian established the result corresponding to theorem 2 for Yang-Mills fields.

Consider a vector bundle E over a Riemannian manifold (M^m, g) and assume E is issued from a principal bundle whose structure group is a compact Lie group G. Yang-Mills fields are connections A on E whose curvature F_A solves

$$d_A^* F_A = 0 \qquad (1.3)$$

where d_A^* is the adjoint to the operator d_A, acting on $\wedge^* M \otimes End \, E$, with respect to the metric g on M and the Killing form of G on the fibers (recall that the Bianchi identity reads instead $d_A F_A = 0$). Yang-Mills fields are the critical points of the Yang-Mills functional $\int |F_A|^2 \, dvol_g$ for perturbations of the form $A + ta$ where $a \in \Gamma(\wedge^1 M \otimes ad \, Q)$. It is proved in [Ti] that, taking a sequence of smooth Yang Mills connections F_{A_n} having a uniformly bounded Yang-Mills energy, one may extract a subsequence (still denoted A_n) such that for some $m-4$ rectifiable closed subset K of M the following holds : in a neighborhood of any point of $M \setminus K$ there exist good choices of gauges such that A_n, expressed in these gauges, converges in C^k topology (for any k) to a limiting form A that defines globally a smooth Yang-Mills connection on $i^* E$, where i is the canonical embedding of $M \setminus K$ in M. Moreover, one has the following convergence in Radon measure :

$$|F_{A_n}|^2 \, dvol_M \rightharpoonup |F_A|^2 \, dvol_M + f(x) \, \mathcal{H}^{m-4} \lfloor K \quad , \qquad (1.4)$$

where f is some non-negative measurable function on K. For $m = 4$, or under the assumption that the measures $|\nabla_{A_n} \nabla_{A_n} F_{A_n}| \, dvol_M$ remain uniformly bounded it is proved by the author in [Ri4] that $f(x)$ is, $\mathcal{H}^{m-4}$ almost everywhere on K, quantized and equal to a sum of energies of Yang-Mills connections over S^4 : the *no neck property* holds. The proof use again the Lorentz space duality $L^{2,\infty} - L^{2,1}$ applied this time to the curvature. These estimates are consequences of $L^{4,2}$ estimates on the connections A for $m = 4$. It is interresting to observe that such an $L^{4,2}$ estimate of the connection plays a crucial role in the resolution by J.Shatah and M. Struwe of the wave map Cauchy problem in 4 dimension for small initial datas in the "natural" space $H^2 \times H^1$ (see [ShS]).

To complete the description of the blow-up phenomena several open questions remain both for harmonic maps and Yang-Mills fields. First, what are the exact nature of the limiting map u and connection A ? Is u still a stationary harmonic map on the whole ball B^m ? Is A still a smooth Yang-Mills connection of some new bundle E_0 over (M, g) ? Can one expect in both cases more regularity than the rectifiability for K ? What is the exact nature of the concentration set K ?

It happens that these two questions are strongly related to eachother : For instance, in the context of harmonic maps, if the weak limiting map u was also a stationary harmonic map, then one would deduce from the monotonicity formula that K is a stationary varifold and therefore inherits nice regularity properties. There is a similar notion of stationary Yang-Mills field (see [Ti]). The limiting A being stationary, as expected, likewise would imply the stationarity of K. These questions in both cases are still widely open and seem to be difficult. It is even unknown, in the general case, whether the limiting map u is weakly harmonic on the whole of B^m. Nevertheless in the case of Yang-Mills, in a work in preparation, T. Tao and G. Tian are proving a *singularity removability* result saying that the weak limit A of <u>smooth</u> Yang-Mills A_n can be extended to a smooth Yang-Mills connection aside from a $\mathcal{H}^{m-4}$ measure zero set.

An important case where these questions have been solved is :

The case of Ω-anti-self-dual Instantons

This notion extends the 4-dimensional notion of instantons to higher dimension. Assuming $m \geq 4$, and given a closed $m - 4$-form Ω, we say that a connection A is an Ω−anti-self-dual connection when the following equation holds

$$- * (F_A \wedge \Omega) = F_A. \tag{1.5}$$

In view of the Bianchi identity $d_A F_A = 0$ and the closedness of Ω it follows that $d_A^* F_A = 0$ that is A is a particular solution to Yang-Mills equations. It is then shown in [Ti] that, if we further assume that the co-mass of Ω is less than 1, then K in (1.4) is a minimizing current in (M, g) calibrated by Ω (i.e. Ω restricted to K coincides with the volume form on K induced by g) and therefore minimizes the area in it's homology class.

A special case of interest arises when (M, g, ω) is a Calabi-Yau 4-fold with a global "holomorphic volume form" θ (i.e. θ is a holomorphic $(4,0)$−form satisfying $\theta \wedge \bar{\theta} = dvol_M$). Let Ω be the form generating the $SU(4)$ holonomy on M (i.e. Ω is a unit holomorphic section of the canonical bundle $\wedge^{(4,0)} T^* M$) and assume M is the product $T \times V^3$ of a 2-torus T with a Calabi Yau 3-fold V^3, for a given $SU(2)$−bundle E, T−invariant solutions to (1.5) are pairs of connections B on E over V^3 and sections ϕ of $End\, E$ that solve a vortex equations (see 6.2.2 in [Ti]). The moduli space of these solutions should be related to the so-called holomorphic Casson invariants (see also R.Thomas work [Th]). The loss of compactness of these solutions arises along T times holomorphic curves in V^3. To relate this moduli space with the holomorphic Casson invariants it is necessary to perturb, in a generic way, the complex structure into a not necessarily integrable one. This generates several non-linear analysis questions related to bubbling and regularity which have to be solved :

i) Show that the weak limits of the vortex equations in the almost complex setting in V^3 have only isolated point singularities located on the pseudo-holomorphic blow-up set (see the extended conjectures on the singular sets of solutions to (1.5) in [Ti]).

ii) Construct solutions to the vortex equations which concentrate on some given choice of pseudo-holomorphic curves.

iii) Show that an arbitrary 1-1 rectifiable cycle (i.e. a 2 dimensional cycle whose tangent plane is invariant under the almost complex structure action) in an almost complex manifold (that may arise as a blow-up set in the above case) is a smooth surface aside from eventually isolated branched points.

Question ii) is studied in a work in progress with F. Pacard. In collaboration with G.Tian, in the direction of iii), the following result was established.

Theorem [RT] *Let (M^4, J) be a smooth almost complex 4-real manifold. Then any 1-1 rectifiable cycle is a smooth pseudo-holomorphic curve aside from isolated branched points.*

2.   Ginzburg-Landau line vortices

2.1.   The strongly repulsive asymptotic

The free Ginzburg-Landau energy on the 3-dimensional ball reads

$$GL(u, A) = \int_{B^3} |d\,u - iAu|^2 + \frac{\kappa^2}{2}(1 - |u|^2)^2 + |dA|^2 \qquad (2.1)$$

where u is a complex function satisfying $|u| \leq 1$ and A is a 1-form on B^3. The study of the above variational problem was initiated by A. Jaffe and C. Taubes in 2 dimensions and by J. Fröhlich and M. Struwe in [FS] in 3 dimensions. The connected components of the zero set of the order parameter u are the so called *vortices* that must generically be lines (for 0 being a regular value of u). The parameter κ plays a crucial role in the theory. Depending on the value of κ, one expects different types of "behaviors" of vortices for minimizing configurations relative to various constraints (boundary datas, topological constraints...) (see [JT] and [Ri3] for a survey on these questions). In particular for large κ, in the so called *strongly repulsive limit*, the vortices of minimizing configurations tend to minimize their length (the first order of the energy is $2\pi \log \kappa$ times the length of the vortices) and repel one another, which is not the case for small κ. The magnetic field dA plays no role in this mechanism ; to simplify the presentation we henforth simply impose that $A = 0$. We are then looking at the functional

$$E_\varepsilon(u) = \int_{B^3} |\nabla u|^2 + \frac{1}{2\varepsilon^2}(1 - |u|^2)^2 = \int_{B_3} e_\varepsilon(u) \qquad (2.2)$$

in the strongly repulsive limit $\varepsilon \to 0$ whose critical points satisfy the non longer gauge invariant Ginzburg-Landau equation

$$\Delta u + \frac{u}{\varepsilon^2}(1 - |u|^2) = 0 \qquad \text{in } \mathcal{D}'(\mathbb{R}^3). \qquad (2.3)$$

Because we are interested in observing finite length vortices, in view of the above remark, we have to restrict to critical points satisfying $E_\varepsilon(u_\varepsilon) = O(\log \frac{1}{\varepsilon})$. Equation (2.3) gives, in particular, that Δu is parallel to u. Therefore, assuming that $|u| > \frac{1}{2}$ in a ball $B_r(x_0)$ and writing $u = |u|e^{i\phi}$, we deduce the nice scalar elliptic equation

$$div(|u|^2 \nabla \phi) = 0 \qquad \text{in } \mathcal{D}'(B_r(x_0)). \qquad (2.4)$$

Assuming $|u_\varepsilon| > 1/2$ on the whole B^3, the compactness of solutions to (2.3), in $W^{1,2}$ say, is reduced to the compactness of the boundary data. It is then clear that, in general, the study of the compactness of solutions u_ε to (2.3), in the limit $\varepsilon \to 0$, involves the study of the compactness of the sets $V_\varepsilon = \{x \; ; \; |u_\varepsilon(x)| < 1/2 \}$ as well as a control of the degree of $u/|u|$ on arbitrary closed curves in the complement of V_ε approximating the vortices. (The number $1/2$ may be of course replaced by an arbitrary number between 0 and 1).

The approximate vortices T_ε

In their study of the 2-dimensional version of the present problem (B^3 replaced by B^2), F. Bethuel, H. Brezis and F. Hélein in [BBH], established that for solutions to (2.3) satisfying a fixed boundary condition from ∂B^2 into S^1 of non-zero degree, V_ε can be covered by a uniformly bounded number of balls of radius $O(\varepsilon)$ around which $u/|u|$ has a uniformly bounded topological degree. Taking then the distribution T_ε given by the sum of the Dirac masses at the center of the balls with the multiplicity given by the surrounding degrees of $u/|u|$, because of the uniform bound of the mass of T_ε, we may extract a converging subsequence of atomic measures $T_{\varepsilon'}$. The compactness of the maps u_ε is a consequence of the convergence of $T_{\varepsilon'}$ via a classical elliptic argument.

Going back to 3 dimensions, the idea remains the same : we introduce an approximation of the vortices and try to prove compactness. More precisely, we consider a minimal 1-dimensional integer current T_ε in the homology class of $H_1(V_\varepsilon, \partial B^3, \mathbb{Z})$ given by the pre-image by u_ε of a regular point in $B^2_{\frac{1}{4}}$. Then, as in 2-dimensions, the key to the compactness of the critical points of E_ε is now the compactness of the familly of 1-dimensional rectifiable current T_ε constructed above.

The energy quantization result exposed in the following subsection shows the compactness of the T_ε.

2.2. Quantization for G-L vortices : the η-compactness

The following energy quantization result introduced by the author in [Ri2] says that there exists an absolute constant $\eta > 0$ such that if the Ginzburg-Landau energy of a critical point in a given ball, suitably renormalized, lies below this number, then there is no vortex passing through the ball of half radius. Precisely we have.

Theorem 5. [LR1] *There exists a positive number* η, *such that for* ε *small enough and for any critical point of* E_ε *in* B_1^3 *satisfying* $|u| \leq 1$, *the following holds. Let* $B_r(x_0) \in B_{\frac{1}{2}}^3$, $r > \varepsilon$ *then*

$$\frac{1}{r} \int_{B_r(x_0)} e_\varepsilon(u_\varepsilon) \leq \eta \ \log \frac{r}{\varepsilon} \qquad \Longrightarrow \qquad |u| > \frac{1}{2} \ in \ B_{\frac{r}{2}}(x_0) \quad .$$

The above result is optimal in the following sense : there are critical points and arbitrary $r > \varepsilon$ such that $\frac{1}{r} \int_{B_r(x_0)} e_\varepsilon(u_\varepsilon) \leq 2\pi \ \log \frac{r}{\varepsilon}$ and $u(x_0) = 0$. Such a result is reminiscent of the $\varepsilon-$regularity result, although one difference here is that one has to handle blowing up energy E_ε which is not a-priori bounded as ε tends to 0.

The $\eta-$compactness was introduced first for minimizers in 3-D in [Ri2] and then obtained for any critical points, still in 3-D, in [LR1]. Regarding dimension 2, the idea of the $\eta-$compactness is implicit in the works before [Ri2]. Making this idea explicit in [Ri5] helped to substantialy simplify the existing proofs in the 2-D case.

It is a striking fact, whose explanation is beyond the scope of this lecture, that the original proof of theorem in [LR1] is based on the same techniques used by the authors to prove the *no neck property* (theorem 3) for harmonic maps. The key of the proof is to obtain a bound independent of ε for the density of energy $\frac{1}{\rho} \int_{B_\rho(x_0)} e_\varepsilon(u)$ (for some ρ between ε and r). The main obstacle in establishing such a bound is to control, independently of ε, the part of the energy in $B_\rho(x_0) \setminus V_\varepsilon$ coming from the vortices T_ε in $B_\rho(x_0)$. This is obtained by controlling, on a generic *good slice* $\partial B_{\rho_1}(x_0) \setminus V_\varepsilon$ ($\rho/2 < \rho_1 < \rho$), the $L^{2,\infty}$ and the $L^{2,1}$ norms of $\nabla\psi$, where ψ is the *vortex potential* $\psi := \Delta^{-1} T_\varepsilon$. Recently in [BBO] F.Bethuel, H.Brezis and G.Orlandi made a nice observation by deriving, from the monotonicity formula associated to the elliptic variational problem E_ε, an L^∞ bound in terms of the local density of energy $\frac{1}{\rho} \int_{B_\rho(x_0)} e_\varepsilon(u)$ for the *vortex potential* ψ in $B_\rho(x_0) \setminus V_\varepsilon$. This bound enabled them to replace in our original proof the use of the $L^{2,\infty} - L^{2,1}$ duality for $\nabla\psi \cdot \nabla\psi$ on a slice by a more standard, in calculus of variation, $L^\infty - L^1$ duality for $\psi \cdot \Delta\psi$ on $B_\rho(x_0)$ itself. As explained in [BBO], this modification is a real improvement in the high dimensional case since it allows to end the proof of the $\eta-$compactness without having to go through an argument based on a *good slice* extraction that would have been certainly more painful for $n \geq 4$. This modification of the duality used to control the energy of the *vortex potential* also allowed C. Wang recently to extend the $\eta-$compactness result from [LR3] for the Ginzburg-Landau heat flow equation from 3 to 4 dimensions.

The η-compactness is a compactness result

Considering, as mentioned above, critical points u_ε to E_ε satisfying $|u| \le 1$ and whose energy are of the order of $\log \frac{1}{\varepsilon}$ ($E_\varepsilon(u_\varepsilon) = O(\log \frac{1}{\varepsilon})$), a Besicovitch covering argument (see [LR1]) combined with the η−compactness above gives

$$V_\varepsilon \subset \cup_{j=1}^{N_\varepsilon} B_\varepsilon(x_j) \qquad \text{with} \qquad N_\varepsilon = O\left(\frac{1}{\varepsilon}\right) \quad .$$

This estimate, combined with the L^∞ bound of ∇u_ε deduced from the equation and the assumption $|u| \le 1$, yields the following fact: $M(T_\varepsilon) = O(1)$, (i.e. the mass of the approximated vortices is uniformly bounded). Combining this estimate with the fact that, by definition, $\partial T_\varepsilon \lfloor B^3 = 0$, as a direct application of Federer-Fleming compactness theorem, we deduce the compactness of the approximated vortices T_ε in the space of integer rectifiable 1-dimensional currents.

Thus, the η−compactness is the compactness of the approximated vortices T_ε, which are <u>rectifiable currents</u> rather than a compactness of <u>maps</u> u_ε. The latter compactness, in $W^{1,p}$ for $p < 3/2$, nevertheless automatically follows from the previous one by the means of classical elliptic arguments for a suitable class of boundary conditions for u_ε (see [LR1] page 219). This class has been recently extended to arbitrary $H^{\frac{1}{2}}(\partial B^3, S^1)$ conditions independent of ε in [BBBO]. This idea of getting convergence of maps going through convergence of currents is in the spirit of the theory of cartesian currents developed by M. Giaquinta, G. Modica and J. Soucek in [GMS].

References

[Be]   F. Bethuel, "On the singular set of stationary harmonic maps" Manuscripta Math., **78** (1993), 417–443.

[BBH]   F. Bethuel, H. Brezis and F. Hélein, "Ginzburg-Landau vortices", Birkhaüser (1994).

[BBO]   F. Bethuel, H. Brezis and G. Orlandi, "Small energy solutions to the Ginzburg-Landau equation" C.R.A.S. Paris, **331** (2000), 10, 763–770.

[BBBO]   F. Bethuel, J. Bourgain, H. Brezis and G. Orlandi, "$W^{1,p}$ estimates for solutions to the Ginzburg-Landau equation with boundary data in $H^{\frac{1}{2}}$" C.R.A.S. Paris, **333** (2001), no 12, 1069–1076.

[CLMS]   R. Coifman, P.-L. Lions, Y. Meyer and S. Semmes, "Compensated Compactness and Hardy Spaces" J. Math. Pure Appl., **72** (1993), 247–286.

[DiT]   W. Ding and G. Tian, "Energy identity for a class of approximate harmonic maps from surfaces" Comm. Anal. Geom., **3** (1995), no.3-4, 543–554.

[DoT]   S.K. Donaldson and R.P. Thomas, "Gauge Theory in higher dimensions" in "*The geometric universe*" (Oxford, 1996), Oxford Univ. Press, 1998, 31–47.

[ES]   J.J. Eells and J.H. Sampson, "Harmonic mappings of Riemannian manifolds" Amer. J. Math., **86**, (1964) 109–160.

[Ev]   L.C. Evans, "Partial regularity for stationary harmonic maps into spheres" Arch. Ration. Mech. Anal., **116**, no2, (1991), 101–113.

[FS]   J. Fröhlich and M. Struwe "Variational problems on vector bundles" Comm. Math. Phys., **131**, (1990), 431–464.

[GMS]   M. Giaquinta, G. Modica and J. Soucek, "Cartesian currents in the calculus of variations" Springer, 1998.

[HLP]   R. Hardt, F.H. Lin and C.C. Poon, "Axially symmetric harmonic maps minimizing a relaxed energy" Comm. Pure App. Math., **45** (1992), no 4, 417–459.

[He]   F. Hélein, "Harmonic maps, Conservation laws and moving frames" Diderot, (1996).

[JT]   A. Jaffe and C. Taubes, "Vortices and Monopoles" Progress in physics, Birkhaüser, (1980).

[Jo]   J. Jost, "Two-dimensional geometric variational problems", Wiley, (1991).

[Li]   F.H. Lin, "Gradient estimates and blow-up analysis for stationary harmonic maps" Ann. Math., **149**, (1999), 785–829.

[LR1]   F.H. Lin and T. Rivière, "A Quantization property for static Ginzburg-Landau Vortices" Comm. Pure App. Math., **54** , (2001), 206–228.

[LR2]   F.H. Lin and T. Rivière, "Energy Quantization for Harmonic Maps" Duke Math. Journal, **111** (2002), 177–193.

[LR3]   F.H. Lin and T. Rivière, "Quantization property for moving Line vortices", Comm. Pure App. Math., **54** (2001), 826–850.

[LW]   F.H. Lin and C.Y. Wang, "Energy identity of harmonic map flows from surfaces at finite singular time", Cal. Var. and P.D.E. **6** (1998), no 4, 369–380.

[Pr]   D. Preiss, "Geometry of measures in $\mathbb{R}^n$: distribution, rectifiability and densities" Ann. of Math., **125** (1987), no 3, 537–643.

[Pa]   T. Parker, "Bubble tree convergence for harmonic maps", J. Diff. Geom., **44** (1996), 545–633.

[Ri1]   T. Rivière, "Everywhere discontinuous Harmonic Maps into Spheres", Acta Matematica, **175** (1995), 197–226.

[Ri2]   T. Rivière, "Line vortices in the $U(1)$-Higgs model" C.O.C.V., **1**, (1996), 77-167.

[Ri3]   T. Rivière, "Ginzburg-Landau vortices : the static model" Séminaire Bourbaki no 868, vol 1999/2000, Asterisque no **276** (2002), 73–103.

[Ri4]   T. Rivière, "Interpolation spaces and energy quantization for Yang-Mills fields" Comm. Anal. Geom., **8** (2002).

[Ri5]   T. Rivière, "Asymptotic analysis for the Ginzburg-Landau equations" Mini-course ETH Zürich 1997, Bolletino UMI, **2-B** (1999), no 8, 537–575.

[RT]   T. Rivière and G. Tian, "The singular set of J-holomorphic maps into algebraic varieties" submitted (2001).

[SaU]   J. Sacks and K. Uhlenbeck, The existence of minimal immersions of 2-spheres, Ann. of Math., **113** (1981), 1–24.

[ScU]   R.Schoen and K.Uhlenbeck, "A regularity theory for harmonic maps" J. Diff. Geom., **17** (1982), 307–335.

[ShS]   J. Shatah and M. Struwe, "The Cauchy problem for wave maps" IMRN, **2002:11** (2002), 555–571.

[Si]   L. Simon, "Rectifiability of the singular set of energy minimizing maps"
         Calc. Var. and P.D.E., **3** (1995), 1–65.

[St]   M. Struwe, "On the evolution of harmonic mappings of Riemannian sur-
         faces" Comm. Math. Helvetici, **60** (1985), 558–581.

[Th]   R. Thomas, "A holomorphic Casson invariant for Calabi-Yau 3-folds, and
         bundles on $K3$ fibrations", J. Diff. Geom, **54** (2000), 367–438.

[Ti]   G. Tian, "Gauge theory and calibrated geometry, I" Ann. Math., **151**
         (2000), 193–268.

ICM 2002 · Vol. III · 209–220

Nonlinear Wave Equations

Daniel Tataru*

Abstract

The analysis of nonlinear wave equations has experienced a dramatic growth in the last ten years or so. The key factor in this has been the transition from linear analysis, first to the study of bilinear and multilinear wave interactions, useful in the analysis of semilinear equations, and next to the study of non-linear wave interactions, arising in fully nonlinear equations. The dispersion phenomena plays a crucial role in these problems. The purpose of this article is to highlight a few recent ideas and results, as well as to present some open problems and possible future directions in this field.

2000 Mathematics Subject Classification: 35L15, 35L70.
Keywords and Phrases: Wave equations, Phase space, Dispersive estimates.

1. Introduction

Consider the constant and variable coefficient wave operators in $\mathbb{R} \times \mathbb{R}^n$,

$$\Box = \partial_t^2 - \Delta_x, \qquad \Box_g = g^{ij}(t,x)\partial_i\partial_j.$$

In the variable coefficient case the summation occurs from 0 to n where the index 0 stands for the time variable. To insure that the equation is hyperbolic in time we assume that the matrix g^{ij} has signature $(1,n)$ and that the time level sets $t = const$ are space-like, i.e. $g^{00} > 0$. We consider semilinear wave equations,

$$\Box u = N(u) \qquad (SLW), \qquad \Box u = N(u, \nabla u) \qquad (GSLW)$$

and quasilinear wave equations,

$$\Box_{g(u)} u = N(u)(\nabla u)^2 \qquad (NLW), \qquad \Box_{g(u,\nabla u)} u = N(u, \nabla u) \qquad (GNLW).$$

To each of these equations we associate initial data in Sobolev spaces

$$u(0) = u_0 \in H^s(\mathbb{R}^n), \qquad \partial_t u(0) = u_1 \in H^{s-1}(\mathbb{R}^n).$$

There are two natural questions to ask: (i) Are the equations locally well-posed in $H^s \times H^{s-1}$? (ii) Are the solutions global, or is there blow-up in finite time?

---

*Department of Mathematics, University of California at Berkeley, Berkeley, CA 94720, USA.
E-mail: tataru@math.berkeley.edu

Local well-posedness.   In a first approximation we define it as follows:

Definition 1. *A nonlinear wave equation is well-posed in $H^s \times H^{s-1}$ if for each $(v_0, v_1) \in H^s \times H^{s-1}$ there is $T > 0$ and a neighborhood V of (v_0, v_1) in $H^s \times H^{s-1}$ so that for each initial data $(u_0, u_1) \in V$ there is an unique solution $u \in C(-T, T; H^s)$, $\partial_t u \in C(-T, T; H^{s-1})$ which depends continuously on the initial data.*

In practice in order to prove uniqueness one often has to further restrict the class of admissible solutions. In most problems, the bound T from below for the life-span of the solutions can be chosen to depend only on the size of the data.

It is not very difficult to prove that all of the above problems are locally well-posed in $H^s \times H^{s-1}$ for large s. The interesting question is what happens when s is small. One indication in this regard is given by scaling. At least in the case when the nonlinear term has some homogeneity, for instance $N(u) = u^p$ or $N(u) = u^p (\nabla u)^q$, one looks for an index α so that all transformations of the form $u(x, t) \to \lambda^\alpha u(\lambda x, \lambda t)$, $\lambda > 0$ leave the equation unchanged. Correspondingly one finds an index $s_0 = \frac{n}{2} - \alpha$ so that the norm of the initial data (u_0, u_1) in the homogeneous Sobolev spaces $\dot{H}^s \times \dot{H}^{s-1}$ is preserved by the above transformations.

Below scaling ($s < s_0$) a small data small time result rescales into a large data large time result. Heuristically one concludes that local well-posedness should not hold. Still, to the author's knowledge there is no proof of this yet.

Conjecture 2. *Semilinear wave equations are ill-posed below scaling.*

This becomes much easier to prove if one strengthens the definition of well-posedness, e.g. by asking for uniformly continuous or C^1 dependence of the solution on the initial data.

If $s = s_0$ then for small initial data local well-posedness is equivalent to global well-posedness. The same would happen for large data if we were to strengthen the definition of well-posedness and ask for a lifespan bound which depends only on the size of the data. This is the only case where this distinction makes a difference.

If $s > s_0$ then a local well-posedness result gives bounds for life-span T_{max} of the solutions in terms of the size of the data,

$$\|(u_0, u_1)\|_{H^s \times H^{s-1}} \leq M \implies T_{max} \gtrsim M^{s_0 - s}.$$

The better localization in time makes the problems somewhat easier to study. However, besides scaling there are also other obstructions to well-posedness. These are related to various concentration phenomena which can occur depending on the precise structure of the equation.

Global well-posedness.   We briefly mention that there is a special case in which the global well-posedness is well understood, namely when the initial data is small, smooth, and decays at infinity. This is not discussed at all in what follows.

Consider first the case when s is above scaling, $s > s_0$, and local well-posedness holds in $H^s \times H^{s-1}$. Then any solution can be continued as long as its size does not blow-up. Hence the goal of any global argument should be to establish a-priori bounds on the $H^s \times H^{s-1}$ norm of the solution. All known results of this type are for problems for which there are either conserved or quasi-conserved positive definite quantities. Such conserved quantities can often be found for equations which are physically motivated or which have some variational structure. For simplicity suppose that there is some index s_c and an energy functional E in $H^{s_c} \times H^{s_c-1}$ which is preserved along the flow. The index s_c needs not be equal to the scaling index s_0. There are three cases to consider:

(i) The subcritical case $s_c > s_0$. Then a local well-posedness result at $s = s_c$ implies the global result for $s \geq s_c$. Furthermore, in recent years there has been considerable interest in establishing global well-posedness also for $s_0 < s < s_c$. This is based on an idea first introduced by Bourgain [5] in a related problem for the Schröedinger equation, and followed up by a number of authors.

(ii) The critical case $s_c = s_0$. Here the energy is not needed for small data, when local and global well-posedness are equivalent. For large data, however, the energy conservation is not sufficient in order to establish the existence of global solutions. In addition, one needs a non-concentration argument, which should say that the energy cannot concentrate inside a characteristic cone.

(iii) The supercritical case, $s_c < s_0$. No global results are known:

Open Problem 3. Are supercritical problems globally well-posed for $s \geq s_0$?

A simple example is the equation (NLW) with $N(u) = |u|^{p-1}u$. The energy is

$$E(u) = \int |u_t|^2 + |\nabla_x u|^2 + \frac{1}{p+1}|u|^{p+1}.$$

Then $s_c = 1$, while $s_0 = \frac{n}{2} - \frac{2}{p-1}$. In $3+1$ dimensions, for instance, $p = 3$ is subcritical, therefore one has global well-posedness in $H^1 \times L^2$. The exponent $p = 5$ is critical and in this case the problem is known to be globally well-posed in $H^1 \times L^2$; the non-concentration argument is due to Grillakis [7]. The exponent $p = 7$ is supercritical.

Blow-up. Not all nonlinear wave equations are expected to have global solutions. Quite the contrary, generic equations are expected to blow up in finite time; only for problems with some special structure it seems plausible that global well-posedness may hold. A simple way to produce blow-up is to look for self-similar solutions, $u(x,t) = t^\gamma u(\frac{x}{t})$. If they exist, self-similar solutions disprove global well-posedness. Because they must respect the scaling of the problem, they are not so useful when trying to disprove local well-posedness.

Another way to produce blow-up solutions is the so-called ode blow-up. In the simplest setting this means looking at one dimensional solutions (say $u(x,t) = u(t)$) which solve an ode and blow up in finite time. Then one can truncate the initial data spatially and still retain the blow-up because of the finite speed of propagation. This is still not very useful for the local problem.

A better idea is to constructs blow-up solutions which are concentrated essentially along a light ray, see Lindblad [10],[11] and Alihnac [1]. In this setup the actual blow-up occurs either because of the increase in the amplitude, in the semilinear case, or because of the focusing of the light rays, in the quasilinear case. As it turns out, the counterexamples of this type are often sharp for the local well-posedness problem.

2.   Semilinear wave equations

Usually, a fixed point argument is used to obtain local results for semilinear equations. We first explain this for the case when $s = s_0$. We define the homogeneous and inhomogeneous solution operators, S and $\Box^{-1}$ by

$$S(u_0, u_1) = u \Longleftrightarrow \{\Box u = 0, \quad u(0) = u_0, \quad \partial_t u(0) = u_1\},$$

$$\Box^{-1} f = u \Longleftrightarrow \{\Box u = f, \quad u(0) = 0, \quad \partial_t u(0) = 0\}.$$

Then the equation (NLW) for instance can be recast as

$$u = S(u_0, u_1) + \Box^{-1} N(u).$$

To solve this using a fixed point argument one needs two Banach spaces X and Y with the correct scaling and the following mapping properties:

$$S : H^s \times H^{s-1} \to X, \quad \Box^{-1} : Y \to X, \quad N : X \to Y.$$

The first two are linear, but the last one is nonlinear. The small Lipschitz constant is always easy to obtain provided the initial data is small and that N decays faster than linear at 0. The solutions given by the fixed point argument are global.

In the case $s > s_0$ the scaling is lost, and with this method one can only hope to get results which are local in time. To localize in time one chooses a smooth compactly supported cutoff function χ which equals 1 near the origin. The fixed point argument is now used for the equation

$$u = \chi S(u_0, u_1) + \chi \Box^{-1} N(u).$$

A solution to this solves the original equation only in an interval near the origin where $\chi = 1$. The modified mapping properties are

$$\chi S : H^s \times H^{s-1} \to X, \quad \chi \Box^{-1} : Y \to X, \quad N : X \to Y.$$

How does one choose the spaces X, Y? One approach is to use the energy estimates for the wave equation and set

$$X = \{u \in L^\infty(H^s), \nabla u \in L^\infty(H^{s-1})\}, \quad Y = L^1(H^{s-1}).$$

The first two mapping properties are trivial. However, if the third holds then we must also have $N : X \to L^\infty(H^{s-1})$. The one unit difference in scaling between L^1 and L^∞ implies that this can only work for $s \geq s_0 + 1$.

What is neglected in the above setup is the dispersive properties of the wave equation. Solutions to the linear wave equation cannot stay concentrated for long time intervals. Instead, they will disperse and decay in time (even though the energy is preserved). In harmonic analysis terms, this is related to the restriction theorem (see [17]) and is a consequence of the nonvanishing curvature of the characteristic set for the wave operator, namely the cone $\xi_0^2 = \xi_1^2 + \cdots + \xi_n^2$. Here ξ stands for the Fourier variable. One way of quantifying the dispersive effects is through the Strichartz estimates. They apply both to the homogeneous and the inhomogeneous equation (see [8] and references therein):

$$S : H^\rho \times H^{\rho-1} \to L^p L^q, \qquad |D|^{1-\rho_1-\rho} \square^{-1} : L^{p_1'} L^{q_1'} \to L^p L^q$$

where (ρ, p, q) and (ρ_1, p_1, q_1) are subject to

$$\frac{1}{p} + \frac{n}{q} = \frac{n}{2} - \rho, \quad \frac{2}{p} + \frac{n-1}{q} \le \frac{n-1}{2}, \quad 2 \le p, q \le \infty, \quad (\rho, p, q) \ne (1, 2, \infty).$$

The worst case in these estimates occurs for certain highly localized approximate solutions to the wave equation, which are called wave packets. A frequency λ wave packet on the unit time scale is essentially a bump function in a parallelepiped of size $1 \times \lambda^{-1} \times (\lambda^{-\frac{1}{2}})^{n-1}$ which is obtained from a $\lambda^{-1} \times (\lambda^{-\frac{1}{2}})^{n-1}$ parallelepiped at time zero which travels with speed 1 in the normal direction. Because of the uncertainty principle, this is the best possible spatial localization which remains coherent up to time 1. Of course one can rescale and produce wave packets on all time scales.

In low dimension $n = 2, 3$ the Strichartz estimates provide a complete set of results for generic equations of both (NLW) and (GNLW) type. Consider the following two examples, of which the second is wrong but almost right:

$$\square u = u^3, \quad n = 3, \quad s = s_0 = \frac{1}{2}, \quad X = L^4, \quad Y = L^{\frac{4}{3}},$$

$$\square u = u \nabla u, \quad n = 3, \quad s_0 = \frac{1}{2}, \quad s = 1 \quad X = |D|^{-1} L^\infty L^2 \cap L^2 L^\infty \quad Y = L^2.$$

For $n \ge 4$, however, the Strichartz estimates no longer provide all the results. The reason is as follows. The worst nonlinear interaction in both (NLW) and (GNLW) occurs for wave packets which travel in the same direction. One can use the Strichartz estimates to accurately describe the interaction of same frequency wave packets. But in the interaction of two wave packets at different frequencies, the low frequency packet is more spread, and only a small portion of it will interact with the high frequency packet. However, unlike in low dimension, the Strichartz estimates do not provide sharp bounds for this smaller part of a wave packet.

A more robust idea due to Bourgain [4] and Klainerman-Machedon [12] is to use the $X^{s,b}$ spaces associated to the wave equation very much in the same way the Sobolev spaces are associated to the Laplacian:

$$\|u\|_{X^{s,b}} = \|(1 + |\xi|)^s (1 + ||\xi_0| - |\xi'||)^b \hat{u}\|_{L^2}.$$

Then one chooses $X = X^{s,\frac{1}{2}}$ and $Y = X^{s-1,-\frac{1}{2}}$. The Strichartz information is not lost since for ρ, p, q as above we have the dual embeddings

$$X^{\rho,\frac{1}{2}+} \subset L^p L^q, \qquad L^{p'} L^{q'} \subset X^{-\rho,-\frac{1}{2}-}.$$

Within the framework of the $X^{s,b}$ spaces one can prove bilinear estimates which provide a better description of the interaction of high and low frequencies, see [6] and references therein. The bilinear estimates are obtained as weighted convolution estimates in the Fourier space, by using the above embeddings, or by combining the two methods. Sometimes even this setup does not suffice and has to be modified further, see [22].

Conjecture 4. *The equation $\Box u = u^p$ is locally well-posed in $H^s \times H^{s-1}$ for $n \geq 4$, $0 \leq s \leq \frac{1}{2}$, $p(\frac{n+1}{4} - s) \leq (\frac{n+5}{4} - s)$.* (see [19] for more details)

The null condition. A natural question to ask is whether there are equations which behave better than generic ones. This may happen if the worst interaction (between parallel wave packets) does not occur in the nonlinearity. A good example is (GNLW) with a quadratic nonlinearity $Q(\nabla u, \nabla u) = q^{ij} \partial_i u \, \partial_j u$. The cancellation condition, called null condition, asserts that

$$q^{ij} \xi_i \xi_j = 0 \qquad \text{in the characteristic set } g^{ij} \xi_i \xi_j = 0.$$

All such null forms are linear combinations of

$$Q_{ij}(\nabla u, \nabla v) = \partial_i u \partial_j v - \partial_i v \partial_j u, \qquad Q_0(u, v) = g^{ij} \partial_i u \partial_j v.$$

Open Problem 5. Study semilinear wave equations corresponding to variable coefficient wave operators for $n \geq 4$ (generic case) or $n \geq 2$ (with null condition).

In the constant coefficient case one can easily use the null condition in the context of the $X^{s,b}$ spaces. This is done using inequalities of the following form:

$$|q_0(\xi, \eta)| \leq c(|p(\xi)| + |p(\eta)| + |p(\xi + \eta)|)$$

respectively

$$|q_{ij}(\xi, \eta)| \leq c|\xi|^{\frac{1}{2}} |\eta|^{\frac{1}{2}} |\xi + \eta|^{\frac{1}{2}} (|p(\xi)|^{\frac{1}{2}} + |p(\eta)|^{\frac{1}{2}} + |p(\xi + \eta)|^{\frac{1}{2}})$$

where by $p(\xi)$ we denote the symbol of the constant coefficient wave operator, given by $p(\xi) = \xi_0^2 - \xi_1^2 - \cdots - \xi_n^2$. Combining this with the embeddings above one can lower the s in the local theory whenever the null condition is satisfied. Unfortunately, this does not always give optimal results. The problem of obtaining improved $L^p L^q$ estimates for null forms has also been explored, see [28][20] [26], but without immediate applications to semilinear wave equations. We limit the following discussion to two of the more interesting models.

Wave maps. These are functions from $\mathbb{R}^n \times \mathbb{R}$ into a complete Riemannian manifold (M, g) which are critical points for

$$I(\phi) = \int_{\mathbb{R}^n \times \mathbb{R}} |\partial_t \phi|_g^2 - |\nabla_x \phi|_g^2 dx \, dt.$$

In local coordinates the equation for wave maps has the form

$$\Box\phi^k = \Gamma_{ij}^k(\phi)Q_0(\phi^i\phi^j)$$

where Γ_{ij}^k are the Riemann-Christoffel symbols. The energy functional is

$$E(u) = \int_{\mathbb{R}^n} |\partial_t\phi|_g^2 + |\nabla_x\phi|_g^2 dx.$$

The scaling index is $s_0 = \frac{n}{2}$ and $s_c = 1$. Local well-posedness for $s > s_c$ can be obtained using the $X^{s,b}$ spaces. For $s = s_c$, using some modified $X^{s,b}$ spaces, local (and therefore small data global) well-posedness was established first in homogeneous Besov spaces $B_{2,1}^{\frac{n}{2}} \times B_{2,1}^{\frac{n}{2}-1}$ in Tataru [25] and then in Sobolev spaces by Tao [21] (for the sphere, $n \geq 2$) and other authors (general target manifold, $n \geq 3$). Large data global well-posedness is false in the supercritical case $n \geq 3$, where self-similar blowup can occur. This leaves open problems in the critical case $n = 2$:

Conjecture 6. *(i) The two dimensional wave maps equation is globally well-posed for small data in $H^1 \times L^2$ for any complete target manifold.*

(ii) The two dimensional wave maps equation is globally well-posed for large data in $H^1 \times L^2$ for "good" target manifolds.

The Yang Mills equations.   Given a compact Lie group $\mathcal{G}$ whose Lie algebra $\mathbf{g}$ admits an invariant inner product $\langle\cdot,\cdot\rangle$ one considers $\mathbf{g}$ valued connection 1-forms $A_j dx^j$ in $\mathbb{R}^n \times \mathbb{R}$. The covariant derivatives of $\mathbf{g}$ valued functions are defined by

$$D_j B = \partial_j B + [A_j, B].$$

The ($\mathbf{g}$ valued) curvature of the connection A is

$$F_{ij} = \partial_i A_j - \partial_j A_i + [A_i, A_j].$$

This is invariant with respect to gauge transformations

$$A_j \to O A_j O^{-1} - \partial_j O O^{-1}, \qquad O \in \mathcal{G}.$$

A Yang-Mills connection is a critical point for the Yang-Mills functional

$$I(A) = \int_{\mathbb{R}^n \times \mathbb{R}} \langle F_{ij}, F^{ij}\rangle dx\, dt$$

where indices are lifted with respect to the Minkovski metric. Then the Yang-Mills equations have the form

$$D^j F_{ij} = 0$$

and the energy functional is

$$E(A) = \int_{\mathbb{R}^n \times \mathbb{R}} \langle F_{ij}, F_{ij}\rangle dx\, dt.$$

A Yang-Mills connection is not a single connection, but instead it is a class of equivalence with respect to the above gauge transformation. In order to view the Yang-Mills equations as semilinear wave equations and solve them one has to fix the gauge, i.e. select a single representative out of each equivalence class. Common gauge choices include: (i) the temporal gauge $A_0 = 0$, (ii) the wave gauge $\partial_j A^j = 0$ and (iii) the Coulomb gauge $\sum_{j=1}^n \partial_j A_j = 0$. To understand the equation better it may help to look first at an oversimplified version, namely

$$\Box u = (u \cdot \nabla_x)u + \nabla_x p, \qquad \nabla_x \cdot u = 0.$$

This exhibits a Q_{ij} type null condition. The scaling index is $s_0 = \frac{n-2}{2}$ and $s_c = 1$. Using the $X^{s,b}$ spaces one can improve the local theory somewhat, but certain more subtle modifications of this are needed in order to handle high-low frequency interactions. In [9] such an approach is used to prove that local well-posedness holds for $s > s_0$, $n \geq 4$.

Open Problem 7. Is the Yang-Mills equation well-posed for $s > s_0$, $n = 2,3$? (Likely not for $n = 2$. For $n = 3$ one can obtain $s > \frac{3}{4}$ using the $X^{s,b}$ spaces.)

Conjecture 8. *(i) The Yang-Mills equation is globally well-posed for small data in $H^{s_0} \times H^{s_0-1}$ for $n \geq 4$.*

(ii) The Yang-Mills equation is globally well-posed for large data in $H^{s_0} \times H^{s_0-1}$ for $n = 4$.

3.   Nonlinear wave equations

We consider (NLW), since (GNLW) reduces to it by differentiation. The fixed point argument in the semilinear case cannot be applied in the nonlinear case, because the wave equation parametrix is not strongly stable with respect to small changes in the coefficients. Instead, one must adopt a different strategy: (i) show that local solutions exist for smooth data, (ii) obtain a-priori bounds for smooth solutions uniformly with respect to initial data in a bounded set in $H^s \times H^{s-1}$ and (iii) prove continuous dependence on the data in a weaker topology, and obtain solutions for $H^s \times H^{s-1}$ data as weak limits of smooth solutions. Steps (i) and (iii) are more or less routine, it is (ii) which causes most difficulties. A good starting point is Klainerman's energy estimate

$$\|\nabla u(t)\|_{H^{s-1}} \lesssim \|\nabla u(0)\|_{H^{s-1}} \exp\left(\int_0^t \|\nabla u(s)\|_{L^\infty} ds\right).$$

This shows that all Sobolev norms of a solution remain bounded for as long as $\|\nabla u\|_{L^1 L^\infty}$ stays bounded. It remains to see how to obtain bounds on $\|\nabla u\|_{L^1 L^\infty}$. The classical approach uses energy estimates and Sobolev embeddings, but, as in the semilinear case, it only yields results one unit above scaling, namely for $s > \frac{n}{2} + 1$.

Better results could be obtained using the Strichartz estimates instead. However, this is very nontrivial as one would have to establish the Strichartz estimates for the operator $\Box_{g(u)}$, which has very rough coefficients. Compounding the difficulty, the argument is necessarily circular,

coefficients $\implies$ Strichartz $\implies$ solution $\implies$ coefficients
regularity   estimates   regularity   regularity

One can get around this with a bootstrap argument of the form

$$\left.\begin{array}{c} \|(u_0, u_1)\|_{H^s \times H^{s-1}} \leq \epsilon \\ \||g(u)|\| \leq 2 \end{array}\right\} \implies \left\{\begin{array}{c} \text{Strichartz estimates for } \Box_{g(u)} \text{ in } [-1, 1] \\ \||g(u)|\| \leq 1 \end{array}\right.$$

where the (possibly nonlinear) triple norm contains the needed information about the metric. Still, apriori there is no clear way to determine exactly how it should be defined. A starting point is to set $\||g(u)|\| = \|\nabla g\|_{L^1 L^\infty}$, but this only leads to partial results. Following partial results independently obtained by Bahouri-Chemin [3],[2] and Tataru [23], [18] and further work of Klainerman-Rodnianski [13], the next result represents the current state of the problem:

Theorem 9. (Smith-Tataru [16]) *The equation (NLW) is locally well-posed in $H^s \times H^{s-1}$ for $s > \frac{n}{2} + \frac{3}{4}$ (n = 2) and $s > \frac{n}{2} + \frac{1}{2}$ (n = 3, 4, 5). In addition, the Strichartz estimates with $q = \infty$ hold for the corresponding wave operator $\Box_{g(u)}$.*

Lindblad's counterexamples correspond to $s = \frac{n+3}{4}$ and show that this result is sharp for $n = 2, 3$. The restriction to $n \leq 5$ is not central to the problem, it can likely be removed with some extra work.

Open Problem 10. Improve the above result in dimension $n \geq 4$.

Wave equation parametrices. In most approaches, the key element in the proof of the Strichartz estimates is the construction of a parametrix for the wave equation. There are many ways to do this for smooth coefficients, however, as the regularity of the coefficients decreases, they start to break down. Let us begin with the classical Fourier integral operator parametrix, used in the work of Bahouri-Chemin:

$$K(x, y) = \int a(x, y, \xi) e^{i\phi(x, y, \xi)} d\xi.$$

The phase ϕ is initialized by $\phi(x, y, \xi) = \xi(x - y)$ when $x_0 = y_0$ and must solve an eikonal equation, while for the amplitude a one obtains a transport equation along the Hamilton flow. The disadvantage is that all spatial localization comes from stationary phase, which seems to require too much regularity for the coefficients.

One way to address the issue of spatial localization is to begin with wave packets, which have the best possible spatial localization on the unit time scale. In the variable coefficient case the frequency λ wave packets are bump functions on curved parallelepipeds of size $1 \times \lambda^{-1} \times (\lambda^{-\frac{1}{2}})^{n-1}$. These parallelepipeds are images of $\lambda^{-1} \times (\lambda^{-\frac{1}{2}})^{n-1}$ parallelepipeds at the initial time, transported along the Hamilton flow for $\Box_g$ corresponding to their conormal direction. Then one can seek approximate solutions for $\Box_g$ as discrete superpositions of wave packets, $u = \sum_T u_T$. It is not too difficult to construct individual wave packets, the more delicate point is to show that the wave packets are almost orthogonal. This approach, which is used in [16], was originally introduced by Smith [14] and used to prove the Strichartz estimates in 2 and 3 dimensions for operators with C^2 coefficients.

Another parametrix with a better built in spatial localization can be obtained by doing a smooth phase space analysis:

$$K(\tilde{y}, y) = \int_C a(x, \xi) e^{i(\phi(y,x,\xi) - \overline{\phi(\tilde{y},x_t,\xi_t)})} dx \, d\xi \, dt \quad \phi(y, x, \xi) = \xi(x - y) + i|\xi|(x - y)^2.$$

Here $(x, \xi) \to (x_t, \xi_t)$ is the Hamilton flow for $\square_g$ on the characteristic cone $C = \{g^{ij}(x)\xi_i\xi_j = 0\}$. One can factor this into a product of three operators, namely an FBI transform, a phase space transport along the Hamilton flow and then an inverse FBI transform. Neglecting the first one, i.e. setting $x = y$ above, produces an operator which is similar to the Fourier integral operators with complex phase. However, it seems to be more useful to keep the Gaussian localizations at both ends. Parametrices of this type were introduced in Tataru [24] and used to prove Strichartz estimates for operators with C^2 coefficients in all dimensions. The C^2 condition was later relaxed in [18] to $\nabla^2 g \in L^1 L^\infty$. Localization and scaling arguments lead also to weaker estimates for operators whose coefficients have less regularity. Such estimates are known to be sharp, see the counterexamples in Smith-Tataru [15].

The null condition. As in the semilinear case, one may ask whether better results can be obtained for equations with special structure. However, unlike the semilinear case, little is known so far. We propose the following

Definition 11. *We say that the equation (GNLW) satisfies the null condition if*

$$\frac{\partial g^{ij}(u, p)}{\partial p_k} \xi_i \xi_j \xi_k = 0 \quad in \ g^{ij}(u, p)\xi_i\xi_j = 0.$$

Conjecture 12. *If the null condition holds then the equation (GNLW) is well-posed in $H^s \times H^{s-1}$ for some $s < \frac{n}{2} + \frac{3}{4}$ $(n = 2)$ respectively for some $s < \frac{n}{2} + \frac{1}{2}$ $(n = 3)$.*

In $3+1$ dimensions a problem which does not quite fit into the above setup but still satisfies some sort of null condition is the Einstein's equations in general relativity. It is similar to the Yang Mills equations in that it has a gauge invariance, and the null condition is only apparent after fixing the gauge. Klainerman-Rodnianski have obtained a different proof of Theorem 9 for this special case of (NLW).

References

[1] Serge Alinhac. *Blowup for nonlinear hyperbolic equations*, Boston: Birkhäuser, 1995. Progress in Nonlinear Differential Equations and their Applications, 17.

[2] Hajer Bahouri and Jean-Yves Chemin, Equations d'ondes quasilineaires et effet dispersif, *Int. Math. Res. Not.*, 1999(21):1141–1178, 1999.

[3] Hajer Bahouri and Jean-Yves Chemin, Equations d'ondes quasilineaires et estimations de Strichartz, *Am. J. Math.*, 121(6):1337–1377, 1999.

[4] Jean Bourgain, Fourier transform restriction phenomena for certain lattice subsets and applications to nonlinear evolution equations. I. Schrödinger equations II. The KdV-equation, *Geom. Funct. Anal.*, 3(3):107–156, 209–262, 1993.

[5] Jean Bourgain, Refinements of Strichartz' inequality and applications to 2D-NLS with critical nonlinearity, *Internat. Math. Res. Notices*, no. 5:253–283, 1998.

[6] Damiano Foschi and Sergiu Klainerman, Bilinear space-time estimates for homogeneous wave equations, *Ann. Sci. École Norm. Sup. (4)*, 33(2):211–274, 2000.

[7] Manoussos Grillakis, Regularity and asymptotic behaviour of the wave equation with a critical nonlinearity, *Ann. of Math.*, 132(3):485–509, 1990.

[8] Markus Keel and Terence Tao, Endpoint Strichartz estimates, *Amer. J. Math.*, 120(5):955-980, 1998.

[9] Sergiu Klainerman and Daniel Tataru, On the optimal local regularity for Yang-Mills equations in R^{4+1}, *J. Amer. Math. Soc.*, 12(1):93–116, 1999.

[10] Hans Lindblad, Counterexamples to local existence for semi-linear wave equations, *Amer. J. Math.*, 118(1):1–16, 1996.

[11] Hans Lindblad, Counterexamples to local existence for quasilinear wave equations, *Math. Res. Lett.*, 5(5):605–622, 1998.

[12] Sergiu Klainerman, Matei Machedon, Space-time estimates for null forms and the local existence theorem, *Comm. Pure Appl. Math.*, 46(9): 1221–1268, 1993.

[13] Sergiu Klainerman and Igor Rodnianski, Improved local well posedness for quasilinear wave equations in dimension three, preprint.

[14] Hart Smith, A parametrix construction for wave equations with $C^{1,1}$ coefficients, *Ann. Inst. Fourier (Grenoble)*, 48(3):797–835, 1998.

[15] Hart Smith and Daniel Tataru, Counterexamples to Strichartz estimates for the wave equation with nonsmooth coefficients, *Math. Res. Lett.*, to appear.

[16] Hart Smith and Daniel Tataru, Sharp local well-posedness results for the nonlinear wave equation. `http://www.math.berkeley.edu/~tataru/nlw.html`

[17] Elias M. Stein, *Harmonic analysis: real-variable methods, orthogonality, and oscillatory integrals*, Princeton University Press, Princeton, NJ, 1993.

[18] Daniel Tataru, Strichartz estimates for operators with nonsmooth coefficients III, to appear, J. Amer. Math. Soc.

[19] Terence Tao, Low regularity semi-linear wave equations *Comm. Partial Differential Equations*, 24(3-4):599–629, 1999.

[20] Terence Tao, Endpoint bilinear restriction theorems for the cone, and some sharp null form estimates, *Math. Z.*, 238(2):215-268, 2001.

[21] Terence Tao, Global regularity of wave maps, II, Small energy in two dimensions, *Comm. Math. Phys.*, 224(2):443–544, 2001.

[22] Daniel Tataru, On the equation $\Box u = |\nabla u|^2$ in $5+1$ dimensions, *Math. Res. Lett.*, 6(5-6):469–485, 1999.

[23] Daniel Tataru, Strichartz estimates for operators with nonsmooth coefficients and the nonlinear wave equation, *Am. J. Math.*, 122(2):349–376, 2000.

[24] Daniel Tataru, Strichartz estimates for second order hyperbolic operators with nonsmooth coefficients, II, *Amer. J. Math.*, 123(3):385–423, 2001.

[25] Daniel Tataru, On global existence and scattering for the wave maps equation, *Amer. J. Math.*, 123(1):37–77, 2001.

[26] Daniel Tataru, Null form estimates for second order hyperbolic operators with

rough coefficients. http://www.math.berkeley.edu/ tataru/nlw.html

[27] Michael E. Taylor, *Pseudodifferential operators and nonlinear PDE.* Birkhäuser, Boston, 1991.

[28] Thomas Wolff, A sharp bilinear cone restriction estimate, *Ann. of Math. (2)*, 153(3):661–698, 2001.

ICM 2002 · Vol. III · 221–231

Affine Maximal Hypersurfaces

Xu-Jia Wang*

Abstract

This is a brief survey of recent works by Neil Trudinger and myself on the Bernstein problem and Plateau problem for affine maximal hypersurfaces.

2000 Mathematics Subject Classification: 35J60, 53A15.
Keywords and Phrases: Affine maximal hypersurfaces, Bernstein problem, Plateau problem.

1. Introduction

The concept of affine maximal surface in affine geometry corresponds to that of minimal surface in Euclidean geometry. The affine Bernstein problem and affine Plateau problem, as proposed in [9,5,7], are two fundamental problems for affine maximal surfaces. We shall describe some recent advances, mostly obtained by Neil Trudinger and myself [21-24], on these two problems.

Given an immersed hypersurface $\mathcal{M} \subset \mathbf{R}^{n+1}$, one defines the *affine metric* (also called the Berwald-Blaschke metric) by $g = |K|^{-1/(n+2)} II$, where K is the Gauss curvature, II is the second fundamental form of $\mathcal{M}$. In order that the metric is positive definite, the hypersurface will always be assumed to be locally uniformly convex, namely it has positive principal curvatures. From the affine metric one has the *affine area functional*,

$$A(\mathcal{M}) = \int_{\mathcal{M}} K^{1/(n+2)}, \tag{1.1}$$

which can also be written as

$$A(u) = \int_{\Omega} [\det D^2 u]^{1/(n+2)} \tag{1.2}$$

if $\mathcal{M}$ is given as the graph of a convex function u over a domain $\Omega \subset \mathbf{R}^n$. The affine metric and affine surface area are invariant under unimodular affine transformations.

---

*Centre for Mathematics and Its Applications, The Australian National University, Canberra, ACT 0200, Australia. E-mail: X.J.Wang@maths.anu.edu.au

A locally uniformly convex hypersurface is called *affine maximal* if it is stationary for the functional A under interior convex perturbation. A convex function is called an affine maximal function if its graph is affine maximal. Traditionally such hypersurfaces were called affine minimal [1,9]. Calabi suggested using the terminology affine maximal as the second variation of the affine area functional is negative [5]. If the hypersurface is a graph of convex function u, then u satisfies the *affine maximal surface equation* (the Euler-Lagrange equation of the functional A),

$$L[u] := U^{ij} w_{ij} = 0, \tag{1.3}$$

where $[U^{ij}]$ is the cofactor of the Hessian matrix $D^2 u$,

$$w = [\det D^2 u]^{-(n+1)/(n+2)}, \tag{1.4}$$

and the subscripts i, j denote partial derivatives with respect to the variables x_i, x_j. Note that for any given i or j, U^{ij}, as a vector field in Ω, is divergence free. The equation (1.3) is a nonlinear fourth order partial differential equation, which can also be written in the short form

$$\Delta_g h = 0, \tag{1.5}$$

where $h = (\det D^2 u)^{-1/(n+2)}$, and Δ_g denotes the Laplace-Beltrami operator with respect to g.

The quantity

$$H_A(\mathcal{M}) = \frac{-1}{n+1} L(u)$$

is called the affine mean curvature of $\mathcal{M}$, and is also invariant under unimodular affine transformations. In particular it is invariant if one rotates the coordinates or adds a linear function to u. The affine mean curvature of the unit sphere is n.

The affine Bernstein problem concerns the uniqueness of entire convex solutions to the affine maximal surface equation, and asks whether an entire convex solution of (1.3) is a quadratic polynomial. The Chern conjecture [9] asserts this is true in dimension two. Geometrically, and more generally, it can be stated as that a Euclidean complete, affine maximal, locally uniformly convex surface in 3-space must be an elliptic paraboloid. Calabi proved the assertion assuming in addition that the surface is affine complete [5], see also [6,7]. A problem raised by Calabi, called the Calabi conjecture in [19], is whether affine completeness alone is enough for the Bernstein theorem. The Chern conjecture was proved true in [21] (see Theorem 3.1 below). The Calabi conjecture was resolved in [22], as a byproduct of our fundamental result that affine completeness implies Euclidean completeness for locally uniformly convex hypersurfaces of dimensions larger than one (Theorem 3.2). See also [14] for a different proof of the Calabi conjecture.

The affine Plateau problem deals with the existence and regularity of affine maximal hypersurfaces with prescribed boundary of which the normal bundles on the boundary coincide with that of a given locally uniformly convex hypersurface. The affine Plateau problem, which had not been studied before, is more complicated when compared with the affine Bernstein problem in 3-space. The first boundary value problem, namely prescribing the solution and its gradient on the boundary, is a special case of the affine Plateau problem. We need to impose two boundary conditions as the affine maximal surface equation is a fourth order equation. We will formulate the Plateau problem as a variational maximization problem and prove the existence and regularity of maximizers to the problem in 3-space [24] (Theorem 5.1). For the existence we need a uniform cone property of locally convex hypersurfaces, proved in [23], which also led us to the proof of the conjecture by Spruck in [20] (Theorem 4.1), concerning the existence of locally convex hypersurfaces of constant Gauss curvature.

Equation (1.3) can be decomposed as a system of two second order partial differential equations, one of which is a linearized Monge-Ampère equation and the other is a Monge-Ampère equation, see (2.6) and (2.7) below. This formulation enables us to establish the regularity for equation (1.3) (Theorem 2.1), using the regularity theory for Monge-Ampère type equations [2,3]. A crucial assumption in Theorem 2.1 is the strict convexity of solutions, which is the key issue for both the affine Bernstein and affine Plateau problems. We succeeded in proving the necessary convexity estimates only in dimension two.

2. A priori estimates

Instead of the homogeneous equation (1.3), we consider here the non-homogeneous (prescribed affine mean curvature) equation

$$L(u) = f \quad \text{in } \Omega, \tag{2.1}$$

where f is a bounded measurable function, and Ω is a normalized convex domain in $\mathbf{R}^n$. A convex domain is called normalized if its minimum ellipsoid, that is the ellipsoid with minimum volume among all ellipsoids containing the domain, is a unit ball.

Let u be a smooth, locally uniformly convex solution of (2.1) which vanishes on $\partial\Omega$. First we need positive upper and lower bounds for the determinant $\det D^2 u$. For the upper bound we have, by constructing appropriate auxiliary function, for any subdomain $\Omega' \subset\subset \Omega$, the estimate

$$\sup_{x \in \Omega'} \det D^2 u(x) \leq C, \tag{2.2}$$

where C depends only on n, $\text{dist}(\Omega', \partial\Omega)$, $\sup_\Omega |Du|$, $\sup_\Omega f$, and $\sup_\Omega |u|$.

For the lower bound we need a key assumption, namely a control on the strict convexity of solutions, which can be measured by introducing the *modulus of convexity*. Let v be a convex function in Ω. For any $y \in \Omega$, $h > 0$, denote

$$S_{h,v}(y) = \{x \in \Omega \mid v(x) = v(y) + Dv(y)(x - y) + h\}.$$

The modulus of convexity of v is a nonnegative function, defined by

$$\rho_v(r) = \inf_{y \in \Omega} \rho_{v,y}(r), \quad r > 0,$$

224 Xu-Jia Wang

where

$$\rho_{v,y}(r) = \sup\{h \geq 0 \mid S_{h,v}(y) \subset B_r(y)\}$$

if there exists $h \geq 0$ such that $S_{h,v}(y) \subset B_r(y)$, otherwise we define $\rho_{v,y}(r) = 0$. We have $\rho_v(r) > 0$ for all $r > 0$ if v is strictly convex.

Let u be a smooth, locally uniformly convex solution of (2.1). Then we have the following lower bound estimate, for any $\Omega' \subset\subset \Omega$,

$$\inf_{x \in \Omega'} \det D^2 u(x) \geq C, \tag{2.3}$$

where C depends on n, dist$(\Omega', \partial\Omega)$, $\sup_\Omega |Du|$, $\inf_\Omega f$, and ρ_u. The proof again can be achieved by introducing an appropriate auxiliary function.

From the a priori estimates (2.2) and (2.3) we then have

Theorem 2.1. *Let $u \in C^4(\Omega) \cap C^0(\overline{\Omega})$ be a locally uniformly convex solution of (2.1). Then for any subdomain $\Omega' \subset\subset \Omega$, we have:*
(i) $W^{4,p}$ estimate,

$$\|u\|_{W^{4,p}(\Omega')} \leq C, \tag{2.4}$$

where $p \in [1, \infty)$, C depends on $n, p, \sup_\Omega |f|$, dist$(\Omega', \partial\Omega)$, $\sup_\Omega |u|$, and ρ_u.
(ii) Schauder estimate,

$$\|u\|_{C^{4,\alpha}(\Omega')} \leq C, \tag{2.5}$$

where $\alpha \in (0,1)$, C depends on $n, \alpha, \|f\|_{C^\alpha(\overline{\Omega})}$, dist$(\Omega', \partial\Omega)$, $\sup_\Omega |u|$, and ρ_u.

Note that the gradient of u is locally controlled by ρ_u, the modulus of convexity of u. To prove Theorem 2.1, we write (2.1) as a second order partial differential system

$$U^{ij} w_{ij} = f \quad \text{in} \ \Omega, \tag{2.6}$$

$$\det D^2 u = w^{-(n+2)/(n+1)} \quad \text{in} \ \Omega, \tag{2.7}$$

where (2.6) is regarded as a second order elliptic equation for w. By (2.2) and (2.3), and the Hölder continuity of linearized Monge-Ampère equation [3], we have the interior a priori Hölder estimate for w. We note that the Hölder continuity in [3] is proved for the homogeneous equation, but the argument there can be easily carried over to the non-homogeneous case under (2.2) and (2.3). By the interior Schauder estimate for the Monge-Ampère equation [2], we obtain the interior a priori $C^{2,\alpha}$ estimate for u. It follows that (2.6) is a linear uniformly elliptic equation with Hölder coefficients. Hence Theorem 2.1 follows.

The control on strict convexity is a key condition in Theorem 2.1. One cannot expect the strict convexity of solutions when $n \geq 3$. Indeed, there are convex solutions to the Monge-Ampère equation

$$\det D^2 u = 1 \tag{2.8}$$

which are not strictly convex, and so not smooth [17]. Note that any non-smooth convex solution of (2.8) can be approximated by smooth ones, and a smooth solution of (2.8) is obviously a solution of (2.1), with $f = 0$.

An interesting problem is to find appropriate conditions to estimate the strict convexity of solutions of (2.1). For the affine Bernstein problem it suffices to prove convexity estimate for solutions vanishing on the boundary. We succeeded only in dimension two, see §5.

3. The affine Bernstein problem

We say a hypersurface $\mathcal{M}$, immersed in $\mathbf{R}^{n+1}$, is Euclidean complete if it is complete under the metric induced from the standard Euclidean metric.

Theorem 3.1. *A Euclidean complete, affine maximal, locally uniformly convex surface in $\mathbf{R}^3$ is an elliptic paraboloid.*

Theorem 3.1 extends Jorgens' theorem [11], which asserts that an entire convex solution of (2.8) in $\mathbf{R}^2$ must be a quadratic function. Jorgens' theorem also leads to the Bernstein theorem for minimal surfaces in dimension two [11]. Jorgens' theorem was extended to higher dimensions by Calabi [4] for $2 \leq n \leq 5$ and Pogorelov [17] for $n \geq 2$. See also [8]. Observe that the Chern conjecture follows from Theorem 3.1 immediately.

The proof of Theorem 3.1 uses the affine invariance of equation (1.3) and the a priori estimates in §2. First note that a Euclidean complete locally uniformly convex hypersurface must be a graph. Suppose the surface in Theorem 3.1 is the graph of a nonnegative convex function u with $u(0) = 0$. For any constant $h > 1$, let T_h be the linear transformation which normalizes the section $S_{h,u}^0 = \{u < h\}$, and let $v_h(x) = h^{-1}u(T_h^{-1}(x))$. By the convexity estimate in dimension two, the modulus of convexity of v_h is independent of h. Hence there is a uniform positive distance from the origin to the boundary $\partial T_h(S_{h,u}^0)$. By Theorem 2.1, we infer that the largest eigenvalue of T_h is controlled by the least one of T_h, which implies that u is defined in the entire $\mathbf{R}^2$. By Theorem 2.1 again, $D^3 v_h(0)$ is bounded. Hence for any given $x \in \mathbf{R}^2$,

$$|D^2 u(x) - D^2 u(0)| \leq Ch^{-1/2} \to 0$$

as $h \to 0$, namely $D^2 u(x) = D^2 u(0)$.

Note that the dimension two restriction is used only for the strict convexity estimate. The affine Bernstein problem was investigated by Calabi in a number of papers [5,6,7]. Using the result that a nonnegative harmonic function (i.e. h in (1.5)) defined on a complete manifold with nonnegative Ricci curvature must be a constant, he proved that, among others, the Bernstein theorem in dimension two, under the additional hypothesis that the surface is also complete under the affine metric.

Instead of the Euclidean completeness as in the Chern conjecture, Calabi asks whether affine completeness alone is sufficient for the Bernstein theorem. This question was recently answered affirmatively in [22]. See also [14] for a different treatment based on the result in [16]. In [22] we proved a much stronger result. That is

Theorem 3.2. *An affine complete, locally uniformly convex hypersurface in* $\mathbf{R}^{n+1}$, $n \geq 2$, *is also Euclidean complete.*

The converse of Theorem 3.2 is not true [12], nor is it for $n = 1$. For the proof, which uses the Legendre transform and Lemma 4.1 below, we refer the reader to [22] for details.

4. Locally convex hypersurfaces with boundary

In this section we present some results in [23], which guarantee the sub-convergence of bounded sequences of locally convex hypersurfaces with prescribed boundary.

Recall that a hypersurface $\mathcal{M} \subset \mathbf{R}^{n+1}$ (not necessarily smooth) is called locally convex if it is a locally convex immersion of a manifold $\mathcal{N}$ and there is a continuous vector field on the convex side of $\mathcal{M}$, transversal to $\mathcal{M}$ everywhere. Let T denote the immersion, namely $\mathcal{M} = T(\mathcal{N})$. For any given point $x \in \mathcal{M}$, $T^{-1}(x)$ may contains more than one point. To avoid confusion when referring to a point $x \in \mathcal{M}$ we understand a pair (x, p) for some point $p \in \mathcal{N}$ such that $x = T(p)$. We say $\omega_x \subset \mathcal{M}$ is a neighborhood of $x \in \mathcal{M}$ if it is the image of a neighborhood of p in $\mathcal{N}$. The r-neighborhood of x, $\omega_r(x)$, is the connected component of $\mathcal{M} \cap B_r(x)$ containing the point x. In [23] we proved the following fundamental lemma for locally convex hypersurfaces.

Lemma 4.1. *Let* $\mathcal{M}$ *be a compact, locally convex hypersurface in* $\mathbf{R}^{n+1}$, $n > 1$. *Suppose the boundary* $\partial \mathcal{M}$ *lies in the hyperplane* $\{x_{n+1} = 0\}$. *Then any connected component of* $\mathcal{M} \cap \{x_{n+1} < 0\}$ *is convex.*

A locally convex hypersurface $\mathcal{M}$ is called convex if it lies on the boundary of the convex closure of $\mathcal{M}$ itself. From Lemma 4.1 it follows that a (Euclidean) complete locally convex hypersurface with at least one strictly convex point is convex, and that a closed, locally convex hypersurface is convex. Lemma 4.1 also plays a key role in the proof of Theorem 3.2.

An application we will use here is the uniform cone property for locally convex hypersurfaces. Let $\mathcal{C}_{x,\xi,r,\alpha}$ denote the cone

$$\mathcal{C}_{x,\xi,r,\alpha} = \{y \in \mathbf{R}^{n+1} \mid |y - x| < r, \ \langle y - x, \xi \rangle \geq \cos \alpha \, |y - x|\}.$$

We say that $\mathcal{C}_{x,\xi,r,\alpha}$ is an inner contact cone of $\mathcal{M}$ at x if this cone lies on the concave side of $\omega_r(x)$. We say $\mathcal{M}$ satisfies the *uniform cone condition* with radius r and aperture α if $\mathcal{M}$ has an inner contact cone at all points with the same r and α.

Lemma 4.2. *Let* $\mathcal{M} \subset B_R(0)$ *be a locally convex hypersurface with boundary* $\partial \mathcal{M}$. *Suppose* $\mathcal{M}$ *can be extended to* $\widetilde{\mathcal{M}}$ *such that* $\partial \mathcal{M}$ *lies in the interior of* $\widetilde{\mathcal{M}}$ *and* $\widetilde{\mathcal{M}} - \mathcal{M}$ *is locally strictly convex. Then there exist* $r, \alpha > 0$ *depending only on* n, R, *and the extended part* $\widetilde{\mathcal{M}} - \mathcal{M}$, *such that the* r-*neighborhood* $\omega_r(x)$ *is convex for any* $x \in \mathcal{M}$, *and* $\mathcal{M}$ *satisfies the uniform cone condition with radius* r *and aperture* α.

In [23] we have shown that if $\partial \mathcal{M}$ is smooth and $\mathcal{M}$ is smooth and locally uniformly convex near $\partial \mathcal{M}$, then $\mathcal{M}$ can be extended to $\widetilde{\mathcal{M}}$ as required in Lemma 4.2. The main point of Lemma 4.2 is that r and α depend only on n, R and the extended part $\widetilde{\mathcal{M}} - \mathcal{M}$. Therefore it holds with the same r and α for a family of locally convex hypersurfaces, which includes all locally uniformly convex hypersurfaces with boundary $\partial \mathcal{M}$, contained in $B_R(0)$, such that its Gauss mapping image coincides with that of $\mathcal{M}$. For any sequence of locally convex hypersurfaces in this family, the uniform cone property implies the sequence converges subsequently and no singularity develops in the limit hypersurface. This property is the key for the existence proof of maximizers to the affine Plateau problem. It also plays a key role for our resolution of the Plateau problem for prescribed constant Gauss curvature (as conjectured in [20]), see [23]. We state the result as follows.

Theorem 4.1. *Let* $\Gamma = (\Gamma_1, \cdots, \Gamma_n) \subset \mathbf{R}^{n+1}$ *be a smooth disjoint collection of closed co-dimension two embedded submanifolds. Suppose* Γ *bounds a locally strictly convex hypersurface* S *with Gauss curvature* $K(S) > K_0 > 0$. *Then* Γ *bounds a smooth, locally uniformly convex hypersurface of Gauss curvature* K_0.

If S is a (multi-valued) radial graph over a domain in S^n which does not contain any hemi-spheres, Theorem 4.1 was established in [10]. Theorem 4.1 has been extended to more general curvature functions in [18].

5. The affine Plateau problem

First we formulate the affine Plateau problem as a variational maximization problem. Let $\mathcal{M}_0$ be a compact, connected, locally uniformly convex hypersurface in $\mathbf{R}^{n+1}$ with smooth boundary $\Gamma = \partial \mathcal{M}_0$. Let $S[\mathcal{M}_0]$ denote the set of locally uniformly convex hypersurfaces $\mathcal{M}$ with boundary Γ such that the image of the Gauss mapping of $\mathcal{M}$ coincides with that of $\mathcal{M}_0$. Then any two hypersurfaces in $S[\mathcal{M}]$ are diffeomorphic. Let $\overline{S}[\mathcal{M}_0]$ denote the set of locally convex hypersurfaces which can be approximated by smooth ones in $S[\mathcal{M}_0]$. Our variational affine Plateau problem is to find a smooth maximizer to

$$\sup_{\mathcal{M} \in \overline{S}[\mathcal{M}_0]} A(\mathcal{M}). \tag{5.1}$$

To study (5.1) we need to extend the definition of the affine area functional to non-smooth convex hypersurfaces. Different but equivalent definitions can be found in [13]. Here we adopt a new definition introduced in [21, 24], which is also more straightforward. Observe that the Gauss curvature K can be extended to a measure on a non-smooth convex hypersurface, and the measure can be decomposed as the sum of a singular part and a regular part, $K = K_s + K_r$, where the singular part K_s is a measure supported on a set of Lebesgue measure zero, and the regular part K_r can be represented by an integrable function. We extend the definition of affine area functional (1.1) to

$$A(\mathcal{M}) = \int_{\mathcal{M}} K_r^{1/(n+2)}. \tag{5.2}$$

The affine area functional is upper semi-continuous [13,15]. See also [21,24] for different proofs.

A necessary condition for the affine Plateau problem is that the Gauss mapping image of $\mathcal{M}_0$ cannot contain any semi-spheres. Indeed if $\mathcal{M}$ is affine maximal such that its Gauss mapping image contains, say, the south hemi-sphere, then the pre-image of the south hemi-sphere is a graph of a convex function u over a domain Ω such that $|Du(x)| \to \infty$ as $x \to \partial\Omega$. Then necessarily $\det D^2 u = \infty$ and so $w = 0$ on $\partial\Omega$. It follows that $w \equiv 0$ in Ω, a contradiction.

Theorem 5.1. *Let $\mathcal{M}_0$ be a compact, connected, locally uniformly convex hypersurface in $\mathbf{R}^3$ with smooth boundary $\Gamma = \partial\mathcal{M}_0$. Suppose the image of the Gauss mapping of $\mathcal{M}_0$ does not contain any semi-spheres. Then there is a smooth maximizer to (5.1).*

To prove the existence we observe that by the necessary condition, there exists a positive constant R such that $\mathcal{M} \subset B_R(0)$ for any $\mathcal{M} \in \overline{S}[\mathcal{M}_0]$. Hence by the uniform cone property, Lemma 4.2, any maximizing sequence in $\overline{S}[\mathcal{M}_0]$ is sub-convergent. The existence of maximizers then follows from the upper semi-continuity of the affine area functional. Note that the existence is true for all dimensions.

To prove the regularity we need to show that
(i) $\mathcal{M}$ can be approximated by smooth affine maximal surfaces; and
(ii) $\mathcal{M}$ is strictly convex.
The purpose of (i) is such that the a priori estimate in Section 2 is applicable. Note that (i) also implies the Bernstein Theorem 3.1 holds for non-smooth affine maximal surfaces.

By the penalty method we proved (i) for all dimensions, using the following classical solvability of the second boundary value problem for the affine maximal surface equation.

Theorem 5.2. *Consider the problem*

$$L(u) = f(x, u) \quad in \ \ \Omega, \tag{5.3}$$
$$u = \varphi \quad on \ \ \partial\Omega,$$
$$w = \psi \quad on \ \ \partial\Omega,$$

where w is given in (1.4), Ω is a uniformly convex domain with $C^{4,\alpha}$ boundary, $0 < \alpha < 1$, f is Hölder continuous, non-decreasing in u, $\varphi, \psi \in C^{4,\alpha}(\overline{\Omega})$, and ψ is positive. Then there is a unique uniformly convex solution $u \in C^{4,\alpha}(\overline{\Omega})$ to the above problem.

To prove Theorem 5.2 we first prove that u satisfies (2.2) and (2.3), and that w is Lipschitz continuous on the boundary, namely $|w(x) - w(y)| \leq C|x - y|$ for any $x \in \Omega$, $y \in \partial\Omega$. Theorem 5.2 is then reduced to the boundary $C^{2,\alpha}$ estimate for the Monge-Ampère equation. The proof for the boundary $C^{2,\alpha}$ estimate involves a delicate iteration scheme. We refer to [24] for details.

The interior $C^{2,\alpha}$ estimate for the Monge-Ampère equation was proved by Caffarelli [2], using a perturbation argument. The boundary $C^{2,\alpha}$ estimate, which also uses a similar perturbation argument, contains substantial new difficulty, as that the sections

$$S^0_{h,v}(y) = \{x \in \Omega \mid v(x) < v(y) + Dv(y)(x-y) + h\}$$

can be normalized for the interior estimate but not for the boundary estimate. We need to prove that $S^0_{h,v}(y)$ has a good shape for sufficiently small $h > 0$ and $y \in \partial\Omega$.

Finally we would like to mention our idea of proving the strict convexity, namely (ii) above. Note that for both the affine Bernstein and affine Plateau problem, the dimension two assumption is only used for the proof of the strict convexity. To prove the strict convexity we suppose to the contrary that $\mathcal{M}$ contains a line segment. Let P be a tangent plane of $\mathcal{M}$ which contains the line segment. Then the contact set F, namely the connected set of $P \cap \mathcal{M}$ containing the line segment, is a convex set. If F has an extreme point which is an interior point of $\mathcal{M}$, by rescaling and choosing appropriate coordinates we obtain a sequence of affine maximal functions which converges to a convex function v, such that $v(0) = 0$ and $v(x) > 0$ for $x \neq 0$ in an appropriate coordinate system, and v is not C^1 at the origin 0. In dimension two this means $\det D^2 v$ is unbounded near 0, which is in contradiction with the estimate (2.2).

If all extreme points of F are boundary points of $\mathcal{M}$, we use the Legendre transform to get a new convex function which is a maximizer of a variational problem similar to (5.1), and satisfies the properties as v above, which also leads to a contradiction.

6. Remarks

We proved the affine Bernstein problem in dimension two. In high dimensions ($n \geq 10$) a counter-example was given in [21], where we proved that the function

$$u(x) = (|x'|^9 + x_{10}^2)^{1/2} \tag{6.1}$$

is affine maximal, where $x' = (x_1, \cdots, x_9)$.

The function u in (6.1) contains a singular point, namely the origin. The graph of u is indeed an *affine cone*, that is all the level sets $S_{h,u} - \{u = h\}$ are affine self-similar, in the sense that there is an affine transformation T_h such that $T_h(S_{h,u}) = S_{1,u}$. The above counter-example shows that there is an affine cone in dimensions $n \geq 10$ which is affine maximal but is not an elliptic paraboloid. We have not been successful in finding smooth counter-examples. Little is known for dimensions $3 \leq n \leq 9$.

For the affine Plateau problem, an interesting problem is whether the maximizer satisfies the boundary conditions. If $\mathcal{M}_0$ is the graph of a smooth, uniformly convex function φ, defined in a bounded domain $\Omega \subset \mathbf{R}^n$, the Plateau problem becomes the first boundary value problem, that is equation (1.3) subject to the boundary conditions:

$$u = \varphi \quad \text{on} \quad \partial\Omega, \tag{6.2}$$

$$Du = D\varphi \quad \text{on} \quad \partial\Omega. \tag{6.3}$$

In this case we also proved the uniqueness of maximizers of (5.1). Obviously the maximizer u satisfies (6.2). Whether u satisfies (6.3) is still unknown. Recall that the Dirichlet problem of the minimal surface equation is solvable for any smooth boundary values if and only if the boundary is mean convex. Therefore an additional condition may be necessary in order that (6.3) is fulfilled.

References

[1] W. Blaschke, Vorlesungen über Differential geometrie, Berlin, 1923.

[2] L.A. Caffarelli, Interior $W^{2,p}$ estimates for solutions of Monge-Ampère equations, Ann. Math., 131 (1990), 135–150.

[3] L.A. Caffarelli and C.E. Gutiérrez, Properties of the solutions of the linearized Monge-Ampère equations, Amer. J. Math., 119(1997), 423–465.

[4] E. Calabi, Improper affine hypersurfaces of convex type and a generalization of a theorem by K. Jörgens, Michigan Math. J., 5(1958), 105–126.

[5] E. Calabi, Hypersurfaces with maximal affinely invariant area, Amer. J. Math. 104(1982), 91–126.

[6] E. Calabi, Convex affine maximal surfaces, Results in Math., 13(1988), 199–223.

[7] E. Calabi, Affine differential geometry and holomorphic curves, Lecture Notes Math. 1422(1990), 15–21.

[8] S.Y. Cheng and S.T. Yau, Complete affine hypersurfaces, I. The completeness of affine metrics, Comm. Pure Appl. Math., 39(1986), 839–866.

[9] S.S. Chern, Affine minimal hypersurfaces, in *minimal submanifolds and geodesics*, Proc. Japan-United States Sem., Tokyo, 1977, 17–30.

[10] B. Guan and J. Spruck, Boundary value problems on S^n for surfaces of constant Gauss curvature. Ann. of Math., 138(1993), 601–624.

[11] K. Jorgens, Über die Lösungen der Differentialgleichung $rt - s^2 = 1$, Math. Ann. 127(1954), 130–134.

[12] K. Nomizu and T. Sasaki, Affine differential geometry, Cambridge, 1994.

[13] K. Leichtweiss, Affine geometry of convex bodies, Johann Ambrosius Barth Verlag, Heidelberg, 1998.

[14] A.M. Li and F. Jia, The Calabi conjecture on affine maximal surfaces, Result. Math., 40(2001), 265–272.

[15] E. Lutwak, Extended affine surface area, Adv. Math., 85(1991), 39–68.

[16] A. Martinez and F. Milan, On the affine Bernstein problem, Geom. Dedicata, 37(1991), 295–302.

[17] A.V. Pogorelov, The muitidimensional Minkowski problems, J. Wiley, New York, 1978.

[18] W.M. Sheng, J. Urbas, and X.-J. Wang, Interior curvature bounds for a class of curvature equations, preprint.

[19] U. Simon, Affine differential geometry, in *Handbook of differential geometry*, North-Holland, Amsterdam, 2000, 905–961.

[20] J. Spruck, Fully nonlinear elliptic equations and applications in geometry, Proc. International Congress Math., Birkhäuser, Basel, 1995, 1145–1152.

[21] N.S. Trudinger and X.-J. Wang, The Bernstein problem for affine maximal hypersurfaces, Invent. Math., 140 (2000), 399–422.

[22] N.S. Trudinger and X.-J. Wang, Affine complete locally convex hypersurfaces, Invent. Math., to appear.

[23] N.S. Trudinger and X.-J. Wang, On locally convex hypersurfaces with boundary, J. Reine Angew. Math., to appear.

[24] N.S. Trudinger and X.-J. Wang, The Plateau problem for affine maximal hypersurfaces, Preprint.

Recent Progress in Mathematical Analysis of Vortex Sheets

Sijue Wu*

Abstract

We consider the motion of the interface separating two domains of the same fluid that moves with different velocity along the tangential direction of the interface. We assume that the fluids occupying the two domains are of constant densities that are equal, are inviscid, incompressible and irrotational, and that the surface tension is zero. We discuss results on the existence and uniqueness of solutions for given data, the regularity of solutions, singularity formation and the nature of solutions after the singularity formation time.

2000 Mathematics Subject Classification: 76B03, 76B07, 76B47, 35Q35, 35J60.
Keywords and Phrases: 2-D incompressible inviscid flow, Birkhoff-Rott equation, Well-posedness, Regularity of solutions.

1. Introduction

Vortex dynamics is of fundamental importance for a wide variety of concrete physical problems, such as lift of airfoils, mixing of fluids, separation of boundary layers, and generation of sounds. In mathematical analysis, one often neglects surface tension and viscosity, when they are small in the real physical problem. This necessitates justifying such simplifications.

In this paper, we consider the motion of the interface separating two domains of the same fluid in R^2 that moves with different velocity along the tangential direction of the interface. We assume that the fluids occupying the two domains separated by the interface are of constant densities that are equal, are inviscid, incompressible and irrotational. We also assume that the surface tension is zero, and there is no external forces. The interface in the aforementioned fluid motion is a so called vortex sheet. We want to study the following problem:

Given a vortex sheet initial data, is there a unique solution to this problem?

*Department of Mathematics, University of Maryland, College Park, MD 20742, USA. E-mail: sijue@math.umd.edu

In general, there are two approaches to the aforementioned problem. One is to solve the initial value problem of the incompressible Euler equation in R^2:

$$\begin{cases} v_t + v \cdot \nabla v + \nabla p = 0 \operatorname{div} v = 0, \\ v(x, y, 0) = v_0(x, y) \end{cases} \qquad (x, y) \in R^2, \ t \geq 0 \qquad (1)$$

where the initial incompressible velocity $v_0 \in L^2_{loc}(R^2)$, in which the vorticity $\omega_0 = \operatorname{curl} v_0$ is a finite Radon measure. Here v is the fluid velocity, p is the pressure, and the density of the fluid is assumed to be one. Notice that a vortex sheet gives a measure valued vorticity supported on the interface. This approach was posed by DiPerna and Majda in 1987 [9]. In 1991, J.M. Delort [8] proved the existence of weak solutions global in time of the 2-D incompressible Euler equation (1) for measure-valued initial vorticity in $H^{-1}_{loc}(R^2)$ that has a distinguished sign. However the problem of uniqueness of the weak solution is still unresolved. In 1963, Yudovich [33] obtained the existence and uniqueness of weak solutions of the 2-D incompressible Euler equation (1) for initially bounded vorticity. The best results on uniqueness upto date are given by Yudovich [34] and Vishik [31] for weak solutions with vorticity in a class slightly larger than L^∞. This does not include vortex sheets, which admit measure-valued vorticity. Examples of weak solutions with the velocity field $v \in L^2(R^2 \times (-T, T))$ that is compactly supported in space-time was constructed by V. Scheffer [29] and later by A. Shnirelman [30]. This gives non-uniqueness of weak solutions in $L^2(R^2 \times (-T, T))$. However non-uniqueness in the physically relevant class of conserved energy $v \in L^\infty([0, \infty), L^2_{loc}(R^2))$ remains open. Numerical evidences of non-uniqueness of weak solutions for vortex sheet data can be found in [25], [18].

Furthermore, weak solutions give little information of the specific nature of the vortex sheet evolution. For instance, does the vorticity remain supported on a curve for a later time given that the initial vorticity is supported on a curve in R^2? Assume further that the free interface between the two fluid domains remains a curve in R^2 at a later time, equation (1) can be reduced to an evolutionary differential-integral equation along the interface. This is the Birkhoff-Rott equation, written explicitly by Birkhoff in [2] and implied in the work of Rott [26]. The second approach uses the Birkhoff-Rott equation as a model for the evolution of the vortex sheet.

2. The Birkhoff-Rott equation

For convenience, we use complex variable $z = x + i y$ to denote a point in R^2. $\bar{z} = x - i y$ denotes the complex conjugate and $f_x = \partial_x f$ is the partial derivative of the function f. H^s indicates Sobolev spaces.

In search of the equation for the evolution of the vortex sheet, we suppose that at time $t \geq 0$ the vorticity is a measure supported on the curve $\Gamma(t)$ given by the complex position $\xi = \xi(s, t)$ in the arclength s, in which $\xi(0, t)$ is the particle path of a reference particle; and on this curve the vorticity density is $\gamma = \gamma(s, t)$. That is the vorticity at time t is $\omega(x,y,t)$ satisfying

$$\iint \phi(x,y)\omega(x,y,t)\,dx\,dy = \int \phi(\xi(s,t))\gamma(s,t)\,ds, \qquad \text{for any } \phi \in C_0^\infty(R^2).$$

From the Biot-Savart law, the velocity field v induced by the vorticity is given by

$$\overline{v}(z,t) = \frac{1}{2\pi i}\int \frac{\gamma(s',t)}{z-\xi(s',t)}\,ds', \qquad \text{for } z \notin \Gamma(t).$$

Notice that the velocity is discontinuous just on $\Gamma(t)$. We define the velocity on the sheet as the average of the velocities at the two sides of the sheet, that is given by the principle value integral:

$$\overline{v}(\xi(s,t),t) = \frac{1}{2\pi i}\,p.v.\int \frac{\gamma(s',t)}{\xi(s,t)-\xi(s',t)}\,ds'. \tag{2}$$

As suggested by the properties of the Euler equation, we assume that the vortex sheet is convected by the average velocity (2), and the vorticity is conserved along the particle path. We arrive at the evolution equation of the vortex sheet:

$$\begin{aligned}
\xi_t(s,t) + a(s,t)\xi_s(s,t) &= v(\xi(s,t),t) \\
\gamma_t(s,t) + \partial_s(a(s,t)\gamma(s,t)) &= 0
\end{aligned} \tag{3}$$

where $a(s,t)$ is a real valued function satisfying $a(0,t)=0$.

A rigorous justification of the equivalence between equation (3) and equation (1) for smooth graphs $\xi(s,t)$ and smooth vortex strength $\gamma(s,t)$ can be found in [19]. It is not hard to extend it to all smooth curves.

Assume $\alpha(s,t) = \int_0^s \gamma(s',t)\,ds'$ defines an increasing function of s, we make a change of variables: $z(\alpha,t) = \xi(s(\alpha,t),t)$, in which $s(\alpha,t)$ is the inverse of $\alpha(s,t)$: $\alpha(s(\alpha,t),t) = \alpha$. We get from equation (3) the Birkhoff-Rott equation

$$\partial_t\overline{z}(\alpha,t) = \frac{1}{2\pi i}\,p.v.\int \frac{1}{z(\alpha,t)-z(\beta,t)}\,d\beta. \tag{4}$$

Notice that $z = z(\alpha,t)$ is a parameterization of the vortex sheet in the circulation variable α, $1/|z_\alpha| = \gamma$ is the vortex strength. A steady solution of (4) is the flat sheet $z = \alpha$.

Equation (4) has been under active investigations over the last four decades. A well-known property of (4) is that perturbations of the flat sheet grow due to the Kelvin-Helmholtz instability, following from a linearization of equation (4) about the flat sheet. For given analytic data, Sulem, Sulem, Bardos and Frisch [28] established the short time existence and uniqueness of solutions in analytic class for 2-D and 3-D vortex sheet evolution. Duchon and Robert [10] obtained the global existence of solutions of equation (4) for a special class of initial data that is close to the flat sheet. However, numerous results show that a vortex sheet can develop a curvature singularity in finite time from analytic data. D.W. Moore [21] was the first to provide analytical evidence that predicts the occurrence and time of singularity formation, which was verified numerically by Meiron, Baker and Orszag [20] and by Krasny [13]. Caflisch and Orellana [3] proved existence almost up to the time of expected singularity formation for analytic data that is close to the flat sheet. Duchon and Robert [10] and Caflisch and Orellana [4] constructed specific examples of solutions of equation (4) where a curvature singularity develops in finite time from analytic data. The example of Caflisch and Orellana [4] has the form $z(\alpha, t) = \alpha + S(\alpha, t) + r(\alpha, t)$, where

$$S(\alpha, t) = \epsilon(1 - i)\{(1 - e^{-t/2 - i\alpha})^{1+\mu} - (1 - e^{-t/2 + i\alpha})^{1+\mu}\}$$

is a solution of the linearized equation in which ϵ is small, $\mu > 0$; $r(\alpha, t)$ is the correction term that is negligible relative to $S(\alpha, t)$ in the sense that $S(\alpha, t) + r(\alpha, t)$ exhibits the same kind of behavior as $S(\alpha, t)$ [4]. Notice that $S(\alpha, t)$ is an analytic function for $t > 0$, but $S(\alpha, 0)$ has an infinite second derivative at $\alpha = 0$ for $\mu \in (0, 1)$. In fact, the $(1 + \nu)$th derivative of $S(\alpha, 0)$ for $\nu > \mu$ becomes infinite at $\alpha = 0$. Now inverting time gives an example $\hat{z}(\alpha, t)$ that is analytic at $t_0 < 0$, but has an infinite second derivative at $\alpha = 0$, $t = 0$. At the singularity formation time $t = 0$, the vortex strength $1/|\hat{z}_\alpha|$ of this example satisfies

$$0 < c \leq 1/|\hat{z}_\alpha| \leq C < \infty \tag{5}$$

for some constants c and C; and $\hat{z}(\alpha, t) \in C^{1+\rho}(R \times [t_0, 0])$ for $0 < \rho < \mu$.

These examples also show that the initial value problem of the Birkhoff-Rott equation (4) is ill-posed in $C^{1+\nu}(R)$, $\nu > 0$, and in Sobolev spaces $H^s(R)$, $s > 3/2$ in the Hadamard sense [4], [10]. Ill-posedness was also proved by Ebin [11] using a different approach. However the existence of solutions in spaces less regular than $C^{1+\nu}(R)$ or $H^s(R)$, and the nature of the vortex sheet at and beyond the singularity time remained unknown analytically in general.

This suggests that we look for solutions of the Birkhoff-Rott equation in the largest possible spaces where the equation makes sense. For the purpose of this paper, we consider functions $z(\alpha, t)$ so that for each fixed time t, both sides of the equation (4) are functions locally in L^2, and on which the L^2-analysis is available. This leads us to consider chord-arc curves, thanks to the work of G. David [7].

Another reason that chord-arc curves are to be considered is due to the numerical calculation of Krasny [14] [15]. Krasny studied the evolution of the vortex sheet beyond singularity using the vortex blob method. He found that the approximating solutions have the form of a spiral beyond singularity. Convergence of the approximating sequence to a weak solution of the Euler equation (1) was proved by J-G. Liu and Z-P. Xin [17], under the assumption that the initial vorticity has a distinguished sign. A special example of chord-arc curves is a logarithmic spiral.

3. Chord-arc curves and some recent results

Let Γ be a rectifiable Jordan curve in R^2 given by $\xi = \xi(s)$ in the arclength s. We say Γ is a chord-arc curve, if there is a constant $M \geq 1$, such that

$$|s_1 - s_2| \leq M|\xi(s_1) - \xi(s_2)|, \qquad \text{for all } s_1, s_2.$$

The infimum of all such constants M is called the chord-arc constant.

For a chord-arc curve $\xi = \xi(s)$, s the arclength, it is proved in [6] that $\xi'(s)$ exists almost everywhere, and there is a choice of the argument function $b \in BMO$, with $\xi'(s) = e^{ib(s)}$. In particular, if the chord-arc constant is close to 1, there is a choice of $b \in BMO$, such that $\|b\|_{BMO}$ is close to 0. Moreover the subset of all those functions b is an open subset of BMO. And if $b \in BMO$, and $\|b\|_{BMO} < 1$, $\xi(s) = \xi_0 + \int_0^s e^{i\,b(s')}\,ds'$ defines a chord-arc curve.

Examples of chord-arc curves include Lipschitz curves and logarithmic spirals $r = \pm e^\theta$, $\theta \in R$, where (r, θ) is the polar coordinates.

A Theorem of G. David [7] states that

Theorem (G. David [7]). *For all chord-arc curves $\Gamma : \xi = \xi(s)$, s the arclength, the corresponding Cauchy integral operator C_Γ, where*

$$C_\Gamma f(s) = p.v. \int \frac{f(s')}{\xi(s) - \xi(s')}\,d\xi(s'),$$

is bounded from $L^2(ds)$ to $L^2(ds)$.

In fact, the result of G. David [7] is stronger than stated above. He proved that the Cauchy integral operator C_Γ is bounded from $L^2(ds)$ to $L^2(ds)$ if and only if Γ is a regular curve. A rectifiable curve Γ is said to be regular if there is a constant M such that for every $r > 0$ and every disc D with radius r, the length of $\Gamma \cap D$ does not exceed Mr. A chord-arc curve is regular but not vice versa.

Now we go back to the Birkhoff-Rott equation. Notice that the Biot-Savart integral representing an incompressible velocity field v in terms of the vorticity ω may be divergent if ω does not vanish fast enough at infinity, even if the velocity field v is well defined, we extend the definition of the Birkhoff-Rott equation (4) by considering the differences of the velocities between any two points:

$$\overline{z_t}(\alpha, t) - \overline{z_t}(\alpha', t) = \frac{1}{2\pi i} p.v. \int_{|\beta| \leq N} \left\{ \frac{1}{z(\alpha, t) - z(\beta, t)} - \frac{1}{z(\alpha', t) - z(\beta, t)} \right\} d\beta$$
$$+ \frac{1}{2\pi i} p.v. \int_{|\beta| > N} \frac{z(\alpha', t) - z(\alpha, t)}{(z(\alpha, t) - z(\beta, t))(z(\alpha', t) - z(\beta, t))}\,d\beta, \tag{6}$$

for all (α, t), (α', t), and some $N > |\alpha| + |\alpha'| + 1$. This admits a larger class of solutions. In particular, the integral on the right hand side of equation (6) is convergent for those similarity solutions considered in [12], [23]-[25], which otherwise give divergent Cauchy integrals in (4) due to the divergent contributions from the vorticities at infinity. It follows from the Theorem of G. David that the integral on the right hand side of the equation (6) is convergent for *a.e.* (α, t), (α', t) and is in $L^\infty([0, T], L^2_{loc}(d\alpha) \times L^2_{loc}(d\alpha'))$ for the solutions considered in Theorem 1 in the following.

Roughly speaking, if a function $z = z(\alpha, t)$ satisfies equation (4), it will also satisfy equation (6). On the other hand, if $z = z(\alpha, t)$ satisfies (6), and the Cauchy integral

$$\frac{1}{2\pi i} \, p.v. \int \frac{1}{z(\alpha, t) - z(\beta, t)} \, d\beta$$

is convergent, then there is a function $c = c(t)$, such that $z(\alpha, t) + c(t)$ satisfies equation (4).

For a local integrable function $f = f(\alpha)$ defined on (a, b), we say f is of bounded local mean oscillation on (a, b) if there exists $\delta_0 > 0$ such that

$$\|f\|_{BMO(a,b),\delta_0} = \sup_{\text{all } I \subset (a,b), |I| \leq \delta_0} \frac{1}{|I|} \int_I |f(\alpha) - f_I| \, d\alpha < \infty,$$

here $f_I = \frac{1}{|I|} \int_I f(\alpha) \, d\alpha$, I is an interval. We say f is analytic on (a, b) if $f \in C^\infty(a, b)$, and for any compact subset K of (a, b), there is a constant $\rho > 0$, such that

$$\sum_{m=o}^{\infty} \frac{\rho^m}{m!} \int_K |\partial_\alpha^m f(\alpha)|^2 \, d\alpha < \infty.$$

Notice that $\ln z$ is multi-valued for complex number z. In the following, $\ln z_\alpha$ refers to one choice of the multi-values. We have the following results concerning the solutions of the Birkhoff-Rott equation (6) (or (4)).

Theorem 1 [32]. *Assume that $z \in H^1([0, T], L^2_{loc}(R)) \cap L^2([0, T], H^1_{loc}(R))$ is a solution of the Birkhoff-Rott equation (6) for $0 \leq t \leq T$, satisfying that*

1. There are constants $m > 0$, $M > 0$, independent of t, such that

$$m|\alpha - \beta| \leq |z(\alpha, t) - z(\beta, t)| \leq M|\alpha - \beta| \qquad \text{for all } \alpha, \ \beta, \ 0 \leq t \leq T. \quad (7)$$

2. For the interval (a, b), there exists $\delta_0 > 0$, independent of t, such that for all $0 \leq t \leq T$,

$$\| \ln z_\alpha(\cdot, t) \|_{BMO(a,b),\delta_0} \leq C(m, M), \quad (8)$$

here $C(m, M)$ is a universal constant depending on m and M. Assume further that $\ln z_\alpha \in L^2([0, T], L^2_{loc}(R))$. Then $z_\alpha \in C((a, b) \times (0, T))$, and for each $t_0 \in (0, T)$, $z_\alpha(\cdot, t_0)$ is analytic on (a, b).

Remark Notice that assumption 2. is satisfied if $\ln z_\alpha \in C([a, b] \times [0, T])$. Assumption 1. is equivalent to assuming that the vortex strength $\gamma = 1/|z_\alpha|$ is bounded away from 0 and ∞, with bounds independent of t, and $z = z(\cdot, t)$ defines a chord-arc curve for each fixed $t \in [0, T]$, with the chord-arc constant independent of t.

Assumption 1. can be relaxed by requiring that $1/|z_\alpha(\alpha,t)|$ is bounded away from 0 and ∞ for $\alpha \in (a,b)$ only, and by assuming some weaker conditions for $\alpha \notin (a,b)$.

A version of Theorem 1 under assumptions that the solution of the Birkhoff-Rott equation (4): $z = z(\alpha,t) \in C^{1+\rho_0}$, for some $\rho_0 > 0$, the vortex strength $c_0 < \gamma(\alpha,t) < C_0$ for some constants $c_0 > 0$, $C_0 > 0$, and the vortex sheets $\Gamma(t)$ are closed Jordan curves is also obtained by Lebeau [16] using an independent approach.

As a consequence of Theorem 1, the example constructed by Caflisch and Orellana [4] will fail to satisfy properties 1 and 2 as stated in Theorem 1 near the singularity after the singularity formation time.

Let Rez and Imz be the real and imaginary parts of the complex number z respectively. Regarding the existence of solutions, we have the following

Theorem 2 [32]. *For any real valued function $w_0 \in H^{\frac{3}{2}}(R)$, there exists $T = T(\|w_0\|_{H^{\frac{3}{2}}}) > 0$, such that the Birkhoff-Rott equation (6) has a solution $z = z(\alpha,t)$ for $0 \le t \le T$, satisfying $\ln z_\alpha \in L^\infty([0,T], H^{\frac{3}{2}}(R)) \cap Lip([0,T], H^{\frac{1}{2}}(R))$ and $Im\{(1+i)\ln z_\alpha(\alpha,0)\} = w_0(\alpha)$, with the properties that there exist constants $m > 0$, $M > 0$ independent of t, such that*

$$m|\alpha - \beta| \le |z(\alpha,t) - z(\beta,t)| \le M|\alpha - \beta|, \qquad \text{for all } \alpha,\ \beta,\ 0 \le t \le T;$$

and there exists $\delta_0 > 0$, independent of t, such that

$$\|\ln z_\alpha(\cdot,t)\|_{BMO(R),\delta_0} \le C(m,M), \qquad for\ 0 \le t \le T,$$

here $C(m,M)$ is the universal constant as in Theorem 1.

Theorem 2 states that if only half of the data $z_\alpha(\alpha,0)$ is given, there is a solution of the Birkhoff-Rott equation (6) for a finite time period. Theorem 2 is a generalization of the existence result of Duchon and Robert [10] to general data.

The following result implies that Theorem 2 is optimal, in the sense that in general, there is no solution of the Birkhoff-Rott equation satisfying properties 1. and 2. as stated in Theorem 1 beyond the initial time $t = 0$ for arbitrarily given data.

Theorem 3 [32]. *Assume that $z \in H^1([0,T], L^2_{loc}(R)) \cap L^2([0,T], H^1_{loc}(R))$ is a solution of the Birkhoff-Rott equation (6) for $0 \le t \le T$, $T > 0$, satisfying the property 1. and property 2. on some interval (a,b) as stated in Theorem 1. Assume further that $\ln z_\alpha \in L^2([0,T], L^2_{loc}(R))$, and $w_0 = Im\{(1+i)\ln z_\alpha(\cdot,0)\}$ is analytic on (a,b). Then $z_\alpha \in C((a,b) \times [0,T))$ and $Re\{(1+i)\ln z_\alpha(\cdot,0)\}$ is also analytic on (a,b).*

4. Open questions

The reason that Theorem 1-3 holds is that under the assumptions 1. and 2. in Theorem 1, the Birkhoff-Rott equation (6) is of "elliptic" type on $(a,b) \times [0,T]$.

It would be interesting to see whether the assumption 2 that requires small local mean oscillation on $\ln z_\alpha(\cdot, t)$ can be removed, or to see whether one can construct a non-smooth solution of equation (6) that violates (8). A good place to start is to construct similarity solutions of the Birkhoff-Rott equation. Similarity solutions are studied numerically in Pullen [23] [25], Pullen and Phillips [24], and analytically in Kambe [12]. However, the similarity solutions in [12], [23]-[25] violate (7). It would also be interesting to relax the assumption 1. by considering vortex strength that is not necessarily bounded away from 0 and infinity on the interval (a, b).

Theorem 1-3 implies that a solution of the Birkhoff-Rott equation is either analytic, or doesn't satisfy properties 1. and 2. in Theorem 1. And there are in general no solutions of reasonable regularity (properties 1. and 2.) beyond the initial time for an arbitrarily given data. Are there solutions of the Birkhoff-Rott equation in spaces of even less regularity? Notice that we can define the solutions of the Birkhoff-Rott equation in the distribution sense by

$$\partial_t \Big(\int \overline{z}(\alpha, t)\eta(\alpha)\, d\alpha \Big) = \frac{1}{4\pi i} \iint \frac{\eta(\alpha) - \eta(\beta)}{z(\alpha, t) - z(\beta, t)}\, d\alpha\, d\beta, \qquad \text{for all } \eta \in C_0^\infty.$$

This admits some even larger classes of solutions $z = z(\alpha, t)$. Is there a solution in the distribution sense for the Birkhoff-Rott equation for any given data? How is it related to the weak solution of the Euler equation?

Theorem 1-3 might as well suggest that the vortex sheet in general fails to be a curve beyond the initial time for general data. Therefore it becomes interesting to study the vortex layers or considering the effects of viscosity. Numerical analysis of the vortex layers can be found in Baker and Shelley [1] etc.

References

[1] G. R. Baker & M. J. Shelley, On the connection between thin vortex layers and vortex sheets, *J. Fluid. Mech.* 215 (1990), 161–194.

[2] G. Birkhoff, Helmholtz and Taylor instability, *Proc. Symp. Appl. Math.* XII, AMS (1962), 55–76.

[3] R.E. Caflisch & O.F. Orellana, Long-time existence for a slightly perturbed vortex sheet, *Comm. Pure Appl. Math* 39 (1986), 807–838.

[4] R.E. Caflisch & O.F. Orellana, Singular solutions and ill-posedness for the evolution of vortex sheets, *SIAM J. Math. Anal.* 20 (1989), 293–307.

[5] R. Coifman & Y. Meyer, Lavrentiev's curves and conformal mappings, *Institut Mittag-Leffler* Report no.5 (1983).

[6] G. David, *Thése de troisiéme cycle, Methématique*, Université de Paris XI.

[7] G. David, Opérateurs intégraux singuliers sur certaines courbes du plan complexe, *Ann. Sci. École. Norm. Sup.(4)* 17, no.1 (1984), 157–189.

[8] J.M. Delort, Existence de nappes de tourbillon en dimension deux, *J. AMS* 4 (1991), 553–586.

[9] R. Diperna & A. Majda, Concentrations in regularizations for 2-D incompressible flow, *Comm. Pure Appl. Math.* 40 (1987), 301–345.

[10] J. Duchon & R. Robert, Global vortex sheet solutions of Euler equations in the plane, *J. Diff. eqns.* 73 (1988), 215–224.

[11] D.G. Ebin, Ill-posedness of the Rayleigh-Taylor and Helmholtz problems for incompressible fluids, *Comm. PDE.* 13 (1988), 1265–1295.

[12] T. Kambe, Spiral vortex solutions of Birkhoff-Rott equation, *Physica D* 37 (1989), 463–473.

[13] R. Krasny, On singularity formation in a vortex sheet by the point-vortex approximation, *J. Fluid Mech.* 167 (1986), 65–93.

[14] R. Krasny, Desingularization of periodic vortex sheet roll-up, *J. Comput. Phys.* 65 (1986), 292–313.

[15] R. Krasny, Computing vortex sheet motion, *Proc. ICM'90, Kyoto, Japan* Vol. II (1991), 1573–1583.

[16] G. Lebeau, Régularité du probléme de Kelvin-Helmholtz pour l'équation d'Euler 2D, *Séminaire: Équations aux Dérivées Partielles, 2000-2001*, Exp. No. II (2001), 12.

[17] J-G. Liu & Z-P. Xin, Convergence of vortex methods for weak solutions to the 2D Euler equations with vortex sheet data, *Comm. Pure Appl. Math.* XLVIII (1995), 611–628.

[18] M.C. Lopes, J. Lowengrub, H.J. Nussenzveig Lopes & Y-X. Zheng, Numerical evidence for nonuniqueness evolution for the 2D incompressible Euler equations: a vortex sheet example, *preprint*.

[19] C. Machioro & M. Pulvirenti, *Mathematical theory of incompressible nonviscous fluids*, Springer, 1994.

[20] D.I. Meiron, G.R. Baker & S.A. Orszag, Analytic structure of vortex sheet dynamics. Part I., *J. Fluid Mech.* 114 (1982), 283.

[21] D.W. Moore, The spontaneous appearance of a singularity in the shape of an evolving vortex sheet, *Proc. Roy. Soc. London Ser. A* 365 (1979), 105–119.

[22] D.W. Moore, Numerical and analytical aspects of Helmholtz instability, *Theoretical and Applied Mechanics, Proc. XVI ICTAM eds. Niordson and Olhoff* (1984), North-Holland, 629–633.

[23] D.I. Pullin, The large scale structure of unsteady self-similar roll-up vortex sheets, *J. Fluid Mech.* 88, part 3 (1978), 401–430.

[24] D.I. Pullin & W.R.C. Phillips, On a generalization of Kaden's problem, *J. Fluid Mech.* 104 (1981), 45–53.

[25] D.I. Pullin, On similarity flows containing two branched vortex sheets, *Mathematical aspects of vortex dynamics, ed. R. Caflisch* (1989), SIAM, 97–106.

[26] N. Rott, Diffraction of a weak shock with vortex generation, *J. Fluid Mech.* 1 (1956), 111–128.

[27] P.G. Saffman, *Vortex dynamics*, Cambridge, 1992.

[28] P. Sulem, C. Sulem, C. Bados& U. Frisch, Finite time analyticity for the two and three dimensional Kelvin-Helmholtz instability, *Comm. Math. Phys.* 80 (1981), 485–516.

[29] V. Scheffer, An inviscid flow with compact support in space-time, *J. Geom. Anal.* 3 (1993), 343–401.

[30] A. Shnirelman, On the non-uniqueness of weak solutions of the Euler equations, *Comm. Pure Appl. Math* L (1997), 1261–1286.

[31] M. Vishik, Incompressible flows of an ideal fluid with vorticity in borderline spaces of Besov type, *Ann. Sci. École Norm. Sup.(4)* 32 no.6 (1999), 769–812.

[32] S. Wu, Recent progress in mathematical analysis of 2-D vortex sheet, preprint.

[33] V. Yudovich, Non-stationary flow of an ideal incompressible liquid, *USSR Comp. Math. and Math. Phys. (English transl.)* 3 (1963), 1407–1457.

[34] V. Yudovich, Uniqueness theorem for the basic nonstationary problem in the dynamics of an ideal, incompressible fluid, *Math. Res. Lett.* 2 (1995), 27–38.

ICM 2002 · Vol. III · 243–252

Quantum Resonances and
Partial Differential Equations

M. Zworski*

Abstract

Resonances, or scattering poles, are complex numbers which mathematically describe meta-stable states: the real part of a resonance gives the rest energy, and its imaginary part, the rate of decay of a meta-stable state. This description emphasizes the quantum mechanical aspects of this concept but similar models appear in many branches of physics, chemistry and mathematics, from molecular dynamics to automorphic forms.

In this article we will will describe the recent progress in the study of resonances based on the theory of partial differential equations.

2000 Mathematics Subject Classification: 35P20, 35P25, 35A27, 47F05, 58J37, 81Q20, 81U.
Keywords and Phrases: Quantum resonances, Scattering theory, Trace formulæ, Microlocal analysis.

1.   Introduction

Eigenvalues of self-adjoint operators appear naturally in quantum mechanics and are in fact what we observe experimentally in many situations. To explain the need for the more subtle notion of *quantum resonances* we consider the following simple example.

Let $V(x)$ be a potential on a bounded interval, as shown in Fig.1. a). If ξ denotes the classical momentum, then the classical energy is given by

$$E = \xi^2 + V(x), \tag{1.1}$$

and the motion of a classical particle can deduced from this equation by considering E as the Hamiltonian. The quantized Hamiltonian is given by

$$P(h) = (hD_x)^2 + V(x), \quad \xi \mapsto hD_x, \quad D_x = \frac{1}{i}\partial_x. \tag{1.2}$$

---

*Department of Mathematics, University of California, Berkeley, CA 94720, USA. E-mail: zworski@math.berkeley.edu

The operator $P(h)$ considered as an unbounded operator on L^2 of a bounded interval (with, say, Dirichlet boundary conditions) has a discrete spectrum, which gets denser and denser as $h \to 0$, which corresponds to getting closer to classical mechanics:

$$P(h)u(h) = E(h)u(h), \quad E(h) \in \mathbf{R}, \quad \int |u(h)|^2 < \infty, \quad u(h)(\pm\pi) = 0. \quad (1.3)$$

The eigenvalues, $E(h)$, are what we (in principle) observe, and the square-integrable eigenfunctions, $u(h)$, are the wave functions.

Now, consider the same example but with a potential $V(x)$ on $\mathbf{R}$, as shown in Fig.1. b). At the energy E, and inside the well created by the potential, the classical motion is the same as in the previous case. Hence, we expect that (at least for h very small) there should exist a quantum state corresponding to the classical one. It is clear in dimension one that for the potential shown in the figure, $P(h)$ given by (1.2) has *no* square integrable eigenfunctions (1.3).

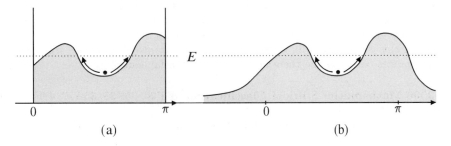

Figure 1: (a) A potential well on a finite interval. (b) A potential on the whole line the same classical picture but a very different quantum picture.

A non-obvious remedy for this is to think of eigenvalues (in the case of Fig.1.a) as the poles of the resolvent of $P(h)$:

$$R(z,h) = (P(h) - z)^{-1}.$$

In the case of Fig1.b), the resolvent is not bounded on $L^2(\mathbf{R})$ for $z > 0$, which corresponds to free motion or scattering. However, under suitable assumptions on $V(x)$ near infinity we have a meromorphic continuation of $R(z,h)$ from $\mathrm{Im}\, z > 0$ to the lower half-plane:

$$R(z,h) \; : \; L^2_{\mathrm{comp}}(\mathbf{R}) \longrightarrow L^2_{\mathrm{loc}}(\mathbf{R}).$$

The complex poles of $R(z,h)$ are the replacement for eigenvalues, are called *resonances*. For a recent presentation of these in the physical literature and for examples of current interest see for instance [7].

In this review, we will restrict ourselves to the case of

$$P(h) = -h^2 \Delta + V(x), \quad x \in \mathbf{R}^n,$$

where the potential $V(x)$ satisfies some decay and analyticity assumptions near infinity, or is simply compactly supported. Then our discussion above applies and resonances are defined as the poles of $R(z, h)$ – see [9] for an attractive new presentation of the meromorphic continuation properties and references. In our convention the resonances are located in the lower half-plane.

We should stress that most of the results hold for very general *black box perturbations* of Sjöstrand-Zworski [22] which allow for a study of diverse problems without going into their specific natures – see [20] for definitions. Also, despite the fact that the motivation presented here came from molecular dynamics, similar issues arise in other settings, from automorpic scattering to electromagnetic (obstacle) scattering – see [8], and [23] respectively.

We suggest the the surveys [12],[20],[31],[33], and [36] for the review of earlier results, and we concentrate on the progress achieved in the last few years. For an account of work based on other methods and motivated by different physical phenomena we refer to [13].

2. Trace formulæ for resonances

Trace formulæ provide one of the most elegant descriptions of the classical-quantum correspondence. One side of a formula is given by a trace of a quantum object, typically derived from a quantum Hamiltonian, and the other side is described in terms of closed orbits of the corresponding classical Hamiltonian. A new general approach based on *quantum monodromy operator* which quantizes the Poincaré map in a natural way was recently given in [24].

In general the spectral (or scattering) side of the formula is given in terms of the trace of a function of the quantum Hamiltonian $f(P(h))$. In the case of self-adjoint problems with discrete sets of eigenvalues the spectral theorem readily provides an expression for tr $f(P(h))$ but the problem becomes subtle for resonances. It has been studied by Lax-Phillips, Bardos-Guillot-Ralston, Melrose, Sjöstrand-Zworski, Guillopé-Zworski – see [34] and references given there.

More recently Sjöstrand [20],[21] introduced *local trace formulæ* for resonances, and that concept was developed further in [18] and [3]. For $P(h)$ given by (1.2) with $V(x)$ decaying sufficiently fast at infinity it can be stated as follows.

Let Ω be an open, simply connected, pre-compact subset of $\{\operatorname{Re} z > 0\} \subset \mathbf{C}$, such that $\Omega \cap \mathbf{R}$ is connected. Suppose that f is holomorphic in a neighbourhood of Ω and that $\psi \in C_c^\infty(\mathbf{R})$ is equal to 1 in $\Omega \cap \mathbf{R}$ and supported in a neighbourhood of $\Omega \cap \mathbf{R}$. Then, denoting the resonance set of $P(h)$ by $\operatorname{Res}(P(h))$,

$$\operatorname{tr}\ ((f\psi)(P(h)) - (f\psi)(-h^2\Delta)) = \sum_{z \in \operatorname{Res}(P(h)) \cap \Omega} f(z) + E_{f,\psi}(h)\,,$$

$$(2.1)$$

$$|E_{f,\psi}(h)| \le h^{-n} C_{\Omega,\psi} \max\{|f(z)|\ :\ z \in \overline{\Omega}_1 \setminus \Omega\,,\ \operatorname{Im} z \le 0\}\,,$$

where Ω_1 is a neighbourhood of Ω (see [3] and [18] for more precise versions).

The basic upper bound on the number of resonances is given as follows:

$$\sharp \operatorname{Res}(P(h)) \cap \Omega = \mathcal{O}(h^{-n})\,,$$

$$(2.2)$$

see [12],[19],[20],[33]. It would consequently appear that the error term in (2.1) is of the same order as the sum over the resonances. However, by choosing the function f so that it is small in $\Omega_1 \setminus \Omega \cap \{\text{Im } z \leq 0\}$, the sum of $f(z)$ can dominate the left hand side of (2.1). Doing that, Sjöstrand has shown that an analytic singularity of $E \mapsto |\{x : V(x) \geq E\}|$ at E_0 gives a lower bound in (2.2) for Ω a neighbourhood of E_0 – see [21] and references given there.

Another application of local trace formulæ techniques is in analysing resonances for *bottles*, that is perturbations of the Euclidean space in which the "size" of the perturbation may grow but which are connected to the euclidean infinity through a fixed "neck of the bottle" – see [21],[18].

3.   Breit-Wigner approximations

In a scattering experiment the physical data is mathematically encoded in the scattering matrix, $S(\lambda, h)$. It is in the behaviour of objects derived from the scattering matrix that we see physical manifestations of abstractly defined resonances. The classical and still central Breit-Wigner approximation provides this connection. The most mathematically tractable case is provided by the *scattering phase* $\sigma(\lambda, h) = \log \det S(\lambda, h)/(2\pi i)$, which is defined for V's with sufficient decay (otherwise relative phase shifts have to be used, see [3]). When λ is close to a resonance $E - i\Gamma$, the Breit-Wigner approximation says that

$$\sigma'(\lambda, h) \sim \frac{1}{\pi} \frac{\Gamma}{(\lambda - E)^2 + \Gamma^2} , \qquad (3.1)$$

that is, if Γ is small, $\sigma(\lambda)$ should change by approximately 1 as λ crosses E. This has been justified rigorously in some situations in whch a given resonance, close to the real axis, is isolated.

In view of (2.2) the number of resonances in fixed regions can be very large and for h small clouds of resonances need to be considered to obtain a correct form of the Breit-Wigner approximation. A formalism for that was introduced by Petkov-Zworski [16],[18], and it was developed further by Bruneau-Petkov [3] and Bony-Sjöstrand [2]. It is closely related to extending the trace formula (2.1) to h dependent Ω's. To formulate it, let us introduce $\omega(z, E) = (1/\pi) \int_E (|\text{Im } z|/|z - \lambda|^2) d\lambda$, $\text{Im } z < 0$, the harmonic measure corresponding to the upper half-plane. Then, for non-critical λ's (that is for λ's for which $\xi^2 + V(x) = \lambda \Rightarrow d_{(x,\xi)}(\xi^2 + V(x)) \neq 0$)

$$\sigma(\lambda + \delta, h) - \sigma(\lambda - \delta, h) = \sum_{\substack{z \in \text{Res}(P(h)) \\ |z - \lambda| < h}} \omega(z, [\lambda - \delta, \lambda + \delta]) + \mathcal{O}(\delta)h^{-n} , \qquad (3.2)$$

$0 < \delta < h/C$. The main point is that δ can be made arbitrarily small and that we only need to include resonances close to λ. If we do not assume that λ is non-critical weaker, but still useful, results can be obtained from factorization of the scattering determinant [18], or more generally, from the the analysis of the phase shift function [3].

One of the consequencies of the development of the Breit-Wigner approximations for clouds of resonances were new estimates on the number of resonances in
small regions, first in [16], and then in greater generality in [1],[3],[18]. If in place
of (2.2) we had a Weyl law with a remainder $\mathcal{O}(h^{1-n})$, then, as for eigenvalues,

$$\sharp\,\{z \in \mathrm{Res}(P(h)) \;:\; |z - \lambda| < \delta\} = \mathcal{O}(\delta)h^{-n}\,, \quad Ch < \delta < 1/C\,. \tag{3.3}$$

It turns out that despite the lack of the Weyl law we still have this estimate for
non-critical λ's.

4.   Resonance expansions of propagators

In the case of discrete spectrum of a self-adjoint operator the propagator,
$\exp(-itP(h)/h)$ can be expanded in terms of the eigenvalues. In fact, our understanding of "state specificity" often comes from such "Fourier decompositions" into
modes. For problems in which decay or escape to infinity are possible such expansions are still expected but just as in the case of trace formulæ far from obvious. For
non-trapping perturbations and in the context of the wave equation the expansions
in terms of resonances were studied in late 60's by Lax-Phillips and Vainberg (see
the references in [30]).

For trapping perturbations the expansions were investigated by Tang-Zworski
[30], and then by Burq-Zworski [5], Christiansen-Zworski [6], Stefanov [26], and
most recently by Nakamura-Stefanov-Zworski [14].

Here, we will recall the general expansion given in [5]: let $\chi \in C_c^\infty(\mathbf{R}^n)$,
$\psi \in C_c^\infty(0, \infty)$, and let chsupp $\psi = [a, b]$. There exists $0 < \delta < c(h) < 2\delta$ and L, so
that we have

$$\chi e^{-itP(h)/h}\chi\psi(P(h)) = \sum_{z \in \Omega(h) \cap \mathrm{Res}(\mathrm{P})} \chi\mathrm{Res}(e^{-it\bullet/h}R(\bullet, h), z)\chi\psi(P(h))$$

$$+ \; \mathcal{O}_{L^2 \to L^2}(h^\infty)\,, \quad \text{for } t > h^{-L}\,, \tag{4.1}$$

$\Omega(h) = (a - c(h), b + c(h)) - i[0, 1/C)$, and where $\mathrm{Res}(f(\bullet), z)$ denotes the residue
of a meromorphic family of operators, f, at z.

The function $c(h)$ depends on the distribution of resonances: roughly speaking we cannot "cut" through a dense cloud of resonances. Even in the very well
understood case of the modular surface [6, Theorem 1] there is, currently at least, a
need for some non-explicit grouping of terms. This is eliminated by the separation
condition [30, (4.4)] which however is hard to verify.

The unpleasant feature of (4.1) is the need for very large times $\sim h^{-L}$ and
the presence of a non-universal parameter, $c(h)$. The former is necessary in this
formulation as one sees by considering the free case $P(h) = -h^2\Delta$. However an
expansion valid for all times is possible in the case when a part of V constitutes a
barrier separating the trapped set in $\{(x, \xi) \;:\; \xi^2 + V(x) \in \mathrm{supp}\,\psi\}$ from infinity [14].
The example shown in Fig.1. b) is of that type. By the trapped set in $\Sigma \subset T^*\mathbf{R}^n$
we mean the set

$$K \cap \Sigma = \{(x, \xi) \in \Sigma \;:\; |\exp(tH_p)(x, \xi)| \not\to \infty\,, \; t \longrightarrow \pm\infty\}\,, \tag{4.2}$$

where H_p is the Hamilton vectorfield of $p = \xi^2 + V(x)$. One of the components comes from the work of Stefanov [27] on making estimates (2.2) more quantitative with constants related to the volume of the trapped set (for yet finer results of that type see Section.6).

5. Separation from the real axis

One of the most striking applications of PDE techniques in the study of resonances is the work of Burq (see [4] and references given there) on the separation of resonances from the real axis, estimates of the cut-off resolvent on the real axis, and the consequent estimates on the time decay of energy.

For non-trapping perturbations we have the following estimate on the truncated meromorphically continued resolvent

$$\|\chi R(z,h)\chi\|_{L^2 \to L^2} \leq C \frac{\exp(C|\operatorname{Im} z|/h)}{h} , \quad \operatorname{Im} z > -Mh \log \frac{1}{h} , \quad \chi \in C_c^\infty(\mathbf{R}^n) ,$$
$$(5.1)$$

for any M, see [11] for the absence of resonances (implicit in (5.1)), and [14] for the (easily derived) estimate.

Since the work of Stefanov-Vodev and Tang-Zworski (see [27],[29],[31] and references given there) we know that in many trapping situations[1], there exist many resonances converging to the real axis[2]. The question then is how close could the resonances approach the real axis or, as it turns out equivalently, how big can the truncated resolvent be on the real axis. Heuristically, tunneling should prevent arbitrary closeness to the real axis, and the separation should be universally given by at least $\exp(-S/h)$.

Tunneling of solutions is made quantitative in PDEs through *Carleman estimates* used classically to show unique continuation of solutions of second order equations. Following the work of Robbiano, and Lebeau-Robbiano, Burq succeeded in applying Carleman estimates to a wide range of resonance problems – see [4] and references given there. In the setting described here his basic result says that if Ω is as in (2.1) then

$$\exists S_1, S_2 , \quad z \in \operatorname{Res}(P(h)) \cap \Omega \implies \operatorname{Im} z > -\exp(-S_1/h) ,$$
$$\|\chi R(z,h)\chi\|_{L^2 \to L^2} \leq \exp(S_2/h), \quad z \in \Omega \cap \mathbf{R}, \quad \chi \in C_c^\infty(\mathbf{R}^n) . \qquad (5.2)$$

In addition, improved estimates are possible if χ is assumed to have support outside the projection of the trapped set:

$$\|\chi R(z,h)\chi\|_{L^2 \to L^2} \leq C/h, \quad z \in \Omega \cap \mathbf{R}, \quad \chi \in C_c^\infty(\mathbf{R}^n \setminus \pi(K)) . \qquad (5.3)$$

That means that away from the interaction region $\pi(K)$ we have, on the real axis, the same estimates as in the non-trapping case (5.1) and that has immediate applications in scattering theory [28].

---

[1] roughly speaking, whenever there exists an elliptic orbit of the H_p-flow

[2] That in specific trapping situations there exist resonances whose distance to the real axis is of order $\exp(-S/h)$ is classical in physics with the most precise mathematical results given by Helffer-Sjöstrand – see [9]. Here we concentrate on very general existence results which guarantee lower bounds on the number of resonances.

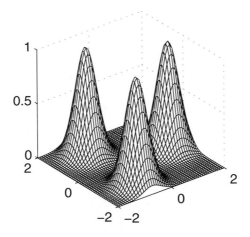

Figure 2: Graph of a potential $V(x)$, for which the classical flow of the Hamiltonian $\xi^2 + V(x)$ is hyperbolic for energies close to 0.5.

6. Resonances in chaotic scaterring

Since the work of Sjöstrand [19] on geometric upper bounds for the number of resonances, it has been expected that for chaotic scattering systems the density of resonances near the real axis can be approximately given by a power law with the power equal to half of the dimension of the trapped set (see (6.1) below). Upper bounds in geometric situations have been obtained in [32] and [35].

An example of a potential in $V \in C_c^\infty(\mathbf{R}^2)$ for which the flow of H_p, $p = \xi^2 + V(x)$, is hyperbolic (and hence scattering exhibits chaotic features) is shown in Fig 6.

A recent numerical study [10] for that potential indicates that the density of resonances satisfies a lower bound related to the dimension of the trapped set – see Fig.6. In a complicated semi-classical situation studied in [10], the dimension is a delicate concept and it may be that different notions of dimension have to be used for upper and lower bounds. That point is emphasized in [25] where numerical data for semi-classical zeta function for several convex obstacles is analyzed.

In the case of *convex co-compact hyperbolic quotients*, $X = \Gamma \backslash \mathbf{H}^2$, studied in [35] the situation is particularly simple as the quantum resonances coincide with the zeros of the zeta-function – see [15]. The notion of the dimension of the trapped set is also clear as it is given by $2(1 + \delta)$. Here $\delta = \dim \Lambda(\Gamma)$ is the dimension of the limit set of Γ, that is the set of accumulation points of the elements of Γ (they are all hyperbolic), $\Lambda(\Gamma) \subset \partial \mathbf{H}^2$.

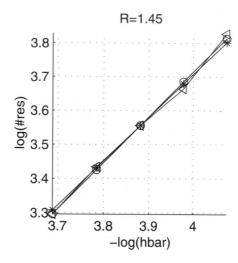

Figure 3: A plot of $\log(N_{res})$ as a function for $-\log(\hbar)$ for the potential shown in in Fig. 6. The value of $\hbar$ ranges from 0.025 to 0.017. Triangles represent numerical data, circles least squares regression, and stars the slope predicted by the conjecture.

Hence we expect that

$$\sum_{|\text{Im } s|\leq r,\ \text{Re } s>-C} m_{\Gamma}(s) \sim r^{1+\delta}, \tag{6.1}$$

where $m_{\Gamma}(s)$ is the multiplicity of the zero of the zeta function of Γ at s. We also changed the traditional convention: here $h^2 s(1-s) = z$ in the notation of previous sections, and $h(\text{Re } s - 1/2)$, $h\text{Im } s$ correspond to Im z, Re z respectively.

An upper bound of this form was established in [35] but a simpler method giving improved upper bounds has been recently presented in [37]. The new method is based on zeta function techniques. What we obtain is a bound in the case of convex co-compact Schottky groups

$$\sum \{m_{\Gamma}(s) : r \leq \text{Im } s \leq r+1, \ \ \text{Re } s > -C_0\} \leq C_1 r^{\delta}, \tag{6.2}$$

where $\delta = \dim \Lambda(\Gamma)$. This improved estimate is a "fractal" version of (3.3): $1 + \delta$ now plays the rôle of n. It is clear that the method works in greater generality and implementing it is part of an ongoing project.

References

[1] J.-F. Bony, *Majoration du nombre de résonances dans des domaines de taille h*, Internat. Math. Res. Notices, **16**(2001), 817–847.

[2] J.-F. Bony and J. Sjöstrand, *Trace formula for resonances in small domains*, J. Funct. Anal., **184**(2001), 402–418.

[3] V. Bruneau and V. Petkov, *Meromorphic continuation of the spectral shift function*, Duke Math. J., to appear.

[4] N. Burq, *Lower bounds on shape resonance width of semi-classical long range Schrödinger operators*, Amer. J. Math., to appear.

[5] N. Burq and M. Zworski, *Resonance expansions in semi-classical propagation*, Comm. Math. Phys., **232**(2001), 1–12.

[6] T. Christiansen and M. Zworski, *Resonance wave expansions: two hyperbolic examples*, Comm. Math. Phys., **212**(2000), 323–336.

[7] P. Gaspard, D. Alonso, and I. Burghardt, in *Advances in Chemical Physics XC*, I. Prigogine and S.A. Rice, eds. John Wiley & Sons, 1995.

[8] L. Guillopé and M. Zworski, *Scattering asymptotics for Riemann surfaces*, Ann. of Math., **129** (1997), 597–660.

[9] A. Lahmar-Benbernou and A. Martinez, *On Helffer-Sjöstrand's theory of resonances*, Int. Math. Res. Not., **13**(2002), 697–717.

[10] K. Lin and M. Zworski, *Quantum resonances in chaotic scattering*, Chem. Phys. Lett., **355**(2002), 201–205.

[11] A. Martinez, *Resonance free domains for non-analytic potentials*, preprint 2001.

[12] R.B. Melrose, *Geometric Scattering Theory*, Cambridge University Press, 1996.

[13] M. Merkli and I.M. Sigal, *A time-dependent theory of quantum resonances*, Comm. Math. Phys., **201** (1999), 549–576.

[14] S. Nakamura, P. Stefanov, and M. Zworski, *Resonance expansions of propagators in the presence of potential barriers*, preprint, 2002.

[15] S.J. Patterson and P. Perry, *Divisor of the Selberg zeta function for Kleinian groups in even dimensions, with an appendix by* C. Epstein, Duke. Math. J., **326**(2001), 321–390.

[16] V. Petkov and M. Zworski, *Breit-Wigner approximation and the distribution of resonances*, Commun. Math. Phys., **204** (1999), 329–351.

[17] V. Petkov and M. Zworski, Erratum to [16], Commum. Math. Phys., **214** (2000), 733–735.

[18] V. Petkov and M. Zworski, *Semi-classical estimates on the scattering determinant*, Annales H. Poincare, **2** (2001), 675–711.

[19] J. Sjöstrand, *Geometric bounds on the density of resonances for semiclassical problems*, Duke Math. J., **60**(1990), 1–57.

[20] J. Sjöstrand, *A trace formula and review of some estimates for resonances*, in *Microlocal analysis and spectral theory* (Lucca, 1996), 377–437, NATO Adv. Sci. Inst. Ser. C, Math. Phys. Sci., 490, Kluwer Acad. Publ., Dordrecht, 1997.

[21] J. Sjöstrand, *Resonances for bottles and related trace formulæ*, Math. Nachr., **221**(2001), 95–149.

[22] J. Sjöstrand and M. Zworski, *Complex scaling and the distribution of scattering poles*, Journal of AMS, **4** (4) (1991), 729–769.

[23] J. Sjöstrand and M. Zworski, *Asymptotic distribution of resonances for convex obstacles*, Acta Math., **183**(1999), 191–253.

[24] J. Sjöstrand and M. Zworski, *Quantum monodromy and semi-classical trace formulæ*, J. Math. Pure Appl.

[25] S. Sridhar, W. Lu, and M. Zworski, *Fractal Weyl laws for chaotic open systems*, in preparation.

[26] P. Stefanov, *Resonance expansions and Rayleigh waves*, Math. Res. Lett., **8**(2001), 105–124.

[27] P. Stefanov, *Sharp upper bounds on the number of resonances near the real axis for trapping systems*, to appear in Amer. J. Math.

[28] P. Stefanov, *Estimates on the residue of the scattering amplitiude*, Asymp. Anal., to appear.

[29] S.H. Tang and M. Zworski, *From quasimodes to resonances*, Math. Res. Lett., **5**(1998), 261–272.

[30] S.H.Tang and M. Zworski, *Resonance expansions of scattered waves*, Comm. Pure Appl. Math., **53** (2000), 1305–1334.

[31] G. Vodev, *Resonances in euclidean scattering*, Cubo Matematica Educacional, **3**(2001), 317–360. http://poincare.math.sciences.univ-nantes.fr/prepub/

[32] J. Wunsch and M. Zworski, *Distribution of resonances for asymptotically Euclidean manifolds*. J. Differential Geom., **55**(2000), 43–82.

[33] M. Zworski, *Counting scattering poles,* in Spectral and scattering theory (Sanda, 1992), 301–331, Lecture Notes in Pure and Appl. Math., **161**, Dekker, New York, 1994.

[34] M. Zworski, *Poisson formulæ for resonances*, Séminaire E.D.P. 1996-1997, École Polytechnique, XIII-1-XIII-12.

[35] M. Zworski, *Dimension of the limit set and the density of resonances for convex co-compact hyperbolic surfaces*, Invent. Math., **136**(1999), 353–409.

[36] M. Zworski, *Resonances in physics and geometry*, Notices Amer. Math. Soc., **46** (1999), 319–328.

[37] M. Zworski, *Density of resonances for Schottky groups*, preprint, 2002.

Section 12. Ordinary Differential Equations and Dynamical Systems

ICM 2002 · Vol. III · 255–264

Non Uniformly Hyperbolic Dynamics: Hénon Maps and Related Dynamical Systems

Michael Benedicks*

Abstract

In the 1960s and 1970s a large part of the theory of dynamical systems concerned the case of uniformly hyperbolic or Axiom A dynamical system and abstract ergodic theory of smooth dynamical systems. However since around 1980 an emphasize has been on concrete examples of one-dimensional dynamical systems with abundance of chaotic behavior (Collet &Eckmann and Jakobson). New proofs of Jakobson's one-dimensional results were given by Benedicks and Carleson [5] and were considerably extended to apply to the case of Hénon maps by the same authors [6]. Since then there has been a considerable development of these techniques and the methods have been extended to the ergodic theory and also to other dynamical systems (work by Viana, Young, Benedicks and many others). In the cases when it applies one can now say that this theory is now almost as complete as the Axiom A theory.

2000 Mathematics Subject Classification: 37D45, 37D25, 37G35, 37H10, 60, 60F05.
Keywords and Phrases: Hénon maps, Strange attractors, Ergodic theory.

1. Introduction

Dynamical systems as a discipline was born in Henri Poincaré's famous treatise of the three body problem. In retrospect arguably one can view as his most remarkable discovery of the *homoclinic phenomenon*. Stable and unstable manifolds of a fixed point or periodic point may intersect at a *homoclinic point* thereby producing a very complicated dynamic behavior—what we now often call chaotic.

In my opinion in the development of theory of dynamical systems—like in the development of all good mathematics—one can clearly see two stages. The first

---

*Department of Mathematics, Royal Institute of Technology, S-100 44 Stockholm, Sweden. E-mail: michaelb@math.kth.se

stage is the understanding of concrete examples and the second stage is generalization. A large part of the second encompasses the introduction of the right concepts
which makes the arguments in the concrete examples into a theory.

One such development starts with the two famous papers by M. L. Cartwright
and J. Littlewood "On non-linear differential equations of the second order I and
II", which was among the first to treat nonlinear differential equations in depth.
Littlewood was astonished by the difficulties that arose in studying these model
problems and called the second of these papers "the monster".

S. Smale gives in "Finding a horseshoe on the beaches of Rio" an entertaining
account of how he was lead to his now ubiquitous horseshoe model for chaotic
dynamics. In fact in his first paper in dynamical systems he made the conjecture
that "chaotic dynamics does not exist" but received a letter from N. Levinson
with a paper clarifying the previous work by Cartwright and Littlewood and which
effectively contained a counterexample to Smale's conjecture. Levinson's paper
contained extensive calculations which Smale found difficult follow and this lead
him to construct a model with minimal complexity but still with the main features
of Levinson's ODE model.

Starting from the horseshoe model, Smale and his group at Berkeley started
to develop the theory of uniformly hyperbolic or Axiom A dynamical systems in
the seventies. One central concept in this theory is that of an axiom A attractor
for a diffeomorphism f of a manifold M.

Let Λ be an invariant set for a diffeomorphism of a manifold M. Λ is said
to be a *hyperbolic set* for f if there is a continuous splitting of the tangent bundle
of M restricted to Λ, $TM|_\Lambda$, which is invariant under the derivative map Df:
$TM|_\Lambda = E^s \oplus E^u$; $Df(E^s) = E^s$; $Df(E^u) = E^u$; and for which there are constants
$C > 0$ and $c > 0$, such that $||Df^n|_{E^s}|| \leq Ce^{-cn}$ and $||Df^n|_{E^u}|| \leq Ce^{-cn}$, for all
$n \geq 0$ and there is a uniform lower bound for the angle between stable and unstable
manifolds: $\mathrm{angle}(E^u(x), E^s(x)) \geq C$, $\forall x \in \Lambda$.

Λ is called an *attractor* (in the sense of Conley) if there is a neighborhood
$U \supset \Lambda$ such that $\overline{f(U)} \subset U$ and $\Lambda = \bigcap_{n=1}^\infty f^n(U)$. This attractor is *topologically
transitive* if there is $x \in \Lambda$ with dense orbit in Λ. If moreover $f|_\Lambda$ is uniformly
hyperbolic Λ is an *Axiom A attractor*.

The ergodic theory of Axiom A attractors was developed by Sinai, Ruelle and
Bowen in the 1970. In particular for an Axiom A attractor they constructed so
called SRB-measures: measures with absolutely continuous conditional measures
on unstable manifolds. This measures μ are also *physical measures* in the sense
of Ruelle since for z_0 in a set of initial points of positive Riemannian measure the
Birkhoff sums $\frac{1}{n} \sum_{j=0}^{n-1} \delta_{f^j z_0} \to \mu$ as $n \to \infty$.

In fact for a topologically transitive Axiom A attractor μ is unique and the
Birkhoff sums converges for a.e. $z_0 \in B$, where B is the *basin of attraction* $B =
\bigcup_{j\geq 0} f^{-j}(U)$. Although the Axiom A theory is quite satisfactory and complete it is
not applicable to very many concrete dynamical systems. There is a general theory
of ergodic theory of smooth dynamical systems due to among others Pesin, Katok,
Ledrappier and others but concrete examples were lacking. Around 1980 the theory
of chaotic one-dimensional maps was started. M. Misiurewicz studied multimodal

maps f of the interval, whose critical set or set of turning points, $\mathcal{C}$, has the property that for all $z_0 \in \mathcal{C}$ and all $j \geq 1$, $\text{dist}(f^j z_0, \mathcal{C}) \geq \delta > 0$ and proved existence of absolutely continuous invariant measures. Then Collet and Eckmann [12] proved abundance, i.e. positive Lebesgue measure, of aperiodic behavior for a family of unimodal maps of the interval and, Jakobson in [20] proved abundance of existence of absolutely continuous invariant measures for the quadratic family. A new proof of Jakobson's theorem was then given by Lennart Carleson and myself in [5] and the methods from this paper were later used by us in [6] to prove aperiodic, chaotic behavior for a class of Hénon maps, which are small perturbations of quadratic maps. The methods of [6] have turned out to be useful for several other dynamical systems and the corresponding ergodic theory has been developed. There have been other accounts of this development, in particular by my collaborators Lai-Sang Young (see e.g. [35]) and Marcelo Viana (see e.g. the proceedings of ICM98 [30]).

2.   Hénon maps

In 1978, M. Hénon proposed as a model for non-linear two-dimensional dynamical systems the map

$$(x, y) \mapsto (1 + y - ax^2, bx) \qquad 0 < a < 2, \ b > 0.$$

He chose the parameters $a = 1.4$ and $b = 0.3$ and proved that $f = f_{a,b}$ has an attractor in the sense of Conley.

He also verified numerically that this Hénon map has sensitive dependence on initial conditions and produced his well-known computer pictures of the attractor. Hénon proposed that this dynamical system should have a "strange attractor" and that it should be more eligible to analysis than the ubiquitous Lorenz system.

In principle most initial points could be attracted to a long periodic cycle. In view of the famous result of S. Newhouse, [24], periodic attractors are topologically generic, so it was not at all a priori clear that the attractor seen by Hénon was not a long stable periodic orbit.

However Lennart Carleson and I, [6], managed to prove that what Hénon conjectured was true—not for the parameters $(a, b) = (1.4, 0.3)$ that Hénon studied—but for small $b > 0$. In fact we managed to prove the following result:

Theorem 1. *There is a constant $b_0 > 0$ such that for all b, $0 < b < b_0$ there is a set A_b of parameters a, such that its one-dimensional Lebesgue measure $|A_b| > 0$ and such that for all $a \in A_b$, $f = f_{a,b}$, has the following properties*

1. *There is an open set $U = U_{a,b}$ such that $\overline{f(U)} \subset U$ and $\Lambda = \bigcap_{n=0}^{\infty} f^n(U) = \overline{W^u(\hat{z})}$, where $W^u(\hat{z})$ is the unstable manifold of the fixed point $\hat{z}$ of f in the first quadrant.*
2. *There is a point $z_0 = z_0(a, b)$ such that $\{f^j(z_0)\}_{j=0}^{\infty}$ is dense on Λ, and there is $c > 0$ such that $|Df^j(z_0)(0, 1)| \geq e^{cj}$, $j = 1, 2, \ldots$.*

Hence Λ is a topological transitive attractor with sensitive dependence on initial conditions. An immediate consequence of Fubinis theorem is that the "good parameter set" $A = \bigcup_{b>0} A_b \times \{b\}$ is (a Cantor set) of positive two-dimensional Lebesgue measure.

The first part of the theorem is easy to prove and the result is true for an open set of parameters, i.e. for a small rectangle close to $a = 2$ and $b = 0$ contained in $\{(a, b) : 0 < a < 2, \ b > 0\}$. The system is dissipative since $|\det D f_{a,b}| = |b| < 1$. In this case applying an argument of Palis and Takens, [28], it follows that a region that is enclosed by pieces of stable and unstable manifolds of the same fixed point $\hat{z}$ is attracted to the unstable manifold $W^u(\hat{z})$. With some additional arguments one can see that also a neighborhood of the closure of the unstable manifold is attracted. However the attractor could a priori be a proper subset of $\overline{W^u(\hat{z})}$.

The second part of the theorem is only true for parameters $(a, b) \in A$. The main ingredient in the proof of the second part of the theorem is the identification of a *critical set* $\mathcal{C}$ for these Hénon attractors. The set $\mathcal{C}$ is countable and located on $W^u(\hat{z})$ but it is natural to expect that the Hausdorff dimension of $\overline{\mathcal{C}}$ is positive.

For each $z_0 \in \mathcal{C}$ the following holds: (i) $|D f^j(z_0)\tau(z_0)| \le e^{-cj} \ \forall j \ge 1$, where $\tau(z_0)$ is the tangent vector of the unstable manifold at z_0; (ii) trough each $z_0 \in \mathcal{C}$ there is a local unstable manifold $W^s(z_0)$, tangent to $W^u(\hat{z})$.

The proof of the theorem is a huge induction in time n. Successively preliminary critical points or sets, *precritical points*, are defined on higher and higher *generations* of the unstable manifold. We roughly say that $z \in W^u(\hat{z})$ is of generation G if $z \in f^G(\gamma) \setminus f^{G-1}(\gamma)$, where γ is the horizontal segment of the local unstable manifold $W^u(\hat{z})$ through the fixed point.

In analogy with the methods from the one-dimensional case of [5] parameters are chosen so that inductively $d(f^j z_0, \mathcal{C}_G) \ge e^{-\alpha j} \ \forall j \le n, \ \forall z_0 \in \mathcal{C}_G$, where $\mathcal{C}_G$ is the set of precritical points of generation $\le G = \theta(b) \cdot n$, where $\theta(b) = C/\log(1/b)$ and $\alpha > 0$ is a suitably chosen numerical constant. Moreover a further parameter selection is made so that, informally speaking, too deep returns close to the critical set do not occur too often. The estimate of the measure of the set, deleted because of this condition, is made by a large deviation argument.

3. The ergodic theory

Existence of SRB-measures. For the set of "good parameters" A of Theorem 1, Lai-Sang Young and I proved in [9], the following result

Theorem 2. *For all $(a, b) \in A$, $f_{a,b}$ has a unique SRB measure supported on the attractor.*

As a consequence it follows by general smooth ergodic theory that there is a set of initial points E of positive two-dimensional Lebesgue measure, a subset of the topological basin B, such that for all $z_0 \in E$ the Birkhoff sums $\frac{1}{n} \sum_{j=0}^{n-1} \delta_{f^j z_0} \to \mu$, weak-$*$ as $n \to \infty$.

Decay of correlation and the central limit theorem. For the same class of Hénon maps as in Theorem 1, L.S. Young and I, [10], managed to prove decay of correlation and a version of the central limit theorem. Our main results may be summarized in the following theorem:

Theorem 3. *Suppose φ and ψ are Hölder observables, i.e. they are functions on the plane that belong to some Hölder class α, $0 < \alpha \leq 1$. Then for some $C > 0$ and $c > 0$*

1. $\left| \int \varphi(f^j(x)) \, \psi(x) \, d\mu - \left(\int \varphi \, d\mu \right) \left(\int \psi \, d\mu \right) \right| \leq C e^{-cj} \qquad \forall j \geq 0;$

2. $\mu \left(\left\{ x : \frac{1}{\sqrt{n}} \left(\sum_{j=0}^{n-1} \varphi(f^j x) - n \int \varphi \, d\mu \right) \leq t \right\} \right) \to \Phi_\sigma(t)$ *as* $n \to \infty$, *where* $\Phi_\sigma(t)$ *is the normal distribution function* $\mathcal{N}(0, \sigma)$.

The methods used to prove this theorem involved the definition of a return set $X = X_u \cap X_s$, where X_u is a set of approximately horizontal long unstable manifolds γ_u and X_s is a set of approximately vertical stable manifolds γ_s, indexed, say, by an arclength coordinate of its intersection with one of the unstable manifolds of X_u. A dynamical tower construction was made and a Markov extension (Markov partition on the tower) was constructed. One of the key estimates concerns the distribution of the return time $R(x) = R_i$ for $x \in X^i$, which is defined as the time the image a partition element X^i returns to the base of the tower. R_i may be defined as the first time such that $f^{R_i}(X^i) \supset (\gamma_u \cap X)$ for all γ_u such that $f^{R_i}(X^i) \cap \gamma_u \neq \emptyset$ and $R(x) \in X$. Our estimate is that there are constants C and $c > 0$ such that for each γ_u, $|\{x \in X \cap \gamma_u : R(x) > t\}| \leq C^{-ct}$. L.S. Young was then able to give a generalization of this setting of dynamical towers, which applies to other dynamical systems. In particular she managed to prove exponential decay of correlation for dissipative billiards ([34]).

The metric basin problem. A natural question that arises in connection with Theorem 2 is for which set of initial points z_0 the Birkhoff sums $n^{-1} \sum_{j=0}^{n-1} \delta_{f^j z_0}$ converge weak-* to the SRB-measure μ. As previously mentioned from the smooth ergodic theory it only follows that this is true for a set of initial points of positive two-dimensional Lebesgue measure.

In [8], Marcelo Viana and I were able to complete the picture: in fact almost all points of the topological basin are generic for the SRB-measure and the basin is foliated by stable manifolds.

Theorem 4. *Let us consider the set of Hénon maps $f_{a,b}$, where $(a, b) \in A$ (the set of good parameters of Theorem 1). Then the following holds*

1. *Through Lebesgue a.e. $z_0 \in B$ there is an infinitely long stable manifold $W^s(z_0)$ that hits the attractor.*

2. *For a.e. initial point $z_0 \in B$, $\frac{1}{n} \sum_{j=0}^{n-1} \delta_{f^j z_0} \to \mu$ weak-* as $n \to \infty$, where μ is the SRB-measure of Theorem 2.*

This work was in fact carried out in the more general setting of the Hénon-like maps of Mora & Viana (see Section 4. below). We were also able to characterize the topological basin in this setting: its boundary is the stable manifold of the fixed point in the third quadrant (this was proved independently by Y. Cao [11]).

4. Other dynamical systems

Hénon-like maps. After a rescaling of the second coordinate, the Hénon maps may be written as $(x, y) \mapsto (1 - ax^2, 0) + \sqrt{b}(y, x)$. More generally Mora & Viana,

[23] con-sided maps of the form $f_a(x, y) = (1 - ax^2, 0) + \psi(x, y)$, where $c_1 b \le |\det(Df(x, y))| \le c_2 b$, $||D(\log|\det Df|)||_{C^1} \le C$ for some c_1, c_2, C and $||\psi||_{C^3} = \mathcal{O}(b^{\frac{1}{2}})$. They managed to carry out the same program as in Theorem 1 for these class of maps, which they called *Hénon-like maps*, and managed to prove prevalence of strange attractors, i.e. that there is a set of positive measure of parameters a so that f_a exhibits a strange attractor.

Mora & Viana used a one-parameter family of maps g_μ, $-\varepsilon < \mu < \varepsilon$, such that g_0 has a homoclinic tangency and proved that there is a positive measure set of parameters μ and a neighborhood U_μ such that $g_\mu^N|_{U_\mu}$ has a Hénon-like strange attractor. This is done following Palis & Takens [27], by proving that there is a linear change of variables and in the parameters Φ_N so that $\Phi_N^{-1} \circ g_\mu^N \circ \Phi_N(\xi, \eta) = (1 - a\xi^2, 0) + \psi(\xi, \eta)$, where ψ satisfies the appropriate estimates of the Hénon-like maps. Note that the consequence of this is not the existence of a global attractor as in Theorem 1 but the existence of a local attractor close to the homoclinic tangency. (A homoclinic tangency does really occur close to Hénon's classical parameters $a = 1.4$ and $b = 0.3$: this was proved by Fornaess and Gavasto, [18].)

Saddle node bifurcations. Another case where methods based on those in [6] turned out to be useful is the case of saddle node bifurcations treated by Diaz, Rocha and Viana in [16]. In that paper they show that when unfolding a one-parameter family with a critical saddle node cycle Hénon like strange attractors appear with positive density at the bifurcation value. Moreover they prove that in an open class of such families the strange attractors are of global type.

This work was continued by M.J. Costa [14], [15], who proved that global strange attractors also appear when destroying a hyperbolic set (horseshoe) by collapsing it with a periodic attractor.

Viana's dynamical systems with multiple expanding directions. In [32], M. Viana studied the dynamical system $f : \mathbf{T} \times I \mapsto \mathbf{T} \times I$ given by the the following skew product $(\theta, x) \mapsto (m\theta \pmod 1, a_0(\theta) - x^2)$, where $a(\theta) = a_0 + \alpha \sin \pi(\theta - \frac{1}{2})$.

If m is a sufficiently large integer (≥ 16 is enough), a_0 is such that $f_{a_0}(x) = a_0 - x^2$ is a Misiurewicz map, i.e. $|f^j(0)| \ge \delta$, $\forall j \ge 1$, and $\alpha > 0$ is sufficiently small, he managed to prove that for a.e. (θ, x) and some constants C and $c > 0$, $|\partial_x f^j(\theta, x)| \ge Ce^{cj}$, $\forall j \ge 1$. In the second part of this paper Viana also considers skew products of Hénon maps driven by circle endomorphisms.

An important difference with the situation in this paper compared to earlier work in the area is that the exponential approach rate condition of an orbit relative to the critical set is no longer satisfied. Instead this is replaced by a statistical property: very deep and very frequent returns to the critical region is very unlikely. The argument is based on an extension of the large deviation argument from [6].

SRB measures for the Viana maps were constructed by J.F. Alves in [1]. An important concept introduced by Alves was the notion of *hyperbolic times*, which are a generalization of the *escape situations* that were considered in [6] and they are also similar to the base in the tower construction in [34], [10].

Infinite-modal maps and flows. Motivated by the study of unfolding of saddle-focus connections for flows in three dimensions Pacifico, Rovella and Viana, [25], studied parameterized families of one-dimensional maps with infinitely many critical points. They prove that for a positive Lebesgue measure set of parameter values the map is transitive and almost all orbits have positive Lyapunov exponent. There has been a considerable amount of work on flows, by Viana, Luzatto, Pumariño, Rodríguez and others, where techniques from [6] have played an important role. For a survey of these and related results I refer to [31].

In a different direction is the recent proof by W. Tucker, [29], of the existence of chaotic behaviour for the Lorenz attractor.

The attractors of Wang and Young. In a recent paper, D. Wang and L.S. Young, [33], carry out a theory of attractors which generalize the Hénon maps in a different direction. They consider maps of a two-dimensional manifold of the form $f(x,y,a,b) = (F(x,y,a,b),0) + b\,(u(x,y,a,b),v(x,y,a,b))$. This class clearly differs from the Hénon-like maps of Mora & Viana. In particular the theory applies to perturbations of certain one-dimensional multimodal maps, with a transversality condition in the parameter dependence. The techniques are analogous to these in [6] but more information on the geometric structure, in particular of the critical set, is achieved. Most previous results are obtained in this setting but also new results, e.g. on some similar dynamical systems and on topological entropy.

5. Random perturbations and stochastic stability

A natural question is how the statistical properties of a dynamical system with chaotic behavior behaves when it is perturbed randomly by some small noise at each iterate. Here we will mainly consider independent additive noise and assume that the underlying ambient space M is either a subset of Euclidean space or a torus but cases of more general manifolds and more general perturbations can also be considered (for this see several papers and books by Y. Kiefer).

Let $f : M \to M$ and suppose that ξ_n, $n \geq 0$, are independent identically distributed random variables with an absolutely continuous probability density supported in a small ball $B(0,\varepsilon)$ around the origin and consider the Markov chain $\{X_n\}_{n=0}^{\infty}$ defined by $X_{n+1} = f(X_n) + \xi_n$. Then there is a stationary transition probability $p_\varepsilon(E|x) = p(X_{n+1} \in E | X_n = x)$ and also at least some stationary measure ν_ε satisfying the fixed point equation $\nu_\varepsilon(E) = \int p_\varepsilon(E|x)\,d\nu_\varepsilon(x)$.

The obvious questions are here whether ν_ε is unique and in that case if ν_ε tends to an invariant measure of the unperturbed system when $\varepsilon \to 0$. This is the problem of *Stochastic Stability*. For the case of Axiom A attractors such results were proved by Y. Kiefer and L.S. Young.

Now such results have also been obtained for the non-uniformly hyperbolic dynamical systems described above. In the case of the quadratic interval maps of [5], L.S. Young and I proved in [9], under suitable assumptions on the density of the perturbations, that ν_ε is unique and $\nu_\varepsilon \to \mu$ weak-*, as $\varepsilon \to 0$, where $d\mu = \varphi\,dx$ is the absolutely continuous invariant measure. V. Baladi and M. Viana, [2], managed to improve this to prove that the density of ν_ε, ψ_ε, converges to φ in L^1-norm.

M. Viana and I have recently proved results on weak-* stochastic stability for the Hénon maps of [6] and the Hénon-like maps of [23] (see [7]). For a recent paper on decay of correlation for *random skew products* of quadratic maps see

Baladi &Benedicks &Maume-Deschampes [3].

6. Open problems and concluding remarks

The most important problem in this general area is the problem of positive Lyapunov exponent for the Standard Map, i.e. the map of the two-dimensional torus defined by $(x, y) \mapsto (2x - y + K \sin 2\pi x, x) \pmod{\mathbf{Z}^2}$.

The general belief is that at least for some parameters K there is at least one ergodic component of positive Lebesgue measure with positive Lyapunov exponent. Nothing is however rigorously known in spite of intensive work by many people. One of the most interesting results by Duarte [17] goes in the opposite direction: for a residual set of parameters K the closure of the elliptic points can have Hausdorff dimension arbitrarily close to 2.

One important difference between the Standard Map and the Hénon maps for the good parameters is that "the critical set" in the Hénon case is rather small. It has a hierarchical structure and conjecturally the Hausdorff dimension of its closure should be $\mathcal{O}(1/(\log(1/b))$. The critical set for the standard map (if possible to define) should have Hausdorff-dimension ≥ 1.

A. Baraviera proved in his recent thesis [4], positive Lyapunov exponent for the Standard Map with parameters driven by an expanding circle endomorphism.

On natural class of problem is to consider are *skew products* of Viana's type, where the parameters are driven by more general maps. One possible choice is to let the driving map be a non-uniformly hyperbolic quadratic map, either a Misiurewicz map or more generally the class of quadratic maps of [5], [6]. A more difficult project would be to study the case when driving map is a circle rotation.

The general picture for dissipative Hénon maps. It is a natural question to consider what happens for other parameters than the ones considered in [6]. One possible scenario is outlined in the following questions, which are much related to J. Palis conjectures, [26], in the C^r-generic setting.

Question 1. Are there for Lebesgue almost every parameter (a, b) in $\{(a, b) \in \mathbf{R}^2 : 0 < a < 2, b > 0\}$ at most finitely many coexisting strange attractors and stable periodic orbits?

If this is true the Newhouse phenomenon of infinitely coexisting sinks and Colli's situation [13] of infinitely many coexisting Hénon-like strange attractors would only appear for a Lebesgue zero set of parameters (a, b)?

Question 2. For the parameter values for which only finitely many Hénon-like attractors or sinks coexists: do the respective basins cover Lebesgue almost all points of the phase space?

Question 3. Is the set of parameters for which the Hénon map $f_{a,b}$ is hyperbolic dense in the parameter space?

This result if true would correspond to the real Fatou conjecture proved fairly recently by Graczyk &Swiatek, [19], and Lyubich [21]. A positive answer to Question 1 would correspond to Lyubich's result that almost all points in the quadratic family is either regular or stochastic, [22].

As can be seen from the above Dynamical Systems is a beautiful mixture of Topology and Analysis. To the hard analysis of Cartwright, Littlewood and Levinson, Smale was able to provide a topological counterpart. Recently again more analytical methods have played an important role. What will be next?

References

[1] J. F. Alves. SRB measures for non-hyperbolic systems with multidimensional expansion. *Ann. Sci. École Norm. Sup.*, 33:1–32, 2000.

[2] V. Baladi and M. Viana. Strong stochastic stability and rate of mixing for unimodal maps. *Ann. Sci. École Norm. Sup. (4)*, 29:483–517, 1996.

[3] Viviane Baladi, Michael Benedicks, and Véronique Maume-Deschamps. Almost sure rates of mixing for i.i.d. unimodal maps. *Ann. Sci. École Norm. Sup. (4)*, 35(1):77–126, 2002.

[4] A. T. Baraviera. *Robust nonuniform hyperbolicity for volume preserving maps.* PhD thesis, IMPA, 2000.

[5] M. Benedicks and L. Carleson. On iterations of $1 - ax^2$ on $(-1, 1)$. *Annals of Math.*, 122:1–25, 1985.

[6] M. Benedicks and L. Carleson. The dynamics of the Hénon map. *Annals of Math.*, 133:73–169, 1991.

[7] M. Benedicks and M. Viana. Random perturbations and statistical properties of certain Hénon-like maps. In preparation.

[8] M. Benedicks and M. Viana. Solution of the basin problem for Hénon-like attractors. *Invent. Math.*, 143:375–434, 2001.

[9] M. Benedicks and L.-S. Young. Absolutely continuous invariant measures and random perturbations for certain one-dimensional maps. *Ergod. Th. & Dynam. Sys.*, 12:13–37, 1992.

[10] M. Benedicks and L.-S. Young. Markov extensions and decay of correlations for certain Hénon maps. *Astérisque*, 261:13–56, 2000.

[11] Yongluo Cao. The nonwandering set of some Hénon map. *Chinese Sci. Bull.*, 44(7):590–594, 1999.

[12] P. Collet and J. P. Eckmann. On the abundance of aperiodic behaviour for maps of the interval. *Comm. Math. Phys.*, 73:115–160, 1980.

[13] E. Colli. Infinitely many coexisting strange attractors. *Ann. Inst. H. Poincaré Anal. Non Linéaire*, 15(5):539–579, 1998.

[14] M. J. Costa. Chaotic behaviour of one-dimensional saddle-node horseshoes. To appear in Discrete and Continous Dynamical Systems.

[15] M.J. Costa. Saddle-node horseshoes giving rise to global Hénon-like attractors. *An. Acad. Brasil. Ciênc.*, 70(3):393–400, 1998.

[16] L. J. Díaz, J. Rocha, and M. Viana. Strange attractors in saddle-node cycles: prevalence and globality. *Invent. Math.*, 125:37–74, 1996.

[17] P. Duarte. Plenty of elliptic islands for the standard family of area preserving maps. *Ann. Inst. H. Poincaré Anal. Non. Linéaire*, 11:359–409, 1994.

[18] J.E. Fornæss and E.A. Gavosto. Tangencies for real and complex Hénon maps: an analytic method. *Experiment. Math.*, 8(3):253–260, 1999.

[19] J. Graczyk and G. Swiatek. Generic hyperbolicity in the logistic family. *Annals of Math.*, 146:1–52, 1997.

[20] M. Jakobson. Absolutely continuous invariant measures for one-parameter families of one-dimensional maps. *Comm. Math. Phys.*, 81:39–88, 1981.

[21] M. Lyubich. Dynamics of quadratic maps I-II. *Acta Math.*, 178:185–297, 1997.

[22] M. Lyubich. Dynamics of quadratic polynomials. III. Parapuzzle and SBR measures. *Astérisque*, (261):xii–xiii, 173–200, 2000. Géométrie complexe et systèmes dynamiques (Orsay, 1995).

[23] L. Mora and M. Viana. Abundance of strange attractors. *Acta Math.*, 171:1–71, 1993.

[24] S. Newhouse. Diffeomorphisms with infinitely many sinks. *Topology*, 13:9–18, 1974.

[25] M. J. Pacifico, A. Rovella, and M. Viana. Infinite-modal maps with global chaotic behavior. *Ann. of Math. (2)*, 148:441–484, 1998. Corrigendum in Annals of Math. 149, 705, 1999.

[26] J. Palis. A global view of Dynamics and a conjecture on the denseness of finitude of attractors. *Astérisque*, 261:335–347, 2000.

[27] J. Palis and F. Takens. Hyperbolicity and the creation of homoclinic orbits. *Annals of Math.*, 125:337–374, 1987.

[28] J. Palis and F. Takens. *Hyperbolicity and sensitive-chaotic dynamics at homoclinic bifurcations.* Cambridge University Press, 1993.

[29] W. Tucker. A rigorous ODE solver and Smale's 14th problem. *Found. Comput. Math.*, 2(1):53–117, 2002.

[30] M. Viana. Dynamics: a probabilistic and geometric perspective. In *Proceedings of the International Congress of Mathematicians, Vol. I (Berlin, 1998)*, volume 1998, 557–578.

[31] M. Viana. Chaotic dynamical behaviour. In *XIth International Congress of Mathematical Physics (Paris, 1994)*, 142–154. Internat. Press, Cambridge, MA, 1995.

[32] M. Viana. Multidimensional nonhyperbolic attractors. *Inst. Hautes Études Sci. Publ. Math.*, 85:63–96, 1997.

[33] Q. Wang and L.S. Young. Strange attractors with one direction of instability. *Comm. Math. Phys.*, 218(1):1–97, 2001.

[34] L.-S. Young. Statistical properties of dynamical systems with some hyperbolicity. *Annals of Math.*, 147:585–650, 1998.

[35] L.S. Young. Developments in chaotic dynamics. *Notices Amer. Math. Soc.*, 45(10):1318–1328, 1998.

ICM 2002 · Vol. III · 265–277

C^1-Generic Dynamics:
Tame and Wild Behaviour

C. Bonatti[*]

Abstract

This paper gives a survey of recent results on the maximal transitive sets of C^1-generic diffeomorphisms.

2000 Mathematics Subject Classification: 37C20, 37D30, 37C29.
Keywords and Phrases: Generic dynamics, Hyperbolicity, Transitivity.

1. Introduction

In order to give a global description of a dynamical system (diffeomorphism or flow) on a compact manifold M, the first step consists in characterizing the parts of M which are, in some sense, indecomposable for the dynamics. This kind of description will be much more satisfactory if these indecomposable sets are finitely many, disjoint, isolated, and not fragile (that is, persistent in some sense under perturbations of the dynamics).

For non-chaotic dynamics, this role can be played by periodic orbits, or by minimal sets. However many (chaotic) dynamical systems have infinitely many periodic orbits and a uncountable number of minimal sets. In order to structure the global dynamics using a smaller number of (larger) sets, we need to relax the notion of indecomposability. A weaker natural notion of topological/dynamical indecomposability is the notion of transitivity.

1.1. Maximal and saturated transitive sets

An invariant compact set K of a diffeomorphism f is *transitive* if there is a point in K whose positive orbit is dense in K. An equivalent definition is the following: for any open subsets U, V of K there is $n > 0$ such that $f^n(U) \cap V \neq \emptyset$.

One easily verifies that the closure of the union of an increasing family of transitive sets is a transitive set. Then Zorn's Lemma implies that any transitive set is contained in a *maximal transitive set* (i.e. maximal for the inclusion).

---

[*]Laboratoire de Topologie UMR 5584 du CNRS, Université de Bourgogne, B.P. 47 870, 21078 Dijon Cedex, France. E-mail: bonatti@u-bourgogne.fr

However, maximal transitive sets are not necessarily disjoint. For this reason, we also consider the stronger notion of *saturated transitive sets*: a transitive set K is saturated if any transitive set intersecting K is contained in K. So two saturated transitive sets are always equal or disjoint.

These notions are motivated by Smale's approach for hyperbolic dynamics, and more specifically for his spectral decomposition theorem (see [40]):

1.2.   Smale's spectral decomposition theorem

For an Axiom A diffeomorphism f on a compact manifold M, the set $\Omega(f)$ of the non-wandering points is the union of finitely many compact disjoint (maximal and saturated) transitive sets Λ_i, called *the basic pieces*, which are uniformly hyperbolic.

If furthermore f as no cycles, there is a *filtration* $\emptyset = M_{k+1} \subset M_k \subset \cdots \subset M_1 = M$ *adapted to* f, that is: the M_i are submanifolds with boundary, having the same dimension as M, and are strictly f-invariant: $f(M_i)$ is contained in the interior $\overset{o}{M}_i$ of M_i. Moreover Λ_i is the *maximal invariant set* of f in $M_i \backslash \overset{o}{M}_{i+1}$, that is $\Lambda_i = \bigcap_{n \in \mathbb{Z}} f^n(M_i \backslash \overset{o}{M}_{i+1})$.

Finally this presentation is *robust*: the same filtration remains adapted to any diffeomorphism g in a C^1-neighborhood of f, and the maximal invariant sets $\Lambda_i(g) = \bigcap_{n \in \mathbb{Z}} g^n(M_i \backslash \overset{o}{M}_{i+1})$ are the basic pieces of g.

1.3.   A global picture of C^1-generic diffeomorphisms

It is known from the sixties (see [2, 39]) that Axiom A diffeomorphisms are not C^1-dense in $Diff^1(M)$ if $dim(M) > 2$ and, of course, general dynamical systems do not admit such a nice global presentation of their dynamics. One would like to give an analogous description for as large as possible a class of diffeomorphisms.

In this paper I will present a collection of works, trying to give a coherent global picture of the dynamics of C^1-generic diffeomorphisms. There are two types of generic behaviours:

- either the manifold contains infinitely many regions having independent dynamical behaviours (we will speak of a *wild diffeomorphism*, and in Section 4. will give examples of such behaviours),
- or one has a description of the dynamics identical to those given by the spectral decomposition theorem: we speak of a *tame diffeomorphism*. In this case the role of basic pieces is played by the *homoclinic classes* (see the definition in Section 2.2.).

An important class of examples of tame dynamics are the robustly transitive dynamics and in Section 3.2. we summarize the known examples of robustly transitive dynamics. The basic pieces of a tame diffeomorphism present a weak form of hyperbolicity called *dominated splitting* and *volume partial hyperbolicity* (see Section 3.3. and 3.4.). In Section 3.5. we try to summarize the dynamical consequences of the dominated splittings.

2. Tame and wild dynamics

2.1. C^1-generic diffeomorphisms

In this paper, we will consider the set of diffeomorphisms $Diff^1(M)$ endowed with the C^1-topology. The choice of the topology comes from the fact that most of the perturbating results (Pugh's closing Lemma [34], Hayashi's connecting Lemma and its generalizations [24, 25, 4, 41]) are only known in this topology.

Recall that a property $\mathcal{P}$ is *generic* if it is verified on a residual subset $\mathcal{R}$ of $Diff^1(M)$ (i.e. $\mathcal{R}$ contains the intersection of a countable family of dense open subsets). In this work we will often use a practical abuse of language; we say:

"Any C^1-generic diffeomorphism verifies $\mathcal{P}$"

instead of:

"There is a residual subset $\mathcal{R}$ of $Diff^1(M)$ such that any $f \in \mathcal{R}$ verifies $\mathcal{P}$."

Let me first recall a famous and classical example, relating Pugh's closing lemma to generic dynamics:

Theorem [34] *Let f be a diffeomorphism on a compact manifold and $x \in \Omega(f)$ be a non-wandering point. There is g arbitrarily C^1-close to f such that x is periodic for g.*

Using a Kupka-Smale argument (genericity of hyperbolicity of the periodic points and the transversality of invariant manifolds) one get:

Corollary *The non-wandering set $\Omega(f)$ of a generic diffeomorphism f is the closure of the set of periodic points of f, which are all hyperbolic.*

2.2. Homoclinic classes

Let f be a diffeomorphism on a compact manifold and p be a hyperbolic periodic point of f of saddle type. Let $W^s(p)$ and $W^u(p)$ denote the invariant manifold of the orbit of p. The *homoclinic class $H(p, f)$* of p is by definition the closure of the transverse intersection points of its invariant manifold:

$$H(p, f) = \overline{W^s(p, f) \pitchfork W^u(p, f)}.$$

The homoclinic class $H(p, f)$ is a transitive set canonically associated to the orbit of the periodic point p.

There is an other way to see the homoclinic class of p: we tell that a periodic point q of saddle type and of same Morse index as p is *homoclinically related to p* if $W^u(q)$ cuts transversally $W^s(p)$ in at least one point and reciprocally $W^s(q)$ cuts transversally $W^u(p)$ in at least one point. The λ-lemma (see [33]) implies that this relation is an equivalence relation and $H(p, f)$ is the closure of the set of periodic points homoclinically related to p.

For Axiom A diffeomorphisms, the homoclinic classes are precisely the basic pieces of Smale's spectral decomposition theorem. However, one easily build examples of diffeomorphisms whose homoclinic classes are not maximal transitive

sets. Moreover, B. Santoro [37] recently build examples of diffeomorphisms on a
3-manifold having periodic points whose homoclinic classes are neither disjoint nor
equal.

2.3.   Homoclinic classes of generic dynamics

Conjectured during a long time, Hayashi's connecting lemma allowed the control the perturbations of the invariant manifolds of the periodic points, opening the door for the understanding of generic dynamics.

Theorem 1 [24]   *Let p and q be two hyperbolic periodic points of some diffeomorphism f. Assume that there is a sequence x_n of points converging to a point $x \in W_{loc}^u(p)$ and positive iterates $y_n = f^{m(n)}(x_n)$, $m(n) \geq 0$, converging to a point $y \in W_{loc}^s(q)$.*

Then there is g, arbitrarily C^1-close to f, such that x and y belong to a same heteroclinic orbit of p and q; in other words:

$$x \in W_{loc}^u(p,g), \; y \in W_{loc}^s(q,g) \text{ and there is } n > 0 \text{ such that } g^n(x) = y.$$

If the periodic points p and q in Theorem 1 belong to a same transitive set, then the sequences x_n and y_n are given by a dense orbit. In [7], using in an essential way Hayashi connecting lemma, we proved:

Theorem 2 *For any C^1-generic diffeomorphism, two periodic orbits belong to the same transitive set if and only if their homoclinic classes coincide.*

Motivated by this result we conjectured:

The homoclinic classes of a generic diffeomorphism coincide with its maximal transitive sets.

We know now that this conjecture, as stated above, is wrong: in [8] (see Section 4.) we show that any manifold M with dimension > 2 admits a non-empty C^1-open subset $U \subset Diff^1(M)$ on which generic diffeomorphisms have an uncoutable family of maximal (an saturated) transitive sets without periodic points.

However, one part of the conjecture is now proved. Generalizations of Hayashi Connecting Lemma (see [25, 4, 41]) recently allowed to show:

Theorem 3 *For any C^1-generic diffeomorphism, the homoclinic class of any periodic point is a maximal (see [4]) and saturated (see [17]) transitive set.*

The proof of this theorem is decomposed in two main steps: first, [4] shows that for a generic diffeomorphism f the homoclinic class of a point p coincides with the intersection of the closure of its invariant manifolds:

$$H(p,f) \overset{def}{=} \overline{W^s(p,f) \pitchfork W^u(p,f)} = \overline{W^s(p,f)} \cap \overline{W^u(p,f)}.$$

Then [17] shows that for a generic diffeomorphism f the closure $\overline{W^u(p,f)}$ is Lyapunov stable (and so admits a base of invariant neighborhoods) and $\overline{W^s(p,f)}$ is Lyapunov stable for f^{-1}. As a consequence a dense orbit of a transitive set T intersecting $H(p,f)$ is capted in arbitrarilly small neighborhoods of $\overline{W^u(p,f)}$ and of $\overline{W^u(p,f)}$, proving that T is contained in $\overline{W^s(p,f)} \cap \overline{W^u(p,f)}$, finishing the proof of the theorem.

2.4.   Tame and wild diffeomorphisms

Using Theorem 2 and the fact that the homoclinic class $H(p, f)$ of a periodic point varies lower semi-continuously with f, [1] shows the existence of a C^1-residual subset $\mathcal{R}$ of diffeomorphisms (or flows), such that the cardinality of the set of homoclinic classes is locally constant on $\mathcal{R}$: for any Kupka-Smale diffeomorphism f let $n(f) \in \mathbb{N} \cup \{\infty\}$ denote the cardinal of the set of different homoclinic classes $H(p, f)$ where p is an hyperbolic periodic point of f; then any $f \in \mathcal{R}$ has a C^1-neighborhood U_f such that any $g \in \mathcal{R} \cap U_f$ verifies $n(g) = n(f)$.

This result induces a natural dichotomy the residual set $\mathcal{R}$:

- a diffeomorphism $f \in \mathcal{R}$ is *tame* if it has finitely many homoclinic classes.
- a diffeomorphism $f \in \mathcal{R}$ is *wild* if it has infinitely many homoclinic classes.

3.   Tame dynamics

3.1.   Filtrations, robust transitivity and generic transitivity

[1] shows that the global dynamics of tame diffeomorphisms admit a good reduction to the dynamics of the transitive pieces (up to reduce the residual set $\mathcal{R}$). Let $f \in \mathcal{R}$ be a tame diffeomorphism, then :

1. as in the Axiom A case, the non-wandering set is the union of finitely many disjoint homoclinic classes $H(p_i, f)$.
2. there is a filtration $\emptyset = M_{k+1} \subset M_k \subset \cdots \subset M_1 = M$ adapted to f such that $H(p_i, f)$ is the maximal invariant set in $M_i \setminus \overset{o}{M}_{i+1}$.
3. moreover (up to reduce the open neighborhood U_f defined above) this filtration holds for any $g \in U_f$, and for $g \in \mathcal{R} \cap U_f$ the maximal invariant set of g in $M_i \setminus \overset{o}{M}_{i+1}$ is the homoclinic class $H(p_{i,g}, g)$.
4. there is a good notion of attractors: either a homoclinic class is a topological attractor (that is, its local basin contains a neighborhood of it) or its stable manifold has empty interior. Then the union of the basin of the attractors of f is a dense open set of M (see [16]).

The item 3 above shows that the transive sets $H(p_i, f)$ are not fragil. In a previous work, [20] introduced the following notion:

Definition 1 *Let f be a diffeomorphism of some compact manifold M. Assume that there is some open set $U \subset M$ and a C^1-neighborhood $\mathcal{V}$ of f such that, for any $g \in \mathcal{V}$, the maximal invariant set $\Lambda_g = \bigcap_{n \in \mathbb{Z}} g^n(\bar{U})$ is a compact transitive set contained in U.*

Then Λ_f is called a robustly transitive set *of f.*

In the definition above, if one has $U = M$ (and so $\Lambda_g = M$ for any $g \in \mathcal{V}$), then f is called a *robustly transitive diffeomorphism*.

This notion is slightly stronger that the property given by the item 3 above; so we have to relax Definition 1: we say that Λ_f is *generically transitive* if, in the notations of Definition 1, the maximal invariant set Λ_g is transitif for g in a residual subset of $\mathcal{V}$.

At this moment, there are no known examples of generic transitive sets which are not robustly transitive. So it is natural to ask if this two notions are equivalent:

$$\text{Generic transitivity} \overset{?}{\Longleftrightarrow} \text{robust transitivity ?}$$

3.2.   Examples of robust transitivity

The Axiom A dynamics are obvious examples of tame dynamics. On compact surfaces, tame diffeomorphisms are, in fact, Axiom A diffeomorphisms, but there are many non-hyperbolic examples in higher dimensions.

Even if this talk is mostly devoted to diffeomorphisms, let us observe that the most famous robustly transitive non-hyperbolic attractor is the Lorenz attractor (geometric model, see [23, 3]) for flows on 3-manifolds. There are many generalizations of this attractor, called singular attractors, for flow on 3-manifolds, see for instance [30]. See also [12] for robust singular attractors in dimension greater ou equal than 4, having a singular point with Morse index (dimension of unstable manifold) greater than 2.

The first example of non-Anosov robustly transitive diffeomorphism is due to Shub [38]: it is a diffeomorphisms on the torus T^4 which is a skew product over an Ansov map on the torus T^2, such that the dynamics on the fibers is dominated by the dynamics on the basis.

Then Mañé [29] built an example of robustly transitive non-hyperbolic diffeomorphism on the torus T^3 by considering a bifurcation of an Anosov map A having 3 real positive different eigenvalues $\lambda_1 < 1 < \lambda_2 < \lambda_3$: he performs a saddle node bifurcation creating a (new) hyperbolic saddle of index 1 (breaking the hyperbolicity) in the weak unstable manifold of a fixed point (of index 2) of the Anosov map A.

Then [6] shows that a diffeomorphism f admits C^1 perturbations which are robustly transitive, if f is:

1. the time one diffeomorphism of any transitive Anosov flow.
2. the product (A, id) where A is some Anosov map and id is the identity map of any compact manifold.

The second case can be easily generalized to any skew product of an Anosov map by rotations of the circle S^1. The same technique also allows [6] to build example of robustly transitive attractors, by perturbating product maps of any hyperbolic attractors by the identity map of some compact manifold.

Each of these previous example was partially hyperbolic (see the definitions in Section 3.3.): they admits a splitting $TM = E^s \oplus E^c \oplus E^u$ where E^s is uniformly contracting and E^u is uniformly expanding, and it was conjectured that partial hyperbolicity was a necessary condition for robust transitivity. Then [13] generalizes Mañé example above and exhibits robustly transitive diffeomorphisms on T^3 having a uniformly contracting 1-dimensional bundle, but no expanding bundle (there is a splitting $TM = E^s \oplus E^{cu}$), and robustly transitive difeomorphisms on T^4 having no hyperbolic subbundles (neither expanding nor contracting): there just admits an invariant dominated splitting $TM = E^{cs} \oplus E^{cu}$).

We do not known what are the manifolds admitting robustly transitive diffeomorphism. For instance:

Conjecture 1 *There is no robuslty transitive diffeomorphism on the sphere S^3.*

This conjecture has been proved in [20] assuming the existence of a codimension 1 (center stable or center unstable) foliation, using Novikov Theorem. Notice that all the known examples of robustly transitive diffeomorphisms on 3-manifolds admits an invariant codimension 1 foliation. However this conjecture remains still open.

3.3. Dominated splitting and partial hyperbolicity: definitions

Let f be a C^1-diffeomorphism of a compact manifold and let $\mathcal{E}$ be an f-invariant compact subset of M. Let $TM_x = E_1(x) \oplus \cdots \oplus E_k(x)$, $x \in \mathcal{E}$, be a splitting of the tangent space at any point of $\mathcal{E}$. This splitting is a *dominated splitting* if it verifies the following properties:

1. For any $i \in \{1, \ldots, k\}$, the dimension $E_i(x)$ is independent of $x \in \mathcal{E}$.
2. The splitting is f_*-invariant (where f_* denots the differential of f): $E_i(f(x)) = f_*(E_i(x))$.
3. There is $\ell \in \mathbb{N}$ such that , for any $x \in \mathcal{E}$, for any $1 \leq i < j \leq k$ and any $u \in E_i(x) \setminus \{0\}, v \in E_j(x) \setminus \{0\}$ one has:

$$\frac{\|f_*^\ell(u)\|}{\|u\|} \leq \frac{\|f_*^\ell(v)\|}{2\|v\|}.$$

Remark 1 - (Continuity) Any dominated splitting on a set $\mathcal{E}$ is continuous and extend in a unique way to the closure $\bar{\mathcal{E}}$.
  - (Extension to a neighborhood) There is a neighborhood U of $\bar{\mathcal{E}}$ on which the maximal invariant set $\Lambda(\bar{U}, f)$ has a dominated splitting extending those on $\mathcal{E}$.
  - (Robust) There is a C^1-neighborhood $\mathcal{U}_f$ of f such that, for any $g \in \mathcal{U}_f$, the maximal invariant set $\Lambda(\bar{U}, g)$ has a dominated splitting varying continuously with g.
  - (Unicity) If $\mathcal{E}$ has a dominated splitting, then there is a (unique) dominated spliting $TM|_\mathcal{E} = E_1 \oplus \cdots \oplus E_k$, called *the finest dominated splitting*, such that any other dominated splitting $F_1 \oplus \cdots \oplus F_l$ over $\mathcal{E}$ is obtained by grouping the E_i in packages.

One of the E_i is *uniformly contracting* if (up to increase ℓ in the definition above) $\frac{\|f_*^\ell(u)\|}{\|u\|} \leq \frac{1}{2}$ for all $x \in \mathcal{E}$ and all $u \in E_i(x) \setminus \{0\}$. In the same way E_i is *uniformly expanding* if $\frac{\|f_*^\ell(u)\|}{\|u\|} \geq \frac{1}{2}$ for all $x \in \mathcal{E}$ and all $u \in E_i(x) \setminus \{0\}$.

An f-invariant compact set K is *hyperbolic* if it has a dominated splitting $TM|_K = E^s \oplus E^u$ where E^s is uniformly contracting and E^u is uniformly expanding.

The compact f-invariant set K is *partially hyperbolic* if it has a dominated splitting and if at least one of the bundles E_i of its finest dominated splitting is uniformly contracting or expanding. Let E^s and E^u be the sum of the uniformly contracting and expanding subbundles, respectively, and let E^c be the sum of the other subbundles. One get a new dominated splitting $E^s \oplus E^c$, $E^c \oplus E^u$ or $E^s \oplus E^c \oplus E^u$, and these bundles are called the stable, central et unstable bundles, respectively.

An f-invariant compact set K is called *volume hyperbolic* if there is a dominated splitting whose extremal bundles E_1 and E_k contracts and expands uniformly the volume, respectively. Notice if one of these bundle has dimension 1, it is uniformly contracting or expanding. In particular, a volume hyperbolic set in dimension 2 is a uniformly hyperbolic set, and in dimension 3 it is partially hyperbolic (having at least one uniformly hyperbolic bundle).

3.4.   Volume hyperbolicity for the robust transitivity

Generalizing previous results by Mañé [28] (in dimension 2) and by [20] in dimension 3, [9] (for robustly transitive set) and [1] for generically transitve sets show:

Theorem 4 *Any robustly (or generically) transitive set is volume hyperbolic.*

Then any robustly transitive set in dimension 2 is a hyperbolic basic set (result of Mañé) and in dimension 3 is partially hyperbolic ([20]). In higher dimension, the dominated splitting may have all the subbundles of dimension greater than 2, so the expansion or contraction of the volume does no more imply the hyperbolicity of the bundle, see the example in [13].

The proof of Theorem 4 has two very different steps (as in [28]). The first one consists in showing that the lake of dominated splitting allows to "mix" the eigenvalues of the periodic orbits, creating an homothecy; a periodic orbit whose differential at the period is an homothecy is (up to a small perturbation) a sink or a source, breaking the transitivity. For that we just perturb the linear cocycle defined by the differential of f, and then we use a Lemma of Franks ([22]) for realizing the linear perturbation as a dynamical perturbation. Let state precisely this result:

Theorem 5 [9]   *Let f be a diffeomorphism of a compact manifold M, and let p be a hyperbolic periodic saddle. Assume that the homoclinic class $H(p, f)$ do not have any dominated splitting . Then, given any $\varepsilon > 0$, there is a periodic point x homoclinically related to p, with the following property:*

Given any neighborhood U of the orbit of x, there is a diffeomorphism g, ε-C^1-close to f, coinciding with f out of U along the orbit of x, such that the differential $g_*^n(x)$ is a homothecy, where n is the periode of x.

The second step consists in proving the uniform contraction and expansion of the volume in the extremal bundles. As in [28], one uses Mañé's Ergodic Closing Lemma to realize a lake of uniform expansion (or uniform expansion) of the volume in the extremal bundle by a periodic orbit z of a C^1-perturbation of f: if furthermore, the differential of this point restricted to the corresponding extremal bundle is an homothecy (as in Theorem 5) one get a sink or a source, breaking the transitivity.

For flows, the existence of singular point lies to additional difficulties. In dimension 3, [31] show that a robustly transitive set K of a flow on a compact 3-manifold is a uniformly hyperbolic set if it does not contain any singular point. If K contains a singular point then all the singular points in K have the same Morse index and K is a singular attractor if this index is 1 and a singular repellor if this index is 2 (see also [18]).

3.5.   Topological description of the dynamics with dominated splittings

The dynamics of diffeomorphisms admitting dominated splitting is already very far to be understood.

In dimension 2, Pujals and Sambarino (see [35, 36]) give a very precise description of C^2-diffeomorphism whose non-wandering set admits a dominated splitting.

- the periods of the non-hyperbolic periodic points is upper bounded.
- $\Omega(f)$ is the union of finitely many normally hyperbolic circles on which a power of f is a rotation, (maybe infinitely many) periodic points contained in a finite family of periodic normally hyperbolic segments and finitely many pairwise disjoint homoclinic classes, each of them containing at most finitely many non-hyperbolic periodic orbits.

This result is close to Mañé 's result, in dimension 1, for C^2-maps far from critical points (see [27]). We hope that this result can be generalized in any dimension, for dynamics having a codimension 1 strong stable bundle:

Conjecture 2 *Let f be a C^2-diffeomorphism and K be a compact locally maximal invariant set of f admitting a dominated splitting $TM|_K = E^s \oplus F$ where F has dimension 1 and E^s is uniformly contracting.*

Then K is the union of finitely many normally hyperbolic circles on which a power of f is a rotation, of periodic points contained in a union of finitely many normally hyperbolic periodic intervals and finitely many pairwize disjoint homoclinic classes each of them containing at most finitely many non-hyperbolic periodic points.

In this direction S. Crovisier [19] obtained some progress in the case where there is a unique non-hyperbolic periodic point.

General dominated splitting cannot avoid wild dynamics: multiplying any diffeomorphism by a uniform contraction and a uniform expansion, we get a normally hyperbolic and partially hyperbolic set. However a dominated splitting give some information of the possible bifurcations and on the index of the periodic point: see [10] which investigate in this direction. In particular a diffeomorphism cannot present any homoclinic tangency if it admits a dominated splitting whose non-hyperbolic bundles are all of dimension 1. We hope that this kind of dominated splitting avoid wild behaviours, but this is unknown, even in dimension 3:

Conjecture 3 *Let M be a compact 3-manifold and denote by $\mathcal{PH}(M)$ the C^1-open set of partially hyperbolic diffeomorphisms of M admitting a dominated splitting $E^s \oplus E^c \oplus E^u$ where all the bundles have dimension 1.*

The open set $\mathcal{PH}(M)$ does not contain any wild diffeomorphism: in other word any generic diffeomorphism in $\mathcal{PH}(M)$ is tame.

For partially hyperbolic diffeomorphism (having a splittin $E^s \oplus E^c \oplus E^u$), Brin and Pesin ([15]) show the existence of unique foliations $\mathcal{F}^s$ and $\mathcal{F}^u$, f-invariant and tangent to E^s and E^u respectively. The dynamics of the strong stable and the strong unstable foliations play an important role for the understanding of the topological and ergodical properties of a partially hyperbolic diffeomorphisms. Let mention two results on these foliations: [21] shows that a dense open subset of partially hyperbolic diffeomorphisms (having strong stable and strong unstable foliaitons) verify the "accessibility property", that is, any two points can be joined by a concatenation of pathes tangent successively to the strong stable or the strong unstable foliations. When the center direction has dimension 1, [11] shows the minimality of at least one of the strong stable or strong unstable foliations for a dense open subset of the robustly transitive systems in $\mathcal{PH}(M)$, where M is a compact 3-manifold.

However there is no general result on the existence of invariant foliations tangent to the central bundle even if it has dimension 1. When a partially hyperbolic diffeomorphism presents an invariant foliation $\mathcal{F}^c$ tangent to the center bundle E^c and which is *plaque expansive*, [26] shows that this foliations is structurally stable: any g close to f admits a foliation $\mathcal{F}_g^c$ topologically conjugated to f and such that (up to this conjugacy of foliation) g is isotopic to f along the center-leaves. This gives a very strong rigidity of the dynamics. This deep result was a key step for the construction of the examples of robustly transitive examples in [38, 29, 6] (there is now new proofs which do not use the stability of the center foliation (see[5])). So an important problem is:

Problem   (1) Does it exist robustly transitive partially hyperbolic diffeomorphisms having an invariant center foliation which is not plaque expansive?

(2) If a transitive partially hyperbolic diffeomorphism admits an invariant center foliation, is it *dynamically coherent*? that is, does it admit invariant center-stable and center unstable foliations which intersect along the center foliation?

(3) If the center bundle is 1-dimensional, is there an invariant center foliation?

4.   Wild dynamics

Very little is known on wild diffeomorphisms: for surfaces, it is not known whether C^1-wild diffeomorphisms exist (recall that the Newhouse phenomenon is a C^2-generic phenomenon, see [32]).

In dimension ≥ 3, the known examples are all of them due to the existence of homoclinic classes which do not admit, in a persistent way, any *dominated splitting* (see the first examples in [7]). Then following the same ideas, [8] present wild diffeomorphisms exhibiting, in a locally generic way, infinitely many hyperbolic and non-hyperbolic non-periodic attractors . The same example will present maximal transitive sets without any periodic orbits. The rest of this section is devoted to a short presentation of these examples:

Consider an open subset $\mathcal{U} \subset Diff^1(M)$ such that for any f in $\mathcal{U}$ there is a periodic point p_f depending continuously on f and verifying:

- For all $f \in \mathcal{U}$ the homoclinic class $H(p_f, f)$ contains two periodic points of different Morse indices, and having each of them a complex (non-real) eigenvalue (this eigenvalue is contracting for one point and expanding for the other).
- For all $f \in \mathcal{U}$ there are two periodic points having the same Morse index as p_f and homoclinically related to p_f such that the jacobian of the derivative of f at the period is strictly greater than one for one off this point and stricly less than one for the other point.

First item means that the homoclinic class $H(p_f, f)$ do not have any dominated splitting, and that this property is robust. So Theorem 5 shows that $H(p_f, f)$ admits periodic points whose derivative can be perturbated in order to get an homothecy. Second item above allows to choose this point having a jacobian (at the period) arbitrarily close to 1. Then a new pertubation allows to get a periodic point whose derivative at the period is the identity. Considering then perturbations of the identity map, we get:

Theorem 6 [8]   *There is a residual part $\mathcal{R}$ of the open set $\mathcal{U}$ defined above, such that any $f \in \mathcal{R}$ admits an infinite family of periodic disks D_n (let t_n denote the period), whose orbits are pairwize disjoint, and verifying the universal following property:*

Given any C^1-open set $\mathcal{O}$ of diffeomorphisms from the disk D^3 to its interior $\overset{o}{D^3}$, there is n such that the restriction of f^{t_n} to the disk D_n is smoothly conjugated to an element of $\mathcal{O}$.

Notice that the set of diffeomorphisms $g : D^3 \to \overset{o}{D^3}$ contains an open subset $\mathcal{U}_0$ verifying the property of $\mathcal{U}$ described above, one get some kind of renormalisation process: there is a residual part of $\mathcal{U}$ containing infinitely many periodic disks D_n containing each of them infinitely many periodic subdisks themself containing infinitely many periodic subdisks and so on... In that way one build a tree such that each branch is a sequence (decreasing for the inclusion), of strictly periodic orbits of disks whose periods go to infinity, and whose radius go to zero. The intersection of this sequence is a Lyapunov stable (and so saturated) transitif compact set, conjugated to an *adding machine* (see for instance [14] for this notion) and so without periodic orbits. The set of the infinite branches of this tree is uncountable, given the following result :

Theorem 7 [8] *Given any compact manifold M of dimension ≥ 3, there is an open subset $\mathcal{V}$ of $Diff^1(M)$ and a residual part $\mathcal{W}$ of $\mathcal{V}$, such that any $f \in \mathcal{W}$ admits an uncountable family of saturated transitif sets without periodic orbits.*

References

[1] F. Abdenur, Generic robustness of spectral decompositions, *preprint IMPA*, (2001)

[2] R. Abraham and S. Smale, Non-genericity of ω-stability, *Global Analysis, vol XIV of Proc. Symp. Pure Path. (Berkeley 1968) Amer. Math Soc.*, (1970).

[3] V.S. Afraimovitch, V.V. Bykov and L.P. Shil'nikov On the appearance and structure of the Lorenz attractor *Dokl. Acad. Sci. USSR*, 234, (1977), 336–339.

[4] M.-C. Arnaud, Création de connexions en topologie C^1, *Ergod. Th. & Dynam. Systems*, 21, (2001), 339–381.

[5] Ch. Bonatti, Dynamique génériques: hyperbolicité et transitivité, *Séminaire Bourbaki* n904, Juin 2002.

[6] Ch. Bonatti and L.J. Díaz, Persistent nonhyperbolic transitive diffeomorphisms. *Ann. of Math.*, (2) 143 (1996), no. 2, 357–396.

[7] Ch. Bonatti and L.J. Díaz, Connexions hétéroclines et généricité d'une infinité de puits ou de sources, *Ann. Scient. Éc. Norm. Sup.*, 4e série, t32, (1999), 135–150.

[8] Ch. Bonatti and L.J. Díaz, On maximal transitive sets of generic diffeomorphisms, *preprint* (2001).

[9] C. Bonatti, L. J. Díaz and E. Pujals, A C^1−generic dichotomy for diffeomorphisms: weak forms of hyperbolicity or infinitely many sinks or sources, to appear at *Annals of Math.*

[10] C. Bonatti, L. J. Díaz, E. Pujals and J. Rocha, Robust transitivity and heterodimensional cycles, To appear in *Asterisque*.

[11] C. Bonatti, L. J. Díaz and R. Ures, Minimality of the strong stable and strong unstable foliations for partially hyperbolic diffeomorphisms, to appear in the *Publ. Math. de l'inst. Jussieu*.

[12] Ch. Bonatti, A. Pumariño and M.Viana, Lorenz attractors with arbitrary expanding dimension, *C.R. Acad Sci Paris*, 1, 325, Serie I, (1997), 863–888.

[13] Ch. Bonatti et M.Viana, SRB measures for partially hyperbolic attractors: the contracting case, *Israel Journal of Math.*,**115**, (2000), 157–193.

[14] J. Buescu and I. Stewart, Liapunov stability and adding machines, *Ergodic Th. & Dyn. Syst.*, **15**(2), (1995), 271–290.

[15] M. Brin et Ya. Pesin , Partially hyperbolic dynamical systems, *Izv. Acad. Nauk. SSSR*, 1, (1974), 177–212.

[16] C. Carballo, C. Morales, Homoclinic classes and finitude of attractors for vector fields on n-manifolds, *Preprint* (2001).

[17] C. Carballo, C. Morales, and M.J. Pacífico, Homoclinic classe for $\mathcal{C}^1$-generic vector fields, to appear at *Erg. The. and Dyn Sys*.

[18] C. Carballo, C. Morales and M.J. Pacífico, Maximal transitive sets with singularities for generic C^1-vector fields, *Boll. Soc. Bras. Mat.* 31, n3, (2000), 287–303.

[19] S. Crovisier, Saddle-node bifurcations for hyperbolic sets, to appear at *Erg. Theor and Dyn Sys*.

[20] L. J. Díaz, E. Pujals and R. Ures, Partial hyperbolicity and robust transitivity *Acta Mathematica* vol. 183, (1999), 1–43.

[21] D. Dolgopyat and A. Wilkinson ,Stable accessibility is C^1-dense, to appear in *Astérisque.*

[22] J. Franks, Necessary conditions for stability of diffeomorphisms, *Trans. A.M.S.*, **158**, (1971), 301-308.

[23] J. Guckenheimer and R.F. Williams, Structural stability of Lorenz attractors, *Publ. Math. IHES* 50, (1979), 59–72.

[24] S. Hayashi, Connecting invariant manifolds and the solution of the C^1-stability and Ω-stability conjectures for flows, *Ann. of Math.*, **145**, (1997), 81–137.

[25] S. Hayashi, A C^1 make or break lemma,*Bol. Soc. Bras. Mat.* 31,(2000), 337–350.

[26] M. Hirsch, C. Pugh, et M. Shub, *Invariant manifolds*, Lecture Notes in Math., 583, Springer Verlag, 1977.

[27] R. Mañé, Hyperbolicity, sinks and measure in one-dimensional dynamics, *Comm. Math. Phys.* 100, (1985), 495–524.

[28] R. Mañé, An ergodic closing lemma, *Annals of Math.* vol. 116, (1982), 503–540

[29] R. Mañé, Contributions to the stability conjecture, *Topology*, 17, (1978), 386–396.

[30] C. Morales, M.J. Pacifico and E. Pujals, Singular hyperbolic systems, *Proc. Amer. Math. Soc.* 127,(1999), 3393–3401.

[31] C. Morales, M.J. Pacifico and E. Pujals, Robust transitive singular sets for 3-flows are partially hyperbolic attractors and repellers, *preprint IMPA* (1999).

[32] S.Newhouse, Diffeomorphisms with infinitely many sinks, *Topology*, **13**, (1974), 9–18.

[33] J. Palis, On Morse Smale dynamical systems, *Topology*, 8, (1969), 385–405.

[34] C. Pugh, The closing lemma, *Amer. Jour. of Math.*, **89**, (1967), 956–1009.

[35] E. Pujals and M. Sambarino, Homoclinic tangencies and hyperbolicity for surface diffeomorphisms. *Ann. of Math.* 151 , no. 3, (2000), 961–1023.

[36] E. Pujals and M. Sambarino, The dynamics of dominated splitting, *Preprint IMPA* (2001).

[37] B. Santoro , Colisão, colapso e explosão de classes holoclínicas, *Thesis PUC Rio de Janeiro* (2001).

[38] M. Shub, Topological transitive diffeomorphism on T^4, *Lect. Notes in Math.*, **206**, 39 (1971).

[39] R. Simon , A 3-dimensional Abraham-Samle example, *Proc. Amer. Math Soc.*, 34, (1972), 629–630.

[40] S. Smale, Differentiable dynamical systems, *Bull. Am. Math. Soc.*, 73, (1967), 747–817.

[41] L. Wen and Z. Xia, C^1 connecting lemmas, *Trans. Amer. Math. Soc.*, 352, (2000), 5213–5230.

ICM 2002 · Vol. III · 279–294

Action Minimizing Solutions of the Newtonian n-body Problem: From Homology to Symmetry

A. Chenciner[*]

(A la mémoire de Nicole Desolneux)

Abstract

An action minimizing path between two given configurations, spatial or planar, of the n-body problem is always a true – collision-free – solution. Based on a remarkable idea of Christian Marchal, this theorem implies the existence of new "simple" symmetric periodic solutions, among which the Eight for 3 bodies, the Hip-Hop for 4 bodies and their generalizations.

2000 Mathematics Subject Classification: 70F07, 70F10, 70F16, 34C14, 34C25.

Keywords and Phrases: n-body problem, Action, Symmetry.

0. Introduction

Finding periodic geodesics on a riemannian manifold as length minimizers in a fixed non-trivial homology or homotopy class is commonplace lore. Advocated by Poincaré [P] as early as 1896, the search for periodic solutions of a given period T of the n-body problem as action minimizers in a fixed non-trivial homology or homotopy class is rendered difficult by the possible existence of collisions due to the relative weakness of the newtonian potential: the action of a solution stays finite even when some of the bodies are colliding. Very few results are available: among them Gordon's characterization of Kepler solutions [G] for 2 bodies in $I\!R^2$, Venturelli's characterization of Lagrange equilateral solutions [V1] for 3 bodies in $I\!R^3$, Arioli, Gazzola and Terracini's characterization of retrograde Hill's orbits [AGT]

---

[*] Astronomie et Systèmes Dynamiques, IMCCE, UMR 8028 du CNRS, 77, avenue Denfert-Rochereau, 75014 Paris, France & Département de Mathématiques, Université Paris VII-Denis Diderot, 16, rue Clisson, 75013 Paris, France. E-mail: chencine@bdl.fr

for the restricted 3-body problem in $I\!\!R^2$. In particular, no truly new solution of the n-body problem was found in this way; indeed, these results confirm the view that the action-minimizing periodic solutions are the "simplest" ones in their class.

The action minimization method has recently been given a new impetus by the replacement of the topological constraints by symmetry ones. This idea was first introduced by the italian school [C-Z][DGM][SeT] as another mean of forcing coercivity of the problem, i.e. forbidding a minimizer to be "at infinity". The bodies were forced to occupy, after half a period, a position symmetrical of the original one with respect to the center of mass of the system. It is proved in [CD] that in a space of even dimension, say $I\!\!R^2$, the minimizers in this symmetry class include relative equilibrium solutions (i.e. solutions which are "rigid body like"); moreover all minimizers are of this form provided a certain "finiteness" hypothesis is verified (see [C3]). Such relative equilibria can occur only for the so called *central configurations* [AC], the most famous of which is Lagrange equilateral triangle.

Recently, a new type of symmetry was considered, which originates in the invariance of the Lagrangian under permutations of equal masses. This has led to the discovery of a whole world of new solutions in the case when all the bodies have the same mass. The most surprising ones are the "choreographies" whose name, given by Carles Simó, fits the beautiful figures they display on the screen in animated computer experiments ([CGMS],[S2]). Referring to my survey article [C3] for a bibliography and a description of the few cases in which existence proofs are available (the Hip-Hop [CV] for 4 bodies in $I\!\!R^3$, the Eight [CM] for 3 bodies in $I\!\!R^2$, Chen's solutions [Ch] for 4 bodies in $I\!\!R^2$), I mainly address here a powerful theorem which solves completely the collision problem for the fixed ends problem in the case of arbitrary masses. This is pertinent because, as we shall see, it allows one to prove the existence of collision-free minimizers under well chosen symmetry constraints. This theorem is the result of the efforts of Richard Montgomery, Susanna Terracini, Andrea Venturelli [V2], and, for the last – fundamental – stone, Christian Marchal [M2] [M3]. I present here a complete proof and, in particular, a simplified version of Marchal's remarkable idea, which avoids numerical computations. I discuss also new applications to minimization under symmetry constraints and open problems.

Notations. By a *configuration* of n bodies in an euclidean space $(E, \langle \rangle)$ we understand an n-tuple $x = (\vec{r}_1, \vec{r}_2, \ldots \vec{r}_n) \in E^n$. The *configuration space* of the n-body problem is the quotient of the set of configurations by the action of translations (see [AC]). It may be identified as in [C3] with the set $\mathcal{X}$ of configurations whose center of mass $\vec{r}_G = (\sum_{i=1}^n m_i)^{-1} \sum_{i=1}^n m_i \vec{r}_i$ is at the origin. It is endowed with the "mass scalar product" $(\vec{r}_1, \ldots, \vec{r}_n) \cdot (\vec{s}_1, \ldots, \vec{s}_n) = \sum_{i=1}^n m_i \langle (\vec{r}_i - \vec{r}_G), (\vec{s}_i - \vec{s}_G) \rangle$. The *non-collision* configurations – the ones such that no two bodies $\vec{r}_i$ coincide – form an open dense subset $\hat{\mathcal{X}}$ of $\mathcal{X}$. The functions $I = x \cdot x$, $J = x \cdot y$, $K = y \cdot y$, defined on the *phase space* $\hat{\mathcal{X}} \times \mathcal{X}$ (whose elements are noted (x, y)) are the basic isometry-invariants of the n-body problem They are respectively the *moment of inertia* of the configuration with respect to its center of mass, its time derivative and twice the *kinetic energy* in a galilean frame which fixes the center of mass. The *potential function* (opposite of the potential energy), the *Hamiltonian* (=total energy) and the *Lagrangian* are respectively defined by

$$U = \sum_{i<j} m_i m_j ||\vec{r}_i - \vec{r}_j||^{-1}, \; H = \frac{1}{2}K - U, \; L = \frac{1}{2}K + U.$$

In terms of the gradient ∇ for the mass metric, the equations of the n-body problem,

$$m_i \ddot{r}_i(t) = \sum_{j \neq i} m_i m_j \frac{\vec{r}_j(t) - \vec{r}_i(t)}{|\vec{r}_j(t) - \vec{r}_i(t)|^3}, \quad i = 1, \ldots, n,$$

can be written $\ddot{x} = \nabla U(x)$. They are the Euler-Lagrange equations of the action, which to a path $x(t)$ associates the real number

$$\mathcal{A}_T\big(x(t)\big) = \int_0^T L\big(x(t), \dot{x}(t)\big) dt.$$

Remark. In the perturbations, we shall not bother about fixing the center of mass because replacing $K = \sum_{i=1}^n m_i ||\vec{v}_i - \vec{v}_G||^2$ by $\sum m_i ||\vec{v}_i||^2$ only increases the action.

1. The fixed-ends problem

Question. Given two configurations, – possibly with collisions – of n point masses in $\mathbb{R}^3$ (resp. $\mathbb{R}^2$) and a positive real number T, does there exist a solution of the Newtonian n-body problem which connects them in the time T ?

A natural way of looking for a solution is to seek for a minimizer of the action $\mathcal{A}_T(x)$ over the space $\Lambda_0^T(x_i, x_f)$ of paths $x(t)$ in the configuration space $\hat{\mathcal{X}}$ which start at time 0 in the configuration x_i and end at time T in the configuration x_f. For the integral to be defined, it is natural to work in the Sobolev space of paths which are square integrable together with their first derivative in the sense of distributions.

The main problem, already mentioned by Poincaré in 1896 (see [P] where he introduces the method in a slightly different context), is that a minimum could well be such that, for a non-empty set of instants (necessarily of measure zero), the system undergoes a collision of two or more bodies, which prevents it form being a true solution (see [C3]). At an isolated collision time, the renormalized configuration is known to be approaching the set of central configurations (the ones which admit homothetic motion [C2]) but very little is understood of these configurations for more than 3 bodies. Continuous families of such configurations could exist (the "finiteness problem") and even if they didn't, there would be no garantee that at collision the renormalized configuration has a limit : it might have one only modulo rotations (the "infinite spin problem"). Nevertheless, we prove the

Theorem. *A minimizer of the action in* $\Lambda_0^T(x_i, x_f)$ *is collision-free on the whole open interval* $]0, T[$. *Hence, the answer to the Question is yes, both in* $\mathbb{R}^3$ *and* $\mathbb{R}^2$.

In the next paragraph, Marchal's idea to prove that isolated collisions do not occur in a minimizer is explained on the Kepler problem. If the finiteness problem is supposed to be solved, it works in the same way for the general n-body problem (surprisingly, the infinite spin problem is irrelevant). We then address the finiteness problem with Terracini's technique of *blow up*, which reduces the problem of isolated collisions to the case of parabolic homothetic solutions; finally we show, following Montgomery and Venturelli, that accumulation of collisions do not occur in a minimizer provided no subclusters collide. The theorem then follows by induction on the number of bodies involved in a collision.

Remark. A similar assertion, based on numerical experiments, was made by Tiancheng Ouyang in Guanajuato (Hamsys, march 2001) but no proof appeared.

2. The Kepler problem as a model for the study of isolated collisions

The case of two bodies contains already many ingredients of the general situation. As is well-known, the 2-body problem is equivalent to the problem of a particle attracted to a fixed center 0, the so-called Kepler problem (or 1-fixed center problem). We call *collision-ejection* a solution in which the particle follows a straight line segment from its initial position $\vec{r}_i$ to the attracting center and (possibly) another straight line segment from the attracting center to its final position $\vec{r}_f$.

A test assertion. *A collision-ejection solution of the Kepler problem does not minimize the action in the Sobolev space* $\Lambda_0^T(\vec{r}_i, \vec{r}_f)$ *of paths joining* $\vec{r}_i$ *to* $\vec{r}_f$.

At least four proofs may be given of the truth of this assertion but only the fourth one using Marchal's idea is robust enough to lead to complete generalization. In the first one, we use the explicit knowledge of the solutions of the 2-body problem [A1] to identify the minimizers with the "direct" arcs of solution, not going "around" the attracting center (this arc is uniquely determined provided $\vec{r}_i$, O and $\vec{r}_f$ do not lie on a line in this order). In the second one, we find a "simple" path without collision (straight line, circle, uniform motion) which has lower action. In the third one, supposing that a minimizer $\vec{r}(t)$ has a collision with the fixed center at time 0, we find a local deformation $\vec{r}_\epsilon(t) = \vec{r}(t) + \epsilon\varphi(t)\vec{s}$, which has lower action and no collision. Such deformations were used by many people, including Susanna Terracini, Gianfausto Dell'Antonio, Richard Montgomery and Christian Marchal. If we chose, with Montgomery, $\varphi(t) = 1$ if $0 \leq t \leq \epsilon^{\frac{3}{2}}$, $\varphi(t) = \epsilon^{-1}(\epsilon^{\frac{3}{2}} + \epsilon - t)$ if $\epsilon^{\frac{3}{2}} \leq t \leq \epsilon^{\frac{3}{2}} + \epsilon$ and $\varphi(t) = 0$ if $t \geq \epsilon^{\frac{3}{2}} + \epsilon$, the gain in action is $c\sqrt{\epsilon}\,(1 + O(\sqrt{\epsilon}\log(1/\sqrt{\epsilon})))$ provided the unit vector $\vec{s}$ is well chosen. We come to the fourth proof, for which we must distinguish two cases according to the dimension of the ambient space.

(i) The case of $\mathbb{R}^3$**.** Let $t \mapsto \vec{r}(t)$ be a collision-ejection solution of the Kepler problem, $\ddot{\vec{r}}(t) = -\vec{r}(t)/|\vec{r}(t)|^3$, such that $\vec{r}(-T') = \vec{r}_i$, $\vec{r}(0) = 0$, $\vec{r}(T) = \vec{r}_f$, $T, T' >$

0. We consider the following family of continuous deformations of $\vec{r}(t)$, parametrized by an element $\vec{s}$ of the unit sphere S^2 in $I\!\!R^3$: if $R'(t) = (1+\frac{t}{T'})\rho$ and $R(t) = (1-\frac{t}{T})\rho$,

$$\vec{r}_{\vec{s}}(t) = \vec{r}(t) + R'(t)\vec{s} \text{ if } -T' \leq t \leq 0, \quad \vec{r}_{\vec{s}}(t) = \vec{r}(t) + R(t)\vec{s} \text{ if } 0 \leq t \leq T.$$

It is a simplification of Marchal's original choice but the idea is the same : to show that the action $\mathcal{A}$ of $\vec{r}(t)$ is strictly bigger than the average $\mathcal{A}_m = \int_{S^2} \mathcal{A}(\vec{r}_{\vec{s}}(t))d\sigma$, where $d\sigma$ denotes the normalized area form, that is the unique rotation invariant probability measure on S^2. This will imply the existence of at least one direction $\vec{s}$ for which $\vec{r}_{\vec{s}}(t)$ has lower action than $\vec{r}(t)$ (because the set of good $\vec{s}$ has positive measure we could choose $\vec{s}$ so that $\vec{r}_{\vec{s}}$ is collision-free but this is irrelevant).

The linearity of the integral and the similar behaviour of ejection and collision allow *to replace in the proof $\vec{r}(t)$ and $\vec{r}_{\vec{s}}(t)$ by their restrictions to the interval $[0,T]$.* Moreover, it follows from the "blow-up" method (see 3.2) that it is enough to consider a *parabolic* solution $\vec{r}(t)$, that is $\vec{r}(t) = \gamma t^{\frac{2}{3}}\vec{c}$, with $\gamma = (9/2)^{\frac{1}{3}}$ if $|\vec{c}| = 1$.

By Fubini theorem applied to the positive integrand,

$$\mathcal{A}_m = \int_{S^2} d\sigma \int_0^T \left(\frac{|\dot{\vec{r}}_{\vec{s}}|^2}{2} + \frac{1}{|\vec{r}_{\vec{s}}|} \right) dt = \int_0^T dt \int_{S^2} \left(\frac{|\dot{\vec{r}}_{\vec{s}}|^2}{2} + \frac{1}{|\vec{r}_{\vec{s}}|} \right) d\sigma, \quad \text{and}$$

$$\mathcal{A}_m - \mathcal{A} = \int_0^T dt \left[\frac{\dot{R}(t)^2}{2} + \int_{S^2} \dot{R}(t)\vec{s} \cdot \vec{r}(t) \, d\sigma \right] + \int_0^T dt \left[\int_{S^2} \frac{d\sigma}{|\vec{r}_{\vec{s}}(t)|} - \frac{1}{|\vec{r}(t)|} \right].$$

The first integral reduces to $\frac{1}{2}\int_0^T \dot{R}(t)^2 dt = \rho^2/2T$ because of the antisymmetry in $\vec{s}$ of the scalar product. The second is the difference in potential resulting from the replacement of the particle $\vec{r}(t)$ by a homogeneous hollow sphere of the same mass and increasing radius $R(t)$. Because of the harmonicity of Newton potential in $I\!\!R^3$, the potential $U_0(\vec{r}, R) := \int \frac{d\sigma}{|\vec{r}-R\vec{s}|} = \int \frac{d\sigma}{|\vec{r}+R\vec{s}|}$ of a homogeneous hollow sphere of radius R is

$$U_0(\vec{r}, R) = \frac{1}{R} \text{ if } |\vec{r}| \leq R, \quad U_0(\vec{r}, R) = \frac{1}{|\vec{r}|} \text{ if } |\vec{r}| \geq R.$$

If 0 enters this sphere at time t_0, $|\vec{r}(t_0)| = R(t_0)$, i.e. $\rho = \gamma t_0^{\frac{2}{3}} + O(t_0^{\frac{5}{3}})$, and

$$\mathcal{A}-\mathcal{A}_m = \frac{\rho^2}{2T}+\int_0^{t_0} \left[\frac{1}{R(t)} - \frac{1}{|\vec{r}(t)|} \right] dt = -\frac{2}{\gamma}t_0^{\frac{1}{3}}+O(t_0^{\frac{4}{3}}) \leq 0 \text{ if } \rho, \text{ hence } t_0, \text{ is small.}$$

(ii) The case of $I\!\!R^2$. The Newtonian potential is not harmonic in $I\!\!R^2$ and this makes things somewhat more complicated. Marchal proposes to replace the sphere by a disk of radius R endowed with the projection $\sigma(\theta, x) = 1/(2\pi R\sqrt{R^2 - x^2})$ (in polar coordinates) of the uniform density on the sphere of the same radius. The potential fonction $U_0(\vec{r}, R)$ of such a disk (total mass 1) may be recovered from the general computation done, via complex function theory, for a thin elliptic plate with a given density which is constant on homothetic ellipses (see [B] and [Ma]):

$$U_0(\vec{r}, R) = \frac{\pi}{2R} \ \ \text{if} \ \ |\vec{r}| \le R, \quad U_0(\vec{r}, R) = \frac{1}{R}\arcsin\left(\frac{R}{|\vec{r}|}\right) \ \ \text{if} \ \ |\vec{r}| \ge R.$$

It does not coïncide any more, but asymptotically, with Newton's potential $1/|\vec{r}|$ of the center of mass outside the disk but it is still constant in the interior and the proof works as well as in the spatial case: as $\arcsin(x) \le x + (\frac{\pi}{2} - 1)x^3$ for $x \ge 0$, the difference in actions between the mean of the modified actions when $\vec{s}$ belongs to the unit disk and the original becomes

$$\mathcal{A}_m - \mathcal{A} = \frac{\rho^2}{2T} + \int_0^{t_0}\left[\frac{\pi}{2R(t)} - \frac{1}{|\vec{r}(t)|}\right]dt + \int_{t_0}^T\left[\frac{1}{R(t)}\arcsin\left(\frac{R(t)}{|\vec{r}(t)|}\right) - \frac{1}{|\vec{r}(t)|}\right]dt$$

$$\le \frac{\rho^2}{2T} + \left[-\frac{\pi T}{2\rho}\log(1 - \frac{t}{T}) - \frac{3}{\gamma}t^{\frac{1}{3}}\right]_0^{t_0} + \int_0^{t_0}(\frac{\pi}{2} - 1)\rho^2(1 - \frac{t}{T})^2\frac{1}{\gamma^3 t^2}dt$$

$$= \frac{\gamma^2 t_0^{\frac{4}{3}}}{2T}\left(1 + O(t_0)\right) + (\frac{\pi}{2} - 3)\frac{1}{\gamma}t_0^{\frac{1}{3}}\left(1 + O(t_0)\right) + (\frac{\pi}{2} - 1)\frac{1}{\gamma}t_0^{\frac{1}{3}} + O\left(t_0^{\frac{4}{3}}\log(\frac{1}{t_0})\right)$$

$$= (\pi - 4)\frac{1}{\gamma}t_0^{\frac{1}{3}} + O\left(t_0^{\frac{4}{3}}\log(\frac{1}{t_0})\right) \le 0 \ \ \text{for } \rho, \text{ hence } t_0, \text{ small.}$$

3. Proof of the theorem

3.1 The induction. We define the following statements about a minimizer $x(t)$:

(I_p) If a collision of p bodies occurs in $x(t)$ for $t \in]0, T[$, it is isolated.

(II_p) No collision of $m \le p$ bodies occurs in $x(t)$ for $t \in]0, T[$.

In 3.2 we prove that (I_p) implies that no collision of p bodies occurs in $]0, T[$, hence that (II_p) and (I_{p+1}) imply (II_{p+1}). In 3.3 we prove that (II_p) implies (I_{p+1}). As (II_1) is empty, it implies (I_2), hence (II_2), etc... up to (II_n) which is the conclusion. If k-collisions are present in x_i or x_f but not j-collisions for $j < k$, the induction proves that $j < k$-collisions are absent. The next step proves that k-collisions, including the ones at the ends, are isolated and everything goes through.

Remark. The induction may succeed because a p-body collision cannot be a limit of q-body collisions with $q > p$. Still, accumulation of collisions involving bodies in different clusters could *a priori* occur, e.g. a sequence of double collisions 23, 12, 34, 23, 13, 24, 23, ... converging to a quadruple collision 1234, or even a converging sequence of such sequences. Induction on the number of bodies in the clusters fortunately avoids having to deal with such problems.

3.2 Elimination of isolated collisions.

3.2.1 The blow-up technique. This technique was introduced by S. Terracini and developped in the thesis of A. Venturelli [V2]. It is based on the homogeneity of the potential (compare [C2]). It allows proving the

Proposition. *If a minimizer $x(t)$ of the fixed ends problem for n-bodies possesses an isolated collision of $p \leq n$ bodies, there is a parabolic (i.e. zero energy) homothetic collision-ejection solution $\overline{x}(t)$ of the p-body problem which is also a minimizer of the fixed ends problem.*

Proof. To keep the exposition as simple as possible, I describe the case of a total collision. In the general case of partial (and possibly simultaneous) collisions, everything goes through in the same way because the blow up sends all bodies not concerned by the collision to infinity (for more details, see [V2]).

Assuming that the collision occurs at $t = 0$, we define $x^\lambda(t) = \lambda^{-\frac{2}{3}} x(\lambda t)$ for $\lambda > 0$. If $x(t)$ is a solution of the n-body problem, so is $x^\lambda(t)$. Moreover, for any path $x(t)$ in $\Lambda_{T_1}^{T_2}(x_i, x_f))$, the path $x^\lambda(t)$ belongs to $\Lambda_{T_1}^{T_2}(x^\lambda(T_1), x^\lambda(T_2))$ and its action is equal to $\lambda^{-\frac{1}{3}}$ times the action of the restriction of $x(t)$ to the interval $[\lambda T_1, \lambda T_2]$. Hence, if $x(t)$ is action minimizer in $\Lambda_{T_1}^{T_2}(x_i, x_f)$, so is x^λ in $\Lambda_{T_1}^{T_2}(x^\lambda(T_1), x^\lambda(T_2))$. Now, Sundman's estimates recalled above imply that, $\{x^\lambda, 0 < \lambda < \lambda_0\}$ is bounded in $H^1([T_1, T_2], \mathcal{X})$, hence weakly compact, so that there exists a sequence $\lambda_n \to 0$ such that x^{λ_n} converges weakly (and hence uniformly) in $H^1([0, T], \mathcal{X})$ to a solution $\overline{x}$. One shows that $\overline{x}$ is made of a parabolic homothetic collision solution followed by a parabolic homothetic ejection solution (the two central configurations involved are a priori distinct). Moreover, it follows from the weak lower semi-continuity of the action that $\overline{x}$ is a minimizer in $\Lambda_{T_1}^{T_2}(\overline{x}(T_1), \overline{x}(T_2))$ (see [V2]).

3.2.2 The mean perturbed action. We shall deal only with the case of $\mathbb{R}^3$ and refer the reader to the Kepler case for the modifications needed in the case of $\mathbb{R}^2$. Thanks to "blow up", we may suppose that our minimizer $x(t)$ is a parabolic homothetic collision-ejection solution $x(t) = (\vec{r}_1(t), \cdots, \vec{r}_p(t)) = x_0 |t|^{\frac{2}{3}}$ of the p-body problem. As in the Kepler case, we may restrict to the ejection part, corresponding to $t \in [0, T]$. One studies deformations of $x(t)$ of the form

$$x_{\vec{s}}^k(t) = \left(\vec{r}_1(t), \ldots, \vec{r}_k(t) + R(t)\vec{s}, \ldots, \vec{r}_p(t) \right),$$

where, as before, $R(t) = (1 - \frac{t}{T})\rho$ with ρ a small positive real number and $\vec{s}$ belongs to the unit sphere. The same computation as in the Kepler case leads to an average action $\mathcal{A}_m^k$ such that

$$\mathcal{A}_m^k - \mathcal{A} \leq \frac{m_k}{2} \frac{\rho^2}{T} + \sum_{j \neq k,\, j \leq p} m_j m_k \int_0^{t_{jk}} \left[\frac{1}{R(t)} - \frac{1}{r_{jk}(t)} \right] dt,$$

where $r_{jk} = |\vec{r}_k - \vec{r}_j|$ and t_{jk} is defined by $r_{jk}(t) = R(t)$ (the inequality sign comes from the fact that the deformations do not keep the center of mass fixed).

As $r_{jk}(t) = c_{jk} t^{\frac{2}{3}}$, one concludes as in the Kepler case that $\mathcal{A}_m^k - \mathcal{A} < 0$.

Remark. We could have dispensed with "blow up" in case similitude classes of central configurations were isolated but certainly not otherwise. This is because, the best control Sundman's theory may give us on the asymptotic behaviour of the colliding bodies is that their moment of inertia I_c with respect to their center of mass and their potential U_c are respectively equivalent to $I_0 t^{\frac{4}{3}}$ and $U_0 t^{-\frac{2}{3}}$ (see [C2]). This implies the existence, for $1 \le j < k \le p$, of $0 < a_{jk} \le b_{jk}$ such that for t small enough, one has $a_{jk} t^{\frac{2}{3}} \le r_{jk}(t) \le b_{jk} t^{\frac{2}{3}}$. It follows that

$$\mathcal{A}_m^k - \mathcal{A} \le \frac{m_k}{2T} b_{jk}^2 t^{\frac{4}{3}} + O(t^{\frac{7}{3}}) - \sum_{j \ne k,\, j \le p} m_j m_k \left(\left[-\frac{1}{a_{jk}} + \frac{3}{b_{jk}} \right] t^{\frac{1}{3}} + o(t^{\frac{1}{3}}) \right).$$

If similitude classes of central configurations are isolated, there is a limit shape and we may take a_{jk} and b_{jk} as close as we wish. Otherwise we cannot conclude.

3.3 The elimination of non-isolated collisions. It remains to prove that (II_p) implies (I_{p+1}). We use energy considerations, an idea which goes back to R. Montgomery and was further developed in Venturelli's thesis [V2].

Proposition. *Let $x(t)$ be a minimizer of the fixed ends problem. If $x(t)$ has no p-body collisions for $p < p_0$, collisions of p_0 bodies are isolated.*

Sketch of proof. I shall give the proof in the case of a total collision (i.e. $p_0 = n$) and then explain what has to be changed in the general case.

(i) Using the behavior of the action under reparametrization, let us prove that the energy stays constant along a minimizer, whatever be the collisions. For this let us consider variations $x_\epsilon(t)$ of the form $x_\epsilon(t) = x(\varphi_\epsilon(t))$ where $t \mapsto \tau = \varphi_\epsilon(t)$ is a differentiable family of diffeomorphisms of $[0, T]$ starting from $\varphi_0(t) \equiv t$:

$$\mathcal{A}(x_\epsilon) = \int_0^T \left(\frac{\|\dot{x}_\epsilon(t)\|^2}{2} + U(x_\epsilon(t)) \right) dt = \int_0^T \left(\frac{1}{\lambda_\epsilon(\tau)} \frac{\|\dot{x}(\tau)\|^2}{2} + \lambda_\epsilon(\tau) U(x(\tau)) \right) d\tau,$$

where $\lambda_\epsilon = dt/d\tau = 1/\dot{\varphi}_\epsilon(\varphi_\epsilon^{-1}(\tau))$. The derivative at $\epsilon = 0$ of $a(\epsilon) = \mathcal{A}(x_\epsilon)$ is

$$\frac{da}{d\epsilon}(0) = \int_0^T \left(\frac{\|\dot{x}(\tau)\|^2}{2} - U(x(\tau)) \right) \delta\lambda(\tau) \, d\tau = \int_0^T H\big(x(\tau), \dot{x}(\tau)\big) \delta\lambda(\tau) d\tau,$$

where $\delta\lambda(\tau) = \frac{d\lambda_\epsilon(\tau)}{d\epsilon}\big|_{\epsilon=0}$. As the variations $\delta\lambda$ satisfy the constraint $\int_0^T \delta\lambda(\tau) d\tau = 0$, which comes from the fact that $\int_0^T \lambda_\epsilon(\tau) d\tau = T$, we get that there exists a real constant c such $H\big(x(\tau), \dot{x}(\tau)\big) = c$ wherever it is defined.

(ii) Let t_0 be an instant at which total collisions accumulate. Let us chose two sequences (a_n) and (b_n) of instants of total collision which converge to t_0 and are such that no total collision occurs in the open intervals $]a_n, b_n[$. The moment of inertia I of the system with respect to its center of mass is equal to zero at each of the instants a_n or b_n and hence has at least one maximum ξ_n in the interval $]a_n, b_n[$. As no partial collision occurs, the motion is regular in each of these intervals and at each such maximum, the second time-derivative $\ddot{I}(\xi_n)$ has to be non positive. But the value $U(\xi_n)$ of the potential function tends to $+\infty$ as $n \to +\infty$, while the energy H stays constant. One then deduces from the Lagrange -Jacobi relation $\ddot{I} = 4H + 2U$ that $\ddot{I}(\xi_n) \to +\infty$, which is a contradiction.

In the general case, when μ is some cluster not containing all the bodies, the energy H_μ of μ is no more constant but one can get from a refinement of the same proof that it is still an absolutely continuous function of time as long as no collision occurs between a body of the cluster and a body of the complementary cluster (see [V2]). This implies that H_μ stays locally bounded and allows the argument of (ii) to work because, by hypothesis, no partial collision occurs in the cluster.

4. Periodic solutions

4.1 Homological or homotopical constraints. Going back to the 1896 Note of Poincaré already alluded to, the idea of constructing periodic solutions of the n-body problem as the "simplest" (action minimizing) ones in a given homology or homotopy class of the configuration space is very natural if one compares to the construction of periodic geodesics as minimizing the length in a non trivial homology or homotopy class. As already noticed by Poincaré, this works beautifully in the so-called "strong force problem", corresponding to a potential in $1/r^2$ or stronger, where each collision path has infinite action [CGMS]. Unfortunately, in the Newtonian case, most of the time minimizers have collisions and hence are not true periodic solutions [M]. This is already true in the planar Kepler problem: it follows from Gordon's work [G] (see also [C3]) that the only minimizers of the action among the loops of a fixed period T whose index in the punctured plane is different from $0, \pm 1$, is an ejection-collision one ! (for an analogue result in the planar three-body problem, see [V1]).

In such cases, solving the fixed ends problem is of no use. Among the cases where minimizers in a fixed homology or homotopy class have no collision are

1) Gordon's theorem for the planar Kepler problem when one fixes the index to ± 1 (resp. when one insists only on the index being different from 0): a minimizer is any elliptic solution of the given period.

2) Venturelli's generalization [V1] of Gordon's theorem to the planar three-body problem whith homogy class fixed in such a way that along a period, each side of the triangle makes exactly one complete turn in the same direction: a minimizer is any elliptic homographic motion of the equilateral triagle, of the given period.

4.2 Symmetry constraints. In order to find "new" solutions as action minimizers, another type of constraints on the loops must be introduced, which somewhat allows using fixed ends type results. We ask the loops to be invariant under the action of a finite group G. An invariant loop is completely defined by its restriction to an interval of time on which G induces no constraint. The restriction to such a "fundamental domain" of a minimizer among G-invariant loops is a minimizer of the fixed ends problems between its extremities. This leads to a new collision problem: a minimizer could well have a collision at the initial or final instant.

(i) Choreographies. We first show, following Andrea Venturelli, the

Theorem. *A minimizer among n-choreographies has no collision.*

Recall that the choreographies are fixed loops under the action of the group $\mathbb{Z}/n\mathbb{Z}$ whose generator cyclically permutes the bodies after one n-th of the period (see [CGMS]); hence a fondamental domain can be chosen as any time interval of length T/n. If there were collisions at the ends, one would get a contradiction with the theorem by just shifting the fundamental domain to the right or to the left. One can prove (using [CD]) that the regular n-gon minimizes in all cases where it minimizes $\tilde{U} = I^{\frac{1}{2}}U$. But this is no more true for $n \geq 6$. So, what is the min ?

(ii) Generalized Hip-Hops. This works also for the "italian" (anti)symmetry:

Theorem. *A minimizer among loops $x(t)$ in $\mathbb{R}^3$ such that $x(t+T/2) = -x(t)$ has no collision. Moreover, it is never a planar solution.*

The last assertion comes from the fact that a relative equilibrium $x(t)$ whose configuration x_0 minimizes $\tilde{U} = I^{\frac{1}{2}}U$ is always a minimizer among the planar (anti)symmetric loops ([CD] and [C3]). But, applied to a variation $z(t) = z_0 \cos \frac{2\pi t}{T}$ normal to the plane of $x(t)$, the Hessian of the action is easily seen [C4] to be

$$d^2\mathcal{A}(x(t))(z(t,z(t)) = I_0^{-\frac{1}{2}}d^2\tilde{U}(x_0)(z_0,z_0)\int_0^T \cos^2\frac{2\pi t}{T}dt,$$

where $I_0 = x_0 \cdot x_0$. Now, results of Pacella and Moeckel [Mo1] say that one can always choose z_0 such that $d^2\tilde{U}(x_0)(z_0,z_0) < 0$. Hence, a relative equilibrium ceases being a minimizer in $\mathbb{R}^3$. This ends the proof because other possible minimizers of the planar problem would have the same action as a relative equilibrium (thanks to A. Venturelli for this remark). In reference to [CV], I propose to call *generalized Hip-Hops* these minimizers. They are the best approximations I can think of in $\mathbb{R}^3$ to the non-existing relative equilibria of non-planar central configurations (recall [AC] that such relative equilibria exist in $\mathbb{R}^4$).

(iii) Eights with less symmetry. As another example, we prove the existence of solutions "of the Eight type" but with less symmetry than the full dihedral group $D_6 = \{s,\sigma|s^6 = 1, \sigma^2 = 1, s\sigma = \sigma s^{-1}\}$ (see [C3]). We consider the subgroups $\mathbb{Z}/6\mathbb{Z} = \{s\}$ and $D_3 = \{s^2,\sigma\}$.

Theorem. *A minimizer among $\mathbb{Z}/6\mathbb{Z}$-invariant loops has no collision. The same is true for a minimizer among D_3-invariant loops.*

Instead of minimizing the action over one twelfth of the period between an Euler configuration at time 0 and an isosceles one at time $T/12$ (see [CM]), one minimizes only over one sixth of the period: in the first case from an isosceles configuration at time t_0 to a symmetric one at time $t_0 + T/6$, in the second one from an Euler configuration at time 0 to another one at time $T/6$. Venturelli's trick of translating the fundamental domain works in the first case where t_0 is arbitrary (a translation of time transforms a minimizer into a minimizer) but not in the second one where, as for the initial D_6-action, an Euler configuration can only occur at times which are integer multiples of $T/6$. To prove the absence of collisions at the initial and the final instant in the second case, we notice that such a collision is necessarily a triple (i.e. total) collision. If this happens, the action of the path is greater than the one of a homothetic ejection solution of equilateral type, a path which is not of the required type, but this is irrelevant here. The conclusion follows because the action of this last path is itself greater than the one of one sixth of the "equipotential model" (see [C3],[CM]). If a minimizer among $\mathbb{Z}/6\mathbb{Z}$ or D_3 symmetric loops possesses the whole D_6 symmetry of the Eight is unknown.

(iv) The P_{12} family. Marchal discovered the P_{12} family, which continues the Eight solution in three-space up to Lagrange equilateral solution, through choreographies in a rotating frame [M1]. It is parametrized by an angle u between 0 and $\frac{\pi}{6}$: the solution labeled by u minimizes the action in fixed time $T/12$ between configurations which are symmetric with respect to a line Δ through the origin which contains body 0 and configurations which are symmetric with respect to a plane P through the origin which contains body 2 and makes angle u with Δ. We shall think of Δ as being horizontal and of P as being vertical (Figure 1).

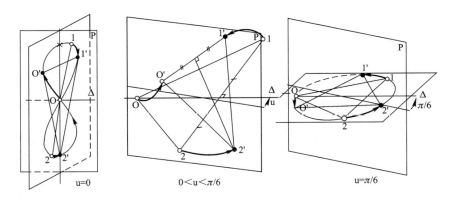

Figure 1 (fixed frame)

For $u = 0$, one gets the Eight in the vertical plane orthogonal to Δ (and hence to P); for $u = \frac{\pi}{6}$, one gets Lagrange solution in the horizontal plane (containing Δ and orthogonal to P). For $\pi/6 \leq u \leq \pi/3$, the minimizer is a horizontal Lagrange solution whose size increases to infinity and action decreases to 0. The x^4-type bifurcation of the minimizer at $u = \pi/6$ was analyzed by Marchal. In a frame rotating around the vertical axis of an angle $-u$ in time $T/12$, one gets a family of D_6-symmetric choreographies of period T between the Eight and twice Lagrange (figure 2).

The relevant action of D_6 on the configuration space of three bodies in $\mathbb{R}^3$ is a direct generalization of the one which leaves the Eight invariant. It is defined as

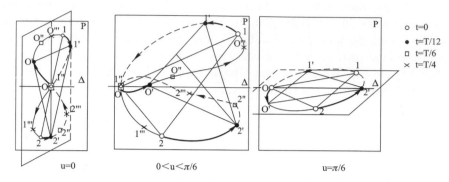

Figure 2 (rotating rame)

follows (the notations are the ones of [C3]):

$$\alpha(s)(\vec{r}_0, \vec{r}_1, \vec{r}_2) = (\Sigma\vec{r}_2, \Sigma\vec{r}_0, \Sigma\vec{r}_1), \quad \beta(s)(t) = t + T/6,$$
$$\alpha(\sigma)(\vec{r}_0, \vec{r}_1, \vec{r}_2) = (\Delta\vec{r}_0, \Delta\vec{r}_2, \Delta\vec{r}_1), \quad \beta(\sigma)(t) = -t,$$

where Σ (resp. Δ) denotes the symmetry with respect to the horizontal plane (resp. to the line Δ).

Thanks to the fact that a minimizer of the fixed ends problem has no collision, the proof boils down to proving that a minimizing path has no collisions at its ends.

(i) Getting rid of triple collisions is easy: one notices firstly that the action of any path undergoing a triple collision is bigger than the action $A_3 = \frac{9}{2}(\frac{\pi^2}{3})^{\frac{1}{3}}T^{\frac{1}{3}}$ (notation of [CM]) of the homothetic solution of the equilateral triangle which goes in the same amount of time T from collision to zero velocity, secondly that this last action is bigger than the one $A_L = \frac{9}{2 \times 3^{\frac{1}{3}}}(\frac{\pi}{3} - u)^{\frac{2}{3}}T^{\frac{1}{3}}$ of a horizontal Lagrange solution which rotates an angle $\frac{\pi}{3} - u$ during the given amount of time. In fact, this last action is even smaller than the action $A_2 = (3\sqrt{2})^{-\frac{2}{3}}A_3$ used in [CM] as long as $u > (\sqrt{2} - 1)\frac{\pi}{6}$, in which case, the absence of any kind of collision in a minimizer follows.

(ii) For double collisions, which we have not yet discarded if $u \le (\sqrt{2}-1)\frac{\pi}{6}$, we provide a local deformation of the path which eliminates the collision and lowers the action. The two cases (collision at initial or final time) are similar, the only difference being the replacement of the symmetry with respect to the line Δ by the symmetry with respect to the plane P in the constraints imposed to the perturbation direction $\vec{v}$. Supposing that a minimizer $x(t) = (\vec{r}_0(t), \vec{r}_1(t), \vec{r}_2(t))$ has a collision between bodies 1 and 2 at the initial time 0, we deform it into $x_\epsilon(t) = x(t) + \epsilon\varphi(t)(0, \vec{v}, -\vec{v})$ where $\varphi(t)$ is chosen as in the third proof for the Kepler case, Using Sundman's estimates on the behavior of the bodies near a double collision, one shows that for a good choice of $\vec{v}$, the action again decreases by $c\sqrt{\epsilon}(1 + O(\sqrt{\epsilon}\log(1/\sqrt{\epsilon})))$.

Remarks. 1) Our argument works for one value of u at a time. As no uniqueness is proved, neither is continuity with u of the family. Such continuity would imply the existence among the family of spatial 3-body choreographies in the fixed frame.

2) A *direct* proof of the existence of the P_{12} family should stem from the following remark: in [CM], the existence of the Eight is deduced from the fact that the action A_2 of a 2-body solution going from collision to zero velocity is (slightly) greater than the action a of an equipotential path for which $I = I_0$ is constant as well as $K = K_0 = U = U_0$. For the Lagrange solution also, $I = I_{\frac{\pi}{6}}, K = K_{\frac{\pi}{6}} = U = U_{\frac{\pi}{6}}$ are constant and $I_{\frac{\pi}{6}} > I_0, \quad K_{\frac{\pi}{6}} = U_{\frac{\pi}{6}} < K_0 = U_0$. In particular, the action of the Lagrange solution is smaller than the one of the equipotential model. When u increases, one should be able to construct a path with the right end conditions whose action is a decreasing function of u and hence is smaller than A_2.

3) The first continuation of the Eight into a family of rotating planar choreographies was given by Michel Hénon [CGMS] who used the same program as in [H]. A third family should exist, rotating around an axis orthogonal to the first two.

5. Related results and open problems

Two global questions seem to be out of reach at the moment: unicity and possible extra symmetries of minimizers.

As an example of the first, numerical evidence by Simo suggests unicity of the Eight but in [CM] we do not even prove that each lobe is convex, only that it is star-shaped (the problem is near the crossing point). This is nevertheless enough to imply that the braid it defines in space time $I\!\!R^2 \times I\!\!R/T\mathbb{Z}$ (equivalent to the homotopy class in the configuration space) is the "Borromean rings", the signature of a truly triple interaction (also noticed in [Ber] in a different context).

As an example of the second one, we do not know if the $\mathbb{Z}/4\mathbb{Z}$-symmetry and the "brake" property of the Hip-Hop solution [CV] follow automatically from minimizing the action among loops such that $x(t+T/2) - -x(t)$ (compare 4.2 (ii)). One is tempted to compare this problem to the celebrated result of Alain Albouy [A2] which states the existence of some symmetry in any central configuration of 4 equal masses (and implies that there is only a finite number of them). But there is Moeckel's numerical example [Mo2] of a central configuration of eight equal masses without any symmetry. And according to [SW], there exists such an example minimizing $\tilde{U}$ for $n = 46$. For more on symmetry, see [V2].

Identifying minimizers, even when one knows that they are collision-free, is usually too difficult a problem (see 4.2 (i) and (ii)). Understanding their stability properties may sometimes be attempted theoretically [Ar],[O], or numerically [S1].

Another type of questions is connected with minimization with mixed constraints: symmetry and homology or homotopy. One can ask, for example, if the Eight is a minimizing choreography in its homology class $(0, 0, 0)$ (each side of the triangle has zero total rotation). An interesting example of mixed conditions may be found in [V2] where generalizations of the Hip-Hop lead to spatial choreographies of 4 equal masses. But, as for most choreographies, no proof was found of the existence of Gerver's "supereight" with four equal masses [CGMS], [C3].

I am indebted to Christian Marchal, Richard Montgomery, David Sauzin, Susanna Terracini and Andrea Venturelli for many illuminating discussions and comments.

References

[A1]　Albouy A. Lectures on the two-body problem, *Classical and Celestial Mechanics: The Recife Lectures, H. Cabral F. Diacu ed.* (in press at Princeton University Press).

[A2]　Albouy A. Symétrie des configurations centrales de quatre corps, *C. R. Acad. Sci. Paris, 320, 217–220 (1995)* & The symmetric central configurations of four equal masses, *Contemporary Mathematics, vol. 198* (1996), 131–135.

[AC]　Albouy A. and Chenciner A., Le problème des n corps et les distances mutuelles, *Inventiones Mathematicæ, 131* (1998), 151–184.

[AGT]　Arioli G., Gazzola F. and Terracini S. Minimization properties of Hill's orbits and applications to some N-body problems, *preprint*, (October 1999).

[Ar]　Arnaud M.C. On the type of certain periodic orbits minimizing the Lagrangian action *Nonlinearity 11* (1998), 143–150.

[B]　Betti E. Teorica delle forze newtoniane e sue applicazioni all' elestrotatica e al magnetismo, *Pisa, Nistri* (1879).

[Ber]　Berger M.A. Hamiltonian dynamics generated by Vassiliev invariants, *Journal of Physics A: Math. Gen. 34* (2001), 1363–1374.

[Ch]　Chen K.C. Action minimizing orbits in the parallelogram four-body problem with equal masses, *Arch. Ration. Mech. Anal., 158, no. 4* (2001), 293–318.

[C1]　Chenciner A. Introduction to the N-body problem, *Ravello summer school* (09-1997), http://www.bdl.fr/Equipes/ASD/person/chenciner/chenciner.htm

[C2]　Chenciner A. Collisions totales, Mouvements complètement paraboliques et réduction des homothéties dans le problème des n corps, *Regular and chaotic dynamics, V.3, 3* (1998), 93–106.

[C3]　Chenciner A. Action minimizing periodic solutions of the n-body problem, *"Celestial Mechanics, dedicated to Donald Saari for his 60th Birthday", A. Chenciner, R. Cushman, C. Robinson, Z.J. Xia ed., Contemporary Mathematics 292* (2002), 71–90.

[C4]　Chenciner A. Simple non-planar periodic solutions of the n-body problem *Proceedings of the NDDS Conference, Kyoto,* (2002).

[CD]　Chenciner A. and Desolneux N. Minima de l'intégrale d'action et équilibres relatifs de n corps, *C.R. Acad. Sci. Paris. t. 326, Série I* (1998), 1209–1212. Corrections in *C.R. Acad. Sci. Paris. t. 327, Série I* (1998), 193 and in [C3].

[CGMS]　Chenciner A., Gerver J., Montgomery R. and Simó C. Simple choreographies of N bodies: a preliminary study, *Geometry, Mechanics and Dynamics, Springer*, (to appear).

[CM]   Chenciner A. and Montgomery R. A remarkable periodic solution of the three body problem in the case of equal masses, *Annals of Math., 152* (2000), 881–901.

[CV]   Chenciner A. and Venturelli A. Minima de l'intégrale d'action du Problème newtonien de 4 corps de masses égales dans $I\!\!R^3$: orbites "hip-hop", *Celestial Mechanics, vol. 77* (2000), 139–152.

[C-Z]   Coti Zelati V. Periodic solutions for N-body type problems, *Ann. Inst. H. Poincaré, Anal. Non Linéaire, v. 7, no. 5* (1990), 477–492.

[DGM]   Degiovanni M., Giannoni F. and Marino A., Periodic solutions of dynamical systems with Newtonian type potentials, *Ann. Scuola Norm. Sup. Pisa Cl. Sci. 15* (1988), 467–494.

[G]   Gordon W.B. A Minimizing Property of Keplerian Orbits, *American Journal of Math. vol. 99, no. 15* (1977), 961–971.

[H]   Hénon M. Families of periodic orbits in the three-body problem, *Celestial Mechanics 10* (1974), 375–388.

[M1]   Marchal C. The family P_{12} of the three-body problem. The simplest family of periodic orbits with twelve symmetries per period, *Fifth Alexander von Humboldt Colloquium for Celestial Mechanics*, (2000).

[M2]   Marchal C. How the method of minimization of action avoids singularities, *Celestial Mechanics and Dynamical Astronomy*, (to appear).

[M3]   Marchal C. Handwritten supplement to the above paper and private discussions.

[Mo1]   Moeckel R. On central configurations, *Math. Z. 205* (1990), 499–517.

[Mo2]   Moeckel R. Some Relative Equilibria of N Equal Masses, N=4,5,6,7; Addendum: N=8; *unpublished paper describing numerical experiments ($\leq$ 1990).*

[M]   Montgomery R. Action spectrum and collisions in the three-body problem, *"Celestial Mechanics, dedicated to Donald Saari for his 60th Birthday"*, A. Chenciner, R. Cushman, C. Robinson, Z.J. Xia ed., *Contemporary Mathematics 292* (2002), 173–184.

[O]   Offin D. *Maslov index and instability of periodic orbits in Hamiltonian systems*, preprint, 2002.

[P]   Poincaré H. Sur les solutions périodiques et le principe de moindre action, *C.R.A.S. t. 123* (1896), 915–918.

[SeT]   Serra E. and Terracini S. Collisionless Periodic Solutions to Some Three-Body Problems, *Arch. Rational Mech. Anal., 120* (1992), 305–325.

[S1]   Simó C. Dynamical properties of the figure eight solution of the three-body problem, *"Celestial Mechanics, dedicated to Donald Saari for his 60th Birthday"*, A. Chenciner, R. Cushman, C. Robinson, Z.J. Xia ed., *Contemporary Mathematics 292*, (2002), 209–228.

A. Chenciner

[S2] Simó C. New families of Solutions in N-Body Problems, *Proceedings of the Third European Congress of Mathematics, C. Casacuberta et al. eds. Progress in Mathematics, 201* (2001), 101–115.

[SW] Slaminka E. & Woerner K. Central configurations and a theorem of Palmore *Celestial Mechanics 48* (1990), 347–355.

[V1] Venturelli A. Une caractérisation variationnelle des solutions de Lagrange du problème plan des trois corps, *C.R. Acad. Sci. Paris, t. 332, Série I* (2001), 641–644.

[V2] Venturelli A. Thèse, *Paris* (to be defended in 2002).

ICM 2002 · Vol. III · 295–303

The Dynamical Systems Approach to the Equations of a Linearly Viscous Compressible Barotropic Fluid

E. Feireisl[*]

Abstract

We develop a dynamical systems theory for the compressible Navier-Stokes equations based on global in time weak solutions. The following questions will be addressed:

- Global existence and critical values of the adiabatic constant;
- dissipativity in the sense of Levinson - bounded absorbing sets;
- asymptotic compactness;
- the long-time behaviour and attractors.

2000 Mathematics Subject Classification: 35Q30, 35A05.
Keywords and Phrases: Compressible Navier-Stokes equations, Long-time behaviour, Weak solutions.

1. Introduction

The long-time behaviour of solutions to the evolutionary equations arising in the mathematical fluid mechanics has been the subject of many theoretical studies. This type of problems is apparently related to the phenomena of turbulence, and there is still a significant gap between many formal "scenarios" and mathematically rigorous results.

The dynamical systems in question are usually related to a system of partial differential equations and, consequently, they are defined in an infinite dimensional phase space. On the other hand, the presence of dissipative terms in the equations due to viscosity results in the existence of global attractors-compact invariant sets attracting uniformly in time all trajectories emanating from a given bounded set. Such a theory is well developed for the incompressible linearly viscous fluids, and

---

[*]Institute of Mathematics, Czech Academy of Sciences, Žitná 25, 115 67 Praha 1, Czech Republic. E-mail: feireisl@math.cas.cz

the reader may consult the monographs of BABIN and VISHIK [1], TEMAM [21] or CONSTANTIN et al. [2] for the recent state of art.

On the other hand, much less seems to be known for the compressible fluids. While there is an existence theory of the weak solutions for the incompressible Navier-Stokes equations due to LERAY [18], its "compressible" counterpart appeared only recently in the work of LIONS [19]. Even in the incompressible case, there is a qualitative difference between the two-dimensional case solved by LADYZHENSKAYA [17], and the three-dimensional case representing one of the most challenging unsolved problems of modern mathematics. It is worth-noting that a similar gap divides the one and more-dimensional problems for the compressible fluids.

The time evolution of the fluid density $\varrho = \varrho(t,x)$ and the velocity $\vec{u} = \vec{u}(t,x)$ is governed by the Navier-Stokes system of equations:

$$\partial_t \varrho + \operatorname{div}(\varrho \vec{u}) = 0, \tag{1.1}$$

$$\partial_t(\varrho \vec{u}) + \operatorname{div}(\varrho \vec{u} \otimes \vec{u}) + \nabla p = \operatorname{div} S + \varrho \vec{f}, \tag{1.2}$$

where p is the pressure, S the viscous stress tensor, and $\vec{f}$ a given external force.

In what follows, we consider linearly viscous (Newtonian) fluids where the viscous stress is related to the velocity by the constitutive law

$$S = \mu\left(\nabla \vec{u} + \nabla \vec{u}^T\right) + \lambda \operatorname{div} \vec{u} \, I, \tag{1.3}$$

where the viscosity coefficients satisfy

$$\mu > 0, \ \lambda + \mu \geq 0. \tag{1.4}$$

Generally speaking, the pressure p depends on the density and the internal energy (temperature) of the fluid. If it is the case, the system (1.1), (1.2) is not closed and should be complemented by the energy equation. Unfortunately, however, the available global existence results for this full system allow for only for small initial data (cf. MATSUMURA and NISHIDA [20]).

There are physically relevant situations when one can assume the flow is barotropic, i.e., the pressure depends solely on the density. This is the case when either the temperature (the isothermal case) or the entropy (the isentropic case) are supposed to be constant. The typical constitutive relation between the pressure and the density then reads

$$p = p(\varrho) = a\varrho^\gamma, \ a > 0, \tag{1.5}$$

where $\gamma = 1$ in the isothermal case, and $\gamma > 1$ represents the adiabatic constant in the isentropic regime. More general and even non-monotone pressure-density constitutive laws arise in nuclear plasma physics (see [4]). In the barotropic regime, the equations (1.1), (1.2) form a closed system and complemented by suitable initial and boundary conditions represent a (at least formally) well-posed problem.

If the problem is posed on a spatial domain $\Omega \subset R^N$, one usually assumes that the fluid adheres completely to the boundary which is mathematically expressed by the no-slip boundary conditions for the velocity:

$$\vec{u}|_{\partial \Omega} = 0. \tag{1.6}$$

Note that for viscous fluids such a condition is in a very good agreement with physical experiment.

In accordance with the deterministic principle, the state of the system at any time $t > t_0$ should be given by the initial conditions

$$\varrho(t_0) = \varrho_I, \quad (\varrho \vec{u})(t_0) = \vec{q}_I. \tag{1.7}$$

The function ϱ_I is non-negative and the momentum $\vec{q}_I$ satisfies the compatibility condition

$$\vec{q}_I = 0 \text{ a.a. on the set } \{\varrho_I = 0\}. \tag{1.8}$$

The reason why we impose the initial conditions for the momentum $\varrho \vec{u}$ rather than for the velocity $\vec{u}$ is that the former quantity is weakly continuous with respect to time while the instantaneous values of the velocity are determined only almost anywhere with respect to time. Clearly such a problem does not arise provided the initial density ϱ_I is strictly positive. However, it is an interesting open problem whether or not this property is preserved at any positive time for any distributional solution of the problem satisfying the natural energy estimates (cf. HOFF and SMOLLER [16]).

2. Finite energy weak solutions and well-posedness

In order to study the long-time behaviour, one should first make sure that the class of objects one deals with is not void. More precisely, one should be able to prove the existence of global-in-time solution for any initial data ϱ_I, $\vec{q}_I$ satisfying some physically relevant hypothesis.

Multiplying the equations of motion by $\vec{u}$ and integrating by parts one deduces the energy inequality

$$\frac{d}{dt} E[\varrho, \vec{u}] + \int_\Omega \mu |\nabla \vec{u}|^2 + (\lambda + \mu) |\text{div } \vec{u}|^2 \, dx \leq \int_\Omega \varrho \vec{f} \cdot \vec{u} \, dx, \tag{2.1}$$

where the total energy E is given by the formula

$$E[\varrho, \vec{u}] = \int_\Omega \varrho |\vec{u}|^2 + P(\varrho) \, dx$$

with

$$P'(\varrho)\varrho - P(\varrho) = p(\varrho).$$

Note that in the most common isentropic case, the function P can be taken in the form

$$P(\varrho) = \frac{a}{\gamma - 1} \varrho^\gamma.$$

The energy inequality is the main (and almost the only one) source of a priori estimates. Accordingly, "reasonble" solutions of the problem (1.1), (1.2) defined on a bounded time interval $I \subset R$ should belong to the class

$$\varrho \geq 0, \ \varrho \in L^\infty(I; L^\gamma(\Omega)), \ \vec{u} \in L^2(I; W_0^{1,2}(\Omega, R^N)). \tag{2.2}$$

The energy inequality (2.1) represents an additional constraint imposed on any solution ϱ, $\vec{u}$ of the problem. Similarly as in the theory of the variational (weak) solutions developed for the incompressible case by Leray, it is not clear if it is satisfied for any weak solution of the problem.

Following DiPERNA and LIONS [3] we shall say that ϱ, $\vec{u}$ is a renormalized solution of the continuity equation (1.1) if the identity

$$\partial_t b(\varrho) + \mathrm{div}(b(\varrho)\vec{u}) + \Big(b'(\varrho)\varrho - b(\varrho)\Big)\mathrm{div}\ \vec{u} = 0 \tag{2.3}$$

holds in the sense of distributions for any function $b \in C^1(R)$ such that

$$b'(\varrho) = 0 \text{ for all } \varrho \geq \mathrm{const}(b). \tag{2.4}$$

Similarly as for the energy inequality, it is not known if any weak solution ϱ, $\vec{u}$ of (1.1) satisfies (2.3).

Definition 2.1 *We shall say that ϱ, $\vec{u}$ is a finite energy weak solution of the problem (1.1 - 1.6) on a set $I \times \Omega$ if the following conditions are satisfied:*
- *The functions ϱ, $\vec{u}$ belong to the function spaces determined in (2.2);*
- *the energy inequality (2.1) is satified in $\mathcal{D}'(I)$;*
- *the continuity equation (1.1) as well as its renormalized version (2.3) hold in $\mathcal{D}'(I \times R^N)$ provided ϱ, $\vec{u}$ were extended to be zero outside Ω;*
- *the momentum equation (1.2) is satisfied in $\mathcal{D}'(I \times \Omega)$.*

The most general available existence result reads as follows:

Theorem 2.1 *Let $\Omega \subset R^N$, $N = 2, 3$ be a bounded Lipschitz domain. Let $I = (0, T)$, and let the initial data ϱ_I, $\vec{q}_I$ satisfy (1.8) together with*

$$\varrho_I \in L^\gamma(\Omega), \ \frac{|\vec{q}_I|^2}{\varrho_I} \in L^1(\Omega).$$

Let $\vec{f}$ be a bounded measurable function of $t \in I$, $x \in \Omega$. Finally, let the pressure $p \in C[0, \infty) \cap C^1(0, \infty)$ be given by a constitutive law

$$p = p(\varrho), \ \frac{1}{a}\varrho^{\gamma-1} - b \leq p'(\varrho) \leq a\varrho^{\gamma-1} + b, \text{ for all } \varrho > 0,$$

where $a > 0$, $b \geq 0$, and

$$\gamma > \frac{N}{2}.$$

Then the problem (1.1 - 1.6) admits a finite energy weak solution ϱ, $\vec{u}$ on $I \times \Omega$ satisfying the initial conditions (1.7).

In his pioneering work, LIONS [19] proved Theorem 2.1 for Ω regular, p monotone, and $\gamma \geq \frac{3}{2}$ if $N = 2$, and $\gamma \geq \frac{9}{5}$ for $N = 3$. The hypotheses concerning γ were relaxed in [9], [12], the case of a general bounded domain Ω treated in [11], and the hypothesis of monotonicity of the pressure removed in [5], [4].

3. Ultimate boundedness

The first issue to be discussed when describing the long-time asymptotics of a given dynamical system is ultimate boundedness or dissipativity. This means there exists an absorbing set bounded in a suitable topology. Here "suitable topology" is of course that one induced by the total energy E. One of possible results in this direction is contained in the following theorem.

Theorem 3.1 *Let $\Omega \subset R^N$, $N = 2, 3$ be a bounded Lipschitz domain. Let $\mathbf{f}$ be a bounded measurable function such that*

$$\text{ess} \sup_{t \in R, x \in \Omega} |\vec{f}(t, x)| \leq F.$$

Let the pressure p be given by the isentropic constitutive law

$$p = a\varrho^\gamma \text{ with } \gamma > 1 \text{ if } N = 2, \ \gamma > \frac{5}{3} \text{ for } N = 3.$$

Finally, set

$$\int_\Omega \varrho \, \mathrm{d}x = m > 0.$$

Then there exists a constant E_∞ depending solely on m and F having the following property:

Given E_I, there exists a time $T = T(E_I)$ such that

$$E[\varrho, \vec{u}](t) \leq E_\infty \text{ for a.a. } t > T$$

provided

$$\text{ess} \limsup_{t \to 0+} E[\varrho, \vec{u}](t) \leq E_I$$

and ϱ, $\vec{u}$ is a finite energy weak solution of the problem (1.1 - 1.6) on $(0, \infty) \times \Omega$.

The proof of Theorem 3.1 can be found in [14] and [7]. The reader will have noticed the "critical" exponent $\gamma > \frac{5}{3}$ which is larger than in the existence Theorem 2.1 and, as a matter of fact, does not include any physically relevant case. Indeed the value $\frac{5}{3}$ happens to be the largest adiabatic constant corresponding to a monoatomic gas.

Under the hypotheses of Theorem 3.1, the dissipative mechanism induced by viscocity is strong enough to prevent any "resonance" phenomena, i.e., the existence of unbounded solutions driven by a bounded external force. Another interesting feature is the existence of periodic solutions provided $\vec{f}$ is periodic in time.

Theorem 3.2 *In addition to the hypotheses of Theorem 3.1, assume that $\vec{f}$ is time periodic, i.e.,*

$$\vec{f}(t + \omega, x) = \vec{f}(t, x) \text{ for a.a. } t \in R, \ x \in \Omega$$

with a certain period $\omega > 0$.

Then there exists at least one finite energy weak solution of the problem (1.1 - 1.6) on $R \times \Omega$ which is ω−periodic in time, i.e.,

$$\varrho(t + \omega) = \varrho(t), \ (\varrho\vec{u})(t + \omega) = (\varrho\vec{u})(t) \text{ for all } t \in R.$$

See [10] for the proof.

4. Asymptotic compactness

The property of asymptotic compactness of a given dynamical system plays a crucial role in the proof of existence of a global (compact) attractor. Here, the main problem is the density component which is bounded only in $L^\gamma(\Omega)$, and, consequently, compact only with respect to the weak topology on this space. Moreover, given the hyperbolic character of the continuity equation, one cannot hope any possible oscillations of the density to be killed at a finite time. However, the amplitude of possible oscillations is decreasing in time uniformly on trajectories emanating from a given bounded set.

Theorem 4.1 *Let $\Omega \subset R^N$, $N = 2,3$ be a bounded Lipschitz domain. Let the pressure p be given by the isentropic constitutive relation*

$$p = a\varrho^\gamma,\ a > 0,\ \gamma > 1\ if\ N = 2,\ \gamma > \frac{5}{3}\ for\ N = 3.$$

Let $\vec{f}_n$ be a sequence of functions such that

$$\operatorname*{ess\ sup}_{t\in R,\ x\in\Omega} |\vec{f}_n(t,x)| \leq F\ independently\ of\ n = 1,2,\dots\ .$$

Finally, let ϱ_n, $\vec{u}_n$ be a sequence of finite energy weak solutions to the problem (1.1 - 1.6) on $(0,\infty) \times \Omega$ such that

$$\int_\Omega \varrho_n\ \mathrm{d}x = m,$$

$$\operatorname*{ess\,lim\,sup}_{t\to 0+} E[\varrho_n, \vec{u}_n](t) \leq E_I$$

independently of $n = 1, 2, \dots$.

Then any sequence of times $t_n \to \infty$ contains a subsequence such that

$$\varrho_n(t_n + t) \to \varrho(t)\ strongly\ in\ L^1(\Omega)\ and\ weakly\ in\ L^\gamma(\Omega),$$

$$(\varrho_n\vec{u}_n)(t_n + t) \to (\varrho\vec{u})(t)\ weakly\ in\ L^1(\Omega, R^N)$$

for any $t \in R$.

Moreover,

$$\int_J \int_\Omega |\varrho_n(t_n + t, x) - \varrho(t,x)|^\gamma\ \mathrm{d}x\mathrm{d}t \to 0,$$

$$\int_J \int_\Omega |(\varrho_n\vec{u}_n)(t_n + t, x) - (\varrho\vec{u})(t,x)|\ \mathrm{d}x\mathrm{d}t \to 0$$

for any bounded interval $J \subset R$. The limit functions ϱ, $\vec{u}$ represent a globally defined (for $t \in R$) finite energy weak solution of the problem (1.1 - 1.6) driven by a force

$$\vec{f} = \lim_{n\to\infty} \vec{f}_n(t_n + \cdot)\ in\ the\ weak\ star\ topology\ of\ the\ space\ L^\infty(R \times \Omega, R^N),$$

and such that

$$\operatorname*{ess\ sup}_{t\in R} E[\varrho, \vec{u}](t) < \infty.$$

The proof of Theorem 4.1 is given in [13] (see also [8]).

5.   The long-time behaviour, attractors

The results presented in Sections 3., 4. allow us to develop a theory of attractors analogous to that one for the incompressible flows (see e.g. TEMAM [21]). Assume, for the sake of simplicity, that the driving force $\vec{f}$ is independent of t. We introduce

$$\mathcal{A} = \{[\varrho_I, \vec{q}_I] \mid \varrho_I = \varrho(0),\ \vec{q}_I = (\varrho\vec{u})(0) \text{ where } \varrho,\ \vec{u} \text{ is a finite energy weak solution}$$

$$\text{of the problem (1.1 - 1.6) on } R \times \Omega \text{ with } E[\varrho, \vec{u}] \in L^\infty(R)\}. \tag{5.1}$$

In other words, the set $\mathcal{A}$ is formed by all globally defined (for $t \in R$) trajectories whose energy is uniformly bounded.

The next result shows that $\mathcal{A}$ is a global attractor in the sense of FOIAS and TEMAM [15].

Theorem 5.1 *Let* $\Omega \subset R^N$, $N = 2,3$ *be a bounded Lipschitz domain. Let* $\vec{f} = \vec{f}(x)$ *be a bounded measurable function independent of time, and let the pressure* p *be given by the isentropic constitutive law*

$$p = a\varrho^\gamma,\ a > 0,\ \gamma > 1 \text{ for } N = 2,\ \gamma > \frac{5}{3} \text{ if } N = 3.$$

Then the set $\mathcal{A}$ *defined by (5.1) is compact in the space*

$$L^\alpha(\Omega) \times L^{\frac{2\gamma}{\gamma+1}}_{weak}(\Omega),$$

and

$$\sup_{[\varrho, \vec{u}] \in \mathcal{B}(E_I)} \left[\inf_{[\varrho_I, \vec{q}_I] \in \mathcal{A}} \left(\|\varrho(t) - \varrho_I\|_{L^\alpha(\Omega)} + \left| \int_\Omega ((\varrho\vec{u})(t) - \vec{q}_I) \cdot \phi \, \mathrm{d}x \right| \right) \right] \to 0$$

for $t \to \infty$

for any $1 \le \alpha < \gamma$ *and any* $\phi \in L^{\frac{2\gamma}{\gamma-1}}(\Omega, R^N)$, *where the symbol* $\mathcal{B}(E_I)$ *stands for the set of all finite energy weak solutions of the problem (1.1 - 1.6) on* $(0, \infty) \times \Omega$ *such that*

$$\operatorname{ess\,lim\,sup}_{t \to 0} E[\varrho, \vec{u}](t) \le E_I.$$

The proof can be found in [6].

To conclude, we give another result concerning the long-time behaviour of solutions on condition that the driving force is a gradient of a scalar potential.

Theorem 5.2 *Let* $\Omega \subset R^N$, $N = 2,3$ *be a bounded Lipschitz domain. Let the pressure* p *be given by the isentropic constitutive relation*

$$p = a\varrho^\gamma,\ a > 0,\ \gamma > \frac{N}{2}.$$

Let the driving force $\vec{f}$ *be of the form*

$$\vec{f} = \nabla F,$$

where $F = F(x)$ be a scalar potential which is globally Lipschitz on $\overline{\Omega}$ and such that the upper level sets

$$[F > k] = \{x \in \Omega \mid F(x) > k\}$$

are connected for any $k \in R$.

Then any finite energy weak solution ϱ, $\vec{u}$ of the problem (1.1 - 1.6) satisfies

$$\varrho(t) \to \varrho_s \ in \ L^{\gamma}(\Omega), \ (\varrho\vec{u})(t) \to 0 \ in \ L^1(\Omega) \ as \ t \to \infty,$$

where ϱ_s is a solution of the static problem

$$a\nabla\varrho_s^{\gamma} = \varrho_s\vec{f} \ on \ \Omega.$$

Acknowledgement The work was supported by Grant No. 201/02/0854 of GA ČR.

References

[1] A.V. Babin and M.I. Vishik. *Attractors of evolution equations*. North-Holland, Amsterdam, 1992.

[2] P. Constantin, C. Foias, and R. an Temam. *Attractors representing turbulent flows*. Mem. Amer. Math. Soc. 53, Providence, 1985.

[3] R.J. DiPerna and P.-L. Lions. Ordinary differential equations, transport theory and Sobolev spaces. *Invent. Math.*, 98 (1989), 511–547.

[4] B. Ducomet, E. Feireisl, H. Petzeltová, and I. Straškraba. Existence global pour un fluide barotrope autogravitant. *C. R. Acad. Sci. Paris, Sér. I.*, 332 (2001), 627–632.

[5] E. Feireisl. Compressible Navier-Stokes equations with a non-monotone pressure law. *J. Differential Equations*, 2002. To appear.

[6] E. Feireisl. Global attractors for the Navier-Stokes equations of three-dimensional compressible flow. *C.R. Acad. Sci. Paris, Sér. I*, 331 (2000), 35–39.

[7] E. Feireisl. Mathematical theory of viscous compressible fluids: Recent development and open problems. 2000. To appear in Mathematical fluid dynamics - Recent results and open questions, Birkhäuser, Basel.

[8] E. Feireisl. Propagation of oscillations, complete trajectories and attractors for compressible flows. *NoDEA*, 2002. To appear.

[9] E. Feireisl. On compactness of solutions to the compressible isentropic Navier-Stokes equations when the density is not square integrable. *Comment. Math. Univ. Carolinae*, 42 (2001), 83–98.

[10] E. Feireisl, Š. Matušů-Nečasová, H. Petzeltová, and I. Straškraba. On the motion of a viscous compressible flow driven by a time-periodic external flow. *Arch. Rational Mech. Anal.*, 149 (1999), 69–96.

[11] E. Feireisl, A. Novotný, and H. Petzeltová. On the domain dependence of solutions to the compressible Navier-Stokes equations of a barotropic fluid. *Math. Meth. Appl. Sci.*, 2002. To appear.

[12] E. Feireisl, A. Novotný, and H. Petzeltová. On the existence of globally defined weak solutions to the Navier-Stokes equations of compressible isentropic fluids. *J. Math. Fluid Dynamics*, 3 (2001), 358–392.

[13] E. Feireisl and H. Petzeltová. Asymptotic compactness of global trajectories generated by the Navier-Stokes equations of compressible fluid. *J. Differential Equations*, 173 (2001), 390–409.

[14] E. Feireisl and H. Petzeltová. Bounded absorbing sets for the Navier-Stokes equations of compressible fluid. *Commun. Partial Differential Equations*, 26 (2001), 1133–1144.

[15] C. Foias and R. Temam. The connection between the Navier-Stokes equations, dynamical systems and turbulence. *In Directions in Partial Differential Equations, Academic Press, New York,* 55–73, 1987.

[16] D. Hoff and J. Smoller. Non-formation of vacuum states for Navier-Stokes equations. Preprint.

[17] O. A. Ladyzhenskaya. *The mathematical theory of viscous incompressible flow.* Gordon and Breach, New York, 1969.

[18] J. Leray. Sur le mouvement d'un liquide visqueux emplissant l'espace. *Acta Math.*, 63 (1934), 193–248.

[19] P.-L. Lions. *Mathematical topics in fluid dynamics, Vol.2, Compressible models.* Oxford Science Publication, Oxford, 1998.

[20] A. Matsumura and T. Nishida. The initial value problem for the equations of motion of compressible and heat conductive fluids. *Comm. Math. Phys.*, 89 (1983), 445–464.

[21] R. Temam. *Infinite-dimensional dynamical systems in mechanics and physics.* Springer-Verlag, New York, 1988.

ICM 2002 · Vol. III · 305–316

Bifurcations without Parameters: Some ODE and PDE Examples

Bernold Fiedler* Stefan Liebscher†

Abstract

In a recent paper the author constructed a continuous map from the configuration space of n distinct ordered points in 3-space to the flag manifold of the unitary group $U(n)$, which is compatible with the action of the symmetric group. This map is also compatible with appropriate actions of the rotation group $SO(3)$. In this paper the author studies the induced homomorphism in $SO(3)$-equivariant cohomology and shows that this contains much interesting information involving representations of the symmetric group.

2000 Mathematics Subject Classification: 34C23, 34C29, 34C37.
Keywords and Phrases: Bifurcation without parameters, Manifolds of equilibria, Normal form, Blow up, Averaging.

1. Applied motivation

In this article we sketch and illustrate some elements of the nonlinear dynamics near equilibrium manifolds. Denoting the equilibrium manifold by $x = 0$, in local coordinates $(x, y) \in \mathbb{R}^n \times \mathbb{R}^k$, we consider systems

$$
\begin{aligned}
\dot{x} &= f(x, y) \\
\dot{y} &= g(x, y)
\end{aligned}
\tag{1.1}
$$

and assume

$$
f(0, y) = g(0, y) = 0, \tag{1.2}
$$

for all y. For simplicity we will only address the cases $k = 1$ of lines of equilibria, and $k = 2$ of equilibrium planes. Sufficient smoothness of f, g is assumed. The occurrence of equilibrium manifolds is infinitely degenerate, of course, in the space of all vector fields (f, g) – quite like many mathematical structures are: equivariance

---

*Free University Berlin, Institute of Mathematics I, Arnimallee 2-6, D-14195 Berlin, Germany. E-mail: fiedler@math.fu-berlin.de

†Free University Berlin, Institute of Mathematics I, Arnimallee 2-6, D-14195 Berlin, Germany. E-mail: liebsch@math.fu-berlin.de, www.math.fu-berlin.de/˜Dynamik/

under symmetry groups, conservation laws, integrability, symplectic structures, and many others. The special case

$$g \;\equiv\; 0 \tag{1.3}$$

in fact amounts to standard bifurcation theory, in the presence of a trivial solution $x = 0$; see for example [5]. Note that condition (1.3), which turns the k-dimensional variable y into a preserved constant parameter, is infinitely degenerate even in our present setting (1.2) of equilibrium manifolds. Due to the analogies of our results and methods with bifurcation theory, we call our emerging theory *bifurcation without parameters*. This terminology emphasizes the intricate dynamics which arises when normal hyperbolicity of the equilibrium manifold fails; see the sections below.

To motivate assumption (1.2), we present several examples. First, consider an "octahedral" graph Γ of $2(m+1)$ vertices $\{\pm 1, \ldots, \pm(m+1)\}$. The graph Γ results from the complete graph by eliminating the "diagonal" edges, which join the antipodal vertices $\pm j$, for $j = 1, \ldots, m+1$. For $m = 1$ we obtain the square, for $m = 2$ the octahedron, and so on. Consider the system

$$\dot{u}_j \;=\; f_j\Big(u_j, \sum_{k \neq \pm j} u_k\Big) \tag{1.4}$$

of oscillators $u_j \in \mathbb{R}^{n'}$ on Γ, additively coupled along the edges by f_j. We assume an antipodal oddness symmetry of the individual oscillator dynamics

$$f_{-j}(-u_j, 0) \;=\; -f_j(u_j, 0). \tag{1.5}$$

As a consequence, the antipode space

$$\Sigma \;:=\; \{u = (u_j)_{j \in \Gamma};\; u_{-j} = -u_j\} \tag{1.6}$$

is invariant under the flow (1.4). Moreover, the flow on Σ completely decouples into a direct product flow of the $m+1$ diagonally antipodal, decoupled pairs

$$\dot{u}_{\pm j} \;=\; f_{\pm j}(u_{\pm j}, 0). \tag{1.7}$$

For the square case $m = 1$, this decoupling phenomenon was first observed in [2]. For more examples see also [3].

An m-plane of equilibria arises from periodic solutions of the decoupled system (1.7). Assume (1.7) possesses time periodic orbits $u_j(t + \varphi_j)$ of equal period $T_j = 2\pi$, for $j = 1, \ldots, m+1$. Choose arbitrary phases $\varphi_j \in S^1$ and let $u_{-j}(t) := -u_j(t)$, $\varphi_{-j} = \varphi_j$. Then

$$u^{\varphi}(t) \;:=\; (u_j(t + \varphi_j))_{j \in \Gamma} \;\in\; \Sigma \tag{1.8}$$

is a 2π-periodic solution of (1.4), (1.7), for arbitrary phases $\varphi \in T^{m+1}$. Eliminating one phase angle φ_{m+1} by passing to an associated Poincaré map, an m-dimensional manifold of fixed points arises, parametrized by the remaining m phase angles. Assuming, in addition to the diagonal oddness symmetry (1.5), equivariance of (1.4) with respect to an S^1-action, the Poincaré map can in fact be obtained as a time-2π map of an autonomous flow within the Poincaré section. In suitable notation, $y = \{\varphi_1, \ldots, \varphi_m\}$, the fixed point manifold then becomes an m-dimensional manifold of equilibria, as presented in (1.1), (1.2) above. For more detailed discussions of this example in the context of bifurcations without parameters see [15, 10, 9].

As a second example of equilibrium manifolds we consider viscous profiles $u = u((\xi - st)/\varepsilon)$ of systems of nonlinear hyperbolic conservation laws and stiff balance laws

$$\partial_t u + \partial_\xi F(u) + \varepsilon^{-1} G(u) \quad = \quad \varepsilon \partial_\xi^2 u. \tag{1.9}$$

Viscous profiles then have to satisfy an ε-independent ODE system

$$\ddot{u} \quad = \quad (F'(u) - s \cdot \mathrm{id})\dot{u} + G(u). \tag{1.10}$$

Standard conservation laws, for example, require $G \equiv 0$. The presence of m conservation laws corresponds to nonlinearities G with range in a manifold of codimension m in u-space. Typically, then, $G(u) = 0$ describes an equilibrium manifold of dimension m of pairs $(u, \dot{u}) = (u, 0)$, in the phase space of (1.10). For an analysis of this example in the context of bifurcations without parameters see [8, 16]. For another example, which relates binary oscillations in central difference discretizations of hyperbolic balance laws with diagonal uncoupling of coupled oscillators, see [11].

We conclude our introductory excursion with a brief summary of some further examples. In [7], lines of equilibria have been observed for the dynamics of models of competing populations. This included a first partial analysis of failure of normal hyperbolicity.

A topologically very interesting example in compact three-dimensional manifolds involves contact structures $\eta(\xi)$ (i.e., nonintegrable plane fields and gradient vector fields $\dot{\xi} = -\nabla V(\xi) \in \eta(\xi)$. See [6] for an in-depth analysis. Examples include mechanical systems with nonholonomic constraints. Notably, level surfaces of regular values of the potential V consist of tori. Under a nondegeneracy assumption, equilibria form embedded circles, that is, possibly linked and nontrivial knots.

For a detailed study of plane Kolmogorov fluid flows in the presence of a line of equilibria with a degeneracy of Takens-Bogdanov type and an additional reversibility symmetry, see [1].

As a caveat we repeat that lines of equilibria, which are transverse to level surfaces of preserved quantities λ do *not* provide bifurcations, without parameters. Rather, $\dot{y} = 0$ for $y := \lambda$ exhibits this problem as belonging to standard bifurcation theory; see (1.3).

2. Sample vector fields and resulting flows

In this section we collect relevant example vector fields (1.1), (1.2) with lines and planes of equilibria $x = 0$; see [10, 9, 1] for further details. We illustrate and comment the resulting flows.

Normally hyperbolic equilibrium manifolds admit a transverse C^0-foliation with hyperbolic linear flows in the leaves. See for example [19], [20] and the ample

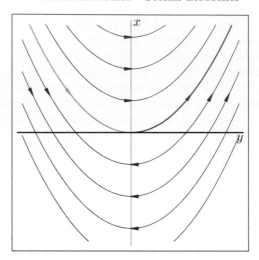

Figure 1: A line of equilibria (y-axis) with a nontrivial transverse eigenvalue zero.

discussion in [4]. As a first example, we therefore consider

$$\begin{aligned} \dot{x} &= xy, \\ \dot{y} &= x. \end{aligned} \tag{2.1}$$

Note the loss of normal hyperbolicity at $x = y = 0$, due to a nontrivial transverse eigenvalue zero of the linearization. Clearly $dx/dy = y$, and the resulting flow lines are parabolas; see Figure 1. For comparison with standard bifurcation theory, where $y = \lambda$, we draw the y-axis of equilibria horizontally.

As a second example, consider

$$\begin{aligned} \dot{x} &= xy, \\ \dot{y} &= \pm x^2. \end{aligned} \tag{2.2}$$

Again, a transverse zero eigenvalue occurs – this time with an additional reflection symmetry $y \mapsto -y$. Dividing by the Euler multiplier x, the reflection becomes a time reversibility. See the left parts of Figure 2 for the resulting flows. Note the resulting integrable, harmonic oscillator case which originates from the elliptic sign $\dot{y} = -x^2$.

As a third example, we consider $x = (x_1, x_2) \in \mathbb{R}^2$, $y \in \mathbb{R}$ with a line $x = 0$ of equilibria and a purely imaginary nonzero eigenvalue $i\omega$ at $x = 0$. Normal-form theory, for example as in [21], then generates an additional S^1-symmetry by the action of $\exp(i\omega t)$ in the x-eigenspace. This equivariance can be achieved, successively, up to Taylor expansions of any finite order. In polar coordinates (r, φ) for x, an example of leading order terms is given by

$$\begin{aligned} \dot{r} &= ry, \\ \dot{y} &= \pm r^2, \\ \dot{\varphi} &= \omega. \end{aligned} \tag{2.3}$$

Since the first two equations in (2.3) coincide with (2.2), the dynamics is then obtained by simply rotating the left parts of Figure 2 around the y-axis at speed w. The right parts of Figure 2 provide three-dimensional views of the effects of higher-order terms which do not respect the S^1-symmetry of the normal forms. In the elliptic case (b), all nonstationary orbits are heteroclinic from the unstable foci, at $y > 0$, to stable foci at $y < 0$. The two-dimensional respective strong stable and unstable manifolds will split, generically, to permit transverse intersections.

Our fourth example addresses Takens-Bogdanov bifurcations without parameters. In suitable rescaled form it reads

$$\ddot{y} + y\dot{y} \;=\; \varepsilon((\lambda - y)\ddot{y} + b\dot{y}^2) \qquad (2.4)$$

with fixed parameters b, λ and ε. The y-axis as equilibrium line is complemented by the two transverse directions $x = (\ddot{y}, \dot{y})$. Note the algebraically triple zero eigenvalue, double in the transverse x-directions, for $\lambda = y = 0$. Two examples of the resulting dynamics for small positive ε are summarized in Figure 3.

The coordinates τ and $\tilde{H}$ in Figure 3 are adapted to the completely integrable case $\varepsilon = 0$. Indeed, obvious first integrals are then given by $\Theta = \ddot{y} + \frac{1}{2}y^2$ and $H = \frac{1}{2}\dot{y}^2 - y\ddot{y} - \frac{1}{3}y^3$. Coordinates are $\tau = \log \Theta$ and $\tilde{H} = \Theta^{-3/2}H$, not drawn to scale. Parameters are $\varepsilon, \lambda > 0$ and, for the hyperbolic case, $-17/12 < b < -1$. For the elliptic case we consider $b > -1$. The equilibrium y-axis, a cusp in (Θ, H) coordinates, transforms to the top (saddles) and bottom (foci) horizontal boundaries $\tilde{H} = \pm\frac{2}{3}\sqrt{2}$, with $y = 0$ relegated to $\tau = -\infty$. Since τ and $\tilde{H}$ are constants of the flow, for $\varepsilon = 0$, they represent slow drifts on the unperturbed periodic motion, for small $\varepsilon > 0$ and $|\tilde{H}| < \frac{2}{3}\sqrt{2}$. The top value $\tilde{H} = +\frac{2}{3}\sqrt{2}$ also represents homoclinics to the saddles, for $\varepsilon = 0$.

Along the focus line $\tilde{H} = -\frac{2}{3}\sqrt{2}$ we observe Hopf bifurcations without parameters, corresponding to $y = \lambda > 0$. The value of b distinguishes elliptic and hyperbolic cases. In addition, lines of saddle-focus heteroclinic orbits and isolated saddle-saddle heteroclinics are generated, for $\varepsilon > 0$, by breaking the homoclinic sheets of the integrable case. Note in particular the infinite swarm of saddle-saddle heteroclinics, in the hyperbolic case.

As a final, fifth example we consider a reversible Takens-Bogdanov bifurcation without parameters:

$$\ddot{y} + (1 - 3y^2)\dot{y} \;=\; ay\ddot{y} + b\dot{y}^2. \qquad (2.5)$$

Here we fix a, b to be small. Again $x = (\ddot{y}, \dot{y})$ denotes the directions transverse to the equilibrium y-axis. Time reversibility generates solutions $-y(-t)$ from solutions $y(t)$. For two examples of the resulting dynamics see Figure 4. Coordinates are the obvious first integrals, for $\varepsilon = 0$, given by $\Theta = \ddot{y} + y - y^3$ and $H = -\ddot{y}y + \frac{1}{2}\dot{y}^2 + \frac{3}{4}y^4 - \frac{1}{2}y^2$. Note the two Takens-Bogdanov cusps, separated by a Hopf point along the lower arc of the equilibrium "triangle". Compare Figures 2, 3. The elliptic Hopf point (b) arises for $a \cdot (a - b) > 0$, whereas $a \cdot (a - b) < 0$ in the hyperbolic case (a). Also note the associated finite and infinite swarms of saddle-saddle heteroclinics, respectively.

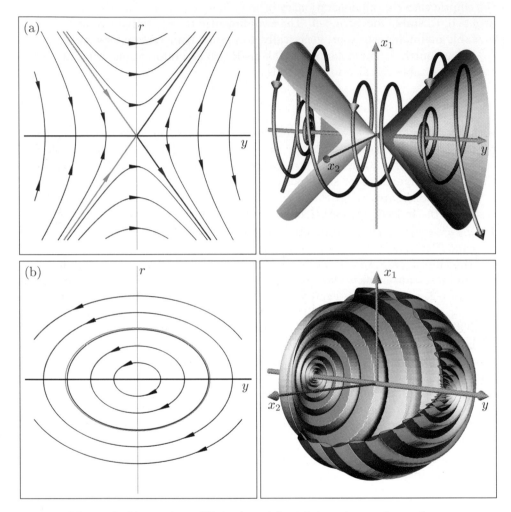

Figure 2: Lines of equilibria (y-axis) with imaginary eigenvalues: Hopf bifurcation without parameters. Case (a) hyperbolic; case (b) elliptic. Red: strong unstable manifolds; green: stable manifolds.

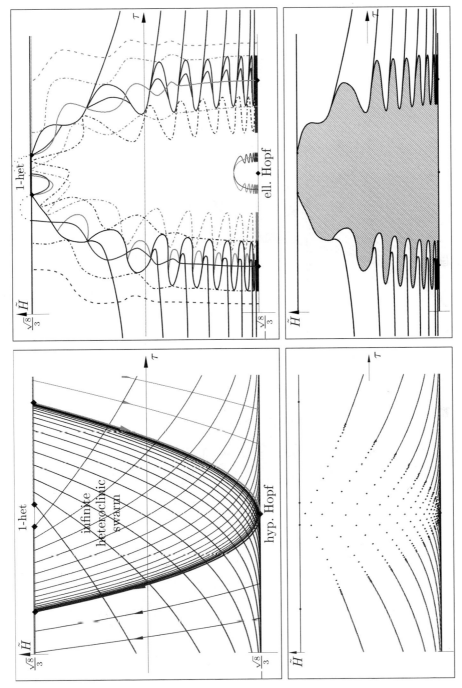

Figure 3: Takens-Bogdanov bifurcations without parameters. Case (a) hyperbolic; case (b) elliptic. Top: stable and unstable manifolds; bottom: invariant sets. For coordinates and fixed parameters see text.

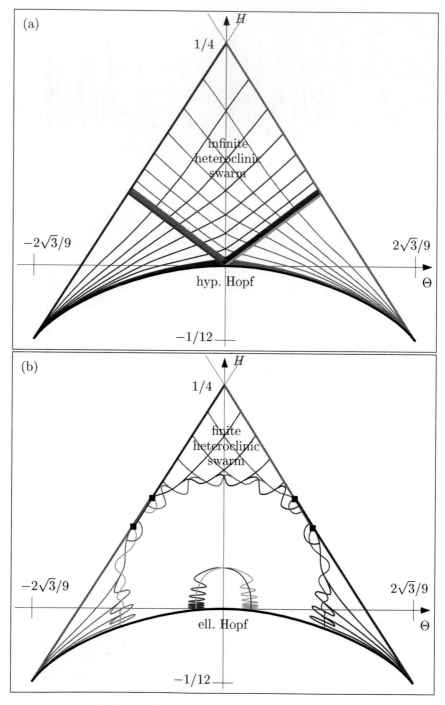

Figure 4: Reversible Takens-Bogdanov bifurcations without parameters. Case (a) hyperbolic; case (b) elliptic. For coordinates
and fixed parameters see text.

3. Methods

Pictures are not proofs. What has been proved, then, and how? We use ingredients involving algebra, analysis, and numerical analysis, as we outline in this section. For further details see [16, 10, 9, 1].

In a first algebraic step, we derive normal forms for vector fields with lines or planes of equilibria, assuming spectral degeneracies of the linearization A in transverse directions. The spectral assumptions on A in fact coincide with those established for parameter-dependent matrix families in standard bifurcation theory. This is reflected in the naming of the five examples of section .

There are more or less standard procedures to derive normal forms of vector fields. By suitable polynomial diffeomorphisms, certain Taylor coefficients of the vector field are eliminated, successively, for higher and higher order. See for example [21] for a systematic choice of normal forms, particularly apt for introducing equivariance of the nonlinear normal-form terms under the action of $\exp(A^T t)$. Normal forms are, however, nonunique in general and other choices are possible.

In the present cases, we adapt the normal-form procedure to preserve the locally flattened equilibrium manifolds. Although the approach in [21] can be modified to accommodate that requirement, it did not provide vector fields convenient for subsequent analysis of the flow. A systematic approach to this combined problem is not known, at present. All examples (2.1)–(2.5) represent truncated normal forms. For the derivation of specific normal forms, for example of (2.4), to any order, and of example (2.5), to third order, see [9], [1], respectively.

Subsequent analysis of the normal-form vector fields is based on scalings, alias blow-up constructions. This is the origin, for example, of the small scaling parameter ε in the Takens-Bogdanov example (2.4). In passing, we note a curious coincidence of two view points for (2.4), concerning the roles of the equilibrium coordinate $y \subset \mathbb{R}$ and the "fixed" real parameter $\lambda \in \mathbb{R}$. First, we may consider λ as a parameter, with a line of equilibria y associated to each fixed λ. Then (2.4) describes the collision of a transverse zero eigenvalue at $y = 0$, as in (2.1), with imaginary Hopf eigenvalues at $y = \lambda > 0$, as in (2.3), as λ decreases through zero. Alternatively, we may consider normal forms for a plane $\mathbf{y} = (y_1, y_2)$ of equilibria with a transverse double zero eigenvalue, at $\mathbf{y} = 0$. It turns out that the two cases coincide, after a scaling blow-up, up to second order in ε, via the correspondence $y = y_1, \lambda = y_2$.

The core of any successful flow analysis in bifurcation theory is an integrable vector field; see again section . The issues of nonintegrable perturbations, by small $\varepsilon > 0$, and of omitted higher order terms, not in normal form, both ensue. Since the underlying integrable dynamics is periodic or homoclinic, in examples (2.3)–(2.5), averaging procedures apply. Indeed, $c > 0$ then introduces a periodically forced, slow flow on first integrals, like (Θ, H), characteristic of $\varepsilon = 0$. We therefore derive an appropriate, but autonomous Poincaré flows, on (Θ, H), such that the associated true Poincaré map can be viewed as a time discretization of first order and step size ε. In the unperturbed periodic region, this amounts to averaging, while the Poincaré flow indicates Melnikov functions at homoclinic or heteroclinic boundaries. The exponential averaging results by Neishtadt [18], for example, then imply that the separatrix splittings in the elliptic Hopf case (b), indicated in Figure 2, are exponentially small in the radius r of the split sphere, for analytic vector fields. See also [12].

Lower bounds of separatrix splittings have not been established, in our settings. This problem is related to the very demanding Lazutkin program of asymptotic expansions for exponentially small separatrix splittings. For recent progress, including the case of Takens-Bogdanov bifurcations for analytic maps, see [13] and the references there. In absence of rigorous lower bounds, our figures indicate only simplest possible splitting scenarios.

While the splitting near elliptic Hopf points are exponentially small, the discretization of the Poincaré flow also exhibits splittings of the unperturbed saddle homoclinic families, which are of first order in the perturbation parameter ε, in example (2.4), or in the small parameters a, b, in example (2.5). Explicit expressions have been derived for the Melnikov functions associated to these homoclinic splittings, in terms of elliptic function in case (2.4), and even of elementary functions in case (2.5). Simplicity and uniqueness of zeros of the Melnikov functions, however, has only been confirmed numerically. While this does not, strictly speaking, match an analytic proof, it still at least supports the validity of the scenarios summarized in Figures 3 and 4.

4. Interpretation and perspective

We indicate some consequences of the above results for the examples of coupled oscillators, viscous shock profiles, and Kolmogorov flows indicated in section . We conclude with a few remarks on the future perspective of bifurcations without parameters.

We first return to the example (1.4)–(1.8) of a coupled oscillator square, $m = 1$. Equilibria y of the Poincaré flow then indicate decoupled antipodal periodic pairs, say with phase difference $y + c$. The case of a transverse zero eigenvalue, (2.1) and Figure 1, then indicates a 50% chance of recovery of decoupling with a stable phase difference $y < 0$, locally, even when the stability threshold $y = 0$ has been exceeded. The hyperbolic Hopf case (2.2), Figure 2 (a), illustrates immediate oscillatory loss of decoupling stability by transverse imaginary eigenvalues. A 100% recovery of decoupling stability, in contrast, occurs at elliptic Hopf points; see (2.2), Figure 2 (b). The exponentially small Neishtadt splitting of separatrices indicates a very delicate variability in the asymptotic phase relations of this recovery, for $t \to \pm\infty$. See [10]. The Takens-Bogdanov cases (2.4), Figures 3 (a), (b) can then be viewed as consequences of a mutual interaction, of a transverse zero eigenvalue with either Hopf case, for recovery of stable decoupling.

In the example (1.9), (1.10) of stiff balance laws, elliptic Hopf bifurcation without parameters as in (2.4), Figure 3 (b), indicates *oscillatory* shock profiles $u(\tau), \tau = (\xi - st)/\varepsilon$. Such profiles in fact contradict the Lax condition, being overcompressive, and violate standard monotonicity criteria. For small viscosities $\varepsilon > 0$, weak viscous shocks in fact turn out unstable, in any exponentially weighted norm, unless they travel at speeds s exceeding all characteristic speeds. The oscillatory profiles can be generated, in fact, by the interaction of inherently non-oscillatory gradient flux functions $F(u)$ with inherently non-oscillatory gradient-like kinetics $G(u)$ in systems of dim ≥ 3. See [16].

The problem of plane stationary Kolmogorov flows asks for stationary solutions of the incompressible Navier-Stokes equations in a strip domain $(\zeta, \eta) \in \mathbb{R} \times [0, 2\pi]$, under periodic boundary conditions in η; see [17]. An η-periodic external force $(F(\eta), 0)$, is imposed, acting in the unbounded ζ-direction. Kolmogorov chose $F(\eta) = \sin \eta$. The Kirchgässner reduction [14] captures all bounded solutions which are nearly homogeneous in ζ, in a center manifold spirit which lets us interpret ζ as "time". The resulting ordinary differential equations in $\mathbb{R}^6$ reduce to $\mathbb{R}^3$, by fixing the values of three first integrals. A line of ζ-homogeneous equilibria appears, in fact, and Kolmogorov's choice corresponds to example (2.5) with $a = b = 0$, to leading orders. In particular note the double reversibility, then, under $y(t) \mapsto \pm y(-t)$ which is generated by

$$
\begin{aligned}
F(\eta) &= -F(\eta + \pi), \qquad \text{and} \\
F(\eta) &= -F(-\eta).
\end{aligned}
\tag{4.1}
$$

As observed by Kolmogorov, an abundance of spatially periodic profiles results. The sample choice $F(\eta) = \sin \eta + c \sin 2\eta$, in contrast, which breaks the first of the symmetries in (4.1), leads to (2.5) with $b = 0 < a$, alias an elliptic reversible Takens-Bogdanov point without parameters; see Figure 4 (b). In particular, the set of near-homogeneous bounded velocity profiles of the incompressible, stationary Navier-Stokes system is then characterized by an abundance of oscillatory heteroclinic wave fronts, which decay to different asymptotically homogeneous ζ-profiles, for $\zeta \to \pm \infty$. The PDE stability of these heteroclinic profiles is of course wide open.

As for perspectives of our approach, we believe to have examples at hand, from sufficiently diverse origin, to justify further development of a theory of bifurcations without parameters. In fact, transverse spectra $\{0, \pm i\omega\}$ and $\{\pm i\omega_1, \pm i\omega_2\}$ still await investigation before we can claim any insight into nonhyperbolicity of even the simple case of an equilibrium plane. This assumes the absence of further structural ingredients like symplecticity, contact structures, symmetries, and the like. Certainly our example collecting activities are far from complete, at this stage.

In addition, we have not addressed the issue of perturbations, so far, which could destroy the equilibrium manifolds by small drift terms. Examples arise, for example, when slightly detuning the basic frequencies $2\pi/T_j$ of our uncoupled oscillators or, much more generally, in the context of multiple scale singular perturbation problems. Feedback and input from our readers will certainly be most appreciated!

References

[1] A. Afendikov, B. Fiedler, and S. Liebscher. Plane Kolmogorov flows, spatial reversibilities, and bifurcation without parameters. in preparation, 2002.

[2] J. Alexander and G. Auchmuty. Global bifurcation of phase-locked oscillators. *Arch. Rational Mech. Anal.*, 93:253–270, 1986.

[3] J. Alexander and B. Fiedler. Global decoupling of coupled symmetric oscillators. In C. Dafermos, G. Ladas, and G. Papanicolaou, editors, *Diff. Equ.*, volume 118 of *Lect. Notes Math.*, New York, 1989. Marcel Dekker Inc.

[4] B. Aulbach. *Continuous and discrete dynamics near manifolds of equilibria*, volume 1058 of *Lect. Notes Math.* Springer, New York, 1984.

[5] S.-N. Chow and J. Hale. *Methods of Bifurcation Theory.* Springer, New York, 1982.

[6] J. Etnyre and R. Ghrist. Gradient flows within plane fields. *Comment. Math. Helv.*, 74:507–529, 1999.

[7] M. Farkas. ZIP bifurcation in a competition model. *Nonlinear Anal., Theory Methods Appl.*, 8:1295–1309, 1984.

[8] B. Fiedler and S. Liebscher. Generic Hopf bifurcation from lines of equilibria without parameters: II. Systems of viscous hyperbolic balance laws. *SIAM J. Math. Anal.*, 31(6):1396–1404, 2000.

[9] B. Fiedler and S. Liebscher. Takens-Bogdanov bifurcations without parameters, and oscillatory shock profiles. In H. Broer, B. Krauskopf, and G. Vegter, editors, *Global Analysis of Dynamical Systems, Festschrift dedicated to Floris Takens for his 60th birthday*, 211–259. IOP, Bristol, 2001.

[10] B. Fiedler, S. Liebscher, and J. Alexander. Generic Hopf bifurcation from lines of equilibria without parameters: I. Theory. *J. Diff. Eq.*, 167:16–35, 2000.

[11] B. Fiedler, S. Liebscher, and J. Alexander. Generic Hopf bifurcation from lines of equilibria without parameters: III. Binary oscillations. *Int. J. Bif. Chaos Appl. Sci. Eng.*, 10(7):1613–1622, 2000.

[12] B. Fiedler and J. Scheurle. *Discretization of Homoclinic Orbits and Invisible Chaos*, volume 570 of *Mem. AMS*. Amer. Math. Soc., Providence, 1996.

[13] V. Gelfreich. A proof of the exponentially small transversality of the separatrices for the standard map. *Commun. math. Phys.*, 201:155–216, 1999.

[14] K. Kirchgässner. Wave solutions of reversible systems and applications. *J. Diff. Eq.*, 45:113–127, 1982.

[15] S. Liebscher. Stabilität von Entkopplungsphänomenen in Systemen gekoppelter symmetrischer Oszillatoren. Diplomarbeit, Freie Universität Berlin, 1997.

[16] S. Liebscher. *Stable, Oscillatory Viscous Profiles of Weak, non-Lax Shocks in Systems of Stiff Balance Laws.* Dissertation, Freie Universität Berlin, 2000.

[17] L. Meshalkin and J. Sinai. Investigation of the stability of the stationary solution of the problem of the motion of some plane incompressible fluid flow (Russian). *Prikl. Math. Mech.*, 25:1140–1143, 1961.

[18] A. Neishtadt. On the separation of motions in systems with rapidly rotating phase. *J. Appl. Math. Mech.*, 48:134–139, 1984.

[19] K. J. Palmer. Exponential dichotomies and transversal homoclinic points. *J. Diff. Eq.*, 55:225–256, 1984.

[20] A. Shoshitaishvili. Bifurcations of topological type of a vector field near a singular point. *Trudy Semin. Im. I. G. Petrovskogo*, 1:279–309, 1975.

[21] A. Vanderbauwhede. Centre manifolds, normal forms and elementary bifurcations. In U. Kirchgraber and H. O. Walther, editors, *Dynamics Reported 2*, pages 89–169. Teubner & Wiley, Stuttgart, 1989.

Asymptotic Behaviour of Ergodic Integrals of 'Renormalizable' Parabolic Flows

G. Forni*

Abstract

Ten years ago A. Zorich discovered, by computer experiments on interval exchange transformations, some striking new power laws for the ergodic integrals of generic non-exact Hamiltonian flows on higher genus surfaces. In Zorich's later work and in a joint paper authored by M. Kontsevich, Zorich and Kontsevich were able to explain conjecturally most of Zorich's discoveries by relating them to the ergodic theory of Teichmüller flows on moduli spaces of Abelian differentials.

In this article, we outline a generalization of the Kontsevich-Zorich framework to a class of 'renormalizable' flows on 'pseudo-homogeneous' spaces. We construct for such flows a 'renormalization dynamics' on an appropriate 'moduli space', which generalizes the Teichmüller flow. If a flow is renormalizable and the space of smooth functions is 'stable', in the sense that the Lie derivative operator on smooth functions has closed range, the behaviour of ergodic integrals can be analyzed, at least in principle, in terms of an Oseledec's decomposition for a 'renormalization cocycle' over the bundle of 'basic currents' for the orbit foliation of the flow.

This approach was suggested by the author's proof of the Kontsevich-Zorich conjectures and it has since been applied, in collaboration with L. Flaminio, to prove that the Zorich phenomenon generalizes to several classical examples of volume preserving, uniquely ergodic, parabolic flows such as horocycle flows and nilpotent flows on homogeneous 3-manifolds.

2000 Mathematics Subject Classification: 37C40, 37E35, 37A17, 37A25, 34D08, 43A85.
Keywords and Phrases: Ergodic integrals, Renormalization dynamics, Cohomological equations, Invariant distributions, Basic currents.

1. Introduction

---
*Department of Mathematics, Northwestern University, Lunt Hall, 2033 Sheridan Road, Evanston, IL, 60208-2730, USA. E-mail: gforni@math.northwestern.edu

A fundamental problem in smooth ergodic theory is to establish quantitative estimates on the asymptotics behaviour of ergodic integrals of smooth functions. For several examples of *hyperbolic* flows, such as geodesic flows on compact manifolds of negative curvature, the asymptotic behaviour of ergodic integrals is described by the *Central Limit Theorem* (Y. Sinai, M. Ratner). In these cases, the dynamical system can be described as an approximation of a 'random' stochastic process, like the outcomes of flipping a coin. Non-hyperbolic systems as not as well understood, with the important exception of toral flows. For generic non-singular area-preserving flows on the 2-torus logarithmic bounds on ergodic integrals of zero-average functions of bounded variation can be derived by the *Denjoy-Koksma inequality* and the theory of continued fractions. For a general ergodic flow, ergodic integrals are bounded for all times for a special class of functions: *coboundaries* with bounded 'transfer' functions (Gottschalk-Hedlund). In the hyperbolic examples and in the case of generic toral flows, a smooth function is a coboundary if and only if it has zero average with respect to all invariant measures.

In this article, we are interested in flows with *parabolic* behaviour. Following A. Katok, a dynamical system is called parabolic if the rate of divergence of nearby orbits is at most polynomial in time, while hyperbolic systems are characterized by exponential divergence. Toral flows are a rather special parabolic example, called *elliptic*, since there is no divergence of orbits. It has been known for many years that typical examples of parabolic flows, such as horocycle flows or generic nilpotent flows are *uniquely ergodic*, but until recently not much was known on the asymptotic behaviour of ergodic averages, with the exception of some polynomial bounds on the speed of convergence in the horocycle case (M. Ratner, M. Burger), related to the polynomial rate of mixing. We have been able to prove, in collaboration with L. Flaminio, that for many examples of parabolic dynamics the behaviour of ergodic averages is typically described as follows.

A smooth flow Φ^X on a finite dimensional manifold M has *deviation spectrum* $\{\lambda_1 > ... > \lambda_i > ... > 0\}$ with multiplicities $m_1, ..., m_i, ... \in \mathbb{Z}^+$ if there exists a system $\{\mathcal{D}_{ij} \,|\, i \in \mathbb{Z}^+, \ 1 \leq j \leq m_i\}$ of linearly independent X-*invariant distributions* such that, for almost all $p \in M$, the ergodic integrals of any smooth function $f \in C_0^\infty(M)$ have an asymptotic expansion

$$\int_0^T f(\Phi^X(t,p))\,dt \ = \ \sum_{i\in\mathbb{N}}\sum_{j=1}^{m_i} c_{ij}(p,T)\,\mathcal{D}_{ij}(f)\,T^{\lambda_i} \ + \ R(p,T)(f)\,, \qquad (1.1)$$

where the real coefficients $c_{ij}(p,T)$ and the distributional remainder $R(p,T)$ have, for almost all $p \in M$, a sub-polynomial behaviour, in the sense that

$$\limsup_{T\to+\infty} \frac{\log\sum_{j=1}^{m_i}|c_{ij}(p,T)|^2}{\log T} = \limsup_{T\to+\infty} \frac{\log\|R(p,T)\|}{\log T} = 0\,. \qquad (1.2)$$

The notion of a deviation spectrum first arose in the work of A. Zorich and in his joint work with M. Kontsevich on non-exact Hamiltonian flows with isolated saddle-like singularities on compact higher genus surfaces. Zorich discovered in numerical experiments on interval exchange transformations an unexpected new phenomenon [8]. He found that, although a generic flow on a surface of genus $g \geq 2$ is uniquely ergodic (H. Masur, W. Veech), for large times the homology classes of return orbits exhibit unbounded polynomial deviations with exponents $\lambda_1 > \lambda_2 > ... > \lambda_g > 0$ from the line spanned in the homology group by the *Schwartzmann's asymptotic cycle*. In his later work [9], [10] and in joint work with M. Kontsevich [6], Zorich was able to explain this phenomenon in terms of conjectures on the Lyapunov exponents of the Teichmüller flow on moduli spaces of holomorphic differentials on Riemann surfaces. Kontsevich and Zorich also conjectured that Zorich's phenomenon is not merely topological, but it extends to ergodic integrals of smooth functions. "There is, presumably, an equivalent way of describing the numbers λ_i. Namely, let f be a smooth function ... Then for a generic trajectory $p(t)$, we expect that the number $\int_0^T f(p(t))dt$ for large T with high probability has size $T^{\lambda_i + o(T)}$ for some $i \in \{1, ..., g\}$. The exponent λ_1 appears for all functions with non-zero average value. The next exponent, λ_2, should work for functions in a codimension 1 subspace of $C^\infty(S)$, etc." [6]

Around the same time, we proved that for a generic non-exact Hamiltonian flow Φ^X on a higher genus surface not all smooth zero average functions are smooth coboundaries [3]. In fact, we found that, in contrast with the hyperbolic case and the elliptic case of toral flows, there are X-invariant distributional obstructions, which are not signed measures, to the existence of smooth solutions of the *cohomological equation $Xu = f$*. This result suggested that Zorich's phenomenon should be related to the presence of invariant distributions other than the (unique) invariant probability measure. In fact, in [4] we were able prove the Kontsevich-Zorich conjectures that the deviation exponents are non-zero and that generic non-exact Hamiltonian flows on higher genus surfaces have a deviation spectrum. Recently, in collaboration with L. Flaminio, we have proved that other classical parabolic examples, such as horocycle flows on compact surfaces of constant negative curvature [1] and generic nilpotent flows on compact 3-dimensional nilmanifolds [2], do have a deviation spectrum, but of countable multiplicity, in contrast with the case of flows on surfaces which have spectrum of finite multiplicity equal to the genus.

We will outline below a general framework, derived mostly from [4] and successfully carried out in [1], [2], for proving that a flow on a *pseudo-homogeneous space* has a deviation spectrum. Our framework is based on the construction of an appropriate *renormalization dynamics* on a moduli space of pseudo-homogeneous structures, which generalizes the Teichmüller flow. A *renormalizable flow* for which the space of smooth functions is *stable* (in the sense of A. Katok), has a deviation spectrum determined by the Lyapunov exponents of a *renormalization cocycle* over a bundle of *basic currents*. Pseudo-homogeneous spaces are a generalization of homogeneous spaces. The motivating non-homogeneous example is given by any punctured Riemann surface carrying a holomorphic differential vanishing only at the punctures. It turns out that renormalizable flows are necessarily parabolic. In fact, the class of renormalizable flows encompasses all parabolic flows which are reasonably well-understood, while not much is known for most non-renormalizable parabolic flows, such as generic geodesic flows on flat surfaces with conical singularities. Our approach unifies and generalizes several classical quantitative equidistribution results such as the Zagier-Sarnak results for periodic horocycles on non-compact hyperbolic surfaces of finite volume [1] or number theoretical results on the asymptotic behaviour of theta sums [2].

2. Renormalizable flows

Let $\mathfrak{g}$ be a finite dimensional real Lie algebra. A $\mathfrak{g}$-*structure* on a manifold M is defined to be a homomorphism τ from $\mathfrak{g}$ into the Lie algebra $\mathcal{V}(M)$ of all smooth vector fields on M. This notion is well-known in the theory of transformation groups (originated in the work of S. Lie) under the name of 'infinitesimal G-transformation group' (for a Lie group G with $\mathfrak{g}$ as Lie algebra). The second fundamental theorem of Lie states that any infinitesimal G-transformation group τ on M can be 'integrated' to yield an essentially unique local G-transformation group. A $\mathfrak{g}$-structure τ will be called *faithful* if τ induces a linear isomorphism from $\mathfrak{g}$ onto $T_x M$, for all $x \in M$. Let τ be a $\mathfrak{g}$-structure. For each element $X \in \mathfrak{g}$, the vector field $X_\tau := \tau(X)$ generates a (partially defined) flow Φ^X_τ on M. Let $E_t(X_\tau) \subset M$ be the closure of the complement of the domain of definition of the map $\Phi^X_\tau(t, \cdot)$ at time $t \in \mathbb{R}$. A faithful $\mathfrak{g}$-structure will be called *pseudo-homogeneous* if for every $X \in \mathfrak{g}$ there exists $t > 0$ such that $E_t(X_\tau) \cup E_{-t}(X_\tau)$ has zero (Lebesgue) measure. A manifold M endowed with a pseudo-homogeneous $\mathfrak{g}$-structure will be called a *pseudo-homogeneous $\mathfrak{g}$-space*. All homogeneous spaces are pseudo-homogeneous.

Let $\mathcal{T}_\mathfrak{g}(M)$ be the space of all pseudo-homogeneous $\mathfrak{g}$-structures on M. The automorphism group $\mathrm{Aut}(\mathfrak{g})$ acts on $\mathcal{T}_\mathfrak{g}(M)$ by composition on the right. The group $\mathrm{Diff}(M)$ acts on $\mathcal{T}_\mathfrak{g}(M)$ by composition on the left. The spaces

$$\mathrm{T}_\mathfrak{g}(M) := \mathcal{T}_\mathfrak{g}(M)/\mathrm{Diff}_0(M) \,, \quad \mathcal{M}_\mathfrak{g}(M) := \mathrm{T}_\mathfrak{g}(M)/\Gamma(M) \,, \tag{2.1}$$

where $\Gamma(M) := \mathrm{Diff}^+(M)/\mathrm{Diff}_0(M)$ is the *mapping class group*, will be called respectively the *Teichmüller space* and the *moduli space* of pseudo-homogeneous $\mathfrak{g}$-structures on M. The group $\mathrm{Aut}(\mathfrak{g})$ acts on the Teichmüller space $\mathrm{T}_\mathfrak{g}(M)$ and on the moduli space $\mathcal{M}_\mathfrak{g}(M)$, since in both cases the action of $\mathrm{Aut}(\mathfrak{g})$ on $\mathcal{T}_\mathfrak{g}(M)$ passes to the quotient.

Let $\mathrm{Aut}^{(1)}(\mathfrak{g})$ be the subgroup of automorphisms with determinant one. An element $X \in \mathfrak{g}$ will be called *a priori renormalizable* if there exists a partially hyperbolic one-parameter subgroup $\{G^X_t\} \subset \mathrm{Aut}^{(1)}(\mathfrak{g})$, $t \in \mathbb{R}$ ($t \in \mathbb{Z}$), in general non-unique, with a single (simple) Lyapunov exponent $\mu_X > 0$ such that

$$G^X_t(X) = e^{t\mu_X} X \,. \tag{2.2}$$

It follows from the definition that the subset of a priori renormalizable elements of a Lie algebra $\mathfrak{g}$ is saturated with respect to the action of $\mathrm{Aut}(\mathfrak{g})$. The subgroup $\{G^X_t\}$ acts on the Teichmüller space and on the moduli space of pseudo-homogeneous $\mathfrak{g}$-structures as a 'renormalization dynamics' for the family of flows Φ^X_τ generated by the vector fields $\{X_\tau \,|\, \tau \in \mathcal{T}_\mathfrak{g}(M)\}$ on M. It will be called a *generalized Teichmüller flow (map)*. A flow Φ^X_τ will be called *renormalizable* if $\tau \in \mathcal{M}_\mathfrak{g}(M)$ is a recurrent point for some generalized Teichmüller flow (map) G_t^X. If μ is a probability G_t^X-invariant measure on the moduli space, then by Poincaré recurrence the flow Φ_τ^X is renormalizable for μ-almost all $\tau \in \mathcal{M}_\mathfrak{g}(M)$.

Let R be an inner product on $\mathfrak{g}$. Every faithful $\mathfrak{g}$ structure τ induces a Riemannian metric R_τ of constant curvature on M. Let ω_τ be the volume form of R_τ. The total volume function $A : \mathcal{T}_\mathfrak{g}(M) \to \mathbb{R}^+ \cup \{+\infty\}$ is $\mathrm{Diff}^+(M)$-invariant and $\mathrm{Aut}^{(1)}(\mathfrak{g})$-invariant. Hence A is well-defined as an $\mathrm{Aut}^{(1)}(\mathfrak{g})$-invariant function on the Teichmüller space and on the moduli space. It follows that the subspace of finite-volume $\mathfrak{g}$-structures has an $\mathrm{Aut}^{(1)}(\mathfrak{g})$-invariant stratification by the level hypersurfaces of the total volume function. Since different hypersurfaces are isomorphic up to a dilation, when studying finite-volume spaces it is sufficient to consider the hypersurface of volume-one $\mathfrak{g}$-structures:

$$\mathrm{T}_\mathfrak{g}^{(1)}(M) := \mathrm{T}_\mathfrak{g}(M) \cap A^{-1}(1) \,, \quad \mathcal{M}_\mathfrak{g}^{(1)}(M) := \mathcal{M}_\mathfrak{g}(M) \cap A^{-1}(1) \,. \qquad (2.3)$$

Let τ be a faithful $\mathfrak{g}$-structure and let $X \in \mathfrak{g}$. If the linear map ad_X on $\mathfrak{g}$ has zero trace, the flow Φ_τ^X preserves the volume form ω_τ and X_τ defines a symmetric operator on $L^2(M, \omega_\tau)$ with domain $C_0^\infty(M)$. If τ is pseudo-homogeneous, by E. Nelson's criterion [7], X_τ is essentially skew-adjoint. It turns out that any a priori renormalizable element $X \in \mathfrak{g}$ is *nilpotent*, in the sense that all eigenvalues of the linear map ad_X are equal to zero, hence the flow Φ_τ^X is volume preserving and parabolic. In all the examples we have considered, the Lie algebra $\mathfrak{g}$ is *traceless*, in the sense that for every element $X \in \mathfrak{g}$, the linear map ad_X has vanishing trace. In this case, any pseudo-homogeneous $\mathfrak{g}$-structure induces a representation of the Lie algebra $\mathfrak{g}$ by essentially skew-adjoint operators on the Hilbert space $L^2(M, \omega_\tau)$ with common invariant domain $C_0^\infty(M)$.

3.   Examples

Homogeneous spaces provide a wide class of examples. Let G be a finite dimensional (non-compact) Lie group with Lie algebra $\mathfrak{g}$ and let $M = G/\Gamma$ be a (compact) homogeneous space. The Teichmüller space $\mathrm{T}_G(M) \subset \mathrm{T}_\mathfrak{g}(M)$ and the moduli space $\mathcal{M}_G(M) \subset \mathcal{M}_\mathfrak{g}(M)$ of all homogeneous G-space structures on M are respectively isomorphic to the Lie group $\mathrm{Aut}(G)$ and to the homogeneous space $\mathrm{Aut}(G)/\mathrm{Aut}(G, \Gamma)$, where $\mathrm{Aut}(G, \Gamma) < \mathrm{Aut}(G)$ is the subgroup of automorphisms which stabilize the lattice Γ. The Teichmüller and moduli spaces $\mathrm{T}_G^{(1)}(M)$ and $\mathcal{M}_G^{(1)}(M)$ of homogeneous volume-one G-space structures are respectively isomorphic to the subgroup $\mathrm{Aut}^{(1)}(G)$ of orientation preserving, volume preserving automorphisms and to the homogeneous space $\mathrm{Aut}^{(1)}(G)/\mathrm{Aut}^{(1)}(G, \Gamma)$.

In the *Abelian* case $\mathfrak{g} = \mathbb{R}^n$, any $X \in \mathbb{R}^n$ is a priori renormalizable. In fact, the group $\mathrm{Aut}^{(1)}(\mathbb{R}^n) = \mathrm{SL}(n, \mathbb{R})$ acts transitively on $\mathbb{R}^n$ and $X_1 = (1, 0, ..., 0)$ is renormalized by the one-parameter group $G_t := \mathrm{diag}(e^t, e^{-t/n}, ..., e^{-t/n}) \subset \mathrm{SL}(n, \mathbb{R})$. Finite volume Abelian homogeneous spaces are diffeomorphic to n-dimensional tori $\mathbb{T}^n$. The generalized Teichmüller flow G_t on the moduli space of all volume-one Abelian homogeneous structures on $\mathbb{T}^n$ is a volume preserving Anosov flow on the finite-volume non-compact manifold $\mathrm{SL}(n, \mathbb{R})/\mathrm{SL}(n, \mathbb{Z})$. Hence, in this case, almost all homogeneous flows are renormalizable, by Poincaré recurrence theorem. The dynamics of the flow G_t has been investigated in depth by D. Kleinbock and G. Margulis in connection with the theory of Diophantine approximations.

In the *semi-simple* case, let $\mathfrak{g} = \mathfrak{sl}(2, \mathbb{R})$ be the unique 3-dimensional simple Lie algebra. There is a basis $\{H, H^{\perp}, X\}$ with commutation relations $[X, H] = H$, $[X, H^{\perp}] = -H^{\perp}$ and $[H, H^{\perp}] = 2X$. The elements H, $H^{\perp}$ are renormalized by the one-parameter group $G_t := \mathrm{diag}(e^t, e^{-t}, 1) \subset \mathrm{Aut}^{(1)}(\mathfrak{g})$, while X is not a priori renormalizable. The unit tangent bundle of any hyperbolic surface S can be identified to a homogeneous $\mathfrak{g}$-space $M := \mathrm{PSL}(2, \mathbb{R})/\Gamma$. The vector fields H, $H^{\perp} \in \mathfrak{g}$ generate the horocycle flows and the vector field X generates the geodesic flow on S. Since G_t is a group of *inner* automorphisms, it is in fact generated by the geodesic vector field X, every point of the moduli space is fixed under G_t. Hence horocycle flows are renormalizable on every homogeneous space $\mathrm{PSL}(2, \mathbb{R})/\Gamma$.

In the *nilpotent*, non-Abelian case, let $\mathfrak{n}$ be the Heisenberg Lie algebra, spanned by elements $\{X, X^{\perp}, Z\}$ such that $[X, X^{\perp}] = Z$ and Z is a generator of the one-dimensional center $Z_{\mathfrak{n}}$. The element X is renormalized by the one-parameter subgroup $G_t := \mathrm{diag}(e^t, e^{-t}, 1) \subset \mathrm{Aut}^{(1)}(\mathfrak{n})$. Since the group $\mathrm{Aut}(\mathfrak{n})$ acts transitively on $\mathfrak{n} \setminus Z_{\mathfrak{n}}$, every $Y \in \mathfrak{n} \setminus Z_{\mathfrak{n}}$ is a priori renormalizable, while the elements of the center are not. A compact nilmanifold modeled over the Heisenberg group N is a homogeneous space $M = N/\Gamma$, where Γ is a co-compact lattice. These spaces are topologically circle bundles over $\mathbb{T}^2$ classified by their Euler characteristic. The moduli space $\mathcal{M}_N^{(1)}(M)$ of volume-one homogeneous $\mathfrak{n}$-structures on M is a 5-dimensional finite-volume non-compact orbifold which fibers over the modular surface $\mathrm{SL}(2, \mathbb{R})/\mathrm{SL}(2, \mathbb{Z})$ with fiber $\mathbb{T}^2$. The generalized Teichmüller flow is an Anosov flow on $\mathcal{M}_N^{(1)}(M)$ [2].

The motivation for our definition of a pseudo-homogeneous space comes from the theory of Riemann surfaces of higher genus. Any holomorphic (Abelian) differential h on a Riemann surface S of genus $g \geq 2$, vanishing at $Z_h \subset S$, induces a (non-unique) pseudo-homogeneous $\mathbb{R}^2$-structure on the open manifold $M_h := S \setminus Z_h$. In fact, the frame $\{X, X^{\perp}\}$ of $TS|M_h$ uniquely determined by the conditions

$$\frac{\sqrt{-1}}{2} \imath_X (h \wedge \bar{h}) = \Im(h) \ , \quad \frac{\sqrt{-1}}{2} \imath_{X^{\perp}} (h \wedge \bar{h}) = -\Re(h) \tag{3.1}$$

satisfies the Abelian commutation relation $[X, X^{\perp}] = 0$ and the homomorphism $\tau_h : \mathbb{R}^2 \to \mathcal{V}(M_h)$ such that $\tau_h(1, 0) = X$, $\tau_h(0, 1) = X^{\perp}$ is a pseudo-homogeneous $\mathbb{R}^2$-structure on M_h. Let $Z \subset S$ be a given subset of cardinality $\sigma \in \mathbb{N}$ and let $\kappa = (k_1, ..., k_{\sigma}) \in (\mathbb{Z}^+)^{\sigma}$ with $\sum k_i = 2g - 2$. Let $\mathcal{H}_{\kappa}(S, Z)$ be the space of Abelian differentials h with $Z_h = Z$ and zeroes of multiplicities $(k_1, ..., k_{\sigma})$. The projection of the set $\{\tau_h \in \mathcal{T}_{\mathbb{R}^2}(M) \, | \, h \in \mathcal{H}_{\kappa}(S, Z)\}$ into the moduli space $\mathcal{M}_{\mathbb{R}^2}(M)$ of pseudo-homogeneous $\mathbb{R}^2$-structures on $M := S \setminus Z$ is isomorphic to a stratum $\mathcal{H}(\kappa)$ of the moduli space of Abelian differentials on S. The flow induced on $\mathcal{H}(\kappa)$ by the one-parameter group of automorphism $G_t = \mathrm{diag}(e^t, e^{-t}) \subset \mathrm{SL}(2, \mathbb{R})$ coincides with the Teichmüller flow on the stratum $\mathcal{H}(\kappa)$.

4.   Cohomological equations

Let $(\mathfrak{g}, R)$ be a finite dimensional Lie algebra endowed with an inner product. Any pseudo-homogeneous $\mathfrak{g}$-structure τ on a manifold M induces a *Sobolev filtration* $\{W_\tau^s(M)\}_{s \geq 0}$ on the space $W_\tau^0(M) := L^2(M, \omega_\tau)$ of square-integrable functions. Let $\triangle_\tau$ be the non-negative *Laplace-Beltrami operator* of the Riemannian metric R_τ on M. The Laplacian is densely defined and symmetric on the Hilbert space $W_\tau^0(M)$ with domain $C_0^\infty(M)$, but it is not in general essentially self-adjoint. In fact, if $\mathfrak{g}$ is traceless, by a theorem of E. Nelson [7], $\triangle_\tau$ is essentially self-adjoint if and only if the representation τ of the Lie algebra $\mathfrak{g}$ on $W_\tau^0(M)$ by essentially skew-adjoint operators induces a unitary representation of a Lie group. Let then $\bar{\triangle}_\tau$ be the *Friederichs extension* of $\triangle_\tau$. The Sobolev space $W_\tau^s(M)$, $s > 0$, is defined as the maximal domain of the operator $(I + \bar{\triangle}_\tau)^{s/2}$ endowed with the norm

$$\|f\|_{s,\tau} := \|(I + \bar{\triangle})^{s/2} f\|_{0,\tau} \,. \tag{4.1}$$

The Sobolev spaces $W_\tau^{-s}(M)$ are defined as the duals of the Hilbert spaces $W_\tau^s(M)$, for all $s > 0$. Let $C_B^0(M)$ be the space of continous bounded functions on M. The pseudo-homogeneous space (M, τ) will be called of *bounded type* if there is a continous (Sobolev) embedding $W_\tau^s(M) \subset C_B^0(M)$ for all $s > \dim(M)/2$. The bounded-type condition is essentially a geometric property of the pseudo-homogeneous structure.

Let $X \in \mathfrak{g}$. Following A. Katok, the space $W_\tau^s(M)$ is called $W_\tau^t(M)$-*stable* with respect to the flow Φ_τ^X if the subspace

$$R^{s,t}(X_\tau) := \{f \in W_\tau^s(M) \,|\, f = X_\tau u \,, \quad u \in W_\tau^t(M)\} \tag{4.2}$$

is closed in $W_\tau^s(M)$. The flow Φ_τ^X will be called *tame* (of degree $\ell > 0$) if $W_\tau^s(M)$ is $W_\tau^{s-\ell}(M)$-stable with respect to Φ_τ^X for all $s > \ell$. In all the examples of §3, generic renormalizable flows are tame. In particular, it is well known that generic toral flows are tame, horocycle flows and generic nilpotent flows on 3-dimensional compact nilmanifolds were proved tame of any degree $\ell > 1$ in [1], [2], generic non-exact Hamiltonian flows on higher genus surfaces were proved tame in [3]. These results are based on the appropriate harmonic analysis: in the homogeneous cases, the theory of unitary representations for the Lie group $SL(2, \mathbb{R})$ [1] and the Heisenberg group [2]; in the more difficult non-homogeneous case of higher genus surfaces, the theory of boundary behaviour of holomorphic functions on the unit disk plays a crucial role [4].

If the Sobolev space $W_\tau^s(M)$ is stable with respect to the flow Φ_τ^X, the closed range $R^{s,l}(X_\tau)$ of the operator X_τ coincides with the distributional kernel $\mathcal{T}^s(X_\tau) \subset W_\tau^{-s}(M)$ of X_τ, which is a space of X_τ-*invariant distributions*. Let X be any smooth vector field on a manifold M. A distribution $\mathcal{D} \in \mathcal{D}'(M)$ is called X-invariant if $X\mathcal{D} = 0$ in $\mathcal{D}'(M)$. Invariant distributions are in bijective correspondence with (homogeneous) one-dimensional *basic currents* for the orbit foliation $\mathcal{F}(X)$ of the flow Φ^X. A one-dimensional basic current C for a foliation $\mathcal{F}$ on M is a continous linear functional on the space $\Omega_0^1(M)$ of smooth 1-forms with compact support such that, for all vector fields Y tangent to $\mathcal{F}$,

$$\imath_Y C \ = \ \mathcal{L}_Y C \ = \ 0 \ (\ \Longleftrightarrow \ \ \imath_Y C \ = \ dC \ = \ 0)\,. \tag{4.3}$$

It follows from the definitions that the one-dimensional current $C := \imath_X \mathcal{D}$ is basic for $\mathcal{F}(X)$ if and only if the distribution $\mathcal{D}$ is X-invariant. Let $\mathcal{I}(X)$ be the space of all X-invariant distributions and $\mathcal{B}(X)$ be the space of all one-dimensional basic currents for the orbit foliation $\mathcal{F}(X)$. The linear map $\imath_X : \mathcal{I}(X) \to \mathcal{B}(X)$ is bijective.

Let (M, τ) be a pseudo-homogeneous space. There is a well-defined Hodge (star) operator and a space $C^0_\tau(M)$ of square-integrable 1-forms on M associated with the metric R_τ. Since the Laplace operator $\triangle_\tau$ extends to $C^0_\tau(M)$ with domain $\Omega^1_0(M)$, it is possible to define, as in the case of functions, a Sobolev filtration $\{C^s_\tau(M)\}_{s\geq 0}$, on the space $C^0_\tau(M)$. The Sobolev spaces $C^{-s}_\tau(M)$ are defined as the duals of the Sobolev spaces $C^s(M)$, for all $s > 0$. Let $\mathcal{B}^s(X_\tau) := \mathcal{B}(X_\tau) \cap C^{-s}_\tau(M)$ be the subspaces of basic currents of Sobolev order $\leq s$ for the orbit foliation $\mathcal{F}(X_\tau)$. The space $\mathcal{B}^s(X_\tau)$ is the image of $\mathcal{I}^s(X_\tau) := \mathcal{I}(X_\tau) \cap W^{-s}_\tau(M)$ under the bijective map $\imath_X : \mathcal{I}(X) \to \mathcal{B}(X)$. In the case of minimal toral flows the space $\mathcal{B}^s(X_\tau)$ is one-dimensional for all $s \geq 0$ (as all invariant distributions are scalar multiples of the unique invariant probability measure). In the parabolic examples we have studied, $\mathcal{B}^s(X_\tau)$ has countable dimension, as soon as $s > 1/2$, for horocycle flows or generic nilpotent flows, while for generic non-exact Hamiltonian flows on higher genus surfaces the dimension is finite for all $s > 0$ and grows linearly with respect to $s > 0$. This finiteness property seems to be an exceptional low dimensional feature.

5.   The renormalization cocycle

The Sobolev spaces $C^s_\tau(M)$ of one-dimensional currents form a smooth infinite dimensional vector bundle over $\mathcal{T}_\mathfrak{g}(M)$. Such bundles can be endowed with a flat connection with parallel transport given locally by the identity maps $C^s_\tau(M) \to C^s_{\tau'}(M)$, for any $\tau \approx \tau' \in \mathcal{T}_\mathfrak{g}(M)$. Since the diffeomorphism group $\mathrm{Diff}(M)$ acts on $C^s_\tau(M)$ by push-forward, we can define (orbifold) vector bundles $C^s_\mathfrak{g}(M)$ over the Teichmüller space $\mathrm{T}_\mathfrak{g}(M)$ or the moduli space $\mathcal{M}_\mathfrak{g}(M)$ of pseudo-homogeneous structures on M. If $X \in \mathfrak{g}$ is a priori renormalizable, a generalized Teichmüller flow (map) G^X_t can be lifted by parallel transport to a 'renormalization cocycle' R^X_t on the bundles of currents $C^s_\mathfrak{g}(M)$ over the Teichmüller space or the moduli space. It follows from the definitions that the sub-bundles $\mathcal{B}^s_\mathfrak{g}(X) \subset C^s_\mathfrak{g}(M)$ with fibers the subspaces of basic currents $\mathcal{B}^s(X_\tau) \subset C^s_\tau(M)$ are R^X_t-invariant. It can be proved that, for any G_t-ergodic probability measure μ on the moduli space, if the flows Φ^X_τ are tame of degree $\ell > 0$ for μ-almost all $\tau \in \mathcal{M}_\mathfrak{g}(M)$, then the sub-bundles $\mathcal{B}^s_\mathfrak{g}(X)$ are μ-almost everywhere defined with closed (Hilbert) fibers of constant rank, for all $s > \ell$.

In the examples considered, with the exception of flows on higher genus surfaces, the Hilbert bundles of basic currents $\mathcal{B}^s_\mathfrak{g}(X)$ are infinite dimensional, and to the author's best knowledge, available Oseledec-type theorems for Hilbert bundles do not apply to the renormalization cocycle. However, the cocycle has a well defined Lyapunov spectrum and an Oseledec decomposition. We are therefore led to formulate the following hypothesis:

$H_1(s)$. The renormalization cocyle R_t^X on the bundle $\mathcal{B}_{\mathfrak{g}}^s(X)$ over the dynamical system (G_t^X, μ) has a Lyapunov spectrum $\{\nu_1 > ... > \nu_k > ... > 0 > ...\}$ and an Oseledec's decomposition

$$\mathcal{B}_{\mathfrak{g}}^s(X) = E_{\mathfrak{g}}^s(\nu_1) \oplus ... \oplus E_{\mathfrak{g}}^s(\nu_k) \oplus ... \oplus N_{\mathfrak{g}}^s , \qquad (5.1)$$

in which the components $E_{\mathfrak{g}}^s(\nu_k)$ correspond to the Lyapunov exponents $\nu_k > 0$, while the component $N_{\mathfrak{g}}^s$ has a non-positive top Lyapunov exponent. Our result on the existence of a deviation spectrum requires an additional technical hypothesis, verified in our examples.

$H_2(s)$. Let $\gamma_\tau^1(p)$ be the one-dimensional current defined by the time $T = 1$ orbit-segment of the flow Φ_τ^X with initial point $p \in M$. (a) The essential supremum of the norm $\|\gamma_\tau^1(p)\|_{\tau,s}$ over $p \in M$ is locally bounded for $\tau \in \mathrm{supp}(\mu) \subset \mathcal{M}_{\mathfrak{g}}(M)$; (b) The orthogonal projections of $\gamma_\tau^1(p)$ on all subspaces $E_\tau^s(\nu_k) \subset \mathrm{C}_{\mathfrak{g}}^{-s}(M)$ are non-zero for μ-almost all $\tau \in \mathcal{M}_{\mathfrak{g}}(M)$ and almost all $p \in M$.

Let $X \in \mathfrak{g}$ be a priori renormalizable and let μ be a G_t^X- invariant Borel probability measure on $\mathcal{M}_{\mathfrak{g}}(M)$, supported on a stratum of bounded-type $\mathfrak{g}$- structures. If the flow Φ_τ^X is tame of degree $\ell > 0$ and the hypoteses $H_1(s)$, $H_2(s)$ are verified for $s > \ell + \dim(M)/2$, for μ- almost $\tau \in \mathcal{M}_{\mathfrak{g}}$, the flow Φ_τ^X has a deviation spectrum with deviation exponents

$$\nu_1/\mu_X > ... > \nu_k/\mu_X > ... > 0 \qquad (5.2)$$

and multiplicities given by the decomposition (5.1) of the renormalization cocyle.

In the homogeneous examples, the Lyapunov spectrum of the renormalization cocycle is computed explicitly in every irreducible unitary representation of the structural Lie group. In the horocycle case, the existence of an Oseledec's decomposition (5.1) is equivalent to the statement that the space of horocycle-invariant distributions is spanned by (generalized) eigenvectors of the geodesic flow, well-known in the representation theory of semi-simple Lie groups as *conical distri-butions* [5]. In the non-homogeneous case of higher genus surfaces, the Oseledec's theorem applies since the bundles $\mathcal{B}_{\mathfrak{g}}^s(X)$ are finite dimensional. We have found in all examples a surprising heuristic relation between the Lyapunov exponents of the renormalization cocycle and the Sobolev regularity of basic currents (or equivalently of invariant distributions): the subspaces $E_\tau^s(\nu_k)$ are generated by basic currents of *Sobolev order* $1 - \nu_k/\mu_X \geq 0$. The Sobolev order of a one-dimensional current C is defined as the infimum of all $s > 0$ such that $C \in \mathrm{C}_\tau^{-s}(M)$.

In the special case of non-exact Hamiltonian flow on higher genus surfaces the Lyapunov exponents of the renormalization cocycle are related to those of the Teichmüller flow. In fact, let S be compact orientable surface of genus $g \geq 2$ and let $\mathcal{H}(\kappa)$ be a stratum of Abelian differentials vanishing at $Z \subset S$. Let $\mathcal{B}_\kappa(X) \subset \mathcal{B}_{\mathbb{R}^2}(X)$ be the measurable bundle of basic currents over $\mathcal{H}(\kappa) \subset \mathcal{M}_{\mathbb{R}^2}(S \setminus Z)$ and let

$H^1_\kappa(S \setminus Z, \mathbb{R})$ be the bundle over $\mathcal{H}(\kappa)$ with fibers isomorphic to the real cohomology $H^1(S \setminus Z, \mathbb{R})$. Since basic currents are closed, there exists a *cohomology map* $j_\kappa : \mathcal{B}_\kappa(X) \to H^1_\kappa(S \setminus Z, \mathbb{R})$ such that, as proved in [4], the restrictions $j_\kappa | \mathcal{B}^s_\kappa(X)$ are surjective for all $s \gg 1$ and, for all $s \geq 1$, there are exact sequences

$$0 \to \mathbb{R} \to \mathcal{B}^{s-1}_\kappa(X) \xrightarrow{\delta_\kappa} \mathcal{B}^s_\kappa(X) \xrightarrow{j_\kappa} H^1_\kappa(S \setminus Z, \mathbb{R}) \ . \tag{5.3}$$

The renormalization cocycle R^X_t on $\mathcal{B}^s_\kappa(X)$ projects for all $s \gg 1$ onto a cocycle on the cohomology bundle $H^1_\kappa(S \setminus Z, \mathbb{R})$, introduced by M. Kontsevich and A. Zorich in order to explain the *homological* asymptotic behaviour of orbits of the flow Φ^X_τ for a generic $\tau \in \mathcal{H}(\kappa) \subset \mathcal{M}_{\mathbb{R}^2}(S \setminus Z)$ [6]. The Lyapunov exponents of the Kontsevich-Zorich cocycle on $H^1_\kappa(S \setminus Z, \mathbb{R})$,

$$\lambda_1 = 1 > \lambda_2 \geq \cdots \geq \lambda_g \geq \overbrace{0 = \cdots = 0}^{\# Z - 1} \geq -\lambda_g \geq \cdots \geq -\lambda_2 > -\lambda_1 = -1 \,, \tag{5.4}$$

are related the Lyapunov exponents of the Teichüller flow on $\mathcal{H}(\kappa)$ [6], [4]. Since the bundle map δ_κ shifts Lyapunov exponents by -1 and, as conjectured in [6] and proved in [4], the Kontsevich-Zorich exponents $\lambda_1 = 1 > \lambda_2 \geq \cdots \geq \lambda_g$ are non-zero, the strictly positive exponents of the renormalization cocycle coincide with the Kontsevich-Zorich exponents. This reduction explains why in the case of non-exact Hamiltonian flows on surfaces the Lyapunov exponents of the Teichmüller flow are related to the deviation exponents for the ergodic averages of smooth functions.

References

[1] L. Flaminio & G. Forni, Invariant distributions and time averages for horocycle flows, preprint.

[2] L. Flaminio & G. Forni, Equidistribution of nilflows and applications to theta sums, in preparation.

[3] G. Forni, Solutions of the cohomological equation for area preserving flows on higher genus surfaces, *Ann. of Math. (2)* 146 (1997), no. 2, 295–344.

[4] G. Forni, Deviations of ergodic averages for area preserving flows on higher genus surfaces, *Ann. of Math. (2)* 154 (2001), no. 1, 1–103.

[5] S. Helgason, A duality for symmetric spaces with applications to group representations, *Advances in Math.* 5 (1970), 1–154.

[6] M. Kontsevich, Lyapunov exponents and Hodge theory, in *The mathematical beauty of physics* (Saclay, 1996), 318–332, Adv. Ser. Math. Phys., 24, World Sci. Publishing, River Edge, NJ, 1997.

[7] E. Nelson, Analytic vectors, *Ann. of Math. (2)* 70 1959, 572–615.

[8] A. Zorich, Asymptotic flag of an orientable measured foliation on a surface, in *Geometric study of foliations* (Tokyo, 1993), 479–498, World Sci. Publishing, River Edge, NJ, 1994.

[9] A.Zorich, Finite Gauss measure on the space of interval exchange transformations. Lyapunov exponents, *Ann. Inst. Fourier* 46 (1996), no. 2, 325–370.

[10] A.Zorich, Deviation for interval exchange transformations, *Ergodic Theory Dynam. Systems* 17 (1997), no. 6, 1477–1499.

ICM 2002 · Vol. III · 327–338

Tangent Bundles Dynamics and Its Consequences[*]

E. R. Pujals[†]

Abstract

We will consider here some dynamics of the tangent map, weaker than hyperbolicity, and we will discuss if these structures are rich enough to provide a good description of the dynamics from a topological and geometrical point of view. This results are useful in attempting to obtain global scenario in terms of generic phenomena relative both to the space of dynamics and to the space of trajectories. Moreover, we will relate these results with the study of systems that remain globally transitive under small perturbations.

2000 Mathematics subject classification: 37C05, 37C10, 37C20, 37C29, 37C70.
Keywords and Phrases: Dynamical systems, Homoclinic bifurcation, Dominated splitting, Partial hyperbolicity, Robust transitivity.

1. Introduction

A long time goal in the theory of dynamical systems is to describe the dynamics of "big sets" (generic or residual, dense, etc) in the space of all dynamical systems.

It was thought in the sixties that this could be realized by the so called hyperbolic ones: systems with the assumption that the tangent bundle over the Limit set $(L(f)$, the accumulation points of any orbit) splits into two complementary subbundles that are uniformly forward (respectively backward) contracted by the tangent map by. The richness of this description would follow from the fact that *the hyperbolic dynamic on the tangent bundle characterizes the dynamic over the manifold* from a geometrical and topological point of view.

Nevertheless, uniform hyperbolicity were soon realized to be a property less universal than it was initially thought: *there are open sets in the space of dynamics which are non-hyperbolic.* After some initial examples of non-density of the hyperbolic systems in the universe of all systems (see [S, AS]), two key aspects were

---

[*]Partially supported by FAPERJ-Brazil and Guggenheim Foundation.

[†]Instituto de Matemática, Universidade Federal do Rio de Janeiro, C. P. 68.530, CEP 21.945-970, Rio de Janeiro, R. J. , Brazil. E-mail: enrique@im.ufrj.br

E. R. Pujals

focused in these examples. On one hand, open sets of non-hyperbolic diffeomorphisms which remain transitive under perturbation (existence of a dense orbit for any system). On the other, residual sets of non-hyperbolic diffeomorphisms, each one exhibiting infinitely many transitive sets. Roughly speaking, it was showed that two kind of different phenomena can appear in the complement of the hyperbolic systems: **a)** *dynamics that robustly can be decomposed into a finite number of closed transitive sets;* **b)** *dynamics that generically exhibit infinitely many disjoint transitive sets.*

The first kind of phenomena occurs in dimension higher or equal than 3 and although the examples are not hyperbolic they exhibit *some kind of decomposition of the tangent bundle into invariant subbundles.* The second one, was obtained by Newhouse (see [N1], [N2], [N3]), who following an early work of the non-density of hyperbolicity for C^2 surface maps, showed that the unfolding of a *homoclinic bifurcation* (non transversal intersection of stable and unstable manifolds of periodic points) leads to a very rich dynamics: residual subsets of open sets of diffeomorphisms whose elements display infinitely many sinks.

These new results naturally pushed some aspects of the theory on dynamical systems in different directions:

1. The study of the dynamical phenomena obtained from homoclinic bifurcations;
2. The characterization of universal mechanisms that could yield to robustly non-hyperbolic behavior;
3. The study and characterization of isolated transitive sets that remain transitive for all nearby system (*robust transitivity*);
4. The dynamical consequences that follows from some kind of the dynamics over the tangent bundle, weaker than the hyperbolic one.

As we will show, these problems are related and they indeed constitute different aspects of the same phenomena. In many cases, such relations provide a conceptual framework, as the hyperbolic theory did for the case of transverse homoclinic orbits.

In the next section, we will discuss the previous aspects for the case of surfaces maps, and in particular we will consider a dynamics in the tangent bundle weaker than hyperbolicity, called dominated splitting. In section 3. we will discuss the problems about the robust transitivity and its relation with other dynamics on the tangent bundle. Finally, in the last section, we will consider the equivalent problems for flows taking into account their intrinsic characteristic that leads to further questions and difficulties do the presence of singularities. We point out that in this survey we will focus more into topological and geometrical aspects of the dynamic rather on the ergodic ones.

Many of the issues discussed here are consequences of works and talks with Martin Sambarino. I also want to thanks Maria J. Pacifico for her help to improve this article.

2. Surfaces maps, homoclinic tangencies and "non-critical" behaviors

After the seminal works of Newhouse, many others were developed in the direction to understand the phenomena that could appear after a bifurcation of homoclinic tangencies (tangent intersection of stable and unstable manifolds of periodic points). In fact, other fundamental dynamic prototype were found in this context, namely the so called cascade of bifurcations, the Hénon-like strange attractor ([BC], [MV]) (even infinitely many coexisting ones [C]), and superexponential growth of periodic points ([K]). Even before these last results, Palis ([PT], [P1]) conjectured that the presence of a homoclinic tangency is a very common phenomenon in the complement of the closure of the hyperbolic ones. In fact, if the conjecture is true, then homoclinic bifurcation could play a central role in the global understanding of the space of dynamics for it would imply that each of these bifurcation phenomena is dense in the complement of the closure of the hyperbolic ones. More precisely, he conjectured that *Every $f \in Diff^r(M^2), r \geq 1$, can be C^r-approximated by a diffeomorphism exhibiting either a homoclinic tangency or by one which is hyperbolic.*

The presence of homoclinic tangencies have many analogies with the presence of *critical points* for one-dimensional endomorphisms. Homoclinic tangecies correspond in the one dimensional setting to preperiodic critical points and it is known that its bifurcation leads to complex dynamics. On the other hand, Mañé (see [M1])showed that for regular and generic one-dimensional endomorphisms, the *absence of critical points is enough to guarantee hyperbolicity*. This result raises the question about the dynamical properties of surface maps exhibiting no homoclinic tangencies. In this direction, first it is proved in that some kind of dynamic over the tangent bundle (weaker than the hyperbolic one) can be obtained in the robust lack of homoclinic tangencies. And later, it is showed that this dynamic on the tangent bundle is rich enough to describe the dynamic on the manifold. More precisely:

Theorem 1 ([PS1]): *Surface diffeomorphisms that can not be C^1-approximated by another exhibiting homoclinic tangencies, has the property that its Limit set has* **dominated splitting**.

An f-invariant set Λ has dominated splitting if the tangent bundle can be decomposed into two invariant subbundles $T_\Lambda M = E \oplus F$, such that:

$$\|Df^n_{/E(x)}\|\|Df^{-n}_{/F(f^n(x))}\| \leq C\lambda^n, \text{ for all } x \in \Lambda, n \geq 0,$$

with $C > 0$ and $0 < \lambda < 1$.

As, dominated splitting prevents the presence of tangencies, we could say that domination plays for surface diffeomorphisms the role that the *non-critical behavior* does for one dimensional endomorphisms.

To have a satisfactory description for this non-critical behavior (existence of a dominated splitting), we should describe its dynamical consequences. A natural question arises: *is it possible to describe the dynamics of a system having dominated splitting?*

The next result gives a positive answer (as satisfactory as in hyperbolic case) when M is a compact surface. More precisely, we give a complete description of the topological dynamics of a C^2 system having a dominated splitting. Actually, first, the dominated decomposition is understood under a generic assumption.

Theorem 2 ([PS1]) :   *Let $f \in Diff^2(M^2)$ and assume that $\Lambda \subset L(f)$ is a compact invariant set exhibiting a dominated splitting such that any periodic point is a hyperbolic periodic point. Then, $\Lambda = \Lambda_1 \cup \Lambda_2$ where Λ_1 is hyperbolic and Λ_2 consists of a finite union of periodic simple closed curves $\mathcal{C}_1, ...\mathcal{C}_n$, normally hyperbolic, and such that $f^{m_i} : \mathcal{C}_i \to \mathcal{C}_i$ is conjugated to an irrational rotation (m_i denotes the period of $\mathcal{C}_i$).*

Using this Theorem and understanding the obstruction for the hyperbolicity assuming domination, we can characterize $L(f)$ without any generic assumption.

Theorem 3 ([PS2]) : *Let $f \in Diff^2(M^2)$ and assume that $L(f)$ has a dominated splitting. Then $L(f)$ can be decomposed into $L(f) = \mathcal{I} \cup \tilde{\mathcal{L}}(f) \cup \mathcal{R}$ such that:*

1. *$\mathcal{I}$ is contained in a finite union of normally hyperbolic periodic arcs.*

2. *$\mathcal{R}$ is a finite union of normally hyperbolic periodic simple closed curves supporting an irrational rotation.*

3. *$f/\tilde{\mathcal{L}}(f)$ is expansive and admits a spectral decomposition (into finitely many homoclinic classes).*

Roughly speaking, the above theorem says that the dynamics of a C^2 diffeomorphism having a dominated splitting can be decomposed into two parts: one where the dynamic consists on periodic and almost periodic motions ($\mathcal{I}$, $\mathcal{R}$) with the diffeomorphism acting equicontinuously, and another one where the dynamics is expansive and similar to the hyperbolic case. Moreover, given a set having dominated decomposition, it is characterized its stable and unstable set, its continuation by perturbation, and their basic pieces (see [PS2]). Let us say also, that in solving the above problem, another kind of differentiable dynamical problem arose: how is affected the dynamics of a system, when its smoothness is improved?

Putting theorem 1 and 2 together, we prove the conjecture of Palis for surface diffeomorphisms in the C^1-topology:

Theorem 4 ([PS1]):   *Let M^2 be a two dimensional compact manifold and let $f \in Diff^1(M^2)$. Then, f can be C^1-approximated either by a diffeomorphism exhibiting a homoclinic tangency or by an Axiom A diffeomorphism.*

Similar arguments, prove that the variation of the topological entropy leads to the unfolding of homoclinic tangencies. Moreover the presence of infinitely many sinks with unbounded period also implies the unfolding of tangencies (see [PS2], [PS4]).

We want to emphasize that theorem 4 and the previous comments are strictly C^1 (on the other hand, theorem 2, and 3 assume that the map is C^2), and nothing is known in the C^2-topology. We would like to understand what happens in the C^2-topology, since many rich dynamical phenomena take place for smooth maps. About this problem we would like to make some remarks.

Recall that for smooth one-dimensional endomorphisms, the absence of critical points was enough to guarantee hyperbolicity. But for surface maps (and due to the lack of understanding) we required C^1-robust absence of tangencies. Many of the tools used in the C^1 case (C^1-closing Lema, perturbation of the tangent map along finite orbits) are unknown in higher topology and even in same particular situation they are also false (see [G], [PS2]). So, taking in mind the scenario for one-dimensional dynamics, instead of ask about the C^r−robust absence of tangencies (for $r \geq 2$), we could try to know what is the two dimensional phenomena whose *presence breaks the domination* and whose *absence guarantee it?* In other words, what are *two dimensional critical points?*

To address these problems we should consider previously some other weaker questions. Observe that any dominated splitting is a continuous one. Is it true the converse, at least generically? To answer this question we would face the following: if there is a continuous invariant splitting over the Limit set for an smooth map (or even assuming an stronger hypothesis: existence of two continuous invariant foliations) can we describe the dynamic of f? It is clear that are dyanmics exhibiting continuous splitting which are not dominated, for instance, maps exhibiting some kind of saddle connection and maps on the torus obtained as $(x, y) \rightarrow (x+\alpha, y+\beta)$. Are those dynamics the unique ones that do not exhibit domination?

On the other hand, splitting dealing with critical behaviors (tangencies or "almost tangencies") are well known in the measure-theoretical setting. This is the case of the non-uniform hyperbolicity (or Pesin theory), where the tangent bundle splits for points a.e. with respect to some invariant measure, and vectors are asymptotically contracted or expanded in a rate that may depend on the base point. Are the invariant measures for smooth maps on surfaces with one non zero Lyapunov exponents, non-uniformly hyperbolic? In other words, one non-zero lyapunov exponent implies that the other is also non-zero? This is not true in general, since there exist ergodic invariant measures with only one non zero Lyapunov exponent: measure supported on invariant circle normally hyperbolic; measure over a non-hyperbolic periodic point; time one map of a Cherry flows; two dimensional version of one dimension phenomena like infinitely renormalizable and absorbing cantor sets. Are those dynamics the unique counterexamples?

Now we would like to say a few words about the proof of Theorem 2. First, it is showed that the local unstable (stable) sets are one dimensional manifolds tangent to the direction F (E respectively). This is achieved by, given an *explicit characterization of the Lyapunov stable sets under the hypothesis of domination* (latter we will give a precise statement). After that, it is proved that the length of the negative iterates of the local unstable (positive for the local stable) manifolds are sumable, and using arguments of distortion hyperbolicity is concluded.

We say that the point x is *Lyapunov stable* (in the future) if given $\epsilon > 0$ there exists $\delta > 0$ such that $f^n(B_\delta(x)) \subset B_\epsilon(f^n(x))$ for any positive integer n. Its characterization is done in any dimension whit the solely assumption that one of the subbundles is one dimensional. To avoid confusion, we call such splitting *codimension one dominated splitting.*

Theorem 5: *Let $f : M \rightarrow M$ be a C^2-diffeomorphism of a finite dimensional compact riemannian manifold M and let Λ be a set having a codimension one dominated splitting. Then there exists a neighborhood V of Λ such that if $f^n(x) \in V$ for any positive integer n and x is Lyapunov stable, one of the following holds:*

   1. $\omega(x)$ *is a periodic orbit,*

   2. $\omega(x)$ *is a periodic curve normally attractive supporting and irrational rota-*

332 E. R. Pujals

tion.

As we said before, this theorem have important consequences related to the direction F. Its local invariant tangent manifold either is dynamically defined (it is a subset of the local unstable set) or there are well understood phenomena: either there are periodic curves normally contractive γ with small length, or there are semi-attracting periodic points, or there are closed invariant curves normally hyperbolic with dynamics conjugated to an irrational rotation. With this characterization in mind, and assuming domination over the whole manifold, it is proved that F is also uniquely integrable.

3. Robust transitivity

As we said in the beginning, in dimension higher or equal than 3, there exist C^r−open set of diffeomorphisms which are transitive and non-hyperbolic ($r \geq 1$). Observe that this phenomena take place in the C^1-topology, fact that it is unknown for one of the dynamics phenomena that we consider in the previous section: the residual sets of infinitely many sinks for surface maps.

The first examples of robust non-hyperbolic systems (examples of robust transitive systems which are not Anosov) were given by M. Shub (see [Sh]), who considered on the 4-torus, skew-products of an Anosov with a Derived of Anosov diffeomorphisms. Then, R. Mañé (see [M]) reduced the dimension of such examples by showing that certain Derived of Anosov diffeomorphisms on the 3-torus are robust transitive. Later, L. Díaz, (see [D1]) constructed examples obtained as a bifurcation of an heteroclinic cycle (cycle involving points of different indices). This last ideas was pushed in [BD] where it was showed a general geometric construction of robust transitive attractors.

All these systems show a *partial hyperbolic* splitting, which allows the tangent bundle to split into Df-invariant subbundles $TM = E^s \oplus E^c \oplus E^u$, where the behavior of vectors in E^s, E^u under iterates of the tangent map is similar to the hyperbolic case, but vectors in E^c may be neutral for the action of the tangent map. On the other hand, recently, it was proved by C. Bonatti and M. Viana that there are opens sets of transitive diffeomorphisms exhibiting a dominated splitting which do not fall into the category of partially hyperbolic ones (see [BV]).

These new situations lead to ask two natural questions: Is there a characterization of robust transitive sets that also gives dynamical information about them? Can we describe the dynamics under the assumption of either partial hyperbolicity or dominated decomposition?

The next result shows that this two questions are extremely related. In fact, some kind of dynamics on the tangent bundle is implied by the robust transitivity (see [M] for surfaces, [DPU] for three dimensional manifolds , and [BDP] for the n.dimensional case):

Theorem 6: *Every robustly transitive set of a C^1-diffeomorphism has dominated splitting whose extremal bundles are uniformly volume contracting or expanding.*

The central idea in this theorem is to show that, in the lack of domination, the eigenspaces of a linear map (obtained by multiplying many bounded linear maps) are very unstable: by small perturbation of each of the factors, one can mix the eigenvalues in order to get a homothety, which will correspond to the creation of either a sink or a source, situation not allowed in the case of the robust transitivity. This last theorem also can be formulated in the following way:

Theorem 7 [BDP]) : *There is a residual subset of C^1-diffeomorphisms such that that for any diffeomorphism in the residual set, it is verified that for any homoclinic class of a periodic point (the closure of the intersection of the stable and unstable manifold of it) either has dominated splitting or it is contained in the closure of infinitely many sources or sinks.*

What about the converse of theorem 6? Is it true that generically a transitive system exhibiting some kind of splitting is robust transitive? On the other hand, all the examples of robust transitivity are based in either a property of the initial system or in a geometrical construction. But, does exist a necessary and sufficient condition among the partial hyperbolic system such that transitivity is equivalent to robust transitivity? *Can this property be characterized in terms of the dynamic of the tangent map?* This is clear for Anosov maps, where transitivity implies robust transitivity, but what about for the non-hyperbolic?. In the direction to understand this problem, in [PS5] was introduced a dynamic on the tangent bundle enough to guarantee robustness of transitivity.

Theorem 8: *Let $f \in Diff^r(M)$ be a transitive partial hyperbolic system verifying that there is $n_0 > 0$ such that for any $x \in M$ there are $y^u(x) \in W_1^{uu}(x)$ and $y^s(x) \in W_1^{ss}(x)$ with:*

1. $|Df_{|E^c(f^m(y^u(x)))}^{n_0}| > 2$ *for any $m > 0$,*

2. $|Df_{|E^c(f^{-m}(y^s(x)))}^{-n_0}| > 2$ *for any $m > 0$,*

then, f is a non-hyperbolic robust transitive system.

Is this property generically necessary?

All these questions naturally push in the direction to understand the dynamic induced by either a partial hyperbolic system or a dominated splitting. In particular, do they exhibit (generically) spectral decomposition, as was showed for a hyperbolic system and for domination on surfaces? Observe that the non-hyperbolicity of these systems is related with the presence of points of different index. Is this (generically) a necessary condition for non-hyperbolicity? In other words, assuming that there are only points of the same index, can one conclude (generically) hyperbolicity? For the case of domination, it is possible to show that if the extremal directions are one-dimensional, then they behave topologically as a hyperbolic one (see [PS4]). Moreover, it is showed that homoclinic classes with codimension one dominated splitting and contractive bundle are generically hyperbolic. But, what happens if the external directions are one-dimensional? And what about the central directions? Of course, this question can be considered in a simpler situation: a partial hyperbolic splitting with only one dimensional central direction. Does the dynamic over the central direction characterize the kind of partial hyperbolic systems?

Many of the questions done for partial hyperbolic systems can be formulated for *Iterated Function Systems*. In same sense, these systems works as a model of partial hyperbolic ones. And its solution, could give an indication how to deal in

334 E. R. Pujals

the general case.

In dimension higher than two, another kind of homoclinic bifurcation breaks the hyperbolicity: the so called heteroclinic cycles (intersection of the stable and unstable manifolds of points of different indices, see [D1] and [D2]). In particular, the unfold of these cycles imply the existence of striking dynamics being the more important, the appearance of non-hyperbolic robust transitive sets (see [KP] also for superexponential growth of periodic point associated to the unfolding of heteoclinic cycles). Moreover, any non-hyperbolic robust transitive sets exhibits generically heteroclinic cycles. In same sense, these cycles play the role for the partial hyperbolic theory as transversal intersection play for the hyperbolic theory.

A similar conjecture as the one for surfaces, was formulated by Palis in any dimension: *Every $f \in Diff^r(M), r \geq 1$, can be C^r-approximated by a diffeomorphism exhibiting either a homoclinic tangency, a heteroclinic cycle or by one which is hyperbolic.*

A similar approach as the one done in dimension 2 could be done: first, try to find the dynamic on the tangent bundle for systems C^1- far from tangencies. About this, in [LW] it was proved a similar result as the one for surfaces: *far from tangencies implies domination.* Does far from heteroclinic cycles imply hyperbolicity? Does this imply that sets with periodic points of different index can not accumulate one on the other? And as we asked before: sets showing dominated decomposition exhibiting points of the same index are generically hyperbolic? These problems are also related with the problems involving tangencies and sinks: can a systems showing infinitely many sinks be approximated by another one showing tangencies? It was showed that this is true for surfaces maps (see [PS3]), and in the case of higher dimension in [PS4] is given a positive answer assuming that the sinks accumulate on a sectional dissipative homoclinic class.

On the other hand, there is a vast works about conservative partial hyperbolic systems describing, in same particular cases, their ergodic properties. The description of the dynamics strongly use the invariance of the volume measure, information that it is not available in the general case that we would like describe. For references about it see the complete review on this subject done by Burns, Pugh , Shub and Wilkinson ([BPSW]). Moreover it is showed in [Bo] and [BoV] some kind of dichotomy (as the one done in theorem 8) for conservative maps in terms of Lyapunov exponents and domination. Also I would like to mention a recent and remarkable work of F. Rodriguez Hertz ([R]) where is proved that many Linear automorphisms on T^4 are stable ergodic, using different kinds of techniques that even if only work in the conservative case, they could be useful to understand the general case.

4. Flows

For flows, a striking example is the Lorenz attractor [Lo], given by the solutions of the polynomial vector field in R^3:

$$X(x,y,z) = \begin{cases} \dot{x} = -\alpha x + \alpha y \\ \dot{y} = \beta x - y - xz \\ \dot{z} = -\gamma z + xy\,, \end{cases} \tag{1}$$

where α, β, γ are real parameters. Numerical experiments performed by Lorenz (for $\alpha = 10, \beta = 28$ and $\gamma = 8/3$) suggested the existence, in a robust way, of a strange attractor toward which tends a full neighborhood of positive trajectories of the above system. That is, the strange attractor could not be destroyed by any perturbation of the parameters. Most important, the attractor contains an equilibrium point $(0,0,0)$, and periodic points accumulating on it, and hence can not be hyperbolic. Notably, only now, three and a half decades after this remarkable work, was it proved [Tu] that the solutions of (1) satisfy such a property for values α, β, γ near the ones considered by Lorenz.

However, already in the mid-seventies, the existence of robust non-hyperbolic attractors was proved for flows introduced in [ABS] and [Gu], which we now call geometric models for Lorenz attractors. In particular, they exhibit, in a robust way, an attracting transitive set with an equilibrium (singularity). Moreover, the properties of this geometrical models, allow one to extract very complete dynamical information. A natural question raises, is such features present for any robust transitive set?

In [MPP] a positive answer for this question is given:

Theorem 9 : C^1 *robust transitive sets with singularities on closed* 3-*manifolds verifies:*

   1. *there are either proper attractors or proper repellers;*

   2. *the eigenvalues at the singularities satisfy the same inequalities as the corresponding ones at the singularity in a Lorenz geometrical model;*

   3. *there are partially hyperbolic with a volume expanding central direction.*

The presence of a singularity prevents these attractors from being hyperbolic. But they exhibit a weaker form of hyperbolicity *singular hyperbolic splitting.* This class of vector fields contains the Axiom Λ systems, the geometric Lorenz attractors and the singular horseshoes in ([LP]), among other systems. Currently, there is a rather satisfactory and complete description of singular hyperbolic vector fields defined on 3-dimensional manifolds (but the panorama in higher dimensions remains open). More precisely, it is proved in a sequel of works that a singular hyperbolic set for flow is K^*.expansive, the periodic orbits are dense in its limit set, and it has a spectral decomposition (see [PP], [K]).

On the other hand, for the case of flows, appears a new kind of bifurcation that leads to a new dynamics distinct from the ones for diffeomorphism: the so called *singular cycles* (cycles involving singularities and periodic orbits, see [BLMP], [Mo], [MP] and [MPP1] for examples of dynamics in the sequel of the unfolding of it). Systems exhibiting this cycles are dense among open set of systems exhibiting a singular hyperbolic splitting. Moreover, recently A. Arroyo and F. Rodriguez Hertz (see [AR]) studying the dynamical consequences of the dominated splitting for the Linear Poincare flow, proved that *any three dimensional flow can be $C^1 - approximated$ either by a flow exhibiting tangency or singular cycle, or by a hyperbolic one.*

References

[ABS]   V. Afraimovich, V. Bykov, and L. Shil'nikov. On the appearance and structure of the Lorenz attractor. *Dokl. Acad. Sci. USSR*, **234**, 336–339, 1977.

[AR]   A. Arroyo, F. Rodriguez Hertz, Homoclinic Bifurcations and Uniform Hyperbolicity for three-dimensional flows, preprint.

[AS]   R. Abraham, S. Smale, Nongenericity of Axiom A and Ω-stability, *Global analysis-Proc. Symp.in Pure Math.*, Vol 14, AMS, Providence, R.I., (1970), 5–8.

[Bo]   J. Bochi, *Genericity of zero Lyapunov exponents*, preprint.

[BC]   M. Benedicks, L. Carleson, The dynamics of the Hénon map, *Annals of Math*, **133** (1991), 73–169.

[BD]   C. Bonatti, L. J. Diaz, Persistence of transitive diffeomorphisms. *Annals of Math*, **143** (1995), 367–396.

[BDP]   C. Bonatti, L. J. Diaz, E. R. Pujals, A C^1−generic dichotomy for diffeomorphisms: weak form of hyperbolicity or infinitely many sinks or sources *Annals of Math*, to appear.

[BLMP]   R. Bamon, R. Labarca, R. Mañ é, M. Pacifico, The explosion of singular cycles, *Publ. Math. IHES*, **78**, 207–232, 1993.

[BV]   C. Bonatti, M. Viana, SRB measures for partially hyperbolic systems whose central direction is mostly contracting. *Israel J. Math.*, **115** (2000), 157–193.

[BoV]   J. Bochi, M. Viana Uniform (projective) hyperbolicity or no hyperbolicity: a dichotomy for generic conservative maps, em Ann. Inst. Henri Poincare (An. nonlin).

[BPSW]   K. Burns, C. Pugh, M. Shub, A. Wilkinson. Recent result about stable ergodicity, *Smooth ergodic theory and its applications* **69** of *Proc. Symp. Pure Math.*, 327–366, 2001.

[C]   E. Colli, Infinitely many coexisting strange attractors, to appear in *Annales de l'Inst. Henri Poincaré, Analyse Nonlinéaire*.

[D1]   L. J. Diaz Robust nonhyperbolic dynamics at heterodimensional cycles. *Ergodic Theory and Dynamical Systems*, **15** (1995), 291–315.

[D2]   L.J.Diaz Persistence of cycles and nonhyperbolic dynamics at heteroclinic bifurcations. *Nonlinearity*, **8** (1995), 693–715.

[DPU]   L. J. Diaz, E. R. Pujals, R. Ures, Partial hyperbolicity and robust transitivity, Acta Mathematica, **183** (1999), 1–43.

[G]   C. Gutierrez, A counter-example to a C^2 closing lemma. *Ergodic Theory Dynam. Systems* **7** (1987), no. 4, 509–530.

[Gu]   J. Guckenheimer. A strange, strange attractor. In J.E. Marsden and M.McCracken, editors, *The Hopf bifurcation theorem and its applications*, 368–381, Springer Verlag, 1976.

[HPS]   M. Hirsch, C. Pugh, M. Shub, Invariant manifolds, *Springer Lecture Notes in Math.*, **583** (1977).

[K]   V. Kaloshin, An extension of the Artin-Mazur theorem. *Ann. of Math.* 2, **150** , 729–741, 1999.

[KP]   V. Kaloshin, E. R. Pujals Superexponential growths of periodic orbits in heteroclinic cyles, *in preparation.*

[Ko]   M. Komuro, Expansive properties of Lorenz attractors, *The theory of dynamical systems and its applications to nonlinear problems (Kyoto, 1984)*, 4–26, World Sci. Publishing, Singapore, 1984.

[Lo]   E. N. Lorenz. Deterministic nonperiodic flow. *J. Atmosph. Sci.*, **20** 130–141, 1963.

[LP]   R. Labarca, M.J. Pacifico. Stability of singular horseshoes *Topology*, **25** 337–352, 1986.

[LW]   Lan Wen, Dominated splitting and tangencies, preprint.

[M]   R. Mañé, Contributions to the stability conjecture. *Topology*, **17** (1978), 386–396.

[M1]   R. Mañé, Hyperbolicity, sinks and measure in one-dymensional dynamics. *Comm. Math. Phys.*, **100**, 495–524, 1985.

[Mo]   C. Morales, Lorenz attractors through Saddle-node bifurcations *Ann. Inst. henri Poincare (An. nonlin)*, **13** 589–617, 1996.

[MP]   C. Morales, E. R. Pujals, Singular strange attractors acroos the boundary of hyperboilic systemes, *Comm. Math. Phys.*, **211**, 527–558, 1997.

[MPP]   C. Morales, M. J. Pacifico, E. R. Pujals, Robust transitive singular sets for 3-flows are partially hyperbolic attractors or repellers, preprint.

[MPP1]   C. Morales, M. J. Pacifico, E. R. Pujals, Strange attractors across the boundary of hyperbolic systems *Comm. Math. Phys.*, **211**, 527–558, 2000.

[MV]   L. Mora, M. Viana, Abundance of strange attractors, *Acta Math.*, **171** (1993), 1–71.

[N1]   S. Newhouse, Non-density of Axiom A(a) on S^2, *Proc. A.M.S. Symp. Pure Math.*, **14** (1970), 191–202.

[N2]   S. Newhouse, Diffeomorphism with infinitely many sinks, *Topology*, **13** (1974), 9–18.

[N3]   S. Newhouse, The abundance of wild hyperbolic sets and nonsmooth stable sets for diffeomorphisms, *Publ. Math. I.H.E.S.*, **50** (1979), 101–151.

[N4]   S. Newhouse, Hyperbolic limit sets, *Trans. A.M.S.*, **167** (1972), 125–150.

[P1]   J. Palis, A global view of dynamics and a conjecture on the denseness of finitude of attractors, to appear in Astérisque.

[PP]   M. J. Pacifico, E. R. Pujals, Dynamical consequences of the singular hyperbolicity, preprint.

[PS1]   E. R. Pujals, M. Sambarino, Homoclinic tangencies and hyperbolicity for surface diffeomorphisms, *Annals of math*, **151** (2000), 961–1023.

[PS2]   E. R. Pujals, M. Sambarino, On homoclinic tangencies, hyperbolicity, creation of homoclinic orbits and variation of entropy, *Nonlinearity*, **13** (2000), 921–926.

[PS3]   E. R. Pujals, M. Sambarino, On the dynamic of dominated splitting, preprint.

[PS4]   E. R. Pujals, M. Sambarino, Codimension one dominated splittings, preprint.

[PS5]   E. R. Pujals, M. Sambarino, Sufficient condition for robust transitivity, preprint.

[PT]   J.Palis and F.Takens *Hyperbolicity and sensitive-chaotic dynamics at homoclinic bifurcations* Cambridge University Press, 1993.

[R]   F. Rodriguez Hertz Stable ergodixity of certain linear automorphisms of the torus, preprint.

[S]   S. Smale, Structurally stable systems are not dense, *Amer. J. Math*, **88** (1966), 491–496.

[Sh]   M. Shub Topologically transitive diffeomorphism of T^4, in *Symposium on Differential Equations and Dynamical Systems (University of Warwick, 1968/69*, 39–40. Lecture Notes in Math., 206. Springer-Verlag, Berlin-NewYork, 1971.

[Tu]   W. Tucker. The Lorenz attractor exists. *C. R. Acad. Sci. Paris*, **328** 1197–1202, 1999.

Applications of Orbit Equivalence to Actions of Discrete Amenable Groups

Daniel J. Rudolph*

Abstract

Since the work of Ornstein and Weiss in 1987 (**Entropy and isomorphism theorems for actions of amenable groups**, *J. Analyse Math.*, **48** (1987)) it has been understood that the natural category for classical ergodic theory would be probability measure preserving actions of discrete amenable groups. A conclusion of this work is that all such actions on nonatomic Lebesgue probability spaces were orbit equivalent. From this foundation two broad developements have been built. First, a full generalization of the various equivalence theories, including Ornstein's isomorphism theorem itself, exists. Fixing the amenable group G and an action of it, one can define a metric-like notion on the full-group of the action, called a size. A size breaks the orbit equivalence class of a single action into subsets, those reachable by a Cauchy sequence (in the size) of full group perturbations. These subsets are the equivalence classes associated with the size. Each size possesses a distinguised "most random" set of classes, the "Bernoulli" classes of the relation. An Ornstein-type theorem can be obtained. Many naturally occuring equivalence relations can be described in this way. Perhaps most interesting, entropy itself can be so described. Second, one can use the characterization of discrete amenable actions as those which are orbit equivalent to a action of $\mathbb{Z}$ to lift theorems from actions of $\mathbb{Z}$ to those of arbitrary amenable groups. The most interesting of these are first, that actions of completely positive entropy (called K-systems for $\mathbb{Z}$ actions) are mixing of all orders (proven jointly with B. Weiss) and that such actions have countable Haar spectrum (proven by Golodets and Dooley). As all ergodic actions are orbit equivalent, only ergodicity is preserved by orbit equivalences in general, but by considering orbit equivalences restricted to be measurable with respect to a sub-σ algebra, many properties relative to that algebra are preserved. This provides the tool for this method to succeed.

2000 Mathematics Subject Classification: 28D15, 37A35.
Keywords and Phrases: Amenable group, Orbit equivalence, Entropy.

---

*Department of Mathematics, University of Maryland, College Park, MD 20742, USA. E-mail: djr@math.umd.edu

1. Definitions and examples of sizes

Our goal in this section is to describe a metric-like notion on the full group of a measure preserving action of an amenable group and show how this leads to various restricted orbit equivalence theories. This work can be found in complete detail in **Restricted Orbit Equivalence for Actions of Discrete Amenable Groups** by D.J. Rudolph and J. Kammeyer, *Cambridge Tracts in Mathematics #146*.

Let $(X, \mathcal{F}, \mu)$ be a fixed nonatomic Lebesgue probability space. Let G be an infinite discrete amenable group. Let $\mathcal{O} \subseteq X \times X$ be an ergodic, measure preserving, hyperfinite equivalence relation. For our purposes, this simply means that $\mathcal{O} = \{(x, T_g(x))\}_{g \in G}$ where $T : G \times X \to G$ (written of course $T_g(x)$) is some ergodic and free, measure preserving action of G on X.

Definition 1.1 *Let G be an infinite countable discrete amenable group. A **G-arrangement** α is any map from $\mathcal{O}$ to G that satisfies:*

  (i) *α is 1-1 and onto, in that for a.e. $x \in X$, for all $g \in G$, there is a unique $x' \in X$ with $\alpha(x, x') = g$. We write $x' = T_g^\alpha(x)$;*
 (ii) *α is measurable and measure preserving, i.e. for all $A \in \mathcal{F}, g \in G$, both $T_g^\alpha(A) \in \mathcal{F}$ and $\mu(T_g^\alpha(A)) = \mu(A)$; and*
(iii) *α satisfies the cocycle equation $\alpha(x_2, x_3)\alpha(x_1, x_2) = \alpha(x_1, x_3)$.*

*As G will not vary for our considerations we will abbreviate this as an **arrangement**. Let $\mathcal{A}$ denote the set of all such arrangements.*

Lemma 1.2 *α is a G-arrangement if and only if there is a measure preserving ergodic free action of G, T, whose orbit relation is $\mathcal{O}$ such that $\alpha(x, T_g(x)) = g$ for all $(x, T_g(x)) \in \mathcal{O}$.*

Thus the vocabulary of G-arrangements on $\mathcal{O}$ is precisely equivalent to the vocabulary of G-actions whose orbits are $\mathcal{O}$. For a G-arrangement α, we write T^α for the corresponding action. For a G-action T, we write α_T for the corresponding G-arrangement.

Definition 1.3 *The **full group** of $\mathcal{O}$ is the group (under composition) Γ of all measure preserving invertible maps $\phi : X \to X$ such that for μ-a.e. $x \in X$, $(x, \phi(x)) \in \mathcal{O}$.*

Definition 1.4 *A **G-rearrangement** of $\mathcal{O}$ is a pair (α, ϕ), where α is a G-arrangement of $\mathcal{O}$ and $\phi \in \Gamma$. As G is fixed for our purposes we will abbreviate this as a **rearrangement**. Let $\mathcal{Q}$ denote the set of all such rearrangements.*

Intuitively, a rearrangement is simply a change (i.e. rearrangement) of an orbit from the arrangement α to the arrangement $\alpha\phi$, where $\alpha\phi(x, x') = \alpha(\phi(x), \phi(x'))$. One can formalize such a rearrangement in three different ways. Set $\mathcal{B}$ to be the set of bijections of G and $\mathcal{B}$ the subgroup of $\mathcal{G}$ fixing the identity. Both are topologized via the product topology on G^G. Notice there is a homomorphism $\hat{H} : \mathcal{B} \to \mathcal{G}$ given by $\hat{H}(q)(g) = q(\mathrm{id})^{-1}q(g)$. The kernel of $\hat{H}$ consist of the left translation maps.

To a rearrangment we can associate a family of functions $q_x^{\alpha,\phi} \in \mathcal{B}$ where

$$q_x^{\alpha,\phi}(g) = \alpha(x, \phi(T_g^\alpha(x))).$$

Now suppose α and β are two arrangements of the orbits $\mathcal{O}$. Regard the first as an initial arrangement and the second as a terminal arrangement. We can associate to this pair and any point x a bijection from G fixing the identity that describes how the arrangement of the orbit has changed:

$$h_x^{\alpha,\beta}(g) = \beta(x, T_g^\alpha(x)).$$

Notice here that $\hat{H}(q_x^{\alpha,\phi}) = h_x^{\alpha,\alpha\phi}$.

Write $h^{\alpha,\beta} : X \to \mathcal{G}$.

The third way to view a rearrangement pair has a symbolic dynamic flavor. For each orbit $\mathcal{O}(x) = \{x'; (x,x') \in \mathcal{O}\}$, a rearrangement (α, ϕ) also gives rise to a natural map $G \to G$ (not a bijection though), given by

$$f_x^{\alpha,\phi}(g) = \alpha(T_g^\alpha(x), \phi(T_g^\alpha(x))).$$

Visually, regarding $\mathcal{O}(x)$ laid out by α as a copy of G, ϕ translates the point at position g to position $f_x^{\alpha,\phi}(g)g$.

There is a natural link between the three functions $h^{\alpha,\alpha\phi}$, $q^{\alpha,\phi}$ and $f^{\alpha,\phi}$ as follows. For any map $f : G \to G$ we define

$$Q(f)(g) = f(g)g \text{ and}$$

$$H(f)(g) = f(g)gf(\mathrm{id})^{-1}.$$

It is an easy calculation that

$$H(f^{\alpha,\phi}) = h^{\alpha,\alpha\phi} \text{ and } Q(f^{\alpha,\phi} = q^{\alpha,\phi}.$$

Let $\{F_i\}$ be a fixed Følner sequence for G. We will describe a number of concepts in terms of the F_i.

We now consider three pseudometrics on the set of rearrangements. These all arise from natural topologies on functions $G \to G$. As G is countable the only reasonable topology is the discrete one, using the discrete 0,1 valued metric. This topologizes G^G as a metrizable space with the product topology. This is the weakest topology for which the evaluations $g : f \to f(g)$ are continuous functions. Notice that H is a continuous map from G^G to itself and the map $h \to h^{-1}$ on $\mathcal{G}$ is continuous.

Define a metric d on $\mathcal{G}$ as follows. List the elements of G as $\{g_1 = \mathrm{id}, g_2, \dots\}$ and let d_0 be the 0,1 valued metric on G. Set

$$d(h_1, h_2) = \sum_i [d_0(h_1(g_i), h_2(g_i)) + d_0(h_1^{-1}(g_i), h_2^{-1}(g_i)))]2^{-(i+1)}.$$

Notice that if h_1, h_2, h_1^{-1}, and h_2^{-1} agree on $g_1, \dots, g_i$ then $d(h_1, h_2) \le 2^{-i}$. On the other hand if $d(h_1, h_2) < 2^{-i}$ then h_1, h_2 and their inverses agree on this list of i terms.

Lemma 1.5 *The metric d on $\mathcal{G}$ gives the restricted product topology and makes $\mathcal{G}$ a complete metric space.*

We can use this to define a complete L^1 metric on arrangements:

$$\|\alpha, \beta\|_1 = \int d(h^{\alpha, \beta}, \mathrm{id}) \, d\mu.$$

As $d(h_1, h_2) = d(h_2^{-1} h_1, \mathrm{id})$ and $(h_x^{\alpha, \beta})^{-1} = h_x^{\beta, \alpha}$ we see that this is a metric.

We can also define a metric similar to d on G^G itself making it a complete metric space by just taking half of the terms in d:

$$d_1(f_1, f_2) = \sum_i d_0(f_1(g_i), f_2(g_i)) 2^{-i}.$$

This also leads to an L^1 metric on G^G-valued functions on a measure space:

$$\|f_1, f_2\|_1 = \int d_1(f_1, f_2) \, d\mu.$$

These two L^1 distances now give us two families of L^1 distances on the full-group, one a metric the other a pseudometric, associated with an arrangement α:

$$\|\phi_1, \phi_2\|_{\mathrm{w}}^\alpha = \int d(h^{\alpha, \alpha\phi_1}, h^{\alpha, \alpha\phi_2}) = \|\alpha\phi_1, \alpha\phi_2\|_1$$

and

$$\|\phi_1, \phi_2\|_{\mathrm{s}}^\alpha = \int d_1(f^{\alpha, \phi_1}, f^{\alpha, \phi_2}) \, d\mu = \|f^{\alpha, \phi_1}, f^{\alpha, \phi_2}\|_1.$$

The **weak** L^1 distance, $\|\cdot, \cdot\|_{\mathrm{w}}^\alpha$, is only a pseudometric but the **strong** L^1 distance, $\|\cdot, \cdot\|_{\mathrm{s}}^\alpha$, is a metric.

To describe the weak*-distance between two arrangements let $G^\star = G \cup \{\star\}$ be the one point compactification of G. Now $(G^\star)^G$ is a compact metric space and hence the Borel probability measures on $(G^\star)^G$, which we write as $\mathcal{M}_1(G^\star)$, are a compact and convex space in the weak* topology. Let $D(\mu_1, \mu_2)$ be an explicit metric giving this topology.

We define the distribution pseudometric between two rearrangements by

$$\|(\alpha, \phi), (\beta, \psi)\|_* = D((f^{\alpha, \phi})^*(\mu), (f^{\beta, \psi})^*(\nu)).$$

We can combine the two L^1-metrics on arrangements and the full group to define a product metric on rearrangements in the form

$$\|(\alpha_1, \phi_1), (\alpha_2, \phi_2)\|_1 = \|\alpha_1, \alpha_2\|_1 + \mu(\{x : \phi_1(x) \neq \phi_2(x)\}).$$

We end this Section by relating this complete L^1-metric on rearrangements to the distribution pseudometric.

We now define the notion of a **size** m on rearrangements (α, ϕ) as a family of pseudometrics m_α on the full-group satisfying some simple relations to the metrics and pseudometrics we just defined.

A size is a function

$$m : \mathcal{Q} \to \mathbb{R}^+$$

such that, if we write

$$m_\alpha(\phi_1, \phi_2) \underset{\text{defn}}{=} m(\alpha\phi_1, \phi_1^{-1}\phi_2),$$

then m satisfies the following three axioms.

Axiom 1 *For each $\alpha \in \mathcal{A}$, m_α is a pseudometric on Γ.*

Axiom 2 *For each $\alpha \in \mathcal{A}$, the identity map*

$$(\Gamma, m_\alpha) \overset{\text{id}}{\to} (\Gamma, \|\cdot, \cdot\|_{\text{w}}^\alpha)$$

is uniformly continuous.

In particular this means that if $m_\alpha(\phi_1, \phi_2) = 0$ then the two arrangements $\alpha\phi_1$ and $\alpha\phi_2$ are identical.

Axiom 3 *m is upper semi-continuous with respect to the distribution metric. That is to say, for every $\varepsilon > 0$, there exists $\delta = \delta(\varepsilon, \alpha, \phi)$, such that if $\|(\alpha, \phi), (\beta, \psi)\|_* < \delta$ then $m(\beta, \psi) < m(\alpha, \phi) + \varepsilon$.*

This last condition implies that if the two measures $(f^{\alpha,\phi})^*(\mu)$ and $(f^{\beta,\psi})^*(\nu)$ are the same, then $m(\alpha, \phi) = m(\beta, \psi)$. Hence the value m is well defined on those measures on G^G which arise as such an image, and we can write

$$m(\alpha, \phi) = m((f^{\alpha,\phi})^*(\mu)).$$

We can now define m-equivalence of two arrangements.

Definition 1.6 *We say α and β are m-equivalent arrangements if there exist ϕ_i which are m_α-Cauchy, ϕ_i^{-1} are m_β-Cauchy and $\alpha\phi_i$ converges in probability to β.*

One can now define m-equivalence of actions on distinct spaces as meaning there are conjugate versions of the actions as arrangments on the same orbit space where the arrangements are m-equivalent in the sense of the definition.

We now give a list of examples to indicate the range of equivalence relations that can be brought under this perspective.

Many examples of sizes have the common feature of being integrals of some pointwise calculation of the distortion of a single orbit. To make this precise we first review some material about bijections of G. Remember that $\mathcal{B}$ is the space of all bijections of the group G with the product topology, $\mathcal{G}$ is the space of bijections fixing id and we metrized both with a complete metric d. The group G can be regarded as a subgroup of $\mathcal{B}$ acting by left multiplication, $(g(g') = gg')$. The map $\hat{H} : \mathcal{B} \to \mathcal{G}$ given by $\hat{H}(q) = qq(\text{id})^{-1}$ is a contraction in d. Also G acting by right multiplication conjugates $\mathcal{B}$ to itself giving an action of G on $\mathcal{B}$. $(T_g(q)(g') = q(g'g)g^{-1}.)$ We view this action by representing an element $q \subset \mathcal{B}$ by a map $f : G \to G$, $f(g) = q(g)g^{-1}$.

Those maps $f \in G^G$ that arise from bijections are a G_δ and hence a Polish space we call F. The map $q \to f$ is obviously a homeomorphism from $\mathcal{B}$ to F. For $f \in F$ let $Q(f)$ be the associated bijection and for $q \in \mathcal{B}$ let $F(q)$ be the associated name in G^G. The action of G on $\mathcal{B}$ in its representation as F is the shift action $\sigma_g(f)(g') = f(g'g)$. Any rearrangement pair (α, ϕ) then gives rise to an ergodic shift invariant measure on this Polish subset of G^G and any ergodic shift invariant measure is an ergodic action of G with a canonical rearrangement pair. The probability measures on a Polish space are weak* Polish and hence the invariant and ergodic measures on this Polish space are weak* Polish.

We will now define a general class of sizes that arise as integrals of valuations made on the bijections $q_x^{\alpha, \phi}$.

Definition 1.7 *A Borel $D : \mathcal{B} \to \mathbb{R}^+$ is called a* **size kernel** *if it satisfies:*

1. $D(q) \geq 0$.
2. $D(\mathrm{id}) = 0$.
3. $D(q(\mathrm{id})^{-1} q^{-1} q(\mathrm{id})) = D(q)$.
4. $D(q_1(\mathrm{id}) q_2 q_1^{-1}(\mathrm{id}) q_1) \leq D(q_1) + D(q_2)$.
5. *For every $\varepsilon > 0$ there is a $\delta > 0$ so that if $D(q) < \delta$ then $d(\mathrm{id}, H(q)) < \varepsilon$.*
6. *The function $\mu \to \int D(q(f)) \, d\mu$ is weak* continuous on the space of shift invariant measures μ on the Polish space F.*

Note. An element of G is regarded as an element of $\mathcal{B}$ acts by left multiplication.

For a size kernel D we define

$$ m^D(\alpha, \phi) = \int D(q_x^{\alpha, \phi}) \, d\mu(x). $$

We call such an m^D an **integral size**.

Example 1 (Conjugacy and Orbit Equivalence)

These first two examples are the extremes of what is possible. For one the equivalence class will be simply the full group orbit and for the other it will be the entire set of arrangements. Both of the pseudometrics $d(q, \mathrm{id})$ and $d(H(q), \mathrm{id})$ are easily seen to be size kernels and so both

$$ \begin{aligned} m^1(\alpha, \phi) &= \|(\alpha, \phi), (\alpha, \mathrm{id})\|_\alpha^s \text{ and} \\ m^0(\alpha, \phi) &= \|(\alpha, \phi), (\alpha, \mathrm{id})\|_\alpha^w \end{aligned} $$

are sizes.

As d makes $\mathcal{B}$ complete, relative to m^1 sequence phi_i is m_α^1 Cauchy iff $\phi_i \to \phi$ in probability. Thus $\alpha \overset{m^1}{\sim} \beta$ iff $\beta = \alpha\phi$ i.e. they differ by an element of the full group and the equivalence class of α is exactly its full group orbit. As for m^0, for any α and β one can construct a sequence of ϕ_i with $\alpha\phi_i \to \beta$ in L^1 with the sequence ϕ_i an m_α Cauchy sequence. Thus all arrangements are m^0 equivalent.

Example 2 (Kakutani Equivalence)

For this example let $G = \mathbb{Z}^n$ and $B_N = [-N, N]^n$ be the standard Følner sequence of boxes centered at $\vec{0}$. We begin with a metric on $\mathbb{Z}^n$ given by

$$\tau(\vec{u}, \vec{v}) = \min\left(\||(\vec{u}/\|\vec{u}\|) - (\vec{v}/\|\vec{v}\|)\|| + \left|\ln(\|\vec{v}\|) - \ln(\|\vec{u}\|)\right|, 1\right)$$

(assuming $\vec{0}/\|\vec{0}\| = \vec{0}$). What is important about τ are the following two properties:

1. τ is a metric on $\mathbb{Z}^n$ bounded by 1 and
2. $\vec{u}$ and $\vec{v}$ are τ close iff the norm of their difference is small in proportion to both of their norms.

For $h \in \mathcal{G}$ set $B_N(h) = \{\vec{v} \in B_N | h(\vec{v}) \in B_N\}$ (those elements of B_N mapped into B_N by h). Now set

$$k(h) = \sup_N \left(\frac{1}{\#B_N}\left(\sum_{\vec{v} \in B_N(h)} \tau(\vec{v}, h(\vec{v})) + \#\{\vec{v} \in B_N | h(\vec{v}) \notin B_N\}\right)\right).$$

Now set $K(q) = k(H(q))$.

Lemma 1.8 *The function K is a size kernel.*

For $d = 1$ standard arguments imply that this size yields even Kakutani equivalence. For $d > 1$ it is leads to an analogous equivalence relation among Katok cross-sections of $\mathbb{R}^d$-actions.

Our last example moves beyond size kernels.

Example 3 (Entropy as a Size)

We discuss this example only for actions of $\mathbb{Z}$ although the ideas extend to general countable amenable groups.

The size at its base will simply be the entropy of the rearrangement itself. We make this precise as follows. The function $g_{(\alpha,\phi)}(x) = \alpha(x, \phi(x))$ takes on countably many values and hence can be regarded as a countable partition $g_{(\alpha,\phi)}$ of X. Set Γ_0^α to be those ϕ for which $g_{(\alpha,\phi)}$ is finite. It is not difficult to see that Γ_0^α is a subgroup and moreover $\Gamma_0^{\alpha\psi} = \psi^{-1}\Gamma_0^\alpha\psi$ as $g_{(\alpha\psi,\psi^{-1}\phi\psi)}(\psi^{-1}(x)) = g_{(\alpha,\phi)}(x)$. It can be shown that the Γ_0^α are all m_α^1 dense in Γ. For $\phi \in \Gamma_0^\alpha$ one can use the entropy of the process $h(T^\alpha, g_{(\alpha,\phi)})$ to start the definition of a size defining

$$e(\alpha, \phi) = \inf_{\phi' \in \Gamma_0^\alpha} h(T^\alpha, g_{(\alpha,\phi')}) + \mu\{x|\phi(x) \neq \phi'(x)\}.$$

Now set the size to be

$$m^e(\alpha, \phi) = e(\alpha, \phi) + m^0(\alpha, \phi).$$

Proposition 1.9 *Two $\mathbb{Z}$-actions are m^e equivalent iff they have the same entropy.*

2. Transference via orbit equivalence

A second natural type of restriction can be placed on an orbit equivalence. Here the interest is in two arrangements α and β of perhaps distinct groups. Suppose $\mathcal{A}$ is an invariant sub σ-algebra for the action T^α of G_1 and β is a G_2 arrangement of the same orbit space. We say the orbit equivalence from α to β is $\mathcal{A}$-measurable if the function $h^{\alpha,\beta}(x, g_1) = \beta(x, T^\alpha_{g_1}(x)$ describing or the orbit is rearranged, is $\mathcal{A}$ measurable for all choices of g_1. Up to conjugacy we can regard all ergodic actions of infinite discrete and amenable groups as residing on the same orbit space, so beyond ergodicity no dynamical property will be preserved by orbit equivalence. On the other hand, many dynamical properties have versions "relative to" an invariant sub σ-algebra and many such properties are indeed invariant under orbit equivalences that are measurable with respect to that sub σ-algebra.

This method first arose in work with B. Weiss showing that actions of discrete amenable groups that have completely positive entropy (cpe), commonly called K-systems, are mixing of all orders. This transference method has been applied to a variety of questions. Here is an outline of the argument to this first result to exhibit the format. The complete argument can be found in **Entropy and mixing for amenable group actions** by D.J. Rudolph and B. Weiss, *Annals of Mathematics*, **151**, (2000)m pp. 1119-1150.

Lemma 2.1 *If T is an action of a discrete amenable group G and $T \times B$, its direct product with a Bernoulli action B of G, is relatively cpe with respect to the Bernoulli second coordinate, then T must be cpe.*

Lemma 2.2 *If $\hat{T}$ and $\hat{S}$ are ergodic actions on the same orbits of two discrete amenable groups G_1 and G_2 and the orbit equivalence between them is $\mathcal{A}$ measurable where $\mathcal{A}$ is a $\hat{T}$ invariant sub σ-algebra, then for any partition P, the conditional entropies $h(T, P|\mathcal{A})$ and $h(S, P|\mathcal{A})$ are equal.*

Let S_i be a list of finite subsets of either G_i. We say the S_i **spread** if any particular $\gamma \neq id$ belongs to at most finitely many of the sets $S_i S_i^{-1}$. If the sets $S_i(x)$ are random sequences of finite sets depending on x, we can again say they are spread if for a.e. x they form a spread sequence. A classical characterization of the K-systems, which we state in a relative form says

Theorem 2.3 *T, a $\mathbb{Z}$-action, is relatively cpe with respect to a sub σ-algebra $\mathcal{A}$ iff for all partitions P and all $\mathcal{A}$ measurable and spread random sequences of sets $S_i(x)$,*

$$\frac{1}{\#S_i}\left[h(\bigvee_{g \in S_i} T_g(P)|\mathcal{A}) - \sum_{g \in S_i} h(T_g(P)|\mathcal{A})\right] \underset{i}{\to} 0$$

in L^1.

That is to say, the translates of P become conditionally ever more independent the more spread they become. Refer to this property as $\mathcal{A}$-relative uniform mixing if it holds for all P.

Lemma 2.4 *If T and S are $\mathcal{A}$-measurably orbit equivalent actions of perhaps distinct groups and T is $\mathcal{A}$-relatively uniformly mixing. Then S is also.*

We now describe the orbit transference proof that cpe actions of discrete amenable groups are always mixing of all orders. Suppose T is a cpe action of the group G. Take $T \times B$ where B is a Bernoulli action of G. This direct product will be ergodic and in fact relatively cpe with respect to the Bernoulli coordinate. Now B is orbit equivalent to a $\mathbb{Z}$ action and this orbit equivalence lifts to an orbit equivalence of $T \times B$ to some ergodic $\mathbb{Z}$ action S. The orbit equivalence will be $\mathcal{A}$ measurable where $\mathcal{A}$ is this Bernoulli coordinate algebra. Now S will still be $\mathcal{A}$ relatively cpe and hence $\mathcal{A}$ relatively uniformly mixing. But now this tells us $T \times B$ is also $\mathcal{A}$ relatively uniformly mixing. Restricting this to partitions that are measurable with respect to the first coordinate tells us T itself is uniformly mixing (without any conditioning) and hence mixing of all orders.

A second and quite significant application of this method, due to Dooley and Golodets, is to show that cpe actions have countable Haar spectrum. In as yet unwritten work, again with B. Weiss, one can show that weakly mixing isometric extensions of Bernoulli actions must be Bernoulli.

This remains an area of very active work. We end on an open question. Consider the known result for $\mathbb{Z}$ actions, that a weakly mixing and isometric extension of a base action that is mixing must itself be mixing. Is this result true for general amenable group actions? To apply the transference method one needs a relativized version of the result for $\mathbb{Z}$ actions. That is to say, one needs to know that a relatively weakly mixing relatively isometric extension of a relatively mixing action is still relatively mixing. What seems an obstacle here is simply the definition of relative mixing over a sub σ-algebra $\mathcal{A}$. Certainly it means that for any sets A and B that

$$\left| E(I_A I_B \circ T^j | \mathcal{A}) - E(I_A | \mathcal{A}) E(I_B \circ T^j | \mathcal{A}) \right| \to 0.$$

The question is, in what sense should it tend to zero. Pointwise convergence behaves well for orbit equivalence but the above relativized question seems answerable only for mean convergence.

Bifurcations and Strange Attractors

Leonid Shilnikov*

Abstract

We reviews the theory of strange attractors and their bifurcations. All
known strange attractors may be subdivided into the following three groups:
hyperbolic, pseudo-hyperbolic ones and quasi-attractors. For the first ones
the description of bifurcations that lead to the appearance of Smale-Williams
solenoids and Anosov-type attractors is given. The definition and the description of the attractors of the second class are introduced in the general case.
It is pointed out that the main feature of the attractors of this class is that
they contain no stable orbits. An etanol example of such pseudo-hyperbolic
attractors is the Lorenz one. We give the conditions of their existence. In
addition we present a new type of the spiral attractor that requires countably
many topological invariants for the complete description of its structure. The
common property of quasi-attractors and pseudo-hyperbolic ones is that both
admit homoclinic tangencies of the trajectories. The difference between them
is due to quasi-attractors may also contain a countable subset of stable periodic orbits. The quasi-attractors are the most frequently observed limit sets
in the problems of nonlinear dynamics. However, one has to be aware that the
complete qualitative analysis of dynamical models with homoclinic tangencies
cannot be accomplished.

2000 Mathematics Subject Classification: 37C29, 37C70, 37C15, 37D45.

1. Introduction

One of the starling discoveries in the XX century was the discovery of dynamical chaos. The finding added a fascinatingly novel type of motions — chaotic
oscillations to the catalogue of the accustomed, in nonlinear dynamics, ones such
as steady states, self-oscillations and modulations. Since then many problems of
contemporary exact and engineering sciences that are modelled within the framework of differential equations have obtained the adequate mathematical description.

---

*Institute for Applied Mathematics, Nizhny Novgorod, Russia.
E-mail:shilnikov@focus.nnov.ru

Meanwhile this also set up a question: the trajectory of what kind are to be connected to dynamical chaos in systems with 3D and higher dimensions of the phase space.

The general classification of the orbits in dynamical systems is due to Poincaré and Birkhoff. As far as chaos is concerned in particular a researcher is interested most of all in non-wandering trajectories which are, furthermore, to be self-liming. In the hierarchy of all orbits of dynamical system the high goes to the set of the unclosed trajectories so-called stable in *Poisson's sense*. The feature of them is that the sequence composed of the differences of the subsequent Poincaré recurrence times of a trajectory (as it gets back into the vicinity of the initial point) may have no upper bound. In the case where the return times of the trajectory are bounded Birkhoff suggested to call such it a *recurrent* one. A partial subclass of recurrent trajectories consists of almost-periodic ones which, by definition, possess a set of almost-periods. Since the spectrum of the set is discrete it follows that the dynamics of a system with the almost-periodic trajectories must be simple. Moreover, the almost-periodic trajectories are also quasi-periodic, the latter ones are associated with the regime of modulation. The closure of a recurrent trajectory is called a *minimal set*, and that of a Poisson stable trajectory is called a *quasi-minimal* one. The last one contains also the continuum of Poisson stable trajectories which are dense in it. The quasi-minimal set may, in addition, contain some closed invariant subsets such as equilibria, periodic orbits, ergodic invariant tori with quasi-periodic covering and so on. All this is the reason why Poincaré recurrent times of a Poisson stable trajectory are unbounded: it may linger in the neighborhoods of the above subsets for rather long times before it passes by the initial point. In virtue of such unpredictable behavior it seems quite reasonable the the the role of dynamical chaos orbits should be assigned to the Poisson stable trajectories.

Andronov was the first who had raised the question about the correspondence between the classes of trajectories of dynamical systems and the types of observable motions in nonlinear dynamics in his work "Mathematical problems of auto-oscillations". Because he was motivated to explain the nature of self-oscillations he repudiated the Poisson stable trajectory forthwith due to their irregular behavior. He expressed the hypothesis that a recurrent trajectory stable in Lyapunov sense would be almost-periodic; he also proposed to Markov to confirm it. Markov proved a stronger result indeed; namely that a trajectory stable both in Poisson and Lyapunov (uniformly) senses would be an almost-periodic one. This means that the Poisson stable trajectories must be unstable in Lyapunov sense to get associated to dynamical chaos. After that it becomes clear that all trajectories in a quasi-minimal set are to be of the saddle type. The importance of such quasi-minimal sets for nonlinear dynamics as the real objects was inferred in the explicit way by Lorenz in 1963. He presented the set of equations, known today as the Lorenz model, that possessed an attracting quasi-minimal set with the unstable trajectory behavior. Later on, such sets got named *strange attractors* after Ruelle and Takens.

Here we arrive at the following problem: how can one establish the existence of the Poisson stable trajectories in the phase space of a system? Furthermore, the applicability of the system as a nonlinear dynamics model requires that such trajectories shall persist under small smooth perturbations. Undoubtedly the second problem is a way complex than that of finding periodic orbits. Below we will yield a list of conditions that guarantee the existence of unstable trajectories stable in Poisson sense.

The most universal criterion of the existence of Poisson stable trajectories is the presence, in a system, of a hyperbolic saddle periodic orbit whose stable and unstable manifolds cross transversally along the homoclinic orbit. This structure implies that the set N of all of the orbits remaining in its small neighborhood consists of only unstable ones. Moreover, the periodic orbits are dense in N, so are the trajectories homoclinic to them, besides the continuum of unclosed trajectories stable in Poisson sense. Generally speaking wherever one can describe the behavior of the trajectories in terms of symbolic dynamics (using either the Bernoulli sub-shift or the Markov topological chains), the proof of the existence of the Poisson stable trajectories becomes trivial. However, the selected hyperbolic sets by themselves are unstable. Nevertheless, their presence in the phase space means the complexity of the trajectory behavior even though they are no part of the strange attractor.

Early sixtieth were characterized the rapid development of the theory of structural stability initiated in the works of Anosov and Smale. Anosov was able to single out the class of the systems for which the hyperbolicity conditions hold in the whole phase space. Such flows and cascade have been named the Anosov systems. Some examples of the Anosov systems include geodesic flows on compact smooth manifolds of a negative curvature [5]. It is well-known that such flow is conservative and its set of non-wandering trajectories coincides with the phase space. Another example of the Anosov diffeomorphism is a mapping of an n-dimensional torus

$$\bar{\theta} = A\theta + f(\theta), \qquad \text{mod } 1, \tag{1}$$

where A is a matrix with integer entries other than 1, $\det |A| = 1$, the eigenvalues of A do not lie on the unit circle, and $f(\theta)$ is a periodic function of period 1.

The condition of hyperbolicity of (1) may be easily verified for one pure class of diffeomorphisms of the kind:

$$\bar{\theta} = A\theta, \qquad \text{mod } 1, \tag{2}$$

which are the algebraic hyperbolic automorphisms of the torus. Automorphism (2) are conservative systems whose set Ω of non-wandering trajectories coincides with the torus $\mathbb{T}^n$ itself.

Conditions of structural stability of high-dimensional systems was formulated by Smale [26]. These conditions are in the following: A system must satisfy both *Axiom A* and the strong transversality condition.

Axiom A requires that:

1A the non-wandering set Ω be hyperbolic;
1B $\Omega = \overline{\text{Per}}$. Here Per denotes the set of periodic points.

Under the assumption of Axiom A the set Ω can be represented by a finite union of non-intersecting, closed, invariant, transitive sets $\Omega_1, \ldots \Omega_p$. In the case of cascades, any such Ω_i can be represented by a finite number of sets having these properties

which are mapped to each other under the action of the diffeomorphism. The sets
$\Omega_1, \dots \Omega_p$ are called *basis sets*.

The basis sets of Smale systems (satisfying the enumerated conditions) may
be of the following three types: attractors, repellers and saddles. Repellers are the
basis sets which becomes attractors in backward time. Saddle basis sets are such
that may both attract and repel outside trajectories. A most studied saddle basis
sets are one-dimensional in the case of flows and null-dimensional in the case of
cascades. The former ones are homeomorphic to the suspension over topological
Markov chains; the latter ones are homeomorphic to simple topological Markov
chains [Bowen [8]].

Attractors of Smale systems are called *hyperbolic*. The trajectories passing
sufficiently close to an attractor of a Smale system, satisfies the condition

$$\mathrm{dist}(\varphi(t,x), A) < ke^{-\lambda t}, \qquad t \geq 0$$

where k and λ are some positive constants. As we have said earlier these attractors
are transitive. Periodic, homo- and heteroclinic trajectories as well as Poisson-
stable ones are everywhere dense in them. In particular, we can tell one more of
their peculiarity: the unstable manifolds of all points of such an attractor lie within
it, i.e., $W_x^s \in A$ where $x \in A$. Hyperbolic attractors may be smooth or non-smooth
manifolds, have a fractal structure, not locally homeomorphic to a direct product
of a disk and a Cantor set and so on.

Below we will discuss a few hyperbolic attractors which might be curious for
nonlinear dynamics. The first example of such a hyperbolic attractor may be the
Anosov torus $\mathbb{T}^n$ with a hyperbolic structure on it. The next example of hyperbolic attractor was constructed by Smale on a two-dimensional torus by means of
a "surgery" operation over the automorphism of this torus with a hyperbolic structure. This is the so-called DA-(*derived from Anosov*) diffeomorphism. Note that the
construction of such attractors is designed as that of minimal sets known from the
Poincaré-Donjoy theory in the case of $\mathbb{C}^1$-smooth vector fields on a two-dimensional
torus.

Let us consider a solid torus $\Pi \in \mathbb{R}^n$, i.e., $\mathbb{T}^2 = \mathbb{D}^2 \times \mathbb{S}^1$ where $\mathbb{D}^2$ is a disk
and $\mathbb{S}^1$ is a circumference. We now expand $\mathbb{T}^2$ m-times (m is an integer) along
the cyclic coordinate on $\mathbb{S}^1$ and shrink it q-times along the diameter of $\mathbb{D}^2$ where
$q \leq 1/m$. We then embed this deformed torus Π_1 into the original one so that its
intersection with $\mathbb{D}^2$ consists of m-smaller disks. Repeat this routine with Π_1 and
so on. The set $\Sigma = \in_{i=1}^{\infty} \Pi_i$ so obtained is called a Witorius-Van Danzig solenoid.
Its local structure may be represented as the direct product of an interval and a
Cantor set. Smale also observed that Witorius-Van Danzig solenoids may have a
hyperbolic structures, i.e., be hyperbolic attractors of diffeomorphisms on solid tori.
Moreover, similar attractors can be realized as a limit of the inverse spectrum of
the expanding cycle map [37]

$$\theta = m\theta, \qquad \mathrm{mod}\ 1.$$

The peculiarity of such solenoids is that they are expanding solenoids. Generally speaking, an expanding solenoid is called a hyperbolic attractor such that its

dimension coincides with the dimension of the unstable manifolds of the points of the attractor. Expanding solenoids were studied by Williams [38] who showed that they are generalized (extended) solenoids. The construction of generalized solenoids is similar to that of minimal sets of limit-quasi-periodic trajectories. Note that in the theory of sets of limit-quasi-periodic functions the Wictorius-Van Danzig solenoids are quasi-minimal sets. Hyperbolic solenoids are called the Smale-Williams solenoids.

We remark also on an example of a hyperbolic attractor of a diffeomorphism on a two-dimensional sphere, and, consequently, on the plane, which was built by Plykin [23]. In fact, this is a diffeomorphism of a two-dimensional torus projected onto a two-dimensional sphere. Such a diffeomorphism, in the simplest case, possesses four fixed points, moreover all of them are repelling.

2. Birth of hyperbolic attractors

Let us now pause to discuss the principal aspects related to the transition from Morse-Smale systems to systems with hyperbolic attractors. The key moment here is a global bifurcation of disappearance of a periodic trajectory. Let us discuss this bifurcation in detail following the paper of Shilnikov and Turaev [34].

Consider a $\mathbb{C}^r$-smooth one-parameter family of dynamical systems X_μ in $\mathbb{R}^n$. Suppose that the flow has a periodic orbit L_0 of the saddle-node type at $\mu = 0$. Choose a neighbourhood U_0 of L_0 which is a solid torus partitioned by the $(n-1)$-dimensional strongly stable manifold $W_{L_0}^{ss}$ into two regions: the node region U^+ where all trajectories tend to L_0 as $t \to +\infty$, and the saddle region U^- where the two-dimensional unstable manifold $W_{L_0}^u$ bounded by L_0 lies. Suppose that all of the trajectories of $W_{L_0}^u$ return to L_0 from the node region U^+ as $t \to +\infty$ and do not lie in W^{ss}. Moreover, since any trajectory of W^u is bi-asymptotic to L_0, $\bar{W}_{L_0}^u$ is compact.

Observe that systems close to X_0 and having a simple saddle-node periodic trajectories close L_0 form a surface B of codimension-1 in the space of dynamical systems. We assume also that the family X_μ is transverse to B. Thus, when $\mu < 0$, the orbit L_0 is split into two periodic orbits, namely: L_μ^- of the saddle type and stable L_μ^+. When $\mu > 0$ L_0 disappears.

It is clear that X_μ is a Morse-Smale system in a small neighbourhood U of the set W^u for all small $\mu < 0$. The non-wandering set here consists of the two periodic orbits L_μ^+ and L_μ^-. All trajectories of $U \backslash W_{L_\mu^-}^s$ tend to L_μ^+ as $t \to +\infty$. At $\mu = 0$ all trajectories on U tend to L_0. The situation is more complex when $\mu > 0$.

The Poincaré map to which the problem under consideration is reduced, may be written in the form

$$
\begin{aligned}
\bar{x} &= f(x, \theta, \mu), \\
\bar{\theta} &= m\theta + g(\theta) + \omega + h(x, \theta, \mu), \qquad \mod 1,
\end{aligned}
\tag{3}
$$

where f, g and h are periodic functions of θ. Moreover, $\|f\|_{\mathbb{C}^1} \to 0$ and $\|h\|_{\mathbb{C}^1} \to 0$ as $\mu \to 0$, m is an integer and ω is a parameter defined in the set $[0, 1)$. Diffeomorphism (3) is defined in a solid-torus $\mathbb{D}^{n-2} \times \mathbb{S}^1$, where $\mathbb{D}^{n-2}$ is a disk $\|x\| < r$, $r > 0$

Observe that (3) is a strong contraction along x. Therefore, mapping (3) is close to the degenerate map

$$\bar{x} = 0,$$
$$\bar{\theta} = m\theta + g(\theta) + \omega, \qquad \text{mod } 1. \qquad (4)$$

This implies that its dynamics is determined by the circle map

$$\bar{\theta} = m\theta + g(\theta) + \omega, \qquad \text{mod } 1, \qquad (5)$$

where $0 \leq \omega < 1$. Note that in the case of the flow in $\mathbb{R}^3$, the integer m may assume the values 0, 1.

Theorem 2.1 *If $m = 0$ and if*

$$\max \|g'(\theta)\| < 1,$$

then for sufficiently small $\mu > 0$, the original flow has a periodic orbit both of which length and period tend to infinity as $\mu \to 0$.

This is the "blue sky catastrophe". In the case where $m = 1$, the closure $\bar{W}^u_{L_0}$ is a two-dimensional torus. Moreover, it is smooth provided that (3) is a diffeomorphism. In the case where $m = -1$ $\bar{W}^u_{L_0}$ is a Klein bottle, also smooth if (3) is a diffeomorphism. In the case of the last theorem $\bar{W}^u_{L_0}$ is not a manifold.

In the case of $\mathbb{R}^n$ ($n \geq 4$) the constant m may be any integer.

Theorem 2.2 *Let $|m| \geq 2$ and let $|m + g'(\theta) > 1$. Then for all $\mu > 0$ suffi-ciently small, the Poincaré map (3) has a hyperbolic attractor homeomorphic to the Smale-Williams solenoid, while the original family has a hyperbolic attractor homeomorphic to a suspension over the Smale-Williams solenoid.*

The idea of the use of the saddle-node bifurcation to produce hyperbolic at-tractors may be extended onto that of employing the bifurcations of an invariant torus. We are not developing here the theory of such bifurcations but restrict ourself by consideration of a modelling situation.

Consider a one-parameter family of smooth dynamical systems

$$\dot{x} = X(x, \mu)$$

which possesses an invariant m-dimensional torus $\mathbb{T}^m$ with a quasi-periodic trajec-tory at $\mu = 0$. Assume that the vector field may be recast as

$$\dot{y} = C(\mu)y,$$
$$\dot{z} = \mu + z^2, \qquad (6)$$
$$\dot{\theta} = \Omega(\mu)$$

in a neighborhood of $\mathbb{T}^m$. Here, $z \in \mathbb{R}^1$, $y \in \mathbb{R}^{n-m-1}$, $\theta \in \mathbb{T}^m$ and $\Omega(0) = (\Omega_1, \ldots, \Omega_m)$. The matrix $C(\mu)$ is stable, i.g., its eigenvalues lie to the left of the imaginary axis in the complex plane. At $\mu = 0$ the equation of the torus is $y = 0$, the equation of the unstable manifold W^u is $y = 0, z > 0$, and that of the strongly unstable manifold W^{ss} partitioning the neighborhood of $\mathbb{T}^m$ into a node and a saddle region, is $z = 0$. We assume also that all of the trajectories of the unstable manifold W^u of the torus come back to it as $t \to +\infty$. Moreover they do not lie in W^{ss}. On a cross-sections transverse to $z = 0$ the associated Poincaré map may be written in the form

$$\begin{aligned}
\bar{y} &= f(y, \theta, \mu), \\
\bar{\theta} &= A\theta + g(\theta) + \omega + h(x, \theta, \mu), \qquad \text{mod } 1,
\end{aligned} \tag{7}$$

where A is an integer matrix, f, g, and h are 1-periodic functions of θ. Moreover $\|f\|_{\mathbb{C}^1} \to 0$ and $\|h\|_{\mathbb{C}^1} \to 0$ as $\mu \to 0$, $\omega = (\omega_1, \cdots, \omega_m)$ where $0 \le \omega_k < 1$.

Observe that the restriction of the Poincaré map on the invariant torus is close to the shortened map

$$\bar{\theta} = A\theta + g(\theta) + \omega, \qquad \text{mod } 1. \tag{8}$$

This implies, in particular, that if (8) is an Anosov map for all ω (for example when the eigenvalues of the matrix A do not lie on the unit circle of the complex plane, and $g(\theta)$ is small), then the restriction of the Poincaré map is also an Anosov map for all $\mu > 0$. Hence, we arrive at the following statement

Proposition 2.1 *If the shortened map is an Anosov map for all small ω, then for all $\mu > 0$ sufficiently small, the original flow possesses a hyperbolic attractor which is topologically conjugate to the suspension over the Anosov diffeomorphism.*

The birth of hyperbolic attractors may be proven not only in the case where the shortened map is a diffeomorphism. Namely, this result holds true if the shortened map is expanding. A map is called expanding of the length if any tangent vector field grows exponentially under the action of the differential of the map. An example is the algebraic map

$$\bar{\theta} = A\theta, \qquad \text{mod } 1,$$

such that the spectrum of the integer matrix A lies strictly outside the unit circle, and any neighboring map is also expanding. If $\|(G'(\theta))^{-1}\| < 1$, where $G = A + g(\theta)$, it follows then that the shortened map

$$\bar{\theta} == \omega + A\theta + g(\theta), \qquad \text{mod } 1, \tag{9}$$

is an expansion for all $\mu > 0$.

Shub [35] established that expanding maps are structurally stable. The study of expanding maps and their connection to smooth diffeomorphisms was continued by Williams [Williams [36]]. Using the result of his work we come to the following result which is analogous to our theorem 2.1, namely

Proposition 2.2 *If $\|(G'(\theta))^{-1}\| < 1$, then for all small $\mu > 0$, the Poincaré map possesses a hyperbolic attractor locally homeomorphic to a direct product of $\mathbb{R}^{m+1}$ and a Cantor set.*

An endomorphism of a torus is called *an Anosov covering* if there exists a continuous decomposition of the tangent space into the direct sum of stable and unstable submanifolds just like in the case of the Anosov map (the difference is that the Anosov covering is not a one-to-one map, therefore, it is not a diffeomorphism). The map (8) is an Anosov covering if, we assume, $|\det A| > 1$ and if $g(\theta)$ is sufficiently small. Thus, the following result is similar to the previous proposition

Proposition 2.3 *If the shortened map (8) is an Anosov covering for all ω, then for all small $\mu > 0$ the original Poincaré map possesses a hyperbolic attractor locally homeomorphic to a direct product of $\mathbb{R}^{m+1}$ and a Cantor set.*

In connection with the above discussion we can ask what other hyperbolic attractors may be generated from Morse-Smale systems?

Of course there are other scenarios of the transition from a Morse-Smale system to a system with complex dynamics, for example, through Ω-explosion, period-doubling cascade, etc. But these bifurcations do not lead explicitly to the appearance of hyperbolic strange attractors.

3. Lorenz attractors

In 1963 Lorenz [18] suggested the model:

$$
\begin{aligned}
\dot{x} &= -\sigma(x - y), \\
\dot{y} &= rx - y - xz, \\
\dot{z} &= -bz + xy
\end{aligned}
\tag{10}
$$

in which he discovered numerically a vividly chaotic behaviour of the trajectories when $\sigma = 10$, $b = 8/3$ and $r = 28$. the important conclusion has been derived from the mathematical studies of the Lorenz model: simple models of nonlinear dynamics may have strange attractors.

Like hyperbolic attractors, periodic as well as homoclinic orbits are everywhere dense in the Lorenz attractor. Unlike hyperbolic attractors the Lorenz attractor is structurally unstable. This is due to the embedding of a saddle equilibrium state with a one-dimensional unstable manifold into the attractor. Nevertheless, under small smooth perturbations stable periodic orbits do not arise. Moreover, it became obvious that such strange attractors may be obtained through a finite number of bifurcations. In particular, in the Lorenz model (due to its specific feature: it has the symmetry group $(x, y, z) \leftrightarrow (-x, -y, z)$) such a route consists of three steps only.

Below we present a few statements concerning the description of the structure of the Lorenz attractor as it was done in [2, 3]. The fact that we are considering only three-dimensional systems is not important, in principle, because the general case where only one characteristic value is positive for the saddle while the others have negative real parts, and the value least with the modulus is real, the result is completely similar to the three-dimensional case. Let B denote the Banach space of $\mathbb{C}^r$-smooth dynamical systems ($r \geq 1$) with the $\mathbb{C}^r$-topology, which are specified on a smooth three-dimensional manifold M. Suppose that in the domain $U \subset B$

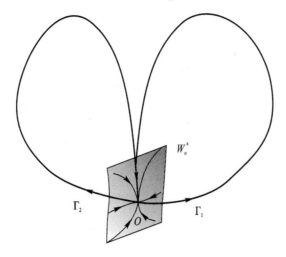

Figure 1: Homoclinic butterfly

each system X has an equilibrium state O of the saddle type. In this case the inequalities $\lambda_1 < \lambda_2 < 0 < \lambda_3$ hold for the roots $\lambda_i = \lambda_i(X)$, $i = 1, 2, 3$ of the characteristic equation at O, and the saddle value $\sigma(X) = \lambda_2 + \lambda_3 > 0$. A stable two-dimensional manifold of the saddle will be denoted by $W^s = W^s(X)$ and the unstable one, consisting of O and two trajectories $\Gamma_{1,2} = \Gamma_{1,2}(X)$ originating from it, by $W^u = W^u(X)$. It is known that both W^s and W^u depend smoothly on X on each compact subset. Here it is assumed that in a certain local map $V = \{(x_1, x_2, x_3)\}$, containing O, X can be written in the form

$$\dot{x}_i = \lambda_i x_i + P_i(x_1, x_2, x_3), \qquad i = 1, 2, 3. \tag{11}$$

Suppose that the following conditions are satisfied for the system $X_0 \subset U$ (see Fig.1):

1. $\Gamma_i(X_0) \subset W^s(X_0)$, $i = 1, 2$, i.e., $\Gamma_i(X_0)$ is doubly asymptotic to O.
2. $\Gamma_1(X_0)$ and $\Gamma_2(X_0)$ approach to O tangentially to each other.

The condition $\lambda_1 < \lambda_2$ implies that the non-leading manifold W_O^{ss} of W_O^s, consisting of O and the two trajectories tangential to the axis x_1 at the point O, divides W_O^s into two open domains: W_+^s and W_-^s. Without loss of generality we may assume that $\Gamma_i(X_0) \subset W_+^s(X_0)$, and hence Γ_i is tangent to the positive semiaxis x_2. Let v_1 and v_2 be sufficiently small neighborhoods of the separatrix "butterfly" $\bar{\Gamma} = \overline{\Gamma_1 \cup O \cup \Gamma_2}$. Let $\mathcal{M}_i$ stand for the connection component of the intersection of $\overline{W_+^s(X_0)}$ with v_i, which contains $\Gamma_i(X_0)$. In the general case $\mathcal{M}_i$ is a two-dimensional $\mathbb{C}^0$-smooth manifold homeomorphic ether to a cylinder or to a Möbius band. The general condition lies in the fact that certain values $A_1(X_0)$ and $A_2(X_0)$, called the separatrix values, should not be equal to zero.

It follows from the above assumptions that X_0 belongs to the bifurcation set B_1^2 of codimension two, and B_1^2 is the intersection of two bifurcation surfaces B_1^1 and B_2^1 each of codimension one, where B_i^1 corresponds to the separatrix loop $\bar{\Gamma}_i = \overline{O \cup \Gamma_i}$. In such a situation it is natural to consider a two-parameter family of dynamical systems $X(\mu)$, $\mu = (\mu_1, \mu_2)$, $|\mu| < \mu_0$, $X(0) = X_0$, such that $X(\mu)$ intersects with B_1^2 only along X_0 and only for $\mu = 0$. It is also convenient to assume that the family $X(\mu)$ is transverse to B_1^2. By transversality we mean that for the system $X(\mu)$ the loop $\Gamma_1(X(\mu))$ "deviates" from $W_+^s(X(\mu))$ by a value of the order of μ_1, and the loop $\Gamma_2(X(\mu))$ "deviates" from $W_+^s(X(\mu))$ by a value of the order of μ_2.

It is known from [30] that the above assumptions imply that in the transition to a system close to X_0 the separatrix loop can generate only one periodic orbit which is of the saddle type. Let us assume, for certainty, that the loop $\Gamma_1(X_0) \cup O$ generates a periodic orbit L_1 for $\mu_1 > 0$ and $\Gamma_2(X_0) \cup O$ generates the periodic orbit L_2 for $\mu_2 > 0$.

The corresponding domain in U, which is the intersection of the stability regions for L_1 and L_2, i.e., i.e., the domain in which the periodic orbits L_1 and L_2 are structurally stable, will be denoted by U_0. A stable manifold of L_i for the system $X \subset U_0$ will be denoted by W_i^s and the unstable one by W_i^u. If the separatrix value $A_i(X_0) > 0$, W_i^u is a cylinder; if $A_i(X_0) < 0$, W_i^u is a Möbius band. Let us note that, in the case where $\mathcal{M}$ is an orientable manifold, W_i^s will also be a cylinder if $A_i(X_0) > 0$. Otherwise it will be a Möbius band. However, in the forthcoming analysis the signs of the separatrix values will play an important role. Therefore, it is natural to distinguish the following three main cases

$$\begin{array}{llll}
\text{Case A (orientable)} & A_1(X_0) > 0, & A_2(X_0) > 0, \\
\text{Case B (semiorientable)} & A_1(X_0) > 0, & A_2(X_0) < 0, \\
\text{Case C (nonorientable)} & A_1(X_0) < 0, & A_2(X_0) < 0.
\end{array}$$

In each of the above three cases the domain U_0 also contains two bifurcation surfaces B_3^1 and B_4^1:

1. In Case A, B_3^1 corresponds to the inclusion $\Gamma_1 \subset W_2^s$ and B_4^1 corresponds to the inclusion $\Gamma_2 \subset W_1^s$;
2. In Case B, B_3^1 corresponds to the inclusion $\Gamma_1 \subset W_1^s$ and B_4^1 corresponds to the inclusion $\Gamma_2 \subset W_1^s$;
3. In Case C, along with the above-mentioned generated orbits L_1 and L_2, there also arises a saddle periodic orbit L_3 which makes one revolution "along" $\Gamma_1(X_0)$ and $\Gamma_2(X_0)$, and if both W_i^u are Möbius bands, $i = 1, 2$, the unstable manifold W_3^u of the periodic orbit L_3 is a cylinder. In this case the inclusions $\Gamma_1 \subset W_2^s$ and $\Gamma_2 \subset W_3^s$ correspond to the surfaces B_3^1 and B_4^1, respectively.

Suppose that B_3^1 and B_4^1 intersect transversely over the bifurcational set B_2^2, see Fig. 2. In a two-parameter family $X(\mu)$ this means that the curves B_3^1 and B_4^1 intersect at some point $\mu^1 = (\mu_{11}, \mu_{12})$. Let us denote a domain lying between B_3^1 and B_4^1 by U_1. Suppose also that for each $X \in U$ there exists a transversal D with the following properties:

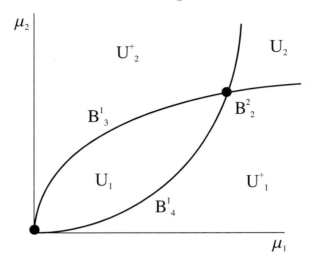

Figure 2: (μ_1, μ_2)-bifurcation diagram

1. The Euclidean coordinates (x, y) can be introduced on D such that

$$D = \Big\{ (x, y) : |x| \leq 1, |y| < 2 \Big\}.$$

2. The equation $y = 0$ describes a connection component S of the intersection $W_O^s \cap D$ such that no ω-semitrajectory that begins on S possesses any point of intersection with D for $t > 0$.

3. The mapping $T_1(X) : D_1 \mapsto D$ and $T_1(X) : D_2 \mapsto D$ are defined along the trajectories of the system X, where

$$D_1 = \Big\{ (x, y) : |x| \leq 1, 0 < y \leq 1 \Big\},$$

$$D_2 = \Big\{ (x, y) : |x| \leq 1, -1 \geq y < 1 \Big\},$$

and $T_i(X)$ is written in the form

$$\begin{aligned} \bar{x} &= f_i(x, y), \\ \bar{y} &= g_i(x, y), \end{aligned} \tag{12}$$

where $f_i, g_i \in \mathbb{C}^r$, $i = 1, 2$.

4. f_i and q_i admit continuous extensions on S, and

$$\lim_{y \to 0} f_i(x, y) = x_i^{**}, \quad \lim_{y \to 0} g_i(x, y) = y_i^{**}, \quad i = 1, 2.$$

5.

$$T_1 D_1 \in Pi_1 = \Big\{ (x, y) : 1/2 \leq x \leq 1, |y| < 2 \Big\},$$

$$T_2 D_2 \in Pi_2 = \Big\{ (x, y) : -1 < x \leq -1/2, |y| < 2 \Big\}.$$

Let $T(X) \equiv T_i(X) \mid D_i$, $(f, g) ==\equiv (f_i, g_i)$ on D_i, $i = 1, 2$.
6. Let us impose the following restrictions on $T(X)$

$$
\left.
\begin{aligned}
&\text{(a)} \quad \|(f_x)\| < 1, \\[1.5ex]
&\text{(b)} \quad 1 - \|(g_y)^{-1}\| \cdot \|f_x\| > 2\sqrt{\|(g_y)^{-1}\| \cdot \|(g_x)\| \cdot \|(g_y)^{-1} \cdot f_y\|}, \\[1.5ex]
&\text{(c)} \quad \|(g_y)^{-1}\| < 1, \\[1.5ex]
&\text{(d)} \quad \|(g_y)^{-1} \cdot f_y\| \cdot \|g_x\| < (1 - \|f_x\|)(1 - \|(g_y)^{-1}\|).
\end{aligned}
\right\} \tag{13}
$$

Hereafter, $\|\cdot\| = \sup\limits_{(x,y)\in D\setminus S} |\cdot|$.

It follows from the analysis of the behavior of trajectories near W_O^s that in a small neighborhood of S the following representation is valid:

$$
\begin{aligned}
f_1 &= x_1^{**} + \varphi_1(x,y)\ y^\alpha, & g_1 &= y_1^{**} + \psi_1(x,y)\ y^\alpha, \\
f_2 &= x_2^{**} + \varphi_2(x,y)(-y)^\alpha, & g_2 &= y_2^{**} + \psi_2(x,y)(-y)^\alpha,
\end{aligned}
\tag{14}
$$

where $\varphi_1, \ldots, \psi_2$ are smooth with respect to x, y for $y \neq 0$, and $T_i(x)$ satisfies estimates (13) for sufficiently small y. Moreover, the limit of φ_1 will be denoted by $A_1(X)$ and that of ψ_2 by $A_2(X)$. The functionals $A_1(X)$ and $A_2(X)$ will be also called the separatrix values in analogy with $A_1(X_0)$ and $A_2(X_0)$ which were introduced above. Let us note that for a system lying in a small neighborhood of the system X all the conditions 1-6 are satisfied near S. Moreover, the concept of orientable, semiorientable and nonorientable cases can be extended to any system $X \in U$. It is convenient to assume, for simplicity, that $A_{1,2}(X)$ do not vanish. It should be also noted that the point P_i with the coordinates (x_i^{**}, y_i^{**}) is the first point of intersection of $\Gamma_i(X)$ with D.

Let us consider the constant

$$
q =
$$

$$
\frac{1 + \|f_x\|\|(g_y)^{-1}\| + \sqrt{1 - \|(g_y)^{-1}\|^2\|(f_x)\| - 4\|(g_y)^{-1}\|\|g_x\|\|(g_y)^{-1}f_y}}{2\|(g_y)^{-1}\|}. \tag{15}
$$

Conditions (15) implies that $q > 1$ and hence all the periodic points will be of the saddle type.

Let Σ denote the closure of the set of points of all the trajectories of the mapping $T(X)$, which are contained entirely in D. Σ is described most simply in the domain U_1. Here the following theorems hold.

Theorem 3.1 *If $X \in U_1$, $T(X) \mid \Sigma$ is topologically conjugated with the Bernoulli scheme (σ, Ω_2) with two symbols.*

Theorem 3.2 *The system $X \in U_2$, has a two-dimensional limiting set Ω, which satisfies the following conditions:*

1. Ω *is structurally unstable.*
2. $[\Gamma_1 \cup \Gamma_2 \cap O] \subset \Omega.$
3. *Structurally stable periodic orbits are everywhere dense in* $\Omega.$
4. *Under perturbations of* X *periodic orbits in* Ω *disappear as a result of matching to the saddle separatrix loops* $\overline{\Gamma_1}$ *and* $\overline{\Gamma_2}.$

Note that in this case the basic periodic orbits will not belong to Ω. In terms of mappings, the properties of Ω can be formulated in more detail. Let us first single out a domain $\tilde{D}$ on D as follows: we assume that in Case A

$$\tilde{D} = \Big\{ (x,y) \in D_1 \cup D_2 \mid y_2(x) < y < y_1(x) \Big\};$$

and in Case B

$$\tilde{D} = \Big\{ (x,y) \in D_1 \cup D_2 \mid y_{12}(x) < y < y_1(x) \Big\};$$

where $y = y_{12}(x)$, $|x| \le 1$, denotes a curve in D whose image lies on the curve $y = y_1(x)$; and finally in Case C

The closure of points of all the trajectories of the mapping $T(X)$, which are entirely contained in $\tilde{D}$, we will denote by $\tilde{\Sigma}$.

Theorem 3.3 *Let* $X \in U_2$. *Then:*

I. $\tilde{\Sigma}$ *is compact, one-dimensional and consists of two connection components in Cases A and C, and of a finite number of connection components in Case B.*

II. $\tilde{D}$ *is foliated by a continuous stable foliation* H^+ *into leaves, satisfying the Lipschitz conditions, along which a point is attracted to* $\tilde{\Sigma}$; *inverse images of the discontinuity line* $S : y = 0$ *(with respect to the mapping* T^k, $k = 1, 2, \ldots$*) are everywhere dense in* $\tilde{D}$.

III. *There exits a sequence of* $T(X)$-*invariant null-dimensional sets* Δ_k, $k \in \mathbb{Z}_+$, *such that* $T(X) \mid \Delta_k$ *is topologically conjugated with a finite topological Markov chain with a nonzero entropy, the condition* $\Delta_k \in \Delta_{k+1}$ *being satisfied, and* $\Delta_k \to \tilde{\Sigma}$ *as* $k \to \infty$.

IV. *The non-wandering set* $\Sigma_1 \in \tilde{\Sigma}$ *is a closure of saddle periodic points of* $T(X)$ *and either* $\Sigma_1 = \tilde{\Sigma}$ *or* $\Sigma_1 = \Sigma^+ \cup \Sigma^-$, *where:*

1. Σ^- *is null-dimensional and is an image of the space* Ω^- *of a certain TMC* (G^-, Ω^-, σ) *under the homeomorphism* $\beta : \Sigma^- \mapsto \Sigma^-$ *which conjugates* $\sigma \mid \Omega$ *and* $T(X) \mid \Sigma^-$;

$$\Sigma^- = \bigcup_{m=1}^{l(X)} \Sigma_m^-, l(X) < \infty,$$

where

$$T(X)\Sigma_m^- = \Sigma_m^-, \qquad \Sigma_{m_1}^- \cap \Sigma_{m_2}^- = \emptyset$$

for $m_1 \ne m_2$ *and* $T(X) \mid \Sigma_m^-$ *is transitive;*

2. Σ^+ *is compact, one-dimensional and*

3. *if* $\Sigma^+ \cap \Sigma^- = \emptyset$, Σ^+ *is an attracting set in a certain neighbourhood;*

4. *if $\Sigma^+ \cap \Sigma^- \neq \emptyset$, then $\Sigma^+ \cap \Sigma^- = \Sigma_m^+ \cap \Sigma_m^-$ for a certain m, and this intersection consists of periodic points of no more than two periodic orbits, and*

(a) *if Σ_m^- is finite, Σ^+ is ω-limiting for all the trajectories in a certain neighbourhood;*

(b) *if Σ_m^- is infinite, Σ^+ is not locally maximal, but is ω-limiting for all the trajectories in $\tilde{D}$, excluding those asymptotic to $\Sigma^- \backslash \Sigma^+$.*

Below we will give the conditions under which the existence of the Lorenz attractor is guaranteed.

Consider a finite-number parameter family of vector field defined by the system of differential equations

$$\dot{x} = X(x, \mu), \tag{16}$$

where $x \in \mathbb{R}^{n+1}$, $\mu \in \mathbb{R}^m$, and $X(x, \mu)$ is a $\mathbb{C}^r$-smooth functions of x and μ. Assume that following two conditions hold

A. System (16) has a equilibrium state $O(0,0)$ of the saddle type. The eigenvalues of the Jacobian at $O(0,0)$ satisfy

$$\mathrm{Re}\lambda_n < \cdots \mathrm{Re}\lambda_2 < \lambda_1 < 0 < \lambda_0.$$

B. The separatrices $\Gamma 1$ and Γ_2 of the saddle $O(0,0)$ returns to the origin as $t \to +\infty$.

Then, for $\mu > 0$ in the parameter space there exists an open set V, whose boundary contains the origin, such that in V system (16) possesses the Lorenz attractor in the following three cases [31]:
Case 1.

A Γ_1 and Γ return to the origin tangentially to each other along the dominant direction corresponding to the eigenvalue λ_1;
B

$$\frac{1}{2} < \gamma < 1, \quad \nu_i > 1, \quad \gamma = -\frac{\lambda_1}{\lambda_0}, \quad \nu_i = -\frac{\mathrm{Re}\lambda_i}{\lambda_0};$$

A The separatrix values A_1 and A_2 (see above) are equal to zero.

In the general case, the dimension of the parameter space is four since we may choose $\mu_{1,2}$, to control the behaviour of the separatrices $\Gamma_{1,2}$ and $\mu_{3,4} = A_{3,4}$. In the case of the Lorenz symmetry, we need two parameters only.
Case 2.

A Γ_1 and Γ belong to the non-leading manifold $W^{ss} \in W^s$ and enter the saddle along the eigen-direction corresponding to the real eigenvector λ_2
B

$$\frac{1}{2} < \gamma < 1, \quad \nu_i > 1, \quad \gamma = -\frac{\lambda_1}{\lambda_0}, \quad \nu_i = -\frac{\mathrm{Re}\lambda_i}{\lambda_0};$$

In the general case, the dimension of the phase space is equal to four. Here, $\mu_{3,4}$ control the distance between the separatrices.
Case 3.

A $\Gamma_{1,2} \notin W^{ss}$;

B $\gamma = 1$;

C $A_{1,2} \neq 0$, and $|A_{1,2}| < 2$.

In this case $m = 3$, $\mu_3 = \gamma - 1$.

In the case where the system is symmetric, all of these bifurcations are of codimension 2.

In A. Shilnikov [27, 28] it was shown that both subclasses (A) and (C) are realized in the Shimizu-Morioka model in which the appearance of the Lorenz attractor and its disappearance through bifurcations of lacunae are explained. Some systems of type (A) were studied by Rychlik [25] and those of type (C) by Robinson [24].

The distinguishing features of strange attractors of the Lorenz type is that they have a complete topological invariant. Geometrically, we can state that two Lorenz-like attractors are topologically equivalent if the unstable manifolds of both saddles behave similarly. The formalization of "similarity" may be given in terms of *kneading invariants* which were introduced by Milnor and Thurston [19] while studying continuous, monotonic mappings on an interval. This approach may be applied to certain discontinuous mappings as well. Since there is a foliation (see above) we may reduce the Poincaré map to the form

$$\begin{aligned} \bar{x} &= F(x,y), \\ \bar{y} &= G(y), \end{aligned} \tag{17}$$

where the right-hand side is, in general, continuous, apart from the discontinuity line $y = 0$, and G is piece-wise monotonic. Therefore, it is natural to reduce (17) to a one-dimensional map

$$\bar{y} = G(y),$$

by using the technique of taking the inverse spectrum, Guckenheimer and Williams [16] showed that a pair of the kneading invariants is a complete topological invariant for the associated two-dimensional maps provided inf $|G'| > 1$.

4. Wild strange attractor

In this section, following the paper by Shilnikov and Turaev [33], we will distinguish a class of dynamical systems with strange attractors of a new type. The peculiarity of such an attractor is that it may contain a wild hyperbolic set. We remark that such an attractor is to be understood as an almost stable, chain-transitive closed set.

Let X be a smooth (C^r, $r \geq 4$) flow in R^n ($n \geq 4$) having an equilibrium state O of *a saddle-focus* type with characteristic exponents $\gamma, -\lambda \pm i\omega, -\alpha_1, \cdots, -\alpha_{n-3}$ where $\gamma > 0$, $0 < \lambda < \text{Re}\,\alpha_j$, $\omega \neq 0$. Suppose

$$\gamma > 2\lambda. \tag{18}$$

This condition was introduced in [21] where it was shown, in particular, that it is necessary in order that no stable periodic orbit could appear when one of the separatrices of O returns to O as $t \to +\infty$ (i.e., when there is *a homoclinic loop*; see also [22]).

Let us introduce coordinates (x, y, z) ($x \in R^1$, $y \in R^2$, $z \in R^{n-3}$) such that the equilibrium state is in the origin, the one-dimensional unstable manifold of O is tangent to the x-axis and the $(n-1)$-dimensional stable manifold is tangent to $\{x = 0\}$. We also suppose that the coordinates $y_{1,2}$ correspond to the leading exponents $\lambda \pm i\omega$ and the coordinates z correspond to the non-leading exponents α.

Suppose that the flow possesses *a cross-section*, say, the surface $\Pi : \{\| y \| = 1, \| z \| \le 1, | x | \le 1\}$. The stable manifold W^s is tangent to $\{x = 0\}$ at O, therefore it is locally given by an equation of the form $x = h^s(y, z)$ where h^s is a smooth function $h^s(0, 0) = 0$, $(h^s)'(0, 0) = 0$. We assume that it can be written in such form at least when $(\| y \| \le 1, \| z \| \le 1)$ and that $| h^s | < 1$ here. Thus, the surface Π is a cross-section for W^s_{loc} and the intersection of W^s_{loc} with Π has the form $\Pi_0 : x = h_0(\varphi, z)$ where φ is the angular coordinate: $y_1 = \| y \| \cos\varphi$, $y_2 = \| y \| \sin\varphi$, and h_0 is a smooth function $-1 < h_0 < 1$. One can make $h_0 \equiv 0$ by a coordinate transformation and we assume that it is done.

We suppose that all the orbits starting on $\Pi \backslash \Pi_0$ return to Π, thereby defining the Poincaré map: $T_+ : \Pi_+ \to \Pi$ $T_- : \Pi_- \to \Pi$, where $\Pi_+ = \Pi \cap \{x > 0\}$ $\Pi_- = \Pi \cap \{x < 0\}$. It is evident that if P is a point on Π with coordinates (x, φ, z), then

$$\lim_{x \to -0} T_-(P) = P^1_-, \; \lim_{x \to +0} T_+(P) = P^1_+,$$

where P^1_- and P^1_+ are the first intersection points of the one-dimensional separatrices of O with Π. We may therefore define the maps T_+ and T_- so that

$$T_-(\Pi_0) = P^1_-, \; T_+(\Pi_0) = P^1_+. \tag{19}$$

Evidently, the region $\mathcal{D}$ filled by the orbits starting on Π (plus the point O and its separatrices) is *an absorbing domain* for the system X in the sense that the orbits starting in $\partial \mathcal{D}$ enter $\mathcal{D}$ and stay there for all positive values of time t. By construction, the region $\mathcal{D}$ is the cylinder $\{\| y \| \le 1, \| z \| \le 1, | x | \le 1\}$ with two glued handles surrounding the separatrices, see Fig.3.

We suppose that the (semi)flow is *pseudohyperbolic* in $\mathcal{D}$. It is convenient for us to give this notion a sense more strong than it is usually done [17]. Namely, we propose the following

Definition. *A semi-flow is called* pseudohyperbolic *if the following two conditions hold:*

A *At each point of the phase space, the tangent space is uniquely decomposed (and this decomposition is invariant with respect to the linearized semi-flow) into a direct sum of two subspaces N_1 and N_2 (continuously depending on the point) such that the maximal Lyapunov exponent in N_1 is strictly less than the minimal Lyapunov exponent in N_2: at each point M, for any vectors $u \in N_1(M)$ and $v \in N_2(M)$*

$$\limsup_{t \to +\infty} \frac{1}{t} \ln \frac{\|u_t\|}{\|u\|} < \liminf_{t \to +\infty} \frac{1}{t} \ln \frac{\|v_t\|}{\|v\|}$$

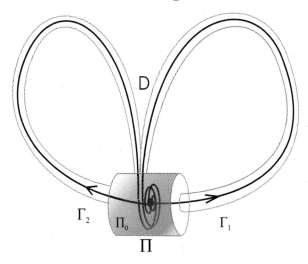

Figure 3: Construction of the wild attractor

where u_t and v_t denote the shift of the vectors u and v by the semi-flow linearized along the orbit of the point M;

B *The linearized flow restricted on N_2 is volume expanding:*

$$V_t \geq const \cdot e^{\sigma t} V_0$$

with some $\sigma > 0$; here, V_0 is the volume of any region in N_2 and V_t is the volume of the shift of this region by the linearized semi-flow.

The additional condition **B** is new here and it prevents of appearance of stable periodic orbits. Generally, our definition includes the case where the maximal Lyapunov exponent in N_1 is non-negative everywhere. In that case, according to condition **A**, the linearized semi-flow is expanding in N_2 and condition **B** is satisfied trivially. In the present paper we consider the opposite case where the linearized semi-flow is *exponentially contracting* in N_1, so condition **B** is essential here.

Note that the property of pseudo-hyperbolicity is stable with respect to small smooth perturbation of the system: according to [17] the invariant decomposition of the tangent space is not destroyed by small perturbations and the spaces N_1 and N_2 depend continuously on the system. Hereat, the property of volume expansion in N_2 is also stable with respect to small perturbations.

Our definition is quite broad; it embraces, in particular, hyperbolic flows for which one may assume $(N_1, N_2) = (N^s, N^u \oplus N_0)$ or $(N_1, N_2) = (N^s \oplus N_0, N^u)$ where N^s and N^u are, respectively, the stable and unstable invariant subspaces and N_0 is a one-dimensional invariant subspace spanned by the phase velocity vector. The geometrical Lorenz model from [2, 3] or [16] belongs also to this class: here N_1 is tangent to the contracting invariant foliation of codimension two and the expansion of areas in a two-dimensional subspace N_2 is provided by the property that the Poincaré map is expanding in a direction transverse to the contracting foliation.

In the present paper we assume that N_1 has codimension three: $dim N_1 = n-3$ and $dim N_2 = 3$ and that the linearized flow (at $t \geq 0$) is exponentially contracting on N_1. Condition **A** means here that if for vectors of N_2 there is a contraction, it has to be weaker than those on N_1. To stress the last statement, we will call N_1 *the strong stable subspace* and N_2 *the center subspace* and will denote them as N^{ss} and N^c respectively.

We also assume that the coordinates (x, y, z) in R^n are such that at each point of $\mathcal{D}$ the space N^{ss} has a non-zero projection onto the coordinate space z, and N^c has a non-zero projection onto the coordinate space (x, y).

Note that our pseudohyperbolicity conditions are satisfied at the point O from the very beginning: the space N^{ss} coincides here with the coordinate space z, and N^c coincides with the space (x, y); it is condition (18) which guarantees the expansion of volumes in the invariant subspace (x, y). The pseudohyperbolicity of the linearized flow is automatically inherited by the orbits in a small neighborhood of O. Actually, we require that this property would extend into the non-small neighborhood $\mathcal{D}$ of O.

According to [17], the exponential contraction in N^{ss} implies the existence of an invariant contracting foliation $\mathcal{N}^{ss}$ with C^r-smooth leaves which are tangent to N^{ss}. As in [3], one can show that the foliation is absolutely continuous. After a factorization along the leaves, the region $\mathcal{D}$ becomes a branched manifold (since $\mathcal{D}$ is bounded and the quotient-semiflow expands volumes it follows evidently that the orbits of the quotient-semiflow must be glued on some surfaces in order to be bounded; cf.[39]).

The property of pseudohyperbolicity is naturally inherited by the Poincaré map $T \equiv (T_+, T_-)$ on the cross-section Π: here, we have:

A* There exists a foliation with smooth leaves of the form $(x, \varphi) = h(z) \mid_{-1 \leq z \leq 1}$, where the derivative $h'(z)$ is uniformly bounded, which possesses the following properties: the foliation is *invariant* in the sense that if l is a leaf, then $T_+^{-1}(l \cap T_+(\Pi_+ \cup \Pi_0))$ and $T_-^{-1}(l \cap T_-(\Pi_- \cup \Pi_0))$ are also leaves of the foliation (if they are not empty sets); the foliation is *absolutely continuous* in the sense that the projection along the leaves from one two-dimensional transversal to another increases or decreases the area in a finite number of times and the coefficients of expansion or contraction of areas are bounded away from zero and infinity; the foliation is *contracting* in the sense that if two points belong to one leaf, then the distance between the iterations of the points with the map T tends to zero exponentially;

B* The quotient maps $\tilde{T}_+$ and $\tilde{T}_-$ are area-expanding.

Statement 3.1 *Let us write the map T as*

$$(\bar{x}, \bar{\varphi}) = g(x, \varphi, z), \quad \bar{z} = f(x, \varphi, z),$$

where f and g are functions smooth at $x \neq 0$ and discontinuous at $x = 0$:

$$\lim_{x \to -0}(g, f) = (x_-, \varphi_-, z_-) \equiv P_-^1, \quad \lim_{x \to +0}(g, f) = (x_+, \varphi_+, z_+) \equiv P_+^1.$$

Let

$$\det \frac{\partial g}{\partial(x, \varphi)} \neq 0. \tag{20}$$

Denote

$$A = \frac{\partial f}{\partial z} - \frac{\partial f}{\partial(x,\varphi)}\left(\frac{\partial g}{\partial(x,\varphi)}\right)^{-1}\frac{\partial g}{\partial z}, \quad B = \frac{\partial f}{\partial(x,\varphi)}\left(\frac{\partial g}{\partial(x,\varphi)}\right)^{-1},$$

$$C = \left(\frac{\partial g}{\partial(x,\varphi)}\right)^{-1}\frac{\partial g}{\partial z}, \qquad\qquad D = \left(\frac{\partial g}{\partial(x,\varphi)}\right)^{-1}.$$

If

$$\lim_{x \to 0} C = 0, \quad \lim_{x \to 0} \| A \| \| D \| = 0, \tag{21}$$

$$\sup_{P \in \Pi \setminus \Pi_0} \sqrt{\| A \| \| D \|} + \sqrt{\sup_{P \in \Pi \setminus \Pi_0} \| B \| \sup_{P \in \Pi \setminus \Pi_0} \| C \|} < 1, \tag{22}$$

then the map has a continuous invariant foliation with smooth leaves of the form $(x, \varphi) = h(z) \mid_{-1 \leq z \leq 1}$ where the derivative $h'(z)$ is uniformly bounded. If, additionally,

$$\sup_{P \in \Pi \setminus \Pi_0} \| A \| + \sqrt{\sup_{P \in \Pi \setminus \Pi_0} \| B \| \sup_{P \in \Pi \setminus \Pi_0} \| C \|} < 1, \tag{23}$$

then the foliation is contracting and if, moreover, for some $\beta > 0$

the functions $A \mid x \mid^{-\beta}$, $D \mid x \mid^{\beta}$, B, C are uniformly bounded and
Hölder continuous,
and $\dfrac{\partial \ln \det D}{\partial z}$ and $\dfrac{\partial \ln \det D}{\partial(x,\varphi)} D \mid x \mid^{\beta}$ are uniformly bounded, $\qquad(24)$

then the foliation is absolutely continuous. The additional condition

$$\sup_{P \in \Pi \setminus \Pi_0} \sqrt{\det D} + \sqrt{\sup_{P \in \Pi \setminus \Pi_0} \| B \| \sup_{P \in \Pi \setminus \Pi_0} \| C \|} < 1 \tag{25}$$

guarantees that the quotient map $\tilde{T}$ expands areas.

It follows from [21, 22] that in the case where the equilibrium state is a saddle-focus, the Poincaré map near $\Pi_0 = \Pi \cap W^s$ is written in the following form under some appropriate choice of the coordinates.

$$(\bar{x}, \bar{\varphi}) = Q_\pm(Y, Z), \quad \bar{z} - R_\pm(Y, Z). \tag{26}$$

Here

$$Y = \mid x \mid^\rho \left(\begin{array}{cc} \cos(\Omega \ln \mid x \mid + \varphi) & \sin(\Omega \ln \mid x \mid + \varphi) \\ -\sin(\Omega \ln \mid x \mid + \varphi) & \cos(\Omega \ln \mid x \mid + \varphi) \end{array} \right) + \Psi_1(x, \varphi, z), \tag{27}$$

$$Z = \Psi_2(x, \varphi, z),$$

where $\rho = \lambda/\gamma < 1/2$ (see (18)), $\Omega = \omega/\gamma$ and, for some $\eta > \rho$,

$$\| \frac{\partial^{p+|q|}\Psi_i}{\partial x^p \partial(\varphi, z)^q} \| = O(|x|^{\eta-p}), \quad 0 \le p+|q| \le r-2 \ ; \tag{28}$$

the functions $Q_\pm$, $R_\pm$ in (26) ("+" corresponds to $x > 0$ - the map T_+, "-" corresponds to $x < 0$ - the map T_-) are smooth functions in a neighborhood of $(Y, Z) = 0$ for which the Taylor expansion can be written down

$$Q_\pm = (x_\pm, \varphi_\pm) + a_\pm Y + b_\pm Z + \cdots, \quad R_\pm = z_\pm + c_\pm Y + d_\pm Z + \cdots \ . \tag{29}$$

It is seen from (26)-(29) that if O is a saddle-focus satisfying (18), then if $a_+ \ne 0$ and $a_- \ne 0$, the map T satisfies conditions (21) and (24) with $\beta \in (\rho, \eta)$. Furthermore, analogues of conditions (20),(22),(23), (25) are fulfilled where the supremum should be taken not over $|x| \le 1$ but it is taken over small x. The map (26),(27),(29) is easily continued onto the whole cross-section Π so that the conditions of the lemma were fulfilled completely. An example is given by the map

$$\bar{x} = 0.9 \, |x|^\rho \cos(\ln |x| + \varphi),$$

$$\bar{\varphi} = 3 \, |x|^\rho \sin(\ln |x| + \varphi), \tag{30}$$

$$\bar{z} = (0.5 + 0.1z \, |x|^\eta) \operatorname{sign} x$$

where $0.4 = \rho < \eta$

As stated above, the expansion of volumes by the quotient-semiflow restricts the possible types of limit behavior of orbits. Thus, for instance, *in $\mathcal{D}$ there may be no stable periodic orbits.* Moreover, *any orbit in $\mathcal{D}$ has a positive maximal Lyapunov exponent.* Therefore, one must speak about a strange attractor in this case.

Beforehand, we recall some definitions and simple facts from topological dynamics. Let $X_t P$ be the time-t shift of a point P by the flow X. For given $\varepsilon > 0$ and $\tau > 0$ let us define as *an (ε, τ)-orbit* as a sequence of points $P_1, P_2, \cdots, P_k$ such that P_{i+1} is at a distance less than ε from $X_t P_i$ for some $t > \tau$. A point Q will be called *(ε, τ)-attainable* from P if there exists an (ε, τ)-orbit connecting P and Q; and it will be called *attainable* from P if, for some $\tau > 0$, it is (ε, τ)-attainable from P for any ε (this definition, obviously, does not depend on the choice of $\tau > 0$). A set C is attainable from P if it contains a point attainable from P. A point P is called *chain-recurrent* if it is attainable from $X_t P$ for any t. A compact invariant set C is called *chain-transitive* if for any points $P \in C$ and $Q \in CC$ and for any $\varepsilon > 0$ and $\tau > 0$ the set C contains an (ε, τ)-orbit connecting P and Q. Clearly, all points of a chain-transitive set are chain-recurrent.

A compact invariant set C is called *orbitally stable*, if for any its neighborhood U there is a neighborhood $V(C) \subseteq U$ such that the orbits starting in V stay in U for all $t \ge 0$. An orbitally stable set will be called *completely stable* if for any its neighborhood $U(C)$ there exist $\varepsilon_0 > 0$, $\tau > 0$ and a neighborhood $V(C) \subseteq U$ such that the (ε_0, τ)-orbits starting in V never leave U. It is known, that a set C is orbitally stable if and only if $C = \bigcap_{j=1}^{\infty} U_j$ where $\{U_j\}_{j=1}^{\infty}$ is a system of embedded open invariant (with respect to the forward flow) sets, and C is completely stable if the sets U_j are not just invariant but they are absorbing domains (i.e.; the orbits starting on ∂U_j enter inside U_j for a time interval not greater than some τ_j; it is clear in this situation that (ε, τ)-orbits starting on ∂U_j lie always inside U_j if ε is sufficiently small and $\tau \geq \tau_j$). Since the maximal invariant set (*the maximal attractor*) which lies in any absorbing domain is, evidently, *asymptotically stable*, it follows that any completely stable set is either asymptotically stable or is an intersection of a countable number of embedded closed invariant asymptotically stable sets.

Definition. *We call the set $\mathcal{A}$ of the points attainable from the equilibrium state O the attractor of the system X.*

This definition is justified by the following theorem.

Theorem 3.4 *The set $\mathcal{A}$ is chain-transitive, completely stable and attainable from any point of the absorbing domain $\mathcal{D}$.*

Let us consider a one-parameter family X_μ of such systems assuming that:
a homoclinic loop of the saddle-focus O exists at $\mu = 0$,
i.e., one of the separatrices (say, Γ_+) returns to O as $t \to +\infty$. In other words, we assume that the family X_μ intersects, at $\mu = 0$, a bifurcational surface filled by systems with a homoclinic loop of the saddle-focus and we suppose that this intersection is *transverse*. The transversality means that when μ varies, the loop splits and if M is the number of the last point of intersection of the separatrix Γ_+ with the cross-section Π at $\mu = 0$ ($P_M^+ \in \Pi_0$ at $\mu = 0$), then the distance between the point P_M^+ and Π_0 changes with a "non-zero velocity" when μ varies. We choose the sign of μ so that $P_M^+ \in \Pi_+$ when $\mu > 0$ (respectively, $P_M^+ \in \Pi_-$ when $\mu < 0$).

Theorem 3.5 *There exists a sequence of intervals Δ_i (accumulated at $\mu = 0$) such that when $\mu \in \Delta_i$, the attractor $\mathcal{A}_\mu$ contains a wild set (non-trivial transitive closed hyperbolic invariant set whose unstable manifold has points of tangency with its stable manifold). Furthermore, for any $\mu^* \in \Delta_i$, for any system C^r-close to a system X_μ^*, its attractor $\mathcal{A}$ also contains the wild set.*

We have mentioned earlier that the presence of structurally unstable (non-transverse) homoclinic trajectories leads to non-trivial dynamics. Using results [11, 13] we can conclude that the systems whose attractors contain structurally non-transverse homoclinic trajectories as well as structurally unstable periodic orbits of higher orders of degeneracies are dense in the given regions in the space of dynamical systems. In particular, the values of μ are dense in the intervals Δ_i for which an attractor of the system contain a periodic orbit of the saddle-saddle type along with its three-dimensional unstable manifold. For these parameter values, the topological dimension of such an attractor is already not less than three. The latter implies that the given class of systems is an example of hyperchaos.

5. Summary

The above listed attractors are, in the ideal, the most suitable images of dynamical chaos. Even though some of them are structurally unstable, nevertheless it is important that no stable periodic orbits appear in the system under small smooth perturbations. Nonetheless, excluding the attractors of the Lorenz type, no others have been ever observed in nonlinear dynamics so far. A research has frequently to deal with the models in which despite the complex behavior of the trajectories appears to be so visually convincing, nevertheless explicit statements regarding the exponential instability of the trajectories in the limit set can be debatable, and therefore should be made with caution. In numeric experiments with such model one finds a positive Lyapunov exponent, a continuous frequency spectrum, fast decaying correlation functions etc., i.e. all the attributes of dynamical chaos seem to get fulfilled so that the presence of the dynamical chaos causes no doubts. However, this "chaotic attractor" may and often do contain countably many stable periodic orbits which have long periods and week and narrow attraction basins. Besides the corresponding stability regions are relatively miniature in the parameter space under consideration and whence those orbits do not reveal themselves ordinarily in numeric simulations except some quite large stability windows where they are clearly visible. If it is the case, the quasi-attractor [4] is a more appropriate term for such chaotic set. The natural cause for this rather complex dynamics is homoclinic tangencies. Today the the systems with homoclinic tangencies are the target of many studies. We briefly outline some valuable facts proven for 3D systems and 2D diffeomorphisms. It will be clear that these results will also hold for the general case where there may be some other peculiarities, as for instance the co-existence of countable sets of saddle periodic orbits of distinct topological types (see [13]).

We suppose that the system possesses an absorbing area embracing the hyperbolic basis set in which the stable and unstable subsets may touch each other. If it is so, such a hyperbolic set is called *wild*. It follows then that either the system itself or a close one will have a saddle periodic orbit with non-transverse homoclinic trajectory along which the stable and unstable manifolds of the cycle have the tangency. In general, the tangency is quadratic. Let the saddle value $|\lambda \gamma|$ be less then 1, where λ and γ are the multipliers of the saddle periodic orbit. This condition is always true when the divergence of the vector field is negative in the absorbing area. Therefore, near the given system there will exist the so-called Newhouse regions [20] in the space of the dynamical system, i.e. the regions of dense structural instability. Moreover, a system in the Newhouse region has countably many stable periodic orbits which cannot principally be separated from the hyperbolic subset. If additionally this hyperbolic set contains a saddle periodic orbit with the saddle value exceeding one, then there will be a countable set of repelling periodic orbits next to it, and whose closure is not separable from the hyperbolic set either. The pictures becomes ever more complex if the divergence of the vector field is sign-alternating in the absorbing area. Such exotic dynamics requires infinitely many continuous topological invariant — *moduli*, needed for the proper description of the system in the Newhouse regions. This result comes from the fact that the systems with the countable set of periodic orbits of arbitrary high degrees of degeneracies are dense in the Newhouse regions [12, 13]. That is why we ought to conclude in a bitter way: the complete theoretical analysis of the models, which admit homoclinic tangencies, including complete bifurcation diagrams and so forth is non realistic.

References

[1] V.S. Afraimovich, V.I. Arnold, Yu.S. Il'yashenko and L.P. Shilnikov [1989], Bifurcation theory in *Dynamical Systems V*, Encyclopedia of Math. Sciences, **5**, Springer-Verlag.

[2] V.S. Afraimovich, V.V. Bykov and L.P. Shilnikov [1977], On the appearance and structure of Lorenz attractor, *DAN SSSR*, **234**, 336–339.

[3] V.S. Afraimovich, V.V. Bykov and L.P. Shilnikov [1983], On the structurally unstable attracting limit sets of Lorenz attractor type, *Tran. Moscow Math. Soc.*, **2**, 153–215.

[4] V.S. Afraimovich and L.P. Shilnikov [1983], Strange attractors and quasi-attractors in *Nonlinear Dynamics and Turbulence* eds. by G.I. Barenblatt, G. Iooss and D.D. Joseph, Pitman, NY, 1–28.

[5] D.Anosov [1967] Geodesic flows on compact Riemannian manifolds of negative curvature, *Trudy Math. Inst. im. V.A.Steklova*, **90**, [in Russian].

[6] D.V. Anosov, S.Kh. Aronson, I.U. Bronshtein and V.Z. Grines [1985], Smooth dynamical systems in *Dynamical Systems*, Encyclopedia of Math. Sciences, **1**, Springer-Verlag.

[7] D.V. Anosov and V.V. Solodov [1991], Hyperbolic sets in *Dynamical systems 9*, Encyclopedia of Math. Sciences, **66**, Springer-Verlag.

[8] R. Bowen [1970], Markov partition for Axiom A diffeomorphisms, *Amer. J. Math.*, **92**(3), 724–747.

[9] R. Bowen [1973], Symbolic dynamics for hyperbolic flows, *Amer. J. Math.*, **95**(2), 429–460.

[10] S.V. Gonchenko and L.P. Shilnikov [1993], On moduli of systems with a structurally unstable homoclinic Poincare curve. *Russian Acad. Sci. Izv. Math.* **41**(3), 417–445.

[11] S.V. Gonchenko, L.P. Shilnikov and D.V. Turaev [1992], On models with a structurally unstable homoclinic Poincaré curve, *Sov. Math.Dokl.*, v.44[2], 422–426.

[12] S.V. Gonchenko, L.P. Shilnikov and D.V. Turaev [1993], On models with non-rough Poincaré homoclinic curves, *Physica D*, v.62, 1–14.

[13] S.V. Gonchenko, D.V. Turaev and L.P. Shilnikov [1993], Dynamical phenomena in multi-dimensional systems with a structurally unstable homoclinic Poincare curve. *Russian Acad. Sci. Dokl. Math.*, **47**(3), 410–415.

[14] S.V. Gonchenko, D.V. Turaev and L.P. Shilnikov [1993], On the existence of Newhouse domains in a neighborhood of systems with a structurally unstable Poincare homoclinic curve (the higher-dimensional case). *Russian Acad. Sci. Dokl. Math.* **47**(?), 268–273.

[15] S.V. Gonchenko, L.P. Shilnikov and D.V. Turaev [1997], On the Newhose regions of two-dimensional diffeomorphisms close to those with a structurally unstable contour. *Proc. Steklov Inst. Math.*, **216**, 76–125.

[16] J. Guckenheimer and R.F. Williams [1979], Structural stability of Lorenz attractors, *Publ. Math. IHES*, **50**, 59–72.

[17] M. Hirsh, C. Pigh and M. Shub [1977], *Invariant manifolds, Lecture Notes in Math.* Springer-Verlag.

[18] E.N. Lorenz [1963] Deterministic non-periodic flow, *J. Atmos. Sci.*, **20**, 130–141.

[19] J. Milnor and R. Thurston [1977] On iterated maps of the interval I ans II. Unpublished notes, Prinseton University Press: Princeton.

[20] S. Newhouse [1979], The abundance of wild hyperbolic sets and nonsmooth stable sets for diffeomorphisms.*Publ. Math. I.H.E.S.* **50**, 101–151.

[21] I.M. Ovsyannikov and L.P. Shilnikov [1986] On systems with a saddle-focus homoclinic curve, *Mat. Sbornik*, **58**, 557–574; English translation in *Math. USSR Sb.* **58**, 557–574 [1987].

[22] I.M. Ovsyannikov and L.P. Shilnikov [1991] Systems with a homoclinic curve of multidimensional saddle-focus type, and spiral chaos *Mat. Sb.* **182**, 1043–1073; English translation in *Math. USSR Sb.* **73**, 415–443 [1992].

[23] R. Plykin, E.A. Sataev and C.V. Shlyachkov [1991], Strange attractors in *Dynamical systems 9*, Encyclopedia of Math. Sciences, **66**, Springer-Verlag.

[24] C. Robinson [1989], Homoclinic bifurcation to a transitive attractor of Lorenz type, *Nonlinearity*, **2**(4), 495–518.

[25] M. Rychlik [1990], Lorenz attractor through Shilnikov type bifurcation, Ergod. Theory and Dyn. Syst., **10**(4), 793–821.

[26] S. Smale [1967] Differentiable dynamical systems,*Bulletin of Amer. Math. Soc.*, **73**, 747–817.

[27] A.L. Shilnikov [1986], Bifurcations and chaos in the Shimizu-Marioka system [In Russian], in *Methods and qualitative theory of differential equations*, Gorky State University, 180–193 [English translation in *Selecta Mathematica Sovietica* **10**, 105–117, [1991]].

[28] A.L. Shilnikov [1993], On bifurcations of the Lorenz attractor in the Shimizu-Marioka system, *Physica* **D62**, 338–346.

[29] L.P. Shilnikov [1965], A case of the existence of a countable set of periodic motions, *Sov.Math.Dokl.*, **6**, 163–166.

[30] L.P. Shilnikov [1970], A contribution to the problem of the structure of an extended neighbourhood of a rough state of saddle-focus type, *Math.USSR Sb.*, **10**, 91–102.

[31] L.P. Shilnikov [1981], The bifurcation theory and quasi-hyperbiloc attractors. *Uspehi Mat. Nauk*, **36**, 240–241.

[32] L.P. Shilnikov [1967], On a Poincare-Birkhoff problem. *Math. USSR Sb.*, **3**, 91–102.

[33] D.V. Turaev and L.P. Shilnikov [1996], An example of a wild strange attractor, *Sbornik: Mathematucs*, **189**(2), 291–314.

[34] D.V. Turaev and L.P. Shilnikov [1995], On a blue sky catastrophy, *Soviet Math. Dokl.*, **342**, N.5, 596–599.

[35] M. Shub [1978], Stabilité globale des systems dynamiques, *Asterisque*, **56**.

[36] R. Williams [1970], Classification of one-dimensional attractors, *Proc. Symp. in Pure Math.*, 361–393.

[37] R.Williams [1967], One-dimensional non-wandering sets, *Topology*, **6**, 473–487.

[38] R.Williams [1974], Expanding attractors, *Publ. Math. IHES*, **43**, 169–203.

[39] R.Williams [1979], The structure of Lorenz attractors, *Publ. Math. IHES*, **50**, 101–152.

Dynamics in Two Complex Dimensions

J. Smillie[*]

Abstract

We describe results on the dynamics of polynomial diffeomorphisms of $\mathbf{C}^2$ and draw connections with the dynamics of polynomial maps of $\mathbf{C}$ and the dynamics of polynomial diffeomorphisms of $\mathbf{R}^2$ such as the Hénon family.

2000 Mathematics Subject Classification: 37F99.
Keywords and Phrases: Hénon diffeomorphism, Julia set, Critical point.

1. Introduction

The subject of this article is part of the larger subject area of higher dimensional complex dynamics. This larger area includes the dynamical study of holomorphic maps of complex projective space, automorphisms of $K3$ surfaces, birational maps, automorphisms of $\mathbf{C}^n$ and higher dimensional Newton's method. Our particular topic of research is polynomial automorphisms of $\mathbf{C}^2$. This area is particularly interesting because of its connections to some fundamental questions of dynamical systems via two real dimensional dynamics and because of its connection to some powerful techniques via one dimensional complex dynamics. I will begin by describing some of these connections. The reader is encouraged to consult [17] for a more thorough discussion of the historical background summarized here.

Over one hundred years after Poincaré observed chaotic behavior in the dynamics of surface diffeomorphisms the problem of creating a comprehensive theory of the dynamics of diffeomorphisms remains unsolved. Though the objective is to create a theory that would apply to diffeomorphisms in any dimension the focus remains on the two dimensional case. On the one hand the chaotic behavior which makes these problems challenging first appears for diffeomorphisms in dimension two, on the other hand there is a sense that if the tools can be developed to solve the problem in dimension two then the higher dimensional problem will be approach-

* Department of Mathematics, Malott Hall, Cornell University, Ithaca, NY, USA. E-mail: smillie@math.cornell.edu

able. There are reasons to believe that if the tools can be developed to thoroughly analyze one specific interesting family of diffeomorphisms then one would be in a good position to attack the general problem. If we were to suggest a family to play the role of a "test case" there is one particular family which stands out. This is the family of diffeomorphisms of $\mathbf{R}^2$ introduced by the French astronomer Hénon:

$$f_{a,b}(x,y) = (a - by - x^2, x).$$

The parameter b is the Jacobian determinant of $f_{a,b}$. When $b \neq 0$ these maps are diffeomorphisms. When $b = 0$ then $f_{a,0}$ is a map with a one dimension range and the behavior of $f_{a,0}$ is essentially that of the quadratic unimodal map $f_a(x) = a - x^2$.

In singling out the Hénon family we are following a well established tradition. This family has appeared often both in the physics and mathematics literature. It has been studied theoretically and numerically.

Virtually all interesting dynamical behavior which is known to occur for two dimensional diffeomorphisms is known to occur in this family. Hénon's original question involved an apparent strange attractor, and this is the first family in which the existence of strange attractors was proved ([11]). For certain parameter values this family exhibits hyperbolic behavior such as the Smale horseshoe ([14]). For other parameters it exhibits persistently nonhyperbolic behavior ([19]).

There is also a great deal that is not understood about the Hénon family. Despite the fact that many different types of dynamic behavior occur it is not known whether the union of these behaviors accounts for a large set of parameter values. There are also open questions about how the complexity of behavior varies with the parameters. When $a \ll 0$ the behavior is non-chaotic. When $a \gg 0$, $f_{a,b}$ exhibits a horseshoe, a model for chaotic behavior. What happens for intermediate values? How is chaos created? (cf [13])

Another reason for looking at the Hénon family is its connection with the one dimensional family of unimodal maps f_a. One dimensional diffeomorphisms exhibit only regular behavior but one dimensional maps exhibit a wealth of chaotic behavior. In contrast to the situation for the Hénon family, the most fundamental questions for the unimodal family have been answered. In the language of [17] the family f_a provides a "qualitatively solvable model of chaos" which is to say that there is a good understanding of attractors, strange and otherwise, for large sets of parameters and there is a good understanding of the transition to chaos.

The quadratic family is distinguished in the family of unimodal maps because it has a natural extension to the complex numbers. In the family $f_a(x) = a - x^2$ both x and a can be taken to be complex. The use of complex methods stands out as a reason for the success of the analysis of the quadratic family and unimodal maps more generally. While there are important results about unimodal maps that do not use complex techniques, these techniques do play a central role in the monotonicity results and in the analysis of attractors for the quadratic family.

Because the Hénon family is also given by polynomial equations it also has a natural complex extension. My first introduction to the importance of the complex

Hénon family was through lectures of J. H. Hubbard in the mid 1980's. Another contributor who brought new ideas to the subject was N. Sibony. Hubbard and his co-authors as well as Fornaess and Sibony and many others have continued to make fundamental contributions to this area and it is not possible to do justice to all of this work in the space provided. I will focus here on work that was carried out jointly with E. Bedford and, in some cases, M. Lyubich over the past 15 years.

2. Basic definitions in one and two variables

The fundamental paper of Friedland and Milnor [15] shows that a natural class of holomorphic diffeomorphisms to consider is the family of polynomial diffeomorphisms of $\mathbf{C^2}$. This class contains the Hénon family and the tools that we use to analyze the Hénon family work equally well for all diffeomorphisms in this class. In studying polynomial maps of $\mathbf{C}$ one focuses on those of degree greater than one because these exhibit chaotic behavior. One way of quantifying chaotic behavior is through the topological entropy, $h_{top}(f)$. In one complex dimension the entropy is the logarithm of the degree so the distinction between degree one and higher degree is the distinction between entropy zero and positive entropy.

For polynomial diffeomorphisms in dimension two the algebraic degree is not a conjugacy invariant and hence not a dynamical invariant. One way to create a conjugacy invariant is to define the following "dynamical degree":

$$d = \lim_{n \to \infty} (\text{algebraic degree } f^n)^{\frac{1}{n}}.$$

It is again true that the topological entropy of a complex diffeomorphism is the logarithm of its dynamical degree, so dynamical degree seems to be the appropriate two dimensional analog of degree. The Hénon diffeomorphisms have the property that the algebraic degree of f^n is 2^n so the dynamical degree is two. Friedland and Milnor show that any diffeomorphism with dynamical degree one is conjugate to an affine or elementary diffeomorphism. They also show that a diffeomorphism with dynamical degree greater than one is, like the Hénon diffeomorphism, conjugate to an explicit diffeomorphism whose actual degree is equal to its dynamical degree. When we refer to the degree of a diffeomorphism we will mean the dynamical degree. We make the standing assumption that all of our polynomial diffeomorphisms have degree greater than one.

Let us review some standard definitions for polynomial maps. Let $f : \mathbf{C} \to \mathbf{C}$ be a polynomial map with degree $d > 1$. The set K is the set of points with bounded orbits. The Julia set, J is the boundary of K. In dimension one all recurrent behavior is contained in K. All chaotic recurrent behavior is contained in J. The ease with which this set can be defined leaves one unprepared for the range of intricate behavior that it exhibits.

Let $f : \mathbf{C^2} \to \mathbf{C^2}$ be a polynomial diffeomorphism with dynamical degree $d > 1$. The set K^+ is the set of points with bounded forward orbits. Following Hubbard we take the set K^- to be the set of points with bounded backward orbits.

The sets $J^{\pm}$ are defined as the boundaries of $K^{\pm}$. The set J is defined to be $J^+ \cap J^-$. In dimension two all chaotic recurrent behavior is contained in J. Thus J seems to be a good analog of the one dimensional Julia set. (In fact there is an alternative analog of J but we will not deal with that here.)

Let p be a periodic saddle point of period n in $\mathbf{C}^2$. Let W_p^u denote the unstable manifold of p. This is the set of points that converge to p under iteration of f^{-1}. Since this definition involves f^{-1} it is less clear what the one variable analog should be. Let us examine the situation more carefully. The set W_p^u is holomorphically equivalent to $\mathbf{C}$. We can find a parameterization $\phi_p : \mathbf{C} \to W_p^u$ which satisfies the functional equation $f^n(\phi_p(z)) = \lambda \cdot z$ where λ is the expanding eigenvalue of Df_p^n. Now if p is a periodic point in $\mathbf{C}$ then the functional equation still makes sense. A function ϕ_p which satisfies this equation is called a linearizing coordinate and this is a good analog of the parameterized unstable manifold in two dimensions.

Hubbard made the key observation that this construction gives a natural way to draw pictures of the sets $W_p^u \cap K^+$ in two variables and a natural way to compare them to the pictures of K in one variable. In both cases we identify a region in $\mathbf{C}$ with the computer screen and choose a color scheme where the color for a pixel corresponding to z is related to the rate of escape of $\phi_p(z)$. The general convention is that points that do not escape (those points in $\phi_p^{-1}(K)$) are colored black. (See [http://www.math.cornell.edu/~dynamics/].)

There is an abstract construction which makes it easier to compare invertible systems such as diffeomorphisms with non-invertible systems. Given a non-invertible system such as $f : \mathbf{C} \to \mathbf{C}$ there is a closely related invertible system called the natural extension. Let us denote this by $\hat{f} : \hat{\mathbf{C}} \to \hat{\mathbf{C}}$. The points in $\hat{\mathbf{C}}$ consist of sequences $(\ldots z_{-1}, z_0, z_1 \ldots)$ such that $f(z_j) = z_{j+1}$. The map $\hat{f}$ acts by shifting such a sequence to the left.

The natural extension gives us a way of justifying the analogy between linearizing coordinates and unstable manifolds. Corresponding to a periodic saddle point p in $\mathbf{C}$ there is a unique periodic point $\hat{p}$ in $\hat{\mathbf{C}}$. Since $\hat{f}$ is invertible we can make sense of the unstable manifold $W_{\hat{p}}^u$ and the linearizing coordinate can be used to parameterize this unstable manifold.

Though $\hat{\mathbf{C}}$ contains "leaves" such as $W_{\hat{p}}^u$ it is a mistake to think of $\hat{\mathbf{C}}$ as a lamination. When f is expanding $\hat{\mathbf{C}}$ is a lamination near the Julia set but the more complicated the dynamics of f, the more degenerate this structure becomes. This complexity arises from recurrent behavior of the critical point for f. This suggests a certain connection between regularity of unstable manifolds in two dimensions and recurrence of critical points in one dimension that we will return to later.

Since points in $\hat{\mathbf{C}}$ have bounded backward orbits, we should think of $\hat{\mathbf{C}}$ as an analog of J^-. Let f_a be an expanding one dimensional map and consider a diffeomorphism $f_{a',b}$ with b small and a' close to a. Hubbard and Oberste-Vorth ([17]) show that J^- is topologically conjugate to the corresponding $\hat{\mathbf{C}}$. When the one dimensional map f is not expanding the relation between $\hat{\mathbf{C}}$ and any particular

J^- should be viewed as metaphorical rather than literal.

3. Potential theory and Pluri-potential theory

A standard construction in potential theory associates to nice sets a measure μ called the harmonic measure or equilibrium measure. The harmonic measure associated to the Julia set turns out to be a measure of dynamical interest. The potential theory construction starts with the Green function. The Green function of K has a dynamical description:

$$G(p) = \lim_{n \to \infty} \frac{1}{d^n} \log^+ |f^n(p)|.$$

The Green function is non-negative and equal to zero precisely on the set K. The harmonic measure μ is obtained by applying the Laplacian to G. The support of μ is the boundary of K which is the set J. The connection between polynomial maps and potential theory first appears in the work of Brolin ([12]). It reappears in a paper of Manning ([18]) and is nicely summarized in [20].

The harmonic measure has connections to entropy and to the connectivity of J. These connections do not play a major role in the one dimensional theory because entropy and connectivity can be approached more directly. In the two dimensional theory these connections are much more important.

The entropy of the measure μ, $h(\mu)$, happens to be $\log d$ which is equal to the topological entropy of the map. The topological entropy dominates the measure theoretic entropy of any invariant measure. A measure for which equality holds is called a measure of maximal entropy. For polynomial maps of $\mathbf{C}$ the measure μ can be characterized as the unique measure of maximal entropy.

The dimension of a measure ν, $\dim_H(\nu)$, is the minimum of the Hausdorff dimensions of subsets of full ν measure. The dimension of the harmonic measure of a planar set is always less than or equal to one. If the set is connected then the dimension is one. For Julia sets the converse is true: the dimension of the harmonic measure is one if and only if J is connected.

The Lyapunov exponent, $\lambda(\mu)$, of f with respect to an ergodic measure measures the rate of growth of tangent vectors under iteration (for a set of full μ measure). The Lyapunov exponent is related to Hausdorff dimension of the measure by the formula:

$$\dim_H(\mu) = h(\mu)/\lambda(\mu).$$

Since $h(\mu)$ is $\log d$, $\lambda(\mu) = \log d$ if and only if J is connected. We will return to this in the next section.

In dimension two we have two rate of escape functions:

$$G^{\pm}(p) = \lim_{n \to \infty} \frac{1}{d^n} \log^+ |f^{\pm n}(p)|.$$

Potential theory in one variable centers on the behavior of the Laplacian. The Laplacian is not holomorphically invariant in two variables but it has a close relative which is. This is the operator dd^c which takes real valued functions to real two forms. The d that appears here is just the exterior derivative and the d^c is a version of the exterior derivative twisted by using the complex structure. In one variable we have:

$$dd^c g = (\triangle g) dx \wedge dy.$$

Not only is dd^c holomorphically natural but it is well defined on complex manifolds of any dimension. Of course, in the two variable context as in the one variable, the functions to which these operators are applied are not smooth and the result has to be interpreted appropriately. The theory connected with the operator dd^c is referred to as pluripotential theory. It was an observation of Sibony that the methods of pluripotential can be profitably applied to the complex Hénon diffeomorphisms.

Define $\mu^{\pm} = \frac{1}{2\pi} dd^c G^{\pm}$. These are dynamically significant currents supported on $J^{\pm}$. Define $\mu = \mu^+ \wedge \mu^-$. This measure μ is the analog of the harmonic measure in one dimension. The following result suggest that "μ" defined above is the dynamical analog as well as the pluripotential theoretic analog.

Theorem 3.1 ([4]) *The measure μ is the unique measure of maximal entropy.*

4. Connectivity and critical points

We want to consider the way in which the dynamical behavior of a polynomial diffeomorphism such as $f_{a,b}$ depends on the parameter. Looking at pictures of $W_p^u \cap K^+$ shows that there are indeed many things that do change. If we want to focus on one fundamental property we might start by looking at connectivity. In one variable the connectivity of the Julia set of f_a defines the Mandelbrot set which is the fundamental object of study for quadratic maps. In two variables there are several notions of connectivity that we could consider. The following has proved useful. We say that f is *stably/unstably connected* if $W_p^{s/u} \cap K^{-/+}$ has no compact components. We can ask about the relation between stable connectivity, unstable connectivity and the connectivity of J. A priori the property of being stably/unstably connected depends on the saddle point p. In fact we show that these properties are independent of p.

Let us look at the situation in one variable. The basic result about connectivity is the following.

Theorem 4.1 (Fatou) *Let f be a polynomial map of* **C**. *Then J is connected if and only if every critical point of f has a bounded orbit.*

The following formula makes a connection between the Lyapunov exponent and critical points ([20]):

$$\lambda(\mu) = \log d + \sum_{\{c_j : f'(c_j) = 0\}} G(c_j).$$

The function G is non-negative and zero precisely on the set K. In light of the theorem above we see that J is connected if and only if the Lyapunov exponent is

$\log d$. This proves an assertion made in Section 1 about the relation between the Lyapunov exponents of μ and the connectivity of J.

In two variables there are two Lyapunov exponents, $\lambda^{\pm}(\mu)$, of f with respect to harmonic measure. The following result establishes the connection between stable/unstable connectivity and these exponents.

Theorem 4.2 ([7]) *We have* $\lambda^{+}(\mu) \geq \log d$; *and* $\lambda^{+}(\mu) = \log d$ *if and only if f is unstably connected. Similarly* $\lambda^{-}(\mu) \leq -\log d$; *and* $\lambda^{-}(\mu) = -\log d$ *if and only if f is stably connected.*

It is clear from this result that neither exponent is zero. Pesin theory shows that stable and unstable manifolds exist for μ almost every point. Let $\mathcal{C}^{u}$ be the set of critical points of the restriction of G^{+} to these unstable manifolds. We define $\mathcal{C}^{s}$ in the corresponding way.

Theorem 4.3 ([7]) *The diffeomorphism f is unstably connected if and only if $\mathcal{C}^{u} = \emptyset$. The diffeomorphism f is stably connected if and only if $\mathcal{C}^{s} = \emptyset$.*

In [6] we prove an analog of the critical point formula where the role of the critical point is played by critical points is played by $\mathcal{C}^{u}$. This formula leads to proofs of the two theorems above.

The following result makes the connection between stable and unstable connectivity and the connectivity of J. Note that in this two variable situation the Jacobian of f enters the picture.

Theorem 4.4 ([7]) *If $|\det Df| < 1$ then f is never stably connected. In this case J is connected if and only if f is unstably connected. If $|\det Df| = 1$ then f is stably connected iff f is unstably connected iff J is connected.*

(The case $|\det Df| > 1$ is analogous to the case $|\det Df| < 1$.) The Jacobian enters the proof through the relation: $\lambda^{+}(\mu) + \lambda^{-}(\mu) = \log|\det Df|$. We see for example that $|\det Df| < 1$ implies that $\lambda^{-}(\mu) < -\log d$ which, by Theorem 4.2 implies that f is unstably disconnected.

Using this result J. H. Hubbard and K. Papadantonakis have developed a computer program that uses the set $\mathcal{C}^{u}$ to draw pictures of the connectivity locus in parameter space. (See [http://www.math.cornell.edu/~dynamics/].)

5. The boundary of the horseshoe locus

Hyperbolic behavior, as exhibited by the horseshoe, is structurally stable. This implies that the set of (a, b) for which $f_{a,b}$ exhibits a horseshoe is open. Let us call this set the horseshoe locus. Standard techniques from dynamical systems can be used to analyze the dynamical behavior inside the horseshoe locus. These techniques break down on the boundary of the horseshoe locus however. By contrast complex techniques from [4], [9] and [10] can be applied on the closure of the horseshoe locus. Thus the analysis of this boundary provides a setting for demonstrating that these techniques derived from complex analysis are not without interest in the real setting.

Let us look at the one dimensional case f_{a}. We say that f_{a} exhibits a horseshoe if $f_a|J_a$ is expanding and topologically conjugate to the one sided two shift. The horseshoe locus here is the set $a > 2$. The boundary of the horseshoe locus is $a = 2$. The map f_2 is the well known example of Ulam and von Neumann. The failure of expansion is demonstrated by the fact that the critical point 0 is in the Julia set, $[-2, 2]$. In fact the critical point maps to the fixed point -2 after two iterates.

The following result describes the failure of hyperbolicity on the boundary of the horseshoe locus for Hénon diffeomorphisms. Note that the property of eventually mapping to the fixed point p in dimension one corresponds to belonging to W_p^s in dimension two.

Theorem 5.1 ([10]). *For $f_{a,b}$ on the boundary of the horseshoe locus there are fixed points p and q so that W_p^s and W_q^u have a quadratic tangency. When $b > 0$ we have $p = q$. When $b < 0$, $p \neq q$.*

The next result gives additional information about the precise nature of the dynamics of maps on the boundary of the horseshoe locus:

Theorem 5.2 (Bedford-Smillie) *For any (a, b) in the boundary of the horseshoe locus the restriction of $f_{a,b}$ to its non-wandering set is conjugate to the full two-shift with precisely two orbits identified. Given (a, b) and (a', b') in the boundary of the horseshoe locus, the restrictions of $f_{a,b}$ and $f_{a',b'}$ to their non-wandering sets are conjugate if and only if b and b' have the same sign.*

There are many techniques which work only for b small. Note that that the result above applies for all values of b including the volume preserving case $b = \pm 1$.

We can ask how the dynamics of $f_{a,b}$ for (a, b) on the boundary of the horseshoe regions $b > 0$ and $b < 0$ compares with the dynamics of f_2 which corresponds to the boundary of the horseshoe region when $b = 0$. The sets $J_{a,b}$ for $b \neq 0$ are totally disconnected while the set J_2 is connected. In particular the inverse limit system $\hat{J}_2$ is not conjugate to either system with $b \neq 0$. This is an example where the insights gained from looking at the inverse limit system need to be interpreted cautiously.

I will touch on the techniques used in the proofs of these theorems. Our fundamental approach to proving these results was to exploit the relationship between the real mapping $f_{a,b} : \mathbf{R}^2 \to \mathbf{R}^2$ and its complex extension $f_{a,b} : \mathbf{C}^2 \to \mathbf{C}^2$. In passing from $\mathbf{C}^2$ to $\mathbf{R}^2$ something may be lost. The first question to ask is how much chaotic behavior do we lose? One way to measure this is through the topological entropy function. If we denote $f_{a,b} : \mathbf{R}^2 \to \mathbf{R}^2$ by $f_\mathbf{R}$ and $f_{a,b} : \mathbf{C}^2 \to \mathbf{C}^2$ by $f_\mathbf{C}$ then we have

$$h_{top}(f_\mathbf{R}) \leq h_{top}(f_\mathbf{C}) = \log 2.$$

If we want to study the real Hénon diffeomorphisms most closely connected to their complex extensions we should focus our attention on those f with $h_{top}(f_\mathbf{R}) = \log 2$. We say that these examples have *maximal entropy*. This is an interesting set to look at. The horseshoe locus is contained in the maximal entropy locus but the maximal entropy locus is larger than the horseshoe locus. The horseshoe locus is open, and the maximal entropy locus is closed. In particular the maximal entropy locus contains the boundary of the horseshoe locus.

For maximal entropy diffeomorphisms the relation between the real and complex dynamics is as close as one could want:

Theorem 5.3 ([4]) $f_\mathbf{R}$ *has maximal entropy if and only if J is contained in* $\mathbf{R}^2$.

This theorem is a consequence of the fact that μ is the unique measure of maximal entropy and the fact that the support of μ is contained in J. The fact that the real and complex dynamics are closely related for this class of maps means that it is a good starting point for applying complex techniques to the real case. It also provides us with useful techniques from harmonic analysis. For example the Green functions of real sets satisfy certain growth conditions, and these translate into conditions insuring expansion and regularity of unstable manifolds. This allows us to show that maximal entropy diffeomorphisms are quasi-expanding ([9]). Quasi-expansion is the two dimensional analog of f having non-recurrent critical points. The exploitation of the properties of quasi-expanding diffeomorphisms leads to the proofs of the Theorems 5.1 and 5.2.

We believe that the connections made so far do not represent the end of the story but only the beginning. We trust that the picture of two dimensional complex dynamics will become clearer with time and, as it does, there will be valuable interactions with the theory of real dynamics.

References

[1] E. Bedford & J. Smillie, *Polynomial diffeomorphisms of* $\mathbf{C}^2$: *currents, equilibrium measures and hyperbolicity*, Inventiones Math. **103** (1991), 69–99.

[2] E. Bedford & J. Smillie, *Polynomial diffeomorphisms of* $\mathbf{C}^2$ *II: stable manifolds and recurrence*, 4 No. 4 Journal of the A.M.S. (1991), 657–679.

[3] E. Bedford & J. Smillie, *Polynomial diffeomorphisms of* $\mathbf{C}^2$ *III: ergodicity, exponents and entropy of the equilibrium measure*, Math. Annalen **294** (1992), 395–420.

[4] E. Bedford, M. Lyubich, and J. Smillie, Polynomial diffeomorphisms of $\mathbf{C}^2$ IV: the measure of maximal entropy and laminar currents, *Inventiones Math.* 112 (1993), 77–125.

[5] E. Bedford, M. Lyubich, and J. Smillie, Distribution of periodic points of polynomial diffeomorphisms of $\mathbf{C}^2$, Inventiones Math. **114** (1993), 277–288.

[6] E. Bedford & J. Smillie, *Polynomial diffeomorphisms of* $\mathbf{C}^2$. *V: Critical points and Lyapunov exponents*, J. Geom. Anal. 8 no. 3, (1998), 349–383.

[7] E. Bedford & J. Smillie, Polynomial diffeomorphisms of $\mathbf{C}^2$. VI: Connectivity of J, *Annals of Mathematics*, 148 (1998), 695–735.

[8] E. Bedford & J. Smillie, Polynomial diffeomorphisms of $\mathbf{C}^2$. VII: Hyperbolicity and external rays, *Ann. Sci. Ecole Norm. Sup. 4 série* 32 (1999), 455–497.

[9] E. Bedford & J. Smillie, Polynomial diffeomorphisms of $\mathbf{C}^2$. VIII: Quasi-expansion, *American Journal of Mathematics* 124 (2002), 221–271.

[10] E. Bedford & J. Smillie, Real polynomial diffeomorphisms with maximal en-

tropy: tangencies, (available at http://www.arXiv.org).

[11] M. Benedicks & L. Carleson, The dynamics of the Hénon map, *Annals of Mathematics*, 133, (1991), 73–179.

[12] H. Brolin, Invariant sets under iteration of rational functions, *Ark. Mat*, 6, (1965), 103–144.

[13] A. de Carvalho & T. Hall, How to prune a horseshoe, *Nonlinearity*, 15 no. 3, (2002), R19–R68.

[14] R. Devaney & Z. Nitecki, Shift automorphisms in the Hénon mapping. Comm. Math. Phys. 67 (1979), no. 2, 137–146.

[15] S. Friedland & J. Milnor, Dynamical properties of plane polynomial automorphisms, *Ergodic Theory Dyn. Syst.* 9, (1989), 67–99 .

[16] J. Hubbard & R. Oberste-Vorth, Hénon mappings in the complex domain. II. Projective and inductive limits of polynomials, in *Real and complex dynamical systems* Kluwer, 1995.

[17] M. Lyubich, The quadratic family as a qualitatively solvable model of chaos. *Notices Amer. Math. Soc.* 47 (2000), no. 9, 1042–1052.

[18] A. Manning, The dimension of the maximal measure for a polynomial map, *Ann. Math.* 119 (1984), 425–430.

[19] S. Newhouse, The abundance of wild hyperbolic sets and nonsmooth stable sets for diffeomorphisms. *Inst. Hautes tudes Sci. Publ. Math.* No. 50 (1979), 101–151.

[20] F. Przytycki, Hausdorff dimension of the harmonic measure on the boundary of an attractive basin for a holomorphic map, *Invent. math.* 80 (1985), 161–179.

ICM 2002 · Vol. III · 383–392

Continuous Averaging in Dynamical Systems

D. Treschev[*]

Abstract

The method of continuous averaging can be regarded as a combination of the Lie method, where a change of coordinates is constructed as a shift along solutions of a differential equation and the Neishtadt method, well-known in perturbation theory for ODE in the presence of exponentially small effects. This method turns out to be very effective in the analysis of one- and multi-frequency averaging, exponentially small separatrix splitting and in the problem of an inclusion of an analytic diffeomorphism into an analytic flow. We discuss general features of the method as well as the applications.

2000 Mathematics Subject Classification: 58F.
Keywords and Phrases: Averaging method, Exponentially small effects, Separatrix splitting.

1. The method

There are several problems in the perturbation theory, of real-analytic ordinary differential equations (ODE), where standard methods do not lead to satisfactory results. We mention as examples the problem of an inclusion of a diffeomorphism into a flow in the analytic set up, and the problem of quantitative description of exponentially small effects in dynamical systems. In this cases one of possible approaches is an application of the continuous averaging method. The method appeared as an extension of the Neishtadt averaging procedure [14]. We begin with the description of the method.

Let us transform the system

$$\dot{z} = \widehat{u}(z), \tag{1.1}$$

by using the change of variables $z \mapsto Z(z,s)$. Here z is a point of the manifold M, $\widehat{u}$ is a smooth vector field on M, s is a non-negative parameter, and the change is

---
[*]Department of Mechanics and Mathematics, Moscow State University, Vorob'evy Gory, Moscow 119899, Russia. E-mail: dtresch@mech.math.msu.su

384   D. Treschev

defined as a shift along solutions of the equation[1]

$$dZ/ds = f(Z,s), \qquad Z(z,0) = z, \quad 0 \le s \le S. \tag{1.2}$$

Let the change $z \mapsto Z$ transform (1.1) to the following system:

$$\dot{Z} = u(Z,s). \tag{1.3}$$

Differentiating (1.3) with respect to s, we have:

$$\dot{f}(Z,s) = u_s(Z,s) + \partial_f u(Z,s) \quad \text{or} \quad u_s = [u,f].$$

Here ∂_f is the differential operator on M, corresponding to the vector field f, the subscript s denotes the partial derivative, and $[\cdot,\cdot]$ is the vector commutator: $[u_1,u_2] = \partial_{u_1}u_2 - \partial_{u_2}u_1$. Putting $f = \xi u$, where ξ is some fixed linear operator, we obtain the Cauchy problem

$$u_s = -[\xi u, u], \qquad u|_{s=0} = \hat{u}. \tag{1.4}$$

We call the system (1.4) averaging. The equation $f = \xi u$ is crucial for our method. The vector field f is usually constructed as a series in the small parameter and not as a result of an application to u of an operator ξ, chosen in advance.

A nonautonomous analog of (1.4) can be easily constructed. If $\hat{u}$ depends explicitly on t then $f = \xi u$ also depends on t and (1.4) should be replaced by the system

$$u_s = (\xi u)_t - [\xi u, u], \qquad u|_{s=0} = \hat{u}(z,t). \tag{1.5}$$

Properties of the averaging system can be illustrated by the following example. Consider the non-autonomous real-analytic system

$$\dot{z} = \varepsilon \hat{u}(z,t), \qquad z \in M. \tag{1.6}$$

Here ε is a small parameter, $\hat{u}$ is 2π-periodic in t. Let us try to weaken the dependence of $\hat{u}$ on time by the change $z \mapsto Z$ (1.2) with $f = \xi u$. We put[2]

$$\xi u(z,t,s) = \sum_{k \in \mathbf{Z}} i\sigma_k u^k(z,s)e^{ikt}, \qquad \sigma_k = \operatorname{sign} k, \tag{1.7}$$

where u^k are Fourier coefficients in the expansion $u(z,t,s) = \sum_{k \in \mathbf{Z}} u^k(z,s)e^{ikt}$.

Equation (1.5) takes the form

$$u_s^k = -|k|u^k + i\varepsilon\sigma_k[u^0,u^k] - 2i\varepsilon\sum_{l+m=k,m<0<l}[u^l,u^m], \tag{1.8}$$
$$u^k|_{s=0} = \hat{u}^k, \qquad k \in \mathbf{Z}.$$

To have an idea of properties of this system, we skip in (1.8) the last term. The equations

$$u_s^k = -|k|u^k + i\varepsilon\,\sigma_k[u^0,u^k], \qquad u^k|_{s=0} = \hat{u}^k, \qquad k \in \mathbf{Z}$$

---

[1]Such method of constructing a change of variables is called the Lie method. The corresponding Hamiltonian version is called the Deprit-Hori method.
[2]Such an operator ξ is called the Hilbert transform.

can be solved explicitly:

$$u^k = e^{-|k|s}\widehat{u}^k \circ g^{i\varepsilon\sigma_k s}, \tag{1.9}$$

where g^s is the time-s shift $z|_{t=0} \mapsto z|_{t=s}$ along solutions of the system $\dot{z} = \widehat{u}^0(z)$.

The complex singularities of the functions $\widehat{u}^k \circ g^\zeta$ of the complex variable ζ prevent an unbounded continuation of the solutions (1.9) to all the set of positive s. Nevertheless, the functions (1.9) can be made exponentially small in ε since s can be chosen of order $\sim 1/\varepsilon$.

If ε is not small, the operator (1.7) can be used to smooth out the dependence of u on t. Indeed, for arbitrarily small $s > 0$ the Fourier coefficients u^k in (1.9) decrease exponentially fast in k even if $\widehat{u}$ is just continuous in t.

Certainly, these hewristic arguments cannot be regarded as a proof of the fact that systems of type (1.8) can be used for averaging or smoothing. Rigorous statements and estimates are based on the majorant method. In this paper we do not go into technical details, but present general ideas and applications.

If the vector field $\widehat{u}$ belongs to some subalgebra χ in the Lie algebra of vector fields on M, it is natural to look for the change $z \mapsto Z$ from the corresponding Lie group of diffeomorphisms. This means that f in (1.2) should be taken from χ. The same remains reasonable in the non-autonomous case. Note that the operator (1.7) is such that if $u(z,t,s) \in \chi$ for any t, the vector field ξu also belongs to χ for any t.

2. Applications

2.1. Fast phase averaging: one-frequency case

Mathematical models of various physical processes use systems of ODE which contain an angular variable changing much faster than other variables in the system. Taking the fast phase as a new time, we can rewrite the equations in the form

$$\dot{z} = \varepsilon\widehat{u}(z,t,\varepsilon), \qquad z \in M, \tag{2.1}$$

where M is the m-dimensional phase space of the system, and ε is a small parameter. The vector field $\widehat{u}$ is assumed to be smooth and to depend on time 2π-periodically.

It is well known that by a change of the variables it is possible to weaken the dependence of the system (2.1) on time. In particular, by using the standard averaging method, it is easy to construct a 2π-periodic in t change of the variables $z \mapsto z_*$ such that the equations (2.1) take the form

$$\dot{z}_* = \varepsilon\widehat{u}^0(z_*) + \varepsilon^2\widehat{u}_*(z_*,\varepsilon) + \varepsilon\widetilde{u}(z_*,t,\varepsilon). \tag{2.2}$$

Here the only term in the right-hand side depending explicitly on time is $\varepsilon\widetilde{u} = O(\varepsilon^K)$. The natural K is arbitrary and $\widehat{u}^0(z) = \frac{1}{2\pi}\int_0^{2\pi}\widehat{u}(z,t,0)\,dt$.

Now suppose that $\widehat{u}$ is real-analytic in z. Poincaré noted in some example that in this case power series in ε presenting a change of variables eliminating time from the equations, exist but diverge: terms at ε^k in these series have the order $k!$. In a general situation this statement has been proved by Sauzin [21].

Neishtadt [14] noted that in this case it is possible to obtain in (2.2)

$$\tilde{u} = O(e^{-\alpha/\varepsilon}), \quad \alpha = \text{const} > 0 \tag{2.3}$$

(ε is assumed to be nonnegative). The method Neishtadt used to prove this assertion is based on a large (of order $1/\varepsilon$) number of successive changes of variables. These changes weaken gradually explicit dependence of the equations on time. Ramis and Schafke [19] obtained analogous results analyzing diverging series, produced by the standard averaging method.

It is known also that in general a constant $A > \alpha$ exists such that it is impossible to construct 2π-periodic in t change $z \mapsto z_*$, such that $\tilde{u} = O(e^{-A/\varepsilon})$. This statement follows, for example, from an estimate of the separatrix splitting rate in Hamiltonian systems of type (2.1) with one and a half degrees of freedom.

In this section we estimate a "maximal" α for which the estimate (2.3) is possible.

Suppose that the manifold M is real-analytic. We fix its complex neighborhood $M_{\mathbf{C}}$ and denote by g^t the phase flow of the averaged system[3] $\dot{z} = \widehat{u}^0(z)$.

Let Q be a compact in M and V_Q its neighborhood in $M_{\mathbf{C}}$. Suppose that for any real s such that $|s| < \alpha$ and for any point $z \in V_Q$ the map g^{is} is analytic at z and moreover, $g^{is}(z) \in M_{\mathbf{C}}$. We define the set

$$U_{Q,\alpha} = \bigcup_{-\alpha < s < \alpha} g^{is}(V_Q).$$

Theorem 1 [24]. *Let the positive constants $\alpha, \rho, \varepsilon_0$ be such that*
(1) $U_{Q,\alpha} \subset M_{\mathbf{C}}$.
(2) The vector field $\widehat{u}$ is analytic in z and C^2-smooth in t, ε on $U_{Q,\alpha} \times \mathbf{T} \times [0, \varepsilon_0]$.
Then for sufficiently small ε_0, there exists a 2π-periodic in t real-analytic in z map $F : V_Q' \times \mathbf{T} \times (0, \varepsilon_0) \to M_{\mathbf{C}}, \quad Q \subset V_Q' \subset V_Q$, such that
(a) The set V_Q' is open in $M_{\mathbf{C}}$,
(b) $F(z, t, \varepsilon) = z_* = z + O(\varepsilon)$,
(c) F transforms (2.1) into (2.2) and the following estimate holds:

$$|\tilde{u}(z, t, \varepsilon)| \le Ce^{-\alpha/\varepsilon}, \qquad z \in V_Q', \quad t \in \Sigma_\rho, \quad \varepsilon \in [0, \varepsilon_0). \tag{2.4}$$

Theorem 1 means in particular, that in the case when components of the field $\widehat{u}$ are entire functions of z, the quantity α in (2.4) can be arbitrary positive number such that for all $s \in [-\alpha, \alpha]$ the maps $z \mapsto g^{is}(z)$ are holomorphic at any point $z \in Q$.

Proof of Theorem 1 is based on the continuous averaging. Namely, we solve the Cauchy problem (1.5), with ξ defined by (1.7). The required change of variables corresponds to the value $s = \alpha/\varepsilon$. The averaging can be performed inside a subalgebra χ in the Lie algebra of all real-analytic vector fields on M. In particular, if the initial vector field $\widehat{u}$ is Hamiltonian then $u_* = u(z, t, \varepsilon, \alpha/\varepsilon)$ is also Hamiltonian, and F is symplectic.

---

[3]It would be more correct to call it by the first approximation averaged system, written with respect to the fast time.

2.2. Averaging: multi-frequency case

Consider a real-analytic slow-fast system

$$\dot{x} = \omega + \varepsilon(\overline{u}(y) + \widehat{u}(x,y,\varepsilon)), \quad \dot{y} = \varepsilon(\overline{v}(y) + \widehat{v}(x,y,\varepsilon)), \qquad x \in \mathbf{T}^n, \ y \in \mathbf{R}^m. \quad (2.5)$$

Average in x of $\widehat{u}$ and $\widehat{v}$ is assumed to be $O(\varepsilon)$. The frequency vector $\omega \in \mathbf{R}^n$ is constant and non-resonant.[4]

We try to weaken the dependence of the right-hand side of the system on the fast variables x by a near-identity change $(x \bmod 2\pi, y) \mapsto (x_\bullet \bmod 2\pi, y_\bullet)$. We put

$$z = \left(\begin{array}{c} x \\ y \end{array}\right), \quad \overline{\varpi} = \left(\begin{array}{c} \omega \\ 0 \end{array}\right), \quad \overline{w} = \left(\begin{array}{c} \overline{u} \\ \overline{v} \end{array}\right), \quad \widehat{w} = \left(\begin{array}{c} \widehat{u} \\ \widehat{v} \end{array}\right).$$

Note that the vector fields $\overline{\varpi}$ and $\overline{w}$ commute. The system (2.5) takes the form

$$\dot{z} = \overline{\varpi} + \varepsilon(\overline{w} + \widehat{w}). \quad (2.6)$$

A1. Diophantine condition. We assume that $n \geq 2$ and the frequency vector ω is Diophantine: there exist $\gamma_0, \gamma > 0$ such that

$$|\langle k, \omega \rangle| \geq \gamma_0 \|k\|^{-\gamma}, \quad \text{for any } k \in \mathbf{Z}^n \setminus \{0\}. \quad (2.7)$$

To formulate the next assumption, we need some definition. Let g^t be the phase flow of the system

$$\dot{z} = \overline{w}(y). \quad (2.8)$$

For any real-analytic function $f(y)$ with values in $\mathbf{C}^{n+m}$ and the vector $k \in \mathbf{Z}^n$ we put $f_k = f(y)e^{i\langle k, x \rangle}$. The function

$$\mathbf{g}_k^s f = e^{-i\langle k, x \rangle} g_*^{-is}(f_k \circ g^{is}), \qquad s \in \mathbf{C}$$

does not depend on x. Here g_*^t is the differential of the map g^t. Note that $\mathbf{g}_k^s$, $s \in \mathbf{R}$ include shifts along g^t in purely imaginary direction. We put

$$\begin{aligned} \Sigma_q &= \{x \in \mathbf{C}^n/(2\pi\mathbf{Z})^n : |\operatorname{Im} x_j| \leq q, \quad j = 1, \ldots, n\}, \\ V_\nu &= \{y \in \mathbf{C}^m : \operatorname{Re} y \in (\overline{\mathcal{D}} + \nu) \subset \mathbf{R}^m, \quad |\operatorname{Im} y_l| \leq \nu, \quad l = 1, \ldots, m\}, \end{aligned}$$

where $\overline{\mathcal{D}}$ is a compact domain and $\overline{\mathcal{D}} + \nu$ is the ν-neighborhood of $\overline{\mathcal{D}}$.

Expand the function $\widehat{w}$ into the Fourier series:

$$\widehat{w}(z, \varepsilon) = \sum_{k \subset \mathbf{Z}^n} \widehat{w}^k(y, \varepsilon) e^{i\langle k, x \rangle}.$$

A2. Analyticity. Let the constant $\alpha > 0$ be such that for any real $s \in [-\alpha, \alpha]$, for any $k \in \mathbf{Z}$ and for any $(x, y) \in \overline{\Sigma}_q \times \overline{V}_\nu = \text{closure}(\Sigma_q) \times \text{closure}(V_\nu)$ the map $(x, y) \mapsto g^{is}(x, y)$ is analytic in x, y and the map $y \mapsto \mathbf{g}_k^s \widehat{w}^k(y)$ is analytic in y.

---

[4]The case of non-constant frequencies $(\omega = \omega(y))$ can be reduced to this one in a small neighbourhood of an unperturbed (may be, resonant) torus $\{y = y^0\}$.

Moreover, suppose that for some real ρ

$$|\mathbf{g}_k^s \widehat{w}^k|_{\overline{V}_\nu \times [0,\varepsilon_0)} \leq \mu \|k\|^{-\rho} e^{-q\|k\|}, \qquad k \neq 0.$$

The function $\widehat{w}^0$ is of order ε and we assume that $|\mathbf{g}_0^s \widehat{w}^0|_{\overline{V}_\nu \times [0,\varepsilon_0)} \leq \varepsilon \mu_0$.

The constants $\mu, \mu_0, \nu, q, \rho, \gamma$ must satisfy some conditions [18]. Here we replace these conditions by more restrictive, but simple ones.

A3. $\mu, \mu_0, \nu, q, \rho, \gamma$ are positive, do not depend on ε, and $\rho > \gamma/(\gamma + 1)$.

Theorem 2 [18]. *Suppose that assumptions* **A1**–**A3** *hold. Then there exists a change of variables*

$$z \mapsto z_\bullet = f(z, \varepsilon), \quad f : \Sigma_{2q/3} \times V_{2\nu/3} \times [0, \varepsilon_0) \to \mathbf{C}^n/(2\pi\mathbf{Z}^n) \times \mathbf{C}^m \qquad (2.9)$$

such that f is analytic in z, smooth in ε, $f(z, \varepsilon) = z + O(\varepsilon)$, and the system (2.6) takes the form

$$\dot{z}_\bullet = \overline{\omega} + \varepsilon(\overline{w}(y_\bullet) + w_\bullet(z_\bullet, \varepsilon)). \qquad (2.10)$$

Let $w_\bullet^0(y_\bullet, \varepsilon)$ be the average in $x_\bullet$ of $w_\bullet$. Then $w_\bullet^0(y_\bullet, 0) = 0$. Moreover,

$$|w_\bullet - w_\bullet^0| \leq C\mu\varepsilon^{\rho/(\gamma+1)} e^{-\overline{q}\varepsilon^{-1/(\gamma+1)}}, \qquad z_\bullet \in \Sigma_{q/2} \times V_{\nu/2}, \qquad (2.11)$$

where C is a constant, not depending on ε and μ,

$$\overline{q} = (1 + \gamma^{-1})(\gamma\gamma_0\alpha q^\gamma)^{1/(\gamma+1)}. \qquad (2.12)$$

If the system (2.6) is Hamiltonian with respect to a certain symplectic structure Ω then (2.10) is also Ω-Hamiltonian and the change (2.9) is Ω-symplectic.

In [18] we also present another theorem which shows that the Fourier series $w_\bullet - w_\bullet^0$ can be divided into 2 parts: one is small and for another we have a sort of control.

Results analogous to Theorem 2 (without estimates for α) are contained in [22, 1, 13].

2.3. Exponentially small separatrix splitting

The phenomenon of exponentially small separatrix splitting was discovered by Poincaré [16]. Intensive quantitative studying of the problem was initiated by papers [7, 10] (see also [2]). The method proposed by Lazutkin with collaborators [12, 11, 5, 6] is based on an analysis of the separatrices in the complex domain. Another method was used in [4], where direct expansions of the Poincaré-Melnikov integral in an additional parameter are analyzed. The resurgent analysis is applyed to these problems in [20].

The main difficulty of the problem is that the traditional Poincaré-Melnikov method can not be applied directly. Indeed, its error has the order of square of the perturbation. Hence, the error considerably exceeds the expected result.

Exponentially small separatrix splitting can be studied with the help of the continuous averaging method. The main idea of to reduce the rate of the perturbation to an exponentially small quantity such that its square is much smaller than the result. Then the Poincaré-Melnikov method is applicable.

It is natural to measure the rate of the separatrix splitting by the area $\mathcal{A}_{\text{lobe}}$ of a lobe domain, bounded by segments $I^{s,u}$ of the stable (s) and unstable (u) separatrix such that I^s and I^u have the same boundary points, these points are homoclinic, and there are no other common points of I^s and I^u.

Consider the system with Hamiltonian

$$\widehat{H}(\widehat{x}, \widehat{y}, t) = \varepsilon(\widehat{y}^2/2 + (1 + 2B\cos t)\cos\widehat{x}), \qquad (2.13)$$

where $\widehat{x}, \widehat{y}$ are canonically conjugated variables and $\varepsilon > 0$ is small. Theorem 1 implies that there exists a symplectic change of coordinates $\widehat{x}, \widehat{y} \mapsto x, y$ which is

i) close to the identity,

ii) 2π-periodic in time,

iii) real-analytic in a complex neighborhood of the separatrices $\Gamma^\pm$ of the system with Hamiltonian $H_0 = \widehat{y}^2/2 + \cos\widehat{x}$

iv) such that the new Hamiltonian function takes the form:

$$H(x, y, t) = \varepsilon\big(H_0(x, y) + \varepsilon H_1(x, y, \varepsilon) + \exp(-c/\varepsilon)H_2(x, y, t, \varepsilon)\big).$$

Here $H_0 = y^2/2 + \cos x$, the constant $c \in [0, \pi/2)$ is arbitrary, the functions H_1, H_2 are real-analytic in x, y in the vicinity of the unperturbed separatrices of the hyperbolic fixed point $(\widehat{x}, \widehat{y}) = (0, 0)$, smooth in $\varepsilon > 0$ and the function H_2 is analytic and 2π-periodic in t.

The ordinary Poincaré-Melnikov theory applied to this system for positive $\tilde{c}$ gives a correct asymptotics of the separatrix splitting, [25]. The following estimate holds:

$$\mathcal{A}_{\text{lobe}} = \frac{8\pi}{\varepsilon}\exp\Big(-\frac{\pi}{2\varepsilon}\Big)\Big(Bf(B^2) + O\big(\frac{\varepsilon}{\log\varepsilon}\big)\Big),$$

where f is an entire real-analytic function, $f(0) = 2$. In [25, 23] numerical values of several Tailor coefficients of f are presented.

Analogous results for the Standard Chirikov Map and for some Hamiltonian systems with 2 degrees of freedom can be found in [23, 26, 15].

2.4. Inclusion of a map into a flow

In this section we consider the following problem: to present a given self-map of a manifold M as the time-one map (the Poincaré map) in some ODE system, generated by a periodic in time vector field.

The problem can be formulated for various classes of maps and the corresponding vector fields. For example, it is possible to consider generic maps and vector fields, reversible ones with respect to some involution, Hamiltonian, preserving a volume, etc. Here we discuss the analytic set up i.e., assuming that the map is real-analytic, we look for its inclusion into a flow generated by a real-analytic vector field. The problem in C^∞ category is much simpler.

The following construction is well-known. Given a diffeomorphism T of a manifold M onto itself consider the direct product $M \times [0,1]$ with the vector field $\partial/\partial t$, where t is the coordinate on $[0,1]$. The map T generates the identification

$$M \times \{0\} \sim M \times \{1\}, \quad (z,0) \sim (T(z),1).$$

This identification converts $M \times [0,1]$ into a manifold $\mathcal{M}$. Let $\pi : M \times [0,1] \to \mathcal{M}$ be the natural projection. The smooth vector field $\partial/\partial t$ generates on the surface $\pi(M \times \{0\}) \subset \mathcal{M}$ the Poincaré map coinciding with T.

This construction does not solve the problem we deal with because in general it is not clear if $\mathcal{M}$ is real-analytically diffeomorphic to $M \times \mathbf{T}^1$. Nevertheless, sometimes this can be proven.

The problem is solved in the symplectic set up for maps which are close to integrable, [3, 8, 9] and for generic maps [27].

Note that all the known proofs in the analytic set up use essentially the Grauert theorem on the inclusion of an analytic manifold into Euclidean space (or modified versions of this theorem). Our result is based on the method of continuous averaging.

First, note that any map which is not isotopic to the identity[5] obviously can not be included into a flow. Let M be an m-dimensional compact real-analytic manifold. Let χ be a closed subalgebra in the Lie algebra $(\mathcal{L}, [\,,])$ of all analytic vector fields on M.

We denote by X the subset of all analytic diffeomorphisms of M obtained as a result of the time-one shift along solutions of a system

$$\dot{z} = u(z,t), \qquad u(\cdot,t) \in \chi, \quad t \in [0,1], \quad z \in M \qquad (2.14)$$

(it is not assumed that $u(z,0) = u(z,1)$). We assume that the vector field u is C^2-smooth with respect to time. This smoothness condition is technical. For example, it can be replaced by continuity in t in Hamiltonian case and in the general one.

Obviously, all diffeomorphisms from X are isotopic to the identity inside X.

Theorem 3 *For any map $T \in X$ there exists a vector field*

$$U = U(z,t), \qquad U(\cdot,t) \in \chi, \quad t \in \mathbf{R}, \quad z \in M$$

which is analytic in z and t, 2π-periodic in t, and such that the time-2π shift along its trajectories coincides with T.

As a corollary we obtain a possibility of the inclusion of analytic maps into analytic flows in general, symplectic, and volume-preserving cases.

The vector field U is obviously not unique.

It can be also proved that if T is reversible with respect to some involution $I : M \to M$, $I^2 = \mathrm{id}_M$ (i.e., $T \circ I = I \circ T^{-1}$), the corresponding vector field U, can be also regarded I-reversible: $U(z,t) = -dI\,U(Iz,-t)$.

---

[5] Two smooth maps $T_j : M' \to M''$, $j = 0,1$ (M' and M'' are manifolds) are called isotopic if there exists a family of maps $\widehat{T}_s : M' \to M''$ of the same smoothness class continuous in the parameter $s \in [0,1]$, such that $\widehat{T}_0 = T_0$ and $\widehat{T}_1 = T_1$. In other words, if T_0 can be continuously deformed into T_1.

Suppose that the map T is close to T_0 ($\mathrm{dist}(T, T_0) = \varepsilon$ in a complex neighborhood of M),[6] where T_0 is included into the flow generated by a periodic analytic vector field U_0. Then the vector field U can be chosen close to U_0 ($|U - U_0| = O(\varepsilon)$ in a smaller complex neighborhood of M). In particular, in the symplectic case if T is close to an integrable map, a Hamiltonian system associated with T also can be chosen close to integrable one and the orders of closeness are the same.

Continuous averaging in the proof is used to smooth out the dependence of the original vector field (2.14) on time, [17].

Acknowledgements. The work was partially supported by Russian Foundation of Basic Research grants 02-01-00400 and 00-15-99269, and by INTAS grant 00-221.

References

[1] Bambusi D., Long time stability of some small amplitude solutions in nonlinear Schrödinger equations, *Comm. Math. Phys.* 189 (1997), 205–226.

[2] Delshams A., Seara T. M., An asymptotic expressions for the splitting of separatrices of the rapidly forced pendulum, *Comm. Math. Phys.* 150 (1992), 433–463.

[3] Douady R., Applications du théoreme des tores invariantes, *Thesis*, Université Paris VII, 1982.

[4] Gallavotti G., Twistless KAM tori, quasi flat homoclinic intersections, and other cancellations in the perturbation series of certain completely integrable hamiltonian systems, A review, Reviews on Mathematical Physics, 6, (1994), 343–411.

[5] Gelfreich V., Reference systems for splitting of separatrices, *Nonlinearity* 10 (1997), 175–193.

[6] Gelfreich V., A proof of the exponentially small transversality of the separatrices for the standard map, *Comm.Math.Phys.* 201 (1999), 155–216.

[7] Holmes P., Marsden J., Scheurle J., Exponentially small splittings of separatrices with applications to KAM theory and degenerate bifurcations, *Contemp. Math.*, 81 (1988), 213–244.

[8] Kuksin S., On the inclusion of an almost integrable analytic symplectomorphism into a Hamiltonian flow, *Russian J. Math. Phys.*, 1, no. 2 (1993), 191–207.

[9] Kuksin S., Pöschel J., On the inclusion of analytic symplectic maps in analytic Hamiltonian flows and its applications, In *Progress in Nonlinear Differential Equations and Their Applications*, V. 12 Ed. Kuksin S., Lazutkin V. F., and Pöschel J. Birkhäuser, 1994, Basel-Boston-Stuttgart, 96–116.

---

[6]The distance can be defined for example as follows:

$$\mathrm{dist}(T, T_0) = \sup_{z \in M'} \rho(T(z), T_0(z)),$$

where M' is a complex neighborhood of M, and ρ is some metric. The choice of the neighborhood and of the metric plays no role due to compactness of the manifold M.

[10] Lazutkin V. F., Splitting of separatrices for standard Chirikov's mapping, VINITI no. 6372-84, 24 Sept. 1984 (in Russian).

[11] Gelfreich V. G., Lazutkin V. F., Tabanov M. B., Exponentially small splitting in Hamiltonian systems, *Chaos*, 1, no. 2 (1991), 137–142.

[12] Gelfreich V. G., Lazutkin V. F., Svanidze N. V., Refined formula to separatrix splitting for the standard map, *Physica D*, 71 (1994), 101–121.

[13] Lochak P., Marco J. P., Sausin D., On the splitting of invariant manifolds in multidimensional near integrable Hamiltonian systems, *Preprint* 1999.

[14] Neishtadt A. I., The separation of motions in systems with rapidly rotating phase, *J. Appl. Math. Mech.*, 48, no. 2 (1984), 133–139.

[15] Novik A., Exponentially small separatrix splitting in some Hamiltonian systems with 2 degrees of freedom, preprint 2002 (in Russian).

[16] Poincaré H., Les métodes nouvelles de la mécanique céleste, V. 1–3. Paris: Gauthier–Villars, 1892, 1893, 1899.

[17] Pronin A., Treschev D., On the inclusion of analytic maps into analytic flows, *Regular and Chaotic Dynamics*, 2, no 2 (1997), 14–24.

[18] Pronin A., Treschev D., *Regular and Chaotic Dynamics*, 5, no 2 (2000), 157–170.

[19] Ramis J. P., Schäfke R., Gevrey separation of fast and slow variables, *Nonlinearity*, 9 (1996), 353–384.

[20] Sauzin D., Résurgence paramétrique et exponentielle petitesse de l'écart des séparatrices du pendule rapidement forcé. *Ann. Inst. Fourier, Grenoble* 45, no. 2 (1995), 453–511.

[21] Sauzin D., Caractère Gevrey des solutions formelles dún problème de moyennisation, *C. R. Acad. Sci. Paris*, T. 315, Serie I, (1992), 991–995.

[22] Simó C., Averaging under fast quasi-periodic forcing in Hamiltonian mechanics: Integrability and chaotic behavior, J. Seimenis ed., NATO Adv. Sci. Inst. Ser. B Phys. 331, Plenum Press, New York (1994), 13–34.

[23] Treschev D., An averaging method for Hamiltonian systems, exponentially close to integrable ones, *Chaos*, 6, no 1 (1996), 6–14.

[24] Treschev D., The continuous averaging method in the problem of separation of fast and slow motions, *Regular and Chaotic Dynamics*, 2, no 3/4 (1997), 9–20 (in Russian).

[25] Treschev D., Separatrix splitting for a pendulum with rapidly oscillating suspension point, *Russian J. of Math. Phys.*, 5, no 1 (1997), 63–98.

[26] Treschev D., Continuous averaging in Hamiltonian systems. in "Hamiltonian systems with three or more degrees of freedom", ed. C.Simo, Kluwer, 1999, NATO ASI Series C, Vol. 533, 244-253.

[27] Trifonov S. I., Analytic diffeomorphisms as monodromy maps of analytic differential equations, *Moscow Univ. Math. Bull.* 41, no 5, (1986), 63–65.

Section 13. Mathematical Physics

ICM 2002 · Vol. III · 395–408

Dirichlet Branes, Homological Mirror Symmetry, and Stability

Michael R. Douglas[*]

Abstract

We discuss mathematics which has come out of the study of Dirichlet branes in superstring theory, focusing on the case of supersymmetric branes in Calabi-Yau compactification. This has led to the formulation of a notion of stability for objects in a derived category, contact with Kontsevich's homological mirror symmetry conjecture, and "hysics proofs" for many of the subsequent conjectures based on it, such as the representation of Calabi-Yau monodromy by autoequivalences of the derived category.

2000 Mathematics Subject Classification: 81T30, 14J32, 81T45.
Keywords and Phrases: Superstring theory, Dirichlet branes, Homological mirror symmetry, Stability.

1.   Introduction

Let M be a complex manifold with Kähler metric, and E a vector bundle on M. The hermitian Yang-Mills (HYM) equations are nonlinear partial equations for a connection on E, written in terms of its curvature two-form F:

$$F^{2,0} \quad = 0, \tag{1.1}$$

$$\omega \cdot F^{1,1} \quad = const. \tag{1.2}$$

The theorems of Donaldson [1] and Uhlenbeck-Yau [2] state that irreducible connections solving these equations are in one-to-one correspondence with μ-stable holomorphic vector bundles E.

In this talk, we will state a conjecture which generalizes this statement, motivated by superstring theory, and some results which follow. In [3] we gave an introduction for mathematicians to the family of problems which comes out of the study of Dirichlet branes in type II string compactification on a Calabi-Yau threefold (henceforth, CY_3), *i.e.* a three complex dimensional Ricci-flat Kähler manifold

---
[*]Dept. of Physics and Astronomy, Rutgers University, Piscataway NJ 08855, USA and IHES, Bures-sur-Yvette 91440, France. E-mail: mrd@physics.rutgers.edu

[4]. These manifolds are of central importance in superstring compactification and their study has led to many results of physical and mathematical importance.

Given a manifold M, one can define a nonlinear sigma model with target space M, using methods of quantum field theory. It has been proven to physicists' satisfaction that for M Calabi-Yau, this sigma model exists and is a $(2,2)$ superconformal field theory (SCFT), which can be used to compactify superstring theory. The space of such theories is parameterized locally by a choice of complex structure on M, and a choice of "stringy Kähler structure." A subset of the stringy Kähler structures are specified by a complexified Kähler class, i.e. a point in $H^2(M, \mathbb{C})/H^2(M, \mathbb{Z})$. However this is only a limit (usually called "large volume limit") in the full moduli space of stringy Kähler structures [5, 6]; we define this space below.

A mirror pair of CY$_3$'s M and W is a pair for which the nonlinear sigma models associated to M and W are the same, up to a nontrivial automorphism of the $(2,2)$ superconformal algebra. This equates observables in the two models, as we discuss below. Physics also provides some techniques for computing these observables, and predicts certain examples of mirror pairs.

Many interesting and mathematically precise conjectures have been made based on this equivalence [7, 8], starting with the work of Candelas *et al* [9] which gave a formula for the number (suitably defined) of rational curves of given degree in M a quintic hypersurface M in $\mathbb{P}^4$, in terms of data about the periods of the holomorphic three-form on its mirror W. This formula has since been proven by Givental [10], by a tour de force computation.

The first conjecture which tried to explain mirror symmetry on a deeper level was made by Kontsevich [11], who proposed that it should be understood as an equivalence between the derived category of coherent sheaves on M, and the Fukaya category on W, a category whose objects are isotopy classes of Lagrangian submanifolds carrying flat connections, and whose morphisms are elements of Floer cohomology.

It took a little while for physicists to appreciate this conjecture. For one thing, the natural physical objects to which it relates, Dirichlet branes, were almost completely unknown when it was made. With the "second superstring revolution," D-branes quickly moved to center stage, but even so the physics relating to this conjecture is rather subtle and took a while to uncover.

Direct contact was made in [12], where it was shown that the derived category of coherent sheaves on M, $D(\text{Coh } M)$, could be obtained as a category of boundary conditions in the B-type topologically twisted sigma model on M. Further developments in this story appear in [13, 14, 15, 16, 17]. This makes contact with the mathematics of equivalences between derived categories in various ways, which we will describe.

On realizing how the derived category arises in string theory, one sees new features which were not present in the standard mathematical treatments. The derived category of coherent sheaves is $\mathbb{Z}$-graded, but it is clear in string theory that the gradings are naturally $\mathbb{R}$-valued, and depend on stringy Kähler data. This was mentioned in [3], and we will describe it in more detail here.

The simplest question about Dirichlet branes is to know which ones exist as physical branes. More precisely, while the general argument leading to the derived category tells us that every B-type brane corresponds to some object in $D(\text{Coh } X)$, not all of these objects are actually physical B-type branes, and we would like to know which ones are.

In the large volume limit, the physical branes are the coherent sheaves associated to solutions of the HYM equations (with certain allowed singularities), and this question was answered by the DUY theorems: the physical branes are the μ-stable sheaves. More precisely, a Yang-Mills connection with $F^{2,0} = 0$ sits in an orbit of the action of the complexified gauge group $G_{\mathbb{C}}$ (say $GL(N, \mathbb{C})$), and one wants to know whether a given orbit of this group contains a solution of (1.2). This question is an infinite dimensional analog of those considered in geometric invariant theory (GIT), and this analogy suggests that this will be true iff the orbit is μ-stable. The DUY theorems justify this expectation.

This answer has the interesting feature that μ-stability and thus the set of stable objects depends on the Kähler class of M, and can change on walls of codimension 1. On the other hand, because the Yang-Mills equations are scale invariant, it does not depend on the overall scale of the Kähler class.

Superstring theory leads to "generalized HYM equations," which reduce in the large volume limit to the HYM equations, but which explicitly introduce a length scale, the "string length." The next step in the discussion would seem to be to study these equations. However, although some results are known in this direction [18, 19], a usable general equation incorporating all stringy effects is not known at this writing. We will comment on this problem in the conclusions.

Rather, what we will do is propose the stringy analog of μ-stability, a necessary and sufficient condition for an object in $D(\text{Coh } X)$ to correspond to a physical boundary condition, called Π-stability. This condition will depend on stringy Kähler moduli, and will describe the generalization of wall crossing and other phenomena in μ-stability to string theory. One can even prove the "easy" half of the DUY theorems, that a Π-unstable object cannot be a physical brane, directly from SCFT.

Despite the direct analogy, the discussion differs dramatically from that made in geometric invariant theory, because the derived category is not an abelian category. Even the most basic elements of the standard discussion need to be rethought. In particular, there is no concept of subobject in the derived category. We will overcome these problems, but by paying a price. Rather than formulating a condition which can be tested on an object E at a point in $\mathcal{M}_K$, we will give a rule which describes the variation of the entire set of stable objects at a point p in $\mathcal{M}_K$, call it Stab_p, to a new set at another point, say $\text{Stab}_{p'}$. This rule depends on the path in $\mathcal{M}_K$ connecting p and p', but conjecturally only on its homotopy class.

This leads to an interesting phenomenon when one traverses a homotopically nontrivial loop in moduli space. Physically, the result of this must be to recover the original list of stable objects. However, the specific objects in the derived category which represent these stable objects can change. In other words, such a loop is naturally associated to an autoequivalence in $D(\text{Coh } X)$. Considerations of mirror symmetry had previously led to conjectures relating loops and monodromies to autoequivalences [20, 21, 22]. Using Π-stability, we can in a sense prove these conjectures, as we will explain.

2. String theory origin of the formalism

We will not try to explain the stringy aspects of the problem in any detail, but just convey the central ideas. The most central idea is the relation between the "physical" string world-sheet, described by SCFT, and the related topologically twisted string or TFT [23]. Given an SCFT, a related TFT is obtained by choosing a world-sheet supercharge, an operator Q in the $N = 2$ algebra satisfying $Q^2 = 0$, and restricting the Hilbert space to its cohomology. In $(2,2)$ SCFT one can make two choices of supercharge, leading to the "A-type" and "B-type" TFT's. In this language, a mirror pair of CY_3's is a pair (M, W) such that their twisted sigma models satisfy $\text{TFT}_A(M) = \text{TFT}_B(W)$ and $\text{TFT}_B(M) = \text{TFT}_A(W)$.

The "stringy" geometry of the CY M is fully encoded in the SCFT, while the TFT only gives a small subset of this. The closed string sector generally corresponds to metric properties, and the A and B-type TFT's respectively encode the stringy Kähler structure and the complex structure. Thus, mirror symmetry can be used to give a definition to the stringy Kähler structure: the stringy Kähler moduli space $\mathcal{M}_K(M)$ is isomorphic to the standard complex structure moduli space $\mathcal{M}_c(W)$.

On general grounds, boundary conditions in an SCFT are objects in a category, whose morphisms $\text{Hom}(E, F)$ are the Q-cohomology of the Hilbert space of open strings stretching from the boundary condition E to the boundary condition F. In $(2,2)$ SCFT, this category will be graded (by "$U(1)$ charge"). The category is not guaranteed to be abelian, but in a sense is generated by a finite set of abelian subcategories (the branes with specified grading).

A priori, each SCFT (i.e. with given complex and Kähler moduli) leads to a different category of boundary conditions. A suitable subset of these descend to TFT, to provide a category of boundary conditions in TFT. The morphisms are called "topological open strings."

In the particular case of the sigma model on M, the simplest B-type boundary conditions are holomorphic submanifolds carrying a holomorphic bundle [23]. If we consider two objects E and F whose support is M and carrying bundles E and F, the operator Q becomes the $\bar{\partial}$ operator coupled to a difference of the gauge connections. Its cohomology, the morphisms $\text{Hom}(E, F)$ are the Dolbeault cohomology, $H^{0,p}(M, E^v \otimes F)$. The grading agrees with the standard grading.

More generally, the role of holomorphic bundles in the discussion of the HYM equations, is played in the string theory generalization by boundary conditions in TFT_B. Thus we need to know if this is the most general boundary condition in $\text{TFT}_B(M)$. In fact, one can find direct evidence that it is not. The main point is the following: the definition of $\text{TFT}_B(M)$ by twisting an SCFT shows that it is independent of the stringy Kähler moduli, and so one must obtain the same TFT_B by twisting any SCFT obtained by deforming in $\mathcal{M}_K(M)$. On the other hand, when one compares the boundary states in examples obtained from various SCFT's, one finds that they are not the same.

Independence of $\text{TFT}_B(M)$ on $\mathcal{M}_K(M)$ can only be recovered by postulating some larger class of boundary conditions which contain the ones produced by twisting any specific SCFT. Since the definition of TFT is made in homological terms, one can attempt to use any of the standard generalizations known in mathematics. Kontsevich's conjecture suggests that we try to formulate boundary conditions which are objects in the derived category of coherent sheaves on X, $D(\text{Coh } X)$. More generally, we would start with any of the various abelian categories A of boundary conditions A in our SCFT, and try to use an object in $D(A)$. In fact this can be done, by more or less repeating the standard mathematical construction.

We still need to justify the claim that this is an appropriate definition of TFT. This involves two non-trivial steps. The first of these is to argue that direct sums of boundary conditions indexed by an integer $\mathbb{Z}$ naturally appear as boundary conditions. One then sees by more or less standard TFT arguments that the operator Q can be modified by turning on maps between terms in the complexes, and that the resulting cohomology is chain maps up to homotopy equivalence. The second non-trivial step is to identify complexes related by quasi-isomorphisms, morphisms which act as the identity on homology.

The justification of the first step, the $\mathbb{Z}$-grading on complexes, is tied to the grading and the structure of the $N = 2$ superconformal algebra [12]. In particular, an A or B-type boundary condition in SCFT naturally carries a grading. However, it turns out that this is not $\mathbb{Z}$-valued but $\mathbb{R}$-valued. This can be used to make complexes, whose terms have grades differing by integers. In fact, it turns out that $D(A)$ is independent of stringy Kähler moduli only as an ungraded category; the gradings depend on the point in $\mathcal{M}_K$, and undergo "flow of gradings." We will describe the consequences of this in the next section.

To a certain extent, one can also justify the second step, quotienting by quasi-isomorphisms, by physical arguments [12, 13, 16]. The basic intuition is that each brane E in string theory comes with an "antibrane" $\bar{E}$, an object whose definition makes clear that a physical configuration containing a E and a E can "annihilate" to a configuration with these two objects removed. This can be formalized in the statement that branes are classified by K theory [24], the additive group completion of the semigroup under direct sum, with deformation equivalent objects identified.

One would prefer to use a finer equivalence relation which allows determining the holomorphic maps between objects (the topological open strings). This requirement can be met by identifying complexes up to quasi-isomorphism; for example we identify the complex $E_0 \longrightarrow E_1$ with cohomology in one degree, with this cohomology. Physically, the intuition behind this is that the brane E_0 and the antibrane $\bar{E}_1$ can "form a bound state," which is the same physical object as the cohomology, so this identification is well motivated. By repeating this construction and moving in $\mathcal{M}_K$, one can justify fairly general identifications [16]. Since the derived category is the universal construction of this type, it is clear that it can be used in this context; we return to this point in the conclusions.

Thus, the boundary conditions in a given SCFT are a subset of the objects in $D(A)$ for any A obtained from an SCFT related by motion in $\mathcal{M}_K$, and we seek a condition which specifies these. Since the problem reduces to the HYM equations in the large volume limit, we are led to look for a condition generalizing μ-stability, which we will describe in the next section.

There is one primary input which can be proven from SCFT and which we can rely on in formulating this condition. Namely, our boundary conditions lead to unitary representations of the $(2,2)$ superconformal algebra. This leads to the general result [25] that the grading φ of a morphism in a SCFT must satisfy $0 \leq \varphi \leq \hat{c} = 3$ in our examples. On the other hand, flow of gradings would generally lead to violations of this bound. This potential contradiction will lead to the stability condition.

3.   Flow of gradings and Π-stability

We now try to state the resulting formalism in a mathematically precise way. We suppose that we start with a CY$_3$, M, with mirror W. Recall that $\mathcal{M}_K(M) \cong \mathcal{M}_c(W)$, the moduli space of complex structures on W. The choice of M determines a "large complex structure limit point" $p_M \in \mathcal{M}_K(M)$. There is furthermore a "topological mirror map," an isomorphism

$$K_0(M) \xrightarrow{m} H_3(W, \mathbb{Z})$$

where $K_0(M)$ is the topological K theory group.

A point p in $\mathcal{M}_c(W)$ determines a holomorphic three-form Ω_p on W, and thus the periods

$$\Pi : \mathcal{M}_c(W) \times H_3(W, \mathbb{Z}) \longrightarrow \mathbb{C}; \qquad \Pi_p(\Sigma) = \int_\Sigma \Omega_p.$$

Composing $\Pi \circ m$, we obtain a character on $K_0(M)$ for each p, which we will also call Π_p. This is usually called the "central charge" in the physics literature.

The periods Π are analytic with singularities on a discriminant locus $\mathcal{D}$ of codimension 1 on which W degenerates. We define $\mathcal{T}_K$ to be the "stringy Teichmüller space," a cover of $\mathcal{M}_K - \mathcal{D}$, on which Π is single valued. A deck transformation on $\mathcal{T}_K$ corresponds to a linear monodromy on Π which preserves the intersection form, i.e. an element of $\mathrm{Sp}(b_3(W), \mathbb{Z})$.

Let Stab$_p$ be the set of stable objects at a point p in $\mathcal{T}_K$, a subset of the objects in $D(\mathrm{Coh}\ X)$. Suppose we know Stab$_p$ at one point p; we will give a rule which, upon following a path $\mathcal{P}$ in $\mathcal{M}_K$ from p to q, determines Stab$_q$.

We assume that at no point x on the path does $\Pi_x(E) = 0$ for an $E \in$ Stab$_x$. Indeed, there are physical arguments [26] that this can only happen on the discriminant locus (and furthermore must happen there).

The basic ingredient in the rule is the assignment of $\mathbb{R}$-valued gradings to all morphisms between stable objects. We denote the grading of a morphism $f \in \mathrm{Hom}(E, F)$ as $\varphi_p(f)$. Given the grading of f between stable objects at a point p, the grading at q obtained by following the path $\mathcal{P}$ is

$$\varphi_q(f) = \varphi_p(f) + \Delta_{p,q}\varphi(F) - \Delta_{p,q}\varphi(E) \tag{3.1}$$

where

$$\Delta_{p,q}\varphi(E) = \frac{1}{\pi}\mathrm{Im}\ \log \frac{\Pi_q(E)}{\Pi_p(E)}, \tag{3.2}$$

resp. for F. We keep track of the branch of the logarithm, so this defines an $\mathbb{R}$-valued shift of the grading $\varphi(f)$. We refer to this as "flow of gradings."

Given the gradings at an initial point p, this determines the gradings at the endpoint of any path. Note that flow preserve the usual additivity of gradings. Also, morphisms from an object to itself do not undergo flow. This implies that Serre duality is still sensible, and (for our CY$_3$'s) we still have $\varphi(f) + \varphi(f^v) = 3$.

We will assume that the gradings satisfy the condition that if $\Pi_p(E)/\Pi_p(F) \in \mathbb{R}$, $\varphi(f_{E \longrightarrow F}) \in \mathbb{Z}$. This is true in string theory and will be true with the "initial conditions" we describe below.

One can equivalently describe the gradings of morphisms by defining a grading on stable objects, an $\mathbb{R}$-valued function $\varphi_p(E)$ which satisfies the constraints $\varphi_p(E[1]) = \varphi_p(E) + 1$ and $\Pi_p(E)/e^{i\pi\varphi_p(E)} \in \mathbb{R}^+$, and varies continuously on $\mathcal{T}_K$. One can then adopt the usual rule that a morphism in $\mathrm{Hom}(E[m], F[n])$ has grading $n - m$, but now with $m, n \in \mathbb{R}$. This tells us that when an object becomes stable, we only need to specify the grading of one morphism involving it, to determine the gradings of all morphisms involving it.

We next consider the distinguished triangles of $D(\mathrm{Coh}\ M)$. With conventional gradings, these take the form

$$(3.3)$$

and involve two morphisms of degree zero, and one of degree one, denoted [1]. After flow of gradings, the gradings of morphisms will typically not be zero or one, but will always satisfy the relation

$$\varphi(u) + \varphi(v) + \varphi(w) = 1. \qquad (3.4)$$

We now distinguish "stable" and "unstable" triangles at each $p \in \mathcal{M}_K$, according to their gradings. In a stable triangle, all three morphisms satisfy the constraints

$$0 \leq \varphi(u), \varphi(v), \varphi(w) \leq 1. \qquad (3.5)$$

If any of these constraints is violated, the triangle is unstable.

As one moves along a path in $\mathcal{T}_K$, the gradings will vary, and the stability of triangles will change, on walls of real codimension 1 in $\mathcal{T}_K$. Given the assumptions we discussed, one can see that when the stability of a triangle changes, two morphisms will have grading zero, and one will have grading one. This can be seen by plotting the central charges of the three objects involved in the complex plane. If we consider the triangle with these vertices, the condition that the distinguished triangle be stable is that this triangle be embedded with positive orientation. A variation of stability is associated to colinear central charges, and a degeneration of this embedding.

The fundamental rule is that when a stable triangle degenerates, the object with largest mass $m(E) = |\Pi(E)|$ becomes unstable. Physically, one says that it "decays into the other two." This is the string theory generalization of "wall-crossing" of μ-stability under variation of Kähler class. As we discussed above, this rule follows directly from CFT, independent of any relation to HYM or other geometric theories. If we grant the claim that the HYM equations emerge as the large volume limit of this formalism, we can say that we have in fact a new derivation of this phenomenon.

Consistency requires us to implement this rule in the opposite direction as well: if an unstable triangle with two stable vertices becomes stable, the third object becomes stable, with grading the same as that of the other two. This determines the grading of all other morphisms it participates in. For example, when the triangle (3.3) becomes stable, the object B becomes stable, with $\varphi(B) = \varphi(A)$; this determines the gradings of all morphisms.

Applying this rule in practice requires keeping track of an infinite set of stable objects and triangles. Some of the issues involved in making sense of this are discussed in [16]. We conjecture that they lead to a unique Stab_p for every $p \in \mathcal{T}_K$, independent of choices of order of application of the rules or other ambiguities, because string theory defines a unique set of B-type branes at each point.

Although we do not have a mathematical proof of consistency, we have some understanding of it, and can point out two nontrivial ingredients involved in this claim. One is the question of whether Stab_q depends on the choice of path taken from p to q. In general, one expects such dependence, but only on the homotopy class of path in $\mathcal{M}_K - \mathcal{D}$, which will disappear on $\mathcal{T}_K$. On the other hand, one can construct simple examples which point to a potential problem with this claim. Namely, one can find two intersecting walls, along one of which one has A decaying to $B + C$, and the other with B decaying to $E + F$. After crossing the second wall, one is not supposed to consider the ABC triangle (since B is not stable), leading to the possibility that following a second path in the same homotopy class would lead to the contradiction that A remains stable. This contradiction does not happen, by an argument given in [16] which invokes the "octahedral axiom" of triangulated categories to show that a third triangle ACF will still predict a decay. This argument also suggests that there are no ambiguities due to rule ordering.

The second ingredient is the physical claim that "the spectrum has a mass gap," in other words that the set $S_p = \{|\Pi_p(E)| : E \in \mathrm{Stab}_p\} \subset \mathbb{R}_{\geq 0}$ at fixed p has a positive lower bound. This assumption prevents infinite chains of brane decay or creation and is physically widely believed to hold in these problems.

4.   Results

We first discuss consequences of the physical argument that $D(A)$ should be the same for the various abelian categories A obtained by explicit SCFT constructions. First, as one $\mathcal{M}_K$ can contain different large volume limits M, M', etc., one infers that $D(\mathrm{Coh}\ M) \cong D(\mathrm{Coh}\ M')$ for all of these limits. In the known physical examples, the various limits are obtained by taking different moment maps in symplectic quotients (the linear sigma model construction [6]), and thus are all birationally equivalent; the simplest picture is that this equivalence holds for any birationally equivalent M and M'. Indeed, this is generally believed to be true (for arbitrary smooth projective varieties) and to some extent has been proven [27, 28].

One can also define and work with boundary conditions at various other "exactly solvable" points in $\mathcal{M}_K$ such as orbifold points [29] and Gepner models [30]. These lead to different abelian categories, some of whose objects can naturally be identified with holomorphic bundles, and some which are not. Physical results on resolution of orbifolds then lead to the general predictions $D(\text{Coh } X) \cong D(\text{Coh}^\Gamma \mathbb{C}^3) \cong D(\text{Mod} - Q_\Gamma)$, where $\Gamma \subset SU(3)$ is a discrete subgroup, $X \to \mathbb{C}^3/\Gamma$ is a crepant resolution, $\text{Coh}^\Gamma M$ are Γ-equivariant sheaves on M, and Q_Γ is the path algebra (with relations) of the McKay quiver for the group Γ [31, 32]. Indeed this is the subject of the generalized McKay correspondence [33], and these facts are shown in [34, 35].

We next turn to discussing Π-stable objects. To get started, one needs to know Stab_p and the gradings at some point p. The simplest choice is a large volume limit, in which the stable objects are coherent sheaves. The gradings which come out of string theory are not quite the standard gradings of sheaf cohomology, but are as follows: a morphism $f \in \text{Ext}^n(E, F)$ between sheafs with support having complex codimension $\text{codim}(E)$ and $\text{codim}(F)$, has grading $\varphi = n + (\text{codim}(F) - \text{codim}(E))/2$. The central charges $\Pi(E)$ are in this limit given by

$$\Pi(E) = \int_M \text{ch}(f_! E) e^{-B - i\omega} \sqrt{\hat{A}(TM)} \quad + O(e^{-\int \omega}), \tag{4.1}$$

where ch is the Chern character, f is an embedding and $f_! E$ is the K-theoretic Gysin map [36].

Dropping the exponentially small terms (world-sheet instanton corrections), one can easily show that Π-stability reduces to a stability condition for the modified HYM equation considered in [18, 19], and in the limit $\omega >> B, F$ to μ-stability [38]. Thus Π-stability does describe B-type branes in string theory without instanton corrections.

Another limit which is available in certain problems (for example, orbifolds $\mathbb{C}^3/\Gamma$ is one in which all central charges Π are real. One can see that in this case, Π-stability reduces to θ-stability in an abelian category [37], and again this can be directly justified from string theory [38]. This simpler description is valid near the limit as long as the central charges for stable objects live in a wedge W and its negation $-W$ in the complex plane.

We now give some stringy examples in the context of sheaves on a projective CY_3 M, say for definiteness the quintic hypersurface in $\mathbb{P}^4$. We follow a path in $\mathcal{M}_K$ parameterized by $J = \int \omega$, decreasing from large J.

The following is a distinguished triangle in $D(\text{Coh } M)$; the numerical superscripts are the gradings of morphisms at large volume;

$$\mathcal{O} \xrightarrow{0} \mathcal{O}(n) \xrightarrow{1/2} \mathcal{O}_\Sigma \xrightarrow{1/2} \mathcal{O}[1] \cdots .$$

All three objects are stable at large volume.

Using $\Pi(\mathcal{O}(n)) \sim \frac{1}{6}(n - iJ)^3$, one finds that $\varphi(\mathcal{O}(n)) - \varphi(\mathcal{O})$ increases as J decreases. When this crosses 1, i.e. $\Pi(\mathcal{O}(n))/\Pi(\mathcal{O})$ is negative real, $\mathcal{O}_\Sigma$ goes unstable.

If one continues towards the stringy regime, eventually $\Pi(\mathcal{O}(n))$ and $\Pi(\mathcal{O})$ become colinear again. At this point, the non-sheaf X_n defined at large volume by

$$\mathcal{O}(n) \xrightarrow{3} \mathcal{O} \longrightarrow X_n \longrightarrow \ldots$$

becomes stable. In [30], a Gepner model boundary state was constructed, which has all the right properties to be this object, confirming the idea that non-coherent sheaves can appear as stable objects.

Another example uses the sequence

$$\mathcal{I}_z \longrightarrow \mathcal{O} \xrightarrow{3/2} \mathcal{O}_z \longrightarrow \mathcal{I}_p[1] \ldots$$

where $\mathcal{O}_z$ is the structure sheaf of a point $z \in M$. Now the ideal sheaf $\mathcal{I}_z$ is unstable at large volume, while the other two objects are stable. Decreasing J can now decrease the 3/2, and eventually the brane $\mathcal{I}_z$ will become stable. In particular, this can happens in the vicinity of the "conifold point," [9] a point near which (choose a local coordinate ψ on $\mathcal{M}_K$),

$$\Pi(\mathcal{O}) \sim \psi.$$

Continuing around this point further decreases the grade of this map to 0, and one finds that the point (or D0) $\mathcal{O}_p$ becomes unstable.

One can then continue back to large volume. Since this is a closed loop, on physical grounds it must act as an autoequivalence on $D(\mathrm{Coh}\ M)$. Furthermore it acts on the K theory class as the mirror Picard-Lefschetz transformation,

$$[X] \longrightarrow [X'] = [X] + \langle X, \mathcal{O} \rangle [\mathcal{O}].$$

These facts and the result we just derived that $\mathcal{O}_z$ has as monodromy image $\mathcal{I}_z$, are enough to prove [16] that this autoequivalence is one conjectured in [20, 21, 22], on grounds of mirror symmetry: X becomes X' defined by the triangle

$$\mathcal{O} \otimes \mathrm{Hom}(\mathcal{O}, X)^* \longrightarrow X \longrightarrow X' \longrightarrow \ldots.$$

On physical grounds, a closed loop $\mathcal{P} \subset \mathcal{M}_K$ must induce an autoequivalence $\mathcal{F}_\mathcal{P}$, such that

$$\mathrm{Stab}_{\mathcal{P}p} \cong \mathcal{F}_\mathcal{P} \mathrm{Stab}_p.$$

Thus we conjecture that Π-stability leads to such an autoequivalence. Granting this, we have derived the particular autoequivalence associated to this loop.

Another equivalence of derived categories which has been obtained from Π-stability [17], is one shown by Bridgeland to describe a flop [27]. It is obtained by going around a point at which the volume of the flopped curve σ vanishes,

$$\Pi_\sigma \sim z = B + iJ.$$

This produces the variations of grading

$$\Delta\varphi(\mathcal{O}_\sigma) = 1; \ \Delta\varphi(\mathcal{O}(-1)_\sigma) = -1.$$

By virtue of

$$\mathcal{O}(-1)_\sigma \longrightarrow \mathcal{O}_\sigma \longrightarrow \mathcal{O}_p \longrightarrow \ldots,$$

all points $p \in \sigma$ (and only these points) become unstable under this variation. On the other hand, if σ is a $(-1,-1)$ curve, one can show that $\dim \mathrm{Ext}^2(\mathcal{O}, \mathcal{O}(-1)) = 2$, and the sequence

$$\mathcal{O}_\sigma \longrightarrow \mathcal{O}(-1)_\sigma \longrightarrow X_p \longrightarrow \ldots,$$

defines "flopped points," which are not sheaves in $D(\mathrm{Coh}\ M)$, but can be shown to be precisely the points on the flopped curve σ' in the flopped CY Coh M'. Again, this transformation on the points extends to a unique equivalence of derived categories.

5.   Conclusions and open questions

To summarize, we conjecture that B-type branes in weakly coupled type II string theory compactified on a CY_3 M, are Π-stable objects in $D(\mathrm{Coh}\ M)$.

Although the statement of this conjecture relies on string theory, one can also regard it as *defining* a natural class of objects which are important in mirror symmetry, the stringy generalization of stable holomorphic bundles (or solutions of the hermitian Yang-Mills equation), so we think it could be of interest to mathematicians. Indeed, by assuming the conjecture and simple physical consistency conditions, one can derive various earlier mathematical conjectures related to mirror symmetry, as we discussed in section 4.

We believe that Π-stability as we formulated it in section 3 is mathematically precise. In any case the first open question is to make it precise, prove its consistency, and see to what extent it uniquely determines the stable objects. Some results in this direction have been announced by Bridgeland [41].

Another application, explored in [13, 16, 17], is to explore "stringy geometry" of Calabi-Yau manifolds by using D-branes as probes [42]. One such problem is to find all "D0-branes" at any $p \in \mathcal{M}_K$. A D0-brane is a family of stable objects E_z with moduli space a CY_3 M_p, such that $D(\mathrm{Coh}\ M_p) \cong D(\mathrm{Coh}\ M)$. In the second example above, one sees that one can have more than one D0 ($\mathcal{I}_z$ and $\mathcal{O}_z$). It might be that in other examples, there is no D0.

A primary application of this conjecture in string theory would be to determine whether Π-stable objects with specified K theory class exist in various regions of $\mathcal{M}_K$. In the large volume limit, the role of μ-stability in this problem was discussed in [43]. For such concrete purposes, it would be better to have a criterion which can be applied at a single point in $\mathcal{M}_K$ and does not require considering the infinite set of stable objects. There are examples (e.g. the Gepner point for the quintic) in which it can be shown that Π-stability does not reduce to stability in a single abelian category, but it could be that the stable objects in any small region could be described as stable objects (in a more conventional sense) in any of a finite set of abelian categories.

The statement of Π-stability is admittedly rather complicated, but this to a large extent only reflects the complexity of its string theory origins. It would be interesting to know to what extent it depends on details of mirror symmetry (such as M a CY_3, and Π given by the periods of the mirror W), and to what extent it can be generalized. The basic definitions make sense for any complex M and any character Π on $K_0(M)$, but we suspect that consistency, and especially the relation between homotopically nontrivial loops and autoequivalences, requires further conditions.

Eventually, one would hope to prove the analog of the DUY theorems. Because general SCFT's on CY_3 are not rigorously defined, this is not yet a mathematically well-posed problem. There are particular cases such as Gepner models which can be rigorously defined using vertex operator algebra techniques [44].

Even proving it to the satisfaction of string theorists seems difficult at present, but might be a reasonable goal. Such problems are much studied by string theorists, under the general rubric of "tachyon condensation." [45] The analogy to HYM at least gives us a guideline for how to proceed in this case. One would need to start with a working definition of boundary CFT, including a large enough space of boundary states, and in which $(2, 2)$ supersymmetry is manifest. One might also start with string field theory as developed in [46].

One could then attempt a direct analog of the Yang-Mills gradient flow method used in [1] to construct solutions. Namely, one would start with an boundary condition formed as a complex of physical boundary conditions. In general this will not be conformal, but will flow to a conformal boundary condition under renormalization group flow. If we start with a large enough class of boundary conditions, presumably all boundary conditions could be obtained this way.

It seems to us that the first and perhaps the major element in this project would be to identify and precisely define the group which plays the role of the complexified gauge group $G_{\mathbb{C}}$ in the HYM theory. One then needs to address the question of how equivalence classes of quasi-isomorphic complexes are actually realized in SCFT. It is not *a priori* obvious or necessary for the conjecture to hold that all such complexes are identified, but only that each class contain a single physical boundary state, which could come about in various ways.

Finally, one also has the mirror, special Lagrangian description of Dirichlet branes. This was also a primary input into formulating the conjecture (particularly [47]) but this ingredient was to some extent subsumed by the SCFT considerations. Nevertheless it is clearly important to make more connections to this side as well.

We believe that the study of Dirichlet branes on Calabi-Yau manifolds is proving to be a worthy continuation of the mirror symmetry story, and will continue to inspire interaction between mathematicians and physicists for some time to come.

I particularly thank my collaborators Paul Aspinwall, Emanuel Diaconescu, Bartomeu Fiol and Christian Römelsberger. I am also grateful to Maxim Kontsevich, Greg Moore, Alexander Polishchuk, Paul Seidel, and Richard Thomas for invaluable discussions.

This work was supported in part by DOE grant DE-FG02-96ER40959.

References

[1] S. Donaldson, Proc. London Math Soc. 50 (1985) 1–26.

[2] K. Uhlenbeck and S.-T. Yau, Commun. Pure App. Math. 39 (1986) S257–S293.

[3] M. R. Douglas, in *Proceedings of the European Congress of Mathematics*, Progress in Mathematics vol. 202, Birkhauser 2001. math.AG/0009209.

[4] S.-T. Yau, *Comm. Pure Appl. Math.* 31 (1978) 339.

[5] P.S. Aspinwall, B.R. Greene, and D.R. Morrison, *Nucl.Phys.* B416 (1994) 414-480; hep-th/9309097.

[6] E. Witten, *Nucl. Phys.* B403 (1993) 159, hep-th/9301042.

[7] D. A. Cox and S. Katz, Mathematical Surveys and Monographs 68, AMS, 1999.

[8] C. Voisin, Mirror Symmetry, SMF/AMS Texts 1, AMS 1999.

[9] P. Candelas, X. C. de la Ossa, P. S. Green, and L. Parkes, *Nucl. Phys.* B359 (1991) 21–74.

[10] A. Givental, in *Topological Field Theory, Primitive Forms, and Related Topics* (Kyoto, 1996), 141-175, Prog. Math. 160, Birkhäuser, 1998.

[11] M. Kontsevich, in *Proceedings of the International Congress of Mathematicians*, pages 120–139, Birkhäuser, 1995, alg-geom/9411018.

[12] M. R. Douglas, *J. Math. Phys.* 42 (2001), 2818; hep-th/0011017.

[13] P. S. Aspinwall and A. E. Lawrence, *JHEP* 08 (2001) 004, hep-th/0104147.

[14] D.-E. Diaconescu, *JHEP* 06 (2001) 016, hep-th/0104200.

[15] C. I. Lazaroiu, Unitarity, D-brane Dynamics and D-brane Categories, hep-th/0102183.

[16] P. S. Aspinwall and M. R. Douglas, *JHEP* 0205 (2002) 031, hep-th/0110071.

[17] P. S. Aspinwall, A Point's Point of View of Stringy Geometry, hep-th/0203111.

[18] M. Marino, R. Minasian, G. Moore and A. Strominger, *JHEP* 0001 (2000) 005, hep-th/9911206.

[19] N. C. Leung, *J. Diff. Geom.* 45 (1997) 514.

[20] M. Kontsevich, 1996, Rutgers Lecture, unpublished.

[21] P. Seidel and R. P. Thomas, *Duke Math. J.* 108 (2001) 37–108, math.AG/0001043.

[22] P. Horja, Derived Category Automorphisms from Mirror Symmetry, math.AG/0103231.

[23] E. Witten, in H. Hofer et al., editors, *The Floer Memorial Volume*, pp. 637–678, Birkhäuser, 1995, hep-th/9207094.

[24] E. Witten, *JHEP* 9812 (1998) 019, hep-th/9810188.

[25] W. Boucher, D. Friedan and A. Kent, *Phys. Lett.* B172:316, 1986

[26] A. Strominger, Nucl.Phys. B451 (1995) 96-108; hep-th/9504090.

[27] T. Bridgeland, Flops and derived categories, math.AG/0009053.

[28] Y. Kawamata, D-equivalence and K-equivalence, math.AG/0205287.

[29] M. R. Douglas and G. Moore, D-branes, Quivers, and ALE Instantons, hep-th/9603167

[30] A. Recknagel and V. Schomerus, *Nucl. Phys.* B531 (1998) 185, hep-th/9712186.

[31] A. V. Sardo Infirri, Partial Resolutions of Orbifold Singularities via Moduli Spaces of HYM-type Bundles, alg-geom/9610004.

[32] M. R. Douglas, B. R. Greene, and D. R. Morrison, *Nucl. Phys.* B506 (1997)

84, hep-th/9704151.

[33] M. Reid, Séminaire Bourbaki (novembre 1999), no. 867, math.AG/9911165.

[34] Y. Ito and H. Nakajima, McKay correspondence and Hilbert schemes in dimension three, math.AG/9803120.

[35] T. Bridgeland, A. King and M. Reid, Mukai implies McKay, math.AG/9908027, submitted to J. Amer. Math. Soc.

[36] R. Minasian and G. Moore, J. High Energy Phys. 11 (1997) 002, hep-th/9710230.

[37] A. D. King, *Quart. J. Math. Oxford* (2), 45 (1994), 515-530.

[38] M. R. Douglas, B. Fiol, and C. Römelsberger, hep-th/0002037.

[39] M. R. Douglas, B. Fiol, and C. Romelsberger, hep-th/0003263.

[40] B. R. Greene, hep-th/9702155.

[41] T. Bridgeland, talk at the 2002 Euroconference in Algebraic Geometry at the Isaac Newton Institute.

[42] M. R. Douglas, in *Nonperturbative Aspects of Strings, Branes and Supersymmetry,* World Scientific (1999), hep-th/9901146.

[43] E. Sharpe, *Adv. Theor. Math. Phys.* 2, 1441 (1999); hep-th/9810064.

[44] Y.-Z. Huang and A. Milas, math.QA/0004039.

[45] A. Sen, *JHEP* 9808 (1998) 012; hep-th/9805170.

[46] N. Berkovits, *Nucl. Phys.* B431 (1994) 258; hep-th/9404162.

[47] D. Joyce, hep-th/9907013.

ICM 2002 · Vol. III · 409–418

Non-Equilibrium Steady States

J.-P. Eckmann*

Abstract

The mathematical physics of mechanical systems in thermal equilibrium is a well studied, and relatively easy, subject, because the Gibbs distribution is in general an adequate guess for the equilibrium state.

On the other hand, the mathematical physics of *non-equilibrium* systems, such as that of a chain of masses connected with springs to two (infinite) heat reservoirs is more difficult, precisely because no such a *priori* guess exists.

Recent work has, however, revealed that under quite general conditions, such states can not only be shown to exist, but are *unique*, using the Hörmander conditions and controllability. Furthermore, interesting properties, such as energy flux, exponentially fast convergence to the unique state, and fluctuations of that state have been successfully studied.

Finally, the ideas used in these studies can be extended to certain stochastic PDE's using Malliavin calculus to prove regularity of the process.

2000 Mathematics Subject Classification: 82C22, 60H15.
Keywords and Phrases: Non-equilibrium statistical mechanics, Stochastic differential equations.

1. The model and results

I report here on work done, in different combinations, together with *Martin Hairer, Luc Rey-Bellet, Claude-Alain Pillet, and Lawrence Thomas*. In it, we considered the seemingly trivial problem of describing the non-equilibrium statistical mechanics of a finite-dimensional non-linear Hamiltonian system coupled to two infinite heat reservoirs which are at *different* temperatures. By this I mean that the stochastic forces of the two heat reservoirs differ. The difficulties in such a problem are related to the absence of an easy *a priori* estimate for the state of the system. We show under certain conditions on the initial data that the system goes to a *unique* non-equilibrium steady state and we describe rather precisely some properties of this steady state. These are,

---

* Section de Mathématiques et Département de Physique Théorique, Université de Genève, 1211 Geneva 4, Switzerland. E-mail: eckmann@physics.unige.ch

- appearance of an energy flux (from the hot to the cold reservoir) whenever the reservoirs are at different temperatures,
- exponential stability of this state,
- fluctuations around this state satisfy the Cohen-Gallavotti fluctuation conjecture.

I first review the construction of the model. Two features need special attention: The modeling of the heat reservoirs and their coupling to the chain, and the nature of the coupling among the masses in the chain. I start with the latter: It is a 1-dimensional chain of n distinct d-dimensional anharmonic oscillators with nearest neighbor coupling. The phase space of the chain is therefore $\mathbf{R}^{2dn}$ and its dynamics is described by a $\mathcal{C}^\infty$ Hamiltonian function of the form

$$H_{\mathrm{S}}(p,q) = \sum_{j=1}^{n} \frac{p_j^2}{2} + \sum_{j=1}^{n} U_j^{(1)}(q_j) + \sum_{i=1}^{n-1} U_i^{(2)}(q_i - q_{i+1}) \equiv \sum_{j=1}^{n} \frac{p_j^2}{2} + V(q) \ , \quad (1.1)$$

where $q = (q_1, \dots, q_n)$, $p = (p_1, \dots, p_n)$, with $p_i, q_i \in \mathbf{R}^d$. We will eventually couple the ends of the chain, $i.e.$, q_1 and q_n, to heat reservoirs. Clearly, for heat conduction to be possible at all we must require that the $U_i^{(2)}$ are non-zero. But, this is not enough and the interaction has to have a minimal strength. A sufficient condition for the main result to hold is: For some $m_2 \geq m_1 \geq 2$, and all sufficiently large $|q|$, we require

$$0 < c_1 \leq \frac{U_i^{(1)}(q)}{(1+|q|)^{m_1}} \leq c_1' \ , \qquad 0 < c_2 \leq \frac{U_i^{(2)}(q)}{(1+|q|)^{m_2}} \leq c_2' \ ,$$

and similar growth conditions on the first and second derivatives. Finally, we require that each of the $(d \times d)$ matrices

$$\nabla_{q_i} \nabla_{q_{i+1}} U_i^{(2)}(q_i - q_{i+1}) \ , \quad i = 1, \dots, n-1 \ , \tag{1.2}$$

is non-degenerate (see [12] for the most general conditions).

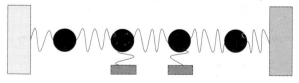

Fig. 1: Model of the chain with the two reservoirs at its ends,
"Hot" at left, "Cold" at right.

Remark 1.1. It seems that relaxing the condition (1.2) poses hard technical problems, although, from a physical point of view, allowing the matrix to be degenerate on hyper-surfaces of codimension ≥ 1 should work. See [9,12] for some possibilities.

Remark 1.2. If $m_2 < m_1$ it seems that the existence of a unique state is jeopardized by the potential appearance of breathers. Indeed, too much energy can then be "stored" in $U^{(1)}$ without being sufficiently "transported" between the oscillators. A more detailed understanding of this problem would be welcome.

Remark 1.3. The nature of the steady state in the limit of an infinite chain $(n \to \infty)$ is a difficult open question.

As a model of a heat reservoir we consider the classical field theory associated with the d-dimensional wave equation. The field φ and its conjugate momentum field π are elements of the real Hilbert space $\mathcal{H} = H_{\mathbf{R}}^1(\mathbf{R}^d) \oplus L_{\mathbf{R}}^2(\mathbf{R}^d)$ which is the completion of $\mathcal{C}_0^\infty(\mathbf{R}^d) \oplus \mathcal{C}_0^\infty(\mathbf{R}^d)$ with respect to the norm defined by the scalar product:

$$\left(\begin{pmatrix} \varphi \\ \pi \end{pmatrix}, \begin{pmatrix} \varphi \\ \pi \end{pmatrix} \right)_{\mathcal{H}} = \int \mathrm{d}x \left(|\nabla \varphi(x)|^2 + |\pi(x)|^2 \right) \equiv 2 H_{\mathrm{B}}(\varphi, \pi) \, , \qquad (1.3)$$

where H_{B} is the Hamiltonian of a bath and the corresponding equation of motion is the ordinary wave equation which we write in the form

$$\begin{pmatrix} \dot{\varphi}(t) \\ \dot{\pi}(t) \end{pmatrix} = \mathcal{L} \begin{pmatrix} \varphi \\ \pi \end{pmatrix} \equiv \begin{pmatrix} 0 & 1 \\ \Delta & 0 \end{pmatrix} \begin{pmatrix} \varphi \\ \pi \end{pmatrix} \, .$$

Finally, we define the coupling between the chain and the heat reservoirs. The reservoirs will be called "L" and "R", the left one coupling to the coordinate q_1 and the right one coupling to the other end of the chain (q_n). Since we consider two heat reservoirs, the phase space of the coupled system, for finite energy configurations, is $\mathbf{R}^{2dn} \times \mathcal{H} \times \mathcal{H}$ and its Hamiltonian will be chosen as

$$H(p, q, \varphi_{\mathrm{L}}, \pi_{\mathrm{L}}, \varphi_{\mathrm{R}}, \pi_{\mathrm{R}}) = H_{\mathrm{S}}(p, q) + H_{\mathrm{B}}(\varphi_{\mathrm{L}}, \pi_{\mathrm{L}}) + H_{\mathrm{B}}(\varphi_{\mathrm{R}}, \pi_{\mathrm{R}})$$
$$+ q_1 \cdot \int \mathrm{d}x \, \varrho_{\mathrm{L}}(x) \nabla \varphi_{\mathrm{L}}(x) + q_n \cdot \int \mathrm{d}x \, \varrho_{\mathrm{R}}(x) \nabla \varphi_{\mathrm{R}}(x) \, .$$
$$(1.4)$$

Here, the $\varrho_i(x) \in L^1(\mathbf{R}^d)$ are charge densities which we assume for simplicity to be spherically symmetric functions. The choice of the Hamiltonian (1.4) is motivated by the dipole approximation of classical electrodynamics. We use the shorthand

$$\phi_i \equiv \begin{pmatrix} \varphi_i \\ \pi_i \end{pmatrix} \, .$$

We set $\alpha_i = \left(\alpha_i^{(1)}, \ldots, \alpha_i^{(d)} \right)$, $i \in \{\mathrm{L}, \mathrm{R}\}$, with

$$\widehat{\alpha}_i^{(\nu)}(k) \equiv \begin{pmatrix} -ik^{(\nu)} \widehat{\varrho}_i(k)/k^2 \\ 0 \end{pmatrix} \, .$$

The "hat" means the Fourier transform $\widehat{f}(k) \equiv (2\pi)^{-d/2} \int \mathrm{d}x \, f(x) e^{-ik \cdot x}$. With this notation the Hamiltonian is

$$H(p, q, \phi_{\mathrm{L}}, \phi_{\mathrm{R}}) = H_{\mathrm{S}}(p, q) + H_{\mathrm{B}}(\phi_{\mathrm{L}}) + H_{\mathrm{B}}(\phi_{\mathrm{R}}) + q_1 \cdot (\phi_{\mathrm{L}}, \alpha_{\mathrm{L}})_{\mathcal{H}} + q_n \cdot (\phi_{\mathrm{R}}, \alpha_{\mathrm{R}})_{\mathcal{H}} \, ,$$

where $H_{\mathrm{B}}(\phi) = \frac{1}{2}\|\phi\|_{\mathcal{H}}^2$. The equations of motions take the form

$$\begin{aligned}
\dot{q}_j(t) &= p_j(t) , \qquad j = 1,\ldots,n , \\
\dot{p}_1(t) &= -\nabla_{q_1} V(q(t)) - (\phi_{\mathrm{L}}(t), \alpha_{\mathrm{L}})_{\mathcal{H}} , \\
\dot{p}_j(t) &= -\nabla_{q_j} V(q(t)) , \qquad j = 2,\ldots,n-1 , \\
\dot{p}_n(t) &= -\nabla_{q_n} V(q(t)) - (\phi_{\mathrm{R}}(t), \alpha_{\mathrm{R}})_{\mathcal{H}} , \\
\dot{\phi}_{\mathrm{L}}(t) &= \mathcal{L}(\phi_{\mathrm{L}}(t) + \alpha_{\mathrm{L}} \cdot q_1(t)) , \\
\dot{\phi}_{\mathrm{R}}(t) &= \mathcal{L}(\phi_{\mathrm{R}}(t) + \alpha_{\mathrm{R}} \cdot q_n(t)) .
\end{aligned} \tag{1.5}$$

The last two equations of (1.5) are easily integrated and lead to

$$\phi_{\mathrm{L}}(t) = e^{\mathcal{L}t}\phi_{\mathrm{L}}(0) + \int_0^t \mathrm{d}s\, \mathcal{L}e^{\mathcal{L}(t-s)}\alpha_{\mathrm{L}} \cdot q_1(s) ,$$

$$\phi_{\mathrm{R}}(t) = e^{\mathcal{L}t}\phi_{\mathrm{R}}(0) + \int_0^t \mathrm{d}s\, \mathcal{L}e^{\mathcal{L}(t-s)}\alpha_{\mathrm{R}} \cdot q_n(s) ,$$

where the $\phi_i(0)$, $i \in \{\mathrm{L},\mathrm{R}\}$, are the initial conditions of the heat reservoirs.

We next assume that the two reservoirs are in thermal equilibrium at inverse temperatures β_{L} and β_{R}. By this I mean that the initial conditions $\Phi(0) \equiv \{\phi_{\mathrm{L}}(0), \phi_{\mathrm{R}}(0)\}$ are random variables distributed according to a Gaussian measure with mean zero and covariance $\langle \phi_i(f)\phi_j(g) \rangle = \delta_{ij}(1/\beta_i)(f,g)_{\mathcal{H}}$. If we assume that the coupling functions $\alpha_i^{(\nu)}$ are in $\mathcal{H}$ for $i \in \{\mathrm{L},\mathrm{R}\}$ and $\nu \in \{1,\ldots,d\}$, then the $\xi_i(t) \equiv \phi_i(0)(e^{-\mathcal{L}t}\alpha_i)$ are d-dimensional Gaussian random processes with mean zero and covariance

$$\langle \xi_i(t)\xi_j(s) \rangle = \delta_{i,j}\frac{1}{\beta_i}C_i(t-s) , \quad i,j \in \{\mathrm{L},\mathrm{R}\} , \tag{1.6}$$

where the $d \times d$ matrices $C_i(t-s)$ are

$$C_i^{(\mu,\nu)}(t-s) = \left(\alpha_i^{(\mu)}, e^{\mathcal{L}(t-s)}\alpha_i^{(\nu)}\right)_{\mathcal{H}} = \frac{1}{d}\delta_{\mu,\nu}\int \mathrm{d}k\, |\widehat{\varrho}_i(k)|^2 \cos(|k|(t-s)) .$$

Finally, we impose a condition on the random force exerted by the heat reservoirs on the chain. We assume that the covariances of the random processes $\xi_i(t)$ with $i \in \{\mathrm{L},\mathrm{R}\}$ satisfy

$$C_i^{(\mu,\nu)}(t-s) = \delta_{\mu,\nu}\lambda_i^2 e^{-\gamma_i|t-s|} , \tag{1.7}$$

with $\gamma_i > 0$ and $\lambda_i > 0$, which can be achieved by a suitable choice of the coupling functions $\varrho_i(x)$, for example

$$\widehat{\varrho}_i(k) = \mathrm{const.}\prod_{m=1}^M \frac{1}{(k^2 + \gamma_i^2)^{1/2}} , \tag{1.8}$$

where all the γ_i are distinct. We continue with the case $M = 1$ for simplicity.

Using (1.7) and enlarging the phase space with auxiliary fields r_i, one eliminates the memory terms (both deterministic and random) of the equations of motion and rewrites them as a system of *Markovian* stochastic differential equations:

$$
\begin{aligned}
dq_j(t) &= p_j(t)dt , & j &= 1,\ldots,n , \\
dp_1(t) &= -\nabla_{q_1} V(q(t))dt + r_L(t)dt , \\
dp_j(t) &= -\nabla_{q_j} V(q(t))dt , & j &= 2,\ldots,n-1 , \\
dp_n(t) &= -\nabla_{q_n} V(q(t))dt + r_R(t)dt , \\
dr_L(t) &= -\gamma_L r_L(t)dt + \lambda_L^2 \gamma_L\, q_1(t)dt - \lambda_L \sqrt{2\gamma_L/\beta_L}\; dw_L(t) , \\
dr_R(t) &= -\gamma_R r_R(t)dt + \lambda_R^2 \gamma_R q_n(t)dt - \lambda_R \sqrt{2\gamma_R/\beta_R}\; dw_R(t) ,
\end{aligned}
\tag{1.9}
$$

which defines a Markov diffusion process on $\mathbf{R}^{d(2n+2)}$.

Theorem 1.4. [3,4,12,1] *There is a constant $\lambda^* > 0$, such that when $|\lambda_L|$, $|\lambda_R| \in (0, \lambda^*)$, the solution of (1.9) is a Markov process which has an absolutely continuous invariant measure μ with a C^∞ density m. This measure is* **unique**, **mixing** *and attracts any other measure at an exponential rate.*

Remark 1.5. On can show even a little more. Let $h_0(\beta)$ be the Gibbs distribution for the case where both reservoirs are at temperature $1/\beta$. If h denotes the density of the invariant measure found in Theorem 1.4, we find that $h/h_0(\beta)$ is in the Schwartz space $\mathcal{S}$ for all $\beta < \min(\beta_L, \beta_R)$. This mathematical statement reflects the intuitively obvious fact that the chain can not get hotter than either of the reservoirs.

Remark 1.6. The restriction on the couplings λ_L, λ_R between the small system and the reservoirs is a condition of stability (against "explosion") of the small system coupled to the heat reservoirs: It is *not* of perturbative nature. Indeed, the reservoirs have the effect of renormalizing the deterministic potential seen by the small system and this potential must be stable. This restricts λ_L and λ_R.

The proof of Theorem 1.4 is based on a detailed study of Eq.(1.9). Let $x = (p, q, r)$ and $r = (r_L, r_R)$. For a Markov process $x(t)$ with phase space X and an invariant measure $\mu(dx)$, ergodic properties may be deduced from the study of the associated semi-group T^t on the Hilbert space $L^2(X, \mu(dx))$. To prove the existence of the invariant measure in Theorem 1.4 one proceeds as follows: Consider first the semi-group T^t on the auxiliary Hilbert space $\mathcal{H}_0 \equiv L^2(X, \mu_0(dx))$, where the reference measure $\mu_0(dx)$ is a generalized Gibbs state for a suitably chosen reference temperature. Our main technical result consists in proving that the generator L of the semi-group T^t on $\mathcal{H}_0$ and its adjoint have compact resolvent. This is proved by generalizing Hörmander's techniques for hypoelliptic operators of "Kolmogorov type" to the problem in unbounded domains described by (1.9). Once this is established, we deduce the existence of a solution to the eigenvalue equation

$(T^t)^* g = g$ in $\mathcal{H}_0$ and this implies immediately the existence of an invariant measure. The original proof [3] was subsequently improved by using more probabilistic techniques [12].

The proof of uniqueness, [4], relies on global controllability of (1.9). In it, one shows that the control equation, in which the noises w_i of (1.9) are replaced by deterministic forces f_i (in the same function space), allows one to reach any given point in phase space in any prescribed time, by choosing the forces f_i adequately. It is here that, at least at the time of this writing, a feature of the problem seems crucial for success:

Remark 1.7. The geometry of the chain: If the chain is not of linear geometry, but with parallel strands, or if the coupling is not of pure nearest neighbor type, uniqueness of the invariant measure does in general not follow from the methods described here. Very simple counterexamples with harmonic chains [14] show that this problem is not easy.

Remark 1.8. We proved in [3, Lemma 3.7] that the density $\varrho = \varrho_T$ is a real analytic function of $\zeta = (T_L - T_R)/(T_L + T_R)$. In particular, this yields the standard perturbative results near equilibrium ($\zeta = 0$).

Question. A fascinating problem is to understand the limit of a chain of infinitely many oscillators, and in particular the nature of heat conduction in this case. I believe that this problem can only be solved if a better understanding of modeling the coupling between the heat bath and the chain can be found.

2. Time-reversal, energy flux, and entropy production

In the wake of the seminal work of Gallavotti and Cohen [6], several authors realized (e.g., [10, 11]) that internal symmetries of stationary non-equilibrium problems lead to an interesting relation for the fluctuations in the stationary state. The model we consider here is no exception, and it is one of the few examples where the Hamiltonian dynamics plays a very nice role.

It will be useful to streamline the notation.[1] The two reservoirs, L and R, are described by the variables $r = (r_L, r_R) \in \mathbf{R}^d \oplus \mathbf{R}^d$. Let Λ be the $(2d \times nd)$ matrix defined by

$$q \cdot \Lambda r = q_1 \cdot \Lambda_L r_L + q_n \cdot \Lambda_R r_R = q_1 \lambda_L r_L + q_n \lambda_R r_R .$$

Define the $(2d \times 2d)$ matrix $\Gamma = \text{diag } \gamma_L \oplus \text{diag } \gamma_R$, let $w = w_L \oplus w_R$ the $2d$-dimensional standard Brownian motion, and finally T the (2×2) diagonal temperature matrix $T = \text{diag}(T_L, T_R)$. It is useful to introduce the change of variables $s = Fr - F^T q$, where $F = \Lambda \Gamma^{-1/2}$. In terms of these variables, one can introduce the effective potential

$$V_{\text{eff}}(q) = V(q) - \tfrac{1}{2} q \cdot \Lambda \Lambda^T q , \tag{2.1}$$

---

[1] This notation generalizes more easily to the case $M > 1$ of (1.8) .

and the "energy" is now $G(s, q, p)$ with

$$G(s, q, p) = \tfrac{1}{2}p^2 + V_{\text{eff}} + \tfrac{1}{2}s \cdot \Gamma s \ . \tag{2.2}$$

Finally, with the adjoint change in the derivatives $\nabla_q \to \nabla_q - F\nabla_s$, the equations of motion (1.9) read

$$
\begin{aligned}
dq &= \nabla_p G dt = p \, dt \ , \\
dp &= -(\nabla_q - F\nabla_s) G dt = -(\nabla_q V_{\text{eff}}(q) - F\Gamma s) dt \ , \\
ds &= -(\nabla_s + F^{\mathrm{T}}\nabla_p) G \, dt - (2T^{1/2}) dw = -(\Gamma s + F^{\mathrm{T}}p) dt - (2T^{1/2}) dw \ .
\end{aligned}
\tag{2.3}
$$

Writing G_p for $\nabla_p G$ and G_q for $\nabla_q G$ (these are vectors with nd components), and G_s for $\nabla_s G$ (this is a vector with $2d$ components), the generator L of the diffusion process takes, in the variables $y = (s, q, p)$, the form

$$L = \nabla_s \cdot T\nabla_s - G_s \cdot \nabla_s + \left(G_p \cdot \nabla_q - G_q \cdot \nabla_p\right) + \left((FG_s) \cdot \nabla_p - G_p \cdot F\nabla_s\right) \ . \tag{2.4}$$

If f is a function on the phase space X, we let

$$S^t f(y) \equiv \left(e^{Lt} f\right)(y) = \int f\left(\xi_y(t)\right) d\mathbf{P}(w) \ .$$

The adjoint L^{T} of L in the space $\mathrm{L}^2(\mathbf{R}^{d(2+2n)})$ is called the Fokker-Planck operator. The density m of the invariant measure is the (unique) normalized solution of the equations $L^{\mathrm{T}} m = 0$.

2.2. The entropy production σ

Using the notation (2.4), we now establish a relation between the energy flux and the entropy production. Since we are dealing with a Hamiltonian setup, the energy flux is defined naturally by the time derivative of the mean evolution S^t of the effective energy, $H_{\text{eff}}(q, p) = p^2/2 + V_{\text{eff}}(q)$. Differentiating, we get from the equations of motion $\partial_t S^t H_{\text{eff}} = S^t L H_{\text{eff}}$, with

$$L H_{\text{eff}} = p \cdot (-\nabla_q V_{\text{eff}} + F\Gamma s) + \nabla_q V_{\text{eff}} \cdot p = p \cdot F\Gamma s \ .$$

We define the total flux by $\Phi = p \cdot F\Gamma s$, and inspection of the definition of F and Γ leads to the identification of the flux at the left and right ends of the chain: $\Phi = \Phi_{\mathrm{L}} + \Phi_{\mathrm{R}}$, with

$$\Phi_{\mathrm{L}} = p_1 \cdot \Lambda_{\mathrm{L}} \Gamma_{\mathrm{L}}^{1/2} s_{\mathrm{L}} \ , \qquad \Phi_{\mathrm{R}} = p_n \cdot \Lambda_{\mathrm{R}} \Gamma_{\mathrm{R}}^{1/2} s_{\mathrm{R}} \ .$$

Note that Φ_{L} is the energy flux from the left bath to the chain, and Φ_{R} is the energy flux from the right bath to the chain. Furthermore, observe that $\langle \Phi \rangle_\mu = 0$, with $\langle f \rangle_\mu \equiv \int \mu(dy) f(y) = \int dy \, m(y) f(y)$, because $\Phi = L H_{\text{eff}}$ and $L^{\mathrm{T}} m = 0$.

Since we have been able to identify the energy flux on the ends of the chain, we can *define* the (thermodynamic) entropy production σ by

$$\sigma = \frac{\Phi_{\mathrm{L}}}{T_{\mathrm{L}}} + \frac{\Phi_{\mathrm{R}}}{T_{\mathrm{R}}} = p \cdot FT^{-1}\Gamma s \ . \tag{2.5}$$

2.3. Time-reversal, generalized detailed balance condition

We next define the "time-reversal" map J by $\big(Jf\big)(s,q,p) = f(s,q,-p)$. This map is the projection onto the space of the s,q,p of the time-reversal of the Hamiltonian flow (on the full phase space of chain plus baths) defined by the original problem (1.5).

Notation. To obtain simple formulas for the entropy production σ we write the (strictly positive) density m of the invariant measure μ as

$$m = Je^{-R}e^{-\varphi} \ , \tag{2.6}$$

where $R = R(s) = \frac{1}{2}s \cdot \Gamma T^{-1}s$. Let L^* denote the adjoint of L in the space $\mathcal{H}_\mu = \mathrm{L}^2(X,\mathrm{d}\mu)$ associated with the invariant measure μ, where $X = \mathbf{R}^{2(2n+2)}$. In terms of the adjoint L^{T} on $\mathrm{L}^2(X,\mathrm{d}s\,\mathrm{d}q\,\mathrm{d}p)$, we have the operator identity

$$L^* = m^{-1}L^{\mathrm{T}}m \ . \tag{2.7}$$

We have the following important symmetry property as suggested by the paper [10].

Theorem 2.9. *Let* $L_\eta = L + \eta\sigma$, *where* $\eta \in \mathbf{R}$. *One has the operator identity*

$$Je^{-J\varphi}(L_\eta)^*e^{J\varphi}J = L_{1-\eta} \ . \tag{2.8}$$

In particular,

$$Je^{-J\varphi}L^*e^{J\varphi}J - L = \sigma \ . \tag{2.9}$$

Remark 2.10. This relation may be viewed as a generalization to non-equilibrium of the detailed balance condition (at equilibrium, one has $JL^*J - L = 0$).

The paper of Gallavotti and Cohen [6] describes fluctuations of the entropy production. It is based on numerical experiments by [5] which were then abstracted to the general context of dynamical systems. In further work, these ideas have been successfully applied to thermostatted systems modeling non-equilibrium problems. In the papers [10] and [11] these ideas have been further extended to non-equilibrium models described by stochastic dynamics. In the context of our model, the setup is as follows: One considers the observable $W(t) = \int_0^t \mathrm{d}\eta\,\sigma\big(\xi_x(\eta)\big)$. By ergodicity, one finds $\lim_{t\to\infty} t^{-1}W(t) = \langle\sigma\rangle_\mu$, for all x and almost all realizations of the Brownian motion $\xi_x(t) = \xi_x(t,\omega)$. The rate function $\widehat{e}$ is characterized by the relation

$$\inf_{y\in I}\widehat{e}(y) = -\lim_{t\to\infty}\frac{1}{t}\log\mathbf{Prob}\left\{\frac{W(t)}{t\langle\sigma\rangle_\mu} \in I\right\} \ .$$

Under suitable conditions it can be expressed as the Legendre transform of the function

$$e(\eta) \equiv -\lim_{t \to \infty} t^{-1} \log \langle e^{-\eta W(t)} \rangle_\mu .$$

Formally, $-e(\eta)$ can be represented as the maximal eigenvalue of L_η. Observing now the relation (2.8), one sees immediately that

$$e(\eta) = e(1 - \eta) . \tag{2.10}$$

Theorem 2.11. [13] *The above relations can be rigorously justified and lead to*

$$\widehat{e}(y) - \widehat{e}(-y) = -y \langle \sigma \rangle_\mu . \tag{2.11}$$

In particular this means that at equal temperatures, when $\langle \sigma \rangle_\mu = 0$, the fluctuations are symmetric around the mean 0, while at unequal temperatures, the odd part is linear in y and proportional to the mean entropy production. Note that when $\langle \sigma \rangle_\mu \neq 0$ this relation describes fluctuations around 0, *not* around the mean! This is the celebrated Gallavotti-Cohen fluctuation theorem.

3. Extensions

The technique for proving uniqueness results presented above can be generalized and applied to many other problems, in particular to certain types of "partially noisy" PDE's (so that now phase space is infinite dimensional). One kind of example must suffice to illustrate the kind of results one can obtain. Consider the stochastic Ginzburg-Landau equation with periodic boundary conditions (written in Fourier components, for $L \gg 1$):

$$du_k = (1 - (k/L)^2) u_k \, dt - \sum_{k_1 \mid k_2 \mid k_3 = k} u_{k_1} u_{k_2} u_{k_3} \, dt + q_k \, dw_k , \quad k \in \mathbf{Z} , \tag{2.12}$$

with $|q_k| \sim k^{-5}$ and where w_k are standard Wiener processes. The point is here that q_k may be zero for all $|k| \leq k_*$.

Theorem 2.12. [2, 8, 7] *The process defined by (2.12) has a unique invariant measure. Any initial condition is attracted exponentially fast to it.*

Acknowledgments. I thank M. Hairer and L. Rey-Bellet for help in preparing this manuscript. This work was supported by the Fonds National Suisse.

References

[1] J.-P. Eckmann & M. Hairer. Non-equilibrium statistical mechanics of strongly anharmonic chains of oscillators. *Comm. Math. Phys.* **212** (2000), 105–164.

[2] J.-P. Eckmann & M. Hairer. Uniqueness of the invariant measure for a stochastic PDE driven by degenerate noise. *Comm. Math. Phys.* **219** (2001), 523–565.

[3] J.-P. Eckmann, C.-A. Pillet & L. Rey-Bellet. Non-equilibrium statistical mechanics of anharmonic chains coupled to two heat baths at different temperatures. *Comm. Math. Phys.* **201** (1999), 657–697.

[4] J.-P. Eckmann, C.-A. Pillet & L. Rey-Bellet. Entropy production in nonlinear, thermally driven Hamiltonian systems. *J. Stat. Phys.* **95** (1999), 305–331.

[5] D. J. Evans, E. G. D. Cohen & G. P. Morriss. Viscosity of a simple fluid from its maximal Lyapunov exponents. *Phys. Rev.* **A43** (1990), 5990–5997.

[6] G. Gallavotti & E. G. D. Cohen. Dynamical ensembles in stationary states. *J. Stat. Phys.* **80** (1998), 3719–3729.

[7] M. Hairer. Exponential mixing properties of stochastic PDEs through asymptotic coupling (2001). Preprint, Geneva.

[8] M. Hairer. Exponential mixing for a stochastic PDE driven by degenerate noise. *Nonlinearity* **15** (2002), 271–279.

[9] V. Jakšić & C.-A. Pillet. Ergodic properties of classical dissipative systems. I. *Acta Math.* **181** (1998), 245–282.

[10] J. Kurchan. Fluctation theorem for stochastic dynamics. *J. Phys.* **A31** (1998), 3719–3729.

[11] J. L. Lebowitz & H. Spohn. A Gallavotti-Cohen-type symmetry in the large deviation functional for stochastic dynamics. *J. Stat. Phys.* **95** (1999), 333–365.

[12] L. Rey-Bellet & L. E. Thomas. Asymptotic behavior of thermal nonequilibrium steady states for a driven chain of anharmonic oscillators. *Comm. Math. Phys.* **215** (2000), 1–24.

[13] L. Rey-Bellet & L. E. Thomas. Exponential convergence to non-equilibrium stationary states in classical statistical mechanics. *Comm. Math. Phys.* **225** (2002), 305–329.

[14] E. Zabey. Diploma, University of Geneva (2001). Unpublished.

ICM 2002 · Vol. III · 419–430

Twisted K-Theory and Loop Groups*

Daniel S. Freed[†]

Abstract

Twisted K-theory has received much attention recently in both mathematics and physics. We describe some models of twisted K-theory, both topological and geometric. Then we state a theorem which relates representations of loop groups to twisted equivariant K-theory. This is joint work with Michael Hopkins and Constantin Teleman.

2000 Mathematics Subject Classification: 19L47, 22E67, 81T45.
Keywords and Phrases: K-theory, Loop groups, Verlinde algebra.

0. Introduction

The loop group of a compact Lie group G is the space of smooth maps $S^1 \to G$ with multiplication defined pointwise. Loop groups have been around in topology for quite some time [Bo], and in the 1980s were extensively studied from the point of view of representation theory [Ka], [PS]. In part this was driven by the relationship to conformal field theory. The interesting representations of loop groups are projective, and with fixed projective cocycle τ there is a finite number of irreducible representations up to isomorphism. Considerations from conformal field theory [V] led to a ring structure on the abelian group $R^\tau(G)$ they generate, at least for *transgressed* twistings. This is the *Verlinde ring*. For G simply connected $R^\tau(G)$ is a quotient of the representation ring of G, but that is not true in general. At about this time Witten [W] introduced a three-dimensional topological quantum field theory in which the Verlinde ring plays an important role. Eventually it was understood that the fundamental object in that theory is a "modular tensor category" whose Grothendieck group is the Verlinde ring. Typically it is a category of representations of a loop group or quantum group.

For the special case of a finite group G the topological field theory is specified by a certain cocycle on G and the category can be calculated explicitly [F1]. We identified it as a category of representations of a Hopf algebra constructed

---

*The author is supported by NSF grant DMS-0072675.
[†]Department of Mathematics, University of Texas, Austin, TX 78712, USA. E-mail: dafr@math.utexas.edu

from G, thus directly linking the Chern-Simons lagrangian and quantum groups. Only recently did we realize that this category has a description in terms of *twisted* equivariant K-theory, and it was natural to guess that the Verlinde ring for arbitrary G has a similar description. Ongoing joint work with Michael Hopkins and Constantin Teleman has confirmed this description. We can speculate further and hope that twisted K-theory provides a construction of the modular tensor category, and perhaps even more of the three-dimensional topological field theory. In another direction the use of K-theory may shed light on Verlinde's formula for certain Riemann-Roch numbers. In any case our result fits well with other uses of K-theory in representation theory [CG], for example in the geometric Langlands program.

The physics motivation for the main theorem is discussed in [F2], [F3]. Here, in §3, we explain the statement of our result in mathematical terms; the proof will appear elsewhere. As background we describe some concrete topological models of twisted K-theory in §1, and give a twisted version of the Chern-Weil construction in §2.

As mentioned above the work I am discussing is being carried out with Michael Hopkins and Constantin Teleman. I thank them for a most enjoyable collaboration.

1. Twistings of K-theory

Let X be a reasonable compact space, say a finite CW complex. Then isomorphism classes of complex vector bundles over X form a semigroup whose group completion is the K-theory group $K^0(X)$. This basic idea was introduced by Grothendieck in the context of algebraic geometry [BS], and was subsequently transported to topology by Atiyah and Hirzebruch [AH]. Vector bundles are local—they can be cut and glued—and in the topological realm this leads to a cohomology theory. In particular, there are groups $K^n(X)$ defined for $n \in \mathbb{Z}$. Historically, K-theory was the first example of a generalized cohomology theory, and it retains the features of ordinary cohomology with one notable exception: the cohomology of a point is nontrivial in all even degrees, as determined by Bott periodicity. There are many nice spaces B which represent complex K-theory in the sense that $K^0(X)$ is the set of homotopy classes of maps from X to B. A particularly nice choice [A1,Appendix], [J] is the space $B = \mathrm{Fred}(H)$ of Fredholm operators on a separable complex Hilbert space H. Thus a map $T\colon X \to \mathrm{Fred}(H)$ determines a K-theory class on X. It is convenient to generalize and allow the Hilbert space H to vary as follows. A *Fredholm complex*[1] is a graded Hilbert space bundle $E^\bullet = E^0 \oplus E^1 \to X$ together with a fiberwise Fredholm map $E^0 \xrightarrow{T} E^1$, and it also represents an element of $K^0(X)$. Another innovation was the introduction by Atiyah and Segal [S2] of the equivariant K-theory groups $K_G^\bullet(X)$ for a compact space X which carries the action of a compact Lie group G. The basic objects are G-equivariant vector bundles $E \to X$, and $K_G^0(X)$ is the group completion of the set of equivalence classes. For example,

---

[1] One [S1] can allow more general topological vector spaces and complexes which are nonzero in degrees other than 0 and 1.

if X is a point then the equivariant K-theory is the representation ring K_G of the compact Lie group G; in general, $K_G(X)$ is a K_G-module.

As a first example of twisted K-theory we consider twisted versions of K_G. A twisting τ is a central extension

$$1 \longrightarrow \mathbb{T} \longrightarrow \tilde{G} \longrightarrow G \longrightarrow 1,$$

where $\mathbb{T} \subset \mathbb{C}$ is the circle group of unit norm complex numbers. Then the twisted representation "ring" K_G^τ is the group completion of equivalence classes of complex representations of $\tilde{G}$ on which the central $\mathbb{T}$ acts by scalar multiplication. Twisted K-theory is *not* a ring, but rather K_G^τ is a K_G-module. A map of twistings, i.e., an isomorphism of central extensions, determines an isomorphism of twisted K-groups. So twisted K-theory is determined up to *noncanonical* isomorphism by the equivalence class of the twisting. For example, $K_{SO(3)} \cong \mathbb{Z}[s]$ is the polynomial ring on a single generator, the 3-dimensional defining representation. Up to equivalence there is a single nontrivial central extension

$$\tau = \{\, 1 \longrightarrow \mathbb{T} \longrightarrow U(2) \longrightarrow SO(3) \longrightarrow 1 \,\}$$

which is induced by the inclusion $\mathbb{Z}/2\mathbb{Z} \hookrightarrow \mathbb{T}$ from the extension

$$1 \longrightarrow \mathbb{Z}/2\mathbb{Z} \longrightarrow SU(2) \longrightarrow SO(3) \longrightarrow 1.$$

Virtual representations of $U(2)$ on which the center acts naturally correspond 1:1 with virtual representations of $SU(2)$ on which the central element acts as -1. Now $K_{SU(2)} \cong \mathbb{Z}[t]$, where t is the defining 2-dimensional representation, and we identify $K_{SO(3)}^\tau \subset K_{SU(2)}$ as the subgroup of odd polynomials in t; the $K_{SO(3)} \cong \mathbb{Z}[s]$-module structure is $s \cdot t = t^3 - t$, by the Clebsch-Gordon rule, and $K_{SO(3)}^\tau$ is a free module of rank one.

More generally, on a G-space X a "cocycle" τ for the equivariant cohomology group $H_G^3(X; \mathbb{Z})$ defines a twisted K-theory group $K_G^\tau(X)$ which is a module over $K_G^0(X)$. A better point of view is that τ is a cocycle, or geometric representative, of a class in $H_G^1(X; A_G)$, where A_G is a group of automorphisms of equivariant K-theory. We will not try to make "automorphism of equivariant K-theory" precise here, but content ourselves with a concrete model, first in the nonequivariant case. Take $\mathrm{Fred}(H)$ to be the classifying space of K-theory. Then the group $A = PGL(H)$ acts as automorphisms by conjugation. Since $\tilde{A} = GL(H)$ is contractible [K], the quotient $A = \tilde{A}/\mathbb{C}^\times$ is homotopy equivalent to $\mathbb{CP}^\infty$, and so $H^1(X; A) \cong H^3(X; \mathbb{Z})$. A twisting τ can be taken to be a principal A-bundle $\pi\colon P \to X$. The action of A on $\mathrm{Fred}(H)$ defines an associated bundle $\mathrm{Fred}(H)_P \to X$, and the twisted K-group $K^\tau(X)$ is the group of homotopy classes of sections of this bundle [A2]. A section is an A-equivariant map $T\colon P \to \mathrm{Fred}(H)$, and as before it is convenient to let the Hilbert space vary. Thus define a *P-twisted Fredholm complex* to be an $\tilde{A}$-equivariant graded Hilbert space bundle $E^\bullet \to P$, where the center $\mathbb{C}^\times \subset \tilde{A}$ acts by scalar multiplication, together with a fiberwise $\tilde{A}$-equivariant Fredholm operator $E^0 \xrightarrow{T} E^1$. Then $(E^\bullet, T)$ represents an element of $K^\tau(X)$.

It is perhaps unsettling that the model is infinite dimensional, but that is unavoidable unless the class of the twisting in $H^3(X; \mathbb{Z})$ is torsion.

There are many other models of twistings and twisted K-theory. For example, *gerbes* are geometric representatives of elements in degree three integral cohomology; see [H], [B], [M] for example. In a Čech description we have a covering $X = \bigcup_i U_i$ of X by open sets and a complex line bundle $L_{ij} \to U_i \cap U_j$ on double intersections. There is further cocycle data on triple intersections. In a model of K-theory on which line bundles act as automorphisms this can be used to define twisted K-theory; see [BCMMS] for example. In fact, the "group" $\mathbb{Z}/2\mathbb{Z} \times \mathbb{CP}^\infty$ of *graded* lines act as automorphisms of K-theory, so there is a larger group of (equivalence classes of) twistings

$$H^1(X; \mathbb{Z}/2\mathbb{Z} \times \mathbb{CP}^\infty) \cong H^1(X; \mathbb{Z}/2\mathbb{Z}) \times H^3(X; \mathbb{Z}). \tag{1}$$

We remark that there is a natural group structure on (1), but it is not the product. In topology these twisted versions of K-theory, at least for torsion twistings, were introduced by Donovan and Karoubi [DK], who also considered real versions. There is another viewpoint and generalization of K-theory using C^*-algebras, and in that context twisted K-theory was discussed by Rosenberg [R] for both torsion and nontorsion twistings. Twisted versions of K-theory have appeared recently in various parts of geometry and index theory, for example in [LU], [AR], [To], [NT], [MMS].

A generalization of the previous model is useful. Here a twisting is a quartet $\tau = (\mathcal{G}, \epsilon, \tilde{\mathcal{G}}, P)$:

$$
\begin{array}{ll}
\mathcal{G} & \text{topological group,} \\
\epsilon \colon \mathcal{G} \to \mathbb{Z}/2\mathbb{Z} & \text{homomorphism termed the } grading, \\
\tilde{\mathcal{G}} \to \mathcal{G} & \text{central extension by } \mathbb{T}, \\
P \to X & \text{principal } \mathcal{G}\text{-bundle.}
\end{array}
\tag{2}
$$

We require the existence of a homomorphism $\tilde{\mathcal{G}} \to GL(H)$ which is the identity on the central $\mathbb{T}$. Let $\tilde{\mathcal{G}}_0 \to \mathcal{G}_0$ be the restriction of the central extension over the subgroup $\mathcal{G}_0 = \epsilon^{-1}(0)$. Then the equivalence class of τ is the obstruction in (1) to restricting/lifting P to a principal $\mathcal{G}_0$ bundle. Our previous construction is the special case $G = PGL(H)$ and ϵ trivial. An element of $K^\tau(X)$ is represented by a $\tilde{\mathcal{G}}$-equivariant $\mathbb{Z}/2\mathbb{Z}$-graded Fredholm complex over P, where the action of $\tilde{\mathcal{G}}$ is compatible with the grading ϵ. If a compact Lie group G acts on X then it is easy to extend this to a model of equivariant twistings and equivariant K-theory.

As an illustration of a nontrivial H^1-twisting, consider $X = pt$ and $\tau = (O(2), \epsilon, O(2) \times \mathbb{T}, O(2))$ with nontrivial grading ϵ. The representation ring of $O(2)$ may be written

$$K_{O(2)} \cong \mathbb{Z}[\sigma, \delta] \,/\, (\sigma(\delta - 1), \delta^2 - 1),$$

where δ is the one-dimensional sign representation and σ the standard two dimensional representation. Then the twisted K-group $K^{\tau}_{O(2)}$ is a module over $K_{O(2)}$ with a single generator t and the relation $\delta \cdot t = t$. There is a nontrivial odd twisted K-group $K^{\tau+1}_{O(2)}$ which is also a module with a single generator u; the relations are $\delta \cdot u = -u$ and $\sigma \cdot u = 0$.

Many topological properties of K-theory, including exact and spectral sequences, have straightforward analogs in the twisted case. This is easiest to see from the homotopy-theoretic view of cohomology theories, and so applies to twisted cohomology theories in general. Computations are usually based on these sequences. A more specialized result is the completion theorem in equivariant K-theory [AS]; its generalization to the twisted case has some new features [FHT]. The Thom isomorphism theorem fits naturally into the twisted theory [DK]. Let $V \to X$ be a real vector bundle of finite rank, which for convenience we suppose endowed with a metric. There is an associated twisting

$$\tau(V) = \bigl(O(n), \epsilon, \mathrm{Pin}^c(n), O(V)\bigr),$$

where $O(V) \to X$ is the orthonormal frame bundle of V and $\epsilon \colon O(n) \to \mathbb{Z}/2\mathbb{Z}$ is the nontrivial grading. The isomorphism class of $\tau(V)$ is the pair of Stiefel-Whitney classes $\bigl(w_1(V), W_3(V)\bigr)$. Denote

$$K^V(X) = K^{\mathrm{rank}(V)+\tau(V)}(X) \tag{3}$$

as the degree-shifted twisted K-theory. Then the Thom isomorphism theorem asserts that the natural map

$$K^{q+V}(X) \longrightarrow K^{q+2\,\mathrm{rank}(V)}(V) \tag{4}$$

is an isomorphism. We remark that if X is a smooth compact manifold, then Poincaré duality identifies $K^{TX}(X)$ with the K-homology group $K_0(X)$. Also, we can define $K^V(X)$ for virtual bundles V.

The Chern character maps twisted K-theory to a twisted version of real cohomology, as we explain next in the context of Chern-Weil theory.

2. A differential-geometric model

For simplicity we work in the nonequivariant context. Our thoughts here were stimulated by reading [BCMMS].

Let $\mathfrak{a} = \mathfrak{pgl}(H)$ denote the Lie algebra of $\tilde{A} = PGL(H)$ and $\tilde{\mathfrak{a}} = \mathfrak{gl}(H)$ the Lie algebra of $A = GL(H)$. They fit into the exact sequence of Lie algebras

$$0 \longrightarrow \mathbb{C} \longrightarrow \tilde{\mathfrak{a}} \longrightarrow \mathfrak{a} \longrightarrow 0.$$

A linear map $L\colon \mathfrak{a} \to \tilde{\mathfrak{a}}$ is a splitting if the composition with $\tilde{\mathfrak{a}} \to \mathfrak{a}$ is the identity. A splitting L determines a closed right-invariant 2-form on A which represents the generator of $H^2(A;\mathbb{Z})$; its value on right-invariant vector fields ξ, η is

$$\bigl[L(\xi), L(\eta)\bigr] - L\bigl([\xi,\eta]\bigr).$$

Let $\pi\colon P \to X$ be a principal A-bundle. We now add two pieces of geometric data:

$$\Theta \in \Omega^1(P; \mathfrak{a}) \qquad\qquad \text{connection on } P \to X,$$

$$L\colon P \to \mathrm{Hom}(\mathfrak{a}, \tilde{\mathfrak{a}}) \qquad A\text{-invariant map into splittings.} \tag{5}$$

Both are sections of affine space bundles over X, so can be constructed using partitions of unity. As usual, define the curvature

$$F_\Theta = d\Theta + \frac{1}{2}[\Theta \wedge \Theta].$$

It is an $\mathfrak{a}$-valued 2-form on P. Introduce the *scalar* 2-form

$$\beta = \left(dL(\Theta) + \frac{1}{2}[L(\Theta) \wedge L(\Theta)] \right) - L(F_A).$$

Then one can check that β is transgressive. In other words, $d\beta = \pi^*\eta$ for a closed scalar 3-form $\eta \in \Omega^3(X)$. The de Rham cohomology class of η in $H^3(X; \mathbb{R})$ represents the image in real cohomology of the isomorphism class of the twisting $P \to X$.

Now let $E^0 \xrightarrow{T} E^1$ be a twisted Fredholm complex over P. Thus $E^i \to P$ are $\tilde{A}$-equivariant Hilbert space bundles, with $\mathbb{C}^\times \subset \tilde{A}$ acting by scalar multiplication, and T is an $\tilde{A}$-equivariant Fredholm map. The $\tilde{A}$ action determines an $\tilde{A}$-invariant partial covariant derivative on $E^\bullet \to P$ along the fibers of $\pi\colon P \to X$. Again we introduce differential-geometric data:

$$\nabla^\bullet \qquad \tilde{A}\text{-invariant extension of the partial covariant derivative.}$$

Such an extension is a section of an affine space bundle over X, so can be constructed via a partition of unity. Introduce a formal parameter u of degree 2 and its inverse u^{-1} of degree -2, and so the graded ring of $\mathbb{R}[[u, u^{-1}]]$-valued differential forms. (One can identify u as a generator of the K-theory of a point.) Following Quillen [Q] define the $\tilde{A}$-invariant superconnection

$$D = \begin{pmatrix} \nabla^0 & uT^* \\ T & \nabla^1 \end{pmatrix}$$

on $E^\bullet \to X$. Its usual curvature $D^2 \in \Omega\big(P; \mathrm{End}\,E \otimes \mathbb{R}[[u, u^{-1}]]\big)^2$ is an A-invariant form of total degree 2. However, it does not descend to the base X. Instead, one can check that the *twisted* curvature

$$F(E^\bullet, T, \nabla^\bullet) := D^2 - \beta \cdot \mathrm{id}$$

is A-invariant and basic, so descends to an element of $\Omega\big(X; \mathrm{End}\,E \otimes \mathbb{R}[[u, u^{-1}]]\big)^2$. (Note that $\mathrm{End}\,E^\bullet \to P$ descends to a graded vector bundle on X, since the center of $\tilde{A}$ acts trivially.) It does not, however, satisfy a Bianchi identity, since β is not closed.

The Chern character form

$$ch(E^\bullet, T, \nabla^\bullet) := \mathrm{Tr}\exp(u^{-1}F) \tag{6}$$

is an element in $\Omega\big(X; \mathbb{R}[[u, u^{-1}]]\big)^0$ of total degree 0. Here we assume favorable circumstances in which the graded trace Tr is finite. For example, if the twisting class is torsion then we can take $E^\bullet \to P$ finite dimensional. Or, if the superconnection comes from a family of elliptic operators as in [Bi] then the graded trace exists. The Chern character form (6) is not closed in the usual sense, but rather

$$(d + u^{-1}\eta)\, ch(E^\bullet, T, \nabla^\bullet) = 0.$$

The differential $d + u^{-1}\eta$ on $\Omega\big(X; \mathbb{R}[[u, u^{-1}]]\big)^\bullet$ computes a twisted version of real cohomology which is the codomain of the Chern character.

This construction works with little change for the more general twistings (2).

The differential geometric model we have outlined not only gives geometric representatives of twisted topological K-theory classes, but also geometric representatives of twisted *differential* K-theory classes. Similarly, the geometric twistings (5) give geometric models for differential cohomology classes. See [HS] for foundations of (untwisted) differential cohomology theories in general, and [L] for the basics differential K-theory.

3. Loop groups

Let G be a compact Lie group. The loop group LG of G is the space of smooth maps $S^1 \to G$. There is a twisted description we use instead. Namely, let $R \to S^1$ be a principal G-bundle and LG_R the group of gauge transformations, i.e., the space of smooth sections of the bundle of groups $G_R \to S^1$ associated to R. A trivialization of R gives an isomorphism $LG_R \cong LG$. Note that R is necessarily trivializable if G is connected, and in general the topological class of R is labeled by a conjugacy class in $\pi_0 G$. Let $G(R) \subset G$ be the union of components which comprise that conjugacy class. The theory of loop groups LG_R is described in [PS], [Ka], and some specific further developments appear in [FF], [We].

One salient feature of loop groups is the existence of nontrivial central extensions

$$1 \longrightarrow \mathbb{T} \longrightarrow LG_R^\tau \longrightarrow LG_R \longrightarrow 1.$$

As in (2) we may also consider gradings $\epsilon\colon LG_R \to \mathbb{Z}/2\mathbb{Z}$. We call the pair $\tau = (LG_R^\tau, \epsilon)$ a *graded central extension* and denote it simply as LG_R^τ. It has an invariant in

$$H^1_G\big(G(R); \mathbb{Z}/2\mathbb{Z}\big) \times H^3_G\big(G(R); \mathbb{Z}\big) \tag{7}$$

as follows. Fix a basepoint $s \in S$ and consider the product

$$P = \mathcal{A}_R \times R_s \tag{8}$$

of the space of connections $\mathcal{A}_R$ on R and the fiber R_s at the basepoint. Then LG_R acts freely on P with quotient the holonomy map

$$\text{hol}: P \longrightarrow G(R).$$

Furthermore, the G action on $G(R)$ by conjugation lifts to the G action on the R_s factor of P, where it commutes with the LG_R action. Thus we have an equivariant twisting

$$\tilde{\tau} = (LG_R, \epsilon, LG_R^\tau, P)$$

whose isomorphism class, called the *level*, lies in (7). (The basepoint is not necessary. It is more natural to replace (8) by $\mathcal{A}_R \times R$ and LG_R by the group of bundle automorphisms which cover rotations in the base S^1.)

As a warm-up to representations of loop groups, recall the Borel-Weil construction of representations of a compact connected Lie group G. Let $T \subset G$ be a maximal torus, and $F = G/T$ the *flag manifold*. Then F admits G-invariant complex structures. Fix one. A character λ of T determines a holomorphic line bundle $L_\lambda \to F$, and the standard construction takes the induced virtual representation of G to be $\oplus_q (-1)^q H^q(F; L_\lambda)$, which is the G-equivariant index of the $\bar{\partial}$ operator. We can use the Dirac operator instead of the $\bar{\partial}$ operator. This has the advantage that no complex structure need be chosen, though it can be described in holomorphic terms as the *ρ-shift* $L_\lambda \longrightarrow L_\lambda \otimes K_F^{1/2}$, where $K_F^{1/2} \to F$ is a square root of the canonical bundle. More generally, if $Z \subset G$ is the centralizer of any subtorus of T, and $F' = G/Z$ the corresponding generalized flag manifold, then there is a Dirac induction map

$$K_Z \cong K_G(F') \longrightarrow K_G. \tag{9}$$

A representation of Z defines a G-equivariant vector bundle on F', and (9) is the equivariant index of the Dirac operator with coefficients in this bundle.

A similar construction works for loop groups. Fix a conjugacy class $C \subset G(R)$. The group $LG_R \times G$ acts transitively on $\text{hol}^{-1}(C)$. Let Z_C denote the stabilizer at some point; it embeds isomorphically into either factor of $LG_R \times G$. Introduce the flag manifold $\mathcal{F}_C = \text{hol}^{-1}(C)/G$; then the loop group LG_R acts transitively $\mathcal{F}_C$ with stabilizer the image of Z_C in LG_R. On the other hand, the image of Z_C in G is the centralizer of an element in C. The geometry of $\mathcal{F}_C$ is similar to that of finite dimensional flag manifolds [F4]. An important special case is $C = \{1\}$ and $R \to S^1$ trivial. Then the flag manifold is the "loop grassmannian" $\mathcal{F} = LG/G$.

There is a Dirac induction for loop groups which generalizes (9). For this we introduce spinors on LG_R and a canonical graded central extension. Consider the principal $SO(\mathfrak{g} \oplus \mathbb{R})$-bundle associated to $R \to S^1$ via the twisted adjoint homomorphism

$$G \longrightarrow SO(\mathfrak{g} \oplus \mathbb{R}),$$
$$g \longmapsto Ad_g \oplus \det Ad_g.$$

It is trivializable, and any trivialization induces an isomorphism $LG_R \to LSO(N)$ for $N = \dim(G) + 1$. There is a canonical graded central extension of $LSO(N)$ from the spin representation, and it pulls back to the desired canonical graded central extension $LG_R^{\sigma_R(G)} \to LG_R$. The spin representation itself defines (projective) spinors on LG_R. For a conjugacy class $C \subset G(C)$ the embedding $i_C \colon Z_C \hookrightarrow LG_R$ induces a graded central extension $Z_C^{i_C^* \sigma_R(G)}$ which may be described instead by the real adjoint representation $\mathfrak{z}_C$, viewed as a Z_C-equivariant vector bundle over a point (see (3)). In this form it carries a degree shift, and so it is natural to also include a degree shift in $\sigma_R(G)$. We do not specify it precisely, but remark that its parity agrees with that of $\dim Z_C$. If G is simple, connected, and simply connected and $R \to S^1$ is trivial, then $\sigma_R(G)$ has degree shift $\dim G$ and the H^3 component of the level in (7) is the dual Coxeter number of G times a generator; the H^1 component vanishes.

For a fixed graded central extension LG_R^τ of LG_R there is a finite set of isomorphism classes of irreducible positive energy representations of LG_R^τ on which the center acts by scalar multiplication. Let $R^\tau(G)$ denote the abelian group generated by these equivalence classes. It is natural to extend this to a $\mathbb{Z}$-graded group with mod 2 periodicity and possibly nontrivial groups in odd degrees. Now we describe Dirac induction for loop groups. As in the first map of (9) a representation of Z_C defines an LG_R-equivariant vector bundle over the flag manifold $\mathcal{F}_C$. However, we are interested in LG_R^τ-equivariant vector bundles, so need to start with a representation of the central extension of Z_C defined by $i_C^* \tau$. Finally, spinors on the flag manifold $\mathcal{F}_C$ may be constructed from spinors on LG_R by "subtracting" spinors on the adjoint representation of Z_C, and this imposes an additional twisting. Altogether, then, Dirac induction is a map

$$K_{Z_C}^{i_C^* \tau - \mathfrak{z}_C} \longrightarrow R^{\tau - \sigma_R(G)}\big(G(R)\big). \tag{10}$$

The *adjoint shift* $\sigma_R(G)$ on the right hand side means that we obtain representations of the fiber product of $LG_R^\tau \to LG_R$ with the inverse of $LG_R^{\sigma_R(G)} \to LG_R$, including a degree shift. For connected, simply connected G it suffices to consider $C = \{1\}$, since then (10) is surjective, but this is not true in general.

The inclusion $\tilde{i}_C \colon C \hookrightarrow G(R)$ induces a pushforward in twisted K-theory (cf. (4)):

$$K_{Z_C}^{i_C^* \tau - \mathfrak{z}_C} \cong K_G^{\tilde{i}_C^* \tilde{\tau} + TC - TG|_C}(C) \longrightarrow K_G^{\tilde{\tau}}\big(G(R)\big). \tag{11}$$

The maps (10) and (11) give, for each conjugacy class C, a correspondence between certain representations of the loop group and a twisted K-theory group. Our main result is

Theorem 1. *These correspondences induce an isomorphism of abelian groups*

$$R^{\tau - \sigma_R(G)}\big(G(R)\big) \longrightarrow K_G^{\tilde{\tau}}\big(G(R)\big). \tag{12}$$

There is a transgression from $H^4(BG;\mathbb{Z})$ to levels (7) with trivial first component. More generally, there is an extension

$$0 \longrightarrow H^4(BG;\mathbb{Z}) \longrightarrow E^4(BG) \longrightarrow H^2(BG;\mathbb{Z}/2\mathbb{Z}) \longrightarrow 0,$$

and elements of $E^4(BG)$ transgress to general levels for all loop groups LG_R simultaneously. If G is connected then any $R \to S^1$ is trivializable, and both sides of (12) have a ring structure for any fixed R. For arbitrary G the sum of each side of (12) over representatives of each topological type of $R \to S^1$ has a ring structure. The multiplication on representations is the *fusion product* [V], [F], [T]; on twisted K-theory it is the pushforward by multiplication $G \times G \to G$ or equivalently the Pontrjagin product in K-homology.

Theorem 2. *If the level of the twisting $\tilde{\tau} + \tau(TG)$ is transgressed, then the isomorphisms* (12) *are compatible with the ring structure.*

Our proof reduces both sides of (12) to the statement for tori, where there is a direct argument.

References

[AR]   A. Adem, Y. Ruan, *Twisted orbifold K-theory*, `math.AT/0107168`.

[A1]   M. F. Atiyah, *K-Theory*, Benjamin, New York, 1967.

[A2]   M. F. Atiyah, *K-theory past and present*, `math.KT/0012213`.

[AH]   M. F. Atiyah, F. Hirzebruch, *Vector bundles and homogeneous spaces*, Proc. Symp. Pure Math. **3** (1961), 7–38.

[AS]   M. F. Atiyah, G. B. Segal, *Equivariant K-theory and completion*, J. Diff. Geom. **3** (1969), 1–18.

[Bi]   J. M. Bismut, *The Atiyah-Singer Index Theorem for families of Dirac operators: two heat equation proofs*, Invent. math. **83** (1986), 91–151.

[BS]   A. Borel, J.-P. Serre, *Le théorème de Riemann-Roch*, Bull. Soc. Math. France **86** (1958), 97–136.

[Bo]   R. Bott, *The space of loops on a Lie group*, Michigan Math. J. **5** (1958), 35–61.

[BCMMS] P. Bouwknegt, A. L. Carey, V. Mathai, M. K. Murray, D. Stevenson, *Twisted K-theory and K-theory of bundle gerbes*, hep-th/0106194.

[B]   J.-L. Brylinski, *Loop spaces, characteristic classes and geometric quantization*, Birkhäuser Boston Inc., Boston, MA, 1993.

[CG]   N. Chriss, V. Ginzburg, *Representation Theory and Complex Geometry*, Birkhäuser, Boston, 1997.

[DK]   P. Donovan, M. Karoubi, *Graded Brauer groups and K-theory with local coefficients*, Inst. Hautes Études Sci. Publ. Math. **38** (1970), 5–25.

[F]   G. Faltings, *A proof for the Verlinde formula*, J. Algebraic Geom. **3** (1994), 347–374.

[FF]   B. L. Feĭgin and E. V. Frenkel, *Affine Kac-Moody algebras and semi-infinite flag manifolds*, Comm. Math. Phys. **128** (1990), 161–189.

[F1]   D. S. Freed, *Higher algebraic structures and quantization*, Commun. Math. Phys. **159** (1994), 343–398, hep-th/9212115.

[F2]   D. S. Freed, *The Verlinde algebra is twisted equivariant K-theory*, Turkish J. Math. **25** (2001), 159–167, math.RT/0101038.

[F3]   D. S. Freed, *K-theory in quantum field theory*, Current Developments in Mathematics 2001, math-ph/0206031.

[F4]   D. S. Freed, *The geometry of loop groups*, J. Diff. Geo. **28** (1988), 223–276.

[FHT]   D. S. Freed, M. J. Hopkins, C. Teleman, *Twisted equivariant K-theory with complex coefficients*, math.AT/0206257.

[H]   N. J. Hitchin, *Lectures on special lagrangian submanifolds*, Winter School on Mirror Symmetry, Vector Bundles and Lagrangian Submanifolds (Cambridge, MA, 1999), Amer. Math. Soc., Providence, RI, 2001, 151–182, math.DG/9907034.

[HS]   M. J. Hopkins, I. M. Singer, *Quadratic functions in geometry, topology, and M-theory* (in preparation).

[J]   K. Jänich, *Vektorraumbündel und der Raum der Fredholm-Operatoren*, Math. Annalen **161** (1965), 129–42.

[Ka]   V. G. Kac, *Infinite dimensional Lie algebras*, Cambridge University Press, Cambridge, 1990.

[K]   N. H. Kuiper, *The homotopy type of the unitary group of Hilbert space*, Topology **3** (1965), 19–30.

[L]   J. Lott, *$\mathbb{R}/\mathbb{Z}$ index theory*, Comm. Anal. Geom. **2** (1994), 279–311.

[LU]   E. Lupercio, B. Uribe, *Gerbes over orbifolds and twisted K-theory*, math.AT/0105039.

[MMS]   V. Mathai, R.B. Melrose, I.M. Singer, *The index of projective families of elliptic operators*, math.DG/0206002.

[M]   M. K. Murray, *Bundle gerbes*, J. London Math. Soc. **54** (1996), 403–416.

[NT]   V. Nistor, E. Troitsky, *An index for gauge-invariant operators and the Dixmier-Douady invariant*, math.KT/0201207.

[PS]   A. Pressley, G. Segal, *Loop Groups*, Oxford University Press, Oxford, 1986.

[Q]   D. Quillen, *Superconnections and the Chern character*, Topology **24** (1985), 89–95.

[R]   J. Rosenberg, *Continuous trace algebras from the bundle theoretic point of view*, Jour. Austr. Math. Soc. **47** (1989), 368–381.

[S1]   G. B. Segal, *Fredholm complexes*, Quart. J. Math. Oxford **21** (1970), 385–402.

[S2]   G. B. Segal, *Equivariant K-theory*, Publ. Math.Inst. Hautes. Études Sci. **34** (1968), 129–151.

[T]   C. Teleman, *Lie algebra cohomology and the fusion rules*, Comm. Math. Phys. **173** (1995), 265–311.

[To]   B. Toen, *Notes on G-theory of Deligne-Mumford stacks*, `math.AG/9912172`.

[V]   E. Verlinde, *Fusion rules and modular transformations in 2D conformal field theory*, Nuclear Phys. B **300** (1988), 360–376.

[We]   R. Wendt, *Weyl's character formula for non-connected Lie groups and orbital theory for twisted affine Lie algebras*, J. Funct. Anal. **180** (2001), 31–65, `math.RT/9909059`.

[W]   E. Witten, *Quantum field theory and the Jones polynomial*, Commun. Math. Phys. **121** (1989), 351–399.

ICM 2002 · Vol. III · 431–443

Mirror Symmetry and Quantum Geometry

Kentaro Hori*

Abstract

Recently, mirror symmetry has been derived as T-duality applied to gauge systems that flow to non-linear sigma models. We present some of its applications to study quantum geometry involving D-branes. In particular, we show that one can employ D-branes wrapped on torus fibers to reproduce the mirror duality itself, realizing the program of Strominger-Yau-Zaslow in a slightly different context. The Floer theory of intersecting Lagrangians plays an essential role.

2000 Mathematics Subject Classification: 81T30, 81T60, 14J32, 53D40, 53D45.
Keywords and Phrases: Mirror symmetry, D-branes, Floer homology.

1. Introduction

Mirror symmetry has played important roles in exploring the quantum modification of geometry in string theory. Things started with the discovery of mirror pairs of Calabi-Yau manifolds [1], with a subsequent application [2] to superstring compactifications. It had an immediate impact on enumerative geometry and motivated various mathematical investigations including the formulation of Gromov-Witten invariants. Another breakthrough was made through the recognition of D-branes as indispensable elements in string theory [3], which was preceded by Konstevich's homological mirror symmetry [4]. Studies and applications of mirror symmetry involving D-branes have enriched our understanding of quantum geometry.

In particular, Strominger-Yau-Zaslow (SYZ) proposed, using the transformation of D-branes under T-duality, that mirror symmetry of Calabi-Yau manifolds is nothing but dualization of special Lagrangian torus fibrations [5]. This provides a very geometric picture of mirror symmetry that has inspired many physicists and mathematicians.

---

*School of Natural Sciences, Institute for Advanced Study, Princeton, NJ 08540, USA; Department of Physics & Mathematics, University of Toronto, Toronto, Ontario M5S 1A7, Canada. E-mail: hori@ias.edu

Recently, more progress has been made via an exact analysis of quantum field theory on the worldsheet [6]. Mirror symmetry is derived as T-duality applied to gauge systems [7] that flow to non-linear sigma models. This, however, turns sigma models into Landau-Ginzburg (LG) models, where the LG potential for the dual fields is generated by the vortex-anti-vortex gas of the high energy gauge system.

What is the relation between this and SYZ, both of which use T-duality applied to torus fibers? Since SYZ employ D-branes wrapped on the torus fibers, it is a natural idea to do the same in the situation of [6]. In this talk, we present some applications of [6] to study the properties of D-branes. In particular, we show that the study of D-branes wrapped on torus fibers indeed reproduces the LG mirror of [6]. The study of Floer homology for intersecting Lagrangians [8] plays an important role. We will also present other aspects of mirror symmetry involving D-branes.

2.   T-Duality and D-Branes

Let us consider a closed string moving on the circle of radius R, which is described by a periodic scalar field $X \equiv X + 2\pi R$ on the worldsheet. The space of states is decomposed into sectors labeled by two conserved charges — the momentum $l \in \mathbb{Z}$ associated with the translation symmetry $X \to X +$constant, and the winding number $m \in \mathbb{Z}$ which counts how many times the string winds around the circle. The ground state in each sector has energy $\frac{1}{2}[(l/R)^2 + (Rm)^2] - \frac{1}{12}$, which is invariant under

$$R \longleftrightarrow \frac{1}{R},$$

and $l \leftrightarrow m$. In fact, the sigma model on the circle of radius R is equivalent to the sigma model on the circle of radius $1/R$. This is called *T-duality*. The exchange of momentum and winding number can be described as the relation between the corresponding currents

$$\partial_t X = \partial_\sigma \widetilde{X}, \quad \partial_\sigma X = \partial_t \widetilde{X}, \tag{2.1}$$

where (t, σ) are the time and space coordinates on the worldsheet and $\widetilde{X} \equiv \widetilde{X} + 2\pi/R$ is the coordinate of the T-dual circle.

Let us now introduce an open string to this system. We need to specify the boundary condition on the scalar field X at the worldsheet boundary, say, at $\sigma = 0$. The Neumann boundary condition $\partial_\sigma X|_{\sigma=0} = 0$ corresponds to the freely moving end point, while the Dirichlet boundary condition $\partial_t X|_{\sigma=0} = 0$ fixes the end point. They describe open strings ending on *D-branes*: the former is for a D1-brane wrapped on the circle while the latter is for a D0-brane at a point of the circle. By the relation (2.1), we see that T-duality exchanges the Neumann and Dirichlet boundary conditions. Thus, T-duality maps a D1-brane wrapped on the circle to a D0-brane at a point of the T-dual circle. The open string end point is charged under the $U(1)$ gauge field on the D-brane. The holonomy $a = \int_{S^1} A$ parametrizes the gauge field configuration on the D1-brane wrapped on the circle. Under T-duality, this parameter is mapped to the position of the D0-brane in the T-dual circle. Therefore, the T-dual of S^1 can be identified as the dual circle $H^1(S^1, U(1))$. This story generalizes to the higher dimensional torus T^n: T-duality inverts the radii of the torus, mapping a Dn-brane wrapped on T^n to a D0-brane at a point of the T-dual torus $\widetilde{T}^n$, where the map provides the identification $\widetilde{T}^n \cong H^1(T^n, U(1))$.

The same story applies to *supersymmetric* theories on the worldsheet, which are obtained by including fermionic fields as the superpartner of the scalar fields. When the target space is a Kähler manifold, the system has $(2,2)$ *supersymmetry*, which is an extended symmetry with four supercharges $Q_+, \overline{Q}_+, Q_-, \overline{Q}_-$ ($\pm$ shows the worldsheet chirality. The supercharges are complex but are related by Hermitian conjugation $\overline{Q}_\pm = Q_\pm^\dagger$). The complex coordinates of the target space are annihilated by $\overline{Q}_+$-variation, but are mapped to the partner Dirac fermions under the $Q_\pm$-variation. The simplest example is the cylinder $\mathbb{C}^\times = \mathbb{R} \times S^1$ with the (flat) product metric parametrized by the radius R of the circle. T-duality applied to the S^1 yields another cylinder $\widetilde{\mathbb{C}}^\times = \mathbb{R} \times \widetilde{S}^1$ with radius $1/R$. This is actually an example of *mirror symmetry*: Q_- and $\overline{Q}_-$ are exchanged under the equivalence. T-duality maps the various D-brane configurations: A D2-brane wrapped on $\mathbb{C}^\times$ is mapped to a D1-brane extending in the $\mathbb{R}$-factor of $\widetilde{\mathbb{C}}^\times$; a D1-brane wrapped on S^1 at a point in the $\mathbb{R}$-direction is mapped to a D0-brane at a point of $\widetilde{\mathbb{C}}^\times$.

Let us next consider a more interesting target space — the two-sphere S^2. It can be viewed as the circle fibration over a segment, and one may ask what happens if T-duality is applied fiber-wise. Since T-duality inverts the radius of the circle, a larger circle is mapped to a smaller circle and a smaller circle is mapped to a larger circle, and one may naïvely expect that the dual geometry is as in Fig. 1.

Since the size of the dual circle blows up towards the two ends, two holes effectively open up and the dual geometry has the topology of a cylinder. This is consistent in one aspect: the conserved momentum associated with the $U(1)$-isometry of S^2 (fiber-rotaion) is mapped to the winding number of the dual system, which is conserved due to the cylindrical topology. However, another aspect is not clear. The winding number is not conserved in the original system because $\pi_1(S^2) = \{1\}$, and this should mean in the dual theory that the momentum is not conserved or the translation symmetry is broken. But how can it be broken? Is it because the metric is secretly not invariant under rotation of the cylinder?

What really happens under T-duality is as follows. It is true that the dual geometry has the topology of a cylinder. However, the dual theory is not just a sigma model but a model called a *Landau-Ginzburg (LG) model*. It has a potential and Yukawa coupling terms determined by a holomorphic function of the target space, called the *superpotential*. Let us parametrize the dual cylinder by a complex coordinate Y which is periodic in the imaginary direction, $Y \equiv Y + 2\pi i$. Then the superpotential of the dual LG model is given by

$$W = e^{-Y} + e^{-t+Y}. \qquad (2.2)$$

Here $t = r - i\theta$ is a complex parameter that corresponds to the data of the original S^2 sigma model: r is the area of the original S^2 and θ determines the B-field (it gives a phase factor $e^{ik\theta}$ to the path-integral measure for a worldsheet mapped to S^2 with degree k). It is this superpotential that breaks the translation symmetry $\text{Im}(Y) \to \text{Im}(Y)+$ constant. This T-duality is a mirror symmetry, as in the example of the cylinder.

This was derived in [6] by exact analysis of quantum field theory on the world-sheet. The derivation applies to the case where the target space is a general toric manifold X. A toric manifold, as S^2 is, can be viewed as a torus fibration over some base manifold, and T-duality sends the sigma model to a LG model. Suppose X is realized as the symplectic quotient of $\mathbb{C}^N$ by the $U(1)^k$ action $(z_i) \mapsto (e^{i \sum_a Q_i^a \lambda_a} z_i)$ with the moment map equation $\sum_i Q_i^a |z_i|^2 = r^a$, and suppose the B-field is such that the path-integral weight is $e^{i \sum_a k_a \theta^a}$ for a map of multi-degree (k_a). Then the dual geometry is an $(N - k)$-dimensional cylinder defined by $\sum_i Q_i^a Y_i = r^a - i\theta^a$ for $Y_i \equiv Y_i + 2\pi i$, and the superpotential is given by $W = \sum_i e^{-Y_i}$. This mirror symmetry explains several observations made earlier [9–13]. The analysis of [6] also includes the derivation of the mirror pairs of Calabi-Yau hypersurfaces or complete intersections in toric manifolds [1, 14]. Furthermore, the method can also be applied to string backgrounds with non-trivial dilaton [15] and H-field [16].

We do not repeat the analysis of [6] in what follows. Instead, we will find some of the consequences of the mirror symmetry, especially on D-branes. We will see that D-brane analysis sheds new light on the duality itself. Recall that, for the circle sigma model, the dual circle was identified as the space of wrapped D1-branes, $\widetilde{S}^1 \cong H^1(S^1, U(1))$. The same will happen here; the dual theory can be rediscovered by looking at the D-branes wrapped on the circle fibers of the toric manifold.

3.   Supersymmetric D-branes

Abstractly, D-branes can be regarded as boundary conditions or boundary interactions on the worldsheet of an open string. We will focus on those preserving a half of the $(2, 2)$ worldsheet supersymmetry. There are two kinds of such D-branes [17]: A-branes preserving the combinations $Q_A = \overline{Q}_+ + Q_-$ and $Q_A^\dagger = Q_+ + \overline{Q}_-$; B-branes preserving $Q_B = \overline{Q}_+ + \overline{Q}_-$ and $Q_B^\dagger = Q_+ + Q_-$. Since mirror symmetry exchanges Q_- and $\overline{Q}_-$, A-branes and B-branes are exchanged under mirror symmetry.

Let us consider the sigma model on a Kähler manifold M. M can be considered as a complex manifold or as a symplectic manifold (with respect to the Kähler form ω). A D-brane wrapped on a cycle γ of M and supporting a unitary gauge field A is an A-brane if γ is a Lagrangian submanifold ($\omega|_\gamma = 0$) and A is flat ($F_A = 0$), while it is a B-brane if γ is a complex submanifold of M and A is holomorphic ($F_A^{(2,0)} = 0$). If we consider a LG model with superpotential W, there is a further condition that the W-image of γ is a straight line parallel to the real axis for A-branes and W is a constant on γ for B-branes [18, 19]. A-branes and B-branes are objects of interest from the point of view of symplectic geometry and complex analytic geometry, respectively. They are exchanged under mirror symmetry.

Of prime interest are the lowest energy states of open strings ending on D-branes, in particular, the supersymmetric ground states which correspond to massless open string modes. The theory of an open string stretched between two A-branes (or two B-branes) has one complex supercharge $Q = Q_A$ (or $Q = Q_B$). In many cases, it obeys the supersymmetry algebra

$$\{Q, Q^\dagger\} = 2H,$$
$$Q^2 = 0, \tag{3.1}$$

where H is the Hamiltonian of the system. Then the system can be regarded as supersymmetric quantum mechanics (with infinitely many degrees of freedom) and the standard method [20] applies. In particular, there is a one-to-one correspondence between the gound states and the Q-cohomology classes. However, in some cases, the above algebra is modified and it can happen that

$$Q^2 \neq 0. \tag{3.2}$$

In such a case, the cohomological characterization of the ground states does not apply. (In fact, there is no supersymmetric ground state.) This does not happen for closed strings and is a new phenomenon peculiar to open strings.

In what follows, we study D-branes in the sigma model on S^2 or more general toric manifolds. In Sec. 4., we study A-branes in S^2 and the mirror B-branes in the LG model. We will see that we can reproduce the mirror duality through the study of D-branes. In Sec. 5., we study B-branes in S^2 and the mirror A-branes in LG.

4. Intersecting Lagrangians and their Mirrors

Let (M, ω) be a Kähler manifold. We study an open string stretched from one A-brane (γ_0, A_0) to another (γ_1, A_1), where γ_i are Lagrangian submanifolds and A_i are flat $U(1)$ connections on them. Classical supersymmetric configurations are the ones mapped identically to the intersection points of γ_0 and γ_1. However, quantum tunneling effects may lift the ground state degeneracy. Only the index $\mathrm{Tr}(-1)^F = \#(\gamma_0 \cap \gamma_1)$ is protected from corrections.

To determine the actual ground state spectrum, one may apply the Morse theory analysis of [20] to the space of open string configurations. The sigma model action defines a Morse function and its critical points are indeed the constant maps to the intersection points of γ_0 and γ_1. Tunneling configurations are holomorphic maps from the strip to M such that the left and the right boundaries are mapped to γ_0 and γ_1 respectively, and the far past and the far future are asymptotic to the constant maps to $\gamma_0 \cap \gamma_1$. This usually leads to a cochain complex that models the original Q-complex. However, the 'coboundary' operator may fail to be nilpotent [8], $\partial^2 \neq 0$, which corresponds to $Q^2 \neq 0$.

As an example, consider two Lagrangian submanifolds in the complex plane $M = \mathbb{C}$ as depicted in Fig. 2.

They intersect at two points q and p, and the constant maps to them are the candidate supersymmetric configurations. The 'cochain' complex has $\mathbb{Z}_2$ grading that distinguishes p and q. There is one tunneling configuration from q to p — the holomorphic map from the strip to the region C. Then the 'coboundary' operator acts as $\partial q = \mathrm{e}^{-A(C)} p$ where $A(C)$ is the area of the region C. Likewise, we find $\partial p = \mathrm{e}^{-A(D)} q$. Then we see that

$$\partial^2 q = \mathrm{e}^{-A(C)} \mathrm{e}^{-A(D)} q = \mathrm{e}^{-A(C \cup D)} q \neq 0. \tag{4.1}$$

The standard proof of $\partial^2 = 0$ does not apply here: there is a one-parameter family of tunneling configurations from q to q, that starts with the composition of C and D at p and ends with the holomorphic disc $C \cup D$. (The family is made of the composition of C and D along the segment $[r,p]$, where $r \in \gamma_1$ moves from p to q.) This is a general phenomenon called "bubbling off of holomorphic discs", which is peculiar to open string systems.

If ∂ happens to be nilpotent, one can define the cohomology group, which is known as the Floer cohomology group $HF((\gamma_0, A_0), (\gamma_1, A_1))$. This is the space of supersymmetric ground states of the open string system.

Let us next consider a LG model with superpotential W on a complex manifold Y. We study an open string stretched from a B-brane Z_0 to another Z_1, where Z_i is a complex submanifold of Y on which W is constant. It is straightforward to show, using the canonical commutation relation, that

$$Q^2 = W|_{Z_1} - W|_{Z_0}, \tag{4.2}$$

and there is no quantum correction to it. Thus, we see that $Q^2 \neq 0$ if the W-values of Z_0 and Z_1 do not agree. If they do agree, the space of supersymmetric ground states is the Q-cohomology group. There is a finite dimensional model of the Q-complex [21, 22]; It consists of anti-holomorphic forms on $Z_0 \cap Z_1$ with values in the exterior powers of $N_{Z_0} \cap N_{Z_1}$, on which the coboundary operator acts as $\bar{\partial} + \partial W \cdot$ (here N_{Z_i} is the normal bundle of Z_i in Y, and $\partial W \cdot$ is the contraction with the holomorphic 1-form ∂W). If Z_0 and Z_1 are points, the complex is non-trivial only if they are the same point, and the cohomology is the exterior power of the tangent space if the point is a critical point of W but vanishes if the point is not a critical point.

Eqn. (4.2), if non-zero, is the mirror counterpart of $Q^2 \neq 0$ for the intersecting Lagrangian systems. Note that we find (4.2) by purely classical analysis, in contrast to the case of A-branes where computation of Q^2 requires the analysis of quantum tunneling effects.

To make it explicit, let us come back to the mirror symmetry between the S^2 sigma model and the LG model with superpotential $W = e^{-Y} + e^{-t+Y}$. We consider A-branes wrapped on the S^1 fibers in Fig. 1 (Left). They are mapped under T-duality to B-branes at points on the Y-cylinder — their $\mathrm{Re}(Y)$ and $\mathrm{Im}(Y)$ coordinates are determined respectively by the location of the S^1 and the holonomy of the $U(1)$ gauge field. One can actually find a detailed map via a field theory analysis [21]. Let $Y = c - ia$ be the location of the mirror D0-brane. Then the area of the disc bounded by the original D1-brane is c and the holonomy of the $U(1)$ connection is $e^{i(a - c\theta/r)}$. Let us now find the condition for $Q^2 = 0$ using the LG description. For two D0-branes at $Y = c_0 - ia_0$ and $Y = c_1 - ia_1$, we find

$$Q^2 = e^{-c_1 + ia_1} + e^{-t + c_1 - ia_1} - e^{-c_0 + ia_0} - e^{-t + c_0 - ia_0}. \tag{4.3}$$

There are two solutions to $Q^2 = 0$: $(c_1, a_1) = (c_0, a_0)$ and $(r - c_0, \theta - a_0)$. On the S^2 side, they correspond to two identical D1-branes (the same location and the same holonomy), and two D1-branes of opposite holonomies such that the interior area of one is equal to the exterior area of the other. The Q-cohomology is non-trivial only if the two points are the same critical point of the superpotential W, which is $\mathrm{e}^{-Y} = \mathrm{e}^{-t/2}$ or $-\mathrm{e}^{-t/2}$ in the present case. In such a case, the cohomology is the exterior power of the tangent space, $\wedge^\bullet \mathbb{C} = \wedge^0 \mathbb{C} \oplus \wedge^1 \mathbb{C}$, which has one bosonic and one fermionic basis vector. In the original S^2 sigma model, $\mathrm{e}^{-c+ia} = \pm \mathrm{e}^{-t/2}$ means that the interior and exterior areas of S^1 are the same and the holonomy is ± 1. Thus, we find that mirror symmetry predicts

$$HF((\gamma_0, A_0), (\gamma_1, A_1)) = \begin{cases} \wedge^\bullet \mathbb{C} & \text{if } \gamma_0 = \gamma_1 \text{ divides } S^2 \text{ into halves} \\ & \quad \text{and } A_0 = A_1 \text{ has holonomy } \pm 1 \\ 0 & \text{otherwise} \end{cases} \qquad (4.4)$$

One can also directly compute the Floer homology group. Let us put the D1-branes in a position as in Fig. 3.

For simplicity, we suppress the θ-angle as well as the $U(1)$ connections on γ_i. The two circles intersect at two points q and p which represent 'cochains' of different degrees (the 'complex' is $\mathbb{Z}_2$ graded). The 'coboundary' operator acts as $\partial q = \mathrm{e}^{-A(D)}p - \mathrm{e}^{-A(F)}p$ and $\partial p = \mathrm{e}^{-A(C)}q - \mathrm{e}^{-A(E)}q$. Thus the square is $\partial^2 q = (\mathrm{e}^{-A(C \cup D)} - \mathrm{e}^{-A(D \cup E)} - \mathrm{e}^{-A(C \cup F)} + \mathrm{e}^{-A(E \cup F)})q$. If we denote the region inside γ_i by D_i, the area of the region outside γ_i is $r - A(D_i)$. We therefore find

$$\partial^2 q = \left[\mathrm{e}^{-A(D_0)} - \mathrm{e}^{-A(D_1)} - \mathrm{e}^{-r+A(D_1)} + \mathrm{e}^{-r+A(D_0)} \right] q \qquad (4.5)$$

This vanishes if and only if $A(D_1) = A(D_0)$ or $A(D_1) = r - A(D_0)$, namely, when the interior area of γ_1 is equal to the interior or exterior area of γ_0. One can also show that the $U(1)$ holonomy of the two should be the same or opposite, respectively. This matches precisely with the LG result. In fact, (4.5) with holonomy included is identical to (4.3) under the map of variables mentioned before. Let us next compute the cohomology. Consider $A(D_0) = A(D_1)$ first. In such a case, $A(C) = A(E)$ and $A(F) - A(D) = A(F \cup C) - A(D \cup C) = r - A(D_1) - A(D_0) = r - 2A(D_0)$. Thus the coboundary operator acts as

$$\partial q = \mathrm{e}^{-A(D)}(1 - \mathrm{e}^{-r+2A(D_0)})p$$
$$\partial p = 0.$$

The cohomology vanishes if $A(D_0) \neq r/2$, while it is non-vanishing if $A(D_0) = r/2$ — each of q and p generates the cohomology in its degree. Note that $A(D_0) = r/2$ is when γ_0 divides S^2 into halves. One can also show that the cohomology is non-vanishing if the holonomy is ± 1. We also find the same conclusion if we start with $A(D_1) = r - A(D_0)$. To summarize, we see that the result (4.4) of mirror symmetry is indeed correct.

We have seen a practical aspect of mirror symmetry in the study of D-branes. Computation of the ground state spectrum of the open string involves highly non-trivial analysis of quantum tunneling effect in the sigma model, while it is done by a simple classical manipulation in the mirror LG model. This story generalizes straightforwardly to more general toric manifolds.

There is another important aspect in the above analysis. We recall that, in the case of the S^1 sigma model, it was enough to analyze the wrapped D1-brane to find the T-dual space; since a D1-brane is T-dual to a D0-brane, its moduli space is equivalent to the space of D0-branes of the T-dual theory, namely the dual space itself. In fact, the same applies here as well. By analysing the D1-branes wrapped on the S^1-fibers, we find the cylinder as the dual space, but we also find the superpotential (up to addition of a constant) through the computation of Q^2, see (4.5). In other words, we can reproduce the mirror symmetry between toric sigma models and LG models by analyzing the D-branes.

This point of view is similar in spirit to Strominger-Yau-Zaslow [5] who proposed, using D-branes, that mirror symmetry of Calabi-Yau manifolds is nothing but dualization of special Lagrangian fibrations. The latter has led, for example, to the topological construction of mirror manifolds [23–25]. In attempts to make it more precise, the treatment of singular fibers constitutes the essential part where quantum corrections are expected to play an important role. (See e.g. [26] for recent progress.) The example considered above includes singular fibers and we have shown how the quantum effect is taken into account. Although we have not dealt with *special* Lagrangian fibrations, we note that the above analysis applied to toric Calabi-Yau yields mirror manifolds consistent with the SYZ program. For example, if we start with the total space of $\mathcal{O}(-1) \oplus \mathcal{O}(-1)$ over $\mathbb{CP}^1$, we obtain the LG model on $(\mathbb{C}^\times)^3$ with superpotential $W = e^{-Y_0}(e^{-Y_1} + e^{-Y_2} + e^{-t-Y_1-Y_2} + 1)$ as the mirror, which in turn is related [18] to the sigma model on the Calabi-Yau hypersurface $e^{-Y_1} + e^{-Y_2} + e^{-t-Y_1-Y_2} + 1 = uv$ in $\mathbb{C}^\times \times \mathbb{C}^\times \times \mathbb{C} \times \mathbb{C} = \{(e^{-Y_1}, e^{-Y_2}, u, v)\}$. This last mirror turns out to be consistent with SYZ topologically [27].

5. Holomorphic Bundles and their Mirrors

Let us consider an open string stretched between B-branes wrapped on M and supporting holomorphic vector bundles E_0 and E_1. The zero mode sector of the open string Hilbert space is identified as the space $\Omega^{0,\bullet}(M, E_0^* \otimes E_1)$ where the supercharge Q acts as the Dolbeault operator. Thus, the space of supersymmetric ground states in the zero mode approximation is the Dolbeault cohomology or $\mathrm{Ext}^\bullet(E_0, E_1)$. In particular, the index is

$$\mathrm{Tr}(-1)^F = \chi(E_0, E_1).$$

In the full theory, some pairs of states of neighboring R-charges could be lifted to non-supersymmetric states. The latter does not happen if $\mathrm{Ext}^p(E_0, E_1)$ is non-zero only for even p (or odd p). An example of such a pair E_0, E_1 is from an *exceptional collection* [29], which is an ordered set of bundles $\{E_i\}$ where $\mathrm{Ext}^p(E_i, E_i) = \delta_{p,0}\mathbb{C}$ while for $i < j$ $\mathrm{Ext}^\bullet(E_j, E_i) = 0$ but $\mathrm{Ext}^p(E_i, E_j)$ can be non-zero only for one value of p. For $\mathbb{CP}^n$, the set of line bundles $\{\mathcal{O}(i)\}_{i=j}^{j+n}$ is an exceptional collection ($\forall j \in \mathbb{Z}$).

Next we consider the LG model with superpotential W which has only non-degenerate critical points $\{p_i\}$. Gradient flows of $\mathrm{Re}(W)$ originating from p_i sweep out a Lagrangian submanifold γ_i whose W-image is a straight horizontal line emanating from the critical value $w_i = W(p_i)$. Thus, the D-brane wrapped on γ_i is an A-brane. Let us consider an open string stretched from γ_i to γ_j. Classical supersymmetric configurations are gradient flows of $-\text{Im}(W)$ from a point in γ_i to a point in γ_j. The index is the number of such gradient flows counted with an appropriate sign. It is

$$\text{Tr}(-1)^F = \#(\gamma_i^- \cap \gamma_j^+)$$

where $\gamma_k^\pm$ is the deformation of γ_k so that the W-image is rotated at w_k by a small angle $\pm\epsilon$. Quantum mechanically, the paths of opposite signs are lifted by instanton effects and $|\#(\gamma_i^- \cap \gamma_j^+)|$ is in fact the number of supersymmetric ground states. One can also quantize the system using the Morse function determined by the LG action. This leads to the LG version of the Floer homology group $\text{HF}_W^\bullet(\gamma_i, \gamma_j)$. (This was studied also by Y.-G. Oh [28].) If $\text{Im}(w_i) > \text{Im}(w_j)$ and there is no critical value between the W images of γ_i and γ_j, then $\#(\gamma_i^- \cap \gamma_j^+)$ is equal to the number S_{ij} of BPS solitons connecting p_i to p_j.

B-branes supporting holomorphic bundles on toric manifolds are mapped under mirror symmetry to A-branes in the LG models. Let us examine the detail in the S^2 sigma model and its mirror LG model $W = \text{e}^{-Y} + \text{e}^{-t+Y}$. The superpotential has two critical points $p_\pm$ with critical values $w_\pm = \pm 2\,\text{e}^{-r/2+i\theta/2}$. We first set $\theta = 0$ so that $w_\pm$ are on the real line. The simplest brane on S^2 is the $U(1)$-bundle with trivial connection, or $\mathcal{O}$ over $\mathbb{CP}^1$. Since we are T-dualizing along the S^1-fibers along which the holonomy is trivial, the dual has to be localized at a constant point in the dual fibers. It is a D1-brane at the horizontal line $\text{Im}(Y) = 0$ whose W-image is a straight line emanating from w_+ — the A-brane γ_+. Another simple B-brane is the D0-brane at a point. Its dual is wrapped on S^1 but remains localized in the horizontal direction. The one at $\text{Re}(Y) = r/2$ has straight W-image and therefore is an A-brane, which we call γ_0.

Introducing n D0-branes to $\mathcal{O}$ means turning on n units of magnetic flux and produces the brane $\mathcal{O}(n)$. Its dual is the combination of γ_+ and $n\gamma_0$. For negative n, $n\gamma_0$ is understood to have the reversed orientation.

Let us now turn on a small positive θ. Then w_+ is slightly above w_- in the imaginary direction, and γ_0 is no longer an A-brane. Instead we find a non-compact A-brane γ_-. See Fig. 5(Left).

γ_- belongs to the homology class $\gamma_+ + \gamma_0$ and therefore is the mirror of $\mathcal{O}(1)$. We note that indeed $\chi(\mathcal{O}, \mathcal{O}(1)) = \#(\gamma_+^- \cap \gamma_-^+) = 2$. If we turn on a small negative θ, then w_+ is slightly below w_-, and we find the A-brane γ_- in the homology class $\gamma_+ - \gamma_0$ (Fig. 5(Right)). Since the orientation of γ_0 is reversed, γ_- is the mirror of $\mathcal{O}(-1)$ for this value of θ.

Note that γ_- for $\theta = +\epsilon$ and γ_- for $\theta = -\epsilon$ are related by the Picard-Lefschetz formula $\gamma_-|_{\theta=+\epsilon} = (-\gamma_- + 2\gamma_+)|_{\theta=-\epsilon}$, where the coefficient 2 is $\#(\gamma_+^- \cap \gamma_-^+)$ at $\theta = +\epsilon$. On the other hand, the mirror bundles $\mathcal{O}(1)$ and $\mathcal{O}(-1)$ appear in the exact sequence

$$0 \longrightarrow \mathcal{O}(-1) \longrightarrow \text{Ext}^0(\mathcal{O}, \mathcal{O}(1)) \otimes \mathcal{O} \longrightarrow \mathcal{O}(1) \longrightarrow 0.$$

In such a case, $\mathcal{O}(-1)$ is said to be the *left mutation* of $\mathcal{O}(1)$ with respect to $\mathcal{O}$ [29]. It was observed in [30, 31] (see also [32]) that mutation of exceptional bundles in certain Fano manifolds is related to the Picard-Lefschetz monodromy that appears in the theory of BPS solitons in sigma models. We can now understand it as a consequence of mirror symmetry in the case where the taget space is a toric manifold. Related work has been done by P. Seidel [33].

6. Concluding Remarks

We have presented some applications of mirror symmetry between toric manifolds and LG models [6], especially to the study of D-branes. In particular, we have seen that D-branes wrapped on torus fibers can tell us about the mirror symmetry itself. Below, we comment on some matters that are not covered here.

The structure of integerable systems in topological string theory, along with the matrix model representation, is an important aspect of quantum geometry. It was first discovered in topological gravity [34, 35], the case where the target space is a point. There are several observations suggesting that it may extend to more general target spaces. Here the mirror LG superpotential is expected to play an important role. For example, for the $\mathbb{CP}^1$ model, $W = e^{-Y} + e^{-t+Y}$ is the Lax operator of the Toda lattice hierarchy [36, 12] under the replacement of Y by a differential operator (see also [37, 38]). For a projective space of higher dimension, the mirror superpotential also plays the role of Lax operator, at least in genus zero [13]. Also, the Virasoro constraint [39] suggests the existence of matrix model representations which in turn are related to integrable systems. Some beautiful story is waiting there to be discovered. The possible role of branes is interesting to explore.

Mirror symmetry has an application to enumerative geometry including holomorphic curves with boundaries. In the case of special Lagrangian submanifolds in Calabi-Yau three-folds, the number of holomorphic curves ending on the submanifolds enters into certain terms in the low energy effective action of the superstring theory. The number of holomorphic discs for a class of special Lagrangians in toric Calabi-Yaus are counted in [40] by computing the space-time superpotential terms in the mirror side. There are several related works including the mathematical tests [41–43] of [40].

The derivation of [6] itself applies only to toric manifolds and submanifolds therein that are realized as vacuum manifolds of *abelian* gauge systems. There are some observations that suggest the form of the LG-type mirror for Grassmann and flag manifolds [13, 44] which are realized as the vacuum manifolds of *non-abelian* gauge systems. It would be a challenging problem to find or derive the mirror of such manifolds and others. The method presented here using D-branes is possibly of some use. Another possible way is to consider compactification of three-dimensional mirror symmetry [45]. This works for the abelian gauge systems [46] and many examples of non-abelian mirror pairs are known in three-dimensional gauge theories (e.g. [47]).

Acknowledgement. I would like to dedicate this manuscript to the memory of Dr. Sung-Kil Yang for whom I have limitless respect. I wish to thank T. Eguchi, A. Iqbal, C. Vafa, S.-K. Yang himself and C.-S. Xiong for collaborations in [12, 13, 6, 18]. I would also like to thank K. Fukaya, Y.-G. Oh, H. Ohta and K. Ono for useful discussions and explanation of their works. This work was supported in part by NSF PHY 0070928.

References

[1] B.R. Greene, M.R. Plesser, Duality in Calabi-Yau moduli space, *Nucl. Phys.* **B338** (1990) 15.

[2] P. Candelas, X.C. De La Ossa, P.S. Green and L. Parkes, A pair of Calabi-Yau manifolds as an exactly soluble superconformal theory, *Nucl. Phys.* **B359** (1991) 21.

[3] J. Polchinski, Dirichlet-Branes and Ramond-Ramond Charges, *Phys. Rev. Lett.* **75** (1995) 4724, hep-th/9510017.

[4] M. Kontsevich, Homological algebra of mirror symmetry, *Proc. ICM* (Zürich, 1994), Birkhäuser, 1995, p120, alg-geom/9411018.

[5] A. Strominger, S.-T. Yau, E. Zaslow, Mirror symmetry is T-duality, *Nucl. Phys.* **B479** (1996) 243, hep-th/9606040.

[6] K. Hori, C. Vafa, Mirror symmetry, hep-th/0002222.

[7] E. Witten, Phases of N = 2 theories in two dimensions, *Nucl. Phys.* **B403** (1993) 159, hep-th/9301042.

[8] K. Fukaya, Y-G. Oh, H. Ohta, K. Ono, Lagrangian Intersection Floer Theory - Anomaly and Obstruction -, preprint.

[9] P. Fendley, K. Intriligator, Scattering and thermodynamics in integrable N=2 theories, *Nucl. Phys.* **B380** (1992) 265, hep-th/9202011.

[10] V.V. Batyrev, Quantum cohomology rings of toric manifolds, *Asterisque* No. 218 (1993) 9, alg-geom/9310004.

[11] A. Givental, Homological geometry and mirror symmetry, *Proc. ICM* (Zürich, 1994) Birkhäuser, 1995, p472; Equivariant Gromov-Witten invariants, alg-geom/9603021.

[12] T. Eguchi, K. Hori and S. K. Yang, Topological sigma models and large N matrix integral, *Int. J. Mod. Phys.* **A10** (1995) 4203, hep-th/9503017.

[13] T. Eguchi, K. Hori, C.-S. Xiong, Gravitational quantum cohomology, *Int. J. Mod. Phys.* **A12** (1997) 1743, hep-th/9605225.

[14] V.V. Batyrev, Dual polyhedra and mirror symmetry for Calabi-Yau hypersurfaces in toric varieties *J. Alg. Geom.* **3** (1994) 493.

[15] K. Hori, A. Kapustin, Duality of the fermionic 2d black hole and N = 2 Liouville theory as mirror symmetry, *JHEP* **0108** (2001) 045 hep-th/0104202; Worldsheet descriptions of wrapped NS five-branes, hep-th/0203147.

[16] D. Tong, NS5 branes, T-duality and worldsheet instantons, hep-th/0204186.

[17] H. Ooguri, Y. Oz, Z. Yin, D-branes on Calabi-Yau spaces and their mirrors, *Nucl. Phys.* **B477** (1996) 407, hep-th/9606112.

[18] K. Hori, A. Iqbal, C. Vafa, D-branes and mirror symmetry, hep-th/0005247.

[19] S. Govindarajan, T. Jayaraman and T. Sarkar, Worldsheet approaches to D-branes on supersymmetric cycles, *Nucl. Phys.* **B580** (2000) 519, hep-th/9907131.

[20] E. Witten, Supersymmetry And Morse Theory, *J. Diff. Geom.* **17** (1982) 661.

[21] K. Hori, Linear models of supersymmetric D-branes, In [48], 111-186, hep-th/0012179.

[22] K. Hori, S. Katz, A. Klemm, R. Pandharipande, R. Thomas, C. Vafa, R. Vakil, E. Zaslow, *a book based on CMI school "Mirror Symmetry"* to appear, Ch. 40: Boundary $\mathcal{N} = 2$ theories.

[23] W.-D. Ruan, Lagrangian torus fibrations and mirror symmetry of Calabi-Yau manifolds, In [48], 385-427, math.DG/0104010.

[24] M. Gross, Topological mirror symmetry, *Invent. Math.* **144** (2001) 75, math.AG/9909015.

[25] D. R. Morrison, Geometric aspects of mirror symmetry, math.AG/0007090.

[26] K. Fukaya, Multivalued Morse theory, asymptotic analysis, and mirror symmetry, preprint.

[27] M. Gross, Examples of special Lagrangian fibrations, In [48], 81-109, math.AG/0012002.

[28] Y.-G. Oh, Floer theory for non-compact Lagrangian submanifolds, preprint.

[29] A. N. Rudakov et al., *London Math. Soc., Lecture Note Series* **148** Cambridge.

[30] E. Zaslow, Solitons and helices: The Search for a math physics bridge, *Commun. Math. Phys.* **175** (1996) 337, hep-th/9408133.

[31] M. Kontsevich, Lectures at the ENS, France 1998 (unpublished lecture notes).

[32] B. Dubrovin, Geometry and analytic theory of Frobenius manifolds, *Proc. ICM* (Berlin, 1998) Doc. Math. 1998, 315.

[33] P. Seidel, Vanishing cycles and mutation, math.SG/0007115; More about vanishing cycles and mutation, In [48], 429-465, math.SG/0010032.

[34] E. Witten, On The Structure Of The Topological Phase Of Two-Dimensional Gravity, *Nucl. Phys.* **B340** (1990) 281; Two-dimensional gravity and intersection theory on moduli space, *Surveys in differential geometry* (Cambridge, MA, 1990) 243–310.

[35] M. Kontsevich, Intersection theory on the moduli space of curves and the matrix Airy function, *Commun. Math. Phys.* **147** (1992) 1–23.

[36] T. Eguchi and S. K. Yang, The Topological CP^1 model and the large N matrix integral, *Mod. Phys. Lett.* **A9** (1994) 2893, hep-th/9407134.

[37] E. Getzler, The Toda conjecture, In [48], 51–79, math.AG/0108108.

[38] A. Okounkov, R. Pandharipande, Gromov-Witten theory, Hurwitz numbers, and Matrix models, I, math.AG/0101147; Gromov-Witten theory, Hurwitz theory, and completed cycles, math.AG/0204305.

[39] T. Eguchi, K. Hori and C. S. Xiong, Quantum cohomology and Virasoro algebra, *Phys. Lett.* **402B** (1997) 71, hep-th/9703086.

[40] M. Aganagic, C. Vafa, Mirror symmetry, D-branes and counting holomorphic discs, hep-th/0012041; M. Aganagic, A. Klemm, C. Vafa, Disk instantons, mirror symmetry and the duality web, *Z. Naturforsch. A* **57** (2002) 1, hep-th/0105045.

[41] S. Katz, C.-C. Liu, Enumerative geometry of stable maps with Lagrangian boundary conditions and multiple covers of the disc, *Adv. Theor. Math. Phys.* 5 (2001) 1–49, math.AG/0103074.

[42] J. Li, Y.-S. Song, Open string instantons and relative stable morphisms, *Adv.*

Theor. Math. Phys. 5 (2001) 67–91, hep-th/0103100.

[43] T. Graber and E. Zaslow, Open-String Gromov-Witten Invariants: Calculations and a Mirror "Theorem", hep-th/0109075.

[44] A. Givental, Stationary phase integrals, quantum Toda lattices, flag manifolds and the mirror conjecture, *Topics in singularity theory*, 103–115, AMS.Transl. Ser. 2, 180, alg-geom/9612001.

[45] K. A. Intriligator and N. Seiberg, Mirror symmetry in three dimensional gauge theories, *Phys. Lett.* **387B** (1996) 513, hep-th/9607207.

[46] M. Aganagic, K. Hori, A. Karch and D. Tong, Mirror symmetry in 2+1 and 1+1 dimensions, *JHEP* **0107** (2001) 022, hep-th/0105075.

[47] J. de Boer, K. Hori, H. Ooguri and Y. Oz, Mirror symmetry in three-dimensional gauge theories, quivers and D-branes, *Nucl. Phys.* **B493** (1997) 101, hep-th/9611063.

[48] K. Fukaya, Y.-G. Oh, K. Ono, G. Tian, editors, *Symplectic Geometry and Mirror Symmetry* (Seoul, 2000) World Scientific, 2001.

ICM 2002 · Vol. III · 445–455

Nonperturbative Localization[*]

S. Jitomirskaya[†]

Abstract

Study of fine spectral properties of quasiperiodic and similar discrete Schrodinger operators involves dealing with problems caused by small denominators, and until recently was only possible using perturbative methods, requiring certain small parameters and complicated KAM-type schemes. We review the recently developed nonperturbative methods for such study which lead to stronger results and are significantly simpler. Numerous applications mainly due to J. Bourgain, M. Goldstein, W. Schlag, and the author are also discussed.

2000 Mathematics Subject Classification: 35, 37, 60, 81.

1. Introduction

Consider an operator acting on on $\ell^2(\mathbb{Z}^d)$ defined by

$$H_\lambda = \Delta + \lambda V, \tag{1.1}$$

where Δ is the lattice tight-binding Laplacian

$$\Delta(n,m) = \begin{cases} 1, & \text{dist}(n,m) = 1, \\ 0, & \text{otherwise,} \end{cases}$$

and $V(n,m) = V_n \delta(n,m)$ is a potential given by $V_n = f(T_1^{n_1} \cdots T_d^{n_d}\theta)$, $\theta \in \mathbb{T}^b$, where $T_i\theta = \theta + \omega_i$, and ω is an incommensurate vector. In certain cases Δ may also be replaced by a long-range Laplacian. Replacing T_i's with other commuting ergodic transformations would give a general framework of ergodic Schrödinger operators ([21]; see Sec. 8. for an example of this kind) but we will mostly focus on the quasiperiodic (QP) operators that have been intensively studied in Physics and Mathematics literature. For another review of some recent developments in this

---

[*]This research was partially supported by NSF grant DMS-0070755.
[†]Department of Mathematics, University of California at Irvine, CA 92697, USA. E-mail:szhitomi@math.uci.edu

area see [7]. The questions of interest are the nature and structure of the spectrum, behaviour of the eigenfunctions, and particularly the quantum dynamics: properties of the time evolution $\Psi_t = e^{itH}\Psi_0$ of an initially localized wave packet Ψ_0.

Of particular importance is the phenomenon of Anderson localization (AL) which is usually referred to the property of having pure point spectrum with exponentially decaying eigenfunctions. A somewhat stronger property of dynamical localization (see Sec. 7.) indicates the insulator behavior, while ballistic transport, which for $d = 1$ follows from the absolutely continuous (ac) spectrum, indicates the metallic behavior.

Operators with ergodic potentials always have spectra (and pure point (p.p.) spectra, understood as closures of the set of eigenvalues) constant for a.e. realization of the potential. Moreover, the p.p. spectrum of operators with ergodic potentials never contains isolated eigenvalues, so p.p. spectrum in such models is dense in a certain closed set. An easy example of an operator with dense pure point spectrum is H_∞ which is operator (1.1) with $\lambda^{-1} = 0$, or pure diagonal. It has a complete set of eigenfunctions, characteristic functions of lattice points, with eigenvalues V_j. H_λ may be viewed as a perturbation of H_∞ for small λ^{-1}. However, since V_j are dense, small denominators $(V_i - V_j)^{-1}$ make any perturbation theory difficult, e.g. requiring intricate KAM-type schemes.

The probabilistic KAM-type scheme was developed by Fröhlich and Spencer [26] for random potentials (V_n are i.i.d.r.v.'s) in the multi-dimensional case, and is called multi-scale analysis. It was significantly modified, improved, and widely applied in the later years by a number of authors, most notably [24]. An alternative method for random localization was found by Aizenman-Molchanov [2] and later further developed by Aizenman and coauthors. While still requiring certain large parameters this method relies on direct estimates of the Green's function rather than a step-by-step perturbation scheme.

For QP potentials none of the above methods work, as, among other reasons, they do not allow rank-one perturbations, nor Wegner-type estimates. The situation here is more difficult and the theory is far less developed than for the random case. With a few exceptions the results are confined to the 1D case, and also 1-frequency case ($b = 1$) has been much better developed than that of higher frequencies.

One might expect that H_λ with λ small can be treated as a perturbation of $H_0 = \Delta$, and therefore have ac spectrum. It is not the case though for random potentials in $d = 1$, where AL holds for all λ. Same is expected for random potentials in $d = 2$ (but not higher). Moreover, in 1D case there is strong evidence (numerical, analytical, as well as rigorous [8]) that even models with very mild stochasticity in the underlying dynamics have point spectrum for all values of λ like in the random case (e.g. $V_n = \lambda f(n^\sigma \alpha + \theta)$ for any $\sigma > 1$). At the same time, for QP potentials one can in many cases show ac spectrum for λ small as well as pure point spectrum for λ large (see below), and therefore there is a metal insulator transition in the coupling constant. It is an interesting question whether quasiperiodic potentials are the only ones with metal-insulator transition in 1D.

2.   Perturbative vs nonperturbative

It is probably fair to say that much of the theory of qusiperiodic operators has been first developed around the almost Mathieu operator, which is

$$H_{\lambda,\omega,\theta} = \Delta + \lambda f(\theta + n\omega) \tag{2.1}$$

acting on $\ell^2(\mathbb{Z})$, with $f : \mathbb{T} \to \mathbb{T}$; $f(\theta) = \cos(2\pi\theta)$. The first KAM-type approaches, in both large and small coupling regimes, were developed for this or similar models [23, 5]. The perturbative proofs of complete Anderson localization for $\lambda > \lambda(\omega)$ large are due to Sinai [41] and Fröhlich-Spencer-Wittwer [27], and both applied to cos-type f. For λ small Chulaevsky-Delyon [19] proved pure ac spectrum using duality and the construction of Sinai [41]. Elliasson (see [25] for a review) developed alternative KAM-type arguments for both large and small λ for the case of real-analytic (actually, somewhat more general) class of f in (2.1).

The common feature of the perturbative *approaches* above is that, besides all of them being rather intricate multi-step procedures, they rely extensively on eigenvalue and eigenfunction parametrization and perturbation arguments.

The common feature of the perturbative *results* in the quasiperiodic setting is that they provide no explicit estimates on how large (or small) the parameter λ should be, and, more importantly, λ clearly depends on ω at least through the constants in the Diophantine characterization of ω.

In contrast, the nonperturbative results allow effective (in many cases even optimal) and, most importantly, independent of ω, estimates on λ. We will take the latter property (uniform in ω estimates on λ) as a definition of a NP result.

Recently developed nonperturbative methods are also quite different from the perturbative ones, in that they do not employ multi-scale schemes: usually only a few (from one to three) sufficiently large scales are involved, do not use the eigenvalue parametrization, and rely instead on direct estimates of the Green's function. They are also significantly less involved, technically. One may think that in these latter respects they resemble the Aizenman-Molchanov method for random localization. It is, however, a superficial similarity, as, on the technical side, they are still closer to and do borrow certain ideas from [26, 24].

Several results that satisfy our definition of nonperturbative appeared prior to the recent developments, and were all related to the almost Mathieu operator (see [32] for a review). In [33, 34] AL was proved for $\lambda > 15$, and existence of p.p. component for $\lambda > 2$. The latter papers, while introducing some of the ingredients of the recent nonperturbative methods, did not take advantage of the positivity of the Lyapunov exponents which proved very important later.

3.   Lyapunov exponents

Here for simplicity we consider the quasiperiodic case, although the definition of the Lyapunov exponents and some of the mentioned facts apply more generally to the 1D ergodic case.

For an energy $E \in \mathbb{R}$ the Lyapunov exponent $\gamma(E)$ is defined as

$$\gamma(E) = \lim_{n \to \infty} \frac{\int_0^1 \ln \|M_k(\theta, E)\| d\theta}{k}, \tag{3.1}$$

where

$$M_k(\theta, E) = \prod_{n=k-1}^{0} \begin{pmatrix} E - \lambda f(\omega n + \theta) & -1 \\ 1 & 0 \end{pmatrix}$$

is the k−step transfer-matrix for the eigenvalue equation $H\Psi = E\Psi$.

We will be interested in the regime when Lyapunov exponents are positive for all energies in a certain interval intersecting the spectrum. It is well known and fairly easy to see that if this condition holds for all $E \in \mathbb{R}$, there is no ac component in the spectrum for a.e. θ (it is actually true for all θ [39]). Positivity of Lyapunov exponents, however, does not imply exponential decay of eigenfunctions (in particular, not for the Liouville ω [3] nor for the resonant $\theta \in \mathbb{T}^b$ [36]).

NP methods, at least in their original form, stem to a large extent from estimates involving the Lyapunov exponents and exploiting their positivity.

The general theme of the results on positivity of $\gamma(E)$, as suggested by perturbation arguments, is that the Lyapunov exponents are positive for large λ. This was first established by Aubry-Andre [1] for the almost Mathieu operator with $\lambda > 2$. Their proof was made rigorous in [3]. Another proof, exploiting the subharmonicity, was given by Herman [31], and applied to trigonometric polynomials f. The lower bound in [31] was in terms of the highest coefficient of the trigonometric polynomial and therefore this did not easily extend to the real analytic case. All the subsequent proofs, however, were also based on subharmonicity. Sorets-Spencer [42] proved that for nonconstant real analytic potentials v on $\mathbb{T}$ ($b = 1$) one has $\gamma(E) > \frac{1}{2} \ln \lambda$ for $\lambda > \lambda(v)$ and all irrational ω. Another proof was given in [11], where this was also extended to the multi-frequency case ($b > 1$) with, however, the estimate on λ dependent on the Diophantine condition on ω. Finally, Bourgain [10] proved that Lyapunov exponents are continuous in ω at every incommensurate ω (for $b > 1$; for $b = 1$ this was previously established in [16]), and that led to the following Theorem which is the strongest result in this general context up to date:

Theorem 1 [10] *Let f be a nonconstant real analytic function on $\mathbb{T}^b$, and H given by (1.1). Then, for $\lambda > \lambda(f)$, we have $\gamma(E) > \frac{1}{2} \ln \lambda$ for all E and all incommensurate vectors ω.*

3.1.   Corollaries of positive Lyapunov exponents

The almost Mathieu operator   On one hand the almost Mathieu operator, while simple-looking, seems to represent most of the nontrivial properties expected to be encountered in the more general case. On the other hand it has a very special feature: the duality (essentially a Fourier) transform maps H_λ to $H_{4/\lambda}$, hence $\lambda = 2$ is the self-dual point. Aubry-Andre [1] conjectured that for this model, for irrational ω a sharp metal-insulator transition in the coupling constant λ occurs at the critical value of coupling $\lambda = 2$: the spectrum is pure point for $\lambda > 2$ and pure ac for $\lambda < 2$. A second, related, conjecture was that the dual of ac spectrum is pure point and vice versa. Both conjectures were modified based on the results of [3, 37, 36]. The first modified conjecture stated pure point spectrum for Diophantine ω and a.e. θ for $\lambda > 2$ and pure ac spectrum for $\lambda < 2$ for all ω, θ. As for the duality, the question, after some prior developments, was resolved in [28] where it was shown that the dual of point spectrum is ac spectrum (the proof applied in a more general context), and it was used (together with [38, 30]) to prove that the spectrum is purely singular continuous at $\lambda = 2$ for a.e. ω, θ.

As with the KAM methods, the almost Mathieu operator was the first model where the positivity of Lyapunov exponents was effectively exploited:

Theorem 2 [32] *Suppose ω is Diophantine and $\gamma(E, \omega) > 0$ for all $E \in [E_1, E_2]$. Then the almost Mathieu operator has Anderson localization in $[E_1, E_2]$ for a.e. θ.*

The condition on θ in [32] was actually explicit (arithmetic) and close to optimal. This, combined with the mentioned results on the Lyapunov exponents for the almost Mathieu operator [31] and duality [28] led to the following corollary:

Corollary 3 *The almost Mathieu operator $H_{\omega, \lambda, \theta}$ has*

1° *[32] for $\lambda > 2$, Diophantine $\omega \in \mathbb{R}$ and almost every $\theta \in \mathbb{R}$, only pure point spectrum with exponentially decaying eigenfunctions.*

2° *[28] for $\lambda = 2$, and a.e. $\omega, \theta \in \mathbb{R}$ purely singular-continuous spectrum.*

3° *[32, 28] for $\lambda < 2$, Diophantine $\omega \in \mathbb{R}$ and a.e. $\theta \in \mathbb{R}$, purely ac spectrum.*

Precise arithmetic descriptions of ω, θ are available. Thus the Aubry-Andre conjecture is settled at least for almost all ω, θ. One should mention, however, that while 1° is almost optimal, both 2° and 3° are expected to hold for all θ and all $\omega \notin \mathbb{Q}$, and such extension remains a challenging problem (see [40]).

The method in [32], while so far the only nonperturbative available allowing precise arithmetic conditions, uses some specific properties of the cosine. It extends to certain other but rather limited situations. A much more robust method was developed by Bourgain-Goldstein [11], which allowed them to extend (a measure-theoretic version of) the above result to the general real analytic as well as the multi-frequency case. Note, that essentially no results were previously available for the multifrequency case, even perturbative.

Theorem 4 [11] *Let f be non-constant real analytic on $\mathbb{T}^b$ and H given by (2.1). Suppose $\gamma(E, \omega) > 0$ for all $E \in [E_1, E_2]$ and a.e. $\omega \in \mathbb{T}^b$. Then for any θ, H has Anderson localization in $[E_1, E_2]$ for a.e. ω.*

Combining this with Theorem 1 one obtains [10] that for $\lambda > \lambda(f)$, H as above satisfies Anderson localization for a.e. ω.

One very important ingredient of the method of [11] is the theory of semi-algebraic sets that allows one to obtain polynomial algebraic complexity bounds for certain "exceptional" sets. Combined with measure estimates coming from the large deviation analysis of $\frac{1}{n} \ln \|M_n(\theta)\|$ (using subharmonic function theory and involving approximate Lyapunov exponents), this theory provides necessary information on the geometric structure of those exceptional sets. Such algebraic complexity bounds also exist for the almost Mathieu operator [32] and are actually sharp albeit trivial in this case due to the specific nature of the cosine.

4.   Without Lyapunov exponents

While having led to significant advances, Lyapunov exponents have obvious limitations, as any method, based on them, is restricted to 1D nearest neighbor Laplacians. It turns out that the above methods can be extended to obtain NP results in certain quasi-1D situations where Lyapunov exponents do not exist.

For the next Theorem let H be an operator (1.1) defined on $\ell^2(S)$ where $S = \mathbb{Z} \times S_0$, is a strip. S_0 here is a finite set with a metric, and $\mathrm{dist}((n, s), (n', s')) = |n - n'| + \mathrm{dist}(s, s')$. Let $V_{(n,s)} = f_s(\theta + n\omega)$, $\theta \in \mathbb{T}$.

Theorem 5 [14] *Assume f_s, $s \in S_0$, are non-constant real analytic functions on $\mathbb{T}$. Then for any $\theta \in \mathbb{T}$ and $\lambda > \lambda(f_s)$, operator H has AL for a.e. ω.*

The following nonperturbative Theorem deals with the case of small coupling:

Theorem 6 [15] *Let H be an operator (2.1), where f is real analytic on $\mathbb{T}$ and ω is Diophantine. Then, for $\lambda < \lambda(f)$, H has purely ac spectrum for a.e. θ.*

We note that an analogue of this Theorem does not hold in the multi-frequency case (see next section). Theorem 6 is a result on non-perturbative localization in disguise as it was obtained using duality [28] from a localization Theorem for a dual model which has in general a long-range Laplacian and was in turn obtained by an extension of the method of [32]. A certain measure-theoretic version of it by the method of [11] allowing non-local Laplacians but leading only to continuous spectrum is also available [6]. Theorem 5 was obtained by an extension of the method of [11]. Both Theorems above rely on large deviations for the quantities of the form $\frac{1}{n} \ln |\det(H - E)_\Lambda|$ and path-determinant expansion for the matrix elements of the resolvent [15]. The methods developed in [15] apply also to certain other situations with long-range Laplacians, for example the kicked rotor model (see Sec. 8.).

5.   Multidimensional case: $d > 1$

As mentined above, there are very few results in the multidimensional lattice case ($d > 1$). Essentially, the only result that existed before the new developments was a perturbative Theorem - an extension by Chulaevsky-Dinaburg [20] of Sinai's [41] method to the case of operator (1.1) on $\ell^2(\mathbb{Z}^d)$ with $V_n = \lambda f(n \cdot \omega)$, $\omega \in \mathbb{R}^d$, where f is a cos-type function on $\mathbb{T}$. Recently, Bourgain [6] obtained this result for real analytic f by a nonperturbative method. Note that since $b = 1$, this avoids most serious difficulties and is therefore significantly simpler than the general multi-dimensional case.

Theorem 7 [20, 6] *For any $\epsilon > 0$ there is $\lambda(f, \epsilon)$, and, for $\lambda > \lambda(f, \epsilon)$, $\Omega(\lambda, f) \subset \mathbb{T}^d$ with $mes(\Omega) < \epsilon$, so that for $\omega \notin \Omega$, operator (1.1) with V_n as above has Anderson localization.*

This should be confronted with the following Theorem of Bourgain [8]

Theorem 8 [8] *Let $d = 2$ and $f(\theta) = \cos 2\pi\theta$ in $H = H_\omega$ defined as above. Then for any λ measure of ω s.t. H_ω has some continuous spectrum is positive.*

Therefore for large λ there will be both ω with complete localization as well as those with at least some continuous spectrum. This shows that nonperturbative *results* do not hold in general in the multi-dimensional case!

A similar (in fact, dual) situation is observed for 1D multi-frequency ($d = 1$; $b > 1$) case at small disorder. One has, by duality:

Theorem 9 [20, 19] *Let H be given by (2.1) with $\theta, \omega \in \mathbb{T}^b$ and f real analytic on $\mathbb{T}^b$. Then for any $\epsilon > 0$ there is $\lambda(f, \epsilon)$ s.t. for $\lambda < \lambda(f, \epsilon)$ there is $\Omega(\lambda, f) \subset \mathbb{T}^b$ with $mes(\Omega) < \epsilon$ so that for $\omega \notin \Omega$, H has purely ac spectrum.*

Theorem 10 [8] *Let $d = 1, b = 2$ and f be a trigonometric polynomial on $\mathbb{T}^2$ with a non-degenerate maximum. Then for any λ measure of ω s.t. H_ω has some point spectrum, dense in a set of positive measure, is positive.*

Therefore, unlike the $b = 1$ case (see Theorem 6), nonperturbative *results* do not hold for absolutely continuous spectrum at small disorder.

6. Perturbative results by NP methods

While the above demonstrates the limitations of the NP results, the nonperturbative *methods* have been applied to significantly simplify the proofs and obtain new perturbative results that previously have been completely beyond reach.

We refer the reader for a description of many such applications to [7, 6]. In particular, new results on the construction of QP solutions in Melnikov problems and nonlinear PDE's, obtained by using certain ideas developed for NP quasi-periodic localization (e.g. the theory of semi-algebraic sets) are presented there.

We will only mention here a theorem by Bourgain-Goldstein-Schlag that is the only one so far treating a "true" $d > 1$ situation. Note that here $d = 2$, and the reasons why it has not yet been extended to higher dimensions are not just technical, but conceptual (there are certain purely arithmetic difficulties).

Theorem 11 [12] *Let $d = b = 2$ and f be real analytic on $\mathbb{T}^2$ such that all functions $f(\theta_1, \cdot)$, $f(\cdot, \theta_2)$, $(\theta_1, \theta_2) \in \mathbb{T}^2$ are nonconstant. Then for any $\epsilon > 0$ there is $\lambda(f, \epsilon)$ s.t. for $\lambda > \lambda(f, \epsilon)$ there is $\Omega(\lambda, f) \subset \mathbb{T}^d$ with $mes(\Omega) < \epsilon$ so that for $\omega \notin \Omega$ operator (1.1) with $V_n = \lambda f(n_1\omega_1, n_2\omega_2)$ has Anderson localization.*

7. Dynamical localization

Anderson localization does not in itself guarantee absense of quantum transport, or nonspread of an initially localized wave packet, as characterised, e.g., by boundedness in time of moments of the position operator [22] ([35] for an example of physical model with coexistence of exponential localization and quantum transport). Considering for simplicity the second moment

$$\langle x^2 \rangle_T = \frac{1}{T} \int_0^T \sum_n |\Psi_t(n)|^2 n^2 dt,$$

we will say that H exhibits dynamical localization (DL) if $\langle x^2 \rangle_T < \text{Const.}$ We will say that the family $\{H_\theta\}_{\theta \in \mathbb{T}^b}$ exhibits strong DL if $\int_{\mathbb{T}^b} d\theta \sup_t \langle x^2 \rangle_t < \text{Const.}$ We note that the results mentioned below will hold with more restrictive definitions of DL (involving the higher moments of the position operator) as well. DL implies p.p. spectrum by RAGE theorem (see, e.g. [21]), so it is a strictly stronger notion.

It turns out that nonperturbative methods allow for such dynamical upgrades as well. For the almost Mathieu operator we have

Theorem 12 [29] *For $\lambda > 2$ and Diophantine ω (as in [32]) strong DL holds.*

While proved in [29] with a slightly more restrictive condition on ω, Theorem 12 holds as stated by a result of [16]. For the results obtained by methods stemming from the approach of [11] one has

Theorem 13 [14] *In Theorems 4,5,7,11 dynamical localization also holds.*

This also applies to other results on nonperturbative localization, e.g. [13].

8. Quantum kicked rotor

The quantum kicked rotor was introduced in [18] as a model in quantum chaos. It is given by the time-dependent Schrodinger equation on $L^2(\mathbb{T})$

$$i\frac{\partial \phi}{\partial t} = a\frac{\partial^2 \phi}{\partial^2 \theta} + ib\frac{\partial \phi}{\partial \theta} + V(t,\theta)\phi$$

where $V(t,\theta) = \kappa \cos 2\pi\theta \sum_{n \in \mathbb{Z}} \delta(t-n)$. It represents quantization of the Chirikov standard map, and a conjecture (e.g.[4]) was that for a.e. a, b the solution ϕ is almost-periodic in time, thus demonstrating "quantum suppression of chaos" (as some chaos is expected for the standard map). Such almost periodicity follows from dynamical localization for the Floquet operator $W : L^2(\mathbb{T}) \to L^2(\mathbb{T})$ defined by $W\phi(t,\theta) = \phi(t+1,\theta)$. W is a unitary operator that in Fourier representation can be written as a product $U \cdot S$ where $S(n,m) = S(n-m)$ is a Töeplitz operator with very fast decay of $S(n)$ and $D(n,m) = D(n)\delta(n,m)$ is a diagonal operator with $D(n) = \exp(2\pi i(T^n x)_2)$, with T being the skew shift of the torus $T(x_1,x_2) = (x_1 + \omega, x_2 + x_1)$, and x_1, x_2, ω determined by a, b. The nonperturbative methods (particularly, the method for skew-shift dynamics localization in [13] and a long-range method [15]) were further developed for this model to obtain

Theorem 14 [9] *For any $\epsilon > 0$ and any fixed b there is $\kappa(\epsilon)$, and for $\kappa < \kappa(\epsilon)$, $\Omega(\kappa) \subset \mathbb{T}$ with $mes(\Omega) < \epsilon$, s.t. for $a \notin \Omega(\kappa)$, operator W has DL.*

Exploiting the multiplicative nature of W one also obtains a nonperturbative counterpart:

Theorem 15 [17] *There is $\kappa_0 > 0$ such that for any b, operator W has dynamical localization for $\kappa < \kappa_0$ and a.e. a.*

This confirms the "quantum suppression of chaos" conjecture for small κ.

References

[1] S. Aubry and G. Andre, Analyticity breaking and Anderson localization in incommensurate lattices. Ann. Israel Phys. Soc. **3**(1980), 133–140.

[2] M. Aizenman and S. Molchanov, Localization at large disorder and at extreme energies. Commun. Math. Phys.**157**(1993),245–278.

[3] J. Avron and B. Simon, Singular continuous spectrum for a class of almost periodic Jacobi matrices. Bull. AMS **6**(1982), 81–85.

[4] J. Bellisard, Noncommutative methods in semiclassical analysis, LNM 1589, 1–64 (1994).

[5] J. Bellisard, R. Lima, and D. Testard, A metal-insulator transition for the almost Mathieu model. Commun. Math. Phys. **88**(1983), 207–234.

[6] J. Bourgain, Green's function extimates for lattice Schrödinger operators and applications. 1–198, Ann. Math. Stud., to appear.

[7] J. Bourgain, New results on the spectrum of lattice Schrödinger operators and applications, Contemp. Math., to appear.

[8] J. Bourgain, On the spectrum of lattice Schrödinger operators woth deterministic potential, I, II, to appear in J. D'Analyse Jerusalem.

[9] J. Bourgain, Estimates on Green's functions, localization and the quantum kicked rotor model, Annals of Math, to appear.

[10] J. Bourgain, Positivity and continuity of the Lyapunov exponent for shifts on $\mathbb{T}^d$ with arbitrary frequency vector and real analytic potential. Preprint.

[11] J. Bourgain, M. Goldstein, On nonperturbative localization with quasiperiodic potential. Annals of Math., **152** (2000), 835–879.

[12] J. Bourgain, M. Goldstein, W. Schlag, Anderson localization on $\mathbb{Z}^2$ with quasiperiodic potential, Acta Math, to appear.

[13] J. Bourgain, M. Goldstein, W. Schlag, Anderson localization for Schrödinger operators on $\mathbb{Z}$ with potentials given by the skew shift, CMP **220**(2001), 583–621.

[14] J. Bourgain, S. Jitomirskaya, Anderson localization for the band model, Lecture Notes in Math. **1745**, 67-79, Springer, Berlin, 2000.

[15] J. Bourgain, S. Jitomirskaya, Absolutely continuous spectrum for 1D quasiperiodic operators, Invent. math., **148**(2002), 453-463.

[16] J. Bourgain, S. Jitomirskaya, Continuity of the Lyapunov exponent for quasiperiodic operators with analytic potential, JSP, to appear.

[17] J. Bourgain, S. Jitomirskaya, Nonperturbative localization for the quantum kicked rotor model,In preparation.

[18] B. Chirikov, F. Israilev, D. Shepelyanskii, Dynamical stochasticity in classical and quantum mechanics. MPR, **2**(1981), 209-267, Harwood Acad..

[19] V. Chulaevsky and F. Delyon, Purely absolutely continuous spectrum for almost Mathieu operators. J. Stat. Phys. **55**(1989), 1279–1284.

[20] V. Chulaevsky and E. Dinaburg, Methods of KAM theory for long range quasi-periodic operators on $\mathbb{Z}^n$, CMP **153**(1993), 559–577.

[21] H. L. Cycon, R. G. Froese, W. Kirsch, B. Simon, Schrödinger Operators. Berlin, Heidelberg, New York: Springer 1987.

[22] R. del Rio, S. Jitomirskaya, Y. Last, B. Simon, Operators with singular continuous spectrum: IV, J. d'Analyse Math. **69**(1996), 153–200.

[23] E. Dinaburg and Ya. Sinai, The one-dimensional Schrödinger equation with a quasi-periodic potential. Funct. Anal. Appl. **9**(1975), 279–289.

[24] H. von Dreifus and A. Klein, A new proof of localization in the Anderson tight binding model. Comm. Math. Phys. **124**(1989), 285–299.

[25] L. H. Eliasson, Reducibility and point spectrum for linear quasi-periodic skew products, Doc. Math, Extra volume ICM 1998 II, 779–787.

[26] J. Fröhlich and T. Spencer, Absence of diffusion in the Anderson tight-binding model for large disorder or low energy, CMP **88**(1983), 151–184.

[27] J. Fröhlich, T. Spencer, P. Wittwer, Localization for a class of one dimensional quasi-periodic Schrödinger operators. CMP **132**(1990), 5–25.

[28] A. Gordon, S. Jitomirskaya, Y. Last, and B. Simon, Duality and singular continuous spectrum in the almost Mathieu equation. Acta Math., **178**(1997), 169–183.

[29] F. Germinet, S. Jitomirskaya, Strong dynamical localization for the almost Mathieu model, Rev. Math. Phys. **13**(2001), 755–765.

[30] B. Helffer and J. Sjöstrand, Semiclassical analysis for Harper's equation. III. Cantor structure of the spectrum. (1989). Mm. Soc. Math. France (N.S.) **39**(1989), 1–124.

[31] M. Herman, Une methode pour minorer les exposants de Lyapunov et quelques exemples montrant le caractere local d'un theoreme d'Arnold et de Moser sur le tore en dimension 2. Comm. Math. Helv. **58**(1983), 453–502.

[32] S. Jitomirskaya, Metal-Insulator transition for the almost Mathieu operator. Annals of Math. **150** (1999), 1159–1175.

[33] S. Jitomirskaya, Anderson localization for the almost Mathieu equation; A nonperturbative proof. Commun. Math. Phys. **165**(1983), 49–58.

[34] S. Jitomirskaya, Anderson localization for the almost Mathieu equation, II: Point spectrum for $\lambda > 2$, Commun. Math. Phys.**168**(1995), 563–570.

[35] S. Jitomirskaya, H. Schulz-Baldes, G. Stolz, Delocalization in random polymer models. Preprint.

[36] S. Jitomirskaya and B. Simon, Operators with singular continuous spectrum, III. Almost periodic Schrödinger operators, Commun. Math. Phys. **165**(1994), 201–205.

[37] Y. Last, A relation between ac spectrum of ergodic Jacobi matrices and the

spectra of periodic approximants. CMP **151**(1993), 183–192.

[38] Y. Last, Zero measure of the spectrum for the almost Mathieu operator, CMP **164**(1994), 421–432.

[39] Y. Last and B. Simon, Eigenfuctions, transfermatrices, and ac spectrum of 1D Schrödinger operators, Invent. Math. **135**(1999), 329–367.

[40] B. Simon, Schrdinger operators in the twenty-first century, Mathematical Physics 2000, Imperial College, London, 283-288.

[41] Ya. Sinai, Anderson localization for one-dimensional difference Schrödinger operator with quasi-periodic potential. J. Stat. Phys. **46**(1987), 861–909.

[42] E. Sorets, T. Spencer, Positive Lyapunov exponents for Schrödinger operators with quasi-periodic potential, CMP **142**(1991), 543–566.

ICM 2002 · Vol. III · 457–466

Mathematical Results Inspired by Physics

Kefeng Liu*

Abstract

I will discuss results of three different types in geometry and topology. (1) General vanishing and rigidity theorems of elliptic genera proved by using modular forms, Kac-Moody algebras and vertex operator algebras. (2) The computations of intersection numbers of the moduli spaces of flat connections on a Riemann surface by using heat kernels. (3) The mirror principle about counting curves in Calabi-Yau and general projective manifolds by using hypergeometric series.

2000 Mathematics Subject Classification: 53D30, 57R91, 81T13, 81T30.
Keywords and Phrases: Localization, Elliptic genera, Moduli spaces, Mirror principle.

1. Introduction

The results I will discuss are all motivated by the conjectures of physicists, without which it is hard to imagine that these results would have appeared. In all these cases the new methods discovered during the process to prove those conjectures often give us many more surprising new results. The common feature of the proofs is that they all depend on localization techniques built upon various parts of mathematics: modular forms, heat kernels, symplectic geometry, and various moduli spaces.

Elliptic genera were invented through the joint efforts of physicists and mathematicians [17]. Actually in Section 2 I will only discuss in detail a vanishing theorem of the Witten genus, which is the index of the Dirac operator on loop space. This is a loop space analogue of the famous Atiyah-Hirzebruch vanishing theorem. It was discovered in the process of understanding the Witten rigidity conjectures for elliptic genera. A loop space analogue of a famous theorem of Lawson-Yau for non-abelian Lie group actions will also be discussed.

Moduli spaces of flat connections on Riemann surfaces have been studied for many years in various subjects of mathematics [2]. The computations of the intersection numbers on such moduli spaces have been among the central problems in

---

*Department of Mathematics, University of California, Los Angeles, CA 90095, USA. E-mail: liu@math.ucla.edu

the subject. In Section 3 I will discuss a very effective way to compute the most interesting intersection numbers by using the localization property of heat kernels. This proves several beautiful formulas conjectured by Witten [38]. We remark that these intersection numbers include those needed for the Verlinde formula.

In Section 4 I discuss some remarkable formulas about counting curves in projective manifolds, in particular in Calabi-Yau manifolds. I will discuss the mirror principle, a general method developed in [27]-[30] to compute characteristic classes and characteristic numbers on moduli spaces of stable maps in terms of hypergeometric series. The mirror formulas from mirror symmetry correspond to the computations of the Euler numbers. Mirror principle computes quite general Hirzebruch multiplicative classes such as the total Chern classes.

2. Elliptic genera

Let M be a compact smooth spin manifold with a non-trivial S^1-action, D be the Dirac operator on M. Atiyah and Hirzebruch proved that in such a situation the index of the Dirac operator $\text{Ind } D = \hat{A}(M) = 0$, where $\hat{A}(M)$ is the Hirzebruch $\hat{A}$-genus [3]. One interesting application of this result is that a $K3$ surface does not allow any non-trivial smooth S^1-action, because it has non-vanishing $\hat{A}$-genus.

Let LM be the loop space of M. LM consists of smooth maps from S^1 to M. There is a natural S^1-action on LM induced by the rotation of the loops, whose fixed points are the constant loops which is M itself. Witten formally applied the Atiyah-Bott-Segal-Singer fixed point formula to the Dirac operator on LM, from which he derived the following formal elliptic operator [36]:

$$D^L = D \otimes \bigotimes_{n=1}^{\infty} S_{q^n} TM = \sum_{n=0}^{\infty} D \otimes V_n \, q^n$$

where q is a formal variable and for a vector bundle E,

$$S_q E = 1 + q E + q^2 S^2 E + \cdots$$

is the symmetric operation and V_n is the combinations of the symmetric products $S^j(TM)$'s by formal power series expansion. So D^L, which is called the Dirac operator on loop space, actually consists of an infinite series of twisted Dirac operators with the pure Dirac operator D as the degree 0 term. The index of D^L, denoted by $\text{Ind } D^L$, is called the Witten genus. The loop space analogue of our vanishing theorem is the following:

Theorem 2.1: ([21]) *Let M be a spin manifold with non-trivial S^1-action. Assume $p_1(M)_{S^1} = n \, \pi^* u^2$ for some integer n, then the Witten genus vanishes:* $\text{Ind } D^L = 0$.

Here $p_1(M)_{S^1}$ is the equivariant first Pontrjagin class and u is the generator of the cohomology group of the classifying space BS^1, and $\pi : M \times_{S^1} ES^1 \to BS^1$ is the natural projection from the Borel construction.

This theorem implies that under the extra condition on the first Pontrjagin class, we have infinite number of elliptic operators with vanishing indices. The condition on the first equivariant Pontrjagin class is equivalent to that the S^1-action preserves the spin structure of LM. If we have a non-abelian Lie group acts on M non-trivially, then for an S^1 subgroup, the condition $p_1(M)_{S^1} = n\,\pi^*u^2$ is equivalent to $p_1(M) = 0$ which implies that LM is spin. As an easy consequence, we get:

Corollary 2.2: *Assume a non-abelian Lie group acts on the spin manifold M non-trivially and $p_1(M) = 0$, then the Witten genus, $\operatorname{Ind} D^L$, vanishes.*

This corollary should be considered as a loop space analogue of a result of Lawson-Yau in [18], which states that if a non-abelian Lie group acts on the spin manifold M non-trivially, then $\operatorname{Ind} D = 0$. Our results motivated Hoehn and Stolz to conjecture that, for a compact spin manifold M with positive Ricci curvature and $p_1(M) = 0$, the Witten genus vanishes. So far all of the known examples have non-abelian Lie group action, therefore our results applies. It should be interesting to see how to combine curvature with modular forms to get vanishing results.

The proof of Theorem 2.1 is an interesting combination of the Atiyah-Bott-Segal-Singer fixed point formula with Jacobi forms. The magic combination of geometry and modular invariance implies the vanishing of the equivariant index of D^L. Similar idea can be used to prove many more rigidity, vanishing and divisibility results for D^L twisted by bundles constructed from loop group representations. Such operators can be viewed as twisted Dirac operators on loop space. See [20] and [21]. In these cases the Kac-Weyl character formulas came into play. If we take the level 1 representations of the loop group of the spin group in our general rigidity theorem, we get the Witten conjectures on the rigidity of elliptic genera [36], which were proved by Taubes [35], Bott-Taubes [8], Hirzebruch [15], Krichever, Landweber-Stong, Ochanine for various cases.

Our method can actually go very far. Recently in [33] we proved rigidity and vanishing theorems for families of elliptic genera and the Witten genus. In [32] we proved similar theorems for foliated manifolds. In [10] such theorems were generalized to orbifolds. More recently in [11] we have proved a far general rigidity theorem for D^L twisted by vertex operator algebra bundles.

If we apply the modular invariance argument to the non-equivariant elliptic genera, we get a general formula which expresses the Hirzebruch L-form in terms of the twisted $\hat{A}$-forms [22]. A 12 dimensional version of this formula, due to Alveraz-Gaume and Witten, called the miraculous cancellation formula, had played important role in the development of string theory. This formula has many interesting mathematical consequences involving the eta-invariants. We refer the reader to [22] and [23].

3. Moduli spaces

Let G be a compact semi-simple Lie group and $\mathcal{M}_u$ be the moduli space of flat connections on a principal flat G-bundle P on a Riemann surface S with boundary, where $u \in Z(G)$ is an element in the center. Here for simplicity we first discuss the case when S has one boundary component, G is simply connected and the moduli space is smooth. A point in $\mathcal{M}_u$ is an equivalence class of flat connection on P with

460 Kefeng Liu

holonomy u around the boundary. In general we let $\mathcal{M}_c$ denote the moduli space of flat connections on P with holonomy around the boundary to be $c \in G$ which is close to u, or equivalently in the conjugacy class of c. The following formula is essentially a refined version of the formula [38] which Witten derived from the path integrals on the space of connections.

Theorem 3.1: ([24], [25]), *We have the following identity:*

$$\int_{\mathcal{M}_u} p(\sqrt{-1}\Omega)e^{\omega_u} = |Z(G)|\frac{|G|^{2g-2}}{(2\pi)^{2N_u}} \cdot \lim_{c \to u}\lim_{t \to 0} \sum_{\lambda \in P_+} \frac{\chi_\lambda(c)}{d_\lambda^{2g-1}}p(\lambda + \rho)e^{-tp_c(\lambda)}.$$

The notations in the above formula are as follows: ω_u is the canonical symplectic form on $\mathcal{M}_u$ induced by Poincare duality on S; $p(\sqrt{-1}\Omega)$ is a Pontrjagin class of the tangent bundle $T\mathcal{M}_u$ of the moduli space associated to the symmetric polynomial p; P_+ is the set of irreducible representations of G identified as a lattice in $\mathcal{T}^*$ which is the dual Lie algebra of the maximal torus T of G; $p_c(\lambda) = |\lambda + \rho|^2 - |\rho|^2$ where $\rho = \frac{1}{2}\sum_{\alpha \in \Delta^+}\alpha$ with respect to the Killing form, and Δ^+ denotes the set of positive roots; χ_λ and d_λ are respectively the character and dimension of λ; $|G|$ denotes the volume of G with respect to the bi-invariant metric induced from the Killing form; $|Z(G)|$ denotes the number of elements in the center $Z(G)$ of G and finally N_u is the complex dimension of $\mathcal{M}_u$.

The starting point for the proof of this theorem is to use the holonomy model of the moduli space and the explicit heat kernel on G. We consider the holonomy map $f : G^{2g} \times O_c \to G$ with $f(x_1, \cdots, y_g; z) = \prod_{j=1}^g [x_j, y_j]z$ where O_c is the conjugacy class through the generic point $c \in G$. It is well-known that the moduli space is given by $\mathcal{M}_c = f^{-1}(e)/G$ where G acts on $G^{2g} \times O_c$ by conjugation.

We have the explicit expression for the heat kernel on G:

$$H(t, x, y) = \frac{1}{|G|}\sum_{\lambda \in P_+} d_\lambda \cdot \chi_\lambda(xy^{-1})e^{-tp_c(\lambda)},$$

where $x, y \in G$ are two points. The key idea is to consider the integral

$$I(t) = \int_{h \in G^{2g} \times O_c} H(t, c, f(h))dh,$$

where dh denotes the induced bi-invariant volume form on $G^{2g} \times O_c$. We compute $I(t)$ in two different ways. First as $t \to 0$, $I(t)$ localizes to an integral on $\mathcal{M}_c$, which is the symplectic volume of $\mathcal{M}_c$ with respect to the canonical symplectic form induced by the Poincare duality on the cohomology groups of S with values in the adjoint Lie algebra bundle. To prove this we used the beautiful observation of Witten [37] that the symplectic volume form of $\mathcal{M}_c$ is the same as the Reidemeister torsion which arises from the Gaussian integral in the heat kernel.

On the other hand the orthogonal relations among the characters of the representations of G easily give us the infinite sum. In summary we have obtained the following more precise version of Witten's beautiful formula for the symplectic volume of the moduli space,

Proposition 3.2: ([24]) *As* $t \to 0$, *we have*

$$\int_{\mathcal{M}_c} e^{\omega_c} = |Z(G)| \frac{|G|^{2g-1}|j(c)|}{(2\pi)^{2N_c}|Z_c|} \sum_{\lambda \in P_+} \frac{\chi_\lambda(c)}{d_\lambda^{2g-1}} e^{-tp_c(\lambda)} + O(e^{-\delta^2/4t}).$$

Here δ is any small positive number, $|Z_c|$ is the volume of the centralizer Z_c of c, $j(c) = \prod_{\alpha \in \Delta^+}(e^{\sqrt{-1}\alpha(C)/2} - e^{\sqrt{-1}\alpha(C)/2})$ is the Weyl denominator, and N_c is the complex dimension of $\mathcal{M}_c$.

To get the intersection numbers from the volume formula, we take derivatives with respect to C where $c = u \exp C$. This is another key observation. By using the relation between the symplectic form on $\mathcal{M}_c$ and that on $\mathcal{M}_u$, and then taking the limits we arrive at the formula in Theorem 3.1. For the details see [24] and [25]. Another easy consequence of the method is that the symplectic volume of $\mathcal{M}_c$ is a piecewise polynomial of degree at most $2g|\Delta^+|$ in $C \in \mathcal{T}$ from which we get certain very general vanishing theorems for those integrals when the degree of the polynomial p is relatively large [25].

Similar results for moduli spaces when S has more boundary components can be obtained in the same way [25]. More precisely, assume S has s boundary components and consider the moduli space of flat connections on the principal G bundle P with holonomy $c_1, \cdots, c_s \in G$ around the corresponding boundaries. Let $\mathcal{M}_\mathbf{c}$ denote the moduli space and $\omega_\mathbf{c}$ denote the canonical symplectic form. Then we have

Theorem 3.3: ([25]) *The following formula holds:*

$$\int_{\mathcal{M}_c} p(\sqrt{-1}\Omega)e^{\omega_c} = |Z(G)| \frac{|G|^{2g-2+s} \prod\limits_{j=1}^{s} j(c_j)}{(2\pi)^{2N_c} \prod\limits_{j=1}^{s} |Z_{c_j}|} \lim_{t \to 0} \sum_{\lambda \in P_+} \frac{\prod\limits_{j=1}^{s} \chi_\lambda(c_j)}{d_\lambda^{2g-2+s}} p(\lambda + \rho)e^{-tp_c(\lambda)}.$$

Here $N_\mathbf{c}$ is the complex dimension of $\mathcal{M}_\mathbf{c}$ and $p(\sqrt{-1}\Omega)$ is a Pontrjagin class of $\mathcal{M}_\mathbf{c}$. By taking derivatives with respect to the c_j's we can get intersection numbers involving the other generators of the cohomology ring of $\mathcal{M}_\mathbf{c}$, as well as the polynomial property. From index formula we know that the integrals in our formulas contain all the information needed for the famous Verlinde formula. Recently the general Verlinde formula has been directly derived along this line of idea [7].

This localization method of using heat kernels can be applied to other general situation like moment maps, from which we derive the non-abelian localization formula of Witten. See [25] for applications to three dimensional manifolds and see [26] for applications involving finite groups and moment maps.

4. Mirror principle

Let X be a projective manifold. Let $\mathcal{M}_{g,k}(d, X)$ denote the moduli space of stable maps of genus g and degree d with k marked points into X. Modulo the obvious equivalence, a point in $\mathcal{M}_{g,k}(d, X)$ is given by a triple $(f; C; x_1, \cdots, x_k)$ where $f : C \to X$ is a degree d holomorphic map and $x_1, \cdots, x_k$ are k points on the genus g curve C. Here $d \in H_2(X,, \mathbf{Z})$ will also be identified as the integral index $(d_1, \cdots, d_n)$ by choosing a basis of $H_2(X,, \mathbf{Z})$ dual to a basis of Kahler classes.

This moduli space may have higher dimension than expected. Even worse, its different components may have different dimensions. To define integrals on such space, we need the virtual fundamental cycle first constructed in [19] and later in [6]. Let us denote by $LT_{g,k}(d, X)$ the virtual fundamental cycle which is a homology class of the expected dimension in $\mathcal{M}_{g,k}(d, X)$.

We first consider the case $k = 0$. Let V be a concavex bundle on X. The notion of concavex bundles was introduced in [27], it is a direct sum of a positive and a negative bundle on X. From a concavex bundle V, we can obtain a sequence of vector bundles V_d^g on $M_{g,k}(d, X)$ by taking either $H^0(C, f^*V)$ or $H^1(C, f^*V)$, or their direct sum. Let b be a multiplicative characteristic class. The main problem of mirror principle is to compute the integral [16]

$$K_d^g = \int_{LT_{g,0}(d,X)} b(V_d^g).$$

More precisely, let λ and $T = (T_1, \cdots,, T_n)$ be formal variables. Mirror principle is to compute the generating series,

$$F(q, \lambda) = \sum_{d, g} K_d^g \, \lambda^g \, e^{d \cdot T}$$

in terms of certain natural explicit hypergeometric series. So far we have rather complete picture for the case of balloon manifolds.

A balloon manifold X is a projective manifold with complex torus action and isolated fixed points. Let $H = (H_1, \cdots, H_n)$ be a basis of equivariant Kahler classes. Then X is called a balloon manifold if $H(p) \neq H(q)$ when restricted to any two fixed points $p, q \in X$, and the tangent bundle T_pX has linearly independent weights for any fixed point $p \in X$. The complex 1-dimensional orbits in X joining every two fixed points in X are called balloons which are copies of $\mathbf{P}^1$. We require the bundle V to have fixed splitting type when restricted to each balloon [27].

Theorem 4.1: ([27]-[30]) *Mirror principle holds for balloon manifolds and concavex bundles.*

In the most interesting cases for the mirror formulas, we simply take characteristic class b to be the Euler class and the genus $g = 0$. The mirror principle implies that mirror formulas actually hold for very general manifolds such as Calabi-Yau complete intersections in toric manifolds and in compact homogeneous manifolds. See [31] and [29] for details. In particular this implies all of the mirror formulas for counting rational curves predicted by string theorists. Actually mirror principle holds even for non-Calabi-Yau and for certain local complete intersections. In [30] we developed the mirror principle for counting higher genus curves, for which the only remaining problem is to find the explicit hypergeometric series. Also our method clearly works well for orbifolds.

As an example, we consider a toric manifold X and genus $g = 0$. Let $D_1, .., D_N$ be the toric invariant divisors, and V be the direct sum of line bundles: $V = \bigoplus_j L_j$ with $c_1(L_j) \geq 0$ and $c_1(X) = c_1(V)$. We denote by $\langle \cdot, \cdot \rangle$ the pairing of homology and cohomology classes. Let b be the Euler class and

$$\Phi(T) = \sum_d K_d^0 \, e^{d \cdot T}$$

where $d \cdot T = d_1 T_1 + \cdots d_n T_n$. Introduce the hypergeometric series

$$HG[B](t) = e^{-H \cdot t} \sum_d \prod_j \prod_{k=0}^{\langle c_1(L_j), d \rangle} (c_1(L_j) - k) \frac{\prod_{\langle D_a, d \rangle < 0} \prod_{k=0}^{-\langle D_a, d \rangle - 1}(D_a + k)}{\prod_{\langle D_a, d \rangle \geq 0} \prod_{k=1}^{\langle D_a, d \rangle}(D_a - k)} \, e^{d \cdot t}$$

with $t = (t_1, \cdots, t_n)$ formal variable.

Corollary 4.2: [29] *There are explicitly computable functions $f(t)$, $g(t) = (g_1(t), \cdots, g_n(t))$, such that*

$$\int_X \left(e^f HG[B](t) - e^{-H \cdot T} e(V) \right) = 2\Phi - \sum_j T_j \frac{\partial \Phi}{\partial T_j}$$

where $T = t + g(t)$.

From this formula we can determine $\Phi(T)$ uniquely. The functions f and g are given by the expansion of $HG[B](t)$. We can also replace V by general concavex bundles [29]. To make our algorithm more explicit, let us consider the the Calabi-Yau quintic, for which we have the famous Candelas formula [9]. In this case $V = \mathcal{O}(5)$ on $X = \mathbf{P}^4$ and the hypergeometric series is:

$$HG[B](t) = e^{H t} \sum_{d=0}^{\infty} \frac{\prod_{m=0}^{5d}(5H + m)}{\prod_{m=1}^{d}(H + m)^5} \, e^{d t},$$

where H is the hyperplane class on $\mathbf{P}^4$ and t is a parameter. Introduce

$$\mathcal{F}(T) = \frac{5}{6}T^3 + \sum_{d>0} K_d^0 \, e^{d T}.$$

The algorithm is to take the expansion in H:

$$HG[B](t) = H\{f_0(t) + f_1(t)H + f_2(t)H^2 + f_3(t)H^3\}.$$

Then the famous Candelas formula can be reformulated as

Corollary 4.3: ([27]) *With $T = f_1/f_0$, we have*

$$\mathcal{F}(T) = \frac{5}{2}\left(\frac{f_1}{f_0} \frac{f_2}{f_0} - \frac{f_3}{f_0} \right).$$

Another rather interesting consequence of mirror principle is the local mirror symmetry which is the case when V is a concave bundle. Local mirror symmetry is called geometric engineering in string theory which is used to explain the stringy origin of the Seiberg-Witten theory. In these cases the hypergeometric series are the periods of elliptic curves which are called the Seiberg-Witten curves. These elliptic curves are the mirror manifolds of the open Calabi-Yau manifolds appeared in the local mirror formulas. For example the total space of the canonical bundle of a del Pezzo surface is an example of open Calabi-Yau manifold covered by the local mirror symmetry. The case $\mathbf{P}^2$ already has drawn a lot of interests in string theory. The case for $\mathcal{O}(-1) \oplus \mathcal{O}(-1)$ on $\mathbf{P}^1$ easily gives the multiple cover formula.

The key ingredients for the proof of the mirror principle consist of the following: linear and non-linear sigma model, Euler data, balloon and hypergeometric Euler data. As explained in [30], these ingredients are independent of the genus of the curves, except the hypergeometric Euler data, which for $g > 0$ is more difficult to find out, while for the genus 0 case it can be easily read out from localization at the smooth fixed points of the moduli spaces which are covers of the balloons. The interested reader is refered to [27]-[30] for details. Our idea is to go to the equivariant setting and to use the localization formula as given in [1] and its virtual version in [14] on two moduli spaces which we called non-linear and linear sigma models. One key observation is the functorial localization formula [27]-[30]. We apply this formula to the equivariant collapsing map between the two sigma models, and to the evaluation maps. One can see [30] for the existence of the collapsing map for arbitrary genus. Hypergeometric series naturally appear from localizations on the linear sigma models and at the smooth fixed points in the moduli spaces.

Euler data is a very general notion, it can include general Gromov-Witten invariants by adding the pull-back classes by the evaluation map ev_j at the marked points. More precisely we can try to compute integrals of the form:

$$K_{d,k}^g = \int_{LT_{g,k}(d,X)} \prod_j ev_j^* \omega_j \cdot b(V_d^g)$$

where $\omega_j \in H^*(X)$. By introducing the generating series with summation over k, we can still get Euler data. One goal of the most general mirror principle is to explicitly compute such series in terms of hypergeometric series.

We remark that the development of the proof of the mirror formulas owes to many people, first to the string theorists Candelas and his collaborators, Witten, Vafa, Warner, Greene, Morrison, Plesser and many others. They used the physical theory of mirror symmetry, and their computations used mirror manifolds and their periods. For the general theory of mirror principle, see [27]-[30]. See also [13], [12] and [5] for different approaches to the mirror formula.

5. Concluding remarks

Localization techniques have been very successful in solving many conjectures from physics. In the meantime string theorists have produced many more exciting new conjectures. We can certainly expect their solutions by using localizations.

Recently several mirror formulas of counting holomorphic discs have been conjectured by Vafa and his collaborators. The boundary of the disc is mapped into a Langrangian submanifolds of the Calabi-Yau. Other related conjectures include the Gopakumar-Vafa conjecture on the higher genus multiple covering formula and the mirror formulas for counting higher genus curves in Calabi-Yau. With Chien-Hao Liu we are trying to extend the mirror principle to these settings. Another exciting conjecture is the S-duality conjecture which includes the Witten conjecture on the equality of the Donaldson invariants with the Seiberg-Witten invariants and the Vafa-Witten conjecture on the modularity of the generating series of the Euler numbers of the moduli spaces of self-dual connections. Some progresses are made by constructing a larger moduli space with circle action, the so-called non-abelian monople moduli spaces. Finally there is the Dijkgraaf-Moore-Verlinde-Verlinde conjecture on the generating series of the elliptic genera of Hilbert schemes. For an approach of using localization, see [34].

References

[1] M. F. Atiyah & R. Bott, The moment map and equivariant cohomology, *Topology* **23** (1984) 1-28.

[2] M. F. Atiyah & R. Bott, The Yang-Mills equations over Riemann surfaces, *Phil. Trans. Roy. Soc. London*, **A308**, 523-615 (1982).

[3] M. F. Atiyah & F. Hirzebruch, Spin manifolds and groups actions, *Essays on topology and Related Topics, Memoirs dédié à Georges de Rham* (ed. A. Haefliger and R. Narasimhan), Springer-Verlag, New York-Berlin (1970), 18-28.

[4] M. F. Atiyah & I. Singer, The index of elliptic operators. III, *Ann. of Math.* (2) **87**(1968), 546–604.

[5] A. Bertram, Another way to enumerate rational curves with torus actions, math.AG/9905159.

[6] K. Behrend & F. Fentachi, The intrinsic normal cone, *Invent. Math.* **128** (1997) 45-88.

[7] J.-M. Bismut & F. Labeurie, Symplectic geometry and the Verlinde formulas, *Surveys in differential geometry: differential geometry inspired by string theory, 97–311*, Int. Press, Boston, MA, 1999.

[8] R. Bott & C. Taubes, On the rigidity theorems of Witten, *J. AMS.* **2** (1989), 137-186.

[9] P. Candelas, X. de la Ossa, P. Green & L. Parkes, A pair of Calabi-Yau manifolds as an exactly soluble superconformal theory. *Nucl. Phys.* **B359** (1991) 21-74.

[10] C. Dong, K. Liu & X. Ma, On orbifold elliptic genus, math.DG/0109005.

[11] C. Dong, K. Liu & X. Ma, Elliptic genus and vertex operator algebras, math.DG/0201135.

[12] A. Gathmann, Relative Gromov-Witten invariants and the mirror formula, math.AG/0009190.

[13] A. Givental, Equivariant Gromov-Witten invariants, alg-geom/9603021.

[14] T. Graber & R. Pandharipande, Localization of virtual classes, alg-geom/9708001.

[15] F. Hirzebruch, T. Berger & R. Jung, *Manifolds and modular forms*, Aspects of

466 Kefeng Liu

Mathematics, E20. Friedr. Vieweg and Sohn, Braunschweig, 1992.

[16] M. Kontsevich, Enumeration of rational curves via torus actions, hep-th/9405035.

[17] P. S. Landweber, *Elliptic Curves and Modular forms in Algebraic Topology*, ed. Landweber P. S., SLNM 1326, Springer, Berlin.

[18] B. Lawson & S.-T. Yau, Scalar curvature, non-abelian group actions, and the degree of symmetry of exotic spheres. *Comment. Math. Helv.* **49** (1974), 232–244.

[19] J. Li & G. Tian, Virtual moduli cycle and Gromov-Witten invariants of algebraic varieties, *J. of Amer. math. Soc.* **11**, no. 1, (1998), 119-174.

[20] K. Liu, On $SL_2(\mathbf{Z})$ and topology, *Math. Research Letter,* **1** (1994), 53-64.

[21] K. Liu, On Modular invariance and rigidity theorems, *J. Diff. Geom.* **41** (1995), 343-396.

[22] K. Liu, Modular invariance and characteristic numbers, *Comm. Math. Phys.,* **174** (1995), no. 1, 29-42.

[23] K. Liu & W. Zhang, Elliptic genus and η-invariant, *International Math. Res. Notices,* no. 8 (1994), 319-327.

[24] K. Liu, Heat kernel and moduli space, *Math. Res. Letter,* **3** (1996), 743-762.

[25] K. Liu, Heat kernel and moduli space II, *Math. Res. Letter,* **4** (1997), 569-588.

[26] K. Liu, Heat kernels, symplectic geometry, moduli spaces and finite groups, *Surveys in differential geometry: differential geometry inspired by string theory,* 527–542, Int. Press, Boston, MA, 1999.

[27] B. Lian, K. Liu & S.-T. Yau, Mirror principle, I, *Asian J. Math.* **1** (1997), 729–763.

[28] B. Lian, K. Liu & S.-T. Yau, Mirror principle, II , **3** (1999), no. 1, 109–146.

[29] B. Lian, K. Liu & S.-T. Yau, Mirror principle, III, *Asian J. Math.* **3** (1999), no. 4, 771–800.

[30] B. Lian, K. Liu & S.-T. Yau, Mirror principle, IV, *Asian J. Math.*

[31] B. Lian, C.-H. Liu, K. Liu & S.-T. Yau, The S^1 fixed points in Quot-schemes and mirror principle computations, math.AG/0111256.

[32] K. Liu, X. Ma & W. Zhang, On elliptic genera and foliations, *Math. Res. Lett.,* **8** (2001), no. 3, 361–376.

[33] K. Liu & X. Ma, On family rigidity theorems I. *Duke Math. J.* **102** (2000), 451–474.

[34] K. Liu, X. Ma & J. Zhou, The elliptic genus of the Hilbert schemes of surfaces, work in progress.

[35] C. Taubes, S^1-actions and elliptic genera, *Comm. Math. Phys.* **122** (1989), 455–526.

[36] E. Witten, The index of the Dirac operator in loop space, in *Elliptic Curves and Modular forms in Algebraic Topology,* Landweber P.S., SLNM 1326, Springer, Berlin, 161–186.

[37] E. Witten, On quantum gauge theory in two dimensions, *Commun. Math. Phys.* **141** (1991), 153–209.

[38] E. Witten, Two-dimensional gauge theory revisited, *J. Geom. Phys.* **9** (1992), 303–368.

ICM 2002 · Vol. III · 467–476

Derivation of the Euler Equations
from Many-body Quantum Mechanics*

Bruno Nachtergaele[†] Horng-Tzer Yau[‡]

(Dedicated to André Verbeure on the occasion of his sixtieth birthday)

Abstract

The Heisenberg dynamics of the energy, momentum, and particle densities for fermions with short-range pair interactions is shown to converge to the compressible Euler equations in the hydrodynamic limit. The pressure function is given by the standard formula from quantum statistical mechanics with the two-body potential under consideration. Our derivation is based on a quantum version of the entropy method and a suitable quantum virial theorem. No intermediate description, such as a Boltzmann equation or semiclassical approximation, is used in our proof. We require some technical conditions on the dynamics, which can be considered as interesting open problems in their own right.

2000 Mathematics Subject Classification: 82C10, 82C21, 82C40.
Keywords and Phrases: Quantum many-body dynamics, Euler equations, Hydrodynamic limit, Quantum entropy method.

1. Introduction

The fundamental laws of non-relativistic microscopic physics are *Newton's* and the *Schrödinger* equation in the classical and the quantum case respectively. These equations are impossible to solve for large systems and macroscopic dynamics is therefore modeled by phenomenological equations such as the *Euler* or the *Navier-Stokes* equations. Although the latter were derived centuries ago from continuum considerations, they are in principle consequences of the microscopic physical laws and should be viewed as secondary equations. It was first observed by Morrey [5] in

---

†Department of Mathematics, University of California, Davis, CA 95616, USA. E-mail: bxn@math.ucdavis.edu

‡Courant Institute, New York University, New York, NY 10012, USA. E-mail: yau@cims.nyu.edu

the fifties that the Euler equations become 'exact' in the Euler limit, provided that the solutions to the Newton's equation are 'locally' in equilibrium. Morrey's original work was far from rigorous and, in particular, the meaning of 'local equilibrium' was not clear. It is nevertheless a very original work which led to the later development of the hydrodynamical limits of interacting particle systems. In terms of a rigorous proof along the lines of Morrey's original argument, however, there has been little progress until the recent work [11]. This long delay is mostly due to a serious lack of tools for analyzing many-body dynamics, in the classical case and even more so in the quantum case.

In this lecture, we will discuss the derivation of the Euler equations from microscopic quantum dynamics. As we want to consider the genuine quantum dynamics for a system with short-range pair interactions, we cannot take a semiclassical limit. Although one-particle quantum dynamics converges to Newtonian dynamics in the limit of infinite mass, this is not the case in the thermodynamic limit, i.e., the heavy-particle limit does not commute with the infinite-number-of-particles limit. This is most clearly seen in the pressure function, for which quantum corrections survive at the macroscopic scale. In fact, one of the conclusions of our work is that under rather general conditions, the pressure function is the only place where the quantum nature of the underlying system, in particular the particle statistics, survives in the Euler limit. At the same time, our derivations also shows that it is the quantum mechanical pressure, without modification, which governs the macroscopic dynamics. A similar result should hold for all systems of macroscopic conservation laws.

The Euler equations have traditionally been derived from the Boltzmann equation both in the classical case and in the quantum case, see Kadanoff and Baym [4] for the quantum case. Since the Boltzmann equation is valid only in very low density regions, these derivations are not satisfactory, especially in the quantum case where the relationship between the quantum dynamics and the Boltzmann equation is not entirely clear. There were, however, two approaches based directly on quantum dynamics. The first was due to Born and Green [1], who used an early version of what was later called the BBGKY hierarchy, together with moment methods and some truncation assumptions. A bit later, Irving and Zwanzig [3] used the Wigner equation, moment methods and truncations to accomplish a similar result. These two approaches rely essentially on the moment method with the Boltzmann equation replaced by the Schrödinger equation. Unlike in the Boltzmann case, where one can do asymptotic analysis to justify this approach, it seems unlikely that this can be done for the Schrödinger dynamics. Therefore, in the present work, we follow a much more direct route to connect the micro- and macroscopic dynamics.

2.   Schrödinger and Euler dynamics

We begin by considering N particles on $\mathbb{R}^3$, evolving according to the Schrödinger equation

$$i\partial_t \psi_t(x_1, \cdots, x_N) = H\psi_t(x_1, \cdots, x_N)$$

where the Hamiltonian is given by

$$H = \sum_{j=1}^{N} \frac{-\Delta_j}{2} + \sum_{1 \leq i < j \leq N} W(x_i - x_j) \,. \qquad (2.1)$$

Here, W is a two-body short-ranged super-stable isotropic pair interaction and $\psi_t(x_1, \cdots, x_N)$ is the wave function of particles at time t. We consider spinless Fermions and thus the state space $\mathcal{H}^N$ is the subspace of antisymmetric functions in $L^2(\mathbb{R}^{3N})$. It is convenient not to fix the total number of particles and to use the second quantization terminology. In fact, it would be extremely cumbersome to work through all arguments without the second quantization formalism. The state space of the particles, called the Fermion Fock space, is thus the direct sum of $\mathcal{H}^N$: $\mathcal{H} := \oplus_{N=0}^{\infty} \mathcal{H}^N$.

Define the annihilation and creation operators a_x and a_x^+ by

$$(a_x \Psi)^N(x_1, \cdots, x_N) = \sqrt{N+1} \Psi^{N+1}(x, x_1, \cdots, x_N),$$

$$(a_x^+ \Psi)^N(x_1, \cdots, x_N) = \frac{1}{\sqrt{N}} \sum_{j=1}^{N} (-1)^{j-1} \delta(x - x_j) \Psi^{N-1}(x_1, \cdots, \widehat{x_j}, \cdots, x_N),$$

where a_x and a_x^+ are operator-valued distributions and, as usual, $\widehat{}$ means "omit".

The annihilation operator a_x is the adjoint of a_x^+ with respect to the standard inner product of the Fock space with Lebesgue measure dx, and

$$[a_x, a_y^+]_+ := a_x a_y^+ + a_x a_y^+ = \delta(x - y) \,,$$

where δ is the delta distribution. The derivatives of these distributions with respect to the parameter x are denoted by ∇a_x and ∇a_x^+. With this notation, we can express the Hamiltonian as $H = H_0 + V$ where $H_0 = \frac{1}{2} \int \nabla a_x^+ \nabla a_x \, dx$ and $V = \frac{1}{2} \int \int dx dy W(x - y) a_x^+ a_y^+ a_y a_x$. It is more convenient to put the Schrödinger equation into the operator form, which is sometimes called the Schrödinger-Liouville equation. Denote the density matrix of the state at time t by γ_t. Only normal states, which can be represented by density matrices, will be considered in the time evolution. Then the Schrödinger equation is equivalent to $i\partial_t \gamma_t = \delta_H \gamma_t$, with $\delta_H \gamma_t := [H, \gamma_t]$. The conserved quantities of the dynamics are the number of particles, the three components of the momentum and the energy. The local densities of these quantities are denoted by $\mathbf{u} = (u^\mu)$, $\mu = 0, \cdots, 4$, and are given by the following expressions:

$$u_x^0 = n_x = a_x^+ a_x,$$

$$u_x^j = p_x^j = -\frac{i}{2}[\nabla_j a_x^+ a_x - a_x^+ \nabla_j a_x], \qquad j = 1, 2, 3,$$

$$u_x^4 = h_x = \frac{1}{2} \nabla a_x^+ \nabla a_x + \frac{1}{2} \int dy W(x - y) a_x^+ a_y^+ a_y a_x. \qquad (2.2)$$

The finite volumes, denoted by Λ, will always by three-dimensional tori and, unless otherwise stated, unbounded observables on Λ will be defined with periodic boundary conditions. E.g., the number of particles in Λ, the total momentum, and the total energy of the particles in Λ, respectively, are defined by

$$N_\Lambda = \int_\Lambda dx\, n_x, \quad P_\Lambda^j = \int_\Lambda dx\, p_x^j, \quad j = 1, 2, 3, \quad H_\Lambda = \int_\Lambda dx\, h_x\,.$$

We slightly generalize the definition of the grand canonical Gibbs states to include a parameter for the total momentum of the system: the Lagrange multiplier $\alpha \in \mathbb{R}^3$. We will work under the assumption that the temperature and chemical potential are in the one-phase region of the phase diagram of the system under consideration, such that the thermodynamic limit is unique. The infinite volume Gibbs states are then given by the following formula:

$$\omega_{\beta,\alpha,\mu}(X) = \lim_{\Lambda \to \mathbb{R}^d} \frac{\mathrm{Tr} X e^{-\beta(H_{0,\Lambda} + V_\Lambda - \alpha \cdot P_\Lambda - \mu N_\Lambda)}}{\mathrm{Tr} e^{-\beta(H_{0,\Lambda} + V_\Lambda - \alpha P_\Lambda - \mu N_\Lambda)}}\,. \tag{2.3}$$

It is convenient to denote the parameters (β, α, μ) by $\boldsymbol{\lambda} = (\lambda^\mu), \mu = 0, \cdots, 4$ with $\lambda^0 = \beta\mu, \lambda^j = \beta\alpha^j, \lambda^4 = \beta$. Define (notice the sign convention)

$$\boldsymbol{\lambda} \cdot \boldsymbol{u} = \sum_{\mu=0}^{3} \lambda^\mu u^\mu - \lambda^4 u^4 \quad \text{and} \quad \langle \boldsymbol{\lambda}, \boldsymbol{u} \rangle_\Lambda = |\Lambda|^{-1} \int_\Lambda dx \boldsymbol{\lambda}(x) \cdot \boldsymbol{u}(x)\,.$$

These notations allow us to give a compact formula for the unique, translation invariant Gibbs state (defined with constant $\boldsymbol{\lambda}$), as well as for the states describing local equilibrium (defined with x-dependent $\boldsymbol{\lambda}$):

$$\omega_{\boldsymbol{\lambda}} = \lim_{\Lambda \to \mathbb{R}^d} e^{|\Lambda|\langle \boldsymbol{\lambda}, \boldsymbol{u} \rangle_\Lambda} / Z_\Lambda(\boldsymbol{\lambda}) \tag{2.4}$$

where $Z_\Lambda(\boldsymbol{\lambda})$ is the partition function given by $Z_\Lambda(\boldsymbol{\lambda}) = \mathrm{Tr} e^{|\Lambda|\langle \boldsymbol{\lambda}, \boldsymbol{u} \rangle_\Lambda}$. If we define the pressure, as a function of the constant vector $\boldsymbol{\lambda}$, by $\psi(\boldsymbol{\lambda}) = \lim_{L \to \infty} |\Lambda|^{-1} \log Z_\Lambda(\boldsymbol{\lambda})$, then

$$\frac{\partial \psi}{\partial \lambda^\mu} = \omega_{\boldsymbol{\lambda}}(u^\mu). \tag{2.5}$$

As the states $\omega_{\boldsymbol{\lambda}}$ are translation invariant, we have $\omega_{\boldsymbol{\lambda}}(u^\mu) = q^\mu$. Explicitly,

$$\rho = \omega_{\boldsymbol{\lambda}}(n_x), \quad \mathfrak{q} = \omega_{\boldsymbol{\lambda}}(p_x), \quad e = \omega_{\boldsymbol{\lambda}}(h_x).$$

Notice that $\mathfrak{q}$ and e are momentum and energy per volume.

Again, we will work under the assumption that these parameters stay in the one-phase region, the limiting Gibbs state is unique and these definitions are unambiguous. Although momentum is preferable as a quantum observable, we also introduce the velocity in order to be able to compare with the classical case. The velocity field $v(x)$ has to be defined as a mean velocity of the particles in a neighborhood of x. Therefore we have $v(x) = \mathfrak{q}(x)/\rho(x)$. We also introduce the energy per particle defined by $\tilde{e} = e/\rho$. The usual Euler equations are written in terms of ρ, v, and $\tilde{e}$.

In order to derive the Euler equations, we need to perform a rescaling. So we shall put all particles in a torus Λ_ε of size ε^{-1} and use $(X, T) = (\varepsilon x, \varepsilon t)$ to denote the macroscopic coordinates. For all equations in this paper periodic boundary conditions are implicitly understood.

The Euler equations are given by

$$\frac{\partial \rho}{\partial T} + \sum_{j=1}^{3} \frac{\partial}{\partial X_j}(\rho v_j) \;=\; 0,$$

$$\frac{\partial (\rho v_k)}{\partial T} + \sum_{j=1}^{3} \frac{\partial}{\partial X_j}[\,\rho v_j v_k\,] + \frac{\partial}{\partial X_k}P(e,\rho) \;=\; 0, \qquad (2.6)$$

$$\frac{\partial (\rho \tilde{e})}{\partial T} + \sum_{j=1}^{3} \frac{\partial}{\partial X_j}[\,\rho \tilde{e} v_j + v_j P(e,\rho)\,] \;=\; 0.$$

These equations are in form identical to the classical ones but all physical quantities are computed quantum mechanically. In particular, $P(e,\rho)$ is the thermodynamic pressure computed from quantum statistical mechanics for the microscopic system. It is a function of X and T only through its dependence on e and ρ. If no velocity dependent forces act between the molecules of the fluid under consideration (we consider only a pair potential), the pressure is independent of the velocity.

Let $\boldsymbol{q} = (q^0, \cdots, q^4)$, related to density, momenta and energy as follows:

$$q^0 = \rho\,, \quad q^i = \rho v^i\,, \quad q^4 = e = \rho\,\tilde{e}. \qquad (2.7)$$

In other words q^1, q^2, q^3, and q^4 are momenta and energy per *volume* instead of *per particle* as in the usual Euler equation (2.6). We rewrite the Euler equations in the following form

$$\frac{\partial q^k}{\partial T} + \sum_{i=1}^{3} \nabla_i^X[A_i^k(q)] = 0\,, \quad k = 0,1,2,3,4. \qquad (2.8)$$

The matrix A is determined by comparison with the Euler equations:

$$A_j^0 = q^j,\; A_j^i = \delta_{ij}P + q_i q_j/q_0,\; A_j^4 = q^j(q_4 + P)/q_0.$$

3.   Local equilibrium

To proceed we need a microscopic description of local equilibrium. Suppose we are given macroscopic functions $\boldsymbol{q}(X)$. We wish to find a local Gibbs state with the conserved quantities given by $\boldsymbol{q}(X)$. The local Gibbs states are states locally in equilibrium. In other words, in a microscopic neighborhood of any point $x \in T^3$ the state is given by a Gibbs state. More precisely, we wish to find a local Gibbs state with the expected values of the energy, momentum, and particle number per unit volume at X given by $\boldsymbol{q}(X)$. To achieve this, we only have to adjust the parameter $\boldsymbol{\lambda}$ at every point X. More precisely, we choose $\boldsymbol{\lambda}(X)$ such that the equation (2.5) holds at every point, i.e.,

$$\frac{\partial \Psi(\boldsymbol{\lambda}(X))}{\partial \lambda^\mu(X)} = q^\mu(X).$$

If we denote the solution to the Euler equation by $q(X,T)$, then we can choose in a similar way a local Gibbs state with given conserved quantities at the time T. Define the local Gibbs state

$$\omega_t^\varepsilon = \frac{1}{c_\varepsilon(t)} \exp\left[\varepsilon^{-3}\langle \boldsymbol{\lambda}(\varepsilon t, \varepsilon \cdot), \boldsymbol{u}\rangle_{\varepsilon^{-1}}\right] \tag{3.1}$$

where $c_\varepsilon(t)$ is the normalization constant. Clearly, we have that $\omega_\varepsilon(t)(u_x^\mu) = q^\mu(\varepsilon x, \varepsilon t)$ to leading order in ε.

In summary, the goal is to show that, in the limit $\epsilon \to 0$, the following diagram commutes:

$$\begin{array}{ccc}
\mathbf{q}(X,0) & \xrightarrow{\text{Euler}} & \mathbf{q}(X,T) \\
\text{local equilibrium} \downarrow & & \uparrow \begin{array}{l}\text{limit } \epsilon \downarrow 0 \text{ of expectation} \\ \text{of locally averaged observ-} \\ \text{ables}\end{array} \\
\gamma_0 & \xrightarrow{\text{Schrödinger}} & \gamma_{\epsilon^{-1}T}
\end{array}$$

As smooth solutions of the Euler equations are guaranteed to exist only up to a finite time [6], say T_0, we will formulate our assumptions on the dynamics of the microscopic system for a finite time interval as well, say $t \in [0, T_0/\varepsilon]$. Note the cutoff assumptions below would hold automatically for lattice models.

4. Assumptions and the main theorem

Our main result is stated in Theorem 4.1 below. First, we state the assumptions of the theorem with some brief comments. There are three kinds of assumptions.

The first category of assumptions could be called *physical* assumptions on the solution of the Euler equations that we would like to obtain as a scaling limit of the underlying dynamics, and on the pair interaction potential of this system.

I. One-phase regime: We assume that the pair potential, W, is C^1 and with support contained in a ball of radius R. Moreover, we assume that W is symmetric under reflections of each of the coordinate axes (e.g., rotation symmetric potentials automatically have this symmetry), and has the usual stability property [12]: there is a constant $B \geq 0$ such that, for all $N \geq 2$, $x_1, \ldots, x_N \in \mathbb{R}^3$,

$$\sum_{1 \leq 1 < j \leq N} W(x_i - x_j) \geq -BN.$$

Of the Fermion system with potential W we assume that there is an open region $D \subset \mathbb{R}^2$, which we will call the *one-phase region*, such that the system has a unique limiting Gibbs state for all values of particle density and energy density $(\rho, e) \in D$.

The solution of the Euler equations we consider, $q(X,T)$, will be assumed to C^1 in X for $T \in [0, T_0]$, and have local particle and energy density in the one-phase region for all times $T \in [0, T_0]$. I.e., $(\rho(X,T), e(X,T)) \in D$, for all $X \in \Lambda_1$ and $T \in [0, T_0]$.

The next category of assumptions is on the local equilibrium states for the Fermion system that we construct and on their time-evolution under the Schrödinger equation. They can be considered conjectures. In fact, these assumptions have not been rigorously proved even for Gibbs states. Although one expects that these assumptions can be proved in the high-temperature and low-density region using some type of cluster expansion methods, this has only been done recently for Bosons in [7]. For the rest of this paper, we shall assume this cutoff assumptions for the solution to the Schrodinger equation that we consider, as well as for the Gibbs states in the one phase regions considered in this paper.

II. Cutoff assumptions: Suppose that γ_t is the solution to the Schrödinger equation with a local equilibrium state as initial condition, constructed with the parameters derived from a solution of the Euler equations (with the appropriate pressure function) that does not leave the one-phase region. We make the following two assumptions about this solution:

1. Finite velocity cutoff assumption: Let $N_p(t) = \mathrm{Tr}\gamma_t a_p^+ a_p$, where $a_p^\#$ is the Fourier transform of $a_x^\#$. Then there is a constant $c > 0$, such that for all $t \leq T_0/\varepsilon$,

$$\varepsilon^d \int dp e^{cp^2} N_p(t) \leq C_{T_0}.$$

2. Non-implosion assumption: There is a constant C_{T_0}, such that for all $t \leq T_0/\varepsilon$,

$$\mathrm{Tr}\gamma_t \varepsilon^d \int dx n_x \left[\int_{|x-y|\leq 2R} n_y dy \right]^2 \leq C_{T_0}, \qquad (4.1)$$

where R is the range of the interaction W.

Finally, we have an assumption on the set of the time-invariant ergodic states of the Fermion system. To state this assumption we need the notion of *relative entropy*, of a normal state γ with respect to another normal state ω. Let γ and ω denote the density matrices of these states. The relative entropy, $S(\gamma \mid \omega)$, is defined by

$$S(\gamma|\omega) = \begin{cases} \mathrm{Tr}\gamma(\log\gamma - \log\omega) & \text{if } \ker\omega \subset \ker\gamma, \\ +\infty & \text{else.} \end{cases}$$

For a pair of translation invariant locally normal states, one can show existence of the relative entropy density [10], defined by the limit

$$s(\gamma|\omega) = \lim_{\varepsilon\downarrow 0} \varepsilon^3 S(\gamma_{\Lambda_\varepsilon} \mid \omega_{\Lambda_\varepsilon}),$$

where $\gamma_{\Lambda_\varepsilon}$ and $\omega_{\Lambda_\varepsilon}$ denote the density matrices of the normal states obtained by restricting γ and ω to the observables localized in $\Lambda_\varepsilon = \varepsilon^{-1}\Lambda_1$. The existence of the limit can be proved under more general conditions on the finite volumes, but this is unimportant for us.

III. Ergodicity assumption ("Boltzmann Hypothesis"): All translation invariant ergodic stationary states to the Schrödinger equation are Gibbs states if they satisfy the following assumptions: *1) the density and energy is in one phase region. 2) The relative entropy density with respect to some Gibbs state is finite.*

We expect that our assumptions hold for the solutions γ_t of the Schrödinger equation that we employ, but it must be said that, at this moment, very little is known. We believe that these are natural conditions. To prove them under rather general conditions or even for a special class of models may be regarded as an important open problem in quantum statistical mechanics.

Theorem 4.1 *Suppose that $\boldsymbol{q}(X,T)$ is a smooth solution to the Euler equation in one phase region up to time $T \leq T_0$. Let ω_t^ε be the local Gibbs state with conserved quantities given by $\boldsymbol{q}(X,T)$. Suppose that the cutoff assumptions and the ergodicity assumption hold. Let γ_t be the solution to the Schrödinger equation and $\gamma_0 = \omega_0^\varepsilon$ (Notice γ_t depends on ε). Then for all $t \leq T_o/\varepsilon$ we have*

$$\lim_{\varepsilon \to 0} s(\gamma_t | \omega_t^\varepsilon) = 0.$$

In other words, ω_t^ε is a solution to the Schrödinger equation in entropy sense. In particular, for any smooth function f on Λ, we have

$$\lim_{\varepsilon \to 0} \varepsilon^3 \int_{\Lambda_\varepsilon} dx\, f(\varepsilon x) \left[\gamma_t(\boldsymbol{u}_x) - \boldsymbol{q}(\varepsilon t, \varepsilon x)\right] = 0.$$

This theorem is a quantum analogue of the classical result by Olla, Varadhan, and Yau [11]. There are a few differences in the assumptions and the strategy followed in their paper with respect to ours. E.g., in the treatment of [11] of the classical case the cut-off assumption 1) was not needed. Instead, the usual quadratic kinetic energy was replaced by one with bounded derivatives with respect to momentum. For Fermion models on a lattice instead of in the continuum, no cut-off assumptions are required. Another difference with the treatment in [11], is that we do not add noise terms to the "native" dynamics. In [11] a weak noise term was added to the Newtonian dynamics in order to be able to prove convergence to local equilibrium. In that paper, the strength of the noise vanishes in the hydrodynamic limit. Here, we do not modify the Heisenberg dynamics in any way, but instead reduce the question of convergence to local equilibrium to the ergodicity property given in Assumption III.

5.   Outline of the proof

The basic structure of our proof follows the relative entropy approach of [11, 13]. The aim is to derive a differential inequality for the relative entropy between the solution to the Schrödinger equation and a time-dependent local Gibbs state constructed to reproduce the solution of the Euler equations. The time derivative of the relative entropy can be expressed as an expectation of the local currents with respect to the solution to the Schrödinger equation: Since γ_t is a solution to the Schrödinger equation, we have for any density matrix ω_t^ε the identity

$$\frac{d}{dt} S(\gamma_t | \omega_t^\varepsilon) = \mathrm{Tr}\,\gamma_t \left\{i\delta_H - \partial_t\right\} \log \omega_t^\varepsilon. \tag{5.1}$$

Using the definition of the local equilibrium states (2.4), we get

$$\frac{d}{dt}s(\gamma_t|\omega_t^\varepsilon) = \varepsilon^3 \mathrm{Tr}\gamma_t \left\{i\delta_H - \partial_t\right\} \left\{\langle \boldsymbol{\lambda}_\varepsilon(t,\cdot), \mathbf{u}\rangle - \log c_\varepsilon(t)\right\}. \tag{5.2}$$

Direct computation yields that $\delta_H \langle \boldsymbol{\lambda}_\varepsilon(t,\cdot), \mathbf{u}\rangle = -\varepsilon \sum_{j=1}^3 \langle \nabla_j \boldsymbol{\lambda}_\varepsilon(t,\cdot), \boldsymbol{\theta}_j(t)\rangle$, to leading order in ε, and where the $\boldsymbol{\theta}_j$ are local observables for the microscopic currents of the conserved quantities. Since we do not know the solution of the Schrödinger equation, γ_t, well enough, this expectation in the RHS of (5.2) cannot be computed or estimated explicitly. The main idea is to bound these expectations in terms of the relative entropy itself, thus obtaining a differential inequality. All these bounds are based on the following inequality, which is a consequence of the variational principle for the relative entropy [9, 10]: for density matrices/states ω and γ, with $\ker \omega = \{0\}$, and for all self-adjoint h, and $\delta > 0$, one has

$$\gamma(h) \le \delta^{-1} \log \mathrm{Tr}\, e^{\delta h + \log \omega} + \delta^{-1} S(\gamma|\omega). \tag{5.3}$$

Most of the work is then to show that the terms resulting from the first term in the RHS of this inequality are sufficiently small. This requires a number of steps. Two essential ingredients are the Euler equations (naturally), and a quantum version of Virial Theorem to relate certain terms in the currents to the thermodynamic pressure as given by quantum statistical mechanics. Next, we explain the main steps in a bit more detail.

Step 1: Replace the local microscopic currents by macroscopic currents. The basic idea in hydrodynamical limit is first to show that the local space time average of the solution is time invariant. From Assumption III, ergodic time invariant states are Gibbs. For Gibbs states, we can replace the local microscopic currents by macroscopic currents.

1a: Construct a commuting version of local conserved quantities. In principle, one should be able to express the macroscopic currents as functions of the conserved quantities. The local conservative quantities, however, are operators which commute only up to boundary terms. Therefore, we either need to prove that the non-commutativity does not affect the meaning of macroscopic currents or we need to construct some commuting version of the local conservative quantities. We follow the second approach and construct a commuting version of local conservative quantities with a method inspired by [2].

1b: Restriction to the one phase region. Since Assumption III can only be applied in the one phase region, we have to introduce suitable cutoff functions. As these, in general, do not commute with the local currents, this is a non-trivial step.

1c: Apply the Virial Theorem. Following the method of [8], we prove and apply a quantum virial theorem to relate the local currents to the pressure.

Step 2: Estimate all errors by local conservative quantities. For this step the entropy inequality (5.3) is crucial.

Step 3: Derive a differential inequality of the entropy with error term given by a large deviation formula. As there is no large deviation theory for non-commuting observables, it is essential that we have expressed everything by commuting objects.

The rest of the argument follows by the standard relative entropy method. The main technical difficulties, in comparison with the classical case, all stem from the non-commutativity of the algebra of observables. Simple inequalities, such as $|A+B| \leq |A|+|B|$ and $|AB| \leq |A|\,|B|$, which are used numerous times in estimates for classical systems, are false for quantum observables. Therefore, all estimates have to be derived without taking absolute values. A full account of our work will appear shortly [14].

Acknowledgments. This material is based upon work supported by the National Science Foundation under Grants # DMS-0070774 (B. N.) and # DMS-9703752 (H.-T. Y.)

References

[1] M. Born, FRS, and H.S Green, *A general kinetic theory of liquids. IV. Quantum mechanics of fluids*, Proc. Roy. Soc. A, **191** (1947) 168.

[2] Conlon, J. G., Lieb, E. H., Yau, H. T., *The Coulomb gas at low temperature and low density*, Comm. Math. Phys. 125 (1989) 153–218.

[3] J.H. Irving and R.W. Zwanzig, *The statistical mechanics theory of transport processes. V. Quantum hydrodynamics*, J. Chem. Phys., **19** (1951) 1173–1180.

[4] L.P. Kadanoff and G. Baym, *Quantum statistical mechanics*, W.A. Benjamin, New York, 1962.

[5] C. B. Morrey: *On the derivation of the equations of hydrodynamics from Statistical Mechanics*, Commun. Pure Appl. Math., **8**, 279–290, (1955).

[6] S. Klainerman and A. Majda, *Compressible and Incompressible Fluids*, Comm. Pure Appl. Math, **35** (1982) 629–651.

[7] G. Gallavotti, J.L. Lebowitz, and V. Mastropietro, *Large deviations in rarefied quantum gases*, preprint mp_arc 01–264, to appear in J. Stat. Phys.

[8] B. Nachtergaele and A. Verbeure, *Groups of canonical transformations and the virial-Noether theorem*, J. Geometry and Physics, **3** (1986) 315–325.

[9] Dénes Petz, *A Variational Expression for the Relative Entropy*, Commun. Math. Phys., **114** (1988) 345–349.

[10] M. Ohya and D. Petz, *Quantum entropy and its use*, Springer Verlag, Berlin-Heidelberg-New York, 1993.

[11] S. Olla, S.R.S. Varadhan, and H.-T. Yau, *Hydrodynamical limit for a Hamiltonian system with weak noise*, Commun. Math. Phys, **155** (1993) 523–560.

[12] David Ruelle, *Statistical Mechanics*, W.A. Benjamin, Reading, MA, 1969.

[13] H.-T. Yau, *Relative entropy and the hydrodynamics of Ginzburg-Landau models*, Lett. Math. Phys, **22** (1991) 63–80.

[14] B. Nachtergaele and H.-T. Yau, *Derivation of the Euler Equations from Quantum Dynamcis*, in preparation.

ICM 2002 · Vol. III · 477–495

Seiberg-Witten Prepotential
from Instanton Counting

Nikita A. Nekrasov*

(*To Arkady Vainshtein on his 60th anniversary*)

Abstract

In my lecture I consider integrals over moduli spaces of supersymmetric gauge field configurations (instantons, Higgs bundles, torsion free sheaves).

The applications are twofold: physical and mathematical; they involve supersymmetric quantum mechanics of D-particles in various dimensions, direct computation of the celebrated Seiberg-Witten prepotential, sum rules for the solutions of the Bethe ansatz equations and their relation to the Laumon's nilpotent cone. As a by-product we derive some combinatoric identities involving the sums over Young tableaux.

1. Introduction

The dynamics of gauge theories is a long and fascinating subject. The dynamics of supersymmetric gauge theories is a subject which is better understood [1] yet may teach us something about the real QCD. The solution of Seiberg and Witten [2] of $\mathcal{N} = 2$ gauge theory using the constraints of special geometry of the moduli space of vacua led to numerous achievements in understanding of the strong coupling dynamics of gauge theory and also in string theory, of which the gauge theories in question arise as low energy limits. The low energy effective Wilsonian action for the massless vector multiplets a is governed by the prepotential $\mathcal{F}$, which receives one-loop perturbative and instanton non-perturbative corrections:

$$\mathcal{F}(a) = \mathcal{F}^{pert}(a; \Lambda) + \mathcal{F}^{inst}(a; \Lambda). \tag{1.1}$$

In spite of the fact that these instanton corrections were calculated in many indirect ways, their gauge theory calculation is lacking beyond two instantons[3][4]. The problem is that the instanton measure seems to get very complicated with the growth of the instanton charge, and the integrals are hard to evaluate.

* Institut des Hautes Etudes Scientifiques, Le Bois-Marie, Bures-sur-Yvette, F-91440, France.
On leave of absence from ITEP, 117259, Moscow, Russia. E-mail: nikita@ihes.fr

The present paper attempts the solution of this problem via the localization technique, proposed long time ago in [5][6][7]. Our method can be explained rather simply in the physical terms. We calculate the vacuum expectation value of certain gauge theory observables. These observables are annihilated by a combination of the supercharges, and their expectation value is not sensitive to various parameters. In particular, one can do the calculation in the ultraviolet, where the theory is weakly coupled and the instantons dominate. Or, one can do the calculation in the infrared, where it is rather simple to relate the answer to the prepotential of the effective low-energy theory. By equating these two calculations we obtain the desired formula.

Remark. We can also formulate our results in a more mathematical language. We study $G \times \mathbf{T}^2$ equivariant cohomology of the (suitably partially compactified) moduli space $\widetilde{\mathcal{M}}_k$ of framed G-instantons on $\mathbb{R}^4$, where G is the gauge group, which acts by rotating the gauge orientation of the instantons at infinity, and $\mathbf{T}^2$ is the maximal torus of $SO(4)$ – the group of rotations of $\mathbb{R}^4$ which also acts naturally on the moduli space. We consider the following quantity:

$$Z(a, \epsilon_1, \epsilon_2; q) = \sum_{k=0}^{\infty} q^k \oint_{\widetilde{\mathcal{M}}_k} 1 \qquad (1.2)$$

where $\oint 1$ denotes the *localization* of 1 in $H^*_{G \times \mathbf{T}^2}(\widetilde{\mathcal{M}}_k)$. The latter takes values in the field of fractions of the ring $H^*_{G \times \mathbf{T}^2}(pt)$ which is identified with the space of $G \times \mathbf{T}^2$ invariant polynomial functions on the Lie algebra of $G \times \mathbf{T}^2$. By the Chevalley theorem the latter is isomorphic to the ring of Weyl invariant functions on the Cartan subalgebra of G and $\mathbf{T}^2$. We denote the coordinates on the Cartan of G by a and the coordinates on the Lie algebra of $\mathbf{T}^2$ by ϵ_1, ϵ_2. In explicit calculations we represent 1 by a cohomologically equal form which allows to replace $\oint 1$ by an ordinary integral:

$$\oint_{\widetilde{\mathcal{M}}_k} 1 = \int_{\widetilde{\mathcal{M}}_k} \exp \omega + \mu_G(a) + \mu_{\mathbf{T}^2}(\epsilon_1, \epsilon_2) \qquad (1.3)$$

where ω is a symplectic form on $\widetilde{\mathcal{M}}_k$, invariant under the $G \times \mathbf{T}^2$ action, and $\mu_G, \mu_{\mathbf{T}^2}$ are the corresponding moment maps.

Our first claim is

$$\boxed{Z(a, \epsilon_1, \epsilon_2; q) = \exp \left(\frac{\mathcal{F}^{inst}(a, \epsilon_1, \epsilon_2; q)}{\epsilon_1 \epsilon_2} \right)}$$

$$(1.4)$$

where the function $\mathcal{F}^{inst}$ is analytic in ϵ_1, ϵ_2 near $\epsilon_1 = \epsilon_2 = 0$.

We also have the following explicit expression for Z in the case $\epsilon_1 = -\epsilon_2 = \hbar$ (in the general case we also have a formula, but it looks less transparent) for $G = SU(N)$

(a simple generalization to SO and Sp cases will be presented in [8]) :

$$Z(a,\hbar,-\hbar;q) = \sum_{\vec{\mathbf{k}}} q^{|\mathbf{k}|} \prod_{(l,i)\neq(m,j)} \frac{a_{lm} + \hbar\,(k_{l,i} - k_{m,j} + j - i)}{a_{lm} + \hbar\,(j - i)}.$$

$$(1.5)$$

Here the sum is over all colored partitions: $\vec{\mathbf{k}} = (\mathbf{k}_1,\ldots,\mathbf{k}_N)$, $\mathbf{k}_l = \{k_{l,1} \geq k_{l,2} \geq \ldots k_{l,n_l} \geq k_{l,n_l+1} = k_{l,n_l+2} = \ldots = 0\}$, and

$$|\vec{\mathbf{k}}| = \sum_{l,i} k_{l,i}\,,$$

and the product is over $1 \leq l, m \leq N$, and $i, j \geq 1$.

Already (1.5) can be used to make rather powerful checks of the Seiberg-Witten solution. But the checks are more impressive when one considers the theory with fundamental matter.

To get there one studies the bundle V over $\widetilde{\mathcal{M}_k}$ of the solutions of the Dirac equation in the instanton background. Let us consider the theory with N_f flavors. It can be shown that the gauge theory instanton measure calculates in this case (cf. [9]):

$$Z(a,m,\epsilon_1,\epsilon_2;q) = \sum_{k} q^{k} \oint_{\widetilde{\mathcal{M}_k}} \mathrm{Eu}_{G\times\mathbf{T}^2\times U(N_f)}(V \otimes M) \qquad (1.6)$$

where $M = \mathbb{C}^{N_f}$ is the flavor space, where the flavor group $U(N_f)$ acts, $m = (m_1,\ldots,m_{N_f})$ are the masses = the coordinates on the Cartan subalgebra of the flavor group Lie algebra, and finally $\mathrm{Eu}_{G\times\mathbf{T}^2\times U(N_f)}$ denotes the equivariant Euler class.

The formula (1.5) generalizes in this case to.

$$Z(a,m,\epsilon_1,\epsilon_2;q) = \sum_{\vec{\mathbf{k}}} \left(q\hbar^{N_f}\right)^{|\mathbf{k}|} \prod_{(l,i)} \prod_{f=1}^{N_f} \frac{\Gamma(\frac{a_l+m_f}{\hbar} + 1 + k_{l,i} - i)}{\Gamma(\frac{a_l+m_f}{\hbar} + 1 - i)}$$
$$\times \prod_{(l,i)\neq(m,j)} \frac{a_{lm} + \hbar\,(k_{l,i} - k_{m,j} + j - i)}{a_{lm} + \hbar\,(j - i)}.$$

$$(1.7)$$

Again, we claim that

$$\mathcal{F}^{inst}(a,m,\epsilon_1,\epsilon_2;q) = \epsilon_1\epsilon_2 \log Z(a,m,\epsilon_1,\epsilon_2;q) \qquad (1.8)$$

is analytic in $\epsilon_{1,2}$.

The formulae (1.5)(1.7) were checked against the Seiberg-Witten solution [10]. Namely, we claim that $\mathcal{F}^{inst}(a,m,\epsilon_1,\epsilon_2)|_{\epsilon_1=\epsilon_2=0} =$ the instanton part of the prepotential of the low-energy effective theory of the $\mathcal{N} = 2$ gauge theory with the gauge group G and N_f fundamental matter hypermultiplets. We have checked this claim by an explicit calculation for up to five instantons, against the formulae in [11]. There is also a generalization of (1.5) to the case of adjoint matter. We shall present it in the main body of the paper.

2. Field theory expectations

In this section we explain our approach in the field theory language. We exploit the fact that the supersymmetric gauge theory on flat space has a large collection of observables whose correlation functions are saturated by instanton contribution in the limit of weak coupling. In addition, in the presence of the adjoint scalar vev these instantons tend to shrink to zero size. Moreover, the observables we choose have the property that the instantons which contribute to their expectation values are localized in space. This solves the problem of the runaway of point-like instantons, pointed out in [5].

2.1. Supersymmetries and twisted supersymmetries

The $\mathcal{N} = 2$ theory has eight conserved supercharges, $Q_\alpha^i, Q_{\dot\alpha}^i$, which transform under the global symmetry group $SU(2)_L \times SU(2)_R \times SU(2)_I$ of which the first two factors belong to the group of spatial rotations and the last one is the R-symmetry group. The indices $\alpha, \dot\alpha, i$ are the doublets of these respective $SU(2)$ factors. The basic multiplet of the gauge theory is the vector multiplet. It is useful to work in the notations which make only $SU(2)_L \times SU(2)_d$ part of the global symmetry group manifest. Here $SU(2)_d$ is the diagonal subgroup of $SU(2)_R \times SU(2)_I$. If we call this subgroup a "Lorentz group", then the supercharges, superspace, and the fermionic fields of the theory will split as follows:

Fermions: $\psi_\mu, \chi_{\mu\nu}^+, \eta$;

Superspace: $\theta^\mu, \bar\theta_{\mu\nu}^+, \bar\theta$;

Superfield: $\Phi = \phi + \theta^\mu \psi_\mu + \frac{1}{2}\theta^\mu \theta^\nu F_{\mu\nu} + \ldots$;

Supercharges: $Q, Q_{\mu\nu}^+, G_\mu$.

The supercharge Q is a scalar with respect to the "Lorentz group" and is usually considered as a BRST charge in the topological quantum field theory version of the susy gauge theory. It is conserved on any four-manifold.

In [12] E. Witten has employed a self-dual two-form supercharge $Q_{\mu\nu}^+$ which is conserved on Kähler manifolds.

Our idea is to use other supercharges G_μ as well. Their conservation is tied up with the isometries of the four-manifold on which one studies the gauge theory. Of course, the idea to regularize the supersymmetric theory by subjecting it to the twisted boundary conditions is very common both in physics [13], and in mathematics [14][15][16].

2.2. Good observables: UV

With respect to the standard topological supercharge Q the observables one is usually interested in are the gauge invariant polynomials $\mathcal{O}^{(0)}_{P,x} = P(\phi(x))$ in the adjoint scalar ϕ, evaluated at space-time point x, and its descendants: $\mathcal{O}^{(k)}_{P,\Sigma} = \int_\Sigma P(\phi + \psi + F)$, where Σ is a k-cycle. Unfortunately for $k > 0$ all such cycles are homologically trivial on $\mathbb{R}^4$ and no non-trivial observables are constructed in such a way. One construct an equivalent set of observables by integration over $\mathbb{R}^4$ of a product of a closed $4-k$-form $\omega = \frac{1}{(4-k)!}\, \omega_{\mu_1 \dots \mu_{4-k}} \theta^{\mu_1} \dots \theta^{\mu_{4-k}}$ and the k-form part of $P(\phi + \psi + F)$:

$$\mathcal{O}_P^\omega = \int d^4 x d^4 \theta\; \omega(x,\theta)\; P(\Phi). \tag{2.1}$$

Again, most of these observables are Q-exact, as any closed k-form on $\mathbb{R}^4$ is exact for $k > 0$.

However, if we employ the rotational symmetries of $\mathbb{R}^4$ and work equivariantly, we find new observables.

Namely, consider the fermionic charge

$$\tilde{Q} = Q + E_a \Omega^a_{\mu\nu} x^\nu G_\mu. \tag{2.2}$$

Here $\Omega^a = \Omega^a_{\mu\nu} x^\nu \partial_\mu$ for $a = 1 \dots 6$ are the vector fields generating $SO(4)$ rotations, and $E \in Lie(SO(4))$ is a formal parameter.

With respect to the charge $\tilde{Q}$ the observables $\mathcal{O}^{(k)}_{P,\Sigma}$ are no longer invariant (except for $\mathcal{O}^{(0)}_{P,0}$ where $0 \in \mathbb{R}^4$ is the origin, left fixed by the rotations.

However, the observables (2.1) can be generalized to the new setup, producing a priori nontrivial $\tilde{Q}$-cohomology classes. Namely, let us take any $SO(4)$-equivariant form on $\mathbb{R}^4$. That is, take an inhomogeneous differential form $\Omega(E)$ on $\mathbb{R}^4$ which depends also on an auxiliary variable $E \in Lie(SO(4))$ which has the property that for any $g \in SO(4)$:

$$g^* \Omega(E) = \Omega(g^{-1} E g) \tag{2.3}$$

where we take pullback defined with the help of the action of $SO(4)$ on $\mathbb{R}^4$ by rotations. Such E-dependent forms are called equivariant forms. On the space of equivariant forms acts the so-called equivariant differential,

$$D = d + \iota_{V(E)} \tag{2.4}$$

where $V(E)$ is the vector field on $\mathbb{R}^4$ representing the infinitesimal rotation generated by E. For equivariantly closed (i.e. D-closed) form $\Omega(E)$ the observable:

$$\mathcal{O}_P^{\Omega(E)} = \int_{\mathbb{R}^4} \Omega(E) \wedge P(\Phi) \tag{2.5}$$

is $\tilde{Q}$-closed.

Any $SO(4)$ invariant polynomial in E is of course an example of the D-closed equivariant form. Such a polynomial is characterized by its restriction onto the Cartan subalgebra of $SO(4)$, where it must be Weyl-invariant. The Cartan subalgebra of $SO(4)$ is two-dimensional. Let us denote the basis in this subalgebra corresponding to the decomposition $\mathbb{R}^4 = \mathbb{R}^2 \oplus \mathbb{R}^2$ into a orthogonal direct sum of two dimensional planes, by (ϵ_1, ϵ_2). Under the identification $Lie(SO(4)) \approx Lie(SU(2)) \oplus Lie(SU(2))$ these map to $(\epsilon_1 + \epsilon_2, \epsilon_1 - \epsilon_2)$.

Let us fix in addition a translationally invariant symplectic form ω on $\mathbb{R}^4$. Its choice breaks $SO(4)$ down to $U(2)$ – the holonomy group of a Kähler manifold. Let us fix this $U(2)$ subgroup. Then we have a moment map:

$$\mu : \mathbb{R}^4 \longrightarrow Lie(U(2))^*, \qquad d\mu(E) = \iota_{V(E)}\omega, \quad E \in Lie(U(2)). \qquad (2.6)$$

And therefore, the form $\omega - \mu(E)$ is D-closed. One can find such euclidean coordinates x^ν, $\nu = 1, 2, 3, 4$ that the form ω reads as follows:

$$\omega = dx^1 \wedge dx^2 + dx^3 \wedge dx^4. \qquad (2.7)$$

The Lie algebra of $U(2)$ splits as a direct sum of one-dimensional abelian Lie algebra of $U(1)$ and the Lie algebra of $SU(2)$. Accordingly, the moment map μ splits as (h, μ^1, μ^2, μ^3). In the x^μ coordinates

$$h = \sum_\mu \left(x^\mu\right)^2, \qquad \mu^a = \tfrac{1}{2}\eta^a_{\mu\nu}x^\mu x^\nu, \qquad (2.8)$$

where $\eta^a_{\mu\nu}$ is the anti-self-dual 't Hooft symbol.

Finally, the choice of ω also defines a complex structure on $\mathbb{R}^4$, thus identifying it with $\mathbb{C}^2$ with complex coordinates z_1, z_2 given by: $z_1 = x^1 + ix^2$, $z_2 = x^3 + ix^4$. For E in the Cartan subalgebra $H = \mu(E)$ is given by the simple formula:

$$H = \epsilon_1|z_1|^2 + \epsilon_2|z_2|^2. \qquad (2.9)$$

After all these preparations we can formulate the correlation function of our interest:

$$Z(a, \epsilon) = \langle \exp\frac{1}{(2\pi i)^2} \int_{\mathbb{R}^4} \left(\omega \wedge \mathrm{Tr}\left(\phi F + \tfrac{1}{2}\psi\psi\right) - H\,\mathrm{Tr}\left(F \wedge F\right)\right) \rangle_a \qquad (2.10)$$

where we have indicated that the vacuum expectation value is calculated in the vacuum with the expectation value of the scalar ϕ in the vector multiplet given by $a \in \mathbf{t}$. More precisely, a will be the central charge of $\mathcal{N} = 2$ algebra corresponding to the W-boson states.

Remarks. 1) Note that the observable in (2.10) makes the widely separated instantons suppressed. More precisely, if the instantons form clusters around points $\vec{r}_1, \ldots, \vec{r}_l$ then they contribute $\sim \exp - \sum_m H(\vec{r}_m)$ to the correlation function.
2) One can expand (2.10) as a sum over different instanton sectors:

$$Z(a, \epsilon) = \sum_{k=0}^{\infty} q^k Z_k(a, \epsilon)$$

where $q \sim \Lambda^{2N}$ is the dynamically generated scale – for us – simply the generating parameter.

3) The supersymmetry guarantees that (2.10) is saturated by instantons. Moreover, the superspace of instanton zero modes is acted on by a finite dimensional version of the supercharge $\tilde{Q}$ which becomes an equivariant differential on the moduli space of framed instantons. Localization with respect to this supercharge reduces the computation to the counting of the isolated fixed points and the weights of the action of the symmetry groups (a copy of gauge group and $U(2)$ of rotations) on the tangent spaces. This localization can be understood as a particular case of the Duistermaat-Heckman formula [17], as (2.10) calculates essentially the integral of the exponent of the Hamiltonian of a torus action (Cartan of G times $\mathbf{T}^2$) against the symplectic measure.

The counting of fixed points can be nicely summarized by a contour integral (see below). This contour integral also can be obtained by transforming the integral over the ADHM moduli space of the observable (2.10) evaluated on the instanton configuration, by adding $\tilde{Q}$-exact terms, as in [7][6]. It also can be derived from Bott's formula [18].

2.3. Good observables: IR

The nice feature of the correlator (2.10) is it simple relation to the prepotential of the low-energy effective theory. In order to derive it let us think of the observable (2.10) as of a slow varying changing of the microscopic coupling constant. If we could completely neglect the fact that H is not constant, then its adding would simply renormalize the effective low-energy scale $\Lambda \to \Lambda e^{-H}$.

However, we should remember that H is not constant, and regard this renormalization as valid up to terms in the effective action containing derivatives of H. Moreover, H is really a bosonic part of the function $\mathcal{H}(x,\theta)$ on the (chiral part of) superspace (in [5] we considered such superspace-dependent deformations of the theory on curved four-manifolds):

$$\mathcal{H}(x,\theta) = H(x) + \tfrac{1}{2}\omega_{\mu\nu}\theta^\mu\theta^\nu.$$

Together these terms add up to the making the standard Seiberg-Witten effective action determined by the prepotential $\mathcal{F}(a;\Lambda)$ to the one with the superspace-dependent prepotential

$$\mathcal{F}(a;\Lambda e^{-\mathcal{H}(x,\theta)}) = \mathcal{F}(a;\Lambda e^{-H}) + \omega\,\Lambda\partial_\Lambda\mathcal{F}(a;\Lambda e^{-H}) + \tfrac{1}{2}\omega^2\left(\Lambda\partial_\Lambda\right)^2\mathcal{F}(a;\Lambda e^{-H}).$$
$$(2.11)$$

This prepotential is then integrated over the superspace (together with the conjugate terms) to produce the effective action.

Now, let us go to the extreme infrared, that is let us scale the metric on $\mathbb{R}^4$ by a very large factor t (keeping ω intact). On flat $\mathbb{R}^4$ the only term which may contribute to the correlation function in question in the limit $t \to \infty$ is the last term in (2.10) as the rest will (after integration over superspace) necessarily contain couplings to the gauge fields which will require some loop diagrams to get non-trivial contractions, which all will be suppressed by inverse powers of t. The last term, on the other hand, gives:

$$Z(a;\epsilon) = \exp - \frac{1}{8\pi^2} \int_{\mathbb{R}^4} \omega \wedge \omega \frac{\partial^2 \, \mathcal{F}(a; \Lambda e^{-H})}{(\partial \log \Lambda)^2} + O(\epsilon) \qquad (2.12)$$

where we used the fact that the derivatives of H are proportional to $\epsilon_{1,2}$. Recalling (2.7)(2.8) the integral in (2.12) reduces to:

$$Z(a; \epsilon_1, \epsilon_2) = \exp \frac{\mathcal{F}^{inst}(a; \Lambda) + O(\epsilon)}{\epsilon_1 \epsilon_2} \qquad (2.13)$$

where

$$\mathcal{F}^{inst}(a; \Lambda) = \int_0^\infty \partial_H^2 \mathcal{F}(a; \Lambda e^{-H}) \, H \mathrm{d}H,$$

thereby explaining our claim about the analytic properties of Z and $\mathcal{F}$.

3. Instanton measure and its localization

3.1. ADHM data

The moduli space $\mathcal{M}_{k,N}$ of instantons with fixed framing at infinity has dimension $4kN$. It has the following convenient description. Take two complex vector spaces V and W of the complex dimensions k and N respectively. These spaces should be viewed as Chan-Paton spaces for $D(p-4)$ and Dp branes in the brane realization of the gauge theory with instantons.

Let us also denote by L the two dimensional complex vector space, which we shall identify with the Euclidean space $\mathbb{R}^4 \approx \mathbb{C}^2$ where our gauge theory lives.

Then the ADHM [19] data consists of the following maps between the vector spaces:

$$V \xrightarrow{\tau} V \otimes L \oplus W \xrightarrow{\sigma} V \otimes \Lambda^2 L \qquad (3.1)$$

where

$$\tau = \begin{pmatrix} B_2 \\ -B_1 \\ J \end{pmatrix}, \qquad \sigma = \begin{pmatrix} B_1 & B_2 & I \end{pmatrix},$$

$$(3.2)$$

$$B_{1,2} \in \text{End}(V), \ I \in \text{Hom}(W, V), \ J \in \text{Hom}(V, W).$$

The ADHM construction represents the moduli space of $U(N)$ instantons on $\mathbb{R}^4$ of charge k as a hyperkähler quotient [20] of the space of operators (B_1, B_2, I, J) by the action of the group $U(k)$ for which V is a fundamental representation, B_1, B_2 transform in the adjoint, I in the fundamental, and J in the anti-fundamental representations.

More precisely, the moduli space of proper instantons is obtained by taking the quadruples $(B_{1,2}, I, J)$ obeying the so-called ADHM equations:

$$\mu_c = 0, \qquad \mu_r = 0, \tag{3.3}$$

where:

$$\begin{aligned}\mu_c &= [B_1, B_2] + IJ, \\ \mu_r &= [B_1, B_1^\dagger] + [B_2, B_2^\dagger] + II^\dagger - J^\dagger J\end{aligned} \tag{3.4}$$

and with the additional requirement that the stabilizer of the quadruple in $U(k)$ is trivial. This produces a non-compact hyperkähler manifold $M_{k,N}$ of instantons with fixed framing at infinity.

The framing is really the choice of the basis in W. The group $U(W) = U(N)$ acts on these choices, and acts on $M_{k,N}$, by transforming I and J in the anti-fundamental and the fundamental representations respectively.

This action also preserves the hyperkähler structure of $M_{k,N}$ and is generated by the hyperkähler moment maps:

$$\mathbf{m}_r = I^\dagger I - JJ^\dagger, \qquad \mathbf{m}_c = JI. \tag{3.5}$$

Actually, $\mathrm{Tr}_W \mathbf{m}_{r,c} = \mathrm{Tr}_V \mu_{r,c}$, thus the central $U(1)$ subgroup of $U(N)$ acts trivially on $M_{k,N}$. Therefore it is the group $G = SU(N)/\mathbb{Z}_N$ which acts non-trivially on the moduli space of instantons.

3.2. Instanton measure

The supersymmetric gauge theory measure can be regarded as an infinite-dimensional version of the equivariant Matthai-Quillen representative of the Thom class of the bundle $\Gamma\left(\Omega^{2,+} \otimes \mathbf{g}_P\right)$ over the infinite-dimensional space of all gauge fields $\mathcal{A}_P$ in the principal G-bundle P (summed over the topological types of P). In physical terms, in the weak coupling limit we are calculating the supersymmetric delta-function supported on the instanton gauge field configurations. In the background of the adjoint Higgs vev, this supersymmetric delta-function is actually an equivariant differential form on the moduli space $M_{k,N}$ of instantons. It can be also represented using the finite-dimensional hyperkähler quotient ADHM construction of $M_{k,N}$ (as opposed to the infinite-dimensional quotient of the space of all gauge fields by the action of the group of gauge transformations, trivial at infinity) [7]:

$$\int \mathcal{D}\phi \mathcal{D}\bar{\phi} \mathcal{D}H \mathcal{D}\vec{\chi} \mathcal{D}\eta \mathcal{D}\Psi \mathcal{D}B \mathcal{D}I \mathcal{D}J \ e^{\tilde{Q}\left(\vec{\chi}\cdot\vec{\mu}(B,I,J) + \Psi\cdot V(\bar{\phi}) + \eta[\phi,\bar{\phi}]\right)} \tag{3.6}$$

where, say:

$$\begin{aligned} &\tilde{Q}B_{1,2} = \Psi_{B_{1,2}}, \quad \tilde{Q}\Psi_{B_{1,2}} = [\phi, B_{1,2}] + \epsilon_{1,2}B_{1,2}, \\ &\tilde{Q}I = \Psi_I, \quad \tilde{Q}\Psi_I = \phi I - Ia, \\ &\tilde{Q}J = \Psi_J, \quad \tilde{Q}\Psi_J = -J\phi + Ja - (\epsilon_1 + \epsilon_2)J, \\ &\tilde{Q}\chi_r = H_r, \ \tilde{Q}H_r = [\phi, \chi_r], \qquad \tilde{Q}\chi_c = H_c, \ \tilde{Q}H_c = [\phi, \chi_c] + (\epsilon_1 + \epsilon_2)\chi_c, \\ &\Psi \cdot V\left(\bar{\phi}\right) = \mathrm{Tr}\left(\Psi_{B_1}[\bar{\phi}, B_1^\dagger] + \Psi_{B_2}[\bar{\phi}, B_2^\dagger] + \Psi_I[\bar{\phi}, I^\dagger] - \Psi_J[\bar{\phi}, J^\dagger] + c.c.\right) \end{aligned} \tag{3.7}$$

(we refer to [7] for more detailed explanations). If the moduli space $M_{k,N}$ was compact and smooth one could interpret (3.6) as a certain topological quantity and apply the powerful equivariant localization techniques [21] to calculate it.

The non-compactness of the moduli space of instantons is of both ultraviolet and of infrared nature. The UV non-compactness has to do with the instanton size, which can be made arbitrarily small. The IR non-compactness has to do with the non-compactness of $\mathbb{R}^4$ which permits the instantons to run away to infinity.

3.3. Curing non-compactness

The UV problem can be solved by relaxing the condition on the stabilizer, thus adding the so-called point-like instantons. A point of the hyperkähler space $\tilde{M}_{k,N}$ with orbifold singularities which one obtains in this way (Uhlenbeck compactification) is an instanton of charge $p \leq k$ and a set of $k - p$ points on $\mathbb{R}^4$:

$$\tilde{M}_{k,N} = M_{k,N} \cup M_{k-1,N} \times \mathbb{R}^4 \cup M_{k-2,N} \times Sym^2(\mathbb{R}^4) \cup \ldots \cup Sym^k(\mathbb{R}^4). \quad (3.8)$$

The resulting space $\tilde{M}_{k,N}$ is a geodesically complete hyperkähler orbifold. Its drawback is the non-existence of the universal bundle with the universal instanton connection over $\tilde{M}_{k,N} \times \mathbb{R}^4$. We actually think that in principle one can still work with this space. Fortunately, in the case of $SU(N)$ gauge group there is a nice space $\widetilde{\mathcal{M}}_{k,N}$ which is obtained from $\tilde{M}_{k,N}$ by a sequence of blowups (resolution of singularities) which is smooth, and after some modification of the gauge theory (noncommutative[22][23][24][25] deformation) becomes a moduli space with the universal instanton. Technically this space is obtained [26] by the same ADHM construction except that now one performs the hyperkähler quotient at the non-zero level of the moment map:

$$\mu_r = \zeta_r \mathbf{1}_V, \qquad \mu_c = 0$$

(one can also make $\mu_c \neq 0$ but this does not give anything new). The cohomology theory of $\widetilde{\mathcal{M}}_{k,N}$ is richer then that of $\tilde{M}_{k,N}$ because of the exceptional divisors. However, our goal is to study the original gauge theory. Therefore we are going to consider the (equivariant) cohomology classes of $\widetilde{\mathcal{M}}_{k,N}$ lifted from $\tilde{M}_{k,N}$.

As we stated in the introduction, we are going to utilize the equivariant symplectic volumes of $\widetilde{\mathcal{M}}_{k,N}$. This is not quite precise. We are going to consider the symplectic volumes, calculated using the closed two-form lifted from $\tilde{M}_{k,N}$. This form vanishes when restricted onto the exceptional variety. This property ensures that we don't pick up anything not borne in the original gauge theory (don't pick up freckle contribution in the terminology of [27]).

The ADHM construction from the previous section gives rise to the instantons with fixed gauge orientation at infinity (fixed framing). The group $G = SU(N)/\mathbb{Z}_N$ acts on their moduli space $\mathcal{M}_{N,k}$ by rotating the gauge orientation. Also, the group of Euclidean rotations of $\mathbb{R}^4$ acts on $\mathcal{M}_{N,k}$. We are going to apply localization techniques with respect to both of these groups.

In fact, it is easier to localize first with respect to the groups $U(k) \times G \times \mathbf{T}^2$ acting on the vector space of ADHM matrices, and then integrate out the $U(k)$ part of the localization multiplet, to incorporate the quotient.

The action of $\mathbf{T}^2$ is free at "infinities" of $\widetilde{\mathcal{M}}_k$, thus allowing to apply localization techniques without worrying about the IR non-compactness. Physically, the integral (2.10) is Gaussian-like and convergent in the IR region.

3.4. Reduction to contour integrals

After all the manipulations as in [7][6] we end up with the following integral[27]:

$$
Z_k(a, \epsilon_1, \epsilon_2) = \frac{1}{k!} \frac{\epsilon^k}{(2\pi i \epsilon_1 \epsilon_2)^k} \oint \prod_{i=1}^{k} \frac{d\phi_i \, Q(\phi_i)}{P(\phi_i) P(\phi_i + \epsilon)} \prod_{1 \le i < j \le k} \frac{\phi_{ij}^2 (\phi_{ij}^2 - \epsilon^2)}{(\phi_{ij}^2 - \epsilon_1^2)(\phi_{ij}^2 - \epsilon_2^2)}
$$
(3.9)

where:

$$
Q(x) = \prod_{j=1}^{N_f} (x + m_j),
$$
(3.10)
$$
P(x) = \prod_{l=1}^{N} (x - a_l),
$$

ϕ_{ij} denotes $\phi_i - \phi_j$ and $\epsilon = \epsilon_1 + \epsilon_2$.

We went slightly ahead of time and presented the formula which covers the case of the gauge theory with N_f fundamental multiplets. In fact, the derivation is rather simple if one keeps in mind the relation to the Euler class of the Dirac zeromodes bundle over the moduli space of instantons, stated in the introduction.

3.5. Classification of the residues

The integrals (3.9) should be viewed as contour integrals. As explained in [6] the poles at $\phi_{ij} = \epsilon_1, \epsilon_2$ should be avoided by shifting $\epsilon_{1,2} \to \epsilon_{1,2} + i0$, those at $\phi_i = a_l$ similarly by $a_l \to a_l + i0$ (this case was not considered in [6] but actually was considered (implicitly) in [7]). The interested reader should consult [28] for more mathematically sound explanations of the contour deformations arising in the similar context in the study of symplectic quotients.

The poles which with non-vanishing contributions to the integral must have $\phi_{ij} \ne 0$, for $i \ne j$, otherwise the numerator vanishes. This observation simplifies the classification of the poles. They are labelled as follows. Let $k = k_1 + k_2 + \ldots + k_N$ be a partition of the instanton charge in N summands which have to be non-negative (but may vanish), $k_l \ge 0$. In turn, for all l such that $k_l > 0$ let Y_l denote a partition of k_l:

$$
k_l = k_{l,1} + \ldots k_{l,\nu^{l,1}}, \qquad k_{l,1} \ge k_{l,2} \ge \ldots \ge k_{l,\nu^{l,1}} > 0.
$$

Let $\nu^{l,1} \ge \nu^{l,2} \ge \ldots \nu^{l,k_{l,1}} > 0$ denote the dual partition. Pictorially one represents these partitions by the Young diagram with $k_{l,1}$ rows of the lengths $\nu^{l,1}, \ldots \nu^{l,k_{l,1}}$. This diagram has $\nu^{l,1}$ columns of the lengths $k_{l,1}, \ldots, k_{l,\nu^{l,1}}$.

In total we have k boxes distributed among N Young tableaux (some of which could be empty, i.e. contain zero boxes). Let us label these boxes somehow (the ordering is not important as it is cancelled in the end by the factor $k!$ in (3.9)). Let us denote the collection of N Young diagrams by $\vec{Y} = (Y_1, \ldots, Y_N)$. We denote by $|Y_l| = k_l$ the number of boxes in the l'th diagram, and by $|\vec{Y}| = \sum_l |Y_l| = k$.

Then the pole of the integral (3.9) corresponding to $\vec{Y}$ is at ϕ_s with s labelling the box (α, β) in the l'th Young tableau (so that $0 \le \alpha \le \nu^{l,\beta}$, $0 \le \beta \le k_{l,\alpha}$) equal to:

$$\vec{Y} \longrightarrow \phi_s = a_l + \epsilon_1(\alpha - 1) + \epsilon_2(\beta - 1). \tag{3.11}$$

3.6. Residues and fixed points

The poles in the integral (3.9) correspond to the fixed points of the action of the groups $G \times \mathbf{T}^2$ on the resolved moduli space $\tilde{\mathcal{M}}_{k,N}$. Physically they correspond to the $U(N)$ (noncommutative) instantons which split as a sum of $U(1)$ noncommutative instantons corresponding to N commuting $U(1)$ subgroups of $U(N)$. The instanton charge k_l is the charge of the $U(1)$ instanton in the l'th subgroup. Moreover, these abelian instantons are of special nature – they are fixed by the group of space rotations. If they were commutative (and therefore point-like) they had to sit on top of each other, and the space of such point-like configurations would have been rather singular. Fortunately, upon the noncommutative deformation the singularities are resolved. The instantons cannot sit quite on top of each other. Instead, they try to get as close to each other as the uncertainty principle lets them. The resulting abelian configurations were classified (in the language of torsion free sheaves) by H. Nakajima [29].

Now let us fix a configuration $\vec{Y}$ and consider the corresponding contribution to the integral over instanton moduli. It is given by the residue of the integral (3.9) corresponding to (3.11):

$$R_{\vec{Y}} = \frac{1}{(\epsilon_1 \epsilon_2)^k} \prod_l \prod_{\alpha=1}^{\nu^{l,1}} \prod_{\beta=1}^{k_{l,\alpha}} \frac{\mathcal{S}_l(\epsilon_1(\alpha - 1) + \epsilon_2(\beta - 1))}{(\epsilon(\ell(s) + 1) - \epsilon_2 h(s))(\epsilon_2 h(s) - \epsilon\ell(s))} \times$$

$$\prod_{l<m} \prod_{\alpha=1}^{\nu^{l,1}} \prod_{\beta=1}^{k_{m,1}} \left(\frac{(a_{lm} + \epsilon_1(\alpha - \nu^{m,\beta}) + \epsilon_2(1 - \beta))(a_{lm} + \epsilon_1\alpha + \epsilon_2(k_{l,\alpha} + 1 - \beta))}{(a_{lm} + \epsilon_1\alpha + \epsilon_2(1 - \beta))(a_{lm} + \epsilon_1(\alpha - \nu^{m,\beta}) + \epsilon_2(k_{l,\alpha} + 1 - \beta))} \right)^2 \tag{3.12}$$

where we have used the following notations: $a_{lm} = a_l - a_m$,

$$\mathcal{S}_l(x) = \frac{Q(a_l + x)}{\prod_{m \ne l}(x + a_{lm})(x + \epsilon + a_{lm})}, \qquad S_l(x) = \frac{Q(a_l + x)}{\prod_{m \ne l}(x + a_{lm})^2}, \tag{3.13}$$

and

$$\ell(s) = k_{l,\alpha} - \beta, \qquad h(s) = k_{l,\alpha} + \nu^{l,\beta} - \alpha - \beta + 1. \tag{3.14}$$

Now, if we set $\epsilon_1 = \hbar = -\epsilon_2$ the formula (3.12) can be further simplified. After some reshuffling of the factors it becomes exactly the summand in (1.7).

3.7. The first three nonabelian instantons

We shall now give the formulae for the first three instanton contributions to the prepotential for the general $SU(N)$ case, with $N_f < 2N$.

We shall work with $\epsilon_1 = \hbar = -\epsilon_2$. It will be sufficient to derive the gauge theory prepotential.

Directly applying the rules (3.9)(3.12) we arrive at the following expressions for the moduli integrals:

$$Z_1 = \frac{1}{\epsilon_1 \epsilon_2} \sum_l S_l,$$

$$Z_2 = \frac{1}{(\epsilon_1 \epsilon_2)^2} \left(\frac{1}{4} \sum_l S_l \left(S_l(+\hbar) + S_l(-\hbar) \right) + \frac{1}{2} \sum_{l \neq m} \frac{S_l S_m}{\left(1 - \frac{\hbar^2}{a_{lm}^2} \right)^2} \right),$$

$$Z_3 = \frac{1}{(\epsilon_1 \epsilon_2)^3} \left(\sum_l \frac{S_l \left(S_l(+\hbar) S_l(+2\hbar) + S_l(-\hbar) S_l(-2\hbar) + 4 S_l(+\hbar) S_l(-\hbar) \right)}{36} \right.$$

$$+ \sum_{l \neq m} \frac{S_l S_m}{4} \left(\frac{S_l(+\hbar)}{\left(1 - \frac{2\hbar^2}{(a_{lm}(a_{lm} + \hbar))} \right)^2} + \frac{S_l(-\hbar)}{\left(1 - \frac{2\hbar^2}{(a_{lm}(a_{lm} - \hbar))} \right)^2} \right)$$

$$\left. + \sum_{l \neq m \neq n} \frac{S_l S_m S_n}{6 \left(\left(1 - \frac{\hbar^2}{a_{lm}^2} \right) \left(1 - \frac{\hbar^2}{a_{ln}^2} \right) \left(1 - \frac{\hbar^2}{a_{mn}^2} \right) \right)^2} \right),$$

$$\text{(3.15)}$$

from which we immediately derive:

$$\mathcal{F}_1 = \sum_l S_l,$$

$$\mathcal{F}_2 = \sum_l \frac{1}{4} S_l S_l^{(2)} + \sum_{l \neq m} \frac{S_l S_m}{a_{lm}^2} + O(\hbar^2),$$

$$\mathcal{F}_3 = \sum_l \frac{S_l}{36} \left(S_l S_l^{(4)} + 2 S_l^{(1)} S_l^{(3)} + 3 S_l^{(2)} S_l^{(2)} \right)$$

$$\text{(3.16)}$$

$$+ \sum_{l \neq m} \frac{S_l S_m}{a_{lm}^4} \left(5 S_l - 2 a_{lm} S_l^{(1)} + a_{lm}^2 S_l^{(2)} \right)$$

$$+ \sum_{l \neq m \neq n} \frac{2 S_l S_m S_n}{3(a_{lm} a_{ln} a_{mn})^2} \left(a_{ln}^2 + a_{lm}^2 + a_{mn}^2 \right) + O(\hbar^2)$$

3.8. Four and five instantons

To collect more "experimental data-points" we have considered the case of the gauge groups $SU(2)$ and $SU(3)$ with fundamental matter. We have computed explicitly the prepotential for four and five instantons and found a perfect agreement (yet a few typos) with the results of [11]. We should stress that this is a non-trivial check. Just as an example, we quote here the expression for $\mathcal{F}_5$:

$$\mathcal{F}_5(a,m) = \frac{\mu_3}{8a^{18}}\left(35a^{12} - 210a^{10}\mu_2 + a^8\left(207\mu_2^2 + 846\mu_4\right)\right.$$

$$\left. -1210a^6\mu_2\mu_4 + a^4\left(1131\mu_4^2 + 3698\mu_3^2\mu_2\right) - 5250a^2\mu_3^2\mu_4 + 4471\mu_3^4\right),$$

where $2a = a_1 - a_2$, $\mu_2 = m_1^2 + m_2^2 + m_3^2$, $\mu_3 = m_1 m_2 m_3$, $\mu_4 = (m_1 m_2)^2 + (m_2 m_3)^2 + (m_1 m_3)^2$.

3.9. Adjoint matter and other matters

So far we were discussing $\mathcal{N} = 2$ gauge theories with matter in the fundamental representations. Now we shall pass to other representations. It is simpler to start with the adjoint representation. The ϵ-integrals (3.9) reflect both the topology of the moduli space of instantons and also of the matter bundle.

The latter is the bundle of the Dirac zero modes in the representation of interest. For the adjoint representation, and on $\mathbb{R}^4$ this bundle can be identified with the tangent bundle to the moduli space of instantons. It has a $U(1)$ symmetry. The equivariant Euler class of the tangent bundle (= the Chern polynomial) is the instanton measure in the case of massive adjoint matter. This reasoning leads to the following ϵ-integral:

$$Z_k = \frac{1}{k!}\left(\frac{(\epsilon_1 + \epsilon_2)(\epsilon_1 + m)(\epsilon_2 + m)}{2\pi i\ \epsilon_1\epsilon_2\ m\ (\epsilon - m)}\right)^k \oint \prod_{i=1}^k \mathrm{d}\phi_i\ \frac{P(\phi_i + m)P(\phi_i + \epsilon - m)}{P(\phi_i)P(\phi_i + \epsilon)}$$

$$\times \prod_{i<j} \frac{\phi_{ij}^2(\phi_{ij}^2 - \epsilon^2)(\phi_{ij}^2 - (\epsilon_1 - m)^2)(\phi_{ij}^2 - (\epsilon_2 - m)^2)}{(\phi_{ij}^2 - \epsilon_1^2)(\phi_{ij}^2 - \epsilon_2^2)(\phi_{ij}^2 - m^2)(\phi_{ij}^2 - (\epsilon - m)^2)}.$$

$$(3.17)$$

Note the similarity of this expression to the contour integrals appearing[6] in the calculations of the bulk contribution to the index of the supersymmetric quantum mechanics with 16 supercharges (similarly, (3.9) is related to the one with 8 supercharges). This is not an accident, of course.

Proceeding analogously to the pure gauge theory case we arrive at the following expressions for the first two instanton contributions to the prepotential (which agrees with [11]):

$$\mathcal{F}_1 = m^2 \sum_l T_l,$$

$$\mathcal{F}_2 = \sum_l \left(-\frac{3m^2}{2}T_l^2 + \frac{m^4}{4}T_l T_l^{(2)}\right)$$

$$+ m^4 \sum_{l \neq n} T_l T_n \left(\frac{1}{a_{ln}^2} - \frac{1}{2(a_{ln} + m)^2} - \frac{1}{2(a_{ln} - m)^2}\right),$$

$$(3.18)$$

where $T_l(x) = \prod_{n \neq l}\left(1 - \frac{m^2}{(x+a_{ln})^2}\right)$, $T_l = T_l(0)$, $T_l^{(n)} = \partial_x^n T_l(x)|_{x=0}$ (cf. [30]).

For generic representation R of the gauge group we should use the equivariant Euler class of the corresponding (virtual) vector bundle $\mathcal{E}_R$ over the moduli space of instantons [8].

3.10. Perturbative part

So far we were calculating the nonperturbative part of the prepotential. It would be nice to see the perturbative part somewhere in our setup, so as to combine the whole expression into something nice.

One way is to calculate carefully the equivariant Chern character of the tangent bundle to $\widetilde{\mathcal{M}}_k$ along the lines sketched in the end of the previous section[8]. The faster way in the $\epsilon_1 + \epsilon_2 = 0$ case is to note that the expression (1.5) is a sum over partition with the universal denominator, which is not well-defined without the non-universal numerator. Nevertheless, let us try to pull it out of the sum.

We get the infinite product (up to an irrelevant constant):

$$\prod_{i,j=1}^{\infty}\prod_{l \neq m}\frac{1}{a_{lm} + \hbar(i-j)} \sim$$

$$\exp - \sum_{l \neq m}\int_0^{\infty}\frac{ds}{s}\frac{e^{-sa_{lm}}}{(e^{\hbar s}-1)(e^{-\hbar s}-1)}. \qquad (3.19)$$

If we regularize this by cutting the integral at $s \to 0$, we get a finite expression, which actually has the form

$$\exp\frac{\mathcal{F}^{pert}(a,\epsilon_1,\epsilon_2)}{\epsilon_1\epsilon_2},$$

with $\mathcal{F}^{pert}$ being analytic in ϵ_1, ϵ_2 at zero. In fact

$$\mathcal{F}^{pert}(a,0,0) = \sum_{l \neq m}\frac{1}{2}a_{lm}^2\log a_{lm} + \text{ambiguous quadratic polynomial in } a_{lm}.$$

The formula (3.19) is a familiar expression for the Schwinger amplitude of a mass a_{lm} particle in the electromagnetic field

$$F \propto \epsilon_1\ dx^1 \wedge dx^2 + \epsilon_2\ dx^3 \wedge dx^4\ . \qquad (3.20)$$

Its appearance be explained in the next section.

Let us now combine $\mathcal{F}^{inst}$ and $\mathcal{F}^{pert}$ into a single ϵ-deformed prepotential

$$\mathcal{F}(a,\epsilon_1,\epsilon_2) = \mathcal{F}^{pert}(a,\epsilon_1,\epsilon_2) + \mathcal{F}^{inst}(a,\epsilon_1,\epsilon_2)$$

where in general we define:

$$\mathcal{F}^{pert}(a, \epsilon_1, \epsilon_2) = \sum_{l \neq m} \int_{\varepsilon}^{\infty} \frac{ds}{s} \frac{e^{-sa_{lm}}}{\sinh\left(\frac{s\epsilon_1}{2}\right) \sinh\left(\frac{s\epsilon_2}{2}\right)} \tag{3.21}$$

with the singular in ε part dropped. We define:

$$\mathcal{Z}(a, \epsilon_1, \epsilon_2; q) = \exp\frac{\mathcal{F}(a, \epsilon_1, \epsilon_2; q)}{\epsilon_1 \epsilon_2}. \tag{3.22}$$

4. τ-function conjecture

This conjecture relates the expansion (1.5) to the dynamics of the Seiberg-Witten curve.

Consider the theory of a free complex chiral fermion ψ, ψ^*,

$$S = \int_{\Sigma} \psi^* \bar{\partial} \psi \tag{4.1}$$

living on the curve Σ:

$$w + \frac{\Lambda^{2N}}{w} = \mathbf{P}(\lambda), \qquad \mathbf{P}(\lambda) = \prod_{l=1}^{N}(\lambda - \alpha_l) \tag{4.2}$$

embedded into the space $\mathbb{C} \times \mathbb{C}^*$ with the coordinates (λ, w). This curve has two distinguished points $w = 0$ and $w = \infty$ which play a prominent role in the Toda integrable hierarchy [31]. Let us cut out small disks D_0 and D_∞ around these two points.

The path integral on the surface Σ with two discs deleted will give a state in the tensor product $\mathcal{H}_0 \otimes \mathcal{H}_\infty^*$ of the Hilbert spaces $\mathcal{H}_0$, $\mathcal{H}_\infty$ associated to ∂D_0 and ∂D_∞ respectively. It can also be viewed as an operator $G_\Sigma : \mathcal{H}_0 \to \mathcal{H}_\infty$.

Choose a vacuum state $|0\rangle \in \mathcal{H}_0$ and its dual $\langle 0| \in \mathcal{H}_\infty^*$ (we use the global coordinate w to identify $\mathcal{H}_0$ and $\mathcal{H}_\infty$). Consider

$$\tau_\Sigma = \left\langle 0 \left| \exp\left(\frac{1}{\hbar} \oint_{\partial D_\infty} S\, J\right) G_\Sigma \exp\left(-\frac{1}{\hbar} \oint_{\partial D_0} S\, J\right) \right| 0 \right\rangle \tag{4.3}$$

where:

$$J =: \psi^* \psi :$$
$$dS = \frac{1}{2\pi i} \lambda \frac{dw}{w} \tag{4.4}$$

and we choose the branch of S near $w = 0, \infty$ such that (cf. [32]) :

$$S = \frac{N}{2\pi i} w^{\mp \frac{1}{N}} + O(\lambda^{-1}).$$

Let us represent Σ as a two-fold covering of the λ-plane. It has branch points at $\lambda = \alpha_l^{\pm}$ where

$$\mathbf{P}(\alpha_l^{\pm}) = \pm 2\Lambda^N.$$

Let us choose the cycles A_l to encircle the cuts between α_l^- and α_l^+. Of course, in $H_1(\Sigma, \mathbb{Z})$, $\sum_l A_l = 0$. Then, we define:

$$a_l = \oint_{A_l} dS .$$

Our final conjecture states:

$$\boxed{\mathcal{Z}(a, \hbar, -\hbar) = \tau_{\Sigma}.}$$

$$(4.5)$$

Note that from this conjecture the fact that $\mathcal{F}_0(a, 0, 0)$ coincides with the Seiberg-Witten expression follows as a consequence of the Krichever universal formula [33]. The remaining paragraph is devoted to the explanation of the motivation behind (4.5).

Let us assume that we are in the domain where $\alpha_l - \alpha_m \gg \Lambda$. Then the surface Σ can be decomposed into two halves $\Sigma_{\pm}$ by N smooth circles C_l which are the lifts to Σ of the cuts connecting α_l^- and α_l^+. The path integral calculating the matrix element (4.3) can be evaluated by the cutting and sewing along the C_l's. The path integral on $\Sigma_{\pm}$ gives a state in

$$\otimes_{l=1}^N \mathcal{H}_{C_l}$$

(its dual). If we first pull the $\oint SJ$ as close to C_l as possible, we shall get the Hilbert space obtained by quantization of the fermions which have $a_l + \frac{1}{2} mod\mathbb{Z}$ moding:

$$\psi(w) \sim \sum_{i \in \mathbb{Z}} \psi_{l,i} w^{a_l + i} \left(\frac{dw}{w}\right)^{\frac{1}{2}} \qquad (4.6)$$

near $C_l \subset \Sigma$. In addition, the states in $\mathcal{H}_{C_l}$ of fixed total $U(1)$ charge are labelled by the partitions $k_{l,i}$. We conjecture, that

$$\left\langle 0 \left| e^{\oint SJ} \prod_{l,i} \psi_{l,k_{l,i}-i} \psi_{l,-i}^* \right| 0 \right\rangle_l \sim \prod_{(l,i)<(m,j)} (a_{lm} + \hbar(k_{l,i} - k_{m,j} + j - i)). \qquad (4.7)$$

It is clear that (4.7) implies (4.5). For $N = 1$ (4.7) is of course a well-known fact (with the coefficient given by $\prod_{i<j} \frac{1}{j-i}$). It gives rise to the formula (which can also be derived using the Schur identities [34]):

$$Z_{N=1}(\hbar, -\hbar) = e^{-\frac{1}{\hbar^2}}$$

which confirms that in spite of the fact that we worked with the resolved moduli space $\cup_k \widetilde{\mathcal{M}}_{k,1} = \cup_k (\mathbb{C}^2)^{[k]}$ the "symplectic" volume we calculated is that of $\cup_k \tilde{M}_{k,1} = \cup_k Sym^k(\mathbb{R}^4)$.

References

[1] V. Novikov, M. Shifman, A. Vainshtein, V. Zakharov, Phys. Lett. **217B** (1989) 103.

[2] N. Seiberg, E. Witten, hep-th/9407087, hep-th/9408099.

[3] N. Seiberg, Phys. Lett. **206B** (1988) 75.

[4] N. Dorey, V.V. Khoze, M.P. Mattis, hep-th/9607066.

[5] A. Losev, N. Nekrasov, S. Shatashvili, hep-th/9711108, hep-th/9801061.

[6] G. Moore, N. Nekrasov, S. Shatashvili, hep-th/9803265.

[7] G. Moore, N. Nekrasov, S. Shatashvili, hep-th/9712241 .

[8] N. A. Nekrasov and friends, to appear.

[9] N. Dorey, V.V. Khoze, M.P. Mattis, hep-th/9706007, hep-th/9708036.

[10] A. Klemm, W. Lerche, S. Theisen, S. Yankielowisz, hep-th/9411048 ;
P. Argyres, A. Faraggi, hep-th/9411057;
A. Hannany, Y. Oz, hep-th/9505074.

[11] G. Chan, E. D'Hoker, hep-th/9906193 ;
E. D'Hoker, I. Krichever, D. Phong, hep-th/9609041;
J. Edelstein, M. Gomez-Reino, J. Mas, hep-th/9904087 ;
J. Edelstein, M. Mariño, J. Mas hep-th/9805172.

[12] E. Witten, hep-th/9403195.

[13] E. Witten, hep-th/9304026 ;
O. Ganor, hep-th/9903110 ;
H. Braden, A. Marshakov, A. Mironov, A. Morozov, hep-th/9812078.

[14] G. Ellingsrud, S.A.Stromme, Invent. Math. **87** (1987) 343–352;
L. Göttche, Math. A.. **286** (1990) 193–207.

[15] M. Konstevich, hep-th/9405035.

[16] A. Givental, alg-geom/9603021.

[17] J. J. Duistermaat, G.J. Heckman, Invent. Math. **69** (1982) 259;
M. Atiyah, R. Bott, "The moment map and equivariant cohomology", Topology **23** No 1 (1984) 1.

[18] R. Bott, J. Diff. Geom. **4** (1967) 311.

[19] M. Atiyah, V. Drinfeld, N. Hitchin, Yu. Manin, Phys. Lett. **65A** (1978) 185.

[20] N.J. Hitchin, A. Karlhede, U. Lindstrom, and M. Rocek, Comm. Math. Phys. **108** (1987) 535.

[21] M. Atiyah, R. Bott, "The Yang-Mills equations over Riemann surfaces", Phil. Trans. Roy. Soc. London **A 308** (1982), 524–615;
E. Witten, hep-th/9204083;
S. Cordes, G. Moore, S. Rangoolam, hep-th/9411210.

[22] A. Connes, "Noncommutative geometry", Academic Press (1994).

[23] A. Connes, M. Douglas, A. Schwarz, JHEP **9802**(1998) 003.

[24] M. Douglas, C. Hull, "D-Branes and the noncommutative torus", JHEP **9802**(1998) 008, hep-th/9711165.

[25] N. Seiberg, E. Witten, hep-th/9908142, JHEP **9909**(1999) 032.

[26] N. Nekrasov, A. S. Schwarz, hep-th/9802068, Comm. Math. Phys. **198** (1998) 689.

[27] A. Losev, N. Nekrasov, S. Shatashvili, hep-th/9908204, hep-th/9911099.

[28] F. Kirwan, "Cohomology of quotients in symplectic and algebraic geometry", Mathematical Notes, Princeton Univ. Press, 1985.

[29] H. Nakajima, "Lectures on Hilbert Schemes of Points on Surfaces"; AMS University Lecture Series, 1999, ISBN 0-8218-1956-9.

[30] T. Hollowood, hep-th/0201075, hep-th/0202197.

[31] K. Ueno, K. Takasaki, Adv. Studies in Pure Math. **4** (1984) 1 ; For an excellent review see, e.g. S. Kharchev, hep-th/9810091.

[32] A. Gorsky, A. Marshakov, A. Mironov, A. Marshakov, hep-th/9802007.

[33] I. Krichever, hep-th/9205110, Comm. Math. Phys. **143** (1992) 415.

[34] I. Macdonald, "Symmetric functions and Hall polynomials", Clarendon Press, Oxford, 1979.

Affine Weyl Group Approach to Painlevé Equations

M. Noumi*

Abstract

An overview is given on recent developments in the affine Weyl group approach to Painlevé equations and discrete Painlevé equations, based on the joint work with Y. Yamada and K. Kajiwara.

2000 Mathematics Subject Classification: 34M55, 39A12, 37K35.
Keywords and Phrases: Painlevé equation, Affine Weyl group, Discrete symmetry.

1. Introduction

The purpose of this paper is to give a survey on recent developments in the affine Weyl group approach to Painlevé equations and discrete Painlevé equations.

It is known that each of the Painlevé equations from P_{II} through P_{VI} admits the action of an affine Weyl group as a group of Bäcklund transformations (see a series of works [16] by K. Okamoto, for instance). Furthermore, the Bäcklund transformations (or the Schlesinger transformations) for the Painlevé equations can already be thought of as discrete Painlevé equations with respect to the parameters. The main idea of the affine Weyl group approach to (discrete) Painlevé systems is to extend this class of Weyl group actions to general root systems, and to make use of them as the common underlying structure that unifies various types of discrete system ([10]). In this paper, we discuss several aspects of affine Weyl group symmetry in nonlinear systems, based on a series of joint works with Y. Yamada and K. Kajiwara.

Before starting the discussion of (discrete) Painlevé equations, we recall some definitions, following the notation of [4]. A (*generalized*) *Cartan matrix* is an integer matrix $A = (a_{ij})_{i,j \in I}$ (with a finite indexing set) satisfying the conditions

$$a_{ii} = 2; \quad a_{ij} \leq 0 \quad (i \neq j); \quad a_{ij} = 0 \Longleftrightarrow a_{ji} = 0. \tag{1.1}$$

*Department of Mathematics, School of Science and Technology, Kobe University, Rokko, Kobe 657-8501, Japan. E-mail: noumi@math.kobe-u.ac.jp

The *Weyl group* $W(A)$ associated with A is defined by the generators s_i $(i \in I)$, called the *simple reflections*, and the fundamental relations

$$s_i^2 = 1, \qquad (s_i s_j)^{m_{ij}} = 1 \quad (i \neq j), \tag{1.2}$$

where $m_{ij} = 2, 3, 4, 6$ or ∞, according as $a_{ij} a_{ji} = 0, 1, 2, 3$ or ≥ 4. When the Cartan matrix $A = (a_{ij})_{i,j=0}^{l}$ is of *affine type* (of type $A_l^{(1)}, B_l^{(1)}, \ldots, D_4^{(3)}$), the corresponding Weyl group is called an *affine Weyl group*.

We fix some notation for the case of type $A_l^{(1)}$ that will be used throughout this paper. The Cartan matrix $A = (a_{ij})_{i,j=0}^{l}$ of type $A_l^{(1)}$ is defied by

$$A = \begin{bmatrix} 2 & -2 \\ -2 & 2 \end{bmatrix} \ (l = 1), \qquad A = \begin{bmatrix} 2 & -1 & & & -1 \\ -1 & 2 & -1 & & \\ & -1 & 2 & \ddots & \\ & & \ddots & \ddots & -1 \\ -1 & & & -1 & 2 \end{bmatrix} \ (l \geq 2). \tag{1.3}$$

The affine Weyl group $W(A_l^{(1)}) = \langle s_0, s_1, \ldots, s_l \rangle$ is defined by the following fundamental relations:

$$\begin{aligned} (l = 1): \quad & s_0^2 = s_1^2 = 1, \\ (l \geq 2): \quad & s_i^2 = 1, \quad s_i s_j = s_j s_i \ (j \neq i, i \pm 1), \quad (s_i s_j)^3 = 1 \ (j = i \pm 1), \end{aligned} \tag{1.4}$$

where we have identified the indexing set with $\mathbb{Z}/(l+1)\mathbb{Z}$. We also define an extension $\widetilde{W}(A_l^{(1)}) = \langle s_0, \ldots, s_l, \pi \rangle$ of $W(A_l^{(1)})$ by adjoining a generator π (rotation of indices) such that $\pi s_i = s_{i+1} \pi$ for all $i = 0, 1, \ldots, l$; we do *not* impose the relation $\pi^{l+1} = 1$.

2. Variations on the theme of P_{IV}

In this section, we present several examples of affine Weyl group action of type $A_2^{(1)}$ to illustrate the role of affine Weyl group symmetry in (discrete) Painlevé equations and related integrable systems.

2.1. Symmetric form of P_{IV}

Consider the following system of nonlinear differential equations for three unknown functions $\varphi_j = \varphi_j(t)$ $(j = 0, 1, 2)$:

$$(N_{\mathrm{IV}}) \qquad \begin{cases} \varphi_0' = \varphi_0(\varphi_1 - \varphi_2) + \alpha_0, \\ \varphi_1' = \varphi_1(\varphi_2 - \varphi_0) + \alpha_1, \\ \varphi_2' = \varphi_2(\varphi_0 - \varphi_1) + \alpha_2, \end{cases} \tag{2.1}$$

where $' = d/dt$ denotes the derivative with respect to the independent variable t, and $\alpha_j = 0$ $(j = 0, 1, 2)$ are parameters. When $\alpha_0 + \alpha_1 + \alpha_2 = 0$, this system provides an integrable deformation of the Lotka-Volterra competition model for three species. When $\alpha_0 + \alpha_1 + \alpha_2 = k \neq 0$, it is essentially the *fourth Painlevé equation*

$$(P_{\mathrm{IV}}) \qquad y'' = \frac{1}{2y}(y')^2 + \frac{3}{2}y^3 + 4ty^2 + (t^2 - \alpha)y + \frac{\beta}{y}. \qquad (2.2)$$

In fact, from $(\varphi_0 + \varphi_1 + \varphi_2)' = k$, we have $\varphi_0 + \varphi_1 + \varphi_2 = kt + c$. Under the renormalization $k = 1$, $c = 0$, system (2.1) can be written as a second order equation for $y = \varphi_0$; it is transformed into P_{IV} with $\alpha = \alpha_2 - \alpha_1$, $\beta = -2\alpha_0^2$ by the change of variables $t \to \sqrt{2}t$, $y \to -y/\sqrt{2}$. In view of this fact, we call (2.1) the *symmetric form* of the fourth Painlevé equation (N_{IV}). This type of representation for P_{IV} was introduced by [19], [1] in the context of nonlinear dressing chains, and by [12] in the study of rational solutions of P_{IV}.

The symmetric form N_{IV} provides a convenient framework for describing the discrete symmetry of P_{IV}. Let $\mathcal{K} = \mathbb{C}(\alpha, \varphi)$ be the field of rational functions in the variables $\alpha = (\alpha_0, \alpha_1, \alpha_2)$ and $\varphi = (\varphi_0, \varphi_1, \varphi_2)$. We define the derivation $' : \mathcal{K} \to \mathcal{K}$ by using formulas (2.1) together with $\alpha_j' = 0$ ($j = 0, 1, 2$); we regard the differential field $(\mathcal{K}, \, ')$ as representing the differential system N_{IV}. In this setting, we say that an automorphism of $\mathcal{K}$ is a *Bäcklund transformation* for N_{IV} if it commutes with the derivation $'$. (A Bäcklund transformation as defined above means a birational transformation of the phase space that commutes with the flow defined by the nonlinear differential system.) As we will see below, N_{IV} has four fundamental Bäcklund transformations that generate the extended affine Weyl group $\widetilde{W} = \langle s_0, s_1, s_2, \pi \rangle$ of type $A_2^{(1)}$. Identifying the indexing set $\{0, 1, 2\}$ with $\mathbb{Z}/3\mathbb{Z}$, we define the automorphisms s_i ($i = 0, 1, 2$) and π of $\mathcal{K}$ by

$$\begin{aligned}
s_i(\alpha_j) &= \alpha_j - \alpha_i a_{ij}, & s_i(\varphi_j) &= \varphi_j + \frac{\alpha_i}{\varphi_i} u_{ij} & (i, j = 0, 1, 2), \\
\pi(\alpha_j) &= \alpha_{j+1}, & \pi(\varphi_j) &= \varphi_{j+1} & (j = 0, 1, 2).
\end{aligned} \qquad (2.3)$$

Here $A = (a_{ij})_{i,j=0}^2$ stands for the Cartan matrix of type $A_2^{(1)}$, and $U = (u_{ij})_{i,j=0}^2$ for the orientation matrix of the Dynkin diagram (triangle) in the positive direction:

$$A = \begin{bmatrix} 2 & -1 & -1 \\ -1 & 2 & -1 \\ -1 & -1 & 2 \end{bmatrix}, \qquad U = \begin{bmatrix} 0 & 1 & -1 \\ -1 & 0 & 1 \\ 1 & -1 & 0 \end{bmatrix}. \qquad (2.4)$$

These automorphisms s_i and π commute with the derivation $'$, and satisfy the fundamental relations

$$s_i^2 = 1, \quad (s_i s_{i+1})^3 = 1, \quad \pi s_i = s_{i+1}\pi \qquad (i = 0, 1, 2) \qquad (2.5)$$

for the generators of $\widetilde{W}(A_2^{(1)})$. Hence we obtain a realization of the extended affine Weyl group $\widetilde{W}(A_2^{(1)})$ as a group of Bäcklund transformations for N_{IV}. Notice that the action of the affine Weyl group $W = \langle s_0, s_1, s_2 \rangle$ on the α-variables is identical to its canonical action on the *simple roots*.

We remark that the affine Weyl group symmetry is deeply related to the structure of special solutions of P_{IV} (with the parameters α_j as in N_{IV}). Along each reflection hyperplane $\alpha_j = n$ $(j = 0, 1, 2; n \in \mathbb{Z})$ in the parameter space, P_{IV} has a one-parameter family of classical solutions expressed in terms of Toeplitz determinants of Hermite-Weber functions; each solution of this class is obtained by Bäcklund transformations from a seed solution at $\alpha_j = 0$ which satisfies a Riccati equation. Also, at each point of the W-orbit of the barycenter $(\alpha_0, \alpha_1, \alpha_2) = (\frac{1}{3}, \frac{1}{3}, \frac{1}{3})$ of the fundamental alcove, it has a rational solution expressed in terms of Jacobi-Trudi determinants of Hermite polynomials.

2.2. q-Difference analogue of P_{IV}

We now introduce a *multiplicative* analogue of the birational realization (2.3) of the extended affine Weyl group $\widetilde{W} = \langle s_0, s_1, s_2, \pi \rangle$ ([5]). Taking the field of rational functions $\mathcal{L} = \mathbb{C}(a, f)$ in the variables $a = (a_0, a_1, a_2)$ and $f = (f_0, f_1, f_2)$, we define the automorphisms s_0, s_1, s_2, π of $\mathcal{L}$ as follows:

$$s_i(a_j) = a_j a_i^{-a_{ij}}, \quad s_i(f_j) = f_j \left(\frac{a_i + f_i}{1 + a_i f_i} \right)^{u_{ij}} \quad (i, j = 0, 1, 2),$$
$$\pi(a_j) = a_{j+1}, \qquad \pi(f_j) = f_{j+1} \qquad (j = 0, 1, 2), \tag{2.6}$$

where a_j are the multiplicative parameters corresponding to the simple roots α_j. These automorphisms again satisfy the fundamental relations for the generators of $\widetilde{W}$. In the following, the $\widetilde{W}$-invariant $a_0 a_1 a_2 = q$ plays the role of the base for q-difference equations. If one parameterizes a_j and f_j as

$$a_j = e^{-\varepsilon^2 \alpha_j / 2}, \quad f_j = -e^{-\varepsilon \varphi_j} \qquad (j = 0, 1, 2) \tag{2.7}$$

with a small parameter ε, one can recover the original formulas (2.3) from (2.6) by taking the limit $\varepsilon \to 0$.

A q-difference analogue of (the symmetric form of) P_{IV} is given by

$$(qP_{\text{IV}}) \quad \begin{cases} T(f_0) = a_0 a_1 f_1 \dfrac{1 + a_2 f_2 + a_2 a_0 f_2 f_0}{1 + a_0 f_0 + a_0 a_1 f_0 f_1}, \\[2mm] T(f_1) = a_1 a_2 f_2 \dfrac{1 + a_0 f_0 + a_0 a_1 f_0 f_1}{1 + a_1 f_1 + a_1 a_2 f_1 f_2}, \\[2mm] T(f_2) = a_2 a_0 f_0 \dfrac{1 + a_1 f_1 + a_1 a_2 f_1 f_2}{1 + a_2 f_2 + a_2 a_0 f_2 f_0}, \\[2mm] T(a_j) = a_j \quad (j = 0, 1, 2), \end{cases} \tag{2.8}$$

where T stands for the discrete time evolution ([5]). Notice that (2.8) implies $T(f_0 f_1 f_2) = (a_0 a_1 a_2)^2 f_0 f_1 f_2 = q^2 f_0 f_1 f_2$; hence one can consistently introduce a time variable t such that $f_0 f_1 f_2 = t^2$. If we consider f_j as functions of t, the discrete time evolution T is identified with the q-shift operator $t \to qt$, so that $T f_j(t) = f_j(qt)$. In this sense, formula (2.8) defines a system of nonlinear q-difference equations, which we call the *fourth q-Painlevé equation* (qP_{IV}).

The time evolution T, regarded as an automorphism of $\mathcal{L}$, commutes with the action of $\widetilde{W}$ that we already described above. Namely, the q-difference system qP_{IV} admits the action of the extended affine Weyl group $\widetilde{W}$ as a group of Bäcklund transformations. Again, by taking the limit as $\varepsilon \to 0$ under the parametrization (2.7), one can show that the q-difference system qP_{IV}, as well as its affine Weyl group symmetry, reproduces the differential system N_{IV}. It is known that qP_{IV} defined above shares many characteristic properties with the original P_{IV}. For example, it has classical solutions expressed by continuous q-Hermite-Weber functions, and rational solutions expressed by of continuous q-Hermite polynomials, analogously to the case of P_{IV} ([5], [7]). We also remark that, when $a_0 a_1 a_2 = 1$, one can regard qP_{IV} as a discrete integrable system which generalizes a discrete version of the Lotka-Volterra equation.

2.3.   Ultra-discretization of P_{IV}

It should be noticed that the discrete time evolution of qP_{IV} is defined in terms of a *subtraction-free* birational transformation; we say that a rational function is subtraction-free if it can be expressed as a ratio of two polynomials with real positive coefficients. Recall that there is a standard procedure, called the *ultra-discretization*, of passing from subtraction-free rational functions to piecewise linear functions ([18], [2], see also [15]). Roughly, it is the procedure of replacing the operations

$$a \cdot b \to A + B, \quad a/b \to A - B, \quad a + b \to \max(A, B). \qquad (2.9)$$

Introducing the variables A_j, F_j $(j = 0, 1, 2)$, from qP_{IV} we obtain the following system of piecewise linear difference equations by ultra-discretization:

$$(uP_{IV}) \begin{cases} T(F_0) = A_0 + A_1 + F_1 + \max(0, A_2 + F_2, A_2 + A_0 + F_2 + F_0) \\ \qquad\quad - \max(0, A_0 + F_0, A_0 + A_1 + F_0 + F_1), \\ T(F_1) = A_1 + A_2 + F_2 + \max(0, A_0 + F_0, A_0 + A_1 + F_0 + F_1) \\ \qquad\quad - \max(0, A_1 + F_1, A_1 + A_2 + F_1 + F_2), \\ T(F_2) = A_2 + A_0 + F_0 + \max(0, A_1 + F_1, A_1 + A_2 + F_1 + F_2) \\ \qquad\quad - \max(0, A_2 + F_2, A_2 + A_0 + F_2 + F_0), \\ T(A_j) = A_j \qquad (j = 0, 1, 2), \end{cases} \qquad (2.10)$$

which we call the *fourth ultra-discrete Painlevé equation* (uP_{IV}). Simultaneously, the affine Weyl group symmetry of qP_{IV} is ultra-discretized as follows:

$$s_i(A_j) = A_j - A_i a_{ij}, \quad s_i(F_j) = F_j + u_{ij}\big(\max(A_i, F_i) - \max(0, A_i + F_i)\big),$$
$$\pi(A_j) = A_{j+1}, \qquad \pi(F_j) = F_{j+1} \qquad (i, j = 0, 1, 2). \qquad (2.11)$$

This time, the extended affine Weyl group $\widetilde{W}$ is realized as a group of piecewise linear transformations on the affine space with coordinates (A, F). We also remark that, when $A_0 + A_1 + A_2 = Q = 0$, uP_{IV} gives rise to an ultra-discrete integrable system. It would be an interesting problem to analyze special solutions of the ultra-discrete system uP_{IV}.

3.   Discrete symmetry of Painlevé equations

In this section, we propose a uniform description of discrete symmetry of the Painlevé equations P_J for $J = $ II, IV, V, VI. We also give some remarks on a generalization of this class of birational Weyl group action to arbitrary root systems.

3.1.   Hamiltonian system H_J

It is known that each Painlevé equation P_J ($J = $ II, III,..., VI) is equivalently expressed as a Hamiltonian system

$$(H_J): \qquad \frac{dq}{dt} = \frac{\partial H}{\partial p}, \quad \frac{dp}{dt} = -\frac{\partial H}{\partial q} \tag{3.1}$$

with a polynomial Hamiltonian $H = H(q, p, t, \alpha) \in \mathbb{C}(t)[q, p, \alpha]$ depending on parameters $\alpha = (\alpha_1, \ldots, \alpha_l)$ (see [3], for instance). Setting $\mathcal{K} = \mathbb{C}(q, p, t, \alpha)$, we define the Poisson bracket $\{\cdot, \cdot\} : \mathcal{K} \times \mathcal{K} \to \mathcal{K}$ and the Hamiltonian vector field $\delta : \mathcal{K} \to \mathcal{K}$ by

$$\{\varphi, \psi\} = \frac{\partial \varphi}{\partial p} \frac{\partial \psi}{\partial q} - \frac{\partial \varphi}{\partial q} \frac{\partial \psi}{\partial p}, \quad \delta(\varphi) = \{H, \varphi\} + \frac{\partial \varphi}{\partial t} \qquad (\varphi, \psi \in \mathcal{K}). \tag{3.2}$$

In this setting, a Bäcklund transformation for H_J is understood as an automorphism $w : \mathcal{K} \to \mathcal{K}$ that commutes with δ. We also say that w is *canonical* if it preserves the Poisson bracket: $w(\{\varphi, \psi\}) = \{w(\varphi), w(\psi)\}$ for any $\varphi, \psi \in \mathcal{K}$.

For each $J = $ II, III,...,VI, it is known that the parameter space for H_J is identified with the Cartan subalgebra of a semisimple Lie algebra, and that an extension of the corresponding affine Weyl group acts on $\mathcal{K}$ as a group of Bäcklund transformations ([16]). A table of fundamental Bäcklund transformations for H_J can be found in [9].

If the Hamiltonian H is chosen appropriately, the affine Weyl group symmetry of H_J for $J = $ II, IV, V, VI can be described in a universal way in terms of root systems. With the notation of [4], the type of the affine root system is specified as follows[1].

H_J	H_{II}	H_{IV}	H_{V}	H_{VI}	
$X_l^{(1)}$	$A_1^{(1)}$	$A_2^{(1)}$	$A_3^{(1)}$	$D_4^{(1)}$	(3.3)

---

[1]In the case of H_{III}, one can use an extension of the affine Weyl group, either of type $C_2^{(1)}$ or of $2A_1^{(1)}$, for describing the same group of Bäcklund transformations. It seems natural to expect that the same principle to be discussed below should apply to H_{III} as well, but we have not completely understood the case of H_{III} yet.

The corresponding Cartan matrix $A = (a_{ij})_{i,j=0}^{l}$ is given by

$$A_1^{(1)}: \qquad A = \begin{bmatrix} 2 & -2 \\ -2 & 2 \end{bmatrix} \qquad A_2^{(1)}: \qquad A = \begin{bmatrix} 2 & -1 & -1 \\ -1 & 2 & -1 \\ -1 & -1 & 2 \end{bmatrix}$$

$$A_3^{(1)}: \quad A = \begin{bmatrix} 2 & -1 & 0 & -1 \\ -1 & 2 & -1 & 0 \\ 0 & -1 & 2 & -1 \\ -1 & 0 & -1 & 2 \end{bmatrix} \qquad D_4^{(1)}: \quad A = \begin{bmatrix} 2 & 0 & -1 & 0 & 0 \\ 0 & 2 & -1 & 0 & 0 \\ -1 & -1 & 2 & -1 & -1 \\ 0 & 0 & -1 & 2 & 0 \\ 0 & 0 & -1 & 0 & 2 \end{bmatrix} \tag{3.4}$$

respectively. For the description of affine Weyl group symmetry, we make use of the following Hamiltonian $H = H(q,p,t,\alpha)$:

$$\begin{aligned} H_{\mathrm{II}}: \qquad & H = \frac{1}{2}p(p - 2q^2 + t) + \alpha_1 q, \\ H_{\mathrm{IV}}: \qquad & H = qp(2p - q - 2t) - 2\alpha_1 p - \alpha_2 q, \\ H_{\mathrm{V}}: \qquad & tH = q(q-1)p(p+t) - (\alpha_1 + \alpha_3)qp + \alpha_1 p + \alpha_2 tq, \\ H_{\mathrm{VI}}: \; t(t-1)H = & q(q-1)(q-t)p^2 - \big\{(\alpha_0 - 1)q(q-1) + \alpha_4(q-1)(q-t) \\ & + \alpha_3 q(q-t)\big\}p + \alpha_1(\alpha_1 + \alpha_2)(q-t). \end{aligned} \tag{3.5}$$

The parameter α_0 is defined so that $\alpha_0 + \alpha_1 + \cdots + \alpha_l = 1$ for $J = $ II, IV, V, and $\alpha_0 + \alpha_1 + 2\alpha_2 + \alpha_3 + \alpha_4 = 1$ for $J = $ VI. (Conventionally, the null root is normalized to be the constant 1.)

3.2. Discrete symmetry of H_J

Our main observation concerning the discrete symmetry of H_J can be summarized as follows.

Theorem 1 *Choose the polynomial Hamiltonian $H \in \mathbb{C}(t)[q,p,\alpha]$ for H_J ($J = $ II, IV, V, VI) as in (3.5). Then there exists a set $\{\varphi_0, \varphi_1, \ldots, \varphi_l\}$ of nonzero elements in the Poisson algebra $\mathcal{R} = \mathbb{C}[q,p,t]$ with the following properties:*
(1) The elements φ_i ($i = 0, 1, \ldots, l$) satisfy the Serre relations of type $X_l^{(1)}$

$$\mathrm{ad}_{\{\}}(\varphi_i)^{-a_{ij}+1}\varphi_j = 0 \qquad (i \neq j), \tag{3.6}$$

where $\mathrm{ad}_{\{\}}(\varphi) = \{\varphi, \cdot\}$ stands for the adjoint action of φ by the Poisson bracket.
(2) For each $i = 0, 1, \ldots, l$, define s_i to be the unique automorphism of $\mathcal{K} = \mathbb{C}(q,p,t,\alpha)$ such that

$$\begin{aligned} s_i(\alpha_j) &= \alpha_j - \alpha_i a_{ij} & (j = 0, 1, \ldots, l), \\ s_i(\psi) &= \exp\Big(\frac{\alpha_i}{\varphi_i}\mathrm{ad}_{\{\}}(\varphi_i)\Big)\psi & (\psi \in \mathcal{R} = \mathbb{C}[q,p,t]). \end{aligned} \tag{3.7}$$

Then these s_i are canonical Bäcklund transformations for H_J. Furthermore, the subgroup $W = \langle s_0, s_1, \ldots, s_l \rangle$ of $\mathrm{Aut}(\mathcal{K})$ is isomorphic to the affine Weyl group $W(X_l^{(1)})$.

Note that, for each $\psi \in \mathcal{R}$, $s_i(\psi)$ is determined as a finite sum

$$s_i(\psi) = \psi + \frac{\alpha_i}{\varphi_i}\{\varphi_i, \psi\} + \frac{1}{2!}\Big(\frac{\alpha_i}{\varphi_i}\Big)^2\{\varphi_i, \{\varphi_i, \psi\}\} + \cdots, \qquad (3.8)$$

since the action of $\mathrm{ad}_{\{\}}(\varphi_i)$ on $\mathcal{R}$ is locally nilpotent. A choice of the generators φ_i $(i = 0, 1, 2, \ldots, l)$ with the properties of Theorem 1 is given as follows:

$$
\begin{aligned}
H_{\mathrm{II}} &: \quad \varphi_0 = -p + 2q^2 + t, \quad \varphi_1 = p. \\
H_{\mathrm{IV}} &: \quad \varphi_0 = -p + \frac{q}{2} + t, \quad \varphi_1 = -\frac{q}{2}, \quad \varphi_2 = p. \\
H_{\mathrm{V}} &: \quad \varphi_0 = p + t, \quad \varphi_1 = tq, \quad \varphi_2 = -p, \quad \varphi_3 = t(1-q). \\
H_{\mathrm{VI}} &: \quad \varphi_0 = q - t, \quad \varphi_1 = 1, \quad \varphi_2 = -p, \quad \varphi_3 = q - 1, \quad \varphi_4 = q.
\end{aligned}
\qquad (3.9)
$$

We remark that, in the case of H_J ($J = \mathrm{II, IV, V}$) of type $A_l^{(1)}$ ($l = 1, 2, 3$), we also have the Bäcklund transformation π corresponding to the diagram rotation; its action is given simply by $\pi(\alpha_j) = \alpha_{j+1}$, $\pi(\varphi_j) = \varphi_{j+1}$.

If we use the polynomials φ_j as dependent variables, the Hamiltonian system H_{IV}, for example, is rewritten as

$$\frac{d\varphi_j}{dt} = 2\varphi_j(\varphi_{j+1} - \varphi_{j+2}) + \alpha_j \qquad (j = 0, 1, 2) \qquad (3.10)$$

with the convention $\varphi_{j+3} = \varphi_j$, from which we obtain the symmetric form N_{IV} by a simple rescaling of the variables. We remark that the polynomials φ_j are the factors of the "leading term" of the Hamiltonian H. Also, in the context of irreducibility of Painlevé equations, the polynomials φ_j are the fundamental *invariant divisors* along the reflection hyperplanes $\alpha_j = 0$ (see [8], [17], for instance). When $\alpha_j = 0$, the specialization of H_J by $\varphi_j = 0$ gives rise to a Riccati equation that reduces to a linear equation of hypergeometric type; for $J = \mathrm{II, IV, V}$ and VI, the differential equations of Airy, Hermite-Weber, Kummer and Gauss appear in this way, respectively.

Apart from differential equations, this class of birational realization of Weyl groups as in Theorem 1 can be formulated for an arbitrary Cartan matrix by means of Poisson algebras (see [13], for the details). In this sense, Bäcklund transformations for Painlevé equations P_J ($J = \mathrm{II, IV, V, VI}$) have a *universal* nature with respect to root systems. In the case where A is of affine type, such a birational realization of the affine Weyl group appears as the symmetry of systems of nonlinear partial differential equations of Painlevé type, obtained by similarity reduction from the principal Drinfeld-Sokolov hierarchy (of modified type). The case of type $A_l^{(1)}$ will be mentioned in the next section. As for the original Painlevé equations, P_{II}, P_{IV} and P_{V} are in fact obtained by similarity reduction from the $(l+1)$-reduced modified KP hierarchy for $l = 1, 2, 3$, respectively. For P_{VI}, an 8×8 Lax pair is constructed in [14] in the framework of the affine Lie algebra $\widehat{\mathfrak{so}}(8)$. This Lax pair is compatible with the affine Weyl group symmetry of Theorem 1. It is not clear, however, how this construction should be understood in relation to the Drinfeld-Sokolov hierarchy of type $D_4^{(1)}$.

4. Painlevé systems with $W(A_l^{(1)})$ symmetry

In this section, we introduce Painlevé systems and q-Painlevé systems with affine Weyl group symmetry of type A; this part can be regarded as a generalization, to higher rank cases, of the variations of P_{IV} discussed in Section 2. In the following, we fix two positive integers M, N, and consider a Painlevé system, as well as its q-version, attached to (M, N).

4.1. Painlevé system of type (M, N)

We investigate the compatibility condition for a system of linear differential equations

$$Nz\partial_z\vec{\psi} = A\vec{\psi}, \quad \partial_{t_m}\vec{\psi} = B_m\vec{\psi} \quad (m = 1, \ldots, M), \tag{4.1}$$

where $\vec{\psi} = (\psi_1, \ldots, \psi_N)^{\mathrm{t}}$ is the column vector of unknown functions, and A, B_m are $N \times N$ matrices, both depending on $(z, t) = (z, t_1, \ldots, t_M)$. We assume that

$$B_m = \sum_{k=0}^{m-1} \mathrm{diag}(\boldsymbol{b}^{(m,k)})\Lambda^k + \Lambda^m, \quad (m = 1, \ldots, M), \tag{4.2}$$

where $\Lambda = \sum_{i=1}^{N-1} E_{i,i+1} + zE_{N,1}$ denotes the cyclic matrix, $E_{ij} = (\delta_{a,i}\delta_{b,j})_{a,b}$ being the matrix units, and $\boldsymbol{b}^{(m,k)} = (b_1^{(m,k)}, \ldots, b_N^{(m,k)})$ are N-vectors depending only on t. Note that the compatibility condition

$$\partial_{t_n}(B_m) - \partial_{t_m}(B_n) + [B_m, B_n] = 0 \qquad (m, n = 1, \ldots, M) \tag{4.3}$$

is the Zakharov-Shabat equation of the N-reduced modified KP hierarchy (restricted to the first M time variables). As for the matrix A, we set

$$A = -\mathrm{diag}(\boldsymbol{\rho}) + \sum_{k=1}^{M} kt_k B_k, \quad \boldsymbol{\rho} = (N-1, N-2, \ldots, 0). \tag{4.4}$$

Then the compatibility condition

$$\partial_{t_m}(A) - Nz\partial_z(B_m) + [A, B_m] = 0 \qquad (m = 1, \ldots, M), \tag{4.5}$$

reduces to the homogeneity condition

$$\sum_{n-1}^{M} nt_n\partial_{t_n}(\boldsymbol{b}^{(m,k)}) = (k-m)\boldsymbol{b}^{(m,k)} \quad (1 \le k \le m \le M) \tag{4.6}$$

for the coefficients of the B matrices. We define the *Painlevé system of type* (M, N) to be the system of nonlinear partial differential equations (4.3) with the similarity constraint (4.6).

We remark that, when $(M, N) = (3, 2), (2, 3), (2, 4)$, this system reduces essentially to the Painlevé equations P_{II}, P_{IV}, P_V, respectively. When $(M, N) = (2, N)$

($N \geq 2$), it corresponds to the higher order Painlevé equation of type $A_{N-1}^{(1)}$ discussed in [11]. Note that the linear problem (4.1) defines a monodromy preserving deformation of linear ordinary differential system of order N on $\mathbb{P}^1$, with one regular singularity at $z = 0$ and one irregular singularity at $z = \infty$.

The Painlevé system of type (M, N) admits the action of the affine Weyl group $W = \langle s_0, s_1, \ldots, s_{N-1} \rangle$ of type $A_{N-1}^{(1)}$ as a group of Bäcklund transformations. Expressing the matrix A as

$$-A = \operatorname{diag}(\boldsymbol{\rho}) - \sum_{k=1}^{M} k B_k = \operatorname{diag}(\varepsilon) + \sum_{k=1}^{M} \operatorname{diag}(\boldsymbol{\varphi}^{(k)})\Lambda^k, \qquad (4.7)$$

we set $\alpha_0 = N - \varepsilon_1 + \varepsilon_N$, $\alpha_i = \varepsilon_i - \varepsilon_{i+1}$ and $\varphi_0 = \varphi_N^{(1)}$, $\varphi_i = \varphi_i^{(1)}$ for $i = 1, \ldots, N-1$. (Note that all the exponents $-\varepsilon_i$ at $z = 0$ are constant.) Then, for each $i = 0, \ldots, N-1$, the Bäcklund transformation s_i is obtained as the compatibility of the gauge transformation

$$s_i \vec{\psi} = G_i \vec{\psi}, \quad G_i = 1 + \frac{\alpha_i}{\varphi_i} F_i \qquad (i = 0, 1, \ldots, N-1), \qquad (4.8)$$

where $F_0 = z^{-1} E_{1,N}$ and $F_i = E_{i+1,i}$ $(i = 1, \ldots, N-1)$. If we regard α_i and $\varphi_i^{(k)}$ as coordinates for the matrix $-A$, the ring $\mathcal{R}$ of polynomials in the φ-variables has a natural structure of Poisson algebra. In these coordinates, the Bäcklund transformation s_i is determined by the universal formula

$$s_i(\alpha_j) = \alpha_j - \alpha_i a_{ij}, \quad s_i(\psi) = \exp\left(\frac{\alpha_i}{\varphi_i}\operatorname{ad}_{\{\}}(\varphi_i)\right)\psi \quad (\psi \in \mathcal{R}). \qquad (4.9)$$

4.2. q-Painlevé system of type (M, N)

We construct the q-Painlevé system of type (M, N) in an analogous way ([7]). With the time variables t_m and the q-shift operators $T_m = T_{q,t_m}$ $(i = 1, \ldots, M)$, we investigate the compatibility condition for a system of linear q-difference equations

$$T_{q^N, z}\vec{\psi} = A\vec{\psi}, \quad T_m \vec{\psi} = B_m \vec{\psi} \quad (m = 1, \ldots, M), \qquad (4.10)$$

where A, B are $N \times N$ matrices depending on (z, t). We assume that

$$B_m = \operatorname{diag}(\boldsymbol{u}^{(m)}) + t_m \Lambda, \quad \boldsymbol{u}^{(m)} = (u_1^{(m)}, \ldots, u_N^{(m)}) \quad (m = 1, \ldots, M), \qquad (4.11)$$

with compatibility condition

$$T_n(B_m)B_n = T_m(B_n)B_m \qquad (m, n = 1, \ldots, M). \qquad (4.12)$$

We consider this condition as the Zakharov-Shabat equation for the N-reduced modified q-KP hierarchy; in this formulation, all the time variables $t_1, \ldots, t_M$ are treated equally. Note that, as to the Euler operator $T = T_1 \cdots T_M$, we have

$$T\vec{\psi} = B_T\vec{\psi}, \quad B_T = T_2 \cdots T_M(B_1) T_3 \cdots T_M(B_2) \cdots T_M(B_{M-1}) B_M. \qquad (4.13)$$

In the linear q-difference system (4.10), we choose the following matrix for A:

$$A = \mathrm{diag}(\boldsymbol{\kappa})^{-1} B_T, \quad \boldsymbol{\kappa} = (q^{N-1}, q^{N-2}, \dots, 1). \tag{4.14}$$

Then the compatibility condition

$$T_m(A)B_m = T_{q^N,z}(B_m)A \qquad (m = 1, \dots, M) \tag{4.15}$$

is equivalent to the homogeneity condition

$$T_1 \cdots T_M(u_i^{(m)}) = u_i^{(m)} \qquad (i = 1, \dots, N; m = 1, \dots, M). \tag{4.16}$$

We define the *q-Painlevé system of type* (M, N) to be the system of nonlinear q-difference equations (4.12) for $M \times N$ unknown functions $u_i^{(m)}$ ($m = 1, \dots, M$; $i = 1, \dots, N$) with the similarity constraint (4.16). This system can be written in the form

$$T_m(u_i^{(n)}) = F_i^{(m,n)}(t, u) \qquad (m, n = 1, \dots, M; i = 1, \dots N); \tag{4.17}$$

in general, these $F_i^{(m,n)}(t, u)$ are complicated rational functions. It turns out, however, that by introducing new variables

$$x_j^i = \frac{1}{t_i} T_{i+1} T_{i+2} \cdots T_M(u_j^{(i)}) \quad (i = 1, \dots, M; j = 1, \dots, N), \tag{4.18}$$

the time evolution of the q-Painlevé system can be described explicitly by means of a birational affine Weyl group action on the x-variables.

4.3.   A birational Weyl group action on the matrix space

For describing the time evolution T_i of the q-Painlevé system, we introduce a birational action of the direct product $\widetilde{W}(A_{M-1}^{(1)}) \times \widetilde{W}(A_{N-1}^{(1)})$ of two extended affine Weyl groups. In the following, we use the notation

$$\widetilde{W}^M = \langle r_0, r_1, \dots, r_{M-1}, \omega \rangle, \quad \widetilde{W}_N = \langle s_0, s_1, \dots, s_{N-1}, \pi \rangle \tag{4.19}$$

for the two extend affine Weyl groups. Introducing two parameters q, p, we take $\mathbb{K} = \mathbb{C}(q, p)$ as the ground field. Let $\mathcal{K} = \mathbb{K}(x)$ be the field of rational functions in the MN variables x_j^i ($1 \le i \le M; 1 \le j \le N$); we regard the x-variables as the canonical coordinates of the affine space of $M \times N$ matrices. For convenience, we extend the indices i, j of x_j^i to $\mathbb{Z}$ by setting $x_j^{i+M} = qx_j^i$, $x_{j+N}^i = px_j^i$.

We define the automorphisms r_k ($k \in \mathbb{Z}/M\mathbb{Z}$), ω, s_l ($l \in \mathbb{Z}/N\mathbb{Z}$), π of $\mathcal{K}$ as follows:

$$r_k(x_j^i) = px_j^{i-1} \frac{P_{j-1}^i}{P_j^i}, \quad r_k(x_j^{i-1}) = p^{-1}x_j^i \frac{P_j^i}{P_{j-1}^i} \quad (i \equiv k \mod M);$$

$$r_k(x_j^i) = x_j^i \quad (i \not\equiv k \mod M); \quad \omega(x_j^i) = x_j^{i+1};$$

$$s_l(x_j^i) = qx_{j+1}^i \frac{Q_j^{i-1}}{Q_j^i}, \quad s_l(x_{j+1}^i) = q^{-1}x_j^i \frac{Q_j^i}{Q_j^{i-1}} \quad (j \equiv l \mod N); \tag{4.20}$$

$$s_l(x_j^i) = x_j^i \quad (j \not\equiv l \mod N); \quad \pi(x_j^i) = x_{j+1}^i,$$

where

$$P_j^i = \sum_{k=1}^{N} \prod_{a=0}^{k-1} x_{j+a}^i \prod_{a=k+1}^{N} x_{j+a}^{i+1}, \quad Q_j^i = \sum_{k=1}^{M} \prod_{a=0}^{k-1} x_j^{i+a} \prod_{a=k+1}^{M} x_{j+1}^{i+a}. \qquad (4.21)$$

Note that all these automorphisms represent subtraction-free birational transformations on the affine space of $M \times N$ matrices.

Theorem 2 *The automorphisms* $r_0, \ldots, r_{M-1}, \omega$ *and* $s_0, \ldots, s_{M-1}, \pi$ *of* $\mathcal{K}$ *defined as above give a realization of the product* $\widetilde{W}^M \times \widetilde{W}_N$ *of extended affine Weyl groups.*

By using this birational action of affine Weyl group $\widetilde{W}^M = \langle r_0, \ldots, r_{M-1}, \omega \rangle$ we define $\gamma_1, \ldots, \gamma_M$ by

$$\gamma_k = r_{k-1} \cdots r_1 \, \omega \, r_M \cdots r_k \qquad (k = 1, \ldots, M). \qquad (4.22)$$

We remark that these elements γ_k generate a free abelian subgroup $L \simeq \mathbb{Z}^M$, and that the extended affine Weyl group $\widetilde{W}^M$ decomposes into the semidirect product $L \rtimes S_M$ of L and the symmetric group of degree M that acts on L by permuting the indices for γ_k.

Theorem 3 *In terms of the variables* x_j^i *defined by* (4.18), *the time evolution of the q-Painlevé system of type* (M, N) *is described by*

$$T_k(x_j^i) = \gamma_k^{-1}(x_j^i) \quad (1 \le i \le M; 1 \le j \le N) \qquad (4.23)$$

for all $k = 1, \ldots, M$, *where* γ_k *is defined by* (4.22) *through the birational action of* $\widetilde{W}^M$ *with* $p = 1$.

This theorem means that the discrete time evolutions T_k $(k = 1, \ldots, M)$ of the q-Painlevé system of type (M, N) coincides with the commuting discrete flows γ_k^{-1} $(k = 1, \ldots, M)$ arising from the affine Weyl group action of $\widetilde{W}^M$ with $p = 1$. Furthermore the q-Painlevé system admits the action of extended affine Weyl group $\widetilde{W}_N$ of type $A_{N-1}^{(1)}$ as a group of Bäcklund transformations. One can show that the fourth q-Painlevé equation qP_{IV} discussed in Section 2 arises from the q-Painlevé system of type $(M, N) = (2, 3)$, consistently with the differential case.

Finally we give some remarks on the ultra-discretization. From the birational action of $\widetilde{W}^M \times \widetilde{W}_N$ with two parameters q, p, we obtain a piecewise linear action of the same group on the space of $M \times N$ matrices, with two parameters Q, P corresponding to q, p. When $P = 0$, the commuting piecewise linear flows $\gamma_k \in \widetilde{W}^M$ may be called the *ultra-discrete Painlevé system of type* (M, N). When $P = Q = 0$, it specializes to an integrable ultra-discrete system; it gives rise to an M-periodic version of the box-ball system.

This class of piecewise linear action is tightly related to the combinatorics of *crystal bases*. The coordinates of the $M \times N$ matrix space can be identified with the coordinates for the tensor product $\mathcal{B}^{\otimes M}$ of M copies of the crystal basis $\mathcal{B}$ for the symmetric tensor representation of GL_N. Under this identification, it turns out that the piecewise linear transformations r_k and s_l with $P = Q = 0$ represent the action of the combinatorial R matrices and the Kashiwara's Weyl group action on $\mathcal{B}^{\otimes M}$, respectively (see [15]).

References

[1] V.E. Adler: Nonlinear chains and Painlevé equations, Physica D **73**(1994), 335–351.

[2] A. Berenstein, S. Fomin and A. Zelevinsky: Parametrization of canonical bases and totally positive matrices, Adv. in Math. **122**(1996), 49–149.

[3] K. Iwasaki, H. Kimura, S. Shimomura and M. Yoshida: *From Gauss to Painlevé—A Modern Theory of Special Functions*, Vieweg, 1991.

[4] V.G. Kac: *Infinite dimensional Lie algebras*, 3rd Edition, Cambridge University Press, 1990.

[5] K. Kajiwara, M. Noumi and Y. Yamada: A study on the fourth q-Painlevé equation, J. Phys. A:Math. Gen. **34**(2001), 8563–8581.

[6] K. Kajiwara, M. Noumi and Y. Yamada: Discrete integrable systems with $W(A_{m-1}^{(1)} \times A_{n-1}^{(1)})$ symmetry, to appear in Lett. Math. Phys. (nlin.SI/0106029).

[7] K. Kajiwara, M. Noumi and Y. Yamada: q-Painlevé systems arising from q-KP hierarchy, preprint (nlin.SI/0112045).

[8] M. Noumi and K. Okamoto: Irreducibility of the second and the fourth Painlevé equations, Funkcial. Ekvac. **40**(1997), 139–163.

[9] M. Noumi, K. Takano and Y. Yamada: Bäcklund transformations and the manifolds of Painlevé systems, to appear in Funkcial. Ekvac.

[10] M. Noumi and Y. Yamada: Affine Weyl groups, discrete dynamical systems and Painlevé equations, Commun. Math. Phys. **199**(1998), 281–295.

[11] M. Noumi and Y. Yamada: Higher order Painlevé equations of type $A_l^{(1)}$, Funkcial. Ekvac. **41**(1998), 483–503.

[12] M. Noumi and Y. Yamada: Symmetries in the fourth Painlevé equations and Okamoto polynomials, Nagoya Math. J. **153**(1999), 53–86.

[13] M. Noumi and Y. Yamada: Birational Weyl group action arising from a nilpotent Poisson algebra, in *Physics and Combinatorics 1999*, Proceedings of the Nagoya 1999 International Workshop (Eds. A.N. Kirillov, A. Tsuchiya and H. Umemura), 287–319, World Scientific, 2001.

[14] M. Noumi and Y. Yamada: A new Lax pair for the sixth Painlevé equation associated with $\widehat{\mathfrak{so}}(8)$, to appear in *Microlocal Analysis and Complex Fourier Analysis*, World Scientific (math-ph/0203030).

[15] M. Noumi and Y. Yamada: Tropical Robinson-Schensted-Knuth correspondence and birational Weyl group actions, preprint(math-ph/0203030).

[16] K. Okamoto: Studies of the Painlevé equations I, Ann. Math. Pura Appl. **146**(1987), 337–381; II, Japan. J. Math. **13**(1987), 47–76; III, Math. Ann. **275**(1986), 221–255; IV, Funkcial. Ekvac. **30**(1987), 305–332.

[17] H. Umemura and H. Watanabe: Solutions of the second and fourth Painlevé equations, I, Nagoya Math. J. **148**(1997), 151–198

[18] T. Tokihiro, D. Takahashi, J. Matsukidaira and J. Satsuma: From soliton equations to integrable cellular automata through a limiting procedure, Phys. Rev. Lett. **76**(1996), 3247–3250.

[19] A.P. Veselov and A.B. Shabat: A dressing chain and the spectral theory of Schrödinger operator, Funct. Anal. Appl. **27**(1993), 81–96.

Weyl Manifolds and Gaussian Thermostats

Maciej P. Wojtkowski*

Abstract

A relation between Weyl connections and Gaussian thermostats is exposed and exploited.

2000 Mathematics Subject Classification: 37D, 53C, 70K, 82C05.

1. Introduction

We consider a class of mechanical dynamical systems with forcing and a thermostating term based on the Gauss' Least Constraint Principle for nonholonomic constraints, [G99]. It was originally proposed as a model for systems out of equilibrium, [HHP87], [GR97], [G99'], [R99].

Let us consider a mechanical system of the form

$$\dot{q} = v, \qquad \dot{v} = -\frac{\partial W}{\partial q} + E,$$

where q, v from $\mathbb{R}^n$ are the configuration and velocity coordinates, $W = W(q)$ is a potential function describing the interactions in the system and $E = E(q)$ is an external field acting on the system. The total energy $H = \frac{1}{2}v^2 + W$ does change because of the effect of the field E. We modify the system by applying the Gauss' Principle to the constraint $H = h$ to obtain the *isoenergetic* dynamical system

$$\dot{q} = v, \qquad \dot{v} = -\frac{\partial W}{\partial q} + E - \frac{\langle E, v \rangle}{v^2} v, \qquad (1.1)$$

with possible singularities where $v = 0$. In the special case when $W = 0$ we have the *isokinetic* dynamics. The thermostating term in the equations (1.1) is called the *Gaussian thermostat*. The numerical discovery, [ECM90] that the Lyapunov

---

*Department of Mathematics, University of Arizona, Tucson, Arizona 85721, USA. E-mail: maciejw@math.arizona.edu

spectrum, at least in the isokinetic case, has the shifted hamiltonian symmetry raised the issue of the mathematical nature of the equations (1.1).

On every energy level $H = h$ the equations (1.1) define a dynamical system. In the isokinetic case the change in h is equivalent to the appropriate rescaling of time and the multiplication of the external field E by a scalar.

Example 1.2.

Let $\mathbb{T}^2$ be the flat torus with coordinates $(x, y) \in \mathbb{R}^2$ and $E = (a, 0)$ be the constant vector field on $\mathbb{T}^2$. The Gaussian thermostat equations on the energy level $\dot{x}^2 + \dot{y}^2 = 1$,

$$\ddot{x} = a\dot{y}^2, \quad \ddot{y} = -a\dot{x}\dot{y},$$

can be integrated and we obtain as trajectories translations of the curve

$$ax = -\ln\cos ay$$

or the horizontal lines. Assuming that E has an irrational direction on $\mathbb{T}^2$ we obtain the following global phase portrait for the isokinetic dynamics. In the unit tangent bundle $S\mathbb{T}^2 = \mathbb{T}^3$ we have two invariant tori A and R with minimal quasiperiodic motions, A contains the unit vectors in the direction of E and it is a global attractor and R contains the unit vectors opposite to E and it is a global repellor. It can be established ([W00]) that these invariant submanifolds are normally hyperbolic so that the phase portrait is preserved under perturbations. This example reveals that Gaussian thermostats, even in the most restricted isokinetic case are not in general hamiltonian with respect to any symplectic structure. In part of the phase space they may contract phase volume and hence have no absolutely continuous invariant measure.

The involution $I(q, v) = (q, -v)$ conjugates the forward and backward in time dynamics, i.e., the system is reversible. Reversibility is close to the hamiltonian property, for instance, when accompanied by enough recurrence it can replace symplecticity in KAM theory,[Se98].

Dettmann and Morriss, [DM96], proved the shifted symmetry of the Lyapunov spectrum, in the case of isokinetic dynamics with a locally potential field E by exposing the locally hamiltonian nature of the equations. For the system (1.1) with $W = 0$ and $E = -\frac{\partial U}{\partial q}$ the change of variables

$$p = e^{-U}v, \quad \frac{dt}{d\tau} = e^{U},$$

brings (1.1) to the form

$$\frac{dq}{d\tau} = \frac{\partial H}{\partial p}, \quad \frac{dp}{d\tau} = -\frac{\partial H}{\partial q}, \quad H = \frac{1}{2}e^{2U}p^2 = \frac{1}{2}v^2.$$

Under the same assumptions, Choquard, [Ch97], found a variational principle, also involving the factor e^{U} which in the physically interesting examples is multivalued, thus making the whole description only local. Liverani and Wojtkowski, [WL98], made the observation that although the form

$$\sum dp \wedge dq = e^{-U} \left(\sum dv \wedge dq - dU \wedge \left(\sum v dq \right) \right),$$

like the coordinate system (p, q) is defined only locally, the globally defined form $\omega = \sum dv \wedge dq - dU \wedge (\sum v dq)$ can be used to develop hamiltonian-like formalism.

The three formulations above ([DM96],[Ch97],[WL98]) apply only to isokinetic dynamics with a locally potential field $E = -\nabla U$. In [W00] a geometric setup was proposed that covers all cases, i.e., isoenergetic and isokinetic, with a potential vector field E and nonpotential as well. We will describe now this setup in detail.

2. Weyl manifolds and W-flows

Let us consider a compact n-dimensional riemannian manifold M and its tangent bundle TM. The metric g will be also denoted by $\langle \cdot, \cdot \rangle$. For a smooth vector field E on M the equations of isokinetic dynamics (the Gaussian thermostat) on the energy level $v^2 = 1$ have the coordinate free form

$$\frac{dq}{dt} = v, \quad \frac{Dv}{dt} = E - \langle E, v \rangle v, \qquad (2..1)$$

where $\frac{D}{dt}$ denotes the covariant derivative, i.e., $\frac{D}{dt} = \nabla_v$ and ∇ is the Levi-Civita connection. We obtain a flow on the unit tangent bundle SM of M.

Let φ be the 1-form associated with the vector field E, i.e., $\varphi(\cdot) = \langle E, \cdot \rangle$. The form φ and the riemannian metric g define a Weyl structure on M, which is a linear symmetric connection $\widehat{\nabla}$ given by the formula (cf. [F70])

$$\widehat{\nabla}_X Y = \nabla_X Y + \varphi(Y) X + \varphi(X) Y - \langle X, Y \rangle E,$$

for any vector fields X, Y on M. The Weyl structure is usually introduced on the basis of the conformal class of g rather than g itself, but in our study we fix the riemannian metric g, which plays the role of a physical space. If we change the metric g to $\widetilde{g} = e^{-2U} g$, then the 1-form φ is replaced by $\widetilde{\varphi} = \varphi + dU$. Hence if the vector field E has a potential, i.e., $E = -\nabla U$ then the Weyl structure coincides with the Levi-Civita connection of the rescaled metric $\widetilde{g}$.

The defining property of the Weyl connection is that it is a symmetric linear connection $\widehat{\nabla}$ such that (cf. [F70])

$$\widehat{\nabla}_X g = 2\varphi(X) g, \qquad (2.2)$$

for any vector field X on M, which is equivalent to the requirement that the parallel transport defined by the linear connection preserves angles. We consider the geodesics of the Weyl connection. They are given by the equations in TM

$$\frac{dq}{ds} = w, \quad \frac{\widehat{D}w}{ds} = 0, \qquad (2.3)$$

where $\frac{\widehat{D}}{ds} = \widehat{\nabla}_w$. These equations provide geodesics with a distinguished parameter s defined uniquely up to scale. It follows from (2.2) and (2.3) that $\frac{d|w|}{ds} = -\varphi(w)|w|$. Assuming that at the initial point $q(0)$ we have $|w| = 1$ we obtain

$$|w| = e^{-\int_{q(0)}^{q(s)} \varphi}.$$

This formula shows that unless the form φ is exact we should not expect the geodesic flow in TM of a Weyl connection to preserve any sphere bundle. We introduce the flow

$$\Phi^t : SM \to SM,$$

which we call the *W-flow* for the field E, by parametrizing the geodesics of the Weyl connection with the arc length given by g. In other words the projection of a trajectory of Φ^t to M is a geodesic of the Weyl connection, t is the arc length parameter defined by the metric g and the trajectory itself is the natural lift of the geodesic to SM. By direct calculation we obtain

Theorem 2.4. *The isoenergetic dynamics*

$$\frac{dq}{dt} = v, \qquad \frac{Dv}{dt} = -\nabla W + E - \frac{\langle E, v \rangle}{v^2} v. \tag{2.5}$$

on the energy level $\frac{1}{2}v^2 + W = h$, reparametrized by the arc length defines a flow on SM which coincides with the W-flow for the vector field

$$\widetilde{E} = \frac{-\nabla W + E}{2(h - W)}.$$

(We assumed implicitly that v^2 does not vanish on the energy level set.) In particular in the isokinetic case on the energy level $v^2 = 1$ we obtain that the equations (2.1) define the W-flow for the field E itself.

This theorem can be interpreted as a generalization of the Maupertuis metric,[A89]. Indeed, in the case of $E = 0$ we have

$$\widetilde{E} = \frac{-\nabla W}{2(h - W)} = -\nabla \left(-\frac{1}{2} \ln(h - W) \right),$$

and hence the Weyl connection for the field $\widetilde{E}$ is the Levi-Civita connection for the Maupertuis metric $(h - W)g$.

In this formulation it becomes transparent how the isoenergetic case differs from the isokinetic. In the former case, if $E = -\nabla U$ has a (local) potential, the vector field $\widetilde{E} = \frac{-\nabla(W+U)}{2(h-W)}$ does not have a potential unless $dW \wedge dU = 0$, i.e., unless W and U are functionally dependent. It fits well with the result of Bonetto,Cohen and Pugh [BCP99] that the isoenergetic dynamics does not in general posses the shifted hamiltonian symmetry of the Lyapunov spectrum. This result implies that the W-flows for nonpotential vector fields are not in general conformally symplectic for any choice of the conformally symplectic structure,[WL98].

3. Jacobi equations, curvature of Weyl connections and linearizations of W-flows

We are interested in studying hyperbolicity of the dynamical system (2.1) on SM. Since it was revealed to be a modification of the geodesic flow of a connection, it is natural to check if the Anosov theory of riemannian geodesic flows can be extended in this direction. The first step in such a study must be the investigation of linearized equations of (2.1). For riemannian geodesic flows a very useful linearization is furnished by the Jacobi equations. The Jacobi equations are valid not only for a Levi-Civita connection but for any symmetric connection. To describe them let us consider a one parameter family of geodesics of the Weyl connection, i.e., a family of solutions of (2.3) parametrized by the parameter u close to zero,

$$q(s,u), w(s,u) = \frac{dq}{ds}, |u| < \epsilon.$$

We introduce the Jacobi field

$$\xi = \frac{dq}{du} \quad \text{and} \quad \widehat{\eta} = \widehat{\nabla}_\xi w.$$

The Jacobi equations read

$$\frac{\widehat{D}\xi}{ds} = \widehat{\eta}, \quad \frac{\widehat{D}\widehat{\eta}}{ds} = -\widehat{R}(\xi, w)w, \tag{3.1}$$

where for any tangent vector fields X, Y,

$$\widehat{R}(X,Y) = \widehat{\nabla}_X \widehat{\nabla}_Y - \widehat{\nabla}_Y \widehat{\nabla}_X - \widehat{\nabla}_{[X,Y]}$$

is the curvature tensor of the Weyl connection.

Let us split the vector field $\widehat{\eta} = \widehat{\eta}_0 + \widehat{\eta}_1$ into the component $\widehat{\eta}_1$ orthogonal to w and the component $\widehat{\eta}_0$ parallel to w. The equations (3.1) can be split accordingly

$$\frac{\widehat{D}\widehat{\eta}_1}{ds} = -\widehat{R}_a(\xi, w)w, \quad \frac{\widehat{D}\widehat{\eta}_0}{ds} = -\widehat{R}_s(\xi, w)w, \tag{3.2}$$

where $\widehat{R}_a(X,Y)$ is the antisymmetric and $\widehat{R}_s(X,Y)$ the symmetric part of the Weyl curvature operator $\widehat{R}(X,Y) = \widehat{R}_a(X,Y) + \widehat{R}_s(X,Y)$ ($\widehat{R}_s$ is called the distance curvature and $\widehat{R}_a$ the direction curvature, cf. [F70]).

We are faced now with two tasks, to derive the linearization of (2.1) from the equations (3.1) and to study the curvature tensor of the Weyl connection.

Together with the family of Weyl geodesics let us consider the respective family of trajectories of the W-low (2.1)

$$q(t,u), v(t,u) = \frac{dq}{dt}, |u| < \epsilon.$$

We define again the Jacobi field by $\xi = \frac{dq}{du}$. Letting $\eta = \nabla_\xi v = \nabla_v \xi$ we can consider (ξ, η) as coordinates in the tangent space of SM, which is described by $\langle v, \eta \rangle = 0$. Further we replace η with χ, the component of $\widehat{\nabla}_\xi v$ orthogonal to v. It can be calculated that

$$\chi = \eta + \langle E, v \rangle \xi - \langle \xi, v \rangle E$$

and hence we can use (ξ, χ) as linear coordinates in the tangent bundle of SM. Note that $\langle v, \eta \rangle = 0$ is equivalent to $\langle v, \chi \rangle = 0$ and in these new coordinates the velocity vector field of the W-flow (2.1) is simply $(v, 0)$. Now the linearization of (2.1) can be written as

$$\frac{\widehat{D}\xi}{dt} = \chi + \varphi(\xi)v, \quad \frac{\widehat{D}\chi}{dt} = -\widehat{R}_a(\xi, v)v + \varphi(v)\chi. \tag{3.3}$$

We will rewrite the equations (3.3) as ordinary linear differential equations with time dependent coefficients. To achieve that we need to choose frames in the tangent spaces of M along the trajectory where we linearize the W-flow. The Weyl parallel transport along a path is a conformal linear mapping and the coefficient of dilation is equal to $e^{-\int \varphi}$. We choose an orthonormal frame $v, e_1, \ldots, e_{n-1}$ in an initial tangent space $T_{q_0}M$ and parallel transport it along a trajectory of our W-flow in the direction $v \in SM$. We obtain the orthogonal frames which we normalize by the coefficient $e^{\int \varphi}$ and denote them by $v(t), e_1(t), \ldots, e_{n-1}(t)$. Let $(\xi_0, \xi_1, \ldots, \xi_{n-1})$ and $(0, \chi_1, \ldots, \chi_{n-1})$ be the components of ξ and χ respectively in these frames. Let further $\widetilde{\xi} = (\xi_1, \ldots, \xi_{n-1}) \in \mathbb{R}^{n-1}$ and $\widetilde{\chi} = (\chi_1, \ldots, \chi_{n-1}) \in \mathbb{R}^{n-1}$. The equations (3.3) will read then

$$\frac{d\xi_0}{dt} = \varphi(\xi - \langle \xi, v \rangle v), \quad \frac{d\widetilde{\xi}}{dt} = -\varphi(v)\widetilde{\xi} + \widetilde{\chi}, \quad \frac{d\widetilde{\chi}}{dt} = -\mathcal{R}\widetilde{\xi} \tag{3.4}$$

where the operator $\mathcal{R}\widetilde{\xi} = \widehat{R}_a(\xi, v)v \in \mathbb{R}^{n-1}$ (the vector $\widehat{R}_a(\xi, v)v$, being orthogonal to v, is considered as an element in $\mathbb{R}^{n-1}$ by the expansion in the basis $e_1(t), \ldots, e_{n-1}(t)$). Note that $\widehat{R}_a(\xi, v)v$ and $\varphi(\xi - \langle \xi, v \rangle v)$ depend $\widetilde{\xi} \in \mathbb{R}^{n-1}$ alone.

The linearized equations (3.4) for $(\widetilde{\xi}, \widetilde{\chi})$ differ from the Jacobi equations in the riemannian case by the presence of the "nonconservative" term $-\varphi(v)\widetilde{\xi}$ in the first equation and the properties of the operator $\mathcal{R}$, which although defined analogously in terms of the curvature tensor is not in general symmetric. The curvature tensor can be calculated directly, [W00], but the result is somewhat cumbersome. However the sectional curvatures $\widehat{K}(\Pi)$ of the Weyl connection in the direction of a plane Π defined as

$$\widehat{K}(\Pi) = \langle \widehat{R}_a(X, Y)Y, X \rangle,$$

for any orthonormal basis $\{X, Y\}$ of Π, are pleasantly transparent

$$\widehat{K}(\Pi) = K(\Pi) - E_\perp^2 - div_\Pi E, \qquad (3.5)$$

where $K(\Pi)$ is the riemannian sectional curvature in the direction of Π, the vector field $E_\perp$ is the component of E orthogonal to Π and $div_\Pi E = \langle \nabla_X E, X \rangle + \langle \nabla_Y E, Y \rangle$, the partial divergence of the vector field E, i.e., the exponential rate of growth of the area in the direction Π under the flow in M of the vector field E.

There is a problem with this sectional curvature. It does not depend on the Weyl connection alone, it is also effected by the choice of the riemannian metric g in the conformal class. However the sign of sectional curvatures is well defined.

If M is 2-dimensional then $E_\perp = 0$. Moreover on a compact manifold M with a Weyl connection there is a unique metric in the conformal class, called the Gauduchon gauge, [Ga84], such that the vector field E is divergence free. We obtain that in dimension 2 the curvature of a Weyl connection with respect to the Gauduchon gauge is equal to the gaussian curvature of the Gauduchon gauge.

Let us summarize our discussion. We have a workable linearization (3.4) of the dynamical system (2.1) and a geometric tool (the sectional Weyl curvature (3.5)) to describe its properties. We are ready to draw conclusions about hyperbolic properties of the W-flows under the assumption of negative Weyl sectional curvature.

4. Hyperbolic properties of W-flows

We obtain the information about hyperbolicity of W-flows studying the quadratic form $\mathcal{J}$ in the tangent spaces of the phase space SM, defined by $\mathcal{J}(\xi, \chi) = \langle \xi, \chi \rangle$. The form $\mathcal{J}$ factors naturally to the quotient bundle (the quotient by the span of the vector field (2.1), i.e., in the (ξ, χ) coordinates the quotient by the span of $(v, 0)$). The quotient space can be represented by the subspace $\langle \xi, v \rangle = 0$, but this subspace is not invariant under the linearization (3.4) of the flow. The form $\mathcal{J}$ in the quotient space is nondegenerate and it has equal positive and negative indices of inertia.

We take the Lie derivative of $\mathcal{J}$ and obtain

$$\frac{d}{dt}\mathcal{J} = \chi^2 - \varphi(v)\mathcal{J} - \widehat{K}(\Pi)\xi^2, \qquad (4.1)$$

where Π is the plane spanned by v and ξ. Because of the middle term, which is absent in the riemannian case, the negativity of the sectional curvature does not guarantee that (4.1) is positive definite. However it does have a weaker property that it is positive where $\mathcal{J} = 0$. We call a flow with this property *strictly $\mathcal{J}$-separated*,[W01]. A strictly $\mathcal{J}$-separated flow has a dominated splitting, [M̃84], i.e., it has a continuous splitting into "weakly stable" and "weakly unstable" subspaces on which the rates of growth are uniformly separated, but which are not necessarily decay and growth respectively. For example the dominated splitting allows exponential growth in the "weakly stable" subspace, but then all of the "weakly unstable" subspace grows with a bigger exponent. It needs to be stressed that the splitting is done in the quotient space, because in contrast to the riemannian/contact case we do not have a priori any invariant subspaces transversal to the flow direction.

It turns out that negative sectional Weyl curvatures guarantee even more hyperbolicity.

Theorem 4.2. *If the sectional curvatures of the Weyl structure are negative everywhere in M then the W-flow is strictly $\mathcal{J}$-separated and hence it has the dominated splitting into the invariant subspaces $\mathcal{E}^+$ and $\mathcal{E}^-$. Moreover there is uniform exponential growth of volume on $\mathcal{E}^+$ and uniform exponential decay of volume on $\mathcal{E}^-$.*

Corollary 4.3. *If the sectional curvatures of the Weyl structure are negative everywhere in M then for any ergodic invariant measure of the W-flow the largest Lyapunov exponent is positive and the smallest Lyapunov exponent is negative.*

We can apply this corollary to an individual periodic orbit and we obtain linear instability. Moreover there are also no repelling periodic orbits under the assumption of negative sectional Weyl curvature.

The 2-dimensional case is special. We have

Theorem 4.4. *For a 2-dimensional compact surface M if the curvature of the Weyl structure is negative, i.e, $\widehat{K} = K - divE < 0$ on M, then the W-flow is a transitive Anosov flow.*

For a locally potential vector field $divE = -\triangle U$ and we get

Corollary 4.5. *If $K < -\triangle U$ on a 2-dimensional surface M then the W-flow is a transitive Anosov flow.*

Corollary 4.6. *If the local potential is harmonic and the Gaussian curvature $K < 0$ on M then the W-flow is a transitive Anosov flow.*

We conclude that in the case of fields given by automorphic forms on surfaces of constant negative curvature, which were studied by Bonetto, Gentile and Mastropietro, [BGM00], the flow is always Anosov. Further it follows from the theory of SRB measures that if such a flow is Anosov then it is also automatically dissipative, i.e., the SRB measure is singular, [W00'].

Note that in this situation we can multiply the vector field E by an arbitrary scalar λ and we still get a transitive Anosov flow. It would be interesting to understand the asymptotics of the SRB measure as $\lambda \to \infty$. Is the limit supported on the union of the integral curves of E? Let us stress that this scenario differs from the perturbative conditions in [Go97], [Gr99], [W00'], where the geodesic curvature of the trajectories cannot be too large. Our trajectories may have arbitrarily large geodesic curvatures and yet they form a transitive Anosov flow.

5. Examples, extensions, open problems and disappointments

A. In view of Theorems 4.2 and 4.4 it is natural to ask, if the negative sectional Weyl curvature is enough to guarantee the Anosov property for the W-flow. It follows immediately from (4.1) and (3.5) that if for every plane Π the sectional Weyl curvatures satisfy

$$\widehat{K}(\Pi) + \frac{1}{4}E_\Pi^2 = K(\Pi) - div_\Pi E - E^2 + \frac{5}{4}E_\Pi^2 < 0,$$

then the W-flow is Anosov. We propose

Conjecture 5.1. *There are manifolds of dimension ≥ 3 and tangent vector fields E such that the Weyl sectional curvatures are negative everywhere but the W-flow is not Anosov.*

It is also plausible that under the assumption of negative sectional curvatures we can obtain W-flows which are nontransitive Anosov flows, as in the examples of Franks and Williams, [FW80].

To resolve these issues we would like to construct examples of Weyl manifolds with negative sectional curvatures which are not small deformations of riemannian metrics of negative sectional curvature. In that direction we found some obstructions.

Proposition 5.2. *There are no Weyl structures with negative sectional curvatures in a small neighborhood of the homogeneous Weyl structure on an n-dimensional torus (as in Example 1.2).*

Conjecture 5.3. *There are no Weyl structures of negative sectional curvature on n-dimensional tori.*

It is so for $n = 2$ since for 2-dimensional manifolds the Weyl curvature is equal to the gaussian curvature of the Gauduchon gauge.

The presence of a negative term in the formula for the Weyl sectional curvature (3.5) gives the impression that it is easier to find manifolds with negative Weyl curvature than with negative riemannian curvature. The following two observations suggest that it is not necessarily so.

If $(M_i, g_i, E_i), i = 1, 2$ are two Weyl manifolds then their cartesian product has a natural Weyl structure, and the Weyl sectional curvature in the direction of the plane spanned by $(E_1, 0)$ and $(0, E_2)$ is either mixed (positive, negative and zero) or always zero (iff $|E_i| = const, i = 1, 2$). Hence just like in the riemannian case, product manifolds cannot have negative sectional curvature.

Secondly, if we look for interesting homogeneous Weyl structures we are confronted with the phenomenon that on symmetric spaces of noncompact type, [H78], there are no homogeneous Weyl structures at all (except for the riemannian metric itself). The only simply connected homogeneous riemannian manifolds with a compact factor and nonpositive sectional curvature are symmetric spaces,[AW76].

Conjecture 5.4. *The only simply connected homogeneous Weyl manifolds with a compact factor and negative Weyl sectional curvatures are riemannian symmetric spaces.*

We obtain a homogeneous Weyl manifold by taking a left invariant metric g on a Lie group and a left invariant vector field E. It is known that on unimodular Lie groups the only left invariant metrics with nonpositive riemannian sectional curvature are metrics with zero curvature, [AW76],[M76]. In contrast, by (3.5) the n-dimensional torus, $n \geq 3$, with a constant vector field E has negative Weyl sectional curvature in the direction of any plane except for the planes that contain E, where it vanishes.

Another instructive example is the 3-dimensional Lie group SOL,[T98], which has mixed riemannian sectional curvatures, but for one of the left invariant fields the Weyl sectional curvature is nonpositive, with some negative curvature.

A weaker version of Conjecture 5.4 is

Conjecture 5.5. *There are no left invariant Weyl structures with negative Weyl sectional curvatures on unimodular Lie groups.*

B. Billiard systems are a natural extension of geodesic flows. Similarly we can consider billard W-flows on Weyl manifolds with boundaries by augmenting the continuous time dynamics with elastic reflections in the boundaries. For example we can remove from a Weyl manifold some subsets ("obstacles"). In case of convex obstacles in a cube (or a flat torus), we obtain Sinai billiards, which have good statistical properties.

We can introduce Weyl strict convexity of the obstacles by requiring that the Weyl geodesics in the exterior of the obstacle can have locally at most one point in common with the obstacle. It can be calculated that this property has the following infinitesimal description. Let N be the field of unit vectors orthogonal to the obstacle and pointing out of it. The riemannian convexity of the obstacle at a point is defined by the positive definiteness of the riemannian shape operator $K\xi_0 = \nabla_{\xi_0} N$, where ξ_0 is from the tangent subspace to the obstacle. Similarly we introduce the operator

$$\widehat{K}\xi_0 = K\xi_0 + \langle N, E \rangle \xi_0, \tag{5.6}$$

which is the orthogonal projection of the " Weyl shape operator" $\widehat{\nabla}_{\xi_0} N$ to the tangent subspace. An obstacle is (strictly) Weyl convex if $\widehat{K}$ is positive (definite) semidefinite.

Assuming that the obstacles are Weyl convex we obtain hyperbolic properties of the billiard W-flows, [W00], in parallel with Section 4.

Two dimensional Lorentz gas with round scatterers of radius r, a constant electric field E and the Gaussian thermostat is a model of this kind. It was studied numerically by Moran and Hoover, [MH87]. Chernov et al,[CELS92], obtained exaustive rigorous results about its SRB measures in the case of small fields and finite horizon. We can prove the uniform hyperbolicity of the model when $r|E| < 1$

and this inequality is sharp. Indeed, the exponential map $F(z) = e^{|E|z}$, takes the trajectories of the W-flow onto the straight lines and being conformal it respects the reflections from the boundary. Hence the mapping F takes the billiard W-flow into a billiard flow, at least locally (globally we get billiards on the Riemann surface of the logarithm). However the image of a disk of radius r under the exponential mapping is strictly convex if and only if $r|E| < 1$. Once the obstacles loose convexity we readily find elliptic periodic orbits which rules out global hyperbolicity, [W00].

C. For obscure reasons in hamiltonian systems of many particles interacting by a pair potential, which are expected to have in general good statistical properties (and do have them in numerical experiments), hyperbolicity in all of the phase space is rarely encountered. The notable exceptions are the Boltzmann-Sinai gas of hard spheres,[S00], and also the one dimensional systems of falling balls, [W98],[W99]. In particular, beyond the 2 dimensional examples of Knauf, [K88],(which have Weyl counterparts, [W00]), we do not know of systems equivalent to geodesic flows on manifolds with negative sectional curvatures.

For systems of particles in an external field the Gaussian thermostat provides additional interactions and the resulting system is not hamiltonian. Examining the simplest examples we find that also in this case the zero and positive Weyl sectional curvatures are common.

For noninteracting particles we get a cartesian product of Weyl manifolds and hence we get zero sectional curvature in some directions, as in A.

When we consider the system of two elastic disks in the 2-dimensional torus, the cylinders that are cut out from the four dimensional configuration space are Weyl convex, but the zero Weyl curvature is enough to allow the presence of local simple attractors of the type of Example 1.2, which rules out global fast mixing and decay of correlations.

For the Lorentz gas of two noninteracting point particles in a 2 dimensional torus with round scaterrers, an external field and the Gaussian thermostat, the calculation of (5.6) shows that the obstacles (products of disks) in the four dimensional torus of configurations are not Weyl convex everywhere.

These examples indicate that in the setup of Weyl geometry there is no more freedom for the occurence of global hyperbolicity then in the riemannian realm. The difficulty in constructing natural examples satisfying the Chaotic Hypothesis of Gallavotti,[G01], seems to be parallel to the scarcity of multidimensional hamiltonian systems with strong mixing properties, [L00].

References

[A89]   V.I. Arnold, *Mathematical Methods of Classical Mechanics*, Springer, 1989.

[AW76]   R. Azencott, E.N. Wilson, *Homogeneous manifolds with negative curvature,I*, Trans. AMS **215** (1976), 323–362.

[BCP98]   F. Bonetto, E.G.D. Cohen, C. Pugh, *On the validity of the conjugate pairing rule for Lyapunov exponents*, J. Statist.Phys. **92** (1998), 587–627.

[BGM00] F. Bonetto, G. Gentile, V. Mastropietro, *Electric fields on a surface of constant negative curvature*, Erg. Th. Dynam. Sys. **20** (2000), 681–696.

[CELS93] N.I. Chernov, G.L. Eyink, J.L. Lebowitz, Ya.G. Sinai, *Steady-state electric conduction in the periodic Lorentz gas*, Commun. Math. Phys. **154** (1993), 569–601.

[Ch98] Ph. Choquard, *Variational principles for thermostated systems*, Chaos **8** (1998), 350–356.

[DM96] C.P. Dettmann, G.P. Morriss, *Hamiltonian formulation of the Gaussian isokinetic thermostat*, Phys. Rev. E **54** (1996), 2495–2500.

[FW80] J.M. Franks, R. Williams, *Anomalous Anosov flows*, Lecture Notes in Math. **819** (1980), 158–174.

[F70] G. B. Folland, *Weyl manifolds*, J. Diff. Geom. **4** (1970), 145–153.

[G99] G. Gallavotti, *Statistical Mechanics*, Springer, 1999.

[G99'] G. Gallavotti, *New methods in nonequilibrium gases and fluids*, Open Sys. Information Dynamics **6** (1999), 101–136.

[GR97] G. Gallavotti, D. Ruelle, *SRB states and nonequilibrium statistical mechanics close to equilibrium*, Commun. Math. Phys. **190** (1997), 279–285.

[Ga84] P. Gauduchon, *La 1-forme de torsion d'une variété hermitienne compacte*, Math. Ann **267** (1984), 495–518.

[Go97] N. Gouda, *Magnetic flows of Anosov type*, Tohoku Math. J. **49** (1997), 165–183.

[Gr99] S. Grognet, *Flots magnetiques en courbure negative*, Erg. Th. Dyn. Syst. **19** (1999), 413–436.

[H78] S. Helgason, *Differential Geometry, Lie Groups and Symmetric Spaces*, Academic Press, 1978.

[K87] A. Knauf, *Ergodic and topological properties of Coulombic potentials*, Commun. Math. Phys. **110** (1987), 89–112.

[L00] C. Liverani, *Interacting particles*, Encyclopedia of Math.Sci. **101** (2000), Springer.

[M̃84] R. Mañé, *Oseledec's theorem from the generic viewpoint*, Proceedings of ICM, Warsaw, 1983 (1984), PWN, Warsaw, 1269–1276.

[Mi86] J. Milnor, *Curvatures of left invariant metrics on Lie groups*, Adv. Math. **21** (1976), 293–329.

[MH87] B. Moran, W.G. Hoover, *Diffusion in periodic Lorentz gas*, J. Stat. Phys. **48** (1987), 709–726.

[R99] D. Ruelle, *Smooth dynamics and new theoretical ideas in nonequilibrium statistical mechanics*, J. Stat.Phys. **95** (1999), 393–468.

[Se98] M.B. Sevryuk, *Finite-dimensional reversible KAM theory*, Physica D **112** (1998), 132–147.

[S00] N. Simányi, *Hard ball systems and semi-dispersive billiards: hypeborlicity and ergodicity*, Encyclopedia of Math.Sci. **101** (2000), Springer, 51–88.

[T98] M. Troyanov, *L'horizon de Sol*, Expo.Math **16** (1998), 441–480.

[W98]   M.P. Wojtkowski, *Hamiltonian systems with linear potential and elastic constraints*, Fundamenta Mathematicae **157** (1998), 305–341.

[W99]   M.P. Wojtkowski, *Complete hyperbolicity in hamiltonian systems with linear potential and elastic collisions*, Rep. Math. Phys. **44** (1999), 301–312.

[W00]   M.P. Wojtkowski, *W-flows on Weyl manifolds and gaussian thermostats*, J. Math. Pures. Appl **79** (2000), 953–974.

[W00']   M.P. Wojtkowski, *Magnetic flows and gaussian thermostats on manifolds of negative curvature*, Fundamenta Mathematicae **163** (2000), 177–191.

[W01]   M.P. Wojtkowski, *Monotonicity, $\mathcal{J}$- algebra of Potapov and Lyapunov exponents*, Proc. Sympos. Pure Math. **169** (2001), 499–521.

[WL98]   M.P. Wojtkowski, C. Liverani, *Conformally symplectic dynamics and symmetry of the Lyapunov spectrum*, Commun. Math. Phys. **194** (1998), 47–60.

Section 14. Combinatorics

ICM 2002 · Vol. III · 527–535

Random Points, Convex Bodies, Lattices

Imre Bárány*

Abstract

Assume K is a convex body in R^d, and X is a (large) finite subset of K. How many convex polytopes are there whose vertices come from X? What is the typical shape of such a polytope? How well the largest such polytope (which is actually conv X) approximates K? We are interested in these questions mainly in two cases. The first is when X is a random sample of n uniform, independent points from K and is motivated by Sylvester's four-point problem, and by the theory of random polytopes. The second case is when $X = K \cap Z^d$ where Z^d is the lattice of integer points in R^d. Motivation comes from integer programming and geometry of numbers. The two cases behave quite similarly.

2000 Mathematics Subject Classification: 52A22, 05A16, 52C07.
Keywords and Phrases: Convex bodies, Lattices, Random samples, Convex polytopes, Limit shape.

1. Sylvester's question

In the 1864 April issue of the Educational Times J. J. Sylvester [26] posed the innocent looking question that read: *"Show that the chance of four points forming the apices of a reentrant quadrilateral is 1/4 if they be taken at random in an indefinite plane."* It was understood within a year that the question is ill-posed. (The culprit is, as we all know by now, the "indefinite plane" without a properly defined probability measure on it.) So Sylvester modified the question: let $K \subset R^2$ be a convex body (that is, a compact, convex set with nonempty interior) and choose four random, independent points uniformly from K, and write $P(K)$ for the probability that the four points form the apices of a reentrant quadrilateral, or, in more modern terminology, that their convex hull is a triangle. How large is $P(K)$, and for what K is $P(K)$ the largest and the smallest? This question became known as Sylvester's four-point problem. It took fifty years to find the answer: Blaschke [16] showed that for all convex bodies $K \subset R^2$

$$P(\text{disk}) \le P(K) \le P(\text{triangle}).$$

---

*Rényi Institute of Mathematics, Hungarian Academy of Sciences, PoB 127, Budapest 1364, Hungary, and Department of Mathematics, University College London, Gower Street, London WC1E 6BT, UK. E-mail: barany@renyi.hu and barany@math.ucl.ac.uk

Assume now, more generally, that $X_n = \{x_1, \ldots, x_n\}$ is a random sample of n uniform, independent points from the convex body K and write $p(n, K)$ for the probability that X_n is in *convex position*, that is, no x_i is in the convex hull of the others. Sylvester's question is just the complementary problem for $n = 4$: $P(K) = 1 - p(4, K)$. The probability $p(n, K)$ has been determined in various special cases (see [22, 17, 9, 27]). The following result from [7] describes the asymptotic behaviour of $p(n, K)$.

Theorem 1.1. *For every convex body $K \subset R^2$ of unit area*

$$\lim_{n \to \infty} n^2 \sqrt[n]{p(n, K)} = \frac{e^2}{4} A^3(K)$$

where $A(K)$ is the supremum of the affine perimeter of all convex sets $S \subset K$.

The affine perimeter, $AP(K)$ can be defined in many ways (see [23]), for instance $AP(K) = \int_{\partial K} \kappa^{1/3} ds$ where κ is the curvature and integration goes by arc-length. (This definition works for smooth convex bodies, the extension for all convex bodies can be found in [23].) Theorem 1 of [6] says that there is a unique convex compact set $K_0 \subset K$ with $AP(K_0) = A(K)$. The proof of Theorem 1.1 gives more than just the asymptotic behaviour of $p(n, K)$, namely, if the random points $x_1, \ldots, x_n$ are in convex position, then their convex hull is, with high probability, very close to K_0. For the precise formulation see [7].

Define $Q(X_n)$ as the collection of all convex polygons spanned by the points of X_n, that is, $P \in Q(X_n)$ iff $P = \text{conv}\{x_{i_1}, \ldots, x_{i_k}\}$ for some k-tuple of points from X_n that is in convex position ($k \geq 3$). Clearly, $Q(X_n)$ is a random collection as it depends on the random sample X_n. How many polygons are there in $Q(X_n)$? The answer is given in [7]. Write $E|Q(X_n)|$ for the expectation of the size of $Q(X_n)$.

Theorem 1.2. *For every convex body $K \subset R^2$ of unit area*

$$\lim_{n \to \infty} n^{-1/3} \log E|Q(X_n)| = 3 \cdot 2^{-2/3} A(K).$$

Further, there is a limit shape to the polygons in $Q(X_n)$, meaning that all but a small fraction of the polygons in $Q(X_n)$ are very close to K_0. We use $\delta(S, T)$ to denote the Haussdorf distance of $S, T \subset R^2$.

Theorem 1.3. *For every convex body $K \subset R^2$ and for every $\varepsilon > 0$*

$$\lim_{n \to \infty} \frac{E|\{P \in Q(X_n) : \delta(P, K_0) > \varepsilon\}|}{E|Q(X_n)|} = 0.$$

We will see in Section 3 that similar phenomena hold for the lattice case. In general, lattice points and random points, in relation to convex bodies, behave very much alike. Quite often one understands in the random case what to expect for lattice points, or the other way around. The proofs are quite different and are omitted in this survey.

2. Higher dimensions

Much less is known in higher dimensions. One reason is that the unicity of the convex subset of K with maximal affine surface area is not known. It is a mystery, for instance, which convex subset of the unit cube in R^3 has maximal affine surface area. But there are other reasons as well, connected to the lack of the multiplicative rule (5.3) from [7]. Yet one can prove the following asymptotic formula [8]. Here $\mathcal{C}^d$ denotes the set of all convex bodies in R^d, and $p(n, K)$ denotes, as before, the probability that the random sample $X_n = \{x_1, \ldots, x_n\}$ from K is in convex position, that is, no x_i is in the convex hull of the others.

Theorem 2.1. *For every $K \in \mathcal{C}^d$, and for all $n \geq n_0$*

$$c_1 < n^{2/(d-1)} \sqrt[n]{p(n, K)} < c_2$$

where n_0, c_1, c_2 are positive constants that depend only on d.

With Vinogradov's convenient $\ll$ or $\ll_d$ notation this says that

$$1 \ll_d n^{2/(d-1)} \sqrt[n]{p(n, K)} \ll_d 1.$$

From this one can estimate the size of $E|Q(X_n)|$ when $K \in \mathcal{C}^d$:

$$n^{(d-1)/(d+1)} \ll_d \log E|Q(X_n)| \ll_d n^{(d-1)/(d+1)}.$$

For comparison let us have a look at lattice polytopes contained in some fixed $K \subset \mathcal{C}^d$. So let Z^d be the lattice of the integers in R^d and consider, for a large integer m, the lattice $\frac{1}{m}Z^d$. Assume K contains n points from this lattice. As m is large, $n = (1 + o(1))m^d \operatorname{Vol} K$. Write $\mathcal{P}_m(K)$ for the collection of all $\frac{1}{m}Z^d$-lattice polytopes contained in K. The next theorem, which follows easily from the results of [12], shows a very strong analogy between $\mathcal{P}_m(K)$ and $Q(X_n)$.

Theorem 2.2. *For every $K \in \mathcal{C}^d$*

$$n^{(d-1)/(d+1)} \ll_d \log |\mathcal{P}_m(K)| \ll_d n^{(d-1)/(d+1)}.$$

The result shows that when $K \in \mathcal{C}^d$ contains n lattice points, these lattice points span (essentially) $\exp\{cn^{(d-1)/(d+1)}\}$ convex polytopes, the same number as in the random case. Lattice points and random points in convex bodies behave similarly: this is the moral.

3. Lattice polygons and limit shape

In the plane Theorem 2.2 can be proved in stronger form (see [5], [6], and [28]):

Theorem 3.1. *For every $K \in \mathcal{C}^2$*

$$\lim n^{-2/3} \log |\mathcal{P}_n(K)| = 3 \sqrt[3]{\frac{\zeta(3)}{4\zeta(2)}} A(K).$$

Here $\zeta(.)$ stands for Riemann ζ function. Note that this result is in complete analogy with Theorem 1.2: just the constant is different. (Also, n is in power $-2/3$ instead of $-1/3$ as K contains $(1 + o(1))n^2 \operatorname{Area} K$ lattice points.) The analogy carries over to Theorem 1.3 as well:

Theorem 3.2. *For every convex body $K \in \mathcal{C}^2$ and for every $\varepsilon > 0$*

$$\lim \frac{|\{P \in \mathcal{P}_n(K)) \colon \delta(P, K_0) > \varepsilon\}|}{|\mathcal{P}_n(K)|} = 0.$$

This shows again that all but a tiny fraction of the polygons in $\mathcal{P}_n(K)$ are very close to K_0. In other words, these polygons have a limit shape. Theorems of this type were first proved by Bárány [5], Vershik [28] (for the case when K is the unit square). Sinai [25] found a different proof which uses probability theory and gives a central limit theorems about how small that tiny fraction of polygons is. This has been generalized by Vershik and Zeitouni [29] to all convex bodies in R^2. A central limit theorem of this type holds for the random sample case as well, see [14] for the precise statement.

4. The integer convex hull

The *integer convex hull*, $I(K)$, of a convex body $K \in \mathcal{C}^d$ is, by definition, the convex hull of the lattice points contained in K:

$$I(K) = \operatorname{conv}(Z^d \cap K).$$

$I(K)$ is clearly a convex polytope. How many vertices does it have? Motivation for the question comes from integer programming, classical enumeration questions (like the circle problem), and from the theory of random polytopes. In integer programming one wants to know that $I(K)$ does not have too many vertices, assuming, say, that K is a nice rational polytope. The latter means that K can be given by m inequalities with integral coefficients; the size of such an inequality is the number of bits necessary to encode it as a binary string. Then the size of the rational polytope is the sum of the sizes of the defining inequalities. Strengthening earlier results by Shevchenko [24], and Hayes and Larman [21], Cook, Hartman, Kannan, and McDiarmid [18] showed that for a rational polytope K of size ϕ

$$f_0(I(K)) \leq 2m^d (12d^2\phi)^{d-1}.$$

Here, as usual, $f_i(P)$ stands for the number of i-dimensional faces of the polytope P. Thus $f_0(P)$ is the number of vertices of P. Most likely, the same inequality holds for all $i = 0, 1, \dots, d-1$:

$$f_i(I(K)) \ll \phi^{d-1}$$

where the implied constant depends on d and m as well.

The above inequality for $f_0(I(K))$ is best possible, as is shown Bárány, Howe, and Lovász in [11]:

Theorem 4.1. *For fixed $d \geq 2$ and for every $\phi > 0$ there exists a rational simplex $P \subset R^d$ of size at most ϕ such that $I(P)$ has $\gg_d \phi^{d-1}$ vertices.*

The construction uses algebraic number theory. It shows further that the estimate $f_i(I(K)) \ll \phi^{d-1}$ for all i is best possible, if true.

What about the integer convex hull of other convex bodies? Balog and Bárány [3] considered case $K = rB^2$ where B^2 is the Euclidean unit ball centered at the origin and r is large and showed that

$$0.3r^{2/3} < f_0(I(rB^2)) < 5.5r^{2/3}.$$

Later Balog and Deshoullier [4] determined the average of $f_0(I(rB^2))$ on an interval $[R, R + H]$ which turned out to be very close to $3.453R^{2/3}$ as R goes to infinity (H has to be large). Bárány and Larman [13] determined the order of magnitude of $f_i(I(rB^d))$. (The method, and the result, apply not only to the unit ball but to smooth enough convex bodies as well.)

Theorem 4.2. *For every $d \geq 2$ and every $i = 0, 1, \ldots, d - 1$*

$$r^{d(d-1)/(d+1)} \ll_d f_i(I(rB^d)) \ll_d r^{d(d-1)/(d+1)}.$$

This result is related to a beautiful theorem of G. E. Andrews [1] stating that a lattice polytope P in R^d with volume $V > 0$ has $\ll_d V^{(d-1)/(d+1)}$ vertices. The above theorem shows that Andrews' estimate is best possible (apart from the constant implied by $\ll_d$). A similar (perhaps less compact) example was given earlier V. I. Arnol'd [2].

This kind of question can be considered in a more general setting. Let G be the group of all isometries of R^d with translations by elements of Z^d factored out. G is a compact topological group with a Haar measure which is a unique invariant probability measure when normalized properly. Assume $g \in G$ is chosen according to this probability measure. Then gK is a random copy of K and we can talk about the expectation of the random variable $f_0(I(gK))$.

For the next result we assume $K \in C^d$ and define the function $u \colon K \to R$ by

$$u(x) = \text{Vol}(K \cap (2x - K)),$$

that is, $u(x)$ is the volume of the intersection of K with K reflected about x. Set, finally, $K(u < t) = \{x \in K \colon u(x) < t\}$. The following is an unpublished result of Bárány and Matoušek:

Theorem 4.3. *Consider all $K \in C^d$ with the ratio of the radii of the smallest circumscribed and the largest inscribed balls to K bounded by D. Then, as $\text{Vol} K$ goes to infinity,*

$$\text{Vol} K(u < 1) \ll E f_0(I(gK)) \ll \text{Vol} K(u < 1)$$

where the constants implied by $\ll$ depend only on d and D.

It follows easily from Minkowski's classical theorem that all vertices of $I(K)$ belong to $K(u < 2^d)$. (This is the first step in proving the upper bound.) It is not hard to see that $\operatorname{Vol} K(u < 2^d) \ll \operatorname{Vol} K(u < 1)$. So the meaning of the theorem is that the average number of vertices of $I(gK)$ is essentially the volume of $K(u < 2^d)$. Probably the same is true for the expected number of i-dimensional faces of $I(gK)$ but there is no proof in sight.

The behaviour of $\operatorname{Vol} K(u < 1)$ is more or less known (from [10], say, but more precise results are known as well): it is of order $(\operatorname{Vol} K)^{(d-1)/(d+1)}$ for smooth enough convex bodies and of order $(\log \operatorname{Vol} K)^{d-1}$ for polytopes, and it is between these bounds for all convex bodies.

We mention further that Theorem 4.3 is quite analogous to a result in the theory of random polytopes. Given $K \in \mathcal{C}^d$, and a random sample of n points, X_n, from K, $K_n = \operatorname{conv} X_n$ is called a *random polytope* on n points. It is shown in [10] that, assuming $\operatorname{Vol} K = n$ (which is the proper scaling for comparison with Theorem 4.3), for all $i = 0, 1, \ldots, d-1$

$$\operatorname{Vol} K(u < 1) \ll E f_i(K_n) \ll \operatorname{Vol} K(u < 1)$$

where the implied constants depend only on dimension.

Note that, unlike Theorem 4.3, this result works for all $i = 0, \ldots, d-1$ (without any condition on the ratio of radii of the circumscribed and inscribed balls). Most likely, Theorem 4.3 also holds for all i, which would make the analogy even more complete.

There is, however, a point here where the analogy breaks down. Let $K \subset R^2$ be the square of area n, so K_n is a random polytope, and $I(gK)$ is the integer hull of a random copy of K. The expectation of $\operatorname{Area}(K \setminus K_n)$ is of order $\log n$ (see [10], say), while the expectation of $\operatorname{Area}(K \setminus I(gK))$ is of order $(\log n)^2$. (The latter result comes again from the unpublished work of Bárány and Matoušek.) The reason is that the boundary of K_n contains no points from X_n apart from its vertices, while the boundary of $I(gK)$ does. A further reason is that what we are measuring here is a metric property, and not a combinatorial one. We think that the same phenomena is bound to happen in higher dimension.

5. Random 0-1 polytopes

Finally we mention a recent development, prompted by a question of K. Fukuda and G. M. Ziegler [30]. They asked how many facets a 0-1 polytope in R^d can have; a 0-1 polytope is a polytope whose vertices only have 0 or 1 coordinates. So such a polytope is the convex hull of a subset of the vertices of the unit cube, Q^d, in R^d. 0-1 polytopes play an important role in combinatorial optimization where the target is, very often, a concise description of the facets of the polytope. This task has turned out to be difficult for several classes of 0-1 polytopes.

Write $G(d)$ for the maximal number of facets a 0-1 polytope can have. It is not hard to see that $2^d \leq G(d) \leq 2d!$. The upper and lower bounds have been improved slightly: the lower bound by a construction of Christoff (see [30]), and the upper bound by Fleiner, Kaibel, and Rote [20].

The vertices of every 0-1 polytope are on a sphere (centered at $(1/2, \ldots, 1/2)$). There is a formula (see for instance [9]) for the expected number of facets of a random polytope with n uniform independent points from the (unit) sphere in R^d. It says that, in the range when $2d < n < 2^d$, the expected number of facets is of order $(\log n/d)^{d/2}$. So if the analogy between random points and lattice points carries over the 0-1 case one should expect $G(d)$ to be of order $d^{d/2}$. This is too much to ask for at the moment, yet the following is true (see [15]).

Theorem 5.1. *There is a constant $c > 0$ such that for all $d \geq 2$*

$$G(d) \gg \left(c \frac{d}{\log d} \right)^{d/4}.$$

The construction giving this estimate is random. Write K_n for the convex hull of n random, uniform, and independent 0-1 vectors. Assume x is a point from Q^d, and define

$$p(x, n) = \text{Prob}[x \in K_n].$$

General principles would tell that, for most $x \in Q^d$, $p(x, n)$ is either close to one or close to zero. To be more specific, set

$$P(t) = \{ x \in Q^d : p(x, n) \geq t \}.$$

The proof of Theorem 5.1 is based on the fact that for all small $\varepsilon > 0$ and large enough d $P(1 - \varepsilon) \subset P(\varepsilon)$, of course, but the drop from $1 - \varepsilon$ to ε is very abrupt: $P(\varepsilon)$ is in a small neighbourhood of $P(1 - \varepsilon)$. This shows that $P(1 - \varepsilon) \subset K_n$ with high probability. But only a tiny fraction of K_n lies outside $P(\varepsilon)$: most of the boundary of $P(\varepsilon)$ is outside K_n. Thus most of the boundary of $P(\varepsilon)$ is cut off by facets of K_n. These facets lie outside $P(1 - \varepsilon)$. Comparing the surface area of $P(\varepsilon)$ with the amount a facet can cut off from it gives the lower bound.

The actual proof is technical, difficult, and makes extensive use a beautiful result of Dyer, Füredi, and McDiarmid [19]. Their target was to determine the threshold $n = n(d)$ such that K_n contains most of the volume of Q^d. As they prove, this happens at $n = (2/\sqrt{e})^d$. Their method describes where $p(x, n)$ drops from one to zero as $d \to \infty$. The analysis carries over for other values of n. In our case higher precision is required as we need a good estimate on how fast $p(x, n)$ drops from one to zero. We were able to control this only where the curvature of the boundary of $P(\varepsilon)$ behaves nicely. This is perhaps the spot where the exponent $d/2$ (for the random spherical polytope) is lost and we only get $d/4$ for K_n.

References

[1]   G. E. Andrews, A lower bound for the volume of strictly convex bodies with many boundary points, *Trans. Amer. Math. Society,* 106 (1963), 270–279.

[2]   V. I. Arnol'd, Statistics of integral lattice polytopes (in Russian), *Funk. Anal. Pril,* 14 (1980), 1–3.

[3]   A. Balog, I. Bárány, On the convex hull of the integer points in a disc, *Discrete Comp. Geometry,* 6 (1992), 39–44.

[4]   A. Balog, J-M. Deshouliers, On some convex lattice polytopes, in: *Number theory in progress,* (K. Győry, ed.), de Gruyter, 1999, 591–606.

[5]   I. Bárány, The limit shape of convex lattice polygons, *Discrete Comp. Geometry,* 13 (1995), 270–295.

[6]   I. Bárány, Affine perimeter and limit shape, *J. Reine Ang. Mathematik,* 484 (1997), 71–84.

[7]   I. Bárány, Sylvester's question: the probability that n points are in convex position, *Annals of Probability,* 27 (1999), 2020–2034.

[8]   I. Bárány, A note on Sylvester's four point problem, *Studia Math. Hungarica,* 38 (2001), 73–77.

[9]   I. Bárány, Z. Füredi, On the shape of the convex hull of random points, *Prob. Theory Rel. Fields,* 77 (1988), 231–240.

[10]   I. Bárány, D. G. Larman, Convex bodies, economic cap coverings, random polytopes, *Mathematika,* 35 (1988), 274–291.

[11]   I. Bárány, R. Howe. L. Lovász, On integer points in polyhedra: a lower bound, *Combinatorica,* 12 (1992), 135–142.

[12]   I. Bárány, A. M. Vershik, On the number of convex lattice polytopes, *GAFA Journal,* 2 (1992), 381–393.

[13]   I. Bárány, D. G. Larman, The convex hull of the integer points in a large ball, *Math. Annalen,* 312 (1998), 167–181.

[14]   I. Bárány, G. Rote, W. Steiger, C. Zhang, A central limit theorem for random convex chains, *Discrete Comp. Geometry,* 23 (2000), 35–50.

[15]   I. Bárány, A. Pór, On 0-1 polytopes with many facets, *Advances in Math.,* (2000), 1–28.

[16]   W. Blaschke, *Vorlesungen über Differenzialgeometrie II. Affine Differenzialgeometrie,* Springer, 1923.

[17]   C. Buchta, On a conjecture of R.E. Miles about the convex hull of random points, *Monatsh. Math.,* 102 (1986), 91–102.

[18]   W. Cook, M. Hartman, R. Kannan, C. McDiarmid, On integer points in polyhedra, *Combinatorica,* 12 (1992), 27–37.

[19]   M. E. Dyer, Z. Füredi, C. McDiarmid, Volumes spanned by random points in the hypercube, *Random Structures and Algorithms,* 3 (1992), 91–106.

[20]   T. Fleiner, V. Kaibel, G. Rote:, Upper bounds on the maximal number of facets of 0/1-polytopes, *European J. Combinatorics,* 21 (2000), 121–130.

[21] A. C. Hayes and D. G. Larman, The vertices of the knapsack polytope, *Discrete App. Math.*, 6 (1983), 135–138.

[22] B. Hostinsky, Sur les probabilités géométriques, *Publ. Fac. Sci. Univ. Brno,* (1925).

[23] E. Lutwak, Extended affine surface area, *Adv. Math.*, 85 (1991), 39–68.

[24] V. N. Shevchenko, On the number of extreme points in linear programming (in Russian), *Kibernetika*, 2 (1981), 133–134.

[25] Ya. G. Sinai, Probabilistic approach to analyze the statistics of convex polygonal curves (in Russian), *Funk. Anal. Pril.*, 28 (1994), 41–48.

[26] J. J. Sylvester, Problem 1491, *The Educational Times, (London)* (April 1864), 1–28.

[27] P. Valtr, The probability that n random points in a triangle are in convex position, *Combinatorica*, 16 (1996), 567–574.

[28] A. M. Vershik, The limit shape for convex lattice polygons and related topics (in Russian), *Funk. Anal. Appl.*, 28 (1994), 16–25.

[29] A. M. Vershik, O. Zeitouni, Large deviations in the geometry of convex lattice polygons, *Israel J. Math.*, 109 (1999), 13–28.

[30] G. M. Ziegler, Lectures on 0/1 polytopes, in: *Polytopes—Combinatorics and Computation* (G. Kalai and G. M. Ziegler, eds.), DMV-Seminars,, Birkhäuser-Verlag, 2000, 1–44.

ICM 2002 · Vol. III · 537–545

Combinatorial Problems in Finite Geometry and Lacunary Polynomials

Aart Blokhuis*

Abstract

We describe some combinatorial problems in finite projective planes and indicate how Rédei's theory of lacunary polynomials can be applied to them.

2000 Mathematics Subject Classification: 05.

1. Introduction

In 1991 I wrote a survey paper called Extremal Problems in Finite Geometries [7]. It concerns among others problems of the following type:

Given a set B of points in a finite projective plane Π, with the property that there is a restricted number of possibilities for the size of the intersection of a line with B. What can be concluded about the size and the structure of B.

The archetypal result is Segre's theorem [27]:

If $\Pi = PG(2,q)$, q odd, and B has at most two points on a line, then $|B| \leq q+1$ (this part is easy), and in case of equality B consists of the points of a conic.

The problem becomes much more difficult when larger intersections are allowed. A subset B of $PG(2,q)$ of size k having at most n points on a line, is called a (k,n)-arc.

A simple counting argument going back to Barlotti [5] gives that if B is a (k,n)-arc, then $k \leq (n-1)(q+1)+1 = nq - q + n$, and equality implies that $n|q$ and all lines intersect B in 0 or n points.

An arc B meeting the above upper bound is called a *maximal arc*. The first non trivial case is $n = 2$, and q is even. In this case $|B| = q+2$ is possible and the maximal arc is called a hyperoval. In fact every $(q+1)$-arc can be extended to a hyperoval by adding one point and classifying or trying to find new hyperovals is one of the very active areas in finite geometry. For a survey on the current situation we refer to the two papers by Hirschfeld and Storme [20, 21].

---

*Eindhoven University of Technology, P.O. Box 513, 5600 MB, Eindhoven, Netherlands. E-mail: aartb@win.tue.nl

For $n = 3$ non-existence of maximal arcs in $PG(2,9)$ was shown by Cossu [16], and later by Thas for general q [29]. Very little is known however in this case. In fact, if $k(q)$ stands for the maximal size of a $(k,3)$-arc in $PG(2,q)$ then it is unknown whether limsup $k(q)/q > 2$ or liminf $k(q) < 3$.

For $n = 4$, and in fact for all $n = 2^a$ and $q = 2^b$, $b > a$ examples are known, most of them due to Denniston [17] and Thas [30, 31].

For odd n it was conjectured in [29] that maximal arcs don't exist. This was finally proved by Ball, Blokhuis and Mazzocca in 1996 [1]. A simplified proof appeared a year later [2].

Now we turn to the case where B intersects every line in at least one point. Such a set is called a *blocking set*. It is called non-trivial if it does not contain a line. Here the classical result is due to Bruen [14]:

A non-trivial blocking set B in a projective plane of order q has size at least $q + \sqrt{q} + 1$, with equality if and only if q is a square, and B is a Baer subplane.

Our understanding of the situation when q is not a square has increased dramatically in the last 10 years, from knowing very little to more or less complete knowledge.

Starting point was the unexpectedly simple proof (in 1993) [8] that a non-trivial blocking set in $PG(2,p)$, p prime, has size at least $3(p+1)/2$.

The proof is based on properties of a certain kind of lacunary polynomial (as introduced and studied by Rédei in [26]) associated to the blocking set. The importance of Rédei's work in this area was realized soon after the appearance of his book in 1970 notably in papers by Bruen and Thas, [15], where his result on the number of directions determined by the graph of a function on a finite field was used.

In contrast to the case that B is an arc, we can still say something if we require that B intersects every line of the plane in at least t points, for some $t > 1$. If q is a square then a natural candidate for B is the union of t disjoint Baer subplanes. These can be found for all appropriate values of t because it is possible to partition $PG(2,q)$ in $q - \sqrt{q} + 1$ disjoint Baer subplanes.

Building on earlier work by Ball [3], Gács, Szőnyi [19] and others Blokhuis, Storme and Szőnyi showed that for $t < q^{1/6}$ a t-fold blocking set in $PG(2,q)$ has size at least $t(q + \sqrt{q} + 1)$ with equality if and only if B is the union of t disjoint Baer subplanes [6].

2. Directions

Let $f : GF(q) \to GF(q)$. Define the set of directions determined by f to be

$$D_f := \left\{ \frac{f(u) - f(v)}{u - v} \mid u, v \in GF(q), u \neq v \right\}.$$

We are interested in functions f for which the set D_f is small. If f is linear, then D_f just consists of the slope of the line defined by the graph of f. Our starting point will be the following important result of Rédei [26], p. 237, Satz 24.

Theorem 2.1 [Rédei, 1970] *Let* $f : GF(q) \to GF(q)$ *be a nonlinear function, where* $q = p^n$, p *prime. Then* $|D_f|$ *is contained in one of the intervals*

$$\left(1 + \frac{q-1}{p^e+1}, \frac{q-1}{p^e-1}\right), e = 1, \ldots, [n/2]; \left(\frac{q+1}{2}, q\right).$$

Examples of functions determining relatively few directions are given by:

1. $f(x) = x^{\frac{q+1}{2}}$, q odd, $|D_f| = (q+3)/2$;
2. $f(x) = x^{p^e}$, where $e|n$, $|D_f| = (q-1)/(p^e-1)$;
3. $f(x) = \mathrm{Tr}(GF(q) \to GF(p^e))(x)$, $|D_f| = (q/p^e) + 1$.

In all the examples $|D_f|$ is contained in one of the Rédei intervals corresponding to a subfield of $GF(q)$, i.e., $e|n$, and the obvious question was whether this could be proved. In [12] Blokhuis, Brouwer and Szőnyi slightly improved Rédei's result but the real progress came with the paper [13] where not only it was proved that only the intervals with $e|n$ occur, but the functions for which $|D_f| < (q+3)/2$ where essentially characterized:

Theorem 2.2 [Ball, Blokhuis, Brouwer, Storme, Szőnyi, 1999] *Let* $f : GF(q) \to GF(q)$, *where* $q = p^h$, p *prime,* $f(0) = 0$. *Let* $N = |D_f|$. *Let* e *(with* $0 \le e \le n$*) be the largest integer such that each line with slope in* D_f *meets the graph of* f *in a multiple of* p^e *points. Then we have one of the following:*

1. $e = 0$ *and* $(q+3)/2 \le N \le q+1$,
2. $e = 1$, $p = 2$ *and* $(q+5)/3 \le N \le q-1$,
3. $p^e > 2$, $e \mid n$, *and* $q/p^e + 1 \le N \le (q-1)/(p^e-1)$,
4. $e = n$ *and* $N = 1$.

Moreover, if $p^e > 3$ *or* $(p^e = 3$ *and* $N = q/3+1)$, *then* f *is a linear map on* $GF(q)$ *viewed as a vector space over* $GF(p^e)$.

Very recently this result has been perfected by Simeon Ball, removing the condition $p^e > 2$ in the third case (and thus getting rid of the second).

When we consider the set B formed by the q points of the graph of f together with the $N = |D_f|$ points on the line at infinity corresponding to the directions determined by f, we get a blocking set. For if l has a slope determined by f then the infinite point of l belongs to B, and if not, then l and its parallels all contain precisely one point of the graph of f.

Conversely, if B is a blocking set in $PG(2,q)$ of size $q + N$, and there is a line intersecting B in N points, then it arises from this construction. The blocking set is then called *of Rédei type*.

As mentioned in the introduction, the smallest non-trivial blocking sets were characterized by Bruen [14] to be Baer subplanes. They are of Rédei type and correspond to the function $x \mapsto x^{\sqrt{q}}$.

If the blocking set B is of Rédei type, then as a consequence of the direction theorem above the structure is very special if $N < (q+3)/2$, or equivalently if $|B| \le \frac{3}{2}(q+1)$. An important step towards showing that this is true in general is the following result for planes of prime order already mentioned in the introduction [8].

Theorem 2.3 [Blokhuis, 1994] *Let B be a blocking set in $PG(2,p)$, p prime, not containing a line. Then*

$$|B| \geq \frac{3}{2}(p+1).$$

The proof is based on properties of lacunary polynomials, introduced and studied by Rédei in [26]. In the same paper it is proved that a blocking set in $PG(2,p^3)$ has size at least $p^3 + p^2 + 1$. Recently Polverino [24] has shown that small blocking sets in $PG(2,p^3)$ are all of Rédei type, and the possible sizes are $p^3 + p^2 + 1$ and $p^3 + p^2 + p + 1$ (corresponding to examples 2 and 3 above). When mentioning possible sizes of blocking sets we will always tacitly assume that they are minimal, so deleting a point destroys the blocking property.

A very interesting and probably feasible problem is to characterize the sets that give equality in the bound for $PG(2,p)$. For all (odd) p the graph of the function

$$f \; : \; x \mapsto x^{(p+1)/2},$$

(the first example) together with it's $(p+3)/2$ directions is an example, and it is the essentially unique one of Rédei type (this was proved already in 1981 by Lovász and Schrijver [23] who also gave an elementary proof of Rédei's result for the case that $q = p$ is prime). Only two examples (of size $3(p+1)/2$) are known that are not of Rédei type, one (with 12 points) in the plane of order 7, it looks like a dual affine plane of order 3. The other (with 21 points) in the plane of order 13 was only found last year by Blokhuis, Brouwer and Wilbrink. Both are unique [11]. In the same paper it is shown that no other examples exist in planes of (prime) order less than 37, and it is extremely unlikely that this is different later on.

Motivated by these results we call blocking sets of size $< \frac{3}{2}(q+1)$ *small*, so small blocking sets only exist in planes of non-prime order. The structure of small blocking sets is restricted by the following theorem of Szőnyi [28]:

Theorem 2.4 [Szőnyi, 1997] *Let B be a (minimal) small blocking set in $PG(2,q)$, where q is a power of the prime p. Then $|B \cap l| = 1 \bmod p$ for every line l.*

For a long time I was convinced, and even conjectured that small blocking sets were necessarily of Rédei type, but this turned out to be false. Nice examples of small non-Rédei type blocking sets were found by Polito and Polverino [25].

The basic idea is very simple. Consider $PG(2,q^s)$. By definition its points and lines are the 1- and 2-dimensional subspaces of $V = V(3,q^s)$, a 3 dimensional vector space over $GF(q^s)$. When we consider V' which is just V as $3s$-dimensional over $GF(q)$ then points and lines correspond to certain s-, and $2s$-dimensional subspaces of V'. Now let W be any $s+1$-dimensional subspace of V'. Let $B(W)$ be the collection of points in $PG(2,q^s)$ for which the corresponding s-space in V' intersects W non-trivially. One readily checks that $B(W)$ is a blocking set (of size at most $(q^{s+1}-1)/(q-1)$), because in the $3s$ dimensional vector space V' an $(s+1)$-space and a $(2s)$-space must intersect in at least a 1-space. Polito and Polverino give examples that are not of Rédei type in all planes $PG(2,p^n)$, $n > 3$. The examples of small blocking sets of Rédei type also fall under this more general construction, by the direction theorem.

Next we consider multiple blocking sets. B is called a t-fold blocking set if every line intersects B in at least t points. If q is a square, then $PG(2,q)$ can be partitioned into Baer subplanes, and taking t of them produces a set with the property that every line intersects it in either t or $t + \sqrt{q}$ points (this makes it a two-intersection set). Again using the theory of lacunary polynomials it can be shown that for small t these are the minimal examples [6]. Our knowledge on the structure of (relatively) small multiple blocking sets is summarized in the following

Theorem 2.5 [Blokhuis, Storme, Szőnyi, 1998] *Let B be a t-fold blocking set in $PG(2,q)$ of size $t(q + 1) + c$. Let $c_2 = c_3 = 2^{-1/3}$ and $c_p = 1$ for $p > 3$.*

1. *If $q = p^{2d+1}$ and $t < q/2 - c_p q^{2/3}/2$ then $c \geq c_p q^{2/3}$.*
2. *If $4 < q$ is a square, $t < q^{1/4}/2$ and $c < c_p q^{2/3}$, then $c \geq t\sqrt{q}$ and B contains the union of t disjoint Baer subplanes.*
3. *If $q = p^2$ and $s < q^{1/4}/2$ and $c < p\lceil \frac{1}{4} + \sqrt{(p+1)/2} \rceil$, then $c \geq t\sqrt{q}$ and B contains the union of t disjoint Baer subplanes.*

What it essentially says is that for $t < q^{1/6}$ a t-fold blocking set has at least the size of t disjoint Baer subplanes, and equality implies that it is just that. In the special case that q is the square of a prime, then the same is true for $t < q^{1/4}/2$.

This result appears to be rather sharp in the following sense: In [4] Ball, Blokhuis and Lavrauw construct a two-intersection set with the same parameters as, but different from the union of $q^{1/4} + 1$ disjoint Baer subplanes, so the above characterization does no longer apply if $t > q^{1/4}$.

The construction is based on the Polito-Polverino idea. So the plane is $PG(2, q^s)$, and $V = V(3, q^s)$ and $V' = V(3s, q)$. Now we take for W an $s + 2$-dimensional subspace of V'. If this has the additional property that intersections with the s-dimensional subspaces of V' corresponding to projective points are at most 1-dimensional, then the corresponding set $B(W)$ is a $(q + 1)$-fold blocking set. To see this note that if L is a $2s$-dimensional subspace of V' corresponding to a line, then $W \cap L$ is at least 2-dimensional, but by assumption W intersects s-spaces corresponding to points in at most 1 dimension, so it has to intersect at least $q + 1$ of them.

The question of whether it is possible to find such subspaces lead to the notion of *scattered subspaces* with respect to spreads. An s-spread in a vector space V is a collection of s-dimensional subspaces partioning the nonzero vectors of V. In order for V to admit an s-spread it is necessary and sufficient that its dimension is a multiple of s. So in the above example the s-spaces in V' corresponding to points of $PG(2, q^s)$ define an s-spread. Given a vector space V together with an s-spread S we say that the subspace W is scattered by S if W intersect each spread element in an at most 1-dimensional subspace. A natural question is what the maximal dimension is of a scattered subspace. Results on this question and related problems can be found in the thesis of Lavrauw [22].

A detailed survey of the many recent results on blocking sets and multiple blocking sets is contained in the paper by Hirschfeld and Storme [21]. Blocking sets of projective planes can also be considered as a special case of the more general concept of covers in hypergraphs, extensively treated in the excellent (but not too recent) survey by Füredi [18].

To conclude this section let me mention two attractive problems (on which no progress has been made in the last 10 years).

The first concerns double blocking sets in $PG(2,p)$. A lower bound due to Blokhuis and Ball gives $|B| \geq 5(p+1)/2$. A trivial example is formed by the union of three lines, of size $3p$. Could it be that this is the minimal size? It is true for $p = 2, 3, 5, 7$, but it might already be false for $p = 11$.

The second question has repeatedly been asked to me by Paul Erdős. Is there a universal constant c (10 say), such that in any plane (or any $PG(2,p)$ say) there is a blocking set with at most c points on every line. In all of the known examples there are some lines with many points of the blocking set. Results by Ughi show that it does not work to use for B the union of a small set of algebraic curves of bounded degree [32], using for instance a union of conics one obtains blocking sets with $c \log(q)$ points on a line. On the other hand, in $PG(2, p^n)$ there is a blocking set with at most $p + 1$ point on every line.

3. Lacunary polynomials

We now turn to the main tool in the recent investigations on (multiple) blocking sets.

Let $f(X) \in GF(q)[X]$ be fully reducible, in other words, $f(X)$ factors into linear factors over $GF(q)$. In [26] Rédei investigates the case $f(X) = X^q + g(X)$ with $\deg(g) < q - 1$, and calls the polynomial lacunary. The problem is to characterize those f where the degree of g is small. As an easy example we prove:

Theorem 3.1 [Rédei, 1970] *Let $f(X) = X^q + g(X)$ be fully reducible in $GF(q)[X]$, where $q = p^n$ is prime. Then either $f(X) \in GF(q)[X^p]$, or $g(X) = -X$ or $\deg(g) \geq (q + 1)/2$.*

Proof: Write $f = s.r$, where s has the same zeroes as f, but with multiplicity one, and r consists of the remaining factors. Then $s \mid X^q - X$, as well as f, so $s \mid X + g$, and $r \mid f' = g'$. Hence $f = s.r \mid (X + g)g'$ so either $(X + g)g' = 0$ or $\deg(g) + \deg(g') \leq \deg(f) = q$.

If $q = p$ is prime, then $g' = 0$ together with $\deg(g) < p$ imply that g is constant.

Much of Rédei's book is devoted to the classification of those f with $\deg(g) = (q + 1)/2$. For us the case $g' = 0$ (and hence $f \in GF(q)[X^p]$ is more interesting however, also for our applications we need to consider polynomials of the form $f(X) = X^q g(X) + h(X)$, where both g and h have degree less than q.

The following theorem summarizes what we know in this case [6]:

Theorem 3.2 [Blokhuis, Storme, Szőnyi, 1998] *Let $f \in GF(q)[X]$, $q = p^n$, p prime, be fully reducible, $f(X) = X^q g(X) + h(X)$, where $(g, h) = 1$. Let $k = \max(\deg(g), \deg(h)) < q$. Let e be maximal such that f is a p^e-th power (so $f \in GF(q)[X^{p^e}]$). Then we have one of the following possibilities:*

1. *$e = n$ and $k = 0$;*
2. *$e \geq 2n/3$ and $k \geq p^e$;*
3. *$2n/3 > e > n/2$ and $k \geq p^{n-e/2} - \frac{3}{2}p^{n-e}$;*

4. $e = n/2$ and $k = p^e$ and $f(X) = a\text{Tr}(bX+c)+d$ or $f(X) = a\text{Norm}(bX+c)+d$
    for suitable constants a, b, c, d. Here Tr and Norm respectively denote the trace
    and norm function from $\mathbf{F}_q$ to $\mathbf{F}_{\sqrt{q}}$;

5. $e = n/2$ and $k \geq p^e \left\lceil \frac{1}{4} + \sqrt{(p^e + 1)/2} \right\rceil$;

6. $n/2 > e > n/3$ and $k \geq p^{(n+e)/2} - p^{n-e} - p^e/2$, or if $3e = n+1$ and $p \leq 3$,
    then $k \geq p^e(p^e + 1)/2$;

7. $n/3 \geq e > 0$ and $k \geq p^e \lceil (p^{n-e} + 1)/(p^e + 1) \rceil$;

8. $e = 0$ and $k \geq (q + 1)/2$;

9. $e = 0$, $k = 1$ and $f(X) = a(X^q - X)$.

It would be very pleasant to have stronger information in the case $n/3 < e < n/2$, this would have very useful applications.

4. The connection

In this section we will illustrate the connection between the direction problem, (multiple) blocking sets, and lacunary polynomials.

Let $f : GF(q) \to GF(q)$ be any map, and let D_f be its set of directions. Consider the auxiliary (Rédei) polynomial

$$R(X, Y) = \prod_{w \in GF(q)} (X - wY + f(w)),$$

introduced by Rédei. Let y and $v \neq w \in GF(q)$. Then $vy - f(v) = wy - f(w)$ if and only if $(f(v) - f(w))/(v - w) = y$. It follows that for $y \notin D_f$ the map $w \mapsto wy - f(w)$ is a bijection, and hence $R(X, y) = \prod_{z \in GF(q)}(X - z) = X^q - X$. Write

$$R(X, Y) = X^q + r_1(Y)X^{q-1} + \cdots + r_{q-1}(Y)X + r_q(Y),$$

where r_i is a polynomial of degree $< i$ in Y (with the exception of r_{q-1} of degree $q - 1$: it is clear that r_i has degree at most i, but the coefficient of Y^i is the i-th elementary symmetric polynomial in the elements of $GF(q)$, so this is 0 for $0 < i \neq q - 1$). If $y \notin D_f$ and $i \neq q - 1$ then $r_i(y) = 0$. So r_i is identically zero for $i \leq |D_f|$. As a consequence we have for $y \in D_f$ that $R(X, y) = X^q + g(X)$ for some g depending on y with $\deg(g) \leq q - |D_f| - 1$. So the Rédei polynomial when specialized for $Y = y \in D_f$ is a lacunary polynomial and information on $\deg(g)$ gives results for $|D_f|$.

The extension of the Rédei polynomial from the graph of a function to point sets in general can be illustrated best in the case of an ordinary blocking set B of $PG(2, q)$. We may coordinatize our plane in such a way that the line at infinity becomes a tangent, containing the point $(1 : 0 : 0)$ of the blocking set.

Let $|B| = q + 1 + d$ then the remaining $q + d$ points have certain affine coordinates (a_i, b_i) and the Rédei polynomial associated to B (in this position) can be defined as:

$$R[X, Y] := \prod(X - a_iY + b_i) = X^{q+d} + r_1(Y)X^{q+d-1} + \cdots + r_{q+d}(Y).$$

For $y, c \in GF(q)$ consider the line $\{(U, V) : V = yU + c\}$. It contains an affine point (a, b) of the blocking set: $c = b - ay$. Hence $R[X, y]$ is divisible by $(X - c)$ for all $c \in GF(q)$, in other words $X^q - X$ divides $R[X, y]$. It follows as before that r_i is identically zero for $i = d + 1, \ldots, q - 2$. As a consequence we obtain the lacunary polynomial

$$f(X) = \prod(X - a_i) = X^q g(X) + h(X),$$

with g of degree d and h of degree at most $d + 1$, and information on f translates to information on B. For the case that $q = p$ is prime it not only gives that $|B| \geq 3(p + 1)/2$, but in case of equality it also gives that each point of the blocking set is on exactly $(p - 1)/2$ tangents, and by classifying the possible polynomials it gives all possibilities for the configuration of the tangents through a particular point. For small p (at most 37) the number of possibilities is sufficiently small to be handled by a computer, and to prove uniqueness of the minimal example (for $p \neq 7, 13$).

References

[1] S. Ball, A. Blokhuis & F. Mazzocca, Maximal arcs in Desarguesian planes of odd order do not exist, *Combinatorica*, 17 (1997), 31–41

[2] S. Ball & A. Blokhuis, An easier proof of the maximal arc conjecture, *Proceedings of the American Mathematical Society*, 126 (1998), 3377-3380.

[3] S. Ball, Multiple blocking sets and arcs in finite planes, *J. London Math. Soc.*, 54 (1996) 581–593.

[4] S. Ball, A. Blokhuis & M. Lavrauw, Linear $(q + 1)$-fold blocking sets in $PG(2, q^4)$, Finite Fields and Applications 6 (2000), 294–301.

[5] A. Barlotti, Un'estensione del teorema di Segre-Kustaanheimo, *Boll. Un. Mat. Ital.*, 10 (1955), 498–506.

[6] A. Blokhuis, L. Storme & T. Szőnyi, Lacunary Polynomials, Multiple Blocking Sets and Baer Subplanes, *J. London Math. Soc.*, 60 (1999), 321-332.

[7] A. Blokhuis, Extremal Problems in Finite Geometries, *Extremal Problems for Finite Sets*, Bolyai Society Mathematical Studies, 3, 111–135.

[8] A. Blokhuis, On the size of a blocking set in $PG(2, p)$, *Combinatorica* 14 (1994), 111–114.

[9] A. Blokhuis, Blocking Sets in Desarguesian Planes, *Combinatorics: Paul Erdős is Eighty, Vol. 2*, János Bolyai Mathematical Society, Budapest, 1994.

[10] A. Blokhuis & A.E. Brouwer, Blocking sets in Desarguesian projective planes. *Bull. London Math. Soc. 18 (1986), 132–134.*

[11] A. Blokhuis, A.E. Brouwer & H.A. Wilbrink, Blocking sets in $PG(2, p)$ for small p, and partial spreads in $PG(3, 7)$, unpublished manuscript (2001).

[12] A. Blokhuis, A.E. Brouwer & T. Szőnyi, The number of directions determined by a function f on a finite field. *J. Comb. Theory, Ser. A* 70 (1995), 349–353.

[13] A. Blokhuis, S. Ball, A.E. Brouwer, L. Storme & T. Szőnyi, On the number of slopes of the graph of a function defined on a finite field, *J. Combin. Theory A*, 86 (1999), 187–196.

[14] A.A. Bruen, Blocking sets in finite projective planes. *SIAM J. Appl. Math.* 21 (1971), 380–392.

[15] A.A. Bruen & J.A. Thas, Blocking Sets, *Geom. Dedicata*, 6 (1977), 193–203.

[16] A. Cossu, Su alcune proprietà dei $\{k; n\}$-archi di un piano proiettivo sopra un corpo finito, *Rend. Mat. e Appl.*, 20 (1961), 271–277.

[17] R.H.F. Denniston, Some maximal arcs in finite projective planes, *J. Combin. Theory*, 6 (1969), 317–319.

[18] Z. Füredi, Matchings and Covers in Hypergraphs, *Graphs and Combinatorics*, 4 (1988), 115–206.

[19] A. Gács & T. Szőnyi, Double blocking sets and Baer subplanes, unpublished manuscript (1995).

[20] J.W.P. Hirschfeld & L. Storme, The packing problem in statistics, coding theory and finite projective spaces. Proceedings of the *Bose Memorial Conference* (Colorado, June 7–11, 1995). *J. Statist. Planning Infer.*, 72 (1998), 355–380.

[21] J.W.P. Hirschfeld & L. Storme, The packing problem in statistics, coding theory and finite projective spaces: update 2001, *manuscript*.

[22] M. Lavrauw, *Scattered subspaces with respect to spreads, and eggs in finite projective spaces* Thesis Technical University Eindhoven (2001).

[23] L. Lovász & A. Schrijver, Remarks on a theorem of Rédei. *Studia Sci. Math. Hungar.* 16 (1981), 449–454.

[24] O. Polverino, Small blocking sets in $PG(2, p^3)$, manuscript.

[25] P. Polito & O. Polverino, On small blocking sets, *Combinatorica* 18 (1998), 133–137.

[26] L. Rédei, *Lückenhafte Polynome über endlichen Körpern*. Birkhäuser Verlag, Basel 1970.

[27] B. Segre, Ovals in a finite projective plane, *Can. J. Math.*, 7 (1955), 414–416.

[28] T. Szőnyi, Blocking sets in Desarguesian affine and projective planes, *Finite Fields Appl.*, 3 (1997), 187–202.

[29] J.A. Thas, Some results concerning $\{(q + 1)(n - 1); n\}$-arcs and $\{(q + 1)(n - 1) + 1; 1\}$-arcs in finite projective planes of order q. *J. Combin. Theory Ser. A*, 19 (1975), 228–232.

[30] J.A. Thas, Construction of maximal arcs and partial geometries, *Geom. Dedicata*, 3 (1974), 61–64.

[31] J.A. Thas, Construction of maximal arcs in translation planes, *Europ. J. Combinatorics*, 1 (1980), 189–192.

[32] E. Ughi, On (k, n)-blocking sets which can be obtained as a union of conics, *Geom. Ded.*, 26 (1988), 241–246.

The Strong Perfect Graph Conjecture[*]

Gérard Cornuéjols[†]

Abstract

A graph is *perfect* if, in all its induced subgraphs, the size of a largest clique is equal to the chromatic number. Examples of perfect graphs include bipartite graphs, line graphs of bipartite graphs and the complements of such graphs. These four classes of perfect graphs will be called *basic*. In 1960, Berge formulated two conjectures about perfect graphs, one stronger than the other. The weak perfect graph conjecture, which states that a graph is perfect if and only if its complement is perfect, was proved in 1972 by Lovász. This result is now known as the perfect graph theorem. The strong perfect graph conjecture (SPGC) states that a graph is perfect if and only if it does not contain an odd hole or its complement. The SPGC has attracted a lot of attention. It was proved recently (May 2002) in a remarkable sequence of results by Chudnovsky, Robertson, Seymour and Thomas. The proof is difficult and, as of this writing, they are still checking the details. Here we give a flavor of the proof. Let us call *Berge graph* a graph that does not contain an odd hole or its complement. Conforti, Cornuéjols, Robertson, Seymour, Thomas and Vušković (2001) conjectured a structural property of Berge graphs that implies the SPGC: Every Berge graph G is basic or has a skew partition or a homogeneous pair, or G or its complement has a 2-join. A *skew partition* is a partition of the vertices into nonempty sets A, B, C, D such that every vertex of A is adjacent to every vertex of B and there is no edge between C and D. Chvátal introduced this concept in 1985 and conjectured that no minimally imperfect graph has a skew partition. This conjecture was proved recently by Chudnovsky and Seymour (May 2002). Cornuéjols and Cunningham introduced 2-joins in 1985 and showed that they cannot occur in a minimally imperfect graph different from an odd hole. Homogeneous pairs were introduced in 1987 by Chvátal and Sbihi, who proved that they cannot occur in minimally imperfect graphs. Since skew partitions, 2-joins and homogeneous pairs cannot occur in minimally imperfect Berge graphs, the structural property of Berge graphs stated above implies the SPGC. This structural property was proved: (i) When G contains the line graph of a bipartite subdivision of a 3-connected graph (Chudnovsky, Robertson, Seymour and Thomas (September 2001)); (ii) When G contains a stretcher (Chudnovsky and Seymour (January

---

[*]This work was supported in part by NSF grant DMI-0098427 and ONR grant N00014-97-1-0196.

[†]GSIA, Carnegie Mellon University, Schenley Park, Pittsburgh, PA 15213, USA. E-mail: gc0v@andrew.cmu.edu

2002)); (iii) When G contains no proper wheels, stretchers or their complements (Conforti, Cornuéjols and Zambelli (May 2002)); (iv) When G contains
a proper wheel, but no stretchers or their complements (Chudnovsky and
Seymour (May 2002)). (ii), (iii) and (iv) prove the SPGC.

2000 Mathematics Subject Classification: 05C17.
Keywords and Phrases: Perfect graph, Odd hole, Strong Perfect Graph
Conjecture, Strong Perfect Graph Theorem, Berge graph, Decomposition, 2-
join, Skew partition, Homogeneous pair.

1. Introduction

In this paper, all graphs are simple (no loops or multiple edges) and finite.
The vertex set of graph G is denoted by $V(G)$ and its edge set by $E(G)$. A *stable
set* is a set of vertices no two of which are adjacent. A *clique* is a set of vertices
every pair of which are adjacent. The cardinality of a largest clique in graph G is
denoted by $\omega(G)$. The cardinality of a largest stable set is denoted by $\alpha(G)$. A
k-coloring is a partition of the vertices into k stable sets (these stable sets are called
color classes). The *chromatic number* $\chi(G)$ is the smallest value of k for which there
exists a k-coloring. Obviously, $\omega(G) \leq \chi(G)$ since the vertices of a clique must be
in distinct color classes of the k-coloring. An *induced subgraph* of G is a graph with
vertex set $S \subseteq V(G)$ and edge set comprising all the edges of G with both ends in S.
It is denoted by $G(S)$. The graph $G(V(G) - S)$ is denoted by $G \setminus S$. A graph G is
perfect if $\omega(H) = \chi(H)$ for every induced subgraphs H of G. A graph is *minimally
imperfect* if it is not perfect but all its proper induced subgraphs are.

A *hole* is a graph induced by a chordless cycle of length at least 4. A hole is
odd if it contains an odd number of vertices. Odd holes are not perfect since their
chromatic number is 3 whereas the size of their largest clique is 2. It is easy to
check that odd holes are minimally imperfect. The complement of a graph G is the
graph $\bar{G}$ with the same vertex set as G, and uv is an edge of $\bar{G}$ if and only if it is not
an edge of G. The odd holes and their complements are the only known minimally
imperfect graphs. In 1960 Berge [3] proposed the following conjecture, known as
the *Strong Perfect Graph Conjecture*.

Conjecture 1.1 (Strong Perfect Graph Conjecture) (Berge [3]) *The only
minimally imperfect graphs are the odd holes and their complements.*

At the same time, Berge also made a weaker conjecture, which states that a
graph G is perfect if and only if its complement $\bar{G}$ is perfect. This conjecture was
proved by Lovász [29] in 1972 and is known as the *Perfect Graph Theorem*.

Theorem 1.2 (Perfect Graph Theorem) (Lovász [29]) *Graph G is perfect if
and only if graph $\bar{G}$ is perfect.*

Proof: Lovász [30] proved the following stronger result.

Claim 1: *A graph G is perfect if and only if, for every induced subgraph H, the number of vertices of H is at most $\alpha(H)\omega(H)$.*

Since $\alpha(H) = \omega(\bar{H})$ and $\omega(H) = \alpha(\bar{H})$, Claim 1 implies Theorem 1.2.

Proof of Claim 1: We give a proof of this result due to Gasparyan [25]. First assume that G is perfect. Then, for every induced subgraph H, $\omega(H) = \chi(H)$. Since the number of vertices of H is at most $\alpha(H)\chi(H)$, the inequality follows.

Conversely, assume that G is not perfect. Let H be a minimally imperfect subgraph of G and let n be the number of vertices of H. Let $\alpha = \alpha(H)$ and $\omega = \omega(H)$. Then H satisfies

$$\omega = \chi(H \backslash v) \text{ for every vertex } v \in V(H)$$

$$\text{and } \omega = \omega(H \backslash S) \text{ for every stable set } S \subseteq V(H).$$

Let A_0 be an α-stable set of H. Fix an ω-coloring of each of the α graphs $H \backslash s$ for $s \in A_0$, let $A_1, \ldots, A_{\alpha\omega}$ be the stable sets occuring as a color class in one of these colorings and let $\mathcal{A} := \{A_0, A_1, \ldots, A_{\alpha\omega}\}$. Let $\mathbf{A}$ be the corresponding stable set versus vertex incidence matrix. Define $\mathcal{B} := \{B_0, B_1, \ldots, B_{\alpha\omega}\}$ where B_i is an ω-clique of $H \backslash A_i$. Let $\mathbf{B}$ be the corresponding clique versus vertex incidence matrix.

Claim 2: *Every ω-clique of H intersects all but one of the stables sets in $\mathcal{A}$.*

Proof of Claim 2: Let $S_1, \ldots, S_\omega$ be any ω-coloring of $H \backslash v$. Since any ω-clique C of H has at most one vertex in each S_i, C intersects all S_i's if $v \notin C$ and all but one if $v \in C$. Since C has at most one vertex in A_0, Claim 2 follows.

In particular, it follows that $\mathbf{A}\mathbf{B}^T = J - I$. Since $J - I$ is nonsingular, $\mathbf{A}$ and $\mathbf{B}$ have at least as many columns as rows, that is $n \geq \alpha\omega + 1$. This completes the proof of Claim 1.

2. Four Basic Classes of Perfect Graphs

Bipartite graphs are perfect since, for any induced subgraph H, the bipartition implies that $\chi(H) \leq 2$ and therefore $\omega(H) = \chi(H)$.

A graph L is the *line graph* of a graph G if $V(L) = E(G)$ and two vertices of L are adjacent if and only if the corresponding edges of G are adjacent.

Proposition 2.1 *Line graphs of bipartite graphs are perfect.*

Proof: If G is bipartite, $\chi'(G) = \Delta(G)$ by a theorem of König [28], where χ' denotes the edge-chromatic number and Δ the largest vertex degree.

If L is the line graph of a bipartite graph G, then $\chi(L) = \chi'(G)$ and $\omega(L) = \Delta(G)$. Therefore $\chi(L) = \omega(L)$. Since induced subgraphs of L are also line graphs of bipartite graphs, the result follows.

Since bipartite graphs and line graphs of bipartite graphs are perfect, it follows from Lovász's perfect graph theorem (Theorem 1.2) that the complements of bipartite graphs and of line graphs of bipartite graphs are perfect. This can also be verified directly, without using the perfect graph theorem. To summarize, in this section we have introduced four basic classes of perfect graphs:

- bipartite graphs and their complements, and
- line graphs of bipartite graphs and their complements.

3. 2-Join

A graph G has a *2-join* if its vertices can be partitioned into sets V_1 and V_2, each of cardinality at least three, with nonempty disjoint subsets $A_1, B_1 \subseteq V_1$ and $A_2, B_2 \subseteq V_2$, such that all the vertices of A_1 are adjacent to all the vertices of A_2, all the vertices of B_1 are adjacent to all the vertices of B_2 and these are the only adjacencies between V_1 and V_2. There is an $O(|V(G)|^2 |E(G)|^2)$ algorithm to find whether a graph G has a 2-join [23].

When G contains a 2-join, we can decompose G into two blocks G_1 and G_2 defined as follows.

Definition 3.1 *If A_2 and B_2 are in different connected components of $G(V_2)$, define block G_1 to be $G(V_1 \cup \{p_1, q_1\})$, where $p_1 \in A_2$ and $q_1 \in B_2$. Otherwise, let P_1 be a shortest path from A_2 to B_2 and define block G_1 to be $G(V_1 \cup V(P_1))$. Block G_2 is defined similarly.*

Next we show that the 2-join decomposition preserves perfection (Cornuéjols and Cunningham [23]; see also Kapoor [27] Chapter 8). Earlier, Bixby [4] had shown that the simpler join decomposition preserves perfection.

Theorem 3.2 *Graph G is perfect if and only if its blocks G_1 and G_2 are perfect.*

Proof: By definition, G_1 and G_2 are induced subgraphs of G. It follows that, if G is perfect, so are G_1 and G_2. Now we prove the converse: If G_1 and G_2 are perfect, then so is G. Let G^* be an induced subgraph of G. We must show

$$(*) \quad \omega(G^*) = \chi(G^*).$$

For $i = 1, 2$, let $V_i^* = V_i \cap V(G^*)$. The proof of $(*)$ is based on a coloring argument, combining $\omega(G^*)$-colorings of the perfect graphs $G(V_1^*)$ and $G(V_2^*)$ (Claim 3) into an $\omega(G^*)$-coloring of G^* (Claim 4). To prove Claim 3, we will use the following results.

Claim 1: (Lovász's Replication Lemma [29]) *Let Γ be a perfect graph and $v \in V(\Gamma)$. Create a new vertex v' adjacent to v and to all the neighbors of v. Then the resulting graph Γ' is perfect.*

Proof of Claim 1: It suffices to show that $\omega(\Gamma') = \chi(\Gamma')$ since, for induced subgraphs, the proof follows similarly. We distinguish two cases. Suppose first that v is contained in some $\omega(\Gamma)$-clique of Γ. Then $\omega(\Gamma') = \omega(\Gamma) + 1$. Since at most one new color is needed in Γ', $\omega(\Gamma') = \chi(\Gamma')$ follows.

Now suppose that v is not contained in any $\omega(\Gamma)$-clique of Γ. Consider any $\omega(\Gamma)$-coloring of Γ and let A be the color class containing v. Then, $\omega(\Gamma \setminus (A - \{v\})) =$

$\omega(\Gamma) - 1$, since every $\omega(\Gamma)$-clique of Γ meets $A - \{v\}$. By the perfection of Γ, the graph $\Gamma \setminus (A - \{v\})$ can be colored with $\omega(\Gamma) - 1$ colors. Using one additional color for the vertices $(A - \{v\}) \cup \{v'\}$, we obtain an $\omega(\Gamma)$-coloring of Γ'. This proves Claim 1.

We say that Γ' is obtained from Γ by *replicating* v. Replication can be applied recursively. We say that v is *replicated k times* if k copies of v are made, including v.

Claim 2: *Let Γ be a graph and uv an edge of Γ such that the vertices u and v have no common neighbor. Let Γ' be the graph obtained from Γ by replicating vertex v into v'. Let H be the graph obtained from Γ' by deleting edge uv'. Then Γ is perfect if and only if H is perfect.*

Proof of Claim 2: If H is perfect, then so is Γ since Γ is an induced subgraph of H.

Conversely, suppose that Γ is perfect and H is not. Let H^* be a minimally imperfect subgraph of H. Let Γ^* be the subgraph of Γ' induced by the vertices of H^*. Since Γ^* is perfect but H^* is not, $V(H^*)$ must contain vertices u and v'. Also $\chi(\Gamma^*) = \chi(H^*)$ and $\omega(\Gamma^*) = \omega(H^*) + 1$. Therefore uvv' is the unique maximum clique in Γ^* and $\omega(H^*) = 2$. The only neighbor of v in H^* is u since otherwise v, v' would be in a clique of cardinality three in H^*. Now v' is a vertex of degree 1 in H^*, a contradiction to the assumption that H^* is minimally imperfect. This proves Claim 2.

For $i = 1, 2$, let $A_i^* = A_i \cap V(G^*)$, $B_i^* = B_i \cap V(G^*)$, $a_i = \omega(A_i^*)$ and $b_i = \omega(B_i^*)$. Let $G_i^* = G_i \setminus (V_i - V_i^*)$ and $\omega \geq \omega(G_i^*)$. In an ω-coloring of G_i^*, let $C(A_i^*)$ and $C(B_i^*)$ denote the sets of colors in A_i^* and B_i^* respectively.

Claim 3: *There exists an ω-coloring of V_i^* such that $|C(A_i^*)| = a_i$ and $|C(B_i^*)| = b_i$. Furthermore, if G_i contains path P_i and*

(i) if P_i has an odd number of edges, then $|C(A_i^*) \cap C(B_i^*)| = \max(0, a_i + b_i - \omega)$,

(ii) if P_i has an even number of edges, then $|C(A_i^*) \cap C(B_i^*)| = \min(a_i, b_i)$.

Proof of Claim 3: First assume that block G_i is induced by $V_i \cup \{p_i, q_i\}$. In G_i^*, replicate p_i $\omega - a_i$ times and q_i $\omega - b_i$ times. By Claim 1, this new graph H is perfect and $\omega(H) = \omega$. Therefore an ω-coloring of H exists. This coloring induces an ω-coloring of V_i^* with $|C(A_i^*)| = a_i$ and $|C(B_i^*)| = b_i$. Now assume that G_i contains path P_i. We consider two cases.

(i) P_i has an odd number of edges.

Let $P_i = x_1, \ldots, x_{2k}$. In G_i^*, replicate vertex x_{2k} into x_{2k}' and remove edge $x_{2k-1} x_{2k}'$. By Claim 2, the new graph is perfect. For i odd, $1 \leq i < 2k$, replicate vertex x_i $\omega - a_i$ times. For i even, $1 < i \leq 2k - 2$, replicate vertex x_i a_i times.

If $a_i + b_i < \omega$, replicate x_{2k} a_i times and replicate x_{2k}' $\omega - a_i - b_i$ times. By Claim 1, this new graph H is perfect. Since $\omega(H) = \omega$, H has an ω-coloring. Note that $|C(A_i^*)| = a_i$ and $|C(B_i^*)| = b_i$ and every vertex of P_i belongs to two cliques of size ω. So the colors that appear in the replicates of x_{2k} are precisely $C(A_i)$. Therefore B_i^* is colored with colors that do not appear in $C(A_i^*)$. Thus $|C(A_i^*) \cap C(B_i^*)| = 0$.

If $a_i + b_i \geq \omega$, replicate x_{2k} $\omega - b_i$ times and remove x'_{2k}. The new graph H is perfect and $\omega(H) = \omega$. Therefore H has an ω-coloring. Again $|C(A_i^*)| = a_i$ and $|C(B_i^*)| = b_i$, and the $\omega - b_i$ colors that appear in the replicates of x_{2k} belong to $C(A_i^*)$. Since these colors cannot appear in $C(B_i^*)$, the number of common colors in $C(A_i^*)$ and $C(B_i^*)$ is $a_i + b_i - \omega$.

(ii) P_i has an even number of edges.

Assume w.l.o.g. that $a_i \leq b_i$. Let $P_i = x_1, \ldots, x_{2k+1}$. In G_i^*, replicate vertex x_i $\omega - a_i$ times for i odd, $1 \leq i \leq 2k - 1$, and replicate vertex x_i a_i times for i even, $1 < i \leq 2k$. Finally, replicate x_{2k+1} $\omega - b_i$ times. By Claim 1, the new graph H is perfect and $\omega(H) = \omega$. In an ω-coloring of H, $|C(A_i^*)| = a_i$ and $|C(B_i^*)| = b_i$ and the colors that appear in the replicates of x_{2k} are precisely $C(A_i^*)$. But then these colors do not appear in the replicates of x_{2k+1} and consequently they must appear in $C(B_i)$. Thus $|C(A_i) \cap C(B_i)| = \min(a_i, b_i)$. This proves Claim 3.

Claim 4: G^* has an $\omega(G^*)$-coloring.

Proof of Claim 4: Let $\omega = \omega(G^*)$. Clearly, $\omega \geq a_1 + a_2$ and $\omega \geq b_1 + b_2$. To prove the claim, we will combine ω-colorings of V_1^* and V_2^*.

If at least one of the sets $A_1^*, A_2^*, B_1^*, B_2^*$ is empty, one can easily construct the desired ω-coloring of G^*. So we assume now that these sets are nonempty. This implies that $\omega \geq \omega(G_1^*)$ and $\omega \geq \omega(G_2^*)$. By Claim 3, there exist ω-colorings of V_i^* such that $|C(A_i^*)| = a_i$ and $|C(B_i^*)| = b_i$. Thus, if A_2^* and B_2^* are in different connected components of $G(V_2^*)$, an ω-coloring of V_1^* can be combined with ω-colorings of the components of $G(V_2^*)$ into an ω-coloring of G^*. So we can assume that both P_1 and P_2 exist. Since G_1 contains no odd hole, every chordless path from A_1 to B_1 has the same parity as P_1. It follows from the definition of 2-join decomposition that P_1 and P_2 have the same parity.

(i) P_1 and P_2 both have an odd number of edges.

Then by Claim 3 (i), there exists an ω-coloring of V_i^* with $|C(A_i^*) \cap C(B_i^*)| = \max(0, a_i + b_i - \omega)$. In the coloring of V_1^*, label by 1 through a_1 the colors that occur in A_1^* and by ω through $\omega - b_1 + 1$ the colors that occur in B_1^*. In the coloring of V_2^*, label by ω through $\omega - a_2 + 1$ the colors that occur in A_2^* and by 1 through b_2 the colors that occur in B_2^*. If this is not an ω-coloring of G^*, there must exist a common color in A_1^* and A_2^* or in B_1^* and B_2^*. But then either $a_1 \geq \omega - a_2 + 1$ or $b_2 \geq \omega - b_1 + 1$, a contradiction.

(ii) P_1 and P_2 both have an even number of edges.

Then by Claim 3 (ii), there exists an ω-coloring of V_i^* with $|C(A_i^*) \cap C(B_i^*)| = \min(a_i, b_i)$. In the coloring of V_1^*, label by 1 through a_1 the colors that occur in A_1^* and by 1 through b_1 the colors that occur in B_1^*. In the coloring of V_2^*, label by ω through $\omega - a_2 + 1$ the colors that occur in A_2^* and by ω through $\omega - b_2 + 1$ the colors that occur in B_2^*. If this is not an ω-coloring of G, there must exist a common color in A_1^* and A_2^* or in B_1^* and B_2^*. But then either $a_1 \geq \omega - a_2 + 1$ or $b_1 \geq \omega - b_2 + 1$, a contradiction.

Corollary 3.3 *If a minimally imperfect graph G has a 2-join, then G is an odd hole.*

Proof: Since G is not perfect, Theorem 3.2 implies that block G_1 or G_2 is not perfect, say G_1. Since G_1 is an induced subgraph of G and G is minimally imperfect, it follows that $G = G_1$. Since $|V_2| \geq 3$, V_2 induces a chordless path. Thus G is a minimally imperfect graph with a vertex of degree 2. This implies that G is an odd hole [32].

We end this section with another decomposition that preserves perfection. A graph G has a *6-join* if $V(G)$ can be partitioned into eight nonempty sets $X_1, X_2, X_3, X_4, Y_1, Y_2, Y_3, Y_4$ with the property that, for any $x_i \in X_i$ ($i = 1, 2, 3$) and $y_j \in Y_j$ ($j = 1, 2, 3$), the graph induced by $x_1, y_1, x_2, y_2, x_3, y_3$ is a 6-hole and these kinds of edges are the only adjacencies between $X = X_1 \cup X_2 \cup X_3 \cup X_4$ and $Y = Y_1 \cup Y_2 \cup Y_3 \cup Y_4$.

Theorem 3.4 (Aossey and Vušković [2]) *No minimally imperfect graph contains a 6-join.*

If G contains a 6-join, define *blocks* G_X and G_Y as follows. G_X is the graph induced by $X \cup \{y_1, y_2, y_3\}$ where $y_j \in Y_j$ ($j = 1, 2, 3$). Similarly G_Y is the graph induced by $Y \cup \{x_1, x_2, x_3\}$ where $x_i \in X_i$ ($i = 1, 2, 3$). It can be shown [1] that G is perfect if and only if its blocks G_X and G_Y are perfect.

4. Skew Partition and Homogeneous Pair

A graph has a *skew partition* if its vertices can be partitioned into four nonempty sets A, B, C, D such that there are all the possible edges between A and B and no edges from C to D. It is easy to verify that the odd holes and their complements do not have a skew partition. Chvátal [6] conjectured that no minimally imperfect graph has a skew partition.

Theorem 4.1 (Skew Partition Theorem) (Chudnovsky and Seymour [13]) *No minimally imperfect graph has a skew partition.*

Chudnovsky and Seymour obtained this result as a consequence of their proof of the SPGC. In order to prove the SPGC, they first proved the following weaker result.

Theorem 4.2 (Chudnovsky and Seymour [12]) *A minimally imperfect Berge graph with smallest number of vertices does not have a skew partition.*

We do not give the proof of this difficult theorem here. Instead, we prove results due to Hoàng [26] on two special skew partitions called T-cutset and U-cutset respectively.

Assume that G is a minimally imperfect graph with skew partition A, B, C, D. Let $a = \omega(A)$, $b = \omega(B)$, $\omega = \omega(G)$ and $\alpha = \alpha(G)$. The vertex sets $A \cup B \cup C$ and $A \cup B \cup D$ induce perfect graphs G_1 and G_2 respectively and both of these graphs contain an ω-clique. Indeed, each vertex of a minimally imperfect graph belongs to ω ω-cliques [32] and, for $u \in C$, these ω-cliques are contained in G_1. For $u \in D$, they are contained in G_2.

Lemma 4.3 (Hoàng [26]) *Let C_i be an ω-coloring of G_i, for $i = 1, 2$. Then C_1 and C_2 cannot have the same number of colors in A.*

Proof: Suppose C_1 and C_2 have the same number of colors in A and assume w.l.o.g. that these colors are $1, 2, \ldots, k$. Let K be the subgraph of G induced by the vertices with colors $1, 2, \ldots, k$ and let $H = G \setminus K$. Since every ω-clique of G is in G_1 or G_2, the largest clique in K has size k and the largest clique in H has size $\omega - k$. The graphs H and K are perfect since they are proper subgraphs of G. Color K with k colors and H with $\omega - k$ colors. Now G is colored with ω colors, a contradiction to the assumption that G is minimally imperfect.

Lemma 4.4 *No ω-clique is contained in $A \cup B$.*

Proof: Suppose that a ω-clique were contained in $A \cup B$. Then any ω-coloring of G_i, for $i = 1, 2$, would contain a colors in A and $b = \omega - a$ colors in B, contradicting Lemma 4.3.

Lemma 4.5 *Every α-stable set intersects $A \cup B$.*

Proof: By Lemma 4.4 applied to the complement graph, no α-stable set is contained in $C \cup D$.

Lemma 4.6 *If some $u \in A$ has no neighbor in C, then there exists an ω-coloring of G_1 with b colors in B.*

proof: Let C_1 be an ω-coloring of G_1 with minimum number k of colors in B and suppose that this number is strictly greater than b. Consider the subgraph H of G_1 induced by the vertices colored with the colors of C_1 that appear in B. The graph $H \cup u$ can be colored with k colors since it is perfect and has no clique of size greater than k. Keeping the other colors of C_1 in $G_1 \setminus (H \cup u)$, we get an ω-coloring of G_1 with fewer colors on B than C_1, a contradiction.

Lemma 4.7 *If some $u \in A$ has no neighbor in C, then every vertex of A has a neighbor in D and every vertex of B has a neighbor in C.*

Proof: By Lemma 4.6, there exists an ω-coloring of G_1 with b colors in B. Thus, by Lemma 4.3, there exists no ω-coloring of G_2 with b colors in B. By Lemma 4.6, this implies that every vertex of A has a neighbor in D.

Suppose that $v \in B$ has no neighbor in C. In the complement graph, u and v are adjacent to all the vertices of C. By Lemma 4.3, $|A| \geq 2$ and $|B| \geq 2$. So $A' = A - u$, $B' = B - v$, $C' = C$, $D' = D \cup \{u, v\}$ form a skew partition. But u has no neighbor in B and v has no neighbor in A, contradicting the first part of the lemma. So every $v \in B$ has a neighbor in C.

A *T-cutset* is a skew partition with $u \in C$ and $v \in D$ such that every vertex of A is adjacent to both u and v.

Lemma 4.8 (Hoàng [26]) *No minimally imperfect graph contains a T-cutset.*

Proof: In the complement, u and v contradict Lemma 4.7.

A *U-cutset* is a skew partition with $u, v \in C$ such that every vertex of A is adjacent to u and every vertex of B is adjacent to v.

Lemma 4.9 (Hoàng [26]) *No minimally imperfect graph contains a U-cutset.*

Proof: In the complement, u and v contradict Lemma 4.7.

We conclude this section with the notion of homogeneous pair introduced by Chvátal and Sbihi [8]. A graph G has a *homogeneous pair* if $V(G)$ can be partitioned into subsets A_1, A_2 and B, such that:

- $|A_1| + |A_2| \geq 3$ and $|B| \geq 2$.
- If a node of B is adjacent to a node of A_1 (A_2) then it is adjacent to all the nodes of A_1 (A_2).

Theorem 4.10 (Chvátal and Sbihi [8]) *No minimally imperfect graph contains a homogeneous pair.*

5. Decomposition of Berge Graphs

A graph is a *Berge graph* if it does not contain an odd hole or its complement. Clearly, all perfect graphs are Berge graphs. The SPGC states that the converse is also true.

Conjecture 5.1 (Decomposition Conjecture) (Conforti, Cornuéjols, Robertson, Seymour, Thomas and Vušković (2001)) *Every Berge graph G is basic or has a skew partition or a homogeneous pair, or G or $\bar{G}$ has a 2-join.*

This conjecture implies the SPGC. Indeed, suppose that the Decomposition Conjecture holds but not the SPGC. Then there exists a minimally imperfect graph G distinct from an odd hole or its complement. Choose G with the smallest number of vertices. G is a Berge graph and it cannot have a skew partition by Theorem 4.2. G cannot have an homogeneous pair by Theorem 4.10. Neither G nor $\bar{G}$ can have a 2-join by Corollary 3.3. So G must be basic by the Decomposition Conjecture. Therefore G is perfect, a contradiction.

Note that there are other decompositions that cannot occur in minimally imperfect Berge graphs, such as 6-joins (Theorem 3.4) or universal 2-amalgams [15] (universal 2-amalgams generalize both 2-joins and homogeneous pairs). These decompositions could be added to the statement of Conjecture 5.1 while still implying the SPGC. However they do not appear to be needed. Paul Seymour commented that homogeneous pairs might not be necessary either. In fact, we had initially formulated Conjecture 5.1 without homogeneous pairs. I added them to the statement to be on the safe side since they currently come up in the proof of the SPGC (see below).

Several special cases of Conjecture 5.1 are known. For example, it holds when G is a Meyniel graph (Burlet and Fonlupt [5] in 1984), when G is claw-free (Chvatal and Sbihi [9] in 1988 and Maffray and Reed [31] in 1999), diamond-free (Fonlupt and Zemirline [24] in 1987), bull-free (Chvátal and Sbihi [8] in 1987), or dart-free (Chvátal, Fonlupt, Sun and Zemirline [7] in 2000). All these results involve special types of skew partitions (such as star cutsets) and, in some cases, homogeneous pairs [8]. A special case of 2-join called augmentation of a flat edge appears in [31]. In 1999, Conforti and Cornuéjols [14] used more general 2-joins to prove Conjecture 5.1 for WP-free Berge graphs, a class of graphs that contains all bipartite graphs and all line graphs of bipartite graphs. This paper was the precursor of a sequence of decomposition results involving 2-joins:

Theorem 5.2 (Conforti, Cornuéjols and Vušković [18]) *A square-free Berge graph is bipartite, the line graph of a bipartite graph, or has a 2-join or a star cutset.*

Theorem 5.3 (Chudnovsky, Robertson, Seymour and Thomas [10]) *If G is a Berge graph that contains the line graph of a bipartite subdivision of a 3-connected graph, then G has a skew partition, or G or $\bar{G}$ has a 2-join or is the line graph of a bipartite graph.*

Given two vertex disjoint triangles a_1, a_2, a_3 and b_1, b_2, b_3, a *stretcher* is a graph induced by three chordless paths, $P^1 = a_1, \ldots, b_1$, $P^2 = a_2, \ldots, b_2$ and $P^3 = a_3, \ldots, b_3$, at least one of which has length greater than one, such that P^1, P^2, P^3 have no common vertices and the only adjacencies between the vertices of distinct paths are the edges of the two triangles. The next result is a real tour-de-force and a key step in the proof of the SPGC.

Theorem 5.4 (Chudnovsky and Seymour [12]) *If G is a Berge graph that contains a stretcher, then G is the line graph of a bipartite graph or G has a skew partition or a homogeneous pair, or G or $\bar{G}$ has a 2-join.*

A *wheel* (H, v) consists of a hole H together with a vertex v, called the *center*, with at least three neighbors in H. If v has k neighbors in H, the wheel is called a *k-wheel*. A *line wheel* is a 4-wheel (H, v) that contains exactly two triangles and these two triangles have only the center v in common. A *twin wheel* is a 3-wheel containing exactly two triangles. A *universal wheel* is a wheel (H, v) where the center v is adjacent to all the vertices of H. A *triangle-free wheel* is a wheel containing no triangle. A *proper wheel* is a wheel that is not any of the above four types. These concepts were first introduced in [14]. The following theorem generalizes an earlier result by Conforti, Cornuéjols and Zambelli [21] and Thomas [35].

Theorem 5.5 (Conforti, Cornuéjols and Zambelli [22]) *If G is a Berge graph that contains no proper wheels, stretchers or their complements, then G is basic or has a skew partition.*

The last step in proving the SPGC is the following difficult theorem.

Theorem 5.6 (Chudnovsky and Seymour [13]) *If G is a Berge graph that contains a proper wheel, but no stretchers or their complements, then G has a skew partition, or G or $\bar{G}$ has a 2-join.*

A monumental paper containing these results is forthcoming [11]. Independently, Conforti, Cornuéjols, Vušković and Zambelli [20] proved that the Decomposition Conjecture holds for Berge graphs containing a large class of proper wheels but, as of May 2002, they could not prove it for all proper wheels. Theorems 5.4, 5.5 and 5.6 imply that Conjecture 5.1 holds, and therefore the SPGC is true.

Corollary 5.7 (Strong Perfect Graph Theorem) *The only minimally imperfect graphs are the odd holes and their complements.*

Conforti, Cornuéjols and Vušković [19] proved a weaker version of the Decomposition Conjecture where "skew partition" is replaced by "double star cutset". A *double star* is a vertex set S that contains two adjacent vertices u, v and a subset of the vertices adjacent to u or v. Clearly, if G has a skew partition, then G has a double star cutset: Take $S = A \cup B$, $u \in A$ and $v \in B$. Although the decomposition result in [19] is weaker than Conjecture 5.1 for Berge graphs, it holds for a larger class of graphs than Berge graphs: By changing the decomposition from "skew partition" to "double star cutset", the result can be obtained for all odd-hole-free graphs instead of just Berge graphs.

Theorem 5.8 (Conforti, Cornuéjols and Vušković [19]) *If G is an odd-hole-free graph, then G is a bipartite graph or the line graph of a bipartite graph or the complement of the line graph of a bipartite graph, or G has a double star cutset or a 2-join.*

One might try to use Theorem 5.8 to construct a polynomial time recognition algorithm for odd-hole-free graphs. Conforti, Cornuéjols, Kapoor and Vušković [17] obtained a polynomial time recognition algorithm for the class of even-hole-free graphs. This algorithm is based on the decomposition of even-hole-free graphs by 2-joins, double star and triple star cutsets obtained in [16].

A useful tool for studying Berge graphs is due to Roussel and Rubio [34]. This lemma was proved independently by Robertson, Seymour and Thomas [33], who popularized it and named it *The Wonderful Lemma*. It is used repeatedly in the proofs of Theorems 5.3-5.6.

Lemma 5.9 (The Wonderful Lemma) (Roussel and Rubio [34]) *Let G be a Berge graph and assume that $V(G)$ can be partitioned into a set S and an odd chordless path $P = u, u', \ldots, v', v$ of length at least 3 such that u, v are both adjacent to all the vertices in S and $G(S)$ is connected. Then one of the following holds:*

(i) An odd number of edges of P have both ends adjacent to all the vertices in S.

(ii) P has length 3 and $\bar{G}(S \cup \{u', v'\})$ contains an odd chordless path between u' and v'.

(iii) P has length at least 5 and there exist two nonadjacent vertices x, x' in S such that $(V(P) \setminus \{u, v\}) \cup \{x, x'\}$ induces a path.

References

[1] C. Aossey, 3PC(.,.)-free Berge graphs are perfect, PhD dissertation, University of Kentucky, Lexington, Kentucky (2000).

[2] C. Aossey and K. Vušković, 3PC(.,.)-free Berge graphs are perfect, working paper, University of Kentucky, Lexington, Kentucky (1999), submitted to *Discrete Mathematics*.

[3] C. Berge, Färbung von Graphen deren sämtliche bzw. deren ungerade Kreise starr sind (Zusammenfassung), *Wissenschaftliche Zeitschrift, Martin Luther Universität Halle-Wittenberg, Mathematisch-Naturwissenschaftliche Reihe* (1961) 114-115.

[4] R.E. Bixby, A composition for perfect graphs, in *Topics on Perfect Graphs* (C. Berge and V. Chvátal eds.), *North-Holland Mathematics Studies 88* North Holland, Amsterdam (1984) 221-224.

[5] M. Burlet and J. Fonlupt, Polynomial algorithm to recognize a Meyniel graph, *Annals of Discrete Mathematics 21* (1984) 225-252.

[6] V. Chvátal, Star-cutsets and perfect graphs, *Journal of Combinatorial Theory B* 39 (1985) 189-199.

[7] V. Chvátal, J. Fonlupt, L. Sun and A. Zemirline, Recognizing dart-free perfect graphs, technical report, Rutgers University (2000).

[8] V. Chvátal and N. Sbihi, Bull-free Berge graphs are perfect, *Graphs and Combinatorics 3* (1987) 127-139.

[9] V. Chvátal and N. Sbihi, Recognizing claw-free Berge graphs, *Journal of Combinatorial Theory B* 44 (1988) 154-176.

[10] M. Chudnovsky, N. Robertson, P. Seymour and R. Thomas, presentation at the Workshop on Graph Colouring and Decomposition, Princeton, September 2001.

[11] M. Chudnovsky, N. Robertson, P. Seymour and R. Thomas, The Strong Perfect Graph Theorem, forthcoming.

[12] M. Chudnovsky and P. Seymour, private communication (January 2002).

[13] M. Chudnovsky and P. Seymour, private communication (May 2002).

[14] M. Conforti and G. Cornuéjols, Graphs without odd holes, parachutes or proper wheels: a generalization of Meyniel graphs and of line graphs of bipartite graphs (1999), submitted to *Journal of Combinatorial Theory B*.

[15] M. Conforti, G. Cornuéjols, G. Gasparyan and K. Vušković, Perfect graphs, partitionable graphs and cutsets, *Combinatorica 22* (2002) 19-33.

[16] M. Conforti, G. Cornuéjols, A. Kapoor and K. Vušković, Even-hole-free graphs, Part I: Decomposition theorem, *Journal of Graph Theory 39* (2002) 6-49.

[17] M. Conforti, G. Cornuéjols, A. Kapoor and K. Vušković, Even-hole-free graphs, Part II: Recognition algorithm, to appear in *Journal of Graph Theory* (2002).

[18] M. Conforti, G. Cornuéjols and K. Vušković, Square-free perfect graphs, preprint (2001), to appear in *Journal of Combinatorial Theory B*.

[19] M. Conforti, G. Cornuéjols and K. Vušković, Decomposition of odd-hole-free graphs by double star cutsets and 2-joins, to appear in the special issue of *Discrete Mathematics* dedicated to the Brazilian Symposium on Graphs, Algorithms and Combinatorics, Fortaleza, Brazil, March 2001.

[20] M. Conforti, G. Cornuéjols, K. Vušković and G. Zambelli, Decomposing Berge graphs containing proper wheels, preprint (April 2001, updated March 2002).

[21] M. Conforti, G. Cornuéjols and G. Zambelli, Decomposing Berge graphs con-

taining no proper wheels, big parachutes or their complements (November 2001).

[22] M. Conforti, G. Cornuéjols and G. Zambelli, Decomposing Berge graphs containing no proper wheels, stretchers or their complements, preprint (May 2002).

[23] G. Cornuéjols and W.H. Cunningham, Composition for perfect graphs, *Discrete Mathematics 55* (1985) 245-254.

[24] J. Fonlupt and A. Zemirline, A polynomial recognition algorithm for perfect K_4-$\{e\}$-free graphs, rapport technique RT-16, Artemis, IMAG, Grenoble, France (1987).

[25] G.S. Gasparyan, Minimal Imperfect Graphs: A Simple Approach, *Combinatorica 16* (1996) 209-212.

[26] C. T. Hoàng, Some properties of minimal imperfect graphs, *Discrete Math. 160* (1996) 165-175.

[27] A. Kapoor, On the structure of balanced matrices and perfect graphs, *PhD Thesis, Carnegie Mellon University* (1994).

[28] D. König, Über Graphen und ihre Anwendung auf Determinantentheorie und Mengenlehre, *Math. Ann. 77* (1916) 453-465.

[29] L. Lovász, Normal Hypergraphs and the Perfect Graph Conjecture, *Discrete Mathematics 2* (1972) 253-267.

[30] L. Lovász, A Characterization of Perfect Graphs, *Journal of Combinatorial Theory B 13* (1972) 95-98.

[31] F. Maffray and B. Reed, A description of claw-free perfect graphs, *Journal of Combinatorial Theory B 75* (1999) 134-156.

[32] M. Padberg, Perfect zero-one matrices, *Math. Programming* 6 (1974) 180-196.

[33] N. Robertson, P. Seymour and R. Thomas, presentation at the Workshop on Graph Colouring and Decomposition, Princeton, September 2001.

[34] F. Roussel and P. Rubio, About skew partitions in minimal imperfect graphs, to appear in *Journal of Combinatorial Theory B.*

[35] R. Thomas, private communication (May 2002).

ICM 2002 · Vol. III · 561–571

Singular Combinatorics*

Philippe Flajolet[†]

Abstract

Combinatorial enumeration leads to counting generating functions presenting a wide variety of analytic types. Properties of generating functions at singularities encode valuable information regarding asymptotic counting and limit probability distributions present in large random structures. "Singularity analysis" reviewed here provides constructive estimates that are applicable in several areas of combinatorics. It constitutes a complex-analytic Tauberian procedure by which combinatorial constructions and asymptotic–probabilistic laws can be systematically related.

2000 Mathematics Subject Classification: 05A15, 05A16, 30B10, 39B05, 60C05, 60F05, 68Q25.
Keywords and Phrases: Combinatorial enumeration, Singularity analysis, Analytic combinatorics, Random structures, Limit probability distributions.

1. Introduction

Large random combinatorial structures tend to exhibit great statistical regularity. For instance, an overwhelming proportion of the graphs of a given large size are connected, and a fixed pattern is almost surely contained in a long random string, with its number of occurrences satisfying central and local limit laws. The objects considered (typically, words, trees, graphs, or permutations) are given by construction rules of the kind classically studied by combinatorial analysts via *generating functions* (abbreviated as GFs). A fundamental problem is then to extract asymptotic information on coefficients of a GF either explicitly given by a formula or implicitly determined by a functional equation. In the univariate case, asymptotic counting estimates are derived; in the multivariate case, moments and limit probability laws of characteristic parameters will be obtained.

*Work supported in part by the IST Programme of the EU under contract number IST-1999-14186 (ALCOM-FT).

[†]Algorithms Project, INRIA-Rocquencourt, 78153 Le Chesnay, France. E-mail: Philippe.Flajolet@inria.fr

In what follows, given a combinatorial class $\mathcal{C}$, we let C_n denote the number of objects in $\mathcal{C}$ of size n and introduce the *ordinary* and *exponential* GF (OGF, EGF),

$$\text{OGF:} \quad C(z) := \sum_{n \geq 0} C_n z^n, \qquad \text{EGF:} \quad \widehat{C}(z) := \sum_{n \geq 0} C_n \frac{z^n}{n!}.$$

Generally, EGFs and OGFs serve for the enumeration of labelled classes (atoms composing objects are distinguished by labels) and unlabelled classes, respectively. One writes $C_n = [z^n]C(z) = n! \, [z^n]\, \widehat{C}(z)$, with $[z^n](\cdot)$ the coefficient extractor.

General rules for deriving GFs from combinatorial specifications have been widely developed by various schools starting from the 1970's and these lie at the heart of contemporary combinatorial analysis. They are excellently surveyed in books of Foata & Schützenberger (1970), Comtet (1974), Goulden & Jackson (1983), Stanley (1986, 1998), Bergeron, Labelle & Leroux (1998). We shall retain here the following simplified scheme relating combinatorial constructions and operations over GFs:

Construction		Labelled case	Unlabelled case
Disjoint union	$\mathcal{F} + \mathcal{G}$	$f(z) + g(z)$	$\widehat{f}(z) + \widehat{g}(z)$
Product	$\mathcal{F} \times \mathcal{G}, \; \mathcal{F} \star \mathcal{G}$	$f(z) \cdot g(z)$	$\widehat{f}(z) \cdot \widehat{g}(z)$
Sequence	$\mathfrak{S}\{\mathcal{F}\}$	$(1 - f(z)))^{-1}$	$(1 - f(z)))^{-1}$
Set	$\mathfrak{P}\{\mathcal{F}\}$	$\exp(f(z))$	$\exp\left(f(z) + \frac{1}{2}f(z^2) + \cdots\right)$
Cycle	$\mathfrak{C}\{\mathcal{F}\}$	$\log(1 - f(z))^{-1}$	$\log(1 - f(z))^{-1} + \cdots$

$$(1.1)$$

Such operations on GFs yield a wide variety of analytic functions, either given explicitly or as solutions to functional equations in the case of recursively defined classes. It is precisely the goal of singularity analysis to provide means for extracting asymptotic informations. What we termed "singular combinatorics" aims at relating combinatorial form and asymptotic-probabilistic form by exploiting *complex-analytic* properties of generating functions. Classical approaches [2] are Tauberian theory and Darboux's method, an offspring of elementary Fourier analysis largely developed by Pólya for his programme of combinatorial chemistry [17]. The path followed here, called "singularity analysis" after [7], consists in developing a systematic correspondence between the local behaviour of a function near its singularities and the asymptotic form of its coefficients. (An excellent survey of central aspects of the theory is offered by Odlyzko in [15].)

2. Basic singularity analysis

Perhaps the simplest coefficient estimate is $[z^n](1 - z)^{-\alpha} \sim n^{\alpha-1}/\Gamma(\alpha)$, a consequence of the binomial expansion and Stirling's formula. For the basic scale

$$\sigma_{\alpha,\beta}(z) = (1 - z)^{-\alpha} \left(\frac{1}{z}\log(1 - z)^{-1}\right)^{\beta},$$

much more is available and one has a fundamental translation mechanism [7]:

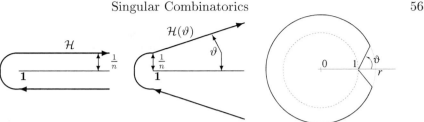

Figure 1: The Hankel contours, $\mathcal{H}$ and $\mathcal{H}(\vartheta)$, and a Δ-domain.

Theorem 1 (Coefficients of the basic scale) *For $\alpha \in \mathbb{C} \setminus \mathbb{Z}_{\leq 0}$, $\beta \in \mathbb{C}$, one has*

$$[z^n]\,\sigma_{\alpha,\beta}(z) \underset{n \to \infty}{\sim} \frac{n^{\alpha-1}}{\Gamma(\alpha)} (\log n)^\beta \,. \tag{2.1}$$

Proof. The estimate is derived starting from Cauchy's coefficient formula,

$$[z^n]f(z) \sim \frac{1}{2i\pi} \int_\gamma f(z)\, \frac{dz}{z^{n+1}},$$

instantiated with $f = \sigma_{\alpha,\beta}$. The idea is then to select for γ a contour $\mathcal{H}$ that is of Hankel type and follows the half-line $(1, +\infty)$ at distance exactly $1/n$ (Fig. 1). This superficially resembles a saddle-point contour, but with the integral normalizing to Hankel's representation of the Gamma function, hence the factor $\Gamma(\alpha)^{-1}$. $\square$

The method of proof is very flexible: it applies for instance to iterated logarithmic terms $(\log\log)$ while providing full asymptotic expansions; see [7] for details.

A remarkable fact, illustrated by Theorem 1, is that larger functions near the singularity $z = 1$ give rise to larger coefficients as $n \to \infty$. This is a general phenomenon under some suitable auxiliary conditions, expressed here in terms of *analytic continuation*: a Δ-domain is an indented disc defined by $(r > 1,\ \vartheta < \pi/2)$

$$\Delta(\vartheta, r) := \{ z \mid |z| < r,\ \vartheta < \mathrm{Arg}(z-1) < 2\pi - \vartheta,\ z \neq 1 \}.$$

Theorem 2 (O-transfer) *With $f(z)$ continuable to a Δ-domain and $\alpha \notin \mathbb{Z}_{\leq 0}$:*

$$f(z) \underset{z \to 1,\ z \in \Delta}{=} O\left(\sigma_{\alpha,\beta}(z)\right) \qquad \implies \qquad [z^n]\,f(z) = O\left([z^n]\,\sigma_{\alpha,\beta}(z)\right).$$

Proof. In Cauchy's coefficient formula, adopt an integration contour $\mathcal{H}(\vartheta)$ passing at distance $1/n$ left of the singularity $z = 1$, then escaping outside of the unit disc within Λ. Upon setting $z = 1 + t/n$, careful approximations yield the result [7]. $\square$

This theorem allows one to transfer error terms in the asymptotic expansion of a function at its singularity (here $z = 1$) to asymptotics of the coefficients. The Hankel contour technique is quite versatile and a statement similar to Theorem 2 holds with $o(\cdot)$-conditions replacing $O(\cdot)$-conditions. The case of α being a negative integer is covered by minor adjustments due to $1/\Gamma(\alpha) = 0$; see [7]. In concrete terms: *Hankel contours combined with Cauchy coefficient integrals accurately "capture" the singular behaviour of a function.*

By Theorems 1 and 2, whenever a function admits an asymptotic expansion near $z = 1$ in the basic scale, one has the implication, with $\sigma \succ \tau \succ \ldots \succ \omega$,

$$f(z) = \lambda\sigma(z) + \mu\tau(z) + \cdots + O(\omega(z)) \quad \Longrightarrow \quad f_n = \lambda\sigma_n + \mu\tau_n + \cdots + O(\omega_n),$$

where $f_n = [z^n]f(z)$. In other words, a *dictionary* translates singular expansions of functions into the asymptotic forms of coefficients. Analytic continuation and validity of functions' expansions outside of the unit circle is a logical necessity, but once granted, application of the method becomes quite mechanical.

For combinatorics, singularities need not be placed at $z = 1$. But since $[z^n]f(z) \equiv \rho^{-n}[z^n]f(z\rho)$, the dictionary can be used for (dominant) singularities that lie anywhere in the complex plane. The case of finitely many dominant singularities can also be dealt with (via composite Hankel contours) to the effect that the translations of local singular expansions get composed additively. In summary one has from function to coefficients:

location+nature of singularity (fn.) $\Longrightarrow$ exponential+polynomial asymptotics (coeff.)

Example 1. *2-Regular graphs* (Comtet, 1974). The class $\mathcal{G}$ of (labelled) 2-regular graphs can be specified as sets of unordered cycles each of length at least 3. Symbolically:

$$\mathcal{G} \cong \mathfrak{P}\{\tfrac{1}{2}\mathfrak{C}_{\geq 3}\{\mathcal{Z}\}\} \quad \text{so that} \quad \widehat{G}(z) = \exp\left(\frac{1}{2}\log(1-z)^{-1} - \frac{z}{2} - \frac{z^2}{4}\right) = \frac{e^{-z/2-z^2/4}}{\sqrt{1-z}}.$$

$\mathcal{Z}$ represents a single atomic node. The function $\widehat{G}(z)$ is singular at $z = 1$, and

$$\mathcal{G}(z) \underset{z\to 1}{\sim} e^{-3/4}(1-z)^{-1/2} \quad \Longrightarrow \quad \frac{G_n}{n!} \underset{n\to\infty}{\sim} \frac{e^{-3/4}}{\sqrt{\pi n}}.$$

This example can be alternatively treated by Darboux's method. $\qquad\square$

Example 2. The *diversity index* of a tree (Flajolet, Sipala & Steyaert, 1990) is the number of non-isomorphic terminal subtrees, a quantity also equal to the size of maximally compact representation of the tree as a directed acyclic graph and related to common subexpression sharing in computer science applications. The mean index of a random binary tree of size $2n+1$ is asymptotic to $Cn/\sqrt{\log n}$, where $C = \sqrt{8\log 2/\pi}$. This results from an exact GF obtained by inclusion-exclusion:

$$K(z) = \frac{1}{2z}\sum_{k\geq 0}\frac{1}{k+1}\binom{2k}{k}\left(\sqrt{1-4z+4z^{k+1}} - \sqrt{1-4z}\right).$$

Singularities accumulate geometrically to the right of $1/4$ while $K(z)$ is Δ–continuable. The unusual singularity type $(1/\sqrt{X\log X})$ precludes the use of Darboux's method. $\quad\square$

Rules like those of Table (1.1) preserve analyticity and analytic continuation. Accordingly, generating functions associated with combinatorial objects described by simple construction rules usually have GFs amenable to singularity analysis. The method is systematic enough, so that an implementation within computer algebra systems is even possible as was first demonstrated by Salvy [18].

3. Closure properties

In what follows, we say that a function is *amenable to singularity analysis*, or "*of S.A. type*" for short, if it is Δ-continuable and admits there a singular expansion in the scale $\mathcal{S} = \{\sigma_{\alpha,\beta}(z)\}$. First, functions of S.A. type include polylogarithms:

Theorem 3 *The generalized polylogarithms* $\mathrm{Li}_{\alpha,k}$ *are of S.A. type, where* $\mathrm{Li}_{\alpha,k}(z) := \sum_{n\geq 1} (\log n)^k n^{-\alpha} z^n$, $k \in \mathbb{Z}_{\geq 0}$.

The proof makes use of the Lindelöf representation

$$\sum_{n\geq 1} \phi(n)(-z)^n = -\frac{1}{2i\pi} \int_{1/2-\infty}^{1/2+\infty} \phi(s) z^s \frac{\pi}{\sin \pi s}\, ds,$$

in a way already explored by Ford [10], with Mellin transform techniques providing validity of the singular expansion in a Δ-domain [5].

Example 3. *Entropy computations.* The GF of $\{\log(k!)\}$ is $(1-z)^{-1} \mathrm{Li}_{0,1}(z)$, which is of S.A. type. The entropy of the binomial distribution, $\pi_{n,k} = \binom{n}{k} p^k (1-p)^{n-k}$, results:

$$H_n := -\sum \pi_{n,k} \log \pi_{n,k} \underset{n\to\infty}{\sim} \frac{1}{2}\log n + \frac{1}{2} + \log\sqrt{2\pi p(1-p)} + \cdots.$$

Such problems are of interest in information theory, where redundancy estimates precisely depend on higher order asymptotic properties (Jacquet-Szpankowski, 1998). Full expansions for functionals of the Bernoulli distribution are also obtained systematically [5]. □

As it is well-known, asymptotic expansions can be integrated while differentiation of asymptotic expansions is permissible for functions analytic in sectors:

Theorem 4 *Functions of S.A. type are closed under differentiation and integration.*

Finally, the Hadamard product of two series, $f \odot g$ is defined as the termwise product: $f(z) \odot g(z) = \sum_n f_n g_n z^n$, if $f(z) - \sum_n f_n z^n$, $g(z) = \sum_n f_n z^n$. Hadamard (1898) proved that singularities get composed multiplicatively. Finer composition properties [4] result from an adaptation of Hankel contours to Hadamard's formula

$$f(z) \odot g(z) = \frac{1}{2i\pi} \int_\gamma f(t) g\left(\frac{w}{t}\right) \frac{dt}{t}.$$

Theorem 5 *Functions of S.A. type are closed under Hadamard product.*

Example 4. *Divide-and-conquer* algorithms solve recursively a problem of size n by splitting it into two subproblems and recombining the partial solutions. Under the assumption of randomness preservation, the expected costs f_n satisfy a "tree recurrence" of the form

$$f_n = t_n + \sum_k \pi_{n,k}\left(f_k + f_{n-k-a}\right), \qquad a \in \{0,1\},$$

where the "toll" sequence t_n usually has a simple form (e.g., $n^\beta, \log n$) and the $\pi_{n,k}$ characterize the stochastic splitting process. The corresponding GFs then satisfy an equation $f(z) = t(z) + \mathcal{L}[f](z)$, where the linear operator $\mathcal{L}$ reflects the splitting probabilities. For instance, binary search trees and the Quicksort algorithm have $\mathcal{L}[f(z)] = 2\int_0^z f(x)dx/(1-x)$. One then has in operator notation $f(z) = (I - \mathcal{L})^{-1}[t](z)$, where the quasi-inverse acts as a "singularity transformer". Closure theorems allow for an asymptotic classification of the cost functions induced by various tolls under various probabilistic models mirrored by the splitting probabilities [4]. □

4. Functional equations

Algebraic functions have expansions at singularities that are expressed by fractional power series (Newton-Puiseux). Consequently, they are of S.A. type with rational exponents; accordingly their coefficient expansions are linear combinations of algebraic elements of the form $\omega^n n^{r/s}$, with ω and algebraic number and $r/s \in \mathbb{Q}$. By Weierstrass preparation, such properties extend to many *implicit* GFs. For instance, GFs of combinatorial families of trees constrained to have degrees in a finite set have a branch point of type $\sqrt{X}$ at their dominant singularity, which in turn corresponds to the "universal" asymptotic form $T_n \sim c\omega^n n^{-3/2}$ for coefficients (Pólya, 1937; Otter 1948; Meir-Moon, 1978).

Order constraints in labelled classes are known to correspond to integral operators and, in the recursive case, there result GFs determined by *ordinary differential equations*. Furthermore, moments (i.e., cumulative values) of additive functionals of combinatorial structures defined by recursion and order constraints have GFs that satisfy *linear differential equations*, for which there is a well-established classification theory going back to the nineteenth century. In particular, in the *Fuchsian case*, singularity analysis applies unconditionally, so that the resulting coefficient estimates are linear combinations of terms $\omega^n n^\alpha (\log n)^k$, with ω, α algebraic and $k \in \mathbb{Z}_{\geq 0}$. This covers a large subset of the class of "holonomic" functions, of which Zeilberger has extensively demonstrated the expressive power in combinatorial analysis [23, 24].

Example 5. *Quadtrees* are a way to superimpose a hierarchical partitioning on sequences of points in d-dimensional space: the first point is taken as the root of the tree and it partitions the whole space in 2^d orthants in which successive points are placed and then made to refine the partition [13]. The problem is expressed by a linear differential equation with coefficients in $\mathbb{C}(z)$. The average cost of finding a point knowing only k of its coordinates is of the asymptotic form $c\, n^\alpha$ with $\alpha = \alpha(k, d)$ an algebraic number of degree d. For instance $k = 1$ and $d = 2$ yield a solution involving a $_2F_1$ hypergeometric function as well as $\alpha = (\sqrt{17} - 3)/2 \doteq 0.56155$, in contrast to an exponent $\frac{1}{2}$ that would correspond to a perfect partitioning, i.e., a regular grid (Flajolet, Gonnet, Puech & Robson, 1993). $\quad\square$

Substitution equations correspond to "balanced structures" of combinatorics. An important rôle in the development of the theory has been played by Odlyzko's analysis [14] of 2-3 trees (such trees have internal nodes of degree 2 and 3 only and leaves are all at the same depth). The OGF satisfies the equation $T(z) = z + T(\tau(z))$, with $\tau(z) := z^2 + z^3$, and has a singularity at $z = 1/\phi$, a fixed point of τ, ($\phi = (1+\sqrt{5})/2$). The singular expansion involves periodic oscillations, corresponding to infinitely many singular exponents having a common real part. Singularity analysis extends to this case and the number of balanced 2-3 trees is found to be of the form $T_n \sim \frac{1}{n}\phi^n \Omega(\log n)$, for some nonconstant smooth periodic function Ω.

A similar problem of *singular iteration* arises in the analysis of the height of binary trees [6]. The GF $y_h(z)$ of trees of height at most h satisfies the Mandelbrot recurrence $y_h = z + y_{h-1}^2$, with $y_0 = z$. The fixed point is the GF of binary trees, that is, of Catalan numbers, $y_\infty = (1 - \sqrt{1 - 4z})/2$ which has its dominant singularity at $1/4$. The analysis of moments of the distribution of height turns out to be equivalent to developing *uniform* approximations to $y_h(z)$ as $z \to 1/4$ and $h \to \infty$ simultaneously, this for z in a Δ-domain. The end result, by singularity analysis and the moment method, is: *the height of a random binary tree with n external nodes when normalized by a factor of $1/(2\sqrt{n})$ converges in distribution to a theta law defined by the density* $4x \sum_{k \geq 1} k^2 (2k^2 x2 - 3)e^{-k^2 x^2}$. The result extends to all simple families of trees in the sense of Meir and Moon and it provides pointwise estimates of the proportion of trees of given height, that is, a *local limit law.*

Generalized digital trees (Flajolet & Richmond, 1992) correspond to a *difference differential equation*, $\partial_z^k \varphi(z) = t(z) + 2e^{z/2} \varphi(z/2)$, whose solution involves basic hypergeometric functions. Catalan sums of the form $\sum_k \binom{2n}{n-k} \nu(k)$, with ν an arithmetical function, arise in the statistics of "order" (also known as Horton-Strahler number) of trees (Flajolet & Prodinger, 1986). Both cases are first subjected to a Mellin transform analysis, which provides the relevant singular expansions. Periodic fluctuations similar to the case of balanced trees then result from singularity analysis.

A notable parallel to the paradigm of generating functions and singularity analysis has been developed by Vallée in a series of papers. In her framework, singularities of certain transfer operators (of Ruelle type) replace singularities of generating functions. See, *e.g.*, [21, 22] for applications to Euclidean algorithms and statistics on sequences produced by a general model of dynamical sources.

5.   Limit laws

One of the important features of singularity analysis, in contrast with Darboux's method or (real) Tauberian theory, is to allow for *uniform* estimates. This makes it possible to analyse asymptotically coefficients of multivariate generating functions, $f(z, u)$, where the auxiliary variable u marks some combinatorial parameter χ. One *first* proceeds to extract $f_n(u) := [z^n]f(z, u)$ by considering $f(z, u)$ as a parameterized family of univariate GFs to which singularity analysis is applied. (The coefficients $f_n(u)$ are, up to normalization, probability generating functions of χ.) A *second* level of inversion is then achieved by the standard continuity theorems for probability characteristic functions (equivalently Fourier transforms). Technically, consideration of a (small) neighbourhood of $u = 1$ is normally sufficient for extracting central limit laws.

Two important cases are those of a smoothly varying singularity and of a smoothly varying exponent. In the first case, $f(z, u)$ has a constant singular exponent α_0 and one has $f(z, u) \sim c(u)(1 - z/\rho(u))^{-\alpha_0}$. Then, uniformity of singularity analysis implies the estimate $f_n(u)/f_n(1) \sim (\rho(1)/\rho(u))^n$. In other words, the probability generating function of χ over objects of size n is analytically similar to the GF of a sum of independent random variables—this situation is described as a *"quasi-powers" approximation*. A Gaussian limit law for χ results from the continuity theorem, with mean and variance that grow in proportion to n. The other case of a smoothly varying exponent is dealt with similarly: one has $f(z, u) \sim c(u)(1 - z/\rho)^{-\alpha(u)}$ implying $f_n(u)/f_n(1) \sim n^{\alpha(u)-\alpha(1)}$; this is once more a quasi-power approximation, but with the parameter now in the scale of $\log n$. (See

Gao & Richmond, 1992, for hybrid cases.)

The technology above builds on early works of Bender [1], continued by Flajolet & Soria [9, 19], and H. K. Hwang [11]. In particular, under general conditions, the following hold: a local limit law expresses convergence to the Gaussian density; speed of convergence estimates result from the Berry-Esseen inequalities; large deviation estimates derive from singularity analysis applied at fixed real values $u \neq 1$.

Example 6. *Polynomials over finite fields.* Consider the family $\mathcal{P}$ of all polynomials with coefficients in the Galois field $\mathbb{F}_q$. A polynomial being determined by its sequence of coefficients, the GF $P(z)$ of all polynomials has a polar singularity. Furthermore, the unique factorization property implies that $\mathcal{P}$ is isomorphic to the class of all multisets ($\mathfrak{M}$) of the irreducible polynomials $\mathcal{I}$: $\mathcal{P} \simeq \mathfrak{M}\{\mathcal{I}\}$. Since taking multisets corresponds to exponentiating singularities of GFs, the singularity of the GF $I(z)$ is logarithmic. By singularity analysis, the number of irreducible polynomials is asymptotic to q^n/n—this is an analogue of the prime number theorem, which was already known to Gauß. The bivariate GF of the number of irreducible factors in polynomials turns out to be of the singular type $(1 - qz)^{-u}$, with a smooth variable exponent, so that: *the number of irreducible factors of a random polynomial over $\mathbb{F}_q$ is asymptotically Gaussian.* This parallels the Erdős-Kac theorem for integers. Similar developments lead to a complete analysis of a major polynomial factorization algorithm (Flajolet, Gourdon & Panario, 2001). □

Movable singularities and exponents occur frequently in the analysis of parameters defined by recursion, leading to algebraic or differential equations, which "normally" admit a smooth perturbative analysis.

Example 7. *Patterns in random strings.* Let Ω be the total number of occurrences of a fixed pattern (as a contiguous block) in a random string over a finite alphabet. For either the Bernoulli model, where letters are independently identically distributed, or the Markov model, the bivariate GF, with z marking the length of the random string and u the number Ω of occurrences, is a *rational function*, as it corresponds to a finite-state device. Perron-Frobenius properties apply for positive u. Therefore the bivariate GF viewed as a function of z has a simple dominant pole at some $\rho(u)$ that is an algebraic (and holomorphic) function of u, for $u > 0$. Quasi-powers approximations therefore hold and the limit law of Ω in random strings of length n is Gaussian. Such facts holds for very general notions of patterns and are developed systematically in Szpankowski's book [20]. □

Example 8. *Non-crossing graphs.* Consider graphs with vertex set the nth roots of unity, constrained to have only non-crossing edges; let the parameter χ be the number of connected components. The bivariate GF $G(z, u)$ is an *algebraic function* satisfying

$$G^3 + (2w^3z^2 - 3w^2z + w - 3)G^2 + (3w^2z - 2w + 3)G + w - 1 = 0.$$

$G(z, 1)$ has a dominant singularity at $\rho(1) = 3/2 - \sqrt{2}$ which gets smoothly perturbed to $\rho(u)$ for u near 1. The singularity type is consistently of the form $(1 - z/\rho(u))^{1/2}$. A central limit law results for the number of components in such graphs (Flajolet–Noy, 1999). Drmota has given general conditions ensuring Gaussian laws for problems similarly modelled by multivariate algebraic functions [3]. □

Example 9. *Profile of quadtrees.* Refer to Example 5. The bivariate GF $f(z, u)$ of node levels in quadtrees satisfies an equation, which, for dimension $d = 3$ reads

$$f(z, u) = 1 + 2^3 u \int_0^z \frac{dx_1}{x_1(1 - x_1)} \int_0^{x_1} \frac{dx_2}{x_2(1 - x_2)} \int_0^{x_2} f(x_3, u)\frac{dx_3}{1 - x_3}.$$

This corresponds to a *linear differential equation* with coefficients in $\mathbb{C}(z, u)$ and a fixed singularity at $z = 1$. The indicial equation is an algebraic one parameterized by u and, when $u \approx 1$, there is a unique largest branch $\alpha(u)$ that determines the dominant regime of the form $(1-z)^{-\alpha(u)}$. This is the case of a movable exponent inducing a central limit law: *The level profile of a d-dimensional quadtree is asymptotically Gaussian.* Such properties are expected in general for models that are perturbations of linear differential equations with a fixed Fuchsian singularity (Flajolet & Lafforgue, 1994). □

Finally, singularity analysis also intervenes by making it possible to "pump" moments of combinatorial distributions. Examples include the height of trees discussed earlier, as well as tree path length (Louchard 1983, Takács 1991) and the construction cost of hashing tables (Flajolet, Poblete & Viola, 1998). The latter problems were first shown in this way to converge to Brownian Excursion Area.

6.   Conclusions

Elementary combinatorial structures are enumerated by generating functions that satisfy a rich variety of functional relations. However, the singular types that are observed are usually somewhat restricted, and *driven by combinatorics*. In simpler cases, the generating functions are explicit combinations of a standard set of special functions. Next, implicitly defined functions (associated with recursion) have singularities that arise from failures of the implicit function theorem and are consequently of the algebraic type, often with exponent $\frac{1}{2}$. Linear differential equations have a well-established classification theory that, in the Fuchsian case, leads to algebraic-logarithmic singularities. In all such cases, the singular expansion is known to be valid outside of the original disc of convergence of the generating function. This means that singularity analysis is *automatically* applicable, and precise asymptotic expansions of coefficients result.

Parameters of combinatorial structures, provided they remain "simple" enough, lead to local deformations (via an auxiliary variable u considered near 1) of the functional relations defining univariate counting generating functions. Under fairly general conditions, such deformations are amenable to perturbation theory and admit of uniform expansions near singularities. Then, since the singularity analysis process preserves uniformity, limit laws result via the continuity theorem for characteristic functions. In this way, the behaviour of a large number of parameters of elementary combinatorial structures becomes predictable. (The theory of functions of several complex variables is thus bypassed. See Pemantle's recent work [16] based on this theory for a global characterization of all the asymptotic regimes involved.)

The generality of the singular approach makes it even possible to discuss *combinatorial schemas* at a fair level of generality [8, 9, 11, 19], the case of polynomial factorization (Ex. 6) being typical. Roughly, combinatorial constructions viewed as "singularity transformers" dictate asymptotic regimes and probabilistic laws. Analytic combinatorics then represents an attractive alternative to probabilistic methods, whenever a strong analytic structure is present—this is the case for most combinatorial problems that are "decomposable" and amenable to the generating function methodology. Very precise asymptotic information on the randomness properties of large random combinatorial objects results from there. This in turn has useful implications in the analysis of many fundamental algorithms and data structures of computer science, following the steps of Knuth's pioneering works [12].

References

[1] Edward A. Bender, *Central and local limit theorems applied to asymptotic enumeration*, Journal of Combinatorial Theory **15** (1973), 91–111.

[2] _____, *Asymptotic methods in enumeration*, SIAM Review **16** (1974), no. 4, 485–515.

[3] Michael Drmota, *Systems of functional equations*, Random Structures & Algorithms **10** (1997), no. 1–2, 103–124.

[4] James A. Fill, Philippe Flajolet, and Nevin Kapur, *Singularity Analysis, Hadamard Products, and Trees Recurrences*, Preprint, 2002.

[5] Philippe Flajolet, *Singularity analysis and asymptotics of Bernoulli sums*, Theoretical Computer Science **215** (1999), no. 1-2, 371–381.

[6] Philippe Flajolet and Andrew M. Odlyzko, *The average height of binary trees and other simple trees*, Journal of Computer and System Sciences **25** (1982), 171–213.

[7] _____, *Singularity analysis of generating functions*, SIAM Journal on Algebraic and Discrete Methods **3** (1990), no. 2, 216–240.

[8] Philippe Flajolet and Robert Sedgewick, *Analytic combinatorics*, 2001, Book in preparation; see also INRIA Research Reports 1888, 2026, 2376, 2956, 3162, 4103.

[9] Philippe Flajolet and Michèle Soria, *General combinatorial schemas: Gaussian limit distributions and exponential tails*, Discrete Mathematics **114** (1993), 159–180.

[10] W. B. Ford, *Studies on divergent series and summability*, 3rd ed., Chelsea, New York, 1960, (From two books originally published in 1916 and 1936.).

[11] Hsien-Kuei Hwang, *Théorèmes limites pour les structures combinatoires et les fonctions arithmetiques*, Ph.D. thesis, École Polytechnique, December 1994.

[12] Donald E. Knuth, *Selected papers on analysis of algorithms*, CSLI Publications, Stanford, CA, 2000.

[13] Hosam M. Mahmoud, *Evolution of random search trees*, John Wiley, New York, 1992.

[14] A. M. Odlyzko, *Periodic oscillations of coefficients of power series that satisfy functional equations*, Advances in Mathematics **44** (1982), 180–205.

[15] _____, *Asymptotic enumeration methods*, Handbook of Combinatorics (R. Graham, M. Grötschel, and L. Lovász, eds.), vol. II, Elsevier, Amsterdam, 1995, 1063–1229.

[16] Robin Pemantle, *Generating functions with high-order poles are nearly polynomial*, Mathematics and computer science (Versailles, 2000), Birkhäuser, Basel, 2000, 305–321.

[17] G. Pólya, *Kombinatorische Anzahlbestimmungen für Gruppen, Graphen und chemische Verbindungen*, Acta Mathematica **68** (1937), 145–254.

[18] Bruno Salvy, *Asymptotique automatique et fonctions génératrices*, Ph. D. thesis, École Polytechnique, 1991.

[19] Michèle Soria-Cousineau, *Méthodes d'analyse pour les constructions combinatoires et les algorithmes*, Doctorat ès sciences, Université de Paris–Sud, Orsay, July 1990.

[20] Wojciech Szpankowski, *Average-case analysis of algorithms on sequences*, John Wiley, New York, 2001.

[21] Brigitte Vallée, *Dynamics of the binary Euclidean algorithm: Functional analysis and operators*, Algorithmica **22** (1998), no. 4, 660–685.

[22] ———, *Dynamical sources in information theory: Fundamental intervals and word prefixes*, Algorithmica **29** (2001), no. 1/2, 262–306.

[23] Jet Wimp and Doron Zeilberger, *Resurrecting the asymptotics of linear recurrences*, Journal of Mathematical Analysis and Applications **111** (1985), 162–176.

[24] Doron Zeilberger, *A holonomic approach to special functions identities*, Journal of Computational and Applied Mathematics **32** (1990), 321–368.

ICM 2002 · Vol. III · 573–586

Finite Metric Spaces—Combinatorics, Geometry and Algorithms

Nathan Linial*

Abstract

Finite metric spaces arise in many different contexts. Enormous bodies of data, scientific, commercial and others can often be viewed as large metric spaces. It turns out that the metric of graphs reveals a lot of interesting information. Metric spaces also come up in many recent advances in the theory of algorithms. Finally, finite submetrics of classical geometric objects such as normed spaces or manifolds reflect many important properties of the underlying structure. In this paper we review some of the recent advances in this area.

2000 Mathematics Subject Classification: Combinatorics, Algorithms, Geometry.
Keywords and Phrases: Finite metric spaces, Distortion, graph, Normed space, Approximation algorithms.

1. Introduction

The constantly intensifying ties between combinatorics and geometry are among the most significant developments in Discrete Mathematics in recent years. These connections are manifold, and it is, perhaps, still too early to fully evaluate this relationship. This article deals only with what might be called *the geometrization of combinatorics*. Namely, the idea that viewing combinatorial objects from a geometric perspective often yields unexpected insights. Even more concretely, we concentrate on finite metric spaces and their embeddings.

To illustrate the underlying idea, it may be best to begin with a practical problem. There are many disciplines, scientific, technological, economic and others, which crucially depend on the analysis of large bodies of data. Technological advances have made it possible to collect enormous amounts of interesting data, and further progress depends on our ability to organize and classify these data so

---

*School of Computer Science and Engineering, Hebrew University, Jerusalem 91904, Israel. E-mail:nati@cs.huji.ac.il

as to allow meaningful and insightful analysis. A case in point is bioinformatics where huge bodies of data - DNA sequences, protein sequences, information about expression levels etc. all await analysis. Let us consider, for example, the space of all proteins. For the purpose of the current discussion, a protein may be viewed as a word in an alphabet of 20 letters (amino acids). Word lengths vary from under fifty to several thousands, the most typical length being several hundred letters. At this writing, there are about half a million proteins whose sequence is known. Algorithms were developed over the years to evaluate the similarity of different proteins, and there are standard computer programs that calculate distances among proteins very efficiently. This turns the collection of all known proteins into a metric space of about half a million elements. Proper analysis of this space is of great importance for the biological sciences. Thus, this huge body of sequence data takes a geometric form, namely, a finite metric space, and it becomes feasible to use geometric concepts and tools in the analysis of this data.

In the combinatorial realm proper, and in the design and analysis of algorithms, similar ideas have proved very useful as well. A graph is completely characterized by its (shortest path, or geodesic) metric. The analysis of this metric provides a lot of useful information about the graph. Moreover, given a graph G, one may modify G's metric by assigning nonnegative *lengths* to G's edges. By varying these edge lengths, a family of finite metrics is obtained, the properties of which reflect a good deal of structural information about G. We mention in passing that there are other useful and interesting geometric viewpoints of graphs. Thus, it is useful to geometrically realize a graph by assigning vectors to the vertices and posit that adjacent vertices correspond to orthogonal vectors. Graphs can encode the intersection patterns of geometric objects. These are all interesting instances of our basic paradigm: In the study of combinatorial objects, and especially graphs, it is often beneficial to develop a perspective from which the graph is perceived geometrically.

Aside from what has already been thus accomplished, this approach holds a great promise. Combinatorics as we know it, is still a very young subject. (There is no official date of birth, and Euler was undoubtedly a giant in our field, but I think that the dawn of modern combinatorics can be dated to the 1930's). Discrete Mathematics stands to gain a lot from interactions with older, better established fields. This geometrization of combinatorics indeed creates clear and tangible connections with various subfields of geometry. So far the study of finite metric spaces has had substantial connections with the theory of finite-dimensional normed spaces, but it seems safe to predict that useful ties with differential geometry will soon emerge. With the possible incorporation of probabilistic tools, now commonplace in combinatorics, we can expect very exciting outcomes.

A good sign for the vitality of this area is the large number of intriguing open problems. We will present here some of those that we particularly like. In a recent meeting (Haifa, March '02), a list of open problems in this area has been collected, see http://www.kam.mff.cuni.cz/˜matousek/haifaop.ps. More extensive surveys of this area can be found in [Mat02] Chapter 15, and [Ind01].

In view of this description, it should not come as a surprise to the reader that this theory is characterized as being

- *Asymptotic:* We are mostly interested in analyzing large, finite metric spaces, graphs and data sets.
- *Approximate:* While it is possible to postulate that the geometric situation agrees perfectly with the combinatorics, it is much more beneficial to investigate the approximate version. This leads to a richer theory that is quantitative in nature. Rather than a binary question whether perfect mimicking is possible or not, we ask *how well* a given combinatorial object can be approximated geometrically.
- *Algorithmic:* Existential results are very important and interesting in this area, but we always prefer it when such a result is accompanied by an efficient algorithm.
- It is mostly *comparative:* There are certain classes of finite metric spaces that we favor. These may have a particularly simple structure or be very well understood. Other, less well behaved spaces are being compared to, and approximated by, these "nice" metrics.

So, how should we compare between two metrics? Let (X, d) and (Y, ρ) be two metric spaces and let $\varphi : X \to Y$ be a mapping between them. We quantify the extent to which φ expands, resp. contracts distances: $\text{expansion}(\varphi) = \sup_{x,y \in X} \frac{\rho(\varphi(x),\varphi(y))}{d(x,y)}$ and $\text{contraction}(\varphi) = \sup_{x,y \in X} \frac{d(x,y)}{\rho(\varphi(x),\varphi(y))}$.

Finally, the main definition is: $\text{distortion}(\varphi) = \text{expansion}(\varphi) \cdot \text{contraction}(\varphi)$.

In other words, we consider the tightest constants $\alpha \geq \beta$ for which $\alpha \geq \frac{\rho(\varphi(x),\varphi(y))}{d(x,y)} \geq \beta$ always holds, and define $\text{distortion}(\varphi)$ as $\frac{\alpha}{\beta}$. We call φ an *isometry* when $\text{distortion}(\varphi) = 1$. This deviates somewhat from the conventional definition, and a map that multiplies all distances by a constant (not necessarily 1) is being considered here as an isometry.

The least distortion with which (X, d) can be embedded in (Y, ρ) is denoted $c_Y(X) = c_Y(X, d)$. If $\mathcal{C}$ is a class of metric spaces, then the infimum of $c_Y(X)$ over all $Y \in \mathcal{C}$ is denoted by $c_{\mathcal{C}}(X)$. When $\mathcal{C}$ is the class of finite-dimensional l_p spaces $\{l_p^n | n = 1, 2, \ldots\}$ we denote $c_{\mathcal{C}}(X)$ by $c_p(X)$.

One of the **major problems** in this area is:

Problem 1. Given a finite metric space (X, d) and a class of metrics $\mathcal{C}$, find the (nearly) best approximation for X by a metric from $\mathcal{C}$. In other words, find a metric space $Y \in \mathcal{C}$ and a map $\varphi : X \to Y$ such that $\text{distortion}(\varphi)$ (nearly) equals $c_{\mathcal{C}}(X)$.

The classes of metric spaces $\mathcal{C}$ for which this problem has so far been studied are: (i) Metrics of normed spaces, especially l_p^n for $\infty \geq p \geq 1$ and $n = 1, 2, \ldots$. (ii) Metrics of special families of graphs, most notably trees, as well as convex combinations thereof.

One more convention: Speaking of l_p, either means infinite dimensional l_p, or, what is often the same, that we do not care about the dimension of the space in which we embed a given metric.

To get a first feeling for this subject, let us consider the smallest nontrivial example. Every 3-point metric embeds isometrically into the plane, but as we show now, the metric of $K_{1,3}$, the 4-vertex tree with a root and three leaves, has no isometric embedding into l_2. Let x, resp. y_i be the image of the root and the leaves of this tree. Since $d(x, y_i) = 1$ and $d(y_i, y_j) = 2$ for all $i \neq j$, it follows that the three points x, y_i, y_j are colinear for every $i \neq j$. Thus, all four points are colinear, leading to a contradiction. It can be shown that the least distorted image of this graph in l_2 is in the plane with $120°$ degree angle among the edges. Below (Section 2.) we present a polynomial-time algorithm that determines $c_2(X)$, the least l_2 distortion for any finite metric (X, d).

Another easy fact which belongs into this warm-up section is that $c_\infty(X) = 1$ for every finite metric (X, d). That is, the space l_∞ space contains an isometric copy of every finite metric space.

Acknowledgment: Helpful remarks on this article by R. Krauthgamer, A. Magen, J. Matoušek, and Yu. Rabinovich are gratefully acknowledged.

2.   Embedding into l_2

This is by far the most developed part of the theory. There are several good reasons for this part of the theory to have attracted the most attention so far. Consider the practical context, where a metric space represents some large data set, and where the major driving force is the search for good algorithms for data analysis. If the data set you need to analyze happens to be a large set of points in l_2, there are many tools at your disposal, from geometry, algebra and analysis. So if your data can be well approximated in l_2, this is of great practical advantage. There is another reason for the special status of l_2 in this area. To explain it, we need to introduce some terminology from Banach space theory. The *Banach-Mazur distance* among two normed spaces X and Y, is said to be $\leq c$, if there is a *linear* map $\varphi : X \to Y$ with distortion$(\varphi) \leq c$. What we are doing here may very well be described as a search for the metric counterpart of this highly developed linear theory. See [MS86] for an introduction to this field and [BL00] for a comprehensive cover of the nonlinear theory. The grandfather of the linear theory is the celebrated theorem of Dvoretzky [Dvo61].

Theorem 1 (Dvoretzky). *For every n and $\epsilon > 0$, every n-dimensional normed space contains a $k = \Omega(\epsilon^2 \cdot \log n)$-dimensional space whose Banach-Mazur distance from l_2 is $\leq 1 + \epsilon$.*

Thus, among embeddings into normed spaces, embeddings into l_2 are the hardest to come by.

We begin our story with an important theorem of Bourgain [Bou85].

Theorem 2. *Every n-point metric space [1] embeds in l_2 with distortion $\leq O(\log n)$.*

Not only is this a fundamental result, Bourgain's proof of the theorem readily translates into an efficient randomized algorithm that finds, for any given finite

---

[1] Here and elsewhere, unless otherwise stated, $n = |X|$, the cardinality of the metric space in question.

(X, d) an embedding in l_2 of distortion $\leq O(\log n)$. The algorithm is so simple that we record it here. Given the metric space (X, d), we map every point $x \in X$ to $\varphi(x)$, an $O(\log^2 n)$-dimensional vector. Coordinates in $\varphi(\cdot)$ correspond to subsets $S \subseteq X$, and the S-th coordinate in $\varphi(x)$ is simply $d(x, S)$, the minimum of $d(x, y)$ over all $y \in S$. To define the map φ, we need to specify, then, the collection of subsets S that we utilize. These sets are selected randomly. Namely, you randomly select $O(\log n)$ sets of size 1, another $O(\log n)$ sets of size 2, of size $4, 8..., \frac{n}{2}$.

In view of Bourgain's Theorem, several questions suggest themselves naturally:

- Is this bound tight? The answer is positive, see Theorem 3.
- Given that $\max c_2(X)$ over all n-point metrics is $\Theta(\log n)$, what about metrics that are closer to l_2? Is there a polynomial-time algorithm to compute $c_2(X, d)$ (That is, the least distortion in an embedding of X into l_2)? Again the answer is affirmative, see below and Theorem 4.
- Are there interesting families of metric spaces for which c_2 is substantially smaller than $\log n$? Indeed, there are, see, e.g., Theorem 5.

So let us proceed with the answers to these questions. *Expanders* are graphs which cannot be disconnected into two large subgraphs by removing relatively few edges. Specifically, a graph G on n vertices is said to be an ϵ-*(edge)-expander* if, for every set S of $\leq n/2$ vertices, there are at least $\epsilon|S|$ edges between S and its complement. It is said to be k-*regular* if every vertex has exactly k neighbors. The theory of expander graphs is a fascinating chapter in discrete mathematics and theoretical computer science. It is not obvious that arbitrarily large k-regular graphs exist with expansion ϵ bounded away from zero. In fact, in the early days of this area, conjectures to the contrary had been made. It turns out, however, that expanders are rather ubiquitous. For every $k \geq 3$, the probability that a randomly chosen k-regular graph has expansion $\epsilon > k/10$ tends to 1 as the number of vertices n tends to ∞. It turns out that the metrics of expander graphs are as far from l_2 as possible. [2]

Theorem 3 ([LLR95], see also [Mat97, LM00]). *Let G be an n-vertex k-regular ϵ-expander graph ($k \geq 3$, $\epsilon > 0$). Then $c_2(G) > c \log n$ where c depends only on k and ϵ.*

Metric geometry is by no means a new subject, and indeed metrics that embed isometrically into l_2 were characterized long ago (see e.g. [Blu70]). This is a special case of the more recent results. Let $\varphi : X \to l_2^n$ be an embedding. The condition that distortion$(\varphi) \leq c$ can be expressed as a system of linear inequalities in the entries of the Gram matrix corresponding to the vectors in $\varphi(X)$. Therefore, the computation of $c_2(X)$ is an instance of *semidefinite quadratic programming* and can be found in polynomial time. [3] This formulation of the problem has, however, other useful consequences. The duality principle of convex programming yields a max-min formula for c_2.

---

[2] We freely interchange between a graph and its (shortest path) metric.

[3] This is not quite accurate. Given an n-point space (X, d) and $\epsilon > 0$, the algorithm can determine $c_2(X, d)$ with relative error $< \epsilon$ in time polynomial in n and $\frac{1}{\epsilon}$.

Theorem 4 ([LLR95]). *For every finite metric space* (X, d),

$$c_2(X, d) = \max \sqrt{\frac{\sum_{i,j:q_{i,j}>0} d^2(i,j)q_{i,j}}{\sum_{i,j:q_{i,j}<0} d^2(i,j)|q_{i,j}|}},$$

where the maximum is over all matrices Q *so that*

1. Q *is positive semidefinite, and*
2. *The entries in every row in* Q *sum to zero.*

Consider the metric of the r-dimensional cube. As shown by Enflo [Enf69], the least distorted embedding of this metric is simply the identity map into l_2^r, which has distortion $\sqrt{r}$. Our first illustration for the power of the quadratic programming method is that we provide a quick elementary proof for this fact, earlier proofs of which required heavier machinery. The rows and columns of the matrix Q are indexed by the 2^r vertices of the r-dimensional cube. The (x, y) entry of Q is: (i) $r - 1$ if $x = y$, (ii) It is -1 if x and y are neighbors (they are represented by two $0, 1$ vectors that differ in exactly one coordinate, and (iii) It is 1 if x and y are antipodal, i.e., they differ in all r coordinates. (iv) All other entries of Q are zero. We leave out the details and only indicate how to prove that Q is positive semidefinite. It is possible to express $Q = (r - 1)I - A + P$, where A is the adjacency matrix of the r-cube and P is the (permutation) matrix corresponding to being antipodal. The eigenfunctions of A are well known, namely, they are the 2^r Walsh functions. The same vectors happen to be also the eigenvectors of Q and all have nonnegative eigenvalues.

As another application of this method (also from [LM00]), here is a quick proof of Theorem 3. It is known [Alo86] that if G is a k-regular ϵ-expander graph and A is G's adjacency matrix, then the second eigenvalue of A is $< k - \delta$ for some δ that depends on k and ϵ, but not on the size of the graph [4]. It is not hard to show that the vertices of a graph with bounded degrees can be paired up so that every two paired vertices are at distance $\Omega(\log n)$. Let P be the permutation matrix corresponding to such a pairing. It is not hard to establish Theorem 3 using the matrix $Q = kI - A + \frac{\delta}{2}(P - I)$. More sophisticated applications of this method will be described below (Theorem 7).

3. Specific families of graph metrics

For various graph families, it is possible find embeddings into l_2 with distortion asymptotically smaller than $\log n$. This often applies as well to graphs with arbitrary nonnegative edge lengths.

3.1. Trees

---

[4] A's first eigenvalue is clearly k. This is the combinatorial analogue of Cheeger's Theorem [Che70] about the spectrum of the Laplacian.

The metrics of trees are quite restricted. They can be characterized through a four-term inequality (e.g. [DL97]). It is also not hard to see that every tree metric embeds isometrically into l_1. They can also be embedded into l_2 with a relatively low distortion.

Theorem 5 (Matoušek [Mat99]). *Every tree on n vertices can be embedded into l_2 with distortion $\leq O(\sqrt{\log\log n})$.*

Bourgain [Bou86] had earlier shown that this bound is attained for complete binary trees. (See [LS] for an elementary proof of this.)

3.2. Planar graphs

It turns out that the metrics of planar graphs have good embedding into l_2. Rao [Rao99] showed:

Theorem 6. *Every planar graph embeds in l_2 with distortion $O(\sqrt{\log n})$.*

A recent construction of Newman and Rabinovich [NR02] shows that this bound is tight.

3.3. Graphs of high girth

The *girth* of a graph is the length of the shortest cycle in the graph. If you restrict your attention (as we do in this section) to graphs in which all vertex degrees are ≥ 3, then it is still a major challenge to construct graphs with very high girth, i.e., having no short cycles. The metrics of such graphs seem far from l_2, so in [LLR95] it was conjectured that $c_2(G) \geq \Omega(g)$ for every graph G of girth g in which all vertex degrees are > 3. There are known examples of n-vertex k-regular expanders whose girth is $\Omega(\log n)$. In view of Theorem 2, such graphs show that this conjecture, if true, is best possible. Recently, the following was shown:

Theorem 7 ([LMN]). *Let G be a k-regular graph $k \geq 3$ with girth g. Then $c_2(G) \geq \Omega(\sqrt{g})$.*

Two proofs of this theorem are given in [LMN]. One is based on the notion of *Markov Type* due to Ball [Bal92]. The underlying idea of this proof is that a random walk on a graph with girth g and all vertex degree ≥ 3 drifts at a constant speed away from its starting point for time $\Omega(g)$. On the other hand, in an appropriately defined class of random walks in Euclidean space, at time T the walk is expected to be only $O(\sqrt{T})$ away from its origin. If we compare between the graph itself and its image under an embedding in l_2, this discrepancy must be accounted for by a metrical distortion. The comparison at time $T = \Theta(g)$ yields a distortion of $\Omega(\sqrt{g})$.

The other proof again employs semidefinite programming, using the matrix $Q = \alpha I - A + \beta B$. Here A is the graph's adjacency matrix, and B is a $0, 1$ matrix where $B_{xy} = 1$ if x and y are at distance $g/2$ in G. The parameters α and β have to satisfy the two conditions from Theorem 4. A key observation is that due to the high girth, B can be expressed as $P_{g/2}(A)$ where P_j is the j-th *Geronimus Polynomial*, a known family of orthogonal polynomials. The proof depends on the distribution of zeros for these polynomials, and other analytical properties that they have.

Our present state of knowledge leads us to ask:

Open Problem 1. How small can $c_2(G)$ be for a a graph G of girth g in which all vertices have degree ≥ 3? The answer lies between $\Omega(\sqrt{g})$ and $O(g)$.

An earlier result of Rabinovich and Raz [RR98] reveals another connection between high girth and distortion. Let φ be a map from a graph of girth g to a graph of smaller *Euler characteristic* ($|E| - |V| + 1$). Then distortion$(\varphi) \geq \Omega(g)$.

4. Algorithmic applications

Among the most pleasing aspects of this field, are the many beautiful applications it has to the design of new algorithms.

4.1. Multicommodity flow and sparsest cuts

Flows in networks are a classical subject in discrete optimization and a topic of many investigations (see [Sch02] for a comprehensive coverage). You are given a *network* i.e., a graph with two specified vertices: The *source s* and the *sink t*. Edges have nonnegative *capacities*. The objective is to ship as much of a given commodity between s and t, subject to two conditions: (i) In every vertex other than s and t, matter is conserved, (ii) The flow through any edge must not exceed the edge capacity. Let the set S *separate* the vertices s and t, i.e., it contains exactly one of them. Define S's *capacity* as the sum of edge capacities over those edges that connect S to its complement. The *Max-flow Min-cut* Theorem states that the largest possible flow equals the minimum such capacity.

Here we consider the *k-commodity* version: Now there are k source-sink pairs $s_i, t_i, i = 1, 2, ..., k$ for the i-th commodity, and the i-th *demand* is $D_i > 0$. We seek to determine the largest $\phi > 0$ for which it is possible to flow $\phi \cdot D_i$ of the i-th commodity between s_i and t_i, simultaneously for all $k \geq i \geq 1$ subject to conditions (i) and (ii) above where in (ii) the *total* flow through an edge should not exceed its capacity. With every subset of the vertices S we associate $\gamma(S) = \frac{\mathrm{cap}(S)}{\mathrm{dem}(S)}$. As before, cap$(S)$ is the sum of the capacities of edges between S and its complement. The denominator dem(S) is $\sum D_i$ over all indices i so that S separates s_i and t_i. It is trivially true that $\phi \leq \gamma(S)$, for every flow and every set S, but unlike the one-commodity case, $\min \gamma(S)$ (the *sparsest cut*) need not equal $\max \phi$. As for the algorithmic perspective, finding $\max \phi$ is a linear program, so it can be computed in polynomial time. However, it is NP-hard to determine the sparsest cut. Also, it is interesting to find out how far $\max \phi$ and $\min \gamma(S)$ can be. Consider the case where the underlying graph is an expander, edges have unit capacities and every pair of vertices form a source-sink pair with a unit demand. It is not hard to see that in this case $\phi \leq O(\frac{\min \gamma(S)}{\log n})$. On the other hand,

Theorem 8 ([LLR95], see also [AR98]). *In the k-commodity problem*

$$\max \phi \geq \Omega(\frac{\min \gamma(S)}{\log k}).$$

We will be able to review the proof in Section 5..

4.2. Graph bandwidth

In this computational problem, we are presented with an n-vertex graph G. It is required to label the vertices with distinct labels from $\{1, \ldots, n\}$ so that the difference between the labels of any two adjacent vertices is not too big. Namely,

$$\mathrm{bw}(G) = \min_{\psi} \max_{xy \in E(G)} |\psi(x) - \psi(y)|,$$

where the minimum is over all $1:1$ maps $\psi : V \to \{1, \ldots, n\}$.

It is NP-hard to compute this parameter, and for many years no decent approximation algorithm was known. However, a recent paper by Feige [Fei00] provides a polylogarithmic approximation for the bandwidth. The statement of his algorithm is simple enough to be recorded here:

1. Compute (a slight modification of) the embedding $\varphi : G \to l_2$ that appears in the proof of Bourgain's Theorem 2.
2. Select a random line l and project $\varphi(G)$ onto it.
3. Label the vertices of G by the order at which their images appear along the line l.

Let $\beta(G) := \max_{x,r} \frac{|B_r(x)|}{r}$ where $B_r(x)$ is the set of those vertices in G at distance $\leq r$ from x. It's easy to see that $\mathrm{bw}(G) \geq \Omega(\beta(G))$ and an interesting feature of Feige's proof is that it shows that $\mathrm{bw}(G) \leq O(\beta(G) \log^c n)$. His paper gives $c = 3.5$ which was later [DV99] improved to $c = 3$.

Open Problem 2. Is it true that $\mathrm{bw}(G) \leq O(\beta(G) \log n)$?

It is not hard to see that this bound would be tight for expanders.

4.3. Bartal's method

The following general structure theorem of Bartal [Bar98] has numerous algorithmic applications:

Theorem 9. *For every finite metric space (X, d) there is a collection of trees $\{T_i \mid i \in I\}$, each of which has X as its set of leaves, and positive weights $\{p_i \mid i \in I\}$ with $\sum_I p_i = 1$. Each of these tree metrics dominates d, i.e., $dist_{T_i}(x, y) \geq d(x, y)$ for every i and every $x, y \in X$. On the other hand, for every $x, y \in X$,*

$$\sum_i p_i \cdot dist_{T_i}(x, y) \leq O(\log n \cdot \log \log n \cdot d(x, y)).$$

Bartal's algorithmic paradigm is a general principle underlying the numerous algorithmic applications of this theorem: Given an algorithmic problem on input a graph or a general metric space (X, d), find a collection of tree metrics T_i and weights p_i as in Theorem 9. Select one of the trees at random, where T_i is selected with probability p_i. Now solve the problem for input T_i. (This description assumes, and this is often the case, that the original optimization problem is NP-hard in general, but feasible for tree metrics.

There are two features of the proof that we'd like to mention:
The trees T_i are *HST*'s. In such trees, edge lengths decrease exponentially as you move from the root toward the leaves. They feature prominently in many recent developments in this area.
The proof makes substantial use of *sparse decompositions* of graphs. Given a graph, one seeks a probability distribution on all partitions of the vertex set, so that (i) Parts have small diameters (ii) Adjacent vertices are very likely to reside in the same part. Such partitions have proved instrumental in the design of many algorithms. In fact, an important tool in Rao's Theorem 6 was an earlier result [KPR93] about the existence of very sparse partitions for the members of any minor-closed families family of graphs.

5.   The mysterious l_1

We know much less about metric embeddings into l_1, and the attempts to understand them give rise to many intriguing open problems. We start by defining the *cut metric* d_S on X where $S \subseteq X$, as follows: $d_S(x, y) = 1$ if x, y are separated by S and is zero otherwise. A simple, but useful observation is that the collection of all n-point metrics in l_1 form a cone $\mathcal{C}$ whose extreme rays are the cut metrics. [5] The book [DL97] provides a coverage of this area.

We are now able to complete the proof of Theorem 8. We retain the terminology of the discussion around that theorem. Linear programming duality yields the following alternative expression for the maximum k-commodity flow problem on $G = (V, E)$:

$$\max \phi = \min \frac{\sum_E d(i, j) \cdot c_{ij}}{\sum_1^k D_j \cdot d(s_j, t_j)}.$$

Here the minimum is over all graphical metrics d on G. Namely, you assign nonnegative *lengths* to G's edges and d is the induced shortest path metric on G's vertices. Now let d be the graphical metric that minimizes this expression. A slight adaptation of Bourgain's embedding algorithm yields an l_1 metric ρ so that $\rho(i, j) \leq d(i, j)$ for all i, j and $\rho(s_j, t_j) \geq \Omega(\frac{d(s_j, t_j)}{\log k})$ for all j. But the minimum of $\frac{\sum_E \rho(i, j) \cdot c_{ij}}{\sum_1^k D_j \cdot \rho(s_j, t_j)}$ over l_1 metrics is attained for ρ a cut metric, since cut metrics are the extreme rays of the cone of l_1 metrics $\mathcal{C}$. This minimum, over cut metrics is simply $\min \gamma(S)$, the sparsest cut value of the network. The conclusion follows.

The identification between l_1 metrics and the cut cone $\mathcal{C}$ makes it desirable to find an algorithm to solve linear optimization problems whose feasible set is this

---

[5]For each n, the n-point metrics in l_1 form a cone $\mathcal{C}_n$, but we suppress the index n.

convex cone. Such an algorithm would solve at one fell swoop a host of interesting (and hard) problems such as max-cut, graph bisection and more. This hope is hard to realize, since the ellipsoid method (e.g. [Sch02]) applies only to convex bodies for which we have efficient *membership and separation oracles*. For the convex cone C, that would mean that we need to efficiently determine whether a given a real symmetric matrix M, represents the metric on n points in l_1. Moreover, if not, we ought to find a hyperplane (in n^2 dimensions) that separates M from C. Unfortunately, these questions are NP-hard (e.g. [DL97]). It becomes, therefore, interesting to *approximate* the cone C. So, can we find another cone that is close to C and for which computationally efficient membership and separation oracles exist? There is a natural candidate for the job. We say that a matrix M is in *square-l_2*, if there are points x_i in l_2 such that $M_{ij} = \|x_i - x_j\|_2^2$. Let S be the collection of all all square-l_2 matrices which are also a metric (i.e. the entries in M also satisfy the triangle inequality). It is not hard to see that $C \subseteq S$, but we ask:

Open Problem 3. What is the smallest $\alpha = \alpha(n)$, such that every $n \times n$ matrix $M \in S$ can be embedded in l_1 with distortion $\leq \alpha$?

It is not hard to see that every finite l_2 metric embeds isometrically into l_1. But what about the opposite direction?

Open Problem 4. Find $\max c_2(X)$ over all (X, d) that are n-point metrics in l_1. As we saw above, for the $n = 2^r$ vertices of the r-cube the answer is $\sqrt{r} = \sqrt{\log n}$. We suspect that this is the extreme case. No example is known where c_2 is asymptotically larger that $\sqrt{\log n}$.

5.1. Dimension reduction

Let us return to the applied aspect of this area. Even when a given metric space can be approximated well in some normed space, the *dimension* of the host space is quite significant. Data analysis and clustering in l_2^N for large N is by no means easy. In fact, practitioners in these areas often speak about the *curse of dimensionality* when they refer to this problem. In l_2 there is a basic result that answers this problem.

Theorem 10 (Johnson Lindenstrauss [JL84]). *Every n-point metric in l_2 can be embedded into l_2^k with distortion $< 1 + \epsilon$ where $k \leq O(\frac{\log n}{\epsilon^2})$.*

Here, again, the proof yields an efficient randomized algorithm. Namely, select a random k-dimensional subspace and project the points to it.
What is the appropriate analogue of this theorem for l_1 metrics?

Open Problem 5. What is the smallest $k = k(n, \epsilon)$ so that every n-point metric in l_1 can be embedded into l_1^k with distortion $< 1 + \epsilon$?

We know very little at the moment, namely $\Omega(\log n) \leq k \leq O(n \log n)$ for constant $\epsilon > 0$. The lower bound is trivial and the upper bound is from [Sch87, Tal90]. Note that if the truth is at the lower bound, then this provides an affirmative answer to Open Problem 4.

5.2. Planar graphs and other minor-closed families

One of the most fascinating problems about l_1 metrics is:

Open Problem 6. Is there is an absolute constant $C > 0$ so that every metric of a planar graph embeds into l_1 with distortion $< C$?

Even more daringly, the same can be asked for every minor-closed family of graphs. Some initial success for smaller graph families has been achieved already [GNRS99].

5.3. Large girth

Is there an analogue of Theorem 7 for embeddings into l_1?

Open Problem 7. How small can $c_1(G)$ be for a a graph G of girth g in which all vertices have degree ≥ 3? Specifically, can $c_1(G)$ stay bounded as g tends to ∞?

6. Ramsey-type theorems for metric spaces

The philosophy of modern Ramsey Theory, (as developed e.g. in [GRS90]) can be stated as follows: Large systems necessarily contain substantial "islands of order". Dvoretzky's Theorem certainly falls into this circle of ideas. But what about the metric analogues?

Open Problem 8. What is the largest $f(\cdot, \cdot)$ so that every n-point metric (X, d) has a subset Y of cardinality $\geq f(n, t)$ with $c_2(Y) \leq t$? (We mean, of course, the metric d restricted to the set Y.)

For t close to 1, the answer is known, namely, $f(n, t) = \Theta(\log n)$. For larger t the behavior is known to be different [BLMN].

References

[Alo86] N. Alon. Eigenvalues and expanders. *Combinatorica*, 6(2):83–96, 1986.

[AR98] Y. Aumann and Y. Rabani. An $O(\log k)$ approximate min-cut max-flow theorem and approximation algorithm. *SIAM J. Comput.*, 27(1):291–301, 1998.

[Bal92] K. Ball. Markov chains, Riesz transforms and Lipschitz maps. *Geom. Funct. Anal.*, 2(2):137–172, 1992.

[Bar98] Yair Bartal. On approximating arbitrary metrices by tree metrics. In *STOC '98 (Dallas, TX)*, 161–168. ACM, New York, 1998.

[BL00] Yoav Benyamini and Joram Lindenstrauss. *Geometric nonlinear functional analysis. Vol. 1.* American Mathematical Society, Providence, RI, 2000.

[BLMN] Y. Bartal, N. Linial, M. Mendel, and A. Naor. On metric Ramsey-type phenomena. Manuscript.

[Blu70]   Leonard M. Blumenthal. *Theory and applications of distance geometry.* Chelsea Publishing Co., New York, 1970.

[Bou85]   J. Bourgain. On Lipschitz embedding of finite metric spaces in Hilbert space. *Israel J. Math.*, 52(1-2):46–52, 1985.

[Bou86]   J. Bourgain. The metrical interpretation of superreflexivity in Banach spaces. *Israel J. Math.*, 56(2):222–230, 1986.

[Che70]   Jeff Cheeger. A lower bound for the smallest eigenvalue of the Laplacian. In *Problems in analysis (Papers dedicated to Salomon Bochner, 1969),* 195–199. Princeton Univ. Press, Princeton, N. J., 1970.

[DL97]   Michel Marie Deza and Monique Laurent. *Geometry of cuts and metrics.* Springer-Verlag, Berlin, 1997.

[DV99]   J. Dunagan and S. Vempala. On Euclidean embeddings and bandwidth minimization. In *Randomization, approximation, and combinatorial optimization (RANDOM-APPROX'99),* 229–240, Berlin, 1999. Springer-Verlag.

[Dvo61]   Aryeh Dvoretzky. Some results on convex bodies and Banach spaces. In *Proc. Internat. Sympos. Linear Spaces (Jerusalem, 1960),* 123–160. Jerusalem Academic Press, Jerusalem, 1961.

[Enf69]   P. Enflo. On the nonexistence of uniform homeomorphisms between L_p-spaces. *Ark. Mat.*, 8:103–105, 1969.

[Fei00]   U. Feige. Approximating the bandwidth via volume respecting embeddings. *J. Comput. System Sci.*, 60(3):510–539, 2000.

[GNRS99]   A. Gupta, I. Newman, Y. Rabinovich, and A. Sinclair. Cuts, trees and l_1 embeddings of graphs. In *40th Annual IEEE Symposium on Foundations of Computer Science,* 230–240, November 1999.

[GRS90]   Ronald L. Graham, Bruce L. Rothschild, and Joel H. Spencer. *Ramsey theory.* John Wiley & Sons Inc., New York, second edition, 1990.

[Ind01]   P. Indyk. Algorithmic applications of low-distortion geometric embeddings. In *42nd Annual IEEE Symposium on Foundations of Computer Science,* 10–33, 2001.

[JL84]   W. B. Johnson and J. Lindenstrauss. Extensions of Lipschitz mappings into a Hilbert space. In *Conference in modern analysis and probability (New Haven, Conn., 1982),* 189–206. Amer. Math. Soc., Providence, RI, 1984.

[KPR93]   P. Klein, S. A. Plotkin, and S. Rao. Excluded minors, network decomposition, and multicommodity flow. In *25th Annual ACM Symposium on Theory of Computing,* 682–690, May 1993.

[LLR95]   N. Linial, E. London, and Y. Rabinovich. The geometry of graphs and some of its algorithmic applications. *Combinatorica,* 15(2):215–245, 1995.

[LM00]   N. Linial and A. Magen. Least-distortion Euclidean embeddings of graphs: products of cycles and expanders. *J. Combin. Theory Ser. B,* 79(2):157–171, 2000.

[LMN]   N. Linial, A. Magen, and A. Naor. Girth and Euclidean distortion. *Geom. Funct. Anal.* To appear.

[LS]   N. Linial and M. Saks. On the Euclidean distortion of complete binary
            trees. *Discrete Comput. Geom.* To appear.

[Mat97]   J. Matoušek. On embedding expanders into l_p spaces. *Israel J. Math.*,
            102:189–197, 1997.

[Mat99]   J. Matoušek. On embedding trees into uniformly convex Banach spaces.
            Israel J. Math., 114:221–237, 1999.

[Mat02]   J. Matoušek. *Lectures on Discrete Geometry.* Springer Verlag, New
            York, 2002. Graduate Texts in Mathematics 212.

[MS86]   V. D. Milman and G. Schechtman. *Asymptotic theory of finite-
            dimensional normed spaces.* Springer-Verlag, Berlin, 1986.

[NR02]   I. Newman and Y. Rabinovich. A lower bound on the distortion of
            embedding planar metrics into Euclidean space. In *Proceedings of the
            18th annual symposium on Computational Geometry*, 94–96. ACM, 2002.

[Rao99]   S. Rao. Small distortion and volume preserving embeddings for planar
            and Euclidean metrics. In *Proceedings of the 15th Annual Symposium
            on Computational Geometry*, 300–306. ACM, 1999.

[RR98]   Y. Rabinovich and R. Raz. Lower bounds on the distortion of embedding
            finite metric spaces in graphs. *Discrete Comput. Geom.*, 19(1):79–94,
            1998.

[Sch87]   Gideon Schechtman. More on embedding subspaces of L_p in l_r^n. *Com-
            positio Math.*, 61(2):159–169, 1987.

[Sch02]   A. Schrijver. *Combinatorial Optimization - Polyhedra and Efficiency.*
            Springer, Heidelberg, 2002.

[Tal90]   Michel Talagrand. Embedding subspaces of L_1 into l_1^N. *Proc. Amer.
            Math. Soc.*, 108(2):363–369, 1990.

ICM 2002 · Vol. III · 587–603

List Colouring of Graphs with at Most $(2 - o(1))\chi$ Vertices

Bruce Reed[*]   Benny Sudakov [†]

Abstract

Ohba has conjectured [9] that if the graph G has $2\chi(G)+1$ or fewer vertices then the list chromatic number and chromatic number of G are equal. In this paper we prove that this conjecture is asymptotically correct. More precisely we obtain that for any $0 < \epsilon < 1$, there exist an $n_0 = n_0(\epsilon)$ such that the list chromatic number of G equals its chromatic number, provided

$$n_0 \leq |V(G)| \leq (2 - \epsilon)\chi(G).$$

2000 Mathematics Subject Classification: 05C15, 05D40.
Keywords and Phrases: Probabilistic Method, Graph coloring, List-chromatic number.

1.   Introduction

Recently, a host of important results on graph colouring have been obtained via the probabilistic method. The first author presented an invited lecture at the 2002 International Congress of Mathematicians surveying a number of these results. The recent monograph [8] provides a more in depth survey of the topic. This paper presents one example of a result proven using the method.

An instance of List Colouring consists of a graph G and a list $L(v)$ of colours for each vertex v of G. We are asked to determine if there is an *acceptable* colouring of G, that is a colouring in which each vertex receives a colour from its list, and no edge has both its endpoints coloured with the same colour. The *list-chromatic number* of G, denoted $\chi^l(G)$ is the minimum integer k such that for every assignment of a

---

[*]CNRS, Paris, France and School of Computer Science, McGill University, Montreal, Canada. E-mail: breed@jeff.cs.mcgill.ca. This research was partially supported by DIMACS and by a CNRS/NSF collaboration grant.

[†]Department of Mathematics, Princeton University, Princeton, NJ 08540, USA and Institute for Advanced Study, Princeton, NJ 08540, USA. E-mail: bsudakov@math.princeton.edu. Research supported in part by NSF grants DMS-0106589, CCR-9987845 and by the State of New Jersey.

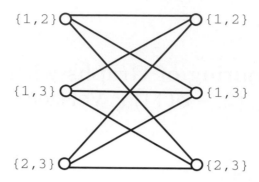

Figure 1: A bipartite graph with list chromatic number three

list $L(v)$ of size at least k to every vertex v of G, there exist an acceptable colouring of G. The list-chromatic number was introduced by Vizing [11], and independently Erdös et al. [3]. This parameter has received a considerable amount of attention in recent years (see, e.g. [4], [1]).

Clearly, by definition, $\chi^l(G) \geq \chi(G)$ because $\chi(G) = k$ precisely if an acceptable colouring exists when each L_v is $\{1, ..., k\}$. However, the converse inequality is not true, e.g. $\chi^l(K_{3,3}) = 3$ as can be easily verified by considering Figure 1. In fact, there are bipartite graphs with arbitrarily high chromatic number (indeed even for bipartite G, $\chi^l(G)$ is bounded from below by a function of the minimum degree which goes to infinity, see [1]). This shows that the gap between $\chi(G)$ and $\chi^l(G)$ can be arbitrarily large. Moreover it shows that $\chi^l(G)$ can not be bound by any function of the chromatic number of G. This gives rise to the following intriguing question in the theory of graph colourings: Find conditions which guarantee the equality of the chromatic and list-chromatic numbers.

There are many conjectures hypothesizing conditions on G which imply that $\chi(G) = \chi^l(G)$. Probably, the most famous of these is the List Colouring Conjecture (see [4]) which states that this is true if G is a line graph. One interesting example of a graph with $\chi = \chi^l$ was obtained in the original paper of Erdös et al. [3]. They proved that if G is complete k-partite graph with each part of size two then $\chi(G) = \chi^l(G) = k$. It took nearly twenty years until Ohba [9] noticed that this example is actually part of much larger phenomenon. He conjectured (cf. [9]) that $\chi(G) = \chi^l(G)$ provided $|V(G)| \leq 2\chi(G) + 1$. This conjecture if it is correct is best possible. Indeed, let G be a complete k-partite graph with $k-1$ parts of size 2 and one part of size 4. Then the number of vertices of G is $2k+2$, the chromatic number is k and it was proved in [9] that list-chromatic number of G is at least $k + 1 > k$.

In his original paper Ohba obtained that $\chi^l(G) = \chi(G)$ for all graphs G with $|V(G)| \leq \chi(G) + \sqrt{2\chi(G)}$. His conjecture was settled for some other special cases in [2]. Recently the result of Ohba was substantially improved by the authors of this paper. In [10] they proved that Ohba's conjecture is true for all graphs G with at most $\frac{5}{3}\chi(G) - \frac{4}{3}$ vertices. In this paper we want to improve this result for large graphs and prove that the conjecture is asymptotically correct. More precisely we obtain the following theorem.

Theorem 1 *For any $0 < \epsilon < 1$, there exist an $n_0 = n_0(\epsilon)$ such that $\chi^l(G) = \chi(G)$ provided $n_0 \leq |V(G)| \leq (2 - \epsilon)\chi(G)$.*

The rest of this paper is organized as follows. In the next section we describe the main steps in the proof of Theorem 1. More precisely, we present our key lemma and show how to deduce from it the assertion of the theorem. We will prove this lemma using probabilistic arguments. In Sections 3 and 4 we discuss the main ideas we are going to use in the proof. We present the details of the proof in Section 5. Finally, the last section of the paper contains some concluding remarks.

2. The key lemma

In this section we present the main steps in the proof of Theorem 1. First we need the following lemma from [10] whose short proof we include here for the sake of completeness.

Lemma 2 *For any integer t, if $\chi^l(G) > t$ then there exist a set of lists $L(v), v \in V(G)$ for which there is no acceptable colouring such that each list has at least t elements and the set $\mathcal{A} = \cup_{v \in V(G)} L(v)$ has size less than $|V(G)|$.*

Proof. Assume $\chi^l(G) > t$ and choose a set of lists $L(v), v \in V(G)$ for which there is no acceptable colouring, in which each list has size at least t and which minimizes $|\mathcal{A}|$.

Now, if $|\mathcal{A}| < |V(G)|$ then we are done. So, we can assume the contrary. We consider the bipartite graph H with bipartition $(\mathcal{A}, V(G))$ and an edge between c and v precisely if $c \in L(v)$. If there is a matching of size $|V|$ in H then this matching saturates V and points out an acceptable colouring for the List Colouring instance in which no colour is used more than once. Since, there is no such acceptable colouring, no such matching exists. Thus there must be a smallest subset B of $\mathcal{A}$ which is not the set of endpoints of a matching in this graph and this set must have at most $|V|$ elements. Clearly, B contains at least two vertices. Now, by the minimality of B there is a matching M in H of size $|B| - 1$ whose endpoints in $\mathcal{A}$ are in B. Further, classical results in matching theory (see e.g. Theorem 1.1.3 of [6]) tell us that if W is the set of endpoints of M in V then for $v \notin W$, we have $L(v) \cap B = \emptyset$.

Let x be any vertex in $G - W$ and replace $L(v)$ by $L(x)$ for every vertex $v \in W$. This yields a new List Colouring Problem in which the total number of colours in all lists is smaller than $|\mathcal{A}|$ (since all the new list are disjoint from B). Therefore by the minimality of our original choice, there exist an acceptable colouring of G for this new Lists Colouring instance. In particular this implies that we can obtain an acceptable colouring of $G - W$ for the original lists $L(v)$. Since no colour in B is used in this colouring, using the colouring of W pointed out by M yields an extension of this colouring to a colouring of G in which no colour of B appears more than once. This contradicts our assumption that there is no acceptable colouring for this instance and proves the lemma. $\qquad\square$

Proof of Theorem 1. Let $0 < \epsilon < 1$ be a fixed constant and let G be a graph satisfying $|V(G)| < (2 - \epsilon)\chi(G)$. We assume that $\chi^l(G) > \chi(G)$ and obtain a contradiction. Since adding an edge between vertices in different colour classes in an optimal colouring of G does not change $\chi(G)$ and can only increase $\chi^l(G)$, we will assume that G is complete $\chi(G)$-partite graph. Thus G has a unique partition into $\chi(G)$ stable sets. We refer to these stable sets as parts rather than colour classes so as to avoid confusion with the colours used in our acceptable colouring of G.

Now by Lemma 2, if $\chi^l(G) > \chi(G)$ then there is an instance of List Colouring on G for which no acceptable colouring exists, in which each list has length at least $\chi(G)$ and such that the size of the union of all lists $L(v)$ is less than $|V(G)|$. This means that in an acceptable colouring at least one colour must be used on more than one vertex. Fortunately, it also implies that for every non-singleton part U there is at least one colour which appears on $L(v)$ for more than half the vertices of U (since each $L(v)$ contains more than half the colours).

Our proof approach is simple. For each non-singleton part U, we choose some colour c_U and colour with c_U all the vertices of U whose list contains c_U (thus we must insist that all the c_U are distinct). We complete the colouring by finding a bijection between the vertices not yet coloured and the colours not yet used so that each such colour is in the list of the vertex with which it is matched. This yields an acceptable colouring in which for each part U there is at most one colour c_U used on more than one vertex of U.

To begin, we consider the case when there is some part U such that some colour appears on all the vertices of U. We show that we can reduce to a smaller problem by using any such colour for c_U. Iteratively repeating this process yields a graph where no such U exists and hence, in particular, there are no parts of size two.

Our choices for the remaining c_U are discussed in the proof of the key Lemma 3 which consists of the analysis of a probabilistic procedure for choosing the remaining c_U. Unfortunately, before discussing this procedure we need to deal with some technical details.

So, to begin we show that we can assume that $\cap_{v \in U} L(v)$ is empty for all parts U of size bigger than 1 in the partition of G. To see this, let U be a part of size at least 2 such that $\cap_{v \in U} L(v) \neq \emptyset$. Then the graph $G - U$ has chromatic number $\chi(G) - 1$ and at most $|V(G)| - 2$ vertices and therefore also satisfies

$$|V(G-U)| \le |V(G)| - 2 \le (2-\epsilon)\chi(G) - 2 = (2-\epsilon)(\chi(G)-1) - \epsilon < (2-\epsilon)\chi(G-U).$$

Note that it also satisfies $\chi^l(G - U) > \chi(G - U) = \chi(G) - 1$ since otherwise we can obtain an acceptable colouring of G from the lists $L(v)$. Indeed, let c be a colour in $\cap_{v \in U} L(v)$. Since $\chi^l(G - U) = \chi(G) - 1$, we know there is an acceptable colouring of $G - U$ from the lists $L(v) - c$. Colouring all vertices in U with c we obtain an extension of this colouring to an acceptable colouring of G from the original lists, a contradiction. Therefore we will consider the graph $G - U$ instead of G and continue this process until we obtain a graph G' and an instance of List Colouring on G' with the following properties.

- G' is $\chi(G')$-partite graph which satisfies $|V(G')| < (2 - \epsilon)\chi(G')$.
- Each list $L'(v)$ has length at least $\chi(G')$ and there is no acceptable colouring of G' from $L'(v)$.
- The size of the union of all lists is less than $|V(G')|$.
- $\cap_{v \in U} L'(v)$ is empty for all parts of size bigger than 1 in the partition of G'.

Since the size of the lists is $\chi(G') > |V(G')|/2$ we obtain that $L'(x) \cap L'(y)$ is non empty for any two vertices $\{x, y\}$ in G'. In particular this implies that in the partition of G' there are no parts of size two. Note that the original graph G has at most $|V(G)|/2$ parts of size ≥ 2 and each time we removed such a part the chromatic number of the remaining graph decreased by one. Therefore we decrease chromatic number of G by at most $|V(G)|/2$ and hence the remaining graph G' should have at least

$$|V(G')| \geq \chi(G') \geq \chi(G) - \frac{|V(G)|}{2} \geq \frac{|V(G)|}{2 - \epsilon} - \frac{|V(G)|}{2} \geq \frac{\epsilon|V(G)|}{4}$$

vertices. So by choosing an appropriate bound on the size of $|V(G)|$ we can make $|V(G')|$ arbitrarily large. This completes our discussion of parts U for which some colour is in $L(v)$ for all vertices v of U. We turn now to the technical details necessary before we present the rest of the ideas needed in the proof.

Let X be the set of all the vertices in the singleton classes in the partition of G'. Pick m to be a sufficiently large integer constant $m = m(\epsilon)$ and let t be an integer which satisfies

$$\frac{t+1}{m} \leq \frac{|X|}{\chi(G')} \leq \frac{t+2}{m}. \tag{2.1}$$

Since in the partition of G' there are no parts of size two, we obtain that $|X| + 3(\chi(G') - |X|) \leq |V(G')| < (2 - \epsilon)\chi(G')$. This implies that $|X| \geq (1 + \epsilon)\chi(G')/2$ and that $m/2 < t \leq m - 2$.

Set $\mathcal{A} = \cup_{v \in G'} L'(v)$. Let H be a bipartite graph with bipartition $(X, \mathcal{A})$ and an edge between c and v precisely when $c \in L'(v)$. Note that the degree of every vertex from X in H is at least $\chi(G') \geq |V(G')|/2 > |\mathcal{A}|/2$. Therefore by well known results on Zarankiewicz's problem (see, e.g., [5], Problem 10.37), H contains a complete bipartite graph with t vertices in X and m vertices in $\mathcal{A}$. Denote the set of vertices from X and $\mathcal{A}$ by $\mathcal{S}_1$ and $\mathcal{C}_1$ respectively and remove them from H. Note that the bound on Zarankiewicz's problem guarantees that we will continue to find a copy of the complete bipartite graph $K_{t,m}$ in H until the minimal degree of a vertex in X is $o(\chi(G')) = o(|\mathcal{A}|)$. Thus in the end we obtain at least $k = (1 - o(1))\chi(G')/m$ disjoint sets of colours $\mathcal{C}_1, \ldots, \mathcal{C}_k$ and also k disjoint sets of singleton partition classes $\mathcal{S}_1, \ldots, \mathcal{S}_k$, such that $\mathcal{C}_i \subset L'(s)$ for every vertex $s \in \mathcal{S}_i$. Denote by $\mathcal{C} = \cup_i \mathcal{C}_i$, by $\mathcal{S} = \cup_i \mathcal{S}_i$ and let C and S be the sizes of $\mathcal{C}$ and $\mathcal{S}$ respectively. Now using (2.1) we can obtain the following inequalities

$$
\begin{aligned}
|X| - S &= |X| - kt \leq \frac{t+2}{m}\chi(G') - (1 + o(1))\frac{t}{m}\chi(G') \\
&= (1 + o(1))\frac{2}{m}\chi(G') = (2 + o(1))k < 3k
\end{aligned}
$$

and

$$
|X| - S = |X| - kt \geq \frac{t+1}{m}\chi(G') - (1 + o(1))\frac{t}{m}\chi(G') = (1 + o(1))\frac{\chi(G')}{m}.
$$

In the above discussion and in particular in the last two inequalities we used that m and t are constants but $|V(G')|$ (and thus also $\chi(G')$) tends to infinity.

Let W be the union of some set of $r = \chi(G') - |\mathcal{C}|$ singleton partition classes which do not belong to $\mathcal{S}$. Such a set W exists, since the number of singleton partition classes outside $\mathcal{S}$ is at least $(1 + o(1))\chi(G')/m \gg r = \chi(G') - km = o(\chi(G'))$. Note that we can obtain an acceptable colouring of W with the lists $L'(v) - \mathcal{C}$ greedily, since the size of $L'(v) - \mathcal{C}$ is equal to r. Let T be the set of r colours used to colour W in one such acceptable colouring. Denote by $G'' = G' - W$ and let $L''(v) = L'(v) - T$ for every vertex $v \in G''$. Then to finish the proof it is enough to show the existence of an acceptable colouring of the G'' from the set of lists $L''(v)$.

By definition, we have that $\chi(G'') = \chi(G') - r = |\mathcal{C}| = C$ and G'' is a complete C partite graph. The number of vertices of G'' satisfies

$$
|V(G'')| = |V(G')| - r < |V(G')| < (2 - \epsilon)\chi(G') = (1 + o(1))(2 - \epsilon)\chi(G'').
$$

So by choosing $\delta = \epsilon/2$ we obtain that $|V(G'')| < (2 - \delta)\chi(G'')$. This together with above discussion implies that G'' satisfies all the condition (1–5) of the next lemma. This lemma guarantees the existence of an acceptable colouring of G'' from the set of lists $L''(v)$ and completes the proof of Theorem 1.

Lemma 3 *Let $0 < \delta < 1$ be a constant and let C, S, k, m, t and n be integers with $m > 6/\delta$, $C = km$, $S = kt$ and $n < (2 - \delta)C$. Suppose, in addition, that m is fixed and n (and hence C) is a sufficiently large function of m. Let G be a complete C-partite graph on n vertices and let $L(v)$ be the set of lists of colours of size C one for each vertex v of G such that the following holds.*

1. *$\cap_{v \in U} L(v) = \emptyset$ for any part U of size bigger than one in the partition of G.*
2. *G contains a set of vertices $\mathcal{S}$ of size S such that the vertices in $\mathcal{S}$ form parts of size one in the partition of G. The set $\mathcal{S}$ is partitioned into k parts $\mathcal{S}_1, \ldots, \mathcal{S}_k$ each of size t.*
3. *G contains no parts of size two and at most $3k$ singleton parts which do not belong to $\mathcal{S}$.*
4. *There exist a set of colours $\mathcal{C}$ of size C and its partition $\mathcal{C}_1, \ldots, \mathcal{C}_k$ into k sets of size m. Such that $\mathcal{C}_i \subset L(s)$ for every vertex $s \in \mathcal{S}_i$. In particular, for any subset of $\mathcal{C}_i$ of size t there exist an acceptable colouring of the vertices of $\mathcal{S}_i$ which uses the colours in this subset.*

5. *The total number of colours in the union of all the lists $L(v)$ is less than n.*

Then there exist an acceptable colouring of G from the lists $L(v)$.

We finish this section with discussion of the proof of Lemma 3. We postpone all the details to the subsequent sections of the paper.

Proof Overview. The proof proceeds as follows:

(I) We choose a random partition of each C_i into two subsets A_i of size t and B_i of size $m - t$ where these choices are made uniformly and independently.

(II) We use the colours in A_i to colour the vertices of S_i which is possible by Condition 4 of the lemma.

(III) We choose a (random) bijection between $\mathcal{B} = \cup_{i=1}^{k} B_i$ and the parts of G not in $\mathcal{S}$ in such a way that, for each part U not in $\mathcal{S}$, U is equally likely to correspond to each colour $c \in \mathcal{B}$. We denote by c_U be the colour corresponding to U.

(IV) For each part U not in $\mathcal{S}$ we colour every vertex v of U for which $c_U \in L(v)$ with the colour c_U.

(V) We match the set V' of vertices not yet assigned a colour with the set of colours not yet used (i.e those colours not in $\mathcal{C}$) so that every vertex is matched with a colour on its list. We colour each vertex of V' with the colour with which it is matched.

If we successfully complete this five step process, we have an acceptable colouring as every colour not in $\mathcal{B}$ appears on at most one vertex, and every colour in $\mathcal{B}$ appears only on a subset of some part, and hence on an independent set of G.

To prove that we can find the colouring in this fashion, we need to describe and analyze our method for choosing the random bijection between the parts and the colours in $\mathcal{C}$ made in Steps I–IV, in order to show that (with positive probability) we can complete the colouring by finding the desired matching in Step V.

A key tool will be Hall's Theorem which states that in a bipartite graph with bipartition (A, B), we can find a matching M such that every vertex of A is the endpoint of an edge of M provided there is no subset X of A such that setting $N(X) = \cup\{N(x) | x \in X\}$ we have $|N(X)| < |X|$.

We remark that although Steps I–III are presented as though they are separate processes performed sequentially, in the more complicated case of our analysis we will need to interleave these processes by first choosing some of the B_i, then choosing the parts with which these colours will be matched, and finally completing Step I and then Step III.

To determine if we can find the desired matching in Step V, we will need to examine the sets $L'(v) = L(v) - \mathcal{C}$ for the vertices of V'. Let H be a bipartite graph with bipartition $(V', \cup_v L'(v))$ and an edge between c and v precisely when $c \in L'(v)$. For each vertex v, we let the weight of v, denoted $w(v)$, be $\frac{1}{|L'(v)|}$. For any set S of vertices we use $W(S)$ to denote the sum of the weights of the vertices in S.

This definition of weight is motivated by the following immediate consequence of Hall's Theorem:

Observation 4 *If we cannot find the desired matching in Step 5 then there exist a subset X of V' such that $W(X) > 1$. In this case $W(V') > 1$ as well.*

Proof. By Hall's Theorem there exist a subset X of V' such that $|N(X)| < |X|$. Then, we obtain that

$$W(V') \geq W(X) = \sum_{x \in X} \frac{1}{|L'(v)|} = \sum_{x \in X} \frac{1}{|N(x)|} \geq \sum_{x \in X} \frac{1}{|N(X)|} \geq \frac{|X|}{|N(X)|} > 1. \qquad \square$$

Thus an analysis of the random parameter $W(V')$ will be crucial to the proof of the lemma. In the next section, by computing the expected value of the parameter, we show that the lemma holds if $n \leq C + S$. In later sections, we complete the proof using a more complicated analysis along the same lines.

3.   The expected value of $W(V')$

For each part U which is not in $\mathcal{S}$, our choices in Steps I and III guarantee that each colour of $\mathcal{C}$ is equally likely to be c_U. Thus, for each vertex v in such a part, the probability that v is in V', i.e., $c_U \notin L(v)$, is $1 - \frac{|L(v) \cap \mathcal{C}|}{C}$. Since $|L(v)| = C$, this is $\frac{|L'(v)|}{C}$. So, we have:

$$\mathbf{E}\big(W(V')\big) = \sum_{v \in V - \mathcal{S}} w(v)\mathbf{Pr}(v \in V') = \sum_{v \in V - \mathcal{S}} \frac{1}{L'(v)} \frac{L'(v)}{C} = \frac{n - S}{C}. \qquad (3.1)$$

So, if $n \leq S + C$, then this expected value is less than or equal to one. Since the probability that a random variable exceeds its expected value is less than one, this implies that we can make the choices in Steps I–IV so that $W(V') \leq 1$, and hence by Observation 4, the desired matching can be found in Step V.

Analyzing the behaviour of the (random) weights of various subsets of V' will allow us extend our proof technique to handle larger values of n. In doing so, the following definitions and observations will prove useful.

We let A be the number of non-singleton parts. Then by Condition 3 of the lemma, A is at least $C - S - 3k$. Since each non singleton colour class has at least three vertices and the total number of classes is C, we obtain $(2-\delta)C \geq n \geq C+2A$, i.e., $A \leq \frac{(1-\delta)C}{2}$. On the other hand, the analysis above shows that we can assume that $n > S+C$ and hence that $S \leq (1-\delta)C$. Thus, $A \geq \delta C - 3k = \delta C - \frac{3}{m}C \geq \frac{\delta C}{2}$. Both these bounds on A will be useful in our analysis. Note also that

$$m - t = \frac{C - S}{k} \geq \frac{C - (1-\delta)C}{k} = \delta\frac{C}{k} = \delta m.$$

4.   Completing the proof: the idea

Let H be a bipartite graph with bipartition $\big(V', \cup_v L'(v)\big)$ and an edge between c and v precisely when $c \in L'(v)$. Our first step will be to check Hall's criterion for a fixed subset of colours K in $\cup_v L'(v)$ and show that the expected number of vertices in $\{v | v \in V', L'(v) \subseteq K\}$ is less than $|K|$.

To begin, we note that for any such v, $w(v) \geq \frac{1}{|K|}$. Therefore, defining the set S_K to be $S_K = \{v | v \in V', L'(v) \subseteq K\}$, we have that

$$\mathbf{E}\big(|S_K|\big) = \sum_{v \in V - S,\, L'(v) \subseteq K} \mathbf{Pr}(v \in V') \leq |K| \sum_{v \in V - S,\, L'(v) \subseteq K} w(v)\mathbf{Pr}(v \in V')$$

$$\leq |K| \sum_{v \in V - S} w(v)\mathbf{Pr}(v \in V') = |K|\mathbf{E}\big(W(V')\big) = |K|\frac{n - S}{C}$$

which, since $n \leq (2 - \delta)C \leq S + A + 3k + (1 - \delta)C$, is at most $|K|(1 + \frac{A + 3k}{C} - \delta)$. On the other hand, this estimate is not good enough to guarantee Hall's criterion, since it still can be greater than $|K|$.

To improve on this bound, we use the fact that no colour c appears on the list of all the vertices of any non-singleton part of G. Note that $k = C/m$, $m > 6/\delta$, the number of non-singleton parts is A and the total number of vertices is at most $n \leq (2 - \delta)C \leq S + A + 3k + (1 - \delta)C$. This altogether implies that for every c,

$$\mathbf{E}\Big(W\big(V' \cap \{v | c \in L'(v)\}\big)\Big) = \mathbf{E}\big(W(V')\big) - \sum_{v \in V - S,\, c \notin L'(v)} w(v)\mathbf{Pr}(v \in V')$$

$$= \mathbf{E}\big(W(V')\big) - \sum_{v \in V - S,\, c \notin L'(v)} \frac{1}{C} \leq \frac{n - S}{C} - \frac{A}{C}$$

$$= \frac{n - S - A}{C} \leq \frac{(1 - \delta)C + 3k}{C}$$

$$= 1 - \delta + \frac{3}{m} \leq 1 - \frac{\delta}{2}. \tag{4.1}$$

Applying this fact for the c in K allows us to improve our bound on $\mathbf{E}\big(|S_K|\big)$. Specifically, we note that summing this bound over all the colours c in K

$$\mathbf{E}\Big(\sum_{c \in K} W\big(V' \cap \{v | c \in L'(v)\}\big)\Big) = \sum_{c \in K} \mathbf{E}\Big(W\big(V' \cap \{v | c \in L'(v)\}\big)\Big) \leq \Big(1 - \frac{\delta}{2}\Big)|K|.$$

Now, each vertex v of S_K contributes $w(v) = 1/|L'(v)|$ to exactly $|L'(v)|$ terms in the first sum in this equation, so its total contribution to the sum is 1. I.e., we have:

$$\mathbf{E}\big(|S_K|\big) \leq \Big(1 - \frac{\delta}{2}\Big)|K|.$$

So, we don't expect any particular set K of colours to provide an obstruction to finding the desired matching in the bipartite graph H in Step V. However, we need to handle all the K at once. In order to do so, we would like to prove that for each K, the size of S_K is highly concentrated around its expected value and hence is greater than $|K|$ only with exponentially small probability. As above, rather than focusing on all the K we actually consider, for each colour c, the weight of the subset V'_c of V' consisting of those v with c on $L'(v)$. There are two major difficulties which complicate our approach.

- some of the parts U can be very large making it impossible for us to obtain the desired concentration results directly (e.g., there could be a part of size exceeding $\frac{n}{3}$).
- If $L'(v)$ is very small then $w(v) = 1/|L'(v)|$ is large and putting v into V' can have a significant effect on the weight of the various V'_c. This makes proving a concentration result directly impossible.

In order to deal with these problems, we proceed as follows:

(A) We colour the "big" parts first, ignoring concentration in our computation and focusing only on the expected weight of the subset of V' intersecting the big parts. We note that by considering the expected overall weight and not focusing on a specific V'_c, we only lose a factor of $\frac{1}{C}$ per part. We will define big parts so that there are $o(1)$ of them, and hence the total loss will not be significant.

(B) We treat v with $|L'(v)|$ small separately using an expected value argument to bound the weight of the vertices in this set.

5.   Completing the proof: the details

In this section we will complete the proof of Lemma 3 using the ideas which have already been discussed above. We choose an integer b so that

$$\frac{\delta^2 C}{40} \leq b(m-t) \leq \frac{\delta^2 C}{20}$$

which is possible because $m \leq \frac{\delta^2 C}{40}$ (this holds, since m and δ are fixed but C tends to infinity) and $m - t > 0$ (in fact it exceeds δm as we remarked at the end of Section 3.). We call the largest $b(m-t)$ parts in our partition of G *big*, and the others *small*. Let *Big* be the union of the vertex sets of the big parts. We will need the following lemma.

Lemma 5 *Every small non-singleton partition class contains at least two v which satisfy:*

$$|L'(v)| > \frac{\delta^3}{80} C.$$

Proof. Let U be a small non-singleton colour class. We already mentioned that every colour of $\mathcal{C}$ is missed by a vertex of U so $\sum_{v \in U} |L'(v)| = \sum_{v \in U} |\mathcal{C} - L(v)| \geq |\mathcal{C}| = C$. Now, since there are less than $n < (2-\delta)C$ colours in total, every $L(v)$ must contain at least δC colours in $\mathcal{C}$ and so the largest $L'(v) = L(v) - \mathcal{C}$ in U has at most $(1-\delta)C$ elements. Thus, the sum of $|L'(v)|$ over the remaining vertices of U is at least δC.

Since there are at least $\frac{\delta^2 C}{40}$ big colour classes, the largest small colour class has at most $\frac{40n}{\delta^2 C} < \frac{80}{\delta^2}$ vertices. So, the second largest $L'(v)$ has size at least $\delta C \cdot (\frac{80}{\delta^2})^{-1}$. This is the desired result. $\qquad\square$

With this auxiliary result in hand, we can now complete the proof. We proceed as follows:

First Process: We randomly choose b of the C_i and a partition of each of these into subsets $\mathcal{A}_i$ of size t and $\mathcal{B}_i$ of size $m - t$ where these choices are all made independently and uniformly. We then choose a uniformly random bijection between the $b(m - t)$ colours in the union of these $\mathcal{B}_i$ and the big parts.

Second Process: We chose a partition of each remaining $\mathcal{C}_i$ into $\mathcal{A}_i$ and $\mathcal{B}_i$ where again these choices are uniform, independent, and independent of all the earlier choices. We then choose a uniformly random bijection between the colours in these $\mathcal{B}_i$ and the small parts.

Denote by c_U the colour which is assigned by the above bijection to the partition class U. Use the colours in $\mathcal{A}_i$ to colour the vertices of $\mathcal{S}_i$ and for each part U not in $\mathcal{S}$ colour every vertex v of U for which $c_U \in L(v)$ with the colour c_U. Let V' be a set of vertices not yet assigned a colour. We set $V'' = V' - Big$ and $V''' = V' \cap Big$.

Note that V''' is determined by our choices in the first process. So, using a computation similar to that in (3.1) we obtain

$$\mathbf{E}\big(W(V''')\big) = \sum_{v \in Big} w(v)\mathbf{Pr}(v \in V''') = \sum_{v \in Big} \frac{1}{L'(v)} \frac{L'(v)}{C} = \frac{|Big|}{C}.$$

Furthermore, by the definition of expectation, there exist at least one set of choices for the first process such that $W(V''') \le \mathbf{E}(W(V''')) = \frac{|Big|}{C}$. We condition on any such set of choices which ensures that this inequality holds. We use **CP** and **CE** for the conditional probability of an event and conditional expectation of a variable for the second process, given this set of choices.

Let $\mathcal{C}'$ be the union of the set of colours in the $\mathcal{C}_i$ which were chosen in the first process. At the end of Section 3 we proved that $m - t$ is at least δm. Therefore $|\mathcal{C}'| = mb = \frac{m}{m-t}b(m - t) \le \delta^{-1} \cdot \frac{\delta^2 C}{20} = \frac{\delta C}{20}$. Hence, we have that for every v in a small part which is not in $\mathcal{S}$,

$$\begin{aligned}
\mathbf{CP}(v \in V'') &\le \frac{|L'(v)|}{C - |\mathcal{C}'|} \le \frac{|L'(v)|}{C}\left(1 + \frac{|\mathcal{C}'|}{C - |\mathcal{C}'|}\right) \le \frac{|L'(v)|}{C}\left(1 + \frac{\delta/20}{1 - \delta/20}\right) \\
&\le \left(1 + \frac{\delta}{10}\right)\frac{|L'(v)|}{C} = \left(1 + \frac{\delta}{10}\right)\mathbf{Pr}(v \in V'').
\end{aligned}$$

Clearly, this implies that for every subset X of the set of vertices $V - \mathcal{S} - Big$ we have

$$\mathbf{CE}\big(W(V'' \cap X)\big) \le \left(1 + \frac{\delta}{10}\right)\mathbf{E}\big(W(V'' \cap X)\big).$$

In particular, for every colour c

$$\mathbf{CE}\Big(W\big(V'' \cap \{v \,|\, c \in L'(v)\}\big)\Big) \le \left(1 + \frac{\delta}{10}\right)\mathbf{E}\Big(W\big(V'' \cap \{v \,|\, c \in L'(v)\}\big)\Big) \quad (5.1)$$

and also

$$\mathbf{CE}\left(W\Big(V'' \cap \Big\{v \,\Big|\, |L'(v)| < \frac{n}{\sqrt{\log n}}\Big\}\Big)\right) \le \quad (5.2)$$

$$\leq \left(1 + \frac{\delta}{10}\right) \mathbf{E}\left(W\left(V'' \cap \left\{v \mid |L'(v)| < \frac{n}{\sqrt{\log n}}\right\}\right)\right).$$

Before we proceed with the proof, we need the following lemma.

Lemma 6 *For every color c the probability that*

$$W\left(V'' \cap \left\{v \mid c \in L'(v), |L'(v)| \geq \frac{n}{\log n}\right\}\right)$$

$$> \mathbf{CE}\left(W\left(V'' \cap \left\{v \mid c \in L'(v), |L'(v)| \geq \frac{n}{\log n}\right\}\right)\right) + \frac{\delta}{20}$$

is $o(n^{-1})$.

Proof. To prove the lemma we need the following variant of a standard large deviation inequality for martingales. Since the proof of this inequality is essentially the same as other proofs which already appeared in the literature (see, e.g., Section 3 of the survey [7]), we will omit it here.

Given a finite set $\{1, 2, \ldots, r\}$, let S_r denotes the set of all $r!$ permutations or linear orders on this set. Let $\mathbf{X} = (X_1, \ldots, X_l)$ be a family of independent random variables, where the random variable X_j takes values in a finite set Ω_j. Thus $\mathbf{X}$ takes values in the set $\Omega = \prod_j \Omega_j$. Let $\pi \in S_r$ be a random permutation independent from $\mathbf{X}$. Suppose that the non-negative real-valued function $h : \Omega \times S_r \to \mathbb{R}$ satisfies the following two conditions for every $(\mathbf{x}, \pi)$.

- For every j, changing the value of a coordinate x_j can change the value of $h(\mathbf{x}, \pi)$ by at most d.
- Swapping any two elements in permutation π can change the value of $h(\mathbf{x}, \pi)$ by at most d.

Denote by $\mathbf{E}h$ the expected value of h. Then for every $t \geq 0$ we have that

$$\mathbf{Pr}\left(|h - \mathbf{E}h| > t\right) \leq e^{-\Omega\left(\frac{t^2}{(r+l)d^2}\right)}.$$

Now fix a color c and define the function h to be

$$h(\mathbf{x}, \pi) = W\left(V'' \cap \left\{v \mid c \in L'(v), |L'(v)| \geq \frac{n}{\log n}\right\}\right),$$

where $(\mathbf{x}, \pi)$ corresponds to the set of random choices for the second process. More precisely, x_i is a random partition of the set $\mathcal{C}_i$ into subsets $\mathcal{A}_i$ and $\mathcal{B}_i$ and π is a random bijection between the colors in these $\mathcal{B}_i$ and the small parts of G. Since we can fix one canonical ordering of these small parts we can assume that π is just a random permutation of the set of colors which is, by definition, independent from the variables x_i.

Next, note that changing the outcome of the variable x_i, i.e., changing one particular $\mathcal{B}_i$ can only affect vertices in at most $m - t$ small parts of G. As we already mentioned in the proof of Lemma 5, each small part contains at most $80/\delta^2$

vertices. Since we considering only vertices v satisfying $|L'(v)| \geq \frac{n}{\log n}$, the weight of such a vertex is at most $w(v) = 1/|L'(v)| \leq \frac{\log n}{n}$. Therefore, changing outcome of one x_i can change the value of h by at most $(m - t)\frac{80}{\delta^2}\frac{\log n}{n} = O\left(\frac{\log n}{n}\right) = d$. Similarly swapping any two colors in π can affect only vertices in two small parts of G. So again this can only change h by at most $d = O\left(\frac{\log n}{n}\right)$.

Since the total number of random variables x_i and also the length of permutation π are bounded by n we have that in our case $(r + l)d^2 \leq O\left(n\left(\frac{\log n}{n}\right)^2\right) = O\left(\frac{\log^2 n}{n}\right)$. Therefore it follows form the above large deviation inequality that

$$\mathbf{Pr}\left(h - \mathbf{E}h > t = \frac{\delta}{20}\right) \leq e^{-\Omega\left(\frac{t^2}{(r+l)d^2}\right)} = e^{-\Omega\left(\frac{(\delta/20)^2}{O(\log^2 n/n)}\right)} = e^{-\Omega\left(\frac{n}{\log^2 n}\right)} = o(n^{-1}).$$

This completes the proof of the lemma.   □

Now, using the fact that the total number of colors is at most n, we deduce from this lemma that with probability $1 - o(1)$ the following holds for every color c

$$W\left(V'' \cap \left\{v \mid c \in L'(v), |L'(v)| \geq \frac{n}{\log n}\right\}\right) \tag{5.3}$$

$$\leq \mathbf{CE}\left(W\left(V'' \cap \left\{v \mid c \in L'(v), |L'(v)| \geq \frac{n}{\log n}\right\}\right)\right) + \frac{\delta}{20}.$$

In addition, we also want to satisfy the following inequality:

$$W\left(V'' \cap \left\{v \mid |L'(v)| < \frac{n}{\sqrt{\log n}}\right\}\right) \tag{5.4}$$

$$\leq \left(1 + \frac{\delta}{10}\right)\mathbf{CE}\left(W\left(V'' \cap \left\{v \mid |L'(v)| < \frac{n}{\sqrt{\log n}}\right\}\right)\right).$$

Since the probability that this last inequality fails is at most $\frac{1}{1+\delta/10} < 1 - o(1)$, there does indeed exist a set of random choices for the second process which satisfies simultaneously (5.3) and (5.4).

Fix any such set of choices. Then, combining the inequalities (5.2) and (5.4) together with the facts that $W(V''') \leq \frac{|Big|}{C}$ and $V' = V'' \cup V'''$ we obtain that

$$W\left(V' \cap \left\{v \mid |L'(v)| < \frac{n}{\sqrt{\log n}}\right\}\right) \leq W\left(V'' \cap \left\{v \mid |L'(v)| < \frac{n}{\sqrt{\log n}}\right\}\right) + W(V''')$$

$$\leq \left(1 + \frac{\delta}{10}\right)^2 \mathbf{E}\left(W\left(V'' \cap \left\{v \mid |L'(v)| < \frac{n}{\sqrt{\log n}}\right\}\right)\right) + \frac{|Big|}{C} \tag{5.5}$$

Note that, by Lemma 5, every small non-singleton partition class contains at least two vertices v such that $|L'(v)| > \frac{\delta^3}{80}C = \Omega(n) > \frac{n}{\sqrt{\log n}}$. Since the number of small non-singleton partition classes is at least $A - \frac{\delta^2}{20}C$ we obtain that

$$\left|(V - \mathcal{S} - Big) \cap \left\{v \mid |L'(v)| < \frac{n}{\sqrt{\log n}}\right\}\right| \leq n - S - |Big| - 2\left(A - \frac{\delta^2}{20}C\right)$$

$$\leq \ (2-\delta)C - S - |Big| - 2A + \frac{\delta^2}{10}C$$

$$= \ (1-\delta)C - |Big| + (C - S - A) - A + \frac{\delta^2}{10}C$$

$$\leq \ (1-\delta)C - |Big| + 3k - A + \frac{\delta^2}{10}C$$

$$\leq \ \left(1 - \frac{4}{5}\delta\right)C - |Big|.$$

Here, in the last inequality we used that $A > \frac{\delta}{2}C > \frac{3}{m}C = 3k$ and $\delta^2 \leq \delta$. Note that a similar computation as in (3.1) shows that for any subset $Y \subseteq V - \mathcal{S}$ the expectation $\mathbf{E}\big(W(V' \cap Y)\big) = \frac{|Y|}{C}$. In particular, for $Y = (V - \mathcal{S} - Big) \cap \{v \,|\, |L'(v)| < \frac{n}{\sqrt{\log n}}\}$ we obtain

$$\mathbf{E}\left(W\left(V'' \cap \left\{v \,\big|\, |L'(v)| < \frac{n}{\sqrt{\log n}}\right\}\right)\right) = \frac{\left|(V - \mathcal{S} - Big) \cap \{v \,|\, |L'(v)| < \frac{n}{\sqrt{\log n}}\}\right|}{C}$$

$$\leq \ 1 - \frac{4}{5}\delta - \frac{|Big|}{C}.$$

Combining this inequality with (5.5) we have

$$W\left(V' \cap \left\{v \,\big|\, |L'(v)| < \frac{n}{\sqrt{\log n}}\right\}\right) \ \leq \ \left(1 + \frac{\delta}{10}\right)^2 \left(1 - \frac{4}{5}\delta - \frac{|Big|}{C}\right) + \frac{|Big|}{C}$$

$$\leq \ \left(1 + \frac{\delta}{4}\right)\left(1 - \frac{4}{5}\delta\right) \leq 1 - \frac{\delta}{2}. \qquad (5.6)$$

This completes our analysis of the weight of vertices with short lists. We now consider the remaining vertices.

As we already mentioned, for every color c and every non-singleton part of G there is at least one vertex v in this part such that $c \notin L(v)$. Since there are at least $A - \frac{\delta^2}{20}C$ small non-singleton parts, a similar computations as in (4.1) shows for every color c that

$$\mathbf{E}\left(W\big(V'' \cap \{v \,|\, c \in L'(v)\}\big)\right) = \frac{\left|(V - \mathcal{S} - Big) \cap \{v \,|\, c \in L'(v)\}\right|}{C}$$

$$\leq \ \frac{n - S - |Big| - (A - \frac{\delta^2}{20}C)}{C}$$

$$\leq \ \frac{(2-\delta)C - S - |Big| - A}{C} + \frac{\delta^2}{20}$$

$$= \ (1-\delta) + \frac{C - S - A}{C} - \frac{|Big|}{C} + \frac{\delta^2}{20}$$

$$\leq \ (1-\delta) + \frac{3k}{C} - \frac{|Big|}{C} + \frac{\delta^2}{20}$$

$$= \ (1-\delta) + \frac{3}{m} - \frac{|Big|}{C} + \frac{\delta^2}{20}$$

$$\leq 1 - \delta + \frac{\delta}{2} + \frac{\delta^2}{20} - \frac{|Big|}{C} \leq 1 - \frac{2}{5}\delta - \frac{|Big|}{C}.$$

Combining this inequality with (5.1) and (5.3) and using the fact that $V' = V'' \cup V'''$ we will have that for every color c

$$W\left(V' \cap \left\{v \middle| c \in L'(v), |L'(v)| \geq \frac{n}{\log n}\right\}\right)$$

$$
\begin{aligned}
&\leq \quad W\left(V'' \cap \left\{v \middle| c \in L'(v), |L'(v)| \geq \frac{n}{\log n}\right\}\right) + W(V''') \\
&\leq \quad \mathbf{CE}\left(W\left(V'' \cap \left\{v \middle| c \in L'(v), |L'(v)| \geq \frac{n}{\log n}\right\}\right)\right) + \frac{\delta}{20} + \frac{|Big|}{C} \\
&\leq \quad \mathbf{CE}\left(W\left(V'' \cap \left\{v \middle| c \in L'(v)\right\}\right)\right) + \frac{\delta}{20} + \frac{|Big|}{C} \\
&\leq \quad \left(1 + \frac{\delta}{10}\right)\mathbf{E}\left(W\left(V'' \cap \left\{v \middle| c \in L'(v)\right\}\right)\right) + \frac{\delta}{20} + \frac{|Big|}{C} \\
&\leq \quad \left(1 + \frac{\delta}{10}\right)\left(1 - \frac{2}{5}\delta - \frac{|Big|}{C}\right) + \frac{\delta}{20} + \frac{|Big|}{C} \\
&\leq \quad \left(1 + \frac{\delta}{10}\right)\left(1 - \frac{2}{5}\delta\right) + \frac{\delta}{20} \leq 1 - \frac{\delta}{4}.
\end{aligned}
\tag{5.7}
$$

Recall that H is a bipartite graph with bipartition $\left(V', \cup_v L'(v)\right)$ and an edge between c and v precisely when $c \in L'(v)$. Let K be any subset of colours in $\cup_v L'(v)$ and denote by $S_K = \{v | v \in V', L'(v) \subseteq K\}$. We complete the proof of the lemma by showing that the graph H satisfies Hall's condition, i.e., $|S_K| \leq |K|$ for every set S_K. Then in Step V we can match all uncoloured vertices in V' with the set of colours yet not used and and produce an acceptable coloring of G.

First, note that any set K of fewer than $\frac{n}{\sqrt{\log n}}$ colours cannot be an obstruction to the existence of the desired matching. Indeed, if $|S_K| > |K|$, then by Observation 4 we have that $W(S_K) > 1$. On the other hand, for every vertex $v \in S_K$ the size of $L'(v)$ is at most $|K| < \frac{n}{\sqrt{\log n}}$. Therefore we obtain a contradiction, since by (5.6)

$$W(S_K) \leq W\left(V' \cap \left\{v \,\middle|\, |L'(v)| < \frac{n}{\sqrt{\log n}}\right\}\right) < 1 - \frac{\delta}{2}.$$

Turning to larger K, we note next that the inequality (5.6) yields:

$$\left|V' \cap \left\{v \,\middle|\, |L'(v)| \leq \frac{n}{\log n}\right\}\right| \leq W\left(V' \cap \left\{v \,\middle|\, |L'(v)| \leq \frac{n}{\log n}\right\}\right)\left(\min_{v, |L'(v)| \leq \frac{n}{\log n}} w(v)\right)^{-1}$$

$$\leq \left(1 - \frac{\delta}{2}\right)\left(\min_{v, |L'(v)| \leq \frac{n}{\log n}} \frac{1}{|L'(v)|}\right)^{-1} \leq \left(1 - \frac{\delta}{2}\right)\frac{n}{\log n} < \frac{n}{\log n}. \tag{5.8}$$

Next, observe that the set of inequalities (5.7) imply that for any set of colours K

$$\left| S_K \cap \left\{ v \,\middle|\, |L'(v)| > \frac{n}{\log n} \right\} \right| = \sum_{v \in S_K, |L'(v)| > \frac{n}{\log n}} w(v) \cdot |L'(v)|$$

$$\leq \sum_{c \in K} \sum_{\{v \mid c \in L'(v), |L'(v)| > \frac{n}{\log n}\}} w(v)$$

$$= \sum_{c \in K} W\left(V' \cap \left\{ v \,\middle|\, c \in L'(v), |L'(v)| \geq \frac{n}{\log n} \right\} \right)$$

$$\leq \left(1 - \frac{\delta}{4} \right) |K|.$$

This, together with the inequality (5.8) yields that any set of colours K of size at least $\frac{n}{\sqrt{\log n}}$ satisfies

$$|S_K| = \left| S_K \cap \left\{ v \,\middle|\, |L'(v)| > \frac{n}{\log n} \right\} \right| + \left| S_K \cap \left\{ v \,\middle|\, |L'(v)| \leq \frac{n}{\log n} \right\} \right|$$

$$\leq \left(1 - \frac{\delta}{4} \right) |K| + \left| V' \cap \left\{ v \,\middle|\, |L'(v)| \leq \frac{n}{\log n} \right\} \right|$$

$$\leq \left(1 - \frac{\delta}{4} \right) |K| + \frac{n}{\log n} < |K|.$$

Thus we obtain that these larger K also do not violate Hall's condition and hence the desired matching of Step V does indeed exist. This completes the proof. $\square$

6.   Concluding remarks

In this paper we proved that for every $\epsilon > 0$ and for every sufficiently large graph G of order n, the list chromatic number of G equals its chromatic number, provided $n \leq (2 - \epsilon)\chi(G)$. A more careful analysys of our methods yields that the value of ϵ in this result can be made as small as $O(1/\log^\eta n)$ for any constant $0 < \eta < 1$. Nevertheless the conjecture of Ohba remains open for graphs with $2\chi(G)$ vertices and it seems one needs new ideas to tackle this problem. Even to show that there is a constant N such that $\chi^l(G) = \chi(G)$ for every graph G with at most $2\chi(G) - N$ vertices, would be very interesting.

In conclusion we would like to propose a related problem, which was motivated by Ohba's conjecture. Let t be an integer and let G be a graph with at most $t\chi(G)$ vertices. Find the smallest constant c_t such that for any such a graph G its list chromatic is bounded by $c_t \chi(G)$. Note that Ohba's conjecture if true, implies that $c_2 = 1$. An additional intriguing question is to determine graphs with $|V(G)| \leq t\chi(G)$ and for which the ratio $\chi^l(G)/\chi(G)$ is maximal. Here the case $t = 2$ gives some indication that a complete multi-partite graph with all parts of size t may have this property.

References

[1] N. Alon, Restricted colorings of graphs, in *Surveys in Combinatorics 1993*, London Math. Soc. Lecture Notes Series 187 (K. Walker, ed.), Cambridge Univ. Press, 1993, 1–33.

[2] H. Enemoto, K. Ohba and J. Sakamoto, Choice numbers of some complete multi-partite graphs, preprint.

[3] P. Erdős, A. L. Rubin and H. Taylor, Choosability in graphs, *Congressus Numerantium* 26 (1979), 125-157.

[4] T. Jensen and B. Toft, *Graph Coloring Problems*, Wiley, New York, 1995.

[5] L. Lovász, *Combinatorial problems and exercises*, North-Holland, Amsterdam, 1993.

[6] L. Lovász and M. Plummer, *Matching Theory*, North Holland, Amsterdam, 1986.

[7] C. McDiarmid, Concentration, in *Probabilistic methods for algorithmic discrete mathematics*, Algorithms and Combinatorics 16, Springer, Berlin, 1998, 195–248.

[8] M. Molloy and B. Reed, *Graph Colouring and the Probabilistic Method*, Springer-Verlag, Berlin, 2002.

[9] K. Ohba, On chromatic-choosable graphs, *J. Graph Theory* 40 (2002), 130–135.

[10] B. Reed and B. Sudakov, List coloring when the chromatic number is close to the order of the graph, *Combinatorica*, to appear.

[11] V. G. Vizing, Coloring the vertices of a graph in prescribed colors (in Russian), em Diskret. Analiz. No. 29, Mctody Diskret. Anal. v. Teorii Kodov i Shem 101 (1976), 3–10.

Hard Constraints and the Bethe Lattice: Adventures at the Interface of Combinatorics and Statistical Physics

Graham R. Brightwell[*]   **Peter Winkler**[†]

Abstract

Statistical physics models with hard constraints, such as the discrete hard-core gas model (random independent sets in a graph), are inherently combinatorial and present the discrete mathematician with a relatively comfortable setting for the study of phase transition.

In this paper we survey recent work (concentrating on joint work of the authors) in which hard-constraint systems are modeled by the space $\text{Hom}(G, H)$ of homomorphisms from an infinite graph G to a fixed finite constraint graph H. These spaces become sufficiently tractable when G is a regular tree (often called a Cayley tree or Bethe lattice) to permit characterization of the constraint graphs H which admit multiple invariant Gibbs measures.

Applications to a physics problem (multiple critical points for symmetry-breaking) and a combinatorics problem (random coloring), as well as some new combinatorial notions, will be presented.

2000 Mathematics Subject Classification: 82B20, 68R10.
Keywords and Phrases: Hard constraints, Bethe lattice, Graph homomorphisms, Combinatorial phase transition.

1. Introduction

Recent years have seen an explosion of activity at the interface of graph theory and statistical physics, with probabilistic combinatorics and the theory of computing as major catalysts. The concept of "phase transition", which a short time ago most graph theorists would barely recognize, has now appeared and reappeared in journals as far from physics as the *Journal of Combinatorial Theory* (Series B).

[*]Department of Mathematics, London School of Economics, Houghton St., London WC2A 2AE England. E-mail: g.r.brightwell@lse.ac.uk
[†]Bell Labs 2C-365, 700 Mountain Ave., Murray Hill NJ 07974-0636, USA. E-mail: pw@lucent.com

Traffic between graph theory and statistical physics is already heavy enough to make a complete survey a book-length proposition, even if one were to assume a readership with knowledge of both fields.

This article is intended for a general mathematical audience, *not* necessarily acquainted with statistical physics, but it is not to serve as an introduction to the field. Readers are referred to texts such as [1, 12, 17, 19] for more background. We will present only a small part (but we hope an interesting one) of the interface between combinatorics and statistical physics, with just enough background in each to make sense of the text. We will focus on the most combinatorial of physical models—those with hard constraints—and inevitably on the authors' own research and related work.

We hope it will be clear from our development that there is an enormous amount of fascinating mathematics to be uncovered by studying statistical physics, quite a lot of which has been or will be connected to graph theory. What follows is only a sample.

2. Random independent sets

In what follows a *graph* $G = \langle V, E \rangle$ consists of a set V (finite or countably infinite) of *nodes* together with a set E of *edges*, each of which is an unordered pair of nodes. We will sometimes permit loops (edges of the form $\{v, v\}$) but multiple edges will not be considered or needed. We write $u \sim v$, and say that u is "adjacent" to v, if $\{u, v\} \in E$; a set $U \subset V$ is said to be *independent* if it contains no edges.

The *degree* of a node u of G is the number of nodes adjacent to u; all graphs considered here will be *locally finite*, meaning that all nodes have finite degree. A *path* in G (of length k) is a sequence $u_0, u_1, \ldots, u_k$ of distinct nodes with $u_i \sim u_{i+1}$; if in addition $u_k \sim u_0$ we have a *cycle* of length $k+1$. If every two nodes of G are connected by a path, G is said to be *connected*.

The plane grid $\mathbb{Z}^2$ is given a graph structure by putting $(i, j) \sim (i', j')$ iff $|i - i'| + |j - j'| = 1$. Let us carve out a big piece of $\mathbb{Z}^2$, say the box $B_n^2 := \{(i, j) \mid -n \leq i, j \leq n\}$. Let I be a uniformly random independent set in the graph B_n^2; in other words, of all sets of nodes (including the empty set) not containing an edge, choose one uniformly at random. What does it look like?

Plate 1 shows such an I (in a rectangular region). Here the nodes are represented by squares, two being adjacent if they have a common vertical or horizontal border segment. The sites belonging to I are colored, the others omitted. It is by no means obvious how to obtain such a random independent set in practice; one cannot simply choose points one at a time subject to the independence constraint. In fact the set in Plate 1 was generated by *Markov chain mixing*, an important and fascinating method in the theory of computing, which has by itself motivated much recent work at the physics-combinatorics interface.

It is common in statistical physics to call the nodes of B_n^2 *sites* (and the edges, *bonds*). Sites in I are said to be *occupied*; one may imagine that each occupied site contains a molecule of some gas, any two of which must be at distance greater than 1.

In the figure, *even* occupied sites (nodes $(i,j) \in I$ for which $i+j \equiv 0 \mod 2$) are indicated by one color, odd sites by another. A certain tendency for colors to clump may be observed; understandably, since occupied sites of the same parity may be as close as $\sqrt{2}$ (in the Euclidean norm) but opposite-parity particles must be at least $\sqrt{5}$ apart.

It stands to reason that if more "particles" were forced into I, then we might see more clumping. Let us weight the independent sets according to size, as follows: a positive real λ, called the *activity* (or sometimes *fugacity*) is fixed, and then each independent set I is chosen with probability proportional to $\lambda^{|I|}$. We call this the "λ-measure". Of course if $\lambda = 1$ we are back to the uniform measure, but if $\lambda > 1$ then larger independent sets are favored.

The λ-measure for $\lambda \neq 1$ is, to a physicist, no less natural than the uniform. In a combinatorial setting, such a measure might arise e.g. if the particles happen to be of two different types, with all "typed" independent sets equiprobable; then the probability that a particular set of sites is occupied is given by the λ-measure with $\lambda = 2$.

Plate 2 shows a random I chosen when $\lambda = 3.787$. The clusters have grown hugely as more particles were packed in. Push λ up just a bit more, to 3.792, and something like Plate 3 is the result: one color (parity) has taken over, leaving only occasional islands of the other.

Something qualitative has changed here, but what exactly? The random independent sets we have been looking at constitute what the physicists call the *hard-core lattice gas model*, or "hard-core model" for short. Readers are referred to the exceptionally readable article [2] in which many nice results are obtained for this model[1]. On the plane grid, the hard-core model has a "critical point" at activity about 3.79, above which the model is said to have experienced a *phase transition*.

3. What is phase transition?

There is no uniformity even among statistical physicists regarding the definition of phase transition; in fact, there is even disagreement about whether the "phases" above are the even-dominated versus odd-dominated configurations at high λ, or the high-λ regime versus the low. Technical definitions involving points of non-analyticity of some function miss the point for us.

The point really is that a slight change in a parameter governing the local behavior of some statistical system, like the hard-core model, can produce a global change in the system, which may be evidenced in many ways. For example, suppose we sampled many independent sets in B_n^2 at some fixed λ, and for each computed the ratio of the number of even occupied sites to the number of odd. For low λ these numbers would cluster around $1/2$, but for high λ they would follow a bimodal distribution; and the larger the box size n, the sharper the transition.

Here's another, more general, consideration. Suppose we look only at independent sets which contain all the even sites on the boundary of B_n^2. For these I

---

[1] Readers, however, are cautioned regarding conducting a web search with key-words "hard-core" and "model".

the origin would be more likely to be occupied than, say, one of its odd neighbors. As n grows, this "boundary influence" will fade—provided λ is low. But when λ is above the critical point, the boundary values tend to make I an even-dominated set, giving any even site, no matter how far away from the boundary, a non-disappearing advantage over any odd one.

Computationally-minded readers might be interested in a third approach. Suppose we start with a fixed independent set I_0, namely the set of all even sites in B_n^2, and change it one site at a time as follows: at each tick of a clock we choose a site u at random. If any of u's neighbors is occupied, we do nothing. Otherwise we flip a biased coin and with probability $\lambda/(1+\lambda)$ we put u in I (where it may already have been), and with probability $1/(1+\lambda)$ we remove it (or leave it out). The result is a Markov chain whose states are independent sets and whose stationary distribution, one can easily verify, is exactly our λ-measure. Thus if we do this for many steps, we will have a nearly perfect sample from this distribution—but how many steps will that take? We *believe* that when λ is below its critical value, only polynomially (in n) steps are required—the Markov chain is said to be *rapidly mixing*; even polylogarithmic, if we count the number of steps per site. But for high λ it appears to take time exponential in n (or perhaps in $\sqrt{n}$) before we can expect to see an odd-dominated independent set. The exact relationship between phase transition and Markov chain mixing is complex and the subject of much study.

All these measures rely on taking limits as the finite box B_n^2 grows; the very nice discovery of Dobrushin, Lanford and Ruelle [9, 15] is that there is a way to understand the phenomenon of phase transition as a property of the infinite plane grid. The idea is to extend the λ-measure to a probability distribution on independent sets on the whole grid, then ask whether the extension is unique.

We cannot extend the definition of the λ-measure directly since $\lambda^{|I|}$ is generally infinite, but we can ask that it behave locally like the finite measure. We say that a probability distribution μ on independent sets in the plane grid is a *Gibbs measure* if for any site u the probability that u is in I, given the sites in $I \cap (\mathbb{Z}^2 \setminus \{u\})$, is $\lambda/(1 + \lambda)$ if the neighborhood of u is unoccupied and, of course, 0 otherwise.

It turns out that Gibbs measures always exist (here, and in far greater generality) but may or may not be unique. When there is more than one Gibbs measure we will say that there is a phase transition. For the hard-core model on $\mathbb{Z}^2$, there is a unique Gibbs measure for low λ; but above the critical value, there is a Gibbs measure in which the even occupied sites are dominant and another in which the odd sites are dominant (all other Gibbs measures are convex combinations of these two). How can you construct these measures? Well, for example, the even measure can be obtained as a limit of λ-measures on boxes whose even boundary sites are forced to be in I. The fact that the boundary influence does not fade (in the high λ case) implies that the even and odd Gibbs measures are different.

We have noted that the critical value of λ for the hard-core model on $\mathbb{Z}^2$ is around 3.79. This is an empirical result and all we mathematicians can prove is that there is at least one critical point, and all such are between 1.1 and some high number. It is believed that, for each d, there is just one critical value λ_d on $\mathbb{Z}^d$. It is also to be expected that λ_d is decreasing in d, but only recently has it been shown that the largest critical value on $\mathbb{Z}^d$ tends to 0 as $d \to \infty$. This result was obtained by David Galvin and Jeff Kahn [11], two combinatorialists, using graph theory, geometry, topology, and lots of probabilistic combinatorics. A consequence of their work is that $\lambda = 1$ is above the critical value(s) for sufficiently large d; this can be stated in a purely combinatorial way: for sufficiently high d and large n, *most* independent sets in B_n^d are dominated by vertices of one parity.

In the next section we explain how we can use graphs to understand models with hard constraints; then, in the section following that, we will switch from $\mathbb{Z}^2$ to a much easier setting, in which we can get our hands on nice Gibbs measures.

4. Hard constraints and graph homomorphisms

We are interested in what are sometimes called "nearest neighbor" hard constraint models, where the constraints apply only to adjacent sites. Each site is to be assigned a "spin" from some finite set, and only certain pairs of spins are permitted on adjacent sites. We can code up the constraints as a finite graph H whose nodes are the spins, and whose edges correspond to spins allowed to appear at neighboring sites. This *constraint graph* H may have some loops; a loop at node $v \in H$ would mean that neighboring sites may both be assigned spin v. We adopt the statistical physics tradition of reserving the letter "q" for the number of spins, that is, the number of nodes in H.

The graph G (e.g. $\mathbb{Z}^2$, above) of sites, usually infinite but always countable and locally finite, is called (by us) the *board*. A legal assignment of spins to the sites of G is nothing more or less than a graph homomorphism from G to H, i.e. a map from the sites of G to the nodes of H which preserves edges. We denote the set of homomorphisms from G to H by $\mathrm{Hom}(G, H)$, and give it a graph structure by putting $\varphi \sim \psi$ if φ and ψ differ at exactly one site of G.

We will often confuse a graph with its set of nodes (or sites). In particular, if U is a subset of the nodes of G then U together with the edges of G contained in U constitute the "subgraph of G induced by U", which we also denote by U.

In the hard-core model, the constraint graph H consists of two adjacent nodes, one of which is looped: a function from a board G to this H is a homomorphism iff the set of sites mapped to the unlooped node is an independent set. Plate 4 shows some constraint graphs found in the literature.

When H is complete and every node is looped as well, there is no constraint and nothing interesting happens.

When H is the complete graph K_q (without loops), homomorphisms to H are just ordinary, "proper" q-colorings of the board. (A proper q-coloring of a graph G is a mapping from the nodes of G to a q-element set in which adjacent nodes are never mapped to the same element.) This corresponds to something called the "anti-ferromagnetic Potts model at zero temperature". In the anti-ferromagnetic Potts model at positive temperature, adjacent sites are merely discouraged (by an energy penalty), not forbidden, from having the same spin; thus this is not a hard constraint model in our terminology. The $q = 2$ case of the Potts model is the famous Ising model.

For a general constraint graph H, we need to elevate the notion of activity to vector status. To each node i of H we assign a positive real activity λ_i, so that H now gets an activity vector $\lambda := (\lambda_1, \ldots, \lambda_q)$. When the board G is finite, each homomorphism $\varphi \in \mathrm{Hom}(G, H)$ is assigned probability proportional to

$$\prod_{v \in G} \lambda_{\varphi(v)} .$$

We can think of λ_i as the degree to which we try to use spin i, when it is available. For example, if we know the spins of the neighbors of site v and consequently, say, spins i, j and k are allowed for v, then the λ-measure forces $\mathrm{Pr}(\varphi(v) = i) = \lambda_i/(\lambda_i + \lambda_j + \lambda_k)$.

When G is infinite, things get a little more complicated. A finite subset (and its induced subgraph) $U \subset G$ will be called a "patch" and its boundary ∂U is the set of sites not in U but adjacent to some site of U. We define $U^+ := U \cup \partial U$. If φ is a function on G, then $\varphi{\upharpoonright}U$ denotes its restriction to the subset U.

We say that μ is a Gibbs measure for λ if: for any patch $U \subset G$, and almost every $\psi \in \mathrm{Hom}(G, H)$,

$$\Pr_{\mu}\left(\varphi{\upharpoonright}U = \psi{\upharpoonright}U \,\middle|\, \varphi{\upharpoonright}(G - U) = \psi{\upharpoonright}(G - U)\right) = \Pr_{U^+}\left(\varphi{\upharpoonright}U = \psi{\upharpoonright}U \,\middle|\, \varphi{\upharpoonright}\partial U = \psi{\upharpoonright}\partial U\right)$$

where "Pr_{U^+}" refers to the finite λ-measure on U^+.

This definition looks messy but it just means that the probability distribution of a random φ inside a patch U depends only on its value on the boundary of U, and is the same as if U and its boundary comprised all of the board. We will see later that when H has a certain nice property, as it does in the case of the hard-core model, it suffices to check the Gibbs condition only on patches consisting of a single site—we call this the one-site condition.

It is a special case of a theorem of Dobrushin [9] that there is always at least one Gibbs measure for any λ on $\mathrm{Hom}(G, H)$; we are concerned with questions about when there is a unique Gibbs measure, and when there is a phase transition (i.e. more than one Gibbs measure).

Let us again look briefly at possible implications for phase transition in the setting of finite boards. Given a finite board G, a constraint graph H and activities λ, we define the *point process* $\mathcal{P}(G, H, \lambda)$ as follows: starting from any element of $\mathrm{Hom}(G, H)$, choose a site u of G uniformly at random, and give it a fresh spin according to the Gibbs condition, so that each 'legal' spin j is chosen with probability proportional to λ_j. The point process is a Markov chain on $\mathrm{Hom}(G, H)$, and it is easy to check that the λ-measure is a stationary distribution (which will be unique provided $\mathrm{Hom}(G, H)$ is connected, a point we will return to later).

Running the point process for sufficiently long will thus generate a random homomorphism according to the λ-measure. However, suppose that the finite board G is a large piece of an infinite board G' exhibiting a phase transition for our λ. Then, if we start with a homomorphism arising from one Gibbs measure on $\mathrm{Hom}(G', H)$ (restricted to G), it is reasonable to expect that the point process will take a long time to reach a configuration resembling that from any other Gibbs measure on $\mathrm{Hom}(G', H)$. Thus it is generally believed that, in some necessarily loose sense, phase transition on an infinite graph corresponds to slow convergence for the point process on finite subgraphs.

5. Cayley trees and branching random walks

Gibbs measures can be elusive and indeed it is generally a difficult task to prove that phase transitions occur on a typical board of interest, like $\mathbb{Z}^d$. In order to get results and intuition physicists sometimes turn to a more tractable board, called by them the Bethe lattice (after Hans Bethe) and by combinatorialists, usually, the Cayley tree.

We denote by $\mathbb{T}^r$ the r-branching Cayley tree, equivalently the unique connected (infinite) graph which is cycle-free and in which every site has degree $r+1$. $\mathbb{T}^r$ is a vastly different animal from $\mathbb{Z}^d$. It is barely connected, falling apart with the removal of any site; its patches have huge boundaries, comparable in size with the patch itself; its automorphism group is enormous. It's surprising that we can learn anything at all about $\mathrm{Hom}(\mathbb{Z}^d, H)$ from $\mathrm{Hom}(\mathbb{T}^r, H)$, and indeed we must be careful about drawing even tentative conclusions in either direction. Basic physical parameters like entropy become dodgy on non-amenable (big-boundary) boards like $\mathbb{T}^r$ and a number of familiar statistical physics techniques become useless. More than making up for these losses, though, are the combinatorial techniques we can use to study $\mathrm{Hom}(\mathbb{T}^r, H)$. There are even situations (e.g. in the study of information dissemination) where $\mathbb{T}^r$ is the natural setting.

We are particularly interested in Gibbs measures on $\mathrm{Hom}(\mathbb{T}^r, H)$ which have the additional properties of being *simple* and *invariant*.

For any site u in a tree T, let $d(u)$ be the number of edges incident with u and let $C_1(u), C_2(u), \ldots, C_{d(u)}(u)$ be the connected components of $T \setminus \{u\}$.

Definition 5.1 *A Gibbs measure μ on $\mathrm{Hom}(T, H)$ is simple if, for any site $u \in T$ and any node $i \in H$, the μ-distributions of*

$$\varphi \restriction C_1(u), \ldots, \varphi \restriction C_{d(u)}(u)$$

are mutually independent given $\varphi(u) = i$.

This condition, which is trivially satisfied by the λ-measure for finite T, would follow from the Gibbs condition itself if fewer than two of the $C_i(u)$'s were infinite.

Definition 5.2 *Let $\mathcal{A}(G)$ be the automorphism group of the board G, and for any subset $S \subset \mathrm{Hom}(G, H)$ and $\kappa \in \mathcal{A}(G)$ let $S \circ \kappa := \{\varphi \circ \kappa : \varphi \in S\}$. We say that a measure μ on $\mathrm{Hom}(G, H)$ is invariant if, for any μ-measurable $S \subset \mathrm{Hom}(G, H)$ and any $\kappa \in \mathcal{A}(G)$, we have $\mu(S \circ \kappa) = \mu(S)$.*

Again, this condition is trivially satisfied for finite G; but for an infinite board with as many automorphisms as $\mathbb{T}^r$, it is quite strong. Later we consider relaxing it slightly. For now, we might well ask, how can we get our hands on any Gibbs measure for $\mathrm{Hom}(\mathbb{T}^r, H)$, let alone a simple, invariant one?

The absence of cycles in $\mathbb{T}^r$ makes it plausible that we can get ourselves a Gibbs measure by building random configurations in $\mathrm{Hom}(\mathbb{T}^r, H)$ one site at a time. We could choose a root $x \in \mathbb{T}^r$, assign it a random spin $i \in H$, then assign the neighbors of i randomly to the $r+1$ children of x; thereafter, each time a site u gets spin j we give its r children random spins from among the neighbors of j.

The process we have described can be thought of as a branching random walk on H. Imagine amoebas staggering from node to adjacent node of H; each time an amoeba steps it divides into r baby amoebas which then move independently at the next time step. Of course, the (usually tiny) constraint graph H is shortly piled high with exponentially many amoebas, but being transparent they happily ignore one another and go on stepping and dividing.

To get started we have to throw the first amoeba onto H where we imagine that its impact will cause it to divide $r+1$ ways instead of the usual r.

Note that the $r = 1$ case is just ordinary random walk, started somewhere on the doubly-infinite path $\mathbb{T}^1$ and run both forward and backward.

To determine what probabilities are used in stepping from one node of H to an adjacent node, we assign a positive real *weight* w_i to each node. For convenience we denote by z_i the sum of the weights of the neighbors of i (including i itself, if there is a loop at i). An amoeba-child born on node i then steps to node j with probability w_j/z_i. If there is a loop at i, the amoeba stays at i with the appropriate probability, w_i/z_i.

Assuming H is connected and not bipartite[2], the random walk (branching or not) will have a stationary distribution π; it is easily verified that π_i is proportional to $w_i z_i$ for each i, and somewhat less easily verified that the mapping $w \mapsto \pi$ is one-to-one provided $\sum w_i$ has been normalized to 1. We use the stationary distribution to pick the starting point for the first amoeba, i.e. to assign a spin to the root of $\mathbb{T}^r$.

Finally, the payoff: not only does this node-weighted branching random walk give us a simple invariant Gibbs measure; it's the *only* way to get one. The following theorem appears in [4] but it is not fundamentally different from characterizations which can be found in Georgii [12] and elsewhere.

Theorem 5.3 *Let H be a fixed connected constraint graph with node-weights w and let r be a positive integer. Then the measure μ induced on $\mathrm{Hom}(\mathbb{T}^r, H)$ by the r-branching w-random walk on H is a simple, invariant Gibbs measure, for some activity λ on H. Conversely, if H, r and λ are given, then every simple, invariant Gibbs measure on $\mathrm{Hom}(\mathbb{T}^r, H)$ is given by the r-branching random walk on H with nodes weighted by some w.*

The proof is actually quite straightforward, and worth including here. Invariance of μ with respect to root-preserving automorphisms of $\mathbb{T}^r$ is trivial, since the random walk treats all children equally; the only issue is whether the selection of root makes a difference. For this we need only check that for two neighboring sites u and v of $\mathbb{T}^r$, μ is the same whether u is chosen as root or v is. But, either way

---

[2] A graph is bipartite if its nodes can be partitioned into two sets neither of which contains an edge. Thus, for example, the existence of a looped node already prevents H from being bipartite.

we may choose $\varphi(u)$ and $\varphi(v)$ as the first two spins and the rest of the procedure is the same; so it suffices to check that for any (adjacent) nodes i and j of H, the probability that $\varphi(u) = i$ and $\varphi(v) = j$ is the same with either root choice. But these two probabilities are

$$\pi_i p_{ij} = z_i w_i \frac{w_j}{z_i} = w_i w_j = z_j w_j \frac{w_i}{z_j} = \pi_j p_{ji}$$

as desired.

To show that μ is simple is, indeed, simple: if we condition on $\varphi(u) = i$ then, using invariance to put the root at u, the independence of φ on the $r+1$ components of $\mathbb{T}^r \setminus \{u\}$ is evident from the definition of the branching random walk.

The activity vector λ for which μ is a Gibbs measure turns out to be given by

$$\lambda_i = \frac{w_i}{z_i^r} \ .$$

Let U be any finite set of sites in $\mathbb{T}^r$, with exterior boundary ∂U. On account of invariance of labeling, we may assume that the root x does not lie in $U^+ = U \cup \partial U$.

Let $g \in \text{Hom}(U^+, H)$; we want to show that the probability that a branching random walk φ matches g on U, given that it matches on ∂U, is the same as the corresponding conditional probability for the λ-measure.

Let T be the subtree of $\mathbb{T}^r$ induced by U^+ and the root x; for any $f \in \text{Hom}(T, H)$,

$$\Pr(\varphi \restriction T = f) = \pi_{f(x)} \cdot \prod_{u \to v} p_{f(u), f(v)}$$

$$= z_{f(x)} w_{f(x)} \prod_{u \to v} \frac{w_{f(v)}}{z_{f(u)}}$$

where $u \to v$ means that v is a child of u in the tree. The factors $z_{f(u)}$ corresponding to sites u in U each occur as denominator r times in the above expression, since each site in U has all of its r successors in T; and of course each $w_{f(u)}$ occurs once as a numerator as well. It follows that if we compare $\Pr(\varphi \restriction T = f)$ with $\Pr(\varphi \restriction T = f')$, where f' differs from f only on U, then the value of the first is proportional to

$$\prod_{u \in U} \frac{w_{f(u)}}{z_{f(u)}^r} = \prod_{u \in U} \lambda_{f(u)}$$

which means that μ coincides with the finite measure, as desired.

Now let us assume that μ is a simple, invariant Gibbs measure on $\text{Hom}(\mathbb{T}^r, H)$ with activity vector λ, with the intent of showing that μ arises from a node weighted branching random walk on H.

We start by constructing a μ-random φ, site by site. Choose a root x of $\mathbb{T}^r$ and pick $\varphi(x)$ from the *a priori* distribution σ of spins of x (and therefore, by invariance, of any other site). We next choose a spin for the child y of x according to the conditional distribution matrix $P = \{p_{ij}\}$ given by

$$p_{ij} := \Pr\left(\varphi(y) = j \,\middle|\, \varphi(x) = i\right) \ ;$$

again, by invariance of μ, P is the same for any pair of neighboring sites. It follows that $\sigma = \sigma \cdot P$, and moreover that P is the transition matrix of a reversible Markov chain, since the roles of x and y can be interchanged.

Next we proceed to the rest of the children of x, then to the grandchildren, etc., choosing each spin conditionally according to all sites so far decided.

We claim, however, that the distribution of possible spins of the non-root v depends only on the spin of its parent u; this is so because μ is simple and all sites so far "spun" are in components of $\mathbb{T}^r \setminus \{u\}$ other than the component containing v. Thus the value of $\varphi(v)$ is given by P for *every* site $v \neq x$, and it follows that μ arises from an r-branching Markov chain with state-space H, starting at distribution σ.

Evidently for any (not necessarily distinct) nodes i, j of H, there will be pairs (u, v) of adjacent sites with $\varphi(u) = i$ and $\varphi(v) = j$ if and only if $i \sim j$ in H. Hence P allows transitions only along edges of H, and there is a unique distribution π satisfying $\pi \cdot P = \pi$; thus $\sigma = \pi$.

It remains only to show that P is a node-weighted random walk, and it turns out that a special case of the Gibbs condition for one-site patches suffices. Let j and j' be nodes of H which have a common neighbor i, and suppose that all of the neighbors of the root x have spin i. Such a configuration will occur with positive probability and according to the Gibbs condition for $U = \{x\}$,

$$\frac{\Pr(\varphi(x) = j')}{\Pr(\varphi(x) = j)} = \frac{\lambda_{j'}}{\lambda_j}$$

but

$$\Pr(\varphi(x) = j) = \frac{\pi_j p_{ji}^{r+1}}{\sum_{k \sim i} \pi_k p_{ki}^{r+1}}$$

and similarly for j', so

$$\frac{\Pr(\varphi(x) = j')}{\Pr(\varphi(x) = j)} = \frac{\pi_{j'} p_{j'i}^{r+1}}{\pi_j p_{ji}^{r+1}} \ .$$

Thus the ratio

$$\frac{p_{j'i}}{p_{ji}} = \left(\frac{\lambda_{j'} \pi_j}{\lambda_j \pi_{j'}} \right)^{\frac{1}{r+1}}$$

is independent of i.

Since P is reversible we have $p_{ij} = \pi_j p_{ji} / \pi_i$, hence

$$\frac{p_{ij'}}{p_{ij}} = \frac{\pi_{j'} p_{j'i}}{\pi_j p_{ji}}$$

is also independent of i, and it follows that P is a node-weighted random walk on H. This concludes the proof of Theorem 5.3.

In view of Theorem 5.3, if we can understand the behavior of the map $w \hookrightarrow \lambda$, we will know, given λ, whether there is a nice Gibbs measure and if so whether there is more than one. The first issue is settled nicely in the following theorem, a proof of which can be found in [4] and requires some topology. A similar result was proved by Zachary [23].

Theorem 5.4 *For every $r \geq 2$, every constraint graph H and every set λ of activities for H, there is a node-weighted branching random walk on $\mathbb{T}^r$ which induces a simple, invariant Gibbs measure on $\mathrm{Hom}(\mathbb{T}^r, H)$.*

It's nice to know that we haven't required so much of our measures that they can fail to exist.

A statistical physics dictum (true in great, but not unlimited, generality) says that there's never a phase transition in dimension 1; that holds here:

Theorem 5.5 *For any connected constraint graph H and any activity vector λ, there is a unique simple invariant Gibbs measure on $\mathrm{Hom}(\mathbb{T}^1, H)$.*

Furthermore, in any dimension, there's always *some* region where the map $w \hookrightarrow \lambda$ is one-to-one:

Theorem 5.6 *For any r and H there is an activity vector λ for which there is only one simple invariant Gibbs measure on $\mathrm{Hom}(\mathbb{T}^r, H)$.*

6. Fertile and sterile graphs

The fascination begins when we hit an H and a λ which boast multiple simple, invariant Gibbs measures. Let us examine a particular case, involving a constraint graph we call the "hinge".

The hinge has three nodes, which we associate with the colors green, yellow and red; all three nodes are looped and edges connect green with yellow, and yellow with red. Thus the only missing edge is green-red, and a $\varphi \in \mathbb{T}^r$ may be thought of as a green-yellow-red coloring of the tree in which no green site is adjacent to a red one.

The hinge constraint in fact corresponds to a discrete version of the Widom-Rowlinson model, in which two gases (whose particles are represented by red and green) compete for space and are not permitted to occupy adjacent sites; see e.g. [3, 21, 22]. When λ_{red} and λ_{green} are equal and large relative to $\lambda_{\mathrm{yellow}}$, the Widom-Rowlinson model tends to undergo a phase transition as one gas spontaneously dominates the other. Plate 5 shows a red-dominated sample from the Widom-Rowlinson model on $\mathbb{Z}^2$, with the unoccupied sites left uncolored instead of being colored yellow.

We can see the phase transition operate on the Cayley tree $\mathbb{T}^2$. If the green, yellow and red nodes are weighted 4, 2 and 1 respectively, λ (normalized to integers) turns out to be $(49, 18, 49)$—equal activity for green and red. How can a random walk which is biased so strongly toward green end up coloring a tree according to a Gibbs measure with symmetric specification? As a clue, let us examine a site u of $\mathbb{T}^2$ which happens to be surrounded by yellow neighbors. To be colored green requires that a certain amoeba stepped from yellow to green, then both of its children returned to yellow. Thus the conditional probability that u is green is proportional to

$$\frac{4}{4+2+1} \cdot \left(\frac{2}{4+2}\right)^2$$

as opposed to

$$\frac{1}{4+2+1} \cdot \left(\frac{2}{2+1}\right)^2$$

for red, but these values are equal.

Clearly the reversed weights 1, 2 and 4 would yield the same activity vector, and in fact a third, symmetric weighting, approximately 6, 7 and 6, does as well. Plate 6 shows pieces of $\mathbb{T}^2$ colored according to these three weightings. Of course the colorings have different proportions and are easily identifiable; checking the stationary distributions for the three random walks, we see that *a priori* a site is colored green with probability about 59% in the first weighting, 30% with the symmetric weighting and only 7% in the reversed weighting. Yet, from a conditional point of view, the three colorings are identical.

It turns out that the hinge is one of seven minimal graphs each of which can produce a phase transition on $\mathbb{T}^r$ for any $r \geq 2$. The graphs are pictured in Plate 7. We say that a graph H is *fertile* if $\text{Hom}(\mathbb{T}^r, H)$ has more than one simple, invariant Gibbs measure for some r and λ; otherwise it is *sterile*. The fertile graphs are exactly those which contain one or more of the seven baby graphs in Plate 7 as an induced subgraph. It turns out that the value of r does not come into play: if the constraint graph is rich enough to produce a phase transition on any $\mathbb{T}^r$, then it does so for all $r \geq 2$. One way to state the result is as follows:

Theorem 6.1 [4] *Fix $r > 1$ and let H be any constraint graph. Suppose that H satisfies the following two conditions:*

(a) *Every looped node of H is adjacent to all other nodes of H;*

(b) *With its loops deleted, H is a complete multipartite graph.*

Then for every activity vector on H, there is a unique invariant Gibbs measure on the space $\text{Hom}(\mathbb{T}^r, H)$.

If H fails either condition (a) or condition (b) then there is a set of activities λ on H for which $\text{Hom}(\mathbb{T}^r, H)$ has at least two simple, invariant Gibbs measures, and therefore λ can be obtained by more than one branching random walk.

The proof of Theorem 6.1 is far too complex to reproduce here, but reasonably straightforward in structure. First, a distinct pair of weightings yielding the same activity vector must be produced for each of the seven baby fertile graphs, and for each $r \geq 2$. Second, it must be demonstrated that if H contains one of the seven as an induced subgraph, then there are weightings (whose restrictions are close to those previously found) which induce phase transitions on $\text{Hom}(\mathbb{T}^r, H)$. Third, a monotonicity argument is employed to show that if H satisfies conditions (a) and (b) of the theorem, then the map from w to λ is injective. Finally, an easy graph-theoretical argument shows that H satisfies (a) and (b) precisely if it does not contain any of the seven baby fertile graphs as an induced subgraph.

7. An application to statistical physics

Theorem 6.1 has many shortcomings, applying as it does only to hard constraint models on the Bethe lattice, and we must also not forget that it considers only the very nicest Gibbs measures. The constraint graph of the hard-core model is sterile, yet it can have multiple Gibbs measures on $\mathbb{T}^r$ (or, as we saw, on $\mathbb{Z}^2$) if we relax the invariance condition.

For $H =$ the hinge, however, and for any r and λ, there are multiple Gibbs measures on $\text{Hom}(\mathbb{T}^r, H)$ if and only if there are multiple simple, invariant Gibbs measures. Like the Ising model, the Widom-Rowlinson model exhibits spontaneous symmetry-breaking; indeed its relationship to the Ising model parallels the relation between nodes and edges of a graph.

One of the nice properties known for the Ising model is that it can exhibit at most one critical point; but the proof of this fact does not work for the Widom-Rowlinson model. Indeed, in [3] the methods above are used to construct a board G for which $\text{Hom}(G, H)$ has three (or more) calculable critical points, with H the hinge. Set $\lambda = \lambda_{\text{green}} = \lambda_{\text{red}}$, fixing $\lambda_{\text{yellow}} = 1$, so that the single parameter λ controls the Widom-Rowlinson model. Then:

Theorem 7.1 *There exist $0 < \lambda_1 < \lambda_2 < \lambda_3$ and an infinite graph G, such that the Widom–Rowlinson model on G with activity λ has a unique Gibbs measure for $\lambda \in (0, \lambda_1] \cup [\lambda_2, \lambda_3]$, and multiple Gibbs measures for $\lambda \in (\lambda_1, \lambda_2) \cup (\lambda_3, \infty)$.*

The board G constructed in [3] is a tree, but not quite a regular one; it is made by dangling seven new pendant sites from each site of $\mathbb{T}^{40}$. Readers are referred to that paper for the calculations, but the intuition is something like this.

For low λ the random coloring of G is mostly yellow, but as λ rises, either green or red tends to take over the interior vertices as in $\mathbb{T}^{40}$. Then comes the third interval, where the septuplets of leaves, wanting to use both green and red, force more yellow on the interior vertices, relieving the pressure and restoring green-red symmetry. Finally the activity becomes so large that the random coloring is willing to give up red-green variety among the septuplets in order to avoid yellow interior vertices, and symmetry-breaking appears once again.

It turns out that multiple critical points can be obtained for the hard-core model in a similar way.

8. Dismantlable graphs

In addition to the fertile and sterile graphs, a second graph dichotomy appears repeatedly in our studies: dismantlable and non-dismantlable graphs. Coincidentally, the term "dismantlable" as applied to graphs was coined by Richard Nowakowski and the second author [16] almost twenty years ago in another context entirely: a pursuit game on graphs.

Two players, a cop $\mathcal{C}$ and a robber $\mathcal{R}$, compete on a fixed, finite, undirected graph H. We will assume that H is connected and has at least one edge, although the concepts make sense even without these assumptions. The cop begins by placing herself at a node of her choice; the robber then does the same. Then the players alternate beginning with $\mathcal{C}$, each moving to an adjacent node. The cop wins if she can "capture" the robber, that is, move onto the node occupied by the robber; $\mathcal{R}$ wins by avoiding capture indefinitely. In doing so $\mathcal{R}$ is free to move (or even place himself initially) onto the same node as the cop, although that would be unwise if the node were looped since then $\mathcal{C}$ could capture him at her next move.

Evidently the robber can win on any loopless graph by placing himself at the same node as the cop and then shadowing her every move; among graphs in which every node is looped, $\mathcal{C}$ clearly wins on paths and loses on cycles of length 4 or more. (In the game as defined in [16, 18], there is in effect a loop at every node of H.)

The graph on which the game is played is said to be *cop-win* if $\mathcal{C}$ has a winning strategy, *robber-win* otherwise. The following structural characterization of cop-win graphs is proved in [16] for the all-loops case, but in fact the proof (which is not difficult, and left here as an exercise) works fine in our more general context.

Let $N(i)$ be the neighborhood of node i in H and suppose there are nodes i and j in H such that $N(i) \subseteq N(j)$. Then the map taking i to j, and every other node of H to itself, is a homomorphism from H to $H \setminus \{i\}$. We call this a *fold* of the graph H. A finite graph H is *dismantlable* if there is a sequence of folds reducing H to a graph with one node (which will necessarily be looped).

Note that dismantlable graphs are easily recognized in polynomial time. Plate 8 shows some dismantlable and non-dismantlable graphs.

The following theorem, from [5], collects a boatload of equivalent conditions.

Theorem 8.1 *The following are equivalent, for finite connected graphs H with at least one edge.*

1. *H is dismantlable.*
2. *H is cop-win.*
3. *For every finite board G, $\mathrm{Hom}(G, H)$ is connected.*
4. *For every board G, and every pair $\varphi, \psi \in \mathrm{Hom}(G, H)$ agreeing on all but finitely many sites, there is a path in $\mathrm{Hom}(G, H)$ between φ and ψ.*
5. *There is some positive integer m such that, for every board G, every pair of sets U and V in G at distance at least m, and every pair of maps $\varphi, \psi \in \mathrm{Hom}(G, H)$, there is a map $\theta \in \mathrm{Hom}(G, H)$ such that θ agrees with φ on U and with ψ on V.*
6. *For every positive integer r, and every pair of maps $\varphi, \psi \in \mathrm{Hom}(\mathbb{T}^r, H)$, there is a site u in $\mathbb{T}^r$ with $\varphi(u) \neq \psi(u)$, a patch U containing u, and a map $\theta \in \mathrm{Hom}(\mathbb{T}^r, H)$ which agrees with ψ on $\mathbb{T}^r \setminus U$ and with φ on u.*
7. *For every board G and activity vector λ, if μ is a measure on $\mathrm{Hom}(G, H)$ satisfying the one-site condition, then μ is a Gibbs measure.*
8. *For every finite board G and activity vector λ, every stationary distribution for the point process $\mathcal{P}(G, H, \lambda)$ is a Gibbs measure.*
9. *For every board G of bounded degree such that $\mathrm{Hom}(G, H)$ is non-empty, there is an activity vector λ such that there is a unique Gibbs measure on $\mathrm{Hom}(G, H)$.*

10. *For every r, there is an activity vector λ such that there is a unique Gibbs measure on $\mathrm{Hom}(\mathbb{T}^r, H)$.*

We will prove here what we think, to a graph theorist, is the most interesting of these equivalences—(i) and (iii). Recall that two maps in $\mathrm{Hom}(G, H)$ are adjacent if they differ on one site of G.

Let us first assume H is dismantlable. If it has only one node the connectivity of $\mathrm{Hom}(G, H)$ is trivial, since it has at most one element. Otherwise there are nodes $i \neq j$ in H with $N(i) \subseteq N(j)$ and we may assume by induction that $\mathrm{Hom}(G, H')$ is connected for $H' := H \setminus \{i\}$.

Define, for φ in $\mathrm{Hom}(G, H)$, the map φ' in $\mathrm{Hom}(G, H')$ (and also in $\mathrm{Hom}(G, H)$) by changing all i's to j's in the image. If α and β are two maps in $\mathrm{Hom}(G, H)$ then there are paths from α to α', α' to β' and β' to β; so $\mathrm{Hom}(G, H)$ is connected as claimed.

For the converse, let H be non-dismantlable, and suppose that nonetheless $\mathrm{Hom}(G, H)$ is connected for all finite boards G; let $q = |H|$ be minimal with respect to these properties.

If there are nodes i and j of H with $N(i) \subseteq N(j)$, then $H := H \setminus \{i\}$ is also non-dismantlable. In this case, we claim that the connectivity of $\mathrm{Hom}(G, H)$ implies connectivity of $\mathrm{Hom}(G, H')$. To see this, define, for $\varphi \in \mathrm{Hom}(G, H)$, the map $\varphi' \in \mathrm{Hom}(G, H')$ by changing all i's to j's in the image as before. If α and β are two maps in $\mathrm{Hom}(G, H')$, then we may connect them by a path $\varphi_1, \ldots, \varphi_t$ in $\mathrm{Hom}(G, H)$; now we observe that the not-necessarily distinct sequence of maps $\varphi'_1, \ldots, \varphi'_t$ connects α and β in $\mathrm{Hom}(G, H')$. This contradicts the minimality of H, so we may assume from now on that there is no pair of nodes $i \neq j$ in H with $N(i) \subseteq N(j)$.

Now let G be the 'weak' square of H, that is, the graph whose nodes are ordered pairs (i_1, i_2) of nodes of H with $(i_1, i_2) \sim (j_1, j_2)$ just when $i_1 \sim j_1$ and $i_2 \sim j_2$. There are two natural homomorphisms from G to H, the projections π_1 and π_2, where $\pi_1(i_1, i_2) = i_1$ and $\pi_2(i_1, i_2) = i_2$; we claim that π_1 is an isolated point of the graph $\mathrm{Hom}(G, H)$, which certainly implies that $\mathrm{Hom}(G, H)$ is disconnected.

If not, there is a map π' taking (say) (i_1, i_2) to $k \neq i_1$ and otherwise agreeing with π_1. Let j_2 be a fixed neighbor of i_2 and j_1 any neighbor of i_1. Then $(i_1, i_2) \sim (j_1, j_2)$, and hence $k \sim j_1$ in H. We have shown that every neighbor of i_1 is also a neighbor of k, contradicting the assumption that no such pair of nodes exists in H. This completes the proof.

For the last part of the proof, there is also a simpler (and smaller) construction that works provided H has at least one loop: see [5] or Cooper, Dyer and Frieze [8].

We have seen that, for a dismantlable constraint graph H, and any board G of bounded degree, there is some λ (which can be taken to depend only on H and the maximum degree of G) such that there is a unique Gibbs measure on $\mathrm{Hom}(G, H)$. Dyer, Jerrum and Vigoda [10] have proved a "rapid mixing" counterpart to this result: given a dismantlable H, and a degree bound Δ, there is some λ such that the point process $\mathcal{P}(G, H, \lambda)$ is rapidly mixing for all finite graphs G with maximum degree at most Δ. Of course, if H is not dismantlable, then no such result can be true as $\mathrm{Hom}(G, H)$ need not be connected.

9. Random colorings of the cayley tree

We have observed that ordinary "proper" q-colorings of a graph G are maps in $\mathrm{Hom}(G, K_q)$; since K_q is sterile, there is never more than one Gibbs measure for $G = \mathbb{T}^r$. However, we see even in the case $q = 2$ that multiple Gibbs measures exist, because each of the two 2-colorings of $\mathbb{T}^r$ determines *by itself* a trivial Gibbs measure, as does any convex combination. However, only the $\frac{1}{2}$, $\frac{1}{2}$ combination is invariant under parity-changing automorphisms of $\mathbb{T}^r$.

All of the Gibbs measures in the $q = 2$ case are, however, simple and invariant under all the parity-preserving automorphisms of the board. Such Gibbs measures are neededed to realize the phase transition for the hard-core model as well, so it is not surprising that it is useful to relax our requirements slightly and to consider these *semi-invariant* simple Gibbs measures.

Fortunately we don't have to throw away all our work on invariant Gibbs measures in moving to semi-invariant ones. Given a constraint graph H on nodes $1, 2, \ldots, q$ which is connected and not bipartite, we form its bipartite "double", denoted $2H$, as follows: the nodes of $2H$ are $\{1, 2 \ldots, q\} \cup \{-1, -2, \ldots, -q\}$ with an edge between i and j just when $i \sim -j$ or $-i \sim j$ in H. Note that $2H$ is loopless; a loop at node i in H becomes the edge $\{-i, i\}$ in $2H$.

A homomorphism ψ from $\mathbb{T}^r$ to $2H$ induces a homomorphism $|\psi|$ to H via $|\psi|(v) = |\psi(v)|$. In the reverse direction, a map φ in $\mathrm{Hom}(\mathbb{T}^r, H)$ may be transformed to a map $\bar{\varphi}$ in $\mathrm{Hom}(\mathbb{T}^r, 2H)$, by putting $\bar{\varphi}(v) = \varphi(v)$ for even sites $v \in \mathbb{T}^r$ and $\bar{\varphi}(v) = -\varphi(v)$ for odd v.

Let $\lambda = (\lambda_1, \ldots, \lambda_n)$ be an activity vector for H and suppose that μ is a simple invariant Gibbs measure on $\mathrm{Hom}(\mathbb{T}^r, H)$ corresponding to λ. From μ we can obtain a simple invariant Gibbs measure $\bar{\mu}$ on $\mathrm{Hom}(\mathbb{T}^r, 2H)$ by selecting φ from μ, and flipping a fair coin to decide between $\bar{\varphi}$ (as defined above) and $-\bar{\varphi}$. Obviously $\bar{\mu}$ yields the activity vector $\bar{\lambda}$ on $2H$ given by $\bar{\lambda}_i = \lambda_{|i|}$. Furthermore, the weights on H which produce μ extend to $2H$ by $w_{-i} = w_i$.

Conversely, suppose ν is a simple invariant Gibbs measure on $\mathrm{Hom}(\mathbb{T}^r, 2H)$ whose activity vector satisfies $\lambda_{-i} = \lambda_i$ for each i. Then the measure $|\nu|$, obtained by choosing ψ from ν and taking its absolute value, is certainly an invariant Gibbs measure on $\mathrm{Hom}(\mathbb{T}^r, H)$ for $\lambda \restriction \{1, \ldots, q\}$, but is it simple?

In fact, if the weights on $2H$ which produce ν do not satisfy $w_{-i} = cw_i$, then $|\nu|$ will fail to be simple. To see this, observe that if the weights are not proportional then there are nodes $i \sim j$ of H such that $p_{-i,-j} \neq p_{i,j}$ in the random walk on $2H$. Suppose that $|\psi|$ is conditioned on the color of the root w of $\mathbb{T}^r$ being fixed at i, and let x and y be distinct neighbors of w. Set $\alpha = \mathrm{Pr}(\psi(w) = i \mid |\psi(w)| = i)$. Then

$$\mathrm{Pr}(|\psi|(x) = j) = (1 - \alpha)p_{-i,-j} + \alpha p_{i,j}$$

but

$$\mathrm{Pr}(|\psi|(x) = j \wedge |\psi|(y) = j) = (1 - \alpha)p_{-i,-j}^2 + \alpha p_{i,j}^2 > \mathrm{Pr}(|\psi|(x) = j)^2$$

so the colors of x and y are not independent given $|\psi|(w)$.

However, we can recover simplicity at the expense of one bit worth of symmetry. Let ν^+ be ν conditioned on $\psi(u) > 0$, and define ν^- similarly. Then $|\nu^+|$ and $|\nu^-|$ are essentially the same as ν^+ and ν^-, respectively, and all are simple; but these measures are only semi-invariant.

On the other hand, suppose μ is a simple semi-invariant Gibbs measure on $\mathrm{Hom}(\mathbb{T}^r, H)$. Let θ be a parity-reversing automorphism of $\mathbb{T}^r$ and define $\mu' := \mu \circ \theta$, so that $\frac{1}{2}\mu + \frac{1}{2}\mu'$ is fully invariant (but generally no longer simple). However, $\nu := \frac{1}{2}\bar{\mu} + \frac{1}{2}(-\overline{\mu'})$ is a simple *and* invariant Gibbs measure on $\mathrm{Hom}(\mathbb{T}^r, 2H)$, thus given by a node-weighted random walk on $2H$. We can recover μ as ν_+, hence:

Theorem 9.1 *Every simple semi-invariant Gibbs measure on* $\mathrm{Hom}(\mathbb{T}^r, H)$ *is obtainable from a node-weighted branching random walk on* $2H$*, with its initial state drawn from the stationary distribution on positive nodes of* $2H$*.*

Suppose, instead of beginning with a measure, we start by weighting the nodes of $2H$ and creating a Gibbs measure as in Theorem 9.1. Suppose the activities of the measure are $\{\lambda_i : i = \pm 1, \ldots, \pm q\}$. By identifying color $-i$ with i for each $i > 0$, we create a measure on H-colorings, but this will not be a Gibbs measure unless it happens that $(\lambda_{-1}, \ldots, \lambda_{-q})$ is proportional to $(\lambda_1, \ldots, \lambda_q)$.

We could assure this easily enough by making the weights proportional as well, e.g. by $w_{-i} = w_i$; then the resulting measure on $\mathrm{Hom}(G, H)$ could have been obtained directly by applying these weights to H, and is thus a fully invariant simple Gibbs measure. To get new, semi-invariant Gibbs measures on $\mathrm{Hom}(G, H)$, we must somehow devise weights for $2H$ such that $w_{-i} \not\propto w_i$ yet $\lambda_{-i} \propto \lambda_i$.

Restated with slightly different notation, simple semi-invariant Gibbs measures are in 1–1 correspondence with solutions to the "fundamental equations"

$$\lambda_i = \frac{u_i}{\left(\sum_{j \sim i} v_j\right)^r} = \frac{v_i}{\left(\sum_{j \sim i} u_j\right)^r}$$

for $i = 1, \ldots, q$. Such a solution will be invariant if $u_i = v_i$ for each i.

Plate 9 illustrates a semi-invariant, but not invariant, simple Gibbs measure for uniform 3-colorings of $\mathbb{T}^3$. Approximate weights of the nodes of $2H = 2K_3$ are given along with part of a sample coloring drawn from this measure. Additional measures may be obtained by permuting the colors or by making all the weights equal (invariant case).

Results for q-colorings of $\mathbb{T}^r$, with $q > 2$ and $r > 1$, are as follows:

When $q < r+1$, all choices of activity vector including the uniform case yield multiple simple semi-invariant Gibbs measures.

When $q > r+1$, there is only one simple semi-invariant Gibbs measure for the uniform activity vector, but multiple simple semi-invariant Gibbs measures for some other choices of activity vector.

The critical case is at $q = r+1$, that is, when the number of colors is equal to the degree of the Cayley tree. Here it turns out that there are multiple simple semi-invariant Gibbs measures for all activity vectors *except* the uniform case, where there is just one.

When $q > r+1$ and the activities are equal, the unique simple semi-invariant Gibbs measure is in fact the only Gibbs measure of any kind. This was conjectured in [6] but proved only for $q > cr$, with fixed $c > 1$; Jonasson [13] has recently, and very nicely, finished the job. Jonasson's result is in a sense a special case of the conjecture that the Markov chain of q-colorings of a finite graph of maximum degree less than q, which progresses by choosing and recoloring sites randomly one at a time, mixes rapidly. So far the best result is Vigoda's [20] which proves this if the maximum degree is at most $6q/11$.

When $q \leq r+1$ there are lots of other Gibbs measures, including ones we call *frozen*. These come about because it is possible for a measure to satisfy the Gibbs condition in a trivial and somewhat unsatisfactory way. For example, suppose we are q-coloring $\mathbb{T}^r$ (with root w) for some $q \leq r+1$, and let ψ be any fixed coloring in which the children of every node exhibit all colors other than the color of the parent. Let μ be the measure which assigns probability 1 to ψ. Then for any finite patch U, which we can assume to be a subtree including the root, the colors on ∂U force the colors on the leaves of U, and we can continue inwards to show that the original coloring $\psi \restriction U$ is the only one consistent with the colors on ∂U. Thus μ satisfies the Gibbs condition trivially, and is also vacuously simple—but not invariant or semi-invariant. We call a Gibbs measure of this type "frozen". A frozen state of $\mathrm{Hom}(\mathbb{T}^2, K_3)$ is illustrated in Plate 10. For more about frozen Gibbs measures the reader is referred to [5].

In a soft constraint model such as the Potts model, frozen Gibbs measures can only occur at zero temperature. Since most of the time statistical physicists are interested only in phases which exist at some positive temperature (and have positive entropy), frozen measures are generally absent from the statistical physics literature. However, they are interesting combinatorially and motivate some definitions in the next section.

10. From statistical physics back to graph theory

We conclude these notes with a theorem and a conjecture in "pure" graph theory, stripped of probability and physics, but suggested by the many ideas which have appeared in earlier sections.

Suppose H is bipartite and we are given some $\varphi \in \mathrm{Hom}(\mathbb{T}^1, H)$, where the sites of $\mathbb{T}^1$ are labeled by the integers $\mathbb{Z}$. Then knowing $\varphi(n)$ even for a very large n tells us something about $\varphi(0)$, namely which "part" of H it is in. We call this phenomenon *long range action*, and define it on Cayley trees as follows: If there is a $\varphi \in \mathrm{Hom}(\mathbb{T}^r, H)$ and a node $i \in H$ such that for any n, no $\psi \in \mathrm{Hom}(\mathbb{T}^r, H)$ agreeing with φ on the sites at distance n from the root can have spin i at the root, we say $\mathrm{Hom}(\mathbb{T}^r, H)$ has long range action.

Theorem 10.1 *If H is k-colorable then $\mathrm{Hom}(\mathbb{T}^{k-1}, H)$ has long range action.*

For example, the coloring described at the end of the previous section, which gives rise to a frozen Gibbs measure, shows that $\mathrm{Hom}(\mathbb{T}^2, K_3)$ (more generally, $\mathrm{Hom}(\mathbb{T}^r, K_{r+1})$) has long range action. We also see from Theorem 8.1 that

$\mathrm{Hom}(\mathbb{T}^r, H)$ has long range action for *no r* if and only if H is dismantlable; of course then H has at least one loop and therefore has infinite chromatic number.

Note that the theorem connects a statement about homomorphisms *from H* to a statement about homomorphisms *to H*. However, it is difficult to see how to turn a k-coloring of H into a suitable map in $\mathrm{Hom}(\mathbb{T}^{k-1}, H)$. Suppose, for instance, that H is the 5-cycle C_5, with nodes represented by the integers modulo 5. We can get a completely frozen map in $\mathrm{Hom}(\mathbb{T}^2, C_5)$ by making sure we use both $i+1$ and $i-1$ on the children of any site of spin i. But what has this map got to do with any 3-coloring of C_5?

The proof of Theorem 10.1, found in [7], uses a vector-valued generalization of coloring to construct the required map in $\mathrm{Hom}(\mathbb{T}^{k-1}, H)$.

We now move from long range action to the familiar notion of connectivity. Theorem 8.1—in fact, the part whose proof is given above—tells us that $\mathrm{Hom}(G, H)$ is connected for any finite G just when H is dismantlable. Suppose we restrict ourselves to boards of bounded degree? If, for example, H is bipartite, $\mathrm{Hom}(K_2, H)$ is already disconnected. If $H = K_q$ then $\mathrm{Hom}(K_q, H)$ is extremely disconnected, consisting of $n!$ isolated maps. By analogy with Theorem 10.1, we should perhaps be able to prove:

Conjecture 10.2 *If H is k-colorable then* $\mathrm{Hom}(G, H)$ *is disconnected for some finite G of maximum degree less than k.*

A proof for $k = 3$ appears in [7] and Lovász [14] has shown that the conjecture holds for $k = 4$ as well. We think that a proof of Conjecture 10.2 would have to capture some basic truths about graphs and combinatorial topology, and fervently hope that some reader of these notes will take up the challenge.

References

[1] R.J. Baxter, *Exactly Solved Models in Statistical Mechanics*, Academic Press, London (1982).

[2] J. van den Berg and J.E. Steif, Percolation and the hard-core lattice gas model, *Stochastic Proc. and their Appls.* **49** (1994), 179–197.

[3] G.R Brightwell, O. Häggström and P. Winkler, Nonmonotonic behavior in hard-core and Widom–Rowlinson models, CDAM Research Report LSE-CDAM-98-13 (June 1998); shorter version to appear in *J. Stat. Physics*.

[4] G.R. Brightwell and P. Winkler, Graph homomorphisms and phase transitions, *J. Comb. Theory (Series B)* **77** (1999), 221–262.

[5] G.R. Brightwell and P. Winkler, Gibbs measures and dismantlable graphs, *J. Comb. Theory (Series B)* **78** (2000), 141–166.

[6] G.R. Brightwell and P. Winkler, Random colorings of a Cayley tree, *Contemporary Combinatorics* (B. Bollobás ed.), Bolyai Society Mathematical Studies series (2002).

[7] G.R. Brightwell and P. Winkler, Graph homomorphisms and long range action, CDAM Research Report LSE-CDAM-20010-7 (2001), London School of Eco-

nomics. To appear in *Graphs, Morphisms and Statistical Physics*, a DIMACS publication.

[8] C. Cooper, M. Dyer and A. Frieze, On Markov chains for randomly H-colouring a graph, *Journal of Algorithms* **39** (2001), 117–134.

[9] R.L. Dobrushin, The description of a random field by means of conditional probabilities and conditions of its regularity, *Thy. of Prob. and its Appls.* **13** #2 (1968), 197–224.

[10] M. Dyer, M. Jerrum and E. Vigoda, Rapidly mixing Markov chains for dismantleable constraint graphs. To appear in *Graphs, Morphisms and Statistical Physics*, a DIMACS publication.

[11] D. Galvin and J. Kahn, On phase transition in the hard-core model on $\mathbb{Z}^d$, preprint (2001).

[12] H.-O. Georgii, *Gibbs Measures and Phase Transitions*, de Gruyter, Berlin (1988).

[13] J. Jonasson, Uniqueness of uniform random colorings of regular trees, preprint (2001), to appear in *Stat. Prob. Letters*.

[14] L. Lovász (2001), private communication.

[15] O.E. Lanford and D. Ruelle, Observables at infinity and states with short range correlations in statistical mechanics, *CMP* **9** (1969), 327–338.

[16] R. Nowakowski and P. Winkler, Vertex-to-vertex pursuit in a graph, *Discrete Math.* **43** (1983), 235–239.

[17] B. Prum and J.C. Fort, *Stochastic Processes on a Lattice and Gibbs Measures*, Kluwer, Dordrecht (1991).

[18] A. Quilliot, Homomorphismes, points fixes, rétractions et jeux de pousuite dans les graphes, les ensembles ordonnés et les espaces métriques, Thése d'Etat, Université de Paris VI, Paris, France (1983).

[19] D. Ruelle, *Statistical Mechanics*, Mathematical Physics Monograph Series, W.A. Benjamin Inc., Reading MA (1969).

[20] E. Vigoda, Improved bounds for sampling colorings, *Proc. 40th Symp. on Foundations of Comp. Sci.*, I.E.E.E. Computer Society, Los Alamitos CA (1999), 51–59.

[21] J.C. Wheeler and B. Widom, Phase equilibrium and critical behavior in a two-component Bethe-lattice gas or three-component Bethe-lattice solution, *J. Chem. Phys.* **52** (1970), 5334–5343.

[22] B. Widom and J.S. Rowlinson, New model for the study of liquid-vapor phase transition, *J. Chem. Phys.* **52** (1970), 1670–1684.

[23] S. Zachary, Countable state space Markov random fields and Markov chains on trees, *Ann. Probab.* **11** (1983), 894–903.

ICM 2002 · Vol. III · 625–634

Face Numbers
of 4-Polytopes and 3-Spheres

Günter M. Ziegler*

Abstract

Steinitz (1906) gave a remarkably simple and explicit description of the set of all f-vectors $f(P) = (f_0, f_1, f_2)$ of all 3-dimensional convex polytopes. His result also identifies the simple and the simplicial 3-dimensional polytopes as the only extreme cases. Moreover, it can be extended to strongly regular CW 2-spheres (topological objects), and further to Eulerian lattices of length 4 (combinatorial objects).

The analogous problems "one dimension higher," about the f-vectors and flag-vectors of 4-dimensional convex polytopes and their generalizations, are by far not solved, yet. However, the known facts already show that the answers will be much more complicated than for Steinitz' problem. In this lecture, we will summarize the current state of knowledge. We will put forward two crucial parameters of *fatness* and *complexity*: Fatness $F(P) := \frac{f_1 + f_2 - 20}{f_0 + f_3 - 10}$ is large if there are many more edges and 2-faces than there are vertices and facets, while complexity $C(P) := \frac{f_{03} - 20}{f_0 + f_3 - 10}$ is large if every facet has many vertices, and every vertex is in many facets. Recent results suggest that these parameters might allow one to differentiate between the cones of f- or flag-vectors of

- connected Eulerian lattices of length 5 (combinatorial objects),
- strongly regular CW 3-spheres (topological objects),
- convex 4-polytopes (discrete geometric objects), and
- rational convex 4-polytopes (whose study involves arithmetic aspects).

Further progress will depend on the derivation of tighter f-vector inequalities for convex 4-polytopes. On the other hand, we will need new construction methods that produce interesting polytopes which are far from being simplicial or simple — for example, very "fat" or "complex" 4-polytopes. In this direction, I will report about constructions (from joint work with Michael Joswig, David Eppstein and Greg Kuperberg) that yield

- strongly regular CW 3-spheres of arbitrarily large fatness,
- convex 4-polytopes of fatness larger than 5.048, and
- rational convex 4-polytopes of fatness larger than $5 - \varepsilon$.

2000 Mathematics Subject Classification: 52B11, 52B10, 51M20.
Keywords and Phrases: Polytopes, Face numbers, Flag-vectors, Tilings.

*Institute of Mathematics, MA 6-2, Technical University of Berlin, D-10623 Berlin, Germany. E-mail: ziegler@math.tu-berlin.de, http://www.math.tu-berlin.de/~ziegler

1. Introduction

Our knowledge about the combinatorics and geometry of 4-dimensional convex polytopes is quite incomplete. This assessment may come as a surprise: After all,

- 3-dimensional polytopes have been objects of intensive study since antiquity,
- properties of convex polytopes are essential to the geometry of Euclidean spaces,
- the *regular* polytopes (in all dimensions) were classified by Schläfli in 1850-52 [20], exactly 150 years ago, and
- modern polytope theory has achieved truly impressive results, in particular since the publication of Grünbaum's volume [9] in 1967, thirty-five years ago.

Moreover, we have a rather satisfactory picture of 3-dimensional convex polytopes by now, where the essential combinatorial and geometric properties of 3-dimensional polytopes were isolated a long time ago. Here we mention the three results that will be most relevant to our subsequent discussion:

1. Steinitz [24] characterized the f-vectors $(f_0, f_1, f_2) \in \mathbb{Z}^3$ of the 3-dimensional convex polytopes: They are the integer points in the 2-dimensional convex polyhedral cone that is defined by the three conditions

$$f_0 - f_1 + f_2 = 2, \qquad f_2 - 4 \leq 2(f_0 - 4), \qquad f_0 - 4 \leq 2(f_2 - 4).$$

   In particular, the f-vectors of the 3-dimensional polytopes are given as *all* the integer points in a rational polyhedral cone. Furthermore, Steinitz' result includes the characterization of the polytopes with extremal f-vectors: The first inequality is tight if and only if the polytope is simplicial, while the second one is tight if and only if the polytope is simple.

2. In 1922, Steinitz [25] published a characterization of the graphs of 3-polytopes: They are all the planar, three-connected graphs on at least 4 nodes.

   In modern terms (as reviewed below) and after some additional arguments the Steinitz theorem may be phrased as follows: Every connected finite Eulerian lattice of length 4 is the face lattice of a rational convex 3-polytope.

3. The famous Koebe–Andreev–Thurston circle-packing theorem [26] implies that every combinatorial type of 3-polytope has a realization with all edges tangent to the unit sphere S^2. Furthermore, the representation is unique up to Möbius transformations; thus, in particular, symmetric graphs/lattices have symmetric realizations. (See Bobenko & Springborn [6] for a powerful treatment of this result, and for references to its involved history.)

This is the situation in dimension 3. The picture in dimension 4 is not only quite incomplete; it is also clear by now that the results for the case of 4-dimensional polytopes will be much more involved. So, it will be a substantial challenge for 2006 (the centennial of Steinitz' little $2\,^1/_2$-page paper [24]) to characterize the closures of the convex cones of f-vectors and flag-vectors for 4-polytopes. This paper sketches some obstacles on the way as well as some efforts that have been undertaken or that should and will be made towards this goal. The obstacles are closely linked to the three results listed above:

1. The geometry of the set of f-vectors of 4-polytopes is rather intricate. It does not consist of all the integer points in its convex hull, its convex hull is not closed, and the cone it spans may be not polyhedral.

2. Moreover, the hierarchy covered by the second Steinitz theorem — connected Eulerian lattices, strongly regular CW spheres, convex polytopes, rational convex polytopes — does not collapse in dimension 4: The set of combinatorial types becomes increasingly restricted in this sequence.

3. Furthermore, the non-existence of edge-tangent representations for many types of 4-polytopes is an obstruction to the "E-construction" (see Section 6) that has recently produced sequences of interesting examples.

Nevertheless, there is hope: The boundary complex of a 4-dimensional polytope is 3-dimensional — thus we are in essence concerned with problems of 3-dimensional combinatorial geometry. That is, 4-dimensional polytopes and their faces can be effectively constructed, handled, and visualized. The tools that we have available in this context include dimensional analogy, Schlegel diagrams (see [29, Lect. 5]), a connection to tilings that will be outlined in Section 7, as well as computational tools (use `Polymake` [8]).

2. f-Vectors and flag-vectors

The combinatorial type of a convex d-dimensional polytope P (d-polytope, for short) is given by its *face lattice* $L(P)$: This is a finite graded lattice of length $d+1$ which is *Eulerian*, that is, every non-trivial interval contains the same number of elements of odd and of even rank (cf. Stanley [22, 23]). Furthermore, for $d > 1$ this lattice is *connected*, that is, the bipartite graph of atoms and coatoms (elements of rank 1 vs. elements of rank d, corresponding to vertices vs. facets) is connected.

The primary numerical data of a polytope, or much more generally of a graded lattice, are the numbers $f_i = f_i(P)$ of i-dimensional faces (i-faces, resp. lattice elements of rank $d + 1$). More generally, one considers the 2^d flag numbers $f_S = f_S(P)$ (for $S \subseteq \{0, 1, \ldots, d - 1\}$) that count the chains of faces with one i-face for each $i \in S$. These are collected to yield the *f-vector* $f(P) := (f_0, f_1, \ldots, f_{d-1}) \in \mathbb{Z}^d$, and the *flag-vector* $\mathrm{flag}(P) := (f_S : S \subseteq \{0, 1, \ldots, d - 1\})$ of the polytope or lattice. In terms of flag numbers, the bipartite graph of atoms and coatoms has $f_0 + f_{d-1}$ vertices and $f_{0,d-1}$ edges.

For general polytopes, the components of the f vector satisfy only one non trivial linear equation, the Euler-Poincaré relation $f_0 - f_1 \pm \cdots + (-1)^{d-1} f_{d-1} = 1 + (-1)^{d-1}$. The flag-vector (with 2^d components, including the f-vector) is highly redundant, due to the linear "generalized Dehn-Sommerville relations" (Bayer & Billera [3]) that allow one to reduce the number of independent components to $F_d - 1$, one less than a Fibonacci number. In particular, for $d = 3$ there is no additional information in the flag-vector, by $f_{01} = f_{12} = 2f_1$, $2f_{02} = f_{012} = 4f_1$. For 4-polytopes, the full flag-vector is determined by

$$\mathrm{flag}(P) := (f_0, f_1, f_2, f_3; f_{03}).$$

(We do *not* delete one of the f_i via the Euler-Poincaré relation, in order to explicitly retain the symmetry for dual polytopes.) As an example, the flag-vector of the 4-simplex is given by $\mathrm{flag}(\Delta_4) = (5, 10, 10, 5; 20)$. The set of all f-resp. flag-vectors of 4-polytopes will be denoted f-Vectors($\mathcal{P}_4$) resp. Flag-Vectors($\mathcal{P}_4$).

The known facts about and partial description of the sets f-Vectors($\mathcal{P}_4$) and Flag-Vectors($\mathcal{P}_4$) have been reviewed in detail in Grünbaum [9, Sect. 10.4], Bayer [2], Bayer & Lee [4, Sect. 3.8], and Höppner & Ziegler [11]. Here we will be concerned only with the *linear* known conditions that are tight at flag(Δ_4), and concentrate of the case of f-vectors rather than flag-vectors.

3. Geometry/Topology/Combinatorics

It pays off to study the f- and flag-vector problems with respect to the following hierarchy of four models—a combinatorial, a topological, and two geometric ones (where the last one includes arithmetic aspects):

Eulerian lattices: Let $\mathcal{L}_4$ be the class of all connected Eulerian lattices of length 5, as defined/reviewed above. (More restrictively, one could require that all intervals of length at least 3 must be connected.)

Cellular spheres: Let $\mathcal{S}_4$ be the class strongly regular cellulations of the 3-sphere, that is, of all regular cell decompositions of S^3 for which any intersection of two cells is a face of both of them (which may be empty). These objects appear as "regular CW 3-spheres with the intersection property" as in Björner et al. [5, pp. 203, 223]; following Eppstein, Kuperberg & Ziegler [7] we call them "strongly regular spheres." The intersection property is equivalent to the fact that the face poset of the cell complex is a lattice.

Convex polytopes: $\mathcal{P}_4$ denotes the combinatorial types of convex 4-polytopes.

Rational convex polytopes: $\mathcal{P}_4^{\mathbb{Q}}$ will denote the combinatorial types of convex 4-polytopes that have a realization with rational (vertex) coordinates.

We have natural inclusions

$$\mathcal{P}_4^{\mathbb{Q}} \subset \mathcal{P}_4 \subset \mathcal{S}_4 \subset \mathcal{L}_4.$$

The first inclusion is strict due to the existence of non-rational 4-polytopes (Richter-Gebert [18]), the second one due to the known examples of non-realizable triangulated 3-spheres, the third since any strongly regular cell decomposition (e.g., a triangulation) of a compact connected 3-manifold without boundary has a connected Eulerian face lattice of length 5.

For each of the four classes we define its *cone of flag-vectors*, that is, the closure of the cone with apex flag(Δ_4) that is spanned by all vectors of the form flag(P) − flag(Δ_4): For each family of combinatorial types we denote by Flag-Vectors($\cdot$) the set of flag-vectors, and by f-Cone($\cdot$) resp. Flag-Cone($\cdot$) the corresponding closures of the cones of f-vectors resp. flag-vectors. So we get the inclusions

$$f\text{-Vectors}(\mathcal{P}_4^{\mathbb{Q}}) \subseteq f\text{-Vectors}(\mathcal{P}_4) \subseteq f\text{-Vectors}(\mathcal{S}_4) \subseteq f\text{-Vectors}(\mathcal{L}_4),$$

and

$$f\text{-Cone}(\mathcal{P}_4^{\mathbb{Q}}) \subseteq f\text{-Cone}(\mathcal{P}_4) \subseteq f\text{-Cone}(\mathcal{S}_4) \subseteq f\text{-Cone}(\mathcal{L}_4),$$

and similarly for flag-vectors — *but can we separate them? Is any of these inclusions strict?* At the moment, that does not appear to be clear, not even if we consider the sets of f- or flag-vectors themselves rather than just the closures of the cones they span!

4. Fatness and complexity

Instead of linear combinations of face numbers or flag numbers (such as the toric h-vector, the cd-index etc. [4]), in the following we will rely on quotients of such. Thus we obtain homogeneous "density" parameters that characterize extremal polytopes. Such a quotient is the *average vertex degree* $\frac{f_{01}}{f_0} = 2\frac{f_1}{f_0}$. However, we will in addition normalize the quotients such that numerator and denominator vanish for the simplex; then every inequality of the form "our parameter $\geq$ constant" translates into a linear inequality that holds with equality at the simplex. Thus instead of the average vertex degree we would use densities like $\delta_0 := \frac{f_1 - 10}{f_0 - 5}$ or $\delta_0 := \frac{f_1 + 2f_3 - 20}{f_0 + f_3 - 10}$. (They are not defined for the simplex.) For both of these densities equality in the valid inequality $\delta_0 \geq 2$ characterizes simple 4-polytopes.

We prefer such density parameters since they provide measures of complexity that are independent of the (combinatorial) "size" of the polytope. For example, they are essentially stable under various operations of "glueing" polytopes or "connected sums" of polytopes (as in [29, p. 274]). In terms of the closed cones of f- and flag-vectors, the density parameters measure "how close we are to the boundary" in terms of a "slope." The following two parameters we call *fatness* and *complexity*:

$$\mathrm{F}(P) := \frac{f_1 + f_2 - 20}{f_0 + f_3 - 10}, \qquad \mathrm{C}(P) := \frac{f_{03} - 20}{f_0 + f_3 - 10}.$$

Both parameters are self-dual: Any polytope and its polar dual have the same fatness and complexity. The definition of fatness given here differs from that in [7] by the additional normalization, which makes the inequalities below come out simpler. Using the generalized Dehn-Sommerville equations, one derives from $f_{023} \geq 3f_{03}$ resp. from $f_{02} \geq 3f_2$ and $f_{13} \geq 3f_1$ that

$$\mathrm{C}(P) \leq 2\mathrm{F}(P) - 2 \quad \text{and} \quad \mathrm{F}(P) \leq 2\mathrm{C}(P) \quad 2,$$

with equality in the first case if and only if all facets of P are simple, and with equality in the second case if and only if P is 2-simple and 2-simplicial. In particular, $\mathrm{C}(P), \mathrm{F}(P) \geq 2$ holds for all polytopes, and more generally for all Eulerian lattices of length 5. Furthermore we see that the two parameters are asymptotically equivalent: Any polytope of high fatness has high complexity, and conversely.

In terms of complexity and fatness, we can e. g. rewrite the flag-vector inequality conjectured by Bayer [2, p. 145] as $\mathrm{F}(P) \geq 2\mathrm{C}(P) - 5$; counter-examples appear in Section 6. Similarly, the conjectured f-vector inequality of Bayer [2, p. 149] becomes self-dual if we add the dual inequality — then we obtain the condition $\mathrm{F}(P) \leq 5$, which again fails on examples given below.

Another important, self-dual inequality reads $f_{03} - 3(f_0 + f_3) > -10$, that is,

$$\mathrm{C}(P) \geq 3.$$

In fact, this is the 4-dimensional case of the condition "$g_2^{\mathrm{tor}}(P) \geq 0$" on the toric h-vector [21] of a polytope; it was proved by Stanley for rational polytopes only, and verified for all convex polytopes by Kalai [13] with a simpler argument based on a rigidity result of Whiteley. It is still not established for the case of strongly regular 3-spheres, or for Eulerian lattices of length 5.

5. The f-Vector cone for 4-polytopes

Theorem (cf. Bayer [2, Sect. 4], Eppstein et al. [7]).
In terms of the homogeneous coordinates $\varphi_0 := \frac{f_0-5}{f_1+f_2-20}$ *and* $\varphi_3 := \frac{f_3-5}{f_1+f_2-20}$, *for*
$(f_1 + f_2 - 20, f_0 - 5, f_3 - 0)$-*space, the five linear inequalities*

  (i) $\varphi_0 + 3\varphi_3 \leq 1$ (with equality for simplicial polytopes),
  (ii) $3\varphi_0 + \varphi_3 \leq 1$ (with equality for simple polytopes),
  (iii) $\varphi_0 \geq 0$ (close-to-equality if there are much fewer vertices than other faces),
  (iv) $\varphi_3 \geq 0$ (close-to-equality if there are much fewer facets than other faces),
  (v) $\varphi_0 + \varphi_3 \leq \frac{2}{5}$ (with equality if $\mathrm{F}(P) = \frac{1}{\varphi_0+\varphi_3} = \frac{5}{2}$, which forces $g_2^{\mathrm{tor}}(P) = 0$)

define a 3-dimensional closed polyhedral cone with apex flag$(\Delta_4) = (0,0,0)$. *It is a cone over a pentagon, which drawn to scale looks as follows:*

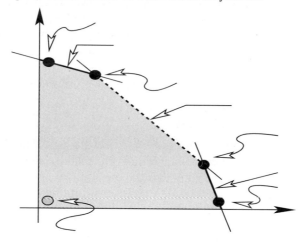

          The five linear inequalities are valid for f-Cone$(\mathcal{P}_4)$, *and they are tight and facet-defining for* f-Cone$(\mathcal{S}_4)$.
          The interesting/crucial parts of this theorem are on the one hand the existence of arbitrarily fat objects, which was established for strongly cellular spheres in [7, Sect. 4] by the M_g-construction (see Section 6), and the last inequality (v), which follows for polytopes from $\mathrm{C}(P) \geq 3$ [2], but whose validity for strongly regular spheres is an open problem.
          Thus *if* $\mathrm{F}(P) \geq \frac{5}{2}$ is valid for all strongly regular 3-spheres, then the five inequalities above give a *complete* linear description of f-Cone$(\mathcal{S}_4)$. On the other hand, *if* fatness is not bounded for (rational) convex 4-polytopes, then the above system is a *complete* description of f-Cone$(\mathcal{P}_4)$ resp. f-Cone$(\mathcal{P}_4^{\mathbb{Q}})$. But if one of the two *if*s fails, then the two cones of f-vectors differ substantially!
          One can attempt to give a similar description for the 4-dimensional cones Flag-Cone$(\mathcal{P}_4)$ and Flag-Cone$(\mathcal{S}_4)$. However, in this case the picture (compare Bayer [2, Sect. 2] and Höppner & Ziegler [11]) is much less complete, yet.
          In both the f-vector and in the flag-vector case the simple polytopes and the simplicial polytopes appear as extreme cases, and they induce facets that meet only at the apex (the simplex). The f- and flag-vectors of simple/simplicial 4-polytopes

and 3-spheres are well-known — a complete picture is given by the g-Theorem (McMullen [15]), which for 4-polytopes was first established by Barnette [1] and for 3-spheres by Walkup [27]).

6. Constructions

In order to prove completeness for linear descriptions of f- or flag-vector cones, one needs to have at one's disposal enough examples or construction techniques for extremal polytopes that go beyond the usual classes of "simple and simplicial" polytopes (neighborly, stacked, random, etc.).

Cubical polytopes. (all of whose proper faces are combinatorial cubes) form a natural class of polytopes. A very specific construction by Joswig & Ziegler [12] produced "neighborly cubical" polytopes as special projections of suitably deformed n-cubes to $\mathbb{R}^4$: These are *rational* cubical 4-polytopes C_4^n with the graph of the n-cube (for $n \geq 4$), hence with flag-vectors

$$\text{flag}(C_4^n) = \big(4, 2n, 3(n-2), n-2; 8(n-2)\big) \cdot 2^{n-2}.$$

Thus we have rational 4-polytopes of fatness $\text{F}(C_4^n)$ arbitrarily close to 5, and complexity $\text{C}(C_4^n)$ arbitrarily close to 8. (One may also show that all cubical polytopes and spheres satisfy $\text{F}(P) < 5$ and $\text{C} < 8$. Indeed, for a polytope of very high fatness and complexity, the facets need to have a very high number of vertices on average).

The E-construction. Eppstein, Kuperberg & Ziegler [7] presented and analyzed a particular 4-dimensional construction that produces interesting example polytopes: Let $P \subset \mathbb{R}^4$ be a simple 4-polytope whose ridges (2-faces) are tangent to S^3; then its polar dual P^Δ is a simplicial edge-tangent polytope. The *E-polytope* of P, obtained as $E(P) := \text{conv}(P \cup P^\Delta)$, is then 2-simple and 2-simplicial. It has fatness

$$\text{F}(E(P)) = \frac{6f_0(P) - 10}{f_0(P) + f_3(P) - 5} < 6.$$

In [7], this construction was used to produce infinite families of 2-simple 2-simplicial polytopes — apparently the first of their kind. It was also used to construct 4-polytopes of fatness larger than 5.048 — currently this is the largest value that has been observed for convex polytopes. (All simple and simplicial polytopes have fatness smaller than 3.) We note, however, that the prerequisites for the E-construction are rather hard to satisfy. Obvious examples where they can be achieved arise from regular convex polytopes. On the other hand, it turned out that, for example, P^Δ cannot be a stacked 4-polytope with more than 6 vertices! Moreover, for most examples the tangency-condition seems to force non-rational coordinates for P, and hence for P^Δ. The analysis in [7] depends on a geometric analysis that puts the Klein model of hyperbolic geometry onto the interior of the 4-ball bounded by S^3.

Thus one may ask: Does the E-construction produce non-rational polytopes? Are there possibly flag-vectors of 2-simple 2-simplicial polytopes that cannot be realized by rational polytopes? While it seems quite reasonable that the E-polytope of the regular 120-cell, with flag-vector $\text{flag}(E(P_{120})) = (720, 3600, 3600, 720; 5040)$, fatness $\mathrm{F}(E(P_{120})) > 5.02$, and 720 biyramids over pentagons as facets, could be non-rational, current investigations (Paffenholz [16]) suggest that E-polytopes are less rigid than one would think at first glance, since in some cases the tangency conditions may be relaxed or dropped.

Fat 3-spheres: The M_g-construction. Based on a covering space argument, Eppstein, Kuperberg & Ziegler [7, Sect. 4] constructed a family of strongly regular CW 3-spheres whose fatness is not bounded at all. The construction starts with a perfect cellulation of the compact connected orientable 2-manifold M_g of genus g with f-vector $(1, 2g, 1)$ and "fatness" $\frac{f_1}{f_0 + f_2} = g$. Then one shows that there is a finite covering $\widetilde{M_g}$ of M_g whose induced cell decomposition (of the same "fatness" g) is strongly regular. Finally, from the standard embedding of $\widetilde{M_g} \times I$ into S^3, where the interval I is subdivided very finely and $\widetilde{M_g} \times I$ gets the product decomposition, one obtains a cellulation of S^3 whose flag vector is dominated by the flag-vector of $\widetilde{M_g} \times I$. This yields strongly regular cell decompositions of S^3 whose f-vector is approximately proportional to $(1, 2g, 1) * (1, 1) = (1, 2g + 1, 2g + 1, 1)$. Thus the resulting spheres have fatness arbitrarily close to $2g + 1$.

Many triangulated 3-spheres. Applied to the fat 3-spheres produced by the M_g-construction, the E-construction yields 3-spheres with substantially more non-simplicial facets than their number of vertices. Thus one obtains (Pfeifle [17]) that on a large number of vertices there are far more triangulated 3-spheres than there are types of simplicial 4-polytopes, thus resolving a problem of Kalai [14].

7. Tilings

There are close connections between d-polytopes ("polyhedral tilings of S^{d-1}") and normal polyhedral tilings of $\mathbb{R}^{d-1}$. In particular, from 4-polytopes one may construct 3-dimensional tilings, for example by starting with a regular tiling of $\mathbb{R}^3$ by congruent tetrahedra and then replacing the tiles by Schlegel diagrams based on a simplex facet. (The converse direction, from tilings of $\mathbb{R}^3$ to 4-polytopes, is non-trivial: It hinges on non-trivial liftability restrictions; see Rybnikov [19].)

Normal tilings are face-to-face tilings of $\mathbb{R}^{d-1}$ by convex polytopes for which the inradii and circumradii of tiles are bounded from below resp. from above — see Grünbaum & Shephard [10, Sect. 3.2]. Of course, all components of an f-vector for tilings would be infinite, but one can try to define ratios, e.g. try to find the "average" number of vertices per tile.

The Euler formula for tilings. Thus, for $\rho > 0$, let $f_i(\rho)$ be the number of all faces of the tiling that intersect the interior of the ball $B^d(\rho)$ of radius ρ around the origin. This yields a regular decomposition of an open d-ball into convex cells; via one-point-compactification (e.g. generated by stereographic projection) by one additional vertex we obtain a regular cell-decomposition of a d-sphere; thus we obtain [28] that for all $\rho > 0$

$$f_0(\rho) - f_1(\rho) + \cdots + (-1)^d f_{d-1}(\rho) = (-1)^{d-1}.$$

In particular, this implies that if the limits $\phi_i := \lim_{\rho \to \infty} \frac{f_i(\rho)}{\sum_j f_j(\rho)}$ exist, then they satisfy $0 \le \phi_i \le \frac{1}{2}$. Furthermore, the existence of the limits ϕ_i is automatic for tilings that are invariant under a full-dimensional lattice of translations, such as the "tilings by Schlegel diagrams" suggested above. In this case the limits ϕ_i are strictly positive, and they satisfy the *Euler formula for tilings*,

$$\phi_0 - \phi_1 + \phi_2 \pm \cdots + (-1)^{d-1}\phi_{d-1} \ = \ 0.$$

For such tilings, we can also define flag-numbers $\phi_S(\rho)$. Then the limit $\rho \to \infty$ of any quotient such as $F(\rho) := \frac{f_1(\rho)+f_2(\rho)}{f_0(\rho)+f_3(\rho)}$ exists, it is positive and finite, and it coincides with

$$F(\mathcal{T}) \ := \ \frac{\phi_1 + \phi_2}{\phi_0 + \phi_3}.$$

One can construct "tilings by Schlegel diagrams" with a high (average) number of vertices per tile, or with a high number of tiles at each vertex. But are both achievable simultaneously? Equivalently, is there is a uniform upper bound on the fatness $F(\mathcal{T})$ of normal tilings of $\mathbb{R}^3$? This is not known, yet, but if fatness is bounded for normal 3-tilings, then it is bounded for 4-polytopes as well.

Fat tilings. Remarkably, there are tilings that have considerably larger fatness than the fattest polytopes we know. In particular, a modified E-construction applied to suitable Schlegel 3-diagrams [29, Lect. 5] of $C_n \times C_n$ and embedding into a cubic tiling one obtains normal lattice-transitive tilings of fatness arbitrarily close to 6.

Acknowledgements. This report has benefitted from joint work and many fruitful discussions with David Eppstein, Michael Joswig, Greg Kuperberg, Julian Pfeifle, Andreas Paffenholz, and Arnold Waßmer. (And thanks to Torsten Heldmann!)

References

[1] D. W. Barnette, *Inequalities for f-vectors of 4-polytopes*, Israel J. Math. **11** (1972), 284–291.

[2] M. M. Bayer, *The extended f-vectors of 4-polytopes*, J. Combinat. Theory, Ser. A **44** (1987), 141–151.

[3] M. M. Bayer and L. J. Billera, *Generalized Dehn-Sommerville relations for polytopes, spheres and Eulerian partially ordered sets*, Inventiones Math. **79** (1985), 143–157.

[4] M. M. Bayer and C. W. Lee, *Combinatorial aspects of convex polytopes*, in: Handbook of Convex Geometry (P. Gruber, J. Wills, eds.), North-Holland, Amsterdam 1993, 485–534.

[5] A. Björner, M. Las Vergnas, B. Sturmfels, N. White, G. M. Ziegler, *Oriented Matroids*, Encyclopedia of Math. **46**, Cambridge Univ. Press, 2nd ed., 1999.

[6] A. I. Bobenko and B. A. Springborn, *Variational principles for circle patterns, and Koebe's theorem*, Report No. 545 (Sfb 288 preprint series), TU Berlin 2002; arXiv: math.GT/0203250, 38.

[7] D. Eppstein, G. Kuperberg, and G. M. Ziegler, *Fat 4-polytopes and fatter 3-spheres*. Preprint, March 2002, 12; arXiv: math.CO/0204007; to appear in: W. Kuperberg Festschrift (A. Bezdek, ed.), Marcel-Dekker 2002.

[8] E. Gawrilow and M. Joswig, *Polymake: A software package for analyzing convex polytopes.* http://www.math.tu-berlin.de/diskregeom/polymake/

[9] B. Grünbaum, *Convex Polytopes*, Interscience, London, 1967. Revised edition (V. Kaibel, V. Klee, G. M. Ziegler, eds.), Springer-Verlag 2002, in preparation.

[10] B. Grünbaum and G. C. Shephard, *Tilings and Patterns*, Freeman, New York, 1987.

[11] A. Höppner and G. M. Ziegler, *A census of flag-vectors of 4-polytopes*, in: Polytopes — Combinatorics and Computation (G. Kalai, G. Ziegler, eds.), *DMV Seminars* **29**, Birkhäuser-Verlag, Basel 2000, 105–110.

[12] M. Joswig and G. M. Ziegler, *Neighborly cubical polytopes*, Discrete Comput. Geometry **24** (2000), 325–344.

[13] G. Kalai, *Rigidity and the lower bound theorem, I*, Inventiones Math. **88** (1987), 125–151.

[14] ——, *Many triangulated spheres*, Discrete Comput. Geometry **3** (1988), 1–14.

[15] P. McMullen, *The numbers of faces of simplicial polytopes*, Israel J. Math. **9** (1971), 559–570.

[16] A. Paffenholz, *Work in progress.* TU Berlin, 2002.

[17] J. Pfeifle, *Work in progress.* TU Berlin, 2002.

[18] J. Richter-Gebert, *Realization Spaces of Polytopes*, Lecture Notes in Mathematics **1643**, Springer-Verlag, Berlin Heidelberg, 1996.

[19] K. Rybnikov, *Stresses and liftings of cell complexes*, Discrete Comput. Geometry **21** (1999), 481–517.

[20] L. Schläfli, *Theorie der vielfachen Kontinuität*, Denkschriften der Schweizerischen naturforschenden Gesellschaft, Vol. 38, 1–237, Zürcher und Furrer, Zürich 1901 (written 1850-1852).

[21] R. P. Stanley, *Generalized h-vectors, intersection cohomology of toric varieties, and related results*, in: Commutative Algebra and Combinatorics (M. Nagata, H. Matsumura, eds.), *Advanced Studies in Pure Mathematics* **11**, Kinokuniya, Tokyo 1987, 187–213.

[22] ——, *A survey of Eulerian posets*, in: "Polytopes: Abstract, convex and computational" (Toronto 1993), Proc. NATO Advanced Study Institute (T. Bisztriczky et al., eds.), Kluwer, Dordrecht 1994, 301–333.

[23] ——, *Enumerative Combinatorics, Volume I*, Second edition, Cambridge Studies in Advanced Mathematics **49**, Cambridge University Press 1997.

[24] E. Steinitz, *Über die Eulerschen Polyederrelationen*, Archiv für Mathematik und Physik **11** (1906), 86–88.

[25] ——, *Polyeder und Raumeinteilungen*, in Encyklopädie der mathematischen Wissenschaften, Geometrie, III.1.2., Heft 9, Kapitel III A B 12, (W. F. Meyer and H. Mohrmann, eds.), B. G. Teubner, Leipzig, 1922, 1–139.

[26] W. P. Thurston, *Geometry and topology of 3-manifolds.* Lecture Notes, Princeton University, Princeton 1977–1978.

[27] D. W. Walkup, *The lower bound conjecture for 3- and 4-manifolds*, Acta Math. **125** (1970), 75–107.

[28] A. Waßmer, *Work in progress.* TU Berlin, 2002.

[29] G. M. Ziegler, *Lectures on Polytopes*, Graduate Texts in Mathematics **152**, Springer-Verlag, New York, 1995. Revised edition, 1998; "Updates, corrections, and more" at www.math.tu-berlin.de/ ziegler.

Section 15. Mathematical Aspects of Computer Science

ICM 2002 · Vol. III · 637–647

How NP Got a New Definition: A Survey of Probabilistically Checkable Proofs[*]

Sanjeev Arora[†]

Abstract

We survey a collective achievement of a group of researchers: the PCP Theorems. They give new definitions of the class NP, and imply that computing approximate solutions to many NP-hard problems is itself NP-hard. Techniques developed to prove them have had many other consequences.

2000 Mathematics Subject Classification: 68Q10, 68Q15, 68Q17, 68Q25.
Keywords and Phrases: Complexity theory, NP, Probabilistically checkable proofs, Approximation algorithms.

1. PCP theorems: an informal introduction

Suppose a mathematician circulates a proof of an important result, say Riemann Hypothesis, fitting several thousand pages. To verify it would take you and your doubting colleagues several years. Can you do it faster? Yes, according to the PCP Theorems. He can rewrite his proof so you can verify it by probabilistically selecting (i.e., using a source of random bits) a constant number of bits —as low as 3 bits—to examine in it. Furthermore, this verification has the following properties: (a) A correct proof will never fail to convince you (that is, no choice of the random bits will make you reject a correct proof) and (b) An incorrect proof will convince you with only negligible probability (2^{-100} if you examine 300 bits). In fact, a stronger assertion is true: if the Riemann hypothesis is false, then you are guaranteed to reject *any string of letters placed before you* with high probability after examining a constant number of bits. (c) This proof rewriting is completely mechanical—a computer could do it—and does not greatly increase its size. (*Caveat:* Before journal editors rush to adopt this new proof verification, we should mention that it currently requires proofs written in a formal axiomatic system —such as Zermelo Fraenkel set theory—since computers do not understand English.)

---

[*]Supported by David and Lucile Packard Fellowship, NSF Grant CCR-0098180, and an NSF ITR Grant.
[†]Department of Computer Science, Princeton University, Princeton, NJ 08544, USA. Email: arora@cs.princeton.edu, Web page: www.cs.princeton.edu/~arora/

This result has a strong ring of implausibility. A mathematical proof is invalid if it has even a single error somewhere. How can this error spread itself all over the rewritten proof, so as to be apparent after we have probabilistically examined a few bits in the proof? (Note that the simple idea of just making multiple copies of the erroneous line everywhere does not work: the unknown mathematician could hand you a proof in which this does not happen, yet that does not make the proof correct.) The methods used to achieve this level of redundancy are reminiscent of the theory of error-correcting codes, though they are novel and interesting in their own right, and their full implications are still being felt (see Section 3.).

1.1. New definition of NP

The PCP Theorems provide interesting new definitions for the complexity class NP. (Clarification: the singular form "PCP Theorem" will refer to a single result NP = PCP$(\log n, 1)$ proved in [3, 2], and the plural form "PCP Theorems" refers to a large body of results of a similar ilk, some predating the PCP Theorem.) Classically, NP is defined as the set of decision problems for which a "Yes" answer has a short certificate verifiable in polynomial time (i.e., if the instance size is n, then the certificate size and the verification time is n^c for some fixed constant c). The following are two examples:

3-SAT = satisfiable boolean formulae of the form AND of clauses of size at most 3, e.g., $(x_1 \vee \neg x_2 \vee x_3) \wedge (\neg x_1 \vee x_2 \vee x_3) \wedge (x_4)$. (The certificate for satisfiability is simply an assignment to the variables that makes the formula true.)

MATH-THEOREM$_{ZFC}$ = set of strings of the form $(T, 1^n)$ where T is a mathematical statement that is a theorem in Zermelo Fraenkel set theory that has a proof n bits long. (The "certificate" for theoremhood is just the proof.)

The famous conjecture P $\neq$ NP —now one of seven Millenium Prize problems in math [19]—says that not every NP problem is solvable in polynomial time. In other words, though the certificate is easy to check, it is not always easy to find.

The PCP Theorem gives a new definition of NP: it is the set of decision problems for which a "Yes" answer has a polynomial-size certificate which can be probabilistically checked using $O(\log n)$ random bits and by examing $O(1)$ (i.e., constant) number of bits in it.

Our earlier claim about proof verification follows from the PCP Theorem, since MATH-THEOREM$_{ZFC}$ is in NP, and hence there is a way to certify a YES answer (namely, theoremhood) that satisfies properties (a) and (b). (Property (c) follows from the constructive nature of the proof of the PCP Theorem in [3, 2].)

Motivated by the PCP Theorems, researchers have proved new analogous definitions of other complexity classes such as PSPACE [22] and PH [43].

1.2. Optimization, approximation, and PCP theorems

The P versus NP question is important because of *NP-completeness* (also, NP-*hardness*). Optimization problems in a variety of disciplines are NP-hard [30], and so if P $\neq$ NP they cannot be solved in polynomial time. The following is one such optimization problem.

MAX-3SAT: Given a 3-CNF boolean formula φ, find an assignment to the variables that maximizes the number of satisfied clauses.

Approximation algorithms represent a way to deal with NP-hardness. An algorithm *achieves an approximation ratio* α for a maximization problem if, for *every* instance, it produces a solution of value at least OPT/α, where OPT is the value of the optimal solution. (For a minimization problem, achieving a ratio α involves finding a solution of cost at most $\alpha\,OPT$.) Note that the approximation ratio is ≥ 1 by definition. For MAX-3SAT we now know a polynomial-time algorithm that achieves an approximation ratio $8/7$ [40].

Though approximation algorithms is a well-developed research area (see [38, 62]), for many problems no good approximation algorithms have been found. The PCP Theorems suggest a reason: for many NP-hard problems, including MAX-CLIQUE, CHROMATIC NUMBER, MAX-3SAT, and SET-COVER, achieving certain reasonable approximation ratios is no easier than computing optimal solutions. In other words, approximation is NP-hard. For instance, achieving a ratio $8/7 - \epsilon$ for MAX-3SAT is NP-hard [37].

Why do the PCP Theorems lead to such results? Details appear in the survey [1] (and [Feige 2002], these proceedings), but we hint at the reason using 3SAT and MAX-3SAT as examples. Cook and Levin [23, 46] showed how to reduce any NP problem to 3SAT, by constructing, for any nondeterministic machine, a 3CNF formula whose satisfying assignments represent the transcripts of accepting computations. Thus it is difficult to satisfy all clauses. Yet it is easy to find assignment satisfying $1 - o(1)$ fraction of the clauses! The reason is that a computation transcript is a very *non-robust* object: changing even a bit affects its correctness. Thus the Cook-Levin reduction does not prove the inapproximability of MAX-3SAT. By providing a more robust representation of a computation, the PCP Theorems overcome this difficulty. We note that MAX-3SAT is a central problem in the study of inapproximability: once we have proved its inapproximability, other inapproximability results easily follow (see [1]; the observation in a weaker form is originally from work on MAX-SNP [52]).

1.3. History and context

PCPs evolved from *interactive proofs*, which were invented by Goldwasser, Micali, and Rackoff [34] and Babai [5] as a probabilistic extension of NP and proved useful in cryptography and complexity theory (see Goldreich's survey [31]), including some early versions of PCPs [29]. In 1990, Lund, Fortnow, Karloff and Nisan [48] and Shamir [59] showed IP=PSPACE, thus giving a new probabilistic definition of PSPACE in terms of interactive proofs. They introduced a revolutionary algebraic way of looking at boolean formulae. In restrospect, this algebraization can also be seen as a "robust" representation of computation. The inspiration to use polynomials came from works on *program checking* [17] (see also [47, 11, 18]). Babai, Fortnow, and Lund [7] used similar methods to give a new probabilistic definition of NEXPTIME, the exponential analogue of NP. To extend this result to NP, Babai, Fortnow, Levin, and Szegedy [8] and Feige, Goldwasser, Lovász, Safra, and Szegedy [26] studied variants of what we now call probabilistically checkable proof systems (Babai et al. called their systems *holographic* proofs).

Feige et al. also proved the first inapproximability result in the PCP area: if any polynomial-time algorithm can achieve a constant approximation ratio for the MAX-CLIQUE problem, then every NP problem is solvable in $n^{O(\log \log n)}$ time. This important result drew everybody's attention to the (as yet unnamed) area of probabilistically checkable proofs. A year later, Arora and Safra [3] formalized and named the class PCP and used it to give a new probabilistic definition of NP. (Babai et al. and Feige et al.'s results were precursors of this new definition.) They also showed that approximating MAX-CLIQUE is NP-hard. Soon, Arora, Lund, Motwani, Sudan, and Szegedy [2] proved the PCP Theorem (see below) and showed that MAX-SNP-hard problems do not have a PTAS if P $\neq$ NP. Since the second paper relied heavily on the still-unpublished first paper, the the PCP theorem is jointly attributed to [3, 2]. For surveys of these developments see [6, 31, 39, 50].

2.   Definitions and results

Now we define the class PCP. We will use "language membership" and "decision problem" interchangeably. A $(r(n), q(n))$-*restricted verifier* for a language L, where r, q are integer-valued functions, is a probabilistic turing machine M that, given an input of size n, checks membership certificates for the input in the following way. The certificate is an array of bits to which the verifier has random-access (that is, it can *query* individual bits of the certificate).

- The verifier reads the input, and uses $O(r(n))$ random bits to compute a sequence of $O(q(n))$ addresses in the certificate.
- The verifier queries the bits at those addresses, and depending upon what they were, outputs "accept" or "reject".
-

$$\forall x \in L \ \exists \text{ certificate } \Pi \text{ s.t. } \Pr[M^{\Pi} \text{accepts}] = 1, \tag{2.1}$$

$$\forall x \notin L \ \forall \text{ certificate } \Pi, \ \Pr[M^{\Pi} \text{accepts}] \leq 1/2 \tag{2.2}$$

(In both cases the probability is over the choice of the verifier's random string.)

$\text{PCP}(r(n), q(n))$ is the complexity class consisting of every language with an $(r(n), q(n))$-restricted verifier. Since NP is the class of languages whose membership certificates can be checked by a deterministic polynomial-time verifier, $\text{NP} = \cup_{c \geq 0} \text{PCP}(0, n^c)$. The PCP Theorem gives an alternative definition: $\text{NP} = \text{PCP}(\log n, 1)$. Other PCP-like classes have been defined by using variants of the definition above, and shown to equal NP (when the parameters are appropriately chosen). We mention some variants and the best results known for them; these are the "PCP Theorems" alluded to earlier.

1. The probability 1 in condition (2.1) may be allowed to be $c < 1$. Such a verifier is said to have *imperfect completeness c*.
2. The probability $1/2$ in condition (2.2) may be allowed to be $s < c$. Such a verifier is said to have *soundness s*. Using standard results on random walks on expanders, it can be shown from the PCP theorem that every NP language

has verifiers with perfect completeness that use $O(k)$ query bits for soundness 2^{-k} (here $k \leq O(\log n)$).

3. The number of query bits, which was $O(q(n))$ above, may be specified more precisely together with the leading constant. The constant is important for many inapproximability results. Building upon past results on PCPs and using fourier analysis, Håstad [37] recently proved that for each $\epsilon > 0$, every NP language has a verifier with completeness $1 - \epsilon$, soundness $1/2$ and only 3 query bits. He uses this to show the inapproximability of MAX-3SAT upto a factor $8/7 - \epsilon$.

4. The *free bit* parameter may be used instead of query bits [27, 15]. This parameter is defined as follows. Suppose the query bit parameter is q. After the verifier has picked its random string, and picked a sequence of q addresses, there are 2^q possible sequences of bits that could be contained in those addresses. If the verifier accepts for only t of those sequences, then we say that the free bit parameter is $\log t$ (note that this number need not be an integer). Samorodnitsky and Trevisan show how to reduce the soundness to $2^{-k^2/4}$ using k free bits [58].

5. *Amortized free bits* may be used [15]. This parameter is $\lim_{s \to 0} f_s / \log(1/s)$, where f_s is the number of free bits needed by the verifier to make soundness $< s$. Håstad [36] shows that for each $\epsilon > 0$, every NP language has a verifier that uses $O(\log n)$ random bits and ϵ amortized free bits. He uses this to show (using a reduction from [26] and modified by [27, 15]) that MAX-CLIQUE is inapproximable upto a factor $n^{1-\delta}$.

6. The certificate may contain not bits but letters from a larger alphabet Σ. The verifier's soundness may then depend upon Σ. In a *p prover 1-round interactive proof system*, the certificate consists of p arrays of letters from Σ. The verifier is only allowed to query 1 letter from each array. Since each letter of Σ is represented by $\lceil \log |\Sigma| \rceil$ bits, the number of *bits* queried may be viewed as $p \cdot \lceil \log |\Sigma| \rceil$. Constructions of such proof systems for NP appeared in [16, 45, 28, 14, 27, 53]. Lund and Yannakakis [49] used these proof systems to prove inapproximability results for SETCOVER and many subgraph maximization problems. The best construction of such proof systems is due to Raz and Safra [54]. They show that for each $k \leq \sqrt{\log n}$, every NP language has a verifier that uses $O(\log n)$ random bits, has $\log |\Sigma| = O(k)$ and soundness 2^{-k}. The parameter p is $O(1)$.

3. Proof of the PCP theorems

A striking feature of the PCP Theorems is that each builds upon the previous ones. However, a few ideas recur. First, note that it suffices to design verifiers for 3SAT since 3SAT is NP-complete and a verifier for any other language can transform the input to a 3SAT instance as a first step. The verifier then expects a certificate for a "yes" answer to be an encoding of a satisfying assignment; we define this next.

For an alphabet Σ let Σ^m denote the set of m-letter words. The *distance* between two words $x, y \in \Sigma^m$, denoted $\delta(x, y)$, is the fraction of indices they differ on. For a set $\mathcal{C} \subseteq \Sigma^m$, let the *minimum distance* of $\mathcal{C}$, denoted min-dist$(\mathcal{C})$, refer to $\min_{x,y\in\mathcal{C};x\neq y} \{\delta(x, y)\}$ and let $\delta(x, \mathcal{C})$ stand for $\min_{y\in\mathcal{C}} \{\delta(x, y)\}$. If min-dist$(\mathcal{C}) = \gamma$, and $\delta(x, \mathcal{C}) < \gamma/2$, then triangle inequality implies there is a unique $y \in \mathcal{C}$ such that $\delta(x, y) = \delta(x, \mathcal{C})$. We will be interested in $\mathcal{C}$ such that min-dist$(\mathcal{C}) \geq 0.5$; such sets are examples of *error-correcting codes* from information theory, where $\mathcal{C}$ is thought of as a map from strings of $\log |\mathcal{C}|$ bits ("messages") to $\mathcal{C}$. When encoded this way, messages can be recovered even if transmitted over a noisy channel that corrupts up to 1/4th of the letters.

The probabilistically checkable certificate is required to contain the encoding of a satisfying assignment using some such $\mathcal{C}$. When presented with such a string, the verifier needs to check, first, that the string is *close* to some codeword, and second, that the (unique) closest codeword is the encoding of a satisfying assignment. As one would expect, the set $\mathcal{C}$ is defined using mathematically interesting objects (polynomials, monotone functions, etc.) so the final technique may be seen as a "lifting" of the satisfiability question to some mathematical domain (such as algebra). The important new angle is "local checkability," namely, the ability to verify global properties by a few random spot-checks. (See below.)

Another important technique introduced in [3] and used in all subsequent papers is *verifier composition*, which composes two verifiers to give a new verifier some of whose parameters are lower than those in either verifier. Verifier composition relies on the notion of a *probabilistically checkable split-encoding*, a notion to which Arora and Safra were led by results in [8]. (Later PCP Theorems use other probabilistically checkable encodings: *linear function codes* [2], and *long codes* [13, 36, 37].) One final but crucial ingredient in recent PCP Theorems is Raz's *parallel repetition theorem* [53].

3.1. Local tests for global properties

The key idea in the PCP Theorems is to design probabilistic local checks that verify global properties of a provided certificate. Designing such local tests involves proving a statement of the following type: if a certain object satisfies some local property "often" (say, in 90% of the local neighborhoods) then it satisfies a global property. Such statements are reminiscent of theorems in more classical areas of math, e.g., those establishing properties of solutions to PDEs, but the analogy is not exact because we only require the local property to hold in most neighborhoods, and not all.

We illustrate with some examples. (A research area called *Property Testing* [55] now consists of inventing such local tests for different properties.) There is a set $\mathcal{C} \subseteq \Sigma^m$ of interest, with min-dist$(\mathcal{C}) \geq 0.5$. Presented with $x \in \Sigma^m$, we wish to read "a few" letters in it to determine whether $\delta(x, \mathcal{C})$ is small.

1. *Linearity test.* Here $\Sigma = GF(2)$ and $m = 2^n$ for some integer n. Thus Σ^m is the set of all functions from $GF(2)^n$ to $GF(2)$. Let $\mathcal{C}_1$ be the set of words that correspond to *linear functions*, namely, the set of $f : GF(2)^n \to GF(2)$ such that $\exists a_1, \ldots, a_n \in GF(2)$ s.t. $f(z_1, z_2, \ldots, z_n) = \sum_i a_i z_i$. The test for linearity involves picking $\overline{z}, \overline{u} \in GF(2)^n$ randomly and accepting iff

$f(\overline{z}) + f(\overline{u}) = f(\overline{z}+\overline{u})$. Let γ be the probability that this test does not accept. Using elementary fourier analysis one can show $\gamma \geq \delta(f, \mathcal{C}_1)/2$ [12] (see also earlier weaker results in [18]).

2. *Low Degree Test.* Here $\Sigma = GF(p)$ for a prime p and $m = p^n$ for some n. Thus Σ^m is the set of all functions from $GF(p)^n$ to $GF(p)$. Let $\mathcal{C}_2$ be the set of words that correspond to *polynomials of total degree* d, namely, the set of $f : GF(p)^n \rightarrow GF(p)$ such that there is a n-variate polynomial g of degree d and $f(z_1, z_2, \ldots, z_n) = g(z_1, z_2, \ldots, , z_n)$. We assume $dn \ll p$ (hence degree is "low"). Testing for closeness to $\mathcal{C}_2$ involves picking random lines. A line has the parametric form $\{(a_1 + b_1t, a_2 + b_2t, \ldots, a_n + b_nt) : t \in GF(p)\}$ for some $a_1, a_2, \ldots, a_n, b_1, b_2, \ldots, b_n \in GF(p)$. (It is a 1-dimensional affine subspace, hence much smaller than $GF(p)^n$.) Note that if f is described by a degree d polynomial, then its restriction to such a line is described by a univariate degree d polynomial in the line parameter t.

- Variant 1: Pick a random line, read its first $d+1$ points to construct a degree d univariate polynomial, and check if it describes f at a randomly chosen point of the line. This test appears in [56] and is similar to another test in [26].

- Variant 2: This test uses the fact that in the PCP setting, it is reasonable to ask that the provided certificate should contain additional useful information to facilitate the test. We require, together with f, a separate table containing a degree d univariate polynomial for the line. We do the test above, except after picking the random line we read the relevant univariate polynomial from the provided table. This has the crucial benefit that we do not have to read $d+1$ separate "pieces" of information from the two tables. If γ is the probability that the test rejects, then $\gamma \geq \min\{0.1, \delta(f, \mathcal{C}_2)/2\}$ (see [2]; which uses [56, 3]).

3. *Closeness to a small set of codewords.* Above, we wanted to check that $\delta(f, \mathcal{C}) < 0.1$, in which case there is a unique word from $\mathcal{C}$ in $\mathrm{Ball}(f, 0.1)$. Proofs of recent PCP Theorems relax this and only require for some ϵ that there is a small set of words $S \subseteq \mathcal{C}$ such that each $s \in S$ lies in $\mathrm{Ball}(f, \epsilon)$. (In information theory, such an S is called a *list decoding* of f.) We mention two important such tests.

For degree d polynomials: The test in Variant 2 works with a stronger guarantee: if β is the probability that the test accepts, then there are $\mathrm{poly}(1/\epsilon)$ polynomials whose distance to f is less than $1 - \epsilon$ provided $p > \mathrm{poly}(nd/\beta\epsilon)$ (see [4], and also [54] for an alternative test).

Long Code test. Here $\Sigma = GF(2)$ and $m = 2^n$ for some integer n. Thus Σ^m is the set of all functions from $GF(2)^n$ to $GF(2)$. Let $\mathcal{C}_3$ be the set of words that correspond to *coordinate functions*, namely,

$$\{f : GF(2)^n \rightarrow GF(2) : \exists i \in \{1, 2, \ldots, n\} \text{ s.t.} f(z_1, z_2, \ldots, z_n) = z_i.\}$$

(This encodes $i \in [1, n]$, i.e., $\log n$ bits of information, using a string of length 2^n, hence the name "Long Code".) The following test works [37], though we do not elaborate on the exact statement, which is technical: Pick $\overline{z}, \overline{w} \in GF(2)^n$

and $\overline{u} \in GF(2)^n$ that is a random vector with 1's in ϵ fraction of the entries. Accept iff $f(\overline{z} + \overline{w}) = f(\overline{z}) + f(\overline{w} + \overline{u})$. (Note the similarity to the linearity test above.)

3.2.   Further applications of PCP techniques

We list some notable applications of PCP techniques. The PCP Theorem is useful in cryptography because many cryptographic primitives involve basic steps that prove Yes/No assertions that are in NP(or even P). The PCP Theorem allows this to be done in a communication-efficient manner. See [42, 51, 10] for some examples. Some stronger forms of the PCP Theorem (specifically, a version involving encoded inputs) have found uses in giving new definitions for polynomial hierarchy [43] and PSPACE [21, 22]. Finally, the properties of polynomials and polynomial-based encodings discovered for use in PCP Theorems have influenced new decoding algorithms for error-correcting codes [35], constructions of pseudorandom graphs called *extractors* [61, 57] and derandomization techniques in complexity theory (e.g. [60]).

References

[1] S. Arora and C. Lund. Hardness of approximations. In [38].

[2] S. Arora, C. Lund, R. Motwani, M. Sudan, and M. Szegedy. Proof verification and intractability of approximation problems. In *Proc. 33rd IEEE Symp. on Foundations of Computer Science*, 13–22, 1992.

[3] S. Arora and S. Safra. Probabilistic checking of proofs: a new characterization of NP. To appear *Journal of the ACM*. Preliminary version in *Proceedings of the Thirty Third Annual Symposium on the Foundations of Computer Science*, IEEE, 1992.

[4] S. Arora and M. Sudan. Improved low degree testing and its applications. *Proceedings of the Twenty Eighth Annual Symposium on the Theory of Computing*, ACM, 1997

[5] L. Babai. Trading group theory for randomness. In *Proc. 17th ACM Symp. on Theory of Computing*, 421–429, 1985.

[6] L. Babai. Transparent proofs and limits to approximations. In *Proceedings of the First European Congress of Mathematicians*. Birkhauser, 1994.

[7] L. Babai, L. Fortnow, and C. Lund. Non-deterministic exponential time has two-prover interactive protocols. *Computational Complexity*, 1:3–40, 1991.

[8] L. Babai, L. Fortnow, L. Levin, and M. Szegedy. Checking computations in polylogarithmic time. In *Proc. 23rd ACM Symp. on Theory of Computing*, 21–31, 1991.

[9] L. Babai and S. Moran. Arthur-Merlin games: a randomized proof system, and a hierarchy of complexity classes. *Journal of Computer and System Sciences*, 36:254-276, 1988.

[10] B. Barak. How to Go Beyond the Black-Box Simulation Barrier. In *Proc. 42nd IEEE FOCS*, 106–115, 2001.

[11] D. Beaver and J. Feigenbaum. Hiding instances in multioracle queries. *Proceedings of the Seventh Annual Symposium on Theoretical Aspects of Computer Science*, Lecture Notes in Computer Science Vol. 415, Springer Verlag, 1990.

[12] M. Bellare, D. Coppersmith, J. Håstad, M. Kiwi and M. Sudan. Linearity testing in characteristic two. *IEEE Transactions on Information Theory* 42(6):1781-1795, November 1996.

[13] M. Bellare, O. Goldreich, and M. Sudan. Free bits and non-approximability–towards tight results. In *Proc. 36th IEEE Symp. on Foundations of Computer Science*, 1995. Full version available from ECCC.

[14] M. Bellare, S. Goldwasser, C. Lund, and A. Russell. Efficient multi-prover interactive proofs with applications to approximation problems. In *Proc. 25th ACM Symp. on Theory of Computing*, 113–131, 1993.

[15] M. Bellare and M. Sudan. Improved non-approximability results. In *Proc. 26th ACM Symp. on Theory of Computing*, 184–193, 1994.

[16] M. Ben-or, S. Goldwasser, J. Kilian, and A. Wigderson. Multi prover interactive proofs: How to remove intractability assumptions. In *Proc. 20th ACM Symp. on Theory of Computing*, 113–121, 1988.

[17] M. Blum and S. Kannan. Designing programs that check their work. *JACM* **42**(1):269-291, 1995.

[18] M. Blum, M. Luby, and R. Rubinfeld. Self-Testing/Correcting with Applications to Numerical Problems. *JCSS* 47(3):549–595, 1993.

[19] Clay Institute. Millenium Prize Problems.

[20] A. Condon. The complexity of the max-word problem and the power of one-way interactive proof systems. *Computational Complexity*, 3:292–305, 1993.

[21] A. Condon, J. Feigenbaum, C. Lund and P. Shor. Probabilistically Checkable Debate Systems and Approximation Algorithms for PSPACE-Hard Functions. *Proceedings of the Twenty Fifth Annual Symposium on the Theory of Computing*, ACM, 1993.

[22] A. Condon, J. Feigenbaum, C. Lund and P. Shor. Random debaters and the hardness of approximating stochastic functions. *SIAM J. Comp.*, 26(2):369-400, 1997.

[23] S. Cook. The complexity of theorem-proving procedures. In *Proc. 3rd ACM Symp. on Theory of Computing*, 151–158, 1971.

[24] P. Crescenzi and V. Kann. A compendium of NP optimization problems. Available from ftp://www.nada.kth.se/Theory/Viggo-Kann/compendium.ps.Z

[25] U. Feige. A threshold of ln n for approximating set cover. *Proceedings of the Twenty Eighth Annual Symposium on the Theory of Computing*, ACM, 1996, 314–318.

[26] U. Feige, S. Goldwasser, L. Lovász, S. Safra, and M. Szegedy. Interactive proofs and the hardness of approximating cliques *Journal of the ACM*, 43(2):268–292, 1996.

[27] U. Feige and J. Kilian. Two prover protocols–low error at affordable rates. In *Proc. 26th ACM Symp. on Theory of Computing*, 172–183, 1994.

[28] U. Feige and L. Lovász. Two-prover one-round proof systems: Their power and their problems. In *Proc. 24th ACM Symp. on Theory of Computing*, 733–741, 1992.

[29] L. Fortnow, J. Rompel, and M. Sipser. On the power of multi-prover interactive protocols. In *Proceedings of the 3rd Conference on Structure in Complexity Theory*, 156–161, 1988.

[30] M. R. Garey and D. S. Johnson. *Computers and Intractability: a guide to the theory of NP-completeness.* W. H. Freeman, 1979.

[31] O. Goldreich. Probabilistic proof systems. *Proceedings of the International Congress of Mathematicians*, Birkhauser Verlag 1994.

[32] O. Goldreich. A Taxonomy of Proof Systems. In *Complexity Theory Retrospective II*, L.A. Hemaspaandra and A. Selman (eds.), Springer-Verlag, New York, 1997.

[33] O. Goldreich, S. Goldwasser and D. Ron. Property Testing and its Connection to Learning and Approximation. *Proceedings of the Thirty Seventh Annual Symposium on the Foundations of Computer Science*, IEEE, 1996, 339–348.

[34] S. Goldwasser, S. Micali, and C. Rackoff. The knowledge complexity of interactive proofs. *SIAM J. on Computing*, 18:186–208, 1989.

[35] V. Guruswami and M. Sudan. Improved decoding of Reed-Solomon and Algebraic-Geometric codes. *IEEE Trans. Inf. Thy.*, **45**:1757-1767 (1999).

[36] J. Håstad. Clique is Hard to Approximate within $n^{1-\epsilon}$. *Acta Mathematica*, **182**:105-142, 1999.

[37] J. Håstad. Some optimal inapproximability results. *Proceedings of the Twenty Eighth Annual Symposium on the Theory of Computing*, ACM, 1997, 1–10.

[38] D. Hochbaum, ed. Approximation Algorithms for NP-hard problems. PWS Publishing, Boston, 1996.

[39] D. S. Johnson. The NP-completeness column: an ongoing guide. *Journal of Algorithms*, 13:502–524, 1992.

[40] H. Karloff and U. Zwick. A 7/8-Approximation Algorithm for MAX 3SAT? In *Proc. IEEE FOCS*, 1997.

[41] R. M. Karp. Reducibility among combinatorial problems. In Miller and Thatcher, editors, *Complexity of Computer Computations*, 85–103. Plenum Press, 1972.

[42] J. Kilian. Improved Efficient Arguments. *Advances in Cryptology - CRYPTO 95* LNCS 963 Coppersmith, D.(ed.), Springer, NY, 311-324, 1995.

[43] M. Kiwi, C. Lund, A. Russell, D. Spielman, and R. Sundaram Alternation in interaction. *Proceedings of the Ninth Annual Conference on Structure in Complexity Theory*, IEEE, 1994, 294–303.

[44] G. Kolata. New shortcut found for long math proofs. *New York Times*, April 7, 1992.

[45] D. Lapidot and A. Shamir. Fully parallelized multi prover protocols for NEXPTIME. In *Proc. 32nd IEEE Symp. on Foundations of Computer Science*, 13–18, 1991.

[46] L. Levin. Universal'nyĭe perebornyĭe zadachi (universal search problems : in Russian). *Problemy Peredachi Informatsii*, 9(3):265–266, 1973.

[47] R. Lipton. New directions in testing. In *Distributed Computing and Cryptography*, J. Feigenbaum and M. Merritt, eds. Dimacs Series in Discrete Mathematics and Theoretical Computer Science, 2. AMS 1991.

[48] C. Lund, L. Fortnow, H. Karloff, and N. Nisan. Algebraic methods for interactive proof systems. *JACM*, 39(4):859–868, 1992.

[49] C. Lund and M. Yannakakis. On the hardness of approximating minimization problems. *JACM*, 41(5):960–981, 1994.

[50] E. W. Mayr, H. J. Promel, and A. Steger. (eds) Lectures on proof verification and approximation algorithms. LNCS Tutorial, Springer Verlag, 1998.

[51] S. Micali. Computationally Sound Proofs. *SIAM J. Comp*, **30**(4), 1253–1298.

[52] C. Papadimitriou and M. Yannakakis. Optimization, approximation and complexity classes. *Journal of Computer and System Sciences*, 43:425–440, 1991.

[53] R. Raz. A parallel repetition theorem. *SIAM J. Comp* 27(3):763-803(1998).

[54] R. Raz and S. Safra. A sub-constant error-probability low-degree test, and a sub-constant error-probability PCP characterization of NP. *Proceedings of the Twenty Eighth Annual Symposium on the Theory of Computing*, ACM, 1997.

[55] D. Ron. Property testing (A Tutorial). In, *Handbook of Randomized Algorithms*, Pardalos, Rajasekaran, Reif, and Rolim, eds., . Kluwer Academic, 2001.

[56] R. Rubinfeld and M. Sudan. Testing polynomial functions efficiently and over rational domains. In *Proc. 3rd Annual ACM-SIAM Symp. on Discrete Algorithms*, 23–32, 1992.

[57] S. Safra, A. Ta-Shma, and D. Zuckerman. Extractors from Reed-Muller codes. In *Proc. IEEE FOCS*, 2001.

[58] A. Samorodnitsky and L. Trevisan. A PCP Characterization of NP with Optimal Amortized Query Complexity. In *Proc. 32nd ACM STOC*, 2000.

[59] A. Shamir. IP = PSPACE. *JACM*, 39(4):869–877, October 1992.

[60] M. Sudan, L. Trevisan and S. Vadhan. Pseudorandom Generators Without the XOR Lemma. *JCSS*, **62**(2):236-266, 2001.

[61] Luca Trevisan. Extractors and Pseudorandom Generators. *JACM*, **48**(4):860-879, 2001.

[62] Vijay Vazirani. Approximation Algorithms. Spring Verlag, 2001.

Approximation Thresholds for Combinatorial Optimization Problems

Uriel Feige[*]

Abstract

An NP-hard combinatorial optimization problem Π is said to have an *approximation threshold* if there is some t such that the optimal value of Π can be approximated in polynomial time within a ratio of t, and it is NP-hard to approximate it within a ratio better than t. We survey some of the known approximation threshold results, and discuss the pattern that emerges from the known results.

2000 Mathematics Subject Classification: 68Q17, 68W25.
Keywords and Phrases: Approximation algorithms, NP-hardness, Thresholds.

1. Introduction

Given an instance I of a combinatorial optimization problem Π, one wishes to find a feasible solution of optimal value to I. For example, in the maximum clique problem, the input instance is a graph $G(V, E)$, a feasible solution is a clique (a set of vertices $S \subset V$ such that for all vertices $u, v \in S$, edge $(u, v) \in E$), the value of the clique is its size, and the objective is to find a feasible solution of maximum value. By introducing an extra parameter k to a combinatorial maximization problem, one can formulate a decision problem of the form "does I have a feasible solution of value at least k?" (or "at most k", for a minimization problem). In this work we consider only combinatorial optimization problems whose decision version is NP-complete. Hence solving these problems is NP-hard, informally meaning that there is no polynomial time algorithm that is guaranteed to output the optimal solution for every input instance (unless P=NP). (For information on the theory of NP-completeness, see [8], or essentially any book on computational complexity.)

One way of efficiently coping with NP-hard combinatorial optimization problems is by using polynomial time approximation algorithms. These algorithms pro-

---

[*]Department of Computer Science and Applied Mathematics, The Weizmann Institute, Rehovot 76100, Israel. E-mail: feige@wisdom.weizmann.ac.il

duce a feasible solution that is not necessarily optimal. The quality of the approximation algorithm is measured by its approximation ratio: the worst case ratio between the value of the solution found by the algorithm and the value of the optimal solution. Numerous approximation algorithms have been designed for various combinatorial optimization problems, applying a diverse set of algorithmic techniques (such as greedy algorithms, dynamic programming, linear programming relaxations, applications of the probabilistic method, and more). See for example [14, 21, 2].

When one attempts to design an approximation algorithm for a specific problem, it is natural to ask whether there are limits to the best approximation ratio achievable. The theory of NP-completeness is useful in this context, allowing one to prove NP-hardness of approximation results. Namely, for some value of ρ, achieving an approximation ratio better than ρ is NP-hard. Of particular interest are *threshold* results. A combinatorial optimization problem Π is said to have an approximation threshold at t if there is a polynomial time algorithm that approximates the optimal value within a ratio of t, and a hardness of approximation result that shows that achieving approximation ratios better than t is NP-hard.

The notion of an approximation threshold is often not fully appreciated, so let us discuss it in more detail. First, let us make the point that thresholds of approximation (in the sense above) need not exist at all. One may well imagine that for certain problems there is a gradual change in the complexity of achieving various approximation ratios: solving the problem exactly is NP-hard, approximating it within a ratio of ρ can be done in polynomial time, and achieving approximation ratios between ρ and 1 is neither in P nor NP-hard, but rather of some intermediate complexity. The existence of an approximation threshold says that nearly all approximation ratios (except for ratios that differ from the threshold only in low order terms) are either NP-hard to achieve, or in P. This is analogous to the well known empirical observation that "most" combinatorial optimization problems that people study turn out to be either in P or NP-hard, with very few exceptions. But note that in the context of approximation ratios, the notion of "most" is well defined. The second point that we wish to make is that an NP-hardness of approximation result is really a polynomial time algorithm. This algorithm reduces instances of 3SAT (or of some other NP-complete language) to instances of Π, and an approximation within a ratio better than ρ for Π can be used in order to solve 3SAT. Hence the threshold is a meeting point between two polynomial time algorithms: the reduction and the approximation algorithm. Here is another way of looking at it. For problem π, we can say that approximation ratio ρ_1 reduces to ρ_2 if there is a polynomial time algorithm that can approximate instances of π within ratio ρ_1 by invoking a subroutine that approximates instances of π within ratio ρ_2. (Each call to the subroutine counts as one time unit.) Two approximation ratios are equivalent if they are mutually reducible to each other. The existence of a threshold of approximation for problem Π says that essentially all approximation ratios for it fall into two equivalent classes (those above the threshold and those below the threshold). A-priori, it is not clear why the number of equivalent classes should be two.

Hardness of approximation results are often (though not always) proved using

"PCP techniques". For more details on these techniques, see for example the survey of Arora [1]. We remark that in all cases the hardness results apply also to the problem of *estimating* the optimal value for the respective problem. An estimation algorithm is a polynomial time algorithm that outputs an upper bound and a lower bound on the value of the optimal solution (without necessarily producing a solution whose value falls within this range). The estimation ratio of the algorithm is the worst case ratio between the upper bound and the lower bound. An approximation algorithm is stronger than an estimation algorithm in a sense that it supports its estimate by exhibiting a feasible solution of the same value. Potentially, designing estimation algorithms is easier than designing approximation algorithms. See for example the remark in Section 2.3.

For many combinatorial optimization problems, approximation thresholds are known. In Section 2. we survey some of these problems. For each such problem, we sketch an efficient algorithm for estimating the optimal value of the solution. The estimation ratios of these algorithms (the analog of approximation ratios) meet the thresholds of approximation for the respective problems (up to low order additive terms), and hence are best possible (unless P=NP). One would expect these estimation algorithms to be the "state of the art" in algorithmic design. However, as we shall see, for every problem above there is a *core* version for which the best possible estimation algorithm is elementary: it bases its estimate only on easily computable properties of the input instance, such as the number of clauses in a formula, without searching for a solution. The core versions that we present often have the property that Hastad [12] characterizes as "non-approximable beyond the random assignment threshold". In these cases (e.g., max 3SAT), the estimates on the value of the optimal solution are derived from analysing the expected value of a random solution.

In contrast, for problems that are known to have a polynomial time approximation scheme (namely, that can be approximated within ratios arbitrarily close to 1), their approximation algorithms do perform an extensive search for a good solution, often using dynamic programming.

The empirical findings are discussed in Section 3.

2.   A survey of some threshold results

In this section we survey some of the known threshold of approximation results. For each problem we show simple lower bounds and upper bounds on the value of the optimal solution. Obtaining any better bounds is NP-hard. (There are certain exceptions to this. For set cover and domatic number the matching hardness result are under the assumption that NP does not have slightly super-polynomial time algorithms. The hardness results for clique and chromatic number assume that NP does not have randomized algorithms that run in expected polynomial time.).

Conventions. For a graph, n denotes the number of vertices, m denotes the number of edges, δ denotes the minimum degree, Δ denotes the maximum degree. For a formula, n denotes the number of variables, m denotes the number of clauses. For each problem we define its inputs, feasible solutions, value of solutions, and objective. We present the known approximation ratios and hardness results, which are given up to low order additive terms. We then present a *core* version of the problem. For the core version, we present an upper bound and a lower bound on the value of the optimal solution. Improving over these bounds (in the worst case) is NP-hard. In most cases, we give hints to the proofs of our claims. We also cite references were full proofs can be found.

2.1.   Max coverage [18, 4]

Input. A collection $S_1, \ldots, S_m$ of sets with $\bigcup_{i=1}^m S_i = \{1, \ldots, n\}$. A parameter k.
Feasible solution. A collection I of k indices.
Value. Number of covered items, namely $|\bigcup_{i \in I} S_i|$.
Objective. Maximize.
Algorithm. Greedy. Iteratively add to I the set containing the maximum number of yet uncovered items, breaking ties arbitrarily.
Approximation ratio. $1 - 1/e$.
Hardness. $1 - 1/e$.
Core. d-regular, r-uniform. Every set is of cardinality d. Every item is in r sets. $k = n/d$.
Upper bound. n.
Lower bound. $(1 - 1/e)n$. (Pick $k = n/d = m/r$ sets at random.)
Hardness of core. $(1 - 1/e + \epsilon)$ for every $\epsilon > 0$, when d, r are large enough.
Remarks. For the special case in which each item belongs to exactly 4 sets, and $k = m/2$, there always is a choice of sets covering at least $15n/16$ items. For very $\epsilon > 0$, a $(1 + \epsilon)15/16$ approximation ratio is NP-hard [16].

2.2.   Min set cover [18, 4]

Input. A collection $S_1, \ldots, S_m$ of sets with $\bigcup_{i=1}^m S_i = \{1, \ldots, n\}$.
Feasible solution. A set of indices I such that $\bigcup_{i \in I} S_i = \{1, \ldots, n\}$.
Value. Number of sets used in the cover, namely, $|I|$.
Objective. Minimize.
Algorithm. Greedy. Iteratively add to I the set containing the maximum number of yet uncovered items, breaking ties arbitrarily.
Approximation ratio. $\ln n$.
Hardness. $\ln n$.
Core. d-regular, r-uniform. Every set is of cardinality d. Every item is in r sets.
Lower bound. n/d.
Upper bound. $(n/d) \ln n$, up to low order terms. (Include every set in I independently with probability $\frac{\ln n}{r}$.)
Hardness of core. $\ln n$.
Remarks. The hardness results for set cover in [4] assume that NP does not have deterministic algorithms that run in slightly super-polynomial time (namely, time $n^{O(\log \log n)}$).

2.3. Domatic number [5]

Input. A graph.
Feasible solution. A domatic partition of the graph. That is, a partition of the vertices of the graph into disjoint sets, where each set is a dominating set. (A dominating set is a set S of vertices that is adjacent to every vertex not in S.)
Value. Number of dominating sets in the partition.
Objective. Maximize.
Algorithm. Let δ be the minimum degree in the graph. Partition the vertices into $(1-\epsilon)(\delta+1)/\ln n$ sets at random, where ϵ is arbitrarily small when $\delta/\ln n$ is large enough. Almost all these sets will be dominating. The sets that are not dominating can be unified with the first of the dominating sets to give a domatic partition. The algorithm can be derandomized to give (a somewhat unnatural) greedy algorithm.
Approximation ratio. $1/\ln n$.
Hardness. $1/\ln n$.
Core. $\delta/\ln n$ is large enough.
Upper bound. $\delta+1$.
Lower bound. $(1-\epsilon)(\delta+1)/\ln n$.
Hardness of core. $(1+\epsilon)/\ln n$, for every $\epsilon > 0$.
Remarks. The hardness results assume that NP does not have deterministic algorithms that run in time $n^{O(\log\log n)}$. The lower bound can be refined to $(1-\epsilon)(\delta+1)/\ln\Delta$, where Δ is the maximum degree in the graph. This is shown using a nonconstructive argument (based on a two phase application of the local lemma of Lovasz). It is not known how to find such a domatic partition in polynomial time.

2.4. k-center [17, 3, 15]

Input. A metric on a set of n points and a parameter $k < n$.
Feasible solution. A set S of k of the points.
Value. Distance between S and point furthest away from S.
Objective. Minimize.
Algorithm. Greedy. Starting with the empty set, iteratively add into S the point furthest away from S, resolving ties arbitrarily.
Approximation ratio. 2.
Hardness. 2.
Core. The n points are vertices of a graph with minimum degree δ, equipped with a shortest distance metric. $k \geq n/(\delta+1)$.
Lower bound. 1. (Because $k < n$.)
Upper bound. 2. (As long as there is a vertex of distance more than 2 from the current set S, every step of the greedy algorithm covers at least $\delta+1$ vertices.)
Hardness of core. 2. (Because it is NP-hard to tell if the underlying graph has a dominating set of size k.)

2.5. Min sum set cover [7]

Input. A collection $S_1,\ldots,S_m$ of sets with $\bigcup_{i=1}^{m} S_i = \{1,\ldots,n\}$.

Feasible solution. A linear ordering. That is, a one to one mapping f from the collection of sets to $\{1, \ldots, m\}$.

Value. The average time by which an item is covered. Namely, $\frac{1}{n} \sum_i \min_{\{j \mid i \in S_j\}} f(j)$.

Objective. Minimize.

Algorithm. Greedy. Iteratively pick the set containing the maximum number of yet uncovered items, breaking ties arbitrarily.

Approximation ratio. 4.

Hardness. 4.

Core. d-regular, r-uniform. Every set is of cardinality d. Every item is in r sets.

Lower bound. $n/2d = m/2r$. (Because every set covers d items.)

Upper bound. $m/(r+1)$. (By ordering the sets at random.)

Hardness of core. $2 - \epsilon$, for every $\epsilon > 0$, when r is large enough.

Remarks. Note that the core has a lower threshold of approximation than the general case.

2.6. Min bandwidth [20]

Input. A graph.

Feasible solution. A linear arrangement. That is, a one to one mapping f from the set of vertices to $\{1, \ldots, n\}$.

Value. Longest stretch of an edge. Namely, the maximum over all edges (i, j) of $|f(i) - f(j)|$.

Objective. Minimize.

Core. The input graph G is a unit length circular arc graph. Namely, the vertices represent arcs of equal length on a circle. Two vertices are connected by an edge if their respective arcs overlap. Let $\omega(G)$ denote the size of the maximum clique in G. (In circular arc graphs, a clique corresponds to a set of mutually intersecting arcs, and $\omega(G)$ can be computed easily.)

Lower bound. $\omega(G) - 1$.

Upper bound. $2\omega(G) - 2$.

Hardness of core. 2.

Remarks. No threshold of approximation is known for the problem on general graphs.

2.7. Max 3XOR [12]

Input. A logical formula with n variables and m clauses. Every clause is the *exclusive or* of three distinct literals.

Feasible solution. An assignment to the n variables.

Value. Number of clauses that are satisfied.

Objective. Maximize.

Algorithm. Gaussian elimination can be used in order to test if the formula is satisfiable. If the formula is not satisfiable, pick a random assignment to the variables.

Approximation ratio. 1/2. (In expectation a random assignments satisfies $m/2$ clauses, whereas no assignment satisfies more than m clauses.)

Hardness. 1/2.

Core. Same as above.
Upper bound. m.
Lower bound. m/2.
Hardness of core. 1/2.

2.8. Max 3AND [12, 22]

Input. A logical formula with n variables and m clauses. Every clause is the *and* of three literals.
Feasible solution. An assignment to the n variables.
Value. Number of clauses that are satisfied.
Objective. Maximize.
Algorithm. Based on semidefinite programming.
Approximation ratio. 1/2.
Hardness. 1/2.
Core. Pairwise independent version. Every variable appears at most once in each clause. For a pair of variables x_i and x_j, let $n_{ij}(0,0)$ ($n_{ij}(0,1)$, $n_{ij}(1,0)$, $n_{ij}(1,1)$, respectively) denote the number of clauses in which they both appear negated (i negated and j positive, i positive and j negated, both positive, respectively). Then for every i and j, $n_{ij}(0,0) = n_{ij}(0,1) = n_{ij}(1,0) = n_{ij}(1,1)$.
Upper bound. m/4. (Consider the pairs of literals in the satisfied clauses. There must be at least three times as many pairs in unsatisfied clauses.)
Lower bound. m/8. (The expected number of clauses satisfied by a random assignment.)
Hardness of core. 1/2.
Remarks. This somewhat complicated core comes up as an afterthought, by analysing the structure of instances that result from the reduction from max 3XOR.

2.9. Max 3SAT [12, 19]

Input. A logical formula with n variables and m clauses. Every clause is the *or* of three literals.
Feasible solution. An assignment to the n variables.
Value. Number of clauses that are satisfied.
Objective. Maximize.
Algorithm. Based on semidefinite programming.
Approximation ratio. 7/8.
Hardness. 7/8.
Core. In every clause, the three literals are distinct.
Upper bound. m.
Lower bound. 7m/8. (The expectation for a random assignment.)
Hardness of core. 7/8.
Remarks. The approximation ratio of the algorithm of [19] is determined using computer assisted analysis.

2.10.   Inapproximable problems

For some problems the best approximation ratios known are of the form $n^{1-\epsilon}$ for every $\epsilon > 0$ (where the range of the objective function is between 1 and n). This is often interpreted as saying that approximation algorithms are almost helpless with respect to these problems. Among these problems we mention finding the smallest maximal independent set [10], max clique [11], and chromatic number [6]. The hardness results in [11, 6] are proved under the assumption that NP does not have expected polynomial time randomized algorithms.

2.11.   Thresholds within multiplicative factors

For some problems, the best approximation ratio is of the form $O(n^c)$, for some $0 < c < 1$, and there is a hardness of approximation result of the form $\Omega(n^{c'})$, where c' can be chosen arbitrarily close to c. We may view these as thresholds of approximation up to a low order multiplicative factor. An interesting example of this sort is the *max disjoint paths* problem [9].

Input. A directed graph and a set S of pairs of terminals $\{(s_1, t_1), \ldots (s_k, t_k)\}$.

Feasible solution. A collection of edge disjoint paths, each connecting s_i to t_i for some i.

Value. Number of pairs of terminals from S that are connected by some path in the solution.

Objective. Maximize.

Algorithm. Greedy. Iteratively add the shortest path that connects some yet unconnected pair from S.

Approximation ratio. $m^{-1/2}$, where m is the number of arcs in the graph.

Hardness. $m^{-1/2+\epsilon}$, for every $\epsilon > 0$.

Core. There is a path from s_1 to t_1.

Upper bound. k. At most all pairs can be connected simultaneously by disjoint paths.

Lower bound. 1. There is a path from s_1 to t_1.

Hardness of core. $1/k$. Here k can be chosen as $m^{1/2-\epsilon}$ for arbitrarily small ϵ.

3.   Discussion

We summarized the pattern presented in Section 2. Given an NP-hard combinatorial optimization problem that has an approximation threshold, we identify a core version of the problem. For the core version we identify certain key parameters. Thereafter, an upper bound and a lower bound on the value of the optimal solution is expressed by a formula involving only these parameters. Even an algorithm that examines the input for polynomial time cannot output better lower bounds or upper bounds on the value of the optimal solution (in the worst case, and up to low order terms), unless P=NP.

Note that if we do not restrict what qualifies as being a key parameter, then the pattern above can be enforced on essentially any problem with an approximation threshold. We can simply take the key parameter to be the output of an approximation algorithm for the same problem. Likewise, if we do not restrict what qualifies as a core version, we can simply take the core version to be all those input instances on which a certain approximation algorithm outputs a certain value.

Hence we would like some restrictions on what may qualify as a key parameter, or as a core version. One option is to enforce a computational complexity restriction. Namely, the core should be such that deciding whether an input instance belongs to the core is computationally easy, for example, computable in logarithmic space. (Note that in this respect, the core that we defined for the max disjoint paths problem may be problematic, because testing whether there is a directed path from s_1 to t_1 is complete for nondeterministic logarithmic space.) Likewise, computing the key parameters should be easy. Another option is to enforce structural restrictions. For example, membership in the core should be invariant over renaming of variables. Ideally, the notions of "core" and "key parameter" should be defined well enough so that we should be able to say that certain classes of inputs do not qualify as being a core (e.g., because the class is not closed under certain operations), and that certain parameters are not legitimate parameters to be used by an estimation algorithm. Also, the definitions should allow in principle the possibility that certain problems have approximation thresholds without having a core.

Hand in hand with suggesting formal definitions to the notion of a core, it would be useful to collect more data. Namely, to find more approximation threshold results, and to identify plausible core versions to these problems. As the case of min bandwidth shows, one may find core versions even for problems for which an approximation threshold is not known.

In this manuscript we mainly addressed the issue of collecting data regarding approximation thresholds, and describing this data using the notion of a core. The issue of uncovering principles that explain why the patterns discussed in this manuscript emerge is left to the reader.

References

[1] S. Arora. "How NP got a new definition: a survey of Probabilistically Checkable Proofs". In *Proceedings of ICM*, 2002.

[2] G. Ausiello, P. Crescenzi, G. Gambosi, V. Kann, A. Marchetti-Spaccamela and M. Protasi. *Complexity and Approximation. Combinatorial optimization problems and their approximability properties.* Springer, 1999.

[3] M. Dyer and A. Frieze. "A simple heuristic for the p-center problem". *Oper. Res. Lett., 3, 1985, 285–288.*

[4] U. Feige. "A Threshold of $\ln n$ for Approximating Set Cover". *Journal of the ACM*, 45(4), 634–652, 1998.

[5] U. Feige, M. Halldorsson, G. Kortsarz and A. Srinivasan. "Approximating the domatic number". Submitted to *SIAM Journal on Computing*. A preliminary version appeared in *Proc. of 32nd STOC, 2000*, 134–143.

[6] U. Feige and J. Kilian. "Zero Knowledge and the Chromatic Number". *Journal of Computer and System Sciences*, 57(2), 187–199, 1998.

[7] U. Feige, L. Lovasz and P. Tetali. "Approximating Min-sum Set Cover". Manuscript, 2002.

[8] M. Garey and D. Johnson. *Computers and Intractability: A Guide to the Theory of NP-Completeness*. W. H. Freeman and Co., San Francisco, 1979.

[9] V. Guruswami, S. Khanna, R. Rajaraman, B. Shepherd, M. Yannakakis. "Near optimal hardness results and approximation algorithms for edge-disjoint paths and related problems". Proceedings of the 31st Annual ACM symposium on Theory of Computing, 1999, 19–28.

[10] M. Halldorsson. "Approximating the minimum maximal independence number". *Inform. Process. Lett. 46 (1993) 169–172.*

[11] J. Hastad. "Clique is hard to approximate within $n^{1-\epsilon}$". *Acta Mathematica*, Vol. 182, 1999, 105–142.

[12] J. Hastad. "Some optimal inapproximability results". *Journal of the ACM*, Vol. 48, 2001, 798–859.

[13] J. Hastad, S. Phillips and S. Safra. "A well characterized approximation problem". *Information Processing Letters*, 47:6, 1993, 301–305.

[14] D. Hochbaum (editor). *Approximation Algorithms for NP-hard Problems*. PWS Publishing Company, 1997.

[15] D. Hochbaum and D. Shmoys. "A best possible approximation algorithm for the k-center problem". *Math. Oper. Res., 10, 1985, 180–184.*

[16] J. Holmerin. "Vertex cover on 4-regular hypergraphs is hard to approximate within $2 - \epsilon$". In *proceedings of the 34th Annual ACM Symposium on Theory of Computing*, 2002, 544–552.

[17] W. Hsu and G. Nemhauser. "Easy and hard bottleneck location problems". *Discrete Applied Math., 1, 1979, 209–216.*

[18] D. Johnson. "Approximation algorithms for combinatorial problems". *J. Comput. System Sci. 9, 1974, 256–278.*

[19] H. Karloff and U. Zwick. "A 7/8-approximation algorithm for max 3SAT?". In *Proceedings of the 38th Annual Symposium on Foundations of Computer Science*, 1997, 406–415.

[20] W. Unger. "The complexity of approximating the bandwidth problem". In *Proceedings of the 39th Annual Symposium on Foundations of Computer Science*, 1998, 82–91.

[21] V. Vazirani. *Approximation Algorithms*. Springer-Verlag, New York, 1999.

[22] U. Zwick. "Approximation algorithms for constraint satisfaction problems involving at most three variables per constraint". *Proceedings of SODA*, 1998, 201–210.

ICM 2002 · Vol. III · 659–672

Hardness as Randomness:
A Survey of Universal Derandomization[*]

Russell Impagliazzo[†]

Abstract

We survey recent developments in the study of probabilistic complexity classes. While the evidence seems to support the conjecture that probabilism can be deterministically simulated with relatively low overhead, i.e., that $P = BPP$, it also indicates that this may be a difficult question to resolve. In fact, proving that probalistic algorithms have non-trivial deterministic simulations is basically equivalent to proving circuit lower bounds, either in the algebraic or Boolean models.

2000 Mathematics Subject Classification: 68Q15, 68Q10, 68Q17, 68W20.
Keywords and Phrases: Probabilistic algorithms, Derandomization, Complexity classes, Pseudo-randomness, Circuit complexity, Algebraic circuit complexity.

1. Introduction

The use of random choices in algorithms has been a suprisingly productive idea. Many problems that have no known efficient deterministic algorithms have fast randomized algorithms, such as primality and polynomial identity testing. But to what extent is this seeming power of randomness real? Randomization is without doubt a powerful algorithm design tool, but does it dramatically change the notion of efficient computation?

To formalize this question, consider BPP, the class of problems solvable by bounded error probabilistic polynomial time algorithms. It is possible that $P = BPP$, i.e., randomness never solves new problems. However, it is also possible that $BPP = EXP$, i.e., randomness is a nearly omnipotent algorithmic tool.

Unlike for $P vs. NP$, there is no consensus intuition concerning the status of BPP. However, recent research gives strong indications that adding randomness does not in fact change what is solvable in polynomial-time, i.e., that $P = BPP$.

---

[*]Research supported by NSF Award CCR-0098197 and USA-Israel BSF Grant 97-00188.
[†]Department of Computer Science, University of California, San Diego, La Jolla, CA 92093-0114, USA. E-mail: russell@cs.ucsd.edu

Surprisingly, the problem is strongly connected to circuit complexity, the question of how many operations are required to compute a function.

A *priori*, possibilities concerning the power of randomized algorithms include:

1. Randomization always helps for intractable problems, i.e., $EXP = BPP$.
2. The extent to which randomization helps is problem-specific. It can reduce complexity by any amount from not at all to exponentially.
3. True randomness is never needed, and random choices can always be simulated deterministically, i.e., $P = BPP$.

Either of the last two possibilities seem plausible, but most consider the first wildly implausible. However, while a strong version of the middle possibility has been ruled out, the implausible first one is still open. Recent results indicate both that the last, $P = BPP$, is both very likely to be the case and very difficult to prove.

More precisely:

1. Either no problem in E has strictly exponential circuit complexity or $P = BPP$. This seems to be strong evidence that, in fact, $P = BPP$, since otherwise circuits can always shortcut computation time for hard problems.
2. Either $BPP = EXP$, or any problem in BPP has a deterministic sub-exponential time algorithm that works on almost all instances. In other words, either randomness solves every hard problem, or it does not help exponentially, except on rare instances. This rules out strong problem-dependence, since if randomization helps exponentially for many instances of *some problem*, we can conclude that it helps exponentially for *all intractible problems*.
3. If $BPP = P$, then either the permanent problem requires super-polynomial algebraic circuits or there is a problem in $NEXP$ that has no polynomial-size Boolean circuit. That is, proving the last possibility requires one to prove a new circuit lower bound, and so is likely to be difficult.

The above are joint work with Kabanets and Wigderson, and use results from many others.

All of these results use the hardness-vs-randomness paradigm introduced by Yao [Yao82]: Use a hard computational problem to define a small set of "pseudo-random" strings, that no limited adversary can distinguish from random. Use these "pseudo-random" strings to replace the random choices in a probabilistic algorithm. The algorithm will not have enough time to distinguish the pseudo-random sequences from truly random ones, and so will behave the same as it would given random sequences.

In this paper, we give a summary of recent results relating hardness and randomness. We explain how the area drew on and contributed to coding theory, combinatorics, and structural complexity theory. We will use a very informal style. Our main objective is to give a sense of the ideas in the area, not to give precise statements of results. Due to space and time limitations, we will be omitting a vast amount of material. For a more complete survey, please see [Kab02].

2. Models of computation and complexity classes

The P vs. BPP question arises in the broader context of the robustness of models of computation. The famous Church-Turing Thesis states that the formal notion of recursive function captures the conceptual notion of computation. While this is not in itself a mathematical conjecture, it has been supported by theorems proving that various ways of formalizing "computability", e.g., Turing Machines and the lambda calculus, are in fact equivalent.

When one considers complexity as well as computability, it is natural to ask if a model also captures the notion of computation time. While it became apparant that exact computation time was model-dependent, simulations between models almost always preserved time up to a polynomial. The time-restricted Church-Turing thesis is that any two reasonable models of computation should agree on time up to polynomials; equivalently, that the class of problems decideable in polynomial time be the same for both models. For many natural models, this is indeed the case, e.g. RAM computation, one-tape Turing machines, multi-tape Turing machines, and Cobham's axioms all define the same class P of poly-time decideable problems.

Probabilistic algorithms for a long time were the main challenge to this time-restricted Church-Turing thesis. If one accepts the notion that making a fair coin flip is a legitimate, finitely realizable computation step, then our model of poly-time computation seems to change. For example, primality testing [SS79, Rab80] and polynomial identity testing [Sch80, Zip79] are now polynomial-time, whereas we do not know any deterministic polynomial-time algorithms. The P vs. BPP question seeks to formalize the question of whether this probabilistic model is actually a counter-example, or whether there is some way to simulate randomness deterministically.

As a philisophical question, the Church-Turing Thesis has some ambiguities. We can distinguish at least two variants: a conceptual thesis that the standard model captures the conceptual notion of computation and computation time, and a physical thesis that the model characterizes the capabilities of physically-implementable computation devices. In the latter interpretation, quantum physics is inherrently probabilistic, so probabilistic machines seem more realistic than deterministic ones as such a characterization.

Recently, researchers have been taking this one step further by studying models for quantum computation. Quantum computation is probably an even more serious challenge to the time-limited Church-Turing thesis than probabilistic computation. This lies beyond the scope of the current paper, except to say that we do not believe that any analagous notion of pseudo-randomness can be used to deterministically simulate quantum algorithms. Quantum computation is intrinsically probabilistic; however, much of its power seems to come from interference between various possible outcomes, which would be destroyed in such a simulation.

2.1. Complexity classes

We assume familiarity with the standard deterministic and non-deterministic computation models (see [Pap94] for background.) To clarify notation, P —

$DTIME(n^{O(1)})$ is the class of decision problems solvable in deterministic poly-nomial time, $E = DTIME(2^{O(n)})$ is the class of such problems decideable in time exponential in the input length, and $EXP = DTIME(2^{n^{O(1)}})$ is the class of prob-lems solvable in time exponential in a polynomial of the input length. NP, NE, and $NEXP$ are the analogs for non-deterministic time. If C_1 and C_2 are complexity classes, we use $Co - C_1$ to denote the class of complements to problems in C_1, and $C_1^{C_2}$ to represent the problems solvable by a machine of the same type as normally accept C_1, but which is also allowed to make oracle queries to a procedure for a fixed language in C_2. (This is not a precise definition, and to make it precise, we would usually have to refer to the definition of C_1. However, it is also usually clear from context how to do this.) The *polynomial hierarchy PH* is the union of $NP, \Sigma_2^P = NP^{NP}, \Sigma_3^P = NP^{\Sigma_2^P}, ...$

A *probabilistic algorithm* running in $t(|x|)$ time is an algorithm A that uses, in addition to its input x, a randomly chosen string $r \in \{0,1\}^{t(n)}$. Thus, $A(x)$ is a probability distribution on outputs $A(x,r)$ as we vary over all strings r. We say that A *recognizes* a language L if for every $x \in L$, $Prob[A(x,r) = 1] > 2/3$ and every $x \notin L$, $Prob[Ax,r = 1] < 1/3$, where probabilities are over the random tape r. BPP is the class of languages recognized by polynomial-time probabilistic algorithms.

The gap between probabilities for acceptance and rejection ensures that there is a statistically significant difference between accepting and rejecting distributions. Setting the gap at $1/3$ is arbitrary; it could be anything larger than inverse polyno-mial, and smaller than $1-$ an inverse exponential, without changing the class BPP. However, it does mean that there are probabilistic algorithms, perhaps even useful ones, that do not accept any language at all. Probabilistic heuristics might clearly accept on some inputs, clearly reject on others, but be undecided sometimes.

To handle this case, we can introduce a stronger notion of simulating prob-abilistic algorithms than solving problems in BPP. Let A be any probabilistic algorithm. We say that a deterministic algorithm B solves the *promise problem* for A if, $B(x) = 1$ whenever $Prob[A(x,r) = 1] > 2/3$ and $B(x) = 0$ whenever $Prob[A(x,r) = 1] < 1/3$. Note that, unlike for BPP algorithms, there may be inputs on which A is basically undecided; for these B can output either 0 or 1. We call the class of promise problems for probabilistic polynomial time machines $Promise - BPP$. Showing that $Promise - BPP \subseteq P$ is at least as strong and seems stronger than showing $BPP = P$. (See [For01, KRC00] for a discussion.)

As happens frequently in complexity, the negation of a good definition for "easy" is not a good definition for "hard". While $EXP = BPP$ is a good formal-ization of "Randomness always helps", $BPP = P$ is less convincing as a translation of "Randomness never helps"; $Promise - BPP \subseteq P$ is a much more robust state-ment along these lines.

$\#P$ is the class of counting problems for polynomail-time verifiable predicates. i.e., For each poly-time predicate $B(x,y)$ and polynomial p, the associated counting problem is: given input x, how many y with $|y| = p(|x|)$ satisfy $B(x,y) = 1$? Valiant showed that computing the permanent of a matrix is $\#P$-complete [Val79], and Toda showed that $PH \subseteq P^{\#P}$ [Toda].

A class that frequently arises in proofs is MA, which consists of languages with probabilistically verifiable proofs of membership. Formally, a language L is in MA if there is a predicate $B(x, y, r)$ in P and a polynomial p so that, if $x \in L$, $\exists y |y| = p(|x|)$ so that $Prob_{r \in_U \{0,1\}^{p(|x|)}}[B(x, y, r) = 1] > 2/3$ and if $x \notin L$, $\forall y, |y| = p(|x|)$, $Prob_{r \in_U \{0,1\}^{p(|x|)}}[B(x, y, r) = 1] < 1/3$. Although MA combines non-determinism and probabilism, there is no direct connection known between derandomizing BPP and derandomizing MA. This is because if $x \in L$, there still may be some poorly chosen witnesses y which are convincing to B about $1/2$ the time. However, derandomizing $Promise - BPP$ also derandomizes MA, because we don't need a strict guarantee.

Lemma 1. *Let $T(n)$ be a class of time-computable functions closed under composition with polynomials. If $Promise - BPP \subseteq NTIME[T(n)]$ then $MA \subseteq NTIME[T(n)]$.*

2.2. Boolean and algebraic circuits

The circuit complexity of a finite function measures the number of primitive operations needed to compute the function. Starting with the input variables, a circuit computes a set of intermediate values in some order. The next intermediate value in the sequence must be computed as a primitive operation of the inputs and previous intermediate values. One or more of the values are labelled as outputs; for one output circuits this is without loss of generality the last value to be computed. The size of a circuit is the number of values computed, and the circuit complexity of a function f, $Size(f)$, is the smallest size of a circuit computing f.

Circuit models differ in the type of inputs and the primitive operations. Boolean circuits have Boolean inputs and the Boolean functions on 1 or 2 inputs as their primitive operations. Algebraic circuits have inputs taking values from a field G and whose primitive operations are addition in G, multiplication in G, and the constants 1 and -1. Algebraic circuits can only compute polynomials. Let f_n represent the function f restricted to inputs of size n We use the notation $P/poly$ to represent the class of functions f so that the Boolean circuit complexity of f_n is bounded by a polynomial in n; we use the notation $AlgP/poly$ for the analagous class for algebraic circuits over the integers.

Circuits are *non-uniform* in that there is no a priori connection between the circuits used to compute the same function on different input sizes. Thus, it is as if a new algorithm can be chosen for each fixed input size. While circuits are often viewed as a combinatorial tool to prove lower bounds on computation time, circuit complexity is also interesting in itself, because it gives a concrete and non-asymptotic measure of computational difficulty.

3. Converting hardness to pseudorandomness

To derandomize an algorithm A, we need to, given x, estimate the fraction of strings r that cause probabilistic algorithm $A(x, r)$ to output 1. If A runs in $t(|x|)$ steps, we can construct an approximately $t(|x|)$ size circuit C which on input

r simulates $A(x, r)$. So the problem reduces to: given a size t circuit $C(r)$, estimate the fraction of inputs on which it accepts. Note that solving this circuit-estimation problem allows us to derandomize $Promise - BPP$ as well as BPP.

We could solve this by searching over all 2^t t-bit strings, but we'd like to be more efficient. Instead, we'll search over a specially chosen small *sample set* $S = \{r_1, ... r_s\}$ of such strings. The average value over $r_i \in S$ of $C(r_i)$ approximate the average over all r's for any small circuit C. This is basically the same as saying that the task of distinguishing between a random string and a member of S is so computationally difficult that it lies beyond the abilities of size t circuits. We call such a sample set *pseudo-random*. Pseudo-random sample sets are usually described as the range of a function called a *pseudo-random generator*. This made sense for the original constructions, which had cryptographic motivations, and where it was important that S could be sampled from very quickly [BM, Yao82]. However, we think the term pseudo-random generator for hardness vs. randomness is merely vestigial, and in fact has misleading connotations, so we will use the term *pseudo-random sample set*.

We want to show the existence of a function with small

Since we want distinguishing members of S to be hard for all small circuits, we need to start with a problem f of high circuit complexity, say $Size(f) \geq t^c$ for some constant $c > 0$. We assume that we have or compute the entire truth table for f.

For the direct applications, we'll obtain f as follows. Start with some function $F \in E$ defined on all input sizes, where F_η is has circuit size at least $H(\eta)$ for a super-polynomial function H. Pick η so that $H(\eta) \geq t^c$ and let $f = F_\eta$. Note that $t^{o(1)} \geq \eta > \log t$. Since $F \in E$, we can construct the truth-table for f in time exponential in η, which means polynomial time in the size of the truth-table, $n = 2^\eta$.

Other applications, in later sections, will require us to be able to use any hard function, not necessarily obtained from a fixed function in E.

We then construct from f the pseudo-random sample set $S_f \subseteq \{0, 1\}^t$. Given the truth table of f, we list the members of S_f in as small a deterministic time as possible. It will almost always be possible to do so in time polynomial in the number of such elements, so our main concern will be minimizing the size of S_f. We then need to show that no t gate circuit can distinguish between members of S_f and truly random sequences. We almost always can do so in a very strong sense: given a test T that distinguishes S_f from the uniform distribution, we can produce a size t^{c-1} size circuit using T as an oracle, C^T, computing f. If such a test were computable in size t, we could then replace the oracle with such a circuit, obtaining a circuit of size t^c computing f, a contradiction.

The simulation is: Choose η. Construct the truth table of $f = F_\eta$. Construct S_f. Run $A(x, r_i)$ for each $r_i \in S_f$. Return the majority answer. In almost all constructions, the dominating term in the simulation's time is the size of S_f. In the most efficient constructions, making the strongest hardness assumption, $H(\eta) \in 2^{\Omega(\eta)}$, [IW97, STV01] obtain constructions with $|S_f| = n^{O(1)} = t^{O(1)}$. This gives us the following theorem:

Theorem 2. *If there is an $F \in E$ with $Size(F_\eta) \in 2^{\Omega(\eta)}$ then $P = BPP$.*

[Uma02] gives an optimally efficient construction for any hardness, not just exponential hardness.

3.1.   The standard steps

The canonical outline for constructing the pseudo-random sample set was first put together in [BFNW93]; however, each of their three steps was at least implicit in earlier papers. Later constructions either improve one of the steps, combine steps, or apply the whole argument recursively. However, a conceptual break-through that changed the way researchers looked at these steps is due to [Tre01] and will be explored in more detail in the next section.

1. Extension and random-self-reduction. Construct from f a function $\hat{f}$ so that, if $\hat{f}$ has a circuit that computes its value correctly on *almost all* inputs, then f has a small circuit that is correct on *all* inputs.
   This is usually done by viewing f as a multi-linear or low-degree polynomial over some field of moderate characteristic (poly in η). Then that polynomial can be extrapolated to define it at non-Boolean inputs, giving the extension $\hat{f}$. If we have a circuit that is almost always correct, we can produce a probabilistic circuit that is always correct as follows. To evaluate $\hat{f}$ at v, pick a point w at random, and evaluate the almost always correct circuit at random points on the line $l = v + x * w$. Since any point is on exactly one line with v, these points are uniform, and chances are the circuit is correct on these points. $\hat{f}$ restricted to l can be viewed as a low-degree polynomial in the single variable x. Thus, we can interpolate this polynomial, and use its value at $x = 0$ to give us the value $\hat{f}(v)$. ([BF90] is the first paper we know with this construction.) The key parameter that influences efficiency for this stage is $\hat{\eta}$, since the size of the truth-table for $\hat{f}$ is $\hat{n} = 2^{\hat{\eta}}$. Ideally, $\hat{\eta} \in O(\eta)$, so that $\hat{n} \in n^{O(1)}$, and we can construct $\hat{f}$ in polynomial-time.

2. Hardness Amplification: From $\hat{f}$, construct a function $\overline{f}$ on inputs of size $\overline{\eta}$ so that, from a circuit that can predict $\overline{f}$ with an ϵ advantage over guessing, we can construct a circuit that computes $\hat{f}$ on almost all inputs.
   The prototypical example of a hardness amplification construction is the exclusive-or lemma [Yao82, Le1]. Here $\overline{f(y_1 \circ y_2 ... \circ y_k)} = \hat{f}(y_1) \oplus \hat{f}(y_2) ... \oplus \hat{f}(y_k)$. Efficiency for this stage is mostly minimizing $\hat{\eta}$. The $\oplus$ construction above is not particularly efficient, so much work went into more efficient amplification.

3. Finding quasi-independent sequences of inputs. Now we have a function whose outputs are almost as good as random bits at fooling a size-limited guesser. However, we need many output bits that look mutually random. In this step, a small sets of input vectors V is constructed so that for $(v_1, ... v_t) \in_U V$, guessing $\overline{f}$ on v_i is hard and in some sense independent of the guess for v_j. Then the sample set will be defined as: $S = \{(\overline{f}(v_1), ... \overline{f}(v_t)) | (v_1, ... v_t) \in V\}$ The classical construction for this step is from [NW94]. This construction starts with a *design*, a family of subsets $D_1, .. D_t \subseteq [1, .. \mu], |D_i| = \overline{\eta}$, and $|D_i \cap$

$D_j| \leqslant \Delta$ for $i \neq j$. Then for each $w \in \{0,1\}^\mu$ we construct $v_1, ...v_t$, where v_i is the bits of w in D_i, listed in order. Intuitively, each v_i is "almost independent" of the other v_j, because of the small intersections. More precisely, if a test predicts $\hat{f}(v_i)$ from the other v_j, we can restrict the parts of w outside D_i. Then each restricted v_j takes on at most 2^Δ values, but we haven't restricted v_i at all. We can construct a circuit that knows these values of $\hat{f}$ and uses them in the predictor.

The size of S_f is 2^μ, so for efficiency we wish to minimize μ. However, our new predicting circuit has size $2^\Delta poly(t)$, so we need $\Delta \in O(\log t)$. Such designs are possible if and only if $\mu \in \Omega(\overline{\eta}^2/\Delta)$. Thus, the construction will be poly-time if we can have $\overline{\eta} = O(\eta) = O(\log t)$.

4. Extractors, Graphs, and Hardness vs. Randomness

As mentioned before, [Tre01] changed our perspective on hardness vs. randomness. We mentioned earlier that it was plausible that nature had truly probabilistic events. But is it plausible that we can physically construct a perfect fair coin? Many physical sources of randomness have imperfections and correlations. From the strong versions of hardness vs. randomness constructions, we can simulate a randomized algorithm without making the assumption that perfect random bits are available. Say we are simulating a randomized algorithm using t perfect random bits. (We don't need to have a time bound for the algorithm). Let T be the set of random sequences on which the algorithm accepts.

Assume we have a physical source outputting n bits, but all we know about it is that no single output occurs more than $2^{-t^{c+1}}$ of the time, i.e., that it has min-entropy at least t^{c+1}. Treating the output of the source as a function f on $\eta = \log n$ bits, we construct the sample set S_f, and simulate the algorithm on the sample set. The min-entropy and a simple counting argument suffices to conclude that most outputs do not have small circuits relative to T. Therefore, most outputs of the source have about the right number of neighbors in T, and so our simulation works with high probability.

This connection has been amazingly fruitful, leading to better constructions of extractors as well as better hardness vs. randomness results.

This construction is also interesting from the point of view of quasi-random graphs. Think globally. Instead of looking at the sample set construction on a single function f, look at it on all possible functions.

This defines a bipartite graph, where on the right side, we have all $2^{2^\eta} = 2^n$ functions on η bits, and on the left side, we have all t bit strings; the edges are between each function f and the members of the corresponding sample set S_f. Let T be any subset of the left side. Then we know that any function f that has many more or fewer than $s|T|/2^t$ neighbors in T has small circuit complexity relative to T. In particular, there cannot be too many such functions. Contrapositively, any large set of functions must have about the right number of neighbors in T. Thus, we get a combinatorially interesting construction of an extremely homogenous bipartite graph from any hardness vs. randomness result.

4.1.   The steps revisited

Once we look at the hardness vs. randomness issue from the point of view of extracting randomness from a flawed source, we can simplify our thoughts about the various steps. Any particular bits, and even most bits, from a flawed random source might be constant, because outputs might tend to be close in Hamming distance. This problem suggests its own solution: Use an error correcting code first. Then any two outputs are far apart, so most bit positions will be random. In fact, in retrospect, what the first two steps of the standard hardness vs. randomness method are doing is error-correcting the function. We do not care very much about rate, unless the rate is not even inverse polynomial. However, we want to be able to correct even if there is only a slight correlation between the recieved coded message and the actual coded message. It is information-theoretically impossible to uniquely decode under such heavy noise, but it is sometimes possible to *list decode*, producing a small set of possible messages. At the end of the hardness amplification stage, this is in fact what we have done to the function.

However, there are some twists to standard error- correction that make the situation unique. Most interestingly, we need decoding algorithms that are super-fast, in that to compute any particular bit of the original message can be done in poly-log time, assuming random access to the bits of the coded message. This kind of *local decodability* was implicit in [AS97], and applied to hardness vs. randomness in [STV01].

In retrospect, much of the effort in hardness-vs-randomness constructions has been in making locally list-decodeable error-correcting codes in an ad hoc manner. [STV01] showed that even natural ways of encoding can be locally list-decodeable. However, there might be some value in the ad hoc approaches. For example, many of the constructions assume the input has been weakly error-corrected, and then do a further construction to increase the amount of noise tolerated. Thus, these constructions can be viewed as error-correction boosters: codes where, given a code word corrupted with noise at a rate of γ, one can recover not the original message, but a message of lower relative noise, i.e. Hamming distance δn from the original message, where $\delta < \gamma$. These might either be known or of interest to the coding community.

5.   Hardness from derandomization

Are circuit lower bounds necessary for derandomization? Some results that suggested they might not be are [IW98] and [Kab01], where average-case derandomization or derandomization vs. a deterministic adversary was possible based on a uniform or no assumption. However, intuitively, the instance could code a circuit adversary in some clever way, so worst-case derandomization based on uniform assumptions seemed difficult. Recently, we have some formal confirmation of this:

Proving worst-case derandomization results automatically prove new circuit lower bounds.

These proofs usually take the contrapositive approach. Assume that a large complexity class has small circuits. Show that randomized computation is unexpectedly powerful as a result, so that the addition of randomness to a class jumps up its power to a higher level in a time hierarchy. Then derandomization would cause the time hierarchy to collapse, contradicting known time hierarchy theorems.

An example of unexpected power of randomness when functions have small circuits is the following result from [BFNW93]:

Theorem 3. *If $EXP \subseteq P/poly$, then $EXP = MA$.*

This didn't lead directly to any hardness from derandomization, because MA is the probabilistic analog of NP, not of P. However, combining this result with Kabanet's easy witness idea ([Kab01]), [IKW01] managed to extend it to $NEXP$.

Theorem 4. *If $NEXP \subseteq P/poly$, then $NEXP = MA$.*

Since as we observed earlier, derandomizing $Promise - BPP$ collapses MA with NP, it does follow that full derandomization is not possible without proving a circuit lower bound for $NEXP$.

Corollary 5. *If $Promise - BPP \subseteq NE$, then $NEXP \nsubseteq P/poly$.*

A very recent unpublished observation of Kabanets and Impagliazzo is that the problem of, given an arithmetic circuit C on n^2 inputs, does it compute the permanent function. is in BPP. This is because one can set inputs to constants to set circuits that should compute the permanent on smaller matrices, and then use the Schwartz-Zippel test ([Sch80, Zip79]) to test that each function computes the expansion by minors of the previous one. Then assume $Perm \in AlgP/poly$. It follows that $PH \subseteq P^{Perm} \subseteq NP^{BPP}$, because one could non-deterministically guess the algebraic circuit for Perm and then verify one's guess in BPP. Thus, if $BPP = P$ (or even $BPP \subseteq NE$) and $Perm \in AlgP/poly$, then $PH \subseteq NE$. If in addition, $NE \subseteq P/poly$, we would have $Co - NEXP = NEXP = MA \subseteq PH \subseteq NE$, a contradiction to the non-deterministic time hierarchy theorems. Thus, if $BPP \subseteq NE$, either $Perm \notin AlgP/poly$ or $NE \nsubseteq P/poly$. In either case, we would obtain a new circuit lower bound.

6. Conclusions

This is an area with a lot of "good news/bad news" results. While the latest results seem pessimistic about finally resolving the P vs. BPP question, the final verdict is still out. Perhaps NE is high enough in complexity that proving a circuit lower bound there would not require a major breakthrough, only persistance. Perhaps derandomization will lead to lower bounds, not the other way around. In any case, derandomization seems to be a nexus of interesting connections between complexity and combinatorics.

References

[ACR98]   A.E. Andreev, A.E.F. Clementi, and J.D.P. Rolim. A new general derandomization method. *Journal of the Association for Computing Machinery*, 45(1):179–213, 1998. (preliminary version in ICALP'96.)

[ACR1]   A. Andreev, A. Clementi and J. Rolim, "Hitting Sets Derandomize BPP", in *XXIII International Colloquium on Algorithms, Logic and Programming (ICALP'96)*, 1996.

[ACR3]   A. Andreev, A. Clementi and J. Rolim, "A new general derandomization method", *J. ACM*, 45(1), 179–213, 1998.

[ACRT]   A. Andreev, A. Clementi, J. Rolim, and L. Trevisan, "Weak random sources, hitting sets, and BPP simulation", *38th FOCS*, 264–272, 1997.

[ALM+98]   S. Arora, C. Lund, R. Motwani, M. Sudan, and M. Szegedy. Proof verification and the hardness of approximation problems. *Journal of the Association for Computing Machinery*, 45(3):501–555, 1998. (preliminary version in FOCS'92.)

[AS97]   S. Arora and M. Sudan. Improved low-degree testing and its applications, In *Proceedings of the Twenty-Ninth Annual ACM Symposium on Theory of Computing*, 485–495, 1997.

[AS98]   S. Arora and S. Safra. Probabilistic checking of proofs: A new characterization of NP. *Journal of the Association for Computing Machinery*, 45(1):70–122, 1998. (preliminary version in FOCS'92.)

[BCW80]   M. Blum, A.K. Chandra, and M.N. Wegman. Equivalence of free Boolean graphs can be tested in polynomial time. *Information Processing Letters*, 10:80–82, 1980.

[BF90]   D. Beaver and J. Feigenbaum. Hiding instances in multioracle queries. In *Proceedings of the Seventh Annual Symposium on Theoretical Aspects of Computer Science*, volume 415 of *Lecture Notes in Computer Science*, 37–48, Berlin, 1990. Springer Verlag.

[BFL91]   L. Babai, L. Fortnow, and C. Lund. Non-deterministic exponential time has two-prover interactive protocols. *Computational Complexity*, 1:3–40, 1991.

[BFNW93]   L. Babai, L. Fortnow, N. Nisan, and A. Wigderson. BPP has subexponential time simulations unless EXPTIME has publishable proofs. *Complexity*, 3:307–318, 1993.

[BFT98]   H. Buhrman, L. Fortnow, and L. Thierauf. Nonrelativizing separations. In *Proceedings of the Thirteenth Annual IEEE Conference on Computational Complexity*, 8–12, 1998.

[BM]   M. Blum and S. Micali. "How to Generate Cryptographically Strong Sequences of Pseudo-Random Bits", *SIAM J. Comput.*, Vol. 13, 850–864, 1984.

[CDGK91]   M. Clausen, A. Dress, J. Grabmeier, and M. Karpinsky. On zero-testing and interpolation of k-sparse multivariate polynomials over finite fields. *Theoretical Computer Science*, 84(2):151–164, 1991.

[CG82]   A.L. Chistov and D.Yu. Grigoriev. Polynomial-time factoring of multivariable polynomials over a global field. *LOMI Preprints*, E-5-82, 1982.

USSR Acad. Sci., Steklov Math. Inst., Leningrad.

[CK97] Z. Chen and M. Kao. Reducing randomness via irrational numbers. In *Proceedings of the Twenty-Ninth Annual ACM Symposium on Theory of Computing*, 200–209, 1997.

[CRS95] S. Chari, P. Rohatgi, and A. Srinivasan. Randomness-optimal unique element isolation with applications to perfect matching and related problems. *SIAM Journal on Computing*, 24(5):1036–1050, 1995.

[For01] L. Fortnow. Comparing notions of full derandomization. In *Proceedings of the Sixteenth Annual IEEE Conference on Computational Complexity*, 28–34, 2001.

[GG99] J. von zur Gathen and J. Gerhard. *Modern Computer Algebra*. Cambridge University Press, New York, 1999.

[GK85] J. von zur Gathen and E. Kaltofen. Factoring multivariate polynomials over finite fields. *Mathematics of Computation*, 45:251–261, 1985.

[GKS90] D.Yu. Grigoriev, M. Karpinsky, and M.F. Singer. Fast parallel algorithms for sparse multivariate polynomial interpolation over finite fields. *SIAM Journal on Computing*, 19(6):1059–1063, 1990.

[GL] O. Goldreich and L.A. Levin. "A Hard-Core Predicate for all One-Way Functions", in *ACM Symp. on Theory of Computing*, 25–32, 1989.

[GLR+91] P. Gemmell, R. Lipton, R. Rubinfeld, M. Sudan, and A. Wigderson. Self-testing/correcting for polynomials and for approximate functions. In *Proceedings of the Twenty-Third Annual ACM Symposium on Theory of Computing*, 32–42, 1991.

[Im] R. Impagliazzo, "Hard-core Distributions for Somewhat Hard Problems", in *36th FOCS*, 538–545, 1995.

[IKW01] R. Impagliazzo, V. Kabanets, and A. Wigderson. In search of an easy witness: Exponential time vs. probabilistic polynomial time. In *Proceedings of the Sixteenth Annual IEEE Conference on Computational Complexity*, 1–11, 2001.

[ISW99] R. Impaglizzo, R. Shaltiel, and A. Wigderson, "Near-Optimal Conversion of Hardness into Pseudo-Randomness", in *40th FOCS*, 181–190, 1999.

[ISW00] R. Impaglizzo, R. Shaltiel, and A. Wigderson, "Extractors and Pseudorandom Generators with optimal seed lengths", in *32nd STOC*, 1–10, 2000.

[IW97] R. Impagliazzo and A. Wigderson. P=BPP if E requires exponential circuits: Derandomizing the XOR Lemma. In *Proceedings of the Twenty-Ninth Annual ACM Symposium on Theory of Computing*, 220–229, 1997.

[IW98] R. Impagliazzo and A. Wigderson. Randomness vs. time: Derandomization under a uniform assumption. In *Proceedings of the Thirty-Ninth Annual IEEE Symposium on Foundations of Computer Science*, 734–743, 1998.

[Kab01] V. Kabanets. Easiness assumptions and hardness tests: Trading time for zero error. *Journal of Computer and System Sciences*, 63(2):236–

252, 2001. (preliminary version in CCC'00.)

[Kab02] V. Kabanets. Derandomization: A brief overview. *Bulletin of the European Association for Theoretical Computer Science*, 76:88–103, 2002. (also available as ECCC TR02-008.)

[Kal85] E. Kaltofen. Polynomial-time reductions from multivariate to bi- and univariate integral polynomial factorization. *SIAM Journal on Computing*, 14(2):469–489, 1985. (preliminary version in STOC'82.)

[Kal92] E. Kaltofen. Polynomial factorization 1987–1991. In I. Simon, editor, *Proceedings of the First Latin American Symposium on Theoretical Informatics*, Lecture Notes in Computer Science, 294–313. Springer Verlag, 1992. (LATIN'92.)

[KL] R. M. Karp and R. J. Lipton, "Turing Machines that Take Advice", *L'Ensignment Mathematique*, 28, 191–209, 1982.

[KRC00] V. Kabanets, C. Rackoff, and S. Cook. Efficiently approximable real-valued functions. *Electronic Colloquium on Computational Complexity*, TR00-034, 2000.

[KS01] A. Klivans and D. Spielman. Randomness efficient identity testing of multivariate polynomials. In *Proceedings of the Thirty-Third Annual ACM Symposium on Theory of Computing*, 216–223, 2001.

[KT90] E. Kaltofen and B. Trager. Computing with polynomials given by black boxes for their evaluations: Greatest common divisors, factorization, separation of numerators and denominators. *Journal of Symbolic Computation*, 9(3):301–320, 1990.

[Len87] A.K. Lenstra. Factoring multivariate polynomials ove algebraic number fields. *SIAM Journal on Computing*, 16:591–598, 1987.

[Le1] L. A. Levin, "One-Way Functions and Pseudorandom Generators", *Combinatorica*, Vol. 7, No. 4, 357–363, 1987.

[LFKN92] C. Lund, L. Fortnow, H. Karloff, and N. Nisan. Algebraic methods for interactive proof systems. *Journal of the Association for Computing Machinery*, 39(4):859–868, 1992.

[Lip91] R. Lipton. New directions in testing. In J. Feigenbaum and M. Merrit, editors, *Distributed Computing and Cryptography*, 191–202. DIMACS Series in Discrete Mathematics and Theoretical Computer Science, Volume 2, AMS, 1991.

[Lov79] L. Lovasz. On determinants, matchings and random algorithms. In L. Budach, editor, *Fundamentals of Computing Theory*. Akademia-Verlag, Berlin, 1979.

[LV98] D. Lewin and S. Vadhan. Checking polynomial identities over any field: Towards a derandomization? In *Proceedings of the Thirtieth Annual ACM Symposium on Theory of Computing*, 438–447, 1998.

[MVV87] K. Mulmuley, U. Vazirani, and V. Vazirani. Matching is as easy as matrix inversion. *Combinatorica*, 7(1):105–113, 1987.

[NW94] N. Nisan and A. Wigderson. Hardness vs. randomness. *Journal of Computer and System Sciences*, 49:149–167, 1994.

[Pap94] C.H. Papadimitriou. *Computational Complexity*. Addison-Wesley,

Reading, Massachusetts, 1994.

[Rab80]   M. O. Rabin. Probabilistic Algorithm for Testing Primality. *Journal of Number Theory*, 12:128–138, 1980.

[RB91]   R.M. Roth and G.M. Benedek. Interpolation and approximation of sparse multivariate polynomials. *SIAM Journal on Computing*, 20(2):291–314, 1991.

[RR97]   A.A. Razborov and S. Rudich. Natural proofs. *Journal of Computer and System Sciences*, 55:24–35, 1997.

[Sch80]   J.T. Schwartz. Fast probabilistic algorithms for verification of polynomial identities. *Journal of the Association for Computing Machinery*, 27(4):701–717, 1980.

[Sha92]   A. Shamir. IP=PSPACE. *Journal of the Association for Computing Machinery*, 39(4):869–877, 1992.

[SS79]   R. Solovay and V. Strassen, A fast Monte Carlo test for primality *SIAM Journal on Computing* 6(1):84–85, 1979.

[STV01]   M. Sudan, L. Trevisan, and S. Vadhan. Pseudorandom generators without the XOR lemma. *Journal of Computer and System Sciences*, 62(2):236–266, 2001. (preliminary version in STOC'99.)

[SU01]   R. Shaltiel and C. Umans. Simple extractors for all min-entropies and a new pseudo-random generator. In *Proceedings of the Forty-Second Annual IEEE Symposium on Foundations of Computer Science*, 648–657, 2001.

[Sud97]   M. Sudan. Decoding of Reed Solomon codes beyond the error-correction bound. *Journal of Complexity*, 13(1):180–193, 1997.

[Toda]   S. Toda, "On the computational power of PP and $\oplus P$", in *30th FOCS*, 514–519, 1989.

[Tre01]   L. Trevisan. Extractors and pseudorandom generators. *Journal of the Association for Computing Machinery*, 48(4):860–879, 2001. (preliminary version in STOC'99.)

[Uma02]   C. Umans. Pseudo-random generators for all hardnesses. In *Proceedings of the Thirty-Fourth Annual ACM Symposium on Theory of Computing*, 2002.

[Val79]   L. Valiant. Completeness classes in algebra. In *Proceedings of the Eleventh Annual ACM Symposium on Theory of Computing*, 249–261, 1979.

[Val92]   L. Valiant. Why is Boolean complexity theory difficult? In M.S. Paterson, editor, *Boolean Function Complexity*, volume 169 of *London Math. Society Lecture Note Series*, 84–94. Cambridge University Press, 1992.

[Yao82]   A.C. Yao. Theory and applications of trapdoor functions. In *Proceedings of the Twenty-Third Annual IEEE Symposium on Foundations of Computer Science*, 80–91, 1982.

[Zip79]   R.E. Zippel. Probabilistic algorithms for sparse polynomials. In *Proceedings of an International Symposium on Symbolic and Algebraic Manipulation (EUROSAM'79)*, Lecture Notes in Computer Science, 216–226, 1979.

Rapid Mixing in Markov Chains

R. Kannan[*]

Abstract

A wide class of "counting" problems have been studied in Computer Science. Three typical examples are the estimation of - (i) the permanent of an $n \times n$ 0-1 matrix, (ii) the partition function of certain $n-$ particle Statistical Mechanics systems and (iii) the volume of an $n-$ dimensional convex set. These problems can be reduced to sampling from the steady state distribution of implicitly defined Markov Chains with exponential (in n) number of states. The focus of this talk is the proof that such Markov Chains converge to the steady state fast (in time polynomial in n).

A combinatorial quantity called conductance is used for this purpose. There are other techniques as well which we briefly outline. We then illustrate on the three examples and briefly mention other examples.

2000 Mathematics Subject Classification: 68W20, 60G50.
Keywords and Phrases: Randomized algorithms, Random walks.

1. Examples

We consider "counting problems", where there is an implicitly defined finite set X and one wishes to compute exactly or approximately $|X|$. In many situations, the approximate counting problem can be reduced to the problem of generating uniformly at random an element of X (the random generation problem). This is often the relatively easier part. Then, the generation problem is solved by devising a Markov Chain with set of states X with uniform steady state probabilities and then showing that this chain "mixes rapidly" - i.e., is close to the steady state distribution after a number of steps which is bounded above by a polynomial in the length of the input. [The proof of rapid mixing is often the challenging part.] We will illustrate the problem settings and scope of the area by means of three examples in this section. Then we will outline some tools for proving rapid mixing and describe very briefly how the tools are applied in some examples. This paper presents a

[*]Department of Computer Science, Yale University, New Haven, CT 06520, USA. E-mail: Kannan@cs.yale.edu

cross-section of methods and results from the area. A more comprehensive survey can be found in [19].

Our first example is the permanent of a $n \times n$ matrix A. Valiant [36] showed that the *exact* computation of the permanent is # P - hard, i.e., every problem in a class of problems called # P is reducible to the exact computation of the permanent of a matrix; thus it is conjectured that it is not solvable in polynomial time. The hardness result holds even for the case with each entry a 0 or a 1 whence the problem is to find $|X|$ where $X = \{\sigma \in S_n : A_{i,\sigma(i)} = 1 \forall i\}$. Note that X here is implicitly defined by A. In the general case, we may think of A as specifying a weight $\prod_i A_{i,\sigma(i)}$ on each σ in X.

As usual, we measure running time as a function of n, a natural parameter of the problem (like the n above) and $1/\epsilon$, where $\epsilon > 0$ is the relative error allowed. Our primary aim is a **polynomial** (in $n, 1/\epsilon$) **time** bounded algorithm; but, we will also discuss methods which help improve the polynomial. A recent breakthrough due to Jerrum, Sinclair, Vigoda [22] gives an approximation algorithm with such a time bound for the permanent (of a matrix with non-negative entries) settling this important open problem.

Our second example starts with the classical problem of computing the volume of a compact convex set in Euclidean $n-$ space $\mathbf{R}^n$. Dyer, Frieze and the author [17] gave polynomial time algorithm for estimating the volume to any specified relative error ϵ. They first reduce the problem to that of drawing a random point from the convex set (with uniform probability density). They then impose a grid on space and do a "coordinate random walk" - from current grid point x in K, pick one of the $2n$ coordinate neighbours y of x at random and go to y if $y \in K$; otherwise, stay at x. Under mild conditions, it is easy to show that the steady state distribution is uniform (over the grid points in the set); they show that in a polynomial (in n) number of steps, we are "close" to the steady state. [The number of states of the chain can be exponentially large.]

Lovász and Simonovits [27] have devised a continuous state space random walk called the "ball walk" which performs better. In this, we choose at the outset a "step size" $\delta > 0$. From the current point x, we pick at random (with uniform density) a point y in a ball of radius δ with x as center. We go to y if it is in K, otherwise, we stay at x.

More generally, we may consider the integration (a "continuous" analog of counting) of a function over a convex set K. Of particular interest are logarithmically-concave (a positive real valued function F is log-concave over a domain if $\log F$ is concave over the domain) functions, since many families of familiar probability density functions like the multi-variate normal are log- concave. One may use the Metropolis version of the random walks for convex set (cf section 1.). Rapid mixing has been proved for this general case too [4].

Our third set of examples concerns the Ising model and other Statistical Mechanics problems. (see [21] and references there). The computational problem arising from the Ising model is the following : we are given a real symmetric $n \times n$ matrix V (the entries of V arise as pairwise interaction energies), a real number B (the external field) and a positive real number β (the temperature). The Ising partition function is defined as

$$Z = Z(V_{ij}, B, \beta) = \sum_{\sigma \in \{-1,+1\}^n} e^{-\beta H(\sigma)}, \quad \text{where } H(\sigma) = -\sum_{i,j} V_{ij}\sigma_i\sigma_j - B\sum_k \sigma_k.$$

Jerrum and Sinclair [21] presented a polynomial time approximation algorithm to compute Z in the case when all V_{ij} are non-negative (called the ferromagnetic case).

Their algorithm for the ferromagnetic case first reduces the problem to the corresponding sampling problem and then more interestingly reduces this sampling problem to another one where we are given a graph (explicitly) $G(V, E)$ with positive edge weights $w(e)$. The problem is to pick a subset of edges of G at random such that the probability of picking a particular subset T is proportional to

$$w(T) = \mu^{|\text{odd}(T)|} \prod_{e \in T} w(e),$$

where μ is a given positive number and $\text{odd}(T)$ denotes the set of odd degree vertices in T. [Note that in this case X is the set of all subsets of edges and we have probabilities $\mathcal{P}$ on X as given above, where $X, \mathcal{P}$ are implicitly defined by giving G, w.]

2. Preliminaries, eigenvalue connection

Most of what we say extends naturally to continuous state space chains (where the set of states is (possibly uncountably) infinite) under mild conditions of measurability, but for ease of notation, here we state it for chains with a finite number of states. If P is the transition probability matrix with P_{xy} denoting the probability of transition from state x to state y, for any natural number t, the matrix power P^t denotes the t−step transition probabilities. All our chains will be connected and aperiodic and thus have steady state probabilities - $\pi(y) = \lim_{t \to \infty} P_{x,y}^t$. ($\pi(y)$ exists and is independent of the start state x). [The notation $\pi(\cdot)$ will be used throughout for steady state probabilities.] We let the vector $p^{(t)} = p^{(0)} P^t$ denote the probabilities at time t where we start with the initial distribution $p^{(0)}$. All our chains will be "time-reversible", i.e., $\pi(x)P_{xy} = \pi(y)P_{yx}$ will be valid for all pairs x, y.

From Linear algebra, we get that the eigenvalues of P are $1 = \lambda_1 > \lambda_2 \geq \lambda_3 \ldots \lambda_N \geq -1$ (where N is the number of states). Standard techniques yield :

Theorem 1. *For a finite time-reversible Markov Chain, with $\pi_0 = \min_x \pi(x)$, for any t,*

$$\sum_x \left| p^{(t)}(x) - \pi(x) \right| \leq \frac{1}{\pi_0} \left[\max(|\lambda_2|, |\lambda_N|) \right]^t.$$

Modifying a Markov Chain by making it stay at the current state with probability $1/2$ and move according to its transition function with probability $1/2$ ensures that $\lambda_N > 0$ while only increasing the (expected) running time by a factor of 2; so in the maximum above, we need only consider λ_2. We call a chain "lazy" if

$P_{xx} \geq \frac{1}{2} \forall x$. We will use the phrase **mixing time** to denote the least positive real τ such that for any $p^{(0)}$, $\sum_x |p^{(\tau)}(x) - \pi(x)| \leq 1/4$. It is known [1] that then for $t \geq \tau \log(1/\epsilon)$, we have $\sum_x |p^{(t)}(x) - \pi(x)| \leq \epsilon$.

If we have a time-reversible Markov Chain on a finite set of states with transition probability matrix P with steady state probabilities $\pi(x)$ and F is a positive real valued function on the states, there is a simple modification of the chain with steady state probabilities - $\pi(x)F(x)/\sum_y F(y)$, called the the **Metropolis** modification. It has transition probabilities - $P'_{xy} = P_{xy}\mathrm{Min}(1, \frac{F(y)}{F(x)})$ for $x \neq y$. This construction is used in many instances including as mentioned in the introduction for sampling according to log-concave functions.

3.   Techniques for proving rapid mixing

3.1.   Conductance

Alon and Milman [3] and Sinclair and Jerrum [34] related λ_2 to a combinatorial quantity called "conductance" (in what may be looked on as a discrete analog of Cheeger's inequality for manifolds). This has turned out to be of great use in practice; often, first proofs of polynomial time convergence use conductance.

For any two subsets S, T of states, the *ergodic flow* from S to T (denoted $Q(S,T)$) is defined as $Q(S,T) = \sum_{x \in S, y \in T} \pi(x)P_{xy}$. The conductance Φ is defined by :

$$\Phi(S) = \frac{Q(S,\bar{S})}{\pi(S)} \qquad \Phi = \min_{S:0<\pi(S)<3/4} \Phi(S).$$

$\Phi(S)$ is the probability of escaping from S to $\bar{S}$ conditioned on starting in S in the steady state; since $p^{(0)}$ may be this distribution, it is intuitively clear that if the conductance of any set is low, then the mixing time is high. More interestingly, [3] and [34] show also a converse.

Theorem 2. *For a time-reversible, lazy, ergodic Markov chain with conductance* Φ, *we have*

$$1 - 2\Phi \leq \lambda_2 \leq 1 - \frac{1}{2}\Phi^2.$$

While conductance has helped bound the mixing time for some complicated chains (including the three examples mentioned in the introduction), it is not a fine enough tool to give the correct bounds for some simple chains. For example, consider the lazy version of the random walk on the 2^n vertices on the $n-$ **cube**, where in each step, one picks at random one of the n neighbours of the current vertex to go to. The mixing time is known to be $O(n \log n)$. Conductance is $\Theta(1/n)$ for this example, yielding only a mixing time of $O(n^3)$ by Theorems (2) and (1).

A striking contrast is the random walk on the vertices of the **cube truncated** by a half-space (i.e., the set of 0-1 vectors satisfying a given linear inequality.) Morris and Sinclair [29] showed that the conductance of this walk is at least $1/p(n)$ for a polynomial $p(\cdot)$.

We now discuss a recent improvement of conductance for chains with a finite number of states from [26], [23]; similar results hold for chains with infinite number of states. In addition to measuring the ergodic flow from S to $\bar{S}$, we now also see if the flow is "well-spread out" in the sense that we "block" a set $B \subseteq \bar{S}$, and then see if $Q(S, \bar{S} \setminus B)$ is still high. We now define for S with $0 < \pi(S) \leq 3/4$,

$$\Psi(S) = \sup_{\alpha \in (0, \pi(S))} \min_{B \subseteq \bar{S}; \ \pi(B) \leq \alpha} \frac{\alpha \, Q(S, \bar{S} \setminus B)}{\pi(S)^2}.$$

It is easy to show that a set B with $\pi(B) \leq \frac{1}{2} Q(S, \bar{S})$, blocks at most $1/2$ of the flow from S to $\bar{S}$, so we have $\Psi(S) \geq \frac{1}{4}\Phi(S)^2$. Thus, an assertion that mixing time is $O(\log(1/\pi_0) \min_S \Psi(S))$ would be at least as strong a result as we get from Theorems (2) and (1). We prove a theorem which implies this assertion; indeed, instead of taking $\min_S \Psi(S)$, the theorem takes an "average" of this quantity over different set sizes. We say that $\psi : [0, 3/4] \to [0, 1]$ is a "blocking conductance function" (b.c.f.) if (the second condition is technical)

$$\forall S, 0 < \pi(S) \leq 3/4, \quad \Psi(S) \geq \psi(\pi(S)) \qquad \text{and} \qquad \psi(t) \leq 2\psi(t') \ \forall 0 \leq t \leq t' \leq \frac{4}{3}t.$$

Theorem 3. *If ψ is a blocking conductance function of a lazy, ergodic, time-reversible, finite Markov chain, with $\pi_0 = \min_x \pi(x)$, then, the mixing time is at most*

$$500 \int_{t=\pi_0}^{3/4} \frac{1}{t\psi(t)} dt.$$

This has been used to improve the analysis of the ball walk for convex sets in [26] and also some other examples in [30]. Also, [5] uses Theorem (3) to argue that the mixing time of the grid lattice, (in a fixed number of dimensions) where some edges have failed according to a standard percolation model is still within a constant of the mixing time of the whole.

3.2. Coupling

Another important technique for proving rapid mixing is "Coupling"[1]. A *coupling* is a stochastic process $(X_t, Y_t), t = 0, 1, 2, \ldots$, where each of $\{X_t, t = 0, 1, \ldots\}$ and $\{Y_t, t = 0, 1, 2, \ldots\}$ is marginally a copy of the chain. [They may be mutually dependent.] So, we run "two copies" of the chain (X_t, Y_t) in tandem. If Y_0 is distributed according to π, the steady state distribution, then, the distribution $p^{(t)}$ of X_t, satisfies

$$\sum_x |p^{(t)}(x) - \pi(x)| \leq \mathbf{Pr}(X_t \neq Y_t).$$

To apply this, one must construct a coupling (X_t, Y_t) for which X_t and Y_t "meet" as fast as possible. This can prove difficult. Path coupling introduced by Bubley and Dyer [8] which we describe now simplifies the task quite a bit. In path coupling, we have an underlying connected directed graph G on the set of states. (G could just be the graph of the Markov Chain.) G defines distances between pairs of states

- namely the length of the shortest path in G. We only need to define a coupling of adjacent pairs of vertices, with the property that for every pair of adjacent (in G) vertices (u, v), the expected distance between the next states of u, v is at most $\beta < 1$. They then show that

Theorem 4. *If D is the diameter (of G), then for any $t > 0$, $\mathbf{Pr}(X_t \neq Y_t) \leq D\beta^t$.*

Propp and Wilson [31] have designed a method they call **Coupling from the Past**. This applies to chains with a partial order on the set of states with a least state $\underline{0}$ and a greatest state $\underline{1}$. They show that running two copies of the Chain backwards - one from $\underline{0}$ and one from $\underline{1}$ - with a coupling satisfying a certain monotonicity condition until they "meet" gives us a good upper bound on the number of steps needed to mix. We refer the reader to [31] for details.

3.3.   Other methods

One way to prove a lower bound on conductance for a chain with a finite set of states X is to construct a family of $|X|^2$ paths - one from each state to each other using as edges the transitions of the Markov Chain, so that no transition is "overloaded" by too many paths. We do not supply here any more details of this technique referred to as the method of "canonical paths" and used by Jerrum and Sinclair [20].

We may look upon the construction of these paths as routing a multi-commodity flow through the network and apply techniques from Network Flows. [33] pursues this. [13] uses different measures of congestion to achieve improved results in some cases and their methods are applied in [15].

Another important method is the use of logarithmic Sobolev inequalities, where, we use (relative) entropy - $\mathrm{Ent}(p^{(t)}) = \sum_x p^{(t)}(x) \log \frac{p^{(t)}(x)}{\pi(x)}$ as the measure of distance. It is known that for ergodic Markov Chains, this distance declines exponentially [12]; i.e., there is a constant $\alpha \in (0, 1)$ such that

$$\mathrm{Ent}(p^{(t)}) \leq \alpha^t \mathrm{Ent}(p^{(0)}).$$

Note that $\mathrm{Ent}(p^{(0)}) \leq \log(1/\pi_0)$. So, it suffices to choose $t = (\log \log \frac{1}{\pi_0} + \log(1/\epsilon))$ $/(1 - \alpha)$ to reduce the entropy to ϵ; the dependence on $1/\pi_0$ is thus better. But we need to determine α which is only known for simple chains. It is known that $\alpha > \lambda_2$, so the most that this method could save over using something like Theorem (1) is the $\log(1/\pi_0)$ factor. [16] and [30] contain several comparisons between the log-Sobolev inequalities, eigenvalue bounds and conductance. [18] uses the log-Sobolev inequality to prove better bounds on the Metropolis version of the coordinate random walk for log-concave functions.

For the random walk on the cube a simple coupling argument, which, moves both X_t and Y_t in the same coordinate, trying to make them equal - shows that mixing time is $O(n \log n)$. Some sophisticated Fourier Transform methods have been used to get much more exact results here and the results are applicable in other contexts too.

A traditional approach to sampling from a probability distribution involves the so-called "Stopping Rules" [1], where one specifies a rule for when to stop the Markov Chain and shows that if we follow the rule, we sample (exactly) from the desired distribution. [2] contain results about the expected time needed for certain stopping rules, which then serves as an upper bound on the number of steps needed to converge.

We also mention two general techniques for deriving convergence rates of a Markov Chain from the knowledge of convergence rates for a simpler-to-analyze chain. The first one is called Comparison and is developed in [11]. The second technique is called Decomposition [32]; here one decomposes the chain into chains on subsets of states and derives a bound on the convergence rate of the whole chain based on the rates for the "sub-chains" and the interconnections between them.

4. Solution of sampling and counting problems

PERMANENT

We consider the permanent of a $n \times n$ 0-1 matrix A. We may define a bipartite graph corresponding to the matrix. Each $\sigma \in S_n$ with $A_{i,\sigma(i)} = 1$ for all i corresponds to a perfect matching in the graph. Let $\mathcal{M}$ be the set of perfect matchings in the graph. Unfortunately, no rapidly mixing Markov Chain with only $\mathcal{M}$ as the set of states is known. Broder [7] first defined the following Markov Chain. We also include the set of "near-perfect" matchings - $\mathcal{M}'$ (a near-perfect matching has $\frac{n}{2} - 1$ edges, no two incident to the same vertex). Transitions of the Markov Chain are as follows: In any current state, M, we pick an edge $e = (u, v)$ of the graph uniformly at random (all edges are equally likely) and

- if $M \in \mathcal{M}_n$ and $e \in M$, move to $M' = M - e$.
- If $M \in \mathcal{M}_{n-1}$ and u and v are both unmatched in M, then move to $M' = M + e$.
- $M \in \mathcal{M}_{n-1}$, u is matched to w in M and v unmatched, then move to $M' = (M + e) - (u, w)$; make a symmetric move if v is matched and u unmatched.
- In all other cases, stay at M.

[20] showed that if A is dense (each row has at least $n/2$ 1's), then the chain above mixes rapidly and in addition that $|\mathcal{M}'| \leq p(n)|\mathcal{M}|$ for a polynomial $P(\cdot)$. Thus, rejection sampling - accept result of a run of the chain if the result is in $\mathcal{M}$ yields a polynomial time sampling procedure.

Jerrum, Sinclair and Vigoda [22] develop an algorithm for the general 0-1 permanent (including the non-dense case). Here is very brief sketch of their algorithm : An edge-weighting w assigns a (positive) real weight $w(e)$ to each edge. For a matching M $w(M) = \prod_{e \in M} w(e)$ is its weight. For a set S of matchings, $w(S) = \sum_{M \in S} w(M)$. Finally, for each pair of vertices (u, v), define $w'(u, v)$ to be the ratio of the weight of all perfect matchings to the weight of all near-perfect matchings which leave u, v unmatched. Then define the "modified weight" $w'(M)$ of a matching M to be $w(M)$ if M is perfect and $w(M)w'(u, v)$ if M leaves u, v unmatched. They first show that a Metropolis version of the above random walk to sample according to $w'(M)$ mixes rapidly. But the w' are not known; they argue that if we start with the complete graph and go through a sequence of graphs, where in each step, we lower the edge weight of a non-edge of G by a factor, then we can successively estimate w' for each edge-weighting (of the complete graph) in the sequence. The final element of the sequence has low enough weights for the non-edges that it gives a good approximation to the permanent.

THE ISING MODEL

Recall the subgraph sampling problem in section 1. Here is the random walk they use. The states of the Markov Chain are the subsets of E. Their chain is the Metropolis version of the following simple Markov Chain whose steady state probabilities are uniform over all subsets of the edges, namely : at any current subset T of E, pick uniformly at random an edge $e \in E$; if $e \in T$, then go to $T' = T - e$, otherwise go to $T' = T + e$. They also make the chain lazy. The proof of a lower bound on conductance relies on a canonical paths argument.

The algorithm that is preferred by physicists is the one due to Swendsen and Wang [35]. This algorithm switches the signs on large blocks of vertices of the graph at once. But while this seems to work well in practice, no proof of rapid mixing is known.

CONVEX SETS, LOG-CONCAVE FUNCTIONS

Consider the ball walk in a convex set K in $\mathbf{R}^n$ with balls of radius δ. We use the notation P_{xy} for the transition probability density from x to y here. The conductance of a (measurable) subset S of K is now defined as

$$\frac{\int_{x \in S} \int_{y \in K \setminus S} \pi(x) P_{xy}}{\min(\pi(S), 1 - \pi(S))}.$$

Let ∂S be the boundary of S interior to K. Since points $x \in S$ on or near ∂S, intuitively have a high $\int_y P_{xy}$, a lower bound on $\mathrm{Vol}_{n-1}(\partial S)$ would seem to imply a lower bound on conductance. This is indeed the case. Lower bounds on $\mathrm{Vol}_{n-1}(\partial S)$ have been the subject of much effort. The most general result known is the following.

Theorem 5. Isoperimetry *Suppose K is a compact convex set in $\mathbf{R}^n$ of diameter d and F is a positive real-valued log-concave function on K. Then for any measurable $S \subseteq K$ with $\int_S F \le (1/2) \int_K F$, and measurable boundary ∂S interior to K, we have*

$$\int_{\partial S} F \ge \frac{2}{d} \int_S F.$$

The theorem was first proved for the case $F \equiv 1$ by Lovász and Simonovits [27] and independently also by Khachiyan and Karzanov [25]. The result was generalized to the case of general log-concave measures F by Applegate and Kannan [4] using the same techniques. We may add an extra factor of $\ln(\int_K F / \int_S F)$ to the right hand side; this was proved independently in [26] and also by Bobkov [6]. The most recent algorithm for computing the volume of convex sets is in [24], where references to earlier papers may be found.

OTHER EXAMPLES

There are many other counting problems on which progress has been made using this method. Again, we are not able to present a comprehensive review here.

A notable result is the one for the truncated cube already mentioned in section 3.. Another example of interest is Contingency Tables - where we are given m, n (positive integers) and the row and column sums of an $m \times n$ matrix A. The problem is to sample uniformly at random from the set of $m \times n$ matrices with non-negative integer entries with these row and column sums. The problem remains open, but there are several partial results [14],[10].

There are many **tiling problems**, where the problem is to pick a random tiling of say a large square in the plane by dominoes of a given shape. These problems arise in Statistical Mechanics. For regular shapes, it is often possible to devise a polynomial time algorithm to count the number exactly. But it is important to devise algorithms with low polynomial time bounds. There has been much progress here - see [28] and references there. Random generation of colorings and independent sets of a graph has received much attention lately due to connections to Statistical Mechanics [9].

Acknowledgment. I thank Ravi Montenegro for suggesting some changes in the manuscript.

References

[1] D. Aldous and J. Fill, Reversible Markov Chains and Random walks on graphs, In preparation. Available at http://stat-www.berkeley.edu/users/aldous/book.html.

[2] D.J. Aldous, L. Lovász and P. Winkler: Mixing times for uniformly ergodic Markov chains, *Stochastic Processes and their Applications* **71** (1997), 165–185.

[3] N. Alon, V. d. Milman, "λ_1, isoperimetric inequalities for graphs and super-concentrators," *Journal of Combinatorial Theory Series B 38*, 73–88, 1985.

[4] D. Applegate, R. Kannan, Sampling and integration of near log-concave functions, *Proceedings of the 23rd ACM Symposium on Theory of Computing*, 156–163, 1991.

[5] I. Benjamini and E. Mosel, On the mixing time of a simple random walk on the super critical percolation cluster, *Probability Theory and Related Fields* (to appear).

[6] S. G. Bobkov, Isoperimetric and analytic inequalities for log-concave probability measures, *Annals of Probability* Vol. 27, No. 4, (1999) 1903–1921.

[7] A.Z. Broder, How hard is it to marry at random? (On the approximation of the permanent), *Proceedings of the 18th ACM Symposium on Theory of Computing*, pp. 50-88, 1986 (Erratum in *Proceedings of the 20th ACM Symposium on Theory of Computing*, 1988, 551).

[8] R. Bubley and M. E. Dyer, Path Coupling: a technique for proving rapid mixing in Markov Chains, *Symposium on Foundations of Computer Science*, (1997) 223–231.

[9] R. Bubley, M. Dyer, C. Greenhill and M. Jerrum, On approximately counting

colorings of a small degree graph, *SIAM Journal on Computing* 29(3) (1999) 834–853.

[10] M. Cryan and M. Dyer, A polynomial-time algorithm to approximately count contingency tables where the number of rows is constant, *Proceedings of the ACM Symposium on Theory of Computing* (2002) 240–250.

[11] P. Diaconis and L. Saloff-Coste, Comparison theorems for reversible Markov Chains, *The Annals of Applied Probability* Vol. 3, No. 3, (1993) 696–730.

[12] P. Diaconis and L. Saloffe-Coste, Logarithmic Sobolev inequalities for finite Markov chains, *Annals of Applied Mathematics* **6** (1996) 695–750.

[13] P. Diaconis, D. Stroock, Geometric bound for eigenvalues of Markov chains, *Annals of Applied Probability 1,* (1991), 36–61.

[14] M. Dyer, R. Kannan and J. Mount, Sampling Contingency tables, *Random Structures and algorithms* **10**(4) (1997) 487–506.

[15] A. M. Frieze, R. Kannan and N. Polson, Sampling from log-concave distributions, *The Annals of Applied Probability* 4 (1994) 812–837.

[16] C. Houdré, Mixed and isoperimetric estimates on the log-Sobolev constant of graphs and Markov chains, *Combinatorica* (2001).

[17] M. E. Dyer, A. Frieze, R. Kannan, A random polynomial time algorithm for approximating the volume of convex bodies, *Journal of the ACM 38,* 1–17, 1991.

[18] A. M. Frieze and R. Kannan : Log-Sobolev inequalities and sampling from log-concave distributions, *Annals of Applied Probability* **9**, (1999) 14–26.

[19] M. Jerrum, Counting, Sampling and integrating : Algorithms and Complexity, in the series *Lectures in Mathematics - ETH, Zürich* (to appear)

[20] M. R. Jerrum, A. Sinclair, Approximating the permanent, *SIAM Journal on Computing18,* 1149–1178, 1989.

[21] M. R. Jerrum, A. Sinclair, "Polynomial-time approximation algorithms for the Ising model," *SIAM Journal on Computing* 22 (1993) 1087–1116.

[22] M. Jerrum, A. Sinclair, E. Vigoda, A polynomial-time approximation algorithm for the permanent of a matrix with non-negative entires, *Electronic Colloquium on Computational Complexity,* ECCC TR00-079, http://www.eccc.uni-trier.de/eccc/

[23] R. Kannan, L. Lovász and R. Montenegro, Rapid mixing using both vertex and edge expansion (in preparation)

[24] R. Kannan, L. Lovász and M. Simonovits, Random walks and an $O^*(n^5)$ volume algorithm, *Random Structures and Algorithms* (1997).

[25] L. Khachiyan, and A. Karzanov, *On the conductance of order Markov chains,* Technical Report DCS 268, Rutgers University, June 1990.

[26] L. Lovász and R. Kannan, Faster mixing via average conductance, *ACM Symposium on Theory of Computing* (1999) 282–287.

[27] L. Lovász and M. Simonovits, Random walks in a convex body and an improved volume algorithm *Random Structures and Algorithms* **4**, 359–412 1993.

[28] M. Luby, D. Randall and A. Sinclair, Markov chain algorithms for planar lattice structures, *SIAM Journal on Computing* **31** (2001) 167–192.

[29] B. Morris and A. Sinclair, Random walks on truncated cubes, *Symposium on*

the Foundations of Computer Science, (1999) 230–240.

[30] R. Montengro and J. B. Son, Edge isoperimetry and rapid mixing on matroids and geometric Markov Chains, *ACM Symposium on Theory of Computing* (2001).

[31] J. Propp and D. Wilson, Exact Sampling with Coupled Markov Chains and Applications to Statistical Mechanics *Random Structures and Algorithms* Vol. 9 (1996), 223–252

[32] N. Madras, D. Randall, Markov Chain decomposition for convergence rate analysis, *Annals of Applied Probability* (to appear).

[33] A. Sinclair, Improved bounds for mixing rates of Markov chains and multi-commodity flow, *Combinatorics, Probability and Computing 1*, 351–370, 1992.

[34] A. J. Sinclair, M.R. Jerrum, "Approximate counting, uniform generation and rapidly mixing Markov chains," *Information and Computation 82*, 93–133, 1989.

[35] R. H. Swendsen, J.S. Wang, "Nonuniversal critical dynamics in Monte Carlo simulations," *Physical Review Letters 58*, 86–88, 1987.

[36] L. G. Valiant, The complexity of computing the permanent, *Theoretical Computer Science 8*, 189–201, 1979.

ICM 2002 · Vol. III · 685–693

$P \neq NP$, Propositional Proof Complexity, and Resolution Lower Bounds for the Weak Pigeonhole Principle

Ran Raz*

Abstract

Recent results established exponential lower bounds for the length of any Resolution proof for the weak pigeonhole principle. More formally, it was proved that any Resolution proof for the weak pigeonhole principle, with n holes and any number of pigeons, is of length $\Omega(2^{n^\epsilon})$, (for a constant $\epsilon = 1/3$). One corollary is that certain propositional formulations of the statement $P \neq NP$ do not have short Resolution proofs. After a short introduction to the problem of $P \neq NP$ and to the research area of propositional proof complexity, I will discuss the above mentioned lower bounds for the weak pigeonhole principle and the connections to the hardness of proving $P \neq NP$.

2000 Mathematics Subject Classification: 68Q15, 68Q17, 03F20, 03D15.
Keywords and Phrases: Lower bounds, Proof theory, Resolution, Pigeonhole principle.

1. Propositional logic

The basic syntactic units (atoms) of propositional logic are Boolean variables $x_1, ..., x_n \in \{0, 1\}$, where the value 0 represents *False* and the value 1 represents *True*. The propositional variables are combined with standard Boolean gates (also called connectives), such as, *AND* (conjunction), *OR* (disjunction), and *NOT* (negation), to form Boolean formulas. Recall that in propositional logic there are no quantifiers.

A *literal* is either an atom (i.e., a variable x_i) or the negation of an atom (i.e., $\neg x_i$). A *clause* is a disjunction of literals. A *term* is a conjunction of literals. A formula f is in *conjunctive-normal-form* (CNF) if it is a conjunction of clauses. A formula f is in *disjunctive-normal-form* (DNF) if it is a disjunction of terms. Since there are standard ways to transform a formula to CNF or DNF (by adding new variables), many times we limit the discussion to CNF formulas or DNF formulas.

*Department of Computer Science, Weizmann Institute for Science, Rehovot 76100, Israel. E-mail: ranraz@wisdom.weizmann.ac.il

A Boolean formula $f(x_1, ..., x_n)$ is a *tautology* if $f(x_1, ..., x_n) = 1$ for every $x_1, ..., x_n$. A Boolean formula $f(x_1, ..., x_n)$ is *unsatisfiable* if $f(x_1, ..., x_n) = 0$ for every $x_1, ..., x_n$. Obviously, f is a tautology if and only if $\neg f$ is unsatisfiable.

Given a formula $f(x_1, ..., x_n)$, one can decide whether or not f is a tautology by checking all the possibilities for assignments to $x_1, ..., x_n$. However, the time needed for this procedure is exponential in the number of variables, and hence may be exponential in the length of the formula f.

2. $P \neq NP$

$P \neq NP$ is the central open problem in complexity theory and one of the most important open problems in mathematics today. The problem has thousands of equivalent formulations. One of these formulations is the following:

Is there a polynomial time algorithm A that gets as input a Boolean formula f and outputs 1 if and only if f is a tautology ?

$P \neq NP$ states that there is no such algorithm.

A related open problem in complexity theory is the problem of $NP \neq Co-NP$. The problem can be stated as follows:

Is there a polynomial time algorithm A that gets as input a Boolean formula f and a string z, and such that: f is a tautology if and only if there exists z s.t.:
 1. The length of z is at most polynomial in the length of f.
 2. $A(f, z) = 1$.

$NP \neq Co-NP$ states that there is no such algorithm. Obviously, $NP \neq Co-NP$ implies $P \neq NP$.

It is widely believed that $P \neq NP$ (and $NP \neq Co - NP$). At this point, however, we are still far from giving a solution for these problems. It is not clear why these problems are so hard to solve.

3.   Propositional proof theory

Propositional proof theory is the study of the length of proofs for different tautologies in different propositional proof systems.

The notion of *propositional proof system* was introduced by Cook and Reckhow in 1973, as a direction for proving $NP \neq co - NP$ (and hence also $P \neq NP$) [6]. A propositional proof system is a polynomial time algorithm $A(f, z)$ such that a Boolean formula f is a tautology if and only if there exists z such that $A(f, z) = 1$ (note that we do not require here that the length of z is at most polynomial in the length of f). We think of the string z as a proof for f in the proof system A. We say that a tautology f is *hard* for a proof system A if any proof z for f in the proof system A is of length super-polynomial in the length of f.

Many times we prefer to talk about unsatisfiable formulas, rather than tautologies, and about refutation systems, rather than proof systems. A *propositional refutation system* is a polynomial time algorithm $A(f, z)$ such that a Boolean formula f is unsatisfiable if and only if there exists z such that $A(f, z) = 1$. We think of the string z as a refutation for f in the refutation system A. We think of a refutation z for f also as a proof for $\neg f$ (and vice versa).

It is easy to see that $NP \neq co - NP$ if and only if for every propositional proof system A there exists a hard tautology, that is, a tautology f with no short proofs. It was hence suggested by Cook and Reckhow to study the length of proofs for different tautologies in stronger and stronger propositional proof systems. It turns out that in many cases these problems are very interesting in their own right and are related to many other interesting problems in complexity theory and in logic, in particular when the tautology f represents a fundamental mathematical principle.

For a recent survey on the main research directions in propositional proof theory, see [2].

4.   Resolution

Resolution is one of the simplest and most widely studied propositional proof systems. Besides its mathematical simplicity and elegance, Resolution is a very interesting proof system also because it generalizes the Davis-Putnam procedure and several other well known proof-search procedures. Moreover, Resolution is the base for most automat theorem provers existing today.

The *Resolution rule* says that if C and D are two clauses and x_i is a variable then any assignment (to the variables $x_1, ..., x_n$) that satisfies both of the clauses, $C \vee x_i$ and $D \vee \neg x_i$, also satisfies the clause $C \vee D$. The clause $C \vee D$ is called the *resolvent* of the clauses $C \vee x_i$ and $D \vee \neg x_i$ on the variable x_i.

Resolution is usually presented as a propositional refutation system for CNF formulas. Since there are standard ways to transform a formula to CNF (by adding new variables), this presentation is general enough. A *Resolution refutation* for a CNF formula f is a sequence of clauses $C_1, C_2, \ldots, C_s$, such that:

1. Each clause C_j is either a clause of f or a resolvent of two previous clauses in the sequence.
2. The last clause, C_s, is the empty clause.

We think of the empty clause as a clause that has no satisfying assignments, and hence a contradiction was obtained.

We think of a Resolution refutation for f also as a proof for $\neg f$. Without loss of generality, we assume that no clause in a Resolution proof contains both x_i and $\neg x_i$ (such a clause is always satisfied and hence it can be removed from the proof). The *length*, or *size*, of a Resolution proof is the number of clauses in it.

We can represent a Resolution proof as an acyclic directed graph on vertices $C_1, \ldots, C_s$, where each clause of f has out-degree 0, and any other clause has two edges pointing to the two clauses that were used to produce it.

It is well known that Resolution is a refutation system. That is, a CNF formula f is unsatisfiable if and only if there exists a Resolution refutation for f. A well-known and widely studied restricted version of Resolution (that is still a refutation system) is called *Regular Resolution*. In a Regular Resolution refutation, along any path in the directed acyclic graph, each variable is resolved upon at most once.

5. Resolution as a search problem

As mentioned above, we represent a Resolution proof as an acyclic directed graph G on the vertices $C_1, \ldots, C_s$. In this graph, each clause C_j which is an original clause of f has out-degree 0, and any other clause has two edges pointing to the two clauses that were used to produce it. We call the vertices of out-degree 0 (i.e., the clauses that are original clauses of f) the *leaves* of the graph. Without loss of generality, we can assume that the only clause with in-degree 0 is the last clause C_s (as we can just remove any other clause with in-degree 0). We call the vertex C_s the *root* of the graph.

We label each vertex C_j in the graph by the variable x_i that was used to derive it (i.e., the variable x_i that was resolved upon), unless the clause C_j is an original clause of f (and then C_j is not labelled). If a clause C_j is labelled by a variable x_i we label the two edges going out from C_j by 0 and 1, where the edge pointing to the clause that contains x_i is labelled by 0, and the edge pointing to the clause that contains $\neg x_i$ is labelled by 1. That is, if the clause $C \vee D$ was derived from the two clauses $C \vee x_i$ and $D \vee \neg x_i$ then the vertex $C \vee D$ is labelled by x_i, the edge from the vertex $C \vee D$ to the vertex $C \vee x_i$ is labelled by 0 and the edge from the vertex $C \vee D$ to the vertex $D \vee \neg x_i$ is labelled by 1. For a non-leaf node u of the graph G, define,

Label(u) = the variable labelling u.

We think of $Label(u)$ as a variable queried at the node u.

Let p be a path on G, starting from the root. Note that along a path p, a variable x_i may appear (as a label of a node u) more than once. We say that the path p evaluates x_i to 0 if $x_i = Label(u)$ for some node u on the path p, and after the last appearance of x_i as $Label(u)$ (of a node u on the path) the path p continues on the edge labelled by 0 (i.e., if u is the last node on p such that $x_i = Label(u)$ then p contains the edge labelled by 0 that goes out from u). In the same way, we say that the path p evaluates x_i to 1 if $x_i = Label(u)$ for some node u on the path p, and after the last appearance of x_i as $Label(u)$ (of a node u on the path) the path p continues on the edge labelled by 1 (i.e., if u is the last node on p such that $x_i = Label(u)$ then p contains the edge labelled by 1 that goes out from u).

For any node u of the graph G, we define $Zeros(u)$ to be the set of variables that the node u "remembers" to be 0, and $Ones(u)$ to be the set of variables that the node u "remembers" to be 1, that is,

Zeros(u) = the set of variables that are evaluated to 0 by every path p from the root to u.

Ones(u) = the set of variables that are evaluated to 1 by every path p from the root to u.

Note that for any u, the two sets $Zeros(u)$ and $Ones(u)$ are disjoint.

The following proposition gives the connection between the sets $Zeros(u)$, $Ones(u)$ and the literals appearing in the clause u. The proposition is particularly interesting when u is a leaf of the graph.

Proposition 1 *Let f be an unsatisfiable CNF formula and let G be (the graph representation of) a Resolution refutation for f. Then, for any node u of G and for any x_i, if the literal x_i appears in the clause u then $x_i \in Zeros(u)$, and if the literal $\neg x_i$ appears in the clause u then $x_i \in Ones(u)$.*

6. The weak pigeonhole principle

The *Pigeonhole Principle* (PHP) is probably the most widely studied tautology in propositional proof theory. The tautology PHP_n is a DNF encoding of the following statement: There is no one to one mapping from $n+1$ pigeons to n holes. The *Weak Pigeonhole Principle* (WPHP) is a version of the pigeonhole principle that allows a larger number of pigeons. The tautology $WPHP_n^m$ (for $m \geq n+1$) is a DNF encoding of the following statement: There is no one to one mapping from m pigeons to n holes. For $m > n+1$, the weak pigeonhole principle is a weaker statement than the pigeonhole principle. Hence, it may have much shorter proofs in certain proof systems.

The weak pigeonhole principle is one of the most fundamental combinatorial principles. In particular, it is used in most probabilistic counting arguments and hence in many combinatorial proofs. Moreover, as observed by Razborov, there are certain connections between the weak pigeonhole principle and the problem of $P \neq NP$ [12]. Indeed, the weak pigeonhole principle (with a relatively large number of pigeons) can be interpreted as a certain encoding of the following statement: There are no small DNF formulas for SAT (where SAT is the satisfiability problem). Hence, in most proof systems, a short proof for certain formulations of the statement "There are no small formulas for SAT" can be translated into a short proof for the weak pigeonhole principle. That is, a lower bound for the length of proofs for the weak pigeonhole principle usually implies a lower bound for the length of proofs for certain formulations of the statement $P \neq NP$. While this doesn't say much about the problem of $P \neq NP$, it does demonstrate the applicability and relevance of the weak pigeonhole principle for other interesting problems.

Formally, the formula $WPHP_n^m$ is expressed in the following way. The underlying Boolean variables, $x_{i,j}$, for $1 \leq i \leq m$ and $1 \leq j \leq n$, represent whether or not pigeon i is mapped to hole j. The negation of the pigeonhole principle, $\neg WPHP_n^m$, is expressed as the conjunction of m *pigeon clauses* and $\binom{m}{2} \cdot n$ *hole clauses*. For every $1 \leq i \leq m$, we have a pigeon clause,

$$(x_{i,1} \vee \ldots \vee x_{i,n}),$$

stating that pigeon i maps to some hole. For every $1 \leq i_1 < i_2 \leq m$ and every $1 \leq j \leq n$, we have a hole clause,

$$(\neg x_{i_1,j} \vee \neg x_{i_2,j}),$$

stating that pigeons i_1 and i_2 do not both map to hole j. We refer to the pigeon clauses and the hole clauses also as pigeon axioms and hole axioms. Note that $\neg WPHP_n^m$ is a CNF formula.

Let G be (the graph representation of) a Resolution refutation for $\neg WPHP_n^m$. Then, by Proposition 1, for any leaf u of the graph G, one of the following is satisfied:

1. u is a pigeon axiom, and then for some $1 \leq i \leq m$, the variables $x_{i,1}, \ldots, x_{i,n}$ are all contained in $Zeros(u)$.

2. u is a hole axiom, and then for some $1 \leq j \leq n$, there exist two different variables $x_{i_1,j}, x_{i_2,j}$ in $Ones(u)$.

7. Resolution lower bounds for the weak pigeonhole principle

There are trivial Resolution proofs (and Regular Resolution proofs) of length $2^n \cdot poly(n)$ for the pigeonhole principle and for the weak pigeonhole principle. In a seminal paper, Haken proved that for the pigeonhole principle, the trivial proof is (almost) the best possible [7]. More specifically, Haken proved that any Resolution proof for the tautology PHP_n is of length $2^{\Omega(n)}$. Haken's argument was further developed in several other papers (e.g., [18, 1, 4]). In particular, it was shown that a similar argument gives lower bounds also for the weak pigeonhole principle, but only for small values of m. More specifically, super-polynomial lower bounds were proved for any Resolution proof for the tautology $WPHP_n^m$, for $m < c \cdot n^2/\log n$ (for some constant c) [5].

For the weak pigeonhole principle with large values of m, there do exist Resolution proofs (and Regular Resolution proofs) which are much shorter than the trivial ones. In particular, it was proved by Buss and Pitassi that for $m > c^{\sqrt{n \log n}}$ (for some constant c), there are Resolution (and Regular Resolution) proofs of length $poly(m)$ for the tautology $WPHP_n^m$ [3]. Can this upper bound be further improved? Can one prove a matching lower bound? A partial progress was made by Razborov, Wigderson and Yao, who proved exponential lower bounds for Regular Resolution proofs, but only when the Regular Resolution proof is of a certain restricted form [17].

The weak pigeonhole principle with large number of pigeons has attracted a lot of attention in recent years. However, the standard techniques for proving lower bounds for Resolution failed to give lower bounds for the weak pigeonhole principle. In particular, for $m \geq n^2$, no non-trivial lower bound was known until very recently.

In the last two years, these problems were completely solved. An exponential lower bound for any Regular Resolution proof was proved in [8], and an exponential lower bound for any Resolution proof was finally proved in [9]. More precisely, it was proved in [9] that for any m, any Resolution proof for the weak pigeonhole principle $WPHP_n^m$ is of length $\Omega(2^{n^\epsilon})$, where $\epsilon > 0$ is some global constant ($\epsilon \approx 1/8$).

The lower bound was further improved in several results by Razborov. The first result [13] presents a proof for an improved lower bound of $\Omega(2^{n^\epsilon})$, for $\epsilon = 1/3$. The second result [14] extends the lower bound to an important variant of the pigeonhole principle, the so called *weak functional pigeonhole principle*, where we require in addition that each pigeon goes to exactly one hole. The third result [15] extends the lower bound to another important variant of the pigeonhole principle, the so called *weak functional onto pigeonhole principle*, where we require in addition that every hole is occupied.

For a recent survey on the propositional proof complexity of the pigeonhole principle, see [16].

8.   Lower bounds for $P \neq NP$

Propositional versions of the statement $P \neq NP$ were introduced by Razborov in 1995 [10] (see also [11]). Razborov suggested to try to prove super-polynomial lower bounds for the length of proofs for these statements in stronger and stronger propositional proof systems. This was suggested as a step for proving the hardness of proving $P \neq NP$. The above mentioned results for the weak pigeonhole principle establish such super-polynomial lower bounds for Resolution.

Let $g : \{0,1\}^d \to \{0,1\}$ be a Boolean function. For example, we can take $g = SAT$, where $SAT : \{0,1\}^d \to \{0,1\}$ is the satisfiability function (or we can take any other NP-hard function). We assume that we are given the truth table of g. Let $t \leq 2^d$ be some integer. We think of t as a large polynomial in d, say $t = d^{1000}$.

Razborov suggested to study propositional formulations of the following statement (in the variables $\vec{Z}$):

$\vec{Z}$ is (an encoding of) a Boolean circuit of size $t \Longrightarrow$
$\vec{Z}$ does not compute the function g.

Note that since the truth table of g is of length 2^d, a propositional formulation of this statement will be of length at least 2^d, and it is not hard to see that there are ways to write this statement as a DNF formula of length $2^{O(d)}$ (and hence, its negation is a CNF formula of that length). The standard way to do that is by including in $\vec{Z}$ both, the (topological) description of the Boolean circuit, as well as the value that each gate in the circuit outputs on each input for the circuit.

In [12], Razborov presented a lower bound for the degree of *Polynomial Calculus* proofs for the weak pigeonhole principle, and used this result to prove a lower bound for the degree of Polynomial Calculus proofs for a certain version of the above statement. Following this line of research, it was proved in [9, 15] (in a similar way) that if t is a large enough polynomial in d (say $t = d^{1000}$) then any Resolution proof for certain versions of the above statement is of length super-polynomial in 2^d, that is, super-polynomial in the length of the statement.

In particular, this can be interpreted as a super-polynomial lower bound for Resolution proofs for certain formulations of the statement $P \neq NP$ (or, more precisely, of the statement $NP \not\subset P/poly$).

It turns out that the exact way to give the (topological) description of the circuit is also important in some cases. This was done slightly differently in [9] and in [15]. In [9], $\vec{Z}$ was used to encode a Boolean circuit of unbounded fan-in, whereas [15] considered Boolean circuits of fan-in 2. It turns out that for the stronger case of unbounded fan-in, the lower bound for the weak pigeonhole principle is enough [9], whereas for the weaker case of fan-in 2 one needs the lower bound for the weak functional onto pigeonhole principle [15] (in fact, this was one of the main motivations to consider the onto functional case). Otherwise, the proof seems to be quite robust in the way the Boolean circuit is encoded.

Acknowledgement. I would like to thank Toni Pitassi for very enjoying collaboration that lead to the results in [8, 9].

References

[1] Beame, P., and Pitassi, T., "Simplified and improved resolution lower bounds," *Foundations of Computer Science*, 1996, 274–282.

[2] Beame, P., and Pitassi, T., "Propositional Proof Complexity: Past, Present, and Future," *Current Trends in Theoretical Computer Science*, 2001, 42–70.

[3] Buss, S., and Pitassi, T., "Resolution and the weak pigeonhole principle," *Lecture Notes in Computer Science, Springer-Verlag*, vol. 1414, 1998, 149–156. (Selected Papers of Computer Science Logic 11th International Workshop, 1997).

[4] Ben-Sasson, E., and Wigderson, A., "Short proofs are narrow–resolution made simple," *Journal of the ACM*, 48(2),2001, 149–168.

[5] Buss, S., and Turan, G., "Resolution proofs of generalized pigeonhole principles," *Theoretical Computer Science*, 62(3), 1988, 311–317.

[6] Cook, S., and Reckhow R., "The relative efficiency of propositional proof systems," *Journal of Symbolic Logic*, 44(1), 1979, 36–50.

[7] Haken, A., "The intractability of resolution," *Theoretical Computer Science*, 39(2-3), 1985, 297–308.

[8] Pitassi, T., and Raz, R., "Regular resolution lower bounds for the weak pigeonhole principle," *Symposium on Theory of Computing*, 2001, 347–355.

[9] Raz, R., "Resolution lower bounds for the weak pigeonhole principle," *Symposium on Theory of Computing*, 2002.

[10] Razborov, A., "Bounded arithmetic and lower bounds in Boolean complexity," *Feasible Mathematics II. Progress in Computer Science and Applied Logic*, vol. 13, 1995, 344–386.

[11] Razborov, A., "Lower bounds for propositional proofs and independence results in Bounded Arithmetic," *Lecture Notes in Computer Science, Springer Verlag*, vol. 1099, 1996, 48–62. (Proc. of the 23rd ICALP).

[12] Razborov, A., "Lower bounds for the polynomial calculus," *Computational Complexity*, 7(4), 1998, 291–324.

[13] Razborov, A., "Improved resolution lower bounds for the weak pigeonhole principle," *Electronic Colloquium on Computational Complexity (ECCC)*, 8(055), 2001.

[14] Razborov, A., "Resolution lower bounds for the weak functional pigeonhole principle," *Electronic Colloquium on Computational Complexity (ECCC)*,

8(075), 2001. (to appear in *Theoretical Computer Science*).

[15] Razborov, A., "Resolution Lower Bounds for Perfect Matching Principles," *Proc. of the 17th IEEE Conference on Computational Complexity*, 2002.

[16] Razborov, A., "Proof Complexity of Pigeonhole Principles," *Developments in Language Theory*, 2001, 100–116.

[17] Razborov, A., Wigderson, A., and Yao, A., "Read-once branching programs, rectangular proofs of the pigeonhole principle, and the transversal calculus," *Symposium on Theory of Computing*, 1997, 739–748.

[18] Urquhart, A., "Hard examples for resolution," *Journal of the ACM*, vol. 34, 1987, 209–219.

Section 16. Numerical Analysis and Scientific Computing

ICM 2002 · Vol. III · 697–706

The Complexity of
Accurate Floating Point Computation

J. Demmel*

Abstract

Our goal is to find accurate and efficient algorithms, when they exist, for evaluating rational expressions containing floating point numbers, and for computing matrix factorizations (like LU and the SVD) of matrices with rational expressions as entries. More precisely, *accuracy* means the relative error in the output must be less than one (no matter how tiny the output is), and *efficiency* means that the algorithm runs in polynomial time. Our goal is challenging because our accuracy demand is much stricter than usual.

The classes of floating point expressions or matrices that we can accurately and efficiently evaluate or factor depend strongly on our model of arithmetic:

1. In the "Traditional Model" (TM), the floating point result of an operation like $a + b$ is $fl(a + b) = (a + b)(1 + \delta)$, where $|\delta|$ must be tiny.
2. In the "Long Exponent Model" (LEM) each floating point number $x = f \cdot 2^e$ is represented by the pair of integers (f, e), and there is no bound on the sizes of the exponents e in the input data. The LEM supports strictly larger classes of expressions or matrices than the TM.
3. In the "Short Exponent Model" (SEM) each floating point number $x = f \cdot 2^e$ is also represented by (f, e), but the input exponent sizes are bounded in terms of the sizes of the input fractions f. We believe the SEM supports strictly more expressions or matrices than the LEM.

These classes will be described by factorizability properties of the rational expressions, or of the minors of the rational matrices. For each such class, we identify new algorithms that attain our goals of accuracy and efficiency. These algorithms are often exponentially faster than prior algorithms, which would simply use a conventional algorithm with sufficiently high precision.

For example, we can factorize Cauchy matrices, Vandermonde matrices, totally positive generalized Vandermonde matrices, and suitably discretized differential and integral operators in all three models much more accurately and efficiently than before. But we provably cannot add $x + y + z$ accurately in the TM, even though it is easy in the other models.

2000 Mathematics Subject Classification: 65F, 65G50, 65Y20, 68Q25.
Keywords and Phrases: Roundoff, Numerical linear algebra, Complexity.

---

*Mathematics Department and Computer Science Division, University of California, Berkeley, CA 94720, USA. E-mail: demmel@cs.berkeley.edu

1. Introduction

We will survey recent progress and describe open problems in the area of accurate floating point computation, in particular for matrix computations. A very short bibliography would include [10, 7, 8, 14, 12, 1, 9, 11, 2].

We consider the evaluation of multivariate rational functions $r(x)$ of floating point numbers, and matrix computations on rational matrices $A(x)$, where each entry $A_{ij}(x)$ is such a rational function. Matrix computations will include computing determinants (and other minors), linear equation solving, performing Gaussian Elimination (GE) with various kinds of pivoting, and computing the singular value decomposition (SVD), among others. Our goals are *accuracy* (computing each solution component with tiny relative error) and *efficiency* (the algorithm should run in time bounded by a polynomial function of the input size).

We consider three models of arithmetic, defined in the abstract, and for each one we try to classify rational expressions and matrices as to whether they can be evaluated or factored accurately and efficiently (we will say "compute(d) accurately and efficiently," or "CAE" for short).

In the Traditional "$1 + \delta$" Model (TM), we have $fl(a \otimes b) = (a \otimes b)(1 + \delta)$, $\otimes \in \{+, -, \times, \div\}$ and $|\delta| \leq \epsilon$, where $\epsilon \ll 1$ is called *machine precision*. It is the conventional model for floating point error analysis, and means that every floating point result is computed with a relative error δ bounded in magnitude by ϵ. The values of δ may be arbitrary real (or complex) numbers satisfying $|\delta| \leq \epsilon$, so that any algorithm proven to CAE in the TM must work for arbitrary real (or complex) number inputs and arbitrary real (or complex) $|\delta| \leq \epsilon$. The size of the input in the TM is the number of floating point words needed to describe it, independent of ϵ.

The Long Exponent (LEM) and Short Exponent (SEM) models, which are implementable on a Turing machine, make errors that may be described by the TM, but their inputs and δ's are much more constrained. Also, we compute the size of the input in the LEM and SEM by counting the number of bits, so that higher precision and wider range take more bits.

This will mean that problems we can provably CAE in the TM are a strict subset of those we can CAE in the LEM, which in turn we conjecture are a strict subset of those we can CAE in the SEM. In all three models we will describe the classes of rational expressions and rational matrices in terms of the factorization properties of the expressions, or of the minors of the matrices.

The reader may wonder why we insist on accurately computing tiny quantities with small relative error, since in many cases the inputs are themselves uncertain, so that one could suspect that the inherent uncertainty in the input could make even the signs of tiny outputs uncertain. It will turn out that in the TM, the class we can CAE appears to be identical to the class where all the outputs are in fact accurately determined by the inputs, in the sense that small relative changes in the inputs cause small relative changes in the outputs. We make this conjecture more precise in section 3 below.

There are many ways to formulate the search for efficient and accurate algorithms [6, 4, 19, 3, 13, 18, 16]. Our approach differs in several ways. In contrast to either conventional floating point error analysis [13] or the model in [6], we ask that even the tiniest results have correct leading digits, and that zero be exact. In [6] the model of arithmetic allows a tiny absolute error in each operation, whereas in TM we allow a tiny relative error. Unlike [6] our LEM and SEM are conventional Turing machine models, with numbers represented as bit strings, and so we can take the cost of arithmetic on very large and very small numbers (i.e. those with many exponent bits) into precise account. For these reasons we believe our models are closer to computational practice than the model in [6]. In contrast to [16], we (mostly) consider the input as given exactly, rather than as a sequence of ever better approximations. Finally, many of our algorithms could easily be modified to explicitly compute guaranteed interval bounds on the output [18].

2. Factorizability and minors

We show here how to reduce the question of accurate and efficient matrix computations to accurate and efficient rational expression evaluation. The connection is elementary, except for the SVD, which requires an algorithm from [10].

Proposition 1 *Being able to CAE the absolute value of the determinant* $|\det(A(x))|$ *is necessary to be able to CAE the following matrix computations on $A(x)$: LU factorization (with or without pivoting), QR factorization, all the eigenvalues λ_i of $A(x)$, and all the singular values of $A(x)$. Conversely, being able to CAE all the minors of $(A(x))$ is sufficient to be able to CAE the following matrix computations on $A(x)$: A^{-1}, LU factorization (with or without pivoting), and the SVD of $A(x)$. This holds in any model of arithmetic.*

Proof First consider necessity. $|\det(A(x))|$ may be written as the product of diagonal entries of the matrices L, U and R in these factorizations, or as the product of eigenvalues or singular values. If these entries or values can be CAE, then so can their product in a straightforward way.

Now consider sufficiency. The statement about A^{-1} is just Cramer's rule, which only needs $n^2 + 1$ different minors. The statement about LU factorization depends on the fact that each nontrivial entry of L and U is a quotient of minors. The SVD is more difficult [10], and depends on the following two step algorithm: (1) Compute a *rank revealing* decomposition $A = X \cdot D \cdot Y$ where X and Y are "well-conditioned" (far from singular in the sense that $\|X\| \cdot \|X^{-1}\|$ is not too large) and (2) use a bisection-like algorithm to compute the SVD from XDY.

We believe that computing $\det(A(x))$ is actually necessary, not just $|\det(A(x))|$. The sufficiency proof can be extended to other matrix computations like the QR decomposition and pseudoinverse by considering minors of matrices like $\begin{bmatrix} I & A \\ A^T & 0 \end{bmatrix}$.

Furthermore, if we can CAE the minors of $C(x) \cdot A(x) \cdot B(x)$, and $C(x)$ and $B(x)$ are well-conditioned, then we can still CAE a number of matrix factorizations, like the SVD. The SVD can be applied to get the eigendecomposition of symmetric matrices, but we know of no sufficient condition for the accurate and efficient calculation of eigenvalues of nonsymmetric matrices.

3. CAE in the traditional model

We begin by giving examples of expressions and matrix computations that we can CAE in the TM, and then discuss what we cannot do. The results will depend on details of the axioms we adopt, but for now we consider the minimal set of operations described in the abstract.

As long as we only do *admissible operations*, namely multiplication, division, addition of like-signed quantities, and addition/subtraction of (exact!) input data $(x \pm y)$, then the worst case relative error only grows very slowly, roughly proportionally to the number of operations. It is when we subtract two like-signed approximate quantities and significant cancellation occurs, that the relative error can become large. So we may ask which problems we can CAE just using only admissible operations, i.e. which rational expressions factor in such a way that only admissible operations are needed to evaluate them, and which matrices have all minors with the same property.

Here are some examples, where we assume that the inputs are arbitrary real or complex numbers. (1) The determinant of a Cauchy matrix $C_{ij} = 1/(x_i + y_j)$ is CAE using the classical expression $\prod_{i<j}(x_j - x_i)(y_j - y_i)/\prod_{i,j}(x_i + y_j)$, as is every minor. In fact, changing one line of the classical GE routine will compute each entry of the LU decomposition accurately in about the same time as the original inaccurate version. (2) We can CAE all minors of sparse matrices, i.e. those with certain entries fixed at 0 and the rest independent indeterminates x_{ij}, if and only if the undirected bipartite graph presenting the sparsity structure of the matrix is *acyclic*; a one-line change to GE again renders it accurate. An important special case are bidiagonal matrices, which arise in the conventional SVD algorithm. (3) The eigenvalue problem for the second centered difference approximation to a Sturm-Liouville ODE or elliptic PDE on a rectangular grid (with arbitrary rectilinear boundaries) can be written as the SVD of an "unassembled" problem $G = D_1 U D_2$ where D_1 and D_2 are diagonal (depending on "masses" and "stiffnesses") and U is *totally unimodular*, i.e. all its minors are ± 1 or 0. Again, a simple change to GE renders it accurate.

In contrast, one can show that it is impossible in the TM to add $x + y + z$ accurately in constant time; the proof involves showing that for *any* algorithm the rounding errors δ and inputs x, y, z can be chosen to have an arbitrarily large relative error. This depends on the δ's being permitted to be arbitrary real numbers in our model.

Vandermonde matrices $V_{ij} = x_i^{j-1}$ are more subtle. Since the product of a Vandermonde matrix and the Discrete Fourier Transform (DFT) is Cauchy, and we can compute the SVD of a Cauchy, we can compute the SVD of a Vandermonde. This fits in our TM model because the roots of unity in the DFT need only be known approximately, and so may be computed in the TM model. In contrast, one can use the result in the last paragraph to show that the inverse of a Vandermonde cannot be computed accurately. Similarly, polynomial Vandermonde matrices with $V_{ij} = P_i(x_j)$, P_i a (normalized) orthogonal polynomial, also permit accurate SVDs, but probably not inverses.

4.   Adding nonnegativity to the traditional model

If we further restrict the domain of (some) inputs to be nonnegative, then much more is possible, $x + y + z$ as a trivial example. A more interesting example are weakly diagonally dominant M-matrices, which arise as discretizations of PDEs; they must be represented as offdiagonal entries and the row sums.

More interesting is the class of *totally positive (TP) matrices*, all of whose minors are positive. Numerous structure theorems show how to represent such matrices as products of much simpler TP matrices. Accurate formulas for the (nonnegative) minors of these simpler matrices combined with the Cauchy-Binet theorem yield accurate formulas for the minors of the original TP matrix, but typically at an exponential cost.

An important class of TP matrices where we can do much better are the TP generalized Vandermonde matrices $G_{ij} = x_i^{\mu_j}$, where the μ_j form an increasing non-negative sequence of integers. $\det(G)$ is known to be the product of $\prod_{i<j}(x_j - x_i)$ and a *Schur function* [15] $s_\lambda(x_i)$, where the sequence $\lambda = (\lambda_j) = (\mu_{n+1-j} - (n-j))$ is called a *partition*. Schur functions are polynomials with nonnegative integer coefficients, so since their arguments x_i are nonnegative, they can certainly be computed accurately. However, straightforward evaluation would have an exponential cost $O(n^{|\lambda|})$, $|\lambda| = \sum_j \lambda_j$. But by exploiting combinatorial identities satisfied by Schur functions along with techniques of divide-and-conquer and memoization, the cost of evaluating the determinant can be reduced to polynomial time $n^2 \prod_j (\lambda_j + 1)^2$. The cost of arbitrary minors and the SVD remains exponential at this time. Note that the λ_i are counted as part of the size of the input in this case.

Here is our conjecture generalizing all the cases we have studied in the TM. We suppose that $f(x_1, ..., x_n)$ is a homogeneous polynomial, to be evaluated on a domain $\mathcal{D}$. We assume that $\mathcal{D} \subseteq \overline{\text{int}\mathcal{D}}$, to avoid pathological domains. Typical domains could be all tuples of the real or complex numbers, or the positive orthant. We say that f satisfies condition (A) (for *Accurate*) if f can be written as a product $f = \prod_m f_m$ where each factor f_m satisfies

- f_m is of the form x_i, $x_i - x_j$ or $x_i + x_j$, or
- $|f_m|$ is bounded away from 0 on $\mathcal{D}$.

Conjecture 1 *Let f and $\mathcal{D}$ be as above. Then condition (A) is a necessary and sufficient condition for the existence of an algorithm in the TM model to compute f accurately on $\mathcal{D}$.*

Note that we make no claims that f can be evaluated efficiently; there are numerous examples where we only know exponential-time algorithms (doing GE with complete pivoting on a totally positive generalized Vandermonde matrix).

5.   Extending the TM

So far we have considered the simplest version of the TM, where (1) we have only the input data, and no additional constants available, (not even integers, let alone arbitrary rationals or reals), (2) the input data is given exactly (as opposed to within a factor of $1 + \delta$), and (3) there is no way to "round" a real number to an integer, and so convert the problem to the LEM or SEM models. We note that in [6], (1) integers are available, (2) the input is rounded, and (3) there is no way to "round" to an integer. Changes to these model assumptions will affect the classes of problems we can solve. For example, if we (quite reasonably) were to permit exact integers as input, then we could CAE expressions like $x - 1$, and otherwise presumably not. If we went further and permitted exact rational numbers, then we could also CAE $9x^2 - 1 = 9(x - \frac{1}{3})(x + \frac{1}{3})$. Allowing algebraic numbers would make $x^2 - 2 = (x - \sqrt{2})(x + \sqrt{2})$ CAE.

If inputs were not given exactly, but rather first multiplied by a factor $1 + \delta$, then we could no longer accurately compute $x \pm y$ where x and y are inputs, eliminating Cauchy matrices and most others. But the problems we could solve with exact inputs in the TM still have an attractive property with inexact inputs: Small relative changes in the inputs cause only a small relative change in the outputs, independent of their magnitudes. The output relative errors may be larger than the input relative error by a factor called a *relative condition number* κ_{rel}, which is at most a polynomial function of $\max(1/\text{rel\_gap}(x_i, \pm x_j))$. Here $\text{rel\_gap}(x_i, \pm x_j) = |x_i \mp x_j|/(|x_i| + |x_j|)$ is the *relative gap* between inputs x_i and $\pm x_j$, and the maximum is taken over all expressions $x_i \mp x_j$ where appearing in $f = \prod_m f_m$. So if all the input differ in several of their leading digits, all the leading digits of the outputs are determined accurately. We note that κ_{rel} can be large, depending on f and $\mathcal{D}$, but it can only be unbounded when a relative gap goes to zero.

If a problem has this attractive property, we say that it possesses a relative perturbation theory. In practical situations, where only a few leading digits of the inputs x_i are known, this property justifies the use of algorithms that try to compute the output as accurately as we do. We state a conjecture very much like the last one about when a relative perturbation theory exists.

Conjecture 2 *Let f and $\mathcal{D}$ be as in the last conjecture. Then condition (A) is a necessary and sufficient condition for f to have a relative perturbation theory.*

6. CAE in the long and short exponent models

Now we consider standard Turing machines, where input floating point numbers $x = f \cdot 2^e$ are stored as the pair of integers (f, e), so the size of x is $\text{size}(x) = \#\text{bits}(f) + \#\text{bits}(e)$. We distinguish two cases, the Long Exponent Model (LEM) where f and e may each be arbitrary integers, and the Short Exponent Model (SEM), where the length of e is bounded depending on the length of f. In the simplest case, when $e = 0$ (or lies in a fixed range) then the SEM is equivalent to taking integer inputs, where the complexity of problems is well understood. This is more generally the case if $\#\text{bits}(e)$ grows no faster than a polynomial function of $\#\text{bits}(f)$.

In particular it is possible to CAE the determinant of an integer (or SEM) matrix each of whose entries is an independent floating point number [5]. This is not possible as far as we know in the LEM, which accounts for a large complexity gap between the two models.

We start by illustrating some differences between the LEM and SEM, and then describe the class of problems that we can CAE in the LEM.

First, consider the number of bits in an expression with LEM inputs can be exponentially larger than the number of bits in the same expression when evaluated with SEM inputs. For example, $\text{size}(x \cdot y) \leq \text{size}(x) + \text{size}(y)$ when x and y are integers, but $\text{size}(x \cdot y) \leq \text{size}(x) \cdot \text{size}(y)$ when x and y are LEM numbers: $(\sum_{i=1}^{n} 2^{e_i}) \cdot (\sum_{i=1}^{n} 2^{e_i'})$ has up to n^2 different bit positions to store, each $2^{e_i + e_j'}$, not $2n$. In other words, LEM arithmetic can encode symbolic algebra, because if e_1 and e_2 have no overlapping bits, then we can recover e_1 and e_2 from the product $2^{e_1} \cdot 2^{e_2} = 2^{e_1 + e_2}$.

Second, the error of many conventional matrix algorithms is typically proportional to the condition number $\kappa(A) = \|A\| \cdot \|A^{-1}\|$. This means that a conventional algorithm run with $O(\log \kappa(A))$ extra bits of precision will compute an accurate answer. It turns out that if $A(x)$ has rational entries in the SEM model, then $\log \kappa(A)$ is at most a polynomial function of the input size, so conventional algorithms run in high precision will CAE the answer. However $\log \kappa(A)$ for LEM matrices can be exponentially larger, so this approach does not work. The simplest example is $\log \kappa(\text{diag}(1, 2^e)) = e = 2^{\#\text{bits}(e)}$. On the other hand $\log \log \kappa(A(x))$ is a lower bound on the complexity of any algorithm, because this is a lower bound on the number of exponent bits in the answer. One can show that $\log \log \kappa(A(x))$ grows at most polynomially large in the size of the input.

Finally, we consider the problem of computing an arbitrary bit in the simple expression $p = \prod_{i=1}^{n}(1 + x_i)$. When the x_i are in the SEM, then p can be computed exactly in polynomial time. However when the x_i are in the LEM, then one can prove that computing an arbitrary bit of p is as hard as computing the permanent, a well-known combinatorially difficult problem. Here is another apparently simple problem not known to even be in NP: testing singularity of a floating point matrix. In the SEM, we can CAE the determinant. But in the LEM, the obvious choice of a "witness" for singularity, a null vector, can have exponentially many bits in it, even if the matrix is just tridiagonal. We conjecture that deciding singularity of an LEM matrix is NP-hard.

So how do we compute efficiently in the LEM? The idea is to use *sparse arithmetic*, or to represent only the nonzero bits in the number. (A long string of 1s can be represented as the difference of two powers of 2 and similarly compressed). In contrast, in the SEM one uses *dense arithmetic*, storing all fraction bits of a number. For example, in sparse arithmetic $2^e + 1$ takes $O(\log e)$ bits to store in sparse arithmetic, but e bits in dense arithmetic. This idea is exploited in practical floating point computation, where extra precise numbers are stored as arrays of conventional floating point numbers, with possibly widely different exponents [17].

Now we describe the class of rational functions that we can CAE in the LEM. We say the rational function $r(x)$ is in factored form if $r(x) = \sum_{i=1}^{n} p_i(x_1, ..., x_k)^{e_i}$, where each e_i is an integer, and $p_i(x_1, ..., x_k)$ is written as an explicit sum of nonzero monomials. We say $\text{size}(r)$ is the number of bits needed to represent it in factored form. Then by (1) computing each monomial in each p_i exactly, (2) computing the leading bits of their sum p_i using sparse arithmetic (the cost is basically sorting the bits), and (3) computing the leading bits of the product of the $p_i^{e_i}$ by conventional rounded multiplication or division, one can evaluate $r(x)$ accurately in time a polynomial in size(r) and size(x). In other words, the class of rational expression that we can CAE are those that we can express in factored form in polynomial space.

Now we consider matrix computations. It follows from the last paragraph that if each minor $r(x)$ of $A(x)$ can be written in factored form of a size polynomial in the size of $A(x)$, then we can CAE all the matrix computations that depend on minors. So the question is which matrix classes $A(x)$ have all their minors (or just the ones needed for a particular matrix factorization) expressible in a factored form no more than polynomially larger than the size of $A(x)$. The obvious way to write $r(x)$, with the Laplace expansion, is clearly exponentially larger than $A(x)$, so it is only specially structured $A(x)$ that will work.

All the matrices that we could CAE in the TM are also possible in the LEM. The most obvious classes of $A(x)$ that we can CAE in the LEM that were impossible in the TM are gotten by replacing all the indeterminates in the TM examples by arbitrary rational expressions of polynomial size. For example, the entries of an M-matrix can be polynomial-sized rational expressions in other quantities. Another class are Green's matrices (inverses of tridiagonals), which can be thought of as discretized integral operators, with entries written as $A_{ij} = x_i \cdot y_j$.

The obvious question is whether A each of whose entries is an independent number in the LEM falls in this class. We conjecture that it does not, as mentioned before.

7. Conclusions and open problems

Our goal has been to identify rational expressions (or matrices) that we can evaluate accurately (or on which we can perform accurate matrix computations), in polynomial time. Accurately means that we want to get a relative error less than 1, and polynomial time means in a time bounded by a polynomial function of the input size.

We have defined three reasonable models of arithmetic, the Traditional Model (TM), the Long Exponent Model (LEM) and the Short Exponent Model (SEM), and tried to identify the classes of problems that can or cannot be computed accurately and efficiently for each model. The TM can be used as a model to do proofs that also hold in the implementable LEM and SEM, but since it ignores the structure of floating point numbers as stored in the computer, it is strictly weaker than either the LEM or SEM. In other words, there are problems (like adding $x + y + z$) that are provably impossible in the TM but straightforward in the other two models.

We also believe that the LEM is strictly weaker than the SEM, in the sense that there appear to be computations (like computing the determinant of a general or even tridiagonal, matrix) that are possible in polynomial time in the SEM but not in the LEM. In the SEM, essentially all problems that can be written down in polynomial space can be solved in polynomial time. For the LEM, only expressions that can be written in *factored form* in polynomial space can be computed efficiently in polynomial time.

A number of open problems and conjectures were mentioned in the paper. We mention just one additional one here: What can be said about the nonsymmetric eigenvalue problem? In other words, what matrix properties, perhaps related to minors, guarantee that all eigenvalues of a nonsymmetric matrix can be computed accurately?

Acknowledgements The author acknowledges Benjamin Diament, Zlatko Drmač, Stan Eisenstat, Ming Gu, William Kahan, Ivan Slapničar, Kresimir Veselič, and especially Plamen Koev for their collaboration over many years in developing this material.

References

[1] J. Barlow and J. Demmel. Computing accurate eigensystems of scaled diagonally dominant matrices. *SIAM J. Num. Anal.*, 27(3):762–791, June 1990.

[2] J. Barlow, B. Parlett, and K. Veselic, editors. *Proceedings of the International Workshop on Accurate Solution of Eigenvalue Problems*, volume 309 of *Linear Algebra and its Applications*. Elsevier, 2001.

[3] L. Blum, F. Cucker, M. Shub, and S. Smale. *Complexity and Real Computation.* Springer, 1997.

[4] L. Blum, M. Shub, and S. Smale. On a theory of computation and complexity over the real numbers: NP-completeness, recursive functions and universal machines. *AMS Bulletin (New Series)*, 21(1):1–46, July 1989.

[5] K. Clarkson. Safe and effective determinant evaluation. In *33rd Annual Symp. on Foundations of Comp. Sci.*, 387–395, 1992.

[6] F. Cucker and S. Smale. Complexity estimates depending on condition and roundoff error. *J. ACM*, 46(1):113–184, Jan 1999.

[7] J. Demmel, Accurate SVDs of structured matrices. *SIAM J. Mat. Anal. Appl.*, 21(2):562–580, 1999.

[8] J. Demmel, M. Gu, S. Eisenstat, I. Slapničar, K. Veselić, and Z. Drmač. Computing the singular value decomposition with high relative accuracy. *Lin. Alg. Appl.*, 299(1–3):21–80, 1999.

[9] J. Demmel and W. Kahan. Accurate singular values of bidiagonal matrices. *SIAM J. Sci. Stat. Comput.*, 11(5):873–912, September 1990.

[10] J. Demmel and P. Koev. Necessary and sufficient conditions for accurate and efficient singular value decompositions of structured matrices. In V. Olshevsky, editor, *Special Issue on Structured Matrices in Mathematics, Computer Science and Engineering*, volume 281 of *Contemporary Mathematics*, 117–145, AMS, 2001.

[11] J. Demmel and K. Veselić. Jacobi's method is more accurate than QR. *SIAM J. Mat. Anal. Appl.*, 13(4):1204–1246, 1992.

[12] I. S. Dhillon. *A New $O(n^2)$ Algorithm for the Symmetric Tridiagonal Eigenvalue/Eigenvector Problem.* PhD thesis, University of California, Berkeley, California, May 1997.

[13] N. J. Higham. *Accuracy and Stability of Numerical Algorithms.* SIAM, Philadelphia, PA, 1996.

[14] P. Koev. *Accurate and efficient computations with structured matrices.* PhD

thesis, University of California, Berkeley, California, May 2002.

[15] I. G. MacDonald. *Symmetric functions and Hall polynomials.* Oxford University Press, 2nd edition, 1995.

[16] M. Pour-El and J. Richards. *Computability in Analysis and Physics.* Springer-Verlag, 1989.

[17] D. Priest. Algorithms for arbitrary precision floating point arithmetic. In P. Kornerup and D. Matula, editors, *Proceedings of the 10th Symposium on Computer Arithmetic*, 132–145, Grenoble, France, June 26-28, 1991. IEEE Computer Society Press.

[18] Reliable Computing (a journal). Kluwer.
www.cs.utep.edu/interval-comp/rcjournal.html

[19] S. Smale. Some remarks on the foundations of numerical analysis. *SIAM Review*, 32(2):211–220, June 1990.

ICM 2002 · Vol. III · 707–716

Computational Modeling of Microstructure*

Mitchell Luskin†

Abstract

Many materials such as martensitic or ferromagnetic crystals are observed to be in metastable states exhibiting a fine-scale, structured spatial oscillation called microstructure; and hysteresis is observed as the temperature, boundary forces, or external magnetic field changes. We have developed a numerical analysis of microstructure and used this theory to construct numerical methods that have been used to compute approximations to the deformation of crystals with microstructure.

2000 Mathematics Subject Classification: 49J45, 65N15, 65N30, 74N10, 74N15, 74N30.
Keywords and Phrases: Microstructure, Martensite, Phase transformation.

1. Introduction

Martensitic crystals are observed to be in metastable states that can be modeled by local minima of the energy [1, 2, 11, 17, 19, 25, 33, 36]

$$\mathcal{E}(y) = \int_\Omega \phi(\nabla y(x), \theta(x))\, dx + \text{interfacial energy} + \text{loading energy}, \qquad (1.1)$$

where $\Omega \subset \mathbb{R}^3$ is the reference configuration of the crystal, $y(x) : \Omega \to \mathbb{R}^3$ is the deformation that may be constrained on the boundary $\partial\Omega$, and $\theta(x) : \Omega \to \mathbb{R}$ is the temperature. The frame-indifferent elastic energy density $\phi(F, \theta) : \mathbb{R}^{3\times 3} \times \mathbb{R} \to \mathbb{R}$ is minimized at high temperature $\theta \geq \theta_T$ on SO(3) and at low temperature $\theta \leq \theta_T$ on the martensitic variants $\mathcal{U} = \mathrm{SO}(3)U_1 \cup \cdots \cup \mathrm{SO}(3)U_N$ where the $U_i \subset \mathbb{R}^{3\times 3}$ are symmetry-related transformation strains satisfying

$$\{\, R_i^T U_1 R_i : R_i \in \mathcal{G} \,\} = \{\, U_1, \ldots, U_N \,\}$$

*This work was supported in part by AFOSR F49620-98-1-0433, by NSF DMS-0074043, by ARO DAAG55-98-1-0335, by the Caltech CIMMS, and by the Minnesota Supercomputer Institute.
†School of Mathematics, University of Minnesota, Minneapolis, Minnesota 55455, USA. E-mail: luskin@math.umn.edu

for the symmetry group $\mathcal{G}$ of the high temperature (austenitic) phase. The loading energy above results from applied boundary forces.

Microstructure occurs when the deformation gradient oscillates in space among the $SO(3)U_i$ to enable the deformation to attain a lower energy than could be attained by a more homogeneous state [1, 25]. The simplest microstructure is a laminate in which the deformation gradient oscillates between $R_iU_i \in SO(3)U_i$ and $R_jU_j \in SO(3)U_j$ for $i \neq j$ in parallel layers of fine scale, but more complex microstructure is observed in nature and is predicted by the theory [2, 25].

We have developed numerical methods for the computation of microstructure in martensitic and ferromagnetic crystals and validated these methods by the development of a numerical analysis of microstructure [4, 6, 12, 14, 16, 22, 25–28]. Related results are given in [9, 10, 15, 21, 24, 31, 32, 34]. For martensitic crystals, we have given error estimates for stable quantities such as nonlinear integrals $\int_\Omega f(x, \nabla y(x))\, dx$ for smooth functions $f(x, F) : \Omega \times \mathbb{R}^{3 \times 3} \to \mathbb{R}$ and for the local volume fractions (Young measure) of the variants $SO(3)U_i$ even though pointwise values of the deformation gradient are not stable under mesh refinement.

To model the evolution of metastable states, we have developed a computational model that nucleates the first order phase change since otherwise the crystal would remain stuck in local minima of the energy as the temperature or boundary forces are varied [8]. Our finite element model for the quasi-static evolution of the martensitic phase transformation in a thin film nucleates regions of the high temperature phase during heating and regions of the low temperature phase during cooling.

Graphical images for the computations of microstructure and phase transformation described in this paper can be found at http://www.math.umn.edu/~luskin and in the cited references. A more extensive description of microstructure and its computation can be found at the above website as well as in the selected references at the end of this paper.

2. Numerical analysis of microstructure

Martensitic crystals typically exist in metastable states for time-scales of technological interest. Many important analytic results have been obtained for mathematical models for martensitic crystals, especially for energy-minimizing deformations with microstructure [1, 2, 11, 20, 30, 35]. These results and concepts for energy-minimizing deformations should also have a role in the analysis of metastability [18, 29]. Similarly, we have developed a numerical analysis of microstructure [4, 6, 12, 14, 16, 22, 25–28] for which results have been obtained primarily for the approximation of energy-minimizing deformations that we think also give insight and some validation for the investigation of metastability by computational methods.

We give here a summary of the numerical analysis of martensitic microstructure that we have developed for temperatures $\theta < \theta_T$ for which the energy density $\phi(F, \theta)$ is minimized on the martensitic variants $\mathcal{U} = SO(3)U_1 \cup \cdots \cup SO(3)U_N$.

We assume that the energy density $\phi(F, \theta)$ is continuous and satisfies near the minimizing deformation gradients $\mathcal{U}$ the quadratic growth condition given by

$$\phi(F,\theta) \geq \mu \, \|F - \pi(F)\|^2 \qquad \text{for all } F \in \mathbb{R}^{3\times 3}, \tag{2.1}$$

where $\mu > 0$ is a constant and $\pi : \mathbb{R}^{3\times 3} \to \mathcal{U}$ is a projection satisfying

$$\|F - \pi(F)\| = \min_{G \in \mathcal{U}} \|F - G\| \qquad \text{for all } F \in \mathbb{R}^{3\times 3}.$$

We also assume that the energy density $\phi(F,\theta)$ satisfies the growth condition for large F given by

$$\phi(F,\theta) \geq C_1 \|F\|^p - C_0 \qquad \text{for all } F \in \mathbb{R}^{3\times 3},$$

where C_0 and C_1 are positive constants independent of $F \in \mathbb{R}^{3\times 3}$ where $p > 3$ to ensure that deformations with finite energy are uniformly continuous.

We can then denote the set of deformations of finite energy by

$$W^\phi = \{\, y \in C(\bar{\Omega}; \mathbb{R}^3) : \int_\Omega \phi(\nabla y(x), \theta) \, dx < \infty \,\},$$

and we can define the set $\mathcal{A}$ of admissible deformations to be

$$\mathcal{A} = \{\, y \in W^\phi : y(x) = y_0(x) \text{ for all } x \in \partial\Omega \,\}. \tag{2.2}$$

Since we assume that the set of admissible deformations $\mathcal{A}$ is constrained on the entire boundary $\partial\Omega$, we can neglect the loading energy in (1.1). We will also set the interfacial energy to be zero in this section so as to consider the idealized model for which the length scale of the microstructure is infinitesimally small. For the theorems below, we assume boundary conditions compatible with a simple laminate mixing QU_i for $Q \in \mathrm{SO}(3)$ with volume fraction λ and U_j with volume fraction $1-\lambda$,

$$y_0(x) = [\lambda Q U_i + (1-\lambda) U_j] \, x \qquad \text{for all } x \in \Omega,$$

where for $a \in \mathbb{R}^3$ and $n \in \mathbb{R}^3$, with a, $n \neq 0$, we have the interface equation [1, 19, 25]

$$QU_i = U_j + a \otimes n.$$

We consider the finite element approximation of the variational problem

$$\inf_{y \in \mathcal{A}} \mathcal{E}(y)$$

given by

$$\inf_{y_h \in \mathcal{A}_h} \mathcal{E}(y_h)$$

where $\mathcal{A}_h$ is a finite-dimensional subspace of $\mathcal{A}$ defined for $h \in (0, h_0]$ for some $h_0 > 0$. The following approximation theorem for the energy has been proven for the P_k or Q_k type conforming finite elements on quasi-regular meshes, in particular for the P_1 linear elements defined on tetrahedra and the Q_1 trilinear elements defined on rectangular parallelepipeds [4, 10, 22, 25–27].

Theorem 2.1. *For each $h \in (0, h_0]$, there exists $y_h \in \mathcal{A}_h$ such that*

$$\mathcal{E}(y_h) = \min_{z_h \in \mathcal{A}_h} \mathcal{E}(z_h) \leq Ch^{1/2}. \tag{2.3}$$

We next define the volume fraction that an admissible deformation $y \in \mathcal{A}$ is in the k-th variant $\mathrm{SO}(3)U_k$ for $k \in \{1, \ldots, N\}$ by

$$\tau_k(y) = \frac{\operatorname{meas} \Omega_k(y)}{\operatorname{meas} \Omega}$$

where

$$\Omega_k(y) = \{\, x \in \Omega : \pi(\nabla y(x)) \in \mathrm{SO}(3)U_k \,\}.$$

The following stability theory was first proven for the orthorhombic to monoclinic transformation ($N = 2$) [26] and then for the cubic to tetragonal transformation ($N = 3$) [22]. The analysis of stability is more difficult for larger N since the additional wells give the crystal more freedom to deform without the cost of additional energy. In fact, for the tetragonal to monoclinic transformation ($N = 4$) [6], the orthorhombic to triclinic transformation ($N = 4$) [16], and the cubic to orthorhombic transformation ($N = 6$) [4] we have shown that there are special lattice constants for which the laminated microstructure is not stable. Error estimates are obtained by substituting the approximation result (2.3) in the following stability results.

In each case for which we have proven the approximation of the microstructure to be stable, we have derived the following basic stability estimate for the approximation of a simple laminate mixing QU_i and U_j which bounds the volume fraction that $y \in \mathcal{A}$ is in the variants $k \neq i, j$

$$\tau_k(y) \leq C \left(\mathcal{E}(y)^{\frac{1}{2}} + \mathcal{E}(y) \right) \qquad \text{for all } k \neq i, j \text{ and } y \in \mathcal{A}. \tag{2.4}$$

For the theorems that follow, we shall assume that the lattice parameters are such that the estimate (2.4) holds.

The following theorem gives estimates for the strong convergence of the projection of the deformation gradient parallel to the laminates (the projection of the deformation gradient transverse to the laminates does not converge strongly [25]), the strong convergence of the deformation, and the weak convergence of the deformation gradient.

Theorem 2.2. *(1) For any $w \in \mathbb{R}^3$ such that $w \cdot n = 0$ and $|w| = 1$, we have the estimate for the strong convergence of the projection of the deformation gradient*

$$\int_{\Omega} |\left(\nabla y(x) - \nabla y_0(x)\right) w|^2 \, dx \leq C \left(\mathcal{E}(y) + \mathcal{E}(y)^{\frac{1}{2}} \right) \qquad \text{for all } y \in \mathcal{A}.$$

(2) We have the estimate for the strong convergence of the deformation

$$\int_{\Omega} |y(x) - y_0(x)|^2 \, dx \leq C \left(\mathcal{E}(y) + \mathcal{E}(y)^{\frac{1}{2}} \right) \qquad \text{for all } y \in \mathcal{A}.$$

(3) For any Lipshitz domain $\omega \subset \Omega$, there exists a constant $C = C(\omega) > 0$ such that we have the estimate for the weak convergence of the deformation gradient

$$\left\| \int_\omega (\nabla y(x) - \nabla y_0(x))\, dx \right\| \leq C \left(\mathcal{E}(y)^{\frac{1}{8}} + \mathcal{E}(y)^{\frac{1}{2}} \right) \qquad \text{for all } y \in \mathcal{A}.$$

For fixed i, j with $i \neq j$ we define a projection operator $\pi_{ij} : \mathbb{R}^{3 \times 3} \to \mathrm{SO}(3)U_i \cup \mathrm{SO}(3)U_j$ by

$$\|F - \pi_{ij}(F)\| = \{\|F - G\| : G \in \mathrm{SO}(3)U_i \cup \mathrm{SO}(3)U_j\} \qquad \text{for all } F \in \mathbb{R}^{3 \times 3},$$

and the operators $\Theta : \mathbb{R}^{3 \times 3} \to SO(3)$ and $\Pi : \mathbb{R}^{3 \times 3} \to \{QU_i, U_j\}$ by the unique decomposition

$$\pi_{ij}(F) = \Theta(F)\Pi(F) \qquad \text{for all } F \in \mathbb{R}^{3 \times 3}.$$

The next theorem shows that the deformation gradients of energy-minimizing sequences must oscillate between QU_i and U_j.

Theorem 2.3. *We have for all $y \in A$ that*

$$\int_\Omega \|\nabla y(x) - \Pi(\nabla y(x))\|^2 \, dx \leq C \left(\mathcal{E}(y) + \mathcal{E}(y)^{\frac{1}{2}} \right).$$

We now present an estimate for the local volume fraction that a deformation $y \in \mathcal{A}$ is near QU_i or U_j. To describe this, we define the sets

$$\omega_\rho^i(y) = \{\, x \in \omega : \Pi(\nabla y(x)) = QU_i \text{ and } \|\nabla y(x) - QU_i\| \leq \rho \,\},$$

$$\omega_\rho^j(y) = \{\, x \in \omega : \Pi(\nabla y(x)) = U_j \text{ and } \|\nabla y(x) - U_j\| \leq \rho \,\},$$

for any subset $\omega \in \Omega$, $\rho > 0$, and $y \in \mathcal{A}$. The next theorem demonstrates that the deformation gradients of energy-minimizing sequences must oscillate with local volume fraction λ near QU_i and local volume fraction $1 - \lambda$ near U_j.

Theorem 2.4. *For any Lipshitz domain $\omega \subset \Omega$ and for any $\rho > 0$, there exists a constant $C = C(\omega, \rho) > 0$ such that for all $y \in \mathcal{A}$*

$$\left| \frac{\mathrm{meas}\, \omega_\rho^i(y)}{\mathrm{meas}\, \omega} - \lambda \right| + \left| \frac{\mathrm{meas}\, \omega_\rho^j(y)}{\mathrm{meas}\, \omega} - (1 - \lambda) \right| \leq C \left(\mathcal{E}(y)^{\frac{1}{8}} + \mathcal{E}(y)^{\frac{1}{2}} \right).$$

We next give an estimate for the weak stability of nonlinear functions of deformation gradients.

Theorem 2.5. *We have for all $f : \Omega \times \mathbb{R}^{3 \times 3} \to \mathbb{R}$ and $y \in \mathcal{A}$ that*

$$\int_\Omega \{f(x, \nabla y(x)) - [\lambda f(x, QU_i) + (1 - \lambda)f(x, U_j)]\}\, dx \leq C\|f\|_\mathcal{V} \left[\mathcal{E}(y)^{\frac{1}{4}} + \mathcal{E}(y)^{\frac{1}{2}} \right]$$

where

$$\|f\|_\mathcal{V}^2 = \int_\Omega \left\{ (\mathrm{ess\, sup}\, \|\nabla_F f(x, F)\|)^2 + |\nabla z_f(x)n|^2 + z_f(x)^2 \right\}\, dx < \infty$$

with $z_f : \Omega \to \mathbb{R}$ defined by

$$z_f(x) = f(x, QU_i) - f(x, U_j) \qquad \text{for all } x \in \Omega.$$

3. A computational model for martensitic phase transformation

We have developed a computational model for the quasi-static evolution of the martensitic phase transformation of a single crystal thin film [8]. Our thin film model [7] includes surface energy, as well as sharp phase boundaries with finite energy. The model also includes the nucleation of regions of the high temperature phase (austenite) as the film is heated through the transformation temperature and nucleation of regions of the low temperature phase (martensite) as the film is cooled. The nucleation step in our algorithm is needed since the film would otherwise not transform.

For our total-variation surface energy model, the bulk energy for a film of thickness $h > 0$ with reference configuration $\Omega_h \equiv \Omega \times (-h/2, h/2)$, where $\Omega \subset \mathbb{R}^2$ is a domain with a Lipschitz continuous boundary $\partial\Omega$, is given by the sum of the surface energy and the elastic energy

$$\kappa \int_{\Omega_h} |D(\nabla u)| + \int_{\Omega_h} \phi(\nabla u, \theta) \, dx, \tag{3.1}$$

where $\int_{\Omega_h} |D(\nabla u)|$ is the total variation of the deformation gradient [7] and κ is a small positive constant.

We have shown in [7] that energy-minimizing deformations u of the bulk energy (3.1) are asymptotically of the form

$$u(x_1, x_2, x_3) = y(x_1, x_2) + b(x_1, x_2)x_3 + \mathrm{o}(x_3^2) \text{ for } (x_1, x_2) \in \Omega, \ x_3 \in (-h/2, h/2),$$

(which is similar to that found for a diffuse interface model [3]) where (y, b) minimizes the thin film energy

$$\mathcal{E}(y, b, \theta) = \kappa \left(\int_\Omega |D(\nabla y|b|b)| + \sqrt{2} \int_{\partial\Omega} |b - b_0| \right) + \int_\Omega \phi(\nabla y|b, \theta) \, dx \tag{3.2}$$

over all deformations of finite energy such that $y = y_0$ on $\partial\Omega$. The map b describes the deformation of the cross-section relative to the film [3]. We denote by $(\nabla y|b) \in \mathbb{R}^{3 \times 3}$ the matrix whose first two columns are given by the columns of ∇y and the last column by b. In the above equation, $\int_\Omega |D(\nabla y|b|b)|$ is the total variation of the vector valued function $(\nabla y|b|b) : \Omega \to \mathbb{R}^{3 \times 4}$.

We describe our finite element approximation of (3.2) by letting the elements of a triangulation τ of Ω be denoted by K and the inter-element edges by e. We denote the internal edges by $e \subset \Omega$ and the boundary edges by $e \subset \partial\Omega$. We define the jump of a function ψ across an internal edge $e \subset \Omega$ shared by two elements $K_1, K_2 \in \tau$ to be

$$[\![\psi]\!]_e = \psi_{e, K_1} - \psi_{e, K_2},$$

where ψ_{e, K_i} denotes the trace on e of $\psi|_{K_i}$, and we define $\psi|_e$ to be the trace on e for a boundary edge $e \subset \partial\Omega$. Next, we denote by $\mathcal{P}_1(\tau)$ the space of continuous, piecewise linear functions on Ω which are linear on each $K \in \tau$ and by $\mathcal{P}_0(\tau)$ the space of piecewise constant functions on Ω which are constant on each $K \in \tau$.

Finally, for deformations $(y, b) \in \mathcal{P}_1(\tau) \times \mathcal{P}_0(\tau)$ and temperature fields $\tilde{\theta} \in \mathcal{P}_0(\tau)$, the energy (3.2) is well-defined and we have that

$$\kappa \left[\int_\Omega |D(\nabla y|b|b)| + \sqrt{2} \int_{\partial\Omega} |b - b_0| \right] + \int_\Omega \phi(\nabla y|b, \tilde{\theta}) \, dx$$

$$= \kappa \left(\sum_{e \subset \Omega} \left| [\![(\nabla y|b|b)]\!]_e \right| |e| + \sqrt{2} \sum_{e \subset \partial\Omega} \left| b|_e - b_0|_e \right| |e| \right) + \sum_{K \in \tau} \phi\big((\nabla y|b, \tilde{\theta})|_K\big) |K|,$$

where $|\cdot|$ denotes the euclidean vector norm, $|e|$ denotes the length of the edge e, $|K|$ is the area of the element K, and

$$\left| [\![(\nabla y|b|b)]\!]_e \right| = \left(\left| [\![\nabla y]\!]_e \right|^2 + 2 \left| [\![b]\!]_e \right|^2 \right)^{1/2}.$$

The above term is not differentiable everywhere, so we have regularized it in our numerical simulations.

Since martensitic alloys are known to transform on a fast time scale, we model the transformation of the film from martensite to austenite during heating by assuming that the film reaches an elastic equilibrium on a faster time scale than the evolution of the temperature, so the temperature $\tilde{\theta}(x, t)$ can be obtained from a time-dependent model for thermal evolution [8]. To compute the evolution of the deformation, we partition the time interval $[0, T]$ for $T > 0$ by $0 = t_0 < t_1 < \cdots < t_{L-1} < t_L = T$ and then obtain the solution $(y(t_\ell), b(t_\ell)) \in \mathcal{A}_\tau$ for $\ell = 0, \ldots, L$ by computing a local minimum for the energy $\mathcal{E}(v, c, \theta(t_\ell))$ with respect to the space of approximate admissible deformations

$$\mathcal{A}_\tau = \{(v, c) \in \mathcal{P}_1(\tau) \times \mathcal{P}_0(\tau) : v = y_0 \text{ on } \partial\Omega\}. \tag{3.3}$$

Since the martensitic transformation strains $\mathcal{U} \subset \mathbb{R}^{3 \times 3}$ are local minimizers of the energy density $\phi(F, \theta)$ for all θ near θ_T, a deformation that is in the martensitic phase will continue to be a local minimum for the bulk energy $\mathcal{E}(v, c, \theta(t))$ for $\theta > \theta_T$. Hence, our computational model will not simulate a transforming film if we compute $(y(t_\ell), b(t_\ell)) \in \mathcal{A}_\tau$ by using an energy-decreasing algorithm with the initial state for the iteration at t_ℓ given by the deformation at $t_{\ell-1}$, that is, if $(y^{[0]}(t_\ell), b^{[0]}(t_\ell)) = (y(t_{\ell-1}), b(t_{\ell-1}))$. We have thus developed and utilized an algorithm to nucleate regions of austenite into $(y(t_{\ell-1}), b(t_{\ell-1})) \in \mathcal{A}_\tau$ to obtain an initial iterate $(y^{[0]}(t_\ell), b^{[0]}(t_\ell)) \in \mathcal{A}_\tau$ for the computation of $(y(t_\ell), b(t_\ell)) \in \mathcal{A}_\tau$.

We used an "equilibrium distribution" function, $P(\theta)$, to determine the probability for which the crystal will be in the austenitic phase at temperature θ and we assume that an equilibrium distribution has been reached during the time between $t_{\ell-1}$ and t_ℓ. The distribution function $P(\theta)$ has the property that $0 < P(\theta) < 1$ and

$$P(\theta) \to 0 \text{ as } \theta \to -\infty \quad \text{and} \quad P(\theta) \to 1 \text{ as } \theta \to \infty.$$

At each time t_ℓ, we first compute a pseudo-random number $\sigma(K, \ell) \in (0, 1)$ on every triangle $K \in \tau$, and we then compute $(y^{[0]}(t_\ell), b^{[0]}(t_\ell)) \in \mathcal{A}_\tau$ by (x_K denotes the barycenter of K):

1. If $\sigma(K, \ell) \leq P\left(\theta(x_K, t_\ell)\right)$ and $\left(\nabla y(x_K, t_{\ell-1}) | b(x_K, t_{\ell-1}), \theta(x_K, t_\ell)\right)$ is in austenite, then set

$$\left(y^{[0]}(t_\ell), b^{[0]}(t_\ell)\right) = \left(y(t_{\ell-1}), b(t_{\ell-1})\right) \text{ on } K.$$

2. If $\sigma(K, \ell) \leq P\left(\theta(x_K, t_\ell)\right)$ and $\left(\nabla y(x_K, t_{\ell-1}) | b(x_K, t_{\ell-1}), \theta(x_K, t_\ell)\right)$ is in martensite, then transform to austenite on K.

3. If $\sigma(K, \ell) > P\left(\theta(x_K, t_\ell)\right)$ and $\left(\nabla y(x_K, t_{\ell-1}) | b(x_K, t_{\ell-1}), \theta(x_K, t_\ell)\right)$ is in austenite, then transform to martensite on K.

4. If $\sigma(K, \ell) > P\left(\theta(x_K, t_\ell)\right)$ and $\left(\nabla y(x_K, t_{\ell-1}) | b(x_K, t_{\ell-1}), \theta(x_K, t_\ell)\right)$ is in martensite, then set

$$\left(y^{[0]}(t_\ell), b^{[0]}(t_\ell)\right) = \left(y(t_{\ell-1}), b(t_{\ell-1})\right) \text{ on } K.$$

We have shown in [8] for a thin film of a CuAlNi alloy in the "tent" configuration that we can compute the nucleation above by setting $y^{[0]}(t_\ell) = y(t_{\ell-1}) \in \mathcal{P}_1(\tau)$ and by updating the piecewise constant $b^{[0]}(t_\ell) \in \mathcal{P}_0(\tau)$ by

$$b^{[0]}(x_K, t_\ell) = \frac{y_{,1}(x_K, t_{\ell-1}) \times y_{,2}(x_K, t_{\ell-1})}{|y_{,1}(x_K, t_{\ell-1}) \times y_{,2}(x_K, t_{\ell-1})|} \qquad \text{on } K$$

to nucleate austenite and

$$b^{[0]}(x_K, t_\ell) = \gamma \, \frac{y_{,1}(x_K, t_{\ell-1}) \times y_{,2}(x_K, t_{\ell-1})}{|y_{,1}(x_K, t_{\ell-1}) \times y_{,2}(x_K, t_{\ell-1})|} \qquad \text{on } K$$

to nucleate martensite.

We then compute $\left(y(t_\ell), b(t_\ell)\right) \in \mathcal{A}_\tau$ by the Polak-Ribière conjugate gradient method with initial iterate $\left(y^{[0]}(t_\ell), b^{[0]}(t_\ell)\right) \in \mathcal{A}_\tau$. We have also experimented with several other versions of the above algorithm for the computation of $b^{[0]}(t_\ell)$. For example, the above algorithm can be modified to utilize different probability functions $P(\theta)$ in elements with increasing and decreasing temperature. We can also prohibit the transformation from austenite to martensite in an element in which the temperature is increasing or prohibit the transformation from martensite to austenite in an element for which the temperature is decreasing.

References

[1] J. Ball & R. James, Fine phase mixtures as minimizers of energy *Arch. Rat. Mech. Anal.* 100 (1987), 13–52.

[2] J. Ball & R. James, Proposed experimental tests of a theory of fine microstructure and the two-well problem, *Phil. Trans. R. Soc. Lond. A* 338 (1992), 389–450.

[3] K. Bhattacharya & R. James, A theory of thin films of martensitic materials with applications to microactuators, *J. Mech. Phys. Solids* 47 (1999), 531–576.

[4] K. Bhattacharya, B. Li, & M. Luskin, The simply laminated microstructure in martensitic crystals that undergo a cubic to orthorhombic phase transformation, *Arch. Rat. Mech. Anal.* 149 (1999), 123–154.

[5] P. Bělík, T. Brule, & M. Luskin, On the numerical modeling of deformations of pressurized martensitic thin films, *Math. Model. Numer. Anal.* 35 (2001), 525–548.

[6] P. Bělík & M. Luskin, Stability of microstructure for tetragonal to monoclinic martensitic transformations. *Math. Model. Numer. Anal. 34* (2000), 663–685.

[7] P. Bělík & M. Luskin, A total-variation surface energy model for thin films of martensitic crystals, *Interfaces and Free Boundaries* 4 (2002), 71–88.

[8] P. Bělík & M. Luskin, A computational model for the indentation and phase transformation of a martensitic thin film, *J.Mech. Phys. Solids* 50 (2002) 1789–1815.

[9] C. Carstensen & P. Plecháč, Numerical analysis of compatible phase transitions in elastic solids, *SIAM J. Numer. Anal.* 37 (2000), 2061–2081.

[10] M. Chipot, C. Collins, & D. Kinderlehrer, Numerical analysis of oscillations in multiple well problems, *Numer. Math.* 70 (1995), 259–282.

[11] M. Chipot & Kinderlehrer, D. Equilibrium configurations of crystals, *Arch. Rat. Mech. Anal.* 103 (1988), 237–277.

[12] C. Collins, D. Kinderlehrer & M. Luskin, Numerical approximation of the solution of a variational problem with a double well potential, *SIAM J. Numer. Anal.* 28 (1991), 321–332.

[13] C. Collins & M. Luskin, The computation of the austenitic-martensitic phase transition, In *Partial Differential Equations and Continuum Models of Phase Transitions* (New York, 1989), vol. 344 of *Lecture Notes in Physics*, Springer-Verlag, pp. 34–50.

[14] C. Collins & M. Luskin, Optimal order error estimates for the finite element approximation of the solution of a nonconvex variational problem, *Math. Comp.* 57 (1991), 621–637.

[15] G. Dolzmann, Numerical computation of rank-one convex envelopes. *SIAM J. Numer. Anal.* 36 (1999), 1621–1635.

[16] Y. Efendiev & M. Luskin, Stability of microstructures for some martensitic transformations, *Mathematical and Computer Modelling* 34 (2000), 1289–1305.

[17] J. Ericksen, Constitutive theory for some constrained elastic crystals, *J. Solids and Structures* 22 (1986), 951–964.

[18] R. James, Hysteresis in phase transformations, In *Proc. ICIAM-95* (1996), K. Kirchgässner, O. Mahrenholtz, and R. Mennicken, Eds., Akademie Verlag, pp. 135–154.

[19] R. James & K. Hane, Martensitic transformations and shape memory materials, *Acta Materialia* 48 (2000), 197–222.

[20] R. Kohn & S. Müller, Surface energy and microstructure in coherent phase transitions, *Comm. Pure and Appl. Math.* 47 (1994), 405–435.

[21] M. Kružik, Numerical approach to double well problems, *SIAM J. Numer. Anal.* 35 (1998), 1833–1849.

[22] B. Li & M. Luskin, Finite element analysis of microstructure for the cubic to tetragonal transformation, *SIAM J. Numer. Anal.* 35 (1998), 376–392.

[23] B. Li & M. Luskin, Theory and computation for the microstructure near the interface between twinned layers and a pure variant of martensite, *Materials*

Science & Engineering A 273 (1999), 237–240.

[24] Z. Li, Rotational transformation method and some numerical techniques for computing microstructures. *Math. Models Methods Appl. Sci.* 8 (1998), 985–1002.

[25] M. Luskin, On the computation of crystalline microstructure. *Acta Numerica* 5 (1996), 191–258.

[26] M. Luskin, Approximation of a laminated microstructure for a rotationally invariant, double well energy density. *Numer. Math.* 75 (1997), 205–221.

[27] M. Luskin & L. Ma, Analysis of the finite element approximation of microstructure in micromagnetics, *SIAM J. Numer. Anal.* 29 (1992), 320–331.

[28] M. Luskin & L. Ma, Numerical optimization of the micromagnetics energy, In *Mathematics in Smart Materials* (1993), SPIE, pp. 19–29.

[29] A. Mielke, F. Theil, & V. I. Levitas, A variational formulation of rate-independent phase transformations using an extremum principle. *Arch. Rat. Mech. Anal.* 162 (2002), 137–177.

[30] S. Müller, Singular perturbations as a selection criterion for periodic minimizing sequences, *Calc. Var.* 1 (1993), 169–204.

[31] R. Nicolaides & N. Walkington, Strong convergence of numerical solutions to degenerate variational problems, *Math. Comp.* 64 (1995), 117–127.

[32] P. Pedregal, On the numerical analysis of non-convex variational problems, *Numer. Math.* 74 (1996), 325–336.

[33] M. Pitteri & G. Zanzotto, *Continuum models for twinning and phase transitions in crystals*, Chapman and Hall, London, 1996.

[34] T. Roubíček, Numerical approximation of relaxed variational problems, *J. Convex Anal.* 3 (1996), 329–347.

[35] V. Šverák, Rank-one convexity does not imply quasiconvexity, *Proc. Royal Soc. Edinburgh* 120A (1992), 185–189.

[36] L. Truskinovsky & G. Zanzotto, Finite-scale microstructures and metastability in one-dimensional elasticity, *Meccanica* 30 (1995), 557–589.

ICM 2002 · Vol. III · 717–726

Adaptive Finite Element Methods for Partial Differential Equations[*]

R. Rannacher[†]

Abstract

The numerical simulation of complex physical processes requires the use of economical discrete models. This lecture presents a general paradigm of deriving a posteriori error estimates for the Galerkin finite element approximation of nonlinear problems. Employing duality techniques as used in optimal control theory the error in the target quantities is estimated in terms of weighted 'primal' and 'dual' residuals. On the basis of the resulting local error indicators economical meshes can be constructed which are tailored to the particular goal of the computation. The performance of this *Dual Weighted Residual Method* is illustrated for a model situation in computational fluid mechanics: the computation of the drag of a body in a viscous flow, the drag minimization by boundary control and the investigation of the optimal solution's stability.

2000 Mathematics Subject Classification: 65N30, 65N50, 65K10.
Keywords and Phrases: Finite element method, Adaptivity, Partial differential equations, Optimal control, Eigenvalue problems.

1. Introduction

Suppose the goal of a simulation is the computation or optimization of a certain quantity $J(u)$ from the solution u of a continuous model with accuracy TOL, by using the solution u_h of a discrete model of dimension N,

$$\mathcal{A}(u) = 0, \qquad \mathcal{A}_h(u_h) = 0.$$

Then, the goal of adaptivity is the optimal use of computing resources, i.e., minimum work for prescribed accuracy, or maximum accuracy for prescribed work. In order to reach this goal, one uses *a posteriori* error estimates

$$|J(u) - J(u_h)| \approx \eta(u_h) := \sum_{K \in \mathbb{T}_h} \rho_K(u_h)\omega_K,$$

---

[*]The author acknowledges the support by the German Research Association (DFG) through SFB 359 'Reactive Flow, Diffusion and Transport'.

[†]Institute of Applied Mathematics, University of Heidelberg, Im Neuenheimer Feld 293/294, D-69120 Heidelberg, Germany. E-mail: rannacher@iwr.uni-heidelberg.de

in terms of the local residuals $\rho_K(u_h)$ of the computed solution and weights ω_K obtained from the solution of a linearized *dual problem*. In the following, we will describe a general optimal control approach to such error estimates in Galerkin finite element methods. For earlier work on adaptivity, we refer to the survey articles [10], [1] and [7]. The contents of this paper is based on material from [5], [6] and [2], where also references to other recent work can be found.

2. Paradigm of a posteriori error analysis

We develop a general approach to a posteriori error estimation for Galerkin approximations of variational problems. The setting uses as little assumptions as possible. Let X be some function space and $L(\cdot)$ a differentiable functional on X. We are looking for stationary points of $L(\cdot)$ determined by

$$L'(x)(y) = 0 \qquad \forall y \in X,$$

and their Galerkin approximation in finite dimensional subspaces $X_h \subset X$,

$$L'(x_h)(y_h) = 0 \qquad \forall y_h \in X_h.$$

For this situation, we have the following general result:

Proposition 1 *There holds the a posteriori error representation*

$$L(x) - L(x_h) = \tfrac{1}{2} L'(x_h)(x - y_h) + R_h, \tag{2.1}$$

for arbitrary $y_h \in X_h$. *The remainder* R_h *is cubic in* $e := x - x_h$,

$$R_h := \tfrac{1}{2} \int_0^1 L'''(x_h + se)(e, e, e)\, s(s-1)\, \mathrm{ds}.$$

Proof We sketch the rather elementary proof. First, we note that

$$L(x) - L(x_h) = \int_0^1 L'(x_h + se)(e)\, \mathrm{ds}$$
$$- \tfrac{1}{2}\{L'(x_h)(e) + L'(x)(e)\} + \tfrac{1}{2} L'(x_h)(e).$$

Since x_h is a stationary point,

$$L'(x_h)(e) = L'(x_h)(x - y_h) + L'(x_h)(y_h - x_h) = L'(x_h)(x - y_h), \quad y_h \in X_h.$$

Finally, using the error representation of the trapezoidal rule,

$$\int_0^1 f(s)\, \mathrm{ds} - \tfrac{1}{2}\{f(0) + f(1)\} = \tfrac{1}{2} \int_0^1 f''(s) s(s-1)\, \mathrm{ds},$$

completes the proof. Notice that the derivation of the error representation (2.1) does not assume the uniqueness of the stationary points. But the a priori assumption $x_h \to x \ (h \to 0)$ makes this result meaningful.

3. Variational equations

We apply the result of Proposition 1 to the Galerkin approximation of *variational equations* posed in some function space V,

$$a(u)(\psi) = 0 \quad \forall \psi \in V. \tag{3.1}$$

Suppose that some functional output $J(u)$ of the solution u is to be computed using a Galerkin approximation in finite dimensional subspaces $V_h \subset V$,

$$a(u_h)(\psi_h) = 0 \quad \forall \psi_h \in V_h. \tag{3.2}$$

The goal is now to estimate the error $J(u) - J(u_h)$. To this end, we employ a formal Euler-Lagrange approach to embed the present situation into the general framework laid out above. Introducing a 'dual' variable z ('Lagrangian multiplier'), we define the Lagrangian functional $\mathcal{L}(u,z) := J(u) - a(u)(z)$. Then, stationary points $\{u,z\} \in V \times V$ of $\mathcal{L}(\cdot,\cdot)$ are determined by the system

$$\mathcal{L}'(u,z)(\varphi,\psi) = \left\{ \begin{array}{c} J'(u)(\varphi) - a'(u)(\varphi,z) \\ -a(u)(\psi) \end{array} \right\} = 0 \quad \forall \{\varphi,\psi\}.$$

The corresponding Galerkin approximation determines $\{u_h, z_h\} \in V_h \times V_h$ by

$$\mathcal{L}'(u_h,z_h)(\varphi_h,\psi_h) = \left\{ \begin{array}{c} J'(u_h)(\varphi_h) - a'(u_h)(\varphi_h,z_h)\rangle \\ -\langle a(u_h)(\psi_h) \end{array} \right\} = 0 \quad \forall \{\varphi_h,\psi_h\}.$$

Set $x := \{u,z\}$, $x_h := \{u_h, z_h\}$, and $L(x) := \mathcal{L}(u,z)$. Then,

$$J(u) - J(u_h) = L(x) + a(u)(z) - L(x_h) - a(u_h)(z_h).$$

Proposition 2 *With the 'primal' and 'dual' residuals*

$$\rho(u_h)(\cdot) := -a(u_h)(\cdot),$$
$$\rho^*(z_h)(\cdot) := J'(u_h)(\cdot) - a'(u_h)(\cdot,z_h),$$

there holds the error identity

$$J(u) - J(u_h) = \tfrac{1}{2}\rho(u_h)(z-\psi_h) + \tfrac{1}{2}\rho^*(z_h)(u-\varphi_h) + \mathcal{R}_h, \tag{3.3}$$

for arbitrary $\varphi_h, \psi_h \in V_h$. The remainder $\mathcal{R}_h$ is cubic in the primal and dual errors $e^u := u - u_h$ and $e^z := z - z_h$.

The evaluation of the error identity (3.3) requires guesses for primal and dual solutions u and z which are usually generated by post-processing from the approximations u_h and z_h, respectively. The cubic remainder term $\mathcal{R}_h$ is neglected. We emphasize that the solution of the dual problem takes only a 'linear work unit' compared to the solution of the generally nonlinear primal problem.

4.   Optimal control problems

Next, we apply Proposition 1 to the approximation of optimal control problems. Let V be the 'state space' and Q the 'control space' for the optimization problem

$$J(u, q) \to \min! \qquad a(u)(\psi) + b(q, \psi) = 0 \quad \forall \psi \in V. \tag{4.1}$$

Its Galerkin approximation uses subspaces $V_h \times Q_h \subset V \times Q$ as follows:

$$J(u_h, q_h) \to \min! \qquad a(u_h)(\psi_h) + b(q_h, \psi) = 0 \quad \forall \psi_h \in V_h. \tag{4.2}$$

For embedding this situation into our general framework, we again employ the Euler-Lagrange approach introducing the Lagrangian functional $\mathcal{L}(u, q, z) := J(u, q) - A(u)(z) - B(q, z)$. Corresponding stationary points $x := \{u, q, z\} \in X := V \times Q \times V$ are determined by the system ('first-order optimality condition')

$$\left. \begin{cases} J'_u(u, q)(\varphi) - a'(u)(\varphi, z) \\ J'_q(u, q)(\chi) - b(\chi, z) \\ -a(u)(\psi) - b(q, \psi) \end{cases} \right\} = 0 \qquad \forall \{\varphi, \chi, \psi\}. \tag{4.3}$$

The Galerkin approximation detrmines $x_h := \{u_h, q_h, z_h\} \in X_h := V_h \times Q_h \times V_h$ in finite dimensional subspace $V_h \subset V$, $Q_h \subset Q$ by

$$\left. \begin{cases} J'_u(u_h, q_h)(\varphi_h) - a'(u_h)(\varphi_h, z_h) \\ J'_q(u_h, q_h)(\chi_h) - b(\chi_h, z_h) \\ -a(u_h)(\psi_h) - b(q_h, \psi_h) \end{cases} \right\} = 0 \qquad \forall \{\varphi_h, \chi_h, \psi_h\}. \tag{4.4}$$

For estimating the accuracy in this discretization, we propose to use the natural 'cost functional' of the optimization problem, i.e., to estimate the error in terms of the difference $J(u, q) - J(u_h, q_h)$. Then, from Proposition 1, we immediately obtain the following result:

Proposition 3 *With the 'primal', 'dual' and 'control' residuals*

$$\rho^*(z_h)(\cdot) := J'_u(u_h, q_h)(\cdot) - a'(u_h)(\cdot, z_h),$$
$$\rho^q(q_h)(\cdot) := J'_q(u_h, q_h)(\cdot) - b(\cdot, z_h),$$
$$\rho(u_h)(\cdot) := -a(u_h)(\cdot) - b(q_h, \cdot),$$

there holds the a posteriori error representation

$$\begin{aligned} J(u, q) - J(u_h, q_h) = \ &\tfrac{1}{2}\rho^*(z_h)(u - \varphi_h) + \tfrac{1}{2}\rho^q(q_h)(q - \chi_h) \\ &+ \tfrac{1}{2}\rho(u_h)(z - \psi_h) + \mathcal{R}_h, \end{aligned} \tag{4.5}$$

for arbitrary φ_h, $\psi_h \in V_h$ and $\chi_h \in Q_h$. The remainder $\mathcal{R}_h$ is cubic in the errors $e^u := u - u_h$, $e^q := q - q_h$, $e^z := z - z_h$.

We note that error estimation in optimal control problems requires only the use of available information from the computed solution $\{u_h, q_h, z_h\}$, i.e., no extra dual problem has to be solve. This is typical for a situation where the discretization error is measured with respect to the 'generating' functional of the problem, i.e. the Lagrange functional in this case. In the practical solution process the mesh adaptation is nested with an outer Newton iteration leading to a successive 'model enrichment'. The 'optimal' solution $\{u_h^{\mathrm{opt}}, q_h^{\mathrm{opt}}\}$ obtained by the adapted discretization may satisfy the state equation only in a rather week sense. If more 'admissibility' is required, we may solve just the state equation with an better discretization (say on a finer mesh) using the computed optimal control q_h^{opt} as data.

5. Eigenvalue problems

Finally, we apply Proposition 1 to the Galerkin approximation of eigenvalue problems. Consider in a (complex) function space V the generalized eigenvalue problem

$$a(u, \psi) = \lambda \, m(u, \psi) \quad \forall \psi \in V, \qquad \lambda \in \mathbb{C}, \ m(u, u) = 1, \tag{5.1}$$

where the form $a(\cdot, \cdot)$ is linear but not necessarily symmetric, and the eigenvalue form $m(\cdot, \cdot)$ is symmetric and positive semi-definit. The Galerkin approximation is defined in finite dimensional subspaces $V_h \subset V$,

$$a(u_h, \psi_h) = \lambda_h m(u_h, \psi_h) \quad \forall \psi_h \in V_h, \qquad \lambda_h \in \mathbb{C}, \ m(u_h, u_h) = 1. \tag{5.2}$$

We want to control the error in the eigenvalues $\lambda - \lambda_h$. To this end, we embed this situation into the general framework of variational equations by introducing the spaces $\mathcal{V} := V \times \mathbb{C}$ and $\mathcal{V}_h := V_h \times \mathbb{C}$, consisting of elements $U := \{u, \lambda\}$ and $U_h := \{u_h, \lambda_h\}$, and the semi-linear form

$$A(U)(\Psi) := \lambda m(u, \psi) - a(u, \psi) + \bar{\mu}\{m(u, u) - 1\}, \quad \Psi = \{\psi, \mu\} \in \mathcal{V}.$$

Then, the eigenvalue problem (5.1) and its Galerkin approximation (5.2) can be written in the compact form

$$A(U)(\Psi) = 0 \qquad \forall \Psi \in \mathcal{V}, \tag{5.3}$$
$$A(U_h)(\Psi_h) = 0 \qquad \forall \Psi_h \in \mathcal{V}_h. \tag{5.4}$$

The error in this approximation will be estimated with respect to the functional

$$J(\Phi) := \mu \, m(\varphi, \varphi),$$

where $J(U) = \lambda$ since $m(u, u) = 1$. The corresponding continuous and discrete dual solutions $Z = \{z, \pi\} \in \mathcal{V}$ and $Z_h = \{z_h, \pi_h\} \in \mathcal{V}_h$ are determined by the problems

$$A'(U)(\Phi, Z) = J'(U)(\Phi) \qquad \forall \Phi \in \mathcal{V}, \tag{5.5}$$
$$A'(U_h)(\Phi_h, Z_h) = J'(U_h)(\Phi_h) \qquad \forall \Phi_h \in \mathcal{V}_h. \tag{5.6}$$

A straightforward calculation shows that these dual problems are equivalent to the *adjoint eigenvalue problems* associated to (5.1) and (5.2),

$$a(\varphi, z) = \pi\, m(\varphi, z) \qquad \forall \varphi \in V, \qquad m(u, z) = 1, \tag{5.7}$$

$$a(\varphi_h, z_h) = \pi_h\, m(\varphi_h, z_h) \quad \forall \varphi_h \in V_h, \quad m(u_h, z_h) = 1. \tag{5.8}$$

Then, application of Proposition 1 yields the following result:

Proposition 4 *With the 'primal' and 'dual' residuals*

$$\rho(u_h, \lambda_h)(\cdot) := a(u_h, \cdot) - \lambda_h\, m(u_h, \cdot),$$
$$\rho^*(z_h, \pi_h)(\cdot) := a(\cdot, z_h) - \pi_h\, m(\cdot, z_h),$$

there holds the a posteriori error representation

$$\lambda - \lambda_h = \tfrac{1}{2}\rho(u_h, \lambda_h)(z - \psi_h) + \tfrac{1}{2}\rho^*(z_h, \pi_h)(u - \varphi_h) - \mathcal{R}_h, \tag{5.9}$$

for arbitrary $\psi_h, \varphi_h \in V_h$, *with the remainder term*

$$\mathcal{R}_h = \tfrac{1}{2}(\lambda - \lambda_h)\, m(v - v_h, z - z_h).$$

We note that in Proposition 4, no assumption about the multiplicity of the approximated eigenvalue λ has been made. In order to make the error representation (5.9) meaningful, we have to use a priori information about the convergence $\{\lambda_h, v_h\} \to \{\lambda, v\}$ as $h \to 0$. The simultaneous solution of primal and dual eigenvalue problems naturally occurs within an optimal multigrid solver of nonsymmetric eigenvalue problems. Further, error estimates with respect to functionals $J(u)$ of eigenfunctions can be derived following the general paradigm. Finally, in solving *stability eigenvalue problems* $\mathcal{A}'(\hat{u})v = \lambda \mathcal{M} v$, we can include the perturbation of the operator $\mathcal{A}'(\hat{u}_h) \approx \mathcal{A}'(\hat{u})$ in the a posteriori error estimate of the eigenvalues.

6. Application in fluid flow simulation

In order to illustrate the abstract theory developed so far, we present some results for the application of 'residual-driven' mesh adaptation for a model problem in computational fluid mechanics, namely 'channel flow around a cylinder' as shown in the figure below. The *stationary* Navier-Stokes system

$$\mathcal{A}(u) := \left\{ \begin{array}{c} -\nu\Delta v + v\cdot\nabla v + \nabla p \\ \nabla\cdot v \end{array} \right\} = 0$$

determines the pair $u := \{v, p\}$ of velocity vector v and scalar pressure p of a viscous incompressible fluid with viscosity ν and normalized density $\rho \equiv 1$. The physical boundary conditions are $v|_{\Gamma_{\text{rigid}}} = 0$, $v|_{\Gamma_{\text{in}}} = v^{\text{in}}$, and $\nu\partial_n v - np|_{\Gamma_{\text{out}}} = 0$, i.e., the flow is driven by the prescribed parabolic inflow v^{in}. The Reynolds number is $\text{Re} = \frac{\bar{U}^2 D}{\nu} = 20$, such that the flow is stationary.

Let the goal of the simulation be the accurate computation of the effective force in the main flow direction imposed on the cylinder, i.e. the so-called 'drag coefficient',

$$J(u) := c_{\text{drag}} = \frac{2}{\max |v^{\text{in}}|^2 D} \int_S n^T (2\nu\tau - pI)e_1 \, ds,$$

where S is the surface of the cylinder, D its diameter, and $\tau = \frac{1}{2}(\nabla v + \nabla v^T)$ the strain tensor. In practice, one uses a volume-oriented representation of c_{drag}.

Here, we cannot describe the standard variational formulation of the Navier-Stokes problem and its Galerkin finite element discretization in detail but rather refer to the literature; see [9], [6], and the references therein.

In the present situation the primal and dual residuals occuring in the a posteriori error representation (3.3) have the following explicit form:

$$\rho(u_h)(z - z_h) := \sum_{K \in \mathbb{T}_h} \left\{ (R_h, z^v - z_h^v)_K + (r_h, z^v - z_h^v)_{\partial K} + (z^p - z_h^p, \nabla \cdot v_h)_K + \dots \right\},$$

$$\rho^*(z_h)(u - u_h) := \sum_{K \in \mathbb{T}_h} \left\{ (R_h^*, v - v_h)_K + (r_h^*, v - v_h)_{\partial K} + (p - p_h, \nabla \cdot z_h^v)_K + \dots \right\},$$

with the cell and edge residuals defined by

$$R_{h|K} := f + \nu \Delta v_h - v_h \cdot \nabla v_h - \nabla p,$$
$$R_{h|K}^* := j + \nu \Delta z_h^v + v_h \cdot \nabla z_h^v - \nabla v_h^T z_h^v + \nabla \cdot v_h z_h^v - \nabla z_h^p,$$
$$r_{h|\Gamma} := \left\{ \begin{array}{ll} \frac{1}{2}[\nu \partial_n v_h - n p_h], & \text{if } \Gamma \not\subset \partial\Omega \\ -\nu \partial_n v_h + n p_h, & \text{if } \Gamma \subset \Gamma_{\text{out}}, \quad (= 0 \text{ else}) \end{array} \right\},$$
$$r_{h|\Gamma}^* := \left\{ \begin{array}{ll} \frac{1}{2}[\nu \partial_n z_h^v + n \cdot v_h z_h^v - z_h^p n], & \text{if } \Gamma \not\subset \partial\Omega \\ -\nu \partial_n z_h^v - n \cdot v_h z_h^v + z_h^p n, & \text{if } \Gamma \subset \Gamma_{\text{out}}, \quad (= 0 \text{ else}) \end{array} \right\},$$

where $[\dots]$ denotes the jump across edges Γ, and '$\dots$' stands for terms representing errors due to boundary and inflow approximation as well as stabilization.

Practical mesh adaptation on the basis of the a posteriori error estimates proceeds as follows: At first, the error functional may have to be regularized according to $\tilde{J}(u) = J(u) + \mathcal{O}(TOL)$. Then, after having computed the primal approximation u_h, the *linear* discrete dual problem is solved:

$$\langle \mathcal{A}'(u_h)^* z_h, \varphi_h \rangle = \tilde{J}'(u_h)(\varphi_h) \quad \forall \varphi_h \in V_h^*. \tag{6.1}$$

The error estimator is localized, $\eta_\omega = \sum_{K \in \mathbb{T}_h} \eta_K$, and approximation of the weights are computed by patch-wise higher-order interpolation: $(z - z_h)_{|K} \approx (I_{2h}^* z_h - z_h)_{|K}$.

Finally, the current mesh is adapted by 'error balancing' $\eta_K \approx \eta_\omega / \#\{K \in \mathbb{T}_h\}$. In the following, we show some results which have been obtained using mesh adaptation on the basis of the Dual Weighted Residual Method ('DWR method').

6.1. Drag computation (from [3])

The drag is computed on meshes generated by the DWR method and by an 'ad hoc' refinement criterion based on smoothness properties of the computed solution.

Table 1: Results for drag computation on adapted meshes (1%-error in bold face).

		Computation of drag		
L	N	c_{drag}	η_{drag}	I_{eff}
4	984	5.66058	$1.1e{-}1$	0.76
5	**2244**	5.59431	$3.1e{-}2$	0.47
6	4368	5.58980	$1.8e{-}2$	0.58
6	7680	5.58507	$8.0e{-}3$	0.69
	∞	5.57953		

Figure 1: Refined meshes by 'ad hoc' strategy (top) and DWR method (bottom)

6.2. Drag minimization (from [4])

The drag coefficient is to be minimized by imposing a pressure drop at the two outlets Γ_i above and below the cylinder. In this case of 'boundary control' the control form is given by $b(q, \psi) := -(q, n \cdot \psi^v)_{\Gamma_1 \cup \Gamma_2}$.

Table 2: Uniform refinement versus adaptive refinement for $\mathrm{Re} = 40$.

Uniform refinement		Adaptive refinement	
N	J_{drag}	N	J_{drag}
10512	3.31321	1572	3.28625
41504	3.21096	4264	3.16723
164928	3.11800	11146	3.11972

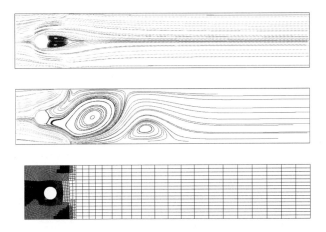

Figure 2: Velocity of the uncontrolled flow (top), controlled flow (middle), corresponding adapted mesh (bottom)

6.3. Stability of optimized flows (from [8])

We want to investigate the stability of the optimized solution $u^{\mathrm{opt}} = \{v^{\mathrm{opt}}, p^{\mathrm{opt}}\}$ by linear stability theory. This is a crucial question since in the present case the optimal solution is obtained by a *stationary* Newton iteration which may converge to physically unstable solutions. In this context, we have to consider the non-symmetric eigenvalue problem for $u := \{v, p\} \in V$ and $\lambda \in \mathbb{C}$:

$$\mathcal{A}'(u^{\mathrm{opt}})u := \left\{ \begin{array}{c} -\nu\Delta v + v^{\mathrm{opt}}{\cdot}\nabla v + v{\cdot}\nabla v^{\mathrm{opt}} + \nabla p \\ \nabla{\cdot}v \end{array} \right\} = \lambda \left\{ \begin{array}{c} v \\ 0 \end{array} \right\}.$$

If the real parts of all eigenvalues are positive, $\mathrm{Re}\,\lambda > 0$, then the (stationary) base flow $\{v^{\mathrm{opt}}, p^{\mathrm{opt}}\}$ is considered as stable (but with respect to possibly only very small perturbations). We find that the optimal solution is at the edge of being unstable.

Figure 3: *Streamlines of real parts of the 'critical' eigenfunction shortly before the Hopf bifurcation and after, depending on the imposed pressure drop*

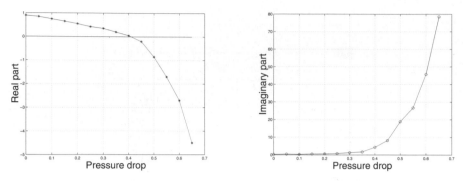

Figure 4: *Real and imaginary parts of the critical eigenvalue as function of the control variable*

References

[1] M. Ainsworth and J. T. Oden, A posteriori error estimation in finite element analysis, *Comput. Methods Appl. Mech. Engrg.*, 142:1–88, 1997.

[2] W. Bangerth and R. Rannacher, Adaptive Methods for Differential Equations, Birkhäuser, Basel, 2002, to appear.

[3] R. Becker, An optimal-control approach to a posteriori error estimation for finite element discretizations of the Navier-Stokes equations, *East-West J. Numer. Math.*, 9:257–274, 2000.

[4] R. Becker, Mesh adaptation for stationary flow control, *J. Math. Fluid Mech.*, 3:317–341, 2001.

[5] R. Becker and R. Rannacher, Weighted a posteriori error control in FE methods, Lecture at ENUMATH-95, Paris, Sept. 18–22, 1995, Preprint 96-01, SFB 359, University of Heidelberg, Proc. *ENUMATH'97* (H. G. Bock *et al.*, eds), 621–637, World Scientific, Singapore, 1998.

[6] R. Becker and R. Rannacher, An optimal control approach to error estimation and mesh adaptation in finite element methods, *Acta Numerica 2000* (A. Iserles, ed.), 1–101, Cambridge University Press, 2001.

[7] K. Eriksson, D. Estep, P. Hansbo, and C. Johnson, Introduction to adaptive methods for differential equations. *Acta Numerica 1995* (A. Iserles, ed.), 105–158, Cambridge University Press, 1995.

[8] V. Heuveline and R. Rannacher, Adaptive finite element discretization of eigenvalue problems in hydrodynamic stability theory, Preprint, SFB 359, Universität Heidelberg, March 2001.

[9] R. Rannacher, *Finite element methods for the incompressible Navier-Stokes equations*, Fundamental Directions in Mathematical Fluid Mechanics (G. P. Galdi, et al., eds), 191–293, Birkhäuser, Basel, 2000.

[10] R. Verfürth, *A Review of A Posteriori Error Estimation and Adaptive Mesh-Refinement Techniques*, Wiley/Teubner, New York Stuttgart, 1996.

High Dimensional Finite Elements for Elliptic Problems with Multiple Scales and Stochastic Data*

C. Schwab[†]

Abstract

Multiple scale homogenization problems are reduced to single scale problems in higher dimension. It is shown that sparse tensor product Finite Element Methods (FEM) allow the numerical solution in complexity independent of the dimension and of the length scale. Problems with stochastic input data are reformulated as high dimensional deterministic problems for the statistical moments of the random solution. Sparse tensor product FEM give a deterministic solution algorithm of log-linear complexity for statistical moments.

2000 Mathematics Subject Classification: 65N30.
Keywords and Phrases: Sparse finite element methods, Homogenization, Stochastic partial differential equations, Wavelets.

1. Introduction

The numerical solution of elliptic problems with multiple scales in a bounded domain D can be achieved either by analytic homogenization [2], [10] through asymptotic analysis or by specially designed Finite Element spaces to capture the fine scales of the problem [8], [7]. In asymptotic theory of homogenization, the fine scale of the solution is averaged and lost in the homogenized limit. Fine scale information can be recovered from so called correctors which must be calculated separately. An alternative which we pursue here is to "unfold" the homogenization problem into a single scale problem in high dimension - so that homogenized and fine scale behaviour are still coupled [3]. As we will show here, this limiting, high-dimensional single scale problem can be solved numerically by sparse tensor product

---

*Research performed in the network "Homogenization and Multiple Scales 2000" (HMS 2000) of the EC and supported by the Swiss Federal government under grant No. BBW 01.0025-1 and by the Swiss National Fund under grant Number SNF 21-58754.99.

†Seminar for Applied Mathematics, ETH Zürich, 8092 Zürich, Switzerland.   E-mail: schwab@sam.math.ethz.ch

FEM in complexity comparable to the optimal one for single scale problems in the physical domain D.

We also show how the same idea can be applied to the fast deterministic calculation of two and M point spatial correlation functions of random solutions to elliptic problems.

2. Homogenization problem

In a bounded domain $D \subset \mathbb{R}^d$ with Lipschitz boundary $\Gamma = \partial D$, we consider the elliptic problem in divergence form

$$-\mathrm{div}(A^\varepsilon(x)\nabla u^\varepsilon) = f(x) \text{ in } D, \ u^\varepsilon = 0 \text{ on } \partial D\,. \qquad (2.1)$$

Problem (2.1) has multiple separated scales in the sense that

$$A^\varepsilon(x) = A\left(x, \left\{\frac{x}{\varepsilon}\right\}_Y\right), \ x \in D \qquad (2.2)$$

where $Y = (0,1)^d$ denotes the unit cell and we denote by $\left[\frac{x}{\varepsilon}\right]_Y$ the unique element in $\mathbb{Z}^d$ such that $x \in \varepsilon(\left[\frac{x}{\varepsilon}\right]_Y + Y)$ and set $\left\{\frac{x}{\varepsilon}\right\}_Y := \frac{x}{\varepsilon} - \left[\frac{x}{\varepsilon}\right]_Y \in Y$. In (2.2), the function $A(x,y) \in L^\infty(D \times Y)_{\mathrm{sym}}^{d \times d}$ is Y-periodic with respect to y and satisfies, for every $(x,y) \in D \times Y$, and some $0 < \alpha < 1$:

$$\forall \xi \in \mathbb{R}^d : \alpha\,|\xi|^2 \le \xi^\top A(x,y)\,\xi \le \alpha^{-1}|\xi|^2\,. \qquad (2.3)$$

Then problem (2.1) admits, for every $f \in L^2(D)$, a unique weak solution:

$$u^\varepsilon \in H_0^1(D): \quad B^\varepsilon(u^\varepsilon, v) = (f, v) \quad \forall v \in H_0^1(D) \qquad (2.4)$$

where $B^\varepsilon(u,v) = \int_D \nabla v \cdot A(x, \frac{x}{\varepsilon})\nabla u\,dx$. As it is well known, as $\varepsilon \to 0$, $u^\varepsilon \to u^0$ in $L^2(D)$ strongly and in $H^1(D)$ weakly, and u^0 is the solution of the homogenized problem

$$-\mathrm{div}(A^0(x)\nabla u^0) = f \text{ in } D, \ u^0 = 0 \text{ on } \partial D\,, \qquad (2.5)$$

and formulas for $A^0(x)$ are available [2, 6]. The lack of strong $H^1(D)$-convergence indicates that in the limit $\varepsilon \to 0$, information on the fine scale of u^ε is lost. It can be recovered by calculating so-called correctors by either differentiating u^ε or by solving a second problem of the type (2.1). Both approaches are not attractive as basis for numerical solution methods: differentiating a numerical approximation of u^0 will reduce convergence rates, and solving (2.1) for correctors amounts to solving a problem akin to the original one.

A variant of the two-scale convergence, originally due to [9, 1], and recently developed in [3], allows to obtain a single scale, "unfolded", limit problem which gives u^0 and essential information on oscillations of u^ε. To describe it, define for every $\varphi \in L^2(D)$ the "unfolding" operator

$$\mathcal{T}_\varepsilon(\varphi)(x,y) := \begin{cases} \varphi\left(\varepsilon\left[\frac{x}{\varepsilon}\right]_Y + \varepsilon y\right) & \text{if } \varepsilon\left(\left[\frac{x}{\varepsilon}\right]_Y + Y\right) \subset D \\ 0 & \text{else} \end{cases} \qquad (2.6)$$

for $x \in D, y \in Y$. Then there holds [3].

Proposition 2.1. *Assume that*

$$\tilde{A}^\varepsilon(x,y) := \mathcal{T}_\varepsilon(A^\varepsilon)(x,y) \to \tilde{A}(x,y) \quad a.e.(x,y) \in D \times Y. \tag{2.7}$$

Then there exists in $u^0 \in H_0^1(D)$ *such that, as* $\varepsilon \to 0$, *the solutions* u^ε *of (2.4) satisfy*

$$u^\varepsilon \underset{H^1(D)}{\rightharpoonup} u^0 \in H_0^1(D) \tag{2.8}$$

and there exists $\phi(x,y) \in L^2(D, H^1_{\mathrm{per}}(Y)/\mathbb{R})$ *such that, as* $\varepsilon \to 0$,

$$\mathcal{T}_\varepsilon(\nabla_x u^\varepsilon) \underset{L^2(D \times Y)^d}{\rightharpoonup} \nabla_x u^0 + \nabla_y \phi \text{ in } L^2(D \times Y)^d. \tag{2.9}$$

The functions $u^0 \in H_0^1(D)$, $\phi \in L^2(D, H^1_{\mathrm{per}}(Y)/\mathbb{R})$ *solve the "unfolded" limiting problem: find* $u^0 \in H_0^1(D)$, $\phi \in L^2(D, H^1_{\mathrm{per}}(Y)/\mathbb{R})$ *such that*

$$
\begin{aligned}
B(u^0, \phi; v, \psi) &:= \\
\int_D \int_Y &\{\nabla_x v + \nabla_y \psi\}^\top \tilde{A}(x,y)\{\nabla_x u^0 + \nabla_y \phi\} dy \, dx = \\
\int_D &f v \, dx \quad \forall v \in H_0^1(D), \ \psi \in L^2(D, H^1_{\mathrm{per}}(Y)/\mathbb{R}).
\end{aligned}
\tag{2.10}
$$

Remark 2.2. i) Problem (2.10) is independent of ε and therefore a single scale problem.

ii) As $\varepsilon \to 0$, $u^\varepsilon(x) \to u^0(x)$ in $L^2(D)$ and $\nabla_x u^\varepsilon \to \nabla_x u^0 + \nabla_y \phi(\cdot, \frac{\cdot}{\varepsilon})$ strongly in $L^2(D, \mathbb{R}^d)$. Therefore, u^0 coincides with the solution of the homogenized problem (2.5) and ϕ retains information on the oscillations of u^ε as $\varepsilon \to 0$. Information on gradients of u^ε is introduced at the price of higher dimension of the limiting problem: while (2.5) is posed on $D \subset \mathbb{R}^d$, (2.10) must be solved on $D \times Y \subset \mathbb{R}^{2d}$.

iii) As (2.10) is derived by weak convergence methods, no information about the boundary behaviour of u^ε is obtained by solving (2.10).

iv) In (2.4), $f = f(x)$ was assumed independent of ε. Loadings of the type $f(x, x/\varepsilon)$ may equally be accommodated.

v) Property (2.7) holds under certain conditions on A^ε, for example if $A(x,y) = A(y)$ or if $A \in L^1(Y, C^0(D))$.

The limit problem (2.10) is well-posed:

Proposition 2.3. *Assume (2.3). Then there is* $C > 0$ *such that*

$$
\begin{aligned}
&\forall u \in H_0^1(D), \ \forall \phi \in L^2(D, H^1_{\mathrm{per}}(Y)/\mathbb{R}) : \\
&B(u, \phi; u, \phi) \geq C\big(\|u\|^2_{H^1(D)} + \|\phi\|^2_{L^2(D, H^1_{\mathrm{per}}(Y))}\big).
\end{aligned}
\tag{2.11}
$$

The proof is immediate if we observe (2.3) and

$$|\nabla_x u + \nabla_y \phi|^2 = |\nabla_x u|^2 + 2\nabla_x u \cdot \nabla_y \phi + |\nabla_y \phi|^2$$

and that, due to $\phi \in L^2(D, H^1_{\mathrm{per}}(Y)/\mathbb{R})$, it holds

$$\int_D \int_Y \nabla_x u \cdot \nabla_y \phi \, dy \, dx = \int_D \nabla_x u \cdot \int_Y \nabla_y \phi \, dy \, dx = 0 \,.$$

The limit problem (2.10) admits the following regularity: if $\tilde{A}(x,y)$, ∂D are smooth, then $f \in H^{-1+k}(D)$ implies

$$u^0 \in H^{1+k}(D), \quad \phi \in H^k(D, C^\infty_{\mathrm{per}}(\overline{Y})) \,. \tag{2.12}$$

3.   Sparse finite element discretization

We discretize the unfolded limiting problem (2.10) by a sparse Finite Element Method (FEM) (e.g. [5] and the references there). To this end, assume D is a bounded Lipschitz polyhedron with straight faces and let $\{\mathcal{T}^D_\ell\}$, $\{\mathcal{T}^Y_\ell\}$ be sequences of nested, quasiuniform meshes in D resp. Y consisting of shape-regular simplices T of meshwidth $h_\ell = O(2^{-\ell})$, and such that the periodic extension of $\mathcal{T}^Y_\ell$ beyond Y is regular.

Let further $p \geq 1$ be a polynomial degree. Then we denote

$$V_D^{\ell,1} = S_0^{p,1}(D, \mathcal{T}^D_\ell) = \{u \in H^1_0(D) : \forall T \in \mathcal{T}^D_\ell : u|_T \in \mathcal{P}_p(T)\} \,,$$

$$V_Y^\ell = S^{p,1}_{\mathrm{per}}(Y, \mathcal{T}^Y_\ell) = \{u \in H^1_{\mathrm{per}}(Y) : \forall T \in \mathcal{T}^Y_\ell : u|_T \in \mathcal{P}_p(T)\}/\mathbb{R} \,,$$

$$V_D^{\ell,0} = S^{p-1,0}(D, \mathcal{T}^D_\ell) = \{u \in L^2(D) : \forall T \in \mathcal{T}^D_\ell : u|_T \in \mathcal{P}_{p-1}(T)\} \,,$$

where $\mathcal{P}_p(T)$ denotes the polynomials of total degree at most p on T.

Since the triangulations are nested, the Finite Element spaces are hierarchical:

$$\begin{aligned} V_D^{0,j} &\subset V_D^{1,j} \subset \cdots \subset V_D^{\ell,j} \subset \cdots \subset H^j(D) \,, \\ V_Y^0 &\subset V_Y^1 \subset \cdots \subset V_Y^\ell \subset \cdots \subset H^1_{\mathrm{per}}(Y) \,. \end{aligned} \tag{3.1}$$

Define orthogonal projections $P_D^{\ell,j} : H^j(D) \to V_D^{\ell,j}$, $j = 0,1$ and $P_Y^\ell : H^1_{\mathrm{per}}(Y) \to V_Y^\ell$ and set $P_D^{-1,j} := 0$, $P_Y^{-1} := 0$. Then we have the increment or detail-spaces

$$\begin{aligned} W_D^{\ell,j} &:= (P_D^{\ell,j} - P_D^{\ell-1,j}) V_D^{\ell,j}, \quad \ell = 0,1,2,\dots \\ W_Y^\ell &:= (P_Y^\ell - P_Y^{\ell-1}) V_Y^\ell, \quad\quad\ \ell = 0,1,2,\dots \end{aligned} \tag{3.2}$$

and, for every $L > 0$, the multilevel decompositions

$$V_D^{L,j} = \bigoplus_{0 \leq \ell \leq L} W_D^{\ell,j}, \qquad V_Y^L = \bigoplus_{0 \leq \ell \leq L} W_Y^\ell, \quad j = 0,1 \tag{3.3}$$

which are orthogonal in $H^j(D)$ resp. $H^1_{\text{per}}(Y)$.

For $L > 0$, define the full tensor product space

$$S^L = V_D^{L,0} \otimes V_Y^L = \bigoplus_{0 \le \ell, \ell' \le L} W_D^{\ell,0} \otimes W_Y^{\ell'} \subset L^2(D, H^1_{\text{per}}(Y)/\mathbb{R}). \tag{3.4}$$

Assume the regularity (2.12) with some $k > 0$.

Then (2.11) shows that the finite element approximation

$$(u^L, \phi^L) \in V_D^{L,1} \times S^L : B(u^L, \phi^L; v^L, \psi^L) = (f, v^L) \quad \forall (v^L, \psi^L) \in V_D^{L,1} \times S^L \tag{3.5}$$

exists, is unique and a tensor product argument show that it satisfies the asymptotic error bounds

$$\begin{aligned} &\|u^0 - u^L\|_{H^1(D)} + \|\phi - \phi^L\|_{L^2(D,H^1(Y))} \\ &\le C\, h_L^{\min(p,k)} \left\{ \|u^0\|_{H^{k+1}(D)} + \|\phi\|_{L^2(D;H^{k+1}(Y))} + \|\phi\|_{H^k(D;H^1(Y))} \right\}. \end{aligned} \tag{3.6}$$

We note that $\dim(V_D^{L,1}) = O(2^{Ld}) = \dim V_Y^L$, whereas $\dim(S^L) = O(2^{2Ld}) = O(\dim(V_D^{L,1})^2)$ as $L \to \infty$, since S^L is a FE-space in $D \times Y \subset \mathbb{R}^{2d}$. Also, note that in (3.6) the full regularity (2.12) of ϕ was not used.

The large number of degrees of freedom in S^L due to high dimension renders the unfolded problem (2.10) impractical for efficient numerical solution. A remedy is to use the sparse tensor product spaces ([5] and the references there)

$$\widehat{S}^L := \bigoplus_{0 \le \ell + \ell' \le L} W_D^{\ell,0} \otimes W_Y^{\ell'} \tag{3.7}$$

which have substantially smaller dimension than the full tensor product spaces (3.4).

The joint regularity (2.12) in x and y of $\phi(x, y)$ allows to retain the error bounds (3.6) with $\widehat{S}^L$ in place of S^L:

Proposition 3.1. *Let $(\widehat{u}^L, \widehat{\phi}^L) \in V_D^{L,1} \times \widehat{S}^L$ denote the sparse FE-solutions which satisfy (3.5) with $\widehat{S}^L$ in place of S^L.*
If $u^0 \in H^{1+k}(D)$, $\phi \in H^k(D, H^{k+1}_{\text{per}}(Y))$, for some $k \ge 1$, then

$$\begin{aligned} &\|u^0 - \widehat{u}^L\|_{H^1(D)} + \|\phi - \widehat{\phi}^L\|_{L^2(D,H^1(Y))} \\ &\le C(L+1)^{\frac{1}{2}} h_L^{\min(p,k)} \left\{ \|u^0\|_{H^{k+1}(D)} + \|\phi\|_{H^k(D;H^{k+1}(Y))} \right\} \end{aligned} \tag{3.8}$$

and the total number of degrees of freedom is bounded by

$$\dim(V_D^{L,1}) + \dim(\widehat{S}^L) \le CL2^{dL}. \tag{3.9}$$

Remark 3.2. i) Since $\dim(V_D^{L,1}) \le C2^{dL}$, computation of $(\widehat{u}, \widehat{\phi}^L)$ in $D \times Y$ requires, up to $L = O(|\log h_L|)$, the same number of degrees of freedom as the FE approximation of u^0 from (2.5) in D. Moreover, the FE approximation of (2.10) does not require the determination of $A^0(x)$. In addition, the convergence rate (3.8) is, up to $(L+1)^{\frac{1}{2}}$, equal to the rate (3.6) for the full tensor product-spaces.

ii) Sparse tensor products of finite elements in one dimension were proposed by Zenger and his students in the 1990ies for the efficient solution of partial differential equations in three dimensions (see [5] and the references there). While allowing similar convergence rates as the full tensor product spaces, extra regularity of the solution (generally not available in non-smooth domains) is required. In (3.7), sparse tensor products of standard finite element spaces in D resp. $Y \subset \mathbb{R}^d$ are taken and realistic regularity of the solution available from the structure of the limit problem (2.10) was used.

iii) The approach generalizes to problems with $M > 2$ scales. The unfolded problem is then posed on a product domain

$$D \times Y_1 \times \cdots \times Y_{M-1} \subset \mathbb{R}^{Md}$$

and an approximation of order $(L+1)^{\frac{M-1}{2}} h_L^{\min(p,k)}$ can be obtained with $O(L^{M-1} 2^{dL})$ degrees of freedom.

iv) Explicit construction of the sparse tensor product-space $\widehat{S}^L$ requires bases for the detail-spaces $W_D^{\ell,0}$, W_Y^ℓ. These are available, for example, via suitable semiorthogonal wavelet bases (e.g. [4]). These bases allow also for optimal preconditioning of the linear system corresponding to (3.5).

v) We discussed here only the diffusion problem (2.1). The results can be generalized to Elasticity, and the Stokes Equations.

vi) If only partial periodicity or patchwise periodic patterns are present, the unfolding approach works equally. See [3] for details.

4. Stochastic data

Let (Ω, Σ, P) be a σ-finite probability space and $A(x) \in L^\infty(D, \mathbb{R}_{\mathrm{sym}}^{d \times d})$ satisfy for every $\xi \in \mathbb{R}^d$

$$\exists \alpha, \beta > 0 : \alpha |\xi|^2 \leq \xi^T A(x) \xi \leq \beta |\xi|^2 \,. \tag{4.1}$$

For a random source term $f \in L^2(\Omega, dP; L^2(D))$, consider the Dirichlet problem

$$L(x, \partial_x) u = -\nabla \cdot A(x) \nabla u(x, \omega) = f(x, \omega) \text{ in } D, \, u = 0 \text{ on } \partial D \tag{4.2}$$

for P-a.e. $\omega \in \Omega$. The random solution $u(x, \omega)$ of (4.2) is searched in the Bochner-space

$$\mathcal{H}_0^1(D) := L^2(\Omega, dP; H_0^1(D)) \cong L^2(\Omega, dP) \otimes H_0^1(D) \,. \tag{4.3}$$

We note that $L^2(\Omega, dP)$, equipped with inner product

$$\langle u, v \rangle = \int_\Omega u(\omega) \, v(\omega) \, dP(\omega) \,, \tag{4.4}$$

is a Hilbert space. The variational form of (4.2) reads: find $u \in \mathcal{H}_0^1(D)$ such that

$$\mathcal{A}(u, v) = \langle f, v \rangle_{\mathcal{L}^2(D)} \quad \forall v \in \mathcal{H}_0^1(D) \tag{4.5}$$

where

$$\mathcal{A}(u, v) = \langle (A \otimes id)(\nabla \otimes id)u, \ (\nabla \otimes id)v \rangle_{\mathcal{L}^2(D)^d}$$

and $\langle f, v \rangle_{\mathcal{L}^2(D)} = \int_D \langle f(x, \cdot), v(x, \cdot) \rangle dx$. The form $\mathcal{A}(\cdot, \cdot)$ in (4.5) is coercive on $\mathcal{H}_0^1(D) \times \mathcal{H}_0^1(D)$, implying the existence of a unique random solution $u(x, \omega)$ of (4.5). Numerical solution of (4.5) that involves a FEM in D and Monte-Carlo in Ω is prohibitively expensive. Alternatively, we might try to compute directly the statistics of $u(x, \omega)$. For example, the mean field $E_u(x) = \int_\Omega u(x, \omega) dP(\omega)$ solves

$$L(x, \partial_x) E_u = E_f \text{ in } D, \ E_u = 0 \text{ on } \partial D. \tag{4.6}$$

For $x, x' \in D$, define the two point correlation function

$$C_u(x, x') = \langle u(x, \cdot), u(x', \cdot) \rangle. \tag{4.7}$$

Then the variance of $u(x, \omega)$ is given by

$$(\text{Var } u(x, \cdot))^2 := (E_u(x))^2 - (C_u(x, x))^2. \tag{4.8}$$

The two-point correlation $C_u(x, x')$ is the solution of a deterministic problem in $D \times D$. Formally

$$L(x, \partial_x) \, L(x', \partial_{x'}) C_u = C_f \text{ in } D \times D, \tag{4.9}$$

and in variational form:

$$C_u \in H_{(0)}^{1,1}(D \times D) : Q(C_u, C_v) = (C_f, C_v) \quad \forall C_v \in H_{(0)}^{1,1}(D \times D) \tag{4.10}$$

where $H_{(0)}^{1,1}(D \times D) := H_0^1(D) \otimes H_0^1(D)$ and

$$Q(C_u, C_v) = \int_{D \times D} \nabla_{xy} C_v \, A(x) \otimes A(y) \, \nabla_{xy} C_u \, dx \, dy, \quad \nabla_{xy} := \nabla_x \otimes \nabla_y.$$

Proposition 4.1. *The two point correlation C_u of the random solution $u(x, \omega)$ is the unique solution of the deterministic problem (4.10) in $D \times D$.*

Hence to get two point correlation functions, Monte-Carlo can be traded for a deterministic problem in high dimensions. The key to its efficient numerical solution lies in the regularity of the solution: if the mean field problem (4.6) satisfies a shift-theorem at order $s > 0$, i.e. $E_f \in H^{-1+s}(D) \Longrightarrow E_u \in H^{1+s}(D)$, then

$$C_f \in H^{-1+s, -1+s}(D \times D) \Longrightarrow C_u \in H_{(0)}^{1+s, 1+s}(D \times D). \tag{4.11}$$

A similar regularity result in weighted spaces of mixed highest derivatives holds if $D \subset \mathbb{R}^2$ has corners [10]. This regularity in spaces of mixed highest derivatives allows to approximate C_u from the sparse tensor product FE spaces

$$\widehat{V}^L := \bigoplus_{0 \leq \ell + \ell' \leq L} W_0^{\ell, 1} \otimes W_D^{\ell', 1} \tag{4.12}$$

of dimension $\dim(\widehat{V}^L) \leq CL2^{dL}$ at a near optimal convergence rate: if $\widehat{C}_u^L \in \widehat{V}^L$ denotes the FE approximation of C_u, it holds [10]

$$\|C_u - \widehat{C}_u^L\|_{H^{1,1}(D\times D)} \leq C \,|\log h_L|^{\frac{1}{2}} \, h_L^{\min(s,p)} \|C_u\|_{H^{s+1,s+1}(D\times D)}. \qquad (4.13)$$

Moreover, using a semiorthogonal wavelet basis of the detail-spaces $W_D^{\ell,1}$ in (3.2), we can design an algorithm which computes $\widehat{C}_u^L$ to the order of the discretization error (4.13) in $O(N_L L^{4d+2})$ operations where $N_L = O(2^{dL})$ denotes the number of degrees of freedom in D (see [10] for details).

As for homogenization problems with multiple scales, M point correlation functions of $u(x,\omega)$ can be approximated at the rate $|\log h_L|^{(M-1)/2} \, h_L^{\min(s,p)}$ with $O(N_L \, L^{M-1})$ degrees of freedom . Let us also remark that high regularity in (4.13) corresponds to strong spatial correlation of $u(x,\omega)$.

In the spatially uncorrelated limit, formally $C_f(x,y) = \delta(x-y)$ and we have for smooth $\partial D, A(x), d \leq 3$:

$$C_u \in H^{1+s,1+s}(D \times D), \ 0 \leq s < 1 - \frac{d}{4},$$

so that only the low convergence rates $|\log h_L|^{\frac{1}{2}} \, h_L^{1-\frac{d}{4}-\varepsilon}$ for $\widehat{C}_u^L$ follow. This is due to the singular support of C_u being the diagonal $\{(x,y) : x = y\}$. The efficient approximation of such C_u is the topic of ongoing research.

References

[1] G. Allaire: Homogenization and Two-scale convergence, SIAM J. Math. Anal. **23**, 1482-1518.

[2] A. Bensoussan, G. Papanicolau and J.L. Lions: Asymptotic Analysis of periodic structures, North Holland 1978.

[3] D. Cioranescu, A. Damlamian and G. Griso: Periodic unfolding and Homogenization, CRAS 2002 (to appear).

[4] W. Dahmen: Wavelet and Multiscale Methods for operator equations, Acta Numerica **6** (1997), 55-228.

[5] M. Griebel, P. Oswald and T. Schiekofer: Sparse grids for boundary integral equations, Numer. Math. **83** (1999), 279-312.

[6] V.V. Jikov, S.M. Kozlov and O.A. Oleinik: Homogenization of Differential Operators and Integral Functionals, Springer Verlag 1994.

[7] A.M. Matache and C. Schwab: Two-scale FEM for Homogenization Problems, Report 2001-06, Seminar for Applied Mathematics, ETH Zürich www.sam.math.ethz.ch/~reports (to appear).

[8] A.M. Matache, I. Babuška and C. Schwab: Generalized p-FEM for Homogenization, Numerische Mathematik, **86** Issue 2 (2000), 319-375.

[9] G. Nguetseng: A general convergence result for a functional related to the theory of homogenization, SIAM J. Math. Anal. **20** (1989), 608-629.

[10] Ch. Schwab and R.A. Todor: Sparse Finite Elements for elliptic problems with stochastic data, Research Report 2002-05, Seminar for Applied Mathematics, ETH Zürich, Switzerland.

ICM 2002 · Vol. III · 735–746

Fast Algorithms for Optimal Control, Anisotropic Front Propagation and Multiple Arrivals

J. A. Sethian*

Abstract

We review some recent work in fast, efficient and accurate methods to compute viscosity solutions and non-viscosity solutions to static Hamilton-Jacobi equations which arise in optimal control, anisotropic front propagation, and multiple arrivals in wave propagation. For viscosity solutions, the class of algorithms are known as "Ordered Upwind Methods", and rely on a systematic ordering inherent in the characteristic flow of information. For non-viscosity multiple arrivals, the techniques hinge on a static boundary value phase-space formulation which again can be solved through a systematic ordering.

2000 Mathematics Subject Classification: 65N06, 65M06, 86A22, 49L25.
Keywords and Phrases: Hamilton-Jacobi equations, Fast marching methods, Ordered upwind methods.

1. Introduction

This paper reviews recent work on algorithms for static Hamilton-Jacobi equations of the form $H(Du, x) = 0$; the solution u depends on $x \in R^n$, and boundary conditions are supplied on a subset of R^n. These equations arise in such areas as wave propagation, optimal control, anisotropic front propagation, medical imaging, optics, and robotic navigation. We develop algorithms to solve these equations remarkably quickly, with the same optimal efficiency as classic algorithms for shortest paths on discrete weighted networks, but extended to continuous Hamilton-Jacobi equations.

The algorithms, which rely on a close examination of the flow of information inherent in static Hamilton-Jacobi equations, are robust, unconditionally stable without time step restriction, and efficient. They are "One-pass" schemes, in that the solution is computed at N grid points in $O(N \log N)$ steps.

*Department of Mathematics, University of California, Berkeley, California, USA. E-mail: sethian@math.berkeley.edu

1.1.   Viscosity vs. non-viscosity solutions

What is meant by a *solution* to $H(Du, x) = 0$? Viscosity solutions [3] provide a unique and well-posed formulation which is linked to the unique viscosity limit of the associated smoothed equation; these are first arrivals in the propagation of information. Fig. 1a shows an example from semiconductor manufacturing in which a beam whose strength is angle-dependent is used to anistropically etch away a metal surface. Fig. 1b shows an optimal control problem to find the shortest exit path for a vehicle with position and direction-dependent speed.

Ion etching in anisotropic front propagation
Fig. 1a

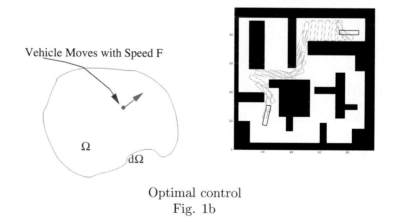

Optimal control
Fig. 1b

Figure 1: Viscosity solutions to static HJ equations

The above are viscosity solutions. However, there are many cases in which later arrivals, or "non-viscosity" solutions, are desirable. Fig. 2a shows the propagation of a wave inwards from a square boundary; the evolving front passes through itself and later arrivals form cusps and swallowtails as they move; Fig. 2b shows multiple arrivals in geophysical wave propagation.

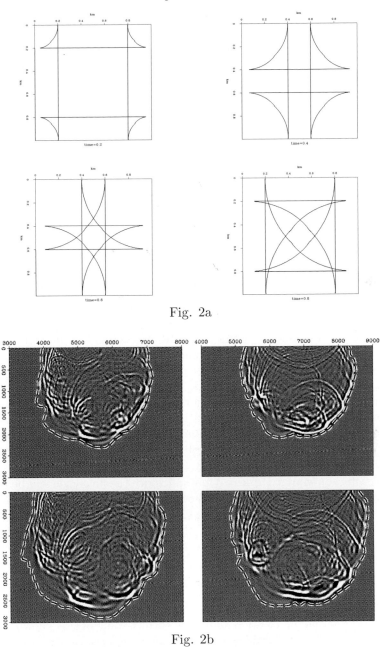

Fig. 2a

Fig. 2b

Figure 2: Non-viscosity solutions

Our goal is to create efficient algorithms which allow us to compute both types of solutions. In the case of viscosity solutions, algorithms are provided by the class of "Ordered Upwind Methods" developed by Sethian and Vladimirsky in [12, 13];

these methods work in physical space and construct the solution in a "One-pass" manner through a careful adherence to a causality inherent in the characteristic flow of the information. In the case of non-viscosity solution, algorithms are provided by the time-independent phase-space formulation developed by Fomel and Sethian [7], which relies on conversion of multiple arrivals into an Eulerian static boundary value problem, which can also be solved very efficiently in a "One-pass" manner which avoids all iteration through a careful ordering procedure. The remainder of this paper is devoted to describing these two classes of algorithms and providing a few computational results.

2.   Fast methods for viscosity solutions

We first discuss "Ordered Upwind Methods" introduced in [12] for computing viscosity solutions.

2.1.   Discrete control: Dijkstra's method

Consider a discrete optimal trajectory problem on a network. Given a network and a cost associated with each node, the global optimal trajectory is the most efficient path from a starting point to some exit set in the domain. Dijkstra's classic algorithm [4] computes the minimal cost of reaching any node on a network in $O(N \log N)$ operations. Since the cost can depend on both the particular node, and the particular link, Dijkstra's method applies to both *isotropic* and *anisotropic* control problems. The distinction is minor for discrete problems, but significant for continuous problems. Dijkstra's method is a "one-pass" algorithm; each point on the network is updated a constant number of times to produce the solution. This efficiency comes from a careful analysis of the direction of information propagation and stems from the optimality principle.

We briefly summarize Dijsktra's method, since the flow logic will be important in explaining our Ordered Upwind Methods. For simplicity, imagine a rectangular grid of size h, where the cost $C_{ij} > 0$ is given for passing through each grid point $x_{ij} = (ih, jh)$. Given a starting point, the minimal total cost U_{ij} of arriving at the node x_{ij} can be written in terms of the minimal total cost of arriving at its neighbors:

$$U_{ij} = \min \left(U_{i-1,j}, U_{i+1,j}, U_{i,j-1}, U_{i,j+1} \right) + C_{ij}. \qquad (2.1)$$

To find the minimal total cost, Dijkstra's method divides mesh points into three classes: *Far* (no information about the correct value of U is known), *Accepted* (the correct value of U has been computed), and *Considered* (adjacent to *Accepted*). The algorithm proceeds by moving the smallest *Considered* value into the *Accepted* set, moving its *Far* neighbors into the *Considered* set, and recomputing all *Considered* neighbors according to formula 2.1. This algorithm has the computational complexity of $O(N \log(N))$; the factor of $\log(N)$ reflects the necessity of maintaining a sorted list of the *Considered* values U_i to determine the next *Accepted* mesh point. Efficient implementation can be obtained using heap-sort data structures.

2.2.   Continuous control: ordered upwind methods

Consider now the problem of continuous optimal control; here, the goal is to find the optimal path from a starting position to an exit set. Dijkstra's method does not converge to the continuous solution as the mesh becomes finer and finer, since (see [11]) it produces the solution to the partial differential equation $\max(|u_x|, |u_y|) = h * C$, where h is the grid size. As h goes to zero, this does not converge to the solution of the continuous Eikonal problem given by $|u_x^2 + u_y^2|^{1/2} = C$. Thus, Dikstra's method cannot be used to obtain a solution to the continuous problem.

2.2.1.   Ordered upwind solvers for continuous isotropic control

In the case of isotropic cost functions in which the cost depends only on position and not on direction, two recent algorithms, first Tsitsiklis's Method [16] and then Sethian's Fast Marching Method [10] have been introduced to solve the problems with the same computational complexity as Dijkstra's method. Both methods exploit information about the flow of information to obtain this efficiency; the causality allows one to build the solution in increasing order, which yields the Dijkstra-like nature of the solutions. Both algorithms result from a key feature of Eikonal equations, namely that their characteristic lines coincide with the gradient lines of the viscosity solution $u(x)$; this allows the construction of one-pass algorithms. Tsitsiklis' algorithm evolved from studying isotropic min-time optimal trajectory problems, and involves solving a minimization problem to update the solution. Sethian's Fast Marching Method evolved from studying isotropic front propagation problems, and involves an upwind finite difference formulation to update the solution. Each method starts with a particular (and different) coupled discretization and each shows that the resulting system can be decoupled through a causality property. We refer the reader to these references for details on ordered upwind methods for Eikonal equations, as well as [13] for a detailed discussion about the similarities and differences between the two techniques.

2.2.2.   Ordered upwind solvers for continuous anisotropic general optimal control

Consider now the full continuous optimal control problem, in which the cost function depends on both position and direction. In [12, 13], Sethian and Vladimirsky built and developed single-pass "Ordered Upwind Methods" for any continuous optimal control problem. They showed how to to produce the solution U_i by recalculating each U_i at most r times, where r depends only the equation and the mesh structure, but not upon the number of mesh points.

Building one-pass Dijkstra-like methods for general optimal control is considerably more challenging than it is for the Eikonal case, since characteristics no longer coincide with gradient lines of the viscosity solution. Thus, characteristics and gradient lines may in fact lie in different simplexes. This is precisely why both Sethian's Fast Marching Method and Tsitsiklis' Algorithm cannot be directly applied in the anisotropic (non-Eikonal) case: it is no longer possible to de-couple the system by computing/accepting the mesh points in the ascending order.

The key idea introduced in [12, 13] is to use the location anisotropy of the cost function to limit of the number of points on the accepted front that must be examined in the update of each Considered point. Consider the anisotropic min-time optimal trajectory problems, in which the speed of motion depends not only on position but also on direction. The value function u for such problems is the viscosity solution of the static Hamilton-Jacobi-Bellman equation

$$\begin{aligned} \max_{a \in S_1} \{(\nabla u(x) \cdot (-a)) f(a, x)\} &= 1, \quad x \in \Omega, \\ u(x) &= q(x), \qquad x \in \partial\Omega. \end{aligned} \tag{2.2}$$

In this formulation, a is the unit vector determining the direction of motion, $f(a, x)$ is the speed of motion in the direction a starting from the point $x \in \Omega$, and $q(x)$ is the time-penalty for exiting the domain at the point $x \in \partial\Omega$. The maximizer a corresponds to the characteristic direction for the point x. If f does not depend on a, Eqn. 2.2 reduces to the Eikonal equation, see [1].

Now, define the anisotropy ratio F_1/F_2, where $0 < F_1 \leq f(a, x) \leq F_2 < \infty$. In [13], two key lemmas were proved:

- **Lemma 1.** *Consider the characteristic passing through $\bar{x} \in \Omega$ and level curve $u(x) = C$, where $q_{max} < C < u(\bar{x})$. The characteristic intersects that level set at some point $\tilde{x}$. If $\bar{x}$ is distance d away from the level set then $\|\tilde{x} - \bar{x}\| \leq d\frac{F_2}{F_1}$.*
- **Lemma 2.** *Consider an unstructured mesh X of diameter h on Ω. Consider a simple closed curve Γ lying inside Ω with the property that for any point x on Γ, there exists a mesh point y inside Γ such that $\|x - y\| < h$. Suppose the mesh point $\bar{x}_i$ has the smallest value $u(\bar{x}_i)$ of all of the mesh points inside the curve. If the characteristic passing through $\bar{x}_i$ intersects that curve at some point $\tilde{x}_i$ then $\|\tilde{x}_i - \bar{x}_i\| \leq h\frac{F_2}{F_1}$.*

Thus, one may use the anisotropy ratio to exclude a large fraction of points on the Accepted Front in the update of any Considered Point; the size of this excluded subset depends on the anisotropy ratio. Building on these results, a fast, Dijkstra-like method was constructed. As before, three of mesh points classes are used. The *Accepted Front* is defined as a set of *Accepted* mesh points, which are adjacent to some not-yet-accepted mesh points. Define the set AF of the line segments $x_j x_k$, where x_j and x_k are adjacent mesh points on the *AcceptedFront*, such that there exists a *Considered* mesh point x_i adjacent to both x_j and x_k. For each *Considered* mesh point x_i one defines the part of AF "relevant to x_i":

$$NF(x_i) = \left\{ (x_j, x_k) \in AF \,|\, \exists \tilde{x} \text{ on } (x_j, x_k) \text{ s.t. } \|\tilde{x} - x_i\| \leq h\frac{F_2}{F_1} \right\}.$$

We will further assume that some consistent upwinding update formula is available: if the characteristic for x_i lies in the simplex $x_i x_j x_k$ then $U_i = K(U_j, U_k, x_i, x_j, x_k)$. For the sake of notational simplicity we will refer to this value as $K_{j,k}$.

1. Start with all mesh points in *Far* ($U_i = \infty$).
2. Move the boundary mesh points ($x_i \in \delta\Omega$) to *Accepted* ($U_i = q(x_i)$).
3. Move all the mesh points x_i adjacent to the boundary into *Considered* and evaluate the tentative value of $U_i = \min_{(x_j, x_k) NF(x_i)} K_{j,k}$.

4. Find the mesh point x_r with the smallest value of U among all the *Considered*.
5. Move x_r to *Accepted* and update the *Accepted Front*.
6. Move the *Far* mesh points adjacent to x_r into *Considered*.
7. Recompute the value for all the *Considered* x_i within the distance $h\frac{F_2}{F_1}$ from x_r. If less than the previous tentative value for x_i then update U_i.
8. If *Considered* is not empty then go to 4).

2.2.3. Analysis and results

This is a "single-pass" algorithm since the maximum number of times each mesh point can be re-evaluated is bounded by the number of mesh points in the $h\frac{F_2}{F_1}$ neighborhood of that point; the method formally has the computational complexity of $O((\frac{F_2}{F_1})^2 M \log(M))$. Convergence of the method to the viscosity solution is proved in [13], and depends on the upwinding update formula $U_i = K(U_j, U_k, x_i, x_j, x_k)$.

As an example, taken from [12], we compute the geodesic distance on the manifold $g(x,y) = .9 \sin(2\pi x)\sin(2\pi y)$ from the origin. This can be shown to be equivalent to solving the static Hamilton-Jacobi equation

$$\|\nabla u(\boldsymbol{x})\| F\left(\boldsymbol{x}, \frac{\nabla u(\boldsymbol{x})}{\|\nabla u(\boldsymbol{x})\|}\right) = 1, \tag{2.3}$$

with speed function F given by $F(x,y,\omega) = \sqrt{\frac{1+g_y^2\cos^2(\omega)+g_x^2\sin^2(\omega)-g_x g_y \sin(2\omega)}{1+g_x^2+g_y^2}}$ where ω is the angle between $\nabla u(x,y)$ and the positive direction of the x-axis. The anisotropy is substantial, since the dependence of F upon ω can be pronounced when ∇g is relatively large. Equidistant contours are shown on the left in Figure 3.

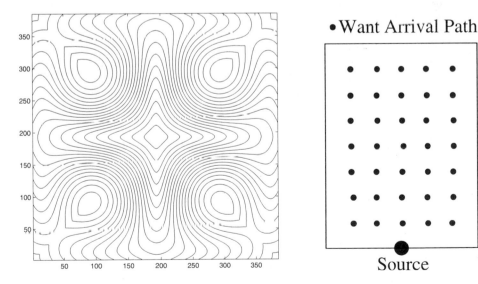

Figure 3: Left: Anistropic front propagation   Right: Arrival paths

3. Fast methods for multiple arrivals

3.1. Computing multiple arrivals

We now consider the problem of multiple arrivals. As an example, consider the two-dimensional Eikonal equation

$$|\nabla u| F(x, y) = 0 \tag{3.1}$$

with $F(x, y)$ given. We imagine a computational domain and a source point, as shown on the right in Figure 3. Suppose the goal is to determine the arrival time and path to each point in the interior from the source point. Here, we are interested not only in the first arrival, but all later arrivals as well.

One popular approach to computing multiple arrivals is to work in phase space, in which the dimensionality of the problem is increased from physical space to include the derivative of the solution as well. There are two approaches to computing these multiple arrivals through a phase space formulation. One is a Lagrangian (ray tracing) approach, in which the phase space characteristic equations are integrated, often from a source point, resulting in a Lagrangian structure which fans out over the domain. Difficulties can occur in either in low ray density zones where there are very few rays or near caustics where rays cross. The other is an Eulerian description of the problem, in either the physical domain or phase space. In recent years, this has led to many fascinating and clever Eulerian PDE-based approaches to computing multiple arrivals, see, for example, [15, 14, 9, 5, 2]. We note that the regularity of the phase space has been utilized previously in theoretical studies on the asymptotic wave propagation [8]. The above phase space approaches to solving for multiple arrivals have two characteristics in common:

- A phase space formulation increases the dimensionality of the problem. In two physical dimensions, the phase space formulation requires three dimensions; in three physical dimensions, the phase space formulation is in five dimensions.
- Given particular sources, the problem is solved with those source location(s) as initial data. Different sources requires re-solving the entire problem.
- The problem is cast as an initial value partial differential equation, and is evolved in time. Time step considerations in regions of high velocity play a role in the stability of the underlying scheme.

3.2. A boundary value formulation

Fomel and Sethian [7] take a different approach. A set of time-independent "Escape Equations" are derived, each of which is an Eulerian boundary value partial differential equation in phase space. Together, they give the exit time, location and derivative of all possible trajectories starting from all possible interior points. Thus, the particular choice of sources is reduced to post-processing. The computational speed depends on whether one wants to obtain results for all possible boundary conditions, or in fact only for a particular subset of possibilities.

3.2.1. Liouville formulation

Briefly (see [7] for details) begin with the static Hamilton-Jacobi equation

$$H(x, \nabla u) = 0, \qquad (3.2)$$

and write the well-known characteristic equations in phase space (x, p), where p corresponds to ∇u (see, for example, [6]). The characteristics must obey

$$\frac{dx}{d\sigma} = \nabla_p H; \qquad \frac{dp}{d\sigma} = -\nabla_x H. \qquad (3.3)$$

Differentiating the function $u(x(\sigma))$, we obtain an additional equation for transporting the function u along the characteristics:

$$\frac{du}{d\sigma} = \nabla u \cdot \frac{dx}{d\sigma} = p \cdot \nabla_p H. \qquad (3.4)$$

Eqns. 3.3,3.4 can be initialized at $\sigma = 0$: $x(0) = x_0$, $p(0) = p_0$, $u(0) = 0$.

One can now convert the phase space approach into a set of Liouville equations. To simplify notation, we denote the phase-space vector (x, p), by y, the right-hand side of system given in Eqn. 3.3 by vector function $R(y)$, and the right-hand side of Eqn. 3.4 by the function $r(y)$. In this notation, the Hamilton-Jacobi system is

$$\frac{\partial y(y_0, \sigma)}{\partial \sigma} = R(y); \qquad \frac{\partial u(y_0, \sigma)}{\partial \sigma} = r(y), \qquad (3.5)$$

and is initialized at $\sigma = 0$ as $y = y_0$ and $u = 0$. This system satisfies

$$\frac{\partial y(y_0, \sigma)}{\partial \sigma} = \nabla_0 y \, R(y_0) , \qquad (3.6)$$

and the transported function u satisfies the analogous equation

$$\frac{\partial u(y_0, \sigma)}{\partial \sigma} = \nabla_0 u \, R(y_0) + r(y_0) , \qquad (3.7)$$

where ∇_0 denotes the gradient with respect to y_0. These are the Liouville equations.

3.2.2. Formulation of escape equations

The key idea in [7] is as follows. Assume a closed boundary $\partial \mathcal{D}$ in the y space that is crossed by every characteristic trajectory originating in $y_0 \in \mathcal{D}$. This defines for every y_0 the function $\sigma = \hat{\sigma}(y_0)$ of the first crossing of the corresponding characteristic with $\partial \mathcal{D}$. Now introduce a differentiable function $\Gamma(y)$ that identifies the boundary, that is, $\Gamma(y) = 0$. In particular, we then have that $\Gamma(y(y_0, \hat{\sigma}(y_0))) = 0$. One can then differentiate with respect to the initial condition y_0 to obtain an escape equation for the parameter $\hat{\sigma}$. Similarly, one can derive escape equations for the position and value, yielding the full set of

$$\underline{\textbf{Escape Equations}} \quad \begin{aligned} 1 + \nabla_0 \hat{\sigma} \cdot R(y_0) &= 0 \\ \nabla_0 \hat{y} \, R(y_0) &= 0 \\ \nabla_0 \hat{u} \cdot R(y_0) + r(y_0) &= 0 \end{aligned}$$

$$(3.8)$$

3.3.   Fast solution of escape equations

Summarizing, rather than compute in physical space, we derive boundary value Escape equations in phase space $y = (x, p)$. All time step considerations are avoided, and one can compute all the arrivals from all possible sources simultaneously. This Eulerian formulation means that the entire domain is covered, even quiet slow zones.

Finally, and most importantly, a constructive, "One-pass" algorithm, similar to the one presented for viscosity solutions, can be designed. Exit time, position, and derivative at the boundary form boundary conditions. We can then systematically march the solution inwards in phase space from the boundary, constructing the solution through an ordering sequence based on the characteristics that ensures computational phase space mesh points need not be revisited more than once.

Consider a square boundary as an example, and suppose we wish to find the time $\widehat{u}(x, z, \theta)$ at which a ray leaving the initial point (x, z) inside the square, initially moving in direction θ, hits the boundary. We assume that the slowness field $n(x, z)$ is given. First, note that the set $\widehat{u}(x, z, \theta) = T$, drawn in x, z, θ space, gives the set of all initial positions and directions which reach the boundary of the square at time T. By the uniqueness of characteristics, the set of all points parameterized by T and given by $\widehat{U}(T) = \{x, z, \theta \mid \widehat{u}(x, z, \theta) = T\}$ sweep out the solution space. Figure 9a shows the solution surfaces $\widehat{u}(x, z, \theta)$ for the collapsing square.

Details on the exact algorithm are given in [7]. As demonstration (see [7]), in Figure 9b, the top pair shows all the arrivals starting from a source at the center of the top wall, together with the slowness field on the right (darker is slower). The bottom pair shows the first arrival and on the amplitude of the displayed arrival (the lighter the tone, the more amplitude).

References

[1] Bellman, R., Introduction to the Mathematical Theory of Control Processes, Academic, New York, 1967.

[2] Benamou, J.D., Big ray tracing:multivalued travel time field computation using viscosity solutions of the eikonal equation, *J. Comp. Phys.*, 128, 463-474, 1996

[3] Crandall, M.G. & Lions, P-L., Viscosity Solutions of Hamilton-Jacobi Equations, *Tran. AMS*, 277, 1-43, 1983.

[4] Dijkstra, E.W., A Note on Two Problems in Connection with Graphs, *Numerische Mathematik*, 1, 269-271, 1959.

[5] Engquist, B., Runborg, O., and Tornberg, A-K., The Segment Projection Method for Geometrical Optics, preprint, 2001.

[6] Evans, L.C., Partial Differential Equations, Amer. Math. Soc., 1998.

[7] Fomel, S., and Sethian, J.A., Fast Phase Space Computation of Multiple Arrivals, *Proc. Nat. Acad. Sci.*, 99, 11, 7329-7334, 2002.

[8] Maslov, V.P., and Fedoriuk, M.V., (1981), Semi-classical approximations in quantum mechanics, Reidel, Dordrecht.

[9] Ruuth, S., Merriman, B., and Osher, S., A Fixed Grid Method for Capturing the Motion of Self-Intersecting Wavefronts and Related PDEs, *J. Comp. Phys.*, 163, 1, 1-21, 2000.

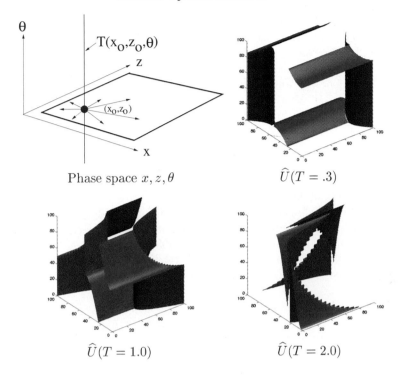

Fig. 9a: Geometry of solution

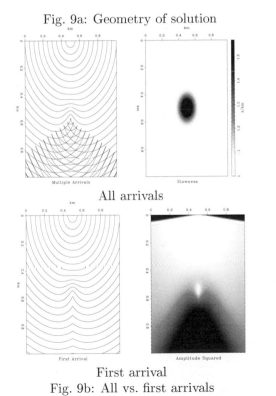

All arrivals

First arrival
Fig. 9b: All vs. first arrivals

[10] Sethian, J.A., A Fast Marching Level Set Method for Monotonically Advancing Fronts, *Proc. Nat. Acad. Sci.*, 93, 4, pp. 1591-1595, February 1996.

[11] Sethian, J.A., Level Set Methods and Fast Marching Methods: Evolving Interfaces in Computational Geometry, Fluid Mechanics, Computer Vision and Materials Sciences, Cambridge University Press, 1999.

[12] Sethian, J.A. & Vladimirsky, A., Ordered Upwind Methods for Static Hamilton-Jacobi Equations, *Proc. Nat. Acad. Sci.*, 98, 20, 11069-11074, 2001.

[13] Sethian,J.A. & Vladimirsky,A., Ordered Upwind Methods for Static Hamilton-Jacobi Equations: Theory and Algorithms, submitted *SINUM*, July 2001.

[14] Steinhoff, J., Fan, M., Wang, L., A New Eulerian Method for the Computation of Propagating Short Acoustic and Electromagnetic Pulses, *J. Comp. Phys.*, 157, 2, 683-706, 2000.

[15] Symes, W.W., A slowness matching finite difference algorithm for traveltimes beyond transmission caustics, Proc., 68th Annual Int. Meeting of the Soc. Expl. Geo., New Orleans, Expanded astract, (1998).

[16] Tsitsiklis, J.N., Efficient Algorithms for Globally Optimal Trajectories, *IEEE Tran. Automatic Control*, 40, 1528-1538, 1995.

ICM 2002 · Vol. III · 747–757

High Resolution Methods for Time Dependent Problems with Piecewise Smooth Solutions

Eitan Tadmor*

Abstract

A trademark of nonlinear, time-dependent, convection-dominated problems is the spontaneous formation of non-smooth macro-scale features, like shock discontinuities and non-differentiable kinks, which pose a challenge for high-resolution computations. We overview recent developments of modern computational methods for the approximate solution of such problems. In these computations, one seeks piecewise smooth solutions which are realized by finite dimensional projections. Computational methods in this context can be classified into two main categories, of local and global methods. Local methods are expressed in terms of point-values (— Hamilton-Jacobi equations), cell averages (— nonlinear conservation laws), or higher localized moments. Global methods are expressed in terms of global basis functions.

High resolution central schemes will be discussed as a prototype example for local methods. The family of central schemes offers high-resolution "black-box-solvers" to an impressive range of such nonlinear problems. The main ingredients here are detection of spurious extreme values, non-oscillatory reconstruction in the directions of smoothness, numerical dissipation and quadrature rules. Adaptive spectral viscosity will be discussed as an example for high-resolution global methods. The main ingredients here are detection of edges in spectral data, separation of scales, adaptive reconstruction, and spectral viscosity.

2000 Mathematics Subject Classification: 65M06, 65M70, 65M12.
Keywords and Phrases: Piecewise smoothness, High resolution, Central schemes, Edge detection, Spectral viscosity.

1. Introduction — piecewise smoothness

*University of California Los-Angeles, Department of Mathematics, Los Angeles, CA, USA 90095, and University of Maryland, Department of Mathematics, Center for Scientific Computation and Mathematical Modeling (CSCAMM) and Institute for Physical Science & Technology (IPST), College Park, MD, USA 20742. E-mail: tadmor@math.umd.edu

A trademark of nonlinear time-dependent convection-dominated problems is the spontaneous formation of non-smooth macro-scale features which challenge high-resolution computations. A prototype example is the formation of shock discontinuities in nonlinear conservation laws,

$$\frac{\partial}{\partial t}u(x,t) + \nabla_x \cdot f(u(x,t)) = 0, \quad u := (u_1, \ldots, u_m)^\top. \tag{1.1}$$

It is well known, e.g., [9], that solutions of (1.1) cease to be continuous, and (1.1) should be interpreted in a weak sense with the derivatives on the left as Radon measures. This requires clarification. If $u(x,t)$ and $v(x,t)$ are two admissible solutions of (1.1) then the following stability estimate is sought (here and below $\alpha, \beta, \ldots$ stand for different generic constants),

$$\|u(\cdot,t) - v(\cdot,t)\| \leq \alpha_t \|u(\cdot,0) - v(\cdot,0)\|. \tag{1.2}$$

Such estimates with different norms, $\|\cdot\|$, are playing a key role in the linear setting — both in theory and computations. For linear hyperbolic systems, for example, (1.2) is responsible for the usual L^2-stability theory, while the stability of parabolic systems is often measured by the L^∞-norm, consult [14]. But for nonlinear conservation laws, (1.2) fails for any L^p-norm with $p > 1$. Indeed, comparing $u(x,t)$ with any fixed translation of it, $v(x,t) := u(x+h,t)$, the L^p version of (1.2) implies

$$\|\Delta_{+h}u(\cdot,t)\|_{L^p(\mathbf{R}^d)} \leq \alpha_t \|\Delta_{+h}u(\cdot,0)\|_{L^p(\mathbf{R}^d)}, \quad \Delta_{+h}u(\cdot,t) := u(\cdot+h,t) - u(\cdot,t).$$

For smooth initial data, however, the bound on the right yields $\|\Delta_{+h}u(\cdot,t)\|_{L^p} \leq \alpha_t|h|$, which in turn, for $p > 1$, would lead to the contradiction that $u(\cdot,t)$ must remain continuous. Therefore, conservation laws cannot satisfy the L^p-stability estimate (1.2) after their finite breakdown time, except for the case $p = 1$. The latter leads to Bounded Variation (BV) solutions, $\|u(\cdot,t)\|_{BV} := \sup\|\Delta_{+h}u(\cdot,t)\|_{L^1}/|h| \leq \alpha_t < \infty$, whose derivatives are interpreted as the Radon measures mentioned above. BV serves as the standard regularity space for admissible solutions of (1.1). A complete BV theory for scalar conservation laws, $m = 1$, was developed Kružkov. Fundamental results on BV solutions of one-dimensional systems, $d = 1$, were obtained by P. Lax, J. Glimm, and others. Consult [2] for recent developments. Relatively little is known for general $(m-1) \times (d-1) > 0$, but cf., [12].

We argue that the space of BV functions is still too large to describe the approximate solutions of (1.1) encountered in computations. Indeed, in such computations one does not 'faithfully' realize arbitrary BV functions but rather, piecewise smooth solutions. We demonstrate this point in the context of scalar approximate solutions, $v^h(x,t)$, depending on a small computational scale $h \sim 1/N$. A typical error estimate for such approximations reads, [4]

$$\|v^h(\cdot,t) - u(\cdot,t)\|_{L^1_{loc}(\mathbf{R})} \leq \|v^h(\cdot,0) - u(\cdot,0)\|_{L^1_{loc}(\mathbf{R})} + \alpha_t h^{1/2}. \tag{1.3}$$

The convergence rate of order $1/2$ is a well understood linear phenomena, which is observed in computations[1]. The situation in the nonlinear case is different. The

---

[1]Bernstein polynomials, $B_N(u)$, provide a classical example of first-order monotone approximation with L^1-error of order $(\|u\|_{BV}/N)^{1/2}$. The general linear setting is similar, with improved rate $\sim h^{r/(r+1)}$ for r-order schemes.

optimal convergence rate for arbitrary BV initial data is still of order one-half, [15], but actual computations exhibit higher-order convergence rate. The apparent difference between theory and computations is resolved once we take into account piecewise smoothness. We can quantify piecewise smoothness in the simple scalar convex case, where the number of shock discontinuities of $u(\cdot, t)$ is bounded by the finitely many inflection points of the initial data, $u(x, 0)$. In this case, the singular support of $u(\cdot, t)$ consists of finitely many points where $\mathcal{S}(t) = \{x \mid \partial_x u(x, t) \downarrow -\infty\}$. Moreover, the solution in between those point discontinuities is as smooth as the initial data permit, [21], namely

$$\sup_{x \in \mathcal{S}_L(t)} |\partial_x^p u(x, t)| \leq e^{pLT} \sup_{x \in \mathcal{S}_L(0)} |\partial_x^p u(x, 0)| + Const_L, \ \mathcal{S}_L(t) := \{x \mid \partial_x u(x, t) \geq -L\}.$$

If we let $d(x, t) := dist(x, \mathcal{S}(t))$ denote the distance to $\mathcal{S}(t)$, then according to [19], the following pointwise error estimate holds, $|v^h(x, t) - u(x, t)| \leq \alpha_t h/d(x, t)$, and integration yields the first-order convergence rate

$$\|v^h(\cdot, t) - u(\cdot, t)\|_{L_{loc}^1(\mathbf{R})} \leq \alpha_t h |\log(h)|. \tag{1.4}$$

There is no contradiction between the optimality of (1.3) and (1.4). The former applies to arbitrary BV data, while the latter is restricted to piecewise smooth data and it is the one encountered in actual computations. The general situation is of course, more complicated, with a host of macro-scale features which separate between regions of smoothness. Retaining the invariant properties of piecewise smoothness in general problems is a considerable challenge for high-resolution methods.

2. A sense of direction

A computed approximation is a finite dimensional realization of an underlying solution which, as we argue above, is viewed as a piecewise smooth solution. To achieve higher accuracy, one should extract more information from the smooth parts of the solution. Macro-scale features of non-smoothness like shock discontinuities, are identified here as barriers for propagation of smoothness, and stencils which discretize (1.1) while crossing discontinuities are excluded because of spurious Gibbs' oscillations. A high resolution scheme should sense the direction of smoothness.

Another sense of directions is dictated by the propagation of information governed by convective equations. Discretizations of such equations fall into one of two, possibly overlapping categories. One category of so-called upwind schemes consists of stencils which are fully aligned with the local direction of propagating waves. Another category of so-called central schemes consists of two-sided stencils, tracing both right-going and left-going waves. A third possibility of stencils which discretize (1.1) 'against the wind' is excluded because of their inherent instability, [14]. A stable scheme should sense the direction of propagation.

At this stage, high resolution stable schemes should compromise between two different sets of directions, where propagation and smoothness might disagree. This require essentially nonlinear schemes, with stencils which adapt their sense of direction according to the computed data. We shall elaborate the details in the context of high-resolution central scheme.

3. Central schemes

We start with a quotation from [14, §12.15], stating "In 1959, Godunov described an ingenious method for one-dimensional problems with shocks". Godunov scheme is in the crossroads between the three major types of local discretizations, namely, finite-difference, finite-volume and finite-element methods. The ingenuity of Godunov's approach, in our view, lies with the evolution of a globally defined approximate solution, $v^h(x, t^n)$, replacing the prevailing approach at that time of an approximate solution which is realized by its discrete gridvalues, $v_\nu(t^n)$. This enables us to pre-process, to evolve and to post-process a globally defined approximation, $v^h(x, t^n)$. The main issue is how to 'manipulate' such piecewise smooth approximations while preserving the desired non-oscillatory invariants.

Godunov scheme was originally formulated in the context of nonlinear conservation laws, where an approximate solution is realized in terms of a first-order accurate, piecewise-constant approximation

$$v^h(x, t^n) := \mathcal{A}_h v(x, t^n) := \sum_\nu \bar{v}_\nu(t^n) 1_{I_\nu}(x), \quad \bar{v}_\nu(t^n) := \frac{1}{|I_\nu|} \int_{I_\nu} v(y, t^n) dy.$$

The cell averages, $\bar{v}_\nu$, are evaluated over the equi-spaced cells, $I_\nu := \{x \setminus |x - x_\nu| \le h/2\}$ of uniform width $h \equiv \Delta x$. More accurate Godunov-type schemes were devised using higher-order piecewise-polynomial projections. In the case of one-dimensional equi-spaced grid, such projections take the form

$$\mathcal{P}_h v(x) = \sum_\nu p_\nu(x) 1_{I_\nu}(x), \qquad p_\nu(x) = v_\nu + v_\nu' \left(\frac{x - x_\nu}{h} \right) + \frac{1}{2} v_\nu'' \left(\frac{x - x_\nu}{h} \right)^2 + \dots.$$

Here, one pre-process the first-order cell averages in order to reconstruct accurate pointvalues, v_ν, and say, couple of numerical derivatives $v_\nu'/h, v_\nu''/h^2$, while the original cell averages, $\{\bar{v}_\nu\}$, should be preserved, $\mathcal{A}_h \mathcal{P}_h v^h = \mathcal{A}_h v^h$. The main issue is extracting information in the direction of smoothness. For a prototype example, let $\Delta_+ \bar{v}_\nu$ and $\Delta_- \bar{v}_\nu$ denote the usual forward and backward differences, $\Delta_\pm v_\nu := \pm(\bar{v}_{\nu\pm1} - \bar{v}_\nu)$. Starting with the given cell averages, $\{\bar{v}_\nu\}$, we set $v_\nu = \bar{v}_\nu$, and compute

$$v_\nu' = mm(\Delta_+ \bar{v}_\nu, \Delta_- \bar{v}_\nu), \quad mm(z_1, z_2) := \frac{sgn(z_1) + sgn(z_2)}{2} \min\{|z_1|, |z_2|\}. \quad (3.1)$$

The resulting piecewise-linear approximation is a second-order accurate, Total Variation Diminishing (TVD) projection, $\|\mathcal{P}_h v^h(x)\|_{BV} \le \|v^h(x)\|_{BV}$. This recipe of so-called minmod numerical derivative, (3.1), is a representative for a large library of non-oscillatory, high-resolution limiters. Such limiters dictate discrete stencils in the direction of smoothness and hence, are inherently nonlinear. Similarly, nonlinear adaptive stencils are used in conjunction with higher-order methods. A description of the pioneering contributions in this direction by Boris & Book, A. Harten, B. van-Leer and P. Roe can be found in [10]. The advantage of dealing with globally defined approximations is the ability to pre-process, to post-process and in particular, to evolve such approximations. Let $u(x, t) = u^h(x, t)$ be the exact solution of (1.1) subject to $u^h(x, t^n) = \mathcal{P}_h v^h(x, t^n)$. The exact solution lies of course outside the finite computational space, but it could be realized in terms of its exact cell averages, $v^h(x, t^{n+1}) = \sum_\nu \bar{v}_\nu(t^{n+1}) 1_{I_\nu}(x)$. Averaging is viewed here a simple post-processing. Two prototype examples are in order.

Integration of (1.1) over control volume $I_\nu \times [t^n, t^{n+1}]$, forms a local stencil which balances between the new averages, $\{\bar{v}_\nu(t^{n+1})\}$, the old ones, $\{\bar{v}_{\nu+k}(t^n)\}$, and the fluxes across the interfaces along $x_{\nu\pm1/2} \times [t^n, t^{n+1}]$. In this case, the solution along these discontinuous interfaces is resolved in terms of Riemann solvers. Since one employs here an exact evolution, the resulting Godunov-type schemes are upwind scheme. The original Godunov scheme based on piecewise constant projection is the forerunner of all upwind schemes. As an alternative approach, one can realize the solution $u(x, t^{n+1})$, in terms of its exact staggered averages, $\{\bar{v}_{\nu+1/2}(t^{n+1})\}$. Integration of (1.1) over the control volume $I_{\nu+1/2} \times [t^n, t^{n+1}]$ subject to piecewise quadratic data given at $t = t^n$, $u(x, t^n) = \mathcal{P}_h v^h(x, t^n)$, yields

$$
\begin{aligned}
\bar{v}_{\nu+1/2}(t^{n+1}) &= \frac{1}{2}\Big(\bar{v}_\nu(t^n) + \bar{v}_{\nu+1}(t^n)\Big) + \\
&+ \frac{1}{8}\Big(v_\nu'(t^n) - v'_{\nu+1}(t^n)\Big) + \frac{\Delta t}{\Delta x}\Big(F_{\nu+1}^{n+1/2} - F_\nu^{n+1/2}\Big), \quad (3.2)
\end{aligned}
$$

where $F_\nu^{n+1/2}$ stands for the averaged flux, $F_\nu^{n+1/2} = \int_{t^n}^{t^{n+1}} f(u(x_\nu, \tau)d\tau/\Delta t$. Thanks to the staggering of the grids, one encounters smooth interfaces $x_\nu \times [t^n, t^{n+1}]$, and the intricate (approximate) Riemann solvers are replaced by simpler quadrature rules. For second-order accuracy, for example, we augment (3.2) with the mid-point quadrature

$$
F_\nu^{n+1/2} = f(v_\nu(t^{n+1/2})), \quad v_\nu(t^{n+1/2}) = v_\nu(t^n) - \frac{\Delta t}{2\Delta x} f(v_\nu(t^n))'. \quad (3.3)
$$

Here, the prime on the right is understood in the usual sense of numerical differentiation of a gridfunction – in this case the flux $\{f(v_\nu(t^n))\}_\nu$. The resulting second-order central scheme (3.3),(3.2) was introduced in [13]. It amounts to a simple predictor-corrector, non-oscillatory high-resolution Godunov-type scheme. For systems, one implements numerical differentiation for each component separately. Discontinuous edges are detected wherever cell-averages form new extreme values, so that $v_\nu'(t^n)$ and $f(v_\nu(t^n))'$ vanish, and (3.3),(3.2) is reduced to the forerunner of all central schemes — the celebrated first-order Lax-Friedrichs scheme, [10]. This first-order stencil is localized to the neighborhood of discontinuities, and by assumption, there are finitely many them. In between those discontinuities, differentiation in the direction of smoothness restores second-order accuracy. This retains the overall high-resolution of the scheme. Consult Figure 1 for example.

Similarly, higher-order quadrature rules can be used in connection with higher-order projections, [3], [11]. A third-order simulation is presented in Figure 2. Finite-volume and finite-element extensions in several space dimensions are realized over general, possibly unstructured control volumes, $\Omega_\nu \times [t^n, t^{n+1}]$, which are adapted to handle general geometries. Central schemes for 2D Cartesian grids were introduced in [6], and extended to unstructured grids in [1]. A similar framework based on triangulated grids for high-resolution central approximations of Hamilton-Jacobi equations was described in [4] and the references therein.

Central schemes enjoy the advantage of simplicity – they are free of (approximate) Riemann solvers, they do not require dimensional splitting, and they apply to arbitrary flux functions[2] without specific references to eigen-decompositions, Jacobians etc. In this context, central schemes offer a "black-box solvers" for an impressive variety of convection-dominated problems. At the same time, the central framework maintain high-resolution by pre -and post-processing in the direction of smoothness. References to diverse applications such as simulations of semiconductors models, relaxation problems, geometrical optics and multiphase computations, incompressible flows, polydisperse suspensions, granular avalanches MHD equations and more can be found at [16].

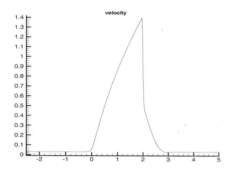

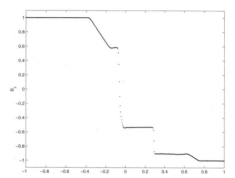

Figure 1: Second-order central scheme simulation of semiconductor device governed by 1D Euler-Poisson equations. Electron velocity in 10^7 cm/s with $N = 400$ cells.

Figure 2: Third-order central scheme simulation of 1D MHD Riemann problem. with $N = 400$ cells. The y-magnetic field at $t = 0.2$.

The numerical viscosity present in central schemes is of order $\mathcal{O}\frac{(\Delta x)^{2r}}{\Delta t}$. It is suitable for the convective regime where $\Delta t \sim \Delta x$, but it is excessive when a small time step is enforced, e.g., due to the presence of diffusion terms. To overcome this difficulty, a new family of central schemes was introduced in [7] and was further refined in [8]. Here, the previous staggered control volumes, $I_{\nu+1/2} \times [t^n, t^{n+1}]$, is replaced by the smaller — and hence less dissipative, $J_{\nu+1/2} \times [t^n, t^{n+1}]$, where $J_{\nu+1/2} := x_{\nu+1/2} + \Delta t \times [a^-, a^+]$ encloses the maximal cone of propagation, $a^\pm \equiv a^\pm_{\nu+1/2} = \binom{\max}{\min}_k \lambda_k^\pm(f_u)$. The fact that the staggered grids are $\mathcal{O}(\Delta t)$ away from each other, yields central stencils with numerical viscosity of order $\mathcal{O}(\Delta x^{2r-1})$. Being independent of Δt enables us to pass to the limit $\Delta t \downarrow 0$. The resulting semi-discrete high-resolution central scheme reads $\dot{\bar{v}}_\nu(t) = -(f_{\nu+1/2}(t) - f_{\nu-1/2}(t))/\Delta x$, with a numerical flux, $f_{\nu+1/2}$, expressed in terms of the reconstructed pointvalues, $v^\pm \equiv v^\pm_{\nu+1/2} = \mathcal{P}_h v^h(x_{\nu+1/2}\pm, t)$,

$$f_{\nu+1/2}(t) := \frac{a^+ f(v^-) - a^- f(v^+)}{a^+ - a^-} + a^+ a^- \frac{v^+ - v^-}{a^+ - a^-}. \qquad (3.4)$$

---

[2]An instructive example is provided by gasdynamics equations with tabulated equations of state.

Instructive examples are provided in Figures 3, 4.

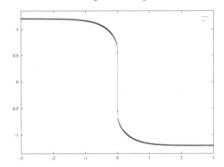

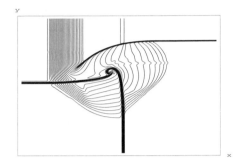

Figure 3: Convection-diffusion eq. $u_t + (u^2)_x/2 = (u_x/\sqrt{1+u_x^2})_x$ simulated by (3.4),(3.1), with 400 cells.

Figure 4: Third-order central scheme for 2D Riemann problem. Density contour lines with 400×400 cells.

4.   Adaptive spectral viscosity methods

Godunov-type methods are based on zeroth-order moments of (1.1). In each time step, one evolves one piece of information per spatial cell — the cell average. Higher accuracy is restored by numerical differentiation in the direction of smoothness. An alternative approach is to compute higher-order moments, where the cell averages, $\bar{v}_\nu$, and say, couple of numerical derivatives, v_ν', v_ν'' are evolved in time. Prototype examples include discontinuous Galerkin and streamline-diffusion methods, where several local moments per computational cell are evolved in time, consult [4] and the references therein. As the number of projected moments increase, so does the size of the computational stencil. At the limit, one arrives at spectral methods based on global projections,

$$v_N(x,t) = \mathcal{P}_N v(x,t) := \sum_{|k|\leq N} \widehat{v}_k(t)\phi_k(x), \quad \widehat{v}_k := \langle v(\cdot,t), \phi_k(\cdot) \rangle.$$

Here, $\phi_k(x)$ are global basis functions, $\phi_k = e^{ikx}, T_k(x), ...$ and $\widehat{v}_k$ are the moments induced by the appropriately weighted, possibly discrete inner-product $\langle \cdot, \cdot \rangle$. Such global projections enjoy superior resolution — the error $\|v(\cdot) - \mathcal{P}_N v(\cdot)\|$ decays as fast as the global smoothness of $v(\cdot)$ permits. With piecewise smooth solutions, however, we encounter first-order spurious Gibbs' oscillations throughout the computational domain. As before, we should be able to pre- and post-process piecewise smooth projections, recovering their high accuracy in the direction of smoothness. To this end, local limiters are replaced by global concentration kernels of the form, [5], $K_N^\eta v_N(x) = \frac{\pi\omega(x)}{N}\sum_{|k|\leq N}\eta\left(\frac{|k|}{N}\right)\widehat{v}_k(\phi_k(x))_x$, where $\eta(\cdot)$ is an arbitrary unit mass $C^\infty[0,1]$ function at our disposal. Detection of edges is facilitated by separation between the $\mathcal{O}(1)$ scale in the neighborhoods of edges and $\mathcal{O}(h^r)$ scales in regions of smoothness, $K_N v_N(x) \sim [v(x)] + \mathcal{O}(Nd(x))^{-r}$. Here, $[v(x)]$ denotes the amplitude of the jump discontinuity at x (– with vanishing amplitudes signaling smoothness), and $d(x)$ is the distance to the singular support of $v(\cdot)$. Two prototype examples in the 2π-periodic setup are in order. With $\eta(\xi) \equiv 1$ one recovers a first-order concentration kernel due to Fejer. In [5] we introduced the concentration kernel, $\eta^{exp}(\xi) :\sim exp\{\beta/\xi(1-\xi)\}$, with exponentially fast decay into regions of smoothness. Performing the minmod limitation, (3.1), $mm\{K_N^1 v_N(x), K_N^{exp} v_N(x)\}$, yields an adaptive, essentially non-oscillatory edge detector with enhanced separation of scales near jump discontinuities. Once macro-scale features of non-smoothness were located, we turn to reconstruct the information in the direction of smoothness. This could be carried out either in the physical space using adaptive mollifiers, $\Psi_{\theta,p}$, or by nonlinear adaptive filters, σ_p. In the 2π periodic case, for example, we set $\Psi * \mathcal{P}_N v(x) := \langle \Psi_{\theta,p}(x - \cdot), \mathcal{P}_N v(\cdot) \rangle$ where Ψ is expressed in terms of the Dirichlet kernel, $D_p(y) := \frac{\sin(p+1/2)\pi y}{2\sin(\pi y/2)}$,

$$\Psi_{\theta,p}(y) = \frac{1}{\theta}\rho\left(\frac{y}{\theta}\right)D_p\left(\left(\frac{y}{\theta}\right)\right), \quad \rho \equiv \rho_\beta(y) := e^{\beta y^2/(y^2-1)}1_{[-1,1]}(y). \qquad (4.1)$$

Mollifiers encountered in applications maintain their finite accuracy by localization, letting $\theta \downarrow 0$. Here, however, we seek superior accuracy by the process of cancellation with increasing $p \uparrow$. To guarantee that the reconstruction is supported in the direction of smoothness, we maximize θ in terms of the distance function, $\theta(x) = d(x)$, so that we avoid crossing discontinuous barriers. Superior accuracy is achieved by the adaptive choice $p :\sim d(x)N$, which yields, [20],

$$|\Psi_{\{d(x),d(x)N\}} * \mathcal{P}_N v(x) - v(x)| \le Const.(d(x)N)^r e^{-\gamma\sqrt{d(x)N}}.$$

The remarkable exponential recovery is due to the Gevrey regularity of $\rho_\beta \in \mathcal{G}_2$ and Figure 5 demonstrates such adaptive recovery with exponential convergence rate recorded in Figure 6. An analogous filtering procedure can be carried out in the dual Fourier space, and as before, it hinges on a filter with an adaptive degree p

$$\Psi * \mathcal{P}_N v(x) := \sum_{|k|\le N} \sigma_p\left(\frac{|k|}{N}\right)\hat{v}_k e^{ikx}, \quad \sigma_p(\xi) = e^{\beta\xi^p/(\xi^2-1)}, \quad p \sim (d(x)N)^{r/r+1}.$$

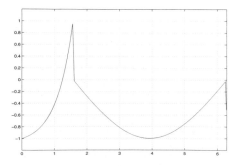
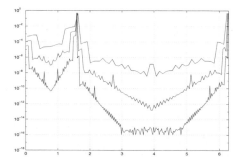

Figure 5: Adaptive reconstruction of the piecewise smooth data from its using $N = 128$-modes using (4.1).

Figure 6: Log error for an adaptive spectral mollifier based on $N = 32$, 64, and 128 modes.

Equipped with this toolkit to process spectral projections of piecewise smooth data, we turn to consider their time evolution. Godunov-type methods are based evolution of cell averages. Cell averaging is dissipative, but projections onto higher-order moments are not. The following example, taken from [18] shows what go wrong with global projections. We consider the Fourier method for 2π-periodic inviscid Burgers' equation, $\partial_t v_N(x,t) + \partial_x S_N(v_N^2(\cdot,t))/2 = 0$. Orthogonality implies that $\|v_N(\cdot,t)\|_{L^2}^2$ is conserved in time, and in particular, that the Fourier method admits a weak limit, $v_N(\cdot,t) \rightharpoonup \bar{v}(t)$. At the same time, $\bar{v}(t)$ is not a strong limit, for otherwise it will contradict the strict entropy dissipation associated with shock discontinuities. Lack of strong convergence indicates the persistence of spurious dispersive oscillations, which is due to lack of entropy dissipation. With this in mind, we turn to discuss the Spectral Viscosity (SV) method, as a framework to stabilize the evolution of global projections without sacrificing their spectral accuracy. To this end one augments the usual Galerkin procedure with high frequency viscosity regularization

$$\frac{\partial}{\partial t} v_N(x,t) + \mathcal{P}_N \nabla_x \cdot f(v_N(\cdot,t)) = -N \sum_{|k| \le N} \sigma\left(\frac{|k|}{N}\right) \widehat{v}_k(t) \phi_k(x), \qquad (4.2)$$

where $\sigma(\cdot)$ is a low-pass filter satisfying $\sigma(\xi) \ge \left(|\xi|^{2s} - \frac{\beta}{N}\right)^+$. Observe that the SV is only activated on the highest portion of the spectrum, with wavenumbers $|k| > \gamma N^{(2s-1)/2s}$. Thus, the SV method can be viewed as a compromise between the stable viscosity approximations corresponding to $s = 0$ but restricted to first order, and the spectrally accurate yet unstable spectral method corresponding to $s = \infty$.

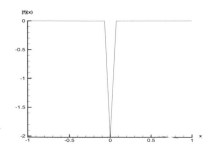 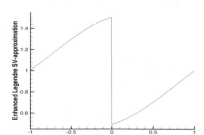

Figure 7: Enhanced detection of edges with $v(x)$ given by $v(x) = sgnx \cdot \cos(x + x \cdot sgnx/2)1_{[-\pi,\pi]}(x)$.

Figure 8: Legendre SV solution of inviscid Burgers' equation. Reconstruction in the direction smoothness.

The additional SV on the right of (4.2) is small enough to retain the formal spectral accuracy of the underlying spectral approximation. At the same time, SV is large enough to enforce a sufficient amount of entropy dissipation and hence, to prevent the unstable spurious Gibbs' oscillations. The original approach was introduced in [18] in conjunction with second-degree dissipation, $s = 1$. Extensions to several

E. Tadmor

space variables, non-periodic expansions, further developments of hyper SV methods with $0 < s < \infty$, and various applications can be found in [17]. We conclude with an implementation of adaptive SV method for simple inviscid Burgers' in Figure 8.

References

[1] P. Arminjon & M.-C. Viallon, Convergence of a finite volume extension of the Nessyahu-Tadmor scheme on unstructured grids for a two-dimensional linear hyperbolic equations, *SIAM J. Numer. Anal.* 36 (1999) 738–771.

[2] S. Bianchini & A. Bressan, Vanishing viscosity solutions to nonlinear hyperbolic systems, Preprint S.I.S.S.A., Trieste 2001.

[3] F. Bianco, G. Puppo & G. Russo, High order central schemes for hyperbolic systems of conservation laws, *SIAM J. Sci. Comput.* 21 (1999) 294–322.

[4] B. Cockburn, C. Johnson, C.-W. Shu & E. Tadmor, *Advanced Numerical Approximation of Nonlinear Hyperbolic Equations*, (A. Quarteroni, ed.), Lecture Notes in Mathematics 1697, Springer 1998.

[5] A. Gelb & E. Tadmor, Detection of edges in spectral data II. Nonlinear enhancement, *SIAM J. Numer. Anal.*, 38 (2000), 1389–1408.

[6] G.-S. Jiang & E. Tadmor, Non-oscillatory central schemes for multidimensional hyperbolic conservation laws *SIAM J. Sci. Comput.*, 19 (1998) 1892–1917.

[7] A. Kurganov & E. Tadmor, New high-resolution central schemes for nonlinear conservation laws and convection-diffusion equations, *J. Comp. Phys.*, 160 (2000), 214–282.

[8] A. Kurganov, S. Noelle & G. Petrova, Semi-discrete central-upwind scheme for hyperbolic conservation laws and Hamilton-Jacobi equations, *SIAM J. Sci. Comput.* 23 (2001), 707–740.

[9] P. D. Lax, *Hyperbolic Systems of Conservation Laws and the mathematical Theory of Shock Waves*, SIAM, 1973.

[10] R. LeVeque, *Numerical Methods for Conservation Laws*, Lecture Notes in Mathematics, Birkhauser, 1992.

[11] D. Levy, G. Puppo & G. Russo, Central WENO schemes for hyperbolic systems of conservation laws *Math. Model. Numer. Anal.* 33 (1999) 547–571.

[12] A. Majda, *Compressible Fluid Flow and Systems of Conservation Laws in Several Space Variables*, Springer, 1984.

[13] H. Nessyahu & E. Tadmor, Non-oscillatory central differencing for hyperbolic conservation laws, *J. Comp. Phys.*, 87 (1990), 408–463.

[14] R. Richtmyer & K. W. Morton, *Difference Methods for Initial-Value Problems*, Interscience, John Wiley, 1967.

[15] F. Şabac, The optimal convergence rate of monotone finite difference methods for hyperbolic conservation laws, *SIAM J. Numer. Anal.*, 34 (1997), 2306–2318.

[16] E. Tadmor, http://www.math.ucla.edu/~tadmor/centralstation/

[17] E. Tadmor, http://www.math.ucla.edu/~tadmor/spectral_viscosity/

[18] E. Tadmor, Convergence of spectral methods for nonlinear conservation laws, *SIAM J. Numer. Anal.*, 26 (1989), 30–44.

[19] E. Tadmor & T. Tang, Pointwise error estimates for scalar conservation laws

with piecewise smooth solutions, *SIAM J. Numer. Anal.*, 36 (1999), 1739–1758.

[20] E. Tadmor & J. Tanner, Adaptive mollifiers – high resolution recovery of piece-wise smooth data from its spectral information, *Found. Comput. Math.*, 2 (2002) 155–189.

[21] E. Tadmor & T. Tassa, On the piecewise regularity of entropy solutions to scalar conservation laws, *Comm. PDEs*, 18 (1993), 1631–1652.

Section 17. Application of Mathematics in the Sciences

ICM 2002 · Vol. III · 761–772

Some Geometric PDEs Related to Hydrodynamics and Electrodynamics

Yann Brenier*

Abstract

We discuss several geometric PDEs and their relationship with Hydrodynamics and classical Electrodynamics. We start from the Euler equations of ideal incompressible fluids that, geometrically speaking, describe geodesics on groups of measure preserving maps with respect to the L^2 metric. Then, we introduce a geometric approximation of the Euler equation, which involves the Monge-Ampère equation and the Monge-Kantorovich optimal transportation theory. This equation can be interpreted as a fully nonlinear correction of the Vlasov-Poisson system that describes the motion of electrons in a uniform neutralizing background through Coulomb interactions. Finally we briefly discuss an equation for generalized extremal surfaces in the 5 dimensional Minkowski space, related to the Born-Infeld equations, from which the Vlasov-Maxwell system of classical Electrodynamics can be formally derived.

2000 Mathematics Subject Classification: 58D05, 35Q, 82D10, 76B.
Keywords and Phrases: Hydrodynamics, Euler equations, Geodesics, System of particles, Monge-Ampère equations, Extremal surfaces, Electrodynamics, Born-Infeld equations.

1. The Euler equations of incompressible fluids

The motion of an incompressible fluid moving in a compact domain D of the Euclidean space $\mathbf{R}^d$ can be mathematically defined as a trajectory $t \to g(t)$ on the set, subsequently denoted by $G(D)$, of all diffeomorphisms of D with unit jacobian determinant. This space can be embedded in the set $S(D)$ of all Borel maps h from D into itself, not necessarily one-to-one, such that

$$\int_D \phi(h(x))dx = \int_D \phi(x)dx$$

for all $\phi \in C(D)$, where dx denotes the Lebesgue measure, normalized so that the measure of D is 1. For the composition rule, $G(D)$ is a group (the identity map I

*CNRS, Université de Nice-Sophia-Antipolis, France (on leave of absence from Université Paris 6) and Institut Universitaire de France. E-mail: brenier@math.unice.fr

being the unity of the group), meanwhile $S(D)$ is a semi-group. Both $G(D)$ and $S(D)$ are naturally embedded in the Hilbert space $H = L^2(D, \mathbf{R}^d)$ of all square integrable mapping from D into $\mathbf{R}^d$ and, therefore, inherit from H a formal Riemannian structure. The equations of geodesics on $G(D)$ turn out to be exactly [AK] the equations of incompressible inviscid fluids introduced by Euler near 1750 [Eu]. The Euler equations play a fundamental role in Fluid Mechanics (for geophysical flow modelling in particular) and their global well-posedness is one of the most challenging problems in the field of nonlinear PDEs. Their mathematical importance is confirmed by the recent publication of several books by Arnold-Khesin [AK], Chemin [Ch], P.-L. Lions [Li], Marchioro-Pulvirenti [MP], as well as by Majda's lecture in the Kyoto ICM [Ma].

From a geometric point of view (different from the usual PDE setting which consists in solving the Euler equations with prescribed initial conditions), it is natural to look for minimizing geodesics between the identity map and prescribed measure preserving maps. More precisely :

Definition 1.1 *Given $h \in G(D)$, find a curve $t \in [0,1] \rightarrow g(t) \in G(D)$ satisfying $g(0) = I$, $g(1) = h$, that minimizes*

$$A_D(g) = \frac{1}{2} \int_0^1 \|g'(t)\|_{L^2}^2 dt = \frac{1}{2} \int_0^1 \int_D |\partial_t g(t,x)|^2 dx dt.$$

The infimum is nothing but $\frac{1}{2}\delta_D^2(I, h)$, where δ_D denotes the geodesic distance on $G(D)$, and any smooth minimizer g must be a smooth solution of the Euler equations (written in "Lagrangian coordinates")

$$g'' \circ g^{-1} = -\nabla p,$$

where $p = p(t,x) \in \mathbf{R}$ is the pressure field and $\nabla p = (\partial_{x_1} p, ..., \partial_{x_d} p)$. The minimization problem will be subsequently called "Shortest Path Problem" (SPP).

The basic local existence and uniqueness theorem for the SPP is due to Ebin and Marsden [EM]. If h and I are sufficiently close in a sufficiently high order Sobolev norm, then there is a unique shortest path. In the large, uniqueness can fail for the SPP. For example, in the case when D is the unit disk, $h(z) = -z$, the SPP has two solutions $g(t,z) = ze^{+i\pi t}$ and $g(t,z) = ze^{-i\pi t}$, where complex notations are used. In 1985, A. Shnirelman [Sh] found, in the case $D = [0,1]^3$, a class of data for which the SPP cannot have a (classical) solution. These data are those of form

$$h(x_1, x_2, x_3) = (H(x_1, x_2), x_3),$$

where H is an area preserving mapping of the unit square, i.e. an element of $G([0,1]^2)$, for which

$$\delta_{[0,1]^3}(I, h) < \delta_{[0,1]^2}(I, H) < +\infty.$$

(This means that, although h is really a two dimensional map, genuinely 3D motions perform better to reach h from I than purely 2D motions.)
Shnirelman also proved [Sh], [Sh2], that $S([0,1]^d)$ is the right completion of $G([0,1]^d)$ for the geodesic distance δ, for all dimension $d \geq 3$. (Notice that $S([0,1]^d)$ is the

L^2 completion of $G([0,1]^d)$ for all $d \geq 2$ [Ne]. So, the case $d = 2$ is very peculiar.) In such situations, a complete existence and uniqueness result for the SPP was obtained in [Br2], *provided* the pressure field is considered as the right unknown and not the path $t \to g(t)$ itself.

Theorem 1.2 *Let* $h \in S([0,1]^3)$ *of form* $h(x_1, x_2, x_3) = (H(x_1, x_2), x_3)$ *with* $H \in S([0,1]^2)$. *Then there is a unique vector-valued measure* $\nabla p(t, x_1, x_2)$ *such that, for each sequence of curves* $t \in [0,1] \to g_n(t) \in G([0,1]^3)$ *labelled by* $n \in \mathbf{N}$ *and satisfying*

$$A_{[0,1]^3}(g_n) \to \frac{1}{2} \delta^2_{[0,1]^3}(I, h), \quad ||g_n(1) - h||_{L^2([0,1]^3)} \to 0,$$

as $n \to \infty$, *then (in the distributional sense)*

$$g_n'' \circ g_n^{-1} \to -\nabla p.$$

In other words, the acceleration field of *all* minimizing sequences converge to $-\nabla p$ which uniquely depends on data h. The proof relies on an appropriate concept of generalized solutions (related to "Young's measures" [Yo], [Ta], [DM], [She]) that describe the oscillatory behaviour of the (g_n) as $n \to +\infty$ and reduces the SPP to a convex minimization problem. (See [Br2] for more details.) More precisely, the associated measures

$$c_n(t, x, a) = \delta(x - g_n(t, a)), \quad m_n(t, x, a) = \partial_t g_n(t, a) \delta(x - g_n(t, a)),$$

have cluster points (c, m) that have the following properties :
1) m is absolutely continuous with respect to c and its vector-valued density $v(t, x, a)$ is $c-$ square integrable;
2) c and v do not depend on x_3 and $v_3 = 0$,
3) c and v solve

$$\partial_t c + \nabla_x.(cv) = 0, \quad \partial_t(cv) + \nabla_x.(cv \otimes v) + c\nabla_x p = 0, \tag{1.1}$$

where the product $c\nabla_x p$ has to be properly defined (in a way related to the work of Zheng and Majda [ZM]). Equations (1.1) are obtained as the optimality equations of the convexified minimization problem. Therefore, it is a priori unclear they have any physical meaning as evolution equations. However, they correspond, up to a change of unknown, to the hydrostatic limit of the Euler equations, obtained from the Euler equations by neglecting the vertical acceleration term, namely:

$$Kg'' \circ g^{-1} = -\nabla p,$$

where K is the singular diagonal matrix $(1,1,0)$. These hydrostatic (or "shear flow") equations are widely used for atmosphere and ocean circulation modelling, as the building block of the so-called "primitive equations". However, they are more singular than the Euler equations and their mathematical analysis is very limited, as discussed in [Li]. Conditional well posedness and derivation from the Euler equations have been established in [Br3] and [Gr].

Remarks

An intriguing question is whether or not the uniqueness of ∇p can be proved by more classical tools even in the case when $H \in G([0,1]^2)$ can be connected to the identity map by a classical shortest path on $G([0,1]^2)$.

Since $S([0,1]^3)$ is the right completion of $G([0,1]^3)$ with respect to the geodesic distance, one could expect the SPP to have a solution in $S([0,1]^3)$ for all data h. This is not true. An example of such a data is $h(x_1, x_2, x_3) = (1 - x_1, x_2, x_3)$. Only generalized flows, as discussed in [Br2], can describe shortest paths in full generality.

Example of generalized solutions

Explicit examples of non trivial generalized shortest paths can be computed either numerically or exactly. Let us just quote a typical example, when D is the cylinder $\{(z,s) = (x_1, x_2, s), \ |z| \le 1, \ 0 \le s \le 1\}$ and $h(z,s) = (-z,s)$. Then, the classical SPP has two distinct solutions $g_+(t,z,s) = (e^{i\pi t}z, s)$ and $g_-(t,z,s) = (e^{-i\pi t}z, s)$, with the same pressure field $p = \pi^2|z|^2/2$, where complex notations are used on the disk $|z| \le 1$. (Notice that there is no motion along the vertical axis s.) Trivial generalized solutions are obtained by mixing these two solutions. However, a non trivial generalized solution exists and can be described as follows. For each fluid particle initially located at (z,s), the elevation s stays unchanged and the initial horizontal position z splits up along a circle of radius $(1 - |z|^2)^{1/2}\sin(\pi t)$, with center $z\cos(\pi t)$, that moves across the unit disk and shrinks down to the point $-z$ as $t = 1$. In addition, each particle is accelerated by the pressure field $p = \pi^2|z|^2/2$, as expected from the theory.

2. Polar factorization of maps and the Monge-Ampère equation

A way to define approximate geodesics on $G = G(D)$ is to introduce a penalty parameter $\epsilon > 0$ and to consider the formal (hamiltonian) dynamical system in the Hilbert space $H = L^2(D, \mathbf{R}^d)$

$$\epsilon^2 \frac{d^2}{dt^2}M + \frac{\delta}{\delta M}\left(\frac{d_H^2(M,G)}{2}\right) = 0, \tag{2.1}$$

where the unknown M is a curve $t \to M(t) \in H$, $\delta/\delta M$ denotes the gradient operator in H, and

$$d_H(M,G) = \inf_{g \in G} ||M - g||_H \tag{2.2}$$

is the distance in H between M and G, where $||.||_H$ is the Hilbert norm of H. This approach is related to Ebin's slightly compressible flow theory [Eb], and is a natural extension of the theory of constrained finite dimensional mechanical systems [RU], [AK]. Notice that the approximate geodesic equation is sensitive only to the L^2 closure of $G(D)$, which is, in the case $D = [0,1]^d$, $d \ge 2$, the entire semi-group $S(D)$ [Ne]. As the penalty parameter ϵ goes to zero, we expect that for appropriate initial data, typically for $M(t = 0) = I$ and $(d/dt)M(t = 0) = v_0$, where v_0 is a smooth divergence free vector field on D tangent to the boundary, the time dependent map M converges to a geodesic curve on G. Because of the classical properties of the distance function in a Hilbert space, for each point $M \in H$ for which there exists a unique closest point $\pi_S(M)$ on $S(D)$, we have

$$\frac{\delta}{\delta M}\left(\frac{d_H^2(M,G)}{2}\right) = M - \pi_S(M). \tag{2.3}$$

Thus, we can formally write the approximate geodesic equation (2.1)

$$\epsilon^2 \frac{d^2}{dt^2}M + M - \pi_S(M) = 0. \tag{2.4}$$

Therefore, it is natural to address the following variational problem, that we call the Closest Point Problem (CPP)

Definition 2.1 *Given $M \in L^2(D, \mathbf{R}^d)$, find $h \in S(D)$ that minimizes*

$$\frac{1}{2}\int_D |M(x) - h(x)|^2 dx.$$

The solution of the CPP is given by the Polar Factorization theorem for maps [Br1]

Theorem 2.2 *Let $M : D \to \mathbf{R}^d$ be an L^2 map such that the probability measure*

$$\rho_M(x) = \int_D \delta(x - M(a))da$$

is a Lebesgue integrable function on D. Then, there exists a unique closest point $\pi_S(M)$ on $S(D)$ and there is a Lipschitz convex function Φ on $\mathbf{R}^d$ such that

$$\pi_S(M)(a) = (\nabla\Phi)(M(a)), \quad a.e. \ a \in D.$$

In addition, Φ is a weak solution, in a suitable sense, of the Monge-Ampère equation

$$\det(\partial_{xx}\Phi(x)) = \rho_M(x).$$

Thus, the Monge-Ampère equation [Ca], which is usually considered as a non variational geometric PDE related to the concept of Gaussian curvature, also is the optimality equation of a variational problem closely linked to the Euler equations of incompressible inviscid fluids. In addition, the Polar Factorization theorem can be seen as a nonlinear version of the Helmholtz-Hodge decomposition theorem for vector fields which asserts that any L^2 vector field on D can be written in a unique way as the (orthogonal) sum of the gradient of a scalar field and a divergence free field tangent to ∂D. Shortly after [Br1], Caffarelli [Ca] established several regularity results for the Polar Factorization. For example, provided D is smooth and strictly convex, any smooth orientation preserving diffeomorphism M of D has a unique Polar Factorization with smooth factors, and $\pi_S(M)$ belongs to $G(D)$. More recently, McCann [Mc] generalized the Polar Factorization theorem when D is a compact Riemannian manifolds.

3. Optimal Transportation Theory

In [Br1], the solution of the CPP problem is based on the Optimal Transportation Theory (OTT). The OTT was introduced by Monge in 1781 [Mo] to solve an engineering problem and renewed by Kantorovich near 1940 [Ka] in the framework of Linear Programing and Probability Theory [RR]. In modern words, this amounts to look for a probability measure μ on a given product measure space $A \times B$, with prescribed projections on A and B, that minimizes

$$\int_{A \times B} c(x,y) d\mu(x,y),$$

where the "cost function" $c \geq 0$ is given on $A \times B$. The CPP roughly corresponds to the case when $A = B = D$, $c(x,y) = |M(x) - y|^2$ and each projection of μ is the (normalized) Lebesgue measure on D. The connexion established in [Br1] between the OTT and the Monge-Ampère equation, enhanced by Caffarelli's regularity theory [Ca], introduced OTT as an active field of research in nonlinear PDEs. Let us first quote the work of Evans-Gangbo [Ev] to solve the original Monge problem with PDE techniques, related to the Eikonal equations, and the recent contributions of Ambrosio, Caffarelli, Feldman, McCann, Trudinger, Wang. (A first attempt was made by Sudakov [Su] with purely probabilistic tools.) Let us next point out the importance of OTT for modelling purposes in Applied Mathematics. First of all, it is fair to say that the OTT and the Monge-Ampère equation were already key ingredients in Cullen and Purser's theory of semi-geostrophic atmospheric flows, which goes back to the early 1980s and preceded our Polar Factorization theorem (see references in [CNP]). Next, Jordan-Kinderlehrer-Otto [JKO], using OTT, established that the heat equation can be seen as a gradient flow for Boltzmann's entropy functional. More systematically, Otto [Ot] showed how the OTT confers a natural Riemannian structure to sets of Probability measures and recognized a large class of dissipative PDEs as gradient flows of various functionals for such Riemannian structures. Examples of such PDEs are porous media equations, lubrication equations, granular flow equations, etc... Let us also mention that OTT has became a powerful tool in Calculus of Variations (through McCann's concept of displacement convexity [Mc]) and Functional Analysis, where all kind of functional inequalities (Minkowski, Brascamp-Lieb, Log Sobolev, Bacry-Emery, etc,...) can be established through OTT arguments, as shown, in particular, by Barthe [Ba], McCann [Mc], Otto, Villani [OV]. Let us finally mention that [BB] has provided for the OTT a formulation different from the Monge-Kantorovich one, by introducing an interpolation variable (which was already present in McCann's concept of displacement convexity). This point of view is useful for both numerical [BB] and theoretical purposes, in particular, by allowing non trivial generalizations of the OTT related to section .

4. Approximate geodesics and Electrodynamics

Let us go back to the approximate geodesic equation (2.4) that can be (formally) written, thanks to the Polar Factorization theorem,

$$\partial_{tt}M(t,a) + (\nabla\phi)(t, M(t,a)) = 0, \quad \det(I - \epsilon^2 \partial_{xx}\phi(t,x)) = \rho_M(t,x) \qquad (4.1)$$

(where $\phi(t,x)$ stands for $\epsilon^{-2}(|x|^2/2 - \Phi(t,x))$). A formal expansion about $\epsilon = 0$ leads, as expected, to the Euler equation (written in Lagrangian coordinates) at the zero order and, at the next order (and exactly as $d = 1$), to

$$\partial_{tt}M(t,a) + (\nabla\phi)(t, M(t,a)) = 0, \quad \epsilon^2 \Delta\phi(t,x) = 1 - \rho_M(t,x), \qquad (4.2)$$

which can be equivalently written as

$$\partial_t f + \xi.\nabla_x f - \nabla_x\phi.\nabla_\xi f = 0, \quad \epsilon^2 \Delta\phi = 1 - \int f d\xi \qquad (4.3)$$

by introducing the "phase density"

$$f(t,x,\xi) = \int_D \delta(x - M(t,a))\delta(\xi - \partial_t M(t,a))da.$$

This system is nothing but the Vlasov-Poisson system that describes the classical non-relativistic motion of a continuum of electrons around a homogeneous neutralizing background of ions through Coulomb interactions.

So, the approximate geodesic equation, which can be written as a "Vlasov-Monge-Ampère" (VMA) system,

$$\partial_t f + \xi.\nabla_x f - \nabla_x\phi.\nabla_\xi f = 0, \quad \det(I - \epsilon^2 \partial_{xx}\phi) = \int f d\xi \qquad (4.4)$$

can be interpreted as a (fully nonlinear) correction of the Vlasov-Poisson system for small values of ϵ. Recently, Loeper [Lo] has shown that the VMA system has local smooth solutions and global weak solutions. Loeper has also proved that the Euler equations and the Vlasov-Poisson system correctly describe the asymptotic behaviour of the VMA system as $\epsilon \to 0$. The asymptotic analysis is based on the so-called modulated energy method already used in [Br5] to derive the Euler equations from the Vlasov-Poisson system.

Notice that, thanks to the substitution of the Monge-Ampère equation (a fully non-linear elliptic PDE) for the classical Poisson equation, the "electric" field $\nabla\phi(t,x)$ is pointwise bounded by the diameter of D divided by ϵ^2, independently on the initial conditions. In particular, point charges do not create unbounded force fields as in classical Electrodynamics.

5. A caricature of Coulomb interaction

The approximate geodesic equation (2.4) can be easily discretized in space by substituting i) for D a discrete set of N "grid" points equally spaced in D, say $A_1, ..., A_N$, ii) for H the euclidean space $\mathbf{R}^{dN}$, iii) for G the discrete set of all sequences $(A_{\sigma_1}, ..., A_{\sigma_N}) \in \mathbf{R}^{dN}$ generated by permutations σ of the first N integers, while keeping unchanged equation (2.4). (Note that such a discretization using permutations cannot be so easily defined for the Euler equations, which formally correspond to the limit case $\epsilon = 0$.) Then $M(t) = (M_1(t), ..., M_N(t))$ can be interpreted as a set of N harmonic oscillators

$$\epsilon^2 \frac{d^2}{dt^2} M_\alpha + M_\alpha - A_{\sigma_\alpha(t)} = 0, \tag{5.1}$$

where the time dependent permutation $\sigma(t)$ is subject to minimize, at all time t, the total potential energy

$$\sum_{\alpha=1}^{N} |M_\alpha(t) - A_{\sigma_\alpha}|^2. \tag{5.2}$$

This system can be seen as a collection of N springs linking each particle M_α to one of the fixed particle A_β according to a dynamical pairing $\beta = \sigma_\alpha(t)$ maintaining the bulk potential energy at the lowest level. There is some ambiguity in the definition of this formal hamiltonian system for which the hamiltonian is given by

$$\frac{1}{2} \sum_{\alpha=1}^{N} |\frac{dM_\alpha}{dt}|^2 + \inf_\sigma \frac{1}{2\epsilon^2} \sum_{\alpha=1}^{N} |M_\alpha - A_{\sigma_\alpha}|^2. \tag{5.3}$$

In particular, $\sigma(t)$ is not uniquely defined at each time t for which several particles have the same position. However, the potential is the sum of a quadratic and a Lipschitz concave functions of M. So its gradient has linear growth at infinity and its second order partial derivatives are locally bounded measures. This is enough, according to recent results by Lions and Bouchut [Bo], [Li2], to ensure that unique global solutions are well defined for Lebesgue almost every initial data $M_\alpha(0)$, $\frac{d}{dt} M_\alpha(0)$, $\alpha = 1, ..., N$. As expected, the limit $N \to +\infty$, $\epsilon \to 0$ (provided N goes fast enough to $+\infty$), leads to the Euler equation, as proven in [Br4]. From the electrostatic point of view, the dynamical system describes a nonlinearly cutoff Coulomb interaction between N electrons (with positions M_α) and a background of N motionless ions (with fixed positions A_α).

6. Generalized extremal surface equations and Electrodynamics

As seen above, the approximate geodesic equation (4.1)—which has been introduced as a natural geometrical approximation to the Euler equations—turns out to be a model for electrostatic interaction with a non-linearly cutoff Coulomb potential. This feature is somewhat reminiscent of the Born-Infeld non-linear theory of the electromagnetic field [BI] (see also [BDLL], [GZ]...). Therefore, one may try to design from similar geometric ideas a non-linearly cutoff theory for classical Electrodynamics. An attempt is made in [Br6]. Instead of considering springs linking two particles of opposite charges we rather consider (with a more space-time oriented point of view) surfaces $(t, s) \rightarrow X(t, s)$ spanning curves $t \rightarrow X_-(t)$ and $t \rightarrow X_+(t)$ followed by two particles of opposite charge, so that $X(s = -1, t) = X_-(t)$ and $X(s = 1, t) = X_+(t)$, $s \in [-1, 1]$ standing for the "interpolation" parameter between the two trajectories. Just by prescribing $(t, s) \rightarrow (t, s, X(t, s))$ to be an extremal surface in the the 5 dimensional Minkowski space (t, s, x_1, x_2, x_3) (with signature $(-+++ +)$), we get the building block of the model. In other words, the individual Action of each surface is

$$\int \sqrt{1 + |\partial_s X|^2 - |\partial_t X|^2 - |\partial_s X \times \partial_t X|^2} dt ds, \qquad (6.1)$$

(which is basically the Nambu-Goto Action of classical string theory). Next, we associate with X a "generalized surface" (ρ, J, E, B) (or more precisely a "cartesian current" in the sense of [GMS]) defined by

$$\rho(t, s, x) = \delta(x - X(t, s)), \quad J(t, s, x) = \partial_t X(t, s) \delta(x - X(t, s)), \qquad (6.2)$$

$$E(t, s, x) = \partial_s X(t, s) \delta(x - X(t, s)), \qquad (6.3)$$

$$B(t, s, x) = \partial_s X(t, s) \times \partial_t X(t, s) \delta(x - X(t, s)) \qquad (6.4)$$

and subject to compatibility conditions

$$\partial_s \rho + \nabla . E = 0, \quad \partial_t \rho + \nabla . J = 0, \quad \partial_t E - \partial_s J - \nabla \times B = 0. \qquad (6.5)$$

In terms of (ρ, J, E, B) the Action of X can be written as

$$K(\rho, J, E, B) = \int \sqrt{\rho^2 - J^2 + E^2 - B^2}. \qquad (6.6)$$

Varying this Action under constraint (6.5) leads to a system of evolution equations for (ρ, J, E, B) (see [Br6] for an explicit form), that we can call "generalized extremal surface equations" (GESE). They enjoy (at least in the simplest cases when the solutions depend on one or two space variables) many interesting properties : hyperbolicity, linear degeneracy of all fields [BDLL], symmetries between t and s, J and E etc... From the GESE, we can derive through various (formal!) limiting process 1) the Born-Infeld and the Maxwell equations, as (ρ, J) are prescribed at $s = -1$ and $s = +1$ (in which case there is no coupling between charged particles and the electromagnetic field), 2) the Vlasov-Born-Infeld and the Vlasov-Maxwell equations as $(E, B) = 0$ is prescribed at $s = -1$ and $s = +1$ (which corresponds to a free boundary condition $\partial_s X = 0$ for an individual surface and yields a full coupling between charged particles and the electromagnetic field). In spite of the possible physical irrelevance of the GESE, their mathematical analysis (global existence, uniqueness, etc...), and the rigorous derivation from them of classical models, such as the Vlasov-Maxwell equations, are, in our opinion, challenging problems in the field of non-linear PDEs.

References

[AK]   V. I. Arnold, B. Khesin, *Topological methods in Hydrodynamics*, Springer Verlag, 1998.

[Ba]   F. Barthe, *On a reverse form of the Brascamp-Lieb inequality*, Invent. Math. *134 (1998) 335–361.*

[BB]   J.-D. Benamou, Y. Brenier, *A computational fluid mechanics solution to the Monge-Kantorovich mass transfer problem*, Numer. Math. *84 (2000) 375–393.*

[BDLL]   G. Boillat. C. Dafermos, P. Lax, T.P. Liu, *Recent mathematical methods in nonlinear wave propagation*, Lecture Notes in Math., *1640*, Springer, Berlin, 1996

[BI]   M. Born, L. Infeld, *Foundations of the new field theory*, Proc. Roy. Soc. London, A *144 (1934) 425–451.*

[Bo]   F. Bouchut, *Renormalized solutions to the Vlasov equation with coefficients of bounded variation*, Arch. Ration. Mech. Anal. *157 (2001) 75–90.*

[Br1]   Y. Brenier, *Polar factorization and monotone rearrangement of vector-valued functions*, Comm. Pure Appl. Math. *44 (1991) 375–417.*

[Br2]   Y. Brenier, *Minimal geodesics on groups of volume-preserving maps*, Comm. Pure Appl. Math. *52 (1999) 411–452.*

[Br3]   Y. Brenier, *Homogeneous hydrostatic flows with convex velocity profiles*, Nonlinearity *12 (1999) 495–512.*

[Br4]   Y. Brenier, *Derivation of the Euler equations from a caricature of Coulomb interaction*, Comm. Math. Phys. *212 (2000) 93–104.*

[Br5]   Y. Brenier, *Convergence of the Vlasov-Poisson system to the incompressible Euler equations*, Comm. Partial Differential Equations *25 (2000) 737–754.*

[Br6]   Y. Brenier, *Lecture notes*, Summer school on Mass transportation problems in kinetic theory and hydrodynamics, Ponta Delgada, Azores, 4–9 september 2000.

[Ca]   L. Caffarelli, *Boundary regularity of maps with convex potentials*, Comm. Pure Appl. Math. *45 (1992) 1141–1151.*

[Ch]   J. -Y. Chemin, *Fluides parfaits incompressibles*, Astérisque *230 (1995).*

[CNP]   M. Cullen, J. Norbury, J. Purser, *Generalised Lagrangian solutions for atmospheric and oceanic flows*, SIAM J. Appl. Math. *51 (1991), 20–31.*

[DM]   R. DiPerna, A. Majda Comm. Math. Phys. *108 (1987), 667–689.*

[Eb]   D. Ebin, *The motion of slightly compressible fluids viewed as a motion with strong constraining force*, Ann. of Math. (2) *105 (1977) 141–200.*

[EM]   D. Ebin, J. Marsden, Ann. of Math. *92 (1970) 102–163.*

[Eu]   L. Euler, *Opera Omnia*, Series Secunda, *12*, 274–361.

[Ev]   L.C. Evans *Partial differential equations and Monge-Kantorovich mass transfer*, Current developments in mathematics, Int. Press, Boston MA 1999.

[GZ]   M. Gaillard, B. Zumino, *Nonlinear electromagnetic self-duality and Legendre transformation*, Duality and supersymmetric theories, Cambridge Univ. Press, Cambridge, 1999.

[GMS]   M. Giaquinta, G. Modica, J. Souček, *Cartesian currents in the calculus* of variations. I. Series of Modern Surveys in Mathematics, 37, Springer-Verlag, Berlin, 1998.

[Gr]   E. Grenier, On the derivation of homogeneous hydrostatic equations, it M2AN Math. Model. Numer. Anal. 33 (1999), no. 5, 965–970.

[JKO]   D. Kinderlehrer, R. Jordan, F. Otto, The variational formulation of the Fokker-Planck equation, SIAM J. Math. Anal. 29 (1998) 1–17.

[Ka]   L.V. Kantorovich, On a problem of Monge, Uspekhi Mat. Nauk. 3 (1948) 225–226.

[Li]   P.-L. Lions, Mathematical topics in fluid mechanics. Vol. 1. Incompressible models, Oxford Lecture Series in Mathematics and its Applications, Oxford University Press, New York, 1996.

[Li2]   P.-L. Lions, Sur les équations différentielles ordinaires et les équations de transport, C. R. Acad. Sci. Paris Sr. I Math. 326 (1998) 833–838.

[Lo]   G. Loeper, On the Vlasov-Monge-Ampère system, preprint, 2002.

[Mc]   R. McCann, A convexity principle for interacting gases, Adv. Math. 128 (1997) 153–179.

[Mc2]   R. McCann, Polar factorization of maps on Riemannian manifolds, Geom. Funct. Anal. 11 (2001) 589–608.

[Ma]   A. Majda, Proceedings of the International Congress of Mathematicians, Kyoto 1990, Springer, 1991.

[MP]   C. Marchioro, M. Pulvirenti, Mathematical theory of incompressible non-viscous fluids, Springer, New York, 1994.

[Mo]   G. Monge, Mém. Math. Phys. Acad. Roy. Sci. Paris (1781), 666–704.

[Ne]   Y. Neretin, Categories of bistochastic measures and representations of some infinite-dimensional groups, Sb. 183 (1992), no. 2, 52–76.

[Ot]   F. Otto, The geometry of dissipative evolution equations: the porous medium equation, Comm. Partial Differential Equations 26 (2001) 101–174.

[OV]   F. Otto, C. Villani, Generalization of an inequality by Talagrand and links with the logarithmic Sobolev inequality, J. Funct. Anal. 173 (2000) 361–400.

[RU]   H. Rubin, P. Ungar, Motion under a strong constraining force, Comm. Pure Appl. Math. 10 (1957) 65–87.

[RR]   L. Rüschendorf, S. T. Rachev, J. of Multivariate Analysis 32 (1990) 48–54.

[She]   V. Shelukhin, Existence theorem in the variational problem for compressible inviscid fluids, Manuscripta Math. 61 (1988) 495–509.

[Sh]   A. Shnirelman, On the geometry of the group of diffeomorphisms and the dynamics of an ideal incompressible fluid, Math. Sbornik USSR 56 (1987) 79–105.

[Sh2]   A. I. Shnirelman, Generalized fluid flows, their approximation and applications, Geom. Funct. Anal. 4 (1994) 586–620.

[Su]   V. N. Sudakov, Proceedings of the Steklov Institute (1979) vol. 141.

[Ta]   L. Tartar, The compensated compactness method applied to systems of conservation laws. Systems of nonlinear PDE, NATO ASI series, Reidel,Dordecht, 1983.

[Yo]   L. C. Young, Lectures on the calculus of variations. Chelsea,New York,

Yann Brenier

1980.

[ZM]   Y. D. Zheng, A. Majda, *Existence of global weak solutions to one-component Vlasov-Poisson and Fokker-Planck-Poisson systems in one space dimension with measures as initial data*, Comm. Pure Appl. Math. 47 (1994) 1365–1401.

Measuring and Hedging
Financial Risks in Dynamical World

Nicole El Karoui*

Abstract

Financial markets have developed a lot of strategies to control risks induced by market fluctuations. Mathematics has emerged as the leading discipline to address fundamental questions in finance as asset pricing model and hedging strategies. History began with the paradigm of zero-risk introduced by Black & Scholes stating that any random amount to be paid in the future may be replicated by a dynamical portfolio. In practice, the lack of information leads to ill-posed problems when model calibrating. The real world is more complex and new pricing and hedging methodologies have been necessary. This challenging question has generated a deep and intensive academic research in the 20 last years, based on super-replication (perfect or with respect to confidence level) and optimization. In the interplay between theory and practice, Monte Carlo methods have been revisited, new risk measures have been back-tested. These typical examples give some insights on how may be used mathematics in financial risk management.

2000 Mathematics Subject Classification: 35B37, 60G44, 60G70, 65C30, 91B20.
Keywords and Phrases: Mathematical finance, Stochastic calculus, Optimization, Monte Carlo methods, Ill-posed problems.

1. Introduction

Financial markets have become an important component of people's life. All sorts of media now provide us with a daily coverage on financial news from all markets around the world. At the same time, not only large institutions but also more and more small private investors are taking an active part in financial trading. In particular, the e-business has led to an unprecedented increase in small investors direct trading. Given the magnitude of the potential impact a financial crisis can

*Centre de Mathématiques Appliquées, Ecole Polytechnique, 91 128 Palaiseau Cedex, France. E-mail: elkaroui@cmapx.polytechnique.fr

have on the real side of the economy, large, and even small, breakdowns have received particularly focal attention. Needless to say, the rapid expansion of financial markets calls upon products and systems designed to help investors to manage their financial risks. The financial risk business now represents more than $15 trillion annually in notional. A large spectrum of simple contracts (futures, options, swaps, etc.) or more exotic financial products (credit derivatives, catastrophe bonds, exotic options, etc.) are offered to private investors who use them to transfer financial risks to specialized financial institutions in exchange for suitable compensation. A classic example is the call option, which provides a protection in case of a large increase in the underlying asset price [1]. More generally, a derivative contract is an asset that delivers a payoff $H(\omega)$ at maturity date, depending upon the scenario ω.

As argued by Merton [19], the development of the financial risk industry would not have been possible without the support of theoretical tools. Mathematics has emerged as the leading discipline to address fundamental questions in financial risk management, as asset pricing models and hedging strategies, based on daily (infinitesimal) risk management. Mathematical finance, which relates to the application of the theory of probability and stochastic processes to finance, in particular the Brownian motion and martingale theory, stochastic control and partial differential equations, is now a field of research in its own right.

2.   The Black & Scholes paradigm of zero-risk

It is surprising that the starting point of financial industry expansion is "the Brownian motion theory and Itô stochastic calculus", first introduced in finance by Bachelier in this PhD thesis (1900, Paris), then used by Black, Scholes and Merton in 1973. Based on these advanced tools, they develop the totally new idea in the economic side that according to an optimal dynamic trading strategy, it is possible for option seller to deliver the contract at maturity without incurring any residual risk. At this stage, any people not familiar with stochastic analysis (as the majority of traders in the bank) may be discouraged. As Foellmer said in Bachelier Congress 2000, it is possible to reduce technical difficulties, and to develop arguments which are essentially probability-free.

A DYNAMICAL UNCERTAIN WORLD

The uncertainty is modelled via a family Ω of scenarios ω, i.e. the possible trajectories of the asset prices in the future. Such paths are described as positive continuous functions ω with coordinates $X_t = \omega(t)$, such that the continuous quadratic variation exists: $[X]_t(\omega) = \lim_n \sum_{t_i \leq t, t_i \in D_n} (X_{t_{i+1}} - X_{t_i})^2$ along the sequence of dyadics partitions D_n. The pathwise version of stochatic calculus yields to Itô's formula,

$$f(t, X_t)(\omega) = f(0, x_0) + \int_0^t f'_x(s, X_s)(\omega)\, dX_s(\omega) + \int_0^t f_t(s, X_s)(\omega)\, dt$$

---

[1] A Call option provides its buyer with the right (and not the obligation) to purchase the risky asset at a pre-specified price (the exercise price) at or before a pre-specified date in the future (maturity date). The potential gain at maturity can therefore be written as $(X_T - K)^+$, where X_T denotes the value of the underlying asset at time T

$$+ \int_0^t \tfrac{1}{2} f_{xx}''(s, X_s)(\omega) d[X]_s(\omega).$$

The second integral is well defined as a Lebesgue-Stieljes integral, while the first exists as Itô's integral, defined as limit of non-anticipating Riemann sums, (where we put $\delta_t = F_x'(t, X_t)$), $\sum_{t_i \leq t, t_i \in D_n} \xi_{t_i}(\omega)(X_{t_{i+1}} - X_{t_i})$.

From a financial point of view, the Itô's integral may be interpreted as the cumulative *gain process* of trading strategies: δ_t is the number of the shares held at time t then the increment in the Riemann sum is the price variation over the period. The non anticipating assumption corresponds to the financial requirement that the investment decisions are based only on the past prices observations. The residual wealth of the trader is invested only in cash, with yield rate (short rate) r_t by time unit. The *self-financing condition* is expressed as:

$$dV_t = r_t(V_t - \delta_t.X_t)dt + \delta_t.dX_t = r_t V_t dt + \delta_t.(dX_t - r_t X_t \, dt), \qquad V_0 = z.$$

HEDGING DERIVATIVES: A SOLVABLE TARGET PROBLEM

Let us come back to the problem of the trader having to pay at maturity T the amount $h(X_T)(\omega)$ in the scenario ω $((X_T(\omega) - K)^+$ for a Call option). This target has to be hedged (approached) in all scenarios by the wealth generated by a self-financing portfolio. The "miraculous" message is that, in Black & Scholes world, a perfect hedge is possible and easily computable, under the additional assumption: the short rate is deterministic and the quadratic variation is absolutely continuous $d[X]_t = \sigma(t, X_t) X_t^2 dt$. The (regular) strictly positive function $\sigma(t, x)$ is a key parameter in financial markets, called the *local volatility*.

Looking for the wealth as a function $f(t, X_t)$, we see that, given Itô's formula and self-financing condition,

$$df(t, X_t) = f_t'(t, X_t)dt + f_x'(t, X_t)dX_t + \tfrac{1}{2} f_{xx}''(t, X_t)X_t^2 \sigma^2(t, X_t) \, dt$$

$$= f(t, X_t)r_t dt + \delta(t, X_t)\big(dX_t - X_t r_t \, dt\big)$$

By identifying the dX_t terms (tanks to assumption $\sigma(t, x) > 0$), $\delta(t, X_t) = f_x'(t, X_t)$, and f should be solution of the following partial differential equation, Pricing PDE in short,

$$f_t'(t, x) + \tfrac{1}{2} f_{xx}''(t, x)x^2 \sigma^2(t, x) + f_x'(t, x)x r_t - f(t, x)r_t = 0, \ f(T, x) = h(x) \qquad (2.1)$$

The derivative price at time t_0 must be $f(t_0, x_0)$, if not, it is easy to generated profit without bearing any risk (arbitrage). That is the *rule of the unique price*, which holds in a liquid market.

The PDE's fundamental solution $q(t, x, T, y)$ $(h(x) = \delta_y(x))$ may be interpreted in terms of Arrow-Debreu "states prices" density, introduced in 1953 by these Nobel Prize winners for a purely theoretical economical point of view and by completely different arguments. The pricing rule becomes: $f(t, x) = \int h(y)q(t, x, T, y)dy$. q is also called *pricing kernel*. When $\sigma(t, x) = \sigma_t$, the pricing kernel is the *log-normal density*, deduced from the Gaussian distribution by an explicit change of variable.

The closed formula for Call option price is the famous [2] Black and Scholes formula, which is known by any practitioner in finance. The impact of this methodology was so important that Black (who already died), Scholes and Merton received the Nobel prize for economics in 1997.

In 1995, B.Dupire [9] give a clever formulation for the dual PDE (one dimensional in state variable) satisfied by $q(t, x, T, y)$ in the variables (T, y). If $C(T, K)$ is the Call price with parameters (T, K) when market conditions are (t_0, x_0), then

$$C'_T(T, K) = \tfrac{1}{2}\sigma^2(T, K)\, K^2 C''_{KK}(T, K) - rK\, C'_K(T, K), \ C(t_0, x_0) = (x_0 - K)^+$$

In short, if $r_t = 0$, $\quad C'_T(T, K) = \tfrac{1}{2}\sigma^2(T, K)\, K^2 C''_{KK}(T, K)$

3.   Model calibration and Inverse problem

In pratice, the main problem is the model calibration. Before discussing that, let me put the problem in a more classical framework. Following P. Lévy, asset price dynamics may be represented through a Brownian motion via stochastic differential equation (SDE)

$$dX_t = X_t(\mu(t, X_t)dt + \sigma(t, X_t)dW_t), \ X_{t_0} = x_0$$

where the Brownian motion W may be viewed as a standardized Gaussian noise with independent increments.

The local expected return $\mu(t, X_t)$ is a trend parameter appearing for the first time in our propose. That is a *key point* in financial risk management. Since this parameter does not appear in the Pricing-PDE, the Call price does not depend on the market trend. It could seem surprising, since the first motivation of this financial product is to hedge the purchaser against underlying rises. By using dynamical hedging strategy, the trader (seller) may be also protected against this unfavorable evolution. For a statistical point of view, this point is very important, because this parameter is very difficult to estimate.

In the B & S model with constant parameters, the volatility square is the variance by time unit of the log return $\ln(X_t) - \ln(X_{t-h}) = R_t$. If the only available information is given by asset price historical data, the statistical estimator to be used is the empirical variance, computed on a more or less long time period, $\widehat{\sigma}^2 = \frac{1}{N-1}\sum_{i=0}^{N-1}(R_{t_i} - \overline{R_N})^2$, where $\overline{R_N} = \frac{1}{N}\sum_{i=0}^{N-1} R_{t_i}$. This estimator is called *historical volatility*.

However, traders are reserved in using this estimator. Indeed, they argue that financial markets are not "statistically" stationary and that past is not enough to

---

[2]In the Black-Scholes model with constant coefficients, the Call option price $C^{BS}(t, x, K, T)$ is given via the Gaussian cumulative function $\mathcal{N}(z) = \int_\infty^z \frac{1}{\sqrt{2\pi}}e^{-\frac{y^2}{2}}\, dy$, and $\theta = T - t$,

$$\begin{cases} \quad C^{BS}(t, x, t + \theta, K) = x\mathcal{N}\big(d_1(\theta, x/K)\big) - K\, e^{-r\theta}\mathcal{N}\big(d_0(\theta, x/K)\big) \\ d_0(\theta, x/K) = \frac{1}{\sigma\sqrt{\theta}}\mathrm{Log}\left(\frac{x}{Ke^{-r\theta}}\right) - \tfrac{1}{2}\sigma\sqrt{\theta}, \quad d_1(\theta, x/K) = d_0(\theta, x/K) + \sigma\sqrt{\theta} \end{cases}$$

Moreover $\Delta(t, x) = \partial_x C^{BS}(t, x, t + \theta, K) = \mathcal{N}\big(d_1(\theta, x/K)\big)$.

explain the future. When it is possible, traders use additional information given by quoted option prices and translate it into volatility parametrization. The *implied volatility* is defined as: $C^{obs}(T,K) = C^{BS}(t_0, x_0, T, K, \sigma^{imp})$.

Moreover, the Δ of the replicating portfolio is $\Delta^{imp} = \partial_x C^{BS}(t_0, x_0, T, K, \sigma^{imp})$.

This strategy is used dynamically by defining the implied volatility and the associated Δ at any renegotiation dates. It was this specific use based on the hedging strategy that may explain the Black & Scholes formula success.

This attractive methodology has been appeared limited: observed implied volatilities are depending on the option parameters (time to maturity and exercise price) (implied volatility surface) in complete contradiction with B&S model assumptions. In particular, the market quotes large movements (heavy tail distribution) higher than in the log-normal framework. The first idea to take into account this empirical observation is to move to a model with local volatility $\sigma(t,x)$. The idea is especially attractive since the Dupire formula (2.2) gives a simple relation between "quoted Call option prices" and local volatility: $\sigma^2(K,T) = 2C'_T(T,K)(K^2 C''_{KK}(T,K))^{-1}$. The local volatility is computable from a continuum of *coherent* (without arbitrage) observed quoted prices. Unfortunately, the option market is typically limited to a relatively few different exercize prices and maturities; a naive interpolation yields to irregularity and instability of the local volatility.

ILL-POSED INVERSE PROBLEM

The problem of determining local volatility can be viewed as a function approximation non linear problem from a finite data set. The data set is the value $C_{i,j}$ of the solution at (t_0, x_0) of Pricing-PDE with boundary conditions $h_{i,j}(x) = (x - K_{i,j})^+$ at maturity T_i. Setting the problem as PDE's inverse problem yields to more robust calibration methods. These ideas appear for the first time in finance in 1997 [16], but the problem is not classical because of the strongly non linearity between option prices and local volatility; the data set is related to a single given initial condition.

Prices adjustment is made through a least square minimization program, including a penalization term related to the local volatility regularity.

$$G(\sigma) = \sum_{i,j} \omega_{i,j} \left(f(t_0, x_0, h_{i,j}, T_i, \sigma(.,.)) - C_{i,j}^{Obs} \right)^2,$$
$$J(\alpha, \sigma) = \alpha ||\nabla \sigma||^2 + G(\sigma) \to \min_\sigma .$$

Existence and uniqueness of solution is only partially solved [4]. Using large deviation theory, the asymptotic in small time of local volatility is expressed in terms of implied volatility: $\sigma^{\text{implied}}(K, t_0)^{-1} = \ln(\frac{K}{x_0})^{-1} \int_{x_0}^{K} \frac{d\xi}{\xi \sigma(\xi, t_0)}$.

Avellaneda & alii [2] have used another penalization criterion based on a stochastic control approach ; the control is the volatility parameter itself constrained to be very close to a prior volatility ($\eta(\sigma) = |\sigma(t, x) - \sigma_0(t, x)|^2$ for instance). The gradient criterion is replaced by $K(\sigma) = U(t_0, x_0, \sigma)$ where $U(T, x, \sigma) = 0$ and

$$U'_t(t, x) + \tfrac{1}{2}\sigma^2(t, x) x^2 U''_{xx}(t, x) + rx\, U'_x(t, x) - U(t, x) + \eta(\sigma(t, x)) = 0.$$

4.   Portfolio, duality and incomplete market

In the previous framework, options market may be entirely explained by underlying prices. In economic theory, it corresponds to *market efficiency*: a security price contains all the information on this particular security. In option world, the observed statistical memory of historical volatility leads naturally to consider stochastic volatility models with specific uncertainty

$$dX_t = X_t\big(\mu(t, X_t, Y_t)dt + \sigma(t, X_t, Y_t)dW_t^1\big), \quad dY_t = \eta(t, X_t, Y_t) + \gamma(t, X_t, Y_t)\,dW_t^2$$

where dW^1 and dW^2 are two correlated Brownian motions. γ is the volatility of the volatility. What does it change ? In fact, everything ! Perfect replication by a portfolio is not possible any more ; the notion of unique price does not exist any longer... But, such a situation is often the general case. What kind of answer may we bring to such a problem?

Super-replication and Robust Hedging

The option problem is still a target problem C_T, to be replicated by a portfolio $V_T(\pi, \delta) = \pi + \int_0^T \sum_i \delta_s^i dX_t^i$ depending on market assets X^i. Constraints (size, sign...) may be imposed on investment decisions (δ_t^i) [3]. Let $\mathcal{V}_T$ be the set of all derivatives, replicable at time T by an admissible portfolio. Their price at t_0 is the value of their replicating portfolio.

Super-replicating C_T is finding the smallest derivative $\widehat{C}_T \in \mathcal{V}_T$ which is greater than C_T in all scenarios. The super-replication price is the price of such a derivative. The $\widehat{C}_T$ replicating portfolio is the C_T *robust hedging*.

There are several ways to characterise the super-replicating portfolio:

1) Dynamic programming on level sets: this is the most direct (but least recent) approach. This method proposed by Soner & Touzi [20] has led the way for original works in geometry by giving a stochastic representation of a class of mean curvature type geometric equations.

2) Duality: this second approach is based on the $\mathcal{V}_T$ "orthogonal space", a set $\mathcal{Q}_T$ of martingale measures to be characterised. The super-replication price is given by

$$\widehat{C}_0 = \sup_{Q \in \mathcal{Q}_T} \mathbf{E}_Q[C_T].$$

We develop this last point, which is at the origin of many works.

Martingale measures

The idea of introducing a dual theory based on probability measures is due to Bachelier (1900), and above all to Harisson & Pliska (1987). The actual and achieved form is due to Delbaen & Schachermayer [7] and to the very active international group in Theoretical Mathematical Finance.

A *martingale measure* is a probability such that: $\forall V_T \in \mathcal{V}_T$, $\mathbf{E}_Q[V_T] = V_0$. Using simple strategies (discrete times, randomly chosen), this property is equivalent to prices of fundamental assets (X_t^i) are $\mathbf{Q}$-(local) martingales: the best X_{t+h}^i-estimated (w.r. to $\mathbf{Q}$) given the past at time t is X_t^i itself. The financial game is fair with respect to martingale measures.

---

[3]For the sake of simplicity, interest rates are assumed to be null

When $\mathcal{V}_T$ contains all possible (path-dependent) derivatives, the market is said to be complete, and the set of martingale measures is reduced to a unique element $\mathbf{Q}$, often called risk-neutral probability. This is the case in the previous framework. Dynamics become $dX_t = X_t\, \sigma(t, X_t)dW_t^Q$ where W^Q is a $\mathbf{Q}$-Brownian motion. This formalism is really efficient as it leads to the following path dependent derivative pricing rule: $\widehat{C}_0 = \mathbf{E}_Q(C_T)$.

Computing the replicating portfolio is more complex. In the diffusion case, the price is a deterministic function of risk factors and the replicating portfolio only depends on partial derivatives. The general case will be mentioned in the paragraph dedicated to Monte-Carlo methods.

INCOMPLETE MARKET

The characterization of the set $\mathcal{Q}_T$ is all the more delicate so since there are many different situations which may lead to market imperfections (non-tradable risks, transaction costs ...).

An abstract theory of super-replication (and more generally of portfolio optimization under constraints) based on duality has been intensively developed. The super-replicating price process is showed [10],[15] to be a super-martingale with respect to any admissible martingale measure. Hence, by the generalization of the Doob-Meyer representation, the super-replicating portfolio is the "$\mathcal{Q}_T$- martingale" part of the super-price Kramkov-decomposition.

Super-replication prices are often too expensive to be used in practice. However, they give an upper bound to the set of possible prices. In the previously described stochastic volatility model, the super-replication price essentially depends on possible values of stochastic volatility:

1. If the set is R^+, then the super-replicating derivative of $h(X_T)$ is $\widehat{h}(X_T)$ where $\widehat{h}$ is the concave envelop of h ; the replicating strategy is the trivial one: buying $\widehat{h}'_x(x_0)$ stocks and holding them till maturity.

2. If the volatility is bounded (up and down relatively to 0), the super-replication price is a (not depending on y) solution of
$$\widehat{h}'_t(t, x) + \tfrac{1}{2}\sup_y(\sigma^2(t, x, y)\widehat{h}''_{xx}(t, x)) = 0, \qquad \widehat{h}(T, x) = h(x).$$

When h is convex, $\widehat{h}(t, x)$ is convex and the super-replication price is the one calculated with the *upper volatility* (in y).

Calibration constraints may be easily taken into account without modifying this framework. We only have to assume that the terminal net cash flows of calibrating derivatives belong to $\mathcal{V}_T$ or equivalently we have to add linear constraints to the dual problem: $\quad \widehat{C}_0^{\text{cal}} = \sup\{\mathbf{E}_Q(C_T)\,; Q \in \mathcal{Q}_T, \mathbf{E}_Q((X_{T_i} - K_{ij})^+) = C_{ij}\}.$

Risk measures

When super-replicating is too expensive, the trader has to measure his market risk exposure. The traditional measure is the variance of the replicating error. But a new criterion, taking into account extreme events, is now used, transforming the risk management at both quantitative and qualitative levels.

VALUE AT RISK

The VaR criterion, corresponding to *the maximal level of losses* acceptable with probability 95%, has taken a considerable importance for several years. Regulation Authorities have required a daily VaR computation of the global risky portfolio from financial institutions. Such a measure is important on the operational point of view, as it affects the provisions a bank has to hold to face market risks. VaR estimation (quantile estimation) and its links with extreme value theory [12] are widely debated in the market, just as by academics.

Moreover, a huge debate has been introduced by academics [1] on the VaR efficiency as risk measure. For instance, its non-additive property enables banks to play with subsidiary creations. This debate has received an important echo from the professional world, which is possibly planning to review this risk measure criterion. Sub-additive and coherent risk measures are an average estimation of losses with respect to a probability family: $\rho(X) = \sup_{Q \in \mathcal{Q}_T} \mathbf{E}_Q(-X)$.

This characterisation has recently been extended to convex risk measures, by adding a penalization term depending on probability density (entropy for instance).

RISK MEASURES AND RESERVE PRICE

A trader willing to relax the super-replication assumption is naturally thinking in terms of potential losses with given probability (level confidence) of his replicating error. It corresponds to the quantile hedging strategy. Other risk measures (quadratic, convex, entropic) may be used. Optimization theory is coming back when looking for the smallest portfolio, generating an acceptable loss. The initial value of this portfolio is called the *reserve price*. Mean-variance and entropic problems have now a complete solution [13]. More surprisingly (because of non-convexity), this also holds for the quantile hedging problem [14]. All these results are in fact sub-products of portfolio optimization in incomplete markets [6] or [8].

5.   New research fields

Monte-Carlo methods

Dual version of super-replication problems, just as new risk measures, underline the interest to compute very well and quickly quantities such as $\mathbf{E}_Q(X)$ and more generally $\sup_{Q \in \mathcal{Q}_T} \mathbf{E}_Q(X)$. For small dimensional diffusions, these quantities may be computed as the solution of some linear PDE (for the expected value) and non-linear PDE (for the sup). However, the computational efficiency falls rapidly with the dimension. That increases the interest for the so-called *probabilistic methods*.

The fundamental idea of Monte-Carlo methods is the computation of $\mathbf{E}_Q(X)$ by simulation, i.e. by drawing a large number ($N \simeq 10^5$) of independent scenarios ω^i and taking the average value of the results $\frac{1}{N} \sum_{i=1}^{N} X(\omega^i)$. Of course, this method does not work very well when being too naive, but convergence may be accelerated by different techniques.

In the finance area, the important quantities are both the price and the sensitivities to different model parameters. Based on integration by parts, efficient methods have been developed to compute in a coherent manner prices and their derivatives [17]. In the case of path-dependent options, the derivative is taken with respect to a Brownian motion perturbation (Malliavin Calculus).

Very original is the actual research on solving, by Monte-Carlo methods, optimization problem expressed as "sup" of a family of expected values (super-replication prices). These solutions are based on dynamic programming, enabling to turn a maximization in expected value into a pathwise maximization. The formulation in terms of Backward SDE's, introduced by Peng and Pardoux in 1987 and in finance[4] in [11],[18] well describes this effect.

In all cases, the problem is to compute conditional expectation by Monte Carlo, using the function approximation theory, or more generally random variable approximation in in the Wiener space (chaos decomposition).

Problems related to the dimension, and statistical modelling

Financial problems are usually multidimensional, but only few *liquid* financial products are depending on multi-assets. Even if the different market actors consider they can have a good knowledge of each individual asset behavior, the question is now to find a multidimensional distribution given each component distribution. This problem is a statistical one, known as the *copula theory*. Copula is a distribution function on $[0,1]^n$ with identity function as marginal. They are useful to give bounds to asset prices. Dynamically, the problem still to be solved is to find the local volatility matrix of multidimensional diffusion given the "Dupire" dynamics of each coordinate.

High dimensional problems arise when computing bank portfolio VaR (with a number of observations less than that of risk factors), or hedging derivatives depending on a large number of underlying assets. Main risk factors may be very different in a Gaussian framework or heavy tail framework (Lévy processes)[5]. Random matrix theory or other asymptotic tools may bring some new ideas to this question.

By presenting the most important tools of the financial risk industry, I have voluntary left apart anything on financial asset statistical modelling, which may be the subject of a whole paper on its own. It is clear that the VaR criterion, the market imperfections are highly dependent on an accurate analysis of the real and historical world [3]. Intense and very innovating research is now developed (High-frequency data, ARCH and GARCH processes, Lévy processes with long memory, random cascades).

6. Conclusion

As a conclusion, applied mathematicians have been highly questioned by problems coming from the financial risk industry. This is a very active world, rapidly evolving, in which theoretical thoughts have often immediate and practical fallout.

---

[4] A BSDE solution is a couple of adapted processes (Y_t, Z_t) such that

$$-dY_t = f(t, Y_t, Z_t)dt - Z_t dW_t, \qquad Y_T = C_T.$$

Thanks to comparison theorem, $\widehat{Y}_t = \sup_{i \in I} Y_t^i$ is the solution of the BSDE with driver $\widehat{f}(t, y, z) = \sup_{i \in I} f^i(t, y, z)$

On the other hand, practical constraints raise new theoretical problems. This paper is far from being an exhaustive view of the financial problems. It is more a subjective view conditioned by my own experience. Many exciting problems, from both theoretical and practical points of view, have not been presented. May active researchers in these fields forgive me.

References

[1] P.Artzner, F.Delbaen, J.M.Eber & D.Heath (1999) Coherent measures of risk. *Mathematical Finance* Vol. 9, no. 3,pp. 203–228.

[2] M. Avellaneda, C. Friedman, R. Holmes & D. Samperi (1997) Calibrating volatility surfaces via entropy. *Applied Mathematical Finance*, March 1997.

[3] O.Barndorff-Nielsen, N.Sheippard (2002) Realised power variation and stochatic volatility models (preprint)

[4] H. Berestycki, J. Busca & I. Florent (2001) Asymptotics and calibration of local volatility models. *Quantitative Finance*, forthcoming.

[5] J.P. Bouchaud, D. Sornette, C. Walter & J.P. Aguilar (1998) Taming large events.. *Int. J. of Theo. and Appl. Finance*, Vol1, no 1, pp. 25–41.

[6] J. Cvitanić & I. Karatzas (1999) On dynamic measures of risk. Finance and Stochastics, Vol. 3, no. 4, pp. 451–482.

[7] F. Delbaen & W. Schachermayer (1994) A General Version of the Fundamental Theorem of Asset Pricing. Mathematische Annalen, Vol. 300, pp. 463–520.

[8] D.Duffie, Dynamic Asset Princing (1997) Academic Press. N.Y.

[9] B. Dupire (1997) Pricing and hedging with smiles. in Mathematics of derivative securities. Dempster and Pliska eds., Cambridge Uni. Press, pp. 103–112.

[10] N. El Karoui & M.C. Quenez (1995) Dynamic programming and pricing in incomplete markets. SIAM J.Control & Optimization, Vol. 33, no 1, pp. 29–66.

[11] N. El Karoui, S. Peng & M.C. Quenez (1997) Backward stochastic differential equations in finance. Mathematical Finance, Vol. 7, no. 1, pp. 1–71.

[12] P. Embrechts, C. Klueppelberg & T. Mikosch (1997) Modelling Extremal Events for Insurance and Finance. Springer book.

[13] M. Frittelli (2000) The minimal entropy martingale measure and the valuation in incomplete markets. Mathematical Finance, Vol. 10, no 1, pp. 39–52.

[14] H. Foellmer & P. Leukert (1999) Quantile hedging. Finance and Stochastics, Vol. 3, no. 3, pp. 251–273.

[15] D.O. Kramkov (1996) Optional decomposition of supermartingales and hedging in incomplete security markets. P.Th.Rel. Fields, Vol. 105, pp. 459–479.

[16] R. Lagnado & S. Osher (1997): A technique for calibrating derivative security pricing models J. of Computational Finance, Vol. 1, no 1, pp. 13–25.

[17] P.-L. Lions, J.-M. Lasry, J. Lebuchoux & E. Fournié (2001) Applications of Malliavin calculus to Monte-Carlo methods in finance II. Finance and Stochastics, Vol. 5, no 2, pp. 201–236.

[18] J. Ma & J. Yong (2000) Forward-Backward SDE and Their Applications.

Lect.Notes in Mathematics.no 1702. Springer

[19] R. Merton (2001) Future Possibilities in Finance Theory and Finance Practice Bachelier Congress 2000, pp. 47–72. Editors Geman & alii, Springer.

[20] H. Soner, H. Mete & N. Touzi (2000) Superreplication under gamma constraints. SIAM J. Control Optim, Vol. 39, no. 1, pp. 73–96 (electronic).

ICM 2002 · Vol. III · 785–794

Exploring the Capability and Limits of the Feedback Mechanism

Lei Guo*

Abstract

Feedback is a most important concept in control systems, its main purpose is to deal with internal and/or external uncertainties in dynamical systems, by using the on-line observed information. Thus, a fundamental problem in control theory is to understand the maximum capability and potential limits of the feedback mechanism. This paper gives a survey of some basic ideas and results developed recently in this direction, for several typical classes of uncertain dynamical systems including parametric and nonparametric nonlinear systems, sampled-data systems and time-varying stochastic systems.

2000 Mathematics Subject Classification: 93B52, 93C40.
Keywords and Phrases: Dynamical systems, Feedback control, Uncertainty, Adaptation, Stabilization.

1. Introduction

Feedback is ubiquitous, and exists in almost all goal-directed behaviors [1]. It is indispensable to the human intelligence, and is important in learning, adaptation, organization and evolution, etc. Feedback is also the most important concept in control, which is a fundamental systems principal when dealing with uncertainties in complex dynamical systems. The uncertainties of a system are usually classified into two types: internal (structure) and external (disturbance) uncertainties, depending on the specific dynamical systems to be controlled. Feedback needs information, and there are also two types of information in a control system: *a priori* information and *posteriori* information. The former is the available information before controlling a system, while the later is the information exhibited by the system dynamic behaviors. It is the *posteriori* information that makes it possible for the feedback to reduce the influences of the uncertainties on control systems. Two of the fundamental questions in control theory are: How much uncertainty can be

*Institute of Systems Science, Chinese Academy of Sciences, Beijing 100080, China. E-mail: Lguo@control.iss.ac.cn

dealt with by feedback ? What are the limits of feedback ? These are conundrums, despite of the considerable progress in control theory over the past several decades.

The existing feedback theory in control systems can be roughly classified into three groups: traditional feedback, robust feedback and adaptive feedback. In the ideal case where the mathematical model can exactly describe the true system, the feedback law that are designed based on the full knowledge of the model may be referred to as traditional feedback. Unfortunately, as is well known, almost all mathematical models are approximations of practical systems, and in many cases there are inevitable large uncertainties in our mathematical descriptions. The primary motivation of robust and adaptive control is to deal with uncertainties by designing feedback laws, and much progress has been made in these two areas. Robust feedback design allows that the true system model is not exactly known but lies in a "ball" centered at a known nominal model with reliable model error bounds(cf. e.g. [2], [3]).

By adaptive feedback we mean the (nonlinear) feedback which captures the uncertain (structure or parameter) information of the underlying system by properly utilizing the measured on-line data. The well-known certainty-equivalence principle in adaptive control is an example of such philosophy. Since an on-line learning mechanism is usually embedded in the structure of adaptive feedback, it is conceivable that adaptive feedback can deal with larger uncertainties than other forms of feedback can do. Over the past several decades, much progress has been made in the area of adaptive control (cf. e.g. [4]–[9]). For linear finite dimensional systems with uncertain parameters, a well-developed theory of adaptive control exists today, both for stochastic systems (cf. [5],[8],[9]) and for deterministic systems with small unmodelled dynamics (cf. [6]). This theory can be generalized to nonlinear systems with linear unknown parameters and with linearly growing nonlinearities (cf. e.g.[12]). However, fundamental difficulties may emerge in the design of stabilizing adaptive controls when these structural conditions are removed. This has motivated a series of studies on the maximum capability (and limits) of the feedback mechanism starting from [10].

To explore the maximum capability and potential limits of the feedback mechanism, we have to place ourselves in a framework that is somewhat different from the traditional robust control and adaptive control. First, the system structure uncertainty may be nonlinear and/or nonparametric, and a known or reliable ball containing the true system, which is centered at a known nominal model, may not be available *a priori*. Second, we need to consider the maximum capability of the whole feedback mechanism (not only a fixed feedback law or a special class of feedback laws). Moreover, we need to answer not only what the feedback can do, but also the more difficult and important question, what the feedback can not do. We shall also work with discrete-time (or sampled-data) feedback laws, as they can reflect the basic causal law as well as the limitations of actuator and sensor in a certain sense, when implemented with digital computers.

In this talk, we will give a survey of some basic ideas and results obtained in the recent few years ([10]–[17]), towards understanding the capability and limits of the feedback mechanism in dealing with uncertainties. For several basic classes of uncertain nonlinear dynamical control systems, we will give some critical values and "Impossibility Theorems" concerning the capability of feedback. The reminder of the paper is organized as follows: the problem formulation will be given in Section 2, and some basic classes of discrete-time parametric and nonparametric nonlinear control systems will be studied in Sections 3 and 4, respectively. Other classes of uncertain systems, including sampled-data systems and time-varying linear systems with hidden Markovian jumps, will be considered in Section 5, and some open problems will be stated in the concluding remarks in Section 6.

2.   Problem formulation

Let $\{u_t, t \geq 0\}$ be an $\mathcal{R}^m$-valued input process of a discrete-time or sampled-data uncertain dynamical control system (whose structure is unknown or not fully known, and is subject to some noise disturbances), and let $\{y_t, t \geq 0\}$ be the corresponding on-line observed $\mathcal{R}^n$-valued output process (see the following figure).

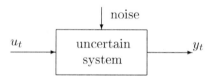

At any moment $t \geq 0$, the input signal u_t is said to be a *feedback signal*, if there is a Lebesgue measurable mapping $h_t(\cdot) : \mathcal{R}^{(t+1)n} \to \mathcal{R}^m$ such that

$$u_t = h_t(y_0, y_1, ..., y_t).$$

In other words, u_t is a function of the *posteriori* information observed up to time t. A *feedback law* u is a sequence of feedback signals, i.e., $u = \{u_t, l \geq 0\}$. Furthermore, the *feedback mechanism* U is defined as the set of all possible feedback laws,

$$U = \{u| \ u \quad \text{is any feedback law}\}.$$

For a complex system whose structure contains uncertainties, it is not a simple (and in fact difficult) problem to find a feedback law u, such that the corresponding output process can achieve a desired goal. This involves questions like: what kind of properties of an uncertain dynamical system can be changed by feedback ? how to construct a satisfactory feedback based on the available information ? More fundamental questions are: how much uncertainty can be deal with by feedback ? What are the maximum capability and limits of the feedback mechanism U ? In the next three sections, we will present some preliminary results towards answering these questions. For further discussion, we need the following definition.

Definition 2.1. *A dynamical control system is said to be globally stabilizable if there exists a feedback law $u \in U$, such that the output process of the system is bounded in the mean square sense, i.e.,*

$$\sup_{t \geq 0} E\|y_t\|^2 < \infty, \quad \text{for any initial value } y_0.$$

3. Parametric nonlinear systems

Consider the following basic discrete-time parametric nonlinear system:

$$y_{t+1} = \theta f(y_t) + u_t + w_{t+1}, \tag{3.1}$$

where y_t, u_t and w_t are the scalar system output, input and noise processes, respectively. For simplicity, we assume that

A1) $\{w_t\}$ is a Gaussian noise process;

A2) θ is an unknown non-degenerate Gaussian random parameter;

A3) The function $f(\cdot)$ is known and has the following growth rate:

$$f(x) \quad \sim \quad Mx^b, \quad \text{as} \quad x \to \infty,$$

where $b \geq 0$, $M > 0$ are constants. Obviously, if $b \leq 1$, then the nonlinear function $f(\cdot)$ has a growth rate which is bounded by linear growth. This case can be easily dealt with by the existing theory in adaptive control (see e.g. [12]). Our prime concern here is to know whether or not the system can be globally stabilized by feedback for any $b > 1$?

The following theorem gives a critical value of b, which characterizes the maximum capability of the feedback mechanism.

Theorem 3.1. *Consider the system (3.1) with Assumptions A1)–A3) holding. Then $b = 4$ is a critical case for feedback stabilizability. In other words,*

(i). If $b \geq 4$, then for any feedback law $u \in U$, there always exists a set D (in the basic probability space) with positive probability such that

$$|y_t| \to \infty, \ \text{on} \ D$$

at a rate faster than exponential.

(ii). If $b < 4$, then the least-squares-based adaptive minimum variance feedback control $u_t = -\theta_t f(y_t)$ where θ_t is the least-squares estimate for θ at time t, can render the system to be globally stable and optimal, with the best rate of convergence:

$$\sum_{t=1}^{T} (y_t - w_t)^2 = O(\log T), \ a.s., \ as \ T \longrightarrow \infty.$$

Remark 3.1. This result is somewhat surprising since the assumptions in our problem formulation have no explicit relationships with the value $b = 4$. We remark that the related results were first found and established in a somewhat general framework in [10]. In particular, the first part (i) was contained in Remark 2.2 in [10], and was later extended to general unknown parameter case in [12] by using a conditional Cramer-Rao inequality. The second part (ii) is a special case of Theorem 2.2 in [10].

Remark 3.2. There are many implications of Theorem 3.1. For example, the limitation of feedback given in Theorem 3.1 (i) is readily applicable to general class of uncertain systems of the form

$$y_{t+1} = f_t(y_t, ..., y_{t-p}, u_t, ..., u_{t-q}) + w_{t+1},$$

as long as it contains the basic class (3.1) as a subclass. Theorem 2.1 can also be used to show the fundamental differences between continuous-time and discrete-time nonlinear adaptive control(see also, [12]). Note that the noise free case can be trivially controlled, regardless of the growth rate of the nonlinear function $f(\cdot)$ (see [11]). This means that the noise effect in (3.1) plays an essential role in the non-stabilizability result of Theorem 3.1 (i): the noise effect gives estimation errors to even the "best" parameter estimates, which are then amplified step by step by the nonlinearity of the system, leading to the final instability of the closed-loop systems, despite of the strong consistency of the parameter estimates ([10]) .

Theorem 3.1 concerns with the case where the unknown parameter θ is a scalar. To see what happens when the number of the unknown parameters increases, let us consider the following polynomial nonlinear regression:

$$y_{t+1} = \theta_1 y_t^{b_1} + \theta_2 y_t^{b_2} + \cdots + \theta_p y_t^{b_p} + u_t + w_{t+1}. \tag{3.2}$$

Again, for simplicity, we assume that

A1)' $b_1 > b_2 > \cdots > b_p > 0$;

A2)' $\{w_t\}$ is a sequence of independent random variable with a common distribution $N(0,1)$;

A3)' $\theta \overset{\Delta}{=} [\theta_1 \cdots \theta_p]^\tau$ is a random parameter with distribution $N(\bar{\theta}, I_p)$.

Now, introduce a characteristic polynomial

$$P(z) = z^{p+1} - b_1 z^p + (b_1 - b_2) z^{p-1} + \cdots + (b_{p-1} - b_p) z + b_p,$$

which plays a crucial role in characterizing the limits of the feedback mechanism as shown by the following "impossibility theorem".

Theorem 3.2. *If there exists a real number $z \in (1, b_1)$ such that $P(z) < 0$, then the above system (3.2) is not stabilizable by feedback. In fact, for any feedback law $u \in U$ and any initial condition $y_0 \in \mathcal{R}^1$, it is always true that*

$$E|y_t|^2 \to \infty, \ as \ t \to \infty$$

at a rate faster than exponential.

Remark 3.3. The proof of Theorem 3.2 can be found in [11], and extensions to non-Gaussian parameter case can be found in [12]. An important consequence of this theorem is that the system (3.2) is not stabilizable by feedback in general, whenever $b_1 > 1$ and the number of unknown parameters p is large(see [11]). This fact implies that the uncertain nonlinear system

$$y_{t+1} = f(y_t) + u_t + w_t$$

with $f(\cdot)$ being unknown but satisfying

$$\|f(x)\| \le c_1 + c_2\|x\|^b, \quad b > 1,$$

may not be stabilizable by feedback in general. This gives us another fundamental limits on feedback in the presence of parametric uncertainties in nonlinear systems, and motivates the study of nonparametric control systems with linear growth conditions in the next section.

4. Nonparametric nonlinear systems

Consider the following first-order nonparametric control system:

$$y_{t+1} = f(y_t) + u_t + w_{t+1}, \quad t \geq 0, \quad y_0 \in \mathcal{R}^1, \tag{4.1}$$

where $\{y_t\}$ and $\{u_t\}$ are the scalar output and input, and $\{w_t\}$ is an "unknown but bounded" noise sequence, i.e. $|w_t| \leq w$, $\forall t$, for some constant $w > 0$. The nonlinear function $f(\cdot) : \mathcal{R}^1 \to \mathcal{R}^1$ is assumed to be completely unknown. We are interested in understanding how much uncertainty in $f(\cdot)$ can be dealt with by feedback. In order to do so, we need to introduce a proper measure of uncertainty first.

Now, define $\mathcal{F} \triangleq \{f : \mathcal{R}^1 \to \mathcal{R}^1\}$ and introduce a quasi-norm on $\mathcal{F}$ as follows:

$$\|f\| \triangleq \lim_{\alpha \to \infty} \sup_{(x,y) \in \mathcal{R}^2} \frac{|f(x) - f(y)|}{|x - y| + \alpha}, \quad \forall f \in \mathcal{F}.$$

Having introduced the norm $\|\cdot\|$, we can then define a ball in the space $(\mathcal{F}, \|\cdot\|)$ centered at its "zero" θ_F with radius L:

$$\mathcal{F}(L) \triangleq \{f \in \mathcal{F} : \|f\| \leq L\}$$

where $\theta_F \triangleq \{f \in \mathcal{F} : \|f\| = 0\}$. It is obvious that the size of $\mathcal{F}(L)$ depends on the radius L, which will be regarded as the measure of the size of uncertainty in our study to follow.

The following theorem establishes a quantitative relationship between the capability of feedback and the size of uncertainty.

Theorem 4.1. *Consider the nonparametric control system (4.1). Then the maximum uncertainty that can be dealt with by feedback is a ball with radius $L = \frac{3}{2} + \sqrt{2}$ in the normed function space $(\mathcal{F}, \|\cdot\|)$, centered at the zero θ_F. To be precise,*

(i) If $L < \frac{3}{2} + \sqrt{2}$, then there exists a feedback law $u \in U$ such that for any $f \in \mathcal{F}(L)$, the corresponding closed-loop control system (4.1) is globally stable in the sense that

$$\sup_{t \geq 0}\{|y_t| + |u_t|\} < \infty, \quad \forall y_0 \in \mathcal{R}^1;$$

(ii) If $L \geq \frac{3}{2} + \sqrt{2}$, then for any feedback law $u \in U$ and any initial value $y_0 \in \mathcal{R}^1$, there always exists some $f \in \mathcal{F}(L)$ such that the corresponding closed-loop system (4.1) is unstable, i.e.,

$$\sup_{t \geq 0} |y_t| = \infty.$$

The proof of the above theorem is given in [14], where it is also shown that once the stability of the closed-loop system is established, it is a relatively easy task to evaluate the control performance.

Remark 4.1. The stabilizing feedback law in Theorem 4.1 (i) can be constructed as follows(see [14]):

$$u_t = \begin{cases} u'_t, & \text{if } |y_t - y_{i_t}| > \epsilon \\ u''_t, & \text{if } |y_t - y_{i_t}| \leq \epsilon \end{cases}$$

where $\epsilon > 0$ is any given threshold. In other words, u_t is a switching feedback based on a stabilizing feedback u'_t and a tracking feedback u''_t, which are defined as follows:

$$u'_t = -\hat{f}_t(y_t) + \frac{1}{2}(\underline{b}_t + \bar{b}_t)$$

where $\hat{f}$ is the nearest neighbor (NN) estimate of f defined by

$$\hat{f}_t(y_t) \overset{\Delta}{=} y_{i_t+1} - u_{i_t}$$

with

$$i_t \overset{\Delta}{=} \underset{0 \le i \le t-1}{\operatorname{argmin}} |y_t - y_i|$$

and where

$$\underline{b}_t = \min_{0 \le i \le t} y_i, \quad \bar{b}_t = \max_{0 \le i \le t} y_i$$

The tracking feedback u''_t is defined

$$u''_t = -\hat{f}_t(y_t) + y^*_{t+1},$$

where $\{y^*_t\}$ is a bounded reference sequence. It is obvious that u_t depends on the observations $\{y_0, y_1, \ldots, y_t\}$ only.

One may try to generalize Theorem 4.1 to the following high-order nonlinear systems $(p \ge 1)$:

$$y_{t+1} = f(y_t, y_{t-1}, \ldots, y_{t-p+1}) + u_t + w_{t+1} \tag{4.2}$$

where $f(\cdot) : R^p \longrightarrow R^1$ is assumed to be completely unknown, but belongs to the following class of Lipschitz functions:

$$\mathcal{F}(L) = \{f(\cdot) : |f(x) - f(y)| \le L\|x - y\|, \forall x, y \in \mathcal{R}^p\}$$

where $L > 0$, $\|x\| = \sum_{i=1}^{p} |x_i|$, $x = (a_1, \ldots, x_p)^\tau \in R^p$. Again, $\{w_t\}$ is a sequence of "unknown but bounded" noises. The following "impossibility theorem" is established in [17].

Theorem 4.2. *If L and p satisfy*

$$L + \frac{1}{2} \ge (1 + \frac{1}{p})(pL)^{\frac{1}{p+1}} \tag{4.3}$$

then there does not exist any globally stabilizing feedback law for the class of uncertain systems (4.2) with $f \in \mathcal{F}(L)$.

It is easy to see that if $p = 1$ then the above inequality (4.3) reduces to $L \ge \frac{3}{2} + \sqrt{2}$, which we know to be a critical value for the capability of feedback by Theorem 4.1. However, when $p > 1$ and (4.3) does not hold, whether or not there exists a stabilizing feedback for the uncertain system (4.2) with $f \in \mathcal{F}(L)$ still remains as an open question.

5.   Other uncertain systems

In this section, we briefly mention some related results on other basic classes of uncertain systems.

Let us first consider the following simple but basic continuous-time system:

$$\dot{x}_t = f(x_t) + u_t, \quad t \geq 0, x_0 \in \mathcal{R}^1. \tag{5.1}$$

The system signals are assumed to be sampled at a constant rate $h > 0$, and the input is assumed to be implemented via the familiar zero-order hold device (i.e., piecewise constant functions):

$$u_t = u_{kh}, \quad kh \leq t < (k+1)h, \tag{5.2}$$

where u_{kh} depends on $\{x_0, x_h, ..., x_{kh}\}$.

The nonlinear function f in (5.1) is assumed to be unknown but belongs to the following class of local Lipschitz (LL) functions:

$$G_c^L = \{f \mid f \text{ is LL and satifies } |f(x)| \leq L|x| + c, \forall x \in R^1\}, \tag{5.3}$$

where $c > 0$ and $L > 0$ are constants. The "slope" L of the unknown nonlinear functions in G_c^L may be regarded as a measure of the size of the uncertainty. Similar to the discrete-time case in Theorem 4.1, L plays a crucial role in the determination of the capability and limits of the sample-data feedback [13].

Theorem 5.1. *Consider the sampled-data control system (5.1)–(5.2). If $Lh > 7.53$, then for any $c > 0$ and any sampled-data control $\{u_{kh}, k \geq 0\}$ there always exists a function $f^* \in G_c^L$, such that the state signal of (5.1)–(5.2) corresponding to f^* with initial point $x_0 = 0$ satisfies $(k \geq 1)$*

$$|x_{kh}| \geq (\frac{Lh}{2})^{k-1} \cdot ch \xrightarrow[k \to \infty]{} \infty.$$

Remark 5.1. This "impossibility theorem" shows that if Lh is larger than a certain value, then there will exist no stabilizing sampled-data feedback. On the other hand, it is easy to show that if $Lh < \log 4$, then a globally stabilizing sampled-data feedback can be constructed (see [13]). An obvious open question here is how to bridge the gap between $\log 4$ and 7.53. Needless to say, Theorem 5.1 gives us some useful quantitative guidelines in choosing properly the sampling rate in practical applications.

Next, we consider the following linear time-varying stochastic model:

$$x_{t+1} = A(\theta_t)x_t + B(\theta_t)u_t + w_{t+1}, \ t \geq 1; \tag{5.4}$$

where $x_t \in \mathcal{R}^n$, $u_t \in \mathcal{R}^m$ and $w_{t+1} \in \mathcal{R}^n$ are the state, input and noise vectors respectively. We assume that

H1) $\{\theta_t\}$ is an unobservable Markov chain which is homogeneous, irreducible and aperiodic, and which takes values in a finite set $\{1, 2, \cdots, N\}$ with transition matrix denoted by $P = (p_{ij})_{NN}$, where by definition $p_{ij} = P\{\theta_t = j | \theta_{t-1} = i\}$.

H2) There exists some $m \times n$ matrix K such that $det\big[(A_i - A_j) - (B_i - B_j)K\big] \neq$ $0, \quad \forall i \neq j, 1 \leq i, j \leq N$, where $A_i \triangleq A(i) \in \mathcal{R}^{nn}$, $B_i \triangleq B(i) \in \mathcal{R}^{nm}$ are the system matrices.

H3) $\{w_t\}$ is a martingale difference sequence which is independent of $\{\theta_t\}$, and satisfies $\sigma I \leq E w_t w_t'$, $\quad E w_t' w_t \leq \sigma_w$, $\quad \forall t$, where σ and σ_w are two positive constants, and the prime superscript represents matrix transpose.

For simplicity of presentation, we denote $S \triangleq \{1, 2, \cdots, N\}$. The following theorem gives a fairly complete characterization of feedback stabilizability for the hidden Markovian model (5.4)[16].

Theorem 5.2. *Let the above Assumptions H1)–H3) hold for the dynamical system (5.4) with hidden Markovian switching. Then the system is stabilizable by feedback if and only if the following coupled algebraic Riccati-like equations have a solution consisting of N positive definite matrices $\{M_i > 0, i \in S\}$:*

$$\sum_j A_j' p_{ij} M_j A_j - \Big(\sum_j A_j' p_{ij} M_j B_j\Big) \Big(\sum_j B_j' p_{ij} M_j B_j\Big)^+ \Big(\sum_j B_j' p_{ij} M_j A_j\Big) - M_i = -I,$$

where $i \in S$ and $(\cdot)^+$ denotes the Moore-Penrose generalized-inverse of the corresponding matrix.

Remark 5.2. Theorem 5.2 shows that the capability of feedback depends on both the structure complexity measured by $\{A_j, B_j, 1 \leq j \leq N\}$ and the information uncertainty measured by $\{p_{ij}, 1 \leq i, j \leq N\}$. To make it more clear in understanding how the capability of feedback depends on both the complexity and uncertainty of the system, we consider the simple scalar variable case with $B(\theta_t) = 1$, where the Markov chain $\{\theta_t\}$ has two states $\{1, 2\}$ only and $p_{12} = p_{21}$. It can be shown by Theorem 5.2 that the system is stabilizable if and only if $CP < 1$, where $C \triangleq (A_2 - A_1)^2$ and $P \triangleq (1 - p_{12})p_{12}$ can be interpreted as measures of the structure complexity (degree of dispersion) and the information uncertainty respectively (see [15] for details).

6. Concluding remarks

For several basic classes of uncertain dynamical control systems, we have given some critical values or equations to characterize the capability and limits of the feedback mechanism, and have shown that "impossibility theorems" hold even for some seemingly simple uncertain dynamical systems. Of course, many important problems still remain open. Examples are as follows:

(i) For general high-dimensional or high order uncertain nonlinear control systems, to find critical conditions characterizing the capability of feedback, at least to find general sufficient conditions under which feedback stabilization is possible in the discrete-time case.

(ii) To characterize the maximum capability of feedback that is designed based on switched linear control models, in dealing with uncertain nonlinear dynamical systems.

(iii) To find a suitable mathematical framework within which the issue of establishing a quantitative relationship among *a priori* information, feedback performance and computational complexity can be addressed adequately.

References

[1] Wiener, N., *Cybernetics, or Control and Communication in the Animal and the Machine,* MIT Press, 1948.

[2] Zames, G., 'Feedback and optimal sensitivity: Model reference transformations, weighted seminorms and approximate inverses', *IEEE Trans.Automat.Contr.,* 23(1981), 301–320.

[3] Zhou, K., J. C. Doyle and K.Glover, *Robust and Optimal Control* , Prentice-Hall, 1996.

[4] Åström, K. J. and B. Wittenmark, *Adaptive Control,* Addison-Wesley, Reading, MA, 2nd ed., 1995.

[5] Chen, H., and L. Guo, *Identification and Stochastic Adaptive Control.* Boston, MA: Birkhäuser, 1991.

[6] Ioannou, P. A., and J. Sun, *Robust Adaptive Control,* Engle-wood Cliffs, NJ: Prentice-Hall, 1996.

[7] Krstić, M., I. Kanellakopoulos and P. V. Kokotović, *Nonlinear and Adaptive Control Design.* New York: John Wiley & Sons, 1995.

[8] Guo, L., and H. F. Chen, 'The Åström-Wittenmark self-tuning regulator revisited and ELS-based adaptive trackers', *IEEE Trans. Automat. Contr.,* 36(7)(1991), 802–812.

[9] Guo, L., 'Self-convergence of weighted least-squares with applications to stochastic adaptive control', *IEEE Trans. Automat. Contr.,* 41(1)(1996), 79–89.

[10] Guo, L., 'On critical stability of discrete-time adaptive nonlinear control', *IEEE Trans. Automat. Contr.,*42(11)(1997),1488–1499.

[11] Xie, L. L., and L. Guo, 'Fundamental limitations of discrete-time adaptive nonlinear control', *IEEE Trans. Automat. Contr.,* 44(9)(1999), 1777–1782.

[12] Xie, L. L., and L. Guo, 'Adaptive control of discrete-time nonlinear systems with structural uncertainties', *AMS/IP Studies in Mathematics, 17(2000),* 49–89.

[13] Xue, F. and L. Guo, 'Stabilizability, uncertainty and the choice of sampling rate', *Proc. IEEE-CDC,* December 2000, Sydney.

[14] Xie, L. L., and L. Guo, 'How much uncertainty can be dealt by feedback?', *IEEE Trans. on Automatic Control, 45*(12)(2000), 2203–2217.

[15] Xue, F., L. Guo and M. Y. Huang, 'Towards understanding the capability of adaptation for time-varying systems', *Automatica, 37*(2001), 1551–1560.

[16] Xue, F. and L. Guo, 'Necessary and sufficient conditions for adaptive stabilizability of jump linear systems', *Communications in Information and Systems, 2*(1),(2001), 205–224.

[17] Zhang, Y. X., and L. Guo, 'A limit to the capability of feedback',*IEEE Trans. Automatic Control,* Vol.47, No.4 (2002),687-692.

A Computer Verification of the Kepler Conjecture

Thomas C. Hales*

Abstract

The Kepler conjecture asserts that the density of a packing of congruent balls in three dimensions is never greater than $\pi/\sqrt{18}$. A computer assisted verification confirmed this conjecture in 1998. This article gives a historical introduction to the problem. It describes the procedure that converts this problem into an optimization problem in a finite number of variables and the strategies used to solve this optimization problem.

2000 Mathematics Subject Classification: 52C17.
Keywords and Phrases: Sphere packings, Kepler conjecture, Discrete geometry.

1. Historical introduction

The Kepler conjecture asserts that the density of a packing of congruent balls in three dimensions is never greater than $\pi/\sqrt{18} \approx 0.74048\ldots$. This is the oldest problem in discrete geometry and is an important part of Hilbert's 18th problem. An example of a packing achieving this density is the face-centered cubic packing (Figure 1).

A packing of balls is an arrangement of nonoverlapping balls of radius 1 in Euclidean space. Each ball is determined by its center, so equivalently it is a collection of points in Euclidean space separated by distances of at least 2. The density of a packing is defined as the lim sup of the densities of the partial packings formed by the balls inside a ball with fixed center of radius R. (By taking the lim sup, rather than lim inf as the density, we prove the Kepler conjecture in the strongest possible sense.) Defined as a limit, the density is insensitive to changes in the packing in any bounded region. For example, a finite number of balls can be removed from the face-centered cubic packing without affecting its density.

Consequently, it is not possible to hope for any strong uniqueness results for packings of optimal density. The uniqueness established by Lemma 2.8 is nearly as

---

*Department of Mathematics, University of Pittsburgh, Thackeray Hall, Pittsburgh PA 15260, USA. E-mail: hales@pitt.edu

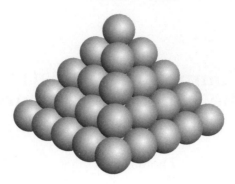

Figure 1: The face-centered cubic packing

strong as can be hoped for. It shows that certain local structures (decomposition stars) attached to the face-centered cubic (fcc) and hexagonal-close packings (hcp) are the only structures that maximize a local density function.

1.1 Hariot and Kepler

The modern mathematical study of close packings can be traced to T. Hariot. Hariot's work—unpublished, unedited, and largely undated—shows a preoccupation with packings of balls. He seems to have first taken an interest in packings at the prompting of Sir Walter Raleigh. At the time, Hariot was Raleigh's mathematical assistant, and Raleigh gave him the problem of determining formulas for the number of cannonballs in regularly stacked piles. Shirley, Hariot's biographer, writes that this study "led him inevitably to the corpuscular or atomic theory of matter originally deriving from Lucretius and Epicurus [25, p.242]."

Kepler became involved in packings of balls through his correspondence with Hariot in the early years of the 17th century. Kargon writes, in his history of atomism in England,

> According to Hariot the universe is composed of atoms with void space interposed. The atoms themselves are eternal and continuous. Physical properties result from the magnitude, shape, and motion of these atoms, or corpuscles compounded from them....
>
> Probably the most interesting application of Hariot's atomic theory was in the field of optics. In a letter to Kepler on 2 December 1606 Hariot outlined his views. Why, he asked, when a light ray falls upon the surface of a transparent medium, is it partially reflected and partially refracted? Since by the principle of uniformity, a single point cannot both reflect and transmit light, the answer must lie in the supposition that the ray is resisted by some points and not others....
>
> It was here that Hariot advised Kepler to abstract himself mathematically into an atom in order to enter 'Nature's house'. In his reply of 2 August 1607, Kepler declined to follow Harriot, ad atomos et vacua. Kepler preferred to think of the reflection-refraction problem in terms of the union of two op-

posing qualities—transparence and opacity. Hariot was surprised. "If those assumptions and reasons satisfy you, I am amazed." [19, p.26]

Despite Kepler's initial reluctance to adopt an atomic theory, he was eventually swayed, and in 1611 he published an essay that explores the consequences of a theory of matter composed of small spherical particles. Kepler's essay was the "first recorded step towards a mathematical theory of the genesis of inorganic or organic form" [28, p.v].

Kepler's essay describes the face-centered cubic packing and asserts that "the packing will be the tightest possible, so that in no other arrangement could more pellets be stuffed into the same container." This assertion has come to be known as the Kepler conjecture. This conjecture was verified with computer assistance in 1998 [15].

1.2   Newton and Gregory

The next episode in the history of this problem is a debate between Isaac Newton and David Gregory. Newton and Gregory discussed the question of how many balls of equal radius can be arranged to touch a given ball. This is the three-dimensional analogue of the simple fact that in two dimensions six pennies, but no more, can be arranged to touch a central penny. This is the kissing-number problem in n-dimensions. In three dimensions, Newton said that the maximum was 12 balls, but Gregory claimed that 13 might be possible. B. L. van der Waerden and Schütte in 1953 showed that Newton was correct [24].

The two-dimensional analogue of the Kepler conjecture is to show that the honeycomb packing in two dimensions gives the highest density. This result was established in 1892 by Thue, with a second proof appearing in 1910 ([26], [27]).

In 1900, Hilbert made the Kepler conjecture part of his 18th problem [16]. The third part of that problem asks, " How can one arrange most densely in space an infinite number of equal solids of given form, e.g. spheres with given radii ..., that is, how can one so fit them together that the ratio of the filled to the unfilled space may be as great as possible?"

1.3   The literature

Past progress toward the Kepler conjecture can be arranged into four categories: (1) bounds on the density, (2) descriptions of classes of packings for which the bound of $\pi/\sqrt{18}$ is known, (3) convex bodies other than balls for which the packing density can be determined precisely, (4) strategies of proof.

Various upper bounds have been established on the density of packings. A list of such bounds appears in [10]. Rogers's bound of 0.7797 is particularly natural [23]. It remained the best available bound for many years.

1.4   Classes of packings

If the infinite dimensional space of all packings is too unwieldy, we can ask if it is possible to establish the bound $\pi/\sqrt{18}$ for packings with special structures.

If we restrict the problem to packings of balls whose centers are the points of a lattice, the packings are described by a finite number of parameters, and the problem becomes much more accessible. Lagrange proved that the densest lattice packing in two dimensions is the familiar honeycomb arrangement [21]. Gauss proved that the densest lattice packing in three dimensions is the face-centered cubic [9]. The enormous list of references in [4] documents the many developments in lattice packings over the past two centuries.

1.5 Other convex bodies

If the optimal packings of balls are too difficult to determine, we might ask whether the problem can be solved for other convex bodies. To avoid trivialities, we restrict our attention to convex bodies whose packing density is strictly less than 1.

The first convex body in Euclidean 3-space that does not tile for which the packing density was explicitly determined is an infinite cylinder [1]. Here A. Bezdek and W. Kuperberg prove that the optimal density is obtained by arranging the cylinders in parallel columns in the honeycomb arrangement.

In 1993, J. Pach exposed the humbling depth of our ignorance when he issued the challenge to determine the packing density for some bounded convex body that does not tile space [22]. (This challenge was met by A. Bezdek [2].)

1.6 Strategies of proof

In 1953, L. Fejes Tóth proposed a program to prove the Kepler conjecture [5]. A single Voronoi cell cannot lead to a bound better than the dodecahedral bound. (The dodecahedral bound is the ratio of the volume of a inscribed ball to the volume of the containing dodecahedron.) L. Fejes Tóth considered weighted averages of the volumes of collections of Voronoi cells. These weighted averages involve up to 13 Voronoi cells. He showed that if a particular weighted average of volumes is greater than the volume of the rhombic dodecahedron, then the Kepler conjecture follows. The Kepler conjecture is an optimization problem in an infinite number of variables. L. Fejes Tóth's weighted-average argument was the first indication that it might be possible to reduce the Kepler conjecture to a problem in a finite number of variables. Needless to say, calculations involving the weighted averages of the volumes of several Voronoi cells are complex.

L. Fejes Tóth made another significant suggestion in [6]. He was the first to suggest the use of computers in the Kepler conjecture. After describing his program, he writes,

> Thus it seems that the problem can be reduced to the determination of the minimum of a function of a finite number of variables, providing a programme realizable in principle. In view of the intricacy of this function we are far from attempting to determine the exact minimum. But, mindful of the rapid development of our computers, it is imaginable that the minimum may be approximated with great exactitude.

A widely publicized attempt to prove the Kepler conjecture was that of Wu-Yi Hsiang [17], [18]. Hsiang's approach can be viewed as a continuation and extension

of L. Fejes Tóth's program. Hsiang's work contains major gaps and errors [3]. A
list of published materials relating to these errors can be found in [10].

2. Structure of the proof

This section describes the structure of the proof of the Kepler Conjecture.

Theorem 2.1. *(The Kepler Conjecture) No packing of congruent balls in Euclidean
three space has density greater than that of the face-centered cubic packing.*

Here, we describe the top-level outline of the proof and give references to the
sources of the details of the proof ([8], [11], [12], [13], [14], [7], [15]).

Consider a packing of congruent balls of unit radius in Euclidean three space.
The density of a packing does not decrease when balls are added to the packing.
Thus, to answer a question about the greatest possible density we may add nonoverlapping balls until there is no room to add further balls. Such a packing will
be said to be *saturated.*

Let Λ be the set of centers of the balls in a saturated packing. Our choice of
radius for the balls implies that any two points in Λ have distance at least 2 from
each other. We call the points of Λ *vertices*. Let $B(x, r)$ denote the ball in Euclidean
three space at center x and radius r. Let $\delta(x, r, \Lambda)$ be the finite density, defined by
the ratio of $A(x, r, \Lambda)$ to the volume of $B(x, r)$, where $A(x, r, \Lambda)$ is defined as the
volume of the intersection with $B(x, r)$ of the union of all balls in the packing. Set
$\Lambda(x, r) = \Lambda \cap B(x, r)$.

The *Voronoi cell* $\Omega(v)$ around a vertex $v \in \Lambda$ is the set of points closer to v than
to any other ball center. The volume of each Voronoi cell in the face-centered cubic
packing is $\sqrt{32}$. This is also the volume of each Voronoi cell in the hexagonal-close
packing.

Let $a : \Lambda \to \mathbb{R}$ be a function. We say that a is *negligible* if there is a constant
C_1 such that for all $x \in \mathbb{R}^3$ and $r \geq 1$, we have

$$\sum_{v \in \Lambda(x,r)} a(v) \leq C_1 r^2.$$

We say that the function a is *fcc-compatible* if for all $v \in \Lambda$ we have the inequality

$$\sqrt{32} \leq \text{vol}(\Omega(v)) + a(v).$$

Lemma 2.2. *If there exists a negligible fcc-compatible function $a : \Lambda \to \mathbb{R}$ for a
saturated packing Λ, then there exists a constant C such that for all $x \in \mathbb{R}^3$ and
$r \geq 1$, we have*

$$\delta(x, r, \Lambda) \leq \pi/\sqrt{18} + C/r.$$

Proof. The numerator $A(x, r, \Lambda)$ of $\delta(x, r, \Lambda)$ is at most the product of the volume
of a ball $4\pi/3$ with the number $|\Lambda(x, r + 1)|$ of balls intersecting $B(x, r)$. Hence

$$A(x, r, \Lambda) \leq |\Lambda(x, r + 1)| 4\pi/3. \tag{2.1}$$

In a saturated packing each Voronoi cell is contained in a ball of radius 2 centered at the *center* of the cell. The volume of the ball $B(x, r + 3)$ is at least the combined volume of Voronoi cells lying entirely in the ball. This observation, combined with fcc-compatibility and negligibility, gives

$$\sqrt{32}|\Lambda(x, r + 1)| \leq \sum_{v \in \Lambda(x, r+1)} (a(v) + \text{vol}(\Omega(v)))$$
$$\leq C_1(r + 1)^2 + \text{vol}\, B(x, r + 3) \tag{2.2}$$
$$\leq C_1(r + 1)^2 + (1 + 3/r)^3 \text{vol}\, B(x, r).$$

Divide through by $\text{vol}\, B(x, r)$ and eliminate $|\Lambda(x, r + 1)|$ between Inequality (2.1) and Inequality (2.2) to get

$$\delta(x, r, \Lambda) \leq \frac{\pi}{\sqrt{18}}(1 + 3/r)^3 + C_1 \frac{(r + 1)^2}{r^3\sqrt{32}}.$$

The result follows for an appropriately chosen constant C.

Remark 2.3. We take the precise meaning of the Kepler Conjecture to be a bound on the essential supremum of the function $\delta(x, r)$ as r tends to infinity. Lemma 2.2 implies that the essential supremum of $\delta(x, r, \Lambda)$ is bounded above by $\pi/\sqrt{18}$, provided a negligible fcc-compatible function can be found. The strategy will be to define a negligible function, and then to solve an optimization problem in finitely many variables to establish that it is fcc-compatible.

The article [8] defines a compact topological space X and a continuous function σ on that space.

The topological space X is directly related to packings. If Λ is a saturated packing, then there is a geometric object $D(v, \Lambda)$ constructed around each vertex $v \in \Lambda$. $D(v, \Lambda)$ depends on Λ only through the vertices in Λ at distance at most 4 from v. The objects $D(v, \Lambda)$ are called *decomposition stars*, and the space of all decomposition stars is precisely X.

Let δ_{tet} be the packing density of a regular tetrahedron. That is, let S be a regular tetrahedron of edge length 2. Let B the part of S that lies within distance 1 of some vertex. Then δ_{tet} is the ratio of the volume of B to the volume of S. We have $\delta_{tet} = \sqrt{8} \arctan(\sqrt{2}/5)$.

Let δ_{oct} be the packing density of a regular octahedron of edge length 2, again constructed as the ratio of the volume of points within distance 1 of a vertex to the volume of the octahedron. We have $\delta_{oct} \approx 0.72$.

Let $pt = -\pi/3 + \sqrt{2}\delta_{tet} \approx 0.05537$.

The following conjecture is made in [8]:

Conjecture 2.4. *The maximum of σ on X is the constant $8\,pt \approx 0.442989$.*

Lemma 2.5. *An affirmative answer to Conjecture 2.4 implies the existence of a negligible fcc-compatible function for every saturated packing Λ.*

Proof. For any saturated packing Λ define a function $a : \Lambda \to \mathbb{R}$ by

$$-\sigma(D(v,\Lambda))/(4\delta_{oct}) + 4\pi/(3\delta_{oct}) = \text{vol}(\Omega(v)) + a(v).$$

Negligibility follows from [8, Prop. 3.14 (proof)]. The upper bound of $8\,pt$ gives a lower bound

$$-8\,pt/(4\delta_{oct}) + 4\pi/(3\delta_{oct}) \le \text{vol}(\Omega(v)) + a(v).$$

The constant on the left-hand side of this inequality equals $\sqrt{32}$, and this establishes fcc-compatibility.

Theorem 2.6. *Conjecture 2.4 is true. That is, the maximum of the function σ on the topological space X of all decomposition stars is $8\,pt$.*

Theorem 2.6, Lemma 2.5, and Lemma 2.2 combine to give a proof of the Kepler Conjecture 2.1.

Let $t_0 = 1.255$ ($2t_0 = 2.51$). This is a parameter that is used for truncation throughout the series of articles on the Kepler Conjecture.

Let $U(v,\Lambda)$ be the set of vertices in Λ at distance at most $2t_0$ from v. From a decomposition star $D(v,\Lambda)$ it is possible to recover $U(v,\Lambda)$ (at least up to Euclidean translation: $U \mapsto U + y$, for $y \in \mathbb{R}^3$). We can completely characterize the decomposition stars at which the maximum of σ is attained.

Theorem 2.7. *Let D be a decomposition star at which the maximum $8\,pt$ is attained. Then the set $U(D)$ of vectors at distance at most $2t_0$ from the center has cardinality 12. Up to Euclidean motion, $U(D)$ is the kissing arrangement of the 12 balls around a central ball in the face-centered cubic packing or hexagonal-close packing.*

2.7 Outline of proofs

To prove Theorems 2.6 and 2.7, we wish to show that there is no counterexample. That is, we wish to show that there is no decomposition star D with value $\sigma(D) > 8\,pt$. We reason by contradiction, assuming the existence of such a decomposition star. With this in mind, we call D a *contravening decomposition star*, if

$$\sigma(D) \ge 8\,pt.$$

In much of what follows we will assume that every decomposition star under discussion is a contravening one. Thus, when we say that no decomposition stars exist with a given property, it should be interpreted as saying that no such contravening decomposition stars exist.

To each contravening decomposition star, we associate a (combinatorial) plane graph. A restrictive list of properties of plane graphs is described in [15, Section 2.3]. Any plane graph satisfying these properties is said to be *tame*. All tame plane graphs have been classified. (There are several thousand, up to isomorphism.) Theorem [15, Th 2.1] asserts that the plane graph attached to each contravening decomposition star is tame. By the classification of such graphs, this reduces the

proof of the Kepler Conjecture to the analysis of the decomposition stars attached
to the finite explicit list of tame plane graphs.

A few of the tame plane graphs are of particular interest. Every decomposition
star attached to the face-centered cubic packing gives the same plane graph (up to
isomorphism). Call it G_{fcc}. Likewise, every decomposition star attached to the
hexagonal-close packing gives the same plane graph G_{hcp}. Let X_{crit} be the set of
decomposition stars D such that the set $U(D)$ of vertices is the kissing arrangement
of the 12 balls around a central ball in the face-centered cubic or hexagonal-close
packing. There are only finitely many orbits of X_{crit} under the group of Euclidean
motions.

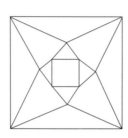 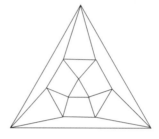

Figure 2: The plane graphs G_{fcc} and G_{hcp}

In [8, Lemma 3.13], the necessary local analysis is carried out to prove the
following local optimality.

Lemma 2.8. *A decomposition star whose plane graph is G_{fcc} or G_{hcp} has score at
most 8 pt, with equality precisely when the decomposition star belongs to X_{crit}.*

In light of this result, we prove 2.6 and 2.7 by proving that any decomposition
star whose graph is tame and not equal to G_{fcc} or G_{hcp} is not contravening

There is one more tame plane graph that is particularly troublesome. It is
the graph G_{pent} obtained from the pictured configuration of twelve balls tangent
to a given central ball (Figure 3). (Place a ball at the north pole, another at the
south pole, and then form two pentagonal rings of five balls.) This case requires
individualized attention. S. Ferguson proves in [7] that if D is any decomposition
star with this graph, then $\sigma(D) < 8\,pt$.

To eliminate the remaining cases, more-or-less generic arguments can be used.
A linear program is attached to each tame graph G. The linear program can be
viewed as a linear relaxation of the nonlinear optimization problem of maximizing
σ over all decomposition stars with a given tame graph G. Because it is obtained
by relaxing the constraints on the nonlinear problem, the maximum of the linear
problem is an upper bound on the maximum of the original nonlinear problem.
Whenever the linear programming maximum is less than $8\,pt$, it can be concluded
that there is no contravening decomposition star with the given tame graph G. This
linear programming approach eliminates most tame graphs.

When a single linear program fails to give the desired bound, it is broken
into a series of linear programming bounds, by branch and bound techniques. For

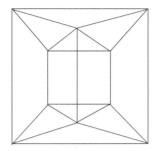

Figure 3: The plane graph G_{pent}

every tame plane graph G other than G_{hcp}, G_{fcc}, and G_{pent}, we produce a series of linear programs that establish that there is no contravening decomposition star with graph G. When every face of the plane graph is a triangle or quadrilateral, this is accomplished in [13]. The general case is completed in the final sections of [15].

References

[1] A. Bezdek and W. Kuperberg, Maximum density space packing with congruent circular cylinders of infinite length, *Mathematica* 37 (1990), 74–80.

[2] A. Bezdek, A remark on the packing density in the 3-space in *Intuitive Geometry*, ed. K. Böröczky and G. Fejes Tóth, *Colloquia Math. Soc. János Bolyai* 63, North-Holland (1994), 17–22.

[3] J. H. Conway, T. C. Hales, D. J. Muder, and N. J. A. Sloane, On the Kepler conjecture, *Math. Intelligencer* 16, no. 2 (1994), 5.

[4] J. H. Conway and N. J. A. Sloane, Sphere packings, lattices and groups, third edition, Springer-Verlag, New York, 1998.

[5] L. Fejes Tóth, *Lagerungen in der Ebene auf der Kugel und im Raum*, Springer, Berlin, first edition, 1953.

[6] L. Fejes Tóth, Regular figures, Pergamon Press, Oxford London New York, 1964.

[7] S. Ferguson, Sphere Packings V, thesis, University of Michigan, 1997.

[8] S. Ferguson, T. Hales, A Formulation of the Kepler Conjecture, preprint math.MG/9811072.

[9] C. F. Gauss, Untersuchungen über die Eigenscahften der positiven ternären quadratischen Formen von Ludwig August Seber, *Göttingische gelehrte Anzeigen*, 1831 Juli 9, also in *J. reine angew. Math.* 20 (1840), 312–320.

[10] T. C. Hales, An Overview of the Kepler Conjecture, preprint math.MG/9811071.

[11] T. C. Hales, Sphere Packings I, Discrete and Computational Geometry, 17 (1997), 1-51.

[12] T. C. Hales, Sphere Packings II, Discrete and Computational Geometry, 18 (1997), 135-149.

[13] T. C. Hales, Sphere Packings III, preprint math.MG/9811075.

[14] T. C. Hales, Sphere Packings IV, preprint math.MG/9811076.

[15] T. C. Hales, The Kepler Conjecture, preprint math.MG/9811078.

[16] D. Hilbert, Mathematische Probleme, *Archiv Math. Physik* 1 (1901), 44–63, also in *Proc. Sym. Pure Math.* 28 (1976), 1–34.

[17] W.-Y. Hsiang, On the sphere packing problem and the proof of Kepler's conjecture, Internat. J. Math 93 (1993), 739-831.

[18] W.-Y. Hsiang, Least Action Principle of Crystal Formation of Dense Packing Type and the Proof of Kepler's Conjecture, World Scientific, 2002.

[19] R. Kargon, Atomism in England from Hariot to Newton, Oxford, 1966.

[20] J. Kepler, The Six-cornered snowflake, Oxford Clarendon Press, Oxford, 1966, forward by L. L. Whyte.

[21] J. L. Lagrange, Recherches d'arithmétique, *Nov. Mem. Acad. Roy. Sc. Bell Lettres Berlin* 1773, in *Œuvres*, vol. 3, 693–758.

[22] W. Moser, J. Pach, Research problems in discrete geometry, DIMACS Technical Report, 93032, 1993.

[23] C. A. Rogers, The packing of equal spheres, *Proc. London Math. Soc.* (3) 8 (1958), 609–620.

[24] K. Schütte and B.L. van der Waerden, Das Problem der dreizehn Kugeln, *Math. Annalen* 125, (1953), 325–334.

[25] J. W. Shirley, *Thomas Harriot: a biography*, Oxford, 1983.

[26] A. Thue, Om nogle geometrisk taltheoretiske Theoremer, *Forandlingerneved de Skandinaviske Naturforskeres* 14 (1892), 352–353.

[27] A. Thue, Über die dichteste Zusammenstellung von kongruenten Kreisen in der Ebene, *Christinia Vid. Selsk. Skr.* 1 (1910), 1–9.

[28] L. L. Whyte, forward to [20].

ICM 2002 · Vol. III · 805–816

Rhythms of the Nervous System: Mathematical Themes and Variations

Nancy Kopell*

Abstract

The nervous system displays a variety of rhythms in both waking and sleep. These rhythms have been closely associated with different behavioral and cognitive states, but it is still unknown how the nervous system makes use of these rhythms to perform functionally important tasks. To address those questions, it is first useful to understood in a mechanistic way the origin of the rhythms, their interactions, the signals which create the transitions among rhythms, and the ways in which rhythms filter the signals to a network of neurons.

This talk discusses how dynamical systems have been used to investigate the origin, properties and interactions of rhythms in the nervous system. It focuses on how the underlying physiology of the cells and synapses of the networks shape the dynamics of the network in different contexts, allowing the variety of dynamical behaviors to be displayed by the same network. The work is presented using a series of related case studies on different rhythms. These case studies are chosen to highlight mathematical issues, and suggest further mathematical work to be done. The topics include: different roles of excitation and inhibition in creating synchronous assemblies of cells, different kinds of building blocks for neural oscillations, and transitions among rhythms. The mathematical issues include reduction of large networks to low dimensional maps, role of noise, global bifurcations, use of probabilistic formulations.

2000 Mathematics Subject Classification: 37N25, 92C20.

1. Introduction

The nervous system creates many different rhythms, each associated with a range of behaviors and cognitive states. The rhythms were first discovered from scalp recordings of humans, and the names by which they are known still come mainly from the electroencephalograph (EEG) literature, which pays attention to

*Boston University, Department of Mathematics and Center for BioDynamics, Boston, MA 02215, USA. E-mail: nk@bu.edu

the frequency and behavioral context of those rhythms, but not to their mechanistic origins. The rhythmic patterns include the alpha (9-11 Hz), beta (12-30 Hz), gamma (30-80 Hz), theta (4-8 Hz), delta (2-4) Hz and slow wave (.5-2 Hz) rhythms. The boundaries of these ranges are rough. More will be said about the circumstances in which some of these rhythms are displayed.

It is now possible to get far more information about the mechanisms behind the dynamics of the nervous system from other techniques, including electrophysiology. The revolutions in experimental techniques, data acquisition and analysis, and fast computation have opened up a broad and deep avenue for mathematical analysis. The general question addressed by those interested in rhythms is: how does the brain make use of these rhythms in sensory processing, sensory-motor coordination and cognition? The mathematical strategy, to be discussed below, is to investigate the "dynamical structure" of the different rhythms to get clues to function. Most of this talk is about dynamical structure, and the mathematical issues surrounding its investigation. I'll return at the end to issues of function.

2. Neuromath

The mathematical framework for the study of brain dynamics are the Hodgkin-Huxley (HH) equations. These are partial differential equations describing the propagation of action potentials in neurons (cells of the nervous system). The equations, which play the same role in neural dynamics that Navier-Stokes does in fluid dynamics, are an elaborate analogy to a distributed electrical circuit. The central equation,

$$Cv' = \sum I_{ion} + \nabla^2 v + \sum I_{syn} + I_{ext}$$

describes conservation of current across a piece of a cell membrane; v is the cross-membrane voltage and the left hand side is the capacitive current. The first sum on the right hand side represents the intrinsically generated ionic currents across the membrane. The term $\nabla^2 v$ represents the spatial diffusion, and I_{ext} the current fed into the cell. $\sum I_{syn}$ represents the currents introduced by coupling from other cells. Thus, these equations can also be used to model networks of interacting neurons, the focus of this talk.

Each of the intrinsic currents I_{ion} is described by Ohm's law: I_{ion} is electromotive force divided by resistance. In this context, one usually uses the concept of "conductance", which is the reciprocal of resistance. The electromotive force depends on the type of charged ion (e.g., Na, K, Ca, Cl) and the voltage of the cell; it has the form $[v - V_{ion}]$, where V_{ion} is the so-called "reversal potential" of that ion.

The dynamics of the conductances are what make the equations so mathematically interesting and rich. In a simple description of the conductance, each ionic current has up to two "gates", which open or close at rates that are dependent on the voltage of the cell. For each such gate, there is then a first order equation of the form

$$x' = [x_\infty(v) - x]/\tau_x(v)$$

where x denotes the fraction of channels of that type that are open at any given time, $x_\infty(v)$ is the steady state value of x for a fixed voltage, and $\tau_x(v)$ is the rate constant of that gating variable. For example, the standard Na current has the form $I_{Na} = \overline{g}m^3h\,[v - V_{Na}]$, where $\overline{g}$ is the maximal conductance, and m and h are gating variables satisfying the above equation for x. The dynamics for m and h differ because $m_\infty(v)$ is an increasing function of v, while $h_\infty(v)$ is a decreasing function; also $\tau_m(v)$ is much smaller than $\tau_h(v)$. The (chemical) synaptic currents have the same form as the intrinsic ones, with the difference that the dependence of the driving force on voltage uses that of the post-synaptic cell, while the conductance depends on the pre-synaptic voltage. That is, I_{syn} has the form $\overline{g}\widehat{x}\,[v - V_{syn}]$, where $\widehat{x}$ satisfies the equation for x above, with v replaced by $\widehat{v}$, the voltage of the cell sending the signal, and V_{syn} is the reversal potential of the synapse. The coupling is said to be excitatory if the current is inward (increases voltage toward the threshold for firing an action potential) or inhibitory if the current is outward (moves voltage away from threshold for firing.)

For a simple version of the HH equations, there are three ionic currents; one of these (Na) creates an inward current leading to an action potential, one (K) an outward current helping to end the action potential, and a leak current (mainly Cl) with no gating variable. The HH equations are not one single set of equations, but a general (and generalizable) form for a family of equations, corresponding to different sets of intrinsic currents (which can depend on position on the neuron), different neuron geometries, and different networks created by interactions of neurons, which may themselves be highly inhomogeneous. Numerical computation has become highly important for observing the behavior of these equations, but does not suffice to understand the behavior, especially to get insight into what the specific ionic currents contribute; this is where the analysis, including simplification, comes in. For an introduction to HH equations, some analysis and some of its uses in models, see [1].

2.1. Some mathematical issues associated with rhythms

It is not possible to analyze the full class of equations in all generality. Our strategy is to look for mathematical structures underlying some classes of behavior observed experimentally; the emphasis is on the role of dynamical systems, as opposed to statistics, though probabilistic ideas enter the analysis.

Our central scientific question here is how rhythms emerge from the "wetware", as modeled by the HH equations. As we will see, different rhythms can be based on different sets of intrinsic currents, different classes of neurons, and different ways of hooking up those cells. There are some behaviors we can see by looking at small networks, and others that do not appear until the networks are large and somewhat heterogeneous. Even in the small networks, there are a multiplicity of different building blocks for the rhythms, with excitation and inhibition playing different roles. Noise appears, and plays different roles from heterogeneity.

Investigators often use simplifications of the HH equations. For example, this talk deals only with "space clamped" cells in which the spatial distribution of each cell is ignored, and the equations become ODEs. (There are circumstances under which this can be a bad approximation, as in [2]). Under some circumstances, the 4-dimensional simplest space-clamped HH equations (one current equation, three gating variables) can be reduced to a one-dimensional equation; thus, networks of neurons can be described by a fraction of the equations that one needs for the full HH network equations. Another kind of reduction replaces the full HH ODEs by maps that follow the times of the spikes. In both cases, there are at least heuristic explanations for why these reductions are often very successful, and hints about how and why the simplifications can be expected to break down.

3. Mathematics and small networks of neurons

3.1. Centrality of inhibition in rhythms

Some kinds of cells coupled by inhibition like to form rhythms and synchronize [3-5]. This is unintuitive, because inhibition to cells can temporarily keep the latter from firing (see below for important exceptions), but mutual inhibition can encourage cells to fire simultaneously.

There are various ways to see this, with methods that are valid in different contexts. For weak coupling, it can be shown rigorously that the full equations reduce to interactions between phases of the oscillators [6]; the particular coupling associated with inhibition can then be shown to be synchronizing (though over many cycles) [7]. If the equations can be reduced to one-dimensional "integrate and fire" models, one can use "spike-response methods" to see the synchronizing effect of inhibitory synapses on timing of spikes . Both of these are described in [6] along with more references.

Another method, which I believe is most intuitive, looks at the ongoing effect of forced inhibition on the voltage of the cells, and how some of the processes are "slaved" to others. This is seen most clearly in the context of another one-dimensional reduction that has become known as the "theta" model, because of the symbols used for the phase of the oscillations [8]. The reduced equations have been shown to be a canonical reduction of equations that are near a saddle-node bifurcation on an invariant circle (limit cycle). Many versions of HH-like models (and some kinds of real neurons) have this property for parameter values near onset of periodic spiking, and they are known as "Type 1" neurons.

The "theta model" has the form

$$\theta' = (1 - \cos\theta) + I(1 + \cos\theta).$$

Here the equation for the phase θ has periodic solutions if the parameter I is positive, and two fixed points (stable, saddle) if I is negative. To understand the effects of forced inhibition, we replace I by a time dependent inhibition given by $I - gs(t)$, where $s(t) = \exp(-t/\tau)$ for $t > 0$, and zero otherwise. With the change of variables $J(t) = 1 - g\exp(-t/\tau)$, this is a 2-D autonomous system. Figures [9] and analysis show that the system has two special orbits, known in the non-standard analysis literature as "rivers" [10], and that almost all of the trajectories feed quickly into one of these, and are repelled from the other. The essential effect is that initial conditions become irrelevant to the outcome of the trajectories. A similar effect works for mutually coupled systems of inhibitory neurons.

The rhythm formed in this way is highly dependent on the time scale of decay of the inhibition for the frequency of the network [11, 12]. These models, and the "fast-firing" inhibitory cells that they represent, can display a large range of frequencies depending on the bias (I_{ext} in HH, the parameter I in the theta model); however, in the presence of a small amount of heterogeneity in parameters, the rhythm falls apart unless the frequency is in the gamma range (30-80 Hz) [4, 13]. This can be understood from spike response methods or in terms of rivers.

The above rhythm is known as ING or interneuron gamma [14, 15]. A variation on this uses networks with fast-firing inhibitory cells (interneurons or I cells) and excitatory cells (pyramidal cells or E-cells). This is called PING (pyramidal interneuron gamma) [14, 15]. Heuristically, it is easy to understand the rhythm: the inhibitory cells are set so they do not fire without input from the E-cells. When the E-cells fire, they cause the I-cells to cross firing threshold and inhibit the E-cells, which fire again when the inhibition wears off. This simple mechanism becomes much more subtle when there is heterogeneity and noise in large networks, which will be discussed later.

3.2.   Excitation and timing maps

The fast-firing cells described above are modeled using only the ionic currents needed to create a spike. Most other neurons have channels to express many other ionic currents as well, with channel kinetics that range over a large span of time constants. These different currents change the dynamical behavior of the cells, and allow such cells to be "Type II", which means that the onset of rhythmic spiking as bias is changed is accompanied by a Hopf bifurcation instead of a saddle node. The type of onset has important consequences for the ability of pair of such cells to synchronize. E.g., models of the fast-firing neuron, if connected by excitatory synapses, do not synchronize, as can be shown from weak coupling or other methods described above (e.g., [7]). However, if the cells are Type II, they do synchronize stably with excitation (and not with inhibition). This was shown by Gutkin and Ermentrout using weak coupling methods [16]. A more specific case study was done by Acker et al. [17], motivated by neurons in the part of the cortex that constitutes the input-output pathways to the hippocampus, a structure of the brain important to learning and recall. These cells are excitatory and of Type II (J.White, in prep.); models of these cells, based on knowledge of the currents that they express, do synchronize with excitatory synapses, and do not with inhibitory synapses.

The synchronization properties of the such cells can be understood from spike-timing functions and maps [17]. Given the HH equations for the cell, one can introduce at any time in the cycle excitation or inhibition whose time course is similar to what the synapse would provide. From this, one can compute how much the next spike is advanced or delayed by this synapse. From such a graph, one can compute a spike-time map which takes the difference in spike times in a single cycle to the difference in the next cycle.

The analysis of such a map is easy, but the process raises deeper mathematical issues. One set of issues concerns what is happening at the biophysical level that gives rise to the Type II bifurcation, which is associated with a particular shape of the spike advance function [18]. Analysis shows that the Type II is associated with slow outward currents or certain slow inward currents that (paradoxically) turn on when the cell is inhibited [16, 17]; this shows how biophysical structure is connected with mathematical structure.

A second set of questions concerns why the high-dimensional coupled HH equations can be well approximated by a 1-D map. (In some parameter ranges, but not all, this is any excellent approximation). The mathematical issues here concern how large subsets of high-dimensional phase space collapse onto what is essentially a one-dimensional space. Ideas similar to those in Section 3.1 are relevant, but with different biophysics creating the collapse of the trajectories. In this case (and others) there are many different ionic currents, with many different time scales, so that a given current can be dominant in some portion of the trajectory and then decrease to zero while others take over; this leads to structure that is more complex than that of the traditional "fast-slow" equations, and which is not nearly as understood. Such reductions to 1-D maps have been used in other investigations of synchrony [19-21] involving multiple cells and multiple kinds of currents.

3.3.   More complex building blocks: Fancier inhibitory cells

So far, I've talked about networks containing fast-firing neurons (inhibitory) or excitatory cells. But there are many different kinds of cells in the nervous system, with intrinsic and synaptic currents that make them dynamically very different from one another. Once there are more currents with more time scales, it is easier to create more rhythms with different frequency. That is, the differences in frequencies often (but not always) come from some time scales in the interacting currents, and cannot be scaled away.

The stellate cell of Section 3.2 is an excellent example of currents creating frequencies; in a wide range of parameters, these cells, even without coupling, form a theta rhythm. Indeed, they are believed to be one of the primary sources of that rhythm in the hippocampus, which is thought by many to use these rhythms in tasks involving learning and recall. As described above, these cells are excitatory, and synchronize when coupled by excitation.

More puzzling are inhibitory cells in the hippocampus that are capable of forming theta rhythms as isolated cells with ionic currents similar to those in the stellate cells. The puzzle is that these cells do not cohere (in models) using inhibitory coupling. (The decay time of inhibition caused by these cells is roughly four times longer than the inhibition caused by the fast-firing cells, but neither fast nor this slower decaying inhibition creates synchrony in models.) So what is providing the coherence seen in the theta rhythm? (The rhythm can be seen in small slices that do not have inputs from other parts of the brain producing theta, so in such a paradigm, the rhythm must be produced locally.)

One suggestion (Rotstein, Kopell, Whittington, in preparation) is that the inhibitory rhythms seen in slice preparations with excitation blocked pharmacologically depend on both kinds of inhibitory cells discussed, the special ones (called O-LM cells [22]) and the others. Simulations show that networks of these cells can have the O-LM cells synchronize and I-cells synchronize at a different phase, to create an inhibitory network with considerably more complexity than interacting fast-firing cells involved in ING. Again, this can be reduced to a low-dimensional map for a minimal network (two O-LM cells, one fast firing I-cell). However, the reduction now requires properties of the currents involved in the O-LM model, including the kinetics of the gating variables.

3.4.   Interaction of rhythms

Another set of mathematical issues is associated with transitions among rhythms. In general, rhythms slower than gamma (e.g., beta, theta and alpha) make use of ionic currents that are active between spikes. These currents are voltage-dependent, so that changes in voltage, in the sub- and super-threshold regimes, can turn on or off these currents. Thus, neuromodulators that change the voltage range of a neuron (e.g., by changing a leak current) can change which other currents are actively expressed. In that way, they can cause a switch from one rhythm to another. For example, models of the alpha rhythm [20] suggest that this rhythm makes use the inhibition-activated "h-current"; this current is effectively off line if the voltage is increased (even below threshold level). Thus, a switch from alpha to a faster rhythm (gamma or beta) can be effected by simply making the E-cells operate in a moderately higher voltage regime.

These switches can be seen in simulations (Pinto, Jones, Kaper, Kopell, in prep.), but are still understood only heuristically. The mathematical issues are associated with reduction of dimension methods. In the regime in which the network is displaying alpha, there are many more variables that are actively changing, notably the gating variables of each of the currents that is important in this rhythm. When there is a switch to gamma and those currents go off line, the phase space becomes effectively smaller. The mathematics here involves understanding how that phase compression takes place.

A related set of mathematical questions concerns rhythms that are "nested", one within another. For example, the theta rhythm often presents as the envelope of a series of faster gamma cycles, and the beta rhythm, at least in some manifestations, occur with the I-cells firing at a gamma rhythm and the E-cells firing at the slower beta rhythm, missing some cycles of the inhibitory rhythm. The gamma/beta switch has been understood from a physiological point of view (see [19] and its references) and has been simulated. The gamma/theta nesting is less understood, though new data and simulations are providing the physiological and heuristic basis for this [22; Rotstein, Kopell, Whittington, in prep.].

4.   Large networks

Though there are many more examples of other building blocks, I'm turning to issues that do not appear in small network analysis. I'm going to go back to a very simple building block, but now put many such together. The simple building block is one E cell, one I-cell, which together can create a gamma rhythm.

4.1. Sparse coupling

We now consider a network with N E-cells and M I-cells, with random coupling from the E-cells to the I-cells and vica versa. Suppose, for example, there is a fixed probability of connection in each direction between any pair of E and I cells. Then the number of inputs to any cell is distributed across the population, leading to heterogeneity of excitation and inhibition. Is it still possible to get coherent gamma rhythms? This can be answered with mathematical analysis using the "theta neuron" model described above [9]. To understand synchrony in E/I networks, it is helpful to understand what each pulse of inhibition does to the population of excitatory cells and vica versa. The part in which both probability and dynamical systems play a large role is the effect of a pulse of inhibition on a population. The "rivers" referred to above in Section 3.1 create synchronization if the inputs to cells have no variance, but with variation in the size of the inputs, there is a spread in the times of the outputs. This can be accurately computed using features of the dynamics and probability theory. Similarly, but with less accuracy, one can compute the the effect of variation of inputs on the spike times of the receiving population due to a pulse of excitation. The results lead to unintutive conclusions, e.g., that increasing the strength of the inhibition (which strengthens the synchronizing effect of the rivers) does not reduce the desynchronizing effects of random connectivity. Furthermore, tight synchrony can be obtained even with extremely sparse coupling provided that variance in the size of the inputs is small.

4.2. Loss of coherence

The above analyses can be put together to understand synchrony of "PING". However, they leave only partially answered many questions about larger networks. One such question, which is central to understanding how the assemblies of neurons are created and destroyed, is the circumstances under which the synchrony falls apart, i.e., what modulations of cells and/or synapses will lead to loss of coherence of the gamma rhythm. The above analysis shows that too large a variation in size of inputs to different cells of the same population can be fatal. Similar phenomena occur with too much variation in drive or intrinsic currents. There are less obvious constraints that are understood from working with smaller networks described above. From those, it is possible to see that ING and PING operate in different parameter regimes: the firing times of the population in ING are governed by the bias of the I-cells (as well as the decay time of the inhibition); in PING, the inhibitory cells are more passive until driven by the E-cells, and the timing comes from bias of the E-cells (as well as decay of inhibition). This means that the mechanism of coherence can switch between ING and PING by changing relative excitability of the two populations. Changing the strengths of the I-E and E-I synapses can also get the population (large or small) out of the regime in which the E-cells synchronize the I-cells, and vica versa.

A more mysterious issue that cannot be addressed within minimal networks is how the size of the sub-populations responding on a given cycle affects the coherence on the next cycle and the numbers of cells participating, especially when there is some heterogeneity in the network. E.g., as the number of inhibitory neurons firing in a cycle changes, it changes the total inhibition to the E-cells, which changes the number of E-cells that are ready to fire when inhibition wears off, and before the next bout of inhibition. If the amount of inhibition gets too small, or inhibition gets too dispersed, the coherence can rapidly die. Without taking into account the trajectories of each of the large number of cells, it is likely that some possibly probabilistic account of the numbers of cells spiking per cycle can give some insight into the dynamical mechanisms surrounding failure of coherence.

Such a reduction has been successfully used in a different setting, involving the long-distance coherence of two populations of heterogeneous cells. In this case, if the populations are each minimal (one E/I pair) for each site, there is 1-D map that describes the synchronization, with the variable the timing between the E and I sites [19] . For large and heterogeneous networks, the synchronization (within some parameter regimes) can be described by a 3-dimensional map, in which the first variable is the time between the first spikes of a cycle in the two E-cell populations, and the others are the fraction of I-cells firing on that cycle in each of the two I-cell populations (McMillen and Kopell, in prep.).

Related work has been done from a different perspective, starting with asynchronous networks and asking how the asynchrony can lose stability [23-25]. Work using multiple time scales to address the formation of "clusters" when synchrony fails is in [26].

4.3. Noise, PING, and frequency control

One of the main differences between ING and PING is the difference in robustness. Small amounts of heterogeneity of any kind make ING coherence fall apart dramatically [4,13]. By contrast, PING is tolerant to large ranges of heterogeneity. The "ping-pong" mechanism of PING is also able to produce frequencies that cover a much wider range than the ING mechanism, which is constrained by loss of coherence to lie in the gamma range of approximately 30-80 Hz [4,13]. Since many versions of gamma seen in experiments are of the PING variety, this raises the question of what constrains the PING rhythms to stay in the gamma frequency range.

A possible answer to this comes from simulations. C. Borgers, D. McMillen and I found that heterogeneity, unless extreme, would not disrupt the PING coherence. However, a very small amount of noise (with fixed amplitude and poisson-distributed times) could entirely destroy coherence of the PING, provided the latter had a frequency below approximately 30 Hz; if the same noise is introduced when the network is in the gamma range, the behavior is only slightly perturbed. Furthermore, the ability to withstand the noise is related to adding some I-I connections, as in ING. A heuristic explanation is that, at low frequencies, the inhibition to the

I-cells (which has a time constant around 10ms) wears off before the excitation from
the E-cells causes these cells to spike. Thus, those cells hang around the threshold
for significant amounts of time, and are therefore vulnerable to being pushed over
threshold by noise. The mathematics has yet to be understood rigorously.

5. Mathematics and clues to function

The mathematical questions are themselves interesting, but the full richness of
the scientific endeavor comes from the potential for understanding how the rhythms
generated by the brain might be used in sensory processing, motor coordination and
cognition. We are still at the outer edges of such an investigation, but there are
many clues from animal behavior, physiology and mathematics. Work done with
EEGs (see, e.g., reviews [27,28]) has shown that many cognitive and motor tasks are
associated with specific rhythms appearing in different parts of the tasks. Gamma
is often associated with attention, awareness and perception, beta with preparation
for motor activies and high-order cognitive tasks, theta with learning and recall and
alpha with quiet awareness (there are several different versions of alpha in different
parts of the brain and found in different circumstances). Work done in whole
animals and in slice preparations are giving clues to the underlying physiology of the
rhythms, and how various neuromodulators change the rhythms, e.g., [14]. Much
of the math done so far has concerned how the networks produce their rhythms
from their ionic currents and connectivity, and has not directly addressed function.
However, the issues of function are starting to be addressed in terms of how the
dynamics of networks affects the computational properties of the latter.

One of the potential functions for these rhythms is the creation of "cell assemblies", temporary sets of neurons that fire synchronously. These assemblies are
believed to be important in distributed processing; they enhance the effect of the
synchronized pulses downstream, and provide a substrate for changes in synapses
that help to encode experience. ("Cells that fire together wire together.") Simulations show, and help to explain, why gamma rhythms have especially good properties for creating cells assemblies, and repressing cells with lower excitatbility or
input [29]. Furthermore, the changes in synapses known to occur during gamma can
facilitate the creation of the beta rhythm (see [19] for references), which appears
in higher-order processing. Mathematical analysis shows that the beta rhythm is
more effective for creating synchrony over distances where the conduction time is
longer. Thus, we can understand the spontaneous gamma-beta switch seen in various circumstances (see [19] and [29]) as creating cell assemblies (during the gamma
portion), using the synaptic changes to get cell assemblies encoded in the beta
rhythm, and then using the beta rhythm to form highly distributed cell assemblies.

The new flood of data, plus the new insights from the mathematics, are opening up many avenues for mathematical research related to rhythms and function.
A large class of such questions concerns how networks that are displaying given
rhythms filter inputs with spatio-temporal structure, and how this affects the changing cell assemblies. This question is closely related to the central and controversial
questions of what is the neural code and how does it operate. These questions will

likely require new techniques to combine dynamical systems and probability, new ways to reduce huge networks to ones amenable to analysis, and new ideas within dynamical systems itself, e.g., to understand switches as global bifurcations; these are large and exciting challenges to the mathematical community.

References

[1] C. Koch and I. Segev, *Methods in Neuronal Modeling*, MIT Press, Cambridge MA., 1998.

[2] P.F. Pinsky and J. Rinzel, Intrinsic and network rhythmogenesis in a reduced Traub model, *J. Comput. Neurosci.* **1** (1994), 39–60.

[3] X.J. Wang and J. Rinzel, Alternating and synchronous rhythms in reciprocally inhibitory model neurons, *Neural Comput.* **4** (1992), 84–97.

[4] J. White, C. Chow, J. Ritt, C. Soto-Trevino and N. Kopell, Synchronization and oscillatory dynamics in heterogeneous, mutually inhibited neurons, *J. Comput. Neurosci.* **5** (1998), 5–16.

[5] D. Terman, N. Kopell and A. Bose, Dynamics of two mutually coupled slow inhibitory neurons, *Physica D* **117** (1998), 241–275.

[6] C. van Vreeswijk, L.F. Abbott, G.B. Ermentrout, When inhibition, not excitation, synchronizes neural firing, *J. Comput. Neurosci.* 1(1994), 313–321.

[7] N. Kopell and G.B. Ermentrout, Mechanisms of phase-locking and frequency control in pairs of coupled neural oscillators, in *Handbook on Dynamical Systems*, vol. 2, *Toward Applications* . Ed. B. Fiedler, Elsevier (2002), 3–54.

[8] G.B. Ermentrout and N. Kopell, Parabolic bursting in an excitable system coupled with a slow oscillation, *SIAM J. Appl. Math.* **46** (1986), 233–253.

[9] C. Borger and N. Kopell, Synchronization in network of excitatory and inhibitory neurons with sparse, random connectivity, to appear in *Neural Comput.*

[10] F. Diener, Proprietes asymptotiques des fleuves, *C.R. Acad. Sci Paris* **302** (1985), 55–58.

[11] C. Chow, J. White, J. Ritt, and N. Kopell, Frequency control in synchronous networks of inhibitory neurons, *J. Comput. Neurosci.*, **5** (1998), 407–420.

[12] M.A. Whittington, R.D. Traub and J. Jefferys, Synchronized oscillation in interneuron networks driven by metabotropic glutamate receptor activation, *Nature* **373** (1995), 612–615.

[13] X.-J. Wang and G. Buzsaki, Gamma oscillation by synaptic inhibition in a hippocampal interneuronal network model, *J. Neurosci.* **16** (1996), 6402–6413.

[14] R.D. Traub, J.G.T. Jefferys and M.A. Whittington, *Fast oscillations in Cortical Circuits*, MIT Press, Cambridge MA, 1999.

[15] M.A. Whittington, R.D. Traub, N. Kopell, G.B. Ermentrout and E.H. Buhl, Inhibition-based rhythms: Experimental and mathematical observation on network dynamics, *Int. J. of Psychophysiology* **38** (2000), 315–336.

[16] G.B. Ermentrout, M. Pascal and B. Gutkin, The effects of spike frequency adaptation and negative feedback on the synchronization of neural oscillators, *Neural Comput.* **13** (2001), 1285–1310.

[17] C. Acker, N. Kopell and J. White, Synchronization of strongly coupled excitatory neurons: relating network behavior to biophysics, to appear in *J. Comput.
Neurosci.*

[18] G.B. Ermentrout, Type I membranes, phase resetting curves and synchrony,
Neural Comput. **8** (1996), 879–1001.

[19] N. Kopell, G.B. Ermentrout, M. Whittington and R.D. Traub, Gamma
rhythms and beta rhythms have different synchronization properties, *Proc.
Nat. Acad. Sci. USA*, **97** (2000), 1867–1872.

[20] S. R. Jones, D. Pinto, T. Kaper and N. Kopell, Alpha-frequency rhythms desyn-
chronize over long cortical distances: a modeling study, *J. Comput. Neurosci.*,
9 (2000), 271–291.

[21] T. LoFaro and N. Kopell, Timing regulation in a network reduced from voltage-
gated equations to a one-dimensional map, *J. Math Biol.* **38** (1999), 479–533.

[22] M.J. Gillies, R.D. Traub, F.E.N LeBeau, C.H. Davies, T. Gloveli, E.H. Buhl
and M.A. Whittington, Stratum oriens interneurons temporally coordinate
atropine-resistant theta oscillations in hippocampal area CA1, to appear in
J. Neurophysiol .

[23] D. Golomb and B. Hansel, The number of synaptic inputs and synchrony of
large sparse neuronal networks, *Neural Comp.* **12** (2000), 1095–1139.

[24] D. Hansel and G. Mato, Existence and stability of persistent states in large
neuronal networks, *Phys. Rev. Lett.* **86** (2001), 4175–4178.

[25] N. Brunel, Dynamics of sparsely connected networks of excitatory and inhibitory spiking neurons, *J. Comput. Neurosci.* **8** (2000), 183–208.

[26] J. Rubin and D. Terman, Geometric analysis of population rhythms in synap-
tically coupled neuronal networks, *Neural Comp.* **12** (2000), 597–645.

[27] S.F Farmer, Rhythmicity, synchronization and binding in human and primate
motor systems, *J. Physiol.* **509** (1998), 3–14.

[28] C. Tallon-Baudry and O. Bertrand, Oscillatory gamma activity in humans and
its role in object representation, *Trends in Cognitive Neurosci.* **3** (1999), 151–
162.

[29] M. Olufsen, MA. Whittington, M. Camperi and N. Kopell, New functions
for the gamma rhythm: Population tuning and preprocessing, to appear in
J.Comput. Neurosci.

ICM 2002 · Vol. III · 817–828

Analysis of Energetic Models for Rate-Independent Materials

Alexander Mielke[*]

Abstract

We consider rate-independent models which are defined via two functionals: the time-dependent energy-storage functional $\mathcal{I} : [0, T] \times X \to [0, \infty]$ and the dissipation distance $\mathcal{D} : X \times X \to [0, \infty]$. A function $z : [0, T] \to X$ is called a solution of the energetic model, if for all $0 \le s < t \le T$ we have

stability: $\quad \mathcal{I}(t, z(t)) \le \mathcal{I}(t, \tilde{z}) + \mathcal{D}(z(t), \tilde{z})$ for all $\tilde{z} \in X$;

energy inequality: $\mathcal{I}(t, z(t)) + \mathrm{Diss}_{\mathcal{D}}(z, [s, t]) \le \mathcal{I}(s, z(s)) + \int_s^t \partial_\tau \mathcal{I}(\tau, z(\tau)) \, \mathrm{d}\tau$.

We provide an abstract framework for finding solutions of this problem. It involves time discretization where each incremental problem is a global minimization problem. We give applications in material modeling where $z \in \mathcal{Z} \subset X$ denotes the internal state of a body. The first application treats shape-memory alloys where z indicates the different crystallographic phases. The second application describes the delamination of bodies glued together where z is the proportion of still active glue along the contact zones. The third application treats finite-strain plasticity where $z(t, x)$ lies in a Lie group.

2000 Mathematics Subject Classification: 74 C 15.
Keywords and Phrases: Energy functionals, Dissipation, Global minimizers, Incremental problems, Bounded variation, Shape-memory alloys, Delamination, Elasto-plasticity.

1 Introduction

Many evolution equations can be written in the abstract form

$$0 \in \partial \Psi(\dot{z}(t)) + \mathrm{D}\mathcal{I}(t, z(t)), \qquad (1.1)$$

where $z \in X$ is the state variable, $\mathcal{I}$ is the energy-storage functional, $\Psi : X \to [0, \infty]$ is a convex dissipation functional, and $\partial \Psi$ means the set-valued subdifferential (see

---

[*]Mathematisches Institut A, Universität Stuttgart, Pfaffenwaldring 57, 70569 Stuttgart, Germany. E-mail: mielke@mathematik.uni-stuttgart.de

[2] for this doubly nonlinear form). Rate-independency is realized by assuming that Ψ is homogeneous of degree 1.

We replace the above differential inclusion by a weaker energetic formulation, which is also more general since it allows for z-dependent dissipation functionals. For given $\mathcal{I} : [0, T] \times X \to [0, \infty]$ and a given dissipation distance $\mathcal{D} : X \times X \to [0, \infty]$ satisfying the triangle inequality, we impose the energetic conditions of *global stability* (S) and the *energy inequality* (E) instead of (1.1). A function $z : [0, T] \to X$ is called a *solution of the energetic model*, if for all $0 \leq s < t \leq T$ we have

(S) $\mathcal{I}(t, z(t)) \leq \mathcal{I}(t, \tilde{z}) + \mathcal{D}(z(t), \tilde{z})$ for all $\tilde{z} \in X$;

(E) $\mathcal{I}(t, z(t)) + \mathrm{Diss}_{\mathcal{D}}(z, [s, t]) \leq \mathcal{I}(s, z(s)) + \int_s^t \partial_\tau \mathcal{I}(\tau, z(\tau)) \, d\tau$.

Here, $\mathrm{Diss}_{\mathcal{D}}(z, [s, t])$ is called the dissipation of z on the interval $[s, t]$ and is defined as the supremum of $\sum_{j=1}^{N} \mathcal{D}(z(t_{j-1}), z(t_j))$ over all $N \in \mathbb{N}$ and all discretizations $s = t_0 < t_1 < \ldots < t_N = t$.

Assuming $\mathcal{D}(z_0, z_1) = \Psi(z_1 - z_0)$, convexity of $\mathcal{I}(t, \cdot)$ and further technical assumptions, this energetic formulation is equivalent to (1.1), see [16]. However, the latter form is more general as it applies to nonconvex problems and it doesn't need differentiability of $t \mapsto z(t)$ nor of $z \mapsto \mathcal{I}(t, z)$. A related energetic approach to equations of the type (1.1) is presented in [20], however, it remains unclear whether that method applies to the rate-independent case.

In Section 2 we discuss the abstract setting in more detail and in Section 3 we provide existence results for solutions for given initial values $z(0) = z_0$. The existence theory is based on time-incremental minimization problems of the form

$$z_k \in \mathrm{argmin}\{ \mathcal{I}(t_k, z) + \mathcal{D}(z_{k-1}, z) \mid z \in X \}$$

and the BV bound for $z : [0, T] \to X$ obtained via the dissipation functional satisfying $\mathcal{D}(z_0, z_1) \geq c_D \|z_0 - z_1\|$. However, one needs additional compactness properties, if X is infinite dimensional. Here we propose a version where $\mathcal{I}$ satisfies coercivity with respect to an embedded Banach space Y, i.e., $\mathcal{I}(t, z) \geq -C_1 + c_1 \|z\|_Y^\alpha$ with $c_1, C_1, \alpha > 0$, where Y is compactly embedded in X.

For the case of $\mathcal{D}$ having the form $\mathcal{D}(z_0, z_1) = \Psi(z_1 - z_0)$ this theory was developed in [16]. The case of general $\mathcal{D}$ can be found in [10].

The flexibility of the energetic formulation allows for applications in continuum mechanics, where $z : \Omega \to Z$ plays the rôle of internal variables in the material occupying the body $\Omega \subset \mathbb{R}^d$. Note that Z may be a manifold containing the internal variables like phase indicators, plastic or phase transformations, damage, polarization or magnetization. By $\mathcal{Z}$ we denote the set of all admissible internal states. The elastic deformation is $\varphi : \Omega \to \mathbb{R}^d$ and $\mathcal{F}$ denotes the set of admissible deformations φ.

Energy storage is characterized via the functional $\mathcal{E} : [0, T] \times \mathcal{F} \times \mathcal{Z} \to \mathbb{R}$, where $t \in [0, T]$ is the (quasi-static) process time, which drives the system via changing loads. In typical material models, $\mathcal{E}$ has the form

$$\mathcal{E}(t, \varphi, z) = \int_\Omega W(x, \mathrm{D}\varphi(x), z(x)) \, dx - \langle \ell_{\mathrm{ext}}(t), \varphi \rangle,$$

where W is the stored-energy density and $\ell_{\mathrm{ext}}(t)$ denotes the external loadings.

Dissipation is characterized by an infinitesimal Finsler metric $\Delta : \Omega \times \mathrm{T}Z \to [0, \infty]$, such that the curve $z : [t_0, t_1] \to \mathcal{Z}$ dissipates the energy

$$\mathrm{Diss}\,(z, [t_0, t_1]) = \int_{t_0}^{t_1} \int_\Omega \Delta(x, z(t, x), \dot{z}(t, x))\,\mathrm{d}x\,\mathrm{d}t.$$

The global dissipation distance $\mathcal{D}(z_0, z_1)$ is then the infimum over all curves connecting z_0 with z_1. The relation to the abstract theory above is obtained by eliminating the elastic deformation via

$$\mathcal{I}(t, z) = \inf\{\,\mathcal{E}(t, \varphi, z) \mid \varphi \in \mathcal{F}\,\} \text{ for } z \in \mathcal{Z} \text{ and } \mathcal{I}(t, z) = +\infty \text{ else.}$$

Obviously, the functional $\mathcal{I}$ is now fairly complicated and it is important to have rather general conditions in the abstract theory.

In Section 4 we illustrate the usefulness of the abstract approach by discussing three quite different applications; however, the theory is used in other areas as well, e.g., in fracture mechanics [4, 3] and in micro-magnetics [8, 19].

Our first model describes phase transformations in shape-memory alloys as discussed in [15, 17, 18, 5]. Here $z : \Omega \to Z$ indicates either the microscopic distribution of the phases or a mesoscopic average of the microscopic distribution. In the first case we choose $Z = Z_p = \{e_1, \dots, e_p\} \subset \mathbb{R}^p$, where e_j denotes the j-th unit vector in $\mathbb{R}^p$ and in the second case we choose $Z = \mathrm{conv}\, Z_p$. In both cases the dissipation distance is given by a volume integral measuring the amount of volume which is transformed into another phase: $\mathcal{D}(z_0, z_1) = \int_\Omega \Delta(z_1(x) - z_0(x))\,\mathrm{d}x$, where $\Delta : \mathbb{R}^p \to [0, \infty[$ is convex and homogeneous of degree 1. This leads naturally to the basic space $X = \mathrm{L}^1(\Omega, \mathbb{R}^p)$ and $\mathcal{Z} = \{\, z \in X \mid z(x) \in Z \text{ a.e.}\,\}$.

Including in $\mathcal{E}$ an interfacial energy proportional to the area of the interfaces between regions of different phases provides a reduced energy $\mathcal{I}$ which is coercive in $Y = \mathrm{BV}(\Omega, \mathbb{R}^p)$, see [9]. For an existence result in the case without interfacial energy we refer to [17].

The second application describes the delamination of a body Ω which is glued together along n hypersurfaces Γ_j, $j = 1, \dots, n$. The internal state $z : \Gamma = \cup_1^n \Gamma_j \to [0, 1]$ denotes the percentage of glue along Γ which remains in effect. The dissipation is given by a material constant $c_\mathcal{D}$ times the destroyed glue, i.e., $\mathcal{D}(z_0, z_1) = c_\mathcal{D} \int_\Gamma z_0(x) - z_1(x)\,\mathrm{d}a(x)$ for $z_1 \leq z_0$ and $\mathcal{D}(z_0, z_1) = +\infty$ else. The basic underlying space is $\mathrm{L}^1(\Gamma)$ and now compactness arises via the trace operator $\mathrm{H}^1(\Omega) \to \mathrm{L}^2(\Gamma)$ which makes the reduced energy functional $\mathcal{I}$ weakly continuous.

The final application is devoted to the modeling of elasto-plasticity with finite strains. There the internal variable $z = (P, p)$ consists of the plastic transformation $P \in \mathrm{SL}(d)$ and hardening parameters $p \in \mathbb{R}^k$. Invariance under previous plastic deformations leads to dissipation metrics which are left-invariant, i.e., $\Delta((P, p), (\dot{P}, \dot{p})) = \Delta((I, p), (P^{-1}\dot{P}, \dot{p}))$. This geometric nonlinearity clearly shows that we need general dissipation distances $\mathcal{D}$ avoiding any linear structure. In single-crystal plasticity Δ is piecewise linear in $P^{-1}\dot{P} \in \mathrm{sl}(d)$ which leads to Banach manifolds and the dissipation metric is then a left-invariant Finsler metric. For applications in this context see [1, 13, 12].

2   Abstract setup of the problem

We start with a Banach space X which is not assumed to be reflexive, since our applications in continuum mechanics (cf. Section 4) naturally lead to spaces of the form $L^1(\Omega, \mathbb{R}^k)$. The first ingredient of the energetic formulation is the *dissipation distance* $\mathcal{D} : X \times X \to [0, \infty]$ satisfying the triangle inequality:

$$\mathcal{D}(z_1, z_3) \leq \mathcal{D}(z_1, z_2) + \mathcal{D}(z_2, z_3) \quad \text{for all } z_1, z_2, z_3 \in X.$$

We don't enforce symmetry, i.e., we allow for $\mathcal{D}(z_0, z_1) \neq \mathcal{D}(z_1, z_0)$ as in Section 4.2. We assume that there is a constant $c_{\mathcal{D}} > 0$ such that $\mathcal{D}(z_0, z_1) \geq c_{\mathcal{D}} \|z_1 - z_0\|_X$ for all $z_0, z_1 \in X$. The latter condition is in fact the one which determines the appropriate function space X for a specific application. Moreover, $\mathcal{D}$ is assumed to be s-weakly lower semicontinuous. (We continue to use the abbreviation s-weak for "sequentially weak".) We call $\mathcal{D}(z_0, z_1)$ the dissipation distance from z_0 to z_1.

For a given curve $z : [0, T] \to X$ we define the total dissipation on $[s, t]$ via

$$\text{Diss}_{\mathcal{D}}(z; [s, t]) = \sup\{ \textstyle\sum_1^N \mathcal{D}(z(\tau_{j-1}), z(\tau_j)) \mid N \in \mathbb{N}, s = \tau_0 < \tau_1 < \cdots < \tau_N = t \}. \quad (2.1)$$

The second ingredient is the energy-storage functional $\mathcal{I} : [0, T] \times X \to [0, \infty]$, which is assumed to be bounded from below and then normalized such that it takes only nonnegative values. Here $t \in [0, T]$ plays the rôle of a (very slow) process time which changes the underlying system via changing loading conditions. For fixed time t, the map $\mathcal{I}(t, \cdot) : X \to [0, \infty]$ is assumed to be s-weakly lower semicontinuous, i.e., $z_j \rightharpoonup z$ implies $\mathcal{I}(t, z) \leq \liminf_{j \to \infty} \mathcal{I}(t, z_j)$. Moreover, we assume that for all z with $\mathcal{I}(t, z) < \infty$ the function $t \mapsto \mathcal{I}(t, z)$ is Lipschitz continuous with $|\partial_t \mathcal{I}(t, z)| \leq C_{\mathcal{I}}$.

Definition 2.1 *A curve $z : [0, T] \to X$ is called a **solution** of the rate-independent model $(\mathcal{D}, \mathcal{I})$, if **global stability (S)** and **energy inequality (E)** holds:*

(S) For all $t \in [0, T]$ and all $\widehat{z} \in X$ we have $\mathcal{I}(t, z(t)) \leq I(t, \widehat{z}) + \mathcal{D}(z(t), \widehat{z})$.

(E) For all t_0, t_1 with $0 \leq t_0 < t_1 \leq T$ we have

$$\mathcal{I}(t_1, z(t_1)) + \text{Diss}_{\mathcal{D}}(z; [t_0, t_1]) \leq \mathcal{I}(t_0, z(t_0)) + \int_{t_0}^{t_1} \partial_t \mathcal{I}(t, z(t)) \, dt.$$

The definition of solutions of (S)&(E) is such that it implies the two natural requirements for evolutionary problems, namely that *restrictions* and *concatenations* of solutions remain solutions. To be more precise, for any solution $z : [0, T] \to E$ and any subinterval $[s, t] \subset [0, T]$, the restriction $z|_{[s,t]}$ solves (S)&(E) with initial datum $z(s)$. Moreover, if $z_1 : [0, t] \to E$ and $z_2 : [t, T] \to E$ solve (S)&(E) on the respective intervals and if $z_1(t) = z_2(t)$, then the concatenation $z : [0, T] \to E$ solves (S)&(E) as well. Under a few additional assumptions, it is shown in [16] that (S) and (E) together imply that, in fact, the energy inequality is in an equality, i.e., for $0 \leq t_0 < t_1 \leq T$ we have

$$\mathcal{I}(t_1, z(t_1)) + \text{Diss}_{\mathcal{D}}(z; [t_0, t_1]) = \mathcal{I}(t_0, z(t_0)) + \int_{t_0}^{t_1} \partial_t \mathcal{I}(t, z(t)) \, dt. \quad (2.2)$$

Rate-independency manifests itself by the fact that the problem has no intrinsic time scale. It is easy to show that z is a solution for $(\mathcal{D}, \mathcal{I})$ if and only if the reparametrized curve $\tilde{z} : t \mapsto z(\alpha(t))$, with $\dot{\alpha} > 0$, is a solution for $(\mathcal{D}, \tilde{\mathcal{I}})$, where $\tilde{\mathcal{I}}(t, z) = \mathcal{I}(\alpha(t), z)$. In particular, the stability (S) is a static concept and the energy estimate (E) is rate-independent, since the dissipation defined via (2.1) is scale invariant like the length of a curve.

The major importance of the energetic formulation is that neither the given functionals $\mathcal{D}$ and $\mathcal{I}(t, \cdot)$ nor the solutions $z : [0, T] \to X$ need to be differentiable. In particular, applications in continuum mechanics often have low smoothness. Of course, under additional smoothness assumptions on $\mathcal{D}$ and $\mathcal{I}$ the weak energetic form (S)&(E) can be replaced by local formulations in the form of differential inclusions like (1.1) ([2, 20]) or variational inequalities. See [16] for a discussion of the implications between these different formulations.

3   Time discretization and existence

The major task is now to develop an existence theory for the initial value problem, i.e., to find a solution in the above sense which additionally satisfies $z(0) = z_0$. In general, we should not expect uniqueness without imposing further conditions like smoothness and uniform convexity of $\mathcal{I}(t, \cdot)$ and $\mathcal{D}$, see [16].

The stability condition (S) can be rephrased by defining the stable sets

$$\mathcal{S}(t) := \{ z \in X \mid \mathcal{I}(t, z) \leq \mathcal{I}(t, \hat{z}) + \mathcal{D}(z, \hat{z}) \text{ for all } \hat{z} \in X \}.$$

Then, (S) simply means $z(t) \in \mathcal{S}(t)$ for all $t \in [0, T]$. The properties of the stable sets turn out to be crucial for deriving existence results.

One of the standard methods to obtain solutions of nonlinear evolution equations is that of approximation by time discretizations. To this end we choose discrete times $0 = t_0 < t_1 < \ldots < t_N = T$ and seek z_k which approximates the solution z at t_k, i.e., $z_k \approx z(t_k)$. Our energetic approach has the major advantage that the values z_k can be found incrementally via minimization problems. Since the methods of the calculus of variations are especially suited for applications in material modeling this will allow for a rich field of applications.

To motivate the following incremental variational problem consider the non-linear parabolic problem $h(\partial_t u) = \operatorname{div}(A \operatorname{D} u) + g$, where we assume $h'(v) \geq 0$. The associated fully implicit incremental problem reads

$$h\left(\tfrac{1}{t_k - t_{k-1}}(u_k - u_{k-1})\right) = \operatorname{div}(A \operatorname{D} u_k) + g(t_k).$$

With $H(v) = \int_0^v h(w) \, dw$ we see that u_k must be a minimizer of the functional

$$\mathcal{J}_k(u_{k-1}; \cdot) : u \mapsto \int_\Omega (t_k - t_{k-1}) H\left(\tfrac{1}{t_k - t_{k-1}}(u - u_{k-1})\right) + \tfrac{1}{2}\langle A \operatorname{D} u, \operatorname{D} u \rangle - g(t_k) u \, dx.$$

In the simplest rate-independent case the function h is given by the signum function which implies $H(v) = |v|$. Hence, the length $t_k - t_{k-1}$ of the k-th time step disappears in the functional $\mathcal{J}_k$. In our more general setting the incremental problem takes the following form:

(IP) For $z_0 \in X$ with $\mathcal{I}(0, z_0) < \infty$ find $z_1, \ldots, z_N \in X$ such that

$$z_k \in \operatorname{argmin}\{\, \mathcal{I}(t_k, z) + \mathcal{D}(z_{k-1}, z) \mid z \in X \,\} \quad \text{for } k = 1, \ldots, N. \tag{3.1}$$

Here "argmin" denotes the set of all minimizers. Using the s-weak lower semi-continuity of $\mathcal{D}$ and $\mathcal{I}$ and the coercivity $\mathcal{I}(t, z) + \mathcal{D}(z_{k-1}, z) \geq c_{\mathcal{D}} \|z - z_{k-1}\|$ we obtain the following result.

Theorem 3.1 *The incremental problem (3.1) always has a solution. Each solution satisfies, for $k = 1, \ldots, N$, the following properties:*

(i) z_k *is stable for time* t_k, *i.e.,* $z_k \in \mathcal{S}(t_k)$;

(ii) $\int_{[t_{k-1}, t_k]} \partial_s \mathcal{I}(s, z_k) \, ds \leq \mathcal{I}(t_k, z_k) - \mathcal{I}(t_{k-1}, z_{k-1}) + \mathcal{D}(z_{k-1}, z_k)$
$$\leq \int_{[t_{k-1}, t_k]} \partial_s \mathcal{I}(s, z_{k-1}) \, ds;$$

(iii) $\mathcal{I}(t_k, z_k) + \sum_{j=1}^{k} \mathcal{D}(z_{j-1}, z_j) \leq \mathcal{I}(0, z_0) + C_{\mathcal{I}} T$;

(iv) $\|z_k\| \leq \|z_0\| + (\mathcal{I}(0, z_0) + C_{\mathcal{I}} T)/c_{\mathcal{D}}$.

The assertions (i) and (ii) are the best replacements for the conditions (S) and (E) in the time-continuous case.

For each discretization $P = \{0, t_1, \ldots, t_{N-1}, T\}$ of the interval $[0, T]$ and each incremental solution $(z_k)_{k=1,\ldots,N}$ of (IP) we define two piecewise constant functions which attain the values z_k at t_k and are constant in-between: Z^P is continuous from the left and $\widehat{Z}^P$ is continuous from the right. Summing the estimates (ii) in Theorem 3.1 over $k = j, \ldots, m$ we find the following two-sided energy estimate.

Corollary 3.2 *Let P be any discretization of $[0, T]$ and $(z_k)_{k=0,\ldots,N}$ a solution of (IP), then for $0 \leq j < m \leq N$ we have the two-sided energy inequality*

$$\mathcal{I}(t_j, Z^P(t_j)) + \int_{t_j}^{t_m} \partial_s \mathcal{I}(s, Z^P(s)) \, ds \leq \mathcal{I}(t_m, Z_P(t_m)) + \operatorname{Diss}_{\mathcal{D}}(Z_P, [t_j, t_m])$$
$$\leq \mathcal{I}(t_j, Z^P(t_j)) + \int_{t_j}^{t_m} \partial_s \mathcal{I}(s, \widehat{Z}_P(s)) \, ds.$$

The existence of solutions can now be established by taking a sequence $(P(l))_{l \in \mathbb{N}}$ of discretizations whose fineness $\delta^{(l)} = \max\{\, t_j^{(l)} - t_{j-1}^{(l)} \mid j = 1, \ldots, N^{(l)} \,\}$ tends to 0. Moreover we assume that the sequence is hierarchical with $P(l) \subset P(l+1)$. The associated solutions of $(\text{IP})^{(l)}$ define $z^{(l)} := Z^{P(l)}$. The construction of a solution of (S)&(E) consists now of two parts.

First we use the dissipation bound (iii) of Theorem 3.1 to obtain an a priori bound in $\operatorname{BV}([0, T], X)$:

$$c_{\mathcal{D}} \int_{[0,T]} \|dz^{(l)}\|_X \leq \operatorname{Diss}_{\mathcal{D}}(z^{(l)}, [0, T]) \leq \mathcal{I}(0, z_0) + C_{\mathcal{I}} T.$$

Then, Part (iv) in Theorem 3.1 and the following additional compactness condition (3.2) allows us to apply a selection principle in the spirit of Helly.

For all $R > 0$ and all $t \in [0, T]$ the sets
$\mathcal{R}_R^t := \{\, z \in X \mid \mathcal{D}(z_0, z) \leq R, \ \mathcal{I}(t, z) \leq R \,\}$ are s-weakly compact. $\qquad (3.2)$

Thus, we can extract a subsequence $(l_n)_{n\in\mathbb{N}}$ such that for all $t \in [0,T]$ the sequence $z^{(l_n)}(t), n \in \mathbb{N}$, converges weakly to a limit $z^{(\infty)}(t)$ with $\mathrm{Diss}_{\mathcal{D}}(z^{(\infty)}, [0,T]) \leq \liminf_{n\to\infty} \mathrm{Diss}_{\mathcal{D}}(z^{(l_n)}, [0,T])$.

Second we need to show that $z^{(\infty)}$ is a solution of (S)&(E). Using Corollary 3.2 it is easy to give conditions which guarantee that $z^{(\infty)}$ satisfies (E) for $t_0 = 0$ and $t_1 = T$, and by (2.2) this is sufficient. To obtain stability of $z^{(\infty)}$ there are essentially two different ways. If additional compactness properties allow us to conclude that the convergence of $z^{(l_n)}(t)$ to $z^{(\infty)}$ also happens in the strong topology, then we are in the good case. Then it suffices to know that the set

$$S_{[0,T]} = \{ (t,z) \in [0,T] \times X \mid z \in \mathcal{S}(t) \} = \cup_{t\in[0,T]}(t,\mathcal{S}(t))$$

is closed in the strong topology. If strong convergence cannot be deduced, one needs to show that $S_{[0,T]}$ is s-weakly closed. This property is quite hard to obtain, since even under nice convexity assumptions on $\mathcal{I}(t,\cdot)$ the sets $\mathcal{S}(t)$ are generally not convex.

The following theorem provides two alternative sets of assumptions which enables us to turn the above construction into a rigorous existence proof.

Theorem 3.3 *Let $\mathcal{D}$ and $\mathcal{I}$ be given as above and satisfy (3.2). If one of the conditions (a) or (b) is satisfied, then for each $z_0 \in X$ with $\mathcal{I}(0,z_0) < \infty$ there is at least one solution $z \in \mathrm{BV}([0,T], X)$ of (S) & (E) with $z(0) = z_0$.*

(a) The set $S_{[0,T]}$ is s-weakly closed and $z \mapsto \partial_t\mathcal{I}(t,z)$ is s-weakly continuous.

(b) The sets $\mathcal{R}_R^t$ in (3.2) are compact, the set $S_{[0,T]}$ is closed, and $z \mapsto \partial_t\mathcal{I}(t,z)$ is continuous (all in the norm topology of X).

Simple nontrivial applications of this theorem with either condition (a) or (b) are as follows: Let $X = \mathrm{L}^1(\Omega)$ with $\Omega \subset \mathbb{R}^d$ bounded and choose the dissipation distance $\mathcal{D}(z_0, z_1) = c_{\mathcal{D}}\|z_1 - z_0\|_X = c_{\mathcal{D}}\int_{\Omega} |z_1(x) - z_0(x)| \,\mathrm{d}x$. As a first case consider

$$\mathcal{I}_1(t,z) = \int_{\Omega} \alpha(x)|z(x)|^{\beta} - g(t,x)z(x)\,\mathrm{d}x + \gamma,$$

where $\alpha(x) \geq \alpha_0 > 0$, $\beta > 1$, and $g \in \mathrm{C}^1([0,T], \mathrm{L}^{\infty}(\Omega))$. The sets $\mathcal{R}_R^t$ are closed convex sets which lie in the intersection of an L^1-ball and an L^{β}-ball. Hence, we obtain the s-weak compactness condition (3.2). Yet, $\mathcal{R}_R^t$ is not strongly compact in $\mathrm{L}^1(\Omega)$. The stable sets for $\mathcal{I}_1$ are given by

$$S_1(t) = \{ z \in \mathrm{L}^1(\Omega) \mid |z(x)|^{\beta-2}z(x) \in [\tfrac{g(t,x)-c_{\mathcal{D}}}{\alpha(x)\beta}, \tfrac{g(t,x)+c_{\mathcal{D}}}{\alpha(x)\beta}] \text{ for a.a. } x \in \Omega \},$$

which shows that they are s-weakly closed since they are convex and closed. Hence, condition (a) is satisfied.

As a second case consider the nonconvex energy functional

$$\mathcal{I}_2(t,z) = \int_{\Omega} \tfrac{1}{2}|\mathrm{D}z(x)|^2 + f(t,x,z(x))\,\mathrm{d}x \text{ for } z \in \mathrm{H}^1(\Omega) \quad \text{and } +\infty \text{ else,}$$

where $f : [0,T] \times \Omega \times \mathbb{R} \to \mathbb{R}$ and $\partial_t f$ are continuous and bounded. Now, $\mathcal{R}_R^t$ is already compact in $\mathrm{L}^1(\Omega)$ since it is closed and contained in an H^1-ball. With these properties, it can be shown that condition (b) of Theorem 3.3 holds.

4 Applications in continuum mechanics

The flexibility of the energetic formulation allows for applications in continuum mechanics. We consider an elastic body which is given through a bounded domain $\Omega \subset \mathbb{R}^d$ with sufficiently smooth boundary. The elastic deformation is given by the mapping $\varphi : \Omega \to \mathbb{R}^d$, and the set of all admissible deformations is denoted by $\mathcal{F}$, which implements the displacement boundary conditions.

The variable $z \in Z$ includes all the internal variables like phase indicators, plastic or phase transformations, damage, polarization or magnetization. A function $z : \Omega \to Z$ gives the internal state of the material, and $\mathcal{Z}$ denotes the set of all admissible internal states. Note that Z may be a manifold with (nonsmooth) boundary. In plasticity we have $Z = \mathrm{SL}(d) \times \mathbb{R}^k$, in phase transformations we let $Z = \{\, z \in [0,1]^k \mid \sum_1^p z^{(j)} = 1 \,\}$, and in micro-magnetism z is the magnetization satisfying $|z(t,x)| = m_0 > 0$. Moreover, below we will also consider an application where z is not defined on all of Ω but at certain parts of the boundary.

Energy storage is characterized via the functional $\mathcal{E} : [0,T] \times \mathcal{F} \times \mathcal{Z} \to \mathbb{R}$ which is the sum of the total elastic energy and the potential energies due to exterior loadings (Gibbs' energy):

$$\mathcal{E}(t, \phi, z) = \int_\Omega W(x, \mathrm{D}\phi(x), z(x))\, \mathrm{d}x - \langle \ell_{\mathrm{ext}}(t), \phi \rangle.$$

Here $t \in [0,T]$ is the (quasi-static) process time which drives the system and the external loads are $\langle \ell_{\mathrm{ext}}(t), \phi \rangle = \int_\Omega f_{\mathrm{ext}}(t,x) \cdot \phi(x)\, \mathrm{d}x + \int_{\Gamma_{\mathrm{tract}}} g_{\mathrm{ext}}(t,x) \cdot \phi(x)\, \mathrm{d}a(x)$.

Dissipation is characterized via the metric $\Delta : \Omega \times \mathrm{T}Z \to [0,\infty]$ such that the curve $z : [0,T] \to \mathcal{Z}$ dissipates the energy

$$\mathrm{Diss}\,(z, [t_0, t_1]) = \int_{t_0}^{t_1} \int_\Omega \Delta(x, z(t,x), \dot{z}(t,x))\, \mathrm{d}x\, \mathrm{d}t \quad \text{on } [t_1, t_2].$$

For each material point $x \in \Omega$, the infinitesimal metric $\Delta(x, \cdot, \cdot) : \mathrm{T}Z \to [0,\infty]$ defines a global distance function $D(x, \cdot, \cdot) : Z \times Z \to [0,\infty]$ and on $\mathcal{Z}$ we obtain the global dissipation distance

$$\begin{aligned} \mathcal{D}(z_0, z_1) &= \int_\Omega D(x, z_0(x), z_1(x))\, \mathrm{d}x \\ &= \inf\{\, \mathrm{Diss}\,(z, [t_0, t_1]) \mid z \in \mathrm{C}^{\mathrm{Lip}}([0,1], \mathcal{Z}), z(0) = z_0, z(1) = z_1 \,\}. \end{aligned}$$

The rate-independent problem for this material model is defined as in the above abstract part, but now the elastic deformation appears as an additional variable, which, however, does not generate any dissipation.

Definition 4.1 *A pair $(\phi, z) : [0,T] \to \mathcal{F} \times \mathcal{Z}$ is called a **solution** of the rate-independent problem associated with $\mathcal{D}$ and $\mathcal{E}$ if the **global stability (S)** and the **energy inequality (E)** hold:*

(S) For all $t \in [0,T]$ and all $(\widehat{\phi}, \widehat{z}) \in \mathcal{F} \times \mathcal{Z}$ we have
$$\mathcal{E}(t, \phi(t), z(t)) \leq \mathcal{E}(t, \widehat{\phi}, \widehat{z}) + \mathcal{D}(z(t), \widehat{z}).$$

(E) For all t_0, t_1 with $0 \leq t_0 < t_1 \leq T$ we have
$$\mathcal{E}(t_1, \phi(t_1), z(t_1)) + \mathrm{Diss}_{\mathcal{D}}(z; [t_0, t_1]) \leq \mathcal{E}(t_0, \phi(t_0), z(t_0)) + \int_{t_0}^{t_1} \partial_t \mathcal{E}(t, \phi(t), z(t))\, \mathrm{d}t.$$

The connection with the above abstract theory is obtained by minimization with respect to the deformations $\phi \in \mathcal{F}$, since the stability condition implies that $\phi(t)$ must be a minimizer of $\mathcal{E}(t, \cdot, z(t))$. We define the associated $\mathcal{I}$ via

$$\mathcal{I}(t, z) = \inf\{\, \mathcal{E}(t, \varphi, z) \mid \varphi \in \mathcal{F} \,\} \text{ for } z \in \mathcal{Z} \quad \text{and } +\infty \text{ else.}$$

While this elimination is suitable for an abstract treatment, the practical approximation of solutions via the incremental approach is better done by keeping the deformation and eliminating the internal variable in each incremental step. In fact, in (IP) we now have to find

$$(\phi_k, z_k) \in \mathrm{argmin}\{\, \mathcal{E}(t_k, \widehat{\phi}, \widehat{z}) + \mathcal{D}(z_{k-1}, \widehat{z}) \mid (\widehat{\phi}, \widehat{z}) \in \mathcal{F} \times \mathcal{Z} \,\}. \tag{4.1}$$

In this minimization problem the internal variable occurs only locally under the integral over Ω and hence can be eliminated pointwise. Defining the local reduced constitutive functions

$$\begin{aligned}
\Psi^{\mathrm{red}}(z_{\mathrm{old}}; x, F) &:= \min\{\, W(x, F, z) + D(x, z_{\mathrm{old}}, z) \mid z \in Z \,\}, \\
Z_{\mathrm{new}}(z_{\mathrm{old}}; x, F) &\in \mathrm{argmin}\{\, W(x, F, z) + D(x, z_{\mathrm{old}}, z) \mid z \in Z \,\},
\end{aligned} \tag{4.2}$$

and the reduced functional $\mathcal{E}^{\mathrm{red}}(z_{\mathrm{old}}; t, \phi) = \int_\Omega \Psi^{\mathrm{red}}(z_{\mathrm{old}}; \mathrm{D}\phi) \,\mathrm{d}x - \langle \ell_{\mathrm{ext}}(t), \phi \rangle$ the solution of (4.1) is equivalent to finding $\phi \in \mathrm{argmin}\{\, \mathcal{E}^{\mathrm{red}}(z_{k-1}; t_k, \widehat{\phi}) \mid \widehat{\phi} \in \mathcal{F} \,\}$ and then letting $z_k = Z_{\mathrm{new}}(z_{k-1}; \mathrm{D}\phi_k)$. For more details we refer to [12].

4.1 Phase transformations in shape-memory alloys

We assume that, in each microscopic point y, an elastic material is free to choose one of p crystallographic phases and that the elastic energy density W is then given by $W_j(\mathrm{D}\phi)$. If the model is made on the mesoscopic level, then the internal variables are phase portions $z^{(j)} \in [0, 1]$ for the j-th phase. We set $Z = \{\, z \in [0, 1]^p \subset \mathbb{R}^p \mid \sum_1^p z^{(j)} = 1 \,\}$ and $X = \mathrm{L}^1(\Omega, \mathbb{R}^p)$. The material properties are described by a mixture function $W : \mathbb{R}^{d \times d} \times Z \to [0, \infty]$, see [11, 17, 5]. The dissipation can be shown to have the form $D(z_0, z_1) = \psi(z_1 - z_0)$ with $\psi(v) = \max\{\, \sigma_m \cdot v \mid m = 1, \ldots, M \,\} \geq C_\psi |v|$, where $\sigma_m \in \mathbb{R}^p$ are thermodynamically conjugated threshold values.

So far we are unable to prove existence results for this model in its full generality. However, the case with only two phases ($p = 2$) has been treated in [17] under the additional assumption that the elastic behavior is linear and both phases have the same elastic tensor. In that case, one sets $z = (\theta, 1 - \theta)$ with $\theta \in [0, 1]$. It can be shown that $\mathcal{I}$ is a quadratic functional in $\theta \subset \mathrm{L}^1(\Omega, [0, 1]) \subset \mathrm{L}^2(\Omega)$. It then follows that the compactness condition (3.2) holds and condition (a) in Theorem 3.3 can be verified using the H-measure to handle the weak convergence of the nonconvex terms.

A microscopic model is treated in [9]. There no phase mixtures are allowed, i.e., we assume $z \in Z_p := \{e_1, e_2, \ldots, e_p\} \subset \mathbb{R}^p$, where e_j is the j-th unit vector. Thus, the functions $z \in \mathcal{Z}$ are like characteristic functions which indicate exactly one phase at each material point. The dissipation is assumed as above, but now the elastic energy contains an additional term measuring the surface area of the interfaces between the different regions:

$$\mathcal{E}(t, \phi, z) = \int_\Omega W(\mathrm{D}\phi, z)\,\mathrm{d}x + \sigma \int_\Omega |\mathrm{D}z| - \langle \ell_{\mathrm{ext}}(t), \phi \rangle,$$

where σ is a positive constant and $\int_\Omega |\mathrm{D}z|$ is $\sqrt{2}$ times the area of all interfaces. Here $\mathcal{Z} = \{ z : \Omega \to Z_p \mid \int_\Omega |\mathrm{D}z| < \infty \}$ and we set $\mathcal{E}(t, \phi, z) = +\infty$ for $z \notin \mathcal{Z}$.

Hence, after minimization with respect to ϕ we still have $\mathcal{I}(t, z) \geq \gamma + \sigma \int_\Omega |\mathrm{D}z|$. This term provides for $\mathcal{R}_R^t$ (cf. (3.2)) an a priori bound in $\mathrm{BV}(\Omega, \mathbb{R}^p)$ and hence we conclude compactness in $X = \mathrm{L}^1(\Omega, \mathbb{R}^p)$. Under the usual additional conditions for the elastic stored-energy densities W_j we obtain for each $z_0 \in \mathcal{Z}$ a solution (ϕ, z) with $\phi \in \mathrm{L}_\mathrm{w}^\infty(]0, T[, \mathrm{W}^{1,2}(\Omega, \mathbb{R}^d))$ and $z \in \mathrm{BV}([0, T], \mathrm{L}^1(\Omega, \mathbb{R}^p)) \cap \mathrm{L}_{\mathrm{w}*}^\infty(]0, T[, \mathrm{BV}(\Omega, \mathbb{R}^p))$ with $z(t) \in \mathcal{Z}$ for all $t \in [0, T]$, see [9].

4.2   A delamination problem

Here we give a simple model for rate-independent delamination and refer to [7] for a better model and the detailed analysis.

Consider a body $\Omega \subset \mathbb{R}^d$ which is given by an open, bounded, and path-connected domain. Assume that the interior of the closure of Ω differs from Ω by a finite set of sufficiently smooth hypersurfaces Γ_j, $j = 1, \ldots, n$. This means that with $\Gamma := \bigcup_{j=1}^n \Gamma_j$ we have $\mathrm{int}(\mathrm{cl}(\Omega)) = \Omega \cup \Gamma$. We assume that the two sides of the body are glued together along these surfaces and that the glue is softer than the material itself. Upon loading, some parts of the glue may break and thus lose its effectiveness. The remaining fraction of the glue which is still effective is denoted by the internal state function $z : \Gamma \to [0, 1]$.

We let $\mathcal{Z} = \{ z : \Gamma \to [0, 1] \mid z \text{ measurable} \} \subset X = \mathrm{L}^1(\Gamma)$. The dissipation distance $\mathcal{D}(z_0, z_1)$ is proportional to the amount of glue that is broken from state z_0 to state z_1:

$$\mathcal{D}(z_0, z_1) = c_\mathcal{D} \int_\Gamma z_0(y) - z_1(y)\,\mathrm{d}a(y) \text{ for } z_0 \geq z_1 \quad \text{and } +\infty \text{ else.}$$

Here we explicitly forbid the healing of the glue by setting $\mathcal{D}$ equal ∞, if $z_0 \not\geq z_1$.

The energy is given by the elastic energy in the body, the elastic energy in the glue, and the potential of the external loadings:

$$\mathcal{E}(t, \phi, z) = \int_\Omega W(\mathrm{D}\phi)\,\mathrm{d}x + \int_\Gamma z(y)Q(y, [\![\phi]\!]_\Gamma(y))\,\mathrm{d}a(y) - \langle \ell_{\mathrm{ext}}(t), \phi \rangle,$$

where for $y \in \Gamma$ the vector $[\![\phi]\!]_\Gamma(y)$ denotes the jump of the deformation ϕ across the interface Γ and $Q(y, \cdot)$ is the potential for the elastic properties of the glue.

For simplicity we assume further that W provides linearized elasticity and Q is quadratic is well, then there is a unique minimizer $\phi = \Phi(t, z) \in \mathrm{H}^1(\Omega, \mathbb{R}^d)$ of $\mathcal{E}(t, \cdot, z)$. It can be shown that the mapping $\Phi(t, \cdot) : \mathcal{Z} \subset \mathrm{L}^1(\Gamma) \to \mathrm{H}^1(\Omega, \mathbb{R}^d)$ is compact, which implies that the functional $\mathcal{I}(t, \cdot) : \mathcal{Z} \to [0, \infty[$ is s-weakly continuous with respect to the L^1-topology on $\mathcal{Z}$. For the latter argument it is essential that z appears only linearly in the definition of $\mathcal{E}(t, \phi, z)$. Theorem (3.3) with condition (a) provides the existence of solutions.

4.3 Elasto-plasticity

The above theory can be applied to linearized elasto-plasticity, see [1, 13]. Here we want to report on recent results concerning elasto-plasticity with finite strain. However, for this application the abstract existence theory is not yet available.

Elasto-plasticity with finite strains is based on the multiplicative decomposition of the deformation gradient $F = \mathrm{D}\phi$ in the form $\mathrm{D}\phi = F_{\text{elast}}P^{-1}$ where the plastic transformation P lies in the Lie group $\mathrm{SL}(d) = \{\, P \in \mathbb{R}^{d \times d} \,|\, \det P = 1 \,\}$. The internal variable has the form $z = (P, p) \in Z$ where $p \in \mathbb{R}^k$ denotes the hardening parameters. We refer to [1, 13, 12] for mechanical motivations and mathematical details. For simplicity, we mention here only the case without hardening where $z = P \in \mathrm{SL}(d) =: Z$ and refer to [6, 12] for more general cases.

The important point in finite-strain elasto-plasticity is that the dissipation distance must be invariant under previous plastic deformations, i.e., $D(QP_0, QP_1) = D(P_0, P_1)$ for all $Q \in \mathrm{SL}(d)$. Equivalently, the infinitesimal metric $\Delta : \mathrm{T}Z \to [0, \infty]$ is left-invariant, i.e., $\Delta(P, \dot{P}) = \Delta(I, P^{-1}\dot{P})$. This implies that the dissipation distance is characterized by a norm $\Delta(I, \cdot)$ on $\mathrm{sl}(d) = \mathrm{T}_I \mathrm{SL}(d)$ and that $D(P_0, P_1)$ behaves logarithmically in $P_0^{-1}P_1$ which introduces strong geometric nonconvexities. So far, even the solution of the incremental problem (IP) is not understood completely. Even in simple cases one has to expect non-attainment in (IP), which leads to the formation of microstructure. The easiest way to see the problems is to study the reduced energy density Ψ^{red} in (4.2). If this density is not quasi-convex, then there are loadings such that (IP) has no solution and relaxation techniques have to be employed, cf. [14].

Acknowledgments This work was partially joint work with Florian Theil, Tomaš Roubíček, Andreas Mainik and Michal Kočvara. The research was partially supported by DFG through the SFB 404 *Multifield Problems in Continuum Mechanics*.

References

[1] C. Carstensen, K. Hackl, and A. Mielke. Non-convex potentials and microstructures in finite-strain plasticity. *Royal Soc. London, Proc. Ser. A*, 458(2018):299–317, 2002.

[2] P. Colli and A. Visintin. On a class of doubly nonlinear evolution equations. *Comm. Partial Differential Equations*, 15(5):737–756, 1990.

[3] G. Dal Maso and R. Toader. A model for quasi-static growth of brittle fractures: existence and approximation results. *Arch. Rat. Mech. Anal.*, 162:101–135, 2002.

[4] G. Francfort and J.-J. Marigo. Revisiting brittle fracture as an energy minimization problem. *J. Mech. Phys. Solids*, 46:1319–1342, 1998.

[5] S. Govindjee, A. Mielke, and G. Hall. The free-energy of mixing for n-variant martensitic phase transformations using quasi-convex analysis. *J. Mech. Physics Solids*, 2002. In press.

[6] K. Hackl, A. Mielke, and D. Mittenhuber. Dissipation distances in multiplicative elasto-plasticity. Preprint June, 2002.

[7] M. Kočvara, A. Mielke, and T. Roubíček. A rate-independent approach to the delamination problem. In preparation, 2002.

[8] M. Kružík. Variational models for microstructure in shape memory alloys and in micromagnetics and their numerical treatment. In A. Ruffing and M. Robnik, editors, *Proceedings of the Bexbach Kolloquium on Science 2000 (Bexbach, 2000)*. Shaker Verlag, 2002.

[9] A. Mainik. Ratenunabhängige Modelle für Phasentransformationen in Formgedächtnislegierungen. Universität Stuttgart. In preparation, 2002.

[10] A. Mainik and A. Mielke. Existence results for rate-independent systems. In preparation, 2002.

[11] A. Mielke. Estimates on the mixture function for multiphase problems in elasticity. In A.-M. Sändig, W. Schiehlen, and W. Wendland, editors, *Multifield problems*, pages 96–103, Berlin, 2000. Springer-Verlag.

[12] A. Mielke. Energetic formulation of multiplicative elasto-plasticity using dissipation distances. Preprint Uni Stuttgart, February 2002.

[13] A. Mielke. Finite elasto-plasticity, Lie groups and geodesics on $SL(d)$. In P. Newton, A. Weinstein, and P. Holmes, editors, *Geometry, Dynamics, and Mechanics*. Springer-Verlag, 2002. In press.

[14] A. Mielke. Relaxation via Young measures of material models for rate-independent inelasticity. Preprint Uni Stuttgart, February 2002.

[15] A. Mielke and F. Theil. A mathematical model for rate-independent phase transformations with hysteresis. In H.-D. Alber, R. Balean, and R. Farwig, editors, *Proceedings of the Workshop on "Models of Continuum Mechanics in Analysis and Engineering"*, 117–129. Shaker-Verlag, 1999.

[16] A. Mielke and F. Theil. On rate-independent hysteresis models. *Nonl. Diff. Eqns. Appl. (NoDEA)*, 2001. To appear.

[17] A. Mielke, F. Theil, and V. Levitas. A variational formulation of rate-independent phase transformations using an extremum principle. *Arch. Rational Mech. Anal.*, 162:137–177, 2002.

[18] T. Roubíček. Evolution model for martensitic phase transformation in shape-memory alloys. *Interfaces Free Bound.*, 2002. In print.

[19] T. Roubíček and M. Kružík. Mircrostructure evolution model in micromagnetics. *Zeits. angew. math. Physik*, 2002. In print.

[20] A. Visintin. A new approach to evolution. *C.R.A.S. Paris*, 332:233–238, 2001.

ICM 2002 · Vol. III · 829–838

Cross-over in Scaling Laws:
A Simple Example from Micromagnetics

Felix Otto*

Abstract

Scaling laws for characteristic length scales (in time or in the model parameters) are both experimentally robust and accessible for rigorous analysis. In multiscale situations cross–overs between different scaling laws are observed. We give a simple example from micromagnetics. In soft ferromagnetic films, the geometric character of a wall separating two magnetic domains depends on the film thickness. We identify this transition from a Néel wall to an Asymmetric Bloch wall by rigorously establishing a cross–over in the specific wall energy.

1. Introduction

Many continuum systems in materials science display pattern formation. These patterns are characterized by one or several length scales. The scaling of these characteristic lengths in the material parameters and/or in time are usually an experimentally robust feature. These scaling laws, and their characterizing exponents, are of interest to theoretical physics since they express a certain universality. At the same time, scaling laws (rather than more detailed features) are ameanable to heuristic and rigorous analysis and thus are a good test for the model and a challenge for mathematics.

Scaling laws and their exponents reflect a scale invariance. In a multiscale model, these scale invariances are broken and only approximately valid in certain parameter and/or time regimes. The cross-over between two scaling laws reflects a change in the dominant physical mechanisms. In studying cross–overs, theoretical analysis may have an advantage over numerical simulation which has to explore many parameter decades and thus has to cope with widely separated length scales.

Together with various collaborators, the author has analyzed scaling laws and their cross-overs in both static (variational) and dynamic models. The dynamic

---

*Department of Applied Mathematics, University of Bonn, Germany. E-mail: otto@iam.uni-bonn.de

models considered were of gradient-flow type and thus endowed with a variational interpretation: steepest descent in a multiscale energy landscape. The examples are

- The branching of domains in uniaxial ferromagnets [1] (with R. Choksi and R. V. Kohn). Strongly uniaxial ferromagnets have only two favored magnetization directions ("up" and "down"). The width of the corresponding domains decreases towards a sample surface perpendicular to the favored axis. We rigorously establish the scaling of the energy in the sample dimensions in support of this behavior. To leading order, the micromagnetic model behaves like a three-dimensional analogue of the Kohn-Müller [10] model for twin branching.

- The period of cross-tie walls in ferromagnetic films [2] (with A. DeSimone, R. V. Kohn and S. Müller). Cross-tie walls are transition layers between domains in ferromagnetic films. They display a periodic structure in the tangential direction. The experimentally observed scaling of the period in the material parameters is not well-understood [9]. In this paper, we present a combination of heuristic and rigorous analysis which reproduces the experimental scaling and thus identifies the relevant mechanism.

- The rate of capillarity-driven spreading of a thin droplet [6] (with L. Giacomelli). Here, the starting point is the lubrication approximation. The scale invariant version of the model is ill-posed and has to be regularized near the contact line, e. g. through allowing finite slippage. In this paper, we rigorously derive a scaling law for the spreading of the droplet in an intermediate time regime. This scaling law depends only logarithmically on the length scale introduced by the regularization, in agreement with a conjecture of de Gennes [5].

- The rate of coarsening in spinodal decomposition [11] (with R. V. Kohn). Spinodal decomposition is usually modelled by a Cahn-Hilliard equation. In the later stages, it is experimentally observed that the phase distribution coarsens in a statistically self-similar fashion. In this paper, we rigorously prove upper bounds for this coarsening process. The exponents are the ones heuristically expected and depend on whether the mobility is degenerate or non-degenerate: $t^{1/4}$ resp. $t^{1/3}$. In [3], we predict a cross-over for almost degenerate mobility due to a change in the coarsening mechanism.

- The first-order correction to the Lifshitz-Slyozov-Wagner theory for Ostwald ripening [7] (with A. Hönig and B. Niethammer). Ostwald ripening describes the late stage of spinodal decomposition in an off-critical mixture (volume fraction of one phase $\phi \ll 1$). The minority phase then consists of several particles immersed in a matrix of the majority phase. The particles are approximately spherical and don't move—the Lifshitz–Slyozov—Wagner theory describes the evolution of the radii distribution. There is a major interest in identifying the next-order correction term in ϕ. We rigorously show that there is a cross-over in the correction term from $\phi^{1/3}$ to $\phi^{1/2}$ depending on the system size.

Our method to rigorously analyze these scaling laws in a multiscale model is based on relating integral quantities (energies, average length scales, dissipation rates...). It is different from the more local method of matched asymptotic expansions. In particular, it differs from the latter by the absence of a specific Ansatz. In order to relate the integral quantities in our *Ansatz-free* approach, we need *interpolation inequalities*. These interpolation inequalities encode the competition of the dominant physical mechanisms in a scale-invariant fashion (e. g. the competition between driving energetics and limiting dissipation or between bulk and surface energy). Hence tools from pure analysis are here employed in a more applied context.

In order to illustrate this set of ideas, we present a simple application.

2. An example from micromagnetics

According to the well-accepted micromagnetic model, the experimentally observed ground-state of the magnetization m is the minimizer of a variational problem. We are interested in transition layers ("walls") between domains in a film of thickness t in the (x_1, x_2)-plane. We assume that the in-plane axis m_2 is favored by the crystalline anisotropy so that domains of magnetization $m = (0, 1, 0)$ or $m = (0, -1, 0)$ form. In order to avoid "magnetic poles", the walls separating such domains are parallel to the x_2-axis. We are interested in their specific energy per unit length in x_2-direction. Hence the admissible magnetizations m are x_2-independent and connect the two end-states

$$m = m(x_1, x_3) \in S^2 \quad \text{for } (x_1, x_3) \in \Omega := (-\infty, \infty) \times (-\tfrac{t}{2}, \tfrac{t}{2})$$
$$\text{and} \quad \lim_{x_1 \to \pm\infty} m_2(x_1, x_3) = \pm 1. \tag{2.1}$$

The specific energy, which is to be minimized, is given by

$$E(m) = d^2 \int_\Omega |\nabla m|^2 \, d^2 x + Q \int_\Omega (m_1^2 + m_3^2) \, d^2 x + \int_{R^2} |\nabla u|^2 \, d^2 x, \tag{2.2}$$

where ∇ refers to the variables $x = (x_1, x_3)$. Here the first term is the "exchange energy", the second term comes from crystalline anisotropy and favors the m_2-axis. The last term is the energy of the stray-field $h_s = -\nabla u$ determined by the static Maxwell equations

$$\nabla \times h_s = 0 \quad \text{and} \quad \nabla \cdot (h_s + m) = 0,$$

which are conveniently expressed in variational form for the potential u

$$\int_\Omega m \cdot \nabla \zeta \, d^2 x = \int_{R^2} \nabla u \cdot \nabla \zeta \, d^2 x \quad \text{for all } \zeta \in C_0^\infty(R^2). \tag{2.3}$$

We see that both "volume charges" ($\nabla \cdot m$ in Ω) and "surface charges" (m_3 on $\partial\Omega$) generate the field h_s and thus are penalized. Since the energy density, i.e. $|\nabla u|^2$, depends on m through (2.3), the problem is non-local. The constraint of unit length, see (2.1), makes the variational problem nonconvex.

The model is already partially non-dimensionalized: The magnetization m and the field $-\nabla u$ are dimensionless, but length is still dimensional. In particular, d has dimensions of length (the "exchange length") and Q is dimensionless (the "quality factor"). Hence the model has two intrinsic length scales (material parameters), namely d and $d/Q^{\frac{1}{2}}$, and one extrinsic length scale (sample geometry), namely t. Despite its simplicity, it is an example of a multiscale model and we expect different regimes depending on the two nondimensional parameters Q and $\frac{t}{d}$.

We will focus on the most interesting regime of "soft" materials (i. e. with low crystalline anisotropy) and thicknesses t close to the exchange length d

$$Q \ll 1 \quad \text{and} \quad Q \ll (\frac{t}{d})^2 \ll Q^{-1}. \tag{2.4}$$

Numerical simulation suggest a cross-over *within* this range [9, Chapter 3.6,Fig. 3.81]:

- For thin films: "Néel walls" (see [9, Chapter 3.6 (C)]), whose geometry is asymptotically characterized by

$$\frac{\partial m}{\partial x_3} \equiv 0 \text{ and } m_3 \equiv 0 \implies m = (\cos\theta(x_1), \sin\theta(x_1), 0). \tag{2.5}$$

- For thick films: "Asymmetric Bloch walls" (see [9, Chapter 3.6 (D)]), whose geometry is asymptotically characterized by

$$-\nabla u \equiv 0 \implies \nabla \cdot m = 0 \text{ in } \Omega \text{ and } m_3 = 0 \text{ on } \partial\Omega$$
$$\implies (m_1, m_3) = (-\frac{\partial\psi}{\partial x_3}, \frac{\partial\psi}{\partial x_1}) \text{ for a } \psi \text{ with } \psi = 0 \text{ on } \partial\Omega. \tag{2.6}$$

This cross-over in the wall geometry is reflected by a cross-over in the scaling of the specific wall energy E. Our proposition rigorously captures this cross-over in energy.

Proposition 1 *In the regime (2.4) we have*

$$\min_{m \text{ satisfies (2.1)}} E(m) \sim \begin{cases} d^2 & \text{for } (\frac{t}{d})^2 \gtrsim \ln\frac{1}{Q} \\ t^2 \frac{1}{\ln\frac{t^2}{Q\,d^2}} & \text{for } (\frac{t}{d})^2 \lesssim \ln\frac{1}{Q} \end{cases}. \tag{2.7}$$

By $\gtrsim$, $\lesssim$ we mean $\geq$ resp. $\leq$ up to a generic universal constant and $\sim$ stands for both $\gtrsim$ and $\lesssim$. This scaling qualitatively agrees with the numerical study of the energy cross-over in the thickness [1] given in [9, Fig 3.79].

Upper bounds are proved by construction. Here we make the Ansatz (2.5), resp. (2.6), and let ourselves be inspired by the physics literature for the details of the construction. The matching *lower* bound in (2.7) states that one cannot beat the Ansatz—at least in terms of energy scaling—by relaxing the geometry assumptions (2.5) or (2.6). Therefore Proposition 1 is a validation of the predicted cross-over in the geometry. We call this type of analysis *Ansatz-free lower bounds*.

---

[1] the x-axis corresponds to $\frac{t}{d}$, the y-axis to $\frac{E}{d\,t}$, and $Q = 0.00025$

3. Proof

The upper bound in Proposition 1 comes from the following two lemmas. We only sketch their proof since our main focus is on lower bounds.

Lemma 1 *For $(\frac{t}{d})^2 \ll Q^{-1}$ there exists an m of the form (2.6) with*

$$E(m) \sim d^2. \tag{3.8}$$

Lemma 2 *For $(\frac{t}{d})^2 \gg Q$ there exists an m of the form (2.5) with*

$$E(m) \sim t^2 \ln^{-1} \frac{t^2}{Q\, d^2}. \tag{3.9}$$

For the lower bound we need to estimate the components m_1 and m_3 by E. In Lemma 3 we control m_3 by the stray-field and exchange energy. More precisely, the stray-field energy penalizes m_3 on $\partial\Omega$ in a weak norm. We interpolate with the $L^2(\Omega)$-control of ∇m to obtain $L^2(\Omega)$-control of m_3. In Lemma 4 we control the vertical average $\overline{m}_1$ of m_1 by stray-field, exchange, and anisotropy energy. More precisely, the penalization of $\nabla \cdot m$ through the stray-field energy yields a penalization of $\frac{d\overline{m}_1}{dx_1}$ in a weak norm. We interpolate with the $L^2(\Omega)$-control of ∇m (exchange) to obtain an estimate on the variation of $\overline{m}_1$. We then interpolate with the $L^2(\Omega)$-control of m_1 (anisotropy) to obtain $L^\infty(\mathbf{R})$-control of $\overline{m}_1$.

Lemma 3 *We have for any m satisfying (2.1)*

$$\int_\Omega m_3^2\, d^2 x \underset{\sim}{<} \left(1 + (\frac{t}{d})^2\right) E(m). \tag{3.10}$$

Lemma 4 *In the regime $(\frac{t}{d})^2 \gg Q$ we have for any m satisfying (2.1)*

$$\sup_{x_1 \in (-\infty, \infty)} \overline{m}_1^2(x_1) \underset{\sim}{<} \left(\frac{1}{t^2} \ln \frac{t^2}{Q\, d^2} + \frac{1}{d^2}\right) E(m). \tag{3.11}$$

Proof of Lemma 1. The construction is due to Hubert [8]. We nondimensionalize length by t, i. e. $t = 1$. One can construct[2] a smooth $\psi \colon \overline{\Omega} \to \mathbf{R}$ with

$$|\nabla\psi|^2 \leq 1 \text{ in } \Omega, \quad \psi = 0 \text{ on } \partial\Omega \text{ and for } |x_1| \gg 1,$$

such that there exists a curve $\gamma \subset \overline{\Omega}$ with

$$\gamma \text{ connects } (0, -\tfrac{1}{2}) \text{ to } (0, \tfrac{1}{2}) \quad \text{and} \quad |\nabla\psi|^2 = 1 \text{ on } \gamma.$$

In line with the Ansatz (2.6), we define $m \colon \Omega \to S^2$ via

$$(m_1, m_3) = (-\frac{\partial\psi}{\partial x_3}, \frac{\partial\psi}{\partial x_1}), \quad m_2 = \left\{ \begin{matrix} - \\ + \end{matrix} \right\} \sqrt{1 - |\nabla\psi|^2} \left\{ \begin{matrix} \text{left} \\ \text{right} \end{matrix} \right\} \text{of } \gamma.$$

---

[2]Indeed, one possible recipe is to start from $\psi(x) = \tfrac{1}{2} - |x|$ and to modify ψ outside of a neighborhood of the curve $\gamma = \left\{ (\tfrac{1}{2}\sqrt{\tfrac{1}{4} - x_2^2}, x_2) \mid x_2 \in [-\tfrac{1}{2}, \tfrac{1}{2}] \right\}$.

834   Felix Otto

Only exchange and anisotropy contribute to the energy:

$$E(m) \sim d^2 + Q,$$

which turns into (3.8) in the regime under consideration.

Proof of Lemma 2. Making the Ansatz (2.5), the energy simplifies to

$$E(m) = d^2 t \int_{-\infty}^{\infty} |\frac{dm}{dx_1}|^2 \, dx_1 + Q t \int_{-\infty}^{\infty} m_1^2 \, dx_1 + \int_{R^2} |\nabla u|^2 \, d^2 x \qquad (3.12)$$

$$\leq t^2 \left\{ \frac{d^2}{t} \int_{-\infty}^{\infty} \frac{1}{1 - m_1^2} (\frac{dm_1}{dx_1})^2 \, dx_1 + \frac{Q}{t} \int_{-\infty}^{\infty} m_1^2 \, dx_1 + \int_{R^2} |\nabla U|^2 \, d^2 x \right\},$$

where U is the harmonic extension [3] of m_1 from $\{x_3 = 0\}$ onto $\mathbf{R}^2$. Hence (3.12) holds for any extension U of m_1. We now have to construct U such that its restriction m_1 satisfies $m_1^2(0) = 1$ in order to allow for the sign change of m_2. $\int_{R^2} |\nabla U|^2 \, d^2 x$ just fails to control the L^∞-norm of U and thus of m_1—the counterexample involves a logarithm which we also use in this construction. The logarithm is cut off at the length scales $\frac{d^2}{t} \ll \frac{t}{Q}$:

$$U(x) = \ln^{-1} \frac{Q d^2}{t^2} \ln \sqrt{\min\{(\frac{Q|x|}{t})^2 + (\frac{Q d^2}{t^2})^2, 1\}}.$$

An elementary calculation shows (3.9) for $m_1(x_1) = U(x_1, 0)$. A more detailed analysis of the reduced variational problem (3.12) is in [4, 13].

Proof of Lemma 3. We rewrite (2.3) as

$$\int_{\Omega} m_3 \frac{\partial \zeta}{\partial x_3} \, d^2 x = \int_{R^2} \frac{\partial u}{\partial x_3} \frac{\partial \zeta}{\partial x_3} \, d^2 x + \int_{R^2} \frac{\partial u}{\partial x_1} \frac{\partial \zeta}{\partial x_1} \, d^2 x + \int_{\Omega} \frac{\partial m_1}{\partial x_1} \zeta \, d^2 x \qquad (3.13)$$

and choose the test function

$$\zeta(x_1, x_3) = \overline{m}_3(x_1) \, \eta(\hat{x}_3) \quad \text{where } x_3 = t \, \hat{x}_3$$

and $\eta \in C_0^\infty(\mathbf{R})$ is chosen such that $\frac{d\eta}{d\hat{x}_3}(\hat{x}_3) = 1$ for $\hat{x}_3 \in (-\frac{1}{2}, \frac{1}{2})$ in order to have

$$\frac{\partial \zeta}{\partial x_3}(x_1, x_3) = \frac{1}{t} \overline{m}_3(x_1) \frac{d\eta}{d\hat{x}_3}(\hat{x}_3) = \frac{1}{t} \overline{m}_3(x_1) \quad \text{for } x_3 \in (-\frac{t}{2}, \frac{t}{2}).$$

Hence the term on the l. h. s. of (3.13) turns into

$$\int_{\Omega} m_3 \frac{\partial \zeta}{\partial x_3} \, d^2 x = \int_{-\infty}^{\infty} \overline{m}_3^2 \, dx_1 \qquad (3.14)$$

and the first term on the r. h. s. of (3.13) is estimated as follows

$$\left| \int_{R^2} \frac{\partial u}{\partial x_3} \frac{\partial \zeta}{\partial x_3} \, d^2 x \right| \lesssim \left(\int_{R^2} (\frac{\partial u}{\partial x_3})^2 \, d^2 x \frac{1}{t} \int_{-\infty}^{\infty} \overline{m}_3^2 \, dx_1 \right)^{\frac{1}{2}}$$

$$\leq \left(\frac{1}{t} E \int_{-\infty}^{\infty} \overline{m}_3^2 \, dx_1 \right)^{\frac{1}{2}}. \qquad (3.15)$$

---

[3] The inequality $\int_{R^2} |\nabla u|^2 \, d^2 x \leq t^2 \int_{R^2} |\nabla U|^2 \, d^2 x$ can best be seen by expressing both integrals in terms of the Fourier transform $\hat{m}_1(k_1)$ of $m_1(x_1)$.

The two remaining terms are also easily dominated:

$$\left| \int_{R^2} \frac{\partial u}{\partial x_1} \frac{\partial \zeta}{\partial x_1} \, d^2x \right| \lesssim \left(\int_{R^2} (\frac{\partial u}{\partial x_1})^2 \, d^2x \; t \int_{-\infty}^{\infty} (\frac{d\overline{m}_3}{dx_1})^2 \, dx_1 \right)^{\frac{1}{2}}$$

$$\leq \left(\int_{R^2} (\frac{\partial u}{\partial x_1})^2 \, d^2x \int_{\Omega} (\frac{\partial m_3}{\partial x_1})^2 \, d^2x \right)^{\frac{1}{2}} \leq \frac{1}{d} E, \quad (3.16)$$

$$\left| \int_{R^2} \frac{\partial m}{\partial x_1} \zeta \, d^2x \right| \lesssim \left(\int_{R^2} (\frac{\partial m}{\partial x_1})^2 \, d^2x \; t \int_{-\infty}^{\infty} \overline{m}_3^2 \, dx_1 \right)^{\frac{1}{2}}$$

$$\leq \left(\frac{t}{d^2} E \int_{-\infty}^{\infty} \overline{m}_3^2 \, dx_1 \right)^{\frac{1}{2}}. \quad (3.17)$$

Collecting (3.14)–(3.17) and using the Cauchy-Schwarz inequality gives

$$\int_{-\infty}^{\infty} \overline{m}_3^2 \, dx_1 \lesssim \left(\frac{1}{t} + \frac{1}{d} + \frac{t}{d^2} \right) E \lesssim \left(\frac{1}{t} + \frac{t}{d^2} \right) E. \quad (3.18)$$

On the other hand, we use Poincaré inequality in the x_3-direction which we integrate over $x_1 \in (-\infty, \infty)$

$$\int_{\Omega} (m_3 - \overline{m}_3)^2 \, d^2x \lesssim t^2 \int_{\Omega} (\frac{\partial m_3}{\partial x_3})^2 d^2x \leq (\frac{t}{d})^2 E. \quad (3.19)$$

Now (3.18) and (3.19) combine as desired into (3.10).

Proof of Lemma 4. In the first step we establish for $0 < \rho \ll \ell$ and $0 \leq \xi_1 - \tilde{\xi}_1 \leq \ell$

$$\left| \frac{1}{\rho} \int_{\xi_1}^{\xi_1 + \rho} \overline{m}_1 \, dx_1 \quad \frac{1}{\rho} \int_{\tilde{\xi}_1 - \rho}^{\tilde{\xi}_1} \overline{m}_1 \, dx_1 \right|^2 \lesssim \left(\frac{1}{t^2} \ln \frac{\ell}{\rho} + \frac{1}{\rho t} \right) E. \quad (3.20)$$

In order to establish (3.20), we construct an appropriate test function ζ for (2.3). We first define ζ on the strip $R \times (-\frac{t}{2}, \frac{t}{2})$ as piecewise linear

$$\zeta(x_1, x_3) = \begin{cases} 0 & \xi_1 + \rho \leq x_1 \\ \frac{1}{\rho}(\xi_1 - x_1 + \rho) & \xi_1 \leq x_1 \leq \xi_1 + \rho \\ 1 & \tilde{\xi}_1 \leq x_1 \leq \xi_1 \\ \frac{1}{\rho}(x_1 - \tilde{\xi}_1 + \rho) & \tilde{\xi}_1 - \rho \leq x_1 \leq \tilde{\xi}_1 \\ 0 & x_1 \leq \tilde{\xi}_1 - \rho \end{cases}. \quad (3.21)$$

ζ is just defined such that

$$\int_{\Omega} m \cdot \nabla \zeta \, d^2x = -\frac{t}{\rho} \int_{\xi_1}^{\xi_1 + \rho} \overline{m}_1 \, dx_1 + \frac{t}{\rho} \int_{\tilde{\xi}_1 - \rho}^{\tilde{\xi}_1} \overline{m}_1 \, dx_1. \quad (3.22)$$

For the r. h. s. of (2.3) we have to extend ζ onto all of R^2: We harmonically extend ζ on the upper and lower half-plane $R \times (\frac{t}{2}, +\infty)$ resp. $R \times (-\infty, -\frac{t}{2})$. We claim

$$\int_{R^2} |\nabla \zeta|^2 \, d^2x \lesssim \ln \frac{\ell}{\rho} + \frac{t}{\rho}. \quad (3.23)$$

This yields the following estimate of the r. h. s. of (2.3)

$$\left| \int_{R^2} \nabla u \cdot \nabla \zeta \, d^2 x \right| \leq \left(\int_{R^2} |\nabla u|^2 \, d^2 x \int_{R^2} |\nabla \zeta|^2 \, d^2 x \right)^{\frac{1}{2}}$$

$$\leq \left(E \left(\ln \frac{\ell}{\rho} + \frac{t}{\rho} \right) \right)^{\frac{1}{2}}. \tag{3.24}$$

Obviously (3.22) & (3.24) yields (3.20).

We now argue in favor of (3.23). On the strip $R \times (-\frac{t}{2}, \frac{t}{2})$ we have

$$\int_{R \times (-\frac{t}{2}, \frac{t}{2})} |\nabla \zeta|^2 \, d^2 x \overset{(3.21)}{=} 2 t \rho (\frac{1}{\rho})^2 \sim \frac{t}{\rho}. \tag{3.25}$$

The Dirichlet integral of the harmonic extension is estimated in terms of its boundary value as follows

$$\int_{R \times (\frac{t}{2}, +\infty)} |\nabla \zeta|^2 \, d^2 x \sim \int_0^\infty \frac{1}{x_3^2} \int_{-\infty}^\infty (\zeta(x_1 + x_3, \frac{t}{2}) - \zeta(x_1, \frac{t}{2}))^2 \, dx_1 \, dx_3,$$

see [12, Théorème 9.4, Théorème 10.2]. Since

$$\int_{-\infty}^\infty (\zeta(x_1 + x_3, \frac{t}{2}) - \zeta(x_1, \frac{t}{2}))^2 \, dx_1 \overset{(3.21)}{\sim} \left\{ \begin{array}{lll} \ell & \ell \underset{\sim}{<} x_3 \\ x_3 & \rho \underset{\sim}{<} x_3 \underset{\sim}{<} \ell \\ \frac{x_3^2}{\rho} & x_3 \underset{\sim}{<} \rho \end{array} \right\},$$

this yields

$$\int_{R \times (\frac{t}{2}, +\infty)} |\nabla \zeta|^2 \, d^2 x \underset{\sim}{<} \ln \frac{\ell}{\rho}. \tag{3.26}$$

Now (3.25) and (3.26) combine into (3.23).

In the second step, we establish for $\ell \gg \frac{d^2}{t}$ and $0 \leq \xi_1 - \tilde{\xi}_1 \leq \ell$

$$|\overline{m}_1(\xi_1) - \overline{m}_1(\tilde{\xi}_1)|^2 \underset{\sim}{<} \left(\frac{1}{t^2} \ln \frac{\ell t}{d^2} + \frac{1}{d^2} \right) E. \tag{3.27}$$

For this, we observe that

$$\left| \frac{1}{\rho} \int_{\xi_1}^{\xi_1 + \rho} \overline{m}_1(x_1) \, dx_1 - \overline{m}_1(\xi_1) \right|^2 \underset{\sim}{<} \rho \int_{-\infty}^\infty (\frac{d\overline{m}_1}{dx_1})^2 \, dx_1$$

$$\leq \frac{\rho}{t} \int_\Omega (\frac{\partial m_1}{\partial x_1})^2 \, d^2 x \leq \frac{\rho}{d^2 t} E,$$

so that together with (3.20) we obtain

$$|\overline{m}_1(\xi_1) - \overline{m}_1(\tilde{\xi}_1)|^2 \underset{\sim}{<} \left(\frac{1}{t^2} \ln \frac{\ell}{\rho} + \frac{1}{\rho t} + \frac{\rho}{d^2 t} \right) E.$$

We now balance the first and last term by choosing $\rho = \frac{d^2}{t} \lesssim \ell$ and so obtain (3.27).

In the last step, we show (3.11) for $\frac{t^2}{Q d^2} \gg 1$. For this we observe that

$$\int_{-\infty}^{\infty} \overline{m}_1^2 \, dx_1 \;\leq\; \frac{1}{t} \int_\Omega m_1^2 \, d^2 x \;\leq\; \frac{1}{Q \, t} E.$$

Hence we obtain together with (3.27) for arbitrary $\xi_1 \in (-\infty, \infty)$

$$
\begin{aligned}
\overline{m}_1(\xi_1)^2 \;&\lesssim\; \frac{1}{\ell} \int_{\xi_1 - \frac{\ell}{2}}^{\xi_1 + \frac{\ell}{2}} \overline{m}_1^2 \, dx_1 + \frac{1}{\ell} \int_{\xi_1 - \frac{\ell}{2}}^{\xi_1 + \frac{\ell}{2}} (\overline{m}_1(\xi_1) - \overline{m}_1(x_1))^2 \, dx_1 \\
&\lesssim\; \left(\frac{1}{Q \, \ell \, t} + \frac{1}{t^2} \ln \frac{\ell \, t}{d^2} + \frac{1}{d^2} \right) E.
\end{aligned}
$$

Choosing $\ell = \frac{t}{Q} \gg \frac{d^2}{t}$, we balance the two first terms and so obtain (3.11).

Proof of Proposition 1. It remains to establish the lower bound. For further reference we remark that by Poincaré's inequality

$$\int_{(-\frac{t}{2}, \frac{t}{2}) \times (-\frac{t}{2}, \frac{t}{2})} (m_i - \overline{m}_i(0))^2 \, d^2 x \;\lesssim\; t^2 \int_\Omega |\nabla m_i|^2 \, d^2 x \;\lesssim\; (\frac{t}{d})^2 E. \qquad (3.28)$$

According to (2.1), we have in particular $\lim_{x_1 \to \pm\infty} \overline{m}_2(x_1) = \pm 1$ and thus there exists an ξ_1 with $\overline{m}_2(\xi_1) = 0$. W. l. o. g. we assume $\xi_1 = 0$ so that $\overline{m}_2(0) = 0$. According to (3.28) we obtain

$$\int_{(-\frac{t}{2}, \frac{t}{2}) \times (-\frac{t}{2}, \frac{t}{2})} m_2^2 \, d^2 x \;\lesssim\; (\frac{t}{d})^2 E. \qquad (3.29)$$

Furthermore, we have according to Lemma 3

$$\int_{(-\frac{t}{2}, \frac{t}{2}) \times (-\frac{t}{2}, \frac{t}{2})} m_3^2 \, d^2 x \;\lesssim\; \left(1 + (\frac{t}{d})^2 \right) E. \qquad (3.30)$$

Since $1 - m_1^2 = m_2^2 + m_3^2$, the estimates (3.29) & (3.30) imply

$$\int_{(-\frac{t}{2}, \frac{t}{2}) \times (-\frac{t}{2}, \frac{t}{2})} (1 - m_1^2) \, d^2 x \;\lesssim\; \left(1 + (\frac{t}{d})^2 \right) E.$$

In view of (3.28), this localizes to

$$1 - \overline{m}_1(0)^2 \;\lesssim\; (\frac{1}{t^2} + \frac{1}{d^2}) E. \qquad (3.31)$$

On the other hand, we have by Lemma 4

$$\overline{m}_1(0)^2 \;\lesssim\; \left(\frac{1}{t^2} \ln \frac{t^2}{Q \, d^2} + \frac{1}{d^2} \right) E \qquad (3.32)$$

provided $(\frac{t}{d})^2 \gg Q$. Combining (3.31) and (3.32), we obtain

$$1 \underset{\sim}{<} \left(\frac{1}{t^2} \ln \frac{t^2}{Q\,d^2} + \frac{1}{d^2} + \frac{1}{t^2} \right) E \sim \left(\frac{1}{t^2} \ln \frac{t^2}{Q\,d^2} + \frac{1}{d^2} \right) E. \qquad (3.33)$$

Since we have by elementary calculus that

$$\frac{1}{t^2} \ln \frac{t^2}{Q\,d^2} \left\{ \begin{matrix} \underset{\sim}{<} \\ \underset{\sim}{>} \end{matrix} \right\} \frac{1}{d^2} \iff \ln \frac{1}{Q} \left\{ \begin{matrix} \underset{\sim}{<} \\ \underset{\sim}{>} \end{matrix} \right\} (\frac{t}{d})^2,$$

(3.33) is equivalent to the lower bound in (2.7).

Acknowledgments. The author thanks A. DeSimone, Weinan E. , R. V. Kohn, and S. Müller for many stimulating discussions on micromagnetics.

References

[1] R. Choksi, R. V. Kohn, F. Otto, Domain branching in uniaxial ferromagnets: a scaling law for minimal energy, *Comm. Math. Phys* **201**, 61–79 (1999).

[2] A. DeSimone, R. V. Kohn, S. Müller, F. Otto, Repulsive interaction of Néel wall, and the internal length scale of the cross-tie wall, submitted.

[3] Weinan E, F. Otto, Thermodynamically driven incompressible fluid mixtures, *J. Chem. Phys.* **107 (23)**, 10177–10184 (1997).

[4] C. Garcia-Cervera, Magnetic domains and magnetic domain walls, Ph-D thesis, New York University (1999).

[5] P. G. de Gennes, Wetting: Statics and dynamics. *Rev. Mod. Phys.* **57** 827–863 (1985).

[6] L. Giacomelli, F. Otto, Droplet Spreading: Intermediate Scaling Law by PDE Methods, *Comm. Pure Appl. Math* **55**, 217–254 (2002).

[7] A. Hönig, B. Niethammer, F. Otto, On first-order corrections to the LSW-theory, submitted.

[8] A. Hubert, Stray-field-free magnetization configurations, *Phys. Status Solidi* **32**, 519–534 (1969).

[9] A. Hubert, R. Schäfer, *Magnetic domains*, Springer (1998).

[10] R. V. Kohn, S. Müller, Surface energy and microstructure in coherent phase transitions, *Comm. Pure Appl. Math.* **47**, 405–435 (1994).

[11] R. V. Kohn, F. Otto, Upper bounds for coarsening rates, will appear in *Comm. Math. Phys.*

[12] J. L. Lions, E. Magenes, *Problèmes aux limites non homogènes*, Dunod (1968).

[13] C. Melcher, The logarithmic tail of Néel walls in thin films, submitted to *Arch. Rat. Mech. Anal.*.

ICM 2002 · Vol. III · 839–849

Mathematical Modelling of the Cardiovascular System

A. Quarteroni[*]

Abstract

In this paper we will address the problem of developing mathematical models for the numerical simulation of the human circulatory system. In particular, we will focus our attention on the problem of haemodynamics in large human arteries.

2000 Mathematics Subject Classification: 93A30, 35Q30, 74F10, 65N30.
Keywords and Phrases: Haemodynamics, Partial differential equations, Finite elements, Fluid structure interaction.

1. Introduction

The simulation of not only the physiological functioning of the blood circulatory system, but also of specific pathological circumstances is of utmost importance since cardiovascular diseases represent the leading cause of death in developed countries, with a tremendous medical, social and economic impact.

In the cardiovascular system, altered flow conditions, such as flow separation, flow reversal, low and oscillatory shear stress areas, are recognised as important factors in the development of arterial diseases. A detailed understanding of the local haemodynamics, the effect of vascular wall modification on flow patterns and its long-term adaptation to surgical procedures can have useful clinical applications. Some of these phenomena are not well understood, making it difficult to foresee short and long term evolution of the disease and the planning of the therapeutic approach. In this context, the mathematical models and numerical simulations can play a crucial role.

Blood flow interacts both mechanically and chemically with the vessel walls. The mechanical coupling requires algorithms that correctly describe the energy transfer between the fluid (typically modelled by the Navier-Stokes equations) and the structure of the vessel wall. On the other hand, the flow equations can be coupled with appropriate models that describe the wall absorption of bio-chemicals (e.g. oxygen, lipids, drugs, etc.) and of their transport, diffusion and kinetics. Numerical simulations of this type may help to understand

---
[*]MOX, Department of Mathematics, Politecnico di Milano, Italy and Institute of Mathematics (IMA), EPFL, Lausanne, Switzerland. E-mail: alfio.quarteroni@epfl.ch

the modifications in bio-chemical exchanges due to an alteration of the flow field caused, for instance, by a stenosis (i.e. a localised narrowing of a vessel lumen, normally due to fat accumulation).

The simulation of large and medium-size arteries is now sufficiently advanced so to envisage the applications of computer models to medical research and, in a medium range, to everyday medical practise. For instance, simulating the flow in a coronary by-pass may help understanding the extent at which its geometry influences the flow and in turn the post-surgery evolution. Also the study of the effects of a vascular prosthesis as well as the study of artificial valve implants are areas which could benefit from a sufficiently accurate simulation of blood flow field.

In this paper we review the principal mathematical steps behind the derivation of the coupled fluid-structure equations which model the blood flow motion in large and medium-sized arteries. Then we mention the way geometrical multiscale models, that combine mathematical models set up in different spatial dimensions, can be conveniently used to simulate the whole circulatory system.

2.   The coupled fluid-structure problem

In this section we will treat the situation arising when the flow in a vessel interacts mechanically with the wall structure. This aspect is particularly relevant for blood flow in large arteries, where the vessel wall radius may vary up to 10% because of the forces exerted by the flowing blood stream.

We will first illustrate a framework for the Navier-Stokes equations in a moving domain which is particularly convenient for the analysis and for the set up of numerical solution methods.

2.1.   The Arbitrary Lagrangian Eulerian (ALE) formulation of the Navier-Stokes equations

Navier-Stokes equations are usually derived according to the *Eulerian* approach where the independent spatial variables are the coordinates of a fixed Eulerian system. When considering the flow inside a portion of a compliant artery, we have to compute the flow solution in a *computational domain* Ω_t varying with time.

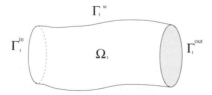

Figure 1: A simple model of a section of an artery. The vessel wall Γ_t^w is moving. The location along the z axis of Γ_t^{in} and Γ_t^{out} are fixed.

The boundary of Ω_t may in general be subdivided into two parts. The first part coincides with the physical fluid boundary, i.e. the vessel wall Γ_t^w (see Fig. 1), which is moving under the effect of the flow field. The other part of $\partial \Omega_t$ corresponds to artificial boundaries which delimit the region of interest from the remaining part of the cardiovascular system.

The "artificial" boundaries are the inlet and outlet (or, using the medical terminology, the proximal and distal) sections, here indicated by Γ_t^{in} and Γ_t^{out}, respectively. The location of these boundaries is fixed a priori. More precisely, Γ_t^{in} and Γ_t^{out} may change with time because of the displacement of Γ_t^w, however they remain planar and their position along the vessel axis is kept fixed. In this case the Eulerian approach becomes impractical.

A possible alternative would be to use the *Lagrangian approach*, where we identify the computational domain on a reference configuration Ω_0, e.g. that at the initial time $t = 0$, and the corresponding domain in the current configuration will be provided by the Lagrangian mapping

$$\Omega_t = \Omega_{\mathcal{L}_t} = \mathcal{L}_t(\Omega_0), \ t > 0, \tag{2.1}$$

which describes the motion of a material particle and whose time derivative is the fluid velocity. Since the fluid velocity at the wall is equal to the wall velocity, the Lagrangian mapping effectively maps Γ_0^w to the correct wall position Γ_t^w at each time t. However, the artificial boundaries in the reference configuration, say Γ_0^{in} and Γ_0^{out}, will now be transported along the fluid trajectories. This is unacceptable, particularly for a relatively large time interval as Ω_t rapidly becomes highly distorted.

A more convenient situation is the one when, even if the wall is moving, one keeps the inlet and outlet boundaries at the same spatial location along the vessel axis. With that purpose, we introduce the *Arbitrary Lagrangian Eulerian* (ALE) mapping $\mathcal{A}_t : \Omega_0 \rightarrow \Omega_{\mathcal{A}_t}$, $\quad \mathbf{Y} \rightarrow \mathbf{x}(t, \mathbf{Y}) = \mathcal{A}_t(\mathbf{Y})$, which provides the spatial coordinates $(t, \mathbf{x})$ in terms of the so-called *ALE coordinates* $(t, \mathbf{Y})$, with the basic requirement that $\mathcal{A}_t$ retrieves, at each time $t > 0$, the desired computational domain, $\Omega_t = \Omega_{\mathcal{A}_t} = \mathcal{A}_t(\Omega_0), t \geq 0$.

The ALE mapping should be continuous and bijective in $\overline{\Omega}_0$. Once given, we may define the *domain* velocity field as

$$\mathbf{w} = \frac{\partial \mathcal{A}_t}{\partial t} \circ \mathcal{A}_t^{-1}, \tag{2.2}$$

where the composition operator applies only to the spatial coordinates. The ALE time derivative of a function $f : I \times \Omega_t \rightarrow \mathbb{R}$, which we denote by $\frac{D^A}{Dt} f$, is defined as

$$\frac{D^A}{Dt} f : I \times \Omega_t \rightarrow \mathbb{R}, \qquad \frac{D^A}{Dt} f = \frac{\partial \widetilde{f}}{\partial t} \circ \mathcal{A}_t^{-1}. \tag{2.3}$$

This definition is readily extended to vector valued functions. The ALE derivative is related to the Eulerian (partial) time derivative by the relation

$$\frac{D^A}{Dt} f = \frac{\partial f}{\partial t} + \mathbf{w} \cdot \boldsymbol{\nabla} f, \tag{2.4}$$

where the gradient is made with respect to the x-coordinates.

The Navier-Stokes equations may be formulated in order to put into evidence the ALE time derivative, obtaining

$$\frac{D^A}{Dt} \mathbf{u} + [(\mathbf{u} - \mathbf{w}) \cdot \boldsymbol{\nabla}] \mathbf{u} + \boldsymbol{\nabla} p - \mathbf{div} \, \mathbf{T}(\mathbf{u}) = \mathbf{f},$$
$$\text{in } \Omega_t, \, t > 0, \tag{2.5}$$

$$\mathrm{div} \, \mathbf{u} = 0,$$

where $\mathbf{T}$ is the Cauchy stress tensor, which for Newtonian fluid is given by $\mathbf{T} = \nu(\nabla\mathbf{u} + \nabla\mathbf{u}^T)$, being ν the blood kinematic viscosity.

When considering small vessels, accounting for non-Newtonian behaviour of blood becomes crucial. In that case the functional dependence of $\mathbf{T}$ on $\mathbf{u}$ becomes more complex, see for instance [18].

2.2. The structure model

The vascular wall has a very complex nature and devising an accurate model for its mechanical behaviour is rather difficult. Its structure is indeed formed by many layers with different mechanical characteristics [8, 11], which are usually in a pre-stressed state. More-over, experimental results obtained by specimens are only partially significant. Indeed, the vascular wall is a living tissue with the presence of muscular cells which contribute to its mechanical behaviour. It may then be expected that the dead tissue used in the laboratory will have different mechanical characteristics than the living one. Moreover, the arterial mechanics depend also on the type of the surrounding tissues, an aspect almost impossible to reproduce in a laboratory. As we have already pointed out, the displacements cannot be considered small (at least in large arteries where the radius may vary up to a few percent during the systolic phase). Consequently, an appropriate model for the structure displace-ment η reads

$$\rho_w \frac{\partial^2 \eta}{\partial t^2} - \mathbf{div}\boldsymbol{\sigma}(\eta, \frac{\partial \eta}{\partial t}) = \mathbf{f}, \quad \text{in } \Omega_t^s, \, t > 0,$$

where Ω_t^s indicates the current configuration and $\boldsymbol{\sigma}$ is the Cauchy stress tensor. The latter may depend on the structure velocity because of viscoelasticity. A full Lagrange formula-tion for the structure on a fixed reference configuration Ω_0^s may be obtained by the usual Lagrange and Piola transformation (see, e.g., [2]). A general framework to derive constitu-tive equations for arterial walls is reported in [11].

It is the role of mathematical modelling to find reasonable simplifying assumptions by which major physical characteristics remain present, yet the problem becomes compu-tationally actractive. In particular, a simpler model may be obtained by considering only displacements in the radial direction and a cylindrical geometry for the vessel. Furthermore if we neglect the geometrical non-linearities (which correspond to assume small displace-ments), as well as the variations along the radial directions (small thickness assumption) we obtain the following "generalised string model" [16, 14] for the evolution of the radial displacement $\eta = R - R_0$,

$$\frac{\partial^2 \eta}{\partial t^2} - a\frac{\partial^2 \eta}{\partial z^2} + b\eta - c\frac{\partial^3 \eta}{\partial t \partial z^2} = H, \quad \text{in } \Gamma_0^w, \, t > 0, \tag{2.6}$$

where $a > 0$ and $b > 0$ are parameters linked to the vessel geometry and mechanical characteristics, $c > 0$ is a viscoelastic parameter and H is a forcing term which depends on the action of the fluid, as we will see in (2.7). More details as well as the derivation of this model are found in the cited references.

Here Γ_0^w is the reference configuration for the structure

$$\Gamma_0^w = \{(r, \theta, z) : r = R_0(z), \, \theta \in [0, 2\pi), \, z \in [0, L]\},$$

where L indicates the length of the arterial element under consideration. In our cylindrical coordinate system (r, θ, z), the z coordinate is aligned along the vessel axes and a plane $z = \bar{z} (= \text{constant})$ defines an *axial section*.

2.3. Coupling with the structure model

We now study the properties of the coupled fluid-structure problem, using for the structure the generalised string model (2.6). We will take $\mathbf{n}$ always to be the outwardly vector normal to the fluid domain boundary. Furthermore we define g as the metric function so that the elemental surface measure $d\sigma$ on Γ_t^w is related to the corresponding measure $d\sigma_0$ on Γ_0^w by $d\sigma = g \, d\sigma_0$.

We will then address the following problem: *For all $t > 0$, find $\mathbf{u}, p, \eta$ such that*

$$\begin{cases} \mathbf{u}, p \quad \text{satisfy problem (2.5)}, \\ \eta \quad \text{satisfies problem (2.6)}, \\ \mathbf{u} \circ \mathcal{A}_t = \dfrac{\partial \eta}{\partial t} \mathbf{e}_r, \quad \text{on } \Gamma_0^w, \\ H = g \dfrac{\rho}{\rho_w h_0} [(p - p_0)\mathbf{n} - \mathbf{T}(\mathbf{u}) \cdot \mathbf{n}] \cdot \mathbf{e}_r \quad \text{on } \Gamma_0^w. \end{cases} \qquad (2.7)$$

Here, p_0 is the pressure acting at the exterior of the vessel, $\mathbf{e}_r$ is the radial unit vector, ρ_w and ρ are the wall and fluid densities, respectively, while h_0 is the wall thickness. The system is complemented by appropriate boundary and initial conditions.

We may then recognise the sources of the coupling between the fluid and the structure models, which are twofold. In view of a possible iterative solution strategy, the fluid solution provides the value of H, which is function of the fluid stresses at the wall. On the other hand, the movement of the vessel wall modifies the geometry on which the fluid equations must be solved, besides providing Dirichlet boundary conditions for the fluid velocity in correspondence to the vessel wall.

Remark 2.1. We may note that the non-linear convective term in the Navier-Stokes equations is crucial to obtain the well-posedness of the coupled problem, because it generates a boundary term which compensates that coming from the treatment of the acceleration term. These two contributions are indeed only present in the case of a moving boundary. See [14] and [1].

2.4. Numerical solution of the coupled fluid-structure problem

In this section we describe an algorithm that at each time-level allows the decoupling of the sub-problem related to the fluid from that related to the vessel wall. As usual, t^k, $k = 0, 1, \ldots$ denotes the k-th discrete time level; $\Delta t > 0$ is the time-step, while v^k is the approximation of the function (either scalar or vector) v at time t^k.

The numerical solution of the fluid-structure interaction problem (2.7) will be carried out by constructing a proper finite element approximation of each sub-problem. In particular, for the fluid we need to devise a finite element formulation suitable for moving domains (or, more precisely, moving grids). In this respect, the ALE formulation will provide an appropriate framework.

To better illustrate the situation we refer to Fig. 2 where we have drawn a 2D fluid structure interaction problem (only the upper portion of the vessel is reported). For the sake of simplicity we have considered only a two-dimensional fluid structure problem, yet the algorithm here presented may be readily extended to more complex situations and three dimensional problems.

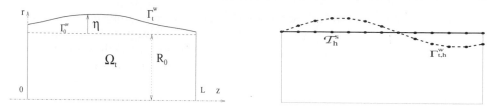

Figure 2: A simple fluid-structure interaction problem. On the left the domain definition and on the right the discretized vessel wall corresponding to a possible value of η_h.

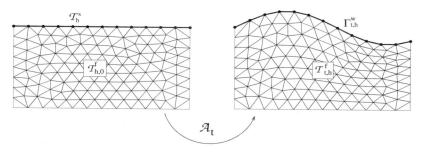

Figure 3: The triangulation used for the fluid problem at each time t is the the image through a map $\mathcal{A}_t$ of a mesh constructed on Ω_0.

The structure on Γ_0^w will be discretized by means of a grid $\mathcal{T}_{t,h}^s$ and employing piecewise linear continuous (P1) finite elements to represent the approximate vessel wall displacement η_h. The position at time t of the discretized vessel wall boundary, corresponding to the discrete displacement field $\eta_h(t)$, is indicated by $\Gamma_{t,h}^w$. Consequently, the fluid domain will be represented at every time by a polygon, which we indicate by $\Omega_{t,h}$. Its triangulation $\mathcal{T}_{t,h}^f$ will be constructed as the image by an appropriate ALE mapping $\mathcal{A}_t$ of a triangulation $\mathcal{T}_{0,h}^f$ of Ω_0, as shown in Fig. 3. Correspondingly, $\Omega_{t,h} = \mathcal{A}_t \Omega_{0,h}$, where $\Omega_{0,h}$ is the discretisation of Ω_0 induced by $\mathcal{T}_{0,h}^f$. The trace of $\mathcal{T}_{0,h}^f$ on Γ_0^w will coincide with the $\mathcal{T}_{t,h}^s$ of the vessel wall, i.e. we here consider *geometrically conforming* finite elements between the fluid and the structure. The possibility of using geometrically non-conforming finite elements has been investigated in [10].

Having at disposal the discrete displacement field η_h^{k+1} at $t = t^{k+1}$ and thus the position of the domain boundary $\partial\Omega_{t^{k+1},h}$, the set up of a map $\mathcal{A}_{t^{k+1}}$ such that $\mathcal{A}_{t^{k+1}}\left(\mathcal{T}_{0,h}^f\right)$

is an acceptable finite element mesh for the fluid domain is not a simple task. However, if we can assume that $\Omega_{t,h}$ is convex for all t and that the displacements are relatively small, the following technique, known as *harmonic extension*, may well serve the purpose. If $\mathbf{X}_h$ indicates the P1 finite element vector space associated to $\mathcal{T}_{0,h}^f$ and $\mathbf{g}_h : \partial\Omega_{0,h} \to \partial\Omega_{t^{k+1},h}$ is the function describing the fluid domain boundary, we build the map by seeking $\mathbf{y}_h \in \mathbf{X}_h$ such that

$$\int_{\Omega_0} \nabla\mathbf{y}_h : \nabla\mathbf{z}_h = 0 \quad \forall\mathbf{z}_h \in \mathbf{X}_h^0, \qquad \mathbf{y}_h = \mathbf{g}_h, \quad \text{on } \partial\Omega_{0,h}, \tag{2.8}$$

and then setting $\mathcal{A}_{t^{k+1}}(\mathbf{Y}) = \mathbf{y}_h(\mathbf{Y}), \quad \forall\mathbf{Y} \in \Omega_{0,h}$. A more general discussion on the construction of the ALE mapping may be found in [5, 12] as well as in [9].

Remark 2.2. Adopting P1 elements for the construction of the ALE map ensures that the triangles of $\mathcal{T}_{h,0}^f$ are mapped into triangles, thus $\mathcal{T}_{h,t}^f$ is a valid triangulation, under the requirement of invertibility of the map (which is assured if the domain is convex and the wall displacements are small).

As for the time evolution, we may adopt a linear time variation within each time slab $[t^k, t^{k+1}]$ by setting

$$\mathcal{A}_t = \frac{t - t^k}{\Delta t}\mathcal{A}_{t^{k+1}} - \frac{t - t^{k+1}}{\Delta t}\mathcal{A}_{t^k}, \quad t \in [t^k, t^{k+1}].$$

Then, the corresponding domain velocity $\mathbf{w}_h$ will be constant on each time slab.

2.4.1.   The iterative algorithm

We are now in the position of describing an iteration algorithm for the solution of the coupled problem. As usual, we assume that all quantities are available at $t = t^k$, $k \geq 0$, provided either by previous calculations or by the initial data and we wish to advance to the new time step t^{k+1}. For ease of notation we here omit the subscript h, with the understanding that we are referring exclusively to finite element quantities.

The algorithm requires to choose a *tolerance* $\tau > 0$, which is used to test the convergence of the procedure, and a *relaxation parameter* $0 < \theta \leq 1$. In what follows, the subscript $j \geq 0$ denotes the sub-iteration counter.

The algorithm reads:

A1 Extrapolate the vessel wall structure displacements and velocity:

$$\eta_{(0)}^{k+1} = \eta^k + \Delta t\dot{\eta}^k, \qquad \dot{\eta}_{(0)}^{k+1} = \dot{\eta}^k.$$

A2 Set $j = 0$.

  A2.1 By using $\eta_{(j)}^{k+1}$ compute the new grid for the fluid domain Ω_t and the ALE map by solving the harmonic extension problem (2.8).

  A2.2 Approximate the Navier-Stokes problem to compute $\mathbf{u}_{(j+1)}^{k+1}$ and $p_{(j+1)}^{k+1}$, using as velocity on the wall boundary the one calculated from $\eta_{(j)}^{k+1}$.

  A2.3 Approximate the structure problem to compute η_*^{k+1} and $\dot{\eta}_*^{k+1}$ using $\mathbf{u}_{(j+1)}^{k+1}$ and $p_{(j+1)}^{k+1}$ to recover the forcing term H.

  A2.4 Unless $\|\eta_*^{k+1} - \eta_{(j)}^{k+1}\|_{L^2(\Gamma_0^w)} + \|\dot{\eta}_*^{k+1} - \dot{\eta}_{(j)}^{k+1}\|_{L^2(\Gamma_0^w)} \leq \tau$, set

$$\eta_{(j+1)}^{k+1} = \theta\eta_{(j)}^{k+1} + (1 - \theta)\eta_*^{k+1}, \quad \dot{\eta}_{(j+1)}^{k+1} = \theta\dot{\eta}_{(j)}^{k+1} + (1 - \theta)\dot{\eta}_*^{k+1},$$

and $j \leftarrow j + 1$. Then return to step 2a.

A3 Set

$$\eta^{k+1} = \eta_*^{k+1}, \qquad \dot{\eta}^{k+1} = \dot{\eta}_*^{k+1}.$$

$$\mathbf{u}^{k+1} = \mathbf{u}_{(j+1)}^{k+1}, \qquad p^{k+1} = p_{(j+1)}^{k+1}.$$

If the algorithm converges, then $\lim_{j \to \infty} \mathbf{u}_{(j)}^{k+1} = \mathbf{u}^{k+1}$ and $\lim_{j \to \infty} \eta_{(j)}^{k+1} = \eta^{k+1}$, where $\mathbf{u}^{k+1}$ and η^{k+1} are the approximate solution of the coupled problem at time step t^{k+1}.

The algorithm entails, at each sub-iteration, the computation of the equation for the structure mechanics, the Navier-Stokes equations and the solution of two Laplace equations (2.8), one for every displacement component. It is therefore quite computationally expensive. Alternatively, less implicit formulations may be adopted, see for instance [13], yet it has been found that for the problem at hand a strong coupling between fluid and structure must be maintained also at discrete level in order to have stable algorithms [12].

3. Multiscale modelling of the cardiovascular system

The cardiovascular system is highly integrated. In many cases, to isolate the part of interest from the rest of the system would require specification of point-wise boundary data on artificial boundary sections. These are difficult to pre-determine. To account for the effect of the global circulatory system when focusing on specific regions we propose to integrate a hierarchy of models operating at different "scales". At the highest level we have the full three-dimensional fluid-structure interaction problem. This will be used where details of local flow fields are needed. At the lowest level, we use lumped parameter models based on the resolution of systems of non-linear ordinary algebraic-differential equations for averaged mass flow and pressure. The latter models are often described by help of an analogy with an electrical circuit, where the voltage represents blood pressure and the current the flow rate. They are well fit to supply the more sophisticated models with the effects of the circulation in small vessels, the capillary bed, the venous system, as well as the action of the heart. A transition between the two extrema could be achieved by convenient one-dimensional models expressed by a first order non-linear hyperbolic system (see Fig. 4). The derivation of the one-dimensional model and a possible numerical implementation may be found in [6, 4, 14].

An analysis of the coupling between fluid-structure models and one dimensional models may be found in [3], while the direct coupling by lumped parameter models and Navier-Stokes model is found in [15, 17], the coupling between lumped parameter models and one-dimensional models is also treated in [7].

Acknowledgements. The research activity here described has been supported by various Swiss and Italian research agencies and isntitutions, in particular the Swiss National Research Fund (FNS), and the Italian CNR, Ministry of Education (MIUR). The author also thanks Luca Formaggia for his contribution to the preparation of this paper.

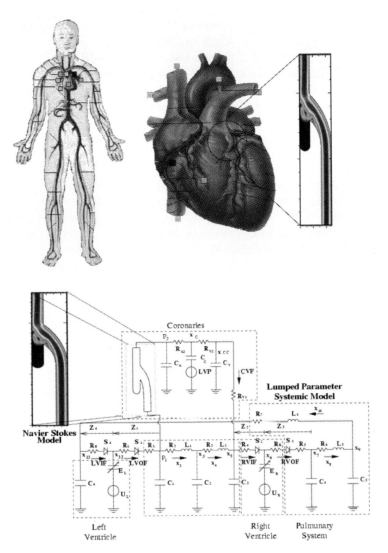

Figure 4: An example of multiscale simulation of blood flow, with the interplay between three-dimensional, one-dimensional and lumped parameters models. On top we show a global model of the circulatory system where a coronary by-pass is being simulated by a Navier-Stokes fluid-structure interaction model. The rest of the circulatory system is described by means of a lumped parameter model, based on the solution of a system of ODEs, is here represented by an electrical circuit analog in the bottom part of the figure.

References

[1] H. Beirão da Veiga. On the existence of strong solutions to a coupled fluid-structure evolution problem. *Arch. Rat. Mech. and Analysis*, 2001. submitted.

[2] P.G. Ciarlet. *Mathematical Elasticity. Volume 1: Three Dimensional Elasticity*, volume 20 of *Studies in Mathematics and its Applications*. North Holland, 1988.

[3] L. Formaggia, J.-F. Gerbeau, F. Nobile, and A. Quarteroni. On the coupling of 3D and 1D Navier-Stokes equations for flow problems in compliant vessels. *Comp. Methods in Appl. Mech. Engng.*, 191:561–582, 2001.

[4] L. Formaggia, D. Lamponi, and A. Quarteroni. One dimensional models for blood flow in arteries. Rapport de Recherche d'Analyse Numerique 03/2002, IMA-MOCS, EPFL, 2002.

[5] L. Formaggia and F. Nobile. A stability analysis for the Arbitrary Lagrangian Eulerian formulation with finite elements. *East-West J. Numer. Math.*, 7:105–131, 1999.

[6] L. Formaggia, F. Nobile, and A. Quarteroni. A one dimensional model for blood flow: application to vascular prosthesis. In I. Babuska, T. Miyoshi, and P.G. Ciarlet, editors, *Mathematical Modeling and Numerical Simulation in Continuum Mechanics*, volume 19 of *Lecture Notes in Computational Science and Engineering*, pages 137–153, Berlin, 2002. Springer-Verlag.

[7] L. Formaggia, F. Nobile, A. Quarteroni, and A. Veneziani. Multiscale modelling of the circulatory system: a preliminary analysis. *Computing and Visualisation in Science*, 2:75–83, 1999.

[8] Y.C. Fung. *Biomechanics: Mechanical Properties of Living Tissues*. Springer-Verlag, New York, 1993.

[9] L. Gastaldi. A priori error estimates for the arbitrary Lagrangian Eulerian formulation with finite elements. *East-West J. Numer. Math.*, 9(2):123–156, 2001.

[10] C. Grandmont and Y. Maday. Nonconforming grids for the simulation of fluid-structure interaction. In *Domain Decomposition Methods, 10 (Boulder, CO, 1997)*, pages 262–270. Amer. Math. Soc., Providence, RI, 1998.

[11] G.A. Holzapfel, T.C. Gasser, and R.W. Ogden. A new constitutive framework for arterial wall mechanics and a comparative study of material models. *Journal of Elasticity*, 61:1–48, 2000.

[12] F. Nobile. *Numerical approximation of fluid-structure interaction problems with application to hemodynamics*. PhD thesis, École Polytechnique Fédérale de Lausanne (EPFL), 2001. Thesis N. 2458.

[13] A. Piperno and C. Farhat. Partitioned procedures for the transient solution of coupled aeroelastic problems. part ii: energy transfer analysis and three dimensional applications. *Comp. Meth. Appl. Mech. Engng.*, 190:3147–3170, 2001.

[14] A. Quarteroni and L. Formaggia. Mathematical modelling and numerical simulation of the cardiovascular system. In N. Ayache, editor, *Modelling of Living Systems*, Handbook of Numerical Analysis (P.G Ciarlet and J.L Lions Eds.). Elsevier, Amsterdam, 2002. (to appear).

[15] A. Quarteroni, S. Ragni, and A. Veneziani. Coupling between lumped and distributed models for blood problems. *Computing and Visualisation in Science*, 4:111–124, 2001.

[16] A. Quarteroni, M. Tuveri, and A. Veneziani. Computational vascular fluid dynamics:

Problems, models and methods. *Computing and Visualisation in Science*, 2:163–197, 2000.

[17] A. Quarteroni and A. Veneziani. Analysis of a geometrical multiscale model based on the coupling of ODE's and PDE's for blood flow simulations. Technical Report 4, MOX, Department of Mathamatics, Politecnico di Milano, Italy, June 2002. submitted to SIAM J. Multiscale Model. Sim.

[18] K.R. Rajagopal. Mechanics of non-newtonian fluids. In G. Galdi and J. Necas, editors, *Recent Developments in Theoretical Fluid Mechanics*. Pitman Research Notes in Mathematics (291) - Longman, 1993.

ICM 2002 · Vol. III · 851–860

Analysis of Some Singular Solutions in Fluid Dynamics[*]

Zhouping Xin[†]

Abstract

Studies on singular flows in which either the velocity fields or the vorticity fields change dramatically on small regions are of considerable interests in both the mathematical theory and applications. Important examples of such flows include supersonic shock waves, boundary layers, and motions of vortex sheets, whose studies pose many outstanding challenges in both theoretical and numerical analysis. The aim of this talk is to discuss some of the key issues in studying such flows and to present some recent progress. First we deal with a supersonic flow past a perturbed cone, and prove the global existence of a shock wave for the stationary supersonic gas flow past an infinite curved and symmetric cone. For a general perturbed cone, a local existence theory for both steady and unsteady is also established. We then present a result on global existence and uniqueness of weak solutions to the 2-D Prandtl's system for unsteady boundary layers. Finally, we will discuss some new results on the analysis of the vortex sheets motions which include the existence of 2-D vortex sheets with reflection symmetry; and no energy concentration for steady 3-D axisymmetric vortex sheets.

2000 Mathematics Subject Classification: 35L70, 35L65, 76N15.
Keywords and Phrases: Singular flows, Shock waves, Vortox sheets, Prandtl's system, Boundary layers.

1. Introduction

Many physically interesting phenomena involve evolutions of singular flows whose velocity fields or vorticity fields change dramatically. Shock waves, vortex sheets, and boundary layer are some of the well-known examples of such singular flows which provide better approximations for significant parts of the flow fields

---

[*]Supported in part by grants from the Research Grants Council of Hong Kong Special Administrative Region CUHK 4219/99D and CUHK 4279/00P.

[†]The Institute of Mathematical Sciences & Department of Mathematics, The Chinese University of Hong Kong, Shatin, N.T., Hong Kong. E-mail: zpxin@ims.cuhk.edu.hk

near the physical boundaries, in the mixing layers and trailing wakes, etc., at high Reynolds numbers. The better understanding of the dynamics of such singular flows is the key in the analysis of general fluid flows governed by the well-known Euler (or Navier-Stokes) systems or their variants for both compressible and incompressible fluids, and has been one of the main focuses for applied analysists for decades. Substantial progress has been made in the past in studying singular flows by either rigorous analysis, or numerical simulations, or asymptotic methods [18]. In particular, a rather complete theory exists for the 1-D shock wave problems, and both the theoretical understanding and numerical methods for 2-D smooth incompressible flows are quite satisfactory. Yet there are still many important issues to be settled, such as, the global existence of weak solutions and asymptotic structures of various approximate solutions generated by either physical considerations or numerical methods for such singular flows.

One of the fundamental problems in the mathematical theory of shock waves for hyperbolic conservation laws is the well-posedness of the multi-dimensional gasdynamical shock waves, for which the celebrated Glimm's method does not apply. Most of the previous studies along this line deal with either short time structural stability of basic fronts or asymptotic analysis and numerical simulations of dynamics of shock fronts. Due to the great complexity and the lack of understanding, it is reasonable for one to begin with some of the physically relevant wave patterns where a lot of experimental data, numerical simulations, and asymptotic results are available. One of the basic model for such studies is the supersonic flow past a pointed body [4], which is one of the fundamental problems in gas dynamics. We first study the stationary supersonic gas flow past an infinite curved and symmetric cone. The flow is governed by the potential equation as well as the boundary conditions on the shock and the surface of the body. This problem has been studied extensively by either physical experiments or numerical simulations. The rigorous analysis starts with the work of Courant and Friedrichs in [4], where they show that if a supersonic flow hits a circular cone with axis being parallel to the velocity of the upstream flow and the vertex angle being less than a critical value, then there appears a circular conical shock attached at the tip of the cone, and the flow field between the shock front and the surface of the body can be determined by solving a boundary value problem of a system of ordinary differential equations. The local existence of supersonic flow past a pointed body has been established recently [1]. Our interest is on the structure of the global solution to this problem. We show that the solution to this problem exists globally in the whole space with a pointed curved shock attached at the tip of the cone and tends to a self-similar solution [2]. Our analysis is based on a global uniform weighted energy estimate for the linearized problem. The method we developed in [2] seems to be quite effective for other multi-dimensional problems. Indeed, similar approach can be used to study unsteady supersonic flows past a curved body for both potential flows and the full Euler system, and we obtain the local existence of shock waves in these cases [3].

Another challenging problem in the mathematical theory of fluid-dynamics is the theoretical foundation of the Prandtl's boundary layer theory [14]. In the presence of physical boundaries, the solutions to the invisicid Euler system cannot be the uniform asymptotic ansatz of the corresponding Navier-Stokes system for large Reynolds number due to the discrepancies between the no-slip boundary conditions for the Navier-Stokes system and the slip boundary condition for the Euler system. Indeed, the physical boundaries will creat vorticity and there is a thin layer (called boundary layer) in which the leading order approximation of the flow velocity is governed by the Prandtl's system for the boundary layers (see (3.1) in Section 3). There are extensive literatures on the theoretical, numerical, and experimental aspects of the Prandtl's boundary layer theory [14]. Yet very little rigorous theory exists for the dynamical boundary layer behavior of the Navier-Stokes solutions for both compressible and incompressible fluids. One of the main difficulties is the well-posedness theory in some standard Hölder or Sobolov spaces for the initial-boundary value problems for the Prandtl's system for boundary layers which is a severely degenerate parabolic-elliptic system, for which the only known existence results are proved locally in the analytic class in [15], except the series of important works of Oleinik who dealt with a class of monotonic data [12]. Indeed, Oleinik considered a plane unsteady flow of viscous incompressible fluid in the presence of an arbitrary injection and removal of the fluid across the boundary. Under the monotonicity assumption (see (3.5) in Section 3), Oleinik proved the well-posedness of local classical solutions to the Prandtl's system [12]. One of the open questions posed in [12] is to prove the global well-posedness of solution for the Prandtl's system under suitable conditions. Recently, we establish a global existence and uniqueness of weak solutions to the 2-D Prandtl's system for unsteady boundary layers in the class considered by Oleinik provided that the pressure is favorable. This is achieved by introducing a viscous splitting method and new weighted total variation estimate [16,17]. See Section 3 and [16,17] for more details.

Finally, we turn to the motion of vortex sheets, which corresponds to a singular inviscid flow where the vorticity field is zero except on lower-dimensional surfaces, the sheets, and can be characterized as inviscid flows with finite local energy and with vorticity fields being finite Radon measures. The study on the existence and structure of solutions for the inviscid Euler system for incompressible fluids with data in such class is of fundamental importance both physically and mathematically. Physically, vortex sheets can be used to model important flows such as high Reynolds number shear layers, and have many engineering applications. Mathematically, the evolution of a vortex sheets gives a classical example of ill-posed problem in the sense of Hadamard, a curvature singularity developes in finite time, and the nature of the solution past singularity formation is of great interest to know. This gives rise to many interesting yet challenging problems. Some of these are: Is there a well-posedness theory of classical weak solutions to the inviscid Euler system with general vortex sheets initial data? What are the structures of the approximate solutions to vortex sheets motions generated by either Navier-Stokes solutions or pratical numerical methods (such as particle method)? Can vorticity concentration and energy defects occur dynamically? etc.. Despite the importance of these problems and past intensive effort in rigorous mathematical analysis, these problems are far from being solved. Better understanding has been achieved before singularity formation in the analytical setting, and studies on global weak solutions and their approximations start with the important works of Diperna-Majda [6]. Delort observes that no vorticity concentration implies that a weak limit in L^2 of an approximate solution sequences is in fact a classical weak solution to the two-dimensional Euler system, and thus proved the first existence of global (in time) classical weak solution to the 2-D incompressible Euler equations with vortex sheets initial data provided that the initial vorticity is of distinguished sign [5]. Similar ideas have been used to study the convergence of approximate solutions generated by either viscous regularization [11] or partical methods [8,9] for vortex sheets with one sign vorticity. In the case that vorticity may change sign, the vortex sheets motion becomes extremely complex after singularity formation. Indeed, many important features of irregular flows seem to be connected with interactions and intertwining of regions of both positive and negative vorticities. Here we consider a mirror-symmetric flow which allows interactions but excludes intertwining of regions of distinguish vorticity, prove that there is no vorticity concentrations for approximate solution of such flow, and thus give the first global (in time) existence of vortex sheets motion with two-sign vorticity. We also consider the 3-D axisymmetric vortex motions. It is well-known that for smooth flows, the analysis for the axisymmetric 3-D Euler system without swirls is almost identitial to that of 2-D incompressible Euler system. However, this parallelness breaks down for the vortex sheets motions. Indeed, we will show that there exist no energy concentrations in approximate solutions to vortex sheets motion with one-sign vorticity for steady axisymmetric 3-D Euler system without swirls [7]. Some partial results on unsteady axisymmetric vortex sheets motion will also be discussed.

2.   The supersonic flow past a pointed body

A projectile moving in the air with supersonic speed, is governed by the inviscid compressible Euler systems

$$\begin{cases} \partial_t \rho + div(\rho u) = 0, \\ \partial_t(\rho u) + div(\rho u \otimes u) + \nabla p = 0, \end{cases} \tag{2.1}$$

where $\rho, u = (u_1, u_2, u_3)$ and p stand for the density, the velocity and the pressure respectively. We will only treat the polytropic gases so that $p = p(\rho) = A\rho^\gamma$ with gas constant $A > 0$ and $1 < \gamma < 3$, γ being the adiabatic exponent.

Suppose that there is a uniform supersonic flow $(u_1, u_2, u_3) = (0, 0, q_0)$ with constant density $\rho_0 > 0$ which comes from negative infinity. Then the flow can be described by the steady Euler system. If we assume further that the flow is irrotational, so that one can introduce a potential function Φ such that $u = \nabla\Phi$. Then the Bernoulli's law implies that $\rho = h^{-1}\left(\frac{1}{2}q_0^2 + h(\rho_0) - \frac{1}{2}|\nabla\Phi|^2\right) \equiv H(\nabla\Phi)$,

where $h(\rho)$ is the specific enthalpy defined by $h'(\rho) = \frac{p'(\rho)}{\rho}$. In this case, (2.1) is reduced to a second order quasilinear equation

$$div(H(\nabla\Phi)\nabla\Phi) = 0, \tag{2.2}$$

which can be verified to be strictly hyperbolic with respect to x_3 if $\partial_3\Phi > c$ with c being the sound speed given by $c^2(\rho) = p'(\rho)$. The flow hits a point body, whose surface is denoted by $m(x_1, x_2, x_3) = 0$. Since no flow can cross the boundary, the natural boundary condition is

$$u \cdot \nabla m \equiv \nabla \Phi \cdot \nabla m = 0 \quad \text{on} \quad m(x_1, x_2, x_3) = 0. \tag{2.3}$$

If the vertex angle of the tangential cone of the pointed body is less than a critical value, it is then expected that a shock front is attached at the tip of the pointed body. Denote by $\mu(x_1, x_2, x_3) = 0$ the equation of the shock front, then the Rankine-Hugoniot conditions become

$$\nabla \mu \cdot [H(\nabla \Phi) \nabla \Phi] = 0, \quad \Phi \ \text{is continuous,} \quad \text{on} \quad \mu(x_1, x_2, x_3) = 0. \tag{2.4}$$

Our aim is to find a solution to this free boundary value problem, (2.2)–(2.4). When the pointed body is small perturbation of a circular cone, the local existence of solution to the problem (2.2)–(2.4) has been established in [1]. Our main goal is to establish a global solution. However, such a global shock wave might not exist in general for arbitrary pointed body due to the possibility of development of new shock waves in the large. Thus we assume further that the pointed body is a curved and symmetric cone. In this case, it will be more convenient to rewrite the problem (2.2)–(2.4) in terms of polar coordinates (r, θ, z) with $r = \sqrt{x_1^2 + x_2^2}$ and $z = x_3$. Assume that the tip of the pointed body locates at the origin, the equation of the surface of the body is $r = b(z)$ with $b(0) = 0$, and the equation of the shock front is $r = S(z)$ with $S(0) = 0$. Set $\Phi = q_0 z + \varphi(r, z)$. Then (2.2)–(2.4) become

$$\left((q_0 + \partial_z \phi)^2 - c^2 \right) \partial_{zz}^2 \phi + \left((\partial_r \phi)^2 - c^2 \right) \partial_{rr}^2 \varphi + 2 \partial_r \varphi (q_0 + \partial_z \varphi) \partial_{rz}^2 \varphi - \frac{c^2}{r} \partial_r \varphi = 0, \tag{2.5}$$

$$-(q_0 + \partial_z \varphi) b'(z) + \partial_r \varphi = 0 \quad \text{on} \quad r = b(z), \tag{2.6}$$

$$-[(q_0 + \partial_z \varphi) H] S'(z) + [\partial_r \varphi H] = 0 \quad \text{on} \quad r = S(z). \tag{2.7}$$

Moreover, the potential $\varphi(r, z)$ is continuous on the shock, so it should satisfy $\varphi(S(z), z) = 0$. Then in [2], we have shown that the problem (2.5)–(2.7) has a globally defined solution as summarized in the following theorem:

Theorem 2.1 *Assume that a curved and symmetric cone is given such that $b(0) = 0$, $b'(0) = b_0$, $b^{(k)}(0) = 0$, $2 \leq k \leq k_1$, and*

$$|z^k \frac{d^k}{dz^k} (b(z) - b_0 z)| \leq \varepsilon_0 \quad \text{for} \quad 0 \leq k \leq k_2, \ z > 0 \tag{2.8}$$

with k_1 and k_2 being some suitable integers. Suppose that a supersonic polytropic flow parallel to the z-axis comes from negative infinite with velocity q_0, and density $\rho_0 > 0$. Then for suitably small ε_0, b_0 and q_0^{-1}, the boundary value problem (2.5)–(2.7) admits a global weak entropy solution with a pointed shock front attached at the origin. Moreover, the location of the shock front and the flow field between the shock and the surface of the body tend to the corresponding ones for the flow past the unperturbed circular cone $r = b_0 z$ with the rate $z^{-\frac{1}{4}}$.

It should be noted that there are no other discontinuities in our solution besides the main shock. Since the deviation of the surface of the body from that of a circular cone is sufficiently small (see (2.8)), any possible compression of the flow will be absorbed by the main shock. This is the mechanism to prevent the formation of any new shocks inside the flow field caused by the perturbation of the body. In particular, our results demonstrate that self-similar solution with a strong shock is structurally stable in a global sense. Indeed, the key element in the proof of Theorem 2.1 is to establish some global uniform weighted energy estimates for the linearized problem of (2.5)–(2.7) around the self-similar solution with a strong shock obtained when the pointed body is a circular cone. This is achieved by a deliberate choice of multipliers which must satisfy a system of ordinary differential inequalities with complicated coefficients due to the structure of the background self-similar solution and the requirement of obtaining global estimates independent of z for the potential function and its derivatives on the boundary as well as its interior of a domain.

When the projectile changes its speed, or it confronts some airstream, then the flow around the projectile will be time-dependent. Thus, we also consider the unsteady supersonic flow past a pointed body. Although our analysis applies to more general case [3], here we will only present the result for two-dimensional polytropic, unsteady, and irrotational flow past a curved wedge. For simplicity in presentation, we also assume that both the wedge and the perturbed incoming flow from infinity are symmetric about x_1-axis. Let $x_2 = b(x_1)$ with $b(0) = 0$ be the equation of the wedge, and $x_2 = S(t, x_1)$ with $S(t, x_1 = 0) = 0$ be the equation of the shock front. Then in terms of the velocity potential function ϕ (so that $u_1 = \partial_1 \phi$ and $u_2 = \partial_2 \phi$), we are looking for solutions to the following initial boundary value problem

$$\partial_t \left(H(-\phi_t + \frac{1}{2}|\nabla\phi|^2) \right) + \Sigma_{i=1}^2 \partial_{x_i} \left(\partial_{x_i} \phi\, H(-\partial_t \phi - \frac{1}{2}|\nabla\phi|^2) \right) = 0,$$

$$t > 0, x_2 > b_1(x_1), \tag{2.9}$$

$$\nabla\phi \cdot (b'(x_1), -1) = 0 \quad \text{on} \quad x_2 = b(x_1), \tag{2.10}$$

$$[H]\partial_t S + [H\,\partial_{x_1}\varphi]\partial_{x_1} S - [H\,\partial_{x_2}\phi] = 0 \quad \text{on} \quad x_2 = S(t, x_1), \tag{2.11}$$

$$S(0, x_1) = S_0(x_1), \quad \phi(0, x_1, x_2) = \phi_0(x_1, x_2), \quad \partial_t \phi(0, x_1, x_2) = 0, \tag{2.12}$$

where $S_0(x_1)$ and $\phi_0(x_1, x_2)$ are suitably small perturbations of the corresponding shock location and potential function respectively for the steady flow. Then we have the following local existence result [3].

Theorem 2.2 *There exist positive constant δ_1 and δ_2 and functions $S(t, x_1)$ and $\phi(t, x_1, x_2)$ defined on the regions $\{(t, x_1)|0 < t < \delta_1, 0 < x_1 < \delta_2\}$ and $\{(t, x_1, x_2)|0 < t < \delta_1, 0 < x_1 < \delta_2, b(x_1) < x_2 < S(t, x_1)\}$ respectively such that $(S(t, x_1), \varphi(t, x_1, x_2))$ solves the problem (2.9)–(2.12).*

3.   Prandtl's system for boundary layers

Consider a plane unsteady flow of viscous incompressible fluid in the presence of an arbitrary injection and removal of the fluid across the boundaries. In this case, the corresponding Prandtl's system takes the form

$$\partial_t u + u\,\partial_x u + v\,\partial_y u + \partial_x p = \nu\,\partial_{yy}^2 u, \quad \partial_x u + \partial_y v = 0, \tag{3.1}$$

in the region, $R = \{(x, y, t)|0 \le x \le L,\ 0 \le y < +\infty,\ 0 < t < T\}$, where ν, L and T are positive constants. The initial and boundary conditions can be imposed as

$$\begin{cases} u|_{t=0} = u_0(x, y),\ u|_{x=0} = u_1(y, t),\ u|_{y=0} = 0,\ v|_{y=0} = v_0(x, t), \\ \lim_{y \to +\infty} u(x, y, t) = U(x, t), \end{cases} \tag{3.2}$$

with $U = U(x, t)$ given as is determined by the corresponding Euler flow. The pressure $p = p(x, t)$ in (3.1) is determined by the Bernoulli's law: $\partial_t U + U\,\partial_x U + \partial_x p = 0$.

It follows from the physical ground that one may assume that

$$U(x, t) > 0,\ u_0(x, y) > 0,\ u_1(y, t) > 0,\ \text{and}\ v_0(x, t) \le 0. \tag{3.3}$$

Due to the degeneracy in the Prandtl's system (3.1), the problem of well-posedness theory of solutions to the problem (3.1)–(3.3) in the standard Hölder space or Sobolev space is quite difficult. In a series of important works by Oleinik and her coauthors [12], they studied this problem under the additional assumption that the data are monotonic in the sense that

$$\partial_y u_0(x, y) > 0,\ \text{and}\ \partial_y u_1(y, t) > 0, \tag{3.4}$$

and prove that there exists a unique local classical smooth solution to the initial-boundary value problem (3.1)–(3.3) provided that the data are monotonic in the sense of (3.4). Here by local we mean that T is small if L is given and fixed, and T is arbitrary if L is small. One of the open problem in [12] is: What are the conditions ensuring the global in time existence and uniqueness of solutions to the problem (3.1)–(3.4) for arbitrarily given L? In [16,17], we study such problem and establish the global (in time) existence and uniqueness of weak solutions to the initial-boundary value problem (3.1)–(3.4) in the case that the pressure is favorable, i.e.,

$$\partial_x p(x, t) \le 0,\ t > 0,\ 0 < x < L. \tag{3.5}$$

More precisely, we have ([16,17]):

Theorem 3.1 *Consider the initial-boundary value problem for the 2-D Prandtl's system, (3.1)–(3.2). Assume that the initial and boundary data satisfy the constraints (3.3), (3.4) and (3.5). Then there exists a unique global bounded weak solutions to the initial-boundary value problem (3.1)–(3.2). Furthermore, these solutions are Lipschitz continuous in both space and time.*

We remark here that the condition (3.5), that the pressure is favorable, is exactly what fluid-dynamists believe for the stability of a laminar boundary layer. This is also consistent with the case of stationary flows [12]. In the case of pressure adverse, i.e., $\frac{\partial p}{\partial x} > 0$, separation of boundary layer may occur, so one would not expect the long time existence of solution to (3.1)–(3.3). Finally, we remark that in the case that (3.5) fails as in many pratical physical situations, short time existence of regular solution is still expected, which has not been established yet.

4.   Vortex sheets motions

Two-dimensional vortex sheets motion corresponds to an inviscid flow whose vorticity is zero except on one-dimensional curves, the sheets. Thus, it is governed by the following Cauchy problem

$$\begin{cases} \partial_t w + u \cdot \nabla w = 0, \ u = K * w, \\ w(t = 0, x) = w_0(x) \in \mathcal{M}(\mathbb{R}^2) \cap H_{loc}^{-1}(\mathbb{R}^2), \end{cases} \tag{4.1}$$

where $u = (u_1, u_2)$ is the velocity field, $w = \nabla^\perp \cdot u$ is the vorticity, and K is the Biot-Sawart kernel, and $\mathcal{M}(\mathbb{R}^2)$ denotes the space of finite Radon measures in $\mathbb{R}^2$. Let $(u^\varepsilon, w^\varepsilon)(w^\varepsilon \equiv \nabla^\perp \cdot u^\varepsilon)$ be a sequence of approximate solution to (4.1) with the following bounds

$$\sup_{0 \le t \le T} \int_{\mathbb{R}^2} |w^2(x, t)| dx \le C_1(T), \quad \sup_{0 \le t \le T} \int_{|x| \le R} |u^\varepsilon(x, t)|^2 dx \le C_2(T, R). \tag{4.2}$$

Then there exist $u \in L^\infty \left(0, T, L^2_{loc}(\mathbb{R}^2)\right)$ and $w \in L^\infty \left(0, T, \mathcal{M}(\mathbb{R}^2)\right)$ with $w = \nabla^\perp \cdot u$ such that

$$w^\varepsilon \rightharpoonup w \text{ in } \mathcal{M}\left([0, T] \times \mathbb{R}^2\right), \ u^\varepsilon \rightharpoonup u \text{ in } L^2_{loc}\left([0, T] \times \mathbb{R}^2\right) \tag{4.3}$$

Two main questions arise: Is (u, w) a classical weak solution to problem (4.1)? Does either vorticity concentration (i.e. $\overline{\lim}_{\varepsilon, r \to 0^+} \int_{B_r} |w^\varepsilon(x, t)| dx > 0$) or energy defects ($\overline{\lim}_{\varepsilon \to 0} \int_{|x| \le R} |u^\varepsilon(x, t)|^2 dx > \int_{|x| \le R} |u(x, t)|^2 dx$ for some $t > 0$, $R > 0$) occur? It is clear from the structure of the 2-D Euler system that no energy defects implies strong L^2-convergence of the velocity field and thus the existence of the classical weak solution to (4.1). A less obvious fact is that no vorticity concentration also implies the weak limit (u, w) being a classical solution to (4.1), which follows from the vorticity formulation of (4.1) as observed by Delort [5]. Then the convergence to a classical weak solution to the Cauchy problem (4.1) is proved for approximate solutions generated by either regularizing the initial data [5], or Navier-Stokes approximation [11], or vortex blob methods [8], or point vortex methods [9] provided that the initial vorticity is of distinguished sign. To study the corresponding issues for flows where interactions of regions of both positive and negative vorticities are allowed, in [10], we study a 2-D mirror-symmetric vortex sheets motion whose vorticity is a integrable perturbation of a non-negative mirror-symmetric radon measure. Here a Radon measure μ is said to be non-negative mirror-symmetric (NMS) if $\mu|_{\mathbb{R}^2_+} \ge 0$ and μ is odd with respect to $x_1 = 0$. Then the main results in [10] can be summarized as

Theorem 4.1 *Assume that $w_0 \equiv \mu_1 + \mu_2$ such that $\mu_1 \in \mathcal{M}_c(\mathbb{R}^2) \cap H_{loc}^{-1}(\mathbb{R}^2) \cap NMS$ and $\mu_2 \in L_c^1(\mathbb{R}^2)$. Then there exists a global (in time) classical solution $(u,w) \in L^\infty\left(0,T,L^2(\mathbb{R}^2)\right) \otimes L^\infty\left(0,T,\mathcal{M}(\mathbb{R}^2)\right)$ to the initial value problem (4.1). Furthermore, this weak solution can be obtained as a limit of either a sequence of smooth inviscid solutions or a sequence of solutions to the Navier-Stokes system.*

It should be noted that this is the only result of existence of classical weak solution to (4.1) involving vorticities with different signs. This is proved by showing $\int_0^T \sup_{x_0 \in \mathbb{R}^2} \int_{B(x_0,\delta)} |w^\varepsilon(y,t)| dy\, dt \to 0$ as $\delta \to 0^+$ uniformly in ε, i.e., no vorticity concentration occur any where. Theorem 4.1 also indicates that interactions of regions of positive and negative vorticities without interwining may not cause concentration in vorticity. There remain many important open problems for the 2-D vortex sheets motion such as the existence of classical weak solution to (4.1) for general vortex sheets initial data, and whether energy defects occur dynamically even in the case of one-sign vorticity.

Finally, we consider 3-D axisymmetric vortex sheets motions. In cylindrical coordinate, (r,θ,z), axisymmetric solutions of 3-D Euler system have the form

$$u(x,t) = u^r(r,z,t)e_r + u^\theta(r,z,t)\, e_\theta + u^z(r,z,t)\, e_z, \quad p(x,t) = p(r,z,t) \qquad (4.4)$$

where $e_r = (\cos\theta, \sin\theta, 0)$, $e_\theta = (-\sin\theta, \cos\theta, 0)$, and $e_z = (0,0,1)$. The axisymmetric flow is said without swirls if $u^\theta \equiv 0$. In this case, the vorticity field is given by $w = \nabla \times u = w^\theta e^\theta$ with $w^\theta = \partial_r u^z - \partial_z u^r$, and $\tilde{D}_t(r^{-1} w^\theta) = 0$ with $\tilde{D}_t = \partial_t + u^r \partial_r + u^z \partial_z$. Thus, for smooth data, the theory for 3-D axisymmetric Euler system without swirls is almost parallel to that of 2-D Euler equation. However, this similarity breaks down for vortex sheets motions. Indeed, in sharp contrast to the 2-D case, we show in [7] that there are no energy defects for a suitable sequence of approximate solutions for 3-D steady axisymmetric vortex sheets motion with one-signed vorticity. Precisely, we have

Theorem 4.2 *Let $(u^\varepsilon, p^\varepsilon)$ be smooth axisymmetric solution to the 3-D steady Euler system: $(u^\varepsilon \cdot \nabla)u^\varepsilon + \nabla p^\varepsilon = f^\varepsilon$, $div\, u^\varepsilon = 0$, $x \in \mathbb{R}^3$, for some given axisymmetric function f^ε with $f^\varepsilon \rightharpoonup f$ weakly in $L^1(\mathbb{R}^3)$. Suppose further that*

$$(w^\varepsilon)^\theta \geq 0, \quad \sup_\varepsilon \int_{\mathbb{R}^3} |w^\varepsilon| dx < +\infty, \quad \sup_\varepsilon \int_{\mathbb{R}^3} |u^\varepsilon|^2 dx < +\infty \qquad (4.5)$$

where $w^\varepsilon = \nabla \times u^\varepsilon = (w^\varepsilon)^\theta\, e_\theta$. Let u be the weak limit of u^ε in $L^2(\mathbb{R}^3)$. Then u is a classical weak solution to $(u \cdot \nabla)u + \nabla p = f$, $div\, u = 0$, $x \in \mathbb{R}^3$. Moreover, there exists a subsequence $\{u^{\varepsilon j}\}$ of $\{u^\varepsilon\}$ such that $u^{\varepsilon j}$ converges to u strongly in $L^2(\mathbb{R}^3)$.

The proof of this theorem is based on a shielding method and the following fact which is valid for both steady and unsteady flows [7].

Theorem 4.3 *Let $\{u^\varepsilon\}$ be a sequence of approximate solutions for 3-D axisymmetric Euler system with general vortex-sheets data generated by either smoothing the initial data or Navier-Stokes approximations. Let A be the support of the defect measure associated with $u^\varepsilon \in L^2([0,T] \times \mathbb{R}^3)$. If $A \neq \phi$, then $A \cap \{(x,t) \in \mathbb{R}^3 \times [0,T] |\ r > 0\} \neq \phi$.*

Theorem 4.3 implies that if there are no energy defects away from the symmetry axis, then strong L^2-convergence takes place. It remains to study whether energy defects occur for unsteady axisymmetric 3-D Euler equations.

References

[1] S. Chen, Existence of stationary supersonic flow past a pointed body, *Arch. Rat. Mech. Aral.*, 156 (2001), 141–181.

[2] S. Chen, Z. P. Xin & H. Yin, Global shock waves for the supersonic flow past a perturbed cone, *Comm. Math. Phys.*, 2002 (to appear).

[3] S. Chen, Z. P. Xin & H. Yin, *Unsteady supersonic flow past a pointed body*, 2002 (Preprint).

[4] R. Courant & K. O. Friedrichs, *Supersonic Flow And Shock Waves*, Interscience Publisher Inc., New York, 1948.

[5] J. M. Delort, Existence de nappes de fourbill on en dimension deux, *J. Amer. Math. Soc.*, 4 (1991), 553–586.

[6] R. J. Diperna & A. Majda, Oscillations and concentrations in weak solutions of the imcompressible fluid equations, *Comm. Math. Phys.*, 108 (1987), 667–689.

[7] Q. S. Jiu & Z. P. Xin, *On strong convergence in vortex-sheets problem for 3-D axisymmetric Euler equations*, 2002 (Preprint).

[8] J. G. Liu & Z. P. Xin, Convergence of vortex methods for weak solutions to the 2-D Euler equations with vortex sheet data, *Comm. Pure Appl. Math.*, 48 (1995), 611–628.

[9] J. G. Liu & Z. P. Xin, Convergence of the point vortex method for 2-D vortex sheet, *Math. Comp.*, 70 (2000), 595–606.

[10] M. C. Lope Filho, H. J. N. Lopes & Z. P. Xin, Existence of vortex sheets with reflection symmetry in two space dimensions, *Arch. Rat. Mech. Anal.*, 158 (2000), 235–257.

[11] A. Majda, Remarks on weak solutions for vortex sheets with a distinguished sign, *Indiane Univ. Math. J.*, 42 (1993), 921–923.

[12] O. A. Oleinik & V. N. Samokhin, *Mathematical Models in Boundary Layer Theory*, Chapman-Hall, 1999.

[13] L. Prandtl, *Uber Flussigkeitsbeuegung bei sehr Kleiner Reibung*, Verbandlung des III Intern. Math-Kongresses, Heidelberg, 1904, 484–491.

[14] H. Schlinchting, *Boundary Layer Theory*, McGraw-Hall, 1987, 7th Edition.

[15] M. Sammartino & R. E. Caflisch, Zero viscosity limit for analytic solutions of Navier-Stokes equation I, *Comm. Math. Phys.*, 192 (1998), 433–461.

[16] Z. P. Xin & L. Zhang, On the global existence of solutions of Prandtl's system, *Advance in Math.*, 2002 (to appear).

[17] Z. P. Xin, L. Zhang & J. Zhao, *Global well-posedness for the two dimensional Prandtl's boundary layer equations*, 2002 (Preprint).

[18] Z. P. Xin, Some current topics in nonlinear conservation laws, *AMS/IP Studies in Advanced Mathematics*, 15 (2000), xiii–xxvii.

ICM 2002 · Vol. III · 861–871

A Numeraire-free and Original Probability Based Framework for Financial Markets[*]

Jia-An Yan[†]

Abstract

In this paper, we introduce a numeraire-free and original probability based framework for financial markets. We reformulate or characterize fair markets, the optional decomposition theorem, superhedging, attainable claims and complete markets in terms of martingale deflators, present a recent result of Kramkov and Schachermayer (1999, 2001) on portfolio optimization and give a review of utility-based approach to contingent claim pricing in incomplete markets.

2000 Mathematics Subject Classification: 60H30, 60G44.
Keywords and Phrases: Expected utility maximization, Fair market, Fundamental theorem of asset pricing (FTAP), Martingale deflator, Minimax martingale deflator, Optional decomposition theorem, Superhedging.

1. Introduction

A widely adopted setting for "arbitrage-free" financial markets is as follows: one models the price dynamics of primitive assets by a vector semimartingale, takes the saving account (or bond) as numeraire, and assumes that there exists an equivalent local martingale measure for the deflated price process of assets. According to the fundamental theorem of asset pricing (FTAP, for short), due to Kreps (1981) and Delbaen and Schachermayer (1994) if the deflated price process is locally bounded, this assumption is equivalent to the condition of "no free lunch with vanishing risk" (NFLVR for short). However, the property of NFLVR is not invariant under a change of numeraire. Moreover, under this setting, the market is "arbitrage-free" only for admissible strategies, the market may allow arbitrage for static trading

---

[*]The work was supported by the 973 project on mathematics of the Ministry of Science and Technology and the knowledge innovation program of the CAS. The author wishes to thank Dr. Jianming Xia for helpful comments.
[†]Academy of Mathematics & System Sciences, Chinese Academy of Sciences, Beijing 100080, China. E-mail: jayan@mail.amt.ac.cn

strategies with short-selling, and a pricing system using an equivalent local martingale measure may not be consistent with the original prices of some primitive assets. In order to remedy these drawbacks, Yan (1998) introduced the numeraire-free notions of "allowable strategy" and fair market. In this paper, we will further present a numeraire-free and original probability based framework for financial markets in a systematic way.

The paper is organized as follows: In Section 2, we introduce the semimartingale model, define the notion of martingale deflator. In Sction 3, we reformulate Kramkov's optional decomposition theorem in terms of martingale deflators, and give its applications to the superhedging of contingent claims and the characterizations of attainable claims and complete markets. In Section 4, we present a recent result of Kramkov and Schachermayer (1999, 2001) on optimal investment and give a review of utility-based approach to contingent claim pricing in incomplete markets.

2.   Semimartingale model and basic concepts

We consider a security market model in which the uncertainty and information structure are described by a stochastic basis $(\Omega, \mathcal{F}, P; (\mathcal{F}_t))$ satisfying the usual conditions with $\mathcal{F}_0$ being trivial. We call P the original (or objective) probability. It models the "real world" probability.

The market consists of d (primitive) assets whose price processes $(S_t^i), i = 1, \cdots, d$ are assumed to be non-negative semimartingales with initial values nonzero. We further assume that the process $\sum_{i=1}^{d} S_t^i$ is strictly positive and that each S_t^i vaniches on $[T^i, \infty)$, where $T^i(\omega) = \inf\{t > 0 : S_t^i(\omega) = 0, \text{ or } S_{t-}^i(\omega) = 0\}$ stands for the ruin time of the company issuing asset i. We will see later that this latter assumption is automatically satisfied for a fair market, since any non-negative supermartingale satisfies this property. In the literature, it was assumed that all primitive assets have strictly positive prices.

Let $S_t = (S_t^1, \cdots, S_t^d)$. Throughout the paper, we will use the following notation:

$$S_t^* = \left(\sum_{i=1}^{d} S_0^i \right)^{-1} \sum_{i=1}^{d} S_t^i.$$

By assumption, S_t^* is a strictly positive semimartingale. In the literature on mathematical finance, one often takes a primitive asset whose price never vanishes as numeraire. In our model, such a primitive numeraire asset may not exist. However, by our assumption on the model, we can always take S_t^* as numeraire.

2.1.   Self-financing strategy

A *trading strategy* is an R^d-valued $\mathcal{F}_t$-predictable process $\theta(t) = (\theta^1(t), \cdots, \theta^d(t))$, which is integrable w.r.t. the semimartingale S_t. Here $\theta^i(t)$ represents the numbers of units of asset i held at time t. The wealth $W_t(\theta)$ at time t of a trading strategy θ is $W_t(\theta) = \theta(t) \cdot S_t$, where $a \cdot b$ denotes the inner product of two vectors

a and b. A trading strategy θ is said to be *self-financing*, if

$$W_t(\theta) = W_0(\theta) + \int_0^t \theta(u)dS_u. \tag{2.1}$$

In this paper we use notation $\int_0^t H_u dX_u$ or $(H.X)_t$ to denote the integral of H w.r.t. X over the interval $(0,t]$. In particular, we have $(H.X)_0 = 0$.

The following theorem concerns a result on stochastic integrals of semimartingales, which represents an important property of self-financing strategies. It was given in Xia and Yan (2002).

Theorem 2.1 *Let X be an R^d-valued semimartingale and H an R^d-valued predictable process. If H is integrable w.r.t. X and*

$$H_t \cdot X_t = H_0 \cdot X_0 + \int_0^t H_s dX_s, \tag{2.2}$$

then for any real-valued semimartingale y, H is integrable w.r.t. yX and

$$y_t(H \cdot X)_t = y_0(H \cdot X)_0 + \int_0^t H_s d(yX)_s. \tag{2.3}$$

As a consequence of Theorem 2.1, we obtain the following

Theorem 2.2 *1) For any given R^d-valued S-integrable predictable process $\theta(t)$ and a real number x there exists a real-valued predictable process $\theta^*(t)$ such that $\{\theta^*(t)1_d + \theta(t)\}$ is a self-financing strategy with initial wealth x, where 1_d is the d-dimensional vector $(1, 1, \cdots, 1)$.*

2) A strategy θ is self-financing if and only if $d\widetilde{W}_t(\theta) = \theta(t)d\widetilde{S}_t$, where $\widetilde{S}_t = S_t(S_t^*)^{-1}, \widetilde{W}_t(\theta) = W_t(\theta)(S_t^*)^{-1}$.

2.2. Fair market and fundamental theorem of asset pricing

Now we consider a finite time horizon T. In Yan (1998), we introduced the notions of allowable strategy and fair market under assumption that all price processes of assets are strictly positive. The following definitions extend these notions to the present model.

Definition 2.1 *A strategy θ is said to be allowable, if it is self-financing and there exists a positive constant c such that the wealth $W_t(\theta)$ at any time t is bounded from below by $-cS_t^*$.*

Definition 2.2 *A market is said to be fair if there exists a probability measure Q equivalent to the original probability measure P such that the deflated price process $(\widetilde{S}_t)$ is a (vector-valued) Q-martingale.*

We call such a Q an *equivalent martingale measure* for the market. Throughout the sequel we denote by $\mathcal{Q}$ the set of all equivalent martingale measures.

If the market is fair, the deflated wealth process of any allowable strategy is a local Q-martingale, and consequently, is also a Q-supermartingale, for all $Q \in \mathcal{Q}$.

By the main theorem in Delbaen and Schachermayer (1994), Yan (1998) obtained an intrinsic characterization of fair markets. This result can be regarded as a numeraire-free version of the FTAP due to Kreps (1981) and Delbaen and Schachermayer (1994). The same result is valid for our more general model.

Theorem 2.3 *The market is fair if and only if there is no sequence (θ_n) of allowable strategies with initial wealth 0 such that $W_T(\theta_n) \geq -\frac{1}{n}S_T^*$ a.s., $\forall n \geq 1,$, and such that $W_T(\theta_n)$ a.s. tends to a non-negative random variable ξ satisfying $P(\xi > 0) > 0$.*

Remark If we take S_t^* as numeraire and consider the market in deflated terms, the condition in Theorem 2.3 is the NFLVR condition introduced in Delbaen and Schachermayer (1994).

2.3. Martingale deflators

In principle, we can take any strictly positive semimartingale as a numeraire, and its reciprocal as a deflator.

Definition 2.3 *A strictly positive semimartingale M_t with $M_0 = 1$ is called a martingale deflator for the market, if the deflated price processes $(S_t^i M_t), i = 1, \cdots, d$ are martingales under the original probability measure P.*

In the literature, such a deflator M is called "state price deflator". Here we propose to name it as "martingale deflator". A martingale deflator M is uniquely determined by its terminal value M_T. In fact, we have $M_t = (S_t^*)^{-1}E[M_T S_T^*|\mathcal{F}_t]$.

In terms of martingale deflators, a market is fair if and only if there exists a martingale deflator for the market.

Assume that the market is fair. We denote by $\mathcal{M}$ the set of all martingale deflators, and denote by $\mathcal{Q}$ the set of all equivalent martingale measures, when S_t^* is taken as numeraire. Note that there exists a one-to-one correspondence between $\mathcal{M}$ and $\mathcal{Q}$. If $M \in \mathcal{M}$, then $\frac{dQ}{dP} = M_T S_T^*$ define an element Q of $\mathcal{Q}$. If $Q \in \mathcal{Q}$, then we can define an element M of $\mathcal{M}$ with $M_T = \frac{dQ}{dP}(S_T^*)^{-1}$. If $\mathcal{M}$ (or $\mathcal{Q}$) contains only one element, the market is said to be complete. Otherwise, the market is said to be incomplete.

We will see in following sections that the use of martingale deflators instead of equivalent martingale measures has some advantages in handling financial problems.

3. Optional decomposition theorem and its applications

The optional decomposition theorem of Kramkov is a very useful tool in mathematical finance. It generalizes the classical Doob-Meyer decomposition theorem for supermartingales. This kind of decomposition was first proved by El Karoui and Quenez (1995), in which the process involved is the value process of a superheding strategy for a contingent claim in an incomplete market modelled by a diffusion process. Kramkov (1996) extended this result to the general semimartingale setting, but under the assumption that the underlying semimartingale is locally bounded and the supermartingale to be decomposed is non-negative and locally bounded. Föllmer and Kabanov (1998) removed any boundedness assumption. But in both papers, the theorem was formulated in the setting that there exists equivalent local martingale measures for the underlying semimartingales.

3.1. Optional decomposition theorem in terms of martingale deflators

Based on Theorem 2.1, Xia and Yan (2002) obtained the following version of the optional decomposition theorem in the equivalent martingale measure setting.

Theorem 3.1 *Let Y be a vector-valued semimartingale with non-negative components. Assume that the set $\mathcal{Q}$ of equivalent martingale measures for Y is nonempty. If X is a local $\mathcal{Q}$-supermartingale, i.e. local $\mathcal{Q}$-supermartingale for all $Q \in \mathcal{Q}$, then there exist an adapted, right continuous and increasing process C with $C_0 = 0$, and a Y-integrable predictable process φ such that*

$$X = X_0 + \varphi.Y - C.$$

Moreover, if X is non-negative, then $\varphi.Y$ is a local $\mathcal{Q}$-martingale.

The following theorem is a reformulation of Theorem 3.1 in terms of martingale deflators.

Theorem 3.2 *Assume that the market is fair. We denote by $\mathcal{M}$ the set of all martingale deflators. Let X be a semimartingale. If XM is a local supermartingale for all $M \in \mathcal{M}$, then there exist an adapted, right continuous and increasing process C with $C_0 = 0$, and an S-integrable predictable process φ such that*

$$X = X_0 + \varphi.S - C.$$

Moreover, if X is non-negative, then $(\varphi.S)M$ is a local martingale for all $M \in \mathcal{M}$.

Proof Let $\widetilde{S}_t = S_t(S_t^*)^{-1}$ and $\widetilde{X}_t = X_t(S_t^*)^{-1}$. Let $\mathcal{Q}$ denote the set of all martingale measures for $\widetilde{S}$. Then $\widetilde{X}$ is a local $\mathcal{Q}$-supermartingale. By Theorem 3.1 we have

$$\widetilde{X} = X_0 + \psi.\widetilde{S} - D,$$

where D is an adapted, right continuous and increasing process with $D_0 = 0$. By Theorem 2.2 there exists a real-valued predictable process $\theta^*(t)$ such that $\{\theta^*(t)1_d + \psi(t)\}$ is a self-financing strategy with initial wealth X_0. Since $\sum_{i=1}^d \widetilde{S}_t^i = \sum_{i=1}^d S_0^i$, we have $\theta^*(t)1_d.\widetilde{S} = 0$. Consequently,

$$X_0 + ((\theta^*1_d + \psi).S)_t = S_t^*(X_0 + ((\theta^*1_d + \psi).\widetilde{S})_t) = S_t^*(X_0 + (\psi.\widetilde{S})_t) = X_t + S_t^* D_t.$$

Put $\varphi = (\theta^* - D_-)1_d + \psi$ and $C = S^*.D$, we got the desired decomposition.

3.2. Superhedging

By a *contingent claim* (or *derivative*) we mean a non-negative $\mathcal{F}_T$-measurable random variable. Let ξ be a contingent claim. In general, one cannot find a self-financing strategy to perfectly replicate ξ. It is natural to raise the question: Does there exist an admissible strategy with the minimal initial value, called superhedging strategy, such that its terminal wealth is no smaller than the claim ξ? Here and henceforth, by an *admissible strategy* we mean a self-financing strategy with non-negative wealth process. For a market with diffusion model, this problem has been solved by El Karoui and Quenez (1995). For a general semimartingale model, it was solved by Kramkov (1996) and Föllmer and Kabanov(1998) using the optional decomposition theorem. The initial value of the superhedging strategy is called the *cost of superhedging* ξ. It can be considered as the "selling price" or "ask price " of ξ.

In a fair market setting, based on the corresponding result of Kramkov (1996), Xia and Yan (2002) proved the following result: if $\sup_{Q\in\mathcal{Q}} E_Q\left[(S_T^*)^{-1}\xi\right] < \infty$, then the cost at time t of superhedging the claim ξ is given by

$$U_t = \text{esssup}_{Q\in\mathcal{Q}} S_t^* E_Q\left[(S_T^*)^{-1}\xi\middle|\,\mathcal{F}_t\right]. \tag{3.1}$$

U is the smallest non-negative $\mathcal{Q}$-supermartingale with $U_T \geq \xi$. In terms of martingale deflators, we can rewrite (3.1) as

$$U_t = \text{esssup}_{M\in\mathcal{M}} M_t^{-1} E\left[M_T\xi\middle|\,\mathcal{F}_t\right]. \tag{3.2}$$

Using the optional decomposition theorem Föllmer & Leukert (2000) showed that the optional decomposition of a suitably modified claim gives a more realistic hedging (called *efficient hedging*) of a contingent claim. This result can be also reformulated in terms of martingale deflators.

3.3.   Attainable claims and completeness of the market

Xia and Yan (2002) introduced the notions of regular and strongly regular strategies. We reformulate them in terms of martingale deflators.

Definition 3.1 *A self-financing strategy ψ is said to be regular (resp. strongly regular), if for some (resp. for all) $M \in \mathcal{M}$, $W_t(\psi)M_t$ is a martingale. A contingent claim is said to be attainable if it can be replicated by a regular strategy.*

By Theorem 3.2, one can easily deduce the following characterizations for attainable claims and complete markets.

Theorem 3.3 *Let ξ be a contingent claim such that $\sup_{M\in\mathcal{M}} E[\xi M_T] < \infty$. Then ξ is attainable (resp. replicable by a strongly regular strategy) if and only if the above supremum is attained by an $M^* \in \mathcal{M}$ (resp. $E[M_T\xi]$ doesn't depend on $M \in \mathcal{M}$).*

Theorem 3.4 *The market is complete if only if any contingent claim ξ dominated by S_T^* is attainable, or equivalently, $E[M_T\xi]$ doesn't depend on $M \in \mathcal{M}$.*

4.   Portfolio optimization and contingent claim pricing

The portfolio optimization and contingent claim pricing and hedging are three major problems in mathematical finance. In a market where assets prices follow an exponential Lévy process, the portfolio optimization problem was studied in Kallsen

(2000). In the general semimartingale model, for utility functions U with effective domains $\mathcal{D}(U) = R_+$, the portfolio optimization problem was completely solved by Kramkov and Schachermayer(1999, 2001), henceforth K-S(1999, 2001). Bellini & Frittelli(2002) and Schachermayer (2002) studied the problem for utility functions U with $\mathcal{D}(U) = R$. The relationship between portfolio optimization and contingent claim pricing was studied in Frittelli(2000) and Goll & Rüschendorf(2001), among others. In what concerning the problem of hedging contingent claims, we refer the reader to Schweizer (2001) for quadratic hedging, Föllmer & Leukert (2000) for efficient hedging, and Delbaen et al. (2001) for exponential hedging.

In this section, under our framework, we will present the main results of K-S(1999, 2001) and give a review of utility-based approach to contingent claim pricing.

4.1. Expected utility maximization

We consider an agent whose objective is to choose a trading strategy to maximize the expected utility from terminal wealth at time T. In the sequel, we only consider such a utility function $U : (0, \infty) \longrightarrow R$, which is strictly increasing, strictly concave, continuously differentiable and satisfies $\lim_{x \downarrow 0} U'(x) = \infty, \lim_{x \to \infty} U'(x) = 0$. We denote by I the inverse function of U'. The conjugate function V of U is defined as

$$V(y) = \sup_{x > 0}[U(x) - xy] = U(I(y)) - yI(y), \; y > 0.$$

For $x > 0$, we denote by $\mathcal{A}(x)$ the set of all admissible strategies θ with initial wealth x. For $x > 0, y > 0$, we put

$$\mathcal{X}(x) = \{W(\theta) : \theta \in \mathcal{A}(x)\}, \qquad \mathcal{X} = \mathcal{X}(1),$$

$$\mathcal{Y} = \{Y \geq 0 : Y_0 = 1, \; YX \text{ is a supermartingale } \forall X \in \mathcal{X}\}, \qquad \mathcal{Y}(y) = y\mathcal{Y},$$

$$\mathcal{C}(x) = \{g \in L^0(\Omega, \mathcal{F}_T, P), 0 \leq g \leq X_T, \text{ for some } X \in \mathcal{X}(x)\}, \qquad \mathcal{C} \hat{=} \mathcal{C}(1),$$

$$\mathcal{D}(y) = \{h \in L^0(\Omega, \mathcal{F}_T, P), 0 \leq h \leq Y_T, \text{ for some } Y \in \mathcal{Y}(y)\}, \qquad \mathcal{D} \hat{=} \mathcal{D}(1).$$

The agent's optimization problem is:

$$\widehat{\psi}(x) = \arg \max_{\psi \in \mathcal{A}(x)} E\left[U(W_T(\psi))\right].$$

To solve this problem we consider two optimization problems (I) and (II):

$$\widehat{X}(x) = \arg \max_{X \subset \mathcal{X}(x)} E\left[U(X_T)\right]; \quad \widehat{Y}(y) = \arg \min_{Y \in \mathcal{Y}(y)} E\left[V(Y_T)\right].$$

Problem (II) is the dual of problem (I). Their value functions are

$$u(x) = \sup_{X \in \mathcal{X}(x)} E\left[U(X_T)\right], \qquad v(y) = \inf_{Y \in \mathcal{Y}(y)} E\left[V(Y_T)\right].$$

The following theorem is the reformulation of the main results of K-S(1999, 2001) under our framework.

Theorem 4.1 *Assume that there is a $\psi \in \mathcal{A}(1)$ such that $W_T(\psi) \geq K$ for a positive constant K (e.g., $S_T^* \geq K$). If $v(y) < \infty, \forall y > 0$, then the value functions $u(x)$ and $v(y)$ are conjugate in the sense that*

$$v(y) = \sup_{x>0}[u(x) - xy], \quad u(x) = \inf_{y>0}[v(y) + xy],$$

and we have:

1. For any $x > 0$ and $y > 0$, both optimization problems (I) and (II) have unique solutions $\widehat{X}(x)$ and $\widehat{Y}(y)$, respectively.

2. If $y = u'(x)$, then $\widehat{X}_T(x) = I(\widehat{Y}_T(y))$ and the process $\widehat{X}(x)\widehat{Y}(y)$ is a martingale.

3. $v(y) = \inf_{M \in \mathcal{M}} E\left[V(yM_T)\right]$.

Proof The proof is almost the same as that in K-S (1999, 2001). We indicate below main differences from K-S (1999, 2001). Obviously, $\mathcal{C}$ and $\mathcal{D}$ are convex sets. By Proposition 3.1 and a slight modification of Lemma 4.2 in K-S(1999), one can show that $\mathcal{C}$ and $\mathcal{D}$ are closed under the convergence in probability. For Items 1 and 2, as in Lemma 3.2 of K-S(1999) and Lemma 1 of K-S(2001), in order to prove the families $(V^-(h))_{h \in \mathcal{D}(y)}$ and $(U^+(g))_{g \in \mathcal{C}(x)}$ are uniformly integrable, we need to use a fact that $\mathcal{C}$ contains a positive constant. In our case, we have indeed $K \in \mathcal{C}$, since by assumption $K \leq W_T(\psi)$ for some $\psi \in \mathcal{A}(1)$. As for Item 3, according to Proposition 1 in K-S(2001) we only need to show $\widehat{\mathcal{D}} = \{M_T : M \in \mathcal{M}\}$ satisfies the following conditions:

- For any $g \in \mathcal{C}$, $\sup_{h \in \widehat{\mathcal{D}}} E[gh] = \sup_{h \in \mathcal{D}} E[gh]$
- $\widehat{\mathcal{D}} \subset \mathcal{D}$, $\widehat{\mathcal{D}}$ is convex and closed under countable convex combinations.

The first condition follows easily from (3.2), the second one is trivial.

4.2.   Utility-based approach to contingent claim pricing

Assume that the market is fair. Let ξ be a contingent claim such that $M_T\xi$ is integrable for some $M \in \mathcal{M}$. We put

$$V_t = (M_t)^{-1}E\left[M_T\xi \,|\, \mathcal{F}_t\right]. \qquad (4.1)$$

If we specify (V_t) as the price process of an asset generated by ξ, then the market augmented with this derivative asset is still fair, because M is still a martingale deflator for the augmented market. So we can define (V_t) as a "fair price process" of ξ. This pricing rule is consistent with the original price processes of primitive assets. However, if the market is incomplete (i.e., the martingale deflator is not unique) we cannot, in general, define uniquely the fair price process of a contingent claim.

In deflated terms, pricing of contingent claims in an incomplete market consists in choosing a reasonable martingale measure. There are several approaches to make such a choice. A well-known one is the so-called "utility-based approach". The basic idea of this approach is as follows. Assume that the representative agent in the market has preference represented by a utility function. In certain cases, the dual optimization problem (II) may produce a so called minimax martingale measure (MMM for short).

Now under our framework we show how the expected utility maximization problem is linked by duality to a martingale deflator. Assume that the solution $\widehat{Y}(y)$ of the dual optimization problem (II) lies in $y\mathcal{M}$. We put $\widehat{M}(y) = y^{-1}\widehat{Y}(y)$. Then $\widehat{M}(y) \in \mathcal{M}$, and we have

$$\widehat{M}(y) = \arg\min_{M\in\mathcal{M}} E\left[V(yM_T)\right].$$

We call $\widehat{M}(y)$ the *minimax martingale deflator*.

The following theorem gives a necessary and sufficient condition for the existence of the minimax martingale deflator.

Theorem 4.2 *Assume that there is a $\psi \in \mathcal{A}(1)$ such that $W_T(\psi) \geq K$ for a positive constant K (e.g., $S_T^* \geq K$), and that $v(y) < \infty$ for all $y > 0$. Let $x > 0$ be the agent's initial wealth and $M^* \in \mathcal{M}$. In order that $M^* \in \mathcal{M}$ is the minimax martingale deflator corresponding to the utility function U if and only if there exist $y > 0$ and $X^* \in \mathcal{X}(x)$ such that $X_T^* = I(yM_T^*)$ and $E[M_T^*X_T^*] = x$. If it is the case, then X^* solves the optimization problem (I).*

Proof We only need to prove the sufficiency of the condition. We have the following inequality

$$U(I(z)) \geq U(w) + z[I(z) - w], \quad \forall w > 0, z > 0.$$

If we replace z and w by yM_T^* and $X_T \in \mathcal{X}(x)$ and take expectation w.r.t. P, we get immediately that $E[U(X_T^*)] \geq E[U(X_T)]$ for all $X \in \mathcal{X}(x)$. This shows that X^* solves the optimization problem (I). On the other hand, since $X_T^* = I(yM_T^*)$ and the assumption $E[M_T^*X_T^*] = x$ implies that M^*X^* is a martingale, by Theorem 3.1, yM^* must solve the optimization problem (II). In particular, M^* is the minimax martingale deflator.

Now assume the minimax martingale deflator $\widehat{M}(y)$ exists. Let ξ be a contingent claim. If we use $\widehat{M}(y)$ to compute a fair price of ξ by (4.1), then it coincides with the fair price of Davis (1997), which is derived through the so-called "marginal rate of substitution" argument. In fact, the Davis' fair price of ξ is defined by

$$\hat{\pi}(\xi) = \frac{E[U'(\widehat{X}_T(x))\xi]}{u'(x)}.$$

Since $y = u'(x)$ and $U'(\widehat{X}_T(x)) = \widehat{Y}(y)$, we have $\hat{\pi}(\xi) = E[\widehat{M}_T(y)\xi]$.

Now we explain the economic meaning of Davis' fair price of a contingent claim. Let ξ be a contingent claim with $E[M_T(y)\xi] < \infty$. Put $\xi_t = (\widehat{M}_t(y))^{-1}E[\widehat{M}_T(y)\xi|\mathcal{F}_t]$. We augment the market with derivative asset ξ, and consider the portfolio maximization problem in the new market. Then it is easy to see that $\widehat{Y}(y)$ is still the solution of the dual optimization problem (II) in the new market. Consequently, the value function v and its conjugate function u remain unchanged. By Theorem 4.1, $\widehat{X}_T(x)$ solves again the optimization problem (I) in the new market. This shows that if the price of a contingent claim is defined by Davis' fair price, no trade on this contingent claim increases the maximal expected utility in comparison to an optimal trading strategy. This fact was observed in Goll and Rüschendorf (2001).

Note that in general the MMM (or minimax martingale deflator) depends on the agent's initial wealth x. This is a disadvantage of the utility-based approach to contingent claim pricing. However, for utility functions $\ln x, \frac{x^p}{p}, -e^{-x}$, where $p \in (-\infty, 1) \setminus \{0\}, \alpha > 0$, the MMM is independent of the agent's initial wealth x. This is due to the fact that the conjugate functions of the above utility functions are $-\ln x - 1, -\frac{p-1}{p} x^{\frac{p}{p-1}}, -x + x \ln x$, respectively, and that $E[dQ/dP] = 1$ for any equivalent martingale measure Q. Under our framework, the situation is a little different: for exponential utility function $U(x) = -e^{-x}$, the minimax martingale deflator depends still on the agent's initial wealth x.

For $U(x) = -e^{-x}$, the corresponding MMM is called the *minimal entropy martingale measure*. We refer the reader to Frittelli (2000), Miyahara (2001) and Xia & Yan (2000) for studies on the subject. If $U(x) = \ln x$, the minimax martingale deflator $\widehat{M}$, if it exists, is nothing but the reciprocal of the wealth process $\widehat{X}(1)$ of the growth optimal portfolio. Yan, Zhang & Zhang (2000) worked out explicit expressions for growth optimal portfolios in markets driven by a jump-diffusion-like process or by a Lévy process. See also Becherer (2001) for a study on the subject.

5.   Concluding remarks

We have introduced a numeraire-free and original probability based framework for financial markets. This framework has the following advantages: Firstly, it permits us to formulate financial concepts and results in a numeraire-free fashion. Secondly, since the original probability models the "real world" probability, one can investigate the martingale deflators by statistical methods using market data. Thirdly, using martingale deflators to deal with problems of pricing and hedging as well as portfolio optimization is sometimes more convenient than the use of equivalent martingale measures. Lastly, our framework includes the traditional one with deflated terms as a particular case. In fact, if the price process of one primitive asset is the constant 1, our framework is reduced to the traditional one.

References

[1] D. Becherer, The numeraire portfolio for unbounded semimartingales, *Finance and Stochastics*, 5(2001), 327–341.

[2] F. Bellni & M. Frittelli, On the existence of minimax martingale measures, *Mathematical Finace*, 12(2002), 1–21.

[3] M. Davis, Option pricing in incomplete markets, In: M. Dempster & S. Pliska (Eds.), *Math. of Derivative Securities*, Cambridge Univ. Press, 1997, 216–226.

[4] F. Delbaen, P. Grandits, Th. Rheinländer, D. Samperi, M. Schweizer, Ch. Stricker, Exponential hedging and entropic penalties, Preprint (2001).

[5] F. Delbaen & W. Schachermayer, A general version of the fundamental theorem of asset pricing, *Math. Annalen.*, 300 (1994), 463–520.

[6] F. Delbaen & W. Schachermayer, The fundamental theorem of asset pricing for unbounded stochastic processes, *Math. Annalen.*, 312(1998), 215–250.

[7] N. El Karoui & M.-C. Quenez, Denamic programming and pricing of contingent claims in an incomp,ete market, *SIAM J. Control Optim.*, 33(1995), 27–60.

[8] M. Frittelli, The minimal entropy martingale measure and the valuation problem in incomplete markets, *Mathematical Finace*, 10(2000), 39–52.

[9] H. Föllmer & Yu. M., Kabanov, Optional decomposition theorem and Lagrange multipliers, *Finance and Stochastics*, 2(1998), 69–81.

[10] H. Föllmer & P. Leukert, Efficient hedges: cost versus shortfall risk, *Finance and Stochastics*, 4(2000), 117–146.

[11] T. Goll & Rüschendorf, Minimax and minmal distance martingale measures and their relationship to portfolio optimization, *Finance and Stochastics*, 5(2001), 557–581.

[12] J. Kallsen, Optimal portfolios for exponential Lévy processes, *Mathematical Methods of Operations Research*, 51 (2000), 357–374.

[13] D. O. Kramkov, Optional decomposition of supermartingales and hedging contingent claims in incomplete security markets, *Probability Theory and Related Fields*, 105 (1996), 459–479.

[14] D. O. Kramkov & W. Schachermayer, The asymptotic elasticity of utility functions and optimal investment in incomplete markets, *The Annals of Applied Probability*, 9 (1999), 904–950.

[15] D. O. Kramkov & W. Schachermayer, Necessary and sufficient conditions in the problem of optimal investment in incomplete markets, Preprint, 2001.

[16] D. M. Kreps, Arbitrage and equilibrium in economies with infinitely many commodities, *J. Math. Econ.* 8 (1981), 15–35.

[17] Y. Miyahara, [Geometric Lévy processes & MEMM] pricing model and related estimation problem, *Asia-Pacific Financial Markets*, 8 (2001), 45–60.

[18] W. Schachermayer, A supermartingale property of the optimal portfolio process, Preprint, 2002.

[19] M. Schweizer, A guided tour through quadratic hedging approaches, in: E. Jouini et al. (Eds.), *Option Pricing, Interest Rates and Risk Management*, Camrige University Press (2001), 58–574.

[20] J. M. Xia & J. A. Yan, The utility maximization approach to a martingale measure constructed via Esscher transform, Preprint (2000).

[21] J. M. Xia & J. A. Yan, Some remarks on arbitrage pricing theory, in: J. Yong (Ed.), *Resent Developments in Mathematical Finance*, World Scientific Publisher, Singapore, 2002, 218–227.

[22] J. A. Yan, A new look at the fundamental theorem of asset pricing, *J. Korean Math. Soc.*, 35(1998), 659–673.

[23] J. A. Yan, Q. Zhang & S. Zhang, Growth optimal portfolio in a market driven by a jump-diffusion-like process or a Lévy process, *Annals of Economics and Finance*, 1 (2000), 101–116.

Section 18. Mathematics Education and Popularization of Mathematics

ICM 2002 · Vol. III · 875–884

Teaching Linear Algebra at University

J.-L. Dorier*

Abstract

Linear algebra represents, with calculus, the two main mathematical subjects taught in science universities. However this teaching has always been difficult. In the last two decades, it became an active area for research works in mathematics education in several countries. Our goal is to give a synthetic overview of the main results of these works focusing on the most recent developments. The main issues we will address concern:

- the epistemological specificity of linear algebra and the interaction with research in history of mathematics
- the cognitive flexibility at stake in learning linear algebra
- three principles for the teaching of linear algebra as postulated by G. Harel
- the relation between geometry and linear algebra
- an original teaching design experimented by M. Rogalski

2000 Mathematics Subject Classification: 97, 01, 15.
Keywords and Phrases: University teaching, Linear algebra, Curriculum, Vector space, Representation, Geometry, Cognitive flexibility, Epistemology.

1. Introduction

In most countries, science-orientated curricula in the first two years at university consist of courses in two main subjects, namely, calculus and linear algebra. The difficulties in these two fields are of different nature. Mathematics education research first developed works on calculus, but in the past 20 years, an increasing number of studies has been carried out about the teaching of linear algebra. One can distinguish roughly two main traditions in the teaching of linear algebra: one focuses on the study of formal vector spaces while the other proposes a more analytical approach based on the study of $\mathbb{R}^n$ and matrix calculus. Between these two orientations, there exist a continuum of teaching designs, in which each pole is more or less dominant. However, the teaching of linear algebra is universally

---

*IUFM de Lyon et équipe DDM, Laboratoire Leibniz, 46, ave. F. Viallet, 38 031 Grenoble Cedex, France. E-mail: Jean-Luc.Dorier@imag.fr

876   J.-L. Dorier

recognised as difficult. Students usually feel that they land on another planet, they
are overwhelmed by the number of new definitions and the lack of connection with
previous knowledge. On the other hand, teachers often feel frustrated and disarmed
when faced with the inability of their students to cope with ideas that they consider
to be so simple. Usually, they incriminate the lack of practice in basic logic and
set theory or the impossibility for the students to use geometrical intuition. These
complaints have a certain validity, but the few attempts at remedying this state of
affairs - with the teaching of Cartesian geometry or/and logic and set theory prior to
the linear algebra course - did not seem to improve the situation substantially. The
aim of this text is to give an account of the main trends in this area of mathematics
education research.

2.   Historical analyses

An epistemological analysis of the history of linear algebra is a way to reveal
some possible sources of students' difficulties as well an inspiration in the design of
activities for students. Several works have been carried out in this direction (see
[4], [6], [8], [17] and [24]). In this paper, we will give an account of only one of the
main result of this type of research. It concerns the last phase of the genesis of the
theory of vector spaces, whose roots can be found in the late nineteenth century,
but really started only after 1930. It corresponds to the axiomatisation of linear
algebra, that is to say a theoretical reconstruction of the methods of solving linear
problems, using the concepts and tools of a new axiomatic central theory. These
methods were operational but they were not explicitly theorised or unified. It is
important to realise that this axiomatisation did not, in itself, allow mathematicians
to solve new problems; rather, it gave them a more universal approach and language
to be used in a variety of contexts (functional analysis, quadratic forms, arithmetic,
geometry, etc.). The axiomatic approach was not an absolute necessity, except for
problems in non-denumerable infinite dimension, but it became a universal way of
thinking and organising linear algebra. Therefore, the success of axiomatisation
did not come from the possibility of reaching a solution to unsolved mathematical
problems, but from its power of generalisation and unification and, consequently, of
simplification in the search for methods for solving problems in mathematics.

As a consequence, one of the most noticeable difficulties encountered in the
learning of unifying and generalising concepts are associated with the pre-existing,
related elements of knowledge or competencies of lower level. Indeed, these need
to be integrated within a process of abstraction, which means that they have to
be looked at critically, and their common characteristics have to be identified, and
then generalised and unified. From a didactic point of view, the difficulty is that
any linear problem within the reach of a first year university student can be solved
without using the axiomatic theory. The gain in terms of unification, generalisation
and simplification brought by the use of the formal theory is only visible to the
expert.

One solution would be to give up teaching the formal theory of vector spaces.
However, many people find it important that students starting university mathe-

matics and science studies get some idea about the axiomatic algebraic structures of which vector space is one of the most fundamental. In order to reach this goal, the question of formalism cannot be avoided. Therefore, students have to be introduced to a certain type of reflection on the use of their previous elements of knowledge and competencies in relation with new formal concepts. This led Dorier, Robert, Robinet and Rogalski to introduce what they called 'meta level activities' (see [5], [6], [9], [11], [21] [22] and [23]). These activities are introduced and maintained by an explicit discourse on the part of the teacher about the significance of the introduced concepts for the general theory, their generalising and unifying character, the change of point of view or a theoretical detour that they offer, the types of general methods they lead to, etc. It hinges on the general attitude of the teacher who induces a constant underlying meta-questioning concerning new possibilities or conceptual gains provided by the use of linear algebra concepts, tools and methods.

3. Cognitive flexibility

One of the main difficulties in learning linear algebra has to do with the variety of languages, semiotic registers of representation, points of view and settings through which the objects of linear algebra can be represented. Students have to distinguish these various ways of representing objects of linear algebra, but they also need to translate from one to another type and, yet, not confuse the objects with their different representations. These abilities could be referred by the general notion of cognitive flexibility. This question is central in several works on the teaching and learning of linear algebra.

Students' difficulties with the formal aspect of the theory of vector spaces are not just a general problem with formalism but mostly a difficulty of understanding the specific use of formalism within the theory of vector spaces and the interpretation of the formal concepts in relation with more intuitive contexts like geometry or systems of linear equations, in which they historically emerged. Various diagnostic studies conducted by Dorier, Robert, Robinet and Rogalski pointed to a single massive obstacle appearing for all successive generations of students and for nearly all modes of teaching, namely, what these authors termed *the obstacle of formalism* (see [6], [7], [10] and [27]).

In [16], Hillel distinguished three basic languages used in linear algebra: the 'abstract language' of the general abstract theory, the 'algebraic language' of the $\mathbb{R}^n$ theory and the 'geometric language' of the two- and three-dimensional spaces. The 'opaqueness' of the representations seems to be ignored by lecturers, who constantly shift the notations and modes of description, without alerting the students in any explicit way. By far, the most confusing case for students is the shift from the abstract to the algebraic representation when the underlying vector space is $\mathbb{R}^n$. In this case, an n-tuple (or a matrix) is represented as another n-tuple (or matrix) relative to another basis. This confusion leads to persistent mistakes in students' solutions related to reading the values of a linear transformation given by a matrix in a basis (see [15]). Parallel with the three languages identified by Hillel, Sierpinska et al. distinguish three modes of thinking that have led to the development of these languages and are necessary for an understanding of the domain: the 'synthetic-geometric', 'analytic-arithmetic', and 'analytic-structural' (see [29]).

In [12], Duval defined *semiotic representations* as productions made by the use of signs belonging to a system of representation which has its own constraints of meaning and functioning. Semiotic representations are, according to him, absolutely necessary in mathematical activity, because its objects cannot be directly perceived and must, therefore, be represented. Moreover, semiotic representations play an essential role in developing mental representations, in accomplishing different cognitive functions (objectification, calculation, etc.), as well as in producing knowledge.

In her work, Pavlopoulou (see [8] pp. 247-252) applied and tested Duval's theory in the context of linear algebra. She distinguished between three registers of semiotic representation of vectors: the graphical register (arrows), the table register (columns of coordinates), and the symbolic register (axiomatic theory of vector spaces). Through several studies, she has shown that the question of registers, especially as regards conversion, is not usually taken into account either in teaching or in textbooks. She also identified a number of student's mistakes that could be interpreted as a confusion between an object and its representation (especially a vector and its geometrical representation) or as a difficulty in converting from one register to another.

The research of Alves-Dias (see [1] and [8] pp. 252-256), an extension of Pavlopoulou's, generalised the necessity of conversions from one semiotic register to another for the understanding of linear algebra to the necessity of 'cognitive flexibility'. Moreover, on the basis of Rogalski's previous work (see [25], [26], [27] and [28]), she focused her study on the question of articulation between the Cartesian and parametric representations of vector subspaces, which is not a mere question of change of register, but deals with more complex cognitive processes involving the use of concepts like rank and duality. Indeed, when a subspace V is represented by Cartesian equations, finding a parametric representation of V mostly consists in finding a set of generators of V, which is not just a change of register, nor an elementary cognitive process, even if it is much easier when the dimension 'd' of V is known. In any case, competencies with regard to the concept of rank and duality are indispensable. Moreover, in order to avoid easy mistakes in calculations or reasoning, it is necessary to be able to have some control over the results obtained. Alves Dias showed that in textbooks and classes, in general, the tasks offered to students are very limited in terms of flexibility. She developed a series of exercises that required the student to mobilise more changes of settings or registers and to exert explicit control via the concepts of rank and duality. Her experimentations demonstrated a variety of difficulties for the students. For instance, students often identified one type of representation exclusively through semiotic characteristics (a representation with x's and y's would be considered as obviously Cartesian) without questioning the meaning of the representation. Concerning the means of control over the validity of the statements by the students and anticipation of results or answers to problems, she found that a theorem like: $dimE = dimKerf + dimImf$, is known and used correctly by many students, but it is very seldom used for those purposes even in cases in which it would immediately bring up a contradiction with the result obtained, or in cases in which it offered valuable information in order to anticipate the correct answer.

In [15], Hillel and Sierpinska stressed that a linear algebra course which is theoretically rather than computationally framed requires a level of thinking that is based on what has been termed by Piaget and Garcia as the 'trans-object level of analysis' which consists in the building of conceptual structures out of what, at previous levels, were individual objects, actions on these objects, and transformations of both the objects and actions (see [18] p. 28). A similar claim was made by Harel in [14], in his assertions that a substantial range of mental processes must be encapsulated into conceptual objects by the time students get to study linear algebra. The difficulty of thinking at the trans-object level leads some students to develop 'defense mechanisms' (to 'survive' the course), consisting in trying to produce a written discourse formally similar to that of the textbook or of the lecture but without grasping the meaning of the symbols and the terminology. This appeared as a major problem for Sierpinska, Dreyfus and Hillel, and the team set out to design an entry into linear algebra that would make this behaviour or attitude less likely to appear in students (see [30]). The designed teaching-learning situations were set in a dynamic geometry environment (Cabri-geometry II) extended by several macro-constructions for the purposes of representing a two-dimensional vector space and its transformations (see [31] and [32]). Further analysis of the students' behaviour in the experimented situations led Sierpinska to postulate certain features of their thinking that could be held partly responsible for their erroneous understandings and difficulties in dealing with certain problems (especially the problem of extending a transformation of a basis to a linear transformation of the whole plane). She proposed that these features be termed "a tendency to think in 'practical' rather than 'theoretical' ways" (see [32]). The distinction between these two ways of thinking was inspired by the Vygotskian notion of scientific, as opposed to spontaneous or everyday concepts. The behaviour of students who were encountering difficulties in the experimentations suggested that their ways of thinking had the features of practical thinking rather than theoretical thinking. In particular they had trouble going beyond the appearance of the graphical and dynamic representations in Cabri that they were observing and manipulating: their relation to these representations was 'phenomenological' rather than 'analytic'. By far the most blatant feature of the students' practical thinking was their tendency to base their understanding of an abstract concept on 'prototypical examples' rather than on its definition. For example, linear transformations were understood as 'rotations, dilations, shears and combinations of these'. This way of understanding made it very difficult for them to see how a linear transformation could be determined by its value on a basis, and consequently, their notion of the matrix of a linear transformation remained at the level of procedure only.

4. Three principles for the teaching of linear algebra

In [14], Harel posits three 'principles' for the teaching of linear algebra, inspired by Piaget's psychological theory of concept development: the *Concreteness Principle*, the *Necessity Principle* and the *Generalisability Principle*.

The Concreteness Principle states, "For students to abstract a mathematical structure from a given model of that structure, the elements of that model must be conceptual entities in the student's eyes; that is to say, the student has mental procedures that can take these objects as inputs". This principle is violated whenever the general concept of vector space is taught as a generalisation from less abstract structures, to students who have not (yet) constructed the elements of these structures as mental entities on which other mental operations can be performed. Starting from the premise that students build their understanding of a concept in a context that is concrete to them, Harel conclude that a sustained emphasis on a geometric embodiment of abstract linear algebra concepts produce a quite solid basis for students' understanding. He insisted, however, that it would be incorrect to conclude that a linear algebra course should start with geometry and build the algebraic concepts through some kind of generalisation from geometry. A teaching experiment built on this premise allowed Harel to observe that when geometry is introduced before the algebraic concepts have been formed, many students remain in the restricted world of geometric vectors, and do not move up to the general case.

The Necessity Principle — For students to learn, they must see an (intellectual, as opposed to social or economic) need for what they are intended to be taught — is based on the Piagetian assumption (which has also been adopted by the Theory of Didactic Situations elaborated by Brousseau in [2]) that knowledge develops as a solution to a problem. If the teacher solves the problems for the students and only asks them to reproduce the solutions, they will learn how to reproduce teacher's solutions, not how to solve problems. Deriving the definition of vector space from a presentation of the properties of $\mathbb{R}^n$ is an example of a violation of the necessity principle.

The last, Generalisability Principle postulated by Harel, is concerned more with didactic decisions regarding the choice of teaching material than with the process of learning itself. "When instruction is concerned with a 'concrete' model, that is a model that satisfies the Concreteness Principle, the instructional activities within this model should allow and encourage the generalisability of concepts." This principle would be violated if the models used for the sake of concretisation were so specific as to have little in common with the general concepts they were aimed at. For example, the notion of linear dependence introduced in a geometric context defined through collinearity or co-planarity is not easily generalisable to abstract vector spaces. Harel's work inspired curriculum reform in the US (see [3]), as well as textbook authors (see [33]).

5.   Geometry and linear algebra

In [19], Robert, Robinet and Tenaud designed and experimented with a geometric entry into linear algebra. The aim was to overcome the obstacle of formalism by giving a more 'concrete' meaning to linear algebra concepts, in particular, through geometrical figures that could be used as metaphors for general linear situations in more elaborate vector spaces. However, as Harel noticed after them in his study mentioned above, the connection with geometry proved to be problematic. Firstly, geometry is limited to three dimensions and therefore some concepts, like rank, for instance, or even linear dependence, have a quite limited field of representation in the geometric context. Moreover, it is not rare that students refer to affine subspaces instead of vector subspaces when working on geometrical examples within linear algebra.

In her work, Gueudet-Chartier (see [13] and [8] pp. 262-264) conducted an epistemological study of the connection between geometry and linear algebra, using the evidence from both historical and modern texts. She found that the necessity of geometric intuition was very often postulated by textbooks or teachers of linear algebra. However, in reality, the use of geometry was most often very superficial. Moreover, some students would use geometrical representations or references in linear algebra, without this always being to their advantage. Indeed, some of them could not distinguish the affine space from the vector space structure; they also often could not imagine a linear transformation that would not be a geometric transformation. In other words, the geometrical reference acted as an obstacle to the understanding of general linear algebra. On the other hand, some very good students were found to use geometric references very rarely. They could operate on the formal level without using geometrical representations. It seems that the use of geometrical representations or language is very likely to be a positive factor, but it has to be controlled and used in a context where the connection is made explicit.

6.   An original teaching experiment

Most of the research conducted in France on the teaching and learning of linear algebra has been more or less directly connected with an experimental course implemented by Rogalski (see [9], [25], [26], [27] and [28]). This course was built on several interwoven and long-term strategies, using meta level activities as well as changes of settings (including intra-mathematical changes of settings), changes of registers and points of view, in order to obtain a substantial improvement in a sufficient number of students. 'Long-term strategy' (see [20]) refers to a type of teaching that cannot be divided into separate and independent modules. The long-term aspect is vital because the mathematical preparation and the changes in the 'didactic contract' (see [2]) have to operate over a period which is long enough to be efficient for the students, in particular as regards assessment. Moreover, the long term strategy refers to the necessity of taking into account the non-linearity of the teaching due to the use of change in points of view, implying that a subject is (re)visited several times in the course of the year.

Rogalski's teaching design has the following main characteristics:

- In order to take into account the specific epistemological nature of the concepts, some activities are introduced, at a favourable and precise time of the teaching, in order to induce a reflection on a 'meta' level.
- A fairly long preliminary phase precedes the actual teaching of elementary

concepts of linear algebra. It prepares the students to understand, through 'meta' activities, the unifying role of these concepts.

- As much as possible, changes of settings and points of view are used explicitly and are discussed.
- Finally, the concept of rank is given a central position in this teaching.

For a long-term teaching design, it is difficult to choose the time suitable for its evaluation, as interference may occur due to students' own organisation of their time and work, in a way that cannot be kept under control. Thus, phenomena of maturing, depending on students' level of involvement (which varies during the year) are difficult to take into account in the evaluation of the teaching. Moreover, such a global teaching design cannot be evaluated by usual comparative analyses, because the differences with the standard course are too important. However, internal evaluations have been conducted, showing several positive effects, even if some questions remain open.

7.   Conclusions

Mathematics education research cannot give a miraculous solution to overcome all the difficulties in learning and teaching linear algebra. Various works have consisted in diagnoses of students' difficulties, epistemological analyses and experimental teaching, offering local remediation. Nevertheless, these works lead to new questions, problems and difficulties. Yet, this should not be interpreted as a failure. Improving the teaching and learning of mathematics cannot consist in one remediation valid for all. cognitive processes and mathematics are far too complex for such an idealistic simplistic view. It is a deeper knowledge of the nature of the concepts, and the cognitive difficulties they enclose, that helps teachers make their teaching richer and more expert; not in a rigid and dogmatic way, but with flexibility. In this sense, in several countries, mathematics education research has influenced curriculum reforms, in an non-formal way, or sometimes very officially, like in North-America through the Linear Algebra Curriculum Study Group (see [3]).

References

[1] M. Alves Dias & M. Artigue, Articulation Problems between Different Systems of Symbolic Representations in Linear Algebra, in *The Proceedings of PME19*, Universidade Federal de Pernambuco, Recife, Brazil, 1995, Volume 2, 34–41.

[2] G. Brousseau, *Theory of Didactic Situations in Mathematics*, Kluwer Academic Publishers, Dordrecht, 1997.

[3] D. Carlson, C. Johnson, D. Lay,. & A. Porter, The Linear Algebra Curriculum Study Group Recommendations for the First Course in Linear Algebra, *College Mathematics Journal*, 24 (1993), 41–46.

[4] J.-L. Dorier, A General Outline of the Genesis of Vector Space Theory, *Historia Mathematica*, 22.3 (1995), 227–261.

[5] J.-L. Dorier, Meta Level in the Teaching of Unifying and Generalizing Concepts in Mathematics, *Educational Studies in Mathematics*, 29.2 (1995), 175–197.

[6] J.-L. Dorier (Ed.), *L'Enseignement de l'Algèbre Linéaire en Question*, La Pensée Sauvage éditions, Grenoble, 1997.

[7] J.-L. Dorier, The Role of Formalism in the Teaching of the Theory of Vector Spaces. *Linear Algebra and its Applications* (275), 1.4 (1998), 141–160.

[8] J.-L. Dorier, *On the Teaching of Linear Algebra*, Kluwer Academic Publishers, Dordrecht, 2000.

[9] J.-L. Dorier, A. Robert, J. Robinet & M. Rogalski, On a Research Program about the Teaching and Learning of Linear Algebra in First Year of French Science University, *International Journal of Mathematical Education in Sciences and Technology*, 31.1 (2000), 27–35.

[10] J.-L. Dorier, A. Robert, J. Robinet & M. Rogalski, The Obstacle of Formalism in Linear Algebra, in J-L. Dorier (Ed.), *On the Teaching of Linear Algebra*, Kluwer Academic Publishers, Dordrecht, 2000, 85–94.

[11] J.-L. Dorier, A. Robert, J. Robinet & M. Rogalski, The Meta Lever, in J-L. Dorier (Ed.), *On the Teaching of Linear Algebra*, Kluwer Academic Publishers, Dordrecht, 2000, 151–176.

[12] R. Duval, *Semiosis et Pensée Humaine. Registres Sémiotiques et Apprentissages Intellectuels*, Peter Lang, Bern, 1995.

[13] G. Gueudet-Chartier, *Rôle du Géométrique dans l'Enseignement et l'Apprentissage de l'Algèbre Linéaire*, Thèse de Doctorat de l'Université Joseph Fourier, Laboratoire Leibniz, Grenoble, 2000.

[14] G. Harel, Principles of Learning and Teaching Mathematics, With Particular Reference to the Learning and Teaching of Linear Algebra: Old and New Observations, in J-L. Dorier (Ed.), *On the Teaching of Linear Algebra*, Kluwer Academic Publishers, Dordrecht, 2000, 177–189.

[15] J. Hillel & A. Sierpinska, On One Persistent Mistake in Linear Algebra, in *The Proceedings PME 18*, University of Lisbon, Portugal, 1994, 65–72.

[16] J. Hillel. Modes of Description and the Problem of Representation in Linear Algebra. in J-L. Dorier (Ed.), *On the Teaching of Linear Algebra*, Kluwer Academic Publishers, Dordrecht, 2000, 191–207.

[17] G. Moore, The Axiomatization of Linear Algebra, *Historia Mathematica*, 22.3, 1995, 262–303.

[18] J. Piaget & R. Garcia, *Psychogenesis and the History of Science, Columbia University Press*, New York, 1983.

[19] A. Robert, J. Robinet & I. Tenaud, *De la Géométrie à l'Algèbre Linéaire* - Brochure 72, IREM de Paris VII, 1987.

[20] A. Robert, Projets Longs et Ingéniorie pour l'Enseignement Universitaire: Questions de Problématique et de Méthodologie. Un exemple: un Enseignement Annuel de Licence en Formation Continue, *Recherches en Didactique des Mathématiques*, 12.2/3 (1992), 181–220.

[21] A. Robert, & J. Robinet, Prise en Compte du Méta en Didactique des Mathématiques, *Recherches en Didactique des Mathématiques*, 16.2 (1996), 145–176.

[22] A. Robert, Outils d'Analyse des Contenus Mathématiques Enseignés au Lycée et l'Université, *Recherches en Didactique des Mathématiques*, 18.2 (1998), 191–230.

[23] A. Robert, Level of Conceptualization and Secondary School Mathematics Education, in J-L. Dorier (Ed.), *On the Teaching of Linear Algebra*, Kluwer Academic Publishers, Dordrecht, 2000, 125–131.

[24] J. Robinet, *Esquisse d'une Genèse des Concepts d'Algèbre Linéaire - Cahier de Didactique des Mathématiques n29*, IREM de Paris VII, 1986.

[25] M. Rogalski, *Un Enseignement de l'Algèbre Linéaire en DEUG A Première Année - Cahier de Didactique des Mathématiques n53*, IREM de Paris 7, 1991.

[26] M. Rogalski. L'Enseignement de l'Algèbre Linéaire en Première Année de DEUG A, *La Gazette des Mathématiciens*, 60 (1994), 39–62.

[27] M. Rogalski, Teaching Linear Algebra: Role and Nature of Knowledge in Logic and Set Theory which Deal with Some Linear Problems, in *The Proceedings PME 20*, Valencia Universidad,. Spain, 1996, Volume 4, 211–218.

[28] M. Rogalski, The Teaching Experimented in Lille, in J-L. Dorier (Ed.), *On the Teaching of Linear Algebra*, Kluwer Academic Publishers, Dordrecht, 2000, 133–149.

[29] A. Sierpinska, A. Defence, T. Khatcherian & L. Saldanha, A propos de trois modes de raisonnement en algèbre linéaire, in J.-L. Dorier (Ed.), *L'Enseignement de l'Algèbre Linéaire en Question*, La Pensée Sauvage éditions, Grenoble, 1997, 249–268.

[30] A. Sierpinska, T. Dreyfus, & J. Hillel, Evaluation of a Teaching Design in Linear Algebra: The Case of Linear Transformations, *Recherches en Didactique des Mathématiques*, 19.1 (1999), 7–40.

[31] A. Sierpinska, J. Trgalova, J. Hillel, & T. Dreyfus, Teaching and Learning Linear Algebra with Cabri. Research Forum paper, in *The Proceedings of PME 23*, Haifa University, Israel, 1999, Volume 1, 119–134.

[32] A. Sierpinska, On Some Aspects of Students' Thinking in Linear Algebra. in J-L. Dorier (Ed.), *On the Teaching of Linear Algebra*, Kluwer Academic Publishers, Dordrecht, 2000, 209–246.

[33] F. Uhlig, *Transform Linear Algebra*, Prentice-Hall, Upper Saddle River, 2002.

ICM 2002 · Vol. III · 885–895

Popularizing Mathematics: From Eight to Infinity

V. L. Hansen*

Abstract

It is rare to succeed in getting mathematics into ordinary conversation without meeting all kinds of reservations. In order to raise public awareness of mathematics effectively, it is necessary to modify such attitudes. In this paper, we point to some possible topics for general mathematical conversation.

2000 Mathematics Subject Classification: 00A05, 00A80, 97B20.
Keywords and Phrases: Mathematics in literature, Symbols, Art and design, Mathematics in nature, Geometry of space, Infinite sums, Mathematics as a sixth sense.

1.   Introduction

"One side will make you grow taller, and the other side will make you grow shorter", says the caterpillar to Alice in the fantasy *Alice's Adventures in Wonderland* as it gets down and crawls away from a marvellous mushroom upon which it has been sitting. In an earlier episode, Alice has been reduced to a height of three inches and she would like to regain her height. Therefore she breaks off a piece of each of the two sides of the mushroom, incidentally something she has difficulties identifying, since the mushroom is entirely round. First Alice takes a bit of one of the pieces and gets a shock when her chin slams down on her foot. In a hurry she eats a bit of the other piece and shoots up to become taller than the trees. Alice now discovers that she can get exactly the height she desires by carefully eating soon from one piece of the mushroom and soon from the other, alternating between getting taller and shorter, and finally regaining her normal height.

Few people link this to mathematics, but it reflects a result obtained in 1837 by the German mathematician Dirichlet, on what is nowadays known as *conditionally convergent infinite series* — namely, that one can assign an arbitrary value to infinite sums with alternating signs of magnitudes tending to zero, by changing the order in

*Department of Mathematics, Technical University of Denmark, Matematiktorvet, Bygning 303, DK-2800 Kgs. Lyngby, Denmark. E-mail: V.L.Hansen@mat.dtu.dk

which the magnitudes are added. Yes, mathematics can indeed be fanciful, and as a matter of fact, *Alice's Adventures in Wonderland* was written, under the pseudonym Lewis Carroll, by an English mathematician at the University of Oxford in 1865. The fantasy about Alice is not the only place in literature where you can find mathematics. And recently, mathematics has even entered into stage plays [18] and movies [19].

2. On the special position of mathematics

When the opportunity arises, it can be fruitful to incorporate extracts from literature or examples from the arts in the teaching and dissemination of mathematics. Only in this way can mathematics eventually find a place in relaxed conversations among laymen without immediately being rejected as incomprehensible and relegated to strictly mathematical social contexts.

Mathematics occupies a special position among the sciences and in the educational system. This position is determined by the fact that mathematics is an *a priori* science building on ideal elements abstracted from sensory experiences, and at the same time mathematics is intimately connected to the experimental sciences, traditionally not least the natural sciences and the engineering sciences. Mathematics can be decisive when formulating theories giving insight into observed phenomena, and often forms the basis for further conquests in these sciences because of its power for deduction and calculation. The revolution in the natural sciences in the 1600s and the subsequent technological conquests were to an overwhelming degree based on mathematics. The unsurpassed strength of mathematics in the description of phenomena from the outside world lies in the fascinating interplay between the concrete and the abstract.

In the teaching of mathematics, and when explaining the essence of mathematics to the public, it is important to get the abstract structures in mathematics linked to concrete manifestations of mathematical relations in the outside world. Maybe the impression can then be avoided that abstraction in mathematics is falsely identified with pure mathematics, and concretization in mathematics just as falsely with applied mathematics. The booklet [3] is a contribution in that spirit.

3. On the element of surprise in mathematics

Without a doubt, mathematics makes the longest-lasting impression when it is used to explain a counter-intuitive phenomenon. The element of surprise in mathematics therefore deserves particular attention.

As an example, most people — even teachers of mathematics — find it close to unbelievable that a rope around the Earth along the equator has to be only 2π metres (i.e., a little more than 6 metres) longer, in order to hang in the air 1 metre over the surface all the way around the globe.

Another easy example of a surprising fact in mathematics can be found in connection with figures of constant width. At first one probably thinks that this property is restricted to the circle. But if you round off an equilateral triangle by substituting each of its sides with the circular arc centred at the opposite corner, one obtains a figure with constant width. This rounded triangle is called a *Reuleaux triangle* after its inventor, the German engineer Reuleaux. It has had several technological applications and was among others exploited by the German engineer Wankel in his construction of an internal combustion engine in 1957. Corresponding figures with constant width can be constructed from regular polygons with an arbitrary odd number of edges. In the United Kingdom, the regular heptagon has been the starting point for rounded heptagonal coins (20p and 50p).

Figures of constant width were also used by the physicist Richard Feynmann to illustrate the dangers in adapting figures to given measurements without knowledge of the shape of the figure. Feynmann was a member of the commission appointed to investigate the possible causes that the space shuttle Challenger on its tenth mission, on 28 January 1986, exploded shortly after take off. A problematic adaption of figures might have caused a leaky assembly in one of the lifting rockets.

4. Mathematics in symbols

Mathematics in symbols is a topic offering good possibilities for conversations with a mathematical touch. Through symbols, mathematics may serve as a common language by which you can convey a message in a world with many different ordinary languages. But only rarely do you find any mention of mathematics in art catalogues or in other contexts where the symbols appear. The octagon, as a symbol for eternity, is only one of many examples of this.

The eternal octagon

Once you notice it, you will find the octagon in very many places - in domes, cupolas and spires, often in religious sanctuaries. This representation is strong in Castel del Monte, built in the Bari province in Italy in the first half of the 1200s for the holy Roman emperor Friedrich II. The castle has the shape of an octagon, and at each of the eight corners there is an octagonal tower [9]. The octagon is also found in the beautiful Al-Aqsa Mosque in Jerusalem from around the year 700, considered to be one of the three most important sites in Islam.

In Christian art and architecture, the octagon is a symbol for the *eighth day* (in Latin, *octave dies*), which gives the day, when the risen Christ appeared to his disciples for the second time after his resurrection on Easter Sunday. In the Jewish counting of the week, in use at the time of Christ, Sunday is the first day of the week, and hence the eighth day is the Sunday after Easter Sunday. In a much favoured interpretation by the Catholic Fathers of the Church, the Christian Sunday is both the first day of the week and the eighth day of the week, and every Sunday is a celebration of the resurrection of Christ, which is combined with the hope of eternal life. Also in Islam, the number eight is a symbol of eternal life.

Such interpretations are rooted in old traditions associated with the number eight in oriental and antique beliefs. According to Babylonian beliefs, the soul wanders after physical death through the seven heavens, which corresponds to the material heavenly bodies in the Babylonian picture of the universe, eventually to reach the eighth and highest heaven. In Christian adaptation of these traditions, the number eight (octave) becomes the symbol for the eternal salvation and fuses with eternity, or infinity. This is reflected in the mathematical symbol for infinity, a figure eight lying down, used for the first time in 1655 by the English mathematician John Wallis.

The regular polyhedra

Polyhedra fascinate many people. A delightful and comprehensive study of polyhedra has been made by Cromwell [4].

Already Pythagoras in the 500s BC knew that there can exist only five types of regular polyhedra, with respectively 4 (tetra), 6 (hexa), 8 (octa), 12 (dodeca) and 20 (icosa) polygonal lateral faces. In the dialogue *Timaeus*, Plato associated the *tetrahedron*, the *octahedron*, the *hexahedron* [cube] and the *icosahedron* with the four elements in Greek philosophy, fire, air, earth, and water, respectively, while in the *dodecahedron*, he saw an image of the universe itself. This has inspired some globe makers to represent the universe in the shape of a dodecahedron.

It is not difficult to speculate as to whether there exist more than the five regular polyhedra. Neither is it difficult to convince people that no kinds of experiments are sufficient to get a final answer to this question. It can only be settled by a mathematical proof. A proof can be based on the theorem that the alternating sum of the number of vertices, edges and polygonal faces in the surface of a convex polyhedron equals 2, stated by Euler in 1750; see e.g. [10].

The cuboctahedron

There are several other polyhedra that are ascribed a symbolic meaning in various cultures. A single example has to suffice.

If the eight vertices are cut off a cube by planes through the midpoints in the twelve edges in the cube one gets a polyhedron with eight equilateral triangles and six squares as lateral faces. This polyhedron, which is easy to construct, is known as the *cuboctahedron*, since dually it can also be constructed from an octahedron by cutting off the six vertices in the octahedron in a similar manner. The cuboctahedron is one of the thirteen semiregular polyhedra that have been known since Archimedes. In Japan cuboctahedra have been widely used as decorations in furniture and buildings. Lamps in the shape of cuboctahedra were used in Japan already in the 1200s, and they are still used today in certain religious ceremonies in memory of the dead [16].

The yin-yang symbol

Yin and yang are old principles in Chinese cosmology and philosophy that represent the dark and the light, night and day, female and male. Originally yin indicated a northern hill side, where the sun does not shine, while yang is the south side of the hill. Everything in the world is viewed as an interaction between yin and yang, as found among others in the famous oracular book *I Ching* (*Book of Changes*), central in the teaching of Confucius (551–479 BC). In this ancient Chinese system of divination, an oracle can be cast by flipping three coins and the oracle is one of sixty-four different hexagrams each composed of two trigrams. The three lines in a trigram are either straight (yang) or broken (yin), thus giving eight different trigrams and sixty-four hexagrams. In Taoism, the eight trigrams are linked to immortality. Yin and Yang as philosophical notions date back to the 400s BC.

The mathematical symbol for yin-yang is a circle divided into two equal parts by a curve made up of two smaller semicircles with their centres on a diameter of the larger circle. The mathematics in this beautiful and very well known symbol is simple, but nevertheless it contains some basic geometrical forms, and thereby offers possibilities for mathematical conversations.

The Borromean link

Braids, knots and links form a good topic for raising public awareness of mathematics, as demonstrated in the CD-Rom [2].

Here we shall only mention the Borromean link. This fascinating link consists of a system of three interlocking rings, in which no pair of rings interlocks. In other words, the three rings in a Borromean link completely falls apart if any one of the rings is removed from the system.

The Borromean link is named after the Italian noble family Borromeo, who gained a fortune by trade and banking in Milano in the beginning of the 1400s. In the link found in the coat of arms of the Borromeo family, the three rings apparently interlock in the way described, but a careful study reveals that the link often has been changed so that it does not completely fall apart if an arbitrary one of the rings is removed from the system. The link in the coat of arms of the Borromeo family is a symbol of collaboration.

5. Mathematics in art and design

Artists, architects and designers give life to abstract ideas in concrete works of art, buildings, furniture, jewellery, tools for daily life, or, by presenting human activities and phenomena from the real world, dynamically in films and television. The visual expressions are realized in the form of paintings, images, sculptures, etc. From a mathematical point of view, all these expressions represent geometrical figures.

For a mathematician, the emphasis is on finding the abstract forms behind the concrete figures. In contrast, the emphasis of an artist or a designer is to realize the abstract forms in concrete figures. At the philosophical level many interesting conversations and discussions about mathematics and its manifestations in the visual arts can take place following this line of thought.

More down to earth, the mathematics of perspective invites many explanations of concrete works of arts [8], and the beautiful patterns in Islamic art inspire discussions on geometry and symmetry [1]. In relation to the works of the Dutch artist Escher, it is possible to enter into conversations about mathematics at a relatively advanced level, such as the Poincaré disc model of the hyperbolic plane [5], [7]. The sculptures of the Australian-British artist John Robinson — such as his wonderful sculpture *Immortality* that exhibits a Möbius band shaped as a clover-leaf knot - offer good possibilities for conversations on knots [2].

In his fascinating book [17], Wilson tells many stories of mathematics in connection with mathematical motifs on postage stamps.

6.   Mathematics in nature

The old Greek thinkers — in particular Plato (427–347 BC) — were convinced that nature follows mathematical laws. This belief has dominated thinking about natural phenomena ever since, as witnessed so strongly in major works by physicists, such as Galileo Galilei in the early 1600s, Isaac Newton in his work on gravitation in *Principia Mathematica* 1687, and Maxwell in his major work on electromagnetism in 1865; cf. [10]. In his remarkable book *On Growth and Form* (1917), the zoologist D'Arcy Wentworth Thompson concluded that wherever we cast our glance we find beautiful geometrical forms in nature. With this point of departure, many conversations on mathematics at different levels are possible. I shall discuss a few phenomena that can be described in broad terms without assuming special mathematical knowledge, but nevertheless point to fairly advanced mathematics.

Spiral curves and the spider's web

Two basic motions of a point in the Euclidean plane are motion in a line, and periodic motion (rotation) around a central point. Combining these motions, one gets spiral motions along *spiral curves* around a *spiral point.* Constant velocity in both the linear motion and the rotation gives an *Archimedean spiral* around the spiral point. Linear motion with exponentially growing velocity and rotation with constant velocity gives a *logarithmic spiral.*

A logarithmic spiral is also known as an *equiangular spiral,* since it has the following characteristic property: *the angle between the tangent to the spiral and the line to the spiral point is constant.*

Approximate Archimedean and logarithmic spirals enter in the spider's construction of its web. First, the spider constructs a Y-shaped figure of threads fastened to fixed positions in the surroundings, and meeting at the centre of the web. Next, it constructs a frame around the centre of the web and then a system of radial threads to the frame of the web, temporarily held together by a non-sticky logarithmic spiral, which it constructs by working its way out from the centre to the frame of the web. Finally, the spider works its way back to the centre along a sticky Archimedean spiral, while eating the logarithmic spiral used during the construction.

Helices, twining plants and an optical illusion

A space curve on a cylinder, for which all the tangent lines intersects the generators of the cylinder in a constant angle, is a *helix*. This is the curve followed by twining plants such as bindweed (right-handed helix) and hops (left-handed helix).

Looking up a long circular cylinder in the wall of a circular tower enclosing a spiral staircase, all the generators of the cylinder appear to form a system of lines in the ceiling of the tower radiating from the central point, and the helix in the banister of the spiral staircase appears as a logarithmic (equiangular) spiral. A magnificent instance of this is found in the spiral staircase in Museo do Popo Galergo, Santiago, Spain, where you have three spiral staircases in the same tower.

Curvature and growth phenomena in nature

Many growth phenomena in nature exhibit curvature. As an example, we discuss the shape of the shell of the primitive cuttlefish nautilus.

The fastest way to introduce the notion of curvature at a particular point of a plane curve is by approximating the curve as closely as possible in a neighbourhood of the point with a circle, called the *circle of curvature*, and then measuring the curvature as the reciprocal of the radius in the approximating circle. If the curve is flat in a neighborhood of the point, or if the point is an inflection point, the approximating circle degenerates to a straight line, the *tangent* of the curve; in such situations, the curvature is set to 0. Finally, the curvature of a curve has a sign: positive if the curve turns to the left (anticlockwise) in a neighbourhood of the point in question, and negative if it turns to the right (clockwise).

Now to the mathematics of a nautilus shell. We start out with an equiangular spiral. At each point of the spiral, take the circle of curvature and replace it with the similar circle centred at the spiral and orthogonal to both the spiral and the plane of the spiral. For exactly one angle in the equiangular spiral, the resulting surface winds up into a solid with the outside of one layer exactly fitting against the inside of the next. The surface looks like the shell of a nautilus.

In the Spring of 2001, I discovered that this unique angle is very close to half the golden angle (the smaller of the two angles obtained by dividing the circumference of the unit circle in golden ratio), so close in fact that it just allows for thickness of the shell. More details can be found on my home page [14].

7. Geometry of space

Questions about geometries alternative to Euclidean geometry easily arise in philosophical discussions on the nature of space and in popularizations of physics. It is difficult material to disseminate, but still, one can come far by immediately incorporating a concrete model of a non-Euclidean geometry into the considerations. In this respect, I find the disc model of *the hyperbolic plane* suggested by Poincaré (1887) particularly useful. In addition to its mathematical uses, it was the inspiration for Escher in his four woodcuts Circle Limit I–IV, ([7], page 180).

In ([13], Chapter 4), I gave a short account of the Poincaré disc model, which has successfully been introduced in Denmark at the upper high-school level, and where you encounter some of the surprising relations in hyperbolic geometry. In particular, the striking difference between tilings with congruent regular n-gons in the Euclidean plane, where you can tile with such n-gons only for $n = 3, 4, 6$, and in the hyperbolic plane, where you can tile with such n-gons for each integer $n \geq 3$.

At a more advanced level, one can introduce a hyperbolic structure on surfaces topologically equivalent to the surface of a sphere with p handles, for each *genus* $p \geq 2$, by pairwise identification of the edges in a regular hyperbolic $4p$-gon; cf. [11].

The appearance of non-Euclidean geometries raised the question of which geometry provides the best model of the physical world. This question was illustrated in an inspired poster designed by Nadja Kutz; cf. [6].

8.   Eternity and infinity

Eternity and, more generally, infinity are notions both expressing the absence of limits. Starting from the classical paradoxes of Zeno, which were designed to show that the sensory world is an illusion, there are good possibilities for conversations about mathematical notions associated with infinity. And one can get far even with fairly difficult topics such as infinite sums; cf. [15].

The harmonic series

A conversation about infinite sums (series) inevitably gets on to the *harmonic* series:

$$\sum_{n=1}^{\infty} \frac{1}{n} = 1 + \frac{1}{2} + \frac{1}{3} + \frac{1}{4} + \cdots + \frac{1}{n} + \cdots .$$

As you know, this series is *divergent*, i.e. the Nth *partial sum*

$$S_N = 1 + \frac{1}{2} + \frac{1}{3} + \frac{1}{4} + \cdots + \frac{1}{N}$$

increases beyond any bound, for increasing N.

The following proof of this fact made a very deep impression on me when I first met it. First notice, that for each number k, the part of the series from $\frac{1}{k+1}$ to $\frac{1}{2k}$ contains k terms, each greater than or equal to $\frac{1}{2k}$, so that

$$\frac{1}{k+1} + \frac{1}{k+2} + \cdots + \frac{1}{2k} \geq \frac{1}{2k} + \frac{1}{2k} + \cdots + \frac{1}{2k} = \frac{1}{2}.$$

With this fact at your disposal, it is not difficult to divide the harmonic series into infinitely many parts, each greater than or equal to $\frac{1}{2}$, proving that the partial sums in the series grow beyond any bound when more and more terms are added.

The alternating harmonic series

The *alternating* harmonic series

$$\sum_{n=1}^{\infty} (-1)^{n-1} \frac{1}{n} = 1 - \frac{1}{2} + \frac{1}{3} - \frac{1}{4} + \cdots + (-1)^{n-1} \frac{1}{n} + \cdots$$

offers great surprises. As shown in 1837 by the German mathematician Dirichlet, this series can be *rearranged*, that is, the order of the terms can be changed, so that the rearrangement is a convergent series with a sum arbitrarily prescribed in advance. The proof goes by observing that the *positive series* (the series of terms with positive sign) increases beyond any upper bound, and that the *negative series* (the series of terms with negative sign) decreases beyond any lower bound, when more and more terms are added to the partial sums in the two series. Now, say that we want to obtain the sum S in a rearrangement of the alternating harmonic series. Then we first take as many terms from the positive series as are needed just to exceed S. Then take as many of the terms from the negative series as needed just to come below S. Continue this way by taking as many terms as needed from the positive series, from where we stopped earlier, just again to exceed S. Then take as many terms as needed from the negative series, from where we stopped earlier, just again to come below S, and so on. Since $\frac{1}{n}$ tends to 0 as n increases, the series just described is convergent with sum S.

The alternating harmonic series is the mathematical idea behind the episode from *Alice's Adventures in Wonderland*, mentioned at the beginning of this paper.

9.   Mathematics as a sixth sense

In 1982, *Dirac's string problem* was presented on a shopping bag from a Danish supermarket chain [12]. This problem illustrates the property of half-spin of certain elementary particles, mathematically predicted by the physicist P.A.M. Dirac in the 1920s for elementary particles such as the electron and the neutron. To convince colleagues sceptical of his theory, Dirac conceived a model to illustrate a corresponding phenomenon in the macroscopical world. This model consists of a solid object (Dirac used a pair of scissors) attached to two posts by loose (or elastic) strings, say with one string from one end of the object and two strings from the other end to the two posts. Dirac then demonstrated that a double twisting of the strings could be removed by passing the strings over and round the object, while he was not able to remove a single twisting of the strings in this way. Rather deep mathematics from topology is needed to explain the phenomenon, and in this case mathematics comes in as a 'sixth sense', by which we 'sense' (understand) counter-intuitive phenomena. An application to the problem of transferring electrical current to a rotating plate, without the wires getting tangled and breaking, was patented in 1971.

10.   Making mathematics visible

In many countries there are increasing demands to make the role and importance of the subjects taught in the school system visible to the general public. If the
curriculum and the methods of teaching are not to stagnate, it is important that
an informed debate takes place in society about the individual subjects in school.

For subjects strongly depending on mathematics, this is a great challenge.
Where mathematicians (and scientists in general) put emphasis on explanations
(proofs), the public (including politicians) are mostly interested in results and consequences. It is doubtless necessary to go part of the way to avoid technical language
in presenting mathematics for a wider audience.

On the initiative of the International Mathematical Union, the year 2000 was
declared *World Mathematical Year*. During the year many efforts were made to
reach the general public through poster campaigns in metros, public lectures, articles in general magazines, etc. and valuable experience was gained. The mathematical year clearly demonstrated the value of an international exchange of ideas
in such matters, and the issue of raising public awareness of mathematics is now on
the agenda all over the world from eight to infinity.

References

[1] S. J. Abas and A.S. Salman, "Symmetries of Islamic geometrical patterns",
    World Scientific, Singapore, 1994.

[2] R. Brown and SUMit software, "Raising Public Awareness of Mathematics
    through Knots", CD-Rom (http://www.cpm.informatics.bangor.ac.uk/), 2001.

[3] M. Chaleyat-Maurel et al., "Mathematiques dans la vie quotidienne", Booklet
    produced in France in connection with World Mathematical Year 2000.

[4] P. R. Cromwell, "Polyhedra", Cambridge Univ. Press, 1997.

[5] M. Emmer, "The Fantastic World of M.C. Escher", Video (50 minutes), ©1980.

[6] EMS-Gallery (design Kim Ernest Andersen), "European Mathematical Society,
    WMY 2000, Poster competition", http://www.mat.dtu.dk/ems-gallery/, 2000.

[7] M.C. Escher, J.L. Locher (Introduction), W.F. Veldhuysen (Foreword), "The
    Magic of M.C. Escher", Thames & Hudson, London, 2000.

[8] J.V. Field, "The Invention of Infinity - Mathematics and Art in the Renais-
    sance", Oxford Univ. Press, 1997.

[9] H. Götze, *Friedrich II and the Love of Geometry*, Mathematical Intelligencer
    17 (1995), No. 4, 48–57.

[10] V.L. Hansen, "Geometry in Nature", A K Peters, Ltd., Wellesley, Mas-
     sachusetts, US, 1993.

[11] V.L. Hansen, *Jakob Nielsen (1890–1959)*, Mathematical Intelligencer 15 (1994),
     No. 4, 44–53.

[12] V.L. Hansen, *The Story of a Shopping Bag*, Mathematical Intelligencer 19
     (1997), No. 2, 50–52.

[13] V.L. Hansen, "Shadows of the Circle - Conic Sections, Optimal Figures and
     Non-Euclidean Geometry", World Scientific, Singapore, 1998.

[14] V.L. Hansen, *Mathematics of a Nautilus shell*, Techn. Univ. of Denmark, 2001.
     See: http://www.mat.dtu.dk/persons/Hansen_Vagn_Lundsgaard/ .

[15] V.L. Hansen, "Matematikkens Uendelige Univers", Den Private Ingeniørfond, Techn. Univ. of Denmark, 2002.

[16] K. Miyazaki, *The Cuboctahedron in the Past of Japan*, 2ε Mathematical Intelligencer 15 (1993), No. 3, 54–55.

[17] R.J. Wilson, "Stamping Through Mathematics", Springer, New York, 2001.

[18] Mathematics in plays: (a) David Auburn: "Proof" (Review Notices AMS, Oct. 2000, pages 1082–1084). (b) Michael Frayn: "Copenhagen". (c) Tom Stoppard: "Hapgood". See also: Opinion, Notices AMS, Feb. 2001, 161.

[19] Mathematics in movies: (a) "Good Will Hunting" (directed by Gus Van Sant), 1997. (b) "A beautiful mind" (directed by Ron Howard), 2001. Based in part on a biography of John Forbes Nash by Sylvia Nasar.

ICM 2002 · Vol. III · 897–906

Reforms of the University Mathematics Education for Non-mathematical Specialties

Shutie Xiao*

Abstract

This article is a part of the report for the research project "Reform of the Course System and Teaching Content of Higher Mathematics (For Non-Mathematical Specialties)" in 1995, supported by the National Ministry of Education. There are thirteen universities participated in this project. The Report not only reflects results of our participants, but also includes valuable opinions of many colleagues in the mathematical education circles.

In this article, after a brief description on the history and reform situation of the higher mathematics education in China, attention concentrates to three aspects. They are: main problems in this field existing; the functions of mathematics accomplishment for college students; these concern course system, teaching and learning philosophy, such as overemphasized specialty education, overlooking to arouse rational thinking and aesthetic conceptions, etc. The last aspect contains a discussion on several important relationships, such as: knowledge impartment and quality cultivation, inherence and modernization of the mathematical knowledge, teacher's guidance and students' initiative, mathematical basic training and mathematical application consciousness and ability cultivation, etc.

2000 Mathematics Subject Classification: 97D30.
Keywords and Phrases: Higher mathematics education.

1. Introduction

This is a part of the report for the research project of " Reforms on the Course System and Teaching Content of Higher Mathematics (For Non-Mathematical Specialties)" in 1995, supported by National Ministry of Education. There are thirteen universities participated in this research project. They worked together cooperatively. Besides, many experts and scholars in mathematical and educational circles

---

*Department of Applied Mathematics, Tsinghua University, South Building 11–5–402, Beijing 100084, China. E-mail: xstwef@mail.tsinghua.edu.cn

had been consulted in many aspects. This essay not only reflects the research results of our participants-universities in more than four years, but also includes valuable opinions of many colleagues in the mathematical education circle.

2. Reform situation of the university mathematics education in China and the main problems existing

For the past half a century, our modern university education, accompanying the progress process of our socialist political and economic development, experiencing the stage of "learning from the Soviet Union in all respects", many times of "educational reforms" and today's "reform and opening to the world", has grown in wave-like development and formed today's scale. In this process, the teaching reform on the university non-mathematics class specialties higher mathematics has almost not been stopped. It can be said that mathematics teaching in this period, generally speaking, adapted itself to the requirements for training professionals in various fields under the planned economic system and made important contributions for the society. In this respect, several generations of educators of mathematical teaching devoted their wisdom and hardworking.

Historically, the focus of the reform of the university mathematical education in China is always concentrated on the problem of the combination of mathematics with practice. In a long period before the reforms and opening to the world, the basic task of the non-mathematics class specialties mathematical teaching is for specialties, or mathematics courses should serve courses of specialties. As mentioned earlier, in the tendency of the extreme situation of putting too much emphasis on specialties, such a saying is natural; but we went even further. In addition, we were in a situation of long-period of isolation. We didn't know the situation of mathematics education in other countries, even were not clear on the reform in the Soviet Union. Without comparison, it is difficult to discover the problems. Moreover, different opinions on such a problem as "the relationship between the teaching of fundamental mathematics and that of specialties," should have been a problem of academic views in the first place and can be discussed and tested; but such opinions were always improperly treated as ideological problems even the political ones. Therefore, it was difficult to express different views. Since the 1980s, our policies of the reform and opening to the world have brought spring for the educational reforms. When we learned of the changes in those decades of years, we realized that we had lagged behind and we should do our best to catch up. At that time, people found that, as a developing country, China was confronted with a series of serious challenges in the socialist market economy. In addition, the rapid appearance of the knowledge economy based on the intelligent resources and innovative competition, we are forced to reconsider problems, such as educational concepts, teaching systems, course systems, teaching content and methods etc., which are formed in the original planned economy.

Viewing from the angle of the university mathematics education, the existing main problems may be summarized as follows:

First, the "specialty education" is overemphasized and thus a one-sided understanding of the role of the university mathematics education "serving specialties" is formed. Such an understanding, as the guiding idea of education, is embodied in every link of teaching. Due to its long duration, the depth and extension of its effects cannot be underestimated. Even now, among quite a large part of teaching cadres and teachers, such an understanding may still work, because in the past the problem of "what is the role of university mathematical courses in the university education" had not been studied properly.

Secondly, as the cultivation target was training "specialists and engineers who could work directly after their graduation," the teaching process was speeded up. Moreover, we have the tradition of putting emphasis on the classroom teaching, and the course of mathematics itself has its own specific strict logical system, the instillation-type of teaching methods can be said to be deep-rooted.

Thirdly, as mathematics teachers of non-mathematical class specialties usually undertake heavy teaching tasks, and in the technological and engineering colleges and universities, the mathematical scientific research not being combined with practice directly is viewed as being isolated from practice, the teachers engaged in teaching basic courses of mathematics have not touched scientific research for a long time. Under this situation it is difficult to improve their professional skills and their teaching levels.

These problems, facing the serious challenges of the 21st century, cause the inadaptability of the university mathematics education to the current education development situation to be more serious. The results are mainly reflected in the following respects.

For students, too specific majors and their one-sided understanding of the role of basic courses lead to their narrow scopes of knowledge (especially knowledge on mathematics), narrow field of vision, lack of creation and all this is usually expressed in insufficient "aftereffects".

For teachers, a single- pattern course content, rigid teaching plans and programs lead to the fact that they could only be responsible for textbooks and exams and it is difficult for them to attend to the cultivation of abilities and qualities of students. The development of teachers' active roles and their growth are both influenced.

For textbooks, most of the content is rather old and their system is stereotyped. There is no mechanism to encourage teachers to edit new teaching textbooks and materials, which gives people an impression of "one thousand people having one face" and results in difficulties for students to grasp mathematical thinking and method and to learn new knowledge in mathematics.

3. Several aspects that should be grasped in doing a good job of our college mathematics education

In recent years, with the rapid development of our economy, higher education in our country has met with an excellent situation of brisk discussions and great reforms. This time of educational reform, whether from the emphasis degree of the government, or from the realistic spirit of policies, the strength of the funds input and the depth and breadth of studying problems, they are all unprecedented. Just as the case in previous educational reforms, the reform of mathematics teaching is still among one of the key problems. What is different from the past is that the reform is not restricted to the long confusing problem of the relationship between mathematics and practice, but rather, the mathematics education is first put under the background of social development and is raised at the height of university quality education. The past is reviewed, the present is examined and the future is planned. In particular, the role of mathematics education in higher education has been clarified. There emerge schemes, new textbooks and new courses on mathematics course content, structure and system reforms that have creative intention and face the epoch. The research and practice on reforming teaching methods of mathematics and introducing new means of teaching have attracted the attention of broad masses of mathematics teachers; but the development is not balanced. There is not enough systematic research and comprehensive experiments. Especially, viewing from the great part of the circle, there are no great changes of the traditional mathematics teaching modes and methods. As for the renewal of the teaching content of mathematics, it is like "walking with the unsteady steps." The main problem still lies in the change of educational ideology and teaching concepts. For the teaching of the basic mathematical courses in non-mathematics class specialties, in our opinion, it is necessary to clarify the following points in understanding.

(1) It is necessary to have a rather comprehensive knowledge of the role of mathematics education in university

In the educational mode in the planned economy, training of professionals in universities was of an upside-down type. The national plan decided the arrangement of specialties, and teaching programs were drawn up according to the requirements of specialties. Courses were arranged according to the procedure of specialty courses - fundamental specialty courses - basic courses. It was emphasized that the latter shall serve the former. For the basic education of mathematics, the excessive emphasis of the aspect of "serving specialties", and the neglect of the inherent unity of mathematics as a rational reasoning system and the specific role of mathematics in the comprehensive quality of students, led to the lack of a comprehensive knowledge of the role of mathematics education in university education. In fact, mathematics is the common foundation for cultivating and training various levels of special professionals. For students of non-mathematics class specialties, the role of university basic mathematics courses lies at least in the following three aspects.

It is the main course for students to grasp mathematical tools. Such a role is very important for students of non-mathematics class specialties and is an important content of "specialty quality." The current problems are as follows: on the one hand, teachers should study how to effectively enable students to grasp and use this tool in the whole course of mathematical teaching and how to lay a foundation in this field in the stage of basic courses; on the other hand, it is necessary to prevent the narrow understanding of "tool", viewing the basic mathematics course only as the tool for serving certain specialty courses, and even the tool specifically for dealing with certain examinations.

It is an important carrier for students to cultivate their rational thinking. What mathematics studies is the model structures of "numerals" and "forms", and what it uses is such rational thinking methods as logic, reasoning and deduction. Large amounts of facts show that it is not a useless theory that is "isolated from practice", but is a thinking creation, which originates from practice and guides practice. The role of such a training of rational thinking could hardly be replaced by other courses. Such a cultivation of rational thinking is of utmost importance for students to improve their quality, to enhance their analytical abilities and to enlighten their creation consciousness. The current problems lie in that we lack the mature teaching experience and excellent textbooks in this respect.

It is a way for students to receive the nurture of beauty sense. The mathematical aesthetics is part of people's quality of the appreciation of beauty. With the development of human civilization and the progress of science, such a fact is being gradually recognized by people. In fact, targets mathematics is striving for — the arrangement of chaos into order, the sublimation of experience to laws and the search for the concise and uniform mathematical expression of motions of substances — are the embodiment of mathematical beauty, and are also man's pursuit for beauty sense. Such a pursuit acts as a subtle influence for the nurture of man's mental world and is always a motive force of innovation. At present we can only say that it is necessary to pay attention to the role of mathematics in aesthetic education. And further search and attempts should be made in its embodiment in teaching and textbooks.

The roles of the three aspects are unified, but in concrete requirements, according to the different kinds of colleges, specialties and students, there should be different sides to be emphasized, so that the comprehensive realization of the knowledge, abilities and quality in mathematics teaching can be expected.

(2) Emphasis should be put on the solution of problems of the renewal of college mathematics course system and content

From the abolition of the imperial civil examination system at the end of Qing Dynasty, the establishment of modern school and up to the liberation of China, the education of science and technology in colleges and universities basically followed modes in Europe and America. For the content of mathematical teaching, before the 1950s, mathematics as a compulsory course in most specialties of science and technology except in few specialties, such as physics, was only limited to simple calculus and differential equations. In the 1950s, the Soviet modes were comprehensively introduced in higher education institutes. A kind of specialty education with the aim of training professional talents according to requirements of the trades was formed. The arrangement of the courses in the specialties was aimed at the requirements of the professional knowledge. Basic courses such as mathematics were required to serve specialty courses. Such a system and guiding ideas have actually continued till now. Though in the 1980s, due to the need to employ computers, linear algebra (taking calculation as the chief content) and numerical methods became compulsory mathematics courses, viewing from the guiding ideas, the main point was still "specialty education" and the changes of teaching content were very limited. Later, there occurred improper evaluations, in addition to the influences of "examination-oriented education" induced by taking the post-graduate entrance examinations, the tendency of one-sided pursuit of the problem-solving techniques was promoted. As a result, the train of thought about reforms became more vague. However, due to the rapid development of the computer technique in the later half of last century, people had more knowledge on the role of modern mathematics. Especially in the recent 30 years, there appears the so-called "modern mathematics technique" which displays its prowess fully in economy and industries. For example, optimization, engineering control, information processing, fuzzy recognition and image reconstruction, etc. they are produced due to the combination of the principles and methods of modern mathematics with computers. They penetrate and are applied in all sectors and trades, and are combined with relevant techniques to form the so-called high and new techniques in those areas. In current developed countries, mathematics is applied to improve the organizational level of economy, and from drawing up macroscopic strategic planning to the storage, distribution and transportation of products and to the market prediction, analysis of finance and insurance businesses, significant progress has been attained. All the facts mentioned above mean that mathematics has turned from part of the traditional natural sciences and engineering techniques to further penetrating into many areas of the modern society and economy and has gradually become one of their indispensable columns. On the other hand, the development of science and technology brings about a series of problems and it is necessary for people to reexamine the relationship between man and society and between man and Nature in a rational way. All this requires that training on the tool property and rationality should be emphasized in college mathematics education. Therefore, it is an urgent task to adjust the basic course system of college mathematics and appropriately renew teaching content.

(3) Reforming the examination-oriented "instillation-type" teaching

In recent years, various kinds of factors have induced a type of examination-oriented teaching. Marks play almost decisive roles for students in many respects such as evaluation in the prizes, graduate entrance examinations and the assignment for jobs. In some occasions, students' marks are the main factor for evaluating teachers. Such a situation causes teachers to teach for exams and students study for exams. In order to raise the average marks of a class, teachers spend a lot of time and energy on the teaching of problem patterns. Accordingly, the "instillation-type" of teaching methods are in a dominating position in our college teaching. Characteristics of such methods are as follows: on the one hand, various kinds of problem patterns are explained in every detail in the classroom and efforts are made to enable students to understand them as soon as they listen to the explanations. In such a way, besides the small amount of classroom information quantity, the psychology of dependence of the students will certainly be enhanced and they will have a bad habit of being lazy in thinking, which will seriously hinder the cultivation of the innovation sense and innovation capability. On the other hand, students are busy problem solving after the class and problems are substituting learning. They rarely read books and pay little attention to the mathematical thinking and mathematical applications, let alone the cultivation of the innovation sense.

(4) Devoting great efforts to the construction of teachers' teams

This is the crux matter of our reform. In recent years, compared to the past, the situation of teachers' teams engaged in non-mathematics class specialty mathematics teaching has turned to the better. Quite a few of young teachers who have obtained Doctor's degrees join the career. In common colleges and universities, proportions of mathematics teachers who have gained Master's degrees are increasing. This is an encouraging phenomenon, indicating that there will be qualified successors to carry on mathematics teaching, but the authorities of colleges and universities have not timely put such a strategic task at a deserved height. For example, knowing clearly that the experienced teachers with high academic levels should be appointed to teach the basic courses, some colleges and universities still classify the teachers according to the order of the postgraduate teaching, senior students teaching, and the basic courses teaching. Such a guiding direction is very unfavorable for cultivating high-quality talents. In addition, viewing from the young teachers' teams, there also exists something worrying. First, their ideas are not so stable; secondly, they cannot deal well with the relationship between scientific research and teaching. In the key universities, there usually exists the atmosphere of paying more attention to scientific research than that to teaching. In common colleges and universities, there always exists the problem of only teaching without scientific research. Therefore, creating certain conditions and asking the young teachers to participate in scientific research are important measures to improve the levels of college mathematics teachers. Thirdly, not enough efforts have been made in conducting education on teacher morality and on superior teaching traditions for the young teachers.

4. Several important relationships in the university mathematics education reforms

The college mathematics education reform is a very careful job. The treatment of many problems in this respect will not only rely on principles, but also be flexible. According to the experience and lessons from the past mathematics education reform, I would like to discuss some problems concerning principles in mathematics teaching reform from the correct treatment of the relationships in several aspects as follows:

(1) Embodying quality education, paying attention to well treating the relationships between knowledge impartment and quality cultivation

In recent years, there are a large amount of discussions on "quality" and "quality education". We assume that quality is a mental and physical attribute expressed when people understand and treat things and events. It is based on the congenital psychological conditions and is gradually formed under the influences of the postnatal environment. Viewing from the angle of education, an individual's quality in a certain aspect is shown in his power of understanding and potential energy in such an aspect. Quality education is a process during which excellent qualities of people are cast through the smelting of systematic knowledge. Any objective existence in the world has its attribute characters of "numerals" and "forms". With the progress of the science of mathematics, people's knowledge of "numerals" and "forms" has enhanced from directly-visual quantity relationships and space forms to the abstract "mathematical structures" and "space concepts" with deeper connotation and wider extension. Such attributes of "numerals" and "forms" in people's understanding things and their comprehension and potential ability in handling the corresponding relationships are obviously a kind of people's quality and we call such a quality mathematical quality. Under the background of the tendency of digitalization and informationalization in the knowledge economy society, the importance of possessing such a quality is very great. In the quality education, viewing from the main body of education (i.e. the main targets of education and corresponding teaching behaviors), the soul of the college mathematics education is exactly the mathematical quality, and is what has been discussed in the previous part — quality property. Viewing from mathematics itself, especially the modern mathematics, its essence is a rational thinking system extracted abstractly from large amounts of objective phenomena. Therefore, in its education, besides displaying how it absorbs nutrition from laws of vivid objective things and provides tools, main focus is the training of the rational thinking techniques and of the mastering mathematical tools. That is the content of the education of mathematical quality. Though it imparts "knowledge" as well, "knowledge" of mathematics acts as a carrier of imparting quality (such a role was always neglected by people in the past), besides one side that it is combined with the material content and thus acts as a tool to serve other disciplines. The quality education in the mathematical teaching is that the teachers put the lively and rational ways of thinking via the knowledge carrier to implement the motivational, psychological and mental guidance for their students.

Mathematics has become a fundamental component of the culture of the contemporary social culture via its ability of tools, rational spirit and sense of beauty. In the society of the 21st century if one has no idea what mathematical technology means, lacks both the rational thinking and the sense of beauty appreciation; then his total quality will be affected. His abilities in insight, judgement and originality would be greatly restricted. Therefore, the accomplishment of mathematical culture is not a kind of "fashion" for one in the ever-increasingly fierce competition among talents of the society; it is indeed a kind of actual necessity for one in his work, study and social communication.

The specific content of the mathematical quality is quite rich, the most outstanding specific characteristics that are generally acknowledged are summarized as follows:

- The sharp consciousness of extracting the attributes of "numerals and forms" of things.
- Thinking mode of using "abstract models and structures" to study things.
- Exploring habits of conducting compact deduction by means of signs and logical systems.

Features of mathematical qualities in these aspects are connected with each other and cannot be separated in the actual educational processes. All college students should be cultivated and educated in such aspects. Of course, the education and cultivation of mathematical qualities are different in emphasis points, depth, and breadth between mathematics major students and non-mathematics class specialties students, and in non-mathematics class specialties, such differences do exist between students of science, engineering, agriculture and medicine and those of humanities majors.

(2) Grasping course systems and content renewal and treating well relationships between the inherence and modernization of the mathematical knowledge

The main body of the textbooks of current basic mathematics courses is mostly mathematics before the 19th century. This is in contrast with other basic textbooks of physics, chemistry and biology that began mostly from the 19th century. Therefore, the problem of the modernization of basic university mathematics textbooks is to be stressed. But we should be cautious in this problem. It is unavoidable to appropriately add the commonly accepted and fundamental content of modern mathematics; but it is necessary to consider another aspect: as we have mentioned above, mathematics is a kind of thinking science and has its own specialties. Its system is constructed by logic and is a structure of series established one layer after another from the bottom to the top. The mathematical knowledge important in the past is the "logical basis" of the current mathematics and throwing away the former will influence the later studies. For example, calculus has a history of over three hundred years, but it still is the foundation stone of modern mathematics and can't be thrown away casually. Such a situation differs from other sciences. For other sciences, if the theories before the 19th century are out-of-date, they can be cancelled, though proceeding and subsequent knowledge has inherent property. The preceding knowledge is not necessarily the direct logical foundation of the subsequent knowledge, so the abandonment of the preceding knowledge will not influence the learning of the subsequent knowledge. In addition, there is still an important reason. Such a part of mathematics content, for example, calculus, has still a rather wide application today. Therefore the idea of "modernization" of the content of college mathematics textbooks is somewhat different from the modernization of other disciplines and thus how much modern content has been listed in the textbook cannot be simply taken as the standard of measurement. We assume that the content modernization of the college mathematics textbooks can mainly include the following aspects. First, the classical mathematics content should be governed by viewpoints and languages of modern mathematics as far as possible; at the same time "roots" in the classical mathematics for some modern mathematics should be appropriately introduced. Secondly, significant results of modern mathematics that have formed the basic part of relevant disciplines should be put into textbooks as far as possible and a poplar introduction should be made. Lastly, it is necessary for students to possess essential basis of modern mathematics necessary for students to further study relevant specialties by themselves.

What is similarly important is to cancel some contents such as reasoning loaded down with trivial details and those calculations that can be done by calculators, and concepts and methods that are relatively old and thus have no development prospect in modern science. In summary, facing an accumulation of nearly two centuries, how to deal with the "metabolism" of the basic content of mathematics is a quite difficult problem. Depending only on "outside extension type" to increase academic hours will not work and efforts should be made on reforms on the structure and content of the course.

(3) Paying attention to the reform of teaching methods and handling well relationships between teachers' guidance and students' initiative

Once the teaching system and content of the course of college mathematics have been determined, teaching methods become the key problem of teaching quality. Obviously, the teaching process of basic mathematics is absolutely not man's repetition of cognition processes for numerals and forms laws, but basic laws of cognition processes should be observed. In addition, the teaching process cannot be the simulation of research process of mathematical problems, but should embody research methods and thinking modes specific for mathematics. The teaching of mathematics includes two sides of teaching and learning and is a comprehensive process in which many teaching activities guided by teachers lead to the initiative of students to study mathematics. The so-called initiative has at least the following meanings. First, students have the interest and dynamic force to learn mathematics. Secondly, students can learn by themselves under the proper guidance of teachers. Thirdly, students can work at and put forward problems independently during the study process. Fourthly, students can use the knowledge their teachers impart and related knowledge gained through self-study after processing and digestion to construct a knowledge system that, even if it is simple and not perfect, actually belongs to themselves.

(4) Emphasizing the practical links of mathematics and paying attention to handling well relationships between mathematical basic training and mathematical application consciousness and ability cultivation

The widespread use of computers and the emergence of a series of powerful mathematical software systems have resulted in profound changes in the roles of mathematics and in mathematical teaching. They make it possible to collect and process large amounts of data, also make mathematical model a means of experiment, and thus greatly promote applications of mathematics in all fields. The combination of mathematical thinking with computers has become an important mode of modern mathematical teaching. For example, since the establishment of the course of "mathematical models" in colleges in the 1980s, there have been hundreds of colleges and universities where this course has been set up, and the course is a favorite of students. Especially the annual national contest of college mathematics model-building attracts students of various specialties and promotes reforms in mathematics teaching. At present, the course "mathematical experiments" aiming at strengthening practice links of mathematics is being tested in several colleges and universities and the preliminary results are encouraging.

The Teaching of Proof

Deborah Loewenberg Ball[*]   Celia Hoyles[†]
Hans Niels Jahnke[‡]   Nitsa Movshovitz-Hadar[§]

Abstract

This panel draws on research of the teaching of mathematical proof, conducted in five countries at different levels of schooling. With a shared view of proof as essential to the teaching and learning of mathematics, the authors present results of studies that explore the challenges for teachers in helping students learn to reason in disciplined ways about mathematical claims.

2000 Mathematics Subject Classification: 97C30, 97C50, 97D20.
Keywords and Phrases: Proof, Didactics of mathematics, Mathematical reasoning.

1.   Introduction

Proof is central to mathematics and as such should be a key component of mathematics education. This emphasis can be justified not only because proof is at the heart of mathematical practice, but also because it is an essential tool for promoting mathematical understanding.

This perspective is not always unanimously accepted by either mathematicians or educators. There have been challenges to the status of proof in mathematics itself, including predictions of the 'death of proof'. Moreover, there has been a trend in many countries away from using proof in the classroom (for a survey see Hanna & Jahnke, 1996).

In contrast to this, the authors of the present paper agree that proof must be central to mathematics teaching at all grades. Nevertheless, there are lessons to be learned from the debates over the role of proof. For many pupils, proof is just a ritual without meaning. This view is reinforced if they are required to write proofs according to a certain pattern or solely with symbols. Much mathematics teaching

---

[*]School of Education, University of Michigan, Ann Arbor, MI 48109-1259, USA. E-mail: dball@umich.edu

[†]Institute of Education, University of London, 20 Bedford Way, London WC1H 0AL, UK. E-mail: choyles@ioe.ac.uk

[‡]Fachbereich Mathematik und Informatik, Universit?t Essen, 45117 Essen, Germany. E-mail: njahnke@uni-essen.de

[§]Department of Education in Science and Technology, Technion-Israel Institute of Technology, Haifa 32000, Israel. E-mail: nitsa@tx.technion.ac.il

in the early grades focuses on arithmetic concepts, calculations, and algorithms, and, then, as they enter secondary school, pupils are suddenly required to understand and write proofs, mostly in geometry. Substantial empirical evidence shows that this curricular pattern is true in many countries.

Needed is a culture of argumentation in the mathematics classroom from the primary grades up all the way through college. However, we need to know more about the difficulties pupils encounter when they are confronted with proof and the challenges faced by teachers who seek to make argumentation central to the mathematics classroom. The epistemological difficulties that confront students in their first steps into proof can be compared to those faced by scientists in the course of developing a new theory. At the beginning, definitions do not exist. It is not clear what has to be proved and what can be presupposed. These problems are interdependent, and researchers (like students) find themselves in danger of circular reasoning. In the infancy of a theory, a proof may serve more to test the credibility or the fruitfulness of an assumption than to establish the truth of a statement. Only later, when the theory has become mature (or the student has come to feel at home in a domain), can a proof play its mathematical function of transferring truth from assumptions to a theorem.

All in all, work is needed in three areas with regard to the teaching of proof. We need (1) a more refined perception of the role and function of proof in mathematics, including studies of the practices of proving in which active mathematicians engage (epistemological analysis), (2) a deeper understanding of the gradual processes and complexities involved in learning to prove (empirical research) and (3) the development, implementation and evaluation of effective teaching strategies along with carefully designed learning environments that can foster the development of the ability to prove in a variety of levels as from the primary through secondary grades and up to college level (design research).

We begin in Section 2 with an analysis of what mathematical proof might involve in the primary grades. Section 3 gives results of a longitudinal study on the development of proving abilities in grades 8 and 9. Section 4 is based upon an empirical investigation of college level teaching and shows how the natural habit of referring to an example can be used as a leverage into the teaching of proof, and section 5 discusses the idea of 'physical mathematics' as an environment for the teaching of proof.

2. What does it take to (teach to) reason in the primary grades?[1]

Although the teaching and learning of mathematical reasoning has often been seen as a focus only beginning in secondary school, calls for improvements in mathematics education in the U.S. have increasingly emphasized the importance of proof and reasoning from the earliest grades (NCTM, 2000, p.56). While some may regard such a focus on reasoning and proof secondary to the main curricular goals

---

[1] Author: Deborah Loewenberg Ball

in mathematics at this level, we consider reasoning to be a basic mathematical skill. Yet what might 'mathematical reasoning' look like with young children, and what might it take for teachers to systematically develop students' capacity for such reasoning? These questions form one strand of our research on the teaching and learning of elementary school mathematics.

We define 'mathematical reasoning' as a set of practices and norms that are collective, not merely individual or idiosyncratic, and that are rooted in the discipline (Ball & Bass, 2000, 2002; Hoover, in preparation). Mathematical reasoning can serve as an instrument of inquiry for discovering and exploring new ideas, a process that we call the *reasoning of inquiry*. Mathematical reasoning also functions centrally in justifying or proving mathematical claims, a process that we call the *reasoning of justification*. It is this latter on which we focus here.

The reasoning of justification in mathematics, as we see it, rests on two foundations. One foundation is an evolving *body of public knowledge* — the mathematical ideas, procedures, methods, and terms that have already been defined and established within a given reasoning community. This knowledge provides a point of departure, and is available for public use by members of that community in constructing mathematical claims and in seeking to justify those claims to others. For professional mathematicians, the base of public knowledge might consist of an axiom system for some mathematical structure simply admitted as given, plus a body of previously developed and publicly accepted knowledge derived from those axioms. Hence, the base of public mathematical knowledge defines the grain size of the logical steps which require no further warrant, that is acceptable within a given context. The second foundation of mathematical reasoning is *mathematical language*—symbols, terms, notation, definitions, and representationsand rules of logic and syntax for their meaningful use in formulating claims and the networks of relationships used to justify them. 'Language' is used here to refer to the entire linguistic infrastructure that supports mathematical communication with its requirements for precision, clarity, and economy of expression. Language is essential for mathematical reasoning and for communicating about mathematical ideas, claims, explanations, and proofs. Some disagreements stem from divergent or unreconciled uses of terminology, whereas others are rooted in substantive and conflicting mathematical claims (Crumbaugh, 1998; Lampert, 1998). The ability to distinguish these requires sensitivity to the nature and role of language in mathematics.

We have been tracing the development of mathematical reasoning in a class of Grade 3 students (ages 8 and 9) across an entire school year using detailed and extensive records of the class: videotapes of the daily lessons, the students' notebooks and tests, interviews with students, and the teacher's plans and notes. By comparing the class's work at different points in time, we are able to discern growth in the students' skills of and dispositions toward reasoning. We offer two brief examples here. Early in the school year, the teacher presented the problem, 'I have pennies (one-cent coins), nickels, (five-cent coins), and dimes (ten-cent coins) in my pocket. Suppose I pull out two coins, what amounts of money might I have?' The children worked to find solutions to this problem: 2, 6, 10, 11, 15, and 20. The teacher asked the students whether they have found <u>all</u> the solutions to the problem, and how they know. Some students seemed uncertain about the question. Other students offered explanations: 'If you keep picking up different coins, you will keep getting the same answers,' 'If you write down the answers and think about it some more until you have them all.' The students believed they had found them all, but it was because they could not find any more. Their empirical reasoning satisfied them. Moreover, they had neither other ideas nor methods for building a logical argument which would allow them to prove that this problem (as worded) had exactly six solutions. They also did not have the mathematical disposition to ask themselves about the completeness of their results when working on a problem with finitely many solutions.

In contrast, consider an episode four months later. Based on their work with simple addition problems, the third graders had developed conjectures about even and odd numbers (e.g., an odd number plus an odd number equals an even number). They generated long lists of examples for each conjecture: $3 + 5 = 8$, $9 + 7 = 16$, $9 + 9 = 18$, and so on. Two girls, amidst this work, argued to their classmates: 'You can't prove that Betsy's conjecture (*odd + odd = even*) always works. Because, um, $\cdots$ numbers go on and on forever, and that means odd numbers and even numbers go on forever, so you couldn't prove that all of them work.' The other children became agitated and one of them pointed out that no other conclusion that the class had reached had met this standard. Pointing to some posted mathematical ideas, the product of previous work, one girl questioned: 'We haven't even tried them with all the numbers there is, so why do you say that those work? We haven't tried those with all the numbers that there ever could be.' And other children reported that they had found many examples, and this showed that the conjecture was true. But some were worried: One student pointed out that there are 'some numbers you can't even pronounce and some numbers you don't even know are there.' A day later, however, challenged by the two girls' claim, the class arrived at a proof. Representing an odd number as a number than can be grouped in twos with one left over, they were able to show that when you add two odd numbers, the two ones left over would form a new group of two, forming an even sum:

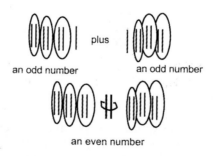

Figure 1: odd + odd = even

This episode illustrates the important role for definitions. Having a shared definition for odd and even numbers enabled these young students to establish a logical argument, based on the structure of the numbers. As one girl explained to her classmates: 'All odd numbers if you circle them by twos, there's one left over, so if you $\cdots$ plus another odd number, then the two ones left over will group together, and it will make an even number.' The definition equipped them to transcend the barrier that 'numbers go on forever' because it afforded them the capacity for quantification over an infinite set. Moreover, this episode shows the students having developed in their ability to construct, inspect, and consider arguments using previously established public mathematical knowledge.

Our research on the nature, structure, and development of mathematical reasoning has made plain that mathematical reasoning can be learned, and has highlighted the important role played by the teacher in developing this capacity. Three domains of work for the teacher have emerged from our analyses. A first concerns the selection of mathematical tasks that create the need and opportunity for substantial mathematical reasoning. The two-coin problem, for instance, did not originally require students to find all the solutions. Asking this transformed an ordinary problem into one that involved the need to reason mathematically about the solution space of the problem. The second domain of teachers' work centers on making mathematical knowledge public and in scaffolding the use of mathematical language and knowledge. Making records of the mathematical work of the class (through student notebooks, public postings, etc.) is one avenue, for it helps to make that work public and available for collective development, scrutiny, and subsequent use. This includes attention to where and in what ways knowledge is recorded, as well as how to name or refer to ideas, methods, problems, and solutions. Making mathematical knowledge and language public also requires moving individuals' ideas into the collective discourse space. A third domain of work, then, concerns the establishment of a classroom culture permeated with serious interest in and respect for others' mathematical ideas. Deliberate attention is required for students to learn to attend and respond to, as well as use, others' solutions or proposals, as a means of strengthening their own understanding and the subsequent contributions they can make to the class's work.

Acknowledgement. The work reported here draws on my research with Hyman Bass, Mark Hoover, Jennifer Lewis, and Ed Wall, as part of the Mathematics Teaching and Learning to Teach Project at the University of Michigan. This research has been supported, in part, by grants from the Spencer Foundation and the National Science Foundation.

3.   The complexity of learning to prove deductively[2]

Deductive mathematical proof offers human beings the purest form of distinguishing right from wrong; it seems so transparently straightforward — yet it is surprisingly difficult for students. Proof relies on a range of 'habits of mind'— looking for structures and invariants, identifying assumptions, organising logical arguments — each of which, individually, is by no means trivial. Additionally these

---

[2]Author: Celia Hoyles

processes have to be coordinated with visual or empirical evidence and mathematical results and facts, and are influenced by intuition and belief, by perceptions of authority and personal conviction, and by the social norms that regulate what is required to communicate a proof in any particular situation (see for example, Clements & Battista, (1992), Hoyles, (1997), Healy, & Hoyles, (2000).

The failure of traditional geometry teaching in schools stemmed at least partly from a lack of recognition of this complexity underlying proof: the standard practice was simply to present formal deductive proof (often in a ritualised two-column format) without regard to its function or how it might connect with students' intuitions of what might be a convincing argument: 'deductivity was not taught as reinvention, as Socrates did, but [that it] was imposed on the learner' (Freudenthal, 1973, p.402). Proving should be part of the problem solving process with students able to mix deduction and experiment, tinker with ideas, shift between representations, conduct thought experiments, sketch and transform diagrams. But what are the main obstacles to achieving this flexible habit of mind?

I present here some examples of geometrical questions that have turned out to be surprisingly difficult— even for high- attaining and motivated students. The analysis forms part of The Longitudinal Proof Project (Hoyles and Kchemann: http://www.ioe.ac.uk/proof/), which is analysing students' learning trajectories in mathematical reasoning over time. Data are collected through annual surveying of high-attaining students from randomly selected schools within nine geographically diverse English regions. Initially 3000 students (Year 8, age 13) from 63 schools were tested in 2000. The same students were tested again in the summer of 2001 using a new test that included some questions from the previous test together with some new or slightly modified questions. The same students will be tested again in June 2002 with the similar aims of testing understandings and development.

Question G1 in both Year 8 and Year 9 (see Fig 1), is concerned with how far students use geometrical reasoning to make decisions in geometry and how far they simply argue from the basis of perception or what 'it looks like' (see Lehrer and Chazan, 1998; Harel and Sowder, 1998). In both cases a geometric diagram is presented, which in the particular case shown, lends support to a conjecture that turns out to be false. Students are asked whether or not they agree with the conjecture and to explain their decision.

Responses to question G1 were coded into 6 broad categories. Surprisingly, a large number of students in both years simply answered on the basis of perception and agreed with the false conjecture with no evidence of progress over the year(Yr 8: 40%, Yr 9: 48%). Additionally 41% in Yr 8 could come up with a correct answer and explain this by reference to an explicit counter-example while only 28% could do this in Yr 9.

Further analysis, however, is thought-provoking. Responses to Yr8G1 showed evidence of three effective strategies: the first to find the most extreme case that obviously shows that the diagonals cannot cross at the centre of the circle; the second to use dynamic reasoning, that is perturbate the diagram in an incremental way, keeping the given properties invariant (e.g., moving one of the vertices round the circumference so the intersection of the diagonals can no longer be at the centre), the third is to focus on the diagonals rather than the quadrilateral and simply to say 'I can find opposite vertices such that the diagonals do not go thru the centre' also evidenced by students who simply drew a diagonals 'cross' without bothering to draw the quadrilateral itself! In answer to Yr9G1, it is harder to find a counter example in a static way as two conditions have to be controlled (neither diagonal can bisect the area) rather than only one (one diagonal must not go thru centre); also it is not possible to find as 'extreme' a counter example as in Y8 (the nearest equivalent is a concave quadrilateral, though here it is still possible to end up with two triangles that look very different but have roughly the same area). The second strategy is also harder in the Year 9 question as the dynamic reasoning has to change an area, not an immediately obvious quantity unlike the coincidence of two points. Clearly avoiding the seduction of perception is only one pitfall in geometrical reasoning.

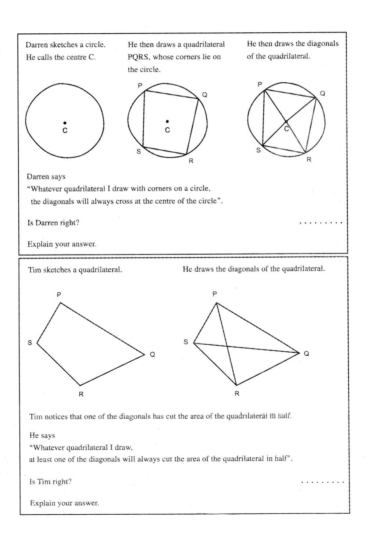

Figure 2: Question Y8G1 (top) and Y9G1: Distinguishing perceptual from geometrical reasoning

We also found that while students did not easily learn over time to reject perception, they did improve in calculation. Both Yr 8 and Yr 9 surveys included a question that required knowledge of certain angle facts (angle on a straight line or at a point, interior angle sum of triangle; angle property of isosceles triangle) and where a 3-step calculation had to be performed to find the size of an angle. We deliberately restricted the task to working with specific numerical values rather than asking students to derive a general relationship as would be required in a standard geometric proof, as this would simply be too hard for our students who have little experience of proving.

First, we note that students made considerable progress in their performance onthe calculation part of the question (from 54% correin Yr 8 to 73% in Yr 9). But on analysing responses to the Yr 9 question, (see Figure 2), where we had asked students to give reasons for each step of their calculation, we discovered that not only did they find it hard to match a step in the calculation to a reason but also they were confused by what it means to give a reason. Many students interpreted 'reasons' in ways that we did not anticipate: that is, as an explanation for the step that they had taken('u is 40 as I took 40 from 360'), or as request to make their plans explicit ('I started with p= 320 as the only thing that I know and I took it from 360 to find u').

Our research is uncovering many more surprises in both student response and progress in proving — in geometry but also in algebra (One of our questions (for 14 year-olds) concerns the sum of odd numbers and shows a remarkably similar spread of responses as those described by Deborah Ball for children age 8/9 years. We know now even more about potential obstacles to 'learning the mathematical game'; but need more systematic work on progress over time. there are no fool-proof approaches and no short cuts or easy solutions.

Acknowledgement. I acknowledge the contribution of Dietmar Kchemann in all the research reported here supported by the Economic and Social Research Council (ESRC), Project number R000237777.

4.   The 'Because for example $\cdots$ 'phenomenon, or transparent pseudo-proofs revisited[3]

This panel is about the teaching of proof in mathematics, or as I interpret it — providing adequate conditions for gaining mathematics knowledge. My presentation is based upon the assumption that mathematics knowledge is *in principle* not different than any other kind of knowledge, although, of course, the nature of the discipline is different. What, then, is knowledge? According to Brook and Stainton (2001), a common, long standing and most plausible answer, given by philosophers to this question, is that in order to be one's knowledge a proposition must comply

---

[3]Author: Nitsa Movshovitz-Hadar

with three necessary (albeit not sufficient) conditions:

(i)   It must be true.
(ii)   One must believe it. And
(iii)   One must have justification for believing it.

Hanna and Jahnke (1993) suggest that in particular for a novice, a preliminary step towards appreciating what it is that is being justified, illumination — namely understanding and believing, is of maximum importance. Bertrand Russell makes an important distinction: Minds do not create truth or falsehood. They create beliefs. What makes a belief true is its correspondence to a fact, and this fact does not in any way involve the mind of the person who holds the belief. This correspondence ensures truth, and its absence entails falsehood. 'Hence we account simultaneously for the two facts that beliefs (a) depend on minds for their *existence*, (b) do not depend on minds for their *truth*$\cdots$ ' (Russell, 1912).

We conclude that for a true mathematical statement, i.e., a theorem, to become one's mathematics knowledge, the learning environment must consist of teaching tools and strategies that support the development of two properties: (a) One's belief in its truth; and (b) one's ability to justify this belief, that is an ability not just to formally prove it, but also to ensure its truth by pointing out its correspondence to facts. Said differently, given a statement p of a mathematical theorem, a learner should be able to relate to two basic questions: (a) 'Do you believe that p?' and, provided the learner's answer to (a) is yes, (b)'Why do you believe that p?'

Quite often students' reply to the earlier question is of the form: 'Yes, because for example $\cdots$ '. Very seldom do the examples that follow, reflect full ability to verify the truth or even a partial understanding of it. For example, 'Yes, the sum of every two even integers is an even integer, because for example 6 plus 8 is 14', does not reflect any insight into the general case, although it does attest to an understanding of the statement, (which *cannot* be said about the reply: 'Yes, for example, because 14 is the sum of 6 and 8'!) The answer: 'Yes, because for example 6, which is 2×3, plus 8, which is 2×4, give 14, which is 2×7', is slightly better but not quite. It ties the belief to some acquaintance with the property of evenness. Although it may be based on deep understanding, it does *not* exhibit more than accepting the general claim as true, possibly due to a message from an external authority. (See also Mason 2001, about warrants and the origins of authority.) To be counted as 'satisfactory' the answer should be something like: 'Yes, because for example 6, which is 2×3, plus 8, which is 2×4, give $2\times(3+4)$ and this IS an even number, as it is a multiple of 2.' This latter one illustrates what we named *a transparent pseudo-proof*.

A transparent proof, is a proof of a particular case which is 'small enough to serve as a concrete example, yet large enough to be considered a non-specific representative of the general case. One can see the general proof through it because nothing specific to the case enters the proof.' Because a transparent proof is not a completely polished proof, this kind of 'proof' was later re-named *Transparent Pseudo-Proof* or as abbreviated: *Transparent P-Proof.* (Movshovitz-Hadar, 1988, 1998).

The delicate pedagogy involved in preparing a transparent p-proof was the focus of my ICME-8 Seville presentation (Movshovitz-Hadar 1998). That paper presents the lessons learned through experimental employment of two slightly different pseudo-proofs, both of them deserving the title 'P-Proof Without Words', yet only one of which — 'transparent'. The 1998 Samose presentation (ibid) included further insight into the notion of transparent p-proof, gained through the preparation of transparent p-proofs as pedagogical tools to be used in first year linear algebra course, at Technion - Israel Institute of Technology.

The study of the impact of using transparent p-proofs went on for four years, and yielded interesting results (Malek, in preparation). Numerous personal interviews of first-year mathematics majors and engineering students taking a linear algebra course, with exposure to transparent p-proofs, yielded clear evidence as to the impact of reading a transparent p-proof, on undergraduate students' ability to write, immediately afterwards, a formal proof of the same claim. A continuing follow-up also yielded comprehensive evidence as to the impact of reading transparent p-proofs, on the (passive) ability to read and comprehend general (formal) proofs, and most important of all, on the (active) ability to compose general proofs and write them in a coherent style.

Consequently, we now strongly advocate, wherever it is appropriate, the use of transparent p-proofs as a pedagogical tool, as it was shown to support both the development of one's belief in the truth of mathematical statements and of one's ability to justify this belief. However, it cannot be overemphasized that extreme care must be taken by the instructor in constructing this tool, be it in verbal-symbolic presentation or in visual-pictorial representation, so that the presentation is indeed of a transparent proof, namely, it does not hang in any way to the specifics of the particular case and hence is readily generalizable. The success of the resulting learning environment in yielding the development of the ability to prove, depends heavily on elaborate and careful preparation of the tools by the instructor.

Acknowledgement. The research work reported here was carried by Aliza Malek under my supervision, and was supported by Technion R & D funds.

5.   Arguments from physics in mathematical proofs[4]

Mathematicians often use arguments from physics in mathematical proofs. Some examples, such as the Dirichlet principle in the variational calculus or Archimedes' use of the law of the lever for determining the volumes of solids, have become famous, and have in fact been regarded by the best mathematicians as elegant proofs, if not necessarily rigorous. It is only natural, then, that several authors, notably Polya (1954) and Winter (1978), have proposed that arguments from physics could and should be used in teaching school mathematics. Besides these publications there are a number of other papers and booklets with examples (see, for example, Tokieda, 1998). Unfortunately, however, this approach to classroom teaching has not been sufficiently explored.

---

[4]Author: Hans Niels Jahnke

The application of physics under discussion goes well beyond the simple physical representation of mathematical concepts, and it is also distinct from drawing general mathematical conclusions by the exploration of a large number of instances. Rather, this approach amounts to using a principle of physics, such as the uniqueness of the centre of gravity, in a proof and treating it as if it were an axiom or a theorem of mathematics.

Let us look at a typical example. The so-called Varignon theorem states that, given an arbitrary quadrangle $ABCD$, the midpoints of its sides W, X, Y, Z form a parallelogram (see figure 3 below). A purely geometrical proof of this result would divide the quadrangle into two triangles and apply a similarity argument.

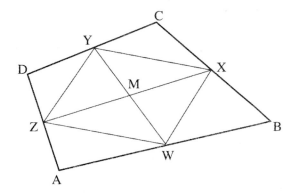

Figure 3: Varignon's theorem

An argument from mechanics, on the other hand, would consider points A, B, C, D as four weights, each of unit mass, connected by rigid but weightless rods. Such a system, with a total mass of 4, has a centre of gravity, and it is this which we need to determine. The two sub-systems AB and CD each have weight 2, and their respective centres of gravity are their midpoints W and Y. Thus, we may replace AB and CD by W and Y loaded with mass 2. Since AB and CD make up the whole system $ABCD$, its centre of gravity is the midpoint M of WY. In the same way we can consider $ABCD$ as made up of BC and DA and its centre of gravity must also be the midpoint of XZ. Since the centre of gravity is unique, this midpoint must be M. This means that M cuts both WY and XZ into equal parts. Thus $WXYZ$, whose diagonals are WY and XZ, is a parallelogram.

The example shows that an argument from physics may

o   provide a more elegant proof
o   reveal the essential features of a complex
     mathematical structure
o   point out more clearly the relevance of a theorem
     to other areas of mathematics or to other scientific disciplines
o   help create a 'holistic' version of a proof, one that can be grasped in its
     entirety, as opposed to an elaborate mathematical argument hard to survey.

Frequently, an argument from physics helps to *generalize* and to arrive at new theorems. Following the lines of our previous argument, for example, we can determine the centre of gravity not only for systems with four masses, but also for those with three, five, six, and so forth. We can also consider three-dimensional configurations and investigate whether we are able to translate the respective statements about the centre of gravity into a purely geometrical theorem.

There are several reasons why this approach to the teaching of proof should be further developed and tested. First, it is unquestionable that, worldwide, we need fresh and possibly more attractive approaches to the teaching of proof. Since using arguments from physics in a proof is an alternative to the established Euclidean routine it might be helpful in motivating teachers to rethink their attitude to proof.

Another reason is that present-day mathematical practice displays a significant emphasis on experimentation, and it is only right that this be reflected in the classroom by a similar emphasis on experimental mathematics. But it would be dangerous from an educational point of view if experimental mathematics were to be represented in the schools only by 'mathematics with computers.' Quite to the contrary: under the heading of experimental mathematics, the curriculum should include a strong component devoted to the classical applications of mathematics to the physical world. In cultivating this type of mathematics, students and teachers should be guided by the question of how mathematics helps to explore and understand the world around us. In this way, the teaching of proof would be embedded in activities of building models, inventing arguments to the question 'why', the study of consequences from assumptions. Working on the border between mathematics and physics, it could be shown that in quite a few cases we cannot only apply mathematics to physics, but, vice versa, can use statements from physics for the derivation of mathematical theorems.

A Canadian and a German group (Gila Hanna, University of Toronto, Hans Niels Jahnke, Universität Essen) study the potentials and pitfalls of this approach in Canadian and German classrooms. Questions investigated concern the feasibility and the acceptance of the approach, given the limited knowledge of physics with students in both countries. It is also asked whether this approach furthers the general understanding of proof and whether the students are aware of the difference between using arguments from physics and the purely empirical appeal to a large number of instances (Hanna & Jahnke, 2002).

References

[1] Ball, D. L. , & Bass, H. (in press). Making mathematics reasonable in school. In J. Kilpatrick, J., Martin, W. G., & Schifter, D. E.(Eds.). (in press). *A research companion to principles and standards for school mathematics.* Reston, VA: National Council of Teachers of Mathematics.

[2] Ball, D. L., and Bass, H. (2000). Making believe: The collective construction of public mathematical knowledge in the elementary classroom. In D. Phillips (Ed.), *Yearbook of the National Society for the Study of Education, Constructivism in Education,* 193–224.Chicago: University of Chicago Press.

[3] Brook, A. and Stainton, R. J., (2001). *Knowledge and mind,* The MIT Press, Cambridge Mass., 29).

[4] Clements, D. H. & Battista, M. T. (1992). Geometry and spatial reasoning. In D. Grouws, (Ed) *Handbook of research on mathematics teaching and learning,* 420–464. New York: Macmillan.

[5] Crumbaugh, C. (1998). *'Yeah, but I thought it would still make a square':* *A study of fourth-graders' disagreement during whole-group mathematics discussion.* Unpublished doctoral dissertation, Michigan State University, East Lansing.

[6] Freudenthal, H. (1973). *Mathematics as an educational task.* Dordrecht: Reidel.

[7] Hanna, G. & Jahnke, N. (1993). Proof and application, *Educational Studies in Mathematics,* 24 (4), 421–438.

[8] Hanna, G. & Jahnke, N. (1996). Proof and proving. In A. Bishop; K. Clements, C. Keitel; J. Kilpatrick; C. Laborde (Eds.), *International handbook of mathematics education,* 877– 908. Dordrecht: Kluwer.

[9] Hanna, G. & Jahnke, N. (2002). Another approach to proof. *Another approach to proof. Zentralblatt fr Didaktik der Mathematik* 34 (1), 1–8.

[10] Harel, G., & Sowder, L. (1998). Students' proof schemes: Results from exploratory studies. In A. H. Schoenfeld, J. Kaput, & E. Dubinsky (Eds.), *Research in collegiate mathematics education.* III, 428. Providence, RI: American Mathematical Society.

[11] Healy, L. & Hoyles, C. (2000). From explaining to proving: a study of proof conceptions in algebra. *Journal for Research in Mathematics Education,* 31, 396–428.

[12] Hoover, M. (in preparation). *What does it take to help students learn to work collectively in mathematics?* Manuscript in preparation, Ann Arbor, MI.

[13] Hoyles C, (1997). The curricular shaping of students' approaches to proof, *For the Learning of Mathematics,* 17 (1), 7–15.

[14] Lampert, M. (1998). *Talking mathematics in school.* Cambridge, MA: Cambridge University Press.

[15] Lehrer, R. and Chazan, D. (Eds), (1998). *Designing learning environments for developing understanding of geometry and space* 3–44. Mahwah, NJ: Lawrence Erlbaum.

[16] Malek, A. (In preparation). *The impact of exposition to transparent pseudo proofs on students' ability to prove theorems in linear algebra* (tentative title). Doctoral dissertation, to be submitted to Technion Graduate School, Fall 2002, Israel.

[17] Mason J. (2001, October). *Talking mathematics in school.* Opening address to QCA Conference, UK, October, 2001.

[18] Movshovitz-Hadar, N. (1988). Stimulating presentations of theorems followed by responsive proofs. *For the Learning of Mathematics,* 8 (2), 12–30.

[19] Movshovitz-Hadar, Nitsa (1997). On striking a balance between formal and informal proofs. In: M. De-Villiers (Ed.): *Proof and Proving — Why, When and How?* Proceedings of Topic Group 8, ICME-VIII - The 8th International Congress on Mathematics Education, Seville, Spain, July 1996. Cen-

trahil, South Africa: AMESA.

[20] Movshovitz-Hadar, N. (1998). *Transparent pseudo-proofs — a bridge to formal proofs.* Presented at and published in the proceedings of the International Conference on the Teaching of Mathematics, University of the Aegean, Samose, Greece, July 3–6, 1998. New York: John Wiley & Sons.

[21] National Council of Teachers of Mathematics. (2000). *Principles and standards for school mathematics.*Reston, VA: National Council of Teachers of Mathematics.

[22] Polya, G. (1954). *Mathematics and plausible reasoning.* Vol.1:*Induction and analogy in mathematics.*Princeton: Princeton University Press.

[23] Polya, G. (1981). Mathematical discovery: *On understanding, learning and teaching problem solving.* New York: John Wiley & Sons.

[24] Russell B. (1912). *Truth and falsehood* (Chapter 12 in The problems of philosophy, Oxford University Press, and Oxford). Reprinted in M. P. Lynch (Ed.), (2001):*The nature of truth, classic and contemporary perspectives,*23–24. Cambridge, MA: MIT Press.

[25] Tokieda, T. F. (1998). Mechanical ideas in geometry. *Mathematical Monthly 105* , 697–703.

[26] Mason J. (2001, October). *Talking mathematics in school.* Opening

[27] Winter, H. (1978). Geometrie vom Hebelgesetz aus – ein Beitrag zur Integration von Physik und Mathematikunterricht der Sekundarstufe I. *Der Mathematikunterricht, 24* (5), 88–125.

Section 19. History of Mathematics

ICM 2002 · Vol. III · 923–934

"Algebraic Truths"
vs
"Geometric Fantasies": Weierstrass' Response to Riemann

U. Bottazzini*

Abstract

In the 1850s Weierstrass succeeded in solving the Jacobi inversion problem for the hyper-elliptic case, and claimed he was able to solve the general problem. At about the same time Riemann successfully applied the geometric methods that he set up in his thesis (1851) to the study of Abelian integrals, and the solution of Jacobi inversion problem. In response to Riemann's achievements, by the early 1860s Weierstrass began to build the theory of analytic functions in a systematic way on arithmetical foundations, and to present it in his lectures. According to Weierstrass, this theory provided the foundations of the whole of both elliptic and Abelian function theory, the latter being the ultimate goal of his mathematical work. Riemann's theory of complex functions seems to have been the background of Weierstrass's work and lectures. Weierstrass' unpublished correspondence with his former student Schwarz provides strong evidence of this. Many of Weierstrass' results, including his example of a continuous non-differentiable function as well as his counter-example to Dirichlet principle, were motivated by his criticism of Riemann's methods, and his distrust in Riemann's "geometric fantasies". Instead, he chose the power series approach because of his conviction that the theory of analytic functions had to be founded on simple "algebraic truths". Even though Weierstrass failed to build a satisfactory theory of functions of several complex variables, the contradiction between his and Riemann's geometric approach remained effective until the early decades of the 20th century.

2000 Mathematics Subject Classification: 01A55, 30-03.
Keywords and Phrases: Abelian integrals, Complex function theory, Jacobi inversion problem, Riemann, Weierstrass.

Introduction

*Dipartimento di matematica, Università di Palermo, via Archirafi 34, 90123 Palermo, Italy. E-mail: bottazzi@math.unipa.it

In 1854 Crelle's *Journal* published a paper on Abelian functions by an unknown school teacher. This paper announced the entry in the mathematical world of a major figure, Karl Weierstrass (1815-1897), who was to dominate the scene for the next forty years to come. His paper presented a solution of Jacobi inversion problem in the hyper-elliptic case. In analogy with the inversion of elliptic integrals of the first kind, Jacobi unsuccessfully attempted a direct inversion of a hyper-elliptic integral of the first kind. This led him to consider multi-valued, "unreasonable" functions having a "strong multiplicity" of periods, including periods of arbitrarily small (non-zero) absolute value. Jacobi confessed he was "almost in despair" about the possibility of the inversion when he realized "by divination" that Abel's theorem provided him with the key for resurrecting the analogy with the inversion of elliptic integrals by considering the sum of a suitable number of (linearly independent) hyper-elliptic integrals instead of a single integral. In his memoir submitted to the Paris Academy in 1826 (and published only in 1841) Abel had stated a theorem which extended Euler's addition theorem for elliptic integrals to more general (Abelian) integrals of the form $\int R(x,y)dx$ in which $R(x,y)$ is a rational function and $y = y(x)$ is an algebraic function defined by a (irreducible) polynomial equation $f(x,y) = 0$. According to Abel's theorem, the sum of any number of such integrals reduces to the sum of a number p of linearly independent integrals and of an algebraic-logarithmic expression (p was later called by Clebsch the genus of the algebraic curve $f(x,y) = 0$). In 1828 Abel published an excerpt of his Paris memoir dealing with the particular (hyper-elliptic) case of the theorem, when $f(x,y) = y^2 - P(x)$, P is a polynomial of degree $n > 4$ having no multiple roots. In this case $p = [(n-1)/2]$, and for hyper-elliptic integrals of the first kind $\int \frac{Q(x)dx}{\sqrt{P(x)}}$ (Q is a polynomial of degree $\leq p-1$) the algebraic-logarithmic expression vanishes ([1], vol. 1, 444-456).

On the basis of Abel's theorem in 1832 Jacobi formulated the problem of investigating the inversion of a system of p hyper-elliptic integrals

$$u_k = \sum_{j=0}^{p-1} \int_0^{x_j} \frac{x^k dx}{\sqrt{P(x)}} \quad (0 \leq k \leq p-1) \quad (\deg P = 2p+1 \text{ or } 2p+2)$$

by studying $x_0, x_1, \cdots, x_{p-1}$ as functions of the variables $u_0, u_1, \cdots, u_{p-1}$. These functions $x_i = \lambda_i(u_0, u_1, \cdots, u_{p-1})$ generalized the elliptic functions to $2p$-periodic functions of p variables. Jacobi's "general theorem" claimed that $x_0, x_1, \cdots, x_{p-1}$ were the roots of an algebraic equation of degree p whose coefficient were single-valued, $2p$-periodic functions of $u_0, u_1, \cdots, u_{p-1}$. Therefore, the elementary symmetric functions of $x_0, x_1, \cdots, x_{p-1}$ could be expressed by means of single-valued functions in C^p. In particular, Jacobi considered the case $p = 2$ ([4], vol. 2, 7-16). His ideas were successfully developed by A. Göpel in 1847 (and, independently of him, J. G. Rosenhain in 1851). The required 4-fold periodic functions of two complex variables were expressed as the ratio of two θ-series of two complex variables obtained by a direct and cumbersome computation. This involved an impressive amount of calculations and could hardly be extended to the case $p > 2$.

Following a completely different route Weierstrass was able to solve the problem for any p. Because of his achievements he was awarded a doctor degree *honoris causa* from the Königsberg University, and two years later he was hired to teach at the Berlin Gewerbeinstitut (later Gewerbeakademie, today Technische Universität). Eventually, in the Fall of 1856 Weierstrass was named *Extraordinarius* at the Berlin University.

1.   Weierstrass' early papers

In the address he gave in 1857 upon entering the Berlin Academy, Weierstrass recognized the "powerful attraction" which the theory of elliptic functions had exerted on him since his student days. In order to become a school teacher in 1839 Weierstrass had entered the Theological and Philosophical Academy of Münster, where he attended for one semester Gudermann's lectures on elliptic functions and became familiar with the concept of uniform convergence which Gudermann had introduced in his papers in 1838. Elliptic functions constituted the subject of Weierstrass' very first paper, an essay he wrote in autumn 1840 for obtaining his *venia docendi*. His starting point was Abel's claim that the elliptic function which is the inverse of the elliptic integral of first kind (sn$u$ in the symbolism Weierstrass took from Gudermann) could be expressed as the ratio of two convergent power series of u, whose coefficients are entire functions of the modulus of the integral. Weierstrass succeeded in proving that sn$u$ (and similarly cn$u$ and dn$u$) could be represented as quotient of certain functions, which he named *Al*-functions in honor of Abel and which he was able to expand in convergent power series.

Working in complete isolation without any knowledge of Cauchy's related results, two years before Laurent, Weierstrass (1841) succeeded in establishing the Laurent expansion of a function in an annulus. In the paper he made an essential use of integrals, and proved the Cauchy integral theorem for annuli ([10], vol. 1, 51-66). In a subsequent paper he stated and proved three theorems on power series. Theorems A) and B) provided estimates (Cauchy inequalities) for the coefficients of a Laurent series in one (and several) complex variables, while Theorem C) was the double series theorem nowadays called after him. As a consequence of it Weierstrass obtained the theorem on uniform differentiation of convergent series ([10], vol. 1, 67-74). Apparently, this paper marked a turning point in Weierstrass' analytic methods for he gave up integrals and choose the power series approach to treat the theory of function of one or more variables on a par. This work was completed by a paper he wrote in spring 1842. There Weierstrass proved that a system of n differential equations

$$\frac{dx}{dt} = G_i(x_1, \cdots, x_n) \ (i = 1, \cdots, n) \ (G_i(x_1, \cdots, x_n) \text{ polynomials})$$

can be solved by a system of n unconditionally and uniformly convergent power series satisfying prescribed initial conditions for $t=0$. In addition, he also showed how the power series

$$x_i = P_i(t - t_0, a_1, \cdots, a_n) \ (i = 1, \cdots, n; t_0, a_1, \cdots, a_n \text{ fixed})$$

convergent in a disk centered at t_0 could be analytically continued outside the disk. Thus, by the early 1840s the essential results of Weierstrass' approach to the theory of analytic functions were already established. His papers, however, remained in manuscript and had no influence on the contemporary development of mathematics.

2.   Abelian functions and integrals

Weierstrass' 1854 paper ([10], vol. 1, 133-152) gave "a short overview" of the work on Abelian functions which he had developed "several years ago" and summarized in the annual report of the Braunsberg Gymnasium for 1848-49. Weierstrass began by considering the polynomial $R(x) = (x - a_0)(x - a_1) \cdots (x - a_{2n})$, with a_i real numbers satisfying the inequalities $a_i > a_{i+1}$. He decomposed $R(x)$ into the factors $P(x) = \prod_{k=1}^{n}(x - a_{2k-1})$, $Q(x) = \prod_{k=1}^{n}(x - a_{2k})$ and considered the system

$$u_m = \sum_{j=1}^{n} \int_{a_{2j-1}}^{x_j} \frac{P(x)}{x - a_{2m-1}} \frac{dx}{2\sqrt{R(x)}} \quad (m = 1, \cdots, n). \tag{2.1}$$

The task Weierstrass gave himself was to "establish in detail" Jacobi's theorem which he considered "the foundations of the whole theory". As Jacobi had remarked, for given values of $x_1, x_2, \cdots, x_n$ the quantities $u_1, u_2, \cdots, u_n$ have infinitely many different values. "Conversely, if the values of $u_1, u_2, \cdots, u_n$ are given, then the values of $x_1, x_2, \cdots, x_n$ as well as the corresponding values of $\sqrt{R(x_1)}, \sqrt{R(x_2)}, \cdots, \sqrt{R(x_n)}$ are uniquely determined". Moreover, "$x_1, x_2, \cdots, x_n$ are roots of a (polynomial) equation of degree n whose coefficient are completely determined, single-valued functions of the variables $u_1, u_2, \cdots, u_n$". Analogously, Weierstrass added, there exists a polynomial function of x, whose coefficients are also single-valued functions of $u_1, u_2, \cdots, u_n$ which gives the corresponding values of $\sqrt{R(x_1)}, \sqrt{R(x_2)}, \cdots, \sqrt{R(x_n)}$ for $x = x_1, x_2, \cdots, x_n$. Every rational symmetric function of $x_1, x_2, \cdots, x_n$ could consequently be regarded as a single-valued function of $u_1, u_2, \cdots, u_n$. Weierstrass considered the product $L(x) = (x - x_1)(x - x_2) \cdots (x - x_n)$ and the $2n+1$ single-valued functions $Al(u_1, u_2, u_n)_m = \sqrt{h_m L(a_m)}$ ($m = 0, \cdots, 2n+1$), with h_m suitable constants, which he called Abelian functions, "since they are the ones which completely correspond to the elliptic functions" to which they reduce when $n = 1$.

He was able to expand his Al-functions in convergent power series and, on the basis of Abel's theorem, he succeeded in establish the "principal property" of such functions, i.e. an addition theorem according to which $Al(u_1 + v_1, u_2 + v_2, \cdots, u_n + v_n)_m$ are rationally expressed in terms of $Al(u_1, u_2, \cdots, u_n)_m$, $Al(v_1, v_2, \cdots, v_n)_m$ and their first-order partial derivatives. Eventually, he determined the algebraic equation whose coefficients were expressed in terms of Al-functions, and whose roots were the quantities $x_1, x_2, \cdots, x_n$ satisfying equations 2.1) for arbitrary $u_1, u_2, \cdots, u_n$. However, as Dirichlet commented, in his paper Weierstrass "gave only partial proofs of his results and lacked the intermediate explanations" ([3], 52).

Two years later Weierstrass resumed this work and published in Crelle's *Journal* the first part of an expanded and detailed version of it ([10], vol. 1, 297-355).

As he had done in his 1854 paper, Weierstrass considered the polynomial $R(x) = A(x - a_1)(x - a_2) \cdots (x - a_{2\rho+1})$, and the analogous product $P(x) = \prod_{j=1}^{\rho}(x - a_j)$, $(j = 1, \cdots, \rho)$ where this time the a_j were any complex numbers such that $a_j \neq a_k$ for $j \neq k$. Instead of equations 2.1) he considered the corresponding system of differential equations

$$du_m = \sum_{j=1}^{\rho} \frac{1}{2} \frac{P(x_j)}{x_j - a_m} \frac{dx}{\sqrt{R(x_j)}} \ (m = 1, \cdots, \rho) \qquad (2.2)$$

and formulated Jacobi inversion problem as the question to find solutions $x_j = x_j(u_1, \cdots, u_\rho)$ of the system 2.2) satisfying the initial conditions $x_j(0, \cdots, 0) = a_j$ $(j = 1, \cdots, \rho)$. In order to obtain the elliptic functions as a special case $(\rho = 1)$, he also gave a slightly different form to his *Al*-functions with respect to his previous paper. Weierstrass succeeded in proving that the solutions $x_j = x_j(u_1, \cdots, u_\rho)$ are single-valued functions of $u_1, \cdots, u_\rho$ in the neighborhood of the origin. They could be considered as the roots of a polynomial equation of degree ρ, whose coefficients were given in terms of *Al*-functions which, for any bounded value of $(u_1, \cdots, u_\rho)$, are single-valued functions expressed as quotient of power series. Then, the symmetric functions of $x_j = x_j(u_1, \cdots, u_\rho)$ have "the character of rational functions". From these results, however, Weierstrass was unable to show that each Abelian function could be represented as the ratio of two everywhere convergent power series. "Here we encounter a problem that, as far as I know, has not yet been studied in its general form, but is nevertheless of particular importance for the theory of functions" ([10], vol. 1, 347).

In the course of his life he returned many times to this problem in an attempt to solve it (see below). Even the factorization theorem for entire functions that Weierstrass was able to establish some 20 years later (see Section 5) can be regarded as an outcome of this research for it provided a positive answer to the problem in the case of one variable ([5], 247). In order to show that his approach permitted one to treat the theory of elliptic and Abelian function on a par, in the concluding part of his 1856 paper Weierstrass presented a detour on elliptic functions, where he summarized the main results he had obtained in 1840. However, the promised continuation of Weierstrass' paper never appeared. Instead, a completely new approach to the theory of Abelian integrals was published by Bernhard Riemann (1826-1866) in 1857 ([7], 88-144) which surpassed by far anything Weierstrass had been able to produce.

In the introductory paragraphs of his paper Riemann summarized the geometric approach to complex function theory he had set up in his 1851 thesis ([7], 3-48). There he defined a complex variable w as a function of $r + iy$ when w varies according to the equation $i\frac{\partial w}{\partial x} = \frac{\partial w}{\partial y}$ "without assuming an expression of w in terms x and y". Accordingly, "by a well known theorem" - Riemann observed without mentioning Cauchy - a function w can be expanded in a power series $\sum a_n(z - a)^n$ in a suitable disk and "can be continued analytically outside it in only one way" ([7], 88). For dealing with multi-valued functions such as algebraic functions and their integrals Riemann introduced one of his deepest achievements, the idea of representing the branches of a function by a surface multiply covering the complex plane (or the Riemann sphere). Thus, "the multi-valued function has only *one* value defined at each point of such a surface representing its branching, and can therefore be regarded as a completely determined (=single-valued) function of position on this surface" ([7], 91). Having introduced such basic topological concepts as cross-cuts and order of connectivity of a surface, Riemann could state the fundamental existence theorem of a complex function on the surface, which he had proved in his dissertation by means of a suitable generalization of the Dirichlet principle. This theorem establishes the existence of a complex function in terms of boundary conditions and behavior of the function at the branch-points and singularities. Then Riemann developed the theory of Abelian functions proper. It is worth noting that, in spite of the fact that both Weierstrass and Riemann gave their paper the same title and used the same wording, they gave it a different meaning. Whereas the former defined Abelian functions to be the single-valued, analytic functions of several complex variables related to his solution of the Jacobi inversion problem, the latter understood Abelian functions to be the integrals of algebraic functions introduced by Abel's theorem. In the first part of his paper Riemann developed a general theory of such functions and integrals on a surface of any genus p, "insofar as this does not depend on the consideration of θ-series" ([7], 100). He was able to classify Abelian functions (integrals) into three classes according to their singularities, to determine the meromorphic functions on a surface, and to formulate Abel's theorem in new terms, thus throwing new light on the geometric theory of birational transformations. The second part of the paper was devoted to the study of θ-series of p complex variables, which express "the Jacobi inverse functions of p variables for an arbitrary system of finite integrals of equiramified, $(2p + 1)$-connected algebraic functions" ([7], 101). In this part Riemann gave a complete solution of Jacobi inversion problem without stating it as a special result. He regarded the work of Weierstrass as a particular case, and mentioned the "beautiful results" contained in the latter's 1856 paper, whose continuation could show "how much their results and their methods coincided". However, after the publication of Riemann's paper Weierstrass decided to withdraw the continuation of his own. Even though Riemann's work "was based on foundations completely different from mine, one can immediately recognize that his results coincide completely with mine" Weierstrass later stated ([10], vol. 4, 9-10). "The proof of this requires some research of algebraic nature". By the end of 1869 he had not been able to overcome all the related "algebraic difficulties". Yet, Weierstrass thought he had succeeded in finding the way to represent any single-valued $2p$-periodic (meromorphic) function as the ratio of two suitable θ-series, thus solving the general inversion problem. However, Weierstrass' paper ([10], vol. 2, 45-48) was flawed by some inaccuracies that he himself later recognized in a letter to Borchardt in 1879 (*ibid.*, 125-133). In particular, Weierstrass (mistakenly) stated that any domain of C^n is the natural domain of existence of a meromorphic function. (This mistake was to be pointed out in papers by F. Hartogs and E.E. Levi in the first decade of the 20th century). By 1857, in his address to the Berlin Academy Weierstrass limited himself to state that "one of the main problems of mathematics" which he decided to investigate was "to give an actual representation" of Abelian functions. He recognized that he had published results "in an incomplete form". "However - Weierstrass continued - it would be foolish if I were to try to think only about solving such a problem, without being prepared by a deep study of the methods that I am to use and without first practicing on the solution of less difficult problems" ([10], vol. 1, 224). The realization of this program became the scope of his University lectures.

3.   Weierstrass' lectures

In response to Riemann's achievements, Weierstrass devoted himself to "a deep study of the methods" of the theory of analytic functions which in his view provided the foundations of the whole building of the theory of both elliptic and Abelian functions. As Poincaré once stated, Weierstrass' work could be summarized as follows: 1) To develop the general theory of functions, of one, two and several variables. This was "the basis on which the whole pyramid should be built". 2) To improve the theory of the elliptic functions and to put them into a form which could be easily generalized to their "natural extension", the Abelian functions. 3) Eventually, to tackle the Abelian functions themselves.

Over the years the aim of establishing the foundations of analytic function theory with absolute rigor on an arithmetic basis became one of Weierstrass' major concerns. From the mid-1860s to the end of his teaching career Weierstrass used to present the whole of analysis in a two-year lecture cycle as follows:

1. Introduction to analytic function theory,
2. Elliptic functions,
3. Abelian functions,
4. Applications of elliptic functions or, alternatively, Calculus of variations.

All of these lectures, except for the introduction to analytic function theory, have been published in Weierstrass' *Werke*. For some twenty years he worked out his theory of analytic functions through continuous refinements and improvements, without deciding to publish it himself. Weierstrass used to present his discoveries in his lectures, and only occasionally communicated them to the Berlin Academy. This attitude, combined with his dislike of publishing his results in printed papers and the fact that he discouraged his students from publishing lecture notes of his courses, eventually gave Weierstrass' lectures an aura of uniqueness and exceptionality.

4.   Conversations in Berlin

In the Fall of 1864, when Riemann was staying in Pisa because of his poor health conditions, the Italian mathematician F. Casorati travelled to Berlin to meet Weierstrass and his colleagues. Rumors about new discoveries made by Weierstrass, combined with lack of publications, motivated Casorati's journey.

"Riemann's things are creating difficulties in Berlin", Casorati recorded in his notes. Kronecker claimed that "mathematicians $\cdots$ are a bit arrogant (*hochmütig*) in using the concept of function". Referring to Riemann's proof of the Dirichlet principle, Kronecker remarked that Riemann himself, "who is generally very precise, is not beyond censure in this regard" ([2], 262).

Kronecker added that in Riemann's paper on Abelian functions the θ-series in several variables "came out of the blue". Weierstrass claimed that "he understood Riemann, because he already possessed the results of his [Riemann's] research". As for Riemann surfaces, they were nothing other than "geometric fantasies". According to Weierstrass, "Riemann's disciples are making the mistake of attributing everything to their master, while many [discoveries] had already been made by and are due to Cauchy, etc.; Riemann did nothing more than to dress them in his manner for his convenience". Analytic continuation was a case in point. Riemann had referred to it in various places but, in Weierstrass's and Kronecker's opinion, nowhere he had treated it with the necessary rigor. Weierstrass observed that Riemann apparently shared the idea that it is always possible to continue a function to any point of the complex plane along a path that avoids critical points (branch-points, and singularities). "But this is not possible", Weierstrass added. "It was precisely while searching for the proof of the general possibility that he realized it was in general impossible". Kronecker provided Casorati with the example of the (lacunary) series

$$\theta_0(q) = 1 + 2 \sum_{n \geq 1} q^{n^2} \tag{4.1}$$

which is convergent for $|q| < 1$, and has the unit circle as a natural boundary. Its unit circle is "entirely made of points where the function is not defined, it can take any value there", Weierstrass observed. He had believed that points in which a function "ceases to be definite" - as was the case of the function $e^{1/x}$ at $x = 0$ because "it can have any possible value" there - "could not form a continuum, and consequently that there is at least one point P where one can always pass from one closed portion of the plane to any other point of it". $\theta_0(q)$ provided an excellent example of this unexpected behavior. This series also played a significant role in Weierstrass' counter-example of a continuous nowhere differentiable function (see Section 6).

5.   Further criticism of Riemann's methods

Apparently, Riemann's theory of complex functions seems to have been the background of Weierstrass' work and lectures. Evidence of this is provided by his (unpublished) correspondence with his former student H. A. Schwarz from 1867 up to 1893. One of the first topics they discussed was Riemann mapping theorem. In his thesis Riemann had claimed that "two given simply connected plane surfaces can always be mapped onto one another in such a way that each point of the one corresponds to a unique point of the other in a continuous way and the correspondence is conformal; moreover, the correspondence between an arbitrary interior point of the one and the other may be given arbitrarily, but when this is done the correspondence is determined completely" ([7], 40). Riemann's proof of the mapping theorem rested on a suitable application of Dirichlet's principle. Because of his criticism of this principle, in Weierstrass' view the Riemann mapping theorem remained a still-open question, worthy of a rigorous answer. Following Weierstrass' suggestion, Schwarz tackled this question after his student days and succeeded in establishing the theorem in particular cases, without resorting to the questionable principle. In a number of papers he gave the solution of the problem of the conformal mapping of an ellipse - or, more generally, of a plane, simply connected figure, with boundaries given by pieces of analytic curves which meet to form non-zero angles - onto the unit disk, by using suitable devices as the lemma and the reflection principle, both named after him ([8], vol. 2, 65-132).

In 1870 Schwarz discovered his alternating method. "With this method - he stated by presenting it in a lecture - all the theorems which Riemann has tried to prove in his papers by means of the Dirichlet principle, can be proved rigorously" ([8], vol. 2, 133). He submitted to Weierstrass an extended version of the paper, and in a letter of July 11, 1870 Schwarz asked him whether he had "objections to raise". Apparently, Weierstrass' answer has been lost. It is quite significant, however, that three days later, on July 14, 1870 Weierstrass presented to the Berlin Academy his celebrated counterexample to the Dirichlet principle ([10], vol. 2, 49-54), and then submitted Schwarz's 1870 paper for publication in the *Monatshefte* of the Academy. Two years later, in a letter of June 20, 1872 Schwarz called Weierstrass' attention to the still widespread idea that a continuous function always is differentiable. As the French mathematician Joseph Bertrand had made this claim in the opening pages of his *Traité*, Schwarz ironically wondered about asking Bertrand to prove that

$$f(x) = \sum_{n \geq 1} \frac{\sin n^2 x}{n^2} \tag{5.1}$$

has a derivative. One month later, on July 18, 1872 Weierstrass presented the Academy with his celebrated example of a continuous, nowhere differentiable function

$$f(x) = \sum_{n \geq 0} b^n \cos a^n x \pi \tag{5.2}$$

where $a=$ is an odd integer, $0 < b < 1$, and $ab > 1 + \frac{3}{2}\pi$. According to Riemann's students, Weierstrass remarked, the very same function 5.1) mentioned by Schwarz had been presented by Riemann in 1861 or perhaps even earlier in his lectures as an example of continuous nowhere differentiable function. "Unfortunately Riemann's proof has not been published", Weierstrass added, and "it is somewhat difficult to prove" that 5.1) has this property, he concluded before producing his own example ([10], vol. 2, 71-74).

Only by the end of 1874 was Weierstrass able to overcome a major difficulty which for a long time had prevented him from building a satisfactory theory of single-valued functions of one variable. This was the proof of the representation theorem of a single-valued function as a quotient of two convergent power series. As he wrote on the same day (December 16, 1874) to both Schwarz and S. Kovalevskaya, this was related to the following question: given an infinite sequence of constants

$\{a_n\}$ with $\lim |a_n| = \infty$ does there always exist an entire, transcendental function $G(x)$ which vanishes at $\{a_n\}$ and only there? He had been able to find a positive answer to it by expressing $G(x)$ as the product $\prod_{n \geq 1} E(x, n)$ of "prime functions"

$$E(x, 0) = 1 + x,$$

$$\cdots$$

$$E(x, n) = (1 + x) \exp(\tfrac{x}{1} + \tfrac{x^2}{2} + \cdots + \tfrac{x^n}{n})$$

which he introduced there for the first time. The "until now only conjectured" representation theorem followed easily. This theorem constituted the core of Weierstrass's 1876 paper on the "systematic foundations" of the theory of analytic functions of one variable ([10], vol. 2, 77-124). In spite of his efforts, however, he was not able to extend his representation theorem to single-valued functions of several variables. "This is regarded as unproved in my theory of Abelian functions" Weierstrass admitted in his letter to Kovalevskaya. (For 2 variables this was done by Poincaré in 1883 and later extended by Cousin in 1895 following different methods from Weierstrass'). Four days later Weierstrass wrote to Schwarz stating that Riemann's (and Dirichlet's) proof of Cauchy integral theorem by means of a double integration process was in his opinion not a "completely methodical" one. On the contrary, a rigorous proof could be obtained by assuming the fundamental concept of analytic element (and its analytic continuation) and by resorting to Poisson integral for the disk, as Schwarz himself had shown in his paper on the integration of the Laplace equation. Criticism of Riemann's ideas and methods were also occasionally expressed by Weierstrass in his letters to Kovalevskaya [6]. On August 20, 1873 he was pleased to quote an excerpt from a letter of Richelot to himself "in which a decisive preference was expressed for the route chosen by Weierstrass in the theory of Abelian functions as opposed to Riemann's and Clebsch's". On January 12, 1875 Weierstrass announced to Kovalevskaya his intention of presenting the essentials of his approach to Abelian functions in a series of letters to Richelot where he hoped "to point out the uniqueness of my method without hesitation and to get into a criticism of Riemann and Clebsch".

Weierstrass openly stated his criticism of Riemann's methods in a often-quoted "confession of faith" he produced to Schwarz on October 3, 1875: "The more I think about the principles of function theory - and I do it incessantly - the more I am convinced that this must be built on the basis of algebraic truths, and that it is consequently not correct when the 'transcendental', to express myself briefly, is taken as the basis of simple and fundamental algebraic propositions. This view seems so attractive at first sight, in that through it Riemann was able to discover so many of the important properties of algebraic functions". Of course, Weierstrass continued, it was not a matter of methods of discovery. It was "only a matter of systematic foundations" ([10], vol. 2, 235). It is worth remarking that Weierstrass added he had been "especially strengthened [in his belief] by his continuing study of the theory of analytic functions of several variable".

6.   Weierstrass' last papers

After Mittag-Leffler, Poincaré and Picard had deeply extended the results of his 1876 paper following "another way" different from his own, Weierstrass felt it necessary to explain his approach to complex function theory and to compare it with those of Cauchy and Riemann. He did this in a lecture that he delivered at the Berlin Mathematical Seminar on May 28, 1884 [11] . Even though "much can be done more easily by means of Cauchy's theorem", Weierstrass admitted, he strongly maintained that the general concept of a single-valued analytic function had to be based on simple, arithmetical operations. His discovery of both continuous nowhere differentiable functions and series having natural boundaries strengthened him in this view. "All difficulties vanish", he stated, "when one takes an arbitrary power series as the foundation of an analytic function" ([11], 3).

Having summarized the main features of his own theory, including in particular the method of analytic continuation, he advanced his criticism of Riemann's general definition of a complex function (*see* Section 2). This was based on the existence of first-order partial derivatives of functions of two real variables, whereas "in the current state of knowledge" the class of functions having this property could not be precisely delimited. Moreover, the existence of partial derivatives required an increasing number of assumptions when passing from one to several complex variables. On the contrary, Weierstrass concluded, his own theory could "easily" be extended to functions of several variables.

A major flaw in Riemann's concept of a complex function had been discovered and published by Weierstrass in 1880. The main theorem of his paper stated that a series of rational functions, converging uniformly inside a disconnected domain may represent different analytic functions on disjoint regions of the domain ([10], vol. 2, 221). Thus, Weierstrass commented, "the concept of a monogenic function of a complex variable does not coincide completely with the concept of dependence expressed by (arithmetic) operations on quantities", and in a footnote he pointed out that "the contrary statement had been made by Riemann" in his thesis. Before proving his theorem Weierstrass discussed an example he had expounded in his lectures "for many years". By combining the theory of linear transformations of elliptic θ-functions with the properties of the lacunary series 4.1), Weierstrass was able to prove that the series

$$F(x) = \sum_{n \geq 0} \frac{1}{x^n + x^{-n}} \qquad (6.1)$$

is convergent for $|x| < 1$, and $|x| > 1$, but "in each region of its domain of convergence it represents a function which cannot be continued outside the boundary of the region" ([10], vol. 2, 211). (It is worth noting that $1 + 4F(x) = \theta_0^2(x)$).

This remark allowed Weierstrass to clarify an essential point of function theory, which deeply related the problem of the analytical continuation of a complex function to the existence of real, continuous nowhere differentiable functions. In order to explain this relation Weierstrass considered the series $\sum_{n \geq 0} b^n x^{a^n}$ which is absolutely and uniformly convergent in the compact disk $|x| \leq 1$, when a is an odd integer, $0 < b < 1$. By a suitable use of his example of a continuous nowhere differentiable function 5.2), he concluded that under the additional condition $ab > 1 + \frac{3}{2}\pi$ the circle $|x| = 1$ reveals to be the natural boundary of the series. Contrary to his habit, in 1886 Weierstrass reprinted this paper in a volume which collected some of his last articles, including a seminal paper where he stated his celebrated "preparation theorem" together with other theorems on single-valued functions of several variables that he used to expound in his lectures on Abelian functions [9].

7. Conclusion

From the 1840s to the end of his life Weierstrass continued to study the theory of Abelian functions, devoting an incredible amount of work to the topic. This theory was the background of many of the results he presented in his papers and lectures, or discussed in his letters to colleagues. In spite of his efforts, however, Weierstrass never succeeded in giving it the complete, rigorous treatment he was looking for. The huge fourth volume of his *Mathematishe Werke* (published posthumously) collects the lectures on Abelian functions he gave in Winter semester 1875-76 and Summer semester 1876. Two thirds of it is devoted to algebraic functions and Abelian integrals, and only the remaining one third to the (general) Jacobi inversion problem. Thus, the editors of the volume, Weierstrass' former students G. Hettner and J. Knoblauch, could aptly state in the preface that the theory of Abelian functions (in Weierstrass' sense) "is sketched only briefly" there. It was not an irony of the history if Weierstrass failed in his pursuit of his his main mathematical goal whereas the machinery that he created to attain it in response to Riemann's "geometric fantasies" became an essential ingredient of modern analysis. The contradiction between Weierstrass' approach, in which all geometric insight was lacking, and Riemann's geometric one remained effective until the early decades of the 20th century, when the theory of functions of several complex variables began to be established in modern terms.

References

[1] N. H. Abel, *Oeuvres complétes*, 2 volls., Christiania 1881.

[2] U. Bottazzini, *The Higher Calculus*, Springer, New York 1986.

[3] P. Dugac, Elements d'analyse de Karl Weiertstrass, *Archive Hist. Ex. Sci.* 10 (1973), 41–176.

[4] C. G. J. Jacobi, *Gesammelte Werke*, Bd. 1-8, Berlin 1881–1891.

[5] *Mathematics of the 19th Century. Geometry. Analytic Function Theory*, A.N. Kolmogorov and A.P. Yushkevich eds., Birkhäuser, Basel 1996.

[6] G. Mittag-Leffler, Weierstrass et Sonja Kowalewsky, *Acta Mathematica* **39** (1923), 133–198.

[7] G. F. B. Riemann, *Gesammelte mathematische Werke*, Leipzig 1876 (2nd ed. 1892; reprint Springer 1990).

[8] H. A. Schwarz, *Gesammelte mathematische Abhandlungen*, Bd.1–2, Berlin 1890.

[9] K. Weierstrass, *Abhandlungen aus der Funktionenlehre*, Berlin 1886.

[10] K. Weierstrass, *Mathematische Werke*, Bd. 1–7, Berlin 1894–1927.

[11] K. Weierstrass, Zur Funtionenlehre, *Acta Mathematica* **45** (1925), 1–10.

ICM 2002 · Vol. III · 935–945

From Quaternions to Cosmology: Spaces of Constant Curvature, ca. 1873–1925

Moritz Epple*

Abstract

After mathematicians and physicists had learned that the structure of physical space was not necessarily Euclidean, it became conceivable that the global topological structure of space was non-trivial. In the context of the late 19th century debates on physical space this speculation gave rise to the problem of classifying spaces of constant curvature from a topological point of view. William Kingdon Clifford, Felix Klein and Wilhelm Killing, the latter of whom devoted a substantial amount of work to the topic in the early 1890s, clearly perceived this problem as relevant for both mathematics and natural philosophy (i.e., physics or cosmology). To some extent, a cosmological interest may even be found among those authors who restated the space form problem in more modern terms in the early 20th century, such as Heinz Hopf.

2000 Mathematics Subject Classification: 01A55, 01A60, 53-03, 57-03.
Keywords and Phrases: 19th century, Geometry, Topology, Cosmology.

1. Scientific contexts of topology

The broader aim of the present paper is to contribute to a better understanding of the emergence of modern topology. From its very beginnings, *analysis situs* or *Topologie*, as Johann Benedikt Listing proposed to call the new field in the 1840s, was perceived as one of the most basic subfields of mathematics. Conceptually independent of many other branches of mathematics, it deserved thorough research in its own right. During the 20th century, this perception became even more pronounced with the gradual growth of structural thinking in mathematics. As is well known, topology — axiomatized in set-theoretical terms following the lead of Felix Hausdorff — became one of three "mother structures" in Bourbaki's architecture of mathematics, making topology into a paradigm field of pure mathematics. Only

---

*Universität Stuttgart, Germany. E-mail: epple@math.uni-bonn.de

in recent decades have the immediate connections of topology with science, and physics in particular, been emphasized in many lines of research.

The purist view of topology also has dominated historical research on the emergence of topology for a long time. However, historians have begun to move beyond a history of topology focusing exclusively or at least predominantly on conceptual developments within pure mathematics. For instance, the importance of Poincaré's interest in celestial mechanics for the development of his qualitative theory of differential equations and of a number of crucial topological ideas (such as the notion of homoclinical points or his "last geometric theorem") has been underlined in several historical studies (see, e.g., [1]). Another area where the interaction of topological research and physics has been investigated thoroughly is the emergence of the theory of Lie groups [2]. Finally, the relations between 19th century studies of vortex motion in ideal fluids by Helmholtz and Thomson, the latter's influential theory of vortex atoms, and the early attempts at a classification of knots and links by Tait and his followers have been studied in detail [3], [4].

This and similar research has made it clear that the gradual formation of topology in the latter half of the 19th century and the first decades of the 20th century involved more than just pure mathematics. In addition to the growing need for topological notions in fields such as (algebraic) function theory or differential geometry, a need for topology was clearly felt in several domains of physics (and maybe even in chemistry). In the following, another major case will be discussed in which mathematical and physical thinking jointly contributed to the emergence of new topological ideas.[1]

2.   The topological space problem

In a well-known series of events ranging from the first mathematical discussions of non-Euclidean geometries to heated public debates in the late 19th century, mathematicians and physicists learned that the most adequate mathematical description of physical space was not necessarily Euclidean. This insight had a wide range of consequences both for the body and for the image of geometric, and indeed mathematical, knowledge (to use a distinction proposed by Yehuda Elkana). One of these consequences was to challenge not only the metric properties of Euclidean space (as a model of physical space) but to question its other properties as well. If there was no *a priori* reason for accepting the axiom of parallels, why should there be *a priori* reasons for accepting, e.g., the topological features of Euclidean space? What were the topological types of the best mathematical descriptions of physical space?

To phrase such a question in modern terms and within our understanding of the relations between geometry and topology sounds anachronistic. Nevertheless, a corresponding problem *was* raised in the terms available to 19th century scientists. Before the establishment of a coherent framework of topological notions, such terms were, in particular, the dimension, the "Zusammenhang" or connectivity, and the continuity of space. While it is well known that the issue of the dimension of space

---

[1]A more detailed treatment of this episode, including full references, will appear in [5].

was in the focus of several 19th century debates, it has less often been emphasized that the properties of connectivity and continuity of space came into question as well. Here I will concentrate on the problem of the connectivity of space.

The notion of "Zusammenhang" was originally introduced by Bernhard Riemann as a tool for distinguishing different types of (Riemann) surfaces in the context of function theory. This notion does not figure prominently in his famous talk *Über die Hypothesen, welche der Geometrie zu Grunde liegen*. There, Riemann introduced the crucial distinction between the "Ausdehnungsverhältnisse" and "Maßverhältnisse" (roughly: topological properties vs. metric properties) of a manifold, but the global topological aspects of manifolds received no special emphasis. This holds in particular for the final sections of his talk which were devoted the geometry of physical space. While acknowledging that there exists a "discrete manifold" of possible "Ausdehnungsverhältnisse" of space, Riemann expressed scepticism about pursuing the global properties of space beyond the issue of dimension: "Questions about the immeasurably large are idle questions for the explanation of nature."

When Riemann's talk reached the scientific public in 1868, another contribution that shaped the later debates on the space problem was on its way. The physicist Hermann v. Helmholtz argued that for epistemological reasons, a crucial assumption in any mathematical description of physical space should be the "free mobility of rigid bodies" of arbitrary size. According to Helmholtz, the existence of freely movable rigid bodies was a precondition for measuring lengths. In mathematical terms, it implied that the classical non-Euclidean geometries were the only possible models of space. Although Helmholtz's argument was soon criticised for technical reasons, his main assumption (not easily stated in precise mathematical terms) was accepted during the 19th century even by many proponents of liberal approaches to the geometry of physical space. With one exception and one crucial modification, this holds for all authors that will be treated below.

The exception is William Kingdon Clifford, the most imaginative follower of Riemann's geometric speculations in Britain. His remarks on a "space theory of matter", according to which all material phenomena might be explained by a time-dependent, wave-like variation of space curvature, are well known. In addition, several of Clifford's writings show a marked interest in different *global* possibilities for manifolds or spaces. In [6], Clifford hinted at a large variety of "algebraic spaces", higher dimensional analogues of Riemann's surfaces. In the same paper, he presented his example of a closed surface embedded in elliptical 3-space, the inherited geometry of which is locally Euclidean. As this example came to play an essential role in the following, let me recall the main line of Clifford's construction.

Identifying points in elliptic space with with one-dimensional subspaces of the quaternions, any given quaternion different from zero induces two isometries of elliptic space by left and right multiplication. Such isometries Clifford called left and right "twists", respectively. (Felix Klein would later term them "Schiebungen", translations.) Every twist possessed a space-filling family of invariant lines, i.e. it moved points along these invariant lines by a constant distance. Any two members of one and the same such family were called "parallels" by Clifford. Next, given any two intersecting lines l and l', Clifford considered the ruled surface generated by all those Clifford (left) parallels to l which met l'. Equivalently, this surface could be described as being generated by all (right) Clifford parallels to l' meeting l. Moreover, there were two commuting one-parameter families of left and right twists inducing isometries of the surface, which had l and l' as invariant lines, respectively. Consequently, the surface had constant curvature zero. In topological terms, the surface was a torus as may be seen from Clifford's description of it as "a finite parallelogram whose opposite sides [given by the lines l and l'] are regarded as identical" [6, p. 193]. Closer inspection shows that the surface is indeed orientable. It is important to keep in mind that Clifford's example was not constructed by endowing the 2-torus with a geometrical structure, but rather as a particular surface embedded in elliptic 3-space arising from the consideration of a particular set of isometries, Clifford's twists or translations.

Several remarks in Clifford's philosophical articles indicate that he was aware of the implications this example had for the problem of giving an adequate mathematical description of physical space: The same local geometry might be tied to spaces that are globally different. Even for spaces of constant curvature one could make different "assumptions [...] about the *Zusammenhang* of space", as he wrote in 1873 [7, p. 387]. Clifford also saw that these differences were of a topological nature. A remark of 1875 may even be read as advocating a more radical kind of 'topologism': "There are many lines of mathematical thought which indicate that distance or quantity may come to be expressed in terms of *position* in the wide sense of the *analysis situs*. And the theory of space-curvature hints at a possibility of describing matter and motion in terms of extension only." [7, p. 289.]

In the 1870s, Cliffords critical remarks about the possibilities of globally different spaces with the same local geometry seem not to have generated resonances within the scientific communities either of physicists or of mathematicians. This changed during the 1880s for reasons that originally had nothing to do with Clifford's ideas. In 1877, the American astronomer Simon Newcomb published a paper on a geometry of space with constant positive curvature (in Kleinian terms: elliptic geometry). In a reaction to this paper, Wilhelm Killing, a student of Weierstrass and mathematics teacher, argued that Newcomb had overlooked the fact that there were actually *two* possible geometries with constant positive curvature that should be discussed: elliptic and spherical space. This prompted Felix Klein, whose earlier contributions on non-Euclidean geometry also had focused on elliptic rather than spherical geometry, to enter into a correspondence with Killing.[2] While Klein pointed out that Killing's remark was fairly obvious from the perspective that Klein had developed, Killing repeatedly emphasized the importance of a theorem (along Helmholtz's line of argument) specifying the full range of geometric spaces compatible with the idea of the free mobility of rigid bodies. According to Killing, there were exactly four such spaces: 3-dimensional Euclidean, hyperbolic, elliptic, and spherical space. Killing was clearly interested in what might be called the foundations of physical geometry as opposed to the framework of projective geometry that

---

[2]Killing's letters to Klein may be found in the Niedersächsische Staats- und Universitätsbibliothek (NSUB) Göttingen, Handschriftenabteilung, Cod. MS Klein 10.

guided Klein. One of Klein's reactions now was to refer to Clifford's flat surface in elliptic space. In his eyes, this example showed that there were many more manifolds satisfying the assumptions that Killing wanted to hold. Killing protested: Clifford's surface did not admit free mobility in the full sense (it did not allow global rotations) and thus was not a "space form satisfying our experience" (Killing to Klein, cf. note 2, 5 October 1880).

It took Killing and Klein several years to sort out their differences. In the end it became clear (not least because of Sophus Lie's additional work on Helmholtz's approach) that the conditions of constant curvature and free mobility in the Helmholtzian sense had to be distinguished. The former was a local, the latter both a local *and* a global property of space. However, Klein pointed out that there was no clear empirical sense which could be given to this latter property — contrary to both Helmholtz's and Killing's intentions. What *might* make sense as an empirical requirement was the free mobility of bodies of finite size, indeed of globally bounded finite size. Of course this restricted condition of free mobility still implied a constant curvature of space. Hence Klein felt justified in posing the following problem, first in a lecture course on non-Euclidean geometry in 1889/1890, then in print: "to enumerate all species of connectivity which may at all occur in closed manifolds of some constant measure of curvature" [8, p. 554]. Obviously, Klein was interested in the global topological differences of such manifolds, not in a finer classification up to isometry. In his paper, he gave a (not quite complete) discussion of the two-dimensional case, emphasizing again Clifford's work. Then he pointed out the general connection between regular tessellations of the standard non-Euclidean spaces of dimension 3 and manifolds of constant curvature. From his own and from Poincaré's work on automorphic functions he knew that this connection lead to quite involved problems. The corresponding section of his paper included an invitation "that the question would be taken up elsewhere". He underlined that the problem was "fundamental for the doctrine of space, inasmuch as we want to start the latter from the condition of free mobility of rigid bodies" [8, p. 564].

Note that Klein here refered to the *restricted* condition of free mobility. By now, Killing accepted Klein's argument that only this latter version of the condition had empirical content, and in the following years he took up the task that Klein had set. One may group his work on what he now called the problem of "Clifford-Klein space forms" (in the following: CK space forms) under three headings: a reformulation of the problem in group-theoretical terms, the construction of new classes of examples, and a discussion of the scientific relevance of spaces of constant curvature. I will return to the two more mathematical aspects in the next section. Here I want to comment on the third.

In both of his relevant publications, Killing included long sections defending a study of CK space forms in the context of the foundations of physical science [9], [10, part 4]. Repeating Klein's argument, Killing advocated an understanding of free mobility in the restricted sense and emphasized that nothing in experience excluded the possibility of space being different from the standard non-Euclidean spaces. In fact, he considered only one possible criticism as requiring a more careful discussion: As yet, neither mechanics nor any other physical theory existed for

CK space forms. But this was just the usual course of science. For the standard non-Euclidean spaces as well, mechanics was just in the process of being developed (Killing himself had made important contributions). In consequence, the primary task was to develop physical theories for CK space forms as well. Only then it would be possible to judge their scientific merits. As a particular phenomenon that mechanics in multiply connected CK space forms might bring up, Killing mentioned anisotropies of the gravitational force between two bodies [10, p. 347]. One may well read this as a hint at a possible local empirical phenomenon that might help in finding out global 'connectivity properties' of physical space.

Killing was not alone in the 1890's in discussing the physical relevance of spaces of constant curvature. In 1899, Klein came into contact with the young and aspiring astronomer Karl Schwarzschild when the latter gave a talk at a large meeting of astronomers discussing "the admissible measure of curvature of space". In this talk, Schwarzschild gave bounds on the radii of curvature of either an elliptic or a hyperbolical universe consistent with astronomical observations of star parallaxes. After the talk and in ensuing correspondence[3], Klein made Schwarzschild aware of the fact that in such a discussion, CK space forms should also be taken into account. Schwarzschild agreed. In the printed version of his talk, he added an appendix in which he briefly discussed whether or not space might actually be a non-standard space of constant curvature. In very intuitive terms, he explained to his readers (the paper was published in an astronomical journal) how one could conceive of astronomical observations suggesting such kinds of spaces: by observing "identical, apparent repetitions of the same world-whole, be it in a Euclidean, elliptic, or hyperbolic space". However, the time was not yet ripe for a full discussion of this possibility: "We may treat the other Clifford-Klein space forms very briefly, the more so since they have not yet been investigated completely even from a mathematical point of view. [...] experience only imposes, in all cases, the condition that their volume has to be larger than that of the visible star system." [11, appendix.]

Killing's mathematical work on CK space forms was reformulated in modern mathematical terms and substantially extended by Heinz Hopf who devoted one of two parts of his dissertation to the problem in 1925. Again I defer a discussion of the mathematical parts of Hopf's work to the next section. However, it must be pointed out that Hopf also shared a cosmological interest in CK space forms with Killing and Klein. When, in 1928, Klein's lectures on non-Euclidean geometry were edited posthumously in a completely rewritten form by Walter Rosemann, Hopf took over the task of writing a new section on spaces of constant curvature (at the time called "homogenous spaces" by him). He closed this section with discussing "the application of geometry to the external world". Here, only "the possibility of homogenous space forms [had] to be taken into consideration", as no empirical data were known that would force one to consider spaces of variable curvature. Of particular value was the "possibility of ascribing to the universe a finite volume, independently of its geometrical structure [...] since the idea of an infinite extent [...] causes various difficulties, for instance in the problem of the distribution of mass." [12, p. 270.] One should note that this was written after the advent of Einstein's

---

[3]See Schwarzschild's letters in NSUB Göttingen, Cod. MS Teubner 44 and Cod. MS Klein 11.

theory of general relativity, and after the development of relativistic cosmology had seriously begun with contributions by Einstein, Schwarzschild, de Sitter, Weyl and others. In this context, constant curvature was no longer a pre-condition of measurement, but rather a consequence of the assumption of a homogenous average distribution of mass throughout the universe.

3. Killing's and Hopf's mathematical contributions

Killing's main mathematical contribution to the problem of CK space forms was its reduction to group theoretical terms. Killing tried to show that in all dimensions n, Klein's problem (see above) was equivalent to finding all finitely or at most countably generated subgroups G of $SL(\mathbb{R}, n+1)$ which for some real parameter $1/k^2$ leave invariant the bilinear form

$$a(x, y) = k^2 x_0 y_0 + x_1 y_1 + ... + x_n y_n , \qquad x, y \in \mathbb{R}^{n+1} ,$$

and satisfy a discontinuity condition that will become clear as we go along [10, p. 322]. Even if it is difficult to follow all details of Killing's argument, its main line is clear. (In the following, modern abbreviations are used to condense Killing's verbal style.) To begin with, if M was a manifold of dimension n with constant curvature $1/k^2$, Killing required the existence of some $r > 0$ such that for all points $P \in M$ there existed a ball $B_r(P) \subset M$ of radius r isometric with a similar ball in Euclidean, hyperbolic or spherical space of the same curvature. This was Killing's way of stating the restricted condition of free mobility. It implied both a kind of completeness of the manifold and the discontinuity condition just mentioned. Using a technique he had learned in a seminar of Weierstrass, Killing translated this into local "coordinates", by which he understood isometric mappings

$$B_r(P) \longrightarrow X_k := \{ x \in \mathbb{R}^{n+1} \mid a(x, x) = k^2 \}$$

mapping P to $\bar{P} := (1, 0, ..., 0)$. (For negative curvature, the condition $x_0 > 0$ was added in the definition of X_k; in the flat case, Killing just considered the hyperplane in $\mathbb{R}^{n+1}$ defined by $x_0 = 1$.) Endowing X_k with the metric d given by

$$k^2 \cos \frac{d(x, y)}{k} = a(x, y) , \qquad x, y \in X_k ,$$

made X_k into a model of the standard Euclidean and non-Euclidean spaces that had been used by Killing in most of his earlier work on these geometries.

Choosing a particular point P as the origin, these "Weierstrassian coordinates" defined a 1-1 correspondence of the bundles of geodesics through P and $\bar{P}$. Using this intuition, Killing extended a local coordinate system around P to a kind of global coordinate system, i.e., a 'mapping' $M \to X_k$, associating a point $Q \in M$ on some geodesic through P with a point $\bar{Q} \in X_k$ on the corresponding geodesic such that the distances between P and Q and between $\bar{P}$ and $\bar{Q}$ were equal. Partly without further argument and partly based on intuitive explanations, Killing assumed that this 'mapping' was in general multi-valued (since in M there might exist closed geodesics), surjective, and locally isometric. If $\bar{Q}_1, \bar{Q}_2 \in X_k$ were two "coordinates" of the same $Q \in M$, then by construction there existed a (local) isometry $K_r(\bar{Q}_1) \to K_r(\bar{Q}_2)$. Killing assumed that this mapping could be uniquely extended to a global isometry $\psi : X_k \to X_k$. He knew that isometries of X_k were induced by elements of the group we denote by $SL(\mathbb{R}, n+1)$, leaving the bilinear form a invariant. Again on intuitive grounds Killing argued that any such ψ was in fact what we would call a covering transformation of M. The collection of all ψ arising in this way formed a discrete subgroup Γ of the isometry group of X_k which had the property that every $\psi \in \Gamma$ moved points by a distance of at least r. M itself was then equivalent to what later was called the quotient space X_k/Γ.

The gaps and intuitive turns in Killing's argument give a striking illustration of the growing need for precise topological arguments in some areas of mathematics at this time. Notions relating to covering spaces or the fundamental group (in his intuitive explanations, Killing repeatedly relied on the consideration of "motions of bodies" along closed geodesics in M) would have helped Killing significantly in securing the vaguer parts of his considerations.

Killing was quite clear that the new problem in group theory was difficult. Accordingly, he was satisfied with describing a few simple Euclidean and spherical space forms. In the flat case, his main example was the analogue of Clifford's surface, the manifold given by $\mathbb{R}^3/\mathbb{Z}^3$. In the case of positive curvature, Killing noticed that in even dimensions, only S^n und $\mathbb{R}P^n$ with their canonical metrics could occur. In dimension 3 he mentioned other possibilities, e.g. $\mathbb{R}P^3/\Gamma$, where Γ is a cyclic group of Clifford's translations.

It was Heinz Hopf who reworked Killing's arguments in a modern framework. In his dissertation of 1925, he presented a completely revised treatment of the problem of CK space forms that made it superfluous to look into the older literature any more [13]. His version of the problem was to classify all geodetically complete Riemannian manifolds of constant curvature in either of two possible senses: One could try to classify the resulting "geometries" (i.e., look for a classification up to isometry) or one might wish to classify just the manifolds carrying these geometries (i.e., look for a classification up to diffeomorphism). In Killing's work, this distinction had never been clearly made.

After a preliminary clarification of the relation between Killing's earlier completeness condition and the weaker condition of geodetic completeness, Hopf gave a new proof of Killing's basic result. In the new setting, this theorem took the form that every geodetically complete Riemannian manifold of constant curvature (again called CK space form by Hopf) was a quotient of Euclidean, hyperbolic or spherical space by a discontinuous group Γ of isometries without fixed points and such that no orbit of Γ had a limit point. The geometric content of Hopf's proof was very similar to Killing's argument – the difference being that Hopf had conceptual tools at his disposal that Killing had missed. Hopf showed that if M was a CK space form in his sense, then every point $P \in M$ still had a neighbourhood that could be mapped isometrically onto a neighbourhood of some point $\bar{P}$ in one of the standard spaces, say X. Using again the resulting 1-1 correspondence of the bundles of geodesics through P and $\bar{P}$, Hopf defined a mapping $X \to M$ (!) of which he showed that it was an isometric covering. As X was simply connected, it was the universal covering space of M. Moreover, the fundamental group $\pi_1(M)$ acted freely and discontinuously in the sense explained above by isometries on X. Therefore, M was isometric to a quotient manifold of the required form.

Instead of multi-valued "coordinates", Hopf could speak of coverings, inverting the direction of the crucial mapping. Relying on the notions of universal covering spaces, covering transformations, local isometries etc., Hopf was able to formulate several steps in his proof as simple arguments by contradiction. For this step into mathematical modernity, Hopf had an important model: Hermann Weyl. The conceptual framework used by Hopf was mainly an adaptation of that outlined in the topological sections of Weyl's monograph *Die Idee der Riemannschen Fläche* of 1913. On the other hand, Weyl himself had pointed out that this framework could be used in discussing manifolds of constant curvature. Using the language of coverings, Weyl showed in an appendix to [14] that every "closed Euclidean space" (i.e., every closed manifold of constant curvature zero) was isometric to a "crystal", i.e., a quotient of Euclidean space by a suitable discrete group of isometries. Weyl's proof only worked in the flat case, and apparently he did not pursue analogous questions for curved manifolds.

Hopf not only used modern topological tools for reformulating the space form problem. He also showed how topology could profit from the geometric ideas involved in this problem. Looked at in this perspective, one point was fairly obvious: CK space forms, constructed as quotient spaces, furnished new examples of manifolds with known fundamental groups. At the time, this was particularly interesting for finite groups; very few examples of manifolds with finite fundamental groups had been known beyond those with cyclic groups. Based on Klein's analysis of the isometry group of elliptic 3-space (which in turn relied on Clifford's ideas), Hopf discussed a series of new 3-dimensional, spherical space forms with finite fundamental groups as well as infinitely many spaces with infinite groups.[4] Moreover, Hopf's research on spaces of constant positive curvature proved to be of decisive importance at a later point in his career: Clifford's parallels provided him with the crucial example of a mapping from S^3 to S^2 that was not homotopic to a constant mapping (lifting a fibration of elliptic 3-space by Clifford parallels to the 2-sheeted covering of elliptic 3-space produced what became known as the Hopf fibration of S^3). Even Hopf's basic invariant for classifying such maps, the linking number of fibres, was derived from an intuitive understanding that the linking behaviour of Clifford's parallels was an obstacle for deforming the Hopf fibration into a constant [16].

4. Conclusions

In these ways, an important strand in the formation of modern manifold topology profited from geometric ideas that had their origin in a 19th century context in which mathematical and cosmological thinking were closely related. If I may use a topological metaphor: Once more it turns out that the fibres of historical develop-

---

[4]Spherical, 3-dimensional space forms with polyhedral fundamental groups had already been added to Killing's examples by the American geometer Woods in 1905.

ments in different scientific fields are intertwined rather than grouped in a locally trivial bundle, the base of which would be some eternal architecture of concepts or structures that only need straightforward elaboration.

From cases like the one I have sketched one might learn that in periods of innovative research, boundaries between different mathematical fields and even between mathematical and physical thinking may tend to blur. In other periods, this may be different. In the present case, it seems (at least at first sight) that the later *solutions* of the Euclidean and the spherical space form problems were found in episodes of autonomous, purely mathematical research. Moreover, the history of the space form problem between Clifford and Hopf reveals a complicated relation between tradition and modernization within mathematics. The analysis of Killing's and Hopf's ways of approaching the space form problem shows that despite the crucial differences due to the non-availability or availability of precise topological notions, Killing's more traditional geometric ideas were taken up in the modern formulations. Another such core of geometric ideas that was handed down to modern topology were Clifford's geometric ideas: in fact *all* later authors discussed here made some use or other of these ideas.

In a period in which the global topological properties of the universe receive new interest among cosmologists, it seems fitting to recall that a century ago, such an interest also motivated some of the topological problems and ideas mathematicians have since become acquainted with. At least in this indirect way, questions about the immeasurably large have *not* been idle questions for the explanation of nature.

References

[1] J. Barrow-Green, *Poincaré and the three body problem*, Providence, RI: AMS/LMS, 1997.

[2] T. Hawkins, *The emergence of the theory of Lie groups*, New York: Springer, 2000.

[3] M. Epple, Topology, matter, and space I, *Archive Hist. Exact Sci.*, 52 (1998), 297–382.

[4] M. Epple, *Die Entstehung der Knotentheorie*, Wiesbaden: Vieweg, 1999.

[5] M. Epple, Topology, matter, and space II; *Archive Hist. Exact Sci.*, (to appear).

[6] W. K. Clifford, Preliminary sketch of biquaternions, in: *Mathematical papers*, London: Macmillan, 1882, 181–200 (originally published 1873).

[7] W. K. Clifford, *Lectures and essays*, vol 1, London: Macmillan, 1901.

[8] F. Klein, Zur Nicht-Euklidischen Geometrie, *Math. Ann.*, 37 (1890), 544–572.

[9] W. Killing, Über die Clifford-Kleinschen Raumformen, *Math. Ann.*, 39 (1891), 257–278.

[10] W. Killing, *Einführung in die Grundlagen der Geometrie*, vol. 2, Paderborn: Schöningh, 1898.

[11] K. Schwarzschild, Über das zulässige Krümmungsmaß des Raumes, *Vierteljahresschrift der Astron. Ges.*, 35 (1900), 337–347.

[12] F. Klein, *Vorlesungen über Nicht-Euklidische Geometrie*, Berlin: Springer, 1928.

[13] H. Hopf, Zum Clifford-Kleinschen Raumproblem, *Math. Ann.*, 95 (1926), 313–339.

[14] H. Weyl, Über die Gleichverteilung von Zahlen mod. Eins, *Math. Ann.*, 77 (1916), 313–352.

[15] H. Samelson, $\pi_3(S^2)$, H. Hopf, W, K. Clifford, F. Klein, in: I. M James (ed.), *History of topology*, Amsterdam: Elsevier, 1999, 575–578.

The Third Approach to the History of Mathematics in China

Anjing Qu*

Abstract

The first approach to the history of mathematics in China led by Li Yan (1892–1963) and Qian Baocong (1892–1974) featured discovering *what* mathematics had been done in China's past. From the 1970s on, Wu Wen-tsun and others shifted this research paradigm to one of recovering *how* mathematics was done in ancient China. Both approaches, however, focus on the same problem, that is mathematics in history. The theme of the third approach is supposed to be *why* mathematics was done. Combining this approach with the former two, the research paradigm will be improved from one of mathematics in history to that of the history of mathematics.

2000 Mathematics Subject Classification: 01A25.
Keywords and Phrases: Chinese mathematics, Research paradigm, Interpolation, Numerical method, Scientific tradition.

1. Introduction

Since the beginning of the last century hundreds of scholars have devoted themselves to the discipline of the history of mathematics in China. Their research has not only thrown light on the various features of traditional Chinese mathematics, but also has led to a better understanding of the diversity of the mathematical sciences.

This research has led to problems, however. Some mathematicians complained that most Chinese historians of mathematics limited their research to ancient China, while it has seemed to other scholars that fresh avenues into the history of traditional Chinese mathematics may, to some extent, have been exhausted. Fewer and fewer young scholars are attracted to the field, and even for some senior historians of mathematics it has been difficult to find exciting new topics to work on.

A similar perception prevailed once before in the 1970s, when many Chinese historians of mathematics had become discouraged about the future. It came as

---
*Northwest University, Department of Mathematics, Xi'an 710069, China. E-mail: qaj@sein.sxgb.com.cn

something of a surprise that, soon after some scholars left to work on other subjects, a fresh upsurge of interest led by Wu Wen-tsun appeared, one that lasted until a few years ago.

The aim of this article is twofold. First, research paradigms adopted by Chinese historians of mathematics will be outlined, followed by a discussion of a new way to approach the history of mathematics in China.

2.   Discovery: the first approach

The question of *what* mathematical science, if any, existed in ancient China was first raised by scholars at the beginning of the 20th century, and was the motivation for those who followed turned their attention to the history of pre-modern Chinese mathematics. There is no doubt that Li Yan (1892–1963) and Qian Baocong (1892–1974), the founders of the subject of the history of mathematics in China, deserve to be named as the representative figures of this movement for discovering *what* mathematics was done in ancient China. Two examples, more or less related to them, can be taken in order to show what the word *discovery* has meant in studies in the history of mathematics in China.

2.1.   Interpolation

In Yi-xing's *Dayan li* (a calendar-making system of 724 AD), a function $f(x)$ designed for calculating the solar equation of center is found as follows:

$$f(x) = \frac{x}{n_1} \times \triangle_1 + (1 - \frac{x}{n_1}) \times \frac{x}{2n_2} \times \triangle^2 \qquad (1)$$

where $0 \leq x < n_1$. While a tropical year is broken into 24 parts (qi), n_1 and n_2 are the lengths in days of two consecutive qi. $\triangle_1$ and $\triangle_2$ are the deviations in du (1 $du = 360°/365.25$) from the mean solar motion to its true one on the intervals n_1 and n_2 respectively. $\triangle^2 = \frac{2n_1n_2}{n_1 + n_2}(\frac{\triangle_1}{n_1} - \frac{\triangle_2}{n_2})$. Suppose $n = n_1 = n_2$, the relation yields to $\triangle^2 = \triangle_1 - \triangle_2$. This special case, formula (1), is found in Liu Zhuo's *Huangji li* (a calendar-making system of 600 AD).

It was Yabuuti Kiyosi (1906–2000) who pointed out for the first time that Liu Zhuo's formula is a quadratic interpolation of equal interval, while Yi-xing's formula is that of unequal interval. Both of them are equivalent to Gauss's interpolation [1].

A decade after Yabuuti's discovery, a more detailed investigation of this topic was made by Li Yan, who demonstrated that Liu Zhuo's quadratic interpolation occupied a leading position among various numerical methods in ancient Chinese mathematical astronomy [2].

Assume that the values of a real function $f(x)$ are given at each of $n + 1$ distinct real values $x_k : f_k(k = 0, 1, 2, \cdots, n)$. The method of finding the values $f(x)$ at x by using these values $f_k = f(x_k)$ is called interpolation. Formulae of interpolation could be stated in many ways, for instance the formulae of Lagrange, Aitken, Newton, Gauss, Stirling, Bessel and Everett, depending on the method you make use of to construct it. In cases where the points of interpolation one chooses are the same, their interpolation functions can be transformed to each other no matter by which kind of interpolation the functions are constructed. As for formula (1), it is easy to verify that

$$f(0) = 0, \quad f(n_1) = \triangle_1, \quad f(n_1 + n_2) = \triangle_1 + \triangle_2.$$

Clearly, $x = 0, n_1$, and $n_1 + n_2$ are three points of interpolation of the function $f(x)$. This result shows that formula (1) is a quadratic interpolation function.

Yabuuti said that formula (1) is equivalent to Gauss's interpolation, while Li Yan considered it to be equivalent to Newton's. Actually, it is neither Gauss's nor Newton's.

What Yabuuti and Li Yan did, as such, was to reveal the fact that formula (1) is a quadratic interpolation function. However, the question of which kind of interpolation Chinese mathematicians made use of to construct formula (1) was left open.

2.2. Remote measurement

In his *Haidao suanjing* (Sea Island Computational Canon, 263 AD), Liu Hui designed nine questions in order to demonstrate the problems of remote measurement. The first of them is a problem concerning how to measure the altitude of an island with two gnomons.

Let HI in Fig.1 be the altitude of an island. AB and CD are two gnomons of equal height. AE and CF are the length of shadows of AB and CD, respectively. Suppose the distance (AC) between the two gnomons is known, a formula for measuring the island's altitude is found in the *Haidao suanjing* as follows:

$$HI = AB + \frac{AB \times AC}{CF - AE} \tag{2}$$

Formula (2) is essential in remote measurement. Other questions in Liu's book are much more complex than this. As many as four gnomons are used in some of them. It is said that there were diagrams drawn by Liu Hui for these questions, but they no longer exist today.

The question of the island's altitude is also known as that of solar altitude. Formula (2) was used to measure the altitude of the sun in the *Zhoubi suanjing* (Zhou Dynasty Canon of Gnomonic Computations, first century BC). A diagram for proving formula (2) is also found in this work. Unfortunately, due to transmission of the text over time, it has been distorted beyond recognition.

For deducing formula (2) or showing its correctness, Qian Baocong added a line DG parallel to the line HE in Fig.1.[3] What Qian did was not unusual for historians of mathematics at that time.

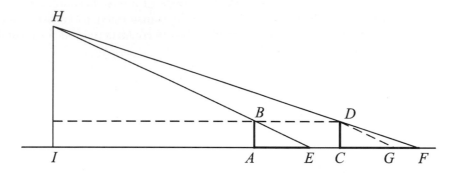

Fig. 1   Altitude of island

However, a new movement in this area of research was heralded in the 1970s when a mathematician who adopted a methodology different to that of Qian's appeared.

3.   Recovery: the second approach

It was Wu Wen-tsun who led studies in the history of mathematics in China on to a second phase, that of recovering *how* mathematics was done in ancient China. Wu explained his approach by criticizing Qian's study on formula (2). Since the parallel line *DG* drawn by Qian in Fig.1 is groundless in traditional Chinese mathematics, his proof of formula (2), as Wu pointed out, should be regarded as a "wrong proof" from the viewpoint of the history of mathematics.

Wu emphasized that demonstrating the correctness of ancient mathematics with modern mathematical notions is by no means the sole purpose of the history of mathematics, and that historians of mathematics should pay more attention to recovering *how* mathematics was actually done in history. He said:

"*Two basic principles of such studies will be strictly observed, viz.:*

"*P1. All conclusions drawn should be based on original texts fortunately preserved up to the present time.*

"*P2. All conclusions drawn should be based on reasoning in the manner of our ancestors in making use of knowledge and in utilizing auxiliary tools and methods available only at that ancient time.*" [4]

Wu, therefore, named his approach recovering mathematical procedure with its original thought.

Let us also take interpolation as an example to show how to recover the history of mathematics in China. How formula (1) was constructed was exactly the topic that Wu's approach focused on. In order to answer this question, the historical background from which the problem of interpolation arose needed to be addressed.

It is well known that formula (1) was invented to solve the problem of irregular solar motion. In the *Chapter of Calendar-making* of the *Tang History*, the evolution of solar theory up to that time was described by the Buddhist monk Yi-xing (683–727) as follows:

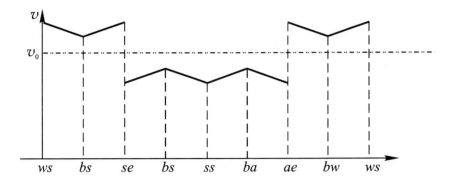

Fig. 2   Pattern of solar speed in the *Huangji li* (600 AD)

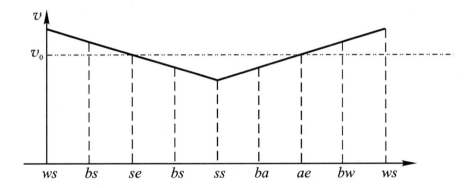

Fig. 3   Pattern of solar speed in the *Dayan li* (724 AD)

In about 560 AD, Zhang Zixin found that solar motion was irregular. From Yi-xing's account, we understand that solar motion in the minds of Liu Zhuo (600 AD) and Yi-xing (724 AD) can be expressed as the patterns in Fig.2 and Fig.3, respectively, where the dotted line v_0 represents the mean velocity of the sun.

Liu's interpolation function is different from Yi-xing's in the division of a year. Liu divided a tropical year into 24 parts of equal length, while Yi-xing divided the ecliptic into 24 parts of 15° each. The small interval was named mean *qi* (*ping qi*) in Liu's division, and ture *qi* (*ding qi*) in Yi-xing's. Since solar motion is irregular, Yi-xing's division is unequally spaced in terms of time. In order to deal with the deviation from the mean motion to the true motion of the sun, interpolation function (1) was constructed by Liu Zhuo and Yi-xing at each *qi*.

The period between the winter solstice (*ws* in Fig.2–4) and the beginning of spring (*bs* in Fig.2–4) consists of 3 *qi*. Let us take this period as an example to demonstrate how formula (1) was constructed by Liu Zhuo and Yi-xing.

In Fig.4, suppose $OM = n_1$, $MN = n_2$, area of $OBCM = \Delta_1$, area of $MEFN = \Delta_2$. Δ_1 and Δ_2 given by observation are deviations of the true motion of the sun from its mean motion on the interval OM and MN, respectively. The doted line OMN stands for the velocity of the mean sun, while the step-like lines BC, EF, and so on stand for the mean velocities of the true sun on each qi.

The idea that Liu Zhou and Yi-xing conceived was how to change the pattern of solar motion from a step pattern to a continuous straight line. The observed data Δ_1 and Δ_2 of the two consecutive qi were used to construct the interpolation function on the former qi so that the slanted lines as a whole could be continuous lines as far as possible. This is a change from a linear to a parabolic interpolation.

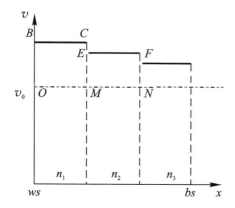 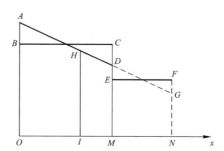

Fig. 4   Linear interpolation   Fig. 5   Quadratic interpolation

As in Fig.5, draw a slanted line AG that crosses the midpoints of lines BC and EF, and intersects line CM at D. Then $AB = DC$, $DE = FG$. Let $x = OI$, then

$$f(x) = \text{area of trapezoid } AOIH$$

is the Chinese interpolation formula (1), $0 \leq x < n_1$.

In order to construct the function $f(x)$, first of all Liu and Yi-xing let the solar speed be changed as an arithmetic sequence from day to day at the interval OM. Then, the arithmetic progression is summed up. The result of the summation is a parabolic function with a variable of days after the initial time at the interval OM. It is the quadratic interpolation function $f(x)$. From the viewpoint of the technique of constructing algorithms, Yi-xing's unequal spaced interpolation is nothing more than Liu Zhuo's equal intervals interpolation.

Since the tropical year is divided into 24 small intervals (qi), and the parabolic interpolation function was constructed on each interval based on the observed data of 24 qi, we call a whole set of the 24 functions of a tropical year *piecewise parabolic interpolation*.

4.   Mathematics in history: original research

During the period of the first movement in the history of Chinese mathematics, *discovery* meant to find out *what* mathematics was done in history. Scholars made their discoveries directly from original historical materials. In the next movement, research discoveries were extended to *recoveries*. The attention of scholars was directed to the question of *how* mathematics was done in history. Recovery work in the history of mathematics is that of rational reconstruction, usually based on indirect historical materials. Recovery, therefore, can be regarded as a kind of indirect discovery.

The paradigm of studies in the history of mathematics in China took shape in the following way: only *discovery*, either direct or indirect, was regarded as original research. Once such a paradigm became the norm in the Chinese history of mathematics community, the model became set for such research.

Obviously, most results gained from the first movement were transformed into the problems to be solved in the second. Problems raised in the first movement became conjectures, while the use of secondary historical materials to "prove" these conjectures was the main trend in the second movement.

This was the research paradigm that historians of mathematics in China followed during the past century. Studies in the history of mathematics were valued according to the viewpoint of this paradigm. It is somewhat equivalent to that of pure mathematics. What they did was to discover or recover mathematics in history.

This picture can help us to clarify the following questions that may puzzle those who are not involved in this field.

First of all, Wu's movement was the consequence of the fact that the research paradigm shifted from *discovery to recovery* after the 1970s. This change was so important that it offered plenty of topics for research to historians of mathematics in the last quarter of the last century.

Secondly, as we have mentioned before, original research in the history of mathematics meant research in original historical materials. For most Chinese scholars, unfortunately, the only original historical materials that they had access to were texts in Chinese. Historical works in Western mathematics or modern mathematics might interest mathematicians or lay readers, and might be widely welcomed, but neither discovery nor recovery could be expected. That meant that these attempts could not be considered as original research. This is the reason why most Chinese history of mathematics was limited to research on ancient China.

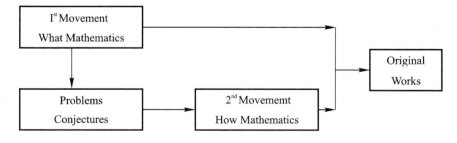

Fig. 6   Studies in the history of mathematics in China: past and present

5.   Why mathematics: the third approach

Discovering *what* mathematics was done is always the basic approach to the history of mathematics. During the period of Li and Qian's movement, it had been taken as the only way to do the history of mathematics in China. Recovering *how* mathematics was done was never seriously accepted as original research until Wu Wen-tsun broke through Li and Qian's paradigm in the 1970s. It helped Chinese historians of mathematics to negotiate through a crisis, and fruitful results have come from this extension of the concept of *original research* in the history of mathematics in China. The model depicted in Fig.6 clarifies the work that Chinese historians of mathematics have done.

The problem now is why are we faced with a crisis once again? How can we step out of such a difficult situation and develop an optimistic future for our field in China? These are vital issues which I have taken upon myself to tackle in the last part of this article.

The reason why the field faces a crisis once again is because the concept of *original work*, either *discovery* or *recovery*, has restricted research on *mathematics in history*. Since the original historical materials that our Chinese scholars can access are limited, they are certain to be exhausted sooner or later.

The answer to achieving our goal of renewed vigor in the field is that the concept of *original work* in the history of mathematics in China has to be extended further. The old research paradigm, therefore, must be improved once again.

Mathematical ideas are the main object of study in the history of mathematics. Hence the history of mathematics, to a great extent, is the history of mathematical thought. When we look at mathematics in history from a historical perspective, an important aspect is often neglected, that is, *why* did mathematics play a part in history? As I have already emphasized, in the past century, two movements dominated studies in the history of mathematics in China namely:

The first approach: *what mathematics* was done.

The second approach: *how mathematics* was done.

It is easy to follow these approaches up with:

A third approach: *why mathematics* was done.

Once the three approaches have merged into a single whole, the research paradigm will shift from mathematics in history to the history of mathematics. *Original research* in the history of mathematics will also be extended in scope.

Actually, *why mathematics* has been regarded as the main purpose of the history of mathematics by leading mathematicians for some time. André Weil at ICM 1978, for instance, presented a plenary speech about *"for whom does one write a general history"* of mathematics. At the end of his presentation, he said *"thus my original question 'why mathematical history?' finally reduces itself to the question 'why mathematics?' which fortunately I do not feel called upon to answer."* [5] Wu Wen-tsun himself also sometimes moved beyond how mathematics in ancient China [6].

6. Why a practical tradition in ancient China

It is often said that, compared with Greek mathematics, Chinese mathematics was characterized by a practical tradition. Many scholars hold that this tradition is the fatal weakness of Chinese mathematical science, one that prevented it from developing into modern science. Some historians of mathematics have argued that certain fundamental factors of the Greek theoretical tradition, such as *proof* and *principle*, can also be detected in the *Nine Chapters of Arithmetic* (1 century BC) and Liu Hui's *Annotations* (263 AD). However, it seems to us that many people are still not convinced.

A scientific tradition represents the principal aspects of scientific methodology, spirit, and style. When we speak of the theoretical tradition of Greek science, we do not mean that there was no applied science at all. What we mean is that compared with this theoretical tradition, the practical tradition was of trivial importance in the development of Greek science.

For a better understanding of the value of Chinese mathematics from a historical perspective, we need to know about the issue of *why mathematics* was done in ancient civilizations, and thus why there was a practical tradition in ancient China is a contact point for this.

In the long history of the Chinese empire, mathematical astronomy was the only subject of the exact sciences that attracted great attention from rulers. In every dynasty, the royal observatory was an indispensable part of the state. Three kinds of expert — mathematicians, astronomers and astrologers — were employed as professional scientists by the emperor. Those who were called mathematicians took charge of establishing the algorithms of the calendar-making systems. Most mathematicians were trained as calendar-makers. Mathematics was thus highly developed for mathematical astronomy besides more general applications, in such areas as indeterminate problems, numerical solutions of algebraic equations, polynomial interpolation and series summation.

Calendar-makers were required to maintain a high degree of precision in prediction. Ceaseless efforts to improve numerical methods were made in order to guarantee that the algorithm could satisfy the precision required for astronomical observation.[7] It was neither necessary nor possible that a geometric model could replace numerical method, which occupied the principal position in Chinese calendar-making systems. The reason for this was that only the numerical method could satisfy the ruler's requirements, that is high accuracy in prediction and computation. As a result numerical analyses won favor over cosmic or geometric model building. As a subject closely related to numerical method, algebra, rather than geometry, became the most developed field of mathematics in ancient China [8]

Science in ancient China was intended primarily to solve concrete problems, such as determining planetary positions. The function of *explaining* natural phenomena never dominated its scientific tradition. What Chinese scientists really cared about was how to solve the problems they faced as accurately as possible.

This is the reason why the practical tradition was chosen in ancient China. From the above description, definitions for the practical and theoretical traditions maybe drawn as follows:

In the practical tradition, science serves to solve concrete problems. Theory is judged by its accuracy of computation. Scientific progress follows the advancement of observation. A theoretical model is always improved to meet the precision requirement step by step.

In the theoretical tradition, on the other hand, science serves to explain natural phenomena. Theory is judged by its function in the explanation. Observation is employed to verify the correctness of the theoretical hypothesis. The old model is always replaced if the new one is more reasonable for the explanation of natural phenomena.

These two traditions differ mainly in their starting and ending points:

In the practical tradition, a model is built up from observation to solve concrete problems. For a more accurate prediction, uninterrupted efforts are made to explore unknown factors, and the related numerical analyses are improved. The more accurate the theory is, the closer the model is to the truth.

In the theoretical tradition, on the other hand, a model is built up from hypotheses that account for natural phenomena. For a better understanding of natural phenomena, the model is revised from time to time on the basis of a new hypothesis. The closer the model is to the truth, the more accurate is the theory.

The attitude to algebraic equations of these two traditions provides a typical example to show us their differences. In the theoretical tradition, mathematicians paid attention to the root formula of the equation. The numerical solution of the equation was seldom taken much notice of by them. The reason is that no matter how effective the numerical method is, its solution is usually an approximate result that does not help to *explain the phenomenon* of the equation.

On the contrary, the root formula is never more important than a numerical solution in the practical tradition. The reason is that even if one could have the exact root from the formula, one has to extract from it a concrete value for application. Hence it seems to mathematicians in the practical tradition that the numerical solution is sufficient. In ancient China, the root formula of equations was not an important subject, although they did know the formula of second-degree equations.

It is believed that modern science, to a large extent, benefits from the heritage of Greek science. Nevertheless, it is hard to say that the theoretical tradition dominates the development of modern science in all aspects. In fact, the practical tradition also plays an important role.

It is obvious that numerical analyses are more frequently used than theoretical hypotheses in modern science. The task for modern scientists is not only to account for natural phenomena, but also to solve concrete problems. The research results arising from the search for solutions to scientific problems have led in two directions: those that are concerned with finding general theorems concerning the problems, and those that are searching for good approximations for solutions. Both explaining natural phenomena and solving concrete problems are the goals that modern scientists strive for. Observations have occupied a substantial position in the development of modern science.

Generally speaking, science in ancient civilizations was often characterized by a distinctive tradition, either the theoretical tradition as in Greece, or the practical tradition as in China. These traditions tended to develop along their own lines. However, the situation in modern times has never been so simple. The diversity of modern science features the blending of the two traditions. It develops in a dualistic mode.

7. Conclusion

The paradigm of the history of mathematics in China has directed the attention of researchers to focus on *mathematics in history*, in particular in ancient China. It is certain that the *what* and *how mathematics* are two approaches for historians of mathematics that will always remain valid. If history continues, they continue.

However, just as Li and Qian's what mathematics approach was replaced by Wu's *how mathematics* approach to become the mainstream of studies in the history of mathematics in China in the last quarter of a century, a new movement is certain to supersede the old one sooner or later. For a historian of mathematics, after the what's and the how's have been figured out, the problem *why* mathematics was done should be addressed.

Following this topic of *why mathematics*, research should shift, to some extent, from mathematics in history to the history of mathematics. Under these circumstances, plenty of new problems are raised for us. Mathematics in ancient China and other old civilizations, for instance, will be placed in the context of the whole history of mathematics. The diversity of mathematics in different civilizations will give us a more distinct picture of the history of mathematics.

Acknowledgements. The author is grateful to Prof. Michio Yano and Prof. Wenlin Li for their invaluable comments on an earlier draft of this article, and to Mr. John Moffett for help with editing. Research for it was supported by the Japan Society for the Promotion of Science (JSPS, P00019).

References

[1] Yabuuti Kiyosi, *Zuitô rekihôshi no kenky? (Research on the History of Calendrical Methods in the Sui and Tang Dynasties)*, Tokyo: Sanseido, 1944, 71–74.

[2] Li Yan, *Zhongsuanjia de neichafa yanjiu* (Studies on the Methods of Interpolation of Ancient Chinese Mathematicians), Beijing: Kexue chubanshe, 1957.

[3] Qian Baocong (ed.), *Suanjing shishu* (The Ten Classical Books in Mathematics), Beijing: Science Press, 1963, 32.

[4] Wu Wen-tsun, Recent Studies of the History of Chinese Mathematics, in *Proceedings of the International Congress of Mathematicians*, 1986. Providence: American Mathematical Society, 1986, 1657.

[5] André Weil, History of Mathematics: Why and How, in *Proceedings of the International Congress of Mathematicians, Helsinki*, 1978. Helsinki: Academia Scientiarum Fennica, 1980, 236.

[6] Wu Wen-tsun, *Mathematics Mechanization*, Beijing: Science Press & Dorrecht: Kluwer Academic Publishers, 2000, 1–66.

[7] Qu Anjing, Numerical Methods in Medieval Chinese mathematical Astronomy, *Journal of Northwest University (Natural Science Edition)*, **28**:2(1998), 99–104.

[8] Qu Anjing, Why Mathematics in Ancient China? in *Matematica e Cultura 2003*, Milan: Springer-Verlag, to appear.

Author Index